AF787092

Electric Systems for Transportation

Electric Systems for Transportation

Editors

Maria Carmen Falvo
Alessandro Ruvio

MDPI • Basel • Beijing • Wuhan • Barcelona • Belgrade • Manchester • Tokyo • Cluj • Tianjin

Editors
Maria Carmen Falvo
University of Rome Sapienza
Italy

Alessandro Ruvio
University of Rome Sapienza
Italy

Editorial Office
MDPI
St. Alban-Anlage 66
4052 Basel, Switzerland

This is a reprint of articles from the Special Issue published online in the open access journal *Energies* (ISSN 1996-1073) (available at: https://www.mdpi.com/journal/energies/special_issues/electric_transportation).

For citation purposes, cite each article independently as indicated on the article page online and as indicated below:

LastName, A.A.; LastName, B.B.; LastName, C.C. Article Title. *Journal Name* **Year**, *Volume Number*, Page Range.

ISBN 978-3-0365-0488-9 (Hbk)
ISBN 978-3-0365-0489-6 (PDF)

© 2021 by the authors. Articles in this book are Open Access and distributed under the Creative Commons Attribution (CC BY) license, which allows users to download, copy and build upon published articles, as long as the author and publisher are properly credited, which ensures maximum dissemination and a wider impact of our publications.
The book as a whole is distributed by MDPI under the terms and conditions of the Creative Commons license CC BY-NC-ND.

Contents

About the Editors

Maria Carmen Falvo received her M.Sc. Degree cum laude in Electrical Engineering in 2002 and her Ph.D. in Electrical Engineering in 2007 at the University of Rome "Sapienza" (Italy). Since 2008, she has been with the Department of Electrical Engineering of the same University, until 2018 as Assistant Professor and now as Associate Professor. Since 2011, she has worked as a professor at the Polytechnic Engineering School of Gijon at University Oviedo (Spain), where she is a teacher for the subjects Electrical Energy for Transport (2011–2012) and Power Systems for Electrical Transportation (2013–2014–2015–2016–2017). She was a visiting professor in 2012 at Instituto de Investigación Tecnológica (IIT) of Universidad Pontificia de Comillas in Madrid (Spain) and, in 2013, at the Electrical and Computer Engineering Department and at the FREEDM Center of North Carolina State University in Raleigh (NC, USA). Her main area of research includes power systems simulation, transmission planning, power systems for electric transport, and smart grids. She is the author of more than 80 papers. She has been an IEEE Senior member since 2013.

Alessandro Ruvio received his M.S. degree in electrical engineering with honors, awarded with the title "Excellent Graduate", and his Ph.D. degree at the Sapienza University of Rome in 2012 and 2019, respectively. He was a Research Assistant at the Engineering Department of Astronautics, Electrical and Energetic at the "Sapienza" University of Rome, from 2013 to 2015. In 2019, he became an Assistant Professor. He is the holder of two Patents. His main research topics have been energy saving in electrical systems for transportation both DC and AC. He is a Member of the Italian Federation of Electrotechnics, Electronics, Automation, Informatics and Telecommunications (AEIT); an assistant on projects financed by Enea, Navy, and Ministry of Economic Development; a reviewer for many technical journals. He is a member of the editorial board of *Electronics*.

Preface to "Electric Systems for Transportation"

Transportation systems play a major role in the reduction of energy consumptions and environmental impact all over the world. The significant amount of energy of transport systems forces the adoption of new solutions to ensure their performance with energy-saving and reduced environmental impact. In this context, technologies and materials, devices and systems, design methods, and management techniques, related to the electrical power systems for transportation are continuously improving thanks to research activities. The main common challenge in all the applications concerns the adoption of innovative solutions that can improve existing transportation systems in terms of efficiency and sustainability.

Maria Carmen Falvo, Alessandro Ruvio
Editors

energies

MDPI

Article

Road Test-Based Electric Bus Selection: A Case Study of the Nanjing Bus Company

Jian Gong [1], Jie He [1,*], Cheng Cheng [1], Mark King [2], Xintong Yan [1], Zhixia He [3] and Hao Zhang [1]

1 School of Transportation, Southeast University, Nanjing 210018, China; timely@topchains.cn (J.G.); gs_chengc@163.com (C.C.); 230198699@seu.edu.cn (X.Y.); andyhao@seu.edu.cn (H.Z.)
2 Centre for Accident Research and Road Safety, Queensland University of Technology, Brisbane 4059, Australia; mark.king@qut.edu.au
3 School of Economics and Management, Southwest Jiaotong University, Chengdu 610031, China; anniehezhixia@163.com
* Correspondence: hejie@seu.edu.cn

Received: 20 February 2020; Accepted: 6 March 2020; Published: 8 March 2020

Abstract: Globally, the use of electric vehicles, and in particular the use of electric buses, has been increasing. The city of Nanjing leads China in the adoption of electric buses, supported by city policies and infrastructure. To lower costs and provide a better service, vehicle selection is crucial, however, existing selection methods are limited. Accordingly, Nanjing Bus Company developed a test method based on road tests to select a bus. This paper presents a detailed description of the test method and a case study of its application. The method included an organization structure, selection of eight test vehicles (four 10 m length, four 8 m length) from four brands (a total of 32 test vehicles), selection of indicators and selection of routes. Data was collected from repeated drives by 65 drivers over an 8-week period. Indicators included power consumption, charging duration, failure duration and driving distance. It is concluded that the road test method designed and conducted by the Nanjing Bus Company provides a good framework for the selection of pure electric buses. Furthermore, subsequent experience with selected buses supports the validity and value of the model.

Keywords: electric bus; vehicle selection; road operation test; sustainable development

1. Introduction

In recent years, the demand for improved transportation systems has been on the rise owing to the enhancements in the quality of individual living. Public transportation plays an increasingly significant role in urban centers worldwide in the terms of its economic, environmental and sustainable capacity [1]. As part of the expansion of urban public transport networks, bus transit systems make up a growing proportion of urban passenger transport. There were 608,600 public buses in China by 2017. At this time, the passenger volume for the bus transit system was approximately 74.5 billion annually, accounting for 58% of the total urban passenger transport volume, which was far more than the rail transport volume [2].

However, concerns regarding air pollution triggered by the emission of traffic exhausts are unavoidable. Consequently, the appeal for the amelioration of air quality to guarantee a certain standard of living led considerable attention to the electrification of urban transportation systems [3]. It is noteworthy that the use of electric buses is not a new concept. The use of battery-powered electric buses can be traced back to the early 2000s, which matured gradually during the mid-2000s, followed by dramatic developments during the past decade [3].

Energy efficient vehicles are being promoted worldwide as an acceptable approach to address vehicle-based air pollution. Some developed countries, like the United States and Sweden, are making efforts to improve energy efficiency and reduce the use of fossil-fuel buses in mass transit systems [4].

Nevertheless, China has been increasing its number of fossil-fuel vehicles rapidly—vehicles with comparatively high emissions and excessive energy usage, which has resulted in severe health, energy and congestion problems [5]. In order to address this dilemma, and keep the pace of the trend of global cities, China has been embracing the use of pure electric buses owing to their characteristics of zero emissions, low noise, superior driving stability and good economic efficiency [6].

The Chinese government has promulgated policies to support this transition, such as financial subsidies for bus-related industries and companies to vigorously promote the use of electric buses [7]. In 2015, Green development has become a national strategy [8], and national ministries have enacted some approaches to managing a mass transit infrastructure in relation to electric vehicles [9].

Vehicle performance indicators are fundamental factors for selecting the proper bus type. However, certain weaknesses of pure electric buses, such as a relatively short endurance mileage and a long charging time, are pending obstacles to be overcome. Therefore, the selection process for pure electric buses is more complicated than it is for common buses.

There are three generally accepted methods used to carry out dynamic behavior testing of new electric vehicles—computer simulation test, test bench simulation, and road testing [10].

Compared with the first two methods, which have shorter test cycles and lower costs and are generally used by manufacturers, road tests are more intuitive to reflect the vehicle properties, as the consequence of their superior capacity to provide accurate results in a setting close to real operational conditions, thus increasing the reliability for bus companies to test vehicle performance.

In regard to the selection of electric vehicles, Zhang et al. concluded that the evaluation and selection of new electric vehicles from certain developed countries like United States, Germany and Japan, relies on relevant research institutions and is based on the advanced vehicle testing activity and fuel cell technology [11], however, it is not the case in China. Although Chinese government has enacted several standards related to pure electric vehicles in terms of basic safety and power systems [12], the selection processes based only on the minimum criteria set out in these standards is inadequate and may not provide sufficient information relevant to actual operation.

In spite of the aforementioned policy deficiency in the standard setting of the pure electric vehicle selection, some research has been conducted in light of the assessment of the pure electric buses, defining some critical performance indicators of the pure electric buses, which can serve as a reference for the selection, e.g., Wang and Zheng used energy consumption per kilometer to evaluate the economic efficiency of pure electric buses [13]. An analysis by Wan et al. argued that that vehicle purchase cost, maintenance cost and energy consumption should be taken into account as well [14].

More recently, Li stated that, in Urumqi, when comparing two different buses and Bus Rapid Transit (BRT) vehicles from different manufacturers, energy consumption and service life should be used [15]. Zhang et al. compared the speed and power consumption of several kinds of electric buses in Kunming and several other cities using road tests [11].

However, there are two main limitations lying in the existing research: (1) The assessment for the pure electric buses is not systematic enough, and thus couldn't provide a clear guidance. (2) The current studies focus on the definition and establishment of theoretical indicators but lack practical validation and specific implementation of the current assessment evaluation model. (3) The environmental impact of electric buses should be emphasized more in the literature, which has been paid increasing public attention these years.

Therefore, the purpose of this paper is to present how the Nanjing Bus Company selects suitable pure electric buses based on a comprehensive selection model. In addition, this paper aims to provide a suitable example of a framework and a method of electric bus selection for other public transportation companies in terms of the environmental indicators.

2. Materials and Methods

2.1. Overview

Nanjing provides a favorable atmosphere for the development of pure electric buses. As a result, it was recognized as a C40 (an international joint urban organization dedicated to climate change) City in 2015 based on the Nanjing New Energy Vehicle Promotion program [16]. Additionally, it won the honorary title of 'National Demonstration City for Transit Metropolis' [17], which is awarded by the Ministry of Transport of the People's Republic of China.

A commercial bus company focuses on profitability in addition to the social benefits that it offers. On the basis of guaranteeing a certain quality of service, the company is required to reduce costs to increase economic efficiency as much as possible. For this purpose, a road operation test was proposed by the Nanjing Bus Company, to determine the most cost-effective vehicles among several bus manufacturers.

The road operation test is a common method used to perform an all-around evaluation of vehicle performance. These tests replicate actual driving conditions and by doing so, differences between tests and reality are minimized. In addition to the characteristics of the vehicle itself, test results can also be influenced by the pavement conditions, traffic conditions, weather, driver behaviors, etc.

2.2. Organizational Structure for the Test

A reasonable organizational structure was designed to define clear duties and labor divisions to further enhance work efficiency. The road test process is generally designed in accordance with the actual operation of buses. Besides the internal work processes of bus companies, road tests require pre-planned arrangements and well-defined organization and coordination of all parts to lay the foundation for efficient testing and accurate results. Based on the operation and evaluation function, the organization structure was divided into three modules—the leadership group, work group, and review group.

2.2.1. Leadership Group

The leadership group was composed of the people overseeing the program and belonged to the Nanjing Bus Company and its subsidiaries and other companies in support roles. The main duties of this group were:

- To monitor and guide the overall operation of the road test;
- To check and approve the implementation plan, and to deal with any possible problem;
- To organize the final review conference and to make a final report, submitted to Nanjing Transportation Bureau.

2.2.2. Working Group

The working group was composed of the various parties involved in road operation, such as operation management, security, maintenance and logistics support sector. The main duties of this group were:

- To conduct the preliminary information survey about electric buses and to determine the outline of operational testing;
- To formulate an implementation plan and the relevant details;
- To supervise subcontractor companies such as Jiangnan and Yangzi to conduct basic data collection and data analysis;
- To make arrangements and to assign vehicles and staff during the road test work;
- To keep track of the progress of the project and to report it to the leadership group on time.

2.2.3. Review Group

The review group was composed of people from the government transport management department, traffic police department, and experts from university and the private sector. The main duties of this group were:

- To comprehensively analyze the relevant data and test results, including investigation of vehicle performance, appearance, and passengers' perceptions and opinions (not reported in this paper);
- To make suggestions after the test and to write a special report on performance difference of buses, to be discussed with all participants.

Every member worked independently but communicated in a detailed manner, sharing feedback and opinions to ensure the successful completion of the test.

2.3. *Test Content*

2.3.1. Test Vehicle Selection

The Nanjing Bus Company selected four common brands of pure electric bus available in the market, each offering two types of buses, whose lengths are approximately 8 m and 10 m. This was done to ensure an equivalent size for the test vehicles.

The test vehicles were divided into two groups based on vehicle length. The first group, Group A, contained buses that were approximately 8 m in length, while Group B contained those that were approximately 10 m in length. All the vehicles were newly manufactured, and they met national standards requirements such as JT/T1026-2016 titled "The general technical conditions of pure electric city buses". The vehicle details are listed in Table 1, and Figures 1–4 show the four brands of buses.

Table 1. Vehicle Brand Involved in the Test.

Type	Brand Name	Model Number	Vehicle Quantity	Length (m)	Electrical Rating (kwh)	Charging Mode
Group A (8 m)	BYD	BYD6810LZEV4	4	8.06	172.8	Slow charge
	Skywell	NJL6859BEV40	4	8.5	93.3	
	Yinlong	GTQ6801BEVBT9	4	8.05	46.4	Fast charge
	Jiankang	NJC6850GBEV2	4	8.5	93.3	
Group B (10 m)	BYD	BYD6100LGEV3	4	10.49	255.4	
	Skywell	NJL6100BEV30	2	10.5	221	Slow charge
		NJL6100BEV37	2		258	
	Yinlong	GTQ6105BEVBT8	4	10.48	81.2	Fast charge
	Jiankang	NJC6105GBEV5	4	10.5	230.4	Slow charge

Figure 1. BYD brand.

Figure 2. Yinlong brand.

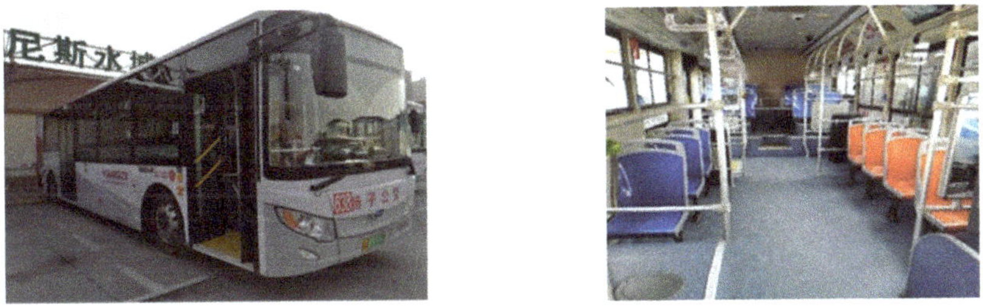

Figure 3. Skywell brand.

Figure 4. Jiankang brand.

2.3.2. Test Environment Selection

The buses were tested on the same route, under the same working conditions, at the same outdoor temperature (adjacent dates), and for days having equivalent humidity, that is, all the tests were conducted during either sunny or rainy days, but not both.

Four representative routes were selected of varying congestion levels, from open and clear to congested, to reflect conditions experienced in normal daily operation. The routes are shown in Figure 5 and the respective details are summarized in Table 2.

Table 2. Routes Involved in the Test.

	Route 1	Route 2	Route 3	Route 4
Bus No.	134W	302W	638W	646W
Length (km)	15.5	15.7	11.6	14.8
Bus stop number	30	33	25	30
Characteristic	No congestion	Congestion	Medium-level congestion	
Description	Traffic conditions are good.	(1) Has over fifty signal lights. (2) Crowded downtown.	Road is narrow.	Far away from the main city.

Figure 5. (**a**) Route 134W; (**b**) Route 302W; (**c**) Route 638W; (**d**) Route 646W.

2.3.3. Staff Selection

The participants, in particular, the drivers, were selected from a group of drivers with a certain level of experience (average driving age = 8–10 years) and no record of accidents to maximize safety. A total of 65 drivers participated in the test, one of whom was on standby. Two drivers were used for each bus. All personnel received the same training at the same time to standardize operational procedures. In addition, these drivers are professional, driving the same route and work a shift every five days, and they are assigned to the their daily driving routes, which can ensure the familiarity with their course of driving and thus reduce the error impact of drivers' behavior caused by the external factors such as the unfamiliarity with different vehicles on the test data.

2.3.4. Road Operation Procedure

Test date: During August and September 2017
Test area: Nanjing, Jiangsu Province, China
Step 1: Establish the organization structure and clarify the duty of the groups.
Step 2: Determine the test conditions and formulate the test process. Vehicle groups A and B were tested during the same period. For both groups, the specific serial number of each brand was assigned to specific test route, though for varying dates and numbers of days, as shown in Table 3.
Step 3: Record data before and after operation for each day.
Step 4: Summarize and analyze data to determine evaluation indexes.
Step 5: Have the experts from the review group evaluate the buses and make a selection.

The road operation test procedure is illustrated in Figure 6.

Table 3. Operation Arrangement.

Type	Brand Name and Model Number	Serial Number	Test Route	Test Date
Group A (8 m)	BYD BYD6810LZEV4	4011	134W	12th Aug. to 30th Sep. Total 50 days
		4012	302W	
		4013	638W	14th Aug. to 30th Sep. Total 48 days
		4014		13th Aug. to 30th Sep. Total 49 days
	Skywell NJL6859BEV40	3801	134W	8th Aug. to 30th Sep. Total 54 days
		3803	302W	
		3800	638	8th Aug. to 7th Sep. and 27th Sep. to 30th Sep. Total 30 days
		3802		8th Aug. to 30th Sep. Total 54 days
	Yinlong GTQ6801BEVBT9	3721	134W	18th Aug. to 21st Sep. and 30th Sep. Total 36 days
		3722	302W	18th Aug. to 28th Sep. Total 42 days
		3723	638W	19th Aug. to 30th Sep. Total 43 days
		3724		19th Aug. to 22nd Aug., 27th Aug. to 9th Sep., 11th Sep. to 25th Sep. and 28th Sep. to 30th Sep. Total 36 days
	Jiankang NJC6850GBEV2	3806	134W	31st Aug. to 30th Sep. Total 31 days
		3807	302W	
		3804	638W	29th Aug. to 30th Sep. Total 33 days
		3805		29th Aug. to 1st Sep., 3rd Sep. to 15th Sep. and 17th Sep. to 30th Sep. Total 31 days
Group B (10 m)	BYD BYD6100LGEV3	4015	134W	12th Aug. to 30th Sep. Total 50 days
		4016	302W	
		4017	646W	14th Aug. to 30th Sep. Total 48 days
		4018		13th Aug. to 30th Sep. Total 49 days
	Skywell NJL6100BEV30 NJL6100BEV37	3809	134W	6th Aug. to 30th Sep. Total 56 days
		3808		
		3810	646W	
		3811	302W	
	Yinlong GTQ6105BEVBT8	3725	134W	18th Aug. to 21st Sep. and 30th Sep. Total 36 days
		3726	302W	18th Aug. to 18th Sep. and 22nd Sep. to 28th Sep. Total 39 days
		3727	646W	19th Aug. to 23rd Sep. and 25th Sep. to 30th Sep. Total 42 days
		3728		19th Aug. to 30th Sep. Total 43 days
	Jiankang NJC6105GBEV5	3813	134W	31st Aug. to 30th Sep. Total 31 days
		3815	302W	
		3812	646W	29th Aug. to 30th Sep. Total 33 days
		3814		29th Aug. to 25th Sep. and 30th Sep. Total 29 days

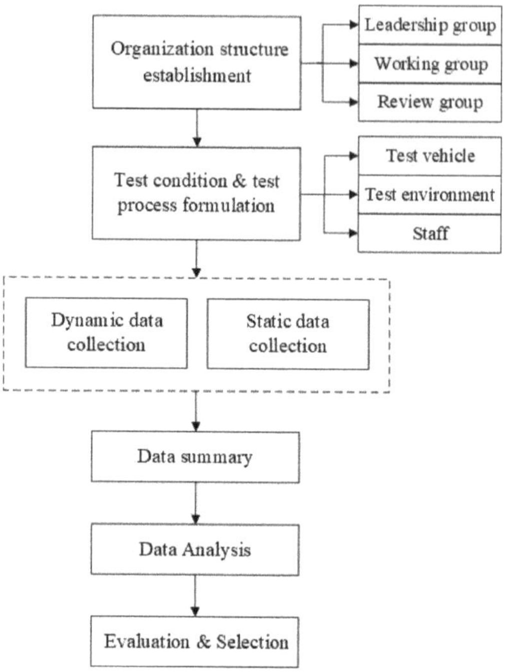

Figure 6. Flow chart of road test operation.

3. Data Collection and Analysis

3.1. Data Collection

Data collection, which had two steps, was performed for the objective data from the vehicle data collectors (vehicle sensors, the equipment installed in the vehicle to collect the data), for charging records and for other car parameters. The steps were as follows:

(1) The first stage of the test was mainly concerned with dynamic data (and various indicators of the vehicle) collected by the vehicle data collectors or determined from dashboard readings while the vehicle was in motion. The dynamic data included driving distance and power consumption.

(2) The second stage of operation test was concerned with static data collection. This included charging and maintenance records, which themselves included charging duration, charging frequency, failure frequency and failure duration.

A data table was completed by the drivers every day they participated in the test and recorded for each bus separately. Table 4 shows an example.

Table 4. Example of Data Collection Table.

Model Number: Skywell-NJL6859BEV40 Serial Number:3801						
Date	**Charging Frequency**	**Charging Duration (min)**	**Power Consumption (kwh)**	**Driving Distance (km)**	**Failure Frequency**	**Failure Duration (s)**
8th Aug.	3	39	140	206	0	0
9th Aug.	1	15	95	143	0	0
1st Sep	3	36	185	275	0	0
2nd Sep	3	40	168	274	0	0

After two months of testing, the data was summarized as shown in Table 5. Because of the confidential business information involved, code names are used for the buses. For example, the code name "V8-1" refers to buses from brand 1 in Group A (about 8 m long) and "V10-1" refers to buses from brand 1 in Group B (about 10 m long).

Table 5. Data Summary Table.

Bus and Route		Charging Frequency	Power Consumption (kwh)	Charging Duration (min)	Driving Distance (km)	Failure Frequency	Failure Duration (s)
V10-1	Total	464	40,061	11,325	40,031	4	8700
	134W	99	10,737	2939	10,793	0	0
	302W	51	8707	1703	7725	0	0
	646W	314	20,617	6683	21,512	4	8700
V8-1	Total	230	31,823	10,987	38,167	6	7800
	134W	75	8817	3124	11,131	2	600
	302W	67	7142	1944	8325	3	0
	638W	88	15,864	5919	18,711	1	7200
V10-2	Total	443	43,391	7506	46,838	1	320
	134W	67	11,195	1466	12,509	0	0
	302W	77	9596	1516	9083	1	320
	646W	299	22,600	4524	25,246	0	0
V8-2	Total	562	29,597	6572	40,284	1	14,400
	134W	147	8655	1839	12,954	0	0
	302W	127	7706	1557	9588	0	0
	638W	288	13,236	3176	17,742	1	14,400
V10-3	Total	829	30,840	8000	29,101	14	148,800
	134W	163	7172	1582	6034	7	48,960
	302W	135	6158	1403	4999	3	4200
	646W	531	17,510	5015	18,068	4	95,640
V8-3	Total	827	23,509	7051	25,051	18	124,380
	134W	196	5747	1713	6795	4	22,140
	302W	164	5945	1595	5810	3	1800
	638W	467	11,817	3743	12,446	11	100,440
V10-4	Total	348	25,389	5834	24,979	5	9900
	134W	60	6440	1217	6610	3	1500
	302W	39	5805	827	5064	2	8400
	646W	249	13,144	3790	13,305	0	0
V8-4	Total	332	17,043	4313	21,294	5	138,120
	134W	78	4445	1118	5958	2	7200
	302W	72	4439	1078	5475	0	0
	638W	182	8159	2117	9861	3	130,920

The various indicators and their data for the road test operation are summarized in Tables 6 and 7.

Table 6. Indicator Data Collected for Group A.

Indicator	V8-1	V8-2	V8-3	V8-4
Total driving distance (km)	38,167	40,284	25,051	21,294
Failure frequency	6	1	18	5
Failure duration (s)	7800	14,400	124,380	138,120
Power consumption (kwh)	31,823	29,597	23,509	17,043
Charging frequency	230	562	827	332
Charging duration (min)	10,987	6572	7051	4313
Average days of operation	49	51	43	32

Table 7. Indicator Data Collected for Group B.

Indicator	V10-1	V10-2	V10-3	V10-4
Total driving distance (km)	40,031	46,838	29,101	24,979
Failure frequency	4	1	14	5
Failure duration (s)	8700	320	148,800	9900
Power consumption (kwh)	40,061	43,391	30,840	25,389
Charging frequency	464	443	829	348
Charging duration (min)	11,325	7506	8000	5834
Average days of operation	49	56	43	32

3.2. Data Analysis

From the perspective of the bus company, economic benefits that can be gained from reducing energy consumption and reducing the time spent on activities or operations are priorities. When use of electric buses first commenced, the utilization rate of pure electric buses was only 50% of that of traditional fuel buses. To maximize the benefits, scheduling and charging mode innovations were implemented, achieving the same utilization rate as a traditional fuel bus. Based on these, the Nanjing Bus Company has processed the indicators in Section 3.1 to indicate four factors that contribute most to electric bus selection:

- *Factor 1: Power consumption per 100 km*

This factor is similar to the 'liters of fuel per hundred kilometers' concept in traditional fuel vehicles. It is the energy consumed by vehicles for a travel distance of 100 km, and it is one of the key economic indicators of pure electric buses. This factor is equal to the power consumption divided by the total driving distance, then multiplied by 100.

- *Factor 2: Charging duration per 100 km*

Factor 2 serves to establish a relationship between battery performance and driving performance. It indicates the battery efficiency and the pros and cons of the charging model used.

- *Factor 3: Daily average driving distance*

This is a basic and necessary index to show vehicle utilization rate, which bus companies value the most, and it is related to operation cost.

- *Factor 4: Failure time per 100 km*

This factor is related to reliability of a vehicle. This factor takes into account frequency and severity of failure, both of which contribute to maintenance costs directly and utilization costs indirectly.

These factors after calculation are given in Tables 8 and 9.

Table 8. Selection Factor Data for Group A.

Indicator	V8-1	V8-2	V8-3	V8-4
Power consumption per 100 km (kwh)	83	73	94	80
Charging duration per 100 km (min)	29	16	28	20
Failure duration per 100 km (s)	20	36	497	649
Daily average driving distance (km)	194	196	146	168

Table 9. Selection Factor Data for Group B.

Indicator	V10-1	V10-2	V10-3	V10-4
Power consumption per 100 km (kwh)	100	93	106	102
Charging duration per 100 km (min)	28	16	27	23
Failure duration per 100 km (s)	22	1	511	40
Daily average driving distance (km)	203	210	170	195

4. Discussion

Based on the analysis of the current situation of domestic pure electric bus, the road operation test is selected as the test method, the composition and responsibility of the road operation test organization are clarified, the test process with the outline of "determine the test conditions—data Collection—data collection" is determined, and the format of relevant record forms is given. According to the evaluation and the discussions of experts and the opinion of the bus company, on the basis of direct and indirect test data, and through observing Tables 8 and 9, it is evident that V8-1 and V8-2 are equally matched in Group A and that V10-2 is the best choice in all aspects, and V10-1 is considered good choice in Group B.

It was confirmed that the manufacturers who provided the V8-1, V8-2, V10-1and the V10-2 were ready to adjust to the conditions in Nanjing regarding the transition to fully electric bus fleets. As a result, the bus company has purchased a batch of pure electric buses of these brands. They have achieved positive operation effects as of the date of this paper.

Also, it should be noted that some limitations still exist in the current study. First, the selection indicators used are primarily objective. The subjective feelings of the drivers and passengers are not considered in the analysis. Second, the differences in battery attenuation and in other aspects of battery performance are ignored for this paper because they meet national standards. Third, the selection process was performed through observation and expert opinion only. Therefore, using different weighting methods, such as an analytic hierarchy process, is suggested to facilitate the selection of more acceptable and justifiable weighting methods [18] and to make the selection method more scientific. Besides, since the buses being completely new, their future performance is uncertain. It would be useful to analyze the performance of the bus throughout its operation life-cycle.

5. Conclusions

The promotion of the bus electrification is in full swing in China, among which pure electric bus, with its advantages of high driving stability, zero emission and outstanding economic performance, has become the focus of many cities. However, it can't be denied that electric buses still have some limitations, e.g., short driving range and long charging time, which are tough to overcome in a short time. Besides, compared with traditional bus, pure electric bus is a relatively new concept, with a more complicated process of its selection, deserving more attention and research.

In this article, a test organized by the Nanjing Bus Company that is based on actual road tests to aid in bus selection was illustrated. Compared with other test methods, road tests were used because they are direct and reliable and they take into account actual transport conditions. The process of the tests can be categorized into three steps: (1) To structure the organization of the testing and selection processes, three groups were introduced: a leadership group, working group, and review group. (2) The test process and test conditions, including bus, route, staff and environmental information were provided for data collection. (3) As for the indicator processing, power consumption per 100 km, charging duration per 100 km, daily average driving distance and failure time per 100 km were determined to be key factors in the selection of a suitable bus.

In general, it is critical for the bus companies to decide to purchase the buses, and thus, the performance, cost, safety, environment and other indicators of electric buses with the support of operation data and the goal of normal operation service are of great significance to be quantified, the method proposed in this paper the bus company applied to make progress in electric bus selection thereby proves the application value, in terms of both process and indicators. In addition, since the test is based on the new buses, it is worth exploring the subsequent performance in the practical operation in the future research.

Author Contributions: Conceptualization, J.G. and J.H.; methodology, J.G. and J.H.; investigation, C.C. and H.Z.; data curation, C.C.; writing—original draft preparation, C.C. and X.Y.; writing—review and editing, M.K. and Z.H. All authors have read and agreed to the published version of the manuscript.

Funding: This research was funded by National Natural Science Foundation of China (Grant No. 51778141, 71874067) and Henan science and technology project (Grant No.182102310733).

Acknowledgments: The authors would like to thank Nanjing Bus Company for providing essential data and support. Their assistance is gratefully acknowledged.

Conflicts of Interest: The authors declare no conflict of interest.

References

1. Mulley, C.; Hensher, D.A.; Rose, J. Do preferences for BRT and LRT vary across geographical jurisdictions? A comparative assessment of six Australian capital cities. *Case Stud. Transp. Policy* **2014**, *4*, 1–9. [CrossRef]
2. Ministry of Transport of the People's Republic of China. *Statistical Communique for the Development of Transportation Industry*; Ministry of Transport of the People's Republic of China: Beijing, China, 2016.
3. Li, J.Q. Battery-electric transit bus developments and operations: A review. *Int. J. Sustain. Transp.* **2016**, *10*, 157–169. [CrossRef]
4. Xylia, M.; Silveira, S. On the road to fossil-free public transport: The case of Swedish bus fleets. *Energy Policy* **2017**, *100*, 397–412. [CrossRef]
5. Kendall, K.; Kendall, M.; Liang, B.; Liu, Z.X. Hydrogen vehicles in China: Replacing the western model. *Int. J. Hydrog. Energy* **2017**, *42*, 30179–30185. [CrossRef]
6. Kühne, R. Electric buses—An energy efficient urban transportation means. *Energy* **2010**, *35*, 4510–4513. [CrossRef]
7. Ministry of Finance of the People's Republic of China. *A Notification on Improvement of Subsidy Policy on Urban Buses Oil Price and Speeding up the Popularization and Application of New Energy Vehicles*; Ministry of Finance of the People's Republic of China: Beijing, China, 2015.
8. Ministry of Transport of the People's Republic of China. *A Notification of the Ministry of Transport on Accelerating the Implement and Application of New Energy Vehicles in the Transportation Industry*; Ministry of Transport of the People's Republic of China: Beijing, China, 2015.
9. Ministry of Transport of the People's Republic of China. *The Assessment Measures of New Energy Bus Application*; Ministry of Transport of the People's Republic of China: Beijing, China, 2015.
10. Wang, W.P.; Wang, B. An introduction of three kinds of testing methods of new energy automobile power system. *Shanghai Auto* **2014**, *2*, 8–11.
11. Zhang, J.; Zhao, L.; Hou, F.; Li, M.; Xu, Y. Technology evaluation of Chinese hybrid electric bus demonstration. *Mitig. Adapt. Starteg. Glob.* **2015**, *20*, 797–815. [CrossRef]
12. Ministry of Transport of the People's Republic of China. *General Technical Conditions for Pure Electric City Bus*; Ministry of Transport of the People's Republic of China: Beijing, China, 2016.
13. Wang, Z.P.; Zhang, Z.J. Study on the evaluation of energy consumption economy for electric vehicles. *Chin. High Technol. Lett.* **2017**, *17*, 171–174.
14. Wan, J.; Ji, J.Z. Analysis on Economy of New Energy Bus Considering Time Value of Funds. *J. Highw. Transp. Res. Dev.* **2015**, *32*, 154–158.
15. Li, Z.Q. New energy bus selection and market opportunity analysis about Urumqi. *Sci. Technol. Ind. Parks* **2017**, *11*, 8.
16. Hua, X. Two Chinese Cities Awarded in PARIS for Efforts to Tackle Climate Change. 2015. Available online: http://news.xinhuanet.com/english/2015-12/04/c_134886361.htm (accessed on 4 December 2015).
17. Nanjing Morning Post. Nanjing Became "A National Demonstration City for Transit Metropolis". 2011. Available online: http://njcb.xhby.net/mp2/pc/c/201711/26/c409288.html (accessed on 26 November 2011).
18. Law, T.H.; Daud, M.S.; Hamid, H.; Haron, N.A. Development of safety performance index for intercity buses: An exploratory factor analysis approach. *Transp. Policy* **2017**, *58*, 46–52. [CrossRef]

© 2020 by the authors. Licensee MDPI, Basel, Switzerland. This article is an open access article distributed under the terms and conditions of the Creative Commons Attribution (CC BY) license (http://creativecommons.org/licenses/by/4.0/).

energies

MDPI

Article

An Extremely High Power Density Asymmetrical Back-to-Back Converter for Aerospace Motor Drive Applications

Yifan Zhang [1], Chushan Li [1,2], David Xu [3], Wuhua Li [1], Jian Zhang [1,*], Hao Ma [1,2] and Xiangning He [1]

1 College of Electrical Engineering, Zhejiang University, Hangzhou 310081, China; zyf1125@zju.edu.cn (Y.Z.); chushan@intl.zju.edu.cn (C.L.); woohualee@zju.edu.cn (W.L.); mahao@zju.edu.cn (H.M.); hxn@zju.edu.cn (X.H.)

2 Zhejiang University–University of Illinois at Urbana-Champaign Institute, Zhejiang University, Haining 314400, China;

3 College of Electrical and Computer Engineering, Ryerson University, Toronto, ON M5B2K3, Canada; dxu@ryerson.ca

* Correspondence: jian_zhang_zju@zju.edu.cn

Received: 26 January 2020; Accepted: 3 March 2020; Published: 10 March 2020

Abstract: Higher-voltage-standard and higher-power-rating aerospace power systems are being investigated intensively in the aerospace industry to address challenges in terms of improving emissions, fuel economy, and also cost. Multilevel converter topologies become attractive because of their higher efficiency under high-voltage and high-switching-frequency conditions. In this paper, an asymmetrical-voltage-level back-to-back multilevel converter is proposed, which consists of a five-level (5L) rectifier stage and a three-level (3L) inverter stage. Based on the comparison, such an asymmetrical back-to-back structure can achieve high efficiency and minimize the converter weight on both rectifier and inverter sides. A compact triple-surface-mounted heatsink structure is designed to realize high density and manufacturable thermal management. This topology and structure are evaluated with a full-rating prototype. According to the evaluation, the achieved power density is 2.61 kVA/kg, which is 30% higher than that of traditional solutions. The efficiency at the rated power of the back-to-back system is 95.8%.

Keywords: more electric aircraft (MEA); multilevel converter; high power density

1. Introduction

The modern aerospace industry is facing challenges in terms of improving emissions, fuel economy, and also cost. The US National Aeronautics and Space Administration (NASA) has laid out ambitious goals for NO_x emissions and fuel consumption for the next three generations of subsonic aircraft, namely N + 1, N + 2, and N + 3 [1,2]. These goals are given in Table 1. The more electric aircraft (MEA) architecture is the state-of-the-art technology that was developed to address those challenges. The reduction of the total full consumption is about 3% [3]. However, such performance is far behind the targets. The essential reason is that, in one typical large commercial aircraft such as Boeing 777, 95% of jet fuel is used for generating propulsion thrust [4]. Hence, a radical propulsion system design is required to meet the abovementioned goals.

The so-called turboelectric distributed propulsion (TeDP) concept, which is now intensely investigated by research agents and companies such as NASA and Airbus [1,5], is considered as the key to meeting those goals. Under this concept, electric generators driven by gas turbine engines supply electrical power to multiple electric fans, which are distributed across the airframe to provide propulsion [2]. The major benefit of this concept is that the separation of gas turbine engines and propulsion fans enables both components to be designed in their optimal points and operated at optimum speeds. It is anticipated that, with the application of this technology coordinated with advanced body design, the fuel consumption can be reduced by 70%.

Table 1. National Aeronautics and Space Administration (NASA)'s goals for future-generation subsonic aircraft [2].

Corners of the Trade Space	N + 1 (2015) Conventional Tube & Wing (relative to B737)	N + 2 (2020) Unconventional Hybrid Wing Body (relative to B777)	N + 3 (2030) Advanced Aerospace Concepts (relative to B737)
Landing and takeoff NO_x emissions	−60%	−75%	−80%
Cruise NO_x emissions	−55%	−70%	−80%
Aircraft fuel/energy consumption	−33%	−50%	−60%

Under the TeDP concept, the total power capacity of the electrical system onboard expands to the megawatt (MW) level. Such high power raises lots of challenges to electrical system designs. Weight is the most critical design factor for all aerospace components, which is usually evaluated with a specific power (unit: kW/kg). It was reported that 1 kg saving will save roughly 1700 t of fuel and 5400 t of CO_2 per year for all air traffic and decrease system costs by US$1000 [6]. Compared to traditional aircraft propulsion systems, TeDP systems require additional high-power generators, distribution cables, and motor drive systems. All these components should achieve extremely high specific power, before TeDP systems can show its significant benefits. The research summarized in [7] showed that the traditional voltage standard of ±270 V for MEA is no longer suitable for this power rating and a higher voltage rating is preferred. In addition, the total weight of the overall electrical system can be optimized [8].

The power electronic converters are the key to enabling components in a TeDP architecture. They realize the power conversion between AC and DC powers. The designs of power electronic converters are quite essential in aerospace electric power systems. For motor drive systems onboard, power electronic parts include two converters: one rectifier and one inverter.

In a traditional MEA architecture, a rectifier usually is realized by a passive equipment called autotransformer rectifier unit (ATRU). Schematics of typical ATRUs can be found in [9,10]. They essentially consist of multiwinding autotransformers plus diode rectifiers. Although ATRUs are proved to be practical in MEA architectures, its power density is rather low because of heavy autotransformers. According to the survey of some commercial products, the specific power of an ATRU is around 1.6–3.3 kW/kg, depending on the power rating [11,12]. In an industrial motor drive system, active front end (AFE) rectifiers are used to achieve a better-quality AC input. By eliminating the use of autotransformers, the total weight of AFE rectifiers can be potentially lower than that of ATRUs. Furthermore, AFE rectifiers can bring potential energy-back capability and recover kinetic energy in some aerospace applications such as E-taxiing [13,14]. Therefore, evaluation and comparison should be carried out between an AFE-rectifier-based power converter and a converter with an ATRU rectifier to show which solution is better for TeDP architectures.

On the other side, two-level (2L) converters are the classic and also the only commercial solution for motor controllers in aerospace, before the TeDP concept emerged. The major reason is that, due to low DC link voltages, 2L converters are sufficient to realize high conversion efficiency. However, following the increase of the DC link voltage rating, 2L converters are no longer suitable for TeDP applications because of their low efficiency and low power density. The comparison in [15] showed that multilevel converters can achieve better efficiency and better power density in higher-power applications. Research activities related to topologies with three-level (3L) [16,17], five-level (5L) [18], and even higher-voltage-level [19] output were investigated. However, these researches did not compare different multilevel converter topologies with different voltage levels to show which topology is more favorable in certain applications. In addition, they did not investigate back-to-back configurations, which has more flexibility in selecting topologies on both rectifier and inverter stages. As a result, comparative work could be done to find the optimal solution based on a certain specification for aerospace applications. With the specification, an optimal design is also worth being investigated to show potential improvements that can be made compared to conventional designs.

In this paper, comparisons are made between different multilevel converter topologies. Based on the comparison results in terms of potential power density and power loss, a high-power-density power converter design for aerospace TeDP applications is demonstrated, which is named as the asymmetrical back-to-back converter topology. It utilizes a 5L converter in a rectifier stage and a 3L converter in an inverter stage. A compact triple-surface-mounted heatsink structure is designed to realize high density and manufacturable thermal management. This paper is organized as follows. In Section 2, the specification of a motor drive system for comparison and design is given and analyzed. In Sections 3 and 4, the topology evaluation and selection as well as the design for thermal solutions are demonstrated. In Section 5, the design is verified by electrical and thermal testing, where the results of the testing are shown. A final evaluation and comparison session are given in Section 6. Section 7 concludes the paper.

2. Introduction of the System Specification

A high-power-rating motor drive system for aerospace applications was assumed in this paper. The specification of the drive system is defined in Table 2. The power demand of the motor was 60 kW. Although in current MEA power systems, a 230-V AC grid is the only choice of power supply for high-power converters. However, in the near future, a 460-V AC grid becomes possible, which makes the DC link voltage up to ±400 V. With the help of this high-voltage DC link, 460-V AC motors can be selected to lower AC cable loss and decrease both cable and motor weights. According to the specification, the nominal AC currents on both the rectifier side and the motor side, the modulation index M, and the power factor were calculated, which is also shown in Table 2. Because that a 200% overloading capability was required by the motor side, 300-A IGBT modules were selected for converters on both sides. The switching frequency was set to be 15 kHz, which is a common selection in converter design for an aerospace motor drive.

Table 2. Specifications of converters for motor control.

Rectifier side	
Input power rating	≈60 kW each
Nominal AC voltage	230–460 V (M = 0.47–0.94)
AC-rated input current	150 A
Motor side (inverter side)	
Motor power rating	≈60 kW each
Inverter apparent power rating	≈80 kVA
Nominal AC voltage	460 V (M = 0.94)
Test-rated output current	100 A (with 200% overcurrent capability)
Converter specification	
IGBT rating	300 A
DC bus voltage	±400 V
Switching frequency	15 kHz

All these specifications present several challenges for the to-be-designed power electronic converter, which are listed as follows:

- High power density and high efficiency: The power electronic converters are the main heavy part of the drive system. Evaluation and comparison should be carried out to show which topology is better in terms of the power modules' size and weight. The efficiency is also important, since it also affects the size of the thermal system. On the other hand, the high temperature created by high-power losses affects the reliability of the converters.

- Density of the cooling system: the additional weights of tubes and liquids make the liquid-cooling system not attractive in aerospace. Air-cooling might be the only choice. Therefore, a compact heatsink design is required to minimize the total weight of the cooling system.
- Filter consideration: The weight and size of a large AC inductor on the grid side also need to be considered. In order to solve this problem, multilevel converters with higher voltage levels are preferred. The trade-off between voltage levels, filters' sizes and weights, and total power losses require to be made.

3. Evaluation of the High-Power-Density Converter Topology

The first step to develop a high-power density converter was to select the optimal converter topology combination for a back-to-back converter. It should potentially have the highest efficiency and power density while minimizing the size and weight of the AC-side filter. In this section, a comparison between different topologies, especially different multilevel converter topologies, was made. Power losses, modules' total sizes and weights, as well as filters' sizes and weights, were considered.

3.1. Topology Selection Consideration

Based on the specifications, several converter topologies, including multilevel converter topologies that are common in industrial drive applications, were selected as the candidate topologies for both the rectifier and the inverter. They were the two-level voltage-source converter (2L-VSC), the two-level H-bridge voltage-source converter (2L-HB), the three-level neutral-point-clamped converter (3L-NPC), the three-level T-type converter (3L-T^{2}C), and the five-level hybrid active neutral-point-clamped converter (5L-HANPC). Their schematics are given in Figure 1. Industrial drive systems based on these topologies are all mature, and products can be found such as Converteam VDM5000 (2L-VSC), Converteam MV7000 (3L-NPC), and ABB ACS 2000 (5L-HANPC) [20]. All these topologies have the same advantage, which is that only one DC source is required for a three-phase circuit. Therefore, a back-to-back configuration at the AFE stage could be built. On the contrary, widely used medium-voltage drive topologies such as cascaded H-bridge converters require multiple isolated DC sources that use zigzag transformers to create. This will significantly increase the system total weight. Another 3L solution is the three-level flying capacitor converter (3L-FC). However, compared to the 3L-NPC, it requires additional flying capacitors, which shows no benefit in terms of total weight. The reason why we considered the 5L-HANPC is that the 5L-HANPC is the only commercially available 5L converter with a single DC source. In addition, with the 5L operation, the filter size may be smaller, and the power loss may potentially be lower.

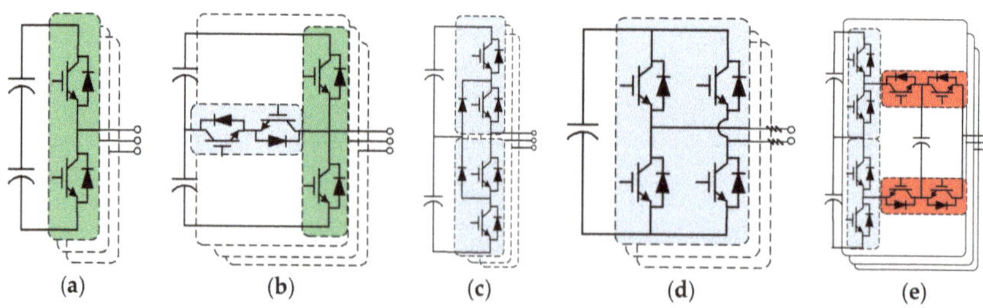

Figure 1. Demonstration of the candidate topologies: (**a**) two-level voltage-source converter (2L-VSC); (**b**) three-level T-type converter (3L-T^{2}C); (**c**) three-level neutral-point clamped converter (3L-NPC); (**d**) two-level H-bridge voltage-source converter (2L-HB); and (**e**) five-level hybrid active neutral-point-clamped converter (5L-HANPC). **Green**: 1700 V IGBT; **blue**: 1200 V IGBT; **red**: 650 V IGBT.

3.2. Device Selection Consideration

After the candidate topologies were determined, IGBT modules were then selected for each topology. In order to guarantee a reliable margin in an aerospace application, a 2/3 voltage derating fact was applied to IGBT modules. It meant that the standard of selection for a ±400 V DC system in aerospace is the same as for a ±600 V DC system in an industrial drive. In other words, this system can also operate under ±600 V DCs, if insulation can be guaranteed. Based on this criterion, the voltage rating of IGBTs in each topology is marked and shown in Figure 1.

Commercially available IGBT modules were considered. In order to guarantee a fair comparison, several types of IGBT modules from the same manufacturer were selected (Infineon in this paper). The results of the selection are given in Table 3. According to Table 3, with the increase of voltage level, the total number of modules to form one three-phase circuit increased. Twelve modules were required to form one three-phase 5L-HANPC circuit, which had the maximum number among all topologies. The total weight of these modules and the minimum heatsink surface area were calculated according to the number of required modules. These two results were quite important for power density evaluation. It is because all these IGBT modules accounted for a large portion of the total weight. On the other hand, the minimum heatsink surface area represented the rectangular baseplate area of all the IGBT modules added together. Although for most converter designs the heatsink surface area is much larger than this minimum area, a well-designed heatsink makes these two numbers similar. Under this condition, the heatsink weight is decided not only by the power loss, but also by how many modules are installed. Based on the comparison given in Table 3, without considering the power loss, the 2L-VSC would be the most suitable topology, since it had the smallest minimum heatsink area as well as the lowest total module weight. On the other hand, the 3L converter was a moderate solution, which showed a similar performance.

Table 3. Selection and comparison of active components for each candidate topology.

Topology	2L-VSC	2L-FB	3L-T^{2}C	3L-NPC	5L-HANPC
Module part number & rating	FF300R17ME4 (1700 V/300 A)	FF300R12ME4-B11 (1200 V/300 A)	FF300R17ME4 (1700 V/300 A) FF300R12KT3P_E (1200 V/300 A)	F3L300R12ME4 (1200 V/300 A)	FF300R12ME4-B11 (1200 V/300 A) FF300R07ME4 (650 V/300 A)
Type of IGBT modules	Half bridge	Half-bridge	Half-bridge & bidirectional switch	Half NPC bridge	Half-bridge
Total number of modules	3	6	6	6	12
Minimum heatsink surface area	454 cm^2	908 cm^2	846 cm^2	908 cm^2	1815 cm^2
Total modules weight	1035 g	2070 g	2010 g	2070 g	4140 g

3.3. Power Loss Comparison

As discussed before, the power loss performance is more critical in designing high-power-density converter systems. In order to evaluate the loss performance for each topology under different load conditions, the power loss on each device was calculated individually. The conduction loss and the switching loss were separately calculated.

A generalized equation to calculate an average conduction loss of an IGBT and a diode in one AC cycle was given by:

$$P_{con} = \frac{\int_{0}^{2\pi} v_{ce}(\theta) i_c(\theta) d_{con}(\theta) \mathrm{d}\theta}{2\pi} \quad (1)$$

where $v_{ce}(\theta)$ represents the on-state voltage drop on the IGBT, $i_{ce}(\theta)$ represents the instantaneous current going through the IGBT when it is turned on, and $d_{con}(\theta)$ indicates the duty cycle of the device. For diodes, $v_f(\theta)$ and $i_f(\theta)$ are used in Equation (1). The voltage drops, instantaneous current, and duty cycle are all variables related to the phase angle θ in one AC cycle. Furthermore, the on-state voltage drop characteristic can be modeled as a constant equivalent voltage source V_0 for both IGBTs and diodes in zero-current-condition series connected with an on-state resistor r_0. Equation (1) can be rewritten as:

$$P_{con} = \frac{\int_0^{2\pi} (V_0 + r_0 \cdot i_c(\theta))i_c(\theta)d_{con}(\theta)\mathrm{d}\theta}{2\pi} \tag{2}$$

The switching loss was calculated under the assumption that the relationships between the switching loss and the on-state current and off-state voltage were linear. From the datasheet, the turn-on energy loss E_{on} and the turn-off energy loss E_{off} of an IGBT and the reverse recovery energy loss E_{rec} of a diode under rated test conditions can be found. Then, by integrating the total energy loss within one output period and then multiplying it with the output frequency, Equation (3) was obtained as:

$$P_{sw} = \frac{\int_0^{2\pi} E_k \frac{i_c(\theta)}{I_{test}} \frac{v_c}{U_{test}} S_{sw}(\theta)\mathrm{d}\theta}{2\pi} f_{out}, S_{sw}(\theta) = \begin{cases} 1 & \text{device has switching loss} \\ 0 & \text{device has no switching loss} \end{cases} \tag{3}$$

where the switching energy loss coefficient E_k represents $E_{on} + E_{off}$ for the IGBT or E_{rec} for the diode, I_{test} and U_{test} are the rated test conditions noted on the datasheet, and $S_{sw}(\theta)$ indicates whether the IGBT or diode has the switching loss or not at a certain phase angle θ in one AC cycle. All topologies were evaluated in both the rated rectifier mode and inverter mode. The results in the inverter mode with the rated conditions are given in Figure 2. According to the results, the 2L-VSC showed a much higher total loss than the other multilevel topologies, because that the 1700 V IGBT module had a poor switching loss performance, which was identified by analyzing the datasheet of devices and the loss distribution. On the other hand, half of IGBT modules in the 3L-T^{2}C were 1700 V-type, and thus the total loss was also too high. Among all the topologies, 5L-HANPC had the lowest switching loss and the lowest total loss. The extremely low switching loss is not only because that the voltage step decreased to 1/4 of the DC link voltage, but also because the 650 V IGBT modules in the 5L-HANPC were the only devices under high-frequency switching after applying a commonly used low-loss modulation scheme. The results in the rectifier mode under the rated conditions are given in Figure 3. It can be found that the tendency kept the same and the 5L-HANPC was the one with the best performance.

Figure 2. Power loss comparison under ±400 V DC in the inverter mode (power factor: 0.76, $M = 0.94$, $f_s = 15$ kHz, and $I_{ac} = 100$ A).

Another comparison was made by fixing the modulation index and power factor in the rated conditions while varying the switching frequency or output current. The results in the inverter mode and the rectifier mode are shown in Figures 4 and 5, respectively, and showed that changing output currents did not affect the tendency since the loss was almost proportional to the output current for all topologies. However, after the switching frequency was decreased to be lower than 7 kHz, the power losses of the 3L-NPC and the 2L-FB became lower than that of the 5L-HANPC. It is because, under low switching frequencies, the conduction loss dominated the total loss and the 5L-HANPC had the highest conduction loss.

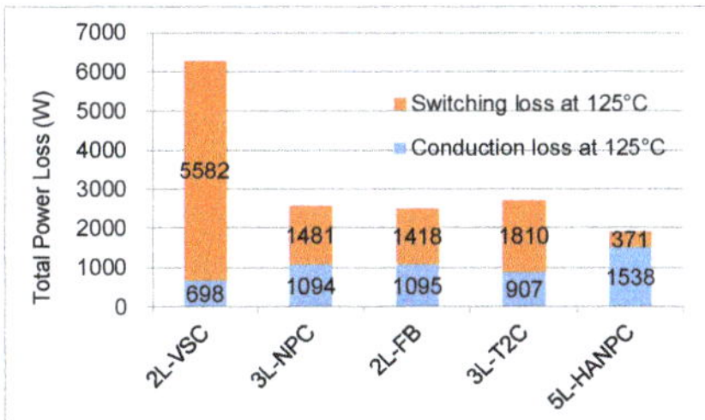

Figure 3. Power loss comparison under ±400 V DC in the rectifier mode (power factor: −1, M = 0.47, f_s = 15 kHz, and I_{ac} =150 A).

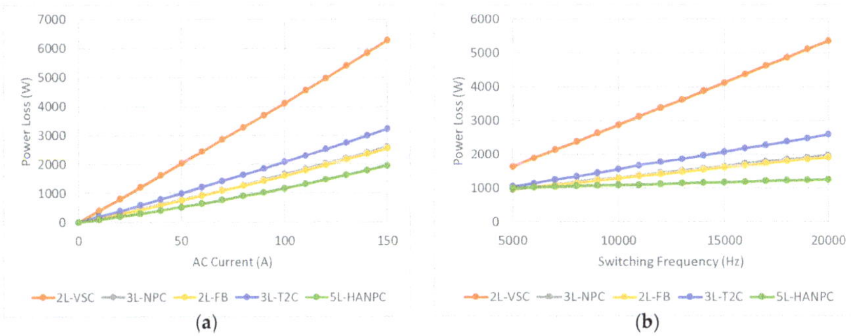

Figure 4. Loss performance comparison in the inverter mode (power factor = 0.76, M = 0.94) under different conditions: (**a**) varied AC currents and a fixed switching frequency (f_s = 15 kHz); and (**b**) varied switching frequencies and a fixed AC current (I_{ac} = 100 A).

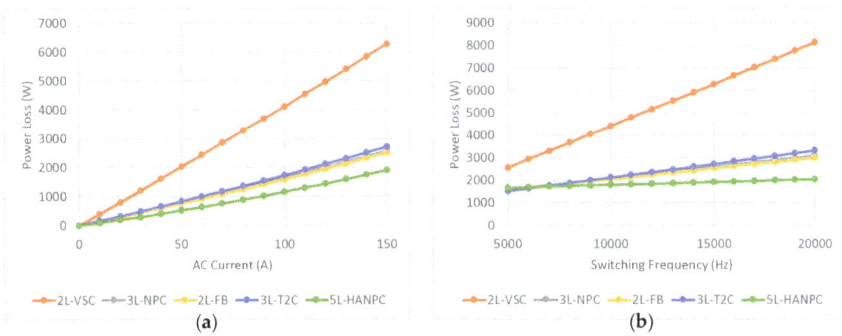

Figure 5. Loss performance comparison in the inverter mode (power factor = −1, M = 0.47) under different conditions: (**a**) varied AC currents and a fixed switching frequency (f_s = 15 kHz); and (**b**) varied switching frequencies and a fixed AC current (I_{ac} = 100 A).

By summarizing the results of the loss comparison, a clear tendency can be discovered. The 2L-VSC was unacceptable because of huge power losses. At the selected operational frequency of 15 kHz, its total loss was 2–3 times higher than those of the other multilevel converter solutions. On the other hand, compared to the 3L-NPC, 2L-FB, and 5L-HANPC, the 3L-T^2C showed no advantages. Its total power loss was 75% higher than those of the 5L-HANPC and 27% higher than those of the 3L-FB and the 2L-FB under the inverter-mode operation. Considering that it also showed no benefit in module weight and total area comparisons according to Table 3, it will not be considered in further evaluation. The 5L-HANPC should be the first choice considering the loss performance, because its total loss was reduced by 30% compared to that of our second choice the 3L-NPC.

3.4. Summary of the Comparison

According to the loss comparison results, the 5L-HANPC was the first choice. Another advantage is that the 5L-HANPC may decrease the volume and weight of an AC inductor, which is necessary for rectifiers. However, according to the comparison in Table 3, more IGBTs were required to install, which in turn increased the surface area of the heatsink and required additional flying capacitors in the 5L-HANPC. On the other hand, by considering both the number of modules and the total loss, we found that the 3L-NPC and the 2L-FB could also be the favorable choices. However, the 2L-FB must connect to motors with three separated stator windings, and thus a specially designed motor was required. As a result, the 3L-NPC could be the best choice, especially for the inverter stage, where no AC-side inductor was required.

Based on the previous discussion, some possible back-to-back converter configurations were discussed. A very interesting fact is that the rectifier stage and the inverter stage did not have to apply converters with the same voltage level. It is because on the rectifier stage, increasing the voltage level can decrease the volume and weight of an AC inductor. On the contrary, no AC-side inductor was required on the inverter stage. Decreasing the voltage level can minimize the number of IGBT modules. Here come three potential solutions as well as one traditional solution used for comparison:

1) Traditional solution: an ATRU and a 2L-VSC;
2) An ATRU and a 3L-NPC;
3) A 3L-NPC and a 3L-NPC;
4) A 5L-HANPC and a 3L-NPC.

Among the three potential solutions, the 3L-NPC should be the only choice for inverter-stage circuits, since it had the highest power density, where no AC inductor was required. For the rectifier stage, a 5L-HANPC, a 3L-NPC, and a mature ATRU circuit are all possible choices. An asymmetrical back-to-back converter topology was proposed and given in Section 4. It aims to optimize the system's total weight on both the rectifier stage and the inverter stage. However, the loss calculation cannot tell which one is better, and only prototype design can clearly show the final winner.

4. Design of a High-Power-Density Converter Prototype

In our research, a 5L-HANPC and 3L-NPC (5L-3L) asymmetrical voltage-level multilevel back-to-back converter prototype was built to evaluate the power density of the power converter. The schematic of this prototype is demonstrated in Figure 6. The specification of the prototype, and the selection results of all main components are given in Table 4. The design of the volume of the passive components followed the criterion, which is for selecting the minimum volume while satisfying requirements of voltage ripples or current ripples.

From the selection results, it can be found that, compared to the DC link capacitor, the flying capacitor required a much larger capacitance. It is because that each flying capacitor was connected in a single phase. A large AC current went through each capacitor directly. Film-type capacitors were applied to both the DC link capacitor and the flying capacitor because of their high reliability, especially because of their open-circuit failure mode. The three-phase AC inductor was designed to have an electrical inductance of 66 μH and a rated input current of 150 A. The inductance was small compared to the traditional voltage source inverter (VSI) design, because the rectifier stage used a 5L converter.

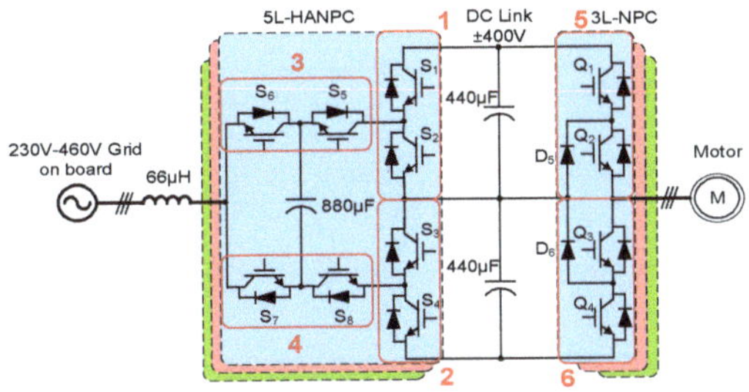

Figure 6. Schematic of a 5L-HANPC and 3L-NPC converter prototype.

Table 4. Specifications and components selection results of the 5L-HANPC and 3L-NPC prototype.

	Specification of the DC link	
DC bus voltage	±400 V	
DC bus capacitor volume	EPCOS (TDK) B32796E2226K	Each half: 440 μF/630 V
	Specification of the rectifier stage	
Grid side line-line AC voltage	$230\ V_{rms}$	
Rated power	60 kW	
Rated input current	150 A	
IGBT type	Main bridge	FF300R12ME4 (1200 V, 300 A)
	Output bridge	FF300R07ME4_B11 (650 V, 300 A)
Flying capacitor volume	EPCOS (TDK) B32778G4117K	Each phase: 880 μF/450 V
AC Inductor	Inductance: 66 μH; current: 150 A	
Switching frequency	15 kHz	
	Specification of the inverter stage	
Rated power	60 kW	
Converter power rating	≈80 kVA	
Rated motor current	100 A	
IGBT type	F3L300R12ME4_B22 (1200 V, 300 A)	
	F3L300R12ME4_B23 (1200 V, 300 A)	
Switching frequency	15 kHz	

The design of a heatsink for the air-cooling system is one of the most challenging parts of the prototype design. According to Table 3, there were 18 IGBT modules; the 3L-NPC was composed of six IGBT modules, and the 5L-HANPC was composed of the other 12 modules. Each converter had a different loss performance. Furthermore, each IGBT module had its own loss. This led to different temperature increases if we used a common heatsink design, where all IGBT modules are mounted on one surface and have an almost identical thermal resistance from the junction to the ambient. On the contrary, an ideal converter thermal system design is to let all IGBTs have the same temperature rise, and thus, the converter can make full use of the cooling system capacity.

In order to design a heatsink structure that can fully exploit the capability of the cooling system, the loss distributions among all IGBT modules under the rated conditions were calculated. The results are given in Table 5. The numbers of modules and switches are kept the same as what are illustrated in Figure 6. It can be found that the high-frequency switches of IGBTs S_5–S_8 in the 5L-HANPC and those of IGBTs in the 3L-NPC had higher power losses. The low-frequency switches of IGBTs S_1–S_4 in the 5L-HANPC had lower power losses. Based on this loss distribution, an innovative heatsink structure was investigated. IGBT modules with higher power loss were placed on the two side surfaces. The IGBT modules with less power loss were placed on the top surface. All fins of the heatsink were arranged to be parallel with the top surface, and thus the two side surfaces could have a better convection area. It is named as triple-surface-mounted heatsink structure, as shown in Figure 7.

Table 5. Theoretical power losses on all IGBT modules.

Topology	Module Number	Switch Number	Power Loss (W)
5L-HANPC	1	S_1, S_2	88.8
	2	S_3, S_4	88.8
	3	S_5, S_6	229.2
	4	S_7, S_8	229.2
3L-NPC	5	Q_1, Q_2, D_5	272.9
	6	Q_3, Q_4, D_6	272.9

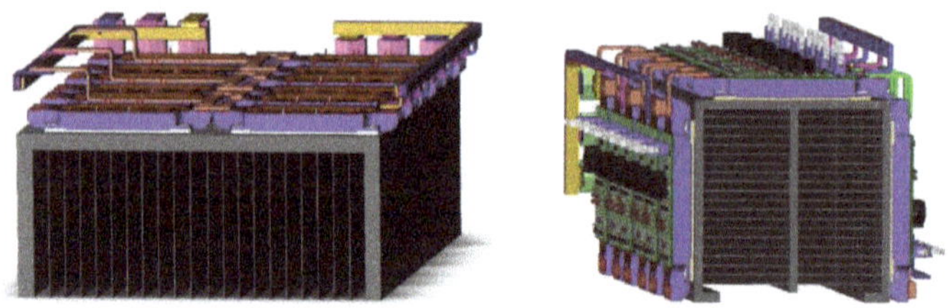

(**a**) (**b**)

Figure 7. Heatsink design: (**a**) traditional planar structure; and (**b**) proposed triple-surface structure.

The thermal performance of the proposed heatsink structure was verified by the finite element analysis (FEA). Two important questions were answered: (1) How large were the optimized areas of the side and the top surfaces? (2) Did the temperature rises on each IGBT module become even? Figure 8 shows the FEA results of the heatsink version I, where the IGBT modules were placed side by side on three surfaces and the width of one surface was equal to the length of one IGBT module. According to the results, the average temperature rises on the heatsink surface did not exceed the upper limit. It meant that heat dissipation capability was strong enough even if the heatsink surface area was just equal to the total bottom side area of the IGBT modules. However, the hottest part was the top surface area near the outlet of the cooling air flow. The temperature difference between the top surface and the side surface was higher than 15 °C. It meant that the top surface area without any fin was hard to dissipate the power loss even if the IGBT modules on the top surface had much less loss.

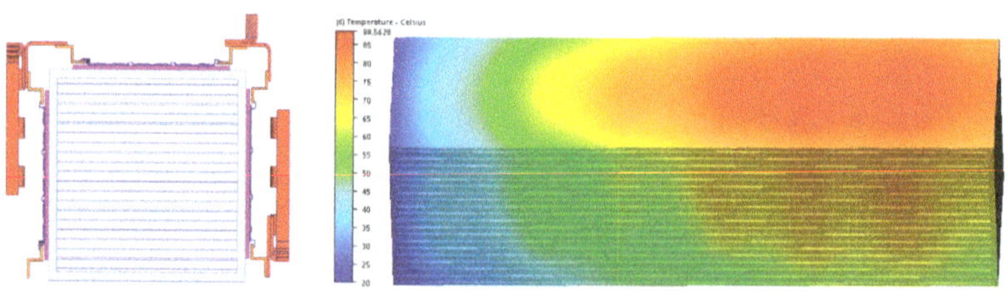

(**a**) (**b**)

Figure 8. (**a**) Heatsink version I; and (**b**) its finite element analysis (FEA) results (maximum temperature on the heatsink surface: 89 °C).

An improved heatsink version II is demonstrated in Figure 9. A thick plate was soldered to the center of the top surface. In this way, part of thermal flow conducted from the top surface to the center plate and was brought out by the airflow. In addition, the bottom side surface, which had a minimum effect on cooling, was removed to further save the total weight. According to the FEA results for version II, the temperature rise on each surface became almost even.

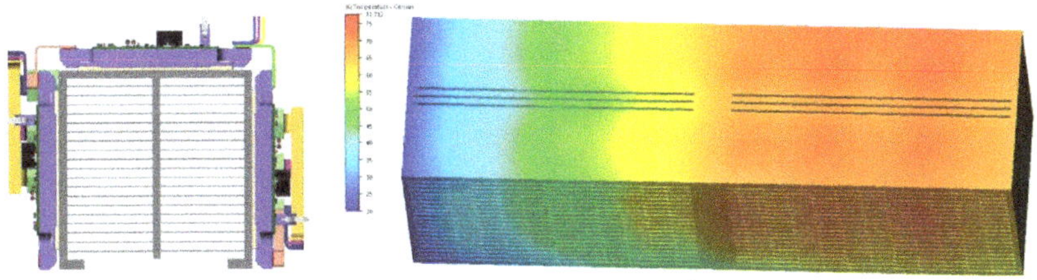

(a) (b)

Figure 9. (**a**) Improved heatsink version II; and (**b**) its FEA results (maximum temperature on the heatsink surface: 78 °C).

The final structure design and the picture of this back-to-back prototype are demonstrated in Figure 10. The switches of IGBTs S_5–S_8 in the 5L-HANPC and those of IGBTs in the 3L-NPC were placed on the two side surfaces of the heatsink version II. The switches of IGBTs S_1–S_4 in the 5L-HANPC were placed on the top surface. The IGBT modules were connected by a copper bar. On the top of the IGBT modules at the two side surfaces, the DC link capacitors and the flying capacitors were fixed by using aluminum stand-offs. The top area was the total surface area of the PCBs for gating, sampling, display, and auxiliary power supply. This structure not only fully made use of the space, but also saved a large amount of weight by decreasing the size of the heatsink and the length of the connecting copper bar compared to structures, where all IGBTs are placed on one surface.

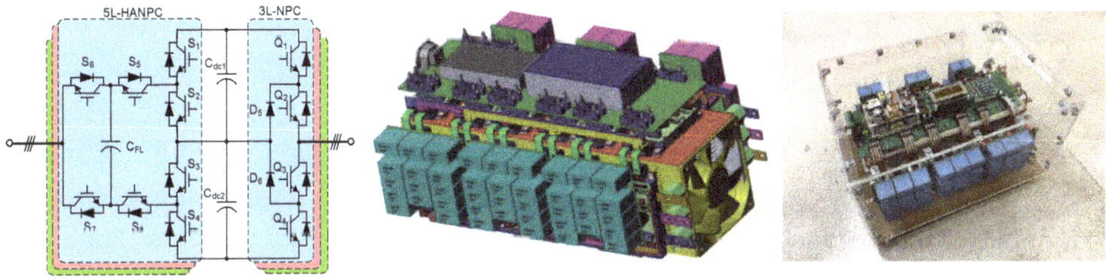

(a) (b) (c)

Figure 10. (**a**) Schematic, (**b**) three-dimensional (3D) structure, and (**c**) photograph demonstration of the converter prototype design.

Table 6 shows the weight distribution of the prototype. The heatsink, IGBT modules, and the AC inductor (not shown in the converter structure) were the main heavy parts. It should be noticed that the flying capacitors with their PCBs used by the 5L-HANPC also contributed to 9.77% of the total weight. This somehow affected the benefit of the 5L-HANPC applied to aerospace.

Table 6. Weight distribution analysis for the converter prototype.

Parts	Weight (kg)	Percentage (%)
Heatsink	9.6	31.27%
IGBT	6.2	20.20%
AC inductor (inductance: 66 μH; current: 150 A)	5.9	19.22%
DC link capacitor	2.0	6.51%
Flying capacitor	3.0	9.77%
Other components	4.0	13.03%
Total weight	30.7	-

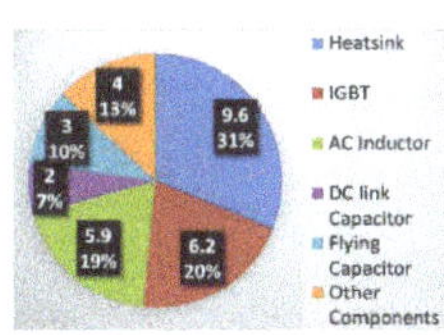

5. Verification of the High-Power-Density Converter Design

After the 5L-3L back-to-back converter prototype was assembled, it was tested under different load conditions to verify two important questions. One was whether the loss estimation was correct or not. Correct loss estimation demonstrated correct topology selection. The other was whether the proposed cooling system structure could dissipate the power loss generated by the converter operation or not.

In order to test the power loss of the converter, each side of the converter was tested in the inverter mode, where the DC link voltage was connected to an 800 V DC source and the AC output was connected to a purely inductive load. The test set is given in Figure 11. The DC-side efficiency was tested by using a power analyzer Norma D 5000 (Fluke, Everett, WA, USA). By using this method, each side of the converter was powered to full-load conditions, while the DC source only provided a small amount of power, which was equal to the power losses on both converters and inductive loads. During the test, by varying the modulation indices M of both sides of the converter, the output current was changed from zero to its maximum point. Then, the total power losses were recorded. The power losses of the inductive loads were estimated with the models described in [21,22]. By deducting the power losses on inductive loads, the power losses on converters were finally derived.

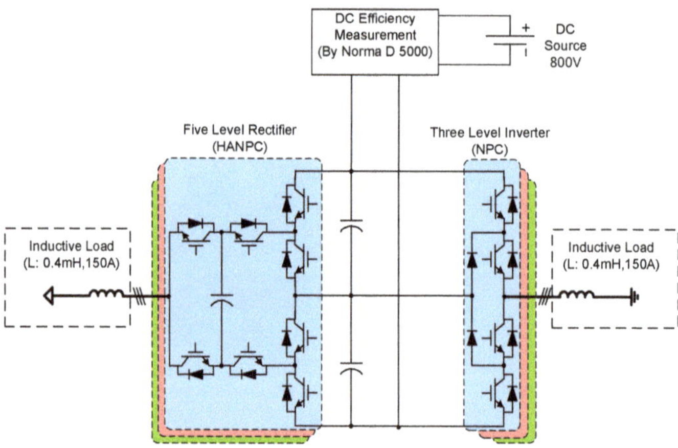

Figure 11. Experimental setup for the full-power test.

Figure 12a shows tested losses and estimated losses for the 3L-NPC under different load conditions. The results showed that the estimation of the conduction loss was quite precise within the whole-load condition. The estimation of the switching loss showed a slightly larger error than that of the conduction loss. Nevertheless, the accuracy was accepted for topology evaluation. Figure 12b shows tested losses and estimated losses for the 5L-HANPC under different load conditions. The results also showed that the estimation of the power loss was correct.

The thermal system performance was verified by investigating the temperature rise on each negative temperature coefficient (NTC) sensor embedded in IGBT modules. There were totally 18 NTC resistors belonging to 18 IGBT modules. The ambient temperature T_a for testing was 20 °C, and the air flow was 120 cubic feet per minute (CFM). The converter was tested under full-load conditions: $V_{dc} = \pm 400$ V; $I_{ac} = 100$ A. The data received from the NTC temperature sensor are plotted in Figure 13. From the results, it can be found that the distributions of the temperature were similar in all three surfaces. On the other hand, the 3L-NPC modules had the highest temperature rise. This is because the power loss per each module in the 3L-NPC was still larger than the power loss per module in the 5L-HANPC.

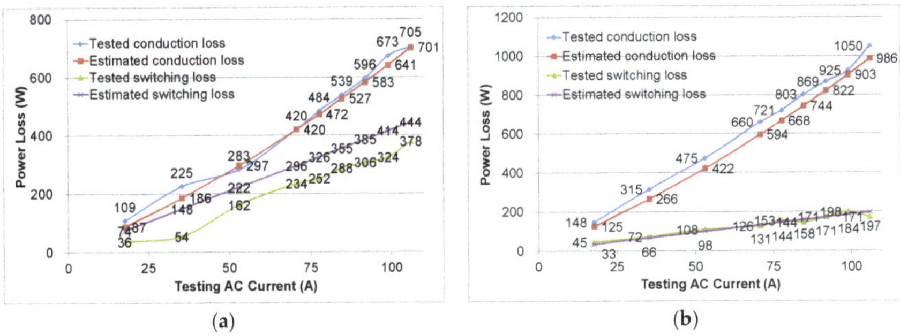

Figure 12. Comparison of tested and estimated power losses for the (**a**) 3L-NPC and (**b**) 5L-HANPC.

Figure 13. Temperature rise on negative temperature coefficient (NTC) sensors of IGBT modules (V_{dc} = ±400 V, I_{ac} = 100 A).

6. Evaluation and Comparison

With the developed 5L-3L back-to-back converter prototype, not only the power density and the efficiency performance of this prototype could be measured, but the other two possible solutions, which were the ATRU/3L-NPC and the 3L-3L back-to-back converter, could also be evaluated by estimating the weight of each main parts and also estimating the power losses according to the calculation derived in Section 3.

Based on the similar commercially available products on the market, the weight of a 60 kW ATRU was aggressively estimated at 20 kg, which is given in Table 7. On the other hand, the AC-input inductor used in the 3L-3L back-to-back converter should be larger than the inductor used in the 5L-3L back-to-back converter prototype. According to the calculation, an inductor with an electrical inductance of 100 µH and an AC current of 150 A was adopted. The weight of this inductor was also estimated and is given in Table 7.

Table 7. Weight estimations for magnetic components used in different configurations.

Parts	Weight (kg)
ATRU (estimated power rating: 60 kW)	20
Inductor (inductance: 60 µH; rated AC current: 150 A)	5.9
Inductor (estimated inductance: 100 µH; rated AC current: 150 A)	8.9

Besides the magnetic components, the weights of the other parts were also estimated. If the 5L-HANPC was replaced with the 3L-NPC on the rectifier part, the weight differences would mainly come from the elimination of six IGBT modules and flying capacitors and from the increased size of the AC Inductor. The weight distribution of each part for the 3L-3L back-to-back converter is shown in column IV of Table 8. If the 5L-HANPC was replaced with the 60 kW ATRU, the heatsink weight could be decreased by half since only the inverter part required the heatsink, and also the heat dissipation roughly decreased by half. All 5L IGBT modules and flying capacitors were eliminated. The weights of other components, including the auxiliary supply, copper bars, and control PCBs, was estimated to decrease 1 kg. By adding the weight of the ATRU, the total weight of the ATRU-3L converter is shown in column II of Table 8.

Table 8. Power density comparison between three possible system configurations.

System Configuration	ATRU-2L	ATRU-3L	5L-3L	3L-3L
Heatsink	-	4.8	9.6	9.6
IGBT	-	2.07	6.2	4.14
AC inductor	-	-	5.9	8.9
DC link capacitor	-	2.0	2.0	2.0
Flying capacitor	-	-	3.0	-
Other components	-	3.0	4.0	4.0
Two-level converter	21.8	-	-	-
ATRU	20.0	20.0	-	-
Total weight	41.8	31.87	30.7	28.64
Power density at 80 kVA	1.91 kVA/kg	2.51 kVA/kg	2.61 kVA/kg	2.79 kVA/kg
Power efficiency at 80 kVA	93.1%	96.0%	95.8%	95.0%

Since according to our testing results the calculation of the power losses matched the tested power losses very well, the calculation was proved to be quite accurate. The estimation of the power losses for each converter also followed the results derived in Figures 4 and 5. The power losses of the ATRU were not directly calculated. From the product brochure [12], the maximum efficiency was 98%.

The final comparisons in terms of total weight, power density, and expected efficiency are given in Table 8. In order to clearly show the benefit of the proposed solutions, the weight performance of the traditional ATRU-2L-VSC solution is also given in Column I of Table 8. The weight of a commercially available 50 kW 2L-VSC inverter product for aerospace with liquid cooling was found to be 21.8 kg. Thus, the total weight of the ATRU-2L-VSC system was 41.8 kg. It should be noticed that the ATRU used in the comparison was also only designed for the liquid-cooling system, which actually is not fully available in this application area. An air-cooled ATRU should have an even heavier weight.

The comparative results showed that the 3L-3L back-to-back converter was a better solution in terms of power density among the three potential solutions. The reasons are as following: Firstly, the heavy ARTU was replaced by the power converter, while at the same time a good balance between the number of active components and the volume of reactive components was found. The highest power density achieved was 2.79 kVA/kg, if we chose the apparent power rating of the machine (80 kVA) as the base power. The power density of the proposed 5L-3L back-to-back converter was 2.61 kVA/kg. It was also quite high, which was 30% higher than that of the traditional solution ATRU-2L converter. By considering the power efficiency, the proposed 5L-3L back-to-back converter showed a better performance compared to the 3L-3L back-to-back converter. The efficiency at the rated power of the back-to-back system, including the rectifier and inverter stages, was 95.8%. This is because of the highly efficient 5L converter. Furthermore, power density improvement can be made for this asymmetrical-voltage-level topology. Since the switching losses of the 5L rectifier stage were pretty low, if the higher switching frequency was selected, it could use smaller flying capacitors and smaller AC side filters, which would in turn decrease the total weight of the system.

7. Conclusions

According to our work, it can be concluded that the power density of a converter for a high-power-rating motor drive system in aerospace can be improved by replacing an ATRU with the power converter. By properly designing the heatsink structure, the multisurface cooling system can achieve considerably low weight and small size. Eventually, the heatsink size and weight were limited by the number and area of IGBT modules, if the converter topology had high efficiency. By evaluating the prototype performance, the finally achieved power density of the 5L-3L back-to-back converter prototype was 2.61 kVA/kg, which was 30% higher than that of the traditional solution. The efficiency could be up to 95.8%. The expected power density of the 3L-3L back-to-back converter prototype was 2.79 kVA/kg.

In the high-power motor drive system for aerospace applications, a multilevel-converter-based system showed advantages including fewer power losses and higher power density compared to transformer-based systems. In fact, in other applications with the TeDP architecture, multilevel converters can also provide advanced solutions with similar benefits. More importantly, multilevel-converter-based solutions can provide more flexibility. For example, a multilevel-converter-based rectifier provides a DC mid-point, which can potentially connect to the ground point in aircraft. Properly designing a modulation scheme can alleviate common-mode problems. In addition, when the motor speed is low, a converter-based rectifier can lower the DC bus voltage to minimize the switching loss of the system. With all these benefits, it is believed that in the near future, multilevel-converter-based rectifier–inverter power units will realize their wide applications in TeDP architectures.

Author Contributions: C.L. and J.Z. proposed the new back-to-back configuration and developed the program to control the converters. Y.Z. prepared the draft of the paper and carried out the experiment. D.X. contributed his expertise to the proposed subject of research and supervised the work when the first author was in Canada. W.L., H.M., and X.H. contributed their expertise to the proposed subject of research and supervised the work in China. All authors have read and agreed to the published version of the manuscript.

Funding: This work is sponsored by the National Nature Science Foundation of China (51807176 and U1834205), the National Key Research and Development Program of China (2017YFE0112400) and the Zhejiang Provincial Key Research and Development Program (2018C01SA150059).

Conflicts of Interest: The authors declare no conflicts of interest.

References

1. Marty, K.B.; Christopher, K.D. *Subsonic Ultra Green Aircraft Research: Phase I Final Report*; National Aeronautics and Space Administration: Washington, DC, USA, 2011.
2. Karen, D.; Patrick, N.; Catherine, J.; Stuart, G.; Mark, H. A review of Turboelectric Distributed Propulsion technologies for aircraft electrical systems. In Proceedings of the 2013 48th International Universities' Power Engineering Conference (UPEC), Dublin, Ireland, 2–5 September 2013; pp. 1–5.
3. Mike, S. 787 no bleed systems: Saving fuel and enhancing operational efficiencies. *Aero Q.* **2007**, *18*, 6–11.
4. Pat, W.; Sergei, B. The More Electric Aircraft: Technology and challenges. *IEEE Electrif. Mag.* **2014**, *2*, 6–12.
5. Frederick, B.; Joseph, P.; Paul, M.; Mark, H.; Graham, D. HTS electrical system for a distributed propulsion aircraft. *IEEE Trans. Appl. Supercond.* **2015**, *25*, 1–5.
6. Xavier, R. New trends and challenges of electrical networks embedded in more electrical aircraft. In Proceedings of the 2011 IEEE International Symposium on Industrial Electronics (ISIE), Gdansk, Poland, 27–30 June 2011; pp. 26–31.
7. Ralph, H.J.; Cheryl, B.; Amy, J.; Rodger, D.; James, F. Overview of NASA Electrified Aircraft Propulsion (EAP) Research for Large Subsonic Transports. In Proceedings of the 53rd AIAA/SAE/ASEE Joint Propulsion Conference, Atlanta, GA, USA, 10–12 July 2017.
8. Kevin, R.A.; Mark, D.G.; Ty, V.M.; Douglas, P.W. Mission Analysis and Aircraft Sizing of a Hybrid-Electric Regional Aircraft. In Proceedings of the 54th AIAA Aerospace Sciences Meeting, San Diego, CA, USA, 4–8 January 2016.
9. Jitendra, S.; Norbert, F.; Joachim, B.; Andreas, A.; Peter, W. High-current variable-voltage rectifiers: State of the art topologies. *IET Power Electron.* **2015**, *8*, 1068–1080.
10. Shahbaz, K.; Gao, Z.H.; Husan, A.; Kashif, H.; Izhar, U.H. Comparative analysis of differential delta configured 18-pulse ATRU. In Proceedings of the 2015 International Conference on Electrical Systems for Aircraft, Railway, Ship Propulsion and Road Vehicles (ESARS),, Aachen, Germany, 3–5 March 2015; pp. 1–6.
11. Latest Solutions to Meet Power Conversion Needs on the More Electric Aircraft. Available online: http://www.craneae.com/ (accessed on 1 January 2020).

12. Thales Avionics Electrical Systems Catalog. Available online: https://www.thalesgroup.com/en/activities/aerospace/electrical-systems (accessed on 1 January 2020).
13. Ganev, E.; Chiang, C.; Fizer, L.; Johnson, E. Electric Drives for Electric Green Taxiing Systems. *SAE Int. J. Aerosp.* **2016**, *9*, 62–73. [CrossRef]
14. Maximilian, T.E.H.; Fabian, K.; Pierre, M.; Ali, E. Regenerative Braking Capability Analysis of an Electric Taxiing System for a Single Aisle Midsize Aircraft. *IEEE Trans. Transp. Electrif.* **2015**, *1*, 298–307.
15. Schweizer, M.; Friedli, T.; Kolar, J.W. Comparative Evaluation of Advanced Three-Phase Three-Level Inverter/Converter Topologies Against Two-Level Systems. *IEEE Trans. Power Electron.* **2013**, *60*, 5515–5527. [CrossRef]
16. Zhang, D.; He, J.; Pan, D. A Megawatt-Scale Medium-Voltage High Efficiency High Power Density "SiC+Si" Hybrid Three-Level ANPC Inverter for Aircraft Hybrid-Electric Propulsion Systems. *IEEE Trans. Ind. Appl.* **2019**, *55*, 5971–5980. [CrossRef]
17. Zhang, D.; He, J.; Pan, D.; Dame, M.; Schutten, M. Development of Megawatt-Scale Medium-Voltage High Efficiency High Power Density Power Converters for Aircraft Hybrid-Electric Propulsion Systems. In Proceedings of the 2018 AIAA/IEEE Electric Aircraft Technologies Symposium (EATS), Cincinnati, OH, USA, 12–14 July 2018; pp. 1–5.
18. Kashihara, Y.; Itoh, J. Performance evaluation among four types of five-level topologies using Pareto front curves. In Proceedings of the Energy Conversion Congress and Exposition (ECCE), Denver, CO, USA, 15–19 September 2013; pp. 1296–1303.
19. Pallo, N.; Foulkes, T.; Modeer, T.; Coday, S.; Pilawa-Podgurski, R. Power-dense multilevel inverter module using interleaved GaN-based phases for electric aircraft propulsion. In Proceedings of the 2018 IEEE Applied Power Electronics Conference and Exposition (APEC), San Antonio, TX, USA, 4–8 March 2018; pp. 1656–1661.
20. Samir, K.; Jose, R.; Bin, W.; Steffen, B.; Marcelo, P. Powering the Future of Industry: High-Power Adjustable Speed Drive Topologies. *Ind. Appl. Mag.* **2012**, *18*, 26–39.
21. Roßkopf, A.; Bär, E.; Joffe, C. Influence of Inner Skin- and Proximity Effects on Conduction in Litz Wires. *IEEE Trans. Power Electron.* **2014**, *29*, 5454–5461. [CrossRef]
22. Muhlethaler, J.; Biela, J.; Kolar, J.W.; Ecklebe, A. Improved Core-Loss Calculation for Magnetic Components Employed in Power Electronic Systems. *IEEE Trans. Power Electron.* **2012**, *27*, 964–973. [CrossRef]

© 2020 by the authors. Licensee MDPI, Basel, Switzerland. This article is an open access article distributed under the terms and conditions of the Creative Commons Attribution (CC BY) license (http://creativecommons.org/licenses/by/4.0/).

energies

MDPI

Article

Optimal Torque Distribution Control of Multi-Axle Electric Vehicles with In-wheel Motors Based on DDPG Algorithm

Liqiang Jin [1], Duanyang Tian [1,*], Qixiang Zhang [1] and Jingjian Wang [2]

1 State Key Laboratory of Automotive Simulation and Control, Jilin University, Changchun 130022, China; jinlq@jlu.edu.cn (L.J.); zhangqx19@mails.jlu.edu.cn (Q.Z.)
2 College of Computer Science and Electronic Engineering, Hunan University, Changsha 410082, China; wangjjhun@163.com
* Correspondence: tiandy19@mails.jlu.edu.cn; Tel.: +86-131-9435-9135

Received: 7 February 2020; Accepted: 10 March 2020; Published: 13 March 2020

Abstract: In order to effectively reduce the energy consumption of the vehicle, an optimal torque distribution control for multi-axle electric vehicles (EVs) with in-wheel motors is proposed. By analyzing the steering dynamics, the formulas of additional steering resistance are given. Aiming at the multidimensional continuous system that cannot be solved by traditional optimization methods, the deep deterministic policy gradient (DDPG) algorithm for deep reinforcement learning is adopted. Each wheel speed and deflection angle are selected as the state, the distribution ratio of drive torque is the optimized action and the state of charge (*SOC*) is the reward. After completing a large number of training for vehicle model, the algorithm is verified under conventional steering and extreme steering conditions. The maximum *SOC* decline of the vehicle can be reduced by about 5% under conventional steering conditions based on the motor efficiency mapused. The combination of artificial intelligence technology and actual situation provides an innovative solution to the optimization problem of the multidimensional state input and the continuous action output related to vehicles or similar complex systems.

Keywords: electric vehicles (EVs); independent-drive technology; deep reinforcement learning (DRL); optimal torque distribution

1. Introduction

The vehicles independently driven by in-wheel motors removes the transmission system of traditional vehicles and the drive torque of each wheel is independently controllable. Besides, the information such as the motor torque and speed can accurately feedback in real-time, so that the transmission efficiency of the vehicle is greatly improved and the layout design becomes more flexible. More importantly, the driving form has significant advantages in terms of stability control, active safety control and energy saving control [1,2], which is a huge attraction for multi-axle heavy vehicles. However, battery technology has always been one of the key issues limiting the development of pure electric vehicles [3]. For heavy vehicles, both the demand and consumption of energy are greater, which means the energy problem is more serious. In the case that the existing battery core technology cannot be solved temporarily, it is necessary to adopt an energy-saving control strategy for the electric vehicle, especially the multi-axle heavy-duty electric vehicle [4].

At present, the energy-saving driving control strategy for electric vehicles is mainly based on three aspects: motor control energy saving, energy feedback and traction control energy saving. The energy-saving of the motor is mainly based on the motor efficiency characteristic curve [5,6], aiming at the optimal system efficiency, and changing the actual working point of each motor by adjusting the front and rear axle torque distribution coefficients to avoid working in the low-efficiency zone, but this method is often only for straight-line driving conditions. Energy feedback mainly refers to regenerative braking technology, which hopes to maximize the recovery of braking energy by using different control strategies during vehicle braking [7–9]. In terms of traction control energy saving, the drive torque and braking torque of each wheel can be controlled independently for electric vehicles. By properly distributing the torque of each wheel, for example, taking the minimum sum of the tire utilization ratios of the driving wheels as the control target [10–12], so as to reduce the energy consumption rate or increase the power of the vehicle [13]. Generally, the optimization method is to turn the torque distribution formula according to vehicle dynamics into the parameter optimization problem under certain constraints [14–16]. However, this kind of method has great limitations in optimizing a multidimensional system.

At present, most of energy-saving control researches are aimed at the straight-line driving conditions evaluated by driving cycles [17] and there are relatively few studies on the vehicle energy-saving control for steering conditions. Compared with two-axle independent drive vehicles, only the two-dimensional optimal torque distribution control between the front and rear axles and between the left and right wheels is needed [18]. Multi-axle electric vehicles need to optimize the multidimensional independent space vector. Meanwhile, there are dynamic and kinematic connections between the wheels, which cannot be solved by traditional optimization algorithms.

The deep deterministic policy gradient (DDPG) [19,20] is an algorithm that improves on the basis of the deep Q network (DQN) [21,22] to solve continuous action problems. In reality, the vehicle is an extremely complex system, and the external environment is dynamic, complex and unknown, which means that it is difficult to simplify it into a fixed expression for quantitative analysis. The DDPG algorithm is highly adaptable and can be optimized for the black-box system in a dynamic environment, which is suitable for solving the practical problems of continuous action.

In the current paper, the four-axle (8 × 8) independent drive electric vehicle is taken as an example to study the torque distribution problem in the steering condition, and a 23-DOF (Degree of Freedom) vehicle dynamics model was built by MATLAB/Simulink (R2015a, MathWorks, Natick, MA, USA). After completing the relevant code of the DDPG algorithm, the data interaction between the algorithm and the vehicle model was realized, and the model was trained enough times through off-line simulation comparing energy consumption of the vehicle under the same conditions, so as to prove that the proposed control algorithm can effectively reduce energy consumption by reasonably distributing the drive torque of each wheel. Under the conventional steering condition and using the motor efficiency map of the current paper, energy consumption of the vehicle can be reduced by up to 5%.

2. Dynamics Model and Energy Analysis

2.1. Model Overview

As the number of axles increases, the dynamics of multi-axle vehicles becomes more complicated. Theoretically, the more the degrees of freedom of the vehicle are considered, the better the simulation effect will be, but the more parameters are actually required to be input, which will affect the results when relevant parameters cannot be obtained. In order to more accurately simulate the impact of vehicle systems and environment on the vehicle during driving, the classical 2-DOF linear model is not used in the vehicle dynamics model. Instead, based on the vehicle system dynamics theory, the differential equations of dynamics and kinematics are derived respectively about vehicle body, wheel and other systems. The suspension part is assumed to be static balance problem, and the tire part is analyzed by "Magic Formula". Finally, the related physical quantities between each system are used to connect the parts into a whole, as shown in Figure 1. Meanwhile, the way of modeling is also suitable for two-axle vehicles, and the simulation accuracy is higher. Based on the dynamics and kinematics equations of each system, the vehicle dynamics model is established by using MATLAB/Simulink. Taking into account 6-DOF of the vehicle body, including longitudinal, lateral, vertical, yaw, pitch, roll, as well as the vertical runout and rotation freedom of each wheel, and steering wheel angle, a total of 23-DOF. In addition, the vehicle adopts the steer-by-wire technology, which can realize all-wheel steering. In the model, according to the fixed relationship between the steering wheel angle and the deflection angle of the right wheel of the first axle and Ackerman steering principle, the S-Function module is built to calculate the actual deflection angle of each wheel, which is directly input into the vehicle dynamic model. The main parameters of the vehicle are shown in Table 1.

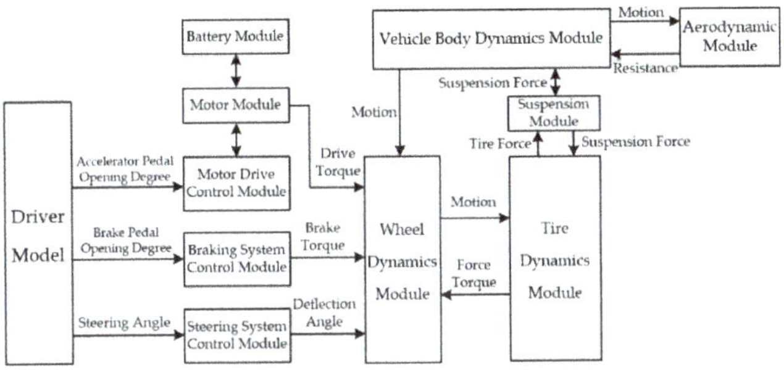

Figure 1. Vehicle dynamics model architecture.

Table 1. Main parameters of the whole vehicle.

Basic Parameters	Value
Total mass of the vehicle (kg)	25,000
Height of the mass (m)	1.20
Wheel rolling radius (m)	0.59
The angle relationship between the steering wheel and the right wheel of the first axle	20:1
1st axle and 2nd axle wheelbase L1 (m)	1.42
2nd axle and 3rd axle wheelbase L2 (m)	2.00
3rd axle and 4st axle wheelbase L3 (m)	1.42
Wheel center distance(m)	2.60
Drive reduction ratio	10.8
Battery rated capacity C_N (Ah)	120
Battery voltage U (V)	900

For electric vehicles with in-wheel motors, due to the complete decoupling of each wheel, in order to achieve electronic differential control, torque control mode is usually adopted for each in-wheel motor [23]. As shown in Figure 2, the drive control architecture is adopted. The total drive torque of the vehicle is obtained by the output of the PID (Proportion Integration Differentiation) controller, and the input of the controller is the deviation of the target speed and the actual speed. In general, the driving torque is evenly distributed to each wheel, so that the speed of wheel will follow according to its stress state. The average distribution mode can ensure the normal driving of vehicles, but it is not the optimal distribution method. Therefore, the optimal distribution mode of drive torque should be proposed, which is the main research content of the current paper.

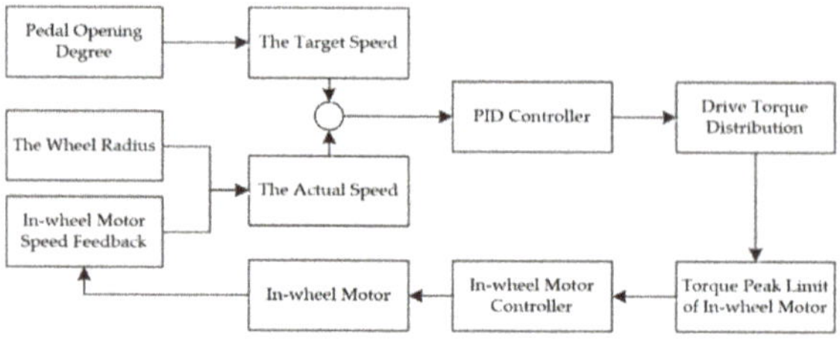

Figure 2. Vehicle drive control.

2.2. Motor and Battery Model

As a high-speed rotating component, the speed characteristic of the motor also determines its high-speed response [24]. In general, the instantaneous response speed of the motor is tens of times faster than that of the wheel, so it can be simplified to a second-order response system [25], whose transfer function is as follows.

$$G(s) = \frac{T_{mi}}{T^*_{mi}} = \frac{1}{2\xi^2 s^2 + 2\xi s + 1}, \tag{1}$$

where T_{mi} is the actual input electromagnetic torque of each in-wheel motor, $T_{mi}{}^*$ is the desired input electromagnetic torque of each in-wheel motor, ξ denotes the damping ratio, which is related to the parameters of the drive motor. According to the response characteristics of PMSM, the value of ξ is 0.001.

At the same time, the motor efficiency map model is adopted. According to the speed and torque of the motor, the working efficiency can be obtained to calculate the corresponding energy loss. The efficiency map of the in-wheel motor used is shown in Figure 3.

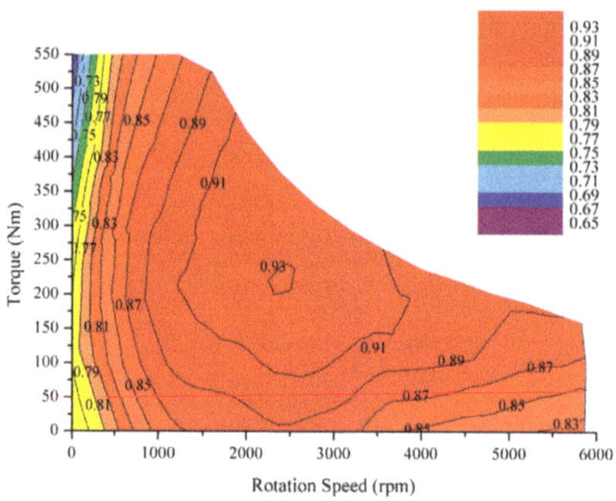

Figure 3. Drive motor efficiency map.

For the battery model, in order to accurately compare the energy consumption, the ampere-hour integral method is adopted to estimate the battery *SOC* [26]. The formula is as follows.

$$SOC = SOC_0 - \frac{1}{C_N}\int \eta I dt = SOC_0 - \frac{1}{C_N}\int \eta \frac{P}{U} dt, \tag{2}$$

where SOC_0 is the initial state of charge and discharge, C_N denotes the battery rated capacity, I is the instantaneous current of the battery, η represents the Coulomb efficiency coefficient, P is the actual working power of the battery, and U is the battery voltage. Generally, without considering the influence of temperature, the battery voltage will decrease with the decrease of *SOC*, but when the battery consumption is between 10% and 90%, the battery voltage variation is relatively small. In order to avoid the impact of the battery voltage change on the *SOC* drop, it is assumed that the battery consumption is always within this range, that is, the battery voltage remains constant.

2.3. Analysis of Steering Energy Consumption

When the vehicle enters the steering condition from the straight driving and the accelerator pedal opening is constant, the vehicle speed will decrease, which indicates that the vehicle driving resistance has increased. The movement of the vehicle is the result of the force from the ground to the vehicle body through the tire. Generally, the force between the tire and the ground is decomposed into longitudinal force and lateral force, and the motion of the vehicle is the result of the combined action. That is, the combined force of the longitudinal force and the lateral force causes the vehicle to generate steering motion. The direction of the resultant force is affected by factors such as drive torque, steering angle, and tire side-slip angle, and in the case of the same drive torque and steering angle, its direction is determined by the tire side-slip angle. When the vehicle turns, the tire force is shown in the Figure 4 below.

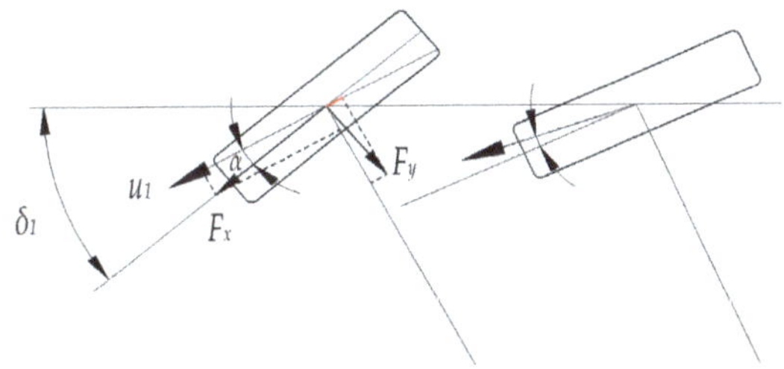

Figure 4. Tire force decomposition diagram.

As shown in Figure 4, δ_1 represents the wheel deflection angle, α is the tire side-slip angle, F_x and F_y denotes the tire longitudinal force and lateral force. Due to δ_1 and α, the lateral force of the wheel will produce a reaction force along the longitudinal axis of the vehicle body, which increases the driving resistance. This explains why the speed of the vehicle will decrease when cornering and the opening of accelerator pedal remains the same, and it also means that if the vehicle wants to maintain the original speed, it needs to consume more energy. By establishing a single-track linear model and assuming that the vehicle moves in a uniform circular motion, the longitudinal force balance equation of the vehicle can be derived as follows.

$$\sum_{i=1}^{4} F_{xi} = F_f + F_a + \mathrm{m}\frac{u^2}{\rho}\left(\frac{l_4}{\mathrm{L}}\sin\alpha_1 + \frac{l_3}{\mathrm{L}}\sin\alpha_2 + \frac{l_2}{\mathrm{L}}\sin\alpha_3 + \frac{l_1}{\mathrm{L}}\sin\alpha_4\right), \tag{3}$$

where F_{xi} is the longitudinal force of each axle, F_f is rolling resistance, F_a denotes air resistance, m is the total mass of the vehicle, u represents the longitudinal velocity, ρ denotes the curvature radius, l_i is the horizontal distance from ith axle to the center of mass, L represents the distance between 1st axle and 4th axle, α_i is the side-slip angle of ith axle. On the left side of the equation is the sum of longitudinal force of each axle and the first two terms on the right are the conventional driving resistance of vehicles. Therefore, the last term is the additional steering resistance caused by the tire slid-slip when the vehicle is steering [27,28], which denoted by F_{af}. If the drive torque of each wheel is changed, the drive force of the outboard wheels is increased and the drive force of the inboard wheels is decreased, then Equation (3) changes as follow.

$$\sum_{i=1}^{4} F_{xi} = F_f + F_a + F_{af} - \sum_{i=1}^{4} \frac{BF_\Delta}{\mathrm{L}}\sin\delta_i, \tag{4}$$

where B is the wheel base, F_{Δ} denotes the change in the drive force, δ_i is the deflection angles of the wheels. With other conditions unchanged, the smaller additional steering resistance, the smaller driving force required by the vehicle, and the less energy consumption. Then it can be seen from Equations (3) and (4) that under certain condition the increase of F_{Δ} is conducive to the reduction of driving resistance. However, as it increases, the tire side-slip angle also increases, which will lead to the increase of the additional steering resistance, so it is not a monotonous change for the total driving resistance. Besides, the speed and deflection angles of wheels also affect the tire side-slip angle, so it is necessary to find the optimal torque distribution ratio at different speeds and steering angle, so as to make the driving resistance of the vehicle minimum.

In addition, the torque distribution of each wheel will also affect the actual working efficiency of the motor. Therefore, the total energy consumption of the vehicle should be taken as the optimization goal, and efficiency of all in-wheel motor is taken into account to achieve dynamic optimization.

3. The DDPG Algorithm

The deep deterministic policy gradient (DDPG) [29,30] is an improved algorithm based on DQN algorithm that can solve the problem of multidimensional continuous action output. This optimization method can operate for continuous action space, and it ignores the specific optimization model, which can complete the black-box learning, focusing on only three concepts [20]: state, action, and reward, and the goal is to get the most cumulative reward.

The selection of DDPG algorithm mainly considers the following points.

(1) The research object of the current paper is the 8 × 8 independent drive electric vehicle, which is equivalent to operating an eight-dimensional independent space vector. It is far different from the two-dimensional optimization problem for 4WD vehicles. The DDPG algorithm is just able to optimize for the problem of multidimensional input and multidimensional continuous output.

(2) The multi-axle vehicle system [31] is complex and difficult to simplify into a fixed expression, whereas DDPG algorithm is more adaptable and capable of learning and optimizing the black-box system.

(3) The actual driving state of the vehicle is constantly changing. In addition to being influenced by the outside, the optimization action at each moment will affect the driving state of the vehicle at the next moment. DDPG algorithm is essentially a kind of reinforcement learning, which can adapt to interact and optimize in a dynamic environment to achieve a better state of adapting to the environment.

In the real word, there is an interaction process between the Agent and its surrounding dynamic environment [32], which can be explained as follows: after the Agent generates an action under a certain state, the environment will give the Agent corresponding reward, and then the Agent enters the next state and will generate the next action. Reinforcement learning is a machine learning model whose modeling goal is to construct the Agent in the environment so that the Agent can always generate actions in the environment to maximize reward. Considering the definition in reinforcement learning, the state of the Agent at time t is s_t, the action under state s_t is a_t, the feedback from the environment is r_t, and the next state the agent enters is s_{t+1}. Corresponding to the content of the current paper, at time t, the vector (w_t, δ_t) composed of the wheel speed (w_t) and deflection angle (δ_t) of each wheel is regarded as s_t. The drive torque distribution ratio of each wheel (p_t) can be regarded as a_t, the vehicle SOC (u_t) can be regarded as r_t. The vector (w_{t+1}, δ_{t+1}) stands for s_{t+1}.

In reinforcement learning, the commonly used optimization objective (R_t) is the expectation of the total future reward at time t, which corresponds to the expectation of battery SOC in the future, as follows.

$$R_t = r_t + \gamma \cdot r_{t+1} + \gamma^2 \cdot r_{t+2} + \ldots = \sum_{i=0}^{+\infty} \gamma^i \cdot r_{t+1}, \tag{5}$$

where γ is a coefficient, $0 < \gamma < 1$, which makes sure that R_t convergence. In order to be able to solve R_t, the above formula can be rewritten as an iterative formula.

$$R_t = r_t + \gamma \cdot R_{t+1}, \tag{6}$$

In the study of Q-learning, if we have the function $Q^* : State \times Action \rightarrow \mathbb{R}$ to represent R_t, and then the optimal action strategy function $\mathcal{A}^*$ can be obtained.

$$\mathcal{A}^*(s_t) = \underset{a_t}{\operatorname{argmax}} Q^*(s_t, a_t), \tag{7}$$

Usually as the environment is poorly understood, Q^* cannot be directly accessed but Deep Neural Network has been proved to be universal function approximator, so it can be used to approximate Q^*. In the current paper, Deep Neural Network is expressed as $Q(s_t, a_t; \Theta)$, where Θ represents the parameter to be solved. In fact, the deep fully-connected neural network is used. Therefore, when Q approaches Q^*, Θ is the optimal parameter Θ^*, and the following equation can be obtained.

$$Q(s_t, a_t; \Theta^*) = r_t + \gamma Q(s_{t+1}, a_{t+1}; \Theta^*), \tag{8}$$

Due to the optimal action strategy function $\mathcal{A}^*$.

$$\mathcal{A}^*(s_t) = \underset{a_t}{\operatorname{argmax}} Q(s_t, a_t; \Theta^*), \tag{9}$$

so the Equation (8) can be expressed as follows.

$$Q(s_t, a_t; \Theta^*) = r_t + \gamma Q(s_{t+1}, \mathcal{A}^*(s_t); \Theta^*), \tag{10}$$

Therefore, the optimization objective of Deep Neural Network can be defined as follows.

$$\begin{gathered} \mathcal{L}(\Theta) = \mathbb{E}_{(s_t, a_t, r_t, s_{t+1}) \sim \mathcal{P}}[((r_t + \gamma \underset{a_t}{\operatorname{argmax}} Q(s_{t+1}, a_t; \Theta)) - Q(s_t, a_t; \Theta))^2] \\ \underset{\Theta}{\operatorname{argmin}} \mathcal{L}(\Theta), \end{gathered} , \tag{11}$$

where $\mathcal{L}(\Theta)$ denotes the optimization objective function with Θ as the independent variable. $\mathbb{E}$ is expectation, and $\mathcal{P}$ represents a probability distribution. The above equation is the optimization objective of DQN algorithm, but the optimization objective is only applicable when a_t is discrete. In the current paper, a_t is the multidimensional continuous space. So, considering an improved algorithm of DQN, DDPG uses Deep Neural Network $\mathcal{A}(s_t; \Phi)$ to approximate the optimal action strategy function $\mathcal{A}^*$, so the optimization objective is as follows.

$$\begin{gathered} \mathcal{L}_1(\Theta) = \mathbb{E}_{(s_t, a_t, r_t, s_{t+1}) \sim \mathcal{P}}\left[((r_t + \gamma Q(s_{t+1}, \mathcal{A}(s_{t+1}; \Phi); \Theta)) - Q(s_t, a_t; \Theta))^2\right] \\ \mathcal{L}_2(\Phi) = \mathbb{E}_{s_t \sim \mathcal{P}}[Q(s_t, \mathcal{A}(s_t; \Phi); \Theta)] \\ \underset{\Theta}{\operatorname{argmin}} \mathcal{L}_1(\Theta) \\ \underset{\Phi}{\operatorname{argmax}} \mathcal{L}_2(\Phi) \end{gathered} , \tag{12}$$

where $\mathcal{L}_1(\Theta)$ represents DQN algorithm optimization target, $\mathcal{L}_2(\Phi)$ denotes the optimization target of approximating the action strategy function $\mathcal{A}^*$. In order to make the optimization process more stable, Φ and Θ in the Equation (12) are replaced with Φ_s and Θ_s corresponding to the soft update parameters.

$$\begin{gathered} \Phi_s = \tau\Phi + (1 - \tau)(\Phi_s) \\ \Theta_s = \tau\Theta + (1 - \tau)(\Theta_s) \end{gathered} , \tag{13}$$

where τ is a coefficient, $0 < \tau < 1$. The expected calculation of $\mathcal{L}_1(\Theta)$ and $\mathcal{L}_2(\Phi)$ can be estimated approximately by Monte Carlo sampling, so the optimization objective is rewritten.

$$\begin{gathered}\mathcal{L}_1(\Theta) = \frac{1}{N}\sum_{i=1}^{N}((r_t^{(i)} + \gamma Q(s_{t+1}^{(i)}, \mathcal{A}(s_{t+1}^{(i)}; \Phi_s); \Theta_s)) - Q(s_t^{(i)}, a_t^{(i)}; \Theta))^2 \\ \mathcal{L}_2(\Phi) = \frac{1}{N}\sum_{i=1}^{N} Q(s_t^{(i)}, \mathcal{A}(s_t^{(i)}; \Phi); \Theta) \\ \underset{\Theta}{\operatorname{argmin}} \mathcal{L}_1(\Theta) \\ \underset{\Phi}{\operatorname{argmax}} \mathcal{L}_2(\Phi)\end{gathered}, \tag{14}$$

where N is the number of the dimension, N = 8, (i) denotes the corresponding wheel number. In fact, stochastic gradient descent algorithm is used to optimize the two optimization targets alternately, and the parameter update method is as follows.

$$\begin{gathered}\Theta^{(t)} = \Theta^{(t-1)} - \alpha_\Theta \nabla \mathcal{L}_1\left(\Theta^{(t-1)}\right) \\ \Phi^{(t)} = \Phi^{(t-1)} + \alpha_\Phi \nabla \mathcal{L}_2\left(\Phi^{(t-1)}\right)\end{gathered}, \tag{15}$$

When the optimal objective is reached, the parameters Θ^* and Φ^* are obtained, corresponding to Deep Neural Network $Q(s_t, a_t; \Theta^*)$ and $\mathcal{A}(s_t; \Phi^*)$. The function $\mathcal{A}(s_t; \Phi^*)$ can output a set of drive torque distribution ratio when the wheel speed and deflection angle are input in real time. The distribution ratio can make the expectation of *SOC* in the future maximum.

The network of a_t is called Actor network, then there are two networks in the algorithm, namely R_t-Q network and Actor network. Actor network is responsible for generating the action, which is the torque distribution ratio of each wheel. R_t-Q network is also commonly referred to as the Critic network, which is used to fit the sum of the system *SOC* for the next n steps, so that Actor network can have a clear optimization target. When the overall algorithm is executed, according to the training logic, Θ in the Q network is updated first, and then as a parameter is input to the Actor network to update Φ, with the aim of minimizing $-$Q. The actual training process is to train Θ and Φ in the two networks, and this process is called joint alternation training.

The overall implementation of architecture design is shown in Figure 5. The DDPG algorithm is directly embedded into the vehicle dynamics model by MATLAB Function to ensure real-time interaction. During the training process, the vehicle system outputs state and reward in real time. A total of 16-dimensional state signal is input to the Actor-network, including eight-dimensional wheel speed and eight-dimensional wheel deflection angle signals, and eight-dimensional wheel torque distribution ratio signal is output. For the Critic network, the same 16-dimensional state signal and eight-dimensional action signal output by the Actor-network are taken as the input to fit the sum of the energy consumption in the next n steps. In addition, the Train function is completed, which contains the logic of the algorithm training process, so that the Actor network and Critic network can update alternately according to the algorithm and complete the corresponding output.

In order to avoid the possible problems of data interaction between the two networks and Train function due to the synchronization of update in the model, all of them are written in a MATLAB Function module and directly called internally. At the same time, taking into account the actual passing ability of the vehicle, and preventing the long-term high torque output of individual motors to reduce the service life, the additional limitation is that the single-axle drive is not allowed in straight-line driving, with the 1st axle and 3rd axle as the main power distribution axle.

In addition, it needs to be clarified that the difference between the application scenario of the current paper and that of the traditional neural network algorithm is that the current action will directly affect the environment at the next moment. If the environment cannot be changed, actually only one step in the overall process is optimized.

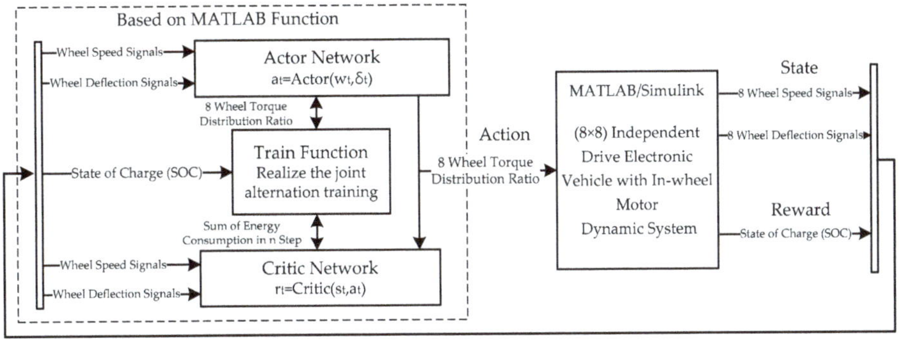

Figure 5. Training process architecture of deep deterministic policy gradient (DDPG) algorithm.

4. Offline Simulation Verification

After the relevant algorithm code is completed and can interact with the vehicle model, the model needs to be trained for a certain number of times first. The purpose is to make the Actor and Critic network update their internal parameters according to the training logic of Train function to adapt to the whole system.

At present, there is no standard cycle condition for the evaluation of vehicle steering energy consumption, which results in the training condition of the model needs to be designed artificially. Different training conditions will affect the final optimization results of the model. The designed training condition should contain enough state samples of the optimized system. At the same time, it should be avoided that due to the influence of training environment, experience with certain type characteristics is particularly abundant, while experience with other type is scarce. At best, experience should have difference and similar experience should be minimized. During neural network training, some unexpected changes are not considered in the current paper, because they are difficult to be included completely. However, in order to avoid related problems, the average distribution as a conservative control scheme was combined with the neural network. By comparing the reward at any time, the control scheme with a higher reward is adopted, so as to ensure that the energy consumption of the vehicle was not lower than the conventional driving mode under any working condition, which is a supportability control strategy.

The state variables in the algorithm are the wheel speed and the wheel deflection angle. Therefore, based on the above principles, the model input of target vehicle speed and steering wheel angle are shown in Figure 6. During training, only the first and second axles were steering axles. Meanwhile, considering the stability problem of the vehicle in high speed, the amplitude of the steering angle decreased after 40 seconds.

According to the training conditions, after completing about 100, 200, ..., 500 times training, data and driving state curves were recorded. Figure 7a shows the change process of vehicle speed after different training times. The change of vehicle speed was little affected by the drive torque distribution and the target vehicle speed could be well followed. Since the optimal torque distribution is equivalent to applying an additional yaw moment for the vehicle, so the yaw rate of the vehicle was increased in each period after distributing, which can be seen in Figure 7b, and it is in line with the actual situation. Figure 7c is a comparison of the *SOC* change after the corresponding training number. It can be seen that the *SOC* decline decreased with the increase of training times. After 500 times of training, the *SOC* decline of this training condition was reduced by about 4.5320%.

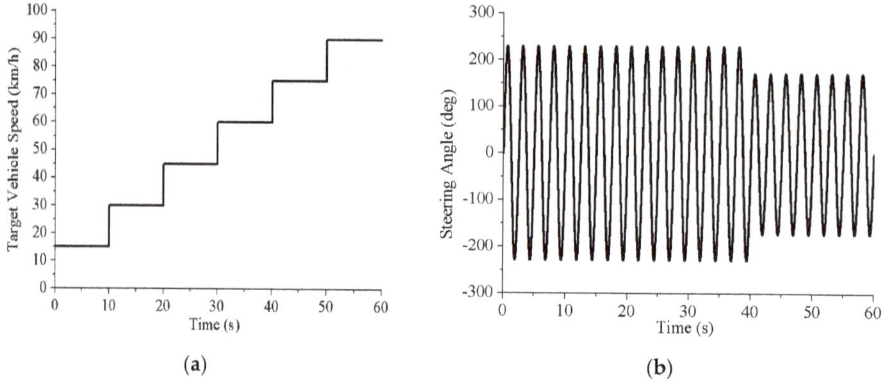

Figure 6. Model input of the training condition: (**a**) target vehicle speed; (**b**) steering wheel angle.

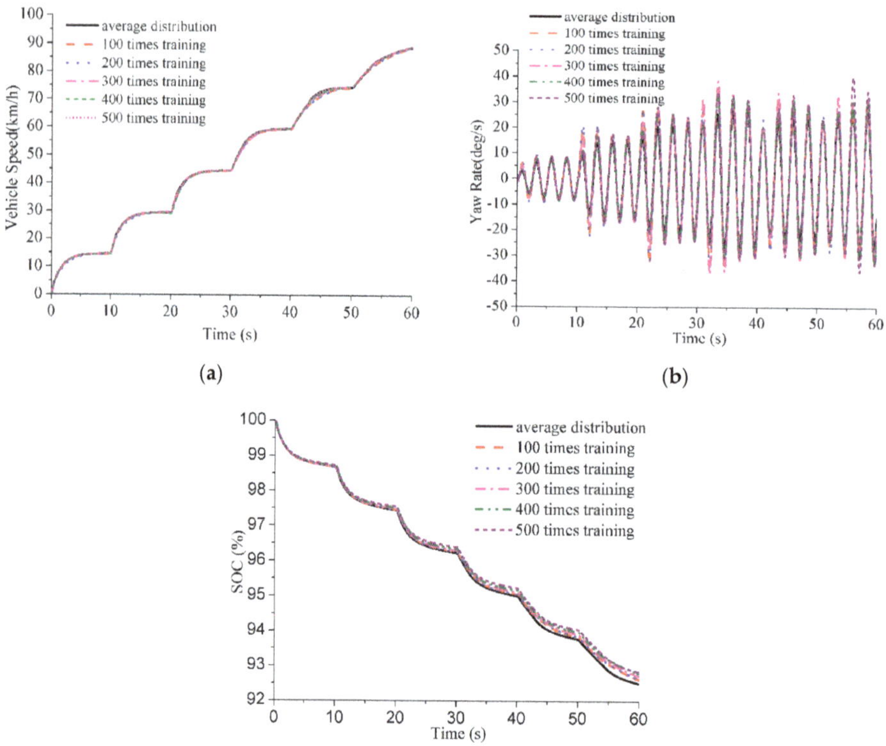

(c)

Figure 7. Changes of vehicle driving parameters and battery state of charge (*SOC*) after training: (**a**). Changes of vehicle speed with training times; (**b**). Changes of yaw rate with training times; (**c**). Changes of *SOC* with training times.

After completing the training, only the parameter matrix in the Actor network is retained and stored into the MATLAB Function, which receive the driving state of the vehicle in real-time and generate the optimal distributing action. In theory, the more training times, the more stable and optimal parameters in the Actor network tend to be, and the better the optimization effect will be. However, with the increase of training times, the rate of optimization return is decreased. Meanwhile, in order to ensure the optimal effect, a fixed simulation step size of 1 millisecond was adopted in the Simulink, while the action was updated every 10 steps by the control algorithm, which led to a significant increase in the computational burden of the model. After completing 400 and 500 times training, and comparing the simulation results, it can be found that the optimization effect was almost the same. Therefore, considering the optimization efficiency, finally the model training was completed for 500 times.

4.1. Conventional Low-Speed Step Steering Condition

The low-speed simulation condition was designed to accelerate the vehicle from the stationary state with a target speed of 30 km/h. At the 20 s, the steering wheel turned about 230° within 1 s, and only the first and second axles were steering axles. Figure 8a shows the actual change in speed of vehicle. It can be seen that after the steering angle change, the vehicle speed decreased slightly, which was caused by the increase of driving resistance. It is consistent with the actual situation. Figure 8b is a detail view of vehicle speed. Compared with the average distribution, the steady-state vehicle speed increased slightly after the optimal distribution of drive torque, but the difference was not significant. Because the redistribution of drive torque led to the reduction of additional steering resistance, the drive torque required to maintain steady state was reduced. It can be seen from Figure 2 that under the condition that the target vehicle speed remained unchanged, the actual vehicle speed increased.

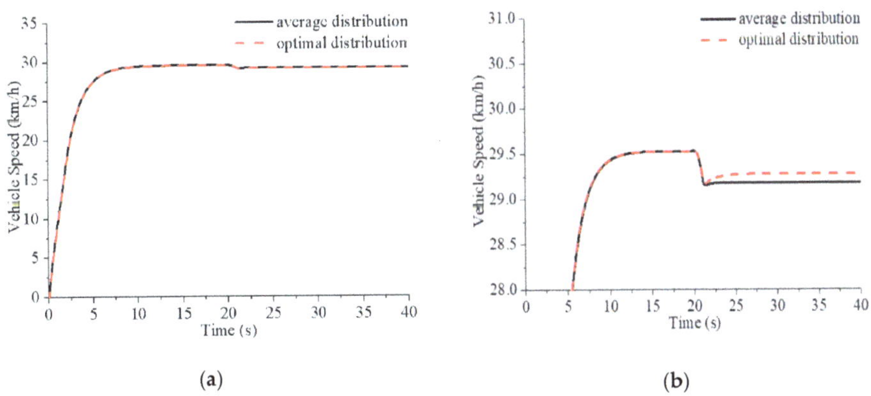

Figure 8. Changes of vehicle speed during low-speed simulation: (**a**). Changes of vehicle speed; (**b**). Partial enlarged drawing.

Figure 9 shows vehicle yaw rate change and the vehicle track comparison respectively. After optimization control, the yaw rate of the vehicle increased by around 1.02%, and the radius of the track was also slightly reduced. From Figures 8 and 9, it can be seen that optimal torque distribution promoted the steering trend, but the influence on the various driving state parameters of the vehicle was not significant, and did not cause the stability problem.

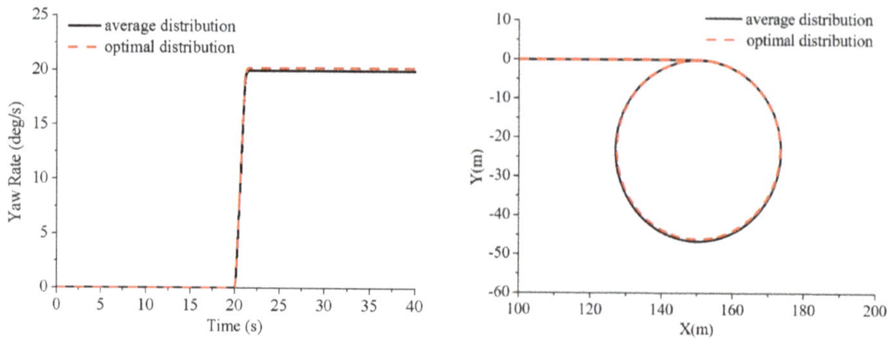

(**a**) (**b**)

Figure 9. Vehicle yaw rate change and the vehicle track comparison: (**a**). Changes of yaw rate; (**b**). Comparison of driving trajectory.

It can be seen from Figure 10a that after adopting torque optimization control, *SOC* decline was significantly reduced and the energy consumption was reduced by about 3.7856% between 0 s and 40 s. However, it included the linear acceleration phase, although the torque was also optimally distributed during straight-line driving, the motor basically worked on the external characteristic curve during acceleration. At the same time, there was no training for the straight-line driving condition, so the optimization effect was not obvious. Then only for the steering phase between 20 s and 40 s, the vehicle energy consumption can be reduced by about 5.112% after optimization.

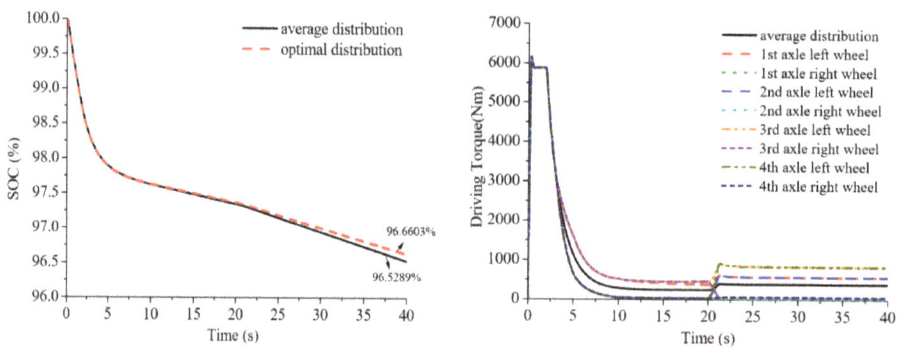

(**a**) (**b**)

Figure 10. Changes of vehicle *SOC* and wheel drive torque after optimization: (**a**). Changes of vehicle *SOC*; (**b**). Changes of wheel drive torque.

Figure 10b shows the change of the drive torque of each wheel. In the linear acceleration phase, the drive torque of the whole vehicle was mainly distributed to 1st axle and 3rd axle, similar to the two-axle drive, which increased the working load of some drive motors and improved overall work efficiency. When steering, the drive torque of the outboard wheel increased, and the drive torque of the inboard wheel decreased. Besides, the drive torque of rear axle of the outboard wheel was relatively larger, because in the same cases, the change of the drive torque of the rear axle had a greater influence on the additional yaw of the whole vehicle, which is more conducive to the reduction of the energy consumption. In addition, the multi-axle vehicle body is longer, resulting in the effect is relatively more obvious. When the vehicle was in steady-state steering, the driving torque of the whole vehicle is about 3107 Nm by average distribution, while the total driving torque is about 2975.4 Nm by optimized distribution, which is relatively reduced by about 4.2356%. Another part of the reduction in energy consumption comes from the improvement of motor working efficiency.

Figure 11 shows the comparison of working point change in the motor efficiency map. The wheel speed and output torque during steady-state steering are respectively derived. Based on the deceleration ratio, the actual working points of each in-wheel motor were calculated. As the relative speed difference between the left and the right wheel was very small, which can be approximately ignored, a point was used to represent the actual working point of each motor when the drive torque was evenly distributed. After the optimal torque distribution control was adopted, the actual working point of each motor was changed. The drive torque of the outboard wheel was increased, and the working efficiency was improved. Though the working efficiency of inboard wheel reduced, its drive torque was small, which led to the overall working efficiency being improved.

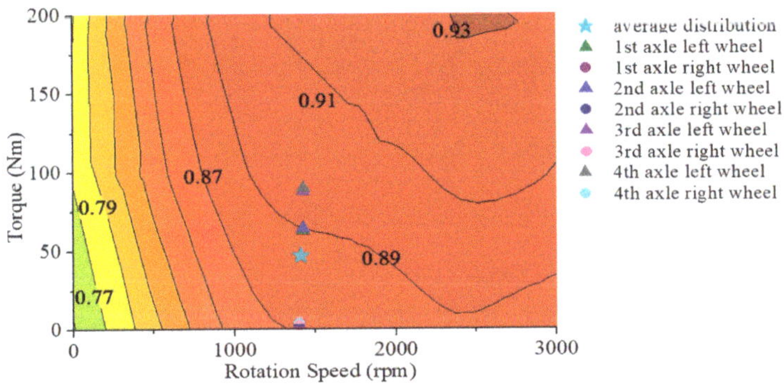

Figure 11. Comparison of motor working points.

4.2. *Conventional High-speed Sinusoidal Steering Condition*

The high-speed simulation condition was designed to accelerate the vehicle from the stationary state with a target speed of 70 km/h. At 20 s, the steering wheel input a sine wave with an amplitude of 110° as shown in Figure 12a. Similarly, 1st axle and 2nd axle were steering axles. Figure 12b,c show changes of the vehicle speed and the yaw rate. Similar to the step steering condition, the change of driving state was not obvious and the peak of yaw rate increased slightly. Figure 12d shows the change of drive torque. Due to the input of the steering wheel constantly changing, the curvature radius of the vehicle driving was also changing. It can be seen from Equation (3) that the additional steering resistance fluctuated accordingly. Therefore, when the driving torque was evenly distributed, the driving torque of each wheel also changed correspondingly. After optimized distribution, the more drive torque was distributed to the wheel of the outboard and rear axles, which promoted the steering of the vehicle. Under the dynamic steering condition, the driving torque of each wheel could follow the changes of system input, which indicates that the optimal control algorithm could adapt to the dynamic environment.

The changes of *SOC* can be seen from Figure 13a. After the optimization control, the *SOC* decline reduced by 2.6213% between 0 s and 40 s. If only comparing the *SOC* change during steering phase, the energy consumption of the vehicle decreased by 4.0482% after optimization as shown in Figure 13b. It was proved that the optimal torque distribution control based on energy consumption could reasonably distribute the drive torque of each wheel and reduce the energy consumption under the dynamic condition. That means the optimization algorithm adopted was not limited to specific working conditions, which can be for any steering conditions, whether static or dynamic. The optimization algorithm could optimize the distribution of driving torque in real time and reduce the vehicle energy consumption. However, the optimization effect was slightly worse than that of low speed test, which was mainly for two reasons. On the one hand, the sine wave input was a dynamic process all the time, but there had to be system inertia in the mechanical system, which may have led to the actual action and control signals not being completely synchronized. Although the effect was relatively small for the electric vehicle with in-wheel motor, it could not guarantee that the drive torque of each wheel was optimal at any time; on the other hand, when the motor worked at a high speed, the high efficiency area on the efficiency map was relatively large, so the optimization effect after the control was slightly lower.

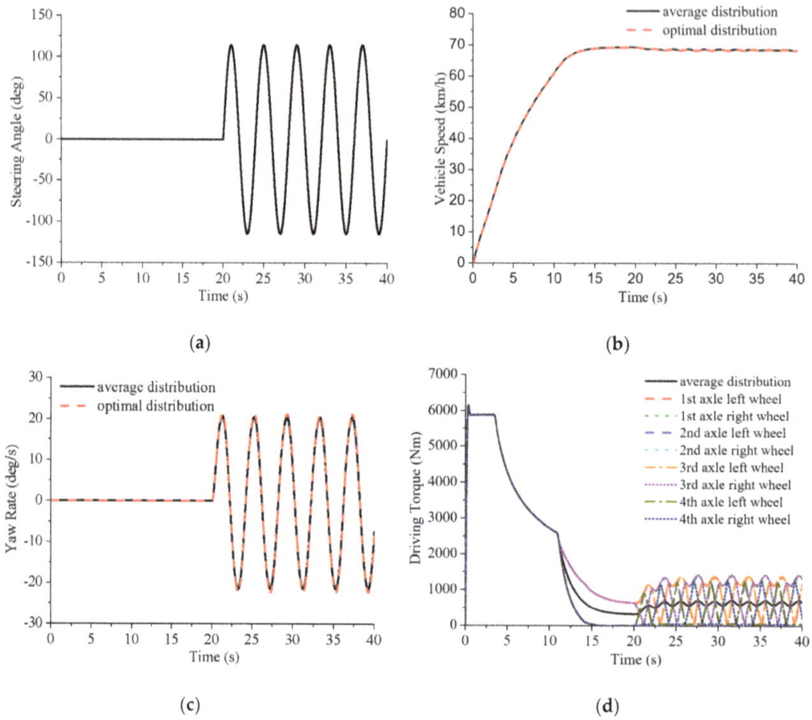

Figure 12. Changes in vehicle driving parameters during high-speed simulation condition: (**a**). Steering wheel angle; (**b**). Vehicle speed; (**c**). Yaw rate; (**d**). Drive torque of each wheel.

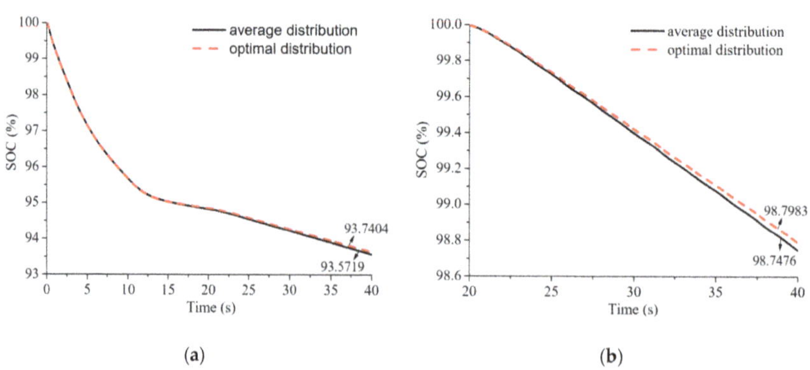

Figure 13. Changes of *SOC* after the optimization control: (**a**). Changes of *SOC* in the whole process; (**b**). Changes of *SOC* in the steering phase.

4.3. Extreme Steering Condition

In order to further reflect the control effect of optimal torque distribution, the extreme steering condition test was carried out. The four-axle reverse phase steering mode was adopted, with the first and second axles deflecting in the opposite direction to the third and fourth axles. The target speed of the vehicle was set to 10 km/h. At 20 s, the right wheel of the first axle deflected about 23° within 2 s, and the deflection angles of other wheels were calculated according to Ackerman steering principle, as shown in Figure 14a. For the change of speed, the vehicle speed after optimal control was still slightly higher than that under average distribution as shown in Figure 14b, which was the same as the previous simulation results. However, when the vehicle was in steady-state steering, the vehicle speed was basically unchanged compared with driving in the straight line, which indicates that the additional steering resistance was relatively small in this working condition.

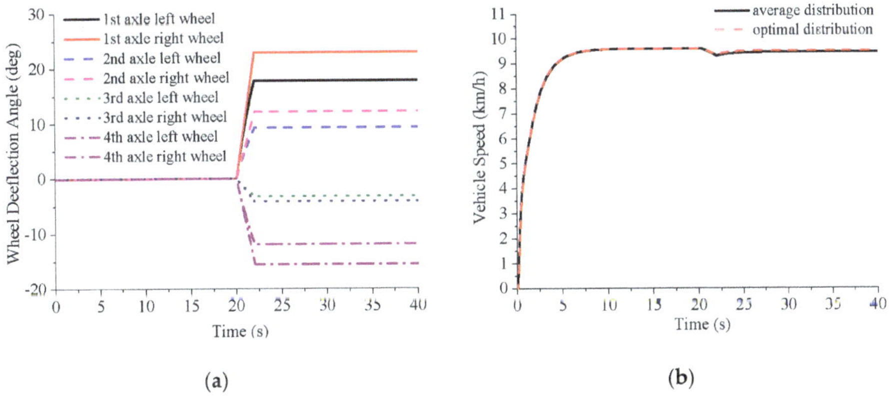

Figure 14. Changes in vehicle driving parameters after the optimization control in reverse phase steering condition: (**a**) wheel deflection angle; (**b**) longitudinal vehicle speed.

As shown in Figure 15, the driving track of the vehicle remained unchanged basically after optimization. The steering radii of the vehicle after average distribution and optimal distribution were 8.1165 m and 8.1053 m respectively, which means that the optimal distribution of drive torque control did not have a great impact on the vehicle trajectory and body posture.

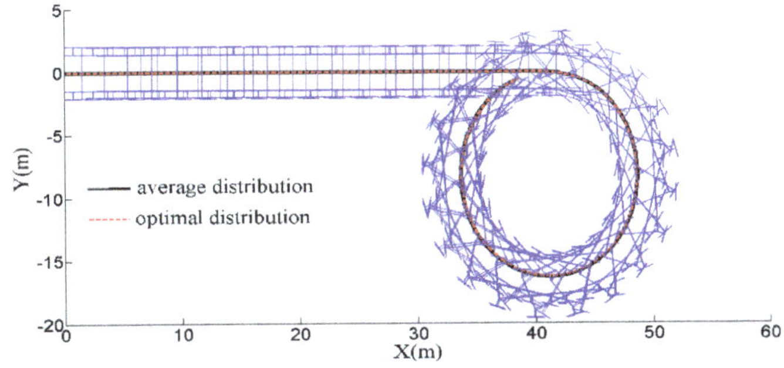

Figure 15. Comparison of vehicle trajectory and body posture under average distribution and optimal distribution.

Figure 16 shows the change of wheel drive torque. 0 s to 20 s was a linear acceleration phase, and the drive torque was distributed between the axles. Since the motor was in the state of low speed and low torque at this stage, in order to improve the overall working efficiency, the driving torque of the vehicle was mainly distributed to the first axle and the third axle to increase the workload of the motor. When entering the steering at 20 s, due to the increase of the driving resistance, the driving torque of the vehicle increased in order to maintain the target speed. However, when the vehicle was in steady-state steering, the drive torque was basically the same as that when the vehicle traveled in a straight line, which was caused by the reduction of driving resistance by the four-axle reverse phase steering. It can be seen that the optimization control made the distribution ratio of the outboard and rear axle wheels increase, which further promoted the reduction of driving resistance, thus achieving the purpose of reducing the driving energy consumption.

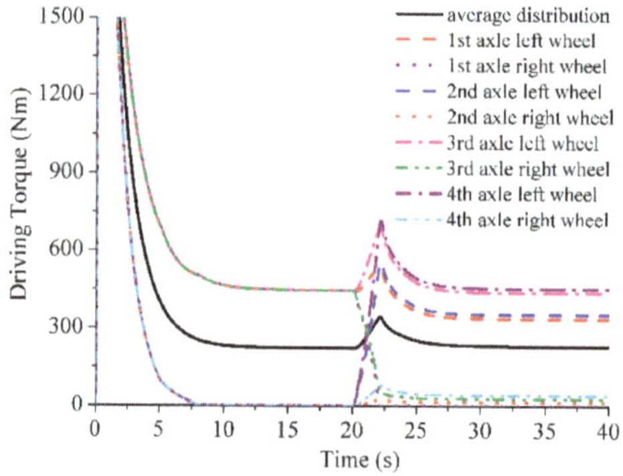

Figure 16. Comparison of wheel drive torque change under average distribution and optimal distribution.

When the vehicle was in steady-state steering, the total required drive torque of the vehicle with the average torque distribution was 1860.0376 Nm, and after the optimal distribution control, it was only 1656.6745 Nm, which was about 10.9332% lower. Then the change of the vehicle *SOC* during the steering phase was compared. The actual energy consumption decreased by about 13.3679%, which was much more obvious than the conclusion obtained by the above that maximum reduction in energy consumption is about 5%. This is mainly because the working efficiency of the motor is extremely low under low speed conditions [33]. Meanwhile, according to the motor efficiency map used in this paper, when the vehicle speed was lower than 30 km/h, the efficiency changed greatly with the torque, so the optimization control effect was better under this working condition. Besides, it was found that when other conditions were the same and four-axle reverse phase steering was adopted, the vehicle demand torque was far less than that when two-axle steering was adopted, sometimes less than half of that. Smaller drive torque led to lower working efficiency, which also led to the more obvious optimization effect.

4.4. Performance Evaluation

It should be emphasized that the optimal distribution of drive torque control can achieve the maximum energy saving effect of about 5% in the conventional steering conditions, but it is only for the motor efficiency map used in the current paper (Figure 3). The motor efficiency map had a great influence on the actual optimization effect. If the high efficiency area of the in-wheel motor was small, the energy saving control effect on the vehicle was obvious. In addition, the selection of algorithm training conditions should be closer to the actual driving state of the vehicle, and enough training times should be ensured to make the parameters in the Actor network tend to the stable and optimal value.

5. Conclusions

(1) Based on the theory of vehicle system dynamics, the dynamic model of an 8 × 8 independent drive electric vehicle is built by MATLAB/Simulink, which contains 23-DOF to more accurately describe the multi-axle vehicle dynamics. On the basis, combining with the analysis of tire force and the mathematical derivation of the single-track linear model, it is concluded that through the reasonable distribution of the driving torque can reduce the additional steering resistance, and then reduce the energy consumption of the vehicle. However, due to the change of the tire side-slip angle and the influence of the motor efficiency, the optimization process is necessarily dynamic.

(2) Considering the research object and content of the current paper, the DDPG algorithm is adopted to optimize the distribution of the drive torque between each wheel to reduce the energy consumption of the vehicle. The formula of DDPG algorithm is derived, and the overall system architecture is designed. The Actor network, Critic network and Train function are completed to interact with the vehicle model with the help of MATLAB Function, and realize the joint alternation training.

(3) Since there is no standard for the evaluation of steering energy consumption, the training condition is designed artificially. After completing 500 times training, the parameter matrix in the Actor-network is stored into the MATLAB Function, which receive the driving state of the vehicle in real-time and generate the optimal distributing action. The low speed, high speed conventional steering and extreme steering simulation tests are carried out respectively. The results show that the vehicle energy consumption can be reduced by about 5% at most under the conventional steering condition with using the motor efficiency map of the current paper, which effectively reduces the energy consumption for the multi-axle electric vehicles with in-wheel motors. Meanwhile, the current paper provides an innovative solution to the vehicle optimization problem of multidimensional state input and multidimensional continuous output.

Author Contributions: Conceptualization, Q.Z. and D.T.; methodology, L.J.; software, D.T. and J.W.; validation, D.T. and L.J.; formal analysis, D.T. and Q.Z.; investigation, D.T.; resources, J.W.; data curation, D.T.; writing—original draft preparation, L.J.; writing—review and editing, D.T., Q.Z. and J.W.; visualization, D.T.; supervision, L.J.; project administration, L.J.; funding acquisition, D.T. All authors have read and agreed to the published version of the manuscript.

Funding: This research was funded by Natural Science Foundation of Jilin Province (Grant No.: 20170101208JC).

Conflicts of Interest: The authors declare no conflict of interest. The funders had no role in the design of the study; in the collection, analyses, or interpretation of data; in the writing of the manuscript, or in the decision to publish the results.

References

1. Chan, C.C.; Chau, K.T. Modern Electric Vehicle Technology. *Power Eng.* **2001**, *16*, 240.
2. Hou, R.F.; Zhai, L.; Sun, T.M.; Hou, Y.H.; Hu, G.X. Steering Stability Control of a Four In-Wheel Motor Drive Electric Vehicle on a Road with Varying Adhesion Coefficient. *IEEE Access* **2019**, *7*, 32617–32627. [CrossRef]
3. Ehsani, M.; Gao, Y.M.; Emadi, A. *Modern Electric, Hybrid Electric, and Fuel Cell Vehicles: Fundamentals, Theory and Design*, 2nd ed.; CRC Press: New York, NY, USA, 2009.
4. Tang, B. Application prospect of heavy-duty multi-axle special transportation vehicle. *Automob. Parts* **2012**, *48*, 30–33.
5. Kim, J. Optimal power distribution of front and rear motors for minimizing energy consumption of 4-wheel-drive electric vehicles. *Int. J. Automot. Technol.* **2016**, *17*, 319–326. [CrossRef]
6. Qian, H.H.; Xu, G.Q.; Yan, J.Y.; Lam, T.L.; Xu, Y.; Xu, K. Energy Management for Four-Wheel Independent Driving Vehicle. In Proceedings of the IEEE/RSJ International Conference on Intelligent Robots and System, Taipei, Taiwan, 18–22 October 2010.

7. Gao, T.M.; Chu, L.; Ehsani, M. Design and Control Principles of Hybrid Braking System for EV, HEV and FCV. In Proceedings of the 2007 IEEE Vehicle Power and Propulsion Conference, Arlington, TX, USA, 9–12 September 2007; pp. 384–391.
8. Li, N.; Zhang, J.Z.; Zhang, S.Y.; Hou, X.; Liu, Y. The influence of accessory energy consumption on evaluation method of braking energy recovery contribution rate. *Energy Convers. Manag.* **2018**, *166*, 545–555. [CrossRef]
9. Chen, Y.; Wang, J.M. Fast and Global Optimal Energy-Efficient Control Allocation with Applications to Over-Actuated Electric Ground Vehicles. *IEEE Trans. Control Syst. Technol.* **2012**, *20*, 1202–1211. [CrossRef]
10. Lenzo, B.; Bucchi, F.; Sorniotti, A.; Frendo, F. On the handling performance of a vehicle with different front-to-rear wheel torque distributions. *Veh. Dyn. Syst.* **2018**, *57*, 1685–1704. [CrossRef]
11. Yamakawa, J.; Watanabe, K. A method of optimal wheel torque determination for independent wheel drive vehicles. *J. Terramech.* **2006**, *43*, 269–285. [CrossRef]
12. Mokhiamar, O.; Abe, M. Simultaneous Optimal Distribution of Lateral and Longitudinal Tire Forces for the Model Following Control. *J. Dyn. Syst. Meas. Control* **2004**, *126*, 753–763. [CrossRef]
13. Park, J.; Jeong, H.; Jang, I.G.; Hwang, S.H. Torque Distribution Algorithm for an Independently Driven Electric Vehicle Using a Fuzzy Control Method. *Energies* **2015**, *8*, 8537–8561. [CrossRef]
14. Dizqah, M.A.; Lenzo, B.; Sorniotti, A.; Gruber, P.; Fallah, S.; De Smet, J. A Fast and Parametric Torque Distribution Strategy for Four-Wheel-Drive Energy-Efficient Electric Vehicles. *IEEE Trans. Ind. Electron.* **2016**, *63*, 4367–4376. [CrossRef]
15. Yu, Z.P.; Zhang, L.J.; Xiong, L. Optimized Torque Distribution Control to Achieve Higher Fuel Economy of 4WD Electric Vehicle with Four In-Wheel Motors. *J. Tongji Univ.* **2005**, *33*, 1355–1361.
16. Fan, J.J.; Mao, M. A Study of Driving Force Distribution Strategy for Three-axles Electric Driving Vehicle Based on Economics. *Veh. Power Technol.* **2007**, *1*, 52–59.
17. Li, B.; Goodarzi, A.; Khajepour, A.; Chen, S.K.; Litkouhi, B. An optimal torque distribution control strategy for four-independent wheel drive electric vehicles. *Veh. Syst. Dyn.* **2015**, *53*, 1172–1189. [CrossRef]
18. Jing, H.H.; Jia, F.J.; Liu, Z.Y. Multi-Objective Optimal Control Allocation for an Over-Actuated Electric Vehicle. *IEEE Access* **2018**, *6*, 4824–4833. [CrossRef]
19. Mnih, V.; Kavukcuoglu, K.; Silver, D.; Graves, A.; Antonoglou, I.; Wierstra, D.; Riedmiller, M. Playing Atari with Deep Reinforcement Learning. *Comput. Sci.* **2013**, arXiv:1312.5602.
20. Lillicrap, T.P.; Hunt, J.J.; Pritzel, A.; Heess, N.; Erez, T.; Tassa, Y.; Silver, D.; Wierstra, D. Continuous control with deep reinforcement learning. *Comput. Sci.* **2015**, *8*, 187–200.
21. Mnih, V.; Kavukcuoglu, K.; Silver, D.; Rusu, A.A.; Veness, J.; Bellemare, M.G.; Graves, A.; Riedmiller, M.; Fidjeland, A.K.; Ostrovski, G.; et al. Human-level control through deep reinforcement learning. *Nature* **2015**, *518*, 529–533. [CrossRef]
22. Silver, D.; Lever, G.; Heess, N.; Degris, T.; Wierstra, D.; Riedmiller, M. Deterministic policy gradient algorithms. In Proceedings of the International Conference on International Conference on Machine Learning, Beijing, China, 21–26 June 2014; pp. 387–395.
23. Jin, L.Q.; Wang, Q.N.; Zhang, H.H.; Wang, J.N. A Study on Differential Technology of In-wheel Motor Drive EV. *Automot. Eng.* **2007**, *29*, 700–704.
24. Tursini, M.; Parasiliti, F.; Zhang, D.Q. Real-time gain tuning of PI controllers for high-performance PMSM drives. *IEEE Trans. Ind. Appl.* **2002**, *38*, 1018–1026. [CrossRef]
25. Harris, T.A.; Kotalas, M.N. *Rolling Bearing Analysis*; CRC Press: Boca Raton, FL, USA, 2010; pp. 124–132.
26. Li, Z.; Lu, L.G.; Ouyang, M.G. Comparison of Methods for Improving SOC Estimation Accuracy through an Ampere-hour Integration Approach. *J. Tsinghua Univ. (Sci. Technol.)* **2010**, *8*, 2193–2196.
27. Zhang, H.H. Research on the Torque Coordinating Control of In-Wheel Motor Driving Electric Vehicle. Ph.D. Thesis, Auto-Body Engineering Department of College of Automobile Engineering of Jilin University, Changchun, China, 2009.
28. Sun, W.; Wang, Q.N.; Wang, J.N. Yaw-moment control of motorized vehicle for energy conservation during cornering. *J. Jilin Univ. (Eng. Technol. Ed.)* **2018**, *48*, 11–19.
29. Kim, S.J.; Kim, H.S.; Kang, D.J. Vibration Control of a Vehicle Active Suspension System Using a DDPG Algorithm. In Proceedings of the IEEE International Conference on Control, Automation and Systems (ICCAS), Daegwallyeong, Korea, 17–20 October 2018.

30. Hou, J.; Li, H.; Hu, J.W.; Zhao, C.; Guo, Y.; Li, S.; Pan, Q. A review of the applications and hotspots of reinforcement learning. In Proceedings of the 2017 IEEE International Conference on Unmanned Systems (ICUS), Beijing, China, 27–29 October 2017.
31. Du, H.; Wang, Z.B.; Wang, Y.; Huang, H. Adaptive Robust Control of Multi-Axle Vehicle Electro-Hydraulic Power Steering System with Uncertain Tire Steering Resistance Moment. *IEEE Access* **2018**, *7*, 5519–5530. [CrossRef]
32. Xu, X.; He, H.G. A Gradient Algorithm for Neural-Network-Based Reinforcement Learning. *Chin. J. Comput.* **2003**, *26*, 227–233.
33. Gu, C.; Liu, H.; Chen, X.B. Torque Distribution Based on Efficiency Optimization of Four-wheel Independent Drive Electric Vehicle. *J. Tongji Univ. (Nat. Sci.)* **2015**, *43*, 1550–1556.

© 2020 by the authors. Licensee MDPI, Basel, Switzerland. This article is an open access article distributed under the terms and conditions of the Creative Commons Attribution (CC BY) license (http://creativecommons.org/licenses/by/4.0/).

energies

MDPI

Article

LCC-S Based Discrete Fast Terminal Sliding Mode Controller for Efficient Charging through Wireless Power Transfer

Naghmash Ali [1], Zhizhen Liu [1,*], Yanjin Hou [2], Hammad Armghan [1], Xiaozhao Wei [1] and Ammar Armghan [3]

1 School of Electrical Engineering, Shandong University, Jinan 250061, China; 14mseenmr@seecs.edu.pk (N.A.); 14mseeharmghan@seecs.edu.pk (H.A.); weixiaozhao@mail.sdu.edu.cn (X.W.)
2 Energy Institute, Qilu University of Technology (Shandong Academy of Sciences), Shandong Provincial Key Laboratory of Biomass Gasification Technology, Jinan 250353, China; houyj@sderi.cn
3 Department of Electrical Engineering, Jouf University, Al-Jawf, Saudi Arabia; aarmghan@ju.edu.sa
* Correspondence: liuzhizhen@sdu.edu.cn

Received: 22 December 2019; Accepted: 12 March 2020; Published: 16 March 2020

Abstract: Compared to the plug-in charging system, Wireless power transfer (WPT) is simpler, reliable, and user-friendly. Resonant inductive coupling based WPT is the technology that promises to replace the plug-in charging system. It is desired that the WPT system should provide regulated current and power with high efficiency. Due to the instability in the connected load, the system output current, power, and efficiency vary. To solve this issue, a buck converter is implemented on the secondary side of the WPT system, which adjusts its internal resistance by altering its duty cycle. To control the duty cycle of the buck converter, a discrete fast terminal sliding mode controller is proposed to regulate the system output current and power with optimal efficiency. The proposed WPT system uses the LCC-S compensation topology to ensure a constant output voltage at the input of the buck converter. The LCC-S topology is analyzed using the two-port network theory, and governing equations are derived to achieve the maximum efficiency point. Based on the analysis, the proposed controller is used to track the maximum efficiency point by tracking an optimal power point. An ultra-capacitor is connected as the system load, and based on its charging characteristics, an optimal charging strategy is devised. The performance of the proposed system is tested under the MATLAB/Simulink platform. Comparison with the conventionally used PID and sliding mode controller under sudden variations in the connected load is presented and discussed. An experimental prototype is built to validate the effectiveness of the proposed controller.

Keywords: wireless power transfer; non-linear; fast terminal sliding mode control; power converters

1. Introduction

Wireless power transfer has been desired since the proposition made by Nikola Tesla about 100 years ago [1]. Due to the recent progress in power electronics technology and advancements in WPT techniques, it is realized that implementing a WPT system is now economical and can be used as a commercial product [2]. Compared to the plug-in charging system, WPT is simpler, reliable, and user-friendly [3]. Companies like Qualcomm, Witricity, and Evatran have developed many commercial products that can be charged wirelessly with good efficiency. Due to such developments, WPT can be used in many industrial applications [4] and in our daily life such as wireless charging of smartphones [5], electric vehicles [6,7], and biomedical implants [8–10].

According to the operating principles, WPT can be broadly divided into four categories, i.e., capacitive wireless power transfer (CWPT), electromagnetic radiation (EMR), acoustic wireless power transfer (AWPT), and resonant inductive power transfer (RIPT) [11,12]. In CWPT, the power is transferred using capacitor plates instead of coils. CWPT is simpler and can be used for both high voltage and low current, but the efficiency decreases when the air-gap between the transmitter and receiver plate increases [13]. In EMR, the power is transferred using microwaves. Although using this operating principle, WPT can achieve long-distance power transfer, this mode has much lower efficiency and has many health hazards due to high power radiation [14]. In optical wireless power transfer (OWPT), which is considered as a subclass of EMR, the same principles for power transfer are used, but the wavelengths are in the visible spectrum [15,16]. At the transmitting side, lasers are used to convert the electrical signal into the optical signal, and at the receiving side, photovoltaic diodes convert the optical signal back into an electrical signal. The advantage of EMR and OWPT techniques is that both techniques have high capability for power beaming. However, due to the conversion steps, almost 40 to 50% of the energy is lost [15]. In AWPT, the power is transferred by propagating energy in the form of sound or vibration waves. At the transmitting side, the electrical signals are converted into pressure waves by a transducer. The waves propagate through a medium and then are collected by the receiving side transducer, which converts it back into electrical signals. The benefit of AWPT is that it can achieve higher power beam directivity than electromagnetic transmitter of the same size and the power is transferred omnidirectional which reduces the losses such as coil misalignment but the power transfer capability and efficiency of the AWPT is very less compared to other WPT systems [12,17,18]. In RIPT, the power transfer takes place between a transmitter coil and a receiver coil using electromagnetic induction. A typical RIPT WPT system is shown in Figure 1, which consists of:

- DC voltage source.
- DC/high-frequency AC inverter.
- Compensation networks.
- Transmitting and receiving coils.
- Rectifier.
- Regulatory circuitry.

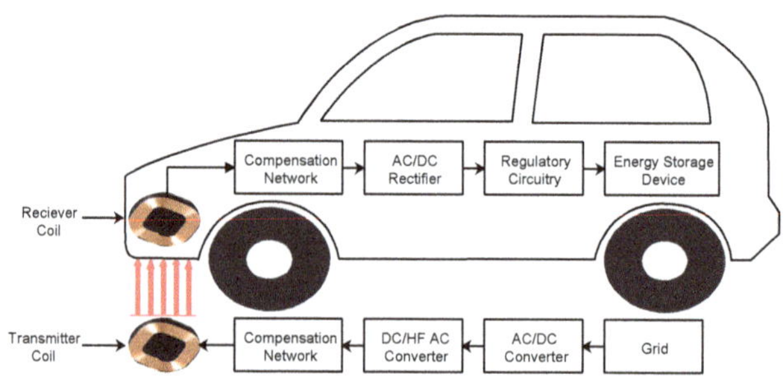

Figure 1. Typical RIPT WPT system.

To transfer the power from the transmitter coil to the receiver coil, the DC power is converted into high-frequency (HF) AC power through an inverter. To cancel out the leakage inductance, improve the system's efficiency, and lower the reactive power transfer in the WPT system, compensation networks are required on both the transmitting and receiving sides. The compensation network on the transmitting side eliminates the phase difference between the voltage and current, which minimizes the reactive power transfer, while on the receiving side, it maximizes the power transfer by improving the efficiency [19,20]. The required system characteristics, i.e., constant voltage or constant current, can also be achieved using suitable compensation networks. Based on the output characteristics, the compensation networks can be broadly divided into four different categories, i.e., series-series (SS), series-parallel (SP), parallel-parallel (PP), and parallel-series (PS) [21]. The equivalent circuit diagrams of these topologies are shown in Figure 2. In PP and PS, the transmitter coil does not transfer power in the absence of the receiver coil, protecting the source. Although it is a safe power transfer, during the misalignment of both coils, the topology cannot transfer high power [22]. The SP topology can transfer high power with constant output voltage, but it depends on the load variation, and the voltage gain is too high [23]. The SS topology is the most commonly used technique as the value of the capacitor is independent of the mutual inductance and load resistance [24]. In the SS topology, the resonant frequency is independent of the coupling coefficient and load conditions. This independence is very important as the coupling coefficient varies with misalignments between the coils, and when charging, the resistance of the battery changes. The problem with using the SS topology is that the output current has an inverse relationship with the duty cycle of the DC-DC converter due to which traditional control methods cannot be used. To solve the problem of this inverse relation, the LCC-S compensation topology was introduced in [25], which can achieve adjustable constant voltage at the input of the DC-DC converter.

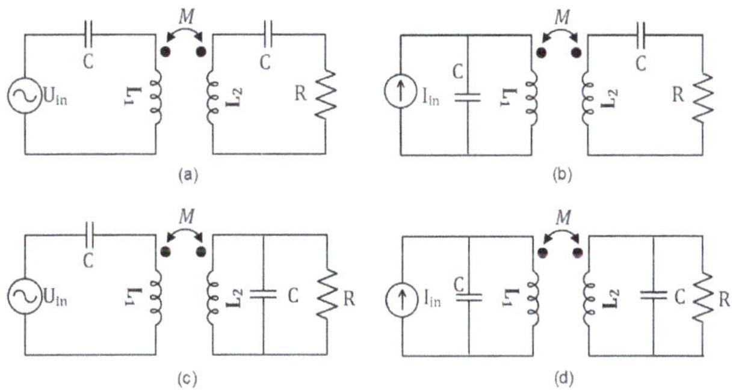

Figure 2. Basic compensation networks (**a**) series-series, (**b**) parallel-series, (**c**) series-parallel, and (**d**) parallel-parallel.

The primary objective of the WPT system is to transfer the energy from the transmitter to charge the energy storage device, i.e., battery, ultra-capacitor, etc. For the simplification of the WPT system design, the energy storage device is considered as a variable load. Furthermore, based on the compensation topology, the WPT system efficiency varies with the load, i.e., the system can achieve the maximum efficiency at a particular resistance value [26]. The objective is to keep the system efficiency high regardless of the load variations. A common approach to solve this issue is to implement a DC-DC converter after the rectifier circuit, which adjust its input resistance by altering the duty cycle of the switch. According to the mentioned approach, researchers have implemented different DC-DC converters such as buck and boost converters [26,27]. By controlling the duty cycle, the input resistance of the buck and boost converter can be altered in the range of $R_L \rightarrow +\infty$ and $0 \rightarrow R_L$, respectively [28]. Conventionally, proportional–integral–derivative (PID) control is the method used to adjust the duty cycle of the DC-DC converter [26,29]. However, due to the linear nature of the PID control, the regulation is limited to a small region. To overcome the shortcomings of PID, the author in [30] proposed a sliding mode control (SMC) for the secondary side DC-DC converter. Due to the non-linear nature of SMC, compared to PID, it is not limited to a small region, but still under the load variations, it exhibits overshoots and has chattering at the equilibrium point. A super-twisting differentiator based high order sliding mode controller (HOSM+STD) was presented in [31]. Compared to the SMC, HOSM+STD has a quicker response during the transition phase, but the controller depends on an optimizing factor "β", which needed to be adjusted for different voltage levels. Otherwise, the response time of the controller will be slow.

Based on the mentioned research work, the LCC-S compensation network based WPT system with a secondary side buck converter is presented in this paper. To control the duty cycle of the buck converter, the discrete fast terminal sliding mode controller (DFTSMC) is proposed to overcome the shortcomings of the SMC. An ultra-capacitor (UC) is connected as the system load, the resistance of which will vary during the charging process. The objective of the paper is to control the duty ratio of the buck converter to maintain the maximum system efficiency during the charging process. Based on the charging requirements of the UC, an efficient control strategy is adopted to ensure that the UC is charged with maximum efficiency. The LCC-S compensation topology is implemented to ensure constant output voltage at the input of the buck converter during the variations in its duty cycle. Depending on the system requirements, the DFTSMC controller regulates the output current or output power under the variations in the connected load.

The paper is structured as follows. Section 2 presents the design and analysis of LCC-S compensation for the WPT system. The relationship of the system efficiency with respect to output load is derived and then transferred to the relationship between the output power and efficiency. The UC charging strategy and the design of the DFTSMC for the buck converter are presented in Section 3. The simulation results of the proposed system and the comparison with other control schemes are discussed in Section 4. Section 5 presents the experimental validation of the proposed system, and Section 6 concludes the paper.

2. Analysis of the LCC-S Compensation Network

The circuit diagram of the proposed system using LCC-S compensation with the buck converter and UC is shown in Figure 3.

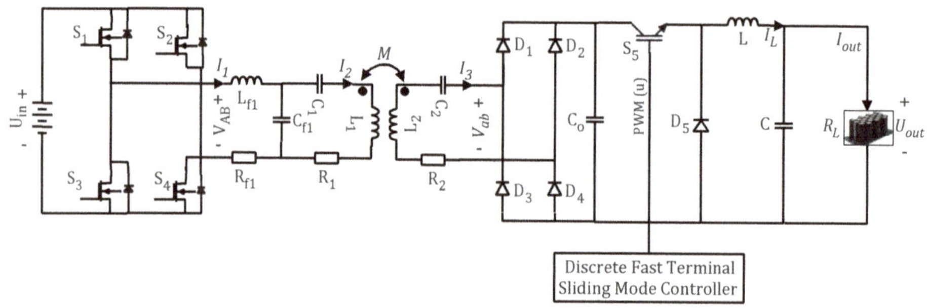

Figure 3. Circuit diagram of the proposed WPT system.

L_{f1} and C_{f1} are the compensation inductor and compensation capacitors for the transmitting side, respectively. L_1 and L_2 are the self-inductances of the primary and secondary coils, and C_1 and C_2 are the series capacitors for the transmitting and receiving side, respectively. V_{AB} is the inverted AC voltage by the high-frequency inverters, and V_{ab} is the output voltage of the receiving coil. M is the mutual inductance between the two coils, and ω_0 is the operating resonant frequency. U_{in} is the input DC voltage, and U_{out} is the output DC voltage. R_{f1}, R_1, and R_2 are the self-resistances of the compensation inductor, primary coil, and secondary coil, respectively. L, C, and S_5 are the inductor, capacitor, and IGBT switch of the buck converter, respectively. R_L is the equivalent resistance of the UC, which varies with its state of charge (SOC).

The equivalent circuit diagram of the proposed system is shown in Figure 4. The R_{eq} is the equivalent resistance of the buck converter, which is equal to:

$$R_{eq} = \frac{1}{u^2} R_L \tag{1}$$

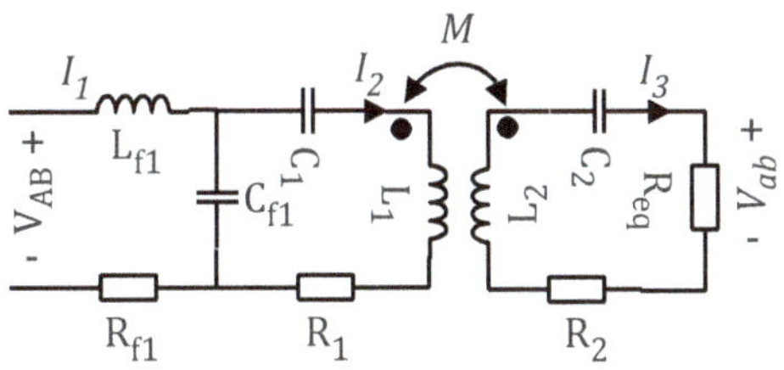

Figure 4. Equivalent diagram of proposed system.

The system can be analyzed using the two-port network theory. To convert the system into a two-port network, the secondary side parameters are transferred to the primary side. The two-port network of the system is shown in Figure 5. Z_r is the transferred impedance of the secondary side, and V_{ab}^* is the voltage across Z_r. Z_r can be calculated as follows,

$$Z_r = \frac{\omega^2 M^2}{R_{eq} + R_2} \tag{2}$$

The system's two-port network can be expressed using the following equations:

$$\begin{cases} V_{AB} = & Z_{11}I_1 + Z_{12}I_2 \\ V_{ab}^* = & Z_{21}I_1 + Z_{22}I_2 \end{cases} \tag{3}$$

Converting Equation (3) into matrix form:

$$\begin{bmatrix} V_{AB} \\ V_{ab}^* \end{bmatrix} = \begin{bmatrix} Z_{11} & Z_{12} \\ Z_{21} & Z_{22} \end{bmatrix} \begin{bmatrix} I_1 \\ I_2 \end{bmatrix} \tag{4}$$

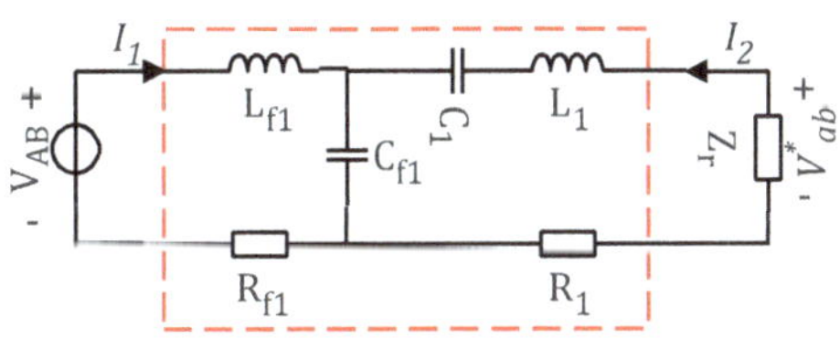

Figure 5. Two port network of the proposed system.

To find the system impedance matrix, two modes are used. In the first mode, shown in Figure 6a, the load is disconnected, which makes $I_2 = 0$. In the second mode, shown in Figure 6b, the input source is disconnected, which makes $I_1 = 0$. Using the two cases, the system impedances are calculated as:

$$Z_{11} = \frac{V_{AB}}{I_1} |_{I_2=0} = R_{f1} + j\left(\omega L_{f1} - \frac{1}{\omega C_{f1}}\right) \tag{5}$$

$$Z_{21} = \frac{V_{ab}^*}{I_1} |_{I_2=0} = \frac{1}{j\omega C_{f1}} \tag{6}$$

$$Z_{22} = \frac{V_{ab}^*}{I_2} |_{I_1=0} = R_1 + j\left(\omega L_1 - \frac{1}{\omega C_{f1}} - \frac{1}{\omega C_1}\right) \tag{7}$$

$$Z_{12} = \frac{V_{AB}}{I_2} |_{I_1=0} = \frac{1}{j\omega C_{f1}} \tag{8}$$

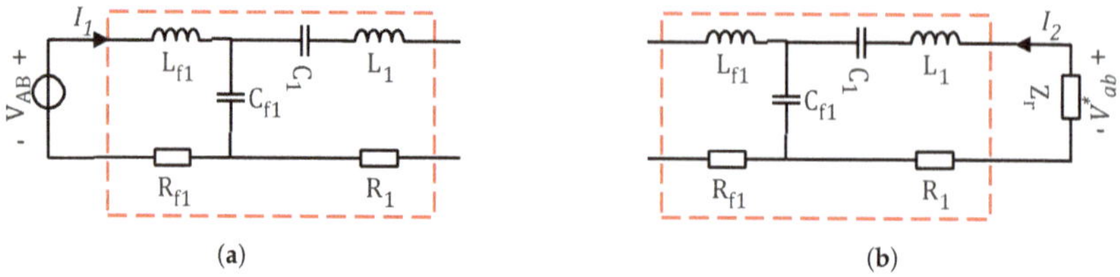

Figure 6. (**a**) Mode 1: Load is disconnected. (**b**) Mode 2: Input source is disconnected.

Using the following equations, the system's parameters are tuned in such a way that the system input voltage and current have zero-phase difference.

$$C_{f1} = \frac{1}{\omega^2 L_{f1}} \tag{9}$$

$$C_1 = \frac{1}{\omega^2 (L_1 - L_{f1})} \tag{10}$$

$$C_2 = \frac{1}{\omega^2 L_2} \tag{11}$$

When the system's parameters satisfy Equations (9)–(11), then Equations (5) to (8) become:

$$\begin{cases} Z_{11} = & R_{f1} \\ Z_{12} = Z_{21} = & j\omega L_{f1} \\ Z_{22} = & R_1 \end{cases} \tag{12}$$

The voltage gain from inverter output voltage to the secondary coil output voltage can be calculated as follows,

$$G_V = \left| \frac{V_{ab}}{V_{AB}} \right| = \left| \frac{V_{ab}}{V^*_{ab}} \right| \left| \frac{V^*_{ab}}{V_{AB}} \right| \tag{13}$$

According to the two-port network theory, $\left| \frac{V^*_{ab}}{V_{AB}} \right|$ can be calculated using the following formula,

$$\left| \frac{V^*_{ab}}{V_{AB}} \right| = \frac{Z_{21} Z_r}{Z_{11}(Z_r + Z_{22}) - Z_{12} Z_{21}} = \frac{\omega^3 L_{f1} M^2}{(R_2 + R_{eq}) \left(R_{f1} \left(\frac{\omega^2 M^2}{R_2 + R_{eq}} + R_1 \right) + \left(\omega L_{f1} \right)^2 \right)} \tag{14}$$

From the circuit diagrams, shown in Figures 5 and 6, V_{ab} and V^*_{ab} can be derived as,

$$V_{ab} = \left(\frac{j\omega M I_2}{R_2 + R_{eq}} \right) R_{eq} \tag{15}$$

$$V^*_{ab} = \frac{\omega^2 M^2 I_2}{R_2 + R_{eq}} \tag{16}$$

Substituting Equations (14)–(16) into Equation (13), the system's voltage gain is derived as,

$$G_V = \frac{\omega^2 L_{f1} M R_{eq}}{(R_2 + R_{eq}) \left(R_{f1} \left(\frac{\omega^2 M^2}{R_2 + R_{eq}} + R_1 \right) + \left(\omega L_{f1} \right)^2 \right)} \tag{17}$$

Similarly, according to the characteristics of the two-port network theory, the inverter output current I_1 and primary coil current I_2 and the current gain G_I can be calculated as:

$$I_1 = \frac{V_{AB}\left(Z_{22}+Z_r\right)}{Z_{11}\left(Z_{22}+Z_r\right)-Z_{12}Z_{21}} = \frac{V_{AB}\left(R_1R_2+R_1R_{eq}+\omega^2M^2\right)}{R_{f1}\left(R_1R_2+R_1R_{eq}+\omega^2M^2\right)+\left(\omega L_{f1}\right)^2\left(R_2+R_{eq}\right)} \tag{18}$$

$$I_2 = \frac{V_{AB}Z_{21}}{Z_{11}\left(Z_{22}+Z_r\right)-Z_{12}Z_{21}} = \frac{V_{AB}\omega L_{f1}(R_2+R_{eq})}{R_{f1}\left(R_1R_2+R_1R_{eq}+\omega^2M^2\right)+\left(\omega L_{f1}\right)^2\left(R_2+R_{eq}\right)} \tag{19}$$

$$G_I = \left|\frac{I_2}{I_1}\right| = \frac{Z_{21}}{Z_r+Z_{22}} = \frac{\omega L_{f1}\left(R_2+R_{eq}\right)}{\omega^2M^2+R_1\left(R_2+R_{eq}\right)} \tag{20}$$

Using Equation (19), the secondary coil current can be derived as,

$$I_3 = \frac{\omega^2 M L_{f1} V_{AB}}{R_{f1}\left(R_1R_2+R_1R_{eq}+\omega^2M^2\right)+\left(\omega L_{f1}\right)^2\left(R_2+R_{eq}\right)} \tag{21}$$

Under the assumption that the system is under resonance condition and there are no conduction losses in the inverter, rectifier, and buck converter, the system efficiency can be calculated as,

$$\eta = \frac{V_{ab}^{*}I_2}{V_{AB}I_1}\frac{R_{eq}}{R_2+R_{eq}} = G_V G_I \frac{R_{eq}}{R_2+R_{eq}} \tag{22}$$

Substituting Equations (17) and (20) into Equation (22), the system's efficiency can be obtained as,

$$\eta = \frac{\omega^4 L_{f1}^2 M^2 R_{eq}}{\left(R_{f1}\left(\frac{\omega^2M^2}{R_2+R_{eq}}+R_1\right)+\left(\omega L_{f1}\right)^2\right)\left(\frac{\omega^2M^2}{R_2+R_{eq}}+R_1\right)\left(R_2+R_{eq}\right)^2} \tag{23}$$

For the system parameters listed in Table 1 and 2, the relationship between the system's efficiency η and load resistance R_{eq} is shown in Figure 7. It can be seen that at a particular resistance R_{op}, i.e., 11 Ω, the system can operate at maximum efficiency. R_{op} can be derived by differentiating Equation (23) with respect to R_{eq}, and R_{op} can be derived as follows,

$$\frac{\partial\eta}{\partial R_{eq}} = 0 \Rightarrow R_{op} = \sqrt{\frac{\left\{R_1R_2R_{f1}+R_2\left(\omega L_{f1}\right)^2+R_{Lf1}\omega^2M^2\right\}\left(R_1R_2+\omega^2M^2\right)}{R_1\left\{R_1R_{Lf1}+\left(\omega L_{f1}\right)^2\right\}}} \tag{24}$$

According to Equation (1), for varying R_L, the buck converter duty cycle u can be used to regulate R_{eq}=R_{op} to make sure that the system operates at maximum efficiency. For the ease of control designing, the optimal resistance R_{op} can be translated into optimal power P_{op}. When the output power of the system is P_{op}, the system will operate at maximum efficiency. Using Equation (17), P_{op} can be obtained as,

$$P_{op} = \frac{V_{ab}^2}{R_{op}} = \left(\frac{1}{R_{op}}\right)\left[\frac{\omega^2 L_{f1} M R_{eq} V_{AB}}{\left(R_2+R_{eq}\right)\left(R_{f1}\left(\frac{\omega^2M^2}{R_2+R_{eq}}+R_1\right)+\left(\omega L_{f1}\right)^2\right)}\right]^2 \tag{25}$$

Using Equation (25), the relationship between the efficiency and the output power is shown in Figure 8. It can be seen that at P_{op}, i.e., 120 Watts, the system efficiency is highest. To track the system output power P_{out} to this maximum efficiency point, the DFTSMC controller is designed in the next section to control the duty cycle u of the secondary side buck converter.

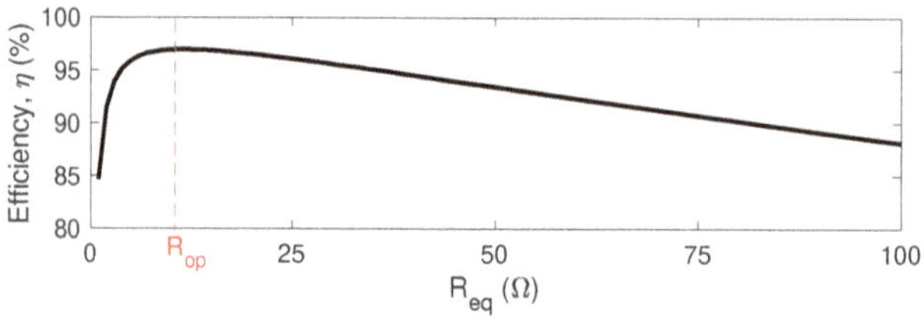

Figure 7. Efficiency, η vs. R_{eq}.

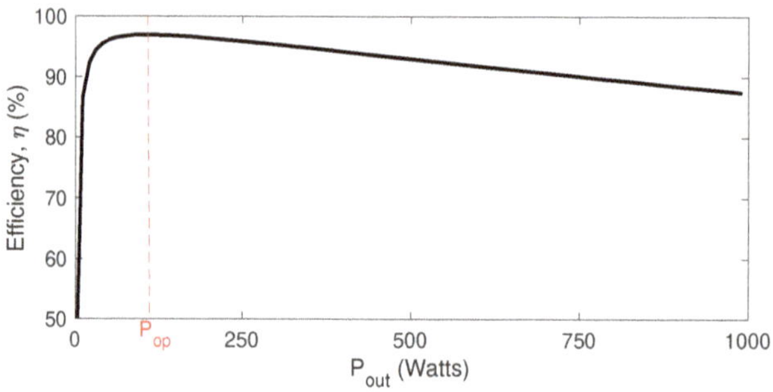

Figure 8. Efficiency, η vs. P_{out}.

3. Discrete Fast Terminal Sliding Mode Controller Design for the Buck Converter

During the charging of the ultra-capacitor, its voltage U_{uc} and current I_{uc} vary in real time due to which R_L constantly varies and can be derived as follows,

$$R_L = \frac{U_{uc}}{I_{uc}} \tag{26}$$

The ultra-capacitor can be charged at maximum efficiency if the system output power is equal to the optimal power, i.e., P_{out}=P_{op}. However, if initially, U_{uc} is too low and it is charged with high power, then it will draw an enormous amount of current, which can damage the system. To solve this issue, the charging of the ultra-capacitor is divided into two stages. The charging strategy is shown in Figure 9. In the first stage, constant current I_{ref} is provided to the ultra-capacitor till its voltage U_{uc} reaches U^r_{uc}, then in the second stage, it is charged with optimal power P_{op} for maximum efficiency.

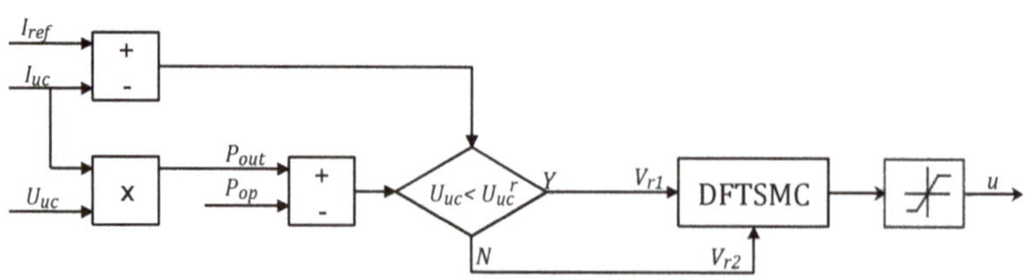

Figure 9. Charging strategy for the ultra-capacitor.

In order to regulate the output current (I_{out}) and output power (P_{out}) to the desired I_{ref} and P_{op}, respectively, a buck converter at the secondary side of the proposed system is used. The required references can be tracked by controlling the duty cycle of the switch $S5$. The circuit diagram of the buck converter is shown in Figure 3. Under the assumption that the buck converter is operating in continuous conduction mode, the following dynamic model is derived.

$$\dot{I}_L = u\frac{V_{ab}}{L} - \frac{U_{out}}{L} \tag{27}$$

$$\dot{U}_{out} = \frac{I_L}{C} - \frac{U_{out}}{R_L C} \tag{28}$$

where I_L and U_{out} are the values of the inductor current and capacitor voltage. $u = [0,1]$ is the duty ratio, used to generate the driving cycle for the switch. For the control design, we will average the model over one switching period. If x_1 is the average value of I_L, x_2 is the average value of U_{out}, and μ is the average value of u, then Equations (27) and (28) take the form,

$$\dot{x}_1 = \mu\frac{V_{ab}}{L} - \frac{x_2}{L} \tag{29}$$

$$\dot{x}_2 = \frac{x_1}{C} - \frac{x_2}{R_L C} \tag{30}$$

By controlling the output voltage of the buck converter, the output current I_{out} and output power P_{out} of the proposed structure can be tracked to the desired references, i.e., I_{ref} and P_{ref}, respectively.

$$V_{r1} = I_{ref} R_L \tag{31}$$

$$V_{r2} = \sqrt{P_{op} R_L} \tag{32}$$

A discrete fast terminal sliding mode controller is designed to generate the required duty ratio for these reference voltages. For this purpose, an error signal is defined.

$$e_1 = x_2 - V_{rj} \tag{33}$$

where (j – 1,2). When j = 1, the controller will track constant current, and for j = 2, the controller will track constant power. By converging the e_1 to zero, we can get our desired result. The dynamic model in Equations (29) and (30) can be rewritten as,

$$\dot{e}_1 = \dot{x}_2 - \dot{V}_{rj} = \frac{x_1}{C} - \frac{x_2}{R_L C} - \dot{V}_{rj} \tag{34}$$

To simplify the expressing, a new variable e_2 is defined as,

$$e_2 = \dot{e}_1 \tag{35}$$

Taking the derivative of e_2 and using Equations (34), (29), and (30), it yields:

$$\dot{e}_2 = \frac{1}{LC}\left(\mu V_{ab} - \left(x_2 - V_{rj}\right) - V_{rj}\right) - \frac{\dot{x}_2}{RC} + \frac{\dot{V}_{rj}}{RC} - \ddot{V}_{rj} \tag{36}$$

The overall dynamic model can be represented as follows,

$$\begin{cases} \dot{e}_1 = & e_2 \\ \dot{e}_2 = & \frac{1}{LC}\left(\mu V_{ab} - e_1 - V_{rj}\right) + \frac{1}{RC}\left(\dot{V}_{rj} - e_2\right) - \ddot{V}_{rj} \end{cases} \tag{37}$$

Using Euler's discretization method, the dynamical model presented in Equation (37) can be discretized as follows,

$$\begin{cases} e_1(k+1) = & e_1(k) + he_2(k) \\ e_2(k+1) = & e_2(k) + \frac{h}{RC}\left(\dot{V}_{rj}(k) - e_2(k)\right) - h\ddot{V}_{rj}(k) + \frac{h}{LC}\left(\mu(k) V_{ab}(k) - e_1(k) - V_{rj}(k)\right) \end{cases} \quad (38)$$

where h is the sampling period. To converge these error signals to zero, a fast terminal sliding surface can be designed as follows,

$$s(k) = e_2(k) + \alpha_1 e_1(k) + \alpha_2 sign(e_1(k)) |e_1(k)|^{\frac{p}{q}} \quad (39)$$

where $0 < \alpha_1 h < 1$, $0 < \alpha_2 h < 1$, and $0 < \frac{p}{q} < 1$ with p and q positive odd integers. As discussed in [32], the sliding mode condition occurs when $s(k+1) = 0$, and Equation (39) becomes,

$$s(k+1) = 0 \Rightarrow \alpha_2 sign(e_1(k+1)) |e_1(k+1)|^{\frac{p}{q}} + e_2(k+1) + \alpha_1 e_1(k+1) = 0 \quad (40)$$

Substituting Equation (38) into (40), it yields,

$$\begin{aligned} 0 =& e_2(k) + \frac{h}{LC}\left(\mu(k) V_{ab}(k) - e_1(k) - V_{rj}(k)\right) + \frac{h}{RC}\left(\dot{V}_{rj}(k) - e_2(k)\right) + \alpha_1\left(e_1(k) + he_2(k)\right) \\ &+ \alpha_2 sign\left(e_1(k) + he_2(k)\right) |e_1(k) + he_2(k)|^{\frac{p}{q}} - h\ddot{V}_{rj}(k) \end{aligned} \quad (41)$$

Finally by solving Equation (41), the DFTSMC law can be obtained as,

$$\begin{aligned} u(k) =& \frac{-LC}{hV_{ab}}\left(e_2(k)\left(1 - \frac{h}{RC} + \alpha_1 h\right) + \frac{h\dot{V}_{rj}(k)}{RC}\right) - \frac{LC}{hV_{ab}}\left(e_1(k)\left(\alpha_1 - \frac{h}{LC}\right) - h\ddot{V}_{rj}(k) - \frac{hV_{rj}}{LC}\right) \\ &- \frac{LC}{hV_{ab}}\left(\alpha_2 sign\left(e_1(k) + he_2(k)\right) |e_1(k) + he_2(k)|^{\frac{p}{q}}\right) \end{aligned} \quad (42)$$

4. Results and Discussion

To verify the performance of the proposed controller, simulations were performed in MATLAB/Simulink using the "Sim Power Systems" toolbox under the abrupt and time-varying fluctuation in load R_L. The uncontrolled rectifier was connected to the load through the buck converter, which was controlled by the DFTSMC. Under variations in "R_L", DFTSMC could regulate the output current I_{out} and output power P_{out}. The specifications of the wireless charging system are shown in Table 1. Using Equations (9)–(11), the parameters of the compensation network were designed and are listed in Table 2. The parameters of the buck converter and DFTSMC are listed in Table 3.

Table 1. Wireless charger parameters.

Symbol	Parameter	Value
V_{in}	Input voltage	80 V
L_1	Transmitting coil inductance	400 µH
R_1	Transmitting coil resistance	~300 mΩ
L_2	Receiving coil inductance	30 µH
R_2	Receiving coil resistance	~120 mΩ
M	Mutual inductance	60 µH
f_s	Switching frequency	40 kHz
P_o	Output power	120 Watts
I_o	Output current	5 A

Table 2. Compensation network parameters.

Symbol	Parameter	Value
L_{f1}	Transmitting side compensation inductance	135 μH
C_{f1}	Transmitting side parallel compensation capacitance	0.117 μF
C_1	Transmitting side series compensation capacitance	0.059 μF
R_{f1}	Transmitting side compensation inductor resistance	200 mΩ
C_2	Receiving side series compensation capacitance	0.52 μF

Table 3. Buck converter and DFTSMC parameters.

Symbol	Parameter	Value
L	Inductor	5 mH
C	Capacitor	500 μF
α_1	Constant	500
α_2	Constant	3000
h	Sampling period	1 ms

According to the charging strategy shown in Figure 9, the DFTSMC was used to generate duty cycle u to regulate the output current to the reference current, i.e., $I_{uc} = I_{ref}$, and the output power to the optimal power $P_{out} = P_{op}$. The current, voltage, and power of the ultra-capacitor are shown in Figure 10.

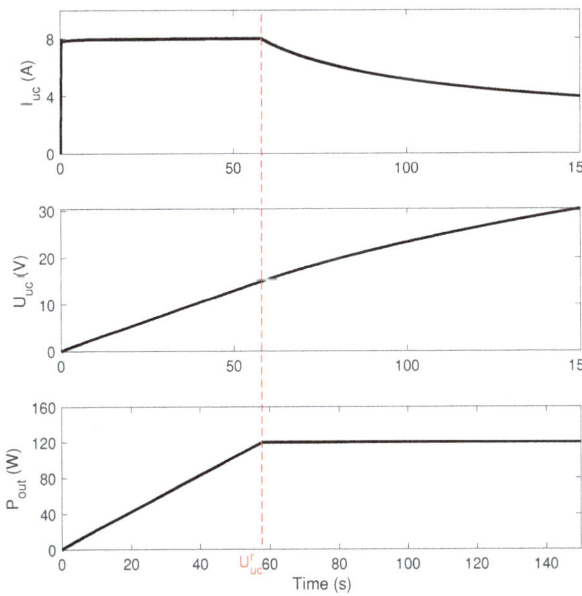

Figure 10. Current, voltage, and power of the ultra-capacitor.

In the first stage, a 5 A current was tracked by the DFTSMC, and then, in the second stage, the output power was regulated to the optimal power, i.e., 120 Watts. It can be seen in Figure 11 that when the DFTSMC regulated P_{out} to P_{op}, the R_{eq} was regulated to R_{op}, i.e., 11 Ω. The duty cycle generated by the DFTSMC, during the charging process, is shown in Figure 12. The efficiency curve is shown in Figure 13, verifying that during the constant power charging stage of the ultra-capacitor, the system operated at a maximum efficiency of about 96%. During the constant power stage, the inverter output voltage and current are shown in Figure 14. According to the voltages and currents in Figure 14, the inverter output power was approximately 129 Watts, which showed that the overall efficiency from the inverter output to the load was approximately 93%.

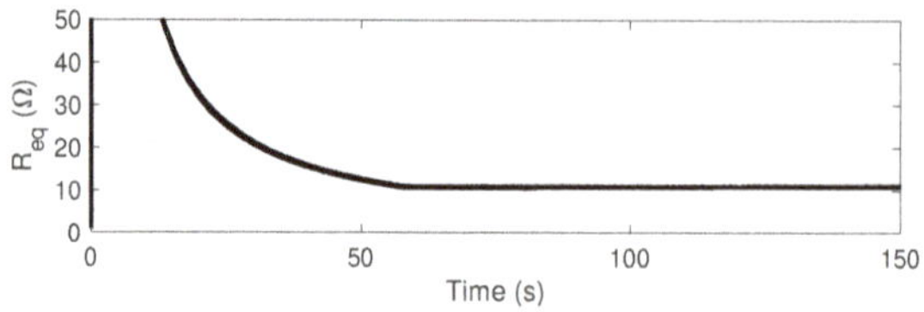

Figure 11. Equivalence resistance of the system.

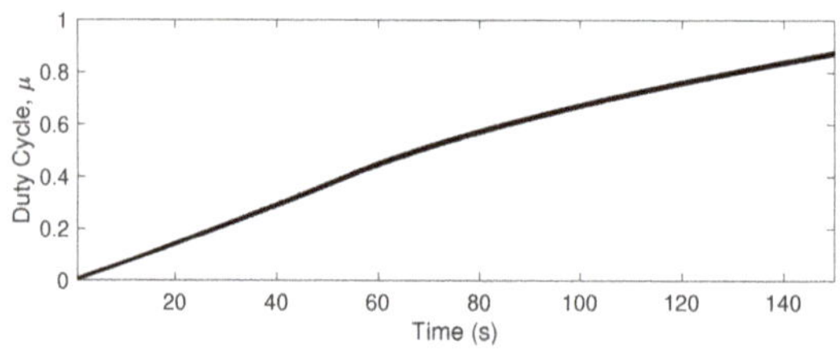

Figure 12. Duty cycle during the charging process.

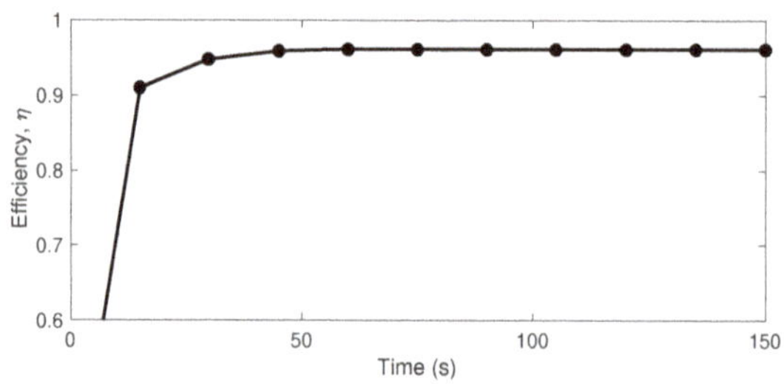

Figure 13. WPT system efficiency during charging process.

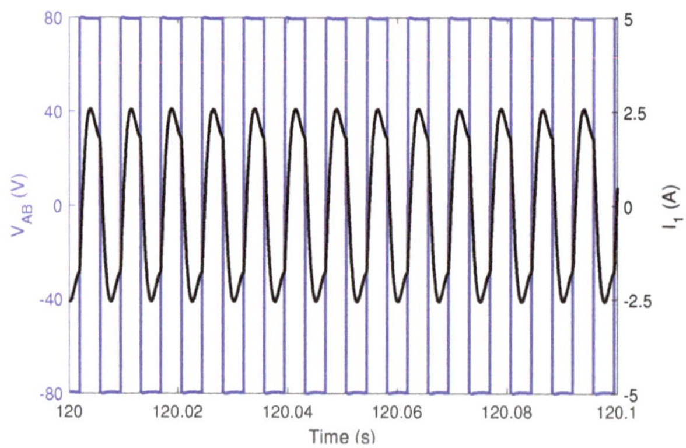

Figure 14. Inverter output voltage and current.

Comparison with Other Control Schemes

To check the robustness of the DFTSMC and compare it with other control schemes such as PID and SMC, load resistance R_L was changed abruptly after every 0.1 s, i.e., with a perturbation frequency of 10 Hz. The value of R_L was initially set at 5 Ω, and then with a fluctuation of 40%, i.e., 2 Ω, it was decremented to 3 Ω and then incremented back to 5 Ω at 0.1 s and 0.2 s, respectively. The perturbation scheme is as follows,

$$R_L = \begin{cases} 5\,\Omega, & t\epsilon\,[0, 0.1)\,s, \\ 3\,\Omega, & t\epsilon\,[0.1, 0.2)\,s, \\ 5\,\Omega, & t\epsilon\,[0.2, 0.3]\,s, \end{cases}$$

Under the mentioned perturbations in R_L, a constant output current of 5 A and constant output power of 120 Watts was tracked by the DFTSMC, sliding mode controller (SMC), and PID controller. Figure 15 shows the regulation of the output current to the referenced current, and Figure 16 shows the convergence of the output power to the referenced power, under the mentioned perturbations in R_L. It can be seen that not only initially, DFTSMC tracked the required current faster, but also under the sudden variations at $t = 0.1$ s and 0.2 s, DFTSMC recovered quickly with respect to PID and SMC with less steady-state error. It can be observed that under these perturbations, although the PID and SMC tracked the required power, compared to DFTSMC, they exhibited initial overshoot, and comparatively, the settling time was large.

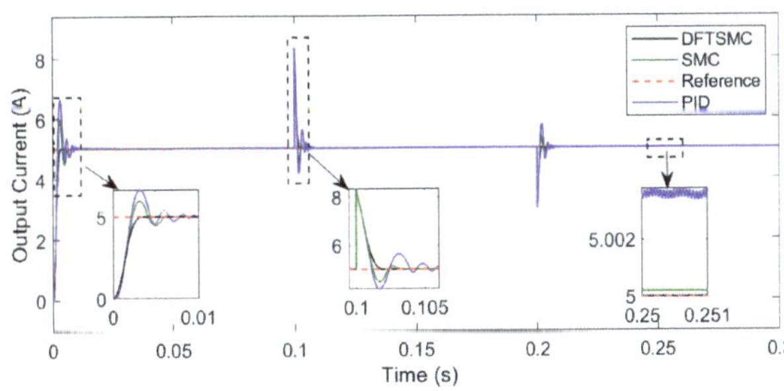

Figure 15. Regulation of output current, I_{out}, during perturbation in R_L.

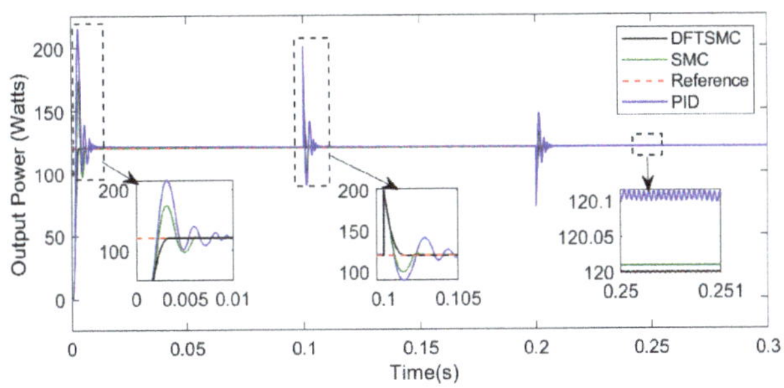

Figure 16. Regulation of output Power, P_{out}, during perturbation in R_L.

5. Experimental Validation

5.1. Experimental Setup

To validate the effectiveness of the proposed system, an experimental platform was fabricated. The topology of the fabricated system was similar to Figure 3, but using DC electronic load instead of the ultra-capacitor. The experiment setup is shown in Figures 17 and 18. The parameters of the experimental setup are consistent with Tables 1–3. The AC/DC rectifier was used to convert the grid AC voltage into DC voltage, and the high-frequency inverter converted the DC voltage into 40 kHz AC voltage. The transmitter and receiver coils were made from tightly wound litz wire with turns of 23 and 10, respectively. The diameter of the transmitter and receiver coils was 29.5 cm and 18.5 cm, respectively, and the gap between the transmitter and receiver coils was 10 cm. The compensation elements such as filter inductance and capacitors were chosen according to the parameters listed in Table 2. The filter inductance was also made from the litz wire, designed in a DD structure to cancel out the cross-coupling effect due to the transmitter coil [33]. The buck converter was connected to the Chroma programmable DC electronic load 63200E, which was configured in constant resistance mode. The DFTSMC controller for the buck converter was implemented using the STM32F334C8 microcontroller, which generated the required PWM signals to track the constant current and power. The output current, output power, and the inverter output current and voltage were observed using the Tektronix MDO3024 Oscilloscope.

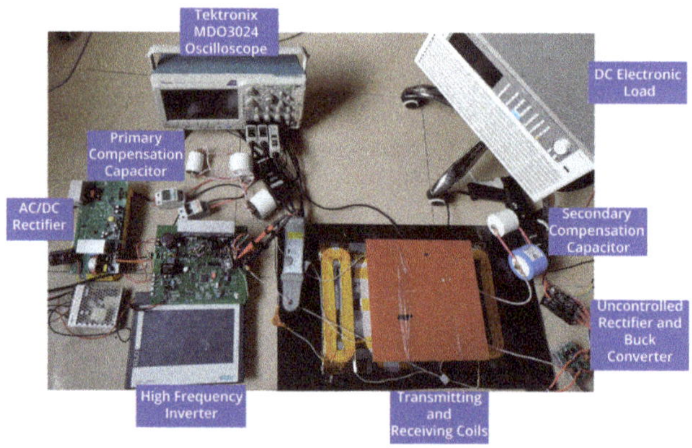

Figure 17. Experimental setup.

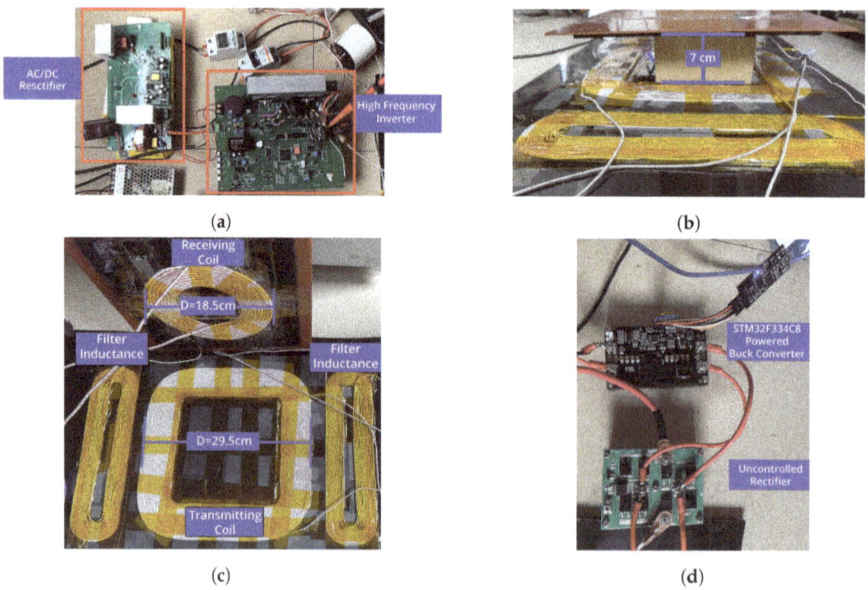

Figure 18. (**a**) AC/DC rectifier and high-frequency inverter. (**b**) Transmitting and receiving coils' gap. (**c**) Transmitting and receiving coils. (**d**) Uncontrolled rectifier and buck converter.

5.2. Results

To check the robustness of the proposed controller, using the programmable DC electronic load, the load resistance was abruptly changed from 5 Ω to 3 Ω and then reverted back from 3 Ω to 5 Ω. During these fluctuations, the controller was used to track the constant current of 5 A and the constant power of 120 Watts.

Figures 19 and 20 show the tracking of the constant current and power, respectively. It can seen that during the fluctuations, the controller quickly recovered and tracked the required reference efficiently. Under the constant power tracking of 120 Watts, Figure 21 shows the inverter output current and voltage. According to Figure 21, the inverter output power was approximately 136 Watts, i.e., the power was transferred from the inverter to the DC Electronic load with an overall efficiency of about 88% and coupling coil efficiency of 95%. The decrease in the overall efficiency was due to losses incurred in the rectifier and buck converter. It can be seen that both the simulation and experimental data exhibited similar behavior, which validated the effectiveness of the controller in real life.

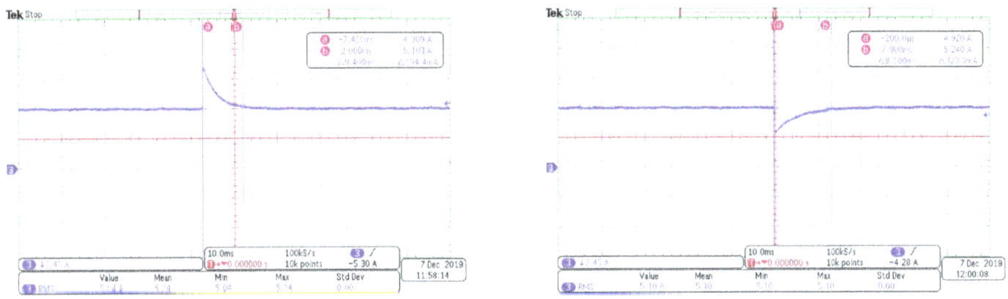

(**a**) Load changed from 5 Ω to 3 Ω (**b**) Load changed from 3 Ω to 5 Ω

Figure 19. Tracking of constant current.

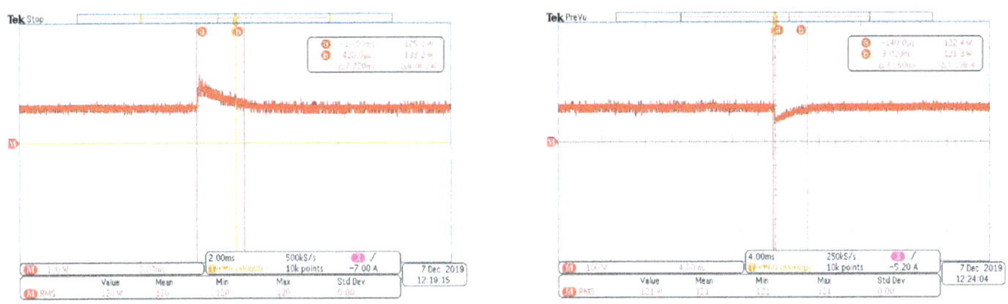

(**a**) Load changed from 5 Ω to 3 Ω (**b**) Load changed from 3 Ω to 5 Ω

Figure 20. Tracking of constant power.

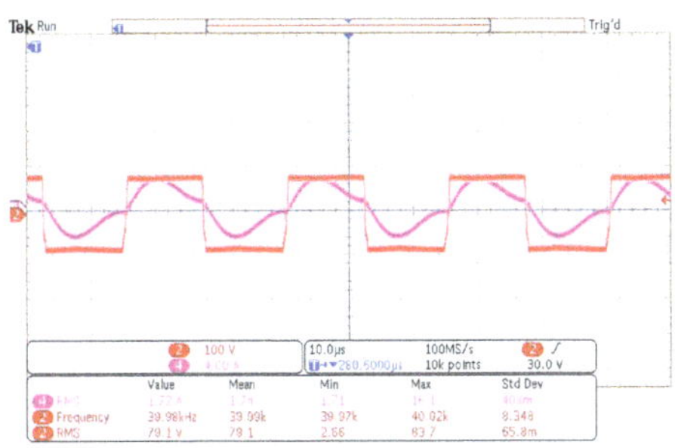

Figure 21. Inverter output voltage and current.

6. Conclusions

In this paper, the LCC-S topology combined with a buck converter at the secondary side was presented. The buck converter was controlled by DFTSMC to regulate the system's output current and power. Using two-port network theory, the system was analyzed, and optimal efficiency equations were derived. Based on the analysis, the proposed system operated at maximum efficiency, when connected with optimal load, or generated optimal power. Using DFTSMC, the duty cycle of the buck converter was generated to track the required optimal power point. The simulation results verified that during charging of the ultra-capacitor, under the constant power stage, the system operated at the maximum efficiency point. The comparison with PID and SMC showed that the proposed controller overall performed better in tracking the desired current and power. Experiments were performed to validate the effectiveness of the proposed controller. The experimental results coincided with the simulation and theoretical analysis. The experimental data verified that under the abrupt perturbation in the load resistance, the controller performed well and tracked the required current and power. The future aim of this study is to model the losses incurred in the inverter, rectifier, and buck converter to analyze the effects on the system efficiency.

Author Contributions: Conceptualization, N.A. and Z.L.; formal analysis, N.A. and Z.L.; funding acquisition, Z.L. and Y.H.; investigation, Z.L. and X.W.; methodology, N.A., X.W. and H.A.; project administration, Z.L. and Y.H.; resources, Z.L. and Y.H.; software, N.A., Y.H. and H.A.; supervision, Z.L.; validation, N.A., Z.L., X.W., H.A. and A.A.; writing, original draft, N.A.; writing, review and editing, N.A., H.A. and A.A. All authors have read and agreed to the published version of the manuscript.

Funding: This research was funded by the Science and Technology Research Project of Shandong Province under Grant No. 2019GGX104080.

Conflicts of Interest: The authors declare no conflict of interest.

Abbreviations

The following abbreviations are used in this manuscript:

WPT	Wireless power transfer
RIPT	Resonant inductive power transfer
CWPT	Capacitive wireless power transfer
EMR	Electromagnetic radiation
AWPT	Acoustic wireless power transfer
OWPT	Optical wireless power transfer
SMC	Sliding mode controller
PID	Proportional-integral-derivative
DFTSMC	Discrete fast terminal sliding mode controller
SOC	State of charge
SS	Series-series
SP	Series-parallel
PS	Parallel-series
PP	Parallel-parallel
LCC	inductor-capacitor-capacitor

References

1. Tesla, N. Apparatus for Transmitting Electrical Energy. US Patent 1,119,732, 1 December 1914.
2. Lu, X.; Wang, P.; Niyato, D.; Kim, D.I.; Han, Z. Wireless Charging Technologies: Fundamentals, Standards, and Network Applications. *IEEE Commun. Surv. Tutor.* **2016**, *18*, 1413–1452. [CrossRef]
3. Barth, H.; Jung, M.; Braun, M.; Schmülling, B.; Reker, U. Concept evaluation of an inductive charging system for electric vehicles. In Proceedings of the 3rd European Conference Smart Grids and E-Mobility, Munich, Germany, 17–18 October 2011.
4. Hui, S.Y.R.; Zhong, W.; Lee, C.K. A Critical Review of Recent Progress in Mid-Range Wireless Power Transfer. *IEEE Trans. Power Electron.* **2014**, *29*, 4500–4511. [CrossRef]

5. Lu, X.; Niyato, D.; Wang, P.; Kim, D.I.; Han, Z. Wireless charger networking for mobile devices: Fundamentals, standards, and applications. *IEEE Wirel. Commun.* **2015**, *22*, 126–135. [CrossRef]
6. Tang, Y.; Chen, Y.; Madawala, U.K.; Thrimawithana, D.J.; Ma, H. A New Controller for Bidirectional Wireless Power Transfer Systems. *IEEE Trans. Power Electron.* **2018**, *33*, 9076–9087. [CrossRef]
7. Lu, F.; Zhang, H.; Mi, C. A Two-Plate Capacitive Wireless Power Transfer System for Electric Vehicle Charging Applications. *IEEE Trans. Power Electron.* **2018**, *33*, 964–969. [CrossRef]
8. RamRakhyani, A.K.; Mirabbasi, S.; Chiao, M. Design and Optimization of Resonance-Based Efficient Wireless Power Delivery Systems for Biomedical Implants. *IEEE Trans. Biomed. Circuits Syst.* **2011**, *5*, 48–63. [CrossRef] [PubMed]
9. Xue, R.; Cheng, K.; Je, M. High-Efficiency Wireless Power Transfer for Biomedical Implants by Optimal Resonant Load Transformation. *IEEE Trans. Circuits Syst. I Regul. Pap.* **2013**, *60*, 867–874. [CrossRef]
10. Mashhadi, I.A.; Pahlevani, M.; Hor, S.; Pahlevani, H.; Adib, E. A New Wireless Power Transfer Circuit for Retinal Prosthesis. *IEEE Trans. Power Electron.* **2019**, *34*, 6425–6439. [CrossRef]
11. Zhang, Y.; Zhao, Z.; Chen, K. Frequency Decrease Analysis of Resonant Wireless Power Transfer. *IEEE Trans. Power Electron.* **2014**, *29*, 1058–1063. [CrossRef]
12. Roes, M.G.; Duarte, J.L.; Hendrix, M.A.; Lomonova, E.A. Acoustic energy transfer: A review. *IEEE Trans. Ind. Electron.* **2012**, *60*, 242–248. [CrossRef]
13. Dai, J.; Ludois, D.C. Wireless electric vehicle charging via capacitive power transfer through a conformal bumper. In Proceedings of the 2015 IEEE Applied Power Electronics Conference and Exposition (APEC), Charlotte, NC, USA, 15–19 March 2015; pp. 3307–3313. [CrossRef]
14. Dai, H.; Liu, Y.; Chen, G.; Wu, X.; He, T.; Liu, A.X.; Ma, H. Safe Charging for Wireless Power Transfer. *IEEE/ACM Trans. Netw.* **2017**, *25*, 3531–3544. [CrossRef]
15. Dickinson, R.M. Wireless power transmission technology state of the art the first Bill Brown lecture. *Acta Astronaut.* **2003**, *53*, 561–570. [CrossRef]
16. Sahai, A.; Graham, D. Optical wireless power transmission at long wavelengths. In Proceedings of the 2011 International Conference on Space Optical Systems and Applications (ICSOS), Santa Monica, CA, USA, 11–13 May 2011; pp. 164–170.
17. Tseng, V. F.-G.; Bedair, S. S.; Lazarus, N. Phased array focusing for acoustic wireless power transfer. *IEEE Trans. Ultrasonics Ferroelectrics Frequency Control.* **2017**, *65*, 39–49. [CrossRef] [PubMed]
18. Awal, M.R.; Jusoh, M.; Sabapathy, T.; Kamarudin, M.R.; Rahim, R.A. State-of-the-art developments of acoustic energy transfer. *Int. J. Antennas Propag.* **2016**, *2016*, 3072528. [CrossRef]
19. Li, S.; Mi, C.C. Wireless power transfer for electric vehicle applications. *IEEE J. Emerg. Sel. Top. Power Electron.* **2015**, *3*, 4–17.
20. Qiu, C.; Chau, K.T.; Ching, T.W.; Liu, C. Overview of wireless charging technologies for electric vehicles. *J. Asian Electr. Veh.* **2014**, *12*, 1679–1685. [CrossRef]
21. Wang, C.S.; Stielau, O.H.; Covic, G.A. Design considerations for a contactless electric vehicle battery charger. *IEEE Trans. Ind. Electron.* **2005**, *52*, 1308–1314. [CrossRef]
22. Kalwar, K.A.; Mekhilef, S.; Seyedmahmoudian, M.; Horan, B. Coil design for high misalignment tolerant inductive power transfer system for EV charging. *Energies* **2016**, *9*, 937. [CrossRef]
23. Cirimele, V.; Freschi, F.; Guglielmi, P. Wireless power transfer structure design for electric vehicle in charge while driving. In Proceedings of the 2014 International Conference on Electrical Machines (ICEM), Berlin, Germany, 2–5 September 2014; pp. 2461–2467.
24. Spanik, P.; Frivaldsky, M.; Drgona, P.; Jaros, V. Analysis of proper configuration of wireless power transfer system for electric vehicle charging. In Proceedings of the ELEKTRO 2016, Strbske Pleso, Slovakia, 16–18 May 2016; pp. 231–237.
25. Geng, Y.; Li, B.; Yang, Z.; Lin, F.; Sun, H. A high efficiency charging strategy for a supercapacitor using a wireless power transfer system based on inductor/capacitor/capacitor (LCC) compensation topology. *Energies* **2017**, *10*, 135. [CrossRef]
26. Li, H.; Li, J.; Wang, K.; Chen, W.; Yang, X. A maximum efficiency point tracking control scheme for wireless power transfer systems using magnetic resonant coupling. *IEEE Trans. Power Electron.* **2014**, *30*, 3998–4008. [CrossRef]

27. Li, B.; Geng, Y.; Lin, F.; Yang, Z.; Igarashi, S. Design of constant voltage compensation topology applied to wpt system for electrical vehicles. In Proceedings of the 2016 IEEE Vehicle Power and Propulsion Conference (VPPC), Hangzhou, China, 17–20 October 2016; pp. 1–6.
28. Li, Z.; Zhu, C.; Jiang, J.; Song, K.; Wei, G. A 3-kW wireless power transfer system for sightseeing car supercapacitor charge. *IEEE Trans. Power Electron.* **2016**, *32*, 3301–3316. [CrossRef]
29. Zhong, W.; Hui, S. Maximum energy efficiency tracking for wireless power transfer systems. *IEEE Trans. Power Electron.* **2014**, *30*, 4025–4034. [CrossRef]
30. Yang, Y.; Zhong, W.; Kiratipongvoot, S.; Tan, S.C.; Hui, S.Y.R. Dynamic improvement of series–series compensated wireless power transfer systems using discrete sliding mode control. *IEEE Trans. Power Electron.* **2017**, *33*, 6351–6360. [CrossRef]
31. Huangfu, Y.; Zhuo, S.; Rathore, A.; Breaz, E.; Nahid-Mobarakeh, B.; Gao, F. Super-twisting differentiator-based high order sliding mode voltage control design for DC-DC buck converters. *Energies* **2016**, *9*, 494. [CrossRef]
32. Xu, J.-X.; Abidi, K. Output Tracking with Discrete-Time Integral Sliding Mode Control. In *Modern Sliding Mode Control Theory: New Perspectives and Applications*; Bartolini, G., Fridman, L., Pisano, A., Usai, E., Eds.; Springer: Berlin/Heidelberg, Germany, 2008; pp. 247–268.
33. Lu, F.; Zhang, H.; Hofmann, H.; Su, W.; Mi, C.C. A dual-coupled LCC-compensated IPT system with a compact magnetic coupler. *IEEE Trans. Power Electron.* **2017**, *33*, 6391–6402. [CrossRef]

© 2020 by the authors. Licensee MDPI, Basel, Switzerland. This article is an open access article distributed under the terms and conditions of the Creative Commons Attribution (CC BY) license (http://creativecommons.org/licenses/by/4.0/).

energies

MDPI

Article

Energy Efficiency Comparison of Hydraulic Accumulators and Ultracapacitors

Jorge Leon-Quiroga [1], Brittany Newell [1], Mahesh Krishnamurthy [2], Andres Gonzalez-Mancera [3] and Jose Garcia-Bravo [1,*]

1 Purdue Polytechnic School of Engineering Technology, Purdue University, West Lafayette, IN 47907, USA; jleonqui@purdue.edu (J.L.-Q.); bnewell1@purdue.edu (B.N.)
2 Department of Electrical and Computer Engineering, Illinois Institute of Technology, Chicago, IL 60616, USA; kmahesh@ece.iit.edu
3 Department of Mechanical Engineering, Universidad de los Andes, Bogota 111711, Colombia; angonzal@uniandes.edu.co
* Correspondence: jmgarcia@purdue.edu

Received: 21 February 2020; Accepted: 20 March 2020; Published: 2 April 2020

Abstract: Energy regeneration systems are a key factor for improving energy efficiency in electrohydraulic machinery. This paper is focused on the study of electric energy storage systems (EESS) and hydraulic energy storage systems (HESS) for energy regeneration applications. Two test benches were designed and implemented to compare the performance of the systems under similar operating conditions. The electrical system was configured with a set of ultracapacitors, and the hydraulic system used a hydraulic accumulator. Both systems were designed to have the same energy storage capacity. Charge and discharge cycle experiments were performed for the two systems in order to compare their power density, energy density, cost, and efficiency. According to the experimentally obtained results, the power density in the hydraulic accumulator was 21.7% higher when compared with the ultracapacitors. Moreover, the cost/power ($/Watt) ratio in the hydraulic accumulator was 2.9 times smaller than a set of ultracapacitors of the same energy storage capacity. On the other hand, the energy density in the set of ultracapacitors was 9.4 times higher, and the cost/energy ($/kWh) ratio was 2.9 times smaller when compared with the hydraulic accumulator. Under the tested conditions, the estimated overall energy efficiency for the hydraulic accumulator was 87.7%, and the overall energy efficiency for the ultracapacitor was 78.7%.

Keywords: efficiency; energy storage systems; electrical power systems; hydraulic power systems; hydraulic accumulator; ultracapacitor

1. Introduction

Energy consumption in the transportation and the industrial sector represented 72% of the total energy consumption during 2018 in the United States. Most of this energy (around 87%) comes from petroleum-based sources and natural gas [1]. In the industrial sector, around 24% of the energy is spent by the agriculture, construction, and mining sectors. Many applications within these sectors use hydraulic equipment and hydraulic machinery, so improving the energy efficiency of the hydraulic systems that are already in use could have a large impact on the reduction of energy consumption and emissions. If the energy efficiency of the hydraulic machinery used in industrial applications is improved by 5%, based on the data provided by the U.S. Energy Department [1], it is possible to estimate an overall annual reduction of 1% in energy consumption in the US.

Over the last 20 years, there has been interest in developing and improving systems for energy regeneration in hydraulic machinery [2–5]. Some of the options for energy storage in energy regeneration devices include flywheels, compressed air, electrical energy storage systems (EESS), and hydraulic energy storage systems (HESS). In the electrical energy regeneration system, an electric accumulator or capacitor is used to store energy [6]. This kind of system is used in electric hybrid machinery. The return line of a boom mechanism in an excavator is connected to a hydraulic motor that is used to move an electric generator that produces electrical energy that is stored in an electric accumulator (ultracapacitor or battery) [2,3,7]. The main benefit of this system is an improvement in energy efficiency, but the complexity that is added to the baseline hydraulic system is evident. Ultracapacitors are mostly used in electric applications that require a high power density such as wind turbines in remote areas [8], energy regeneration and engine size reduction in rubber wheeled gantry cranes [9], and even biomedical applications [10].

The hydraulic energy regeneration system is similar to its electric counterpart. Instead of having an electric accumulator (ultracapacitor or battery), the hydraulic energy regeneration system uses a hydraulic accumulator that works as the energy storage device [6]. One of the main benefits when using a hydraulic regenerative system is the relative ease of installation, since the baseline application already uses a hydraulic system. Moreover, no complex power controls are needed, which is a significant advantage. On the other hand, the energy storage density in hydraulic accumulators can be a drawback when compared to a traditional electric system. Hydraulic regenerative systems have been studied for applications in relief valves [11], where the flow in the return line of the valve is used to charge a hydraulic accumulator. Alternatives like digital hydraulics have been studied to improve energy efficiency in hydraulic systems [12,13]. In these studies, a network of valves is used to change the flow rate in the actuator and to reconfigure the system in order to use the flow from assistive loads to move actuators with resistive loads.

2. Relevance of This Work

Most of the previous studies regarding EESS and HESS have focused on the characteristics of each energy storage system separately. Some studies have focused on the study of the performance of ultracapacitors [14–17] and others have focused on hydraulic accumulators. [18,19], but very little research has been done to compare the two storage systems (hydraulic accumulators and ultracapacitors) side by side. The main purpose of this study was to have a direct comparison of the performance characteristics of ultracapacitors and hydraulic accumulators when used as energy storage devices. Previous studies concerning EESS and HESS for hybrid applications have not compared the benefits and drawbacks of both systems under similar operating conditions; these studies have been focused on each system individually. For this work, an experimental procedure was developed for measuring the charging and discharging cycles of a hydraulic accumulator. Likewise, a test bench using ultracapacitors of similar energy storage capabilities was tested. The estimated energy capacity of each system was modeled with the equations shown in the following section. The experimental procedure and experimental equipment are also described in that section.

The results of this research can be used to stablish a control strategy to optimize systems that use hydraulic accumulators as energy storage systems or as a design strategy to select one over the other.

3. Test Bench Description

The main purpose of the test benches developed in this study was to compare an electrical and a hydraulic energy storage system under similar operating conditions in order to determine the efficiency, power density, energy density, and cost by energy capacity. The main characteristics of the hydraulic accumulator and the ultracapacitor used are shown below in Tables 1 and 2.

Table 1. Characteristics of the hydraulic accumulator used in this study.

Hydraulic Accumulator	
Manufacturer	**Parker Hannifin**
Reference	A2N0058D1K
Mass (kg)	4.53
Vol. Capacity (cm^3)	950
Max. Pressure (Bar)	207

Table 2. Characteristics of the ultracapacitor used in this study.

Ultracapacitor	
Manufacturer	**Maxwell Technologies**
Reference	BCAP0050 P270 S01
Mass (g)	12.2
Energy Capacity (mWh)	50.6
Max. Voltage (V)	2.7
Max. Current (A)	6.1

To compare both devices in a similar way, two test benches were designed and built. Both test benches were designed to measure charge and discharge response of the systems, which in a hydraulic system are the pressure and flow rate and in the electric system are the voltage and current.

3.1. Hydraulic Test Bench

The test bench for the hydraulic system was designed to measure the flow of energy through the accumulator while charging and discharging. The main components of the system were the battery, the electric motor, the hydraulic pump, and the hydraulic accumulator. The battery was the main source of power for the system, and it was connected to the electric motor with a DC/AC inverter. The electric motor was used to drive the hydraulic pump, which moved the fluid from the reservoir to the hydraulic accumulator. The hydraulic testbench and the DC/AC inverter are shown in Figures 1 and 2, respectively. The schematic representation of the test bench that was designed and constructed for the hydraulic accumulator is presented in Figure 3.

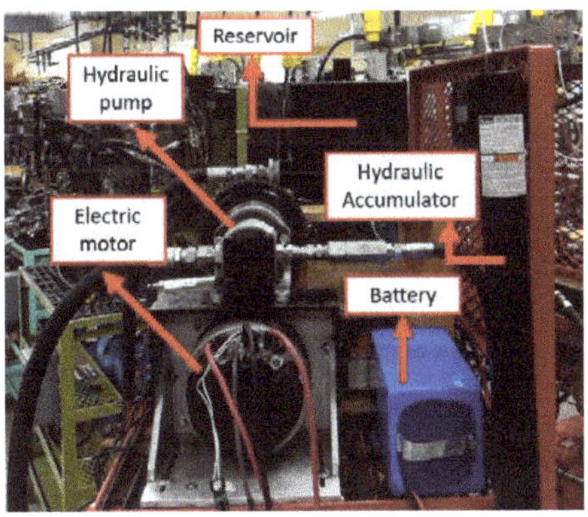

Figure 1. Hydraulic test bench.

Figure 2. DC/AC inverter used in the hydraulic test bench.

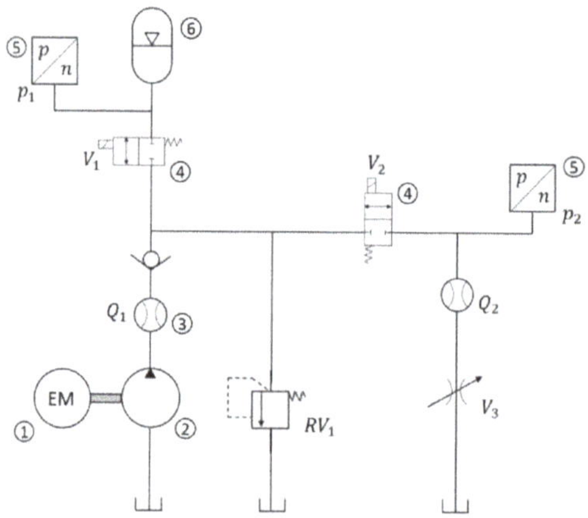

Figure 3. Schematic representation for the hydraulic accumulator test bench.

A list of the components used in the test bench is presented in Table 3.

Table 3. List of components used in the hydraulic test bench.

Index	Device	Reference	Characteristics
1	Electric motor	Motenergy 0907	Max. Speed: 5000 rpm Peak torque: 38 Nm
2	Hydraulic pump	GP-F20-12-P-A	Disp: 12 cm^3/rev Max. Flow: 40 L/min Max. Pressure: 252 Bar
3	Flow meter	FlowTech FSC 375	Max. Pressure: 6 kpsi Max. Flow: 26.45 L/min
4	V_1 and V_2	Hydraforce 12 V DC NC solenoid valve	Max. Flow: 56.7 L/min Max. Pressure: 3 kpsi
5	Pressure gage	Wika A-10	Max. Pressure: 5 kpsi Signal output: 4 to 20 mA
6	Hydraulic accumulator	Parker A2N0058D1K	Capacity: 58 in^3 Max. Pressure: 3 kpsi Precharge pressure: 1 kpsi

The test bench could be operated in either the charging or discharging modes. During charging mode, the electric AC motor was turned on and was used to move the hydraulic pump. Valve V_1 was switched on, allowing flow to go through the accumulator while valve V_2 was closed. The relief valve RV1 was closed unless a relief pressure of 3000 psi was reached. A schematic representation of the test bench during charging mode is presented in Figure 4. The red lines show the high-pressure lines.

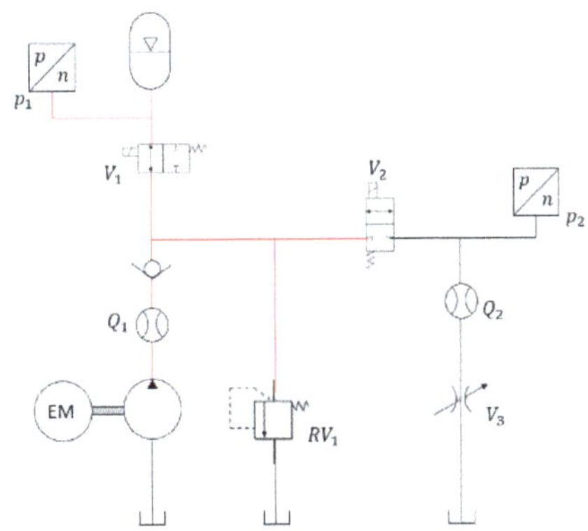

Figure 4. Hydraulic test bench during charging mode.

During charging mode, the measured variables were the current and voltage from the battery connected to the electric AC motor powering the pump and the pressure and flow rate going to the accumulator. The battery and the AC motor were connected through a DC/AC inverter. With these variables, it was possible to calculate the power input from the battery, the power going to the accumulator, the consumption of electric energy, and the energy saved in the accumulator. The speed of the electric motor was changed with a motor controller, which was connected between the battery and the electric motor. The schematic representation of the electric system used to power the pump is shown in Figure 5. The technical data of the battery, the inverter, and the electric motor are shown in Table 4.

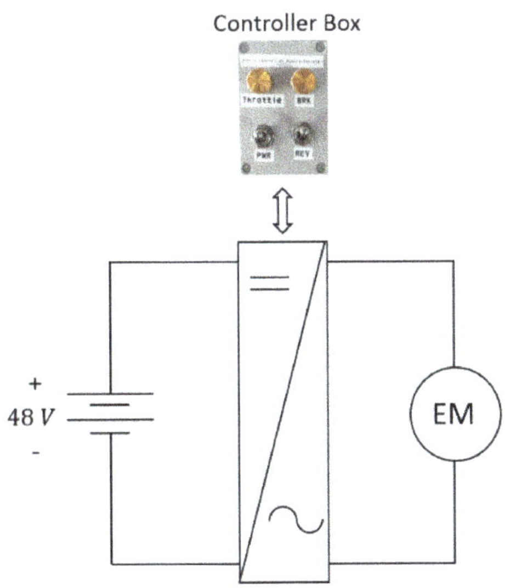

Figure 5. Schematic representation of the electric system.

Table 4. Technical data of the electric system.

Device	Reference	Characteristics
Electric motor	Motenergy 0907	Max. Speed: 5000 rpm Peak torque: 38 Nm Continuous current of 80 Amps AC Inductance phase to phase: 0.1 millihenry
Inverter	KEB48600	Max. Power: 6 kW Max. voltage: 48 V Max. Current: 125 A
Battery	SUN-CYCLE LiFePO4 48 V 24 Ah	Max. Voltage: 48 V Max. discharge current: 60 A Weight: 9.8 kg

To measure the efficiency of the hydraulic accumulator, a similar test for the discharge cycle was developed. In the discharge test, the output load was simulated with a variable orifice, V_3. The load applied to the hydraulic system increased when V_3 was progressively closed. A no-load condition was simulated when the valve was completely open. A schematic representation of the system during discharge mode is presented in Figure 6.

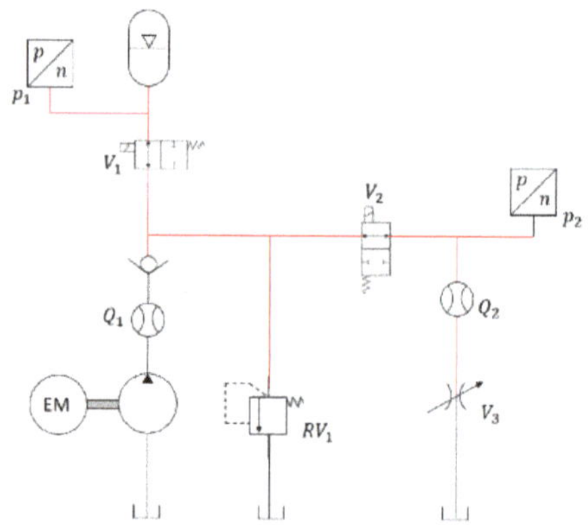

Figure 6. Hydraulic test bench during discharging mode.

To discharge the accumulator, valves V_1 and V_2 were open, while the orifice was open just between 0% and 10%. The measured variables during the discharge mode were the pressure and the flow rate just before the orifice; with these variables, it was possible to calculate the instantaneous hydraulic power used from the accumulator. The output power could be calculated by multiplying the flow rate and the pressure, and the efficiency of the hydraulic accumulator could be derived. Equations (1) and (2) were used to calculate the power input and power output in the accumulator, and Equation (3) was used to calculate the efficiency, the nomenclature shown in Table 5 describes the variables used. The results for a single test are shown in Figure 7 as a reference.

$$P_{In} = Q_{In} p_{Acc,In} \tag{1}$$

$$P_{Out} = Q_{Out} p_{Acc,Out} \tag{2}$$

$$\eta_{Acc} = \frac{\int Q_{Out} p_{Acc,Out} dt}{\int Q_{In} p_{Acc,In} dt} \tag{3}$$

Table 5. Variables used for determining the instantaneous efficiency of the accumulator.

Variable	Description
P_{In} (W)	Power charging the accumulator
P_{Out} (W)	Power discharging the accumulator
Q_{In} (gal/min)	Volumetric flow charging the accumulator
Q_{Out} (gal/min)	Volumetric flow discharging the accumulator
$p_{Acc,In}$ (psi)	Pressure charging the accumulator
$p_{Acc,Out}$ (psi)	Pressure discharging the accumulator
η_{Acc}	Efficiency of the accumulator

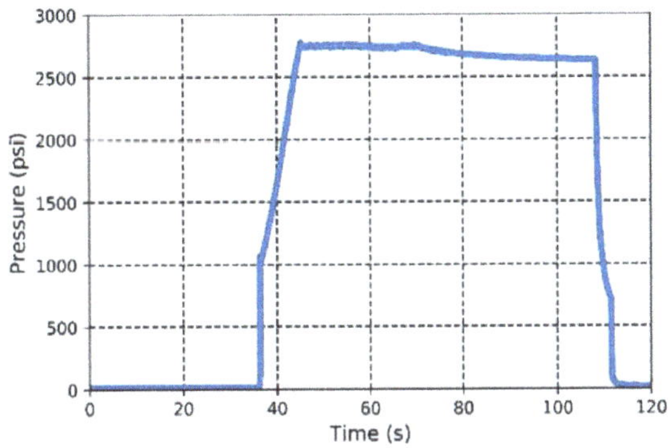

Figure 7. Pressure in the accumulator during the charging and discharging process.

The results for pressure and flow rate in the accumulator during charging are shown in Figure 8. This plot shows a sudden rise in pressure when V_1 was opened and a steady increase in pressure once the pre-charge pressure was reached. It took approximately 17 seconds to fully charge the accumulator to its maximum pressure of 2800 psi. The flow rate plot showed a sudden increase in flow rate and a decrease in flow rate once the pre-charge pressure was reached.

The results for pressure p_2 and flow rate Q_2 during discharging are shown in Figure 9. The discharge time for the accumulator was approximately 3 seconds with the valve 25% open.

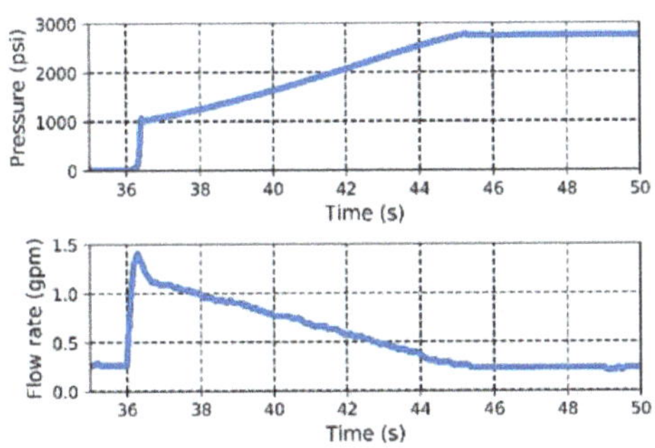

Figure 8. Pressure and flow rate while charging.

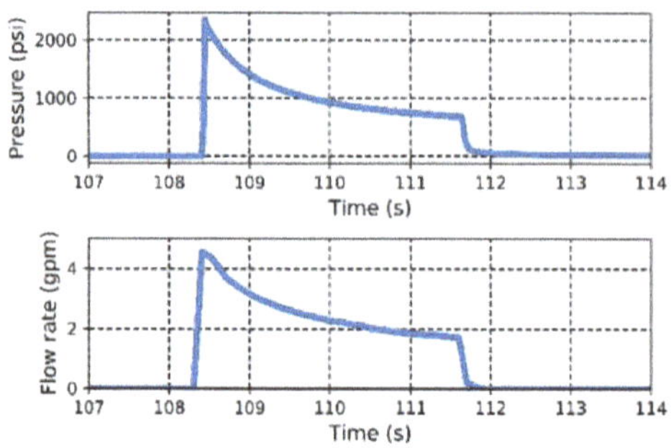

Figure 9. Pressure and flow rate while discharging.

Several experiments like the one described in this section were carried out to calculate the efficiency and performance of the hydraulic accumulator. Load conditions were changed in all the experiments by changing the orifice area. A list of the experiments is shown in Table 6. The 100% value for the orifice area was 11.7 mm^2.

Table 6. List of experiments performed in the hydraulic test bench.

Orifice Area While Charging	Orifice Area While Discharging	Orifice Area While Charging	Orifice Area While Discharging
3.1%	3.1%	12.5%	3.1%
	6.2%		6.2%
	9.4%		9.4%
	12.5%		12.5%
	25.0%		25.0%
	100.0%		100.0%
6.2%	3.1%	25.0%	3.1%
	6.2%		6.2%
	9.4%		9.4%
	12.5%		12.5%
	25.0%		25.0%
	100.0%		100.0%
9.4%	3.1%	100.0%	3.1%
	6.2%		6.2%
	9.4%		9.4%
	12.5%		12.5%
	25.0%		25.0%
	100.0%		100.0%

3.2. Electric Test Bench

To compare the hydraulic accumulator with the ultracapacitor set, an electric test bench was designed and constructed. A schematic representation of the test bench is presented in Figure 10.

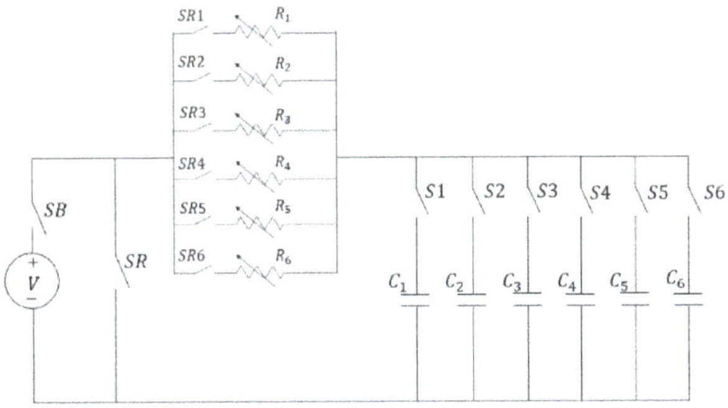

Figure 10. Schematic representation of the test bench used for ultracapacitors.

The ultracapacitors and the rheostats are shown in Figures 11 and 12, respectively.

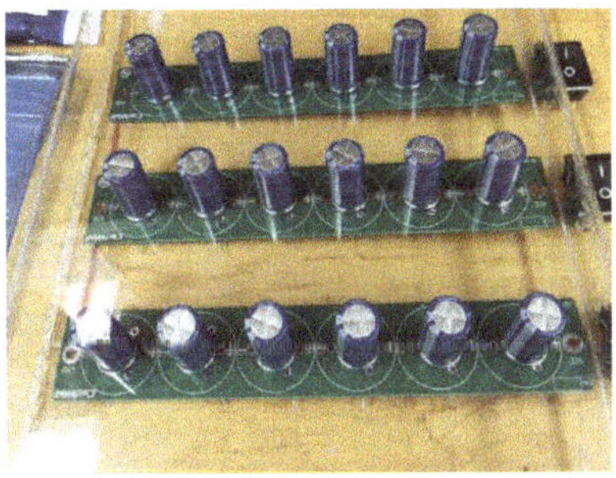

Figure 11. Ultracapacitors used in the electric test bench.

Figure 12. Resistor bank used in the electric test bench.

This test bench was designed to measure the flow of energy through the ultracapacitors during charging and discharging. The resistance or load was made using a bank of resistors connected in parallel, which made the value of the resistance variable and easy to adjust manually. The ultracapacitor arrangement had six cells connected in series in a balancing board and six of these boards connected in parallel, so the total number of ultracapacitors used was 36. This number was selected based on calculations for the energy storage capacity for hydraulic accumulators and ultracapacitors. The selection of 36 ultracapacitors made the energy storage capacity comparable between the two systems. Switches S1 to S6 were used to activate the boards connected in parallel.

As mentioned previously, the number of ultracapacitors used in the test bench was based on the theoretical calculations for energy storage capacity in both systems. Energy stored in a hydraulic accumulator can be calculated with the following equation:

$$E_{acc} = -\int_{v_0}^{v_f} p dv \tag{4}$$

In Equation (4), E_{acc} is the total energy storage capacity of the hydraulic accumulator, p is the pressure, v_0 is the initial volume, and v_f is the final volume. The charging and discharging process of the accumulator can be assumed as adiabatic, and the polytropic index of nitrogen can be assumed as 1.4, according to Rabie [20], so the relationship between pressure and the volume in the hydraulic accumulator can be expressed as follows:

$$pv^n = p_0 v_0^n \tag{5}$$

Plugging Equation (5) into Equation (4) to obtain Equation (6):

$$E_{acc} = \int_{V_o}^{V_f} p_0 v_0^n v^{-n} dv$$

$$E_{acc} = p_0 v_0^n \frac{v^{1-n}}{1-n}\bigg|_{v_0}^{v_f}$$

$$E_{acc} = \frac{p_0 v_0^n}{1-n}\left[v_f^{1-n} - v_0^{1-n}\right] \tag{6}$$

The final compressed volume can be expressed as a function of the maximum pressure in the accumulator.

$$p_{max} v_f^n = p_0 v_0^n$$

$$v_f = \left(\frac{p_0}{p_{max}}\right)^{1/n} v_0 \tag{7}$$

The final equation for energy in the accumulator can be obtained by plugging Equation (7) into Equation (6). Equation (8) is the expression for the energy in the hydraulic accumulator:

$$E_{acc} = \frac{p_0 v_0}{n-1}\left[\left(\frac{p_0}{p_{max}}\right)^{\frac{1-n}{n}} - 1\right] \tag{8}$$

In Equation (8), p_0 is the precharge pressure in the hydraulic accumulator, v_0 is the initial gas volume, p_{max} is the maximum pressure, and n is the ideal gas constant. The values used for the calculation of energy capacity in the hydraulic accumulator are shown in Table 7.

Table 7. Estimated energy capacity of the hydraulic accumulator.

Variable	Value
p_0 (psi)	1000
p_{max} (psi)	3000
n	1.4
v_0 (liter)	1
E_{acc} (Wh)	1.77

The energy storage capacity of the ultracapacitor arrangement needs to be approximately equal to the energy estimated from Equation (8) for the systems to be comparable. The energy in an ultracapacitor can be calculated with Equation (9):

$$E_{ult} = \frac{1}{2} C V_{ut}^2 \tag{9}$$

In Equation (9), C is the capacitance and V_{ut} is the voltage of the ultracapacitors. The capacitance and the voltage of the arrangement can be calculated as function of the number of cells in series (N_C) and the number of boards in parallel (N_B) with Equations (10) and (11):

$$C = \frac{N_B}{N_C} C_{cell} \tag{10}$$

$$V_{ut} = N_C V_{cell} \tag{11}$$

The total energy in the ultracapacitor arrangement can be calculated with Equation (12):

$$E_{ult} = \frac{1}{2} C_{cell} V_{cell}^2 (N_B N_C) \tag{12}$$

The energy capacity of the ultracapacitors is close to the energy capacity in an accumulator with six cells connected in series in a single board and six boards connected in parallel. The estimated energy storage capacity of the ultracapacitor arrangement is shown in Table 8.

Table 8. Estimated energy capacity of the ultracapacitors.

Variable	Value
C_{cell} (F)	50
V_{cell} (V)	2.7
N_B	6
N_C	6
E_{ult}(Wh)	1.82

As mentioned previously, this test bench was designed to measure the flow of energy through the ultracapacitors while charging and discharging. During charging mode, switch SB was on (see Figure 10), while switch SR was off. The current in the circuit depended on the value of the resistance. During discharge mode, switch SB was turned off and switch SR was turned on, which allowed the current to flow from the ultracapacitors to the rheostats, where the energy stored was dissipated as heat. During the experiments, the measured variables included voltage across the ultracapacitors and the current flowing through them. With these variables, the instantaneous power could be calculated, and the energy stored in the ultracapacitors could be estimated. The results for a charge and discharge experiment are shown in Figure 13 to illustrate the output.

Sixty experiments (30 for charge and 30 for discharge) like the one described in this section were made to calculate the efficiency and performance of the ultracapacitor arrangement. In the first round of experiments, just one board with six cells was connected, and five different values of resistance were tested. For the second round of experiments, two boards were connected and five different values of resistance were tested, this process was carried out until six boards were connected and tested with five different values of resistance. The same procedure was applied for discharge. The details for the tests performed in the electric testbench are summarized in Table 9; the number of boards connected, the equivalent capacitance (C), and the equivalent resistance (R) are presented in the table. A detailed explanation of the results is included in the next sections.

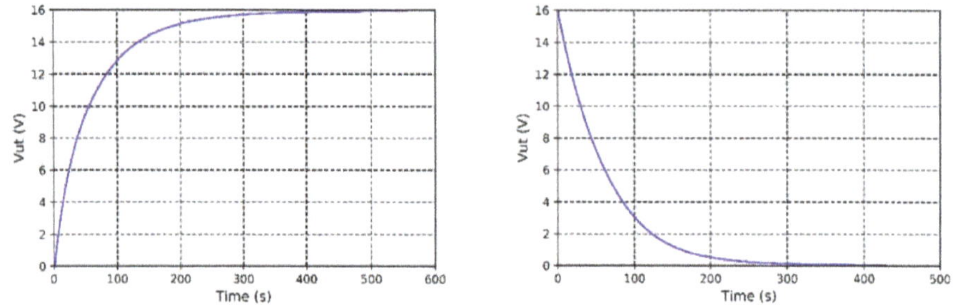

(**a**) (**b**)

Figure 13. Results for charging (**a**) and discharging (**b**).

Table 9. Information of the tests performed in the ultracapacitor test bench.

Number of Boards Connected	C (F)	R (Ω)	Number of Boards Connected	C (F)	R (Ω)
		7.7			7.7
		6			6
1	8.33	4.3	4	33.32	4.3
		2.1			2.1
		1.6			1.6
		7.7			7.7
		6			6
2	16.66	4.3	5	41.65	4.3
		2.1			2.1
		1.6			1.6
		7.7			7.7
		6			6
3	24.99	4.3	6	50	4.3
		2.1			2.1
		1.6			1.6

4. Results for the Hydraulic Accumulator

Instantaneous power in the accumulator could be calculated as the product of pressure and flow rate. These variables were measured during charging and discharging. The results for energy calculation and energy efficiency for the hydraulic accumulator are shown in Table 10. The energy cycle efficiency is the ratio between the total energy released while discharging and the total energy stored while charging.

Table 10. Results for energy calculation in the hydraulic accumulator.

Variable	Value
Total Energy Stored While Charging	1.308 ± 0.003 Wh
Total energy released while discharging	1.147 ± 0.005 Wh
Energy cycle efficiency	87.7 ± 0.6%

The efficiency of the hydraulic system in transferring power from the shaft of the hydraulic pump to the hydraulic accumulator is shown in Table 10. For vehicular application, the kinetic energy of the wheels would move the shaft of a hydraulic pump, which would move the fluid to the hydraulic accumulator in order to absorb the kinetic energy of the vehicle. In the test bench, the wheel was replaced by an electric system. In previous studies, it has been demonstrated that using hydraulic accumulators in vehicle drivetrains can have a positive impact in the efficiency of a vehicle. Wang et al. [19] demonstrated the advantages of a simulated drivetrain for a light passenger vehicle, where, although the energy used for the simulated drive cycle was better using the pure electric drivetrain, the acceleration performance was better for the hydraulic drivetrain thanks to its higher power density. Moreover, Hui et al. [18] studied the effect of using a hydraulic accumulator for extending the state of charge of a battery when hybridizing an electric drivetrain with a hydraulic regeneration with positive results due to the high efficiency of hydraulic accumulators. The power in the hydraulic pump shaft was calculated as the product of the shaft torque and the rotational speed. The torque was estimated based on the pressure at outlet of the pump. The pressure and the torque were correlated according to the next expression.

$$T = \frac{D\Delta p}{\eta_m} \tag{13}$$

In Equation (13), η_m is the mechanical efficiency of the pump. The efficiency is a function of the pressure and the flow rate in the system, so the efficiency changes throughout the experiment. The pressure and the flow rate were measured, and with the datasheets provided by the manufacturer of the pump, it was possible to estimate the efficiency and the torque at any operating conditions. Thus, at any time in the experiment, the input power in the hydraulic pump could be estimated, as could the power going to the accumulator. Then, the instantaneous efficiency of the system could be obtained. At any instant of the experiment, the flow rate and the pressure could be measured. At the same instant, the power input could be measured. With this information, it was possible to create an efficiency map for the hydraulic accumulator, which is shown in Figure 14. The same map without the data points is shown in Figure 15. The map between datapoints was estimated with linear interpolation using the Matlab function griddata [21].

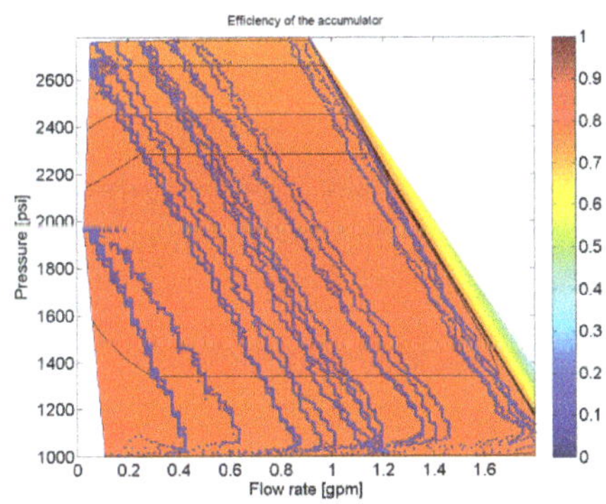

Figure 14. Instantaneous efficiency of the hydraulic accumulator with data points.

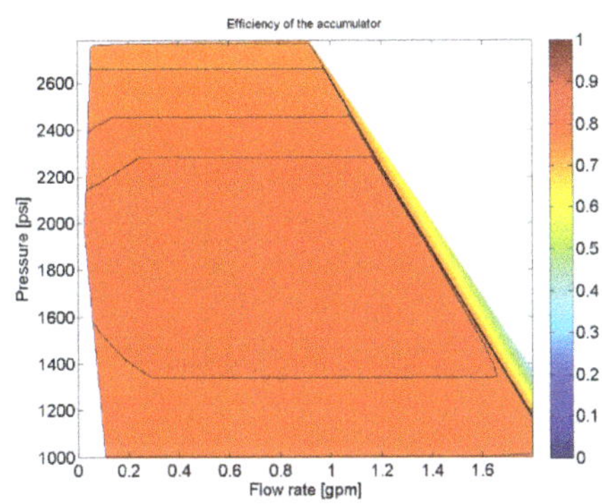

Figure 15. Instantaneous efficiency of the hydraulic accumulator using a gear pump.

The map presented in Figures 14 and 15 is important in identifying operating conditions that would be optimal for a system like the one proposed in this study. According to these plots, the highest efficiency was around 80% and was obtained for flow rates of approximately 1 gpm and pressures of approximately 1800 psi.

The current and voltage of the electric system were also measured during the experiment. A similar map of efficiency could be made for the conversion of electric power to hydraulic power. The map is shown in Figure 16.

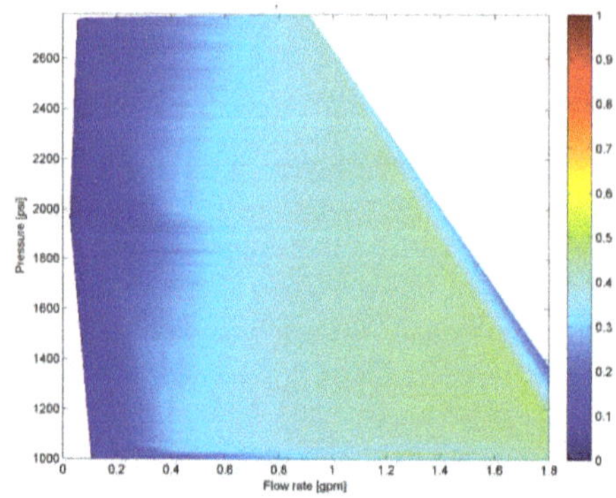

Figure 16. Instantaneous efficiency for conversion of electric power to hydraulic power.

The maximum efficiency was around 50%, relatively low because this was the efficiency of converting the electric power taken from the battery to hydraulic power in the accumulator. The electric power from the battery had to be converted into AC power in the inverter. After that, the AC power was converted into mechanical power by the electric motor. This mechanical power was used to move the shaft of the hydraulic pump, and the hydraulic pump moved the fluid from the reservoir to the accumulator through the hydraulic system, which had some power losses due to friction.

Another important aspect to consider was that the current of the battery was very high during the charging process, which was not the most efficient way to use the battery. The results for the battery for one of the charging experiments are shown in Figure 17.

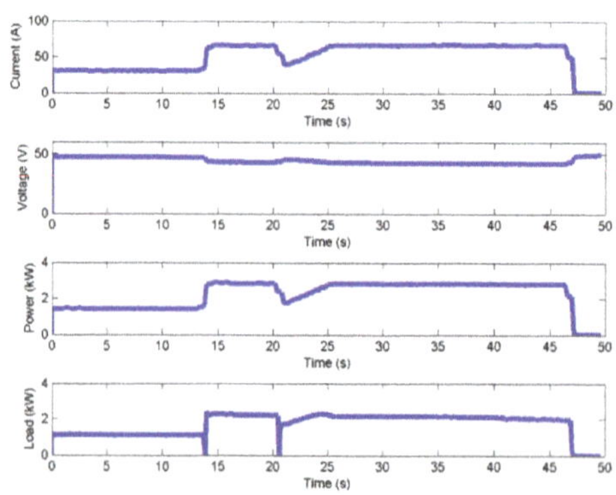

Figure 17. Experimental results for current, voltage, power, and load.

The maximum current during this experiment was 66 A, which was much higher than the continuous current recommended for the efficient operation of this battery, according to the technical data presented in Table 4. The efficiency results for a system with a model of a piston pump were estimated. The model of the piston pump was made by using commercially available datasheets of different piston pumps and then estimated with interpolation for a piston pump with a volumetric displacement of 0.73 in^3/rev, which was 11.9 cc/rev. The results of the numerical estimation are shown in Figure 18.

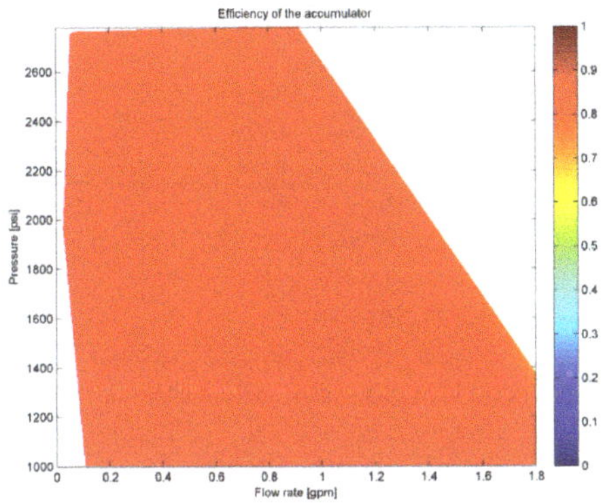

Figure 18. Instantaneous efficiency of the hydraulic accumulator using a piston pump (estimated).

The estimated results of Figure 16 show that the instantaneous efficiency of the system using a model of a piston pump could be improved, mostly because axial piston pumps had higher efficiencies. The results for the efficiency including the electric motor are shown in Figures 19 and 20. These figures show the efficiency of conversion of electric power to mechanical power in the electric system using a gear pump (experimental) and a piston pump (estimated).

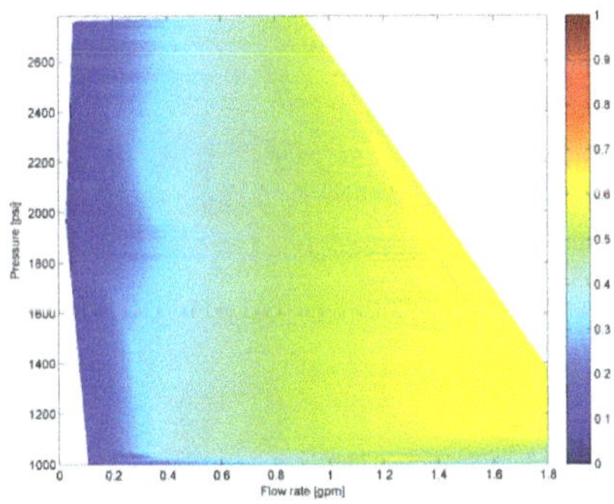

Figure 19. Efficiency of the electric system in the charging process using a gear pump.

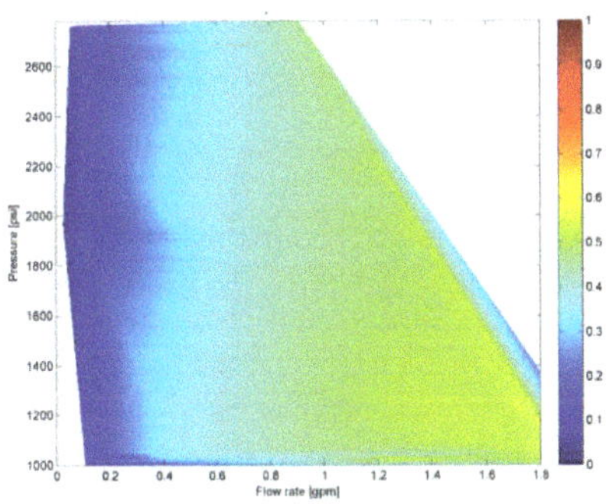

Figure 20. Efficiency of the electric system in the charging process using a piston pump (estimated).

From Figures 19 and 20, it can be observed that the electric system worked better when using a gear pump. The difference in the results was due to the input torque needed to turn the shaft of the hydraulic pump, which in this case is lower for the gear pump than for the piston pump used. The overall efficiency of the system was highly dependent on pump efficiency.

5. Results for the Ultracapacitors

The tests in the ultracapacitor test bench were made by changing the number of boards connected and the resistance used in the circuit. For each of the six possible board configurations, five different values of resistance were used. Starting with one board of six ultracapacitors, the charge and discharge tests were performed five times, and each time, the resistance selected was different. In total, thirty experiments were conducted for charging and thirty experiments for discharging. The results during charge and discharge cycles are presented below.

From Figure 21, it can be seen that the maximum voltage level of the system was reached faster when fewer boards were used—that is, when fewer ultracapacitors were energized. In Figures 21 and 22, each line represents one experiment for one value of resistance.

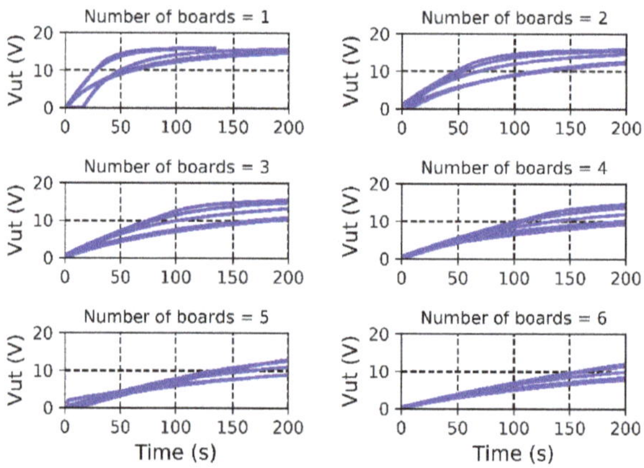

Figure 21. Voltage while charging according to the number of boards connected.

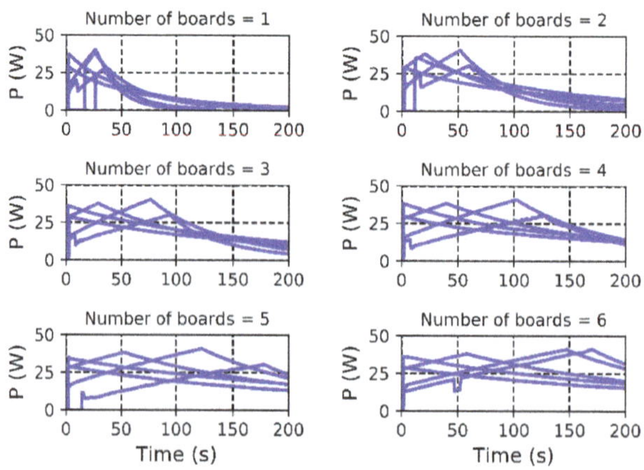

Figure 22. Power while charging according to the number of boards connected.

Now the same results are presented, but they are presented according to the number of the test (Figures 23 and 24). Test number 1 had the highest value of resistance, and test number 5 had the lowest value of resistance. The individual lines represent the number of boards connected in the electric system. The time to reach maximum voltage value was lower at a lower resistance. The value of the resistance used in each test is presented in Table 11.

Table 11. Resistance value for the tests.

Test Number	Resistance Value
Test 1	7.7 Ω
Test 2	6.0 Ω
Test 3	4.3 Ω
Test 4	2.1 Ω
Test 5	1.6 Ω

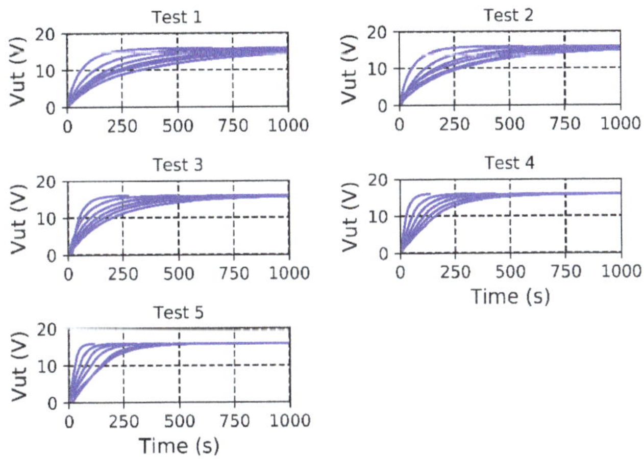

Figure 23. Voltage while charging according to the value of resistance.

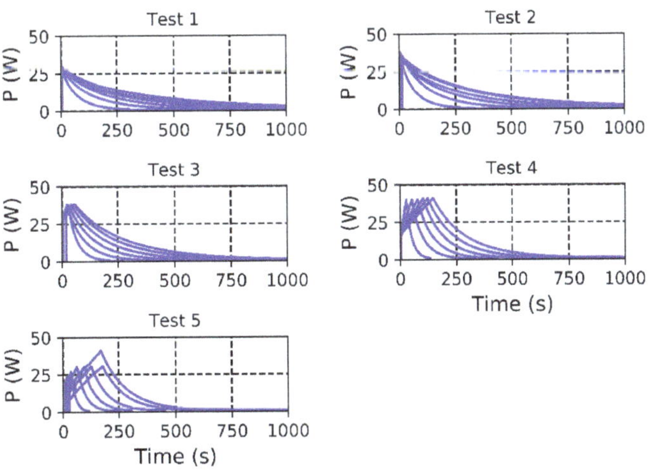

Figure 24. Power while charging according to the value of resistance.

A discharging experiment was conducted for each charging experiment. The results according to the number of boards are presented on Figures 25 and 26, and the results according the number of the test are presented in Figures 27 and 28.

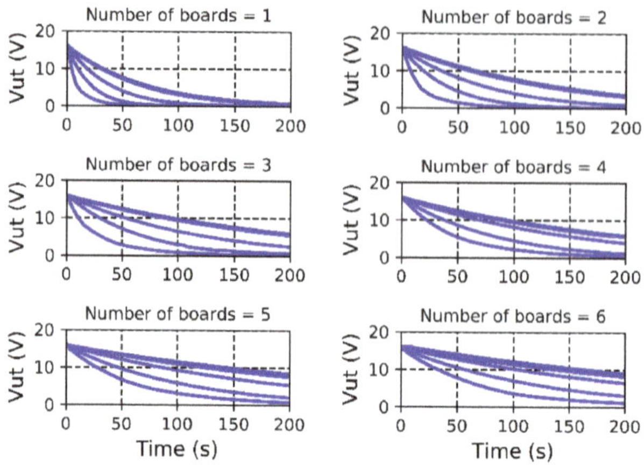

Figure 25. Voltage while discharging according to the number of boards connected.

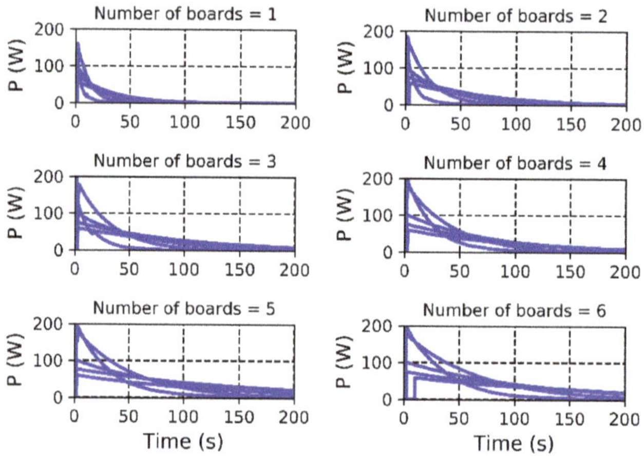

Figure 26. Power while discharging according to the number of boards connected.

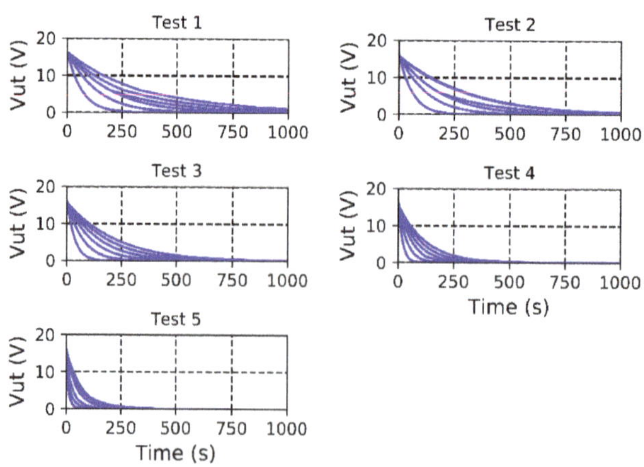

Figure 27. Voltage while discharging according to the number of the test.

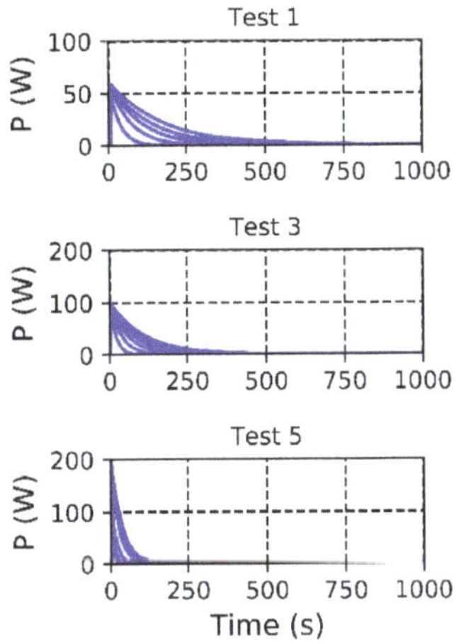

Figure 28. Power while discharging grouped according to the number of the test.

Using results presented in Figures 21–28, it was possible calculate the energy stored in the ultracapacitors and the efficiency. The results for energy calculations in the ultracapacitors are presented in Table 12.

Table 12. Results for energy calculation in the ultracapacitors.

Variable	Number of Boards Connected in Parallel					
	1		2		3	
	Value	Error	Value	Error	Value	Error
Energy stored while charging (Wh)	0.31	0.01	0.63	0.02	0.92	0.03
Energy stored while discharging (Wh)	0.25	0.01	0.50	0.01	0.76	0.02
Energy efficiency	77.46	2.33	77.2	2.7	80.75	3.3
Variable	**Number of Boards Connected in Parallel**					
	4		5		6	
	Value	Error	Value	Error	Value	Error
Energy saved while charging (Wh)	1.24	0.04	1.61	0.05	1.89	0.06
Energy stored while discharging (Wh)	0.97	0.03	1.29	0.04	1.47	0.05
Energy efficiency	78.99	3.91	80.4	4.73	77.74	5.42

6. Comparative Evaluation

The results for energy density and power density are presented in Table 13. Three energy systems were compared: a battery, which is the principal source of energy, the ultracapacitor, and the hydraulic accumulator.

Table 13. Energy density and power density.

Energy Storage System	Energy/Vol (Wh/m^3)	Energy/Mass (Wh/kg)	Cost/Energy (US$/Wh)
Battery	195144	115.2	0.45
Ultracapacitor	2539.7	2.72	138.67
Accumulator	1227	0.29	404.68
Energy Storage System	**Power/Vol (kW/m^3)**	**Power/Mass (kW/kg)**	**Cost/Power (US$/kW)**
Battery	325.24	0.192	270.83
Ultracapacitor	2588	2.21	217
Accumulator	7548	2.69	75

The radar plot presented in Figure 29 better illustrates these results. The radar plot shows the results in a scale from 0 to 10 for the three energy storage systems used in this study. The "*Score*" value for each component of the radar plot was determined using this equation:

$$Score = \frac{System\ Value}{Best\ Value} \cdot 10 \tag{14}$$

In Equation (14), the "*System Value*" is the value of each system: battery, ultracapacitor, or accumulator, and "*Best Value*" is the best value among the three energy storage systems. This "*Best Value*" depends on the specific characteristic to be studied. For instance, in energy/volume, the "*Best Value*" would be the highest value among the three systems, but in cost/power, the "*Best Value*" will be lowest.

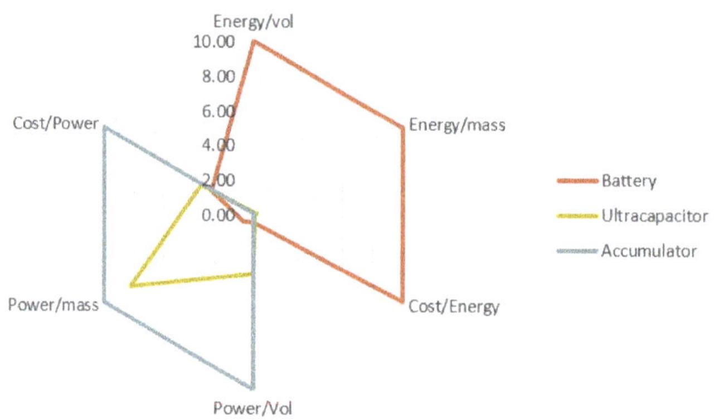

Figure 29. Radar plot comparing three energy storage systems.

From the radar plot, it is possible to see that the energy side of the plot is mostly covered by the battery. Neither of the other two systems could compete against the battery in terms of energy storage. On the other hand, the hydraulic accumulator dominates the power part of the plot. This means that the hydraulic accumulator is suited for high power applications. A comparison between the hydraulic accumulator and the ultracapacitor is shown in Figure 30.

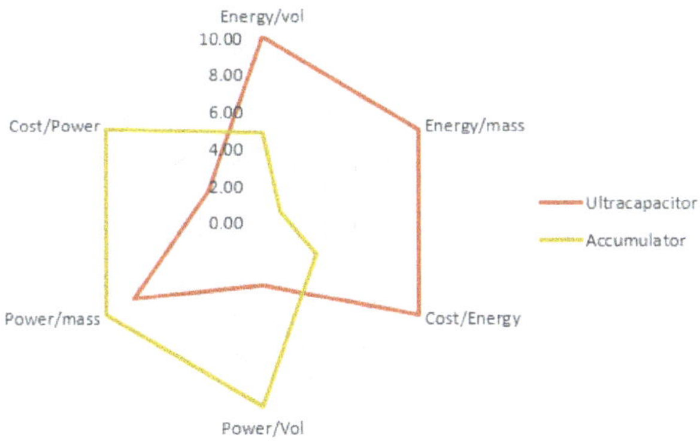

Figure 30. Radar plot comparing ultracapacitors and hydraulic accumulators.

From Figure 29, it is possible to see that the hydraulic accumulator is more suitable than the ultracapacitors in the power segment of the plot, while the ultracapacitor is better in the energy segment of the plot. However, the battery was better than these two systems for storing energy. It is important to note that the energy efficiency for the ultracapacitor was around 78.7%, and the energy efficiency of the hydraulic accumulator was 87.7%. This improvement in energy efficiency and the better power density compared with ultracapacitors could be a determining factor in choosing a hydraulic system over an electric system for a specific application when needing to rapidly charge or discharge energy storage devices, such as in the case for regenerative breaking.

The net cost of each system is presented in Table 14, but the information of this table is not enough to make an effective cost analysis of the energy storage systems considered in this study. It is necessary to consider the results presented in Table 14 and in the radar plots of Figures 29 and 30.

Table 14. Cost of each system.

Energy Storage System	Cost
Battery	390 USD
Ultracapacitors	210 USD
Hydraulic accumulator	391 USD

From the comparative results presented in the radar plots of Figures 29 and 30, it is possible to see that regarding cost/energy ratio, the battery was far better than ultracapacitors and hydraulic accumulators. The cost per unit of energy in the battery was 0.45 USD/Wh, the cost per unit of energy in the ultracapacitors was 138.7 USD/Wh, and the cost per unit of energy in the hydraulic accumulator was 404.7 USD/Wh. Regarding the cost per unit of power, the best system was the hydraulic accumulator with 75 USD/kW, the ultracapacitor had a cost/power ratio of 217 USD/kW, and the battery had a cost/power ratio of 270.8 USD. These results demonstrate the potential of hydraulic accumulators for applications that require a high power density. Nevertheless, it is necessary to consider how expensive an accumulator could be if it is required to store energy for a large period, because its cost per unit of energy is not the best.

7. Conclusions

Two test benches were designed and built to test two different energy storage systems: a hydraulic accumulator and an ultracapacitor with identical energy capacities. The energy efficiency under the test conditions for the hydraulic accumulator was 87.7%, and the energy efficiency of the ultracapacitor was 78.7%.

The efficiency map from this study can be used to determine a control strategy for a regenerative system with hydraulic accumulators. This efficiency map can be replicated for different hydraulic pumps by using numerical models for a pump. In addition, the analysis of the efficiency map for a piston pump shows that a hydraulic accumulator would be more efficient if a piston pump is used instead of a gear pump.

This study also shows that energy segments of the radar plot were dominated by the ultracapacitor, while the power segments were dominated by the hydraulic accumulator. It is interesting to note that there were segments in the radar plot that were not covered by either of the three energy storage systems. In other words, none of the systems showed a good score in cost/power and energy/volume. This means that energy storage systems with high energy density can provide high power but at a high cost. Moreover, there was no system with a high score in power/volume and cost/energy, which means that the energy storage systems with good power density can be used to store energy but at a high cost.

The higher energy efficiency in the hydraulic accumulator and the better power density compared with ultracapacitors could be determining factors in choosing a hydraulic system over an electric system for a specific application, where there is a need to rapidly charge or discharge energy storage devices, such as in the case of regenerative breaking.

Author Contributions: Conceptualization, J.L.-Q., J.G.-B. and M.K.; methodology, J.L.-Q., J.G.-B. and M.K.; software, J.L.-Q.; validation, J.L.-Q., B.N., M.K., A.G.-M., and J.G.-B.; formal analysis, J.L.-Q., and J.G.-B.; investigation, J.L.-Q., M.K., A.G.-M., and J.G.-B.; resources, B.N., and J.G.-B.; data curation, J.L.-Q.; writing—original draft preparation, J.L.-Q.; writing—review and editing, J.L.-Q., B.N., M.K., A.G.-M., and J.G.-B.; visualization, J.L.-Q., B.N., M.K., A.G.-M., and J.G.-B.; supervision, B.N., M.K., A.G.-M., and J.G.-B.; project administration, B.N., and J.G.-B.; funding acquisition, B.N., and J.G.-B. All authors have read and agreed to the published version of the manuscript.

Funding: This research was funded by the Purdue Polytechnic, office of the vice dean of research, and the school of Engineering Technology at Purdue University.

Acknowledgments: Special thanks for Sun Hydraulics for providing testing components.

Conflicts of Interest: The authors declare no conflict of interest.

References

1. U.S. Energy Information Administration. *Monthly Energy Review*; U.S. Energy Information Administration: Washington, DC, USA, 2019.
2. Lin, T.; Wang, Q.; Hu, B.; Gong, W. Development of hybrid powered hydraulic construction machinery. *Autom. Constr.* **2010**, *19*, 11–19. [CrossRef]
3. Kwon, T.S.; Lee, S.W.; Sul, S.K.; Park, C.G.; Kim, N.I.; Kang, B.I.; Hong, M.S. Power Control Algorithm for Hybrid Excavator with Supercapacitor. *IEEE Trans. Ind. Appl.* **2010**, *46*, 1447–1455. [CrossRef]
4. Xiao, Q.; Wang, Q.; Zhang, Y. Control strategies of power system in hybrid hydraulic excavator—ScienceDirect. *Autom. Constr.* **2018**, *17*, 361–367. [CrossRef]
5. Wang, T.; Wang, Q.; Lin, T. Improvement of boom control performance for hybrid hydraulic excavator with potential energy recovery—ScienceDirect. *Autom. Constr.* **2013**, *30*, 161–169. [CrossRef]
6. Lin, T.; Chen, Q.; Ren, H.; Huang, W.; Chen, Q.; Fu, S. Review of boom potential energy regeneration technology for hydraulic construction machinery. *Renew. Sustain. Energy Rev.* **2017**, *79*, 358–371. [CrossRef]
7. Lin, T.; Huang, W.; Ren, H.; Fu, S.; Liu, Q. New compound energy regeneration system and control strategy for hybrid hydraulic excavators. *Autom. Constr.* **2016**, *68*, 11–20. [CrossRef]
8. Tan, Y.; Ciufo, P.; Meegahapola, L.; Muttaqi, K.M. Enhanced Frequency Response Strategy for a PMSG-Based Wind Energy Conversion System Using Ultracapacitor in Remote Area Power Supply Systems. *IEEE Trans. Ind. Appl.* **2016**, *53*, 549–558. [CrossRef]
9. Kim, S.M.; Sul, S.K. Control of Rubber Tyred Gantry Crane with Energy Storage Based on Supercapacitor Bank. *IEEE Trans. Power Electron.* **2006**, *21*, 1420–1427. [CrossRef]
10. Pandey, A.; Allos, F.; Hu, A.P.; Budgett, D. Integration of supercapacitors into wirelessly charged biomedical sensors. In Proceedings of the 2011 6th IEEE Conference on Industrial Electronics and Applications, Beijing, China, 21–23 June 2011; pp. 56–61. [CrossRef]

11. Lin, T.; Zhou, S.; Chen, Q.; Fu, S. A Novel Control Strategy for an Energy Saving Hydraulic System With Near-Zero Overflowing Energy-Loss. *IEEE Access* **2018**, *6*, 33810–33818. [CrossRef]
12. Wang, B. High Speed On-Off Valve Self-adapting Clamping System. *J. Appl. Sci.* **2014**, *14*, 279–284. [CrossRef]
13. Andruch, J.; Lumkes, J.H. Regenerative Hydraulic Topographies using High Speed Valves. *SAE Tech. Pap. Ser.* **2009**. [CrossRef]
14. Fouda, M.; Elwakil, A.S.; Radwan, A.G.; Allagui, A. Power and energy analysis of fractional-order electrical energy storage devices. *Energy* **2016**, *111*, 785–792. [CrossRef]
15. Hartley, T.T.; Veillette, R.J.; Adams, J.L.; Lorenzo, C.F. Energy storage and loss in fractional-order circuit elements. *IET Circuits Devices Syst.* **2015**, *9*, 227–235. [CrossRef]
16. Allagui, A.; Freeborn, T.J.; Elwakil, A.S.; Maundy, B.J. Reevaluation of Performance of Electric Double-layer Capacitors from Constant-current Charge/Discharge and Cyclic Voltammetry. *Sci. Rep.* **2016**, *6*, 38568. [CrossRef] [PubMed]
17. Freeborn, T.J. Estimating supercapacitor performance for embedded applications using fractional-order models. *Electron. Lett.* **2016**, *52*, 1478–1480. [CrossRef]
18. Hui, S.; Lifu, Y.; Junqing, J.; Yanling, L. Control strategy of hydraulic/electric synergy system in heavy hybrid vehicles. *Energy Convers. Manag.* **2011**, *52*, 668–674. [CrossRef]
19. Wang, F. Comparison between Hydraulic Hybrid and Electric Hybrid Passenger Vehicles using ADVISOR 2004. In Proceedings of the 52nd National Conference on Fluid Power, Las Vegas, NV, USA, 23–25 March 2011; pp. 31–40.
20. Rabie, M.G. *Fluid Power Engineering*; McGraw-Hill: New York, NY, USA, 2009; Volume 28.
21. Interpolate 2-D or 3-D Scattered Data—MATLAB Griddata. Available online: https://www.mathworks.com/help/matlab/ref/griddata.html (accessed on 15 January 2020).

© 2020 by the authors. Licensee MDPI, Basel, Switzerland. This article is an open access article distributed under the terms and conditions of the Creative Commons Attribution (CC BY) license (http://creativecommons.org/licenses/by/4.0/).

energies

MDPI

Article

Comparison of SOC Estimation between the Integer-Order Model and Fractional-Order Model Under Different Operating Conditions

Guoqing Jin [1], Lan Li [2], Yidan Xu [2], Minghui Hu [2,*], Chunyun Fu [2,*] and Datong Qin [2]

1 Chongqing Changan Automobile Co., Ltd., Chongqing 400023, China; jingq@changan.com.cn
2 State Key Laboratory of Mechanical Transmissions, School of Automotive Engineering, Chongqing University, Chongqing 400044, China; lilan@cqu.edu.cn (L.L.); eden_xyd@163.com (Y.X.); dtqin@cqu.edu.cn (D.Q.)
* Correspondence: hu_ming@cqu.edu.cn (M.H.); fuchunyun@cqu.edu.cn (C.F.)

Received: 20 February 2020; Accepted: 26 March 2020; Published: 7 April 2020

Abstract: Accurate estimation of the state of charge (SOC) is an important criterion to prevent the batteries from being over-charged or over-discharged, and this assures an electric vehicle's safety and reliability. To investigate the effect of different operating conditions on the SOC estimation results, a dual-polarization model (DPM) and a fractional-order model (FOM) are established in this study, taking into account the prediction accuracy and structural complexity of a battery model. Based on these two battery equivalent circuit models (ECMs), a hybrid Kalman filter (HKF) algorithm is adopted to estimate the SOC of the battery; the algorithm comprehensively utilizes the ampere-hour (Ah) integration method, the Kalman filter (KF) algorithm, and the extended Kalman filter (EKF) algorithm. The SOC estimation results of the DPM and FOM, under the dynamic stress test (DST), federal urban driving schedule (FUDS), and hybrid pulse power characteristic (HPPC) cycle conditions, are compared and analyzed through six sets of experiments. Simulation results show that the SOC estimation accuracy of both the models is high and that the errors are within the range of ±0.06. Under any operating conditions, the SOC estimation error, based on the FOM, is always lower than the SOC estimation error of the DPM, but the adaptability of the FOM is not as high as that of the DPM.

Keywords: lithium-ion battery; dual-polarization model; fractional-order model; SOC estimation; hybrid Kalman filter

1. Introduction

Lithium-ion power batteries are widely used in electric vehicles (EVs), owing to their advantages of high energy density, low self-discharge rate, long cycle life, and no memory effect [1]. To ensure the safe, efficient, and stable operation of the power batteries, it is essential to manage the batteries effectively [2]. It is known that the state of charge (SOC) serves as an important indicator to characterize the remaining battery capacity. Therefore, an accurate SOC estimation is the basis for preventing over-charge and over-discharge, and for equilibrium processing. Accurate SOC estimation is the core of an effective battery management system (BMS) [3–5].

An effective battery model is a prerequisite for estimating the SOC of a battery. A model that is not effective directly reduces the accuracy of the SOC estimation algorithm, and could even cause the estimation algorithm to diverge directly in severe cases. There are three main types of working lithium-ion battery models: the black-box models, the electrochemical models, and the equivalent circuit models (ECMs) [6]. These three types of models describe the characteristics of the lithium-ion batteries from different detail levels [7]. The black-box models are similar to a linear or nonlinear mapping function. This function reflects the characteristics of battery voltage response, whereas ignores the internal mechanism of the battery and has no physical existence. The electrochemical models contain many equations and parameters, but the simulation accuracy of the battery under complex working conditions is low. The ECMs are used to simulate the external operating characteristics of the battery by the matching of electronic components and are widely used in battery SOC estimation [8]. The pseudo-two-dimensional electrochemical mechanism model proposed by Doyle et al. [9] is often used as a full-order reference mechanism model, and it is also used to evaluate and test other simplified mechanism models. Wang et al. [10] established a nonlinear black-box battery model, and the verification under federal urban driving schedule (FUDS) operating conditions showed that the relative error of voltage was within 3.8%. Plett et al. [11] developed the commonly used ECMs in detail, including internal resistance models, models considering hysteresis effects, etc. Kim et al. [12] proposed a hybrid battery model consisting of a KiBaM model and a dual-polarization model (DPM), which can simultaneously describe the external dynamic characteristics and the recovery effect of the battery. Hu et al. [13] used a second-order fractional-order model to predict the terminal voltage under FUDS cycling conditions, and the average relative error does not exceed 0.1%, which proved the high accuracy of the model.

At present, the SOC estimation methods used both locally and internationally include: the ampere-hour (Ah) integration method [14], electrochemical impedance spectroscopy (EIS) method [15], open-circuit voltage (OCV) method [16], internal resistance method [1], particle filter (PF) [17], Kalman filter (KF) [18], fuzzy logic (FL) [19], artificial neural network (ANN) [20], support vector machine (SVM) [21], and relevance vector machine (RVM) [22]. The authors in [23] critically reviewed the existing SOC estimation methods in the past five years and introduced the basic principles and main disadvantages of various methods. Among these methods, the KF is an optimized autoregressive data filtering algorithm [24], which utilizes the principle of minimum mean square error to achieve an optimal state estimation for a complex dynamic system. Not only does the KF correct the initial error of the system but it also effectively suppresses the noise in the actual measurement process [8]. This feature makes the KF stand out among the current SOC estimation models. Therefore, a variety of improved algorithms have also been derived, such as the extended Kalman filter (EKF) [25], the adaptive unscented Kalman filter (AUKF) [26], and the central difference Kalman filter (CDEF) [27]. In 2004, Gregory L Plett [11] employed the EKF algorithm to perform the battery state and parameters' estimation based on the ECM. He proposed an EKF algorithm as the core control method, which was supported and improved by many researchers [28–30]. Xu et al. [31] proposed a fractional-order model (FOM) for SOC estimation. Compared to the integer-order model, the accuracy of the SOC estimation is significantly improved with the FOM. Lai et al. [32] combined the Ah integration method and the EKF to estimate the SOC based on multi-model global identification. The results proved the robustness of the algorithm. Xu et al. [33] employed the double Kalman filter (DKF) algorithm to estimate the SOC of a lithium-ion battery based on the temperature-dependent DPM. After verification, the battery SOC estimation error could be kept within the range of $\pm$ 0.004 under different temperature conditions.

At present, most works in the existing literature focus on the SOC estimation algorithm, and the effects of operating conditions on SOC estimation are barely considered. In this study, we studied the influence of different operating conditions on SOC estimation based on different battery models. Simulation results show that the SOC estimation accuracy of the DPM and the FOM is satisfactory, and the errors are within the range of $\pm$0.06. Under any operating condition, the SOC estimation error of the FOM is always less than that of the DPM, but the adaptability of the FOM is not as good as that of the DPM.

This paper is organized as follows: first, we establish a DPM and a FOM. Second, we apply a mixed-swarm-based cooperative particle swarm optimization (MCPSO) algorithm to identify the battery parameters. The accuracy of the model is verified by using dynamic stress test (DST) operating condition. Third, a hybrid Kalman filter (HKF) algorithm is used to estimate the SOC of the battery, comprehensively utilizing the Ah integration method, KF, and EKF. Finally, the SOC estimation results of the DPM and the FOM under DST, FUDS, and hybrid pulse power characteristic (HPPC) cycling conditions are analyzed by comparing six sets of experiments.

2. Establishment of Lithium-Ion Battery Model

The ECMs (integer-order models and FOMs) are the most widely used type of battery models in various battery-related research because of their clear physical representation, ease of mathematical analysis, and simple parameter identification [34]. Among these models, considering the trade-off between prediction accuracy and structural complexity, the DPM stands out from all integer-order models [35]. However, the integer-order models cannot accurately reflect the electrochemical reactions inside the battery. Therefore, in [36], the authors have used fractional-order impedance elements to improve the integer-order model further. This is because, from the perspective of EIS, a circuit composed of fractional-order impedance elements can better fit the impedance characteristics of a lithium-ion battery, and thus has better applications in battery principle analysis, battery modeling, and state estimation. To investigate the effects of different operating conditions on the battery SOC estimation, the SOC estimation experiments under different operating conditions were conducted in this study, based on the DPM and FOM.

Figure 1a shows the structure of the DPM, and Figure 1b shows the structure of the FOM, where U_d represents the battery terminal voltage, U_{OCV} stands for the OCV, the current is denoted by I, R_0 indicates the Ohmic internal resistance, the polarization internal resistances are represented by R_1 and R_2, C_1 and C_2 stand for the polarization capacitances, the constant phase elements are denoted by CPE_1 and CPE_2, and W indicates the Wahlberg element.

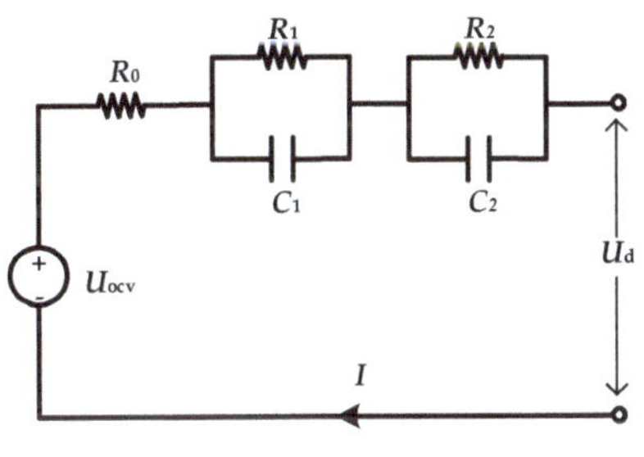

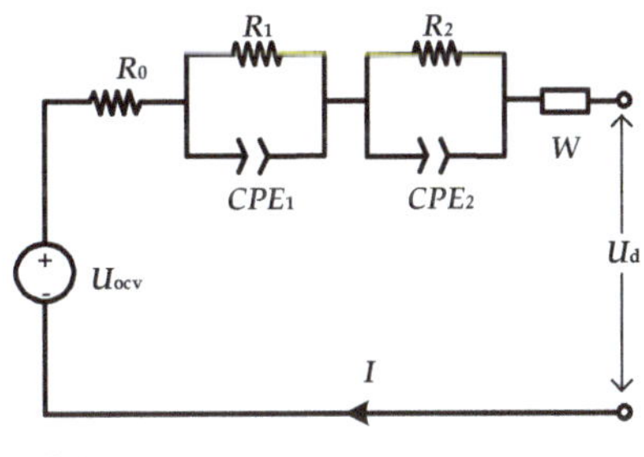

Figure 1. Structural diagram of equivalent circuit models (ECMs).

2.1. Establishment of Lithium-Ion Battery DPM

The terminal voltage of the battery is shown as follows:

$$U_d = U_{OCV} + R_0 I + U_1 + U_2 \tag{1}$$

The changes in the rates of voltages U_1 and U_2 are $\dot{U}_1$ and $\dot{U}_2$, respectively, which can be expressed by Equations (2) and (3):

$$\dot{U}_1 = -\frac{U_1}{R_1 C_1} + \frac{I}{C_1} \tag{2}$$

$$\dot{U}_2 = -\frac{U_2}{R_2 C_2} + \frac{I}{C_2} \tag{3}$$

The definition of SOC is presented in Equations (4) as follows:

$$SOC(t) = SOC(t_0) + \int_{t_0}^{t} \frac{\eta I(\tau)}{Q} d\tau \tag{4}$$

where the values of SOC at time t and t_0 are denoted by $SOC(t)$ and $SOC(t_0)$, respectively, Q stands for the maximum available capacity, the charging and discharging efficiency is represented by η. Supposing ΔT indicates the sampling time, discretizing Equations (1), (2), and (4) as follows:

$$U_1(k) = \exp(\frac{-\Delta T}{R_1C_1})U_1(k-1) + R_1I(k)[1-\exp(\frac{-\Delta T}{R_1C_1})] \tag{5}$$

$$U_2(k) = \exp(\frac{-\Delta T}{R_2C_2})U_2(k-1) + R_2I(k)[1-\exp(\frac{-\Delta T}{R_2C_2})] \tag{6}$$

$$SOC(k) = SOC(k-1) + \frac{\eta\Delta T}{Q}I(k) \tag{7}$$

The parameters to be identified are:

$$\theta = [R_0\ R_1\ C_1\ R_2\ C_2\ U_{OCV}\ Q] \tag{8}$$

2.2. Establishment of Lithium-Ion Battery FOM

The transfer function is presented in Equation (9) for the lithium-ion battery FOM as follows:

$$\frac{U_{OCV}(s) - U_d(s)}{I(s)} = \frac{R_1}{1+R_1Z_1^{-1}} + \frac{R_2}{1+R_2Z_2^{-1}} + Z_W + R_0 \tag{9}$$

where $Z_1 = (CPE_1S^{\alpha})^{-1}$ and $Z_2 = \left(CPE_1S^{\beta}\right)^{-1}$ indicate the impedances of CPE_1 and CPE_2 element, respectively. The impedance of the Wahlberg element is denoted by $Z_W = (WS^{\gamma})^{-1}$.

Assuming the system input is $u = I(t)$, and the output is $y = U_{OCV}(t) - U_d(t)$, then the system model is denoted by a fractional calculus equation, as shown in Equation (10).

$$\begin{aligned}
&(WD^{\gamma} + WR_1CPE_1D^{\alpha+\gamma} + WR_2CPE_2D^{\beta+\gamma} + WR_1CPE_1R_2CPE_2D^{\alpha+\beta+\gamma})y(t) \\
&= [R_1CPE_1D^{\alpha} + R_2CPE_2D^{\beta} + (R_0+R_1+R_2)WD^{\gamma} + \\
&R_1CPE_1R_2CPE_2D^{\alpha+\beta} + (R_0+R_2)WR_1CPE_1D^{\alpha+\gamma} + \\
&(R_0+R_1)WR_2CPE_2D^{\beta+\gamma} + R_0WR_1CPE_1R_2CPE_2D^{\alpha+\beta+\gamma}]u(t) + u(t)
\end{aligned} \tag{10}$$

Here, the fractional-orders of CPE_1 and CPE_2 are represented by α and β, and the fractional-order of the Wahlberg element is denoted by γ. The parameters D^{γ}, D^{α}, D^{β}, $D^{\alpha+\gamma}$, $D^{\beta+\gamma}$, $D^{\alpha+\beta}$, $D^{\alpha+\beta+\gamma}$ indicate the fractional-order operators.

The definition of SOC of the FOM is given by:

$$D^1SOC(t) = -\frac{\eta}{C_n}u(t) \tag{11}$$

where C_n is the rated capacity of the battery; η stands for the battery coulomb efficiency.

By using calculations as presented in [13], we can transform Equation (10) into a first-order difference equation as follows:

$$y(k) = -\frac{A(1)}{A(0)}y(k-1) + \frac{1+B(0)}{A(0)}u(k) + \frac{B(1)}{A(0)}u(k-1) \tag{12}$$

Discretizing Equation (11):

$$SOC(k) = SOC(k-1) - \frac{\eta T}{C_n}u(k) \tag{13}$$

The FOMs of lithium-ion batteries can be established using Equations (12) and (13), and the parameters that the model needs to identify are as follows:

$$\theta = [R_0 \quad R_1 \; CPE_1 \; R_2 \; CPE_2 \; W \; \alpha \; \beta \; \gamma] \tag{14}$$

2.3. *Model Parameter Identification*

2.3.1. Model Parameter Identification

The equipment employed in our experimentation included a battery testing device (BTS-5 V 100 A), an incubator (HL404C), a charging and discharging facility, and a computer. The sampling time of the battery test system was set to 0.1 s. Figure 2 shows the configuration of the battery testing system. The A123 ternary lithium-ion soft pack batteries were selected as the experimental objects, and their specifications can be seen in Table 1.

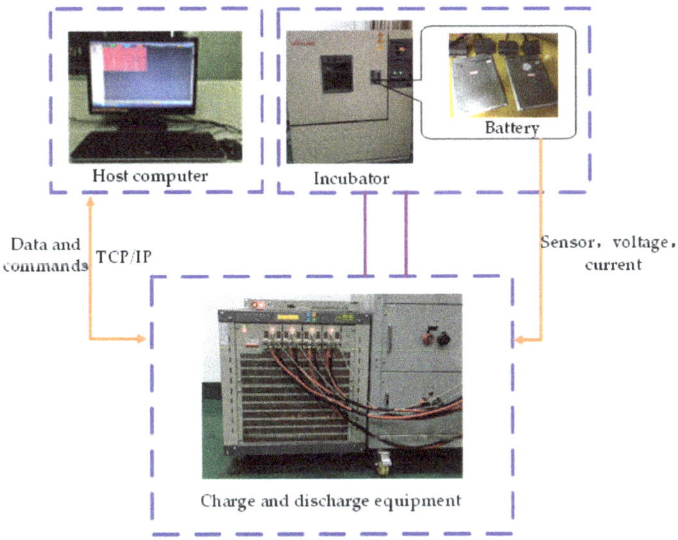

Figure 2. Configuration of the battery testing bench.

Table 1. Specification of the battery.

Rated Capacity	Charging Cut-Off Voltage	Discharging Cut-Off Voltage	Charging Cut-Off Current
25 Ah	4.20 V	2.50 V	1.25 A

The OCV is an important parameter of the lithium-ion battery, and it usually has a relatively fixed corresponding relationship with the SOC. By testing the OCV at different SOC points, the SOC-OCV curve of the battery can be drawn. Generally, if the lithium-ion battery is left in the open state for a sufficiently long period, the measured battery terminal voltage can be approximately considered as the battery's OCV [37]. According to the Empirical Formula (15) and the fitted data of Table 2, Figure 3 shows the connection between OCV and SOC [11]. The fitted parameters for the OCV-SOC curve are shown in Table 3.

$$U_{OCV}(SOC) = C_0 + C_1 SOC + C_2 \frac{1}{SOC} + C_3 \ln(SOC) + C_4 \ln(1 - SOC) \tag{15}$$

Table 2. Open-circuit voltage-state of charge (OCV-SOC) fitting curve parameter table.

SOC	10%	20%	30%	40%	50%	60%	70%	80%	90%
OCV/V	3.47	3.54	3.61	3.64	3.67	3.73	3.83	3.92	4.03

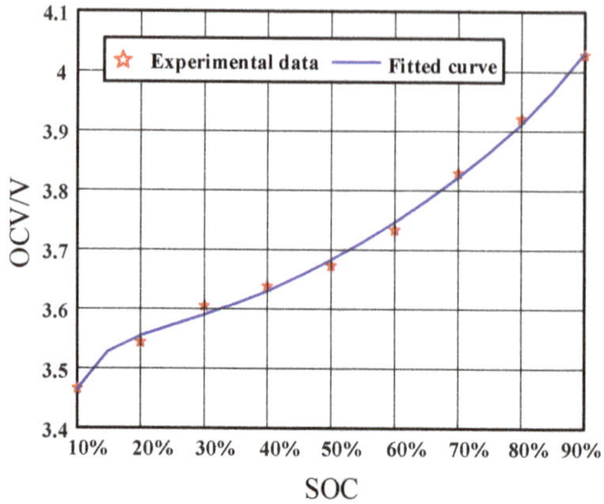

Figure 3. Fitted curve of open-circuit voltage-state of charge (OCV-SOC).

Table 3. Fitted parameters for the OCV-SOC curve.

C_0	C_1	C_2	C_3	C_4
2.951	0.998	−0.052	−0.400	−0.086

To simulate cells operating in EVs and obtain their characteristics, static capacity test (SCT), DST, FUDS, and HPPC tests were performed at an environmental temperature of 25 °C. Later, the MCPSO algorithm [13] was employed to identify the parameters of the above-mentioned DPM and FOM in the time domain. The results of parameter identification are shown in Tables 4 and 5.

Table 4. Dual-polarization model (DPM) model parameter identification results.

R_0	R_1	C_1	R_2	C_2	Q
0.049	0.005	3089.2	0.003	552.129	24.994

Table 5. Fractional-order model (FOM) parameter identification results.

R_0	R_1	CPE_1	R_2	CPE_2	W	α	β	γ
0.012	0.008	19.714	17.808	51.103	155.786	0.839	0.213	0.167

2.3.2. Model Accuracy Verification

In this study, the DST test cycle conditions are used to verify the model accuracy for both the DPM and the FOM. The battery current is used as the model input variable so that the corresponding model voltage output can be measured and plotted. Figure 4 shows the current profile of DST operating condition. The comparison chart and error chart of the DPM output voltage and measured voltage are shown in Figure 5. The comparison chart and error chart of the FOM output voltage and measured voltage are shown in Figure 6.

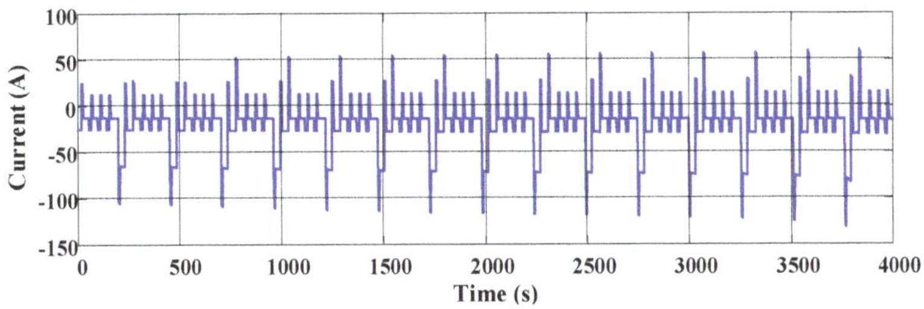

Figure 4. Current profile of the dynamic stress test (DST) test cycle condition.

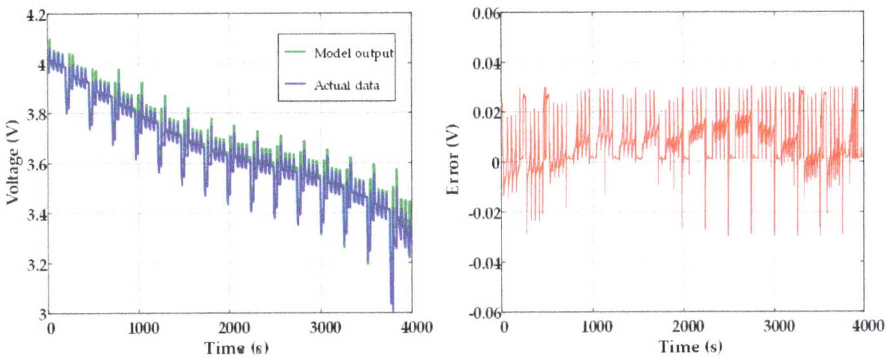

(a) Comparison between output voltage and measured voltage under DST test cycle

(b) Error between output voltage and measured voltage under DST test cycle

Figure 5. DST working condition verification of the dual-polarization model (DPM).

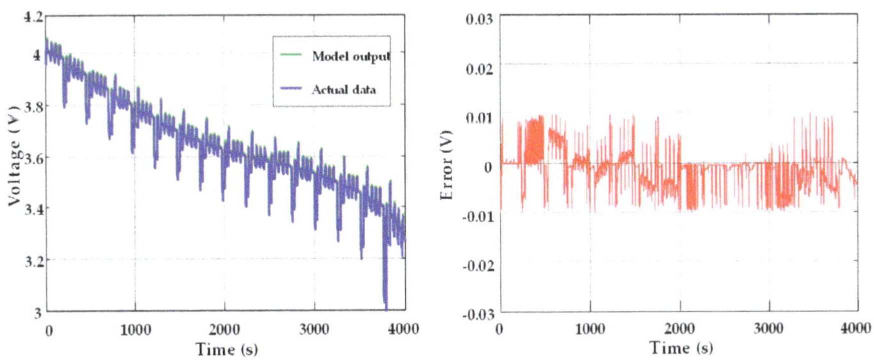

(a) Comparison between output voltage and measured voltage under DST test cycle

(b) Error between output voltage and measured voltage under DST test cycle

Figure 6. The DST working condition verification of the fractional-order model (FOM).

Figures 5 and 6 show that the output voltage curves and measured voltage curves of the DPM and the FOM are highly fitted. In addition, the voltage error of the DPM did not exceed 40 mV, and the voltage error of the FOM was kept within 20 mV. This shows that the established model yielded high accuracy under DST working conditions. However, the accuracy of the FOM was higher than that of the DPM. The quantitative results are presented for comparison in Table 6.

Table 6. RMSEs and relative errors for different SOC ranges.

Model	DPM	FOM
RMSE	12.20 mV	8.19 mV
Relative mean error	0.391%	0.156%

3. SOC Estimation Based on HKF Algorithm

The SOC estimation accuracy of the lithium-ion battery is essential for the BMS. It is reflected in other functions, including charging and discharging control, balance management, safety management, and fault diagnosis cannot be achieved without the high-precision SOC estimation. In this paper, an HKF algorithm based on an established DPM and the FOM is proposed for SOC estimation. The algorithm comprehensively utilizes the merits of the Ah integration method, the KF algorithm, and the EKF algorithm, with verified effectiveness under different operating conditions.

3.1. KF Algorithm and EKF Algorithm

For a discrete linear system, the KF algorithm can be used to improve the estimation accuracy of the system state variables. Recursive formulas used in the algorithm are shown in Table 7. However, the KF algorithm is only effective for linear systems. The DPM and the FOM of lithium-ion batteries established in this paper both are nonlinear models, challenging the usefulness of the KF algorithm. The EKF algorithm can linearize a nonlinear system, thus introducing approximation in the process, and inducing inevitable model errors. However, the algorithm poses certain advantages, such as being simple and fast in its implementation. Hence, the EKF algorithm has been extensively used to deal with the estimation problems for nonlinear models, and its recursive formula is shown in Table 8.

Table 7. The Kalman filter (KF) algorithm recursive formula.

Prior estimation:	$x(k/k-1) = A_k x(k-1/k-1) + B_k u(k)$ $P(k/k-1) = A_k P(k-1/k-1) A_k{}^T + Q_\omega$
Posterior estimation:	$Kg(k) = P(k/k-1) C_k{}^T / [C_k P(k/k-1) C_k{}^T + R_v]$ $x(k/k) = x(k/k-1) + Kg(k)[y(k) - (C_k x(k/k-1) + D_k u(k))]$ $P(k/k) = [I - Kg(k) C_k] P(k/k-1)$

Table 8. The extended Kalman filter (EKF) algorithm recursive formula.

Prior estimation:	$\hat{x}_{k/k-1} = f(\hat{x}_{k-1/k-1}, u_{k-1})$ $P_{k/k-1} = E_{k-1} P_{k-1/k-1} E_{k-1}{}^T + Q_\omega$
Posterior estimation:	$Kg_k = P_{k/k-1} F_k{}^T / (F_k P_{k/k-1} F_k{}^T + R_v)$ $\hat{x}_{k/k} = \hat{x}_{k/k-1} + Kg_k (y_k - g(\hat{x}_{k/k-1}, u_k))$ $P_{k/k} = [I - Kg_k F_k] P_{k/k-1}$

3.2. SOC Estimation Based on HKF

Figure 7 shows the proposed HKF-based SOC estimation algorithm. This algorithm uses a two-layer filtering arrangement to proceed with SOC estimation. In the first layer, the error between the measured voltage (denoted by Vce) and the battery model output voltage (i.e., V) is utilized as the input to an EKF. The output from the EKF provides feedback to the established battery model to obtain a correctional SOC estimate, SOC_{EKF}. The objective of the EKF is to dispose the nondeterminacy resulted from modeling defects. Thus, the estimation performance of SOC is improved. The second layer uses the KF to deal with the cumulative error caused by the Ah integration method. The specific way is that the error difference between SOC_{EKF} and the SOC estimate came from the Ah integration method (namely SOC_{Ah}) is sent to a KF, then the output from the KF provides feedback to the Ah integration algorithm, producing a further correctional SOC estimate value, which is the output SOC. Therefore, the accuracy of SOC estimation can be further improved by the second filter.

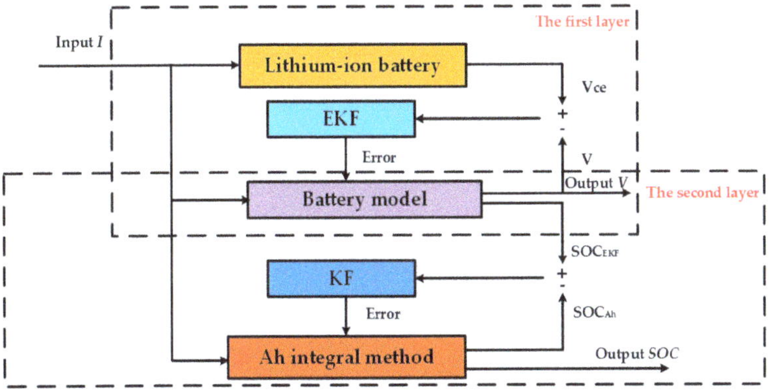

Figure 7. Schematic diagram of the hybrid Kalman filter (HKF) algorithm.

To perform related calculations of the HKF, the DPM is converted into a state-space expression by Equations (5)–(7):

$$\begin{cases} x_{k+1} = Ax_k + Bu_k + \omega_k \\ y_k = Cx_k + Du_k + v_k \end{cases} \tag{16}$$

Among those, the state variable of the system is set to $x = [SOC\ U_1\ U_2]$; the input is $u = I$, and the output is $y = U$. The definitions of A, B, C, and D are shown as follows:

$$A = \begin{bmatrix} 1 & 0 & 0 \\ 0 & \exp\left(\frac{-t}{R_1C_1}\right) & 0 \\ 0 & 0 & \exp\left(\frac{-t}{R_2C_2}\right) \end{bmatrix} \tag{17}$$

$$B = \begin{bmatrix} \frac{\eta t}{Q} \\ R_1\left(1 - \exp\left(\frac{t}{R_1C_1}\right)\right) \\ R_2\left(1 - \exp\left(\frac{-t}{R_2C_2}\right)\right) \end{bmatrix} \tag{18}$$

$$C = \begin{bmatrix} \frac{dOCV}{dSOC} & 1 & 1 \end{bmatrix} \tag{19}$$

$$D = R_0 \tag{20}$$

According to the FOM, the continuous state-space function can be expressed by Equations (21)–(23).

$$D^{\alpha}V_1(t) = -\frac{1}{R_1CPE_1}V_1(t) + \frac{1}{CPE_1}I(t) \tag{21}$$

$$D^{\beta}V_2(t) = -\frac{1}{R_2CPE_2}V_2(t) + \frac{1}{CPE_2}I(t) \tag{22}$$

$$D^{\gamma}V_W(t) = -\frac{1}{W}I(t) \tag{23}$$

According to the definition, the system state SOC of the FOM can be denoted as Equation (24):

$$D^1SOC(t) = -\frac{\eta}{C_n}I(t) \tag{24}$$

Supposing $x(t) = [V_1(t)\ V_2(t)\ V_W(t)\ SOC(t)]$ is state variable of the system, $u = I(t)$ is the input of the system, and $y = U_d(t)$ is the output from the system. Therefore, the pseudo-system state equation of the FOM in this paper is built as follows:

$$\begin{cases} D^N x(t+1) = Ax(t) + Bu(t) \\ y(t) = f[x(t)] + Cx(t) + Du(t) \end{cases} \quad (25)$$

where, the definitions of A, B, C, and D are shown in Equations (26)–(29):

$$A = \begin{bmatrix} -\frac{1}{R_1 CPE_1} & 0 & 0 & 0 \\ 0 & -\frac{1}{R_2 CPE_2} & 0 & 0 \\ 0 & 0 & 0 & 0 \\ 0 & 0 & 0 & 0 \end{bmatrix} \quad (26)$$

$$B = \begin{bmatrix} \frac{1}{CPE_1} \\ \frac{1}{CPE_2} \\ -\frac{1}{W} \\ -\frac{\eta}{C_n} \end{bmatrix} \quad (27)$$

$$C = [-1 \ \ -1 \ \ -1 \ \ 0] \quad (28)$$

$$D = R_0 \quad (29)$$

U_{OCV} and SOC are functionally related. The fractional-order calculus and EKF principle are used to discretize Equation (25) as follows:

$$\begin{cases} x_{k+1} = A_k x_k + B_k u_k \\ y_k = C_k x_k + D_k u_k \end{cases} \quad (30)$$

where A_k, B_k, C_k and D_k are defined as shown in Equations (31)–(34).

$$A_k = \begin{bmatrix} \alpha - T^{\alpha} \frac{1}{R_1 CPE_1} & 0 & 0 & 0 \\ 0 & \beta - T^{\beta} \frac{1}{R_2 CPE_2} & 0 & 0 \\ 0 & 0 & \gamma & 0 \\ 0 & 0 & 0 & 1 \end{bmatrix} \quad (31)$$

$$B_k = \begin{bmatrix} \frac{T^{\alpha}}{CPE_1} \\ \frac{T^{\beta}}{CPE_2} \\ -\frac{T^{\gamma}}{W} \\ -\frac{\eta T}{C_n} \end{bmatrix} \quad (32)$$

$$C_k = [-1 \ \ -1 \ \ -1 \ \ \frac{df(SOC_k)}{dSOC}] \quad (33)$$

$$D_k = R_0 \quad (34)$$

3.3. Comparison of SOC Estimation Results under Different Working Conditions

The test data in this section were based on the lithium-ion power battery test platform and the related basic characteristic tests. The main parameters are shown in Table 1. Besides, the tests in this section were performed at the ambient temperature of 25 °C. The initial SOC values in all the experiments were set to 90%, and the commissioning and verification of the SOC estimation under the HKF for the DPM and the FOM were completed under the parameters set, as explained. The experimental scheme is shown in Table 9. The basic characteristic tests performed as required for the battery SOC estimation algorithm included the DST cycle test, the FUDS cycle test, and the HPPC cycle test. Each working condition test refers to the relevant test methods in the *USABC Electric Vehicle Battery Test Procedures Manual* and the *Freedom CAR Battery Test Manual for Power-Assist Hybrid Electric*

Vehicles. The specific experimental steps are given in Table 10. The sampling frequency was 10 Hz. The charging and discharging current and battery terminal voltage response of each test are shown in Figure 8.

Table 9. Experiments under different models and different test conditions.

Test Objects	DPM			FOM		
Test number	1	2	3	4	5	6
Parameter settings	DST	FUDS	HPPC	DST	FUDS	HPPC

Table 10. The specific experimental steps of the dynamic stress test (DST), federal urban driving schedule (FUDS), and hybrid pulse power characteristic (HPPC) test conditions.

Test Condition	Test Steps
DST	1. Set the temperature of the incubator to 25 °C, and let the experimental battery shelve for an hour 2. 1 C constant current charging to charge cut-off voltage 4.2 V 3. 4.2 V constant voltage charging to charge cut-off current 1.25 A 4. Battery shelved for an hour 5. Running the current profile of DST test cycle condition developed with reference to *USABC Electric Vehicle Battery Test Procedures Manual*
FUDS	1. Set the temperature of the incubator to 25 °C, and let the experimental battery shelve for an hour 2. 1 C constant current charging to charge cut-off voltage 4.2 V 3. 4.2 V constant voltage charging to charge cut-off current 1.25 A 4. Battery shelved for an hour 5. Running the current profile of FUDS test cycle condition developed with reference to *USABC Electric Vehicle Battery Test Procedures Manual*
HPPC	1. Set the temperature of the incubator to 25 °C, and let the experimental battery shelve for an hour 2. 1 C constant current charging to charge cut-off voltage 4.2 V 3. 4.2 V constant voltage charging to charge cut-off current 1.25 A 4. Battery shelved for an hour 5. Discharging 10% of battery capacity (10% SOC) at 1 C constant current 6. Battery shelved for an hour 7. Discharging at 1 C constant current for 10 s, battery shelved for 40 s, charging at 0.75 C constant current for 10 s 8. Repeat steps 5~7 nine times

Considering that the HKF needs to complete the prior estimation of the SOC by the Ah integration method, the sampling time interval is particularly important. Here we chose to collect the current and voltage signal every 0.1 s. It should be noted that, for most devices, the sampling interval cannot be completely and constantly controlled; that is, the sampling time interval fluctuates slightly around 0.1 s.

The above experimental data are battery current and voltage values. To compare the effects of DPM and FOM on SOC estimation, it was necessary to obtain reference values for SOC changes. Considering that under the experimental conditions, the voltage and current acquisition accuracy were very high, and the battery's initial SOC and maximum available capacity were known, the sufficiently accurate SOC change curve could be obtained by the Ah integration method. The curve found by integration was used as a reference value for the SOC estimation results of the two models, and the simulation results are shown in Figures 9–11.

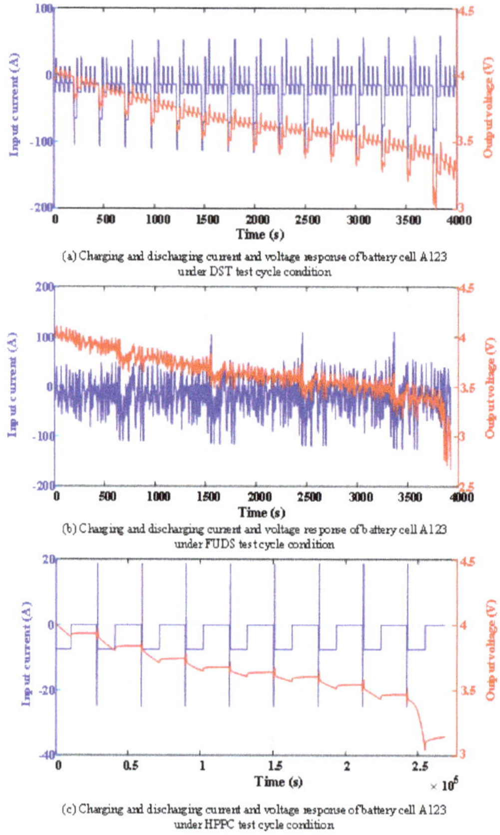

Figure 8. Charging and discharging current and voltage response of battery cell A123 under different test cycle conditions.

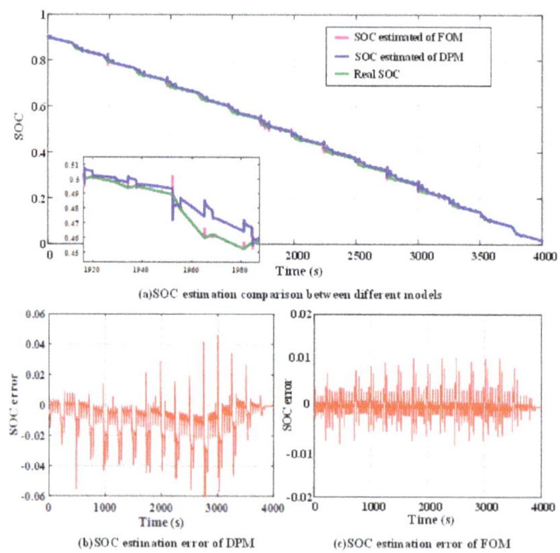

Figure 9. SOC estimation results under the DST test cycle condition.

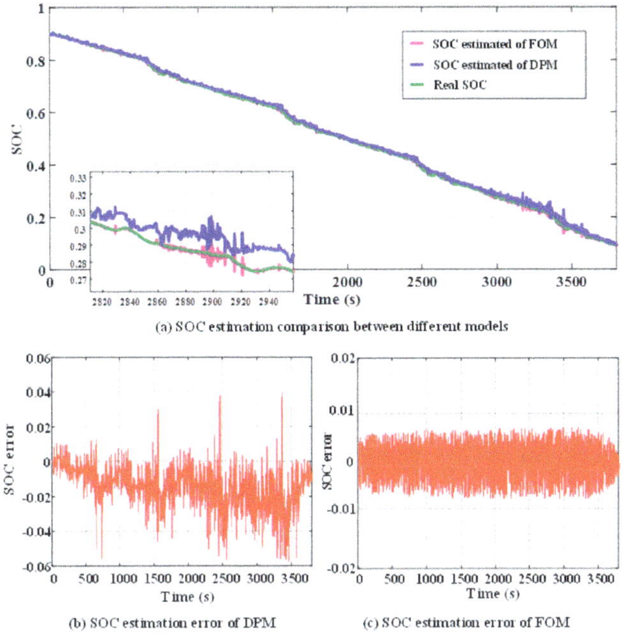

Figure 10. SOC estimation results under the federal urban driving schedule (FUDS) test cycle condition.

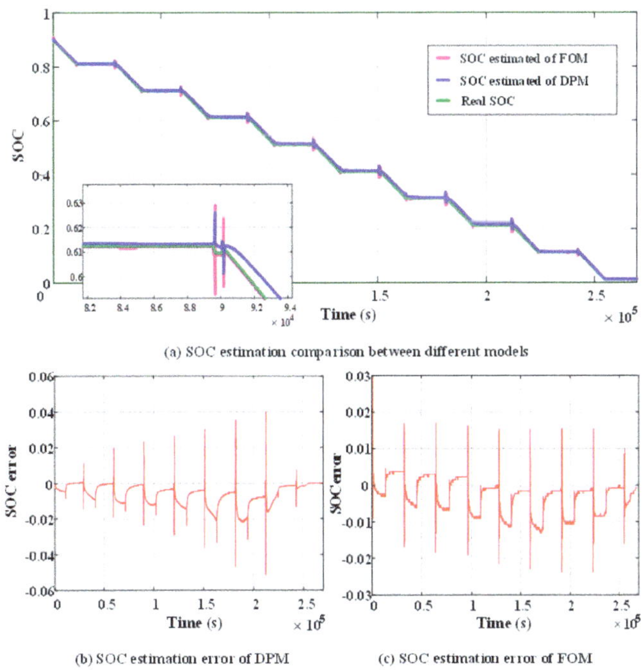

Figure 11. SOC estimation results under the hybrid pulse power characteristic (HPPC) test cycle condition.

As can be seen from Figure 9, the SOC estimation error based on the DPM fluctuated within a range of ±0.06 under the DST test cycle condition, and the SOC estimation error based on the FOM remained within a range of ±0.01. Figure 10 shows that the error of SOC estimation based on the DPM fluctuated within the range of ±0.06 under the FUDS test cycle condition, and the error of SOC estimation based on the FOM can fluctuate within ±0.01. It is clear from Figure 11 that under the condition of HPPC, the error of SOC estimation based on the DPM fluctuated within ±0.05 range, and the SOC estimation error based on the fractional-order model remained fluctuating within ±0.03 range. The results of SOC estimation errors under different models and different test conditions are summarized in Table 11. It can be observed that irrespective of the working conditions applied, the SOC estimation error of the FOM was always less than the estimation error of the DPM, which benefits from the more accurate simulation of the battery impedance characteristics achieved by the fractional-order model. This illustrates that the higher the accuracy of the selected model, the lower the SOC estimation error.

Table 11. SOC estimation errors using different models under different test conditions.

Battery Models	DPM			FOM		
Index	RMSE	Maximum Error	Average Error	RMSE	Maximum Error	Average Error
DST	0.0186	0.06	0.0117	0.0124	0.01	0.0034
FUDS	0.0189	0.06	0.0123	0.0125	0.01	0.0039
HPPC	0.0177	0.05	0.0098	0.0146	0.03	0.0047

From Figure 9b, Figure 10b, and Figure 11b, it could be seen that in the DPM, the frequency of current change in the DST mode and FUDS mode was nearly the same, and the error of SOC estimation was within ± 0.06 range; the SOC estimation error could be maintained within ±0.05 range under HPPC test cycle condition. The maximum fluctuation of the SOC estimation error between different operating conditions was 0.01. From Figure 9c, Figure 10c, and Figure 11c, the SOC estimation error of the FOM in the DST and FUDS test cycle conditions fluctuated within ±0.01 range. Under the HPPC test cycle condition, the SOC estimation result was relatively poor, but the SOC estimation error could also remain within a range of ±0.03. The maximum fluctuation of the SOC estimation error between different operating conditions was 0.02. This shows that the adaptability of the FOM was not as good as that of the DPM. As can be seen from Figures 9a and 11a, in most cases, the SOC simulation values of the two models basically fluctuated slightly around the measured values, and only fluctuated slightly larger when the voltage and current change suddenly. Moreover, the fluctuation range of the FOM was larger than that of the DPM, but it immediately converged near the theoretical value, which had better convergence than the DPM. In summary, although the SOC estimation accuracy of the FOM was higher, the adaptability of working conditions was not as good as that of the DPM.

4. Conclusions

To investigate the effect of different operating conditions on the SOC estimation results, a DPM and a FOM were established in this study, taking into account the prediction accuracy and structural complexity of a battery model. Further, based on these two battery ECMs, an HKF was used to estimate the SOC of the battery, utilizing the Ah integration method, KF, and EKF. The SOC estimation results of the DPM and the FOM under DST, FUDS, and HPPC test cycle conditions were compared and analyzed through six sets of experiments. Simulation results showed that the SOC estimation accuracy of both models was satisfactory, and the errors were within the range of ± 0.06. Under any operating condition, the SOC estimation error of the FOM was always less than that of the DPM, but the adaptability of the FOM was not as good as that of the DPM. In future work, we consider combining several typical working conditions to form a new working condition, then identify the parameters of the battery models to make them more adaptable in different operating conditions.

Author Contributions: Conceptualization, M.H.; methodology, M.H.; validation, G.J.; formal analysis, G.J., L.L., and Y.X.; investigation, G.J., L.L., Y.X., and D.Q.; writing—original draft, G.J. and L.L.; writing—review and editing, M.H. and C.F.; supervision, M.H. and C.F.; project administration, L.L. and Y.X.; funding acquisition, D.Q., M.H. and C.F. All authors have read and agreed to the published version of the manuscript.

Funding: This work was supported by the National Key R&D Program of China [No. 2018YFB0106102], and the Major Program of Chongqing Municipality [No. cstc2018jszx-cyztzxX0007].

Conflicts of Interest: The authors declare no conflict of interest. The funders had no role in the design of the study; in the collection, analyses, or interpretation of data; in the writing of the manuscript; or in the decision to publish the results.

Abbreviations

SOC	state of charge
DPM	dual-polarization model
FOM	fractional-order model
ECM	equivalent circuit model
HKF	hybrid Kalman filter
Ah	ampere-hour
KF	Kalman filter
EKF	extended Kalman filter
DST	dynamic stress test
FUDS	federal urban driving schedule
HPPC	hybrid pulse power characteristic
EVs	electric vehicles
BMS	battery management system
EIS	electrochemical impedance spectroscopy
OCV	open-circuit voltage
PF	particle filter
FL	fuzzy logic
ANN	artificial neural network
SVM	support vector machine
RVM	relevance vector machine
AUKF	adaptive unscented Kalman filter
CDKF	central difference Kalman filter
DKF	double Kalman filter
MCPSO	mixed-swarm-based cooperative particle swarm optimization
SCT	static capacity test

References

1. Piller, S.; Perrin, M.; Jossen, A. Methods for state-of-charge determination and their applications. *J. Power Sources* **2001**, *96*, 113–120. [CrossRef]
2. Stan, A.-I.; Swierczynski, M.; Stroe, D.-I.; Teodorescu, R.; Andreasen, S.J. Lithium ion battery chemistries from renewable energy storage to automotive and back-up power applications—An overview. In Proceedings of the 2014 International Conference on Optimization of Electrical and Electronic Equipment (OPTIM), Bran, Romania, 22–24 May 2014; pp. 713–720.
3. Xiong, R.; Cao, J.; Yu, Q.; He, H.; Sun, F. Critical Review on the Battery State of Charge Estimation Methods for Electric Vehicles. *IEEE Access* **2018**, *6*, 1832–1843. [CrossRef]
4. Ng, K.S.; Moo, C.-S.; Chen, Y.-P.; Hsieh, Y.-C. Enhanced coulomb counting method for estimating state-of-charge and state-of-health of lithium-ion batteries. *Appl. Energy* **2009**, *86*, 1506–1511. [CrossRef]
5. Chaoui, H.; Ibe-Ekeocha, C.C. State of Charge and State of Health Estimation for Lithium Batteries Using Recurrent Neural Networks. *IEEE Trans. Veh. Technol.* **2017**, *66*, 8773–8783. [CrossRef]
6. Seaman, A.; Dao, T.-S.; McPhee, J. A survey of mathematics-based equivalent-circuit and electrochemical battery models for hybrid and electric vehicle simulation. *J. Power Sources* **2014**, *256*, 410–423. [CrossRef]

7. Farmann, A.; Sauer, D.U. Comparative study of reduced order equivalent circuit models for on-board state-of-available-power prediction of lithium-ion batteries in electric vehicles. *Appl. Energy* **2018**, *225*, 1102–1122. [CrossRef]
8. Xiong, R.; Sun, F.; Chen, Z.; He, H. A data-driven multi-scale extended Kalman filtering based parameter and state estimation approach of lithium-ion olymer battery in electric vehicles. *Appl. Energy* **2014**, *113*, 463–476. [CrossRef]
9. Doyle, M.; Fuller, T.; Newman, J. Modeling of galvanostatic charge and discharge of the lithium/polymer/insertion cell. *J. Electrochem. Soc.* **1993**, *140*, 1526–1533. [CrossRef]
10. Junping, W.; Quanshi, C.; Binggang, C. Support vector machine based battery model for electric vehicles. *Energ Convers Manag.* **2006**, *47*, 858–864. [CrossRef]
11. Plett, G.L. Extended Kalman filtering for battery management systems of LiPB-based HEV battery packs: Part 2. Modeling and identification. *J. Power Sources* **2004**, *134*, 262–276. [CrossRef]
12. Kim, T.; Qiao, W. A Hybrid Battery Model Capable of Capturing Dynamic Circuit Characteristics and Nonlinear Capacity Effects. *IEEE Trans. Energy Convers.* **2011**, *26*, 1172–1180. [CrossRef]
13. Hu, M.; Li, Z.; Li, Y.; Li, S.; Fu, C.; Qin, D. Lithium-ion battery modeling and parameter identification based on fractional theory. *Energy* **2018**, *165*, 153–163. [CrossRef]
14. Deng, Y.; Hu, Y.; Cao, Y. An improved algorithm of SOC testing based on open-circuit voltage-ampere hour method. *Commun. Comput. Inf. Sci.* **2014**, *463*, 258–267.
15. Yang, Q.X.; Xu, J.; Cao, B.G.; Li, X.Q. A simplified fractional order impedance model and parameter identification method for lithium-ion batteries. *PLoS ONE* **2017**, *12*, e0172424. [CrossRef] [PubMed]
16. Lee, S.; Lee, J.; Kim, J.; Cho, B.H. State-of-charge and capacity estimation of lithium-ion battery using a new open-circuit voltage versus state-of-charge. *J. Power Sources* **2008**, *185*, 1367–1373. [CrossRef]
17. Orchard, M.E.; Hevia-Koch, P.; Zhang, B.; Tang, L. Risk Measures for Particle-Filtering-Based State-of-Charge Prognosis in Lithium-Ion Batteries. *IEEE Trans. Ind. Electron.* **2013**, *60*, 5260–5269. [CrossRef]
18. Mastali, M.; Vazquez-Arenas, J.; Fraser, R.; Fowler, M.; Afshar, S.; Stevens, M. Battery state of the charge estimation using Kalman filtering. *J. Power Sources* **2013**, *239*, 294–307. [CrossRef]
19. Zenati, A.; Desprez, P.; Razik, H. Estimation of the SOC and the SOH of li-ion batteries, by combining impedance measurements with the fuzzy logic inference. In Proceedings of the IECON 2010 - 36th Annual Conference on IEEE Industrial Electronics Society, Glendale, AZ, USA, 7–10 November 2010; pp. 1773–1778.
20. Weigert, T.; Tian, Q.; Lian, K. State-of-charge prediction of batteries and battery–supercapacitor hybrids using artificial neural networks. *J. Power Sources* **2011**, *196*, 4061–4066. [CrossRef]
21. Hansen, T.; Wang, C.-J. Support vector based battery state of charge estimator. *J. Power Sources* **2005**, *141*, 351–358. [CrossRef]
22. Zhang, Y.; Guo, B. Online capacity estimation of lithium-ion batteries based on novel feature extraction and adaptive multi-kernel relevance vector machine. *Energies* **2015**, *8*, 12439–12457. [CrossRef]
23. Rivera-Barrera, J.P.; Muñoz-Galeano, N.; Sarmiento-Maldonado, H.O. Soc estimation for lithium-ion batteries: Review and future challenges. *Electronics (Switzerland)* **2017**, *6*, 102. [CrossRef]
24. Sun, F.; Hu, X.; Zou, Y.; Li, S. Adaptive unscented Kalman filtering for state of charge estimation of a lithium-ion battery for electric vehicles. *Energy* **2011**, *36*, 3531–3540. [CrossRef]
25. Plett, G.L. Extended Kalman filtering for battery management systems of LiPB-based HEV battery packs: Part 3. State and parameter estimation. *J. Power Sources* **2004**, *134*, 252–261. [CrossRef]
26. Meng, J.; Luo, G.; Gao, F. Lithium Polymer Battery State-of-Charge Estimation Based on Adaptive Unscented Kalman Filter and Support Vector Machine. *IEEE T Power Electr.* **2016**, *31*, 2226–2238. [CrossRef]
27. He, H.; Zhang, Y.; Xiong, R.; Wang, C. A novel Gaussian model based battery state estimation approach: State-of-Energy. *Appl. Energy* **2015**, *151*, 41–48. [CrossRef]
28. Xiong, R.; He, H.; Sun, F.; Zhao, K. Evaluation on State of Charge Estimation of Batteries With Adaptive Extended Kalman Filter by Experiment Approach. *IEEE Trans. Veh. Technol.* **2013**, *62*, 108–117. [CrossRef]
29. Schwunk, S.; Armbruster, N.; Straub, S.; Kehl, J.; Vetter, M. Particle filter for state of charge and state of health estimation for lithium–iron phosphate batteries. *J. Power Sources* **2013**, *239*, 705–710. [CrossRef]
30. Chen, C.; Xiong, R.; Shen, W. A Lithium-Ion Battery-in-the-Loop Approach to Test and Validate Multiscale Dual H Infinity Filters for State-of-Charge and Capacity Estimation. *IEEE Trans. Power Electron.* **2018**, *33*, 332–342. [CrossRef]

31. Xu, J.; Mi, C.C.; Cao, B.; Cao, J. A new method to estimate the state of charge of lithium-ion batteries based on the battery impedance model. *J. Power Sources* **2013**, *233*, 277–284. [CrossRef]
32. Lai, X.; Qiao, D.; Zheng, Y.; Zhou, L. A fuzzy state-of-charge estimation algorithm combining ampere-hour and an extended Kalman filter for li-ion batteries based on multi-model global identification. *Appl. Sci. (Switz.)* **2018**, *8*, 2028. [CrossRef]
33. Xu, Y.D.; Hu, M.H.; Fu, C.Y.; Cao, K.B.; Su, Z.; Yang, Z. State of Charge Estimation for Lithium-Ion Batteries Based on Temperature-Dependent Second-Order RC Model. *Electronics* **2019**, *8*, 1012. [CrossRef]
34. Widanage, W.D.; Barai, A.; Chouchelamane, G.H.; Uddin, K.; McGordon, A.; Marco, J.; Jennings, P. Design and use of multisine signals for Li-ion battery equivalent circuit modelling. Part 2: Model estimation. *J. Power Sources* **2016**, *324*, 61–69. [CrossRef]
35. Hu, X.; Li, S.; Peng, H. A comparative study of equivalent circuit models for Li-ion batteries. *J. Power Sources* **2012**, *198*, 359–367. [CrossRef]
36. Ma, Y.; Zhou, X.; Li, B.; Hong, C. Fractional Modeling and SOC Estimation of Lithium-ion Battery. *J. Autom. Engl.* **2016**, *3*, 281–287.
37. Pattipati, B.; Balasingam, B.; Avvari, G.V.; Pattipati, K.R.; Bar-Shalom, Y. Open circuit voltage characterization of lithium-ion batteries. *J. Power Sources* **2014**, *269*, 317–333. [CrossRef]

© 2020 by the authors. Licensee MDPI, Basel, Switzerland. This article is an open access article distributed under the terms and conditions of the Creative Commons Attribution (CC BY) license (http://creativecommons.org/licenses/by/4.0/).

energies

MDPI

Article

Electric Vehicle–Grid Integration with Voltage Regulation in Radial Distribution Networks

Chong Cao [1,*], Zhouquan Wu [1] and Bo Chen [1,2]

1 Department of Electrical and Computer Engineering, Michigan Technological University, 1400 Townsend Dr, Houghton, MI 49931, USA; wzhouqua@mtu.edu (Z.W.); bochen@mtu.edu (B.C.)
2 Department of Mechanical Engineering-Engineering Mechanics, Michigan Technological University, 1400 Townsend Dr, Houghton, MI 49931, USA
* Correspondence: chongcao@mtu.edu; Tel.: +1-(906)-370-5135

Received: 10 March 2020; Accepted: 31 March 2020; Published: 8 April 2020

Abstract: In this paper, a vehicle–grid integration (VGI) control strategy for radial power distribution networks is presented. The control schemes are designed at both microgrid level and distribution level. At the VGI microgrid level, the available power capacity for electric vehicle (EV) charging is optimally allocated for charging electric vehicles to meet charging requirements. At the distribution grid level, a distributed voltage compensation algorithm is designed to recover voltage violation when it happens at a distribution node. The voltage compensation is achieved through a negotiation between the grid-level agent and VGI microgrid agents using the alternating direction method of multipliers. In each negotiation round, individual agents pursue their own objectives. The computation can be carried out in parallel for each agent. The presented VGI control schemes are simulated and verified in a modified IEEE 37 bus distribution system. The simulation results are presented to show the effectiveness of the VGI control algorithms and the effect of algorithm parameters on the convergence of agent negotiation.

Keywords: vehicle–grid integration; distribution network voltage regulation; alternating direction method of multipliers

1. Introduction

Plug-in electric vehicles (PEVs) have become a practical option for reducing global greenhouse emissions and fossil fuel depletion. However, PEVs also bring challenges to the operation of the power grid if the penetration of PEVs increases. Some of our previous studies [1–3] focus on PEV charging scheduling and optimization within a microgrid, such as via load shaping, charging cost minimization, etc. Study [4] uses the real-time simulation method to validate the PEV charging control algorithm in a VGI microgrid. Large-scale PEV charging activities bring more challenges to power distribution grids. Papers [5–7] use deterministic and stochastic approaches to analyze the PEV charging impact on the distribution networks, including overloading, transformer aging, voltage drop, frequency deviation, and network operating costs. Investigations have been conducted to mitigate some of the aforementioned grid challenges. For example, Cao, et al. [8] formulate the PEV charging activities in a distribution grid as a generalized Nash equilibrium problem. Without violating the node and substation power limits, a Nikaido–Isoda-based control algorithm is developed to minimize individual customers' PEV charging costs. Wang et. al, [9] develop a fully distributed consensus-based large-scale PEV charging coordination algorithm in a power distribution grid. The objectives of this development are to minimize the charging power loss and maximize the PEV power for vehicle-to-grid services. In [10], the authors further provide a dual-level consensus-based electric vehicle charging control scheme for distribution grid frequency regulation. The upper-level control aims to minimize the frequency deviation, and the lower-level control aims to minimize the frequency regulation cost and battery degradation.

In a traditional radial distribution grid, power is delivered from the head node to the end node through the feeder line. Our previous study [5] proposes that the reason for the voltage drop is the excessive high load peaks in a distribution grid. The study provides intuitive approaches for PEV charging, load shifting, and curtailment based on the time of use (ToU) and direct load control (DLC) demand response. Though on-board tap changers (OLTC) [11] and capacitor banks [12] are widely used in distribution grids for voltage regulation, the OLTC is usually used to regulate a relatively large-area network and only monitors voltage at specific nodes. The capacitor banks, though they react quickly, are not installed throughout the entire network. As a result, these devices lack the flexibility for voltage regulation in distribution networks. Some researchers have studied the possibility of utilizing the re-active power operation of the PEV charger for voltage regulation. For example, the authors in [13] propose a vehicle-to-grid reactive power support strategy in cooperation with a high penetration of distributed generation to provide the distribution grid voltage-regulation service. Paper [14] introduces four operation modes of the PEV charger, which include the combination of charging/discharging and inductive/capacitive operation. Though a bi-directional charger capable of reactive power operation is conceptually feasible [15,16], existing on-board chargers on the market may not have this functionality. In fact, the charging system testing data from [17] show that the on-board chargers of the major PEV models are unidirectional, and the related power factors are stable between 0.95–0.99.

In this paper, we aim to study the capability of PEVs to regulate voltage in a VGI distribution grid. A two-level control system was developed to find a balance between the PEV charging requirement and the distribution grid voltage recovery requirement. The contributions of this paper include:

1. A distributed multi-agent negotiation algorithm is developed to recover voltage violation in a distribution network with the alternating direction method of multipliers (ADMM) [18]. In this negotiation algorithm, a series of PEV charging power curtailment decisions are made through a negotiation process. The negotiation agents are the computation and communication units that perform objective pursuit and conduct bargaining in the negotiation. The negotiation agents in this study include multiple VGI microgrid agents who wish to operate the microgrid at a critical charging point for PEVs, and a grid-level agent who wants to recover voltage violation. The negotiation process aims to find a balance between the voltage compensation and the PEV charging requirement.
2. The proposed "capacity of curtailment" (CoC)-based optimal VGI microgrid control algorithm is an improvement of the previously developed iterative PEV curtailment control algorithm [19]. The algorithm determines the microgrid PEV charging power capacity either from PEVs' maximum power demand or base on the distribution grid voltage compensation requirement. With a nonlinear optimization technique, the new algorithm allocates the limited PEV charging power to maximize the infimum of the vehicle CoC value set in a VGI microgrid.
3. The effects of negotiation parameters, such as the penalty factor and the proximal factor, on the negotiation convergence and convergence speed are studied. The selection range of these parameters is provided based on our simulation testing.

The rest of the paper is organized as follows: Section 2 provides an overview of a distribution-level VGI system. Section 3 presents the development of a CoC-based optimal VGI microgrid control algorithm. Section 4 gives the formulation of the voltage regulation negotiation in a distributed manner. Section 5 shows the simulation results of a use case study. Section 6 concludes the paper.

2. Overview of Distribution-Level VGI Control

A radial distribution system is the simplest and the most commonly seen power distribution system. When massive electric vehicle charging activities present in a radial distribution system, the system is at the risk of aggravated power loss and excessive voltage drop. A distributed PEV charging management scheme is designed to provide emergent voltage recovery in a distribution grid and PEV charging power allocation at local microgrids. Figure 1 shows the control and communication among the components of a distribution-level VGI system. Over the feeder branch, a grid-level agent takes charge of bus voltage monitoring and voltage regulation. This consists of modules for bus voltage monitoring, voltage–load variation relation analysis, ADMM optimization, and the distributed negotiation service. At each bus, the buildings and PEV charging stations are grouped to form a microgrid-level VGI system. A microgrid agent is composed of the PEV charging control module that is responsible for dispatching the limited charging power to individual PEVs and the ADMM optimization module, which coordinates with other agents during the process of voltage recovery negotiation.

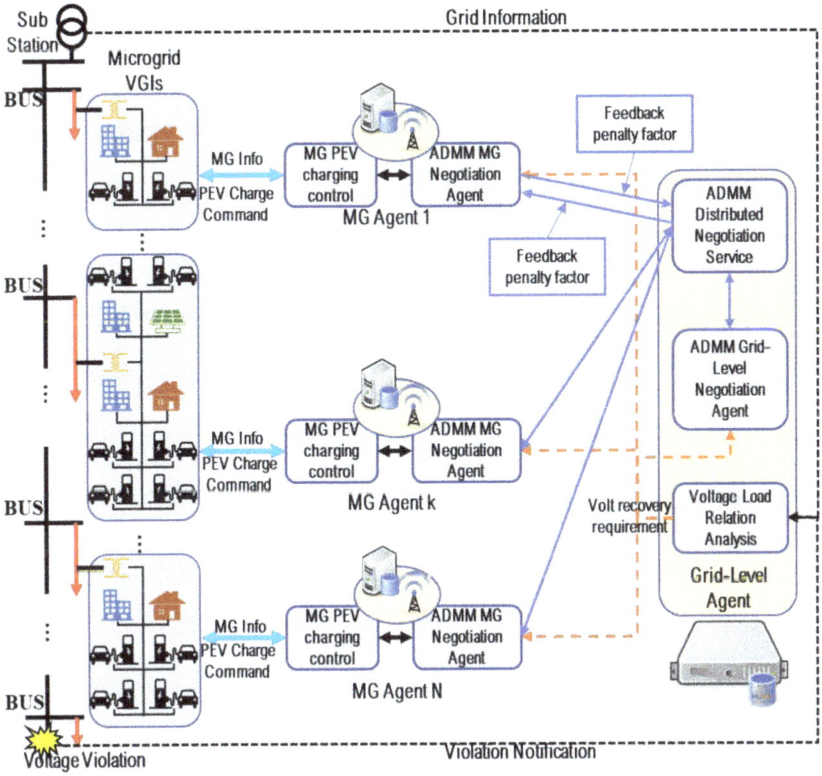

Figure 1. The physical components and control architecture of a distribution-level VGI system.

When the grid-level agent senses a voltage violation at a bus, it sends a notification to all the VGI microgrid agents. Each microgrid individually finds and decreases the microgrid PEV charging power to a critical charging power point. This critical charging power point is the lowest microgrid PEV charging power capacity that meets the PEV state of charge (SOC) requirements. If the voltage violation remains, a voltage–load relation analysis is conducted by the grid-level agent. The coefficients of the voltage–load relation, which reflects the impact of the load change of individual microgrids on voltage variation at current violation bus, are broadcasted to all the agents in the network. A negotiation is initiated among all the involved microgrid agents and the grid-level agent. The negotiation uses the ADMM-based method to find out the balance between the load curtailment amount of each microgrid and the voltage recovery requirement. Figure 2 shows the PEV charging power capacity determination process before the microgrid PEV charging control can be applied.

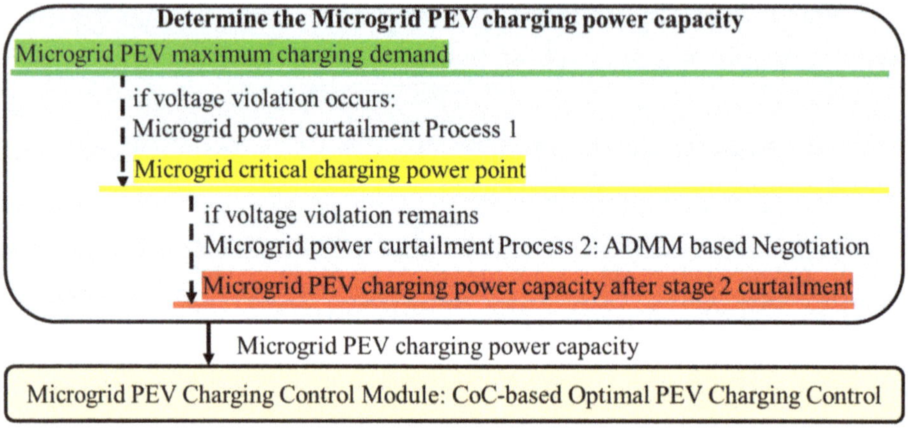

Figure 2. The determination of PEV charging power capacity in microgrids.

3. Microgrid-Level VGI Control

In this section, a CoC-based optimal VGI microgrid control algorithm is presented, which aims to allocate limited PEV charging power to better meet the PEV charging requirements in a microgrid. The PEV charging process is analyzed based on the steady-state PEV charging test results [17]. The process is divided into two stages: constant power charging and constant voltage charging. In the constant power charging stage, the PEV on-board charger can provide charging power control by specifying the input AC current [20]. When the PEV battery voltage reaches a certain level, the charging enters a constant voltage charging stage in which the PEV charging power decreases gradually. The constant power charging stage is the major controllable period that smart charging schemes are applied to. The VGI microgrid control algorithm is designed with following assumptions.

1. The target SOC, departure time, and vehicle type are specified by the PEV charging customers at charging stations.
2. All charging PEVs are enrolled in the smart charging control program.
3. The Electric Vehicle Supply Equipment (EVSE) can detect the PEV connectivity, the PEV SOC and charging stage. In addition, four control states are defined for an EVSE. State ($s_i = 0$): there is no PEV connected to the EVSE; State 1 ($s_i = 1$): a PEV is connected to the EVSE and is in charging stage 1; State 2 ($s_i = 2$): a PEV is connected to the EVSE and is in the charging stage 2; State 3 (($s_i = 3$): reserved for temporary usage.

3.1. The CoC-Based Optimal PEV Charging Control

The concept of capacity of curtailment is defined to evaluate the capability of a PEV to allow charging power curtailment as shown below:

$$\mathrm{CoC} = \left(e^{-(\mathrm{T_{Dep}}-t)} + 1\right)\left(\frac{\mathrm{SOC_t}}{100} \cdot \mathrm{cap_{batt}} + \mathrm{I_{batt}}(\mathrm{T_{Dep}} - t) - \frac{\mathrm{SOC_{Tar}}}{100} \cdot \mathrm{cap_{batt}}\right) \tag{1}$$

where $\mathrm{SOC_t}$ and $\mathrm{SOC_{Tar}}$ represent the current and target state of charge, respectively. $\mathrm{I_{batt}}$ denotes the battery charging current at the DC terminal. $\mathrm{T_{Dep}}$ is the PEV departure time and $\mathrm{cap_{batt}}$ is the vehicle battery capacity in amp-hours. A positive value of COC means that the target SOC can be reached before the departure time with the current charging power rate. A time-related weighting factor $e^{-(T_{Dep}-1)}$ boosts the CoC value when a vehicle approaches its departure time.

The objective of CoC-based optimal VGI microgrid control algorithm is to curtail the charging power of PEVs with higher CoC values and leave the power capacity for the PEVs with lower CoC values when the total charging power of a microgrid is limited. In other word, the set of CoC values of the controllable PEVs in a microgrid is considered as a collection of time-varied functions of the PEV charging power: $\mathbf{CoC} = \left\{CoC_i(P_{EV,i}), |i \in \mathbf{I}_{(s_i=1)}\right\}$, where the set $\mathbf{I}_{(s_i=1)}$ represents the set of controllable PEVs. The objective is to calculate the proper charging powers for the controllable PEVs to maximize the infimum of the CoC set as shown in Equation (2).

$$\begin{gathered} \max_{P_{EV,i}, i \in \mathbf{I}_{(s_i=1)}} \\ \text{s.t. } y \leq CoC_i(P_{EV,i});\ P_{EV,i} \in \phi_{EVSE,i};\ \text{and} \sum_{i \in \mathbf{I}} P_{EV,i} \leq P_{cap} \end{gathered} \tag{2}$$

where y is the infimum, which is the largest value that is smaller than all the CoC values. P_{cap} is the limited charging power of the microgrid. s_i represents the EVSE state. $\phi_{EVSE,i}$ is the feasible range of power consumption of EVSE i, which can be represented as Equation (3)

$$\phi_{EVSE,i} = \begin{cases} \{0\} & if\ s_i = 0 \\ \left\{\left[P_{J1772,min}, P_{EV,i,max}\right]\right\} & if\ s_i = 1 \\ \left\{P_{EV,i}\right\} & if\ s_i = 2 \\ \left\{P_{EV,i,max}\right\} & if\ s_i = 3 \end{cases} \tag{3}$$

where $P_{J1772,min}$ is the minimum PEV charging rate when the AC minimum charging current is 6 A. $P_{EV,i,max}$ denotes the maximum charging power of the PEV charger.

The CoC-based optimal VGI microgrid control algorithm presents the control design for different available power capacities. If the available charging power is greater than the charging power demand, all the PEVs in charging stage 1 will be charged at the maximum power of the charging stations and the PEVs in charging stage 2 will be charged at the required power for constant voltage charging. If the available charging power is not enough for all connected PEVs even at the minimum charging power, the control scheme temporarily shuts off some of PEVs in charging stage 1. If the available charging power is between the maximum and minimum charging power demands, the optimization is performed to reallocate the charging power to individual PEVs in charging stage 1. If a PEV is being charged at maximum charging power but still has the lowest CoC value, this PEV is considered as an uncontrollable load.

3.2. Microgrid PEV Charging Power Capacity

Before dispatching power to the PEVs, the microgrid needs to determine the PEV charging power capacity, P_{Cap}. If there is no distribution voltage violation, all the PEVs obtain as much power as their maximum demands. When voltage violation occurs, a microgrid curtails its PEV charging power capacity to a value, P_{CP}, named as "The critical charging power point". This P_{CP} is the minimum power capacity that can meet the PEV charging SOC requirement through the microgrid PEV charging control algorithm. The critical charging power point can be obtained by using Equation (4)

$$\begin{gathered} P_{CP} = \min\left(P_{Cap}\right) \\ \text{s.t. } CoC_i\left(P_{EV,i}\right) \geq 0 \end{gathered} \tag{4}$$

where all the $P_{EV,i}$ are the outputs of the PEV charging control algorithm when applying P_{Cap}. If the first process of curtailment does not resolve the voltage violation, a second PEV charging power capacity curtailment process is needed. The further power curtailment is defined as ΔP_{Cap}, which is determined through a distributed negotiation process.

The CoC-based Optimal VGI Microgrid Control Algorithm

Obtain charging information:

- Obtain PEV charging stage information.
- Obtain EVSE states from a state set $S = \{s_i | i \in \mathbf{I}\}$ based on the charging stage information
- Obtain vehicle current/target SOCs: $SOC_{ins,i}$, $SOC_{Tar,i}$
- Obtain vehicle departure time, $T_{Dep,i}$

Obtain the PEV charging power capacity, P_{Cap}:

if no voltage violation occurs

$P_{Cap} = \sum_{s_i=1} P_{EV,i,max} + \sum_{s_i=2} P_{EV,i}$;

elseif voltage violation occurs

Curtailment Process 1: $P_{Cap} = P_{cp}$;

if voltage violation remains at **Process 1**

Curtailment Process 2: $P_{Cap} = P_{cp} - \Delta P_{Cap}$; ΔP_{Cap} is the power curtailment decision of the negotiation.

Enter the Optimal PEV Charging Control

- Obtain the uncontrolled charging load $P_{un} = \sum_{s_i=2} P_{EV,i}$, the controllable load $P_{con} = P_{Cap} - P_{un}$
- Attain the maximum PEV charging demand: $D_{EV} = \sum_{s_i=1} P_{EV,i,max} + \sum_{s_i=2} P_{EV,i}$

if $P_{con} \geq D_{EV} - \sum_{s_i=2} P_{EV,i}$

$P_{EV,i} = P_{EV,i,max}; s_i = 1; P_{EV,i} = P_{EV,i}; s_i = 2;$

elseif $P_{con} < \text{length}(s_i(s_i = 1)) * P_{J1772,min}$

Calculate $CoC_{(i|s_i=1)}$ with $P_{EV,i} = P_{J1772,min}$

Sort $\mathbf{CoC}_{(i|s_i=1)}$ set in ascending order.

Calculate $n_s = \text{integer}\left(\frac{P_{con}}{P_{J1772,min}}\right)$

Perform optimization defined in Equation (2) among the first n_s number of PEVs.

Shut off charging of rest stage 1 PEVs – set to state 0

else

Perform optimization defined in Equation (2)

While $(P_{EV,i=\mathbf{I}(\min(\mathbf{CoC}))} == P_{EV,i,max})$

set $s_i = 3$;

$P_{con} = P_{con} - P_{EV,i}$;

Perform optimization defined in Equation (2)

Exit

4. Voltage Compensation in a Distribution-Level VGI System

A voltage compensation process is divided into two parts. In the first part, the voltage–load variation relation analysis is conducted by the grid-level agent. With this relation, the microgrid agents and the grid-level agent negotiate and find out the further power curtailment of the microgrids and the adjusted voltage compensation target.

4.1. The Voltage–Load Variation Relationship Analysis

The relationship between the voltage change and the load variation of the nodes is analyzed at the grid-level agent with a Jacobi iterative method. This relationship is useful for the microgrid agents to determine their PEV charging power curtailment and the grid-level agent to adjust its voltage compensation target in performing the voltage regulation. Figure 3 shows a distribution feeder branch from a substation BUS 0 as a power source to the end bus—BUS N. The distribution line section parameters are simplified as an impedance model R + jX. The cross-coupling effects of inductance and shunt capacitance from different phases are not considered.

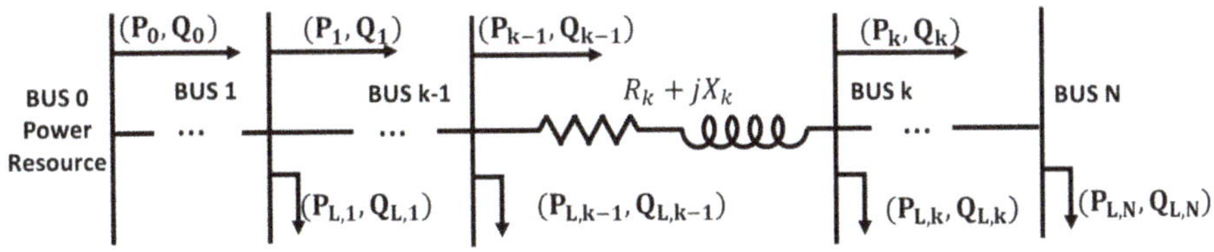

Figure 3. The adjacent buses in a distribution feeder branch.

Based on the results deduced in [12], the active and reactive supply power at a random BUS k can be related to the power loss on the distribution line and the grid parameters of BUS k − 1, as shown in Equations (5) and (6). The voltage relationship between BUS k − 1 and BUS k can also be derived as Equation (7):

$$P_k = P_{k-1} - P_{Loss,k} - P_{L,k} = P_{k-1} - R_k \frac{P_{k-1}^2 + Q_{k-1}^2}{|V_{k-1}|^2} - P_{L,k} \quad (5)$$

$$Q_k = Q_{k-1} - Q_{Loss,k} - Q_{L,k} = Q_{k-1} - X_k \frac{P_{k-1}^2 + Q_{k-1}^2}{|V_{k-1}|^2} - Q_{L,k} \quad (6)$$

$$|V_k|^2 = |V_{k-1}|^2 - 2(R_k P_{k-1} + X_k Q_{k-1}) + \frac{\left(R_k^2 + X_k^2\right)\left(P_{k-1}^2 + Q_{k-1}^2\right)}{|V_{k-1}|^2} \quad (7)$$

The square of the downstream root-mean-square (RMS) voltage square can be represented as a function of its adjacent upper-stream supply power and the square of RMS voltage. It is also related to the distribution line impedance between the two buses. Fazio et. al, [21] provide a proof that a random bus parameter variation $\left[\Delta P_k,\ \Delta Q_k,\ \Delta|V_k|^2\right]$ can be estimated as a linear combination of all buses' loads along the distribution feeder and the source voltage variation square. By using Taylor expansion and the chain rule, the voltage–load variation relationship is deduced; the square of RMS voltage variation for a random BUS k can be represented as Equation (8).

$$\Delta|V_K|^2 = \sum_{k=1}^{N}\left[c_{Q,k}, c_{P,k}\right]\left[\Delta P_{L,k}, \Delta Q_{L,k}\right]^T + c_{V_0}\Delta|V_0|^2 \quad (8)$$

The coefficients $\left[c_{P,k}, c_{Q,k}\right]$ compose the weighting factor vector that represents the effect of load variation at BUS k on the BUS K RMS voltage square. The values of the coefficients are deduced from [21]. $\Delta|V_0|^2$ denotes the power source voltage fluctuation. c_{V_0} represents the weighting factor between the variation of power source voltage and the BUS k voltage. Considering the power factor of the PEV charging in each microgrid as pf_k, the reactive PEV charging power curtailment of a microgrid can be represented as $\Delta Q_{Cap,k} = \frac{\sqrt{1-pf_k^2}}{pf_k}\Delta P_{Cap,k}$. Assuming that the voltage source V_0 does not fluctuate, the distribution voltage improvement at BUS K about power curtailment at each node along the distribution feeder line then can be simplified and reformulated as

$$c_0\Delta|V_k|^2 + \sum_{k=1}^{N} c_k \Delta P_{Cap,k} = 0, \text{ where } c_0 = -1 \quad (9)$$

where the parameter $c_k = \left[c_{P,k}, c_{Q,k}\right]\left[1, \frac{\sqrt{1-pf_k^2}}{pf_k}\right]^T$. The relationship (Equation (9)) is broadcasted to all the voltage recovery participants—the microgrid agents and the grid-level agent.

4.2. The ADMM-Based Voltage Compensation Negotiation

With the voltage–load variation relationship information, a distributed negotiation is triggered to compensate the voltage violation at BUS K in a distribution feeder branch amongst all the agents in the feeder branch. The voltage compensation negotiation aims to maximally recover the voltage back to the allowable range with least sacrifice of PEV charging requirements.

The microgrid agents want to minimize further power curtailment from their critical charging power point. The objective function of microgrid agents are defined in Equation (10).

$$f_k(\Delta P_{Cap,k}) = \left\|\Delta P_{Cap,k}\right\|_2^2;\ k = 1, \ldots, N \tag{10}$$

The objective of the grid-level agent is to meet the distribution voltage compensation target as much as possible. Define the distribution RMS voltage square compensation target as $\Delta|V_K|_T^2 = |V_K|_T^2 - |V_K|_{ins}^2$, which is the difference between the target RMS voltage square and the instant RMS voltage square at the violation bus. The value of the target RMS voltage is selected to be a little higher than the distribution voltage lower bound, 0.95 p.u. The objective function of the grid-level agent is represented as Equation (11)

$$f_0(\Delta|V_K|^2) = \left\|\Delta|V_K|^2 - \Delta|V_K|_T^2\right\|_2^2 \tag{11}$$

The voltage compensation problem is converted to a coordination problem that balances the objectives of the distribution grid voltage requirement and the PEV charging requirements in multiple microgrids as shown in Equation (12).

$$\begin{aligned} &\min_{\mathbf{x}}\left[f_0(\Delta|V_K|^2) + \textstyle\sum_{k=1}^N f_k(\Delta P_{Cap,k})\right] \\ &\text{s.t: } c_0\Delta|V_K|^2 + \sum_{k=1}^{N} c_k \Delta P_{Cap,k} = 0 \end{aligned} \tag{12}$$

This minimization problem is solved using ADMM iteratively. The augmented Lagrangian is expressed as Equation (13) in the first step.

$$\begin{aligned} L_\rho\left(\left[\Delta|V_K|^2, \mathbf{\Delta P_{Cap}}\right], \lambda\right) = f_0(\Delta|V_K|^2) + \textstyle\sum_{k=1}^N f_k(\Delta P_{Cap,k}) + \\ \lambda\left(c_0\Delta|V_K|^2 + \sum_{k=1}^{N} c_k \Delta P_{Cap,k}\right) + \frac{\rho}{2}\left\|c_0\Delta|V_K|^2 + \sum_{k=1}^{N} c_k \Delta P_{Cap,k}\right\|_2^2 \end{aligned} \tag{13}$$

$\mathbf{\Delta P_{Cap}} = \left\{\Delta P_{Cap,k}|k = 1, \ldots, N\right\}$ is a vector of microgrid PEV charging power curtailments. The symbol λ is the Lagrangian multiplier. ρ is called the penalty parameter. The optimal solution set of the voltage compensation and the microgrid PEV charging curtailments $\left\{\Delta|V_K|^2,\ \mathbf{\Delta P_{Cap}}\right\}$ can be found through an iterative optimization process as shown in Equation (14).

$$\begin{aligned} \left[\Delta|V_K|^2, \mathbf{\Delta P_{Cap}}\right]^{m+1} &:= \underset{[\Delta|V_K|^2, \mathbf{\Delta P_{Cap}}]}{\operatorname{argmin}}\ L_\rho\left(\left[\Delta|V_K|^2, \mathbf{\Delta P_{Cap}}\right], \lambda^m\right); \\ \lambda^{m+1} &= \lambda^m + \rho\left(c_0\Delta|V_K|^2 + \sum_{k=1}^{N} c_k \Delta P_{Cap,k}\right); \end{aligned} \tag{14}$$

Due to the separability of the minimization objective (Equation (12)), the update of decision variable can be conducted by each negotiation participant in parallel using the Jacobian type of method. This method fixes variables that are not directly related to a microgrid to the last iteration decisions, therefore simplifies the calculation. Furthermore, a proximal term $\frac{\phi}{2}\left(\|\Delta|V_K|^2 - \Delta\left|V_K^m\right|^2\|_2^2 + \sum_{k=1}^N \|\Delta P_{Cap,k} - \Delta P_{Cap,k}^m\|_2^2\right)$ is added to strengthen the convexity of the augmented Lagragian and accelerate the convergence. The distributed solution of the distribution voltage compensation problem can be achieved through the following iterations.

Distributed Voltage Compensation Negotiation Process

Initialization

- ○ Set $m = 0$ as the negotiation iteration counter.
- ○ Set the initial values of $\Delta P^{m=0}_{Cap,k} = 0,\ k = 1, \ldots, N$, for all the microgrid agents, and $\Delta\left|V^{m=0}_{K}\right|^2 = \Delta|V_K|^2_T$ for the grid-level agent
- ○ Set the Lagrangian multiplier $\lambda^{m=0}$, penalty factor ρ and the proximal factor ϕ
- ○ Calculate the Residual = $c_0\Delta\left|V^{m=0}_{K}\right|^2 + \sum_{k=1}^{N} c_k \Delta P^{m=0}_{Cap,k}$

While |Residual | > ϵ

- **For each microgrid**, k = 1 . . . , N
  $\Delta P^{m+1}_{Cap,k} = \underset{\Delta P_{Cap,k}\in[0,P_{CP,k}]}{\operatorname{argmin}} \left(\|\Delta P_{Cap,k}\|_2^2 + \frac{\rho}{2}\| c_0\Delta\left|V^{m}_{K}\right|^2 + c_k\Delta P_{Cap,k} + \sum_{l\neq k} c_l \Delta P^{m}_{Cap,l} - \frac{\lambda^m}{\rho}\|_2^2 + \right.$
  $\left.\frac{\phi}{2}\|\Delta P_{Cap,k} - \Delta P^{m}_{Cap,k}\|_2^2\right);$
- **For the grid-level agent**
  $\Delta\left|V^{m+1}_{K}\right|^2 = \underset{x_0}{\operatorname{argmin}}\left(\|\Delta|V_K|^2 - \Delta|V_K|^2_T\|_2^2 + \frac{\rho}{2}\| c_0\Delta|V_K|^2 + \sum_{k=1}^{N} c_k \Delta P^{m}_{Cap,k} - \frac{\lambda^m}{\rho}\|_2^2 + \right.$
  $\left.\frac{\phi}{2}\|\Delta|V_K|^2 - \Delta\left|V^{m}_{K}\right|^2\|_2^2\right);$
  Then update Lagrangian Multiplier
  Residual = $c_0\Delta\left|V^{m}_{K}\right|^2 + \sum_{k=1}^{N} c_k \Delta P^{m}_{Cap,k}$
  $\lambda^{m+1} = \lambda^m - \rho(\text{Residual})$

Exit

In the negotiation process shown in Figure 4, each participant agent places an initial value for its decision variable, $\Delta\left|V^{m=0}_{K}\right|^2$ for the grid-level agent and $\Delta P^{m}_{Cap,k}$, $k = 1, \ldots, N$, for each of the microgrid agents, and notifies all other participant agents. Then, with the knowledge of all other agents' proposed decisions, and the voltage–load variation relationship coefficients, c_k, each participant agent optimizes the augmented Lagrangian with the proximal, considering its own decision as the only variable. Due to the separability of the objective in Equation (12), the optimization conducted by each negotiation participant is closed to its target value with the constraint of the voltage–load variation relationship. In addition, all components of the individual optimization functions are square errors, which are quadratic. This ensures the strict convexity of these functions and guarantees that a unique vector of minimizers can be found in each round of negotiation. The grid-level agent collects all the optimization results from all negotiation participants. The residual is calculated and used to judge if the negotiation agreement is obtained. If the residual is beyond a pre-specified range, a new round of negotiation starts with the updated Lagrangian multiplier, λ^{m+1}. The negotiation iteration continues until the residual converges to a small enough value. Notice that the selection of the initial values, such as the initial Lagrangian multiplier, λ, the penalty factor ρ, the proximal factor ϕ and VGI microgrid load curtailments can all have the effects on the negotiation convergence. The investigation of these impacts is conducted to ensure the performance of the control scheme.

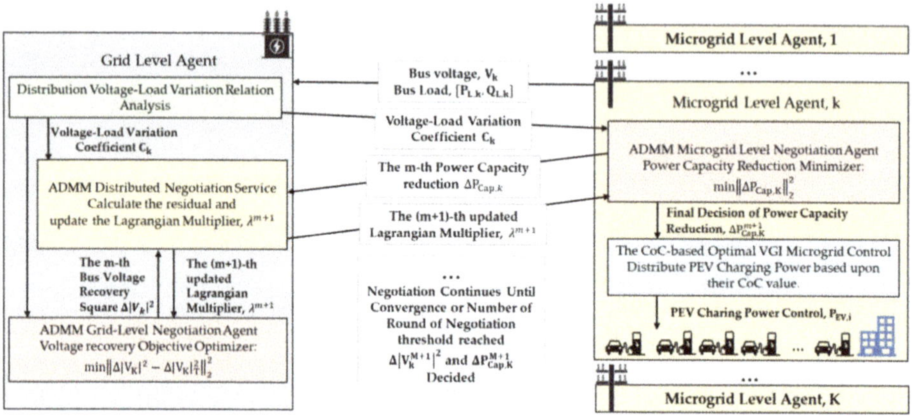

Figure 4. The development and message flow of the distributed voltage compensation negotiation and VGI microgrid control.

5. Use Case Study

To validate the presented voltage compensation algorithm, the longest branch of the IEEE 37 Node Test feeder [22] from Node 701–Node 741 was selected to simulate a mid-level distribution feeder branch with commercial loads. The transmission voltage level of 230 kV is transformed in distribution substation to 4.8 kV in the feeder. Different combinations of commercial building load profiles obtained from [23] were used as uncontrollable load of microgrids connected to each node. In addition to building loads, microgrids on node 702, 730, 708, 734, 738 and 741 had PEV chargers. Figure 5 shows the distribution grid topology of the use case. The building loads and the number of EVSEs under each node are listed in Table 1. A total of 120 EVSEs were installed in six microgrids to charge three types of PEVs. The PEV battery and charging parameters are listed in Table 2. The power factors of the building load and PEV charger were 0.95 and 0.98, respectively.

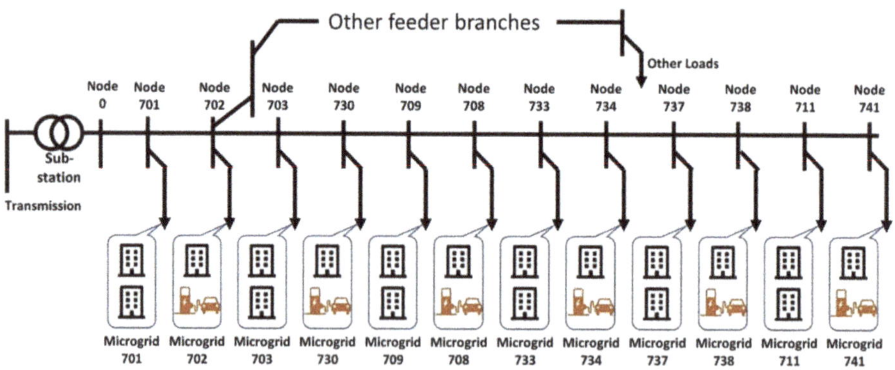

Figure 5. The distribution grid topology in the use case study.

Table 1. The building and number of EVSEs in microgrids.

MG	Building Types	EVSE Units
701	Medium office, Strip mall	0
702	Supermarket	20
703	Small restaurant, Retail store	0
730	Large office	20
709	Supermarket	0
708	Supermarket, Warehouse	20
733	Strip mall, Retail store	0
734	Small and medium office	20
737	Restaurant, Retail store	0
738	Multiple small offices	20
711	Primary school, small offices	0
741	Warehouse	20

Table 2. PEV battery and charging parameters [17].

PEV Type	Battery Capacity	Maximum Charge Rate	Minimum Charge Rate
Nissan Leaf 2013	66.2 Ah	6.6 kW	1.44 kW
BMW i3 2014	60.0 Ah	7.2 kW	1.44 kW
Chevy Volt 2013	45.0 Ah	3.1 kW	1.44 kW

We simulated the voltage compensation at 10:00 am and assumed that all the EVSEs were connected with PEVs. This resulted in 120 PEVs (31 Nissan Leaf 2013s, 20 BMW i3 2014s, and 69 Chevy Volt 2013s) in the simulation. These PEVs were randomly allocated to EVSEs. The initial and target SOCs of the PEVs were also randomly generated. Table 3 shows the uncontrollable load and maximal PEV charging power demand at 10:00 am. The variation and effect from other feeder branches are not considered. In addition, it was assumed that the grid was balanced, and the node loads were all connected to the same phase on the feeder branch.

One distribution voltage violation was simulated at Node 741 with 0.933 per unit when all microgrids chose to charge PEVs at maximum charging power. After all microgrids curtailed PEV charging power to critical point, the voltage violation still existed at Node 741 with 0.948 per unit. In this situation, a negotiation among the grid-level agent and microgrid agents to further curtail PEV charging power for the voltage compensation is triggered. Firstly, the voltage–load variation relationship coefficients were found to be coef = [0.2396, 0.4420, 0.7242, 1.0718, 1.1869, 1.3710, 1.5548, 1.8788, 2.2462, 2.4738, 2.6975, 2.9188], which were used as the grid voltage compensation reference. Each VGI microgrid wishes to retain a charging power close to its PEV critical charging power point. On the other hand, the grid-level agent wishes to compensate the voltage at Node 741 back to 0.9505 per unit. The negotiation process is shown in Figure 4.

Table 3. Uncontrollable load and PEV charging power demand at 10:00 am in each node along the distribution feeder branch.

Microgrid	Building Demand (kW)	MG PEV Charging Demand (kW)	MG Critical Charging Power Point (kW)
701	93.32	0	0
702	129.6	112.8	65.1
703	97.76	0	0
730	74.36	119.2	66.0
709	129.6	0	0
708	66.35	113.4	55.8
733	62.22	0	0
734	101.4	109.3	50.5
737	96.87	0	0
738	99.15	117.5	57.5
711	84.87	0	0
741	69.43	123.3	62.0

Figures 6 and 7 show the selection of negotiation parameters, such as the penalty factor ρ and the proximal factor ϕ, and their effects on the residual convergence. It is shown in Figure 6 that a larger penalty factor ρ can accelerate the negotiation. However, the negotiation oscillates when the penalty factor exceeds a certain value. The introduction of a proximal term helps to improve the convergence of the negotiation process. Figure 7 shows simulation results for different values of the proximal factor ϕ when the penalty factor $\rho = 0.2$. It is observed that the proximal term enhances the possibility of convergence of the negotiation. However, an excessively large proximal factor results in divergence. After a number of simulations, an empirical conclusion was that it is relatively safe to choose a penalty factor that is smaller than 1 and a proximal factor that is 5–10 times larger than the penalty factor.

With the penalty factor, $\rho = 0.2$ and the proximal factor, $\phi = 1$, the negotiation shows a fast convergence speed within 30 rounds. Figure 8 shows the update sequence of the microgrid power curtailment and the grid level-agent decisions in the negotiation process. The negotiation finally reaches an agreement that the VGI microgrids' charging power curtailments are 2.44k W (node 702), 5.37 kW (node 730), 6.79 kW (node 708), 9.19 kW (node 734), 12.02 kW (node 738), and 14.16 kW (node 741) from their critical charging points. On the other hand, the grid-level agent adjusts its $\Delta|V_{741}|^2$ requirement from 109.35×10^3 to 104.49×10^3.

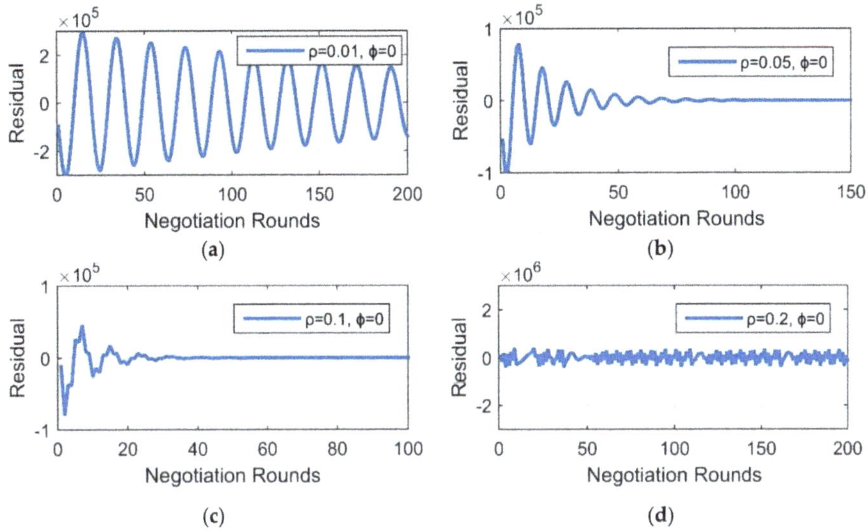

Figure 6. The negotiation process with different values of penalty factor ρ. Sub-figure (**a**) shows a very slow convergence rate with $\rho = 0.01$, Sub-figure (**b**,**c**) show faster convergence within the range $0.05 \leq \rho \leq 0.1$, but (**c**) demonstrates an even better convergence rate. Sub-figure (**d**) show negotiation does not converge when ρ increases and reaches 0.2.

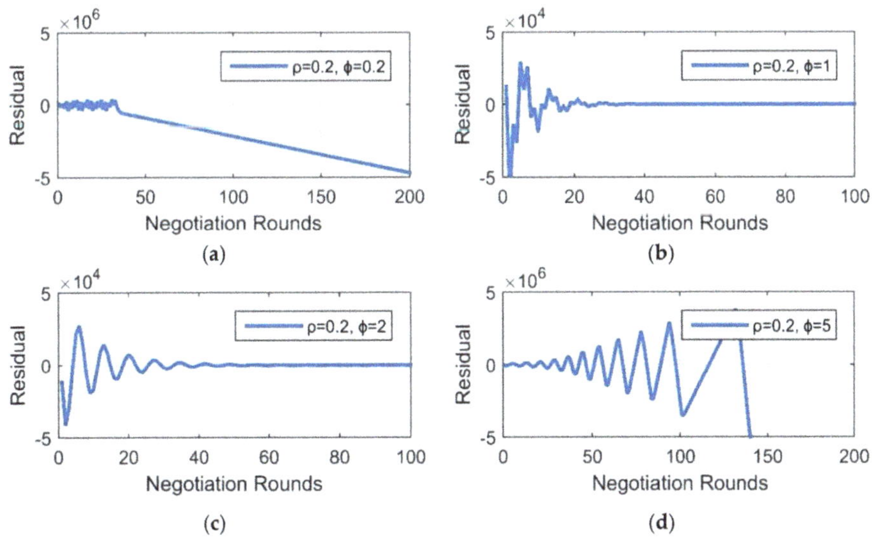

Figure 7. The negotiation process with different values of proximal factor ϕ. Adding proper proximal factor can increase the robustness of the negotiation. Sub-figure (**a**) shows adding a small Proximal Factor, $\phi = 0.2$, does not help negotiation with a large penalty factor $\rho = 0.2$ to converge. Sub-figure (**b**,**c**) show that if the proximal factor, ϕ, is within proper range, the robustness of negotiation increases. (**d**) demonstrates an exceedingly large proximal factor, ϕ, breaks the negotiation convergence.

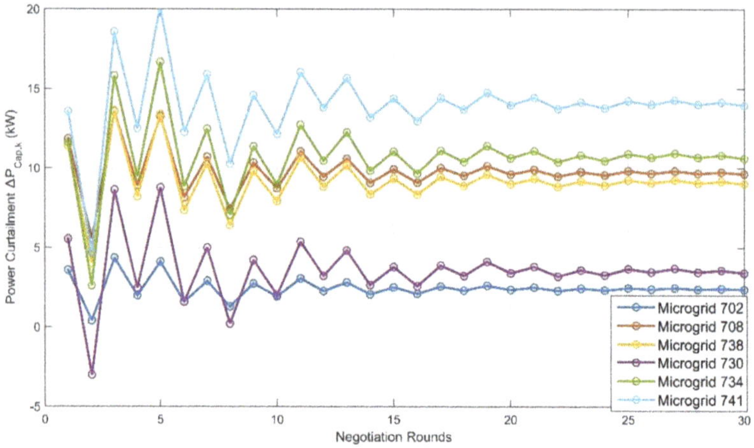

(**a**)

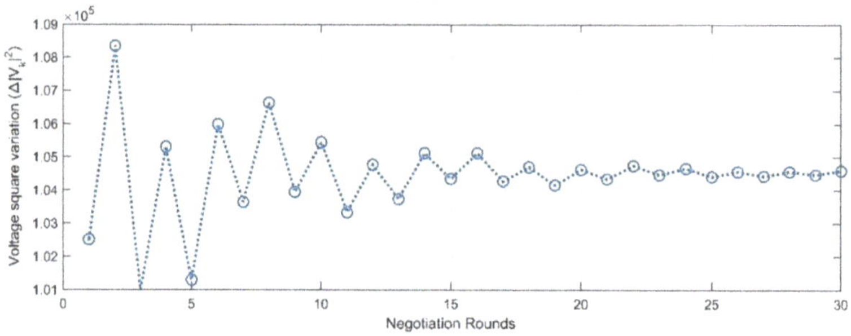

(**b**)

Figure 8. Updates of microgrid power curtailment $\Delta P_{Cap,k}$ and the voltage square variation $\Delta|V_{741}|^2$ during negotiation. (**a**) shows the power capacity curtailment evolvements of different VGI Microgrids along the distribution line during the negotiation process. (**b**) shows the voltage square difference caused by change of VGI Microgrids power capacity change during the negotiation process.

Figure 9 shows the PEV charging power and CoC comparison in each microgrid before and after the voltage compensation negotiation. The charging power of some PEVs is further reduced and results in the CoC value falling below zero. It was also found that VGI microgrids that are close to the voltage violation node curtail more power than the VGI microgrids that are far away from the voltage violation node. The reason is that the load variation near the voltage violation node has a larger effect on the voltage regulation.

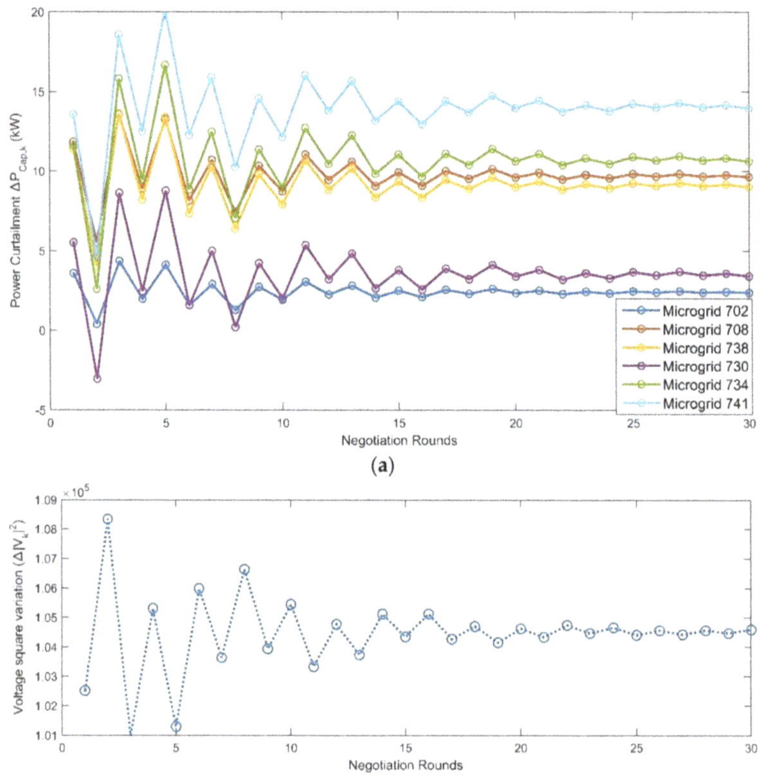

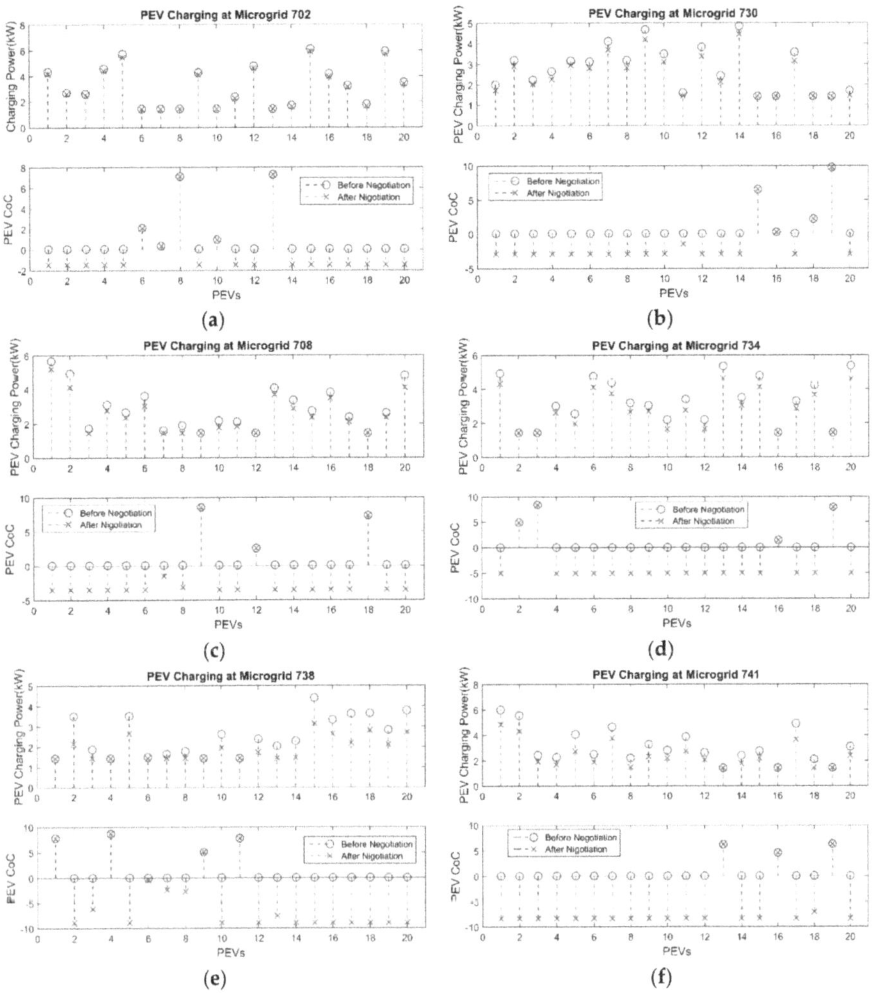

Figure 9. The negotiation effects on PEV charging in the microgrids. (**a**–**f**) show PEV Charging power and CoC value comparison before and after the negotiation process in different VGI Microgrids along the distribution lines.

The simulation result testifies to the effectiveness of the distribution voltage compensation. Figure 10 shows the distribution voltage of each node along the distribution feeder line. The comparison is shown for the PEV charging scenarios: maximal power demand charging, charging at the VGI microgrid PEV critical charging power point, and charging after voltage compensation negotiation. The voltage compensation negotiation successfully raises the Node 741 voltage to around 0.9504 per unit.

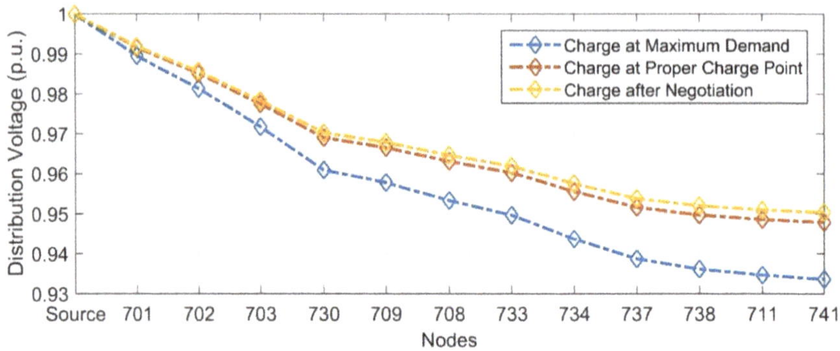

Figure 10. Voltage along the distribution feeder branch in different charging scenarios at 10:00 am.

6. Conclusions

This paper presents a distributed VGI control to realize voltage compensation service in a distribution network. The control scheme design is composed of two levels–the microgrid level and the distribution grid level. At the top level—the distribution grid level—an ADMM-based distributed voltage compensation negotiation among the multiple VGI microgrid agents and a grid-level agent is triggered when a voltage violation occurs. This distribution grid agent aims to maintain its distribution voltage level above the lowest threshold by sacrificing the microgrids' power capacity. On the other side, each VGI microgrid agent wishes to minimize its power capacity curtailment to lower the impact to the PEV charging activities with in the VGI microgrid. Though having conflicting objectives, the negotiation coordinates the interests of all agents as an entity and finally reaches an agreement that all agents can accept. After determining the power capacity curtailment for each VGI microgrid, the CoC-based optimal microgrid VGI control algorithm aims to reasonably dispatch limited power to the charging PEVs and results in an average PEV charging satisfaction at the lower level, the VGI microgrid. We used a case study to simulate the application of the proposed algorithm in a VGI distribution grid to prove its effectiveness and advantages. First, the distributed algorithm allows each agent to pursue its own objective under the coordination of negotiation. This greatly reduces the computation burden of a single control unit compared to a centralized control design. Secondly, the two-level control design decouples the distribution grid-level control and single PEV charging control at the microgrid level. Each VGI microgrid becomes a relatively independent entity that controls the PEV charging activities within its range. This organization increases the scalability of the control scheme. However, the scenario presented in this paper applies certain simplifications. For example, distributed generation, like from renewable energy sources, was not considered. The possible utilization of a four-quadrant power inverter and large energy storage were also not in consideration. All these factors may provide both additional flexibility and uncertainty for the VGI control design in the distribution grid. We are going to consider improving the proposed control algorithm to better adapt to the factors above in future study.

Author Contributions: Conceptualization, C.C., Z.W. and B.C.; Methodology, C.C., Z.W. and B.C.; Software, C.C.; Validation, C.C.; Formal Analysis, C.C.; Investigation, C.C.; Resources, C.C.; Data curation, C.C.; Writing—original draft preparation, C.C.; Writing—review and editing, C.C. and B.C.; Visualization, C.C.; Supervision, B.C.; Project administration, B.C.; Funding acquisition, B.C. All authors have read and agreed to the published version of the manuscript.

Funding: This research received no external funding.

Conflicts of Interest: The authors declare no conflict of interest.

References

1. Cao, C.; Cheng, M.; Chen, B. Optimal Scheduling of PEV Charging/Discharging in Microgrids with Combined Objectives. *SGRE* **2016**, *7*, 115–130. [CrossRef]
2. Wang, L.; Cao, C.; Chen, B. Grid-Tied Single-Phase Bi-Directional PEV Charging/Discharging Control. *SAE Int. J. Passeng. Cars Electron.* **2016**, *9*, 275–285. [CrossRef]
3. Wang, L.; Cao, C.; Chen, B. Model-Based Micro-Grid Modeling and Optimal PEV Charging Control. In Proceedings of the 2016 12th IEEE/ASME International Conference on Mechatronic and Embedded Systems and Applications (MESA), Auckland, New Zealand, 29–31 August 2016; pp. 1–6.
4. Cao, C.; Wang, L.; Chen, B.; Harper, J.; Bohn, T.; Dobrzynski, D.; Hardy, K. Real-Time Modeling to Enable Hardware-in-the-Loop Simulation of Plug-In Electric Vehicle-Grid Interaction. In Proceedings of the 13th ASME/IEEE MESA, Cleveland, OH, USA, 6–9 August 2017.
5. Cao, C.; Wang, L.; Chen, B. Mitigation of the Impact of High Plug-in Electric Vehicle Penetration on Residential Distribution Grid Using Smart Charging Strategies. *Energies* **2016**, *9*, 1024. [CrossRef]
6. Nour, M.; Ramadan, H.; Ali, A.; Farkas, C. Impacts of Plug-in Electric Vehicles Charging on Low Voltage Distribution Network. In Proceedings of the 2018 International Conference on Innovative Trends in Computer Engineering (ITCE), Egypt, Aswan, 19–21 February 2018; pp. 357–362.
7. Leou, R.; Su, C.; Lu, C. Stochastic Analyses of Electric Vehicle Charging Impacts on Distribution Network. *IEEE Trans. Power Syst.* **2014**, *29*, 1055–1063. [CrossRef]
8. Cao, C.; Chen, B. Generalized Nash Equilibrium Problem Based Electric Vehicle Charging Management in Distribution Networks. *Int. J. Energy Res.* **2018**, *42*, 4584–4596. [CrossRef]
9. Wang, L.; Chen, B. Distributed control for large-scale plug-in electric vehicle charging with a consensus algorithm. *Int. J. Electr. Power Energy Syst.* **2019**, *109*, 369–383. [CrossRef]
10. Wang, L.; Chen, B. Dual-Level Consensus-Based Frequency Regulation Using Vehicle-to-Grid Service. *Electr. Power Syst. Res.* **2019**, *167*, 261–276. [CrossRef]
11. Ranamuka, D.; Agalgaonkar, A.P.; Muttaqi, K.M. Dynamic Adjustment of OLTC Parameters Using Voltage Sensitivity While Utilizing DG for Volt/VAr Support. In Proceedings of the 2014 IEEE PES General Meeting, Gaylord National Resort & Convention Center On The Potomac, National Harbor, MD, USA, 27–31 July 2014; pp. 1–5.
12. Baran, M.; Wu, F.F. Optimal Sizing of Capacitors Placed on a Radial Distribution System. *IEEE Trans. Power Deliv.* **1989**, *4*, 735–743. [CrossRef]
13. Azzouz, M.A.; Shaaban, M.F.; El-Saadany, E.F. Real-Time Optimal Voltage Regulation for Distribution Networks Incorporating High Penetration of PEVs. *IEEE Trans. Power Syst.* **2015**, *30*, 3234–3245. [CrossRef]
14. Wu, X.; Li, L.; Zou, J.; Zhang, C. EV based voltage regulation in line distribution grid. In Proceedings of the 2016 IEEE International Instrumentation and Measurement Technology Conference Proceedings, Taipei, Taiwan, 23–26 May 2016; pp. 1–6.
15. Kesler, M.; Kisacikoglu, M.C.; Tolbert, L.M. Vehicle-to-Grid Reactive Power Operation Using Plug-In Electric Vehicle Bidirectional Offboard Charger. *IEEE Trans. Ind. Electron.* **2014**, *61*, 6778–6784. [CrossRef]
16. Kisacikoglu, M.C.; Ozpineci, B.; Tolbert, L.M. EV/PHEV Bidirectional Charger Assessment for V2G Reactive Power Operation. *IEEE Trans. Power Electron.* **2013**, *28*, 5717–5727. [CrossRef]
17. Idaho National Laboratory. Advanced Vehicle Testing Activity. Available online: https://avt.inl.gov/content/charging-system-testing/vehicle-charging-system-testing (accessed on 15 June 2019).
18. Boyd, S.; Parikh, N.; Chu, E.; Peleato, B.; Eckstein, J. Distributed Optimization and Statistical Learning via the Alternating Direction Method of Multipliers. *Found. Trends Mach. Learn.* **2011**, *3*, 1–122. [CrossRef]
19. Cao, C.; Chen, B. The Hardware-in-the-Loop Simulation of Vehicle-Grid Integration in a Distribution Grid (Accepted). In Proceedings of the 2019 IEEE PES General Meeting, Atlanta, GA, USA, 4–8 August 2019.
20. Hybrid – EV Committee SAE. *SAE Electric Vehicle and Plug in Hybrid Electric Vehicle Conductive Charge Coupler - J1772*; SAE International: Warrendale, PA, USA, 2017.
21. Di Fazio, A.R.; Russo, M.; Valeri, S.; De Santis, M. Sensitivity-Based Model of Low Voltage Distribution Systems with Distributed Energy Resources. *Energies* **2016**, *9*, 801. [CrossRef]

22. IEEE Power & Energy Society. IEEE PES AMPS DSAS Test Feeder Working Group/resources. Available online: http://sites.ieee.org/pes-testfeeders/resources/ (accessed on 20 April 2019).
23. Office of Energy Efficiency & Renewable Energy. Commercial and Residential Hourly Load Profiles for all TMY3 Locations in the United States. Available online: https://openei.org/datasets/dataset/commercial-and-residential-hourly-load-profiles-for-all-tmy3-locations-in-the-united-states (accessed on 15 February 2019).

© 2020 by the authors. Licensee MDPI, Basel, Switzerland. This article is an open access article distributed under the terms and conditions of the Creative Commons Attribution (CC BY) license (http://creativecommons.org/licenses/by/4.0/).

energies

Article

Assessment of the Worthwhileness of Efficient Driving in Railway Systems with High-Receptivity Power Supplies

Alejandro Cunillera [1], Adrián Fernández-Rodríguez [1], Asunción P. Cucala [1], Antonio Fernández-Cardador [1] and Maria Carmen Falvo [2,*]

1 Institute for Research in Technology, ICAI School of Engineering, Comillas Pontifical University, 23 Alberto Aguilera Street, 28015 Madrid, Spain; Alejandro.Cunillera@iit.comillas.edu (A.C.); Adrian.Fernandez@iit.comillas.edu (A.F.-R.); paloma.cucala@iit.comillas.edu (A.P.C.); Antonio.Fernandez@icai.comillas.edu (A.F.-C.)

2 DIAEE—Electrical Engineering, University of Rome Sapienza, via delle Sette Sale 12b, 00184 Rome, Italy

* Correspondence: mariacarmen.falvo@uniroma1.it

Received: 16 March 2020; Accepted: 6 April 2020; Published: 10 April 2020

Abstract: Eco-driving is one of the most important strategies for significantly reducing the energy consumption of railways with low investments. It consists of designing a way of driving a train to fulfil a target running time, consuming the minimum amount of energy. Most eco-driving energy savings come from the substitution of some braking periods with coasting periods. Nowadays, modern trains can use regenerative braking to recover the kinetic energy during deceleration phases. Therefore, if the receptivity of the railway system to regenerate energy is high, a question arises: is it worth designing eco-driving speed profiles? This paper assesses the energy benefits that eco-driving can provide in different scenarios to answer this question. Eco-driving is obtained by means of a multi-objective particle swarm optimization algorithm, combined with a detailed train simulator, to obtain realistic results. Eco-driving speed profiles are compared with a standard driving that performs the same running time. Real data from Spanish high-speed lines have been used to analyze the results in two case studies. Stretches fed by 1 × 25 kV and 2 × 25 kV AC power supply systems have been considered, as they present high receptivity to regenerate energy. Furthermore, the variations of the two most important factors that affect the regenerative energy usage have been studied: train motors efficiency ratio and catenary resistance. Results indicate that the greater the catenary resistance, the more advantageous eco-driving is. Similarly, the lower the motor efficiency, the greater the energy savings provided by efficient driving. Despite the differences observed in energy savings, the main conclusion is that eco-driving always provides significant energy savings, even in the case of the most receptive power supply network. Therefore, this paper has demonstrated that efforts in improving regenerated energy usage must not neglect the role of eco-driving in railway efficiency.

Keywords: railway transport; eco-driving; energy efficiency; optimization algorithm; power systems

1. Introduction

Railways are considered an inexpensive, fast and safe transport mode. Furthermore, this transport mode is energy-efficient and its contribution to global warming is less severe compared with others [1]. Despite that, its increasing activity throughout the world [2] makes it necessary to continue working on the energy efficiency of rail transport to achieve a sustainable transport industry without losing sight of service quality improvements.

Eco-driving, also named speed profile optimization or train trajectory optimization, is one of the most important measures with which to achieve significant energy reductions in the energy consumption of railway operations. It consists of obtaining a way to drive a train on a journey to fulfil a target running time with minimal energy consumption. Eco-driving has the advantage of being a short-term action that requires low investments, while other measures, such as improving the infrastructure [3–12] or rolling stock [13–17], usually require a significant investment and long/mid-term actions.

The eco-driving problem has been studied since 1968, when Ichikawa applied the Pontryagin's maximum principle to a greatly simplified train dynamics model, to derive the optimal control of a train [18]. Since then, numerous researchers have contributed to the eco-driving field with their proposals.

Control theory [18,19] shows that the optimal speed profile for a train running from two stations in a flat track consists of a sequence of four regimes: maximum traction, cruising, coasting and maximal braking. If the line geometry is more complex [20,21] or the train has regenerative braking [22], the most energy-efficient speed profile results as a smart combination of these four regimes.

In the literature, different optimization techniques have been proposed to find the right combination of the efficient driving regimes. These techniques can be divided into two groups [23]: analytical and numerical methods.

Typically, analytical methods apply the Pontryagin maximum principle to obtain the optimality necessary conditions, and using these results, apply different techniques to find the optimal speed profile. These techniques are: constructive algorithms [20–22,24–26], dynamic programming [27–31], sequential quadratic programming [30,32] and the Lagrange multiplier method over the discretized problem [33]. Other analytical methods are based on transforming the optimal control problem into a non-linear problem and solving it directly [34,35].

Analytical methods are, in most cases, fast procedures that can produce the optimal solution. However, the complexity of the problem must be reduced to apply these methods, due to the requirements for obtaining the analytical solution. This leads to simplifications in the train and line models and in the operational restrictions that the solution must comply.

On the contrary, the use of numerical methods does not demand simplifications in the train and line model, and any model to observe the requirements of passengers' comfort or operation characteristics can be included. In recent years, these methods have received growing attention aimed toward solving the eco-driving problem, because of the improvement of computational performance; in most cases, they are more computationally expensive than analytical procedures.

Different numerical methods have been applied to optimize the energy consumption of train driving: direct search algorithms [36,37], brute force [38], Monte Carlo simulation [39], artificial neural networks [40,41] and nature inspired algorithms [29,42–47]. Among these techniques, nature inspired computational intelligence is one of the most common methods applied to solve the speed profile optimization problem. These techniques provide a framework that can be easily implemented and are independent of the specificities of the problem. Moreover, they can be used in combination with complex train dynamics models that can be easily substituted by other different models when the characteristics of studied railway line change.

Many nature inspired computational intelligence techniques have been proposed in the literature to solve different problems [48]. When looking in detail at the algorithms applied to solve the train eco-driving problem, the following are present: the genetic algorithm (GA) [29,42,47,49–54], the multi-population genetic algorithm (MPGA) [55,56], GA with fuzzy parameters [57–60], differential evolution [46], ant colony optimization [29,61,62], simulated annealing [45,63], the indicator based evolutionary algorithm (IBEA) [64], the non-dominated sorting genetic algorithm II (NSGA-II) [43,44], multi-objective particle swarm optimization (MOPSO) [44,65] and dynamic versions of NSGA-II and MOPSO [66,67].

Apart from the specific technique applied to obtain the efficient train driving, the literature shows consistently that the use of eco-driving provides important energy savings [68]. Typically, trains on mainline railways are manually driven and the whole driving application is made by means of driver advisory systems (DASs). Several works have reported applications of eco-driving in DASs [69–73], wherein energy saving between 7% and 22% have been obtained. On the other hand, nowadays many urban railways are usually driven automatically by means of automatic train operation (ATO) systems. In these systems, efficient driving commands must be programmed to perform eco-driving speed profiles. Several authors reported applications of eco-driving in ATO systems, wherein energy savings between 6% and 18% have been obtained [44,65,74–76].

Savings obtained thanks to eco-driving are mainly due to the substitution of the use of braking by coasting periods. Besides, acceleration phase can optimized to reach higher energy efficiency. Nowadays, most trains are equipped with regenerative braking that allows recovering the kinetic energy during deceleration phases. In DC systems, regenerated energy cannot be totally used because it is necessary to have another train in the system consuming the amount of energy regenerated (except for trains which are equipped with on-board accumulation devices [77]). However, the use of reversible substations [5,6,8,9] allows them to use all the energy regenerated in a DC system by sending the regenerated energy not used by other trains to the utility grid. In modern high-speed systems the electrification is AC and substations allow the flow of energy in both directions (consumption or generation); thus, as in the previous case, it is possible to use all the regenerated energy produced (except electrical losses).

In this context of high use of regenerated energy, a question arises: is it worthwhile to minimize the energy consumption of a train, or will any driving over the same running time provide similar energy consumption?

The research objectives of this work are, on one hand, to assess the effectiveness of eco-driving under different scenarios of regenerated energy receptivity, and on the other hand, to see whether there is a scenario wherein eco-driving presents no relevant efficiency, providing an answer to the previous question.

The studies presented in this paper are based on real data from Spanish high-speed lines. The studied lines have been chosen because they are AC electrified, which offers high regenerated energy usage. The lines selected are Madrid–Barcelona, which is fed by a 2 × 25 kV power supply system, and Madrid–Sevilla, which is fed by a 1 × 25 kV power supply system. In the case study, two driving strategies are compared: an optimized eco-driving strategy and a standard driving strategy. The standard driving strategy consists of maintaining a holding speed during the journey to meet the running time. On the other hand, the efficient eco-driving consists of two driving commands: a holding without braking command and a coasting command before the braking up to the arrival station.

A MOPSO (multi objective particle swarm optimization) algorithm [78] combined with a detailed train simulation model [47] has been used in this piece of research to generate not only the efficient driving but also the standard driving for different target running times. This algorithm has the advantage of generating in a single run a set of non-dominated solutions, also known as Pareto front. A non-dominated solution is one that cannot be improved at the same time in all the problem's objectives. In the eco-driving problem, the objectives are the minimization of energy consumption and running time. As these objectives are conflicting, the result of the algorithm is Pareto front of solutions, where each solution is a speed profile with the minimum energy consumption for each possible running time. This way, MOPSO allows one to compare results easily in a wide range of running time values. MOPSO algorithm has demonstrated its suitability to the eco-driving problem in [44,66,67].

The usable regenerated energy that a train can produce depends mainly on two factors: the efficiency ratio of the train motor (that measures the efficiency of the train as a kinetic energy accumulator), and the transmission losses produced at catenary [30,79,80]. For this reason, the influences of different realistic values of these factors are analyzed in this paper to provide insights into the benefits provided by eco-driving on these types of lines.

This article is organized as follows: Section 2 describes the simulation model used to evaluate the different speed profiles. Section 3 describes the different driving strategies studied. The manual driving model is described in Section 4. Section 5 introduces the nature-inspired algorithm used to obtain the energy-efficient speed profiles. The results obtained and the discussion are shown in Section 6, and Section 7 presents the conclusions of this article.

2. Train Simulation Model

In this work, an accurate simulation model is used to calculate the speed profiles of a train and its consumption measured at the pantograph and at substations. It is divided into three modules: train, line and manual driving [47]. The train simulator is a software developed by Comillas Pontifical University researchers. This simulator is a desktop application for Windows coded in an object-oriented environment using the C++/CLI programming language. A validation against real on-board registered measurements was performed for high-speed trains on commercial services and night tests on the Madrid–Barcelona line, demonstrating its accuracy. Simulation results taken from this software showed an average error of 1.2% in running time and 0.4% in energy consumption compared with on-board measurements [47]. This accurate simulator has been applied for the design of efficient driving and tested in collaboration with Renfe and Adif.

The train module includes mass, length, running resistance, adhesion coefficient and the engine characteristics. These are introduced through the maximum traction and braking effort curves, which depend on the train speed. In this work, the traction and braking efficiencies are taken as equal and non-dependent on the train speed. This module includes also the energy consumption of the on-board auxiliary systems and a minimum constant braking effort that is applied along neutral zones in order to feed the auxiliary systems when no braking curve is affecting the train.

The physical characteristics of the line—track geometry (grades, grades transitions and track curves and their transitions), speed limits, tunnels, stopping points, stations, neutral zones, electrical substations and catenary features—are introduced in the line module.

The manual driving module calculates the traction and braking force needed at every time step, taking into account track data, speed limits, the position and speed of the train and stopping points. The driving module takes as input driving commands that can be easily interpreted by a human driver and generates speed profiles. In this work, the human driving is emulated by using a PI (proportional–integral) controller, in which the engine effort is modelled as a variable that represents the proportion of traction or brake that is demanded with respect to the maximum values for the current train speed.

2.1. Train Dynamics Model

At every time step the train position $s(t)$, speed $v(t)$ and acceleration $a(t)$ are determined by using Newton's second law Equations (1)–(5),

$$\frac{ds}{dt} = v(t), \tag{1}$$

$$\frac{dv}{dt} = a(t), \tag{2}$$

$$a(t) = \frac{F_m(\mathrm{t}) - F_b(\mathrm{t}) - F_r(v) - F_g(s)}{\rho m}, \tag{3}$$

$$s(t_0) = s_0,\ s\left(t_f\right) = s_f, \tag{4}$$

$$v(t_0) = v_0,\ v\left(t_f\right) = v_f, \tag{5}$$

where m is the train mass, ρ is the dimensionless rotating mass factor, $F_r\ (v)$ is the running resistance and $F_g(s)$ are the forces that affect the train due to the track geometry. $F_m(t)$ is the engine traction force and $F_b(t)$ is the braking force. t_0, t_f, s_0, s_f, v_0 and v_f are the initial and final times, positions and speeds of the train, respectively.

The engine traction force $F_m(t)$ is bounded by a maximum traction curve that depends on the speed of the train. The braking force $F_b(t)$ has two components: the electrical or regenerative and the pneumatic brakes. The regenerative brake is limited by a maximum electrical braking curve; if the train needs to brake it will use pneumatic brake to complement the electrical when the electrical one is saturated. The manual driving module computes both forces at every simulation time step to control the train dynamics considering the driving commands while obeying speed limits and stopping the train at the stopping points. In neutral zones, a minimum braking effort is applied to feed the auxiliary systems using regenerative energy.

The running resistance $F_r(v)$ depends quadratically on the train speed and it is modelled by the Davis formula (Equation (6)):

$$F_r(v) = A + Bv + T_f C v^2 \tag{6}$$

where A, B and C are the Davis formula coefficients. T_f is a factor that models the impact of tunnels on the running resistance and it modifies the quadratic term. It is equal to one when the train is outside a tunnel and greater otherwise.

The considered forces on the train due to the track geometry, $F_g(s)$, are those due to the track grades and curves:

$$F_g(s) = mgp(s) + mg\frac{K}{R(s)} \tag{7}$$

where m is the mass of the train; g is the gravity acceleration; $p(s)$ and $R(s)$ are the averages of the grades and curve radius respectively affecting the train in the position s. K is a constant that depends on the gauge of the track which units are the same as $R(s)$ (meters). The gauge in the considered lines is standard UIC (Internal Union of Railways) gauge, 1.435 m, and $K = 600$. Notice that the force due to curves is modelled as an equivalent grade.

2.2. Energy Consumption Model

The simulation model calculates the power consumed at the pantograph and at the electrical substations at every time step as a function of the train speed and the engine traction or braking effort by using Equations (8) and (9)

$$P_{pantograph} = \frac{F_m v}{\eta_t} + P_{aux} - \eta_b \min(F_b, F_{max}(v))v, \tag{8}$$

$$P_{substations} = P_{pantograph} + r(s)\left(\frac{P_{pantograph}}{V cos\varphi}\right)^2 \tag{9}$$

where η_t is the engine traction efficiency, η_b is the engine regenerative brake efficiency, and both are considered constant. P_{aux} is the power consumed by the auxiliary systems; $F_{max}(v)$ is the maximum regenerative brake force at speed v. $r(s) = \hat{r}|s - s_{ss}|$ is the resistance of the catenary that depends on the distance between the train and the electrical substation, which is located at position s_{ss}. In this article the influence of the lineal resistance $\hat{r}$ on the energy consumption is studied. V is the nominal line voltage and $cos\varphi$ is the power factor, and they are assumed to be constant. If a train is in a neutral zone, the power consumed at the electrical substations is zero. The energy consumption at the pantograph and substations can be obtained by integrating Equations (8) and (9) in time, respectively. This section may be divided by subheadings. It should provide a concise and precise description of the experimental results and their interpretation, and the experimental conclusions that can be drawn.

3. Driving Strategies and Commands

In this work the performances of two driving strategies are compared in terms of energy savings. In particular, a holding speed without braking and final coasting phase eco-driving strategy is compared against a standard driving strategy consisting in holding a constant speed with braking. All the regenerated energy at the pantograph is assumed to be returned to the AC power grid. The dependence of the energy consumption measured at the pantograph and electrical substations on the engine traction and brake efficiency and on the losses at the catenary is analyzed.

For the eco-driving strategy, the journey is divided in two phases. In the first phase, the driver must hold a certain speed by applying traction. If braking is necessary to maintain the constant speed, coasting is applied instead, and thus, the speed of the train will increase. In the second phase, the driver must apply coasting until the train reaches its braking curve to stop at the station.

For the standard strategy, a cruise speed is determined for the whole journey. The driver has to maintain that speed until reaching the braking curve. To drive at a constant speed, braking can be applied if necessary. Any driving strategy has to be described in easily-interpretable commands so that a driver can apply them. The driving commands are defined by using a command vector. It has two components. The first one contains the position s_c that corresponds to the final point of the holding speed phase. The holding speed v_c is introduced in the second component. The final coasting phase is performed starting at s_c up to the braking curve in the holding speed without braking and the final coasting driving strategy. Notice that speed limits of the line are observed by the driving model braking the train when necessary.

4. Driving Model

A closed-loop PI controller is used to emulate the manual driving by controlling the train dynamics. It takes into account the maximum acceleration and service deceleration of the train, the jerk limitation, the maximum tractive effort, the speed limits and the energy that has to be regenerated by braking in neutral zones. The control loop minimizes the difference between an objective speed v_{obj} and the train speed v. At every time step the objective speed is computed in terms of the driving commands and the track speed limits. To do so, the braking curves to the next ceiling speed reduction, $v_b(s)$, and to the next stop at a station, $v_{fb}(s)$, are calculated by using Equations (10) and (11).

$$v_b(s) = \sqrt{v_{next}^2 + 2 \cdot d_{service} \cdot (s_{next} - s)}, \tag{10}$$

$$v_{fb}(s) = \sqrt{2 \cdot d_{service} \cdot (s_{end} - s)}, \tag{11}$$

where v_{next} is the next ceiling speed of the track, $d_{service}$ is the service deceleration of the train, s is the position of the train, s_{next} is the position of the next ceiling speed corrected with the train length if necessary and s_{end} is the next stopping point.

The objective speed v_{obj} is therefore computed in terms of the maximum admissible speed v_{max}, which is the minimum between the next braking curve and the ceiling speed at the position of the train $v_{ceiling}(s)$,

$$v_{max}(t) = \min\left(v_{ceiling}(s), v_b(s), v_{fb}(s)\right), \tag{12}$$

$$\begin{gathered} v_{obj}(t) = \min(v_{max}(t), v_c) \quad if\ s \le s_c \\ v_{obj}(t) = 0 \quad if\ s_c < s \end{gathered} \tag{13}$$

The control variable U_{reg} is defined as the traction or brake demand of the PI controller. To calculate it, at every time step the proportional and integral control components are calculated as described in Equations (14)–(18). For the standard driving the control variable is computed using Equation (17), while for the eco-driving strategy Equation (18) is used instead.

$$e(t) = v_{obj}(t) - v(t) \tag{14}$$

$$U_P(t) = K_P \cdot e(t) \tag{15}$$

$$U_I(t) = \frac{K_P}{K_I} \cdot \frac{(e(t) + e(t - \Delta t))}{2} + U_I(t - \Delta t) \tag{16}$$

$$U_{reg\ (std)}(t) = U_P(t) + U_I(t) \tag{17}$$

$$\begin{aligned} U_{reg\ (eco)}(t) &= U_P(t) + U_I(t) \quad if\ v(t) \geq v_{max}(t)\ or\ U_P(t) + U_I(t) \geq 0 \\ U_{reg\ (eco)}(t) &= 0 \quad if\ v(t) < v_{max}(t)\ and\ U_P(t) + U_I(t) < 0 \end{aligned} \tag{18}$$

where $e(t)$ is the difference between the objective and the train speeds, whose absolute value has to be minimized by the control. $U_P(t)$ and $U_I(t)$ are the proportional and integral components of the control variable $U_{reg}(t)$. K_P and K_I are the proportional and the integral gains, respectively. Δt is the simulation time-step. If the control variable is positive, the control demands tractive effort, while if it is negative, it demands braking. If it is zero, the control demands coasting. In the eco-driving strategy (Equation (18)), the control demand is set to coasting when the train speed is below the maximum admissible speed and the control would demand braking under the current condition.

Once that the control variable has been obtained, the engine maximum traction and brake limits; comfort constraints—maximum acceleration and jerk; and the presence of neutral zones, have to be taken into account. Strong accelerations are considered to be unpleasant for passengers, so the maximum acceleration and deceleration have to be controlled. $F_{maxAcc}(s, v)$ and $F_{maxDec}(s, v)$ model the forces corresponding to the maximum admissible acceleration and deceleration respectively with corrections for the running and track geometry resistances. Equations (19) and (20) are used for calculating these forces.

$$F_{maxAcc}(s, v) = a_{max} \cdot \rho \cdot m + \left(F_r(v) + F_g(s)\right) \tag{19}$$

$$F_{maxDec}(s, v) = d_{max} \cdot \rho \cdot m + \left(F_r(v) + F_g(s)\right) \tag{20}$$

where a_{max} and d_{max} are the maximum admissible acceleration and deceleration rates in absolute values, respectively.

The rate of change of the train acceleration, jerk, can also affect passengers' comfort [49,81–83]. The maximum and minimum admissible effort due to the jerk limitation, $F_{maxJerk}(t)$ and $F_{minJerk}(t)$, are calculated by using Equations (21) and (22).

$$F_{maxJerk}(t) = F_m(t - \Delta t) - F_b(t - \Delta t) + (j_{max} \cdot \rho \cdot m) \cdot \Delta t \tag{21}$$

$$F_{minJerk}(t) = F_m(t - \Delta t) - F_b(t - \Delta t) - (j_{max} \cdot \rho \cdot m) \cdot \Delta t \tag{22}$$

where j_{max} is the maximum admissible jerk.

The total traction or braking effort demanded to the train after applying the comfort restrictions and engine maximum traction and brake effort, $F_{TB}(t)$, is computed by evaluating Equation (23) or Equation (24).

$$\begin{gathered} F_{TB}(t) = \min\left(U_{reg}(t), F_{max}(v), F_{maxAcc}(s, v), F_{maxJerk}(t)\right) \\ if\ U_{reg}(t) \geq 0\ and\ s \notin \left(s_{NZ_{start}^k}, s_{NZ_{end}^k}\right)\ for\ k \in \{1, 2, \ldots, z\}, \end{gathered} \tag{23}$$

$$\begin{gathered} F_{TB}(t) = \max\left(U_{reg}(t), F_{maxDec}(s, v), F_{minJerk}(t)\right) \\ if\ U_{reg}(t) < 0\ and\ s \notin \left(s_{NZ_{start}^k}, s_{NZ_{end}^k}\right)\ for\ k \in \{1, 2, \ldots, z\}. \end{gathered} \tag{24}$$

If the train is on a neutral zone, it has to use the regenerative brake to maintain the comfort and auxiliary systems; therefore, the total braking effort to be applied is

$$\begin{gathered} F_{TB}(t) = \max\left(\min\left(U_{reg}(t), F_{NZ}(v)\right), F_{maxDec}(s, v), F_{minJerk}(t)\right) \\ if\ s \in \left(s_{NZ_{start}^k}, s_{NZ_{end}^k}\right), \quad k \in \{1, 2, \ldots, z\} \end{gathered} \tag{25}$$

where $F_{NZ}(v)$ is the minimum force required for keeping the auxiliary systems on; $s_{NZ_{start}^k}$ and $s_{NZ_{end}^k}$ are the initial and final points of the k-th neutral zone; and z is the total number of neutral zones in the considered line.

Finally, the values of the tractive and brake forces to be applied in the train are obtained from the total traction or braking effort simply by considering that if $F_{TB}(t)$ is positive the train will apply traction and it will brake if $F_{TB}(t)$ is negative, as described in Equations (25) and (26).

$$\begin{array}{ll} F_m(t) = F_{TB}(t), & if\ F_{TB}(t) \geq 0, \\ F_m(t) = 0 & if\ F_{TB}(t) < 0, \end{array} \tag{26}$$

$$\begin{array}{ll} F_b(t) = 0 & if\ F_{TB}(t) \geq 0, \\ F_b(t) = -F_{TB}(t) & if\ F_{TB}(t) < 0. \end{array}$$

5. Eco-Driving Optimization Algorithm

A multi-agent based optimization method called MOPSO (multi-objective particle swarm optimization algorithm) [78] is used to generate energy-efficient speed profiles [44]. Energy consumption/running time Pareto fronts are obtained for different values of the engine traction and brake efficiencies and the linear resistance of the catenary.

This nature-inspired algorithm imitates the behavior of a swarm of insects looking for food. These insects are considered particles that move iteratively within a search space. In this case, each particle corresponds to a speed profile and the running time and energy consumption have to be optimized. A fitness value is assigned to each particle. In biology terms, fitness stands for a degree of adaptability of an insect or how close it is to an abundant source of food. To optimize its fitness, each particle position is updated at every iteration by computing its velocity as a function of its past position with best fitness, *pbest*, and the global best position found by the whole swarm, *gbest*. The latter variable accounts for the information shared between the particle ensemble.

To initialize the algorithm, a set of particles with random positions and velocities is generated. At every iteration i, the non-dominated solutions, which are those particles for which there are not any calculated solutions with lower running time and energy consumption, are stored in an archive A. The non-dominated solutions in A are sorted in decreasing order by using a crowing distance (CD) operator [44]. This way, the low-density zones of the Pareto front are given more priority. *pbest* of each particle is updated and *gbest* is drawn randomly from the archive, giving priority to the solutions with higher crowding distance, which correspond to the solutions at the top of A.

The positions and velocities of the particles at the next iteration are generated by using Equations (27) and (28),

$$\hat{vp}_j(i) = w\hat{vp}_j(i-1) + c_1 r_1\left(pbest_j - \hat{xp}_j(i-1)\right) + c_2 r_2\left(gbest - \hat{xp}_j(i-1)\right) \tag{27}$$

$$\hat{xp}_j(i) = \hat{xp}_j(i-1) + \hat{vp}_j(i) \tag{28}$$

where $\hat{xp}_j$, $\hat{vp}_j$ are the position and velocity of the j-th particle; w is an inertia constant that weights the previous velocities; c_1 and c_2 are two social factor constants that weight the distance to $pbest_j$; and $gbest$, r_1 and r_2 are random numbers drawn uniformly between 0 and 1.

After the position of all the particles in the current iteration has been updated, the new non-dominated solutions are included in the archive and all the newly dominated solutions are deleted. *pbest* is updated for all the solutions in the archive. The solutions in A are sorted in decreasing order of CD, and finally, the *gbest* is drawn randomly from A, giving a higher probability to those solutions with higher CD. This procedure is repeated until a certain number of iterations I is reached. Figure 1 shows the MOPSO algorithm flowchart.

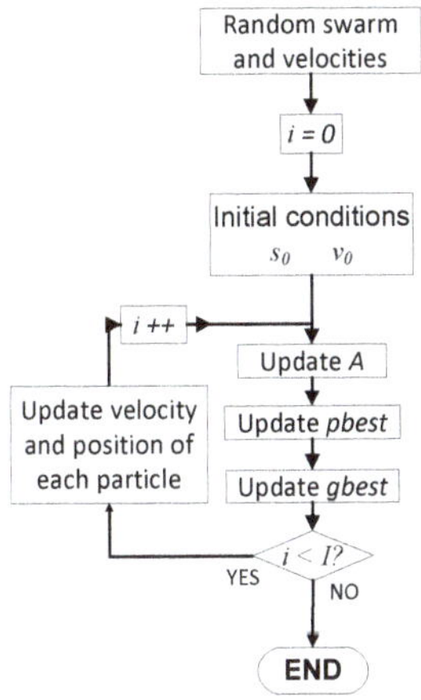

Figure 1. MOPSO algorithm flowchart.

6. Case Study and Results

In this section, the holding speed without braking with final coasting eco-driving strategy is compared against the standard holding speed driving, and the difference in energy consumption is quantified. In both cases, there is a single cruise speed and all the energy regenerated by the electrical brake is assumed to be returned to the power grid. The dependence of the energy consumption (measured at pantograph and at substations) on the engine efficiency and catenary linear resistance is analyzed. The nature-inspired optimization algorithm introduced in Section 5 is used for obtaining the Pareto front of speed profiles for the Talgo-Bombardier class 102 train running on two sections of Spanish high-speed lines. Track grades, curves, speed limits, tunnels, neutral zones and the positions of electrical substations are considered in the line models.

The first section analyzed is an 85.4 km long section between Calatayud and Zaragoza in the Madrid–Barcelona line with a 2 × 25 kV power supply system. The second section is a 38.9 km long line section between Puertollano and Ciudad Real in the Madrid–Sevilla line, with a 1 × 25 kV power supply system. The train has two 8 MW engines of 200 kN of maximum traction effort. The maximum traction curve is shown in Figure 2. The power consumed by the auxiliary systems is 325 kW and in any neutral zone the train has to regenerate enough energy to maintain these systems on, as presented in Equation (24). The running resistance curve is presented in Figure 3. The empty mass of the train is 324 t, and its length is 200 m. Furthermore, apart from the train's empty mass, 50 t-worth of passengers have been taken into account.

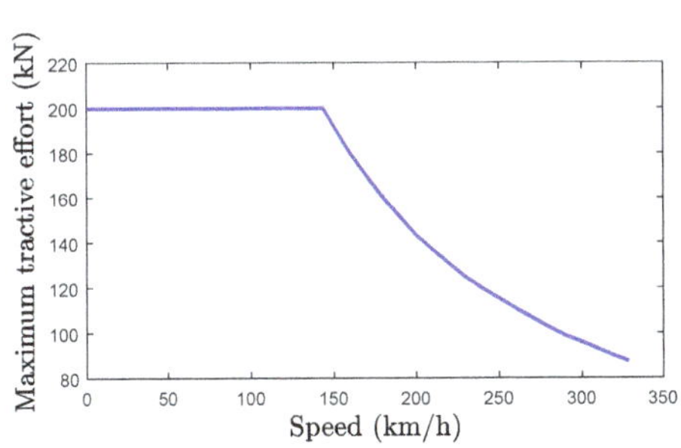

Figure 2. Maximum tractive effort curve for the considered Talgo-Bombardier class 102 train.

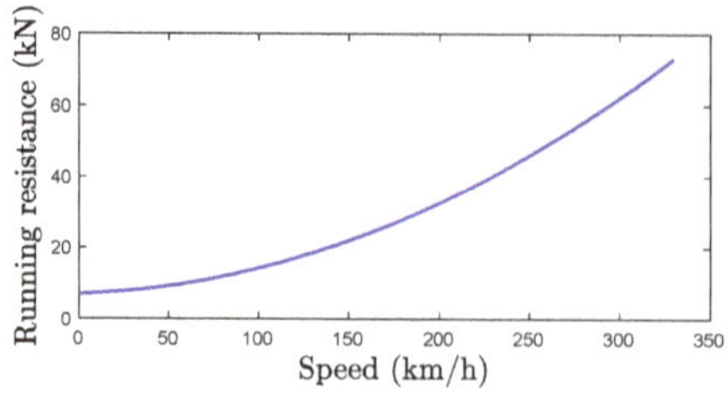

Figure 3. Running resistance for the Talgo-Bombardier class 102 train. It is modelled by using Equation (6).

Regarding the comfort constraints, the maximum and the service decelerations are 0.4 m/s^2, the maximum acceleration is 0.67 m/s^2 and the maximum admissible jerk is, in absolute value, 0.7 m/s^3. The power constant is considered to be $cos\varphi = 1$. The social factors c_1 and c_2 in the MOPSO algorithm are equal to 2 and the inertia constant w is 0.2.

6.1. *Calatayud–Zaragoza Case Study*

Figure 4 shows the height profile of the line section between Calatayud and Zaragoza. As can be seen, the considered section is mostly downhill. There are three neutral zones, which are represented in magenta in the lower part of the figure.

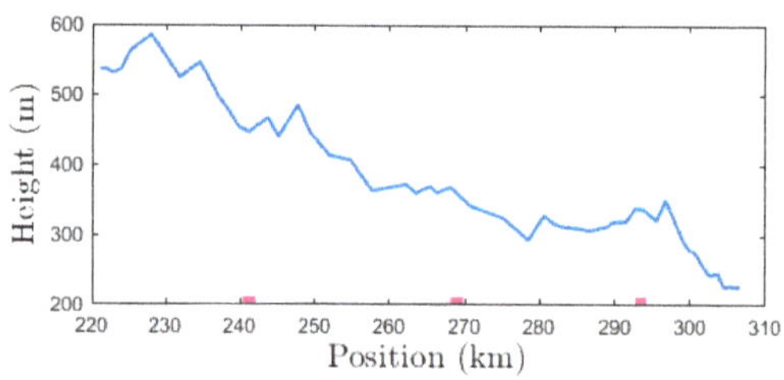

Figure 4. Height profile of the Calatayud–Zaragoza section. Neutral zones are represented in magenta.

In this subsection, the impact of the efficiency ratio of the motors on the energy consumption is analyzed for both driving strategies. The traction and brake engine efficiencies are considered equal and constant within a range between 85% and 97%. The influence of the linear resistance of the catenary on the energy consumption is also studied. In the case of the Calatayud–Zaragoza section, the catenary configuration is 2 × 25 kV, and different values of its linear resistance, between $21{\cdot}10^{-6}$ Ω/m and $37{\cdot}10^{-6}$ Ω/m, are tested. The MOPSO algorithm was used to generate speed profiles for both standard and eco-driving strategies. For each driving strategy the non-dominated speed profiles in terms of energy consumption and running time define the Pareto front. Figure 4 shows Pareto fronts for the net consumption measured at the electrical substations for the extreme values of the considered interval for the efficiency ratio of the motors and for the linear resistance of the catenary. The Pareto fronts corresponding to the eco-driving strategy are represented in blue, while those corresponding to the standard driving strategy are represented in red.

As shown in Figure 5, the consumed energy decreases as the engine efficiency increases. The energy consumption increases for higher values of the linear resistance, since more regenerated energy is dissipated at the catenary. For the fastest speed profilesm the difference in energy consumption is significantly lower than for the rest. These speed profiles are close to the flat-out speed profile; thus, the final coasting phase is very short, and therefore has little relevance in terms of energy-saving, so both driving strategies produce similar results in terms of energy consumption. It can also be observed that the eco-driving strategy produces faster and less energy-consuming speed profiles than the standard driving strategy, as all the solutions for the latter one are dominated by those produced by the former. Table 1 is focused on the analysis for the commercial running time (24 min), varying the efficiency ratio and the linear resistance. It shows the difference (in %) in net energy consumption at the electrical substations between both driving strategies. Results shown in this table are especially relevant for the infrastructure administrator.

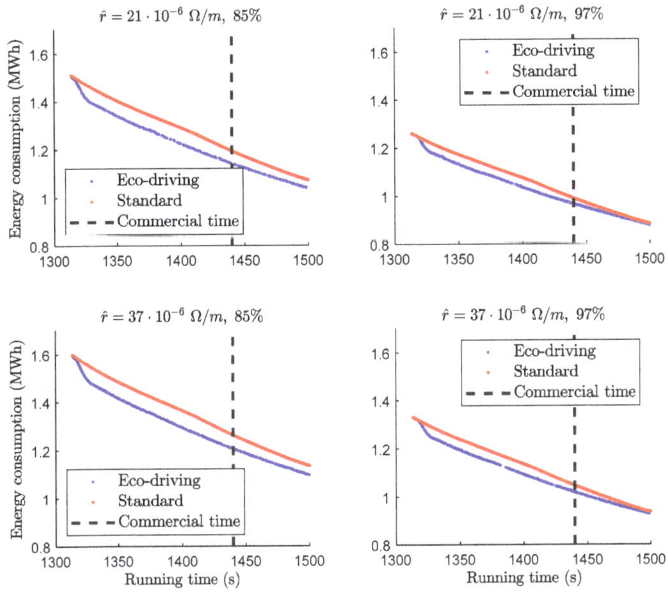

Figure 5. Pareto fronts for the net consumption at the electrical substations for the considered extreme values of the engine traction and braking efficiency (85% and 97%) and linear resistance of the catenary ($21 \cdot 10^{-6}$ Ω/m and $37 \cdot 10^{-6}$ Ω/m), for the Calatayud–Zaragoza section.

Table 1. Difference (in %) in net energy consumption at the electrical substations between the eco-driving and standard driving strategies for different values of the energy traction and braking efficiency and the linear resistance in the Calatayud–Zaragoza section for the commercial running time (24 min).

		Engine Traction and Braking Efficiency (%)				
		85.00	**88.00**	**91.00**	**94.00**	**97.00**
Linear resistance of the catenary ($\times 10^{-6}$ Ω/m)	21	4.54%	3.89%	3.56%	2.99%	2.48%
	24	4.56%	4.08%	3.58%	3.10%	2.51%
	28	4.59%	4.13%	3.62%	3.12%	2.54%
	32	4.63%	4.17%	3.68%	3.17%	2.64%
	37	4.69%	4.23%	3.78%	3.30%	2.76%

In Table 1, two trends can be observed. The difference in energy consumption decreases as the linear resistance of the catenary decreases. It decreases even more significantly when the efficiencies of the engine brake and traction are higher. The energy saving at the electrical substations between the eco-driving strategy and the standard driving is 2.47–4.69%.

Obviously, the measured energy consumption at the pantograph does not depend on the linear resistance of the catenary. Figure 6 shows the dependence of the energy savings (measured at the pantograph) on the energy efficiency.

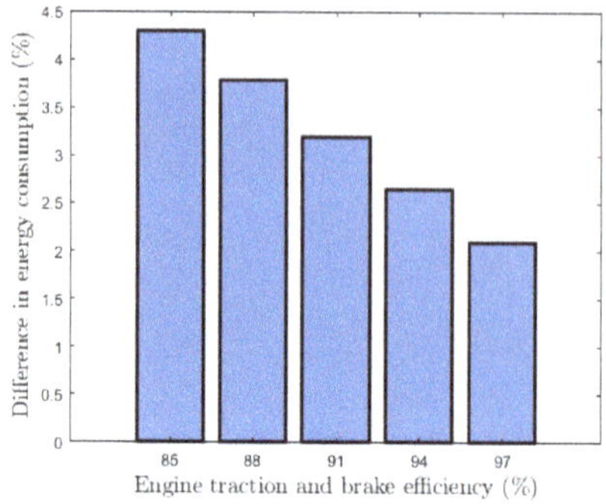

Figure 6. Difference in energy consumption at the pantograph (in %) between the considered driving strategies for different values of the engine traction and brake efficiency in Calatayud–Zaragoza section.

It may be observed that for the considered engine efficiency interval, there is an approximately linear relation between the difference in energy consumption and the engine efficiency. The eco-driving strategy produces speed profiles with lower energy consumption than the standard driving strategy. Results shown in Figure 6 are especially relevant for train operators. The energy saving measured at pantograph of eco-driving with respect to standard driving is between 2.1% and 4.3%.

In Figure 7 the difference in energy consumption at the substations is analyzed for each linear resistance of catenary. Results are shown for different values of the engine efficiency and running time. A grey vertical dashed line is used to highlight the commercial running time, which is 24 min.

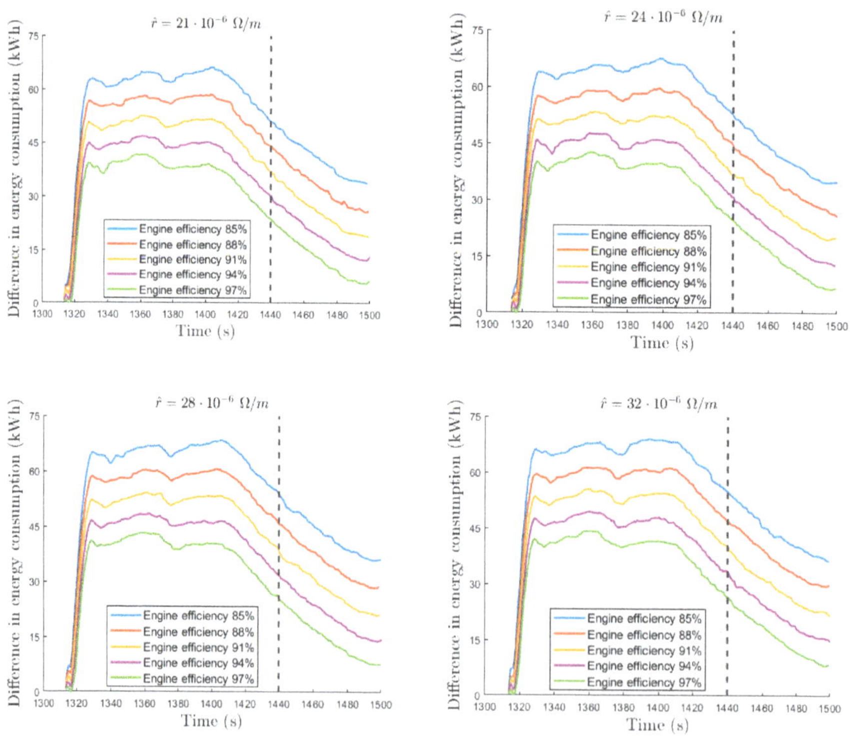

Figure 7. *Cont.*

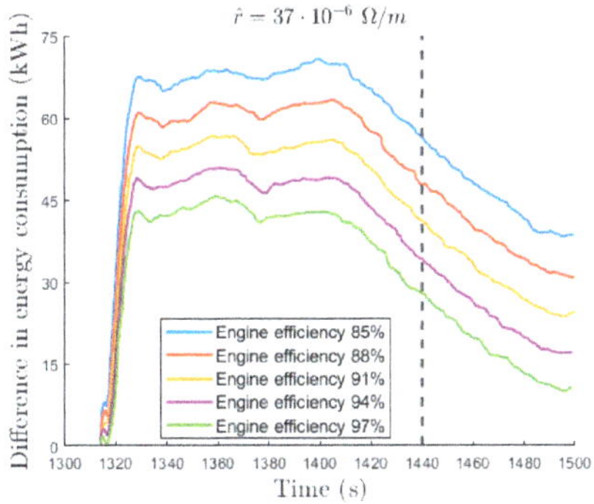

Figure 7. Difference in energy consumption (in kWh) between the considered driving strategies at the electrical substations for the Calatayud–Zaragoza section for different values of the catenary linear resistance $\hat{r}$.

Three different scenarios can be distinguished. Close to the flat-out running time, there is a large increase in the difference in energy consumption between the eco-driving and the standard driving strategies; this is due to the action of the final coasting phase in the former driving strategy. Next, there is a scenario in which the difference is approximately constant. For the slowest speed profiles, the difference in energy consumption is significantly lower, as can be seen in Figure 5, and the difference in energy consumption between the considered driving strategies becomes smaller as the running time increases, as the coasting phase will be shorter and therefore its impact will be less significant. If the dependence of the results on the energy braking and traction efficiency is considered, it can be observed that for a fixed running time the difference in energy consumption seems to depend linearly on the difference in energy efficiency. For the same value of the engine efficiency, the difference in energy consumption increases slightly with the linear resistivity of the catenary.

Last, the speed profiles obtained for the commercial time with the two considered driving strategies are presented. Table 2 shows the command vectors for these two speed profiles. Notice that the considered line section ends at kilometric point (k.p.) 306.7 at Zaragoza station (stopping point). The coasting phase in eco-driving starts 18 km before the stopping point.

Table 2. Command vectors for the commercial time speed profiles obtained with the two considered driving strategies for Calatayud–Zaragoza section.

	Eco-Driving Strategy	Standard Driving Strategy
Cruise speed (km/h)	253.0	249.2
Final position of the cruise phase (km)	288.7	Braking curve

In Figures 8 and 9, the speed profiles are shown. The train speed is represented in blue, the speed limits are represented by red lines and the track altitude is shown in green. The holding speed is represented by an orange dashed line.

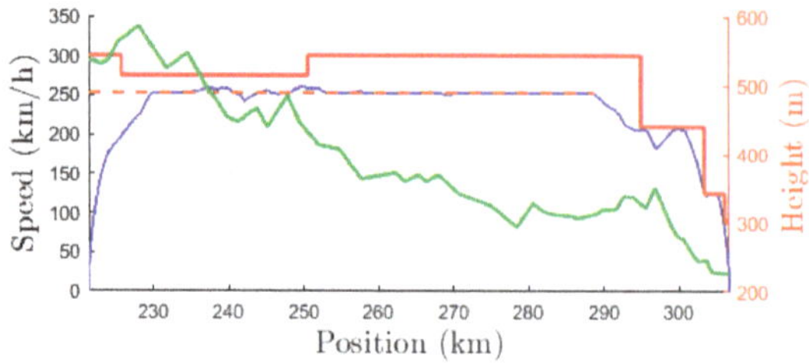

Figure 8. Eco-driving speed profile for the commercial time in the Calatayud–Zaragoza section.

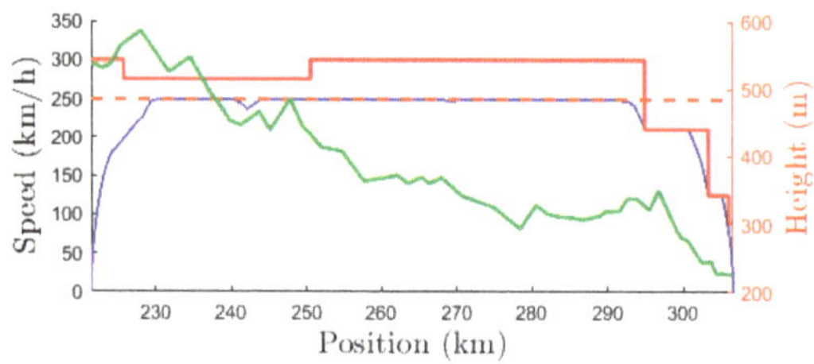

Figure 9. Standard driving speed profile for the commercial time in the Calatayud–Zaragoza section.

Figure 8 shows that in the holding speed regimes there are few points in which the train goes faster than the cruise speed in the eco-driving strategy, as there are few steep downhill grades. The most significant difference between the two speed profiles in Figures 8 and 9 is after position 288.7 km, in which the train is coasting in Figure 8, while in Figure 9 it is holding its speed until it reaches the braking curve. Near position 300 km the train has to brake due to the presence of a speed limit.

6.2. Puertollano–Ciudad Real Case Study

Finally, results are shown for the second case study, in which eco-driving speed profiles were designed for the considered Talgo-Bombardier class 102 train in a 38.9 km long line section between Puertollano and Ciudad Real. This line section has a catenary configuration of 1 × 25 kV, and three different values of its linear resistance between $54 \cdot 10^{-6}$ Ω/m and $110 \cdot 10^{-6}$ Ω/m are considered in this study. In Figure 10 the height profile of the considered line section is represented in blue, while the two neutral zones are represented in magenta in the lower part of the figure. In general terms, this line section is flatter than the Calatayud–Zaragoza line section.

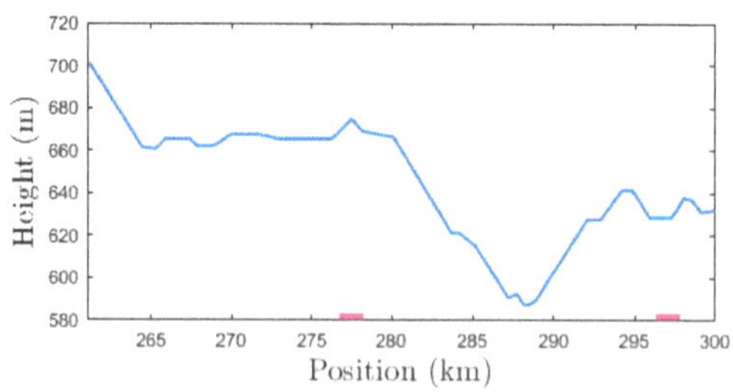

Figure 10. Height profile of the line section between Puertollano and Ciudad Real. Neutral zones are represented in magenta.

Again, speed profiles were generated for the two considered driving strategies by using the MOPSO algorithm. The obtained Pareto fronts are represented in Figure 11, in which they are sorted in terms of the net consumption at the substations. Results are shown for the extreme values of the considered engine efficiency interval, 85% and 97% and for catenary linear resistances $54 \cdot 10^{-6}$ Ω/m, $70 \cdot 10^{-6}$ Ω/m and $110 \cdot 10^{-6}$ Ω/m. The standard and eco-driving speed profiles are represented in red and blue, respectively. The commercial time, which is 13 min, is highlighted by a vertical grey dashed line.

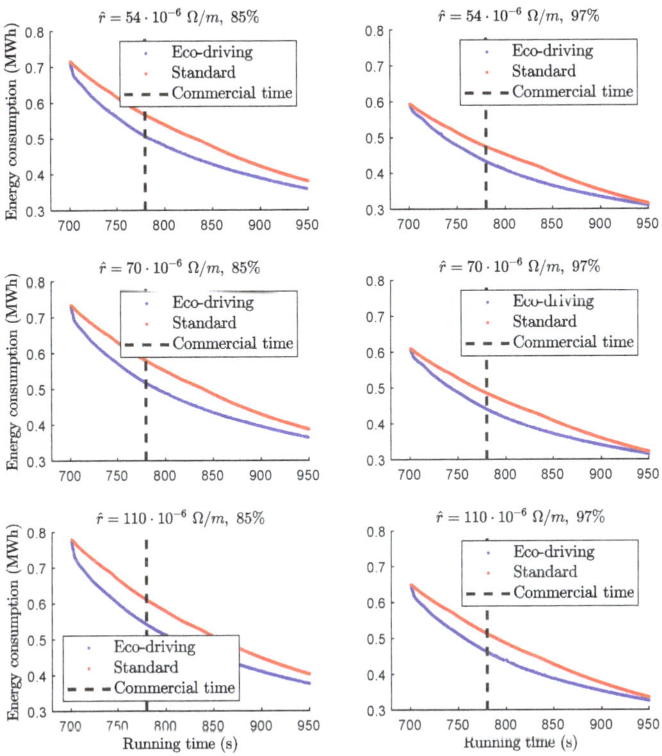

Figure 11. Pareto fronts for the net consumption at the electrical substations for the extreme values of the engine traction and braking efficiency (85% and 97%) and the three considered values of the linear resistance of the catenary ($54 \cdot 10^{-6}$ Ω/m, $70 \cdot 10^{-6}$ Ω/m and $110 \cdot 10^{-6}$ Ω/m).

Results presented in Figure 11 are similar to those in Figure 5, although this time the difference in energy consumption between two solutions with the same running time is usually larger than in the Calatayud–Zaragoza case. For a fixed running time, the energy consumption is larger when the engine is more inefficient and when the catenary has a larger linear resistance. The speed profiles in the holding speed without a braking Pareto curve dominate those in the holding speed with a braking Pareto curve.

Table 3 shows the difference in energy consumption at the substations between the two considered driving strategies for the commercial running time in the Puertollano–Ciudad Real section. Results are shown for different values of the engine efficiency and the linear resistance of the catenary.

Table 3. Difference (in %) in energy consumption at the electrical substations between the holding speed with braking and the holding speed without braking and final coasting driving strategies for different values of the energy traction and braking efficiency and the linear resistance in the Puertollano–Ciudad Real section for the commercial running time (13 min).

		Engine Traction and Braking Efficiency (%)				
		85.00	**88.00**	**91.00**	**94.00**	**97.00**
Linear resistance of the catenary ($\times 10^{-6}$ Ω/m)	54	11.86%	11.33%	10.81%	10.28%	9.70%
	70	12.07%	11.62%	11.12%	10.40%	10.15%
	110	12.55%	12.30%	11.78%	11.47%	11.07%

Once again, the lower the linear resistance is, the lower the difference in energy consumption is. Moreover, the difference in consumption between two speed profiles produced by using the two considered strategies is higher when the train engine is less efficient. The difference in energy consumption between the considered driving strategies for the commercial running time is between 9.70% and 12.55%.

Figure 12 shows the energy consumption measured at the pantograph for the commercial time speed profiles. As can be observed, for an engine efficiency of 85%, the eco-driving strategy can be used for achieving a 10.8% energy saving with respect to the standard driving strategy, while for an engine efficiency of 97% the energy saving between the two strategies is 8.3%.

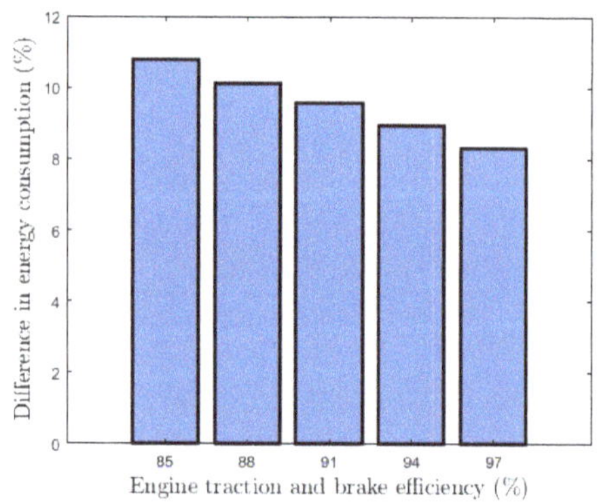

Figure 12. Difference in energy consumption at the pantograph (in %) between the considered driving strategies for different values of the engine traction and brake efficiency in Puertollano–Ciudad Real section.

The difference in energy consumption between the considered driving strategies is analyzed in Figure 13 for the considered linear resistances of the catenary and engine efficiencies and for different running times.

The difference in energy consumption grows up to a maximum close to the commercial running time. The commercial running time (13 min, that is 780 s) is highlighted by means of a grey dashed line. The more-or-less constant scenario in Figure 7 is at this point concave; this indicates that the shape of the difference in energy consumption depends on the height profile of the track. This time, the difference in energy consumption is larger when considering different values of the linear resistivity, as there is a larger difference between those values. Again, the difference in consumption seems to decrease linearly with the engine efficiency.

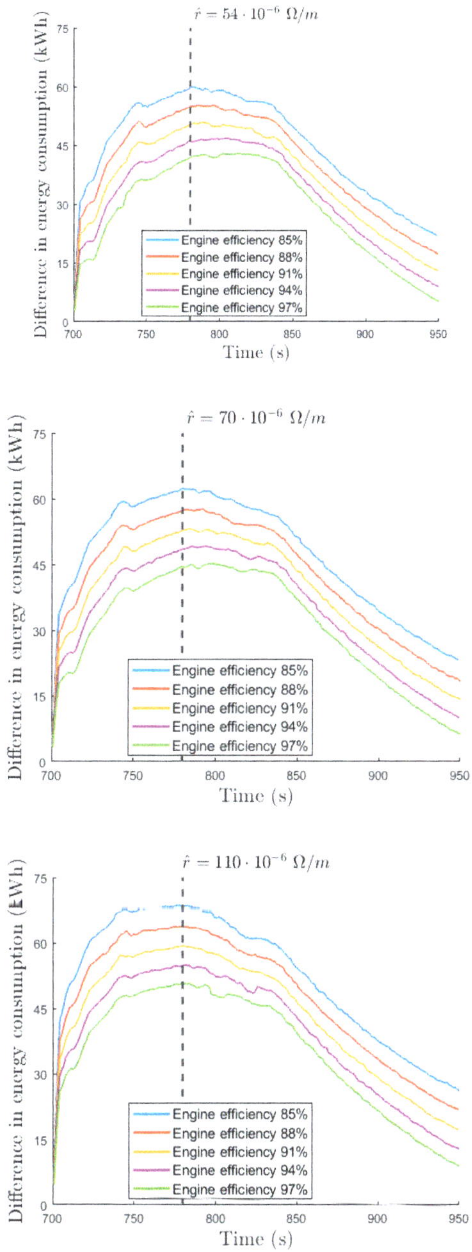

Figure 13. Difference in energy consumption (in kWh) between the considered driving strategies at the electrical substations for the Puertollano–Ciudad Real section for different running times. Different values of the catenary linear resistance $\hat{r}$ and different values of the engine efficiency are considered.

In Table 4 the command vectors of the two driving speed profiles that satisfy the commercial running time are presented. The eco-driving final coasting phase is approximately 10 km long and its cruise speed is 17.4 km/h higher. Figures 14 and 15 show the commercial speed profiles obtained for the eco-driving and standard driving strategies, respectively. The train speed is represented in blue, the speed limits are represented by red lines and the track altitude is shown in green. The holding speed is represented by an orange dashed line.

Table 4. Command vectors for the commercial time speed profiles obtained with the two considered driving strategies for the Puertollano–Ciudad Real section.

	Eco-Driving Strategy	Standard Driving Strategy
Cruise speed (km/h)	236.5	219.1
Final position of the cruise phase (km)	287.1	Braking curve

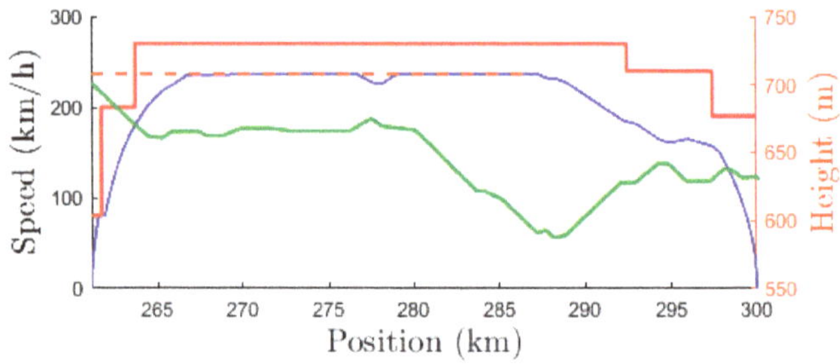

Figure 14. Eco-driving speed profile for the commercial time in the Puertollano–Ciudad Real line section.

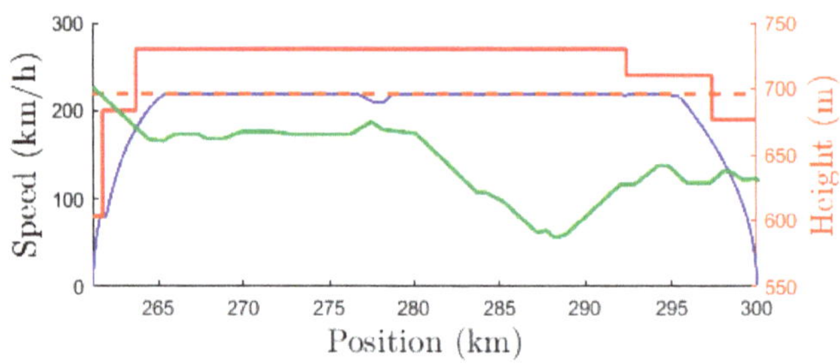

Figure 15. Standard driving speed profile for the commercial time in the Puertollano–Ciudad Real line section.

In this case, the train coasts approximately in 25% of the total length of the line section. In comparison to Figures 8 and 9, in this section the role of the final coasting phase is more important. This explains the fact that the difference in energy consumption between the considered driving strategies in percentage is much higher for the Puertollano–Ciudad Real section.

7. Discussion

The studies developed in this paper have been carried out with the objective of assessing the effectiveness of eco-driving under different electrical scenarios and to answer the question about whether or not eco-driving is useful in high-receptivity networks. The results obtained suggest that the answer to the question is that it is useful to apply eco-driving. Besides, it has to be taken into account that the value of energy saving because of efficient driving will depend on the efficiency of the electrical chain (catenary loses and motor efficiency) and on the running time.

The cases studied in Section 6 are two cases of an alternate current power supply system with high receptivity. These systems are, by nature, much more effective using regenerate energy than direct current systems. In alternate current systems, the energy generated by train braking can be used by other trains or sent back to the distribution network for other uses. However, there is part of the regenerated energy that will be lost anyway because of the catenary losses and the motor's efficiency. Saving part of the lost energy is the main benefit of eco-driving, and the results support this. It has been shown that, the lower the energy lost in the motor and catenary, the lower the difference between efficient non-efficient driving.

The results obtained in Puertollano–Ciudad Real show energy savings produced by eco-driving according to the typical values between 10% and 15% in the literature [68]. On the other hand, the results obtained in Calatayud–Zaragoza present values that are further reduced. One reason to explain this is the track profile. Although the average grade in both cases is downhill, the difference in altitude between the end and start of the journey is higher in Calatayud–Zaragoza. The presence of steeper downhills leads to lower energy consumption results and lower differences between different speed profiles.

Catenary losses can also explain why the energy reduction figures obtained by the eco-driving strategy in the Calatayud–Zaragoza stretch are lower than those obtained in the Puertollano–Ciudad Real stretch (the 2 × 25 kV system is more efficient than the 1 × 25 kV system because of lower rate of power losses).

The influence of the commercial running time cannot be neglected. In both cases, it can be observed that there are three scenarios of energy saving due to eco-driving depending on the running time. Close to the flat-out running time, the energy saving that can be obtained by efficient driving is zero because there is no possibility of modifying the driving to obtain the minimum running time. From there, there is a small interval of running time wherein the energy saving provided by eco-driving increases as the running time increases. After that, there is a zone of running time values in which the energy saving rate is approximately constant. This zone typically contains the commercial running time for the journey. Finally, for the largest running times, the difference in energy consumption between efficient and non-efficient driving is gradually reduced.

Therefore, a second order reason for the reduced energy saving figures in Calatayud–Zaragoza is that the commercial running time of this journey is higher than usual and it is located in the running time interval in which the energy savings provided by eco-driving are reduced.

Despite this, the minimum energy saving provided by eco-driving is relevant (2.5%) in a scenario of 97% motor efficiency. This motor efficiency is very high compared with its typical values, which are between 85% and 90%. Therefore, taking into account the typical motor efficiency, the minimum energy saving provided by eco-driving can reach more that 3.5%.

With these results, it can be said that high network receptivity to braking energy can mitigate the effect of eco-driving, but it is still significant.

Obviously, results can vary in a field-test because of simulation error. However, as explained in Section 2, the value of simulation error in energy consumption is 0.4%. This error can slightly quantitatively modify the results obtained but cannot qualitatively modify them and the trends obtained.

8. Conclusions

The basis of eco-driving strategy is the substitution of the use of braking with coasting periods. Trains equipped with regenerative braking allow for recovering the kinetic energy during deceleration phases, which can be used by other trains or sent back to the utility grid in AC power supply systems or in DC systems equipped with reversible substations. In this context of high use of regenerated energy, it is necessary to analyze whether eco-driving design based on coast commands has an impact on energy savings. In this work, an eco-driving strategy has been compared to a standard driving strategy. The former one consists on holding a cruise speed without braking (if braking is needed to maintain that speed, it is not applied and the speed increases) and coasting before braking. The latter one consists of holding the cruise speed and braking if it is necessary to do so. A comparison has been performed in two real case studies by simulating a train on two high-speed Spanish lines. One of them is a section of the Madrid–Barcelona high-speed line, which is fed by a 2 × 25 kV power supply system. The other one is a section of the Madrid–Sevilla high-speed line, which is fed by a 1 × 25 kV power supply system. The influence of the regenerated energy losses at the catenary—which depend on its linear resistance, and the engine traction and braking efficiency, which are considered constant and equal—on the energy consumption of the speed profiles generated for both driving strategies has been studied. To generate the driving commands, a nature-inspired algorithm called MOPSO has been used for both driving strategies. Speed profiles have been obtained by simulating driving commands. For each driving strategy and for some values of the linear resistance and engine efficiency, the Pareto curves consisting of the non-dominated speed profiles in terms of energy consumption and running time have been computed. The difference in energy consumption between the considered driving strategies has been calculated for different values of the catenary resistance, engine efficiency and running time, with a special focus on the commercial time. For the commercial running time, the eco-driving strategy produces energy savings with respect to the standard driving strategy between 2.48% and 4.69% at the pantograph for the first case study and between 9.7% and 12.55% for the second.

For a specific running time, the holding speed without braking and final coasting eco-driving strategy produces solutions that consume less energy than the holding speed with braking standard driving strategy. The difference in energy consumption between the considered driving strategies is larger when the losses at the catenary are more significant and when the engine is less efficient.

It was found that the holding speed without braking and final coasting driving strategy produces speed profiles more energy-efficient than the standard driving strategy, mostly due to the final coast phase. For slower speed profiles, the energy consumption is much lower, as the cruise speed is lower; the final coast phase takes place more close to the brake curve at the end of the section; and then the difference in energy consumption between the considered driving strategies is smaller.

Finally, it can be concluded that, although the power supply network presents a high receptivity to braking energy, the eco-driving design can produce significant energy savings.

Author Contributions: A.C.: investigation, writing—original draft preparation. A.F.-R.: investigation, software, writing—original draft preparation. A.P.C.: conceptualization, methodology, investigation, writing—review and editing. A.F.-C.: conceptualization, methodology, investigation, writing—review and editing. M.C.F.: writing—review and editing. All authors have read and agreed to the published version of the manuscript.

Funding: This research received no external funding.

Conflicts of Interest: The authors declare no conflict of interest.

References

1. IEA. Energy Consumption and CO_2 Emissions. In *UIC Railway Handbook 2017*; IEA: Paris, France, 2017.
2. *UNIFE/Roland Berger World Rail Market Study: Forecast 2016 to 2021*; UNIFE: Brussels, Belgium, 2016.
3. Bae, C.H. A simulation study of installation locations and capacity of regenerative absorption inverters in DC 1500 V electric railways system. *Simul. Model. Pract. Theory* **2009**, *17*, 829–838. [CrossRef]
4. Barrero, R.; Mierlo, J.V.; Tackoen, X. Energy savings in public transport. *IEEE Veh. Technol. Mag.* **2008**, *3*, 26–36. [CrossRef]
5. Cornic, D. Efficient recovery of braking energy through a reversible dc substation. In Proceedings of the Electrical Systems for Aircraft, Railway and Ship Propulsion, Bologna, Italy, 19–21 October 2010; pp. 1–9.
6. López-López, Á.J.; Pecharromán, R.R.; Fernández-Cardador, A.; Cucala, A.P. Assessment of energy-saving techniques in direct-current-electrified mass transit systems. *Transp. Res. Part C Emerg. Technol.* **2014**, *38*, 85–100. [CrossRef]
7. Lv, X.; He, B.; Li, Y. Inbound Energy-Saving Slope Design Method of High-Speed Railway. In Proceedings of the ICTE 2013: Safety, Speediness, Intelligence, Low-Carbon, Innovation, Chengdu, China, 19–20 October 2013.
8. Roch-Dupré, D.; Cucala, A.P.; Pecharromán, R.; López-López, Á.J.; Fernández-Cardador, A. Evaluation of the impact that the traffic model used in railway electrical simulation has on the assessment of the installation of a Reversible Substation. *Int. J. Electr. Power Energy Syst.* **2018**, *102*, 201–210. [CrossRef]
9. Roch-Dupré, D.; López-López, Á.J.; Pecharromán, R.R.; Cucala, A.P.; Fernández-Cardador, A. Analysis of the demand charge in DC railway systems and reduction of its economic impact with Energy Storage Systems. *Int. J. Electr. Power Energy Syst.* **2017**, *93*, 459–467. [CrossRef]

10. Fernández-Rodríguez, A.; Fernández-Cardador, A.; Cucala, A.P.; Falvo, M.C. Energy Efficiency and Integration of Urban Electrical Transport Systems: EVs and Metro-Trains of Two Real European Lines. *Energies* **2019**, *12*, 366. [CrossRef]
11. Falvo, M.C.; Lamedica, R.; Bartoni, R.; Maranzano, G. Energy management in metro-transit systems: An innovative proposal toward an integrated and sustainable urban mobility system including plug-in electric vehicles. *Electr. Power Syst. Res.* **2011**, *81*, 2127–2138. [CrossRef]
12. Iannuzzi, D.; Pagano, E.; Tricoli, P.; Iannuzzi, D.; Pagano, E.; Tricoli, P. The Use of Energy Storage Systems for Supporting the Voltage Needs of Urban and Suburban Railway Contact Lines. *Energies* **2013**, *6*, 1802–1820. [CrossRef]
13. Arboleya, P.; Coto, M.; González-Morán, C.; Arregui, R. On board accumulator model for power flow studies in DC traction networks. *Electr. Power Syst. Res.* **2014**, *116*, 266–275. [CrossRef]
14. Beusen, B.; Degraeuwe, B.; Debeuf, P. Energy savings in light rail through the optimization of heating and ventilation. *Transp. Res. Part Transp. Environ.* **2013**, *23*, 50–54. [CrossRef]
15. Bombardier. *Aeroefficient Optimised Train Shaping*; Bombardier: Hennigsdorf, Germany, 2010.
16. Kondo, M.; Miyabe, M.; Manabe, S. Development of a High Efficiency Induction Motor and the Estimation of Energy Conservation Effect. *Q. Rep. RTRI* **2014**, *55*, 138–143. [CrossRef]
17. Matsuoka, K.; Kondo, M. Energy Saving Technologies for Railway Traction Motors. *IEEJ Trans. Electr. Electron. Eng.* **2014**, *5*, 278–284. [CrossRef]
18. Ichikawa, K. Application of Optimization Theory for Bounded State Variable Problems to the Operation of Train. *Bull. JSME* **1968**, *11*, 857–865. [CrossRef]
19. Howlett, P. An optimal strategy for the control of a train. *J. Aust. Math. Soc. Ser. B* **1990**, *31*, 454–471. [CrossRef]
20. Howlett, P.G.; Pudney, P.J.; Vu, X. Local energy minimization in optimal train control. *Automatica* **2009**, *45*, 2692–2698. [CrossRef]
21. Liu, R.; Golovitcher, I.M. Energy-efficient operation of rail vehicles. *Transp. Res. Part Policy Pract.* **2003**, *37*, 917–932. [CrossRef]
22. Khmelnitsky, E. On an optimal control problem of train operation. *IEEE Trans. Autom. Control* **2000**, *45*, 1257–1266. [CrossRef]
23. Yang, X.; Li, X.; Ning, B.; Tang, T. A Survey on Energy-Efficient Train Operation for Urban Rail Transit. *IEEE Trans. Intell. Transp. Syst.* **2016**, *17*, 2–13. [CrossRef]
24. Albrecht, A.R.; Howlett, P.G.; Pudney, P.J.; Vu, X. Energy-efficient train control: From local convexity to global optimization and uniqueness. *Automatica* **2013**, *49*, 3072–3078. [CrossRef]
25. Su, S.; Li, X.; Tang, T.; Gao, Z. A Subway Train Timetable Optimization Approach Based on Energy-Efficient Operation Strategy. *IEEE Trans. Intell. Transp. Syst.* **2013**, *14*, 883–893. [CrossRef]
26. Yang, J.; Jia, L.; Lu, S.; Fu, Y.; Ge, J. Energy-Efficient Speed Profile Approximation: An Optimal Switching Region-Based Approach with Adaptive Resolution. *Energies* **2016**, *9*, 762. [CrossRef]
27. Albrecht, T.; Binder, A.; Gassel, C. Applications of real-time speed control in rail-bound public transportation systems. *IET Intell. Transp. Syst.* **2013**, *7*, 305–314. [CrossRef]
28. Ichikawa, S.; Miyatake, M. Energy Efficient Train Trajectory in the Railway System with Moving Block Signaling Scheme. *IEEJ J. Ind. Appl.* **2019**, *8*, 586–591. [CrossRef]
29. Lu, S.; Hillmansen, S.; Ho, T.K.; Roberts, C. Single-Train Trajectory Optimization. *IEEE Trans. Intell. Transp. Syst.* **2013**, *14*, 743–750. [CrossRef]
30. Miyatake, M.; Ko, H. Optimization of Train Speed Profile for Minimum Energy Consumption. *IEEJ Trans. Electr. Electron. Eng.* **2010**, *5*, 263–269. [CrossRef]
31. Miyatake, M.; Matsuda, K. Energy Saving Speed and Charge/Discharge Control of a Railway Vehicle with On-board Energy Storage by Means of an Optimization Model. *IEEJ Trans. Electr. Electron. Eng.* **2009**, *4*, 771–778. [CrossRef]
32. Gu, Q.; Tang, T.; Cao, F.; Song, Y. Energy-Efficient Train Operation in Urban Rail Transit Using Real-Time Traffic Information. *IEEE Trans. Intell. Transp. Syst.* **2014**, *15*, 1216–1233. [CrossRef]
33. Rodrigo, E.; Tapia, S.; Mera, J.; Soler, M. Optimizing Electric Rail Energy Consumption Using the Lagrange Multiplier Technique. *J. Transp. Eng.* **2013**, *139*, 321–329. [CrossRef]

34. Wang, Y.; De Schutter, B.; van den Boom, T.J.J.; Ning, B. Optimal trajectory planning for trains—A pseudospectral method and a mixed integer linear programming approach. *Transp. Res. Part C Emerg. Technol.* **2013**, *29*, 97–114. [CrossRef]
35. Wang, Y.; De Schutter, B.; van den Boom, T.J.J.; Ning, B. Optimal trajectory planning for trains under fixed and moving signaling systems using mixed integer linear programming. *Control Eng. Pract.* **2014**, *22*, 44–56. [CrossRef]
36. Wong, K.K.; Ho, T.K. Coast control for mass rapid transit railways with searching methods. *IEE Proc. Electr. Power Appl.* **2004**, *151*, 365–376. [CrossRef]
37. De Cuadra, F.; Fernandez, A.; de Juan, J.; Herrero, M.A. *Energy-Saving Automatic Optimisation of Train Speed Commands Using Direct Search Techniques*; Wessex Institute of Technology: Southampton, UK, 1996; Volume 10.
38. Zhao, N.; Roberts, C.; Hillmansen, S.; Tian, Z.; Weston, P.; Chen, L. An integrated metro operation optimization to minimize energy consumption. *Transp. Res. Part C Emerg. Technol.* **2017**, *75*, 168–182. [CrossRef]
39. Tian, Z.; Weston, P.; Zhao, N.; Hillmansen, S.; Roberts, C.; Chen, L. System energy optimisation strategies for metros with regeneration. *Transp. Res. Part C Emerg. Technol.* **2017**, *75*, 120–135. [CrossRef]
40. Acikbas, S.; Soylemez, M.T. Coasting point optimisation for mass rail transit lines using artificial neural networks and genetic algorithms. *IET Electr. Power Appl.* **2008**, *2*, 172–182. [CrossRef]
41. Chuang, H.-J.; Chen, C.-S.; Lin, C.-H.; Hsieh, C.-H.; Ho, C.-Y. Design of Optimal Coasting Speed for MRT Systems Using ANN Models. *IEEE Trans. Ind. Appl.* **2009**, *45*, 2090–2097. [CrossRef]
42. Bocharnikov, Y.V.; Tobias, A.M.; Roberts, C. Reduction of train and net energy consumption using genetic algorithms for Trajectory Optimisation. In Proceedings of the IET Conference on Railway Traction Systems (RTS 2010), Birmingham, UK, 13–15 April 2010; pp. 1–5.
43. Carvajal-Carreño, W.; Cucala, A.P.; Fernández-Cardador, A. Optimal design of energy-efficient ATO CBTC driving for metro lines based on NSGA-II with fuzzy parameters. *Eng. Appl. Artif. Intell.* **2014**, *36*, 164–177. [CrossRef]
44. Domínguez, M.; Fernández-Cardador, A.; Cucala, A.P.; Gonsalves, T.; Fernández, A. Multi objective particle swarm optimization algorithm for the design of efficient ATO speed profiles in metro lines. *Eng. Appl. Artif. Intell.* **2014**, *29*, 43–53. [CrossRef]
45. Keskin, K.; Karamancioglu, A. Energy-Efficient Train Operation Using Nature-Inspired Algorithms. *J. Adv. Transp.* **2017**, *2017*, 12. [CrossRef]
46. Kim, Y.-G.; Jeon, C.-S.; Kim, S.-W.; Park, T.-W. Operating speed pattern optimization of railway vehicles with differential evolution algorithm. *Int. J. Automot. Technol.* **2013**, *14*, 903–911. [CrossRef]
47. Sicre, C.; Cucala, A.P.; Fernández, A.; Lukaszewicz, P. Modeling and optimizing energy-efficient manual driving on high-speed lines. *IEEJ Trans. Electr. Electron. Eng.* **2012**, *7*, 633–640. [CrossRef]
48. Yang, X.-S. *Nature-Inspired Optimization Algorithms*; Elsevier: Amsterdam, The Netherlands, 2014; ISBN 978-0-12-416745-2.
49. Chang, C.S.; Sim, S.S. Optimising train movements through coast control using genetic algorithms. *Electr. Power Appl. IEE Proc.* **1997**, *144*, 65–73. [CrossRef]
50. Lechelle, S.A.; Mouneimne, Z.S. *OptiDrive: A Practical Approach for the Calculation of Energy-Optimised Operating Speed Profiles*; IET: London, UK, 2010; p. 23.
51. Li, X.; Lo, H.K. An energy-efficient scheduling and speed control approach for metro rail operations. *Transp. Res. Part B Methodol.* **2014**, *64*, 73–89. [CrossRef]
52. Wong, K.K.; Ho, T.K. Dynamic coast control of train movement with genetic algorithm. *Int. J. Syst. Sci.* **2004**, *35*, 835–846. [CrossRef]
53. Wong, K.K.; Ho, T.K. Coast control of train movement with genetic algorithm. In Proceedings of the 2003 Congress on Evolutionary Computation, CEC '03, Canberra, Australia, 8–12 December 2003; Volume 2, pp. 1280–1287.
54. Yang, L.; Li, K.; Gao, Z.; Li, X. Optimizing trains movement on a railway network. *Omega* **2012**, *40*, 619–633. [CrossRef]
55. Wei, L.; Qunzhan, L.; Bing, T. Energy Saving Train Control for Urban Railway Train with Multi-population Genetic Algorithm. In Proceedings of the International Forum on Information Technology and Applications, IFITA '09, Chengdu, China, 15–17 May 2009; Volume 2, pp. 58–62.

56. Huang, Y.; Ma, X.; Su, S.; Tang, T. Optimization of Train Operation in Multiple Interstations with Multi-Population Genetic Algorithm. *Energies* **2015**, *8*, 14311–14329. [CrossRef]
57. Bocharnikov, Y.V.; Tobias, A.M.; Roberts, C.; Hillmansen, S.; Goodman, C.J. Optimal driving strategy for traction energy saving on DC suburban railways. *IET Electr. Power Appl.* **2007**, *1*, 675–682. [CrossRef]
58. Cucala, A.P.; Fernández, A.; Sicre, C.; Domínguez, M. Fuzzy optimal schedule of high speed train operation to minimize energy consumption with uncertain delays and driver's behavioral response. *Eng. Appl. Artif. Intell.* **2012**, *25*, 1548–1557. [CrossRef]
59. Hwang, H.-S. Control strategy for optimal compromise between trip time and energy consumption in a high-speed railway. *IEEE Trans. Syst. Man Cybern. Part Syst. Hum.* **1998**, *28*, 791–802. [CrossRef]
60. Sicre, C.; Cucala, A.P.; Fernández-Cardador, A. Real time regulation of efficient driving of high speed trains based on a genetic algorithm and a fuzzy model of manual driving. *Eng. Appl. Artif. Intell.* **2014**, *29*, 79–92. [CrossRef]
61. Ke, B.-R.; Lin, C.-L.; Yang, C.-C. Optimisation of train energy-efficient operation for mass rapid transit systems. *IET Intell. Transp. Syst.* **2012**, *6*, 58–66. [CrossRef]
62. Yan, X.; Cai, B.; Ning, B.; ShangGuan, W. Online distributed cooperative model predictive control of energy-saving trajectory planning for multiple high-speed train movements. *Transp. Res. Part C Emerg. Technol.* **2016**, *69*, 60–78. [CrossRef]
63. Xie, T.; Wang, S.; Zhao, X.; Zhang, Q. Optimization of Train Energy-Efficient Operation Using Simulated Annealing Algorithm. In *Intelligent Computing for Sustainable Energy and Environment*; Li, K., Li, S., Li, D., Niu, Q., Eds.; Communications in Computer and Information Science; Springer: Berlin/Heidelberg, Germany, 2013; pp. 351–359. ISBN 978-3-642-37104-2.
64. Chevrier, R.; Pellegrini, P.; Rodriguez, J. Energy saving in railway timetabling: A bi-objective evolutionary approach for computing alternative running times. *Transp. Res. Part C Emerg. Technol.* **2013**, *37*, 20–41. [CrossRef]
65. Fernández-Rodríguez, A.; Fernández-Cardador, A.; Cucala, A.P.; Domínguez, M.; Gonsalves, T. Design of Robust and Energy-Efficient ATO Speed Profiles of Metropolitan Lines Considering Train Load Variations and Delays. *IEEE Trans. Intell. Transp. Syst.* **2015**, *16*, 2061–2071. [CrossRef]
66. Fernández-Rodríguez, A.; Fernández-Cardador, A.; Cucala, A.P. Balancing energy consumption and risk of delay in high speed trains: A three-objective real-time eco-driving algorithm with fuzzy parameters. *Transp. Res. Part C Emerg. Technol.* **2018**, *95*, 652–678. [CrossRef]
67. Fernández-Rodríguez, A.; Fernández-Cardador, A.; Cucala, A.P. Real time eco driving of high speed trains by simulation-based dynamic multi-objective optimization. *Simul. Model. Pract. Theory* **2018**, *84*, 50–68. [CrossRef]
68. Douglas, H.; Roberts, C.; Hillmansen, S.; Schmid, F. An assessment of available measures to reduce traction energy use in railway networks. *Energy Convers. Manag.* **2015**, *106*, 1149–1165. [CrossRef]
69. Albrecht, A.; Howlett, P.; Pudney, P.; Vu, X.; Zhou, P. The two-train separation problem on non-level track—Driving strategies that minimize total required tractive energy subject to prescribed section clearance times. *Transp. Res. Part B Methodol.* **2018**, *111*, 135–167. [CrossRef]
70. Coleman, D.; Howlett, P.; Pudney, P.; Xuan, V.; Yee, R. *Coasting Boards vs. Optimal Control*; IET: London, UK, 2010; pp. 1–5.
71. Franke, R.; Terwiesch, P.; Meyer, M. An algorithm for the optimal control of the driving of trains. In Proceedings of the 39th IEEE Conference on Decision and Control, Sydney, NSW, Australia, 12–15 December 2000; Volume 3, pp. 2123–2128.
72. Johnson, P.; Brown, S. A simple in-cab schedule advisory system to save energy and improve on-time performance. In Proceedings of the IET Conference on Railway Traction Systems (RTS 2010), Birmingham, UK, 13–15 April 2010; p. 15.
73. Scheepmaker, G.M.; Goverde, R.M.P. The interplay between energy-efficient train control and scheduled running time supplements. *J. Rail Transp. Plan. Manag.* **2015**, *5*, 225–239. [CrossRef]
74. González-Gil, A.; Palacin, R.; Batty, P.; Powell, J.P. A systems approach to reduce urban rail energy consumption. *Energy Convers. Manag.* **2014**, *80*, 509–524. [CrossRef]
75. Pudney, P.; Howlett, P. Optimal driving strategies for a train journey with speed limits. *ANZIAM J.* **1994**, *36*, 38–49. [CrossRef]

76. Domínguez, M.; Fernández, A.; Cucala, A.P.; Cayuela, L.P. Computer-aided design of ATO speed commands according to energy consumption criteria. *WIT Trans. Built Environ.* **2008**, *103*, 183–192.
77. Dominguez, M.; Fernandez, A.; Cucala, A.P.; Blanquer, J. Efficient Design of Automatic Train Operation Speed Profiles with On Board Energy Storage Devices. *WIT Trans. Built Environ.* **2010**, *114*, 509–520.
78. Coello, C.A.C.; Pulido, G.T.; Lechuga, M.S. Handling multiple objectives with particle swarm optimization. *IEEE Trans. Evol. Comput.* **2004**, *8*, 256–279. [CrossRef]
79. Scheepmaker, G.M.; Goverde, R.M.P. Energy-efficient train control including regenerative braking with catenary efficiency. In Proceedings of the 2016 IEEE International Conference on Intelligent Rail Transportation (ICIRT), Birmingham, UK, 23–25 August 2016; pp. 116–122.
80. Falvo, M.C.; Sbordone, D.; Fernández-Cardador, A.; Cucala, A.P.; Pecharromán, R.R.; López-López, A. Energy savings in metro-transit systems: A comparison between operational Italian and Spanish lines. *Proc. Inst. Mech. Eng. Part F J. Rail Rapid Transit* **2016**, *230*, 345–359. [CrossRef]
81. Chang, C.S.; Xu, D.Y.; Quek, H.B. Pareto-optimal set based multiobjective tuning of fuzzy automatic train operation for mass transit system. *Electr. Power Appl. IEE Proc.* **1999**, *146*, 577–583. [CrossRef]
82. Chang, C.S.; Xu, D.Y. Differential evolution based tuning of fuzzy automatic train operation for mass rapid transit system. *Electr. Power Appl. IEE Proc.* **2000**, *147*, 206–212. [CrossRef]
83. Wang, Y.; Ning, B.; Cao, F.; De Schutter, B.; van den Boom, T.J.J. A survey on optimal trajectory planning for train operations. In Proceedings of the 2011 IEEE International Conference on Service Operations, Logistics and Informatics, Beijing, China, 10–12 July 2011; pp. 589–594.

© 2020 by the authors. Licensee MDPI, Basel, Switzerland. This article is an open access article distributed under the terms and conditions of the Creative Commons Attribution (CC BY) license (http://creativecommons.org/licenses/by/4.0/).

energies

MDPI

Article

Examination of EV Abilities to Provide Vehicle-to-Home Service in Low Voltage Installation

Paweł Kelm *, Rozmysław Mieński, Irena Wasiak and Katarzyna Wojciechowska

Institute of Electrical Power Engineering, Lodz University of Technology, 90-924 Łódź, Poland; rozmyslaw.mienski@p.lodz.pl (R.M.); irena.wasiak@p.lodz.pl (I.W.); kwojciechowska@miratrans.pl (K.W.)

* Correspondence: pawel.kelm@p.lodz.pl

Received: 11 March 2020; Accepted: 7 April 2020; Published: 10 April 2020

Abstract: This paper deals with the application of an electric vehicle (EV) motor inverter and its batteries as an energy storage device supporting the operation of home electrical installation. This additional functionality of EV is called a Vehicle-to-Home (V2H) service. Two kind of services are considered: a peak shaving and an emergency power supply. The simulation model developed in the PSCAD program is presented. It allows for the examination of the EV battery control and operation during EV driving and parking. Additionally, it allows an evaluation of the availability of home installation for the V2H service. Control algorithms enabling the implementation of discussed work options are presented. Results of simulations are presented illustrating the EV control and operation in different operational modes.

Keywords: ancillary services; EV motor inverter; electrical installation; on-board charger; vehicle-to-home; home energy management system

1. Introduction

The dynamic technological development of electric vehicles (EV) and different support instruments introduced in various countries according to EU directives encourage individual customers to invest in EVs. The goal of the European Union is to reduce the greenhouse gas emissions into the atmosphere by 2050 by 60% compared to 1990. One of the assumptions to achieve the above objective is halving the number of vehicles with conventional drive in the urban transport by 2030 and total elimination of such vehicles from cities by 2050 [1].

There are three main types of EVs available on the market: hybrid vehicles (HEV), hybrid vehicles with the possibility of charging a battery pack (PHEV) and fully electric vehicles (BEV) [2,3]. The BEV vehicles are the subject of this paper. Due to the high energy density, reasonable battery life and acceptable recharging speed the lithium-ion batteries are used to power this type of vehicle [4].

The increasing number of EVs may bring a significant impact on the supply network. The most negative aspect for the network is an uncontrolled EV charging process, especially if it takes place during a time overlapping with high power demand. The increase of peak load due to mass charging may lead to branch congestion and large voltage drops. This problem is described in the literature. The assessment framework of a plug-in electric vehicles (PEV) impact on the supplying network is proposed in [5] taking into account the great uncertainty in many crucial assumptions and parameters (e.g., EV battery charging power and time). In [6,7] the rating of electricity demand profile is assessed on the basis of information about EV user behaviors. Furthermore, in [8] the diagrams of medium voltage (MV) network loading are presented for a scenario where 52% of all used cars are EVs. It is shown that dump charging can cause peak a power increase by about 100% and controllable charging is necessary to avoid network disruptions. Similarly, in [9,10] the impact of the increased number of EVs on the operation of low voltage (LV) power network by both controlled and uncontrolled EVs charging is analyzed. Different charging strategies are tested and compared in [11–13]. The concept of using renewables, in particular the solar panels, in charging infrastructure is discussed in [14].

The impact of EVs on the supply network significantly depends on EV user behaviors. If EV is used for commercial purposes or to travel for longer distances, it requires the infrastructure of fast and superfast chargers. In that case the location of charging points must be based on factors such as EV ranges, road infrastructure, investment costs, etc. In certain cases, there will be a need to reinforce a supply network and so the planning is crucial to minimize additional costs. The methodology for planning minimal fast charging infrastructure deployment is evaluated and depicted in [15].

If the user exploits EV on relatively short distances, for example for commuting, the EV can be charged at home in the evening. At present this is the most typical scenario and it will be discussed further. While the EV is connected to the home electrical installation, the energy stored in EV battery can be used to support the installation operation and improve the reliability of supply. Such an EV service, called the vehicle-to-home (V2H), is especially useful in the case of supply network failure or when the EV user wants to reduce energy costs by using an EV battery to supply loads in the period of high energy prices.

Benefits from using EV for additional services have already been indicated in the literature [16]. It is shown that through demand side management (DSM) the V2H may support the local loads during severe supply network loading or outages, hence alleviating energy demand on the network and its reliability requirements. In [17] an active management model integrated with generation curtailment mechanism and the charging/discharging management of plug-in EVs was built for the local distribution system. In [18–20] PEVs are used for the backup power during an outage and on the basis of residential energy usage data, the duration of backup power that could be achieved is identified. Similarly, in [21,22] EVs are used as energy storage for a local smart grid to reduce energy costs. In [23] the exploitation of vehicle-to-grid (V2G) services for power quality improvement is discussed. The optimal charge/discharge schedule for an EV plugged into a smart building is presented in [24,25] taking into account time-varying electricity pricing schemes.

At present the V2H and V2G services require an off-board bi-directional inverter which generate high additional costs. In addition, the size of such an inverter is not without significance, especially in the context of limited space in garages or parking lots. In this paper, authors analyze the usage of an on-board motor inverter (instead of off-board) for V2H services. It means that the motor inverter is used for both: motor propulsion and ancillary (V2G/V2H) services. Different approaches to the adaptation of motor converters for the ancillary services are available in the literature. The review of the drivelines in all-electric vehicles is presented in [26]. The topology of an on-board integrated battery charger using an asymmetrical nine-phase machine is described in [27]. In [28], a single-phase integrated charger, using the EV propulsion machine and its traction converter, is introduced.

The presented literature concentrates on indicating the possible benefits of the V2H services for EV users and the supplying network operators. However, to implement the additional functionalities of EV some technical issues should be examined including the control of EV motor inverter. In this paper the authors present an original simulation model, which can be a useful tool for both: the examination of EV motor inverter for V2H services and for the adaptation of home installation for such services.

The paper is arranged as follows. In Section 2 the EV user behavior scenarios are identified to justify the possibilities and significance of EV additional functionalities. Then in Section 3 we describe the simulation model. In the next Section, we present the simulation results for the chosen case study. The paper finishes with some conclusions.

2. EV User Behavior Scenario and V2H Charging Strategy

The availability of EV to be used for V2H services depends on EV user behaviors, which are unpredictable to a large extent. However, some typical user behaviors can be identified on the basis of factors describing the way of EV exploitation, i.e.,:

- the distance, which a user drives during one day;
- the place, where a user charges their electric car;
- the time, when a user charges his electric car.

In this paper, the following scenario of EV exploitation is assumed. The EV is used in urban and non-urban areas for commuting. The vehicle is mainly charged on a private property. The most probable choice is charging in the user's private garage in the evening, less probable is charging in the workplace during the day.

Currently the typical charging solutions may include: AC on-board charger (rectifier), or off-board DC charger, or on-board motor converter (also used for motor propulsion, braking energy regeneration and battery charging). In a further consideration, only the controlled AC charging, with the use of motor inverter, will be dealt with.

The control strategy applied defines charging and control parameters [29] and requires infrastructure. A vehicle used for commuting is parked most of the day and can be connected to the electric installation. Hence, it is justified to extend the charging control strategy in order to use the EV battery as energy storage for the EV user needs.

The control of the charging process can be based on the tariff signal and the power limit value set by the distribution system operator (DSO). This strategy is beneficial for both the EV user and the network operator. The EV user reduces energy costs and the DSO receives network management opportunities (such as network load control).

The utilization of energy storage devices by end-users has been known for a long time and reported in the literature [30–33]. The typical application concerns a peak-shaving. When a distinction is made in tariffs between peak and off-peak hours, it is profitable for an end-users to reduce the peak demand. The EV battery is charged during off-peak hours, when energy is cheaper, and it is discharged during peak load periods. For EV users the reduction of peak power demand means decreasing demand charges. It should be noted that the peak demand occurs usually in the evening time period, so, according to the described scenario, while the EV is parked and its battery is available for ancillary service.

Another useful application of energy storage in LV installations is the supply of loads during dips and short interruptions. Its purpose is lowering the costs of losses incurred by end-users due to such disturbances in the network.

According to authors the infrastructure for the V2H services requires both the EV motor inverter and the building installation equipped with a home energy management system (HEMS) (Figure 1). HEMS is responsible for the calculation of the starting and ending time of charging or discharging processes. It is based on the following parameters:

- current energy cost;
- battery state of charge (SOC);
- current energy consumption at home;
- EV user needs (time when EV user exploits their car).

The vehicle should be connected to the AC power network via dedicated EVSE (Electric Vehicle Suppling Equipment) i.e., socket outlet, connecting cable and wallbox. Communication between EV and EVSE must be bi-directional.

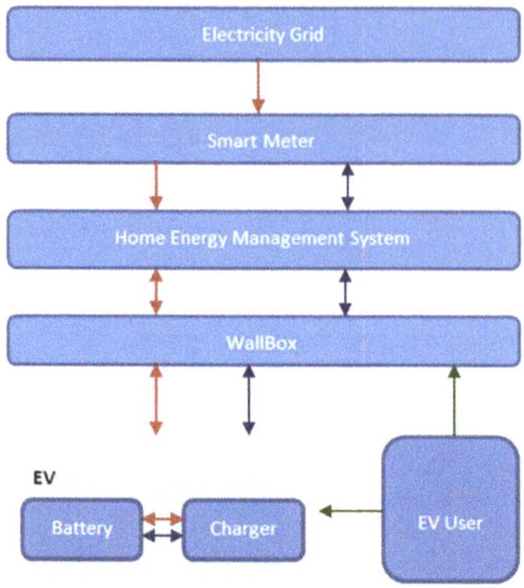

Figure 1. Vehicle-to-home (V2H) charging scheme.

3. Description of EV Simulation Model

3.1. Asumptions

During driving, the EV motor inverter provides power to its motor or is used for regenerative braking. While an EV is parked, the motor inverter is connected to 3-phase LV home installation and is used to charge the batteries or for the V2H services such as energy storage operation or uninterruptable power supply (UPS). This functionality requires an appropriate control system and procedures to run depending on the function being performed: drive, energy storage or UPS (when the installation is disconnected from the network). The controller is located on-board the vehicle. Its operation modes and settings should include all limitations regarding all the cooperating devices: battery, vehicle engine, electric loads and the supplying feeder.

During driving the EV driver controls the power fed to the EV motor within the limits of its rated power (using accelerator and brake pedals).

While EV is parked and connected to the home installation, the battery can be charged or discharged according to the user's decision or predefined profile. The amount of power drawn from or returned to the customer installation is limited by EV motor rated power, technical parameters of home installation and should be managed by the HEMS. Additionally, it should be noted that every EV is equipped with the on-board battery management systems (BMS). It is responsible for battery protection, monitoring and controlling its state. The HEMS cannot and should not have priority over BMS so it cannot jeopardize the battery safety. However, HEMS has access to the additional information (the energy tariffs, current energy consumption at home or other EV user preferences). That data can be used to effectively manage services like the peak shaving or planning the charging schedule.

The UPS option requires fully autonomous operation of the controller in order to balance an active and reactive power in the user's home installation (islanded operation). The amount of power drawn from or returned to the customer installation is limited by EV motor rated power and technical parameters of the home installation.

The objective of the presented model is to create a simulation tool to examine and evaluate the EV capability for V2H services. In particular, the model can be used to:

- evaluate the possibilities of EV inverter construction in terms of the ability to provide additional functionalities for achieving technical and economic benefits for EV users;
- evaluate the home electric installation with regard to its abilities for the islanded operation with the use of an EV battery system;
- verify the EV control algorithms and the selection of settings according to EV user's requirements.

3.2. Model Structure

The presented EV simulation model, for which the general scheme is shown in Figure 2, was built in the PSCAD program [34]. The model encompasses the following modules: part of LV distribution network (NETWORK), EV user electric home installation supplied from the network (EV OWNER HOME INSTALLATION) and the on-board EV infrastructure (EV).

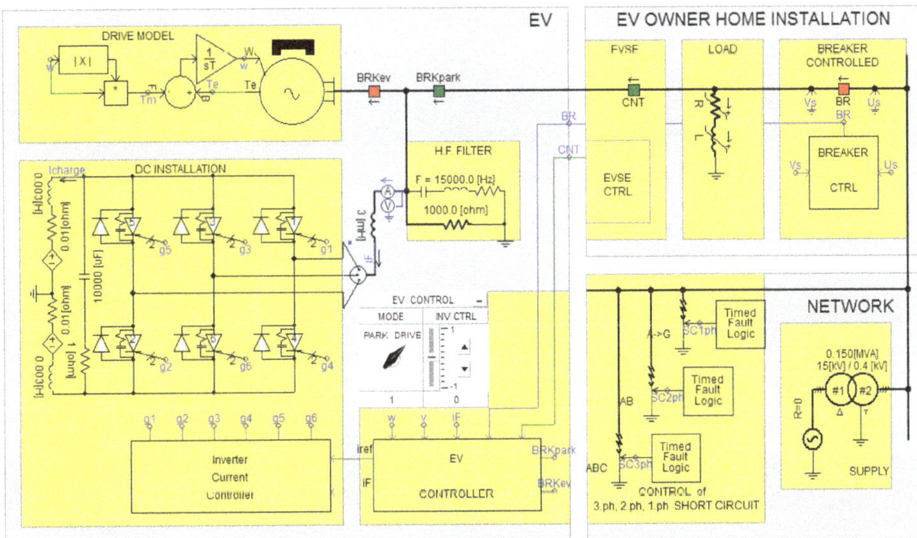

Figure 2. Schematic diagram of an electric vehicle (EV) and home installation simulator mode.

3.2.1. NETWORK Module

The network module consists of two blocks: SUPPLY and CONTROL of 3.ph, 2.ph, 1.ph SHORT CIRCUIT. The first block include ideal 3-phase voltage source of nominal voltage 15 kV and 15/0.4 kV distribution transformer supplying LV busbars (SUPPLY). The second allow the emulation of different, asymmetrical faults in the network and illustrate their effects at the point of coupling with end-user installation. The beginning and duration of each fault can be set using Time Fault Logic blocks.

3.2.2. EV OWNER INSTALLATION Module

User installation module is represented by R, L load block (LOAD), which through a BR switch is connected to the transformer LV busbars. The value of load impedance can be changed with constant tgφ. The EVSE terminal has electrical and control circuits. The CNT switch is controlled by the EVSE CTRL block.

The BR switch is controlled by the BRAKER CTRL block. In this block the supplying network voltage vector v_n is monitored and if the voltage exceeds permissible values, the switch is opened.

The controller can close the breaker only when the voltage vectors $\boldsymbol{v_n}$ and $\boldsymbol{v_h}$, on both sides of the switch, expressed by Equations (1) and (2), fulfil the synchronization requirements.

$$\boldsymbol{v_h} = V_h e^{j(\omega_h t+\varphi_h)} \tag{1}$$

$$\boldsymbol{v_n} = V_n e^{j(\omega_n t+\varphi_n)} \tag{2}$$

where:

$\boldsymbol{v_h}$, V_h, ω_h, φ_h—instantaneous value, module, angular velocity and phase angle of voltage vector in end-user (home) installation in α, β coordinates.

$\boldsymbol{v_n}$, V_n, ω_n, φ_n—instantaneous value, module, angular velocity and phase angle of supplying network voltage vector in α, β coordinates.

3.2.3. EV Module

The EV module consists of four parts: a drive module (DRIVE MODEL), a DC circuit (DC INSTALLATION), a harmonic filter (H.F. FILTER) and a central controller (EV CONTROLLER). They are described below.

DRIVE MODEL is based on the equation of rotary motion dynamics:

$$J\frac{dw}{dt} = T_e - T_m(w) \tag{3}$$

where:

J—moment of machine inertia brought to the machine shaft,
w—instantaneous value of the angular velocity of the machine rotor,
T_e, T_m—electromagnetic torque and mechanical torque in the function of w.

The EV engine is represented by the module of a synchronous machine with permanent magnets. The integral part imitates the car inertia brought to the machine shaft. The input of this block is the sum of torques: electromagnetic driving T_e depended on the current generated by the inverter and provided to the vehicle engine and mechanical T_m braking the car's movement. The output is the value of the machine's angular velocity w. It is assumed that the resistance to the movement and so the torque T_m are proportional to the angular velocity $\boldsymbol{w}$ of the car.

DC INSTALLATION represents EV battery and DC/AC motor inverter. The battery is represented by the DC source. The inverter is built as three-phase (6T) bridge of gate turn-off (GTO) thyristors. Firing pulses to thyristors are generated using pulse-width modulation (PWM) technique in the block of Inverter Current Controller. Hysteresis control is applied, in which the inverter AC current $\boldsymbol{i_{inv}}$ reproduces the reference current vector $\boldsymbol{i_{ref}}$ determined in the block of EV CONTROLLER.

In the EV module passive filter (H.F.FILTER) and the remaining R, L, C elements visible in the simulator diagram are used to reduce current harmonics of high frequencies resulting from thyristor switches operation.

Two EV operation modes are modelled in the simulator: Driving and Parking. They can be selected by the position of MODE switch in EV CONTROL panel. For the first mode the switch should be set to the DRIVE position. It results in opening BRKpark switch and closing BRKev switch. With BRKev closed the EV inverter is connected to the stator of the synchronous machine and the EV is ready to drive. For the parking mode the position PARK at the control panel should be selected—it causes BRKpark switch to close. As a result, the connection between the EV inverter and EVSE (user installation) is established.

3.3. EV Control during Driving and Parking

The operation of EV devices is managed by an EV CONTROLLER block. For each operation mode it determines the value of reference current vector $\boldsymbol{i}_{ref}$ applied to the inverter. In the DRIVING mode the vector module is the function of control signal INV CTRL setting (representing accelerator and brake pedals). Its angular speed is equal to the synchronous machine angular velocity w.

In this case the vector of stator current is practically equal to $\boldsymbol{i}_{ref}$ vector, i.e., it rotates with the same angular velocity as the permanent magnet magnetic flux of the rotor. This results in the electromagnetic torque T_e driving or braking the vehicle depending on sign of the INV CTRL signal.

In the PARKING mode, after closing the switch CNT, all control and power circuits of the car are connected to the EVSE (part of "EV USER INSTALLATION" module). The EV can perform a chosen function (charging or ancillary services). Two control options are possible depending on the state of the supply network:

- normal network conditions—EV can be charged or used as energy storage system (V2G);
- emergency network conditions—EV is used as UPS system.

In the first option the inverter transmits active power equal to the value set on the INV CTRL slider (in this case the INV CTRL setting can be determined by the Home Energy Management System). For positive signal value the current is in phase with the supplying voltage, which means that the EV batteries are charging. If the signal has a negative value, the current is in opposite phase to the voltage, which means energy is supplied to the consumer installation. EV CONTROLLER block determines reference current vector according to the formula:

$$\boldsymbol{i}_{ref} = \frac{P + jQ}{1.5(|\boldsymbol{v}_h|)^2} \boldsymbol{v}_h \tag{4}$$

where:

P, Q—active and reactive power transmitted by the inverter.

Equation (4) is based on the definition of instantaneous power in the machine stator. Its derivation is provided in Appendix A.

The second option (UPS) is activated if there is a fault in the supply network. It results in the BR switch opening. In this case the EV function is to supply the consumer loads with voltage close to nominal. The reference current values are determined to ensure active and reactive power balance within the consumer installation. Simultaneously the voltage in the home installation is compared with the supplying network voltage in block BRAKE CTRL. If the network fault is eliminated the switch BR can be reclosed. Then, the EV CONTROLLER module comes back to performing the function it did before the disturbance.

4. Simulation

In order to demonstrate the performance of the developed model, the following scenarios were simulated in sequence, in single calculation cycle, i.e.,:

- driving state, in which the EV is driving (accelerate and brake several times);
- parking state, in which the vehicle is connected to the EVSE and operates as energy storage system;
- parking state, in which the vehicle is connected to the EVSE, the battery energy storage operation is interrupted by faults in the supply network causing the EV operate as UPS system.

Time constant representing the inertia of the car was reduced in such a way that all states could be examined in one simulation and presented in one graph, Figure 3. The states of all switches included in the simulator diagram and the waveforms of quantities enabling the evaluation of the modelled devices work in transient states were recorded. Individual operation states are presented in Figures 4–8, respectively.

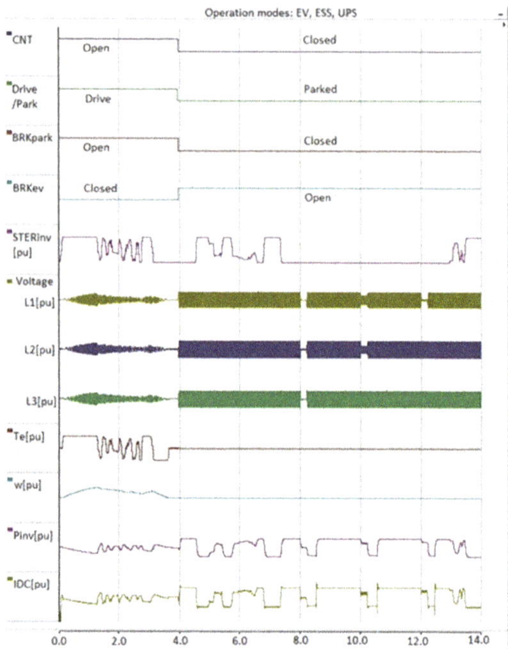

Figure 3. Main parameters describing processes during all EV operation states, breakers positions, voltages, currents, engine torque and speed.

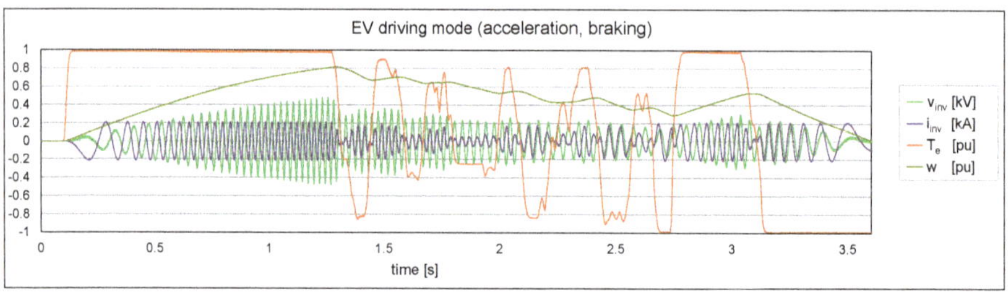

(**a**)

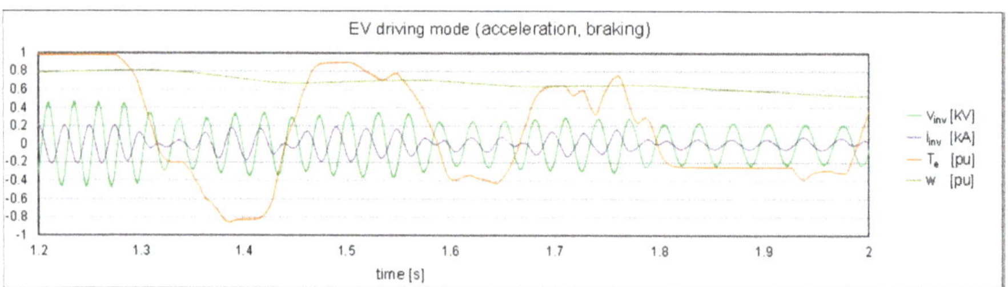

(**b**)

Figure 4. (**a**) EV driving state (accelerating and braking); (**b**) The zoom of the upper diagram, period (1.2–2) s.

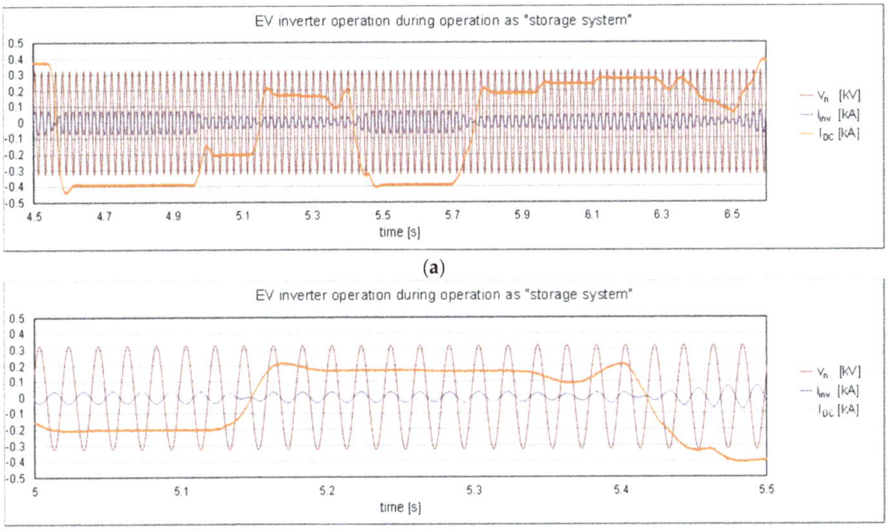

(a)

(b)

Figure 5. (**a**) EV operates as energy storage system in normal network state. (**b**) The zoom of the upper diagram, period (5–5.5) s.

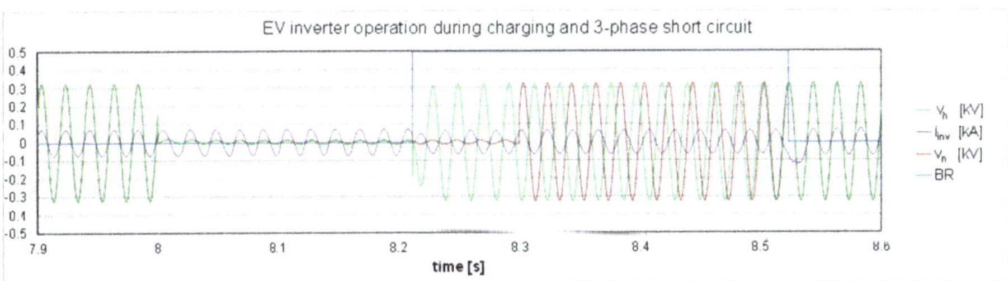

Figure 6. Inverter voltage and current and the network voltage during charging, 3-phase short circuit, island operation and again charging. The BR signal equeal "0" means that BR breaker is closed.

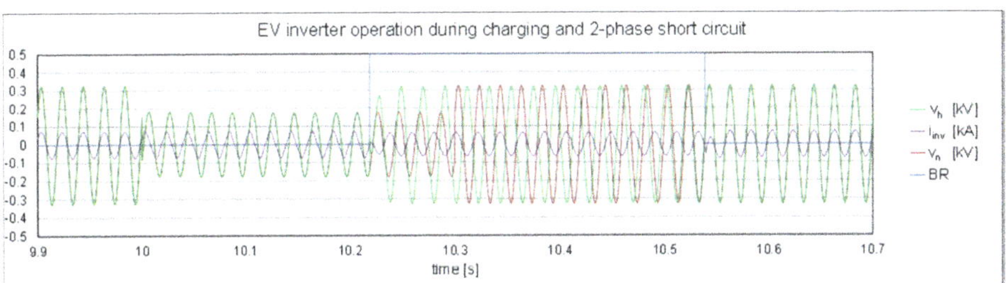

Figure 7. Inverter voltage and current and the network voltage during charging, 2-phase short circuit, island operation and again charging. The BR signal equal "0" means that BR breaker is closed.

In the first scenario (driving state) it was assumed that the vehicle accelerates and brakes to adapt its speed (w) to the road traffic conditions, see Figure 4. During the EV acceleration, electromagnetic torque (T_e) of the EV motor reaches its maximum value. When the road conditions prevent further acceleration (the driver releases the accelerator pedal), the electromagnetic torque reverses and the vehicle begins to decelerate. During the acceleration, when the invertor current (i_{inv}) and voltage (v_{inv})

are in opposite phases, the energy is transferred from batteries to the engine. Conversely, during regenerative braking, the engine operates as a generator, thus the inverter current and voltage are in phase, the energy is transferred from engine to the batteries.

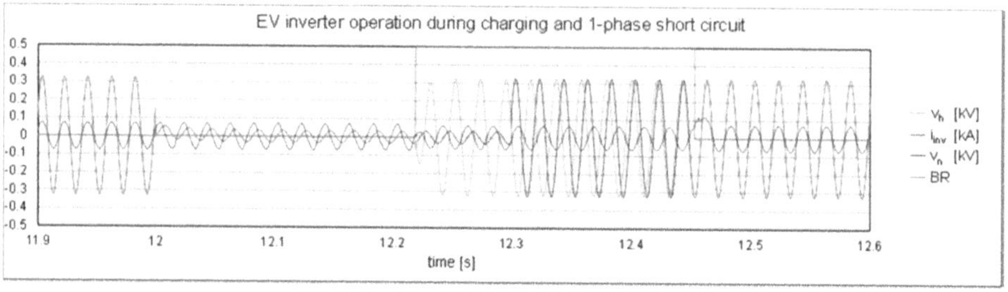

Figure 8. Inverter voltage and current and the network voltage during charging, 1-phase short circuit, island operation and again charging. The BR signal equal "0" means that BR breaker is closed.

During the next scenario the vehicle is parked and connected to the supply network, see Figure 5. The bi-directional power flow is enabled, thus the execution of V2H services can be applied. This service can be controlled by the home energy management system on the basis of factors such as energy prices, user requirements, SOC of EV battery or home installation rated power. The voltages of the traction invertor (v_{inv}), the voltage in the home installation (v_h) and the voltage in the supply network (v_n) are equal. The inverter current (i_{inv}) changes (amplitude, phase) so the power can be transferred from or to the supply network. In Figure 5 the phase of inverter current corresponds to the battery current (I_{DC}). When the on-board motor inverter operates as the battery charger (I_{DC} is positive) the voltage and invertor current are in phase. When energy is transferred to the customer installation (battery discharging, I_{DC} is negative) the phase between current and voltage is shifted by 180 degrees.

The next three drawings correspond to the third scenario, i.e.,: the 3-phase short circuit—Figure 6, the 2-phase short circuit—Figure 7 and the single-phase short circuit—Figure 8. In all the figures, the order of events is the same. The simulation starts with the EV battery being charged. The voltages of the motor invertor, home installation and supplying network are the same. The motor invertor current and voltage are in phase. Then, the fault in the supply network occurs, which results in v_{inv}, v_n and v_h voltages decreasing, respective to the type of short circuit. In the response to the network fault, the protection controller trips the BR breaker. The vehicle inverter changes control strategy from battery charger to UPS and the home installation starts the islanded operation. One can notice that the home installation voltage reaches nominal value immediately. Moreover, the amplitude of invertor current changes respective to the customers loads. During island operation the EV inverter operates autonomously. When the fault in the supply network is removed, the synchronization process can be started. When the voltages in the home installation and in the supply network are in phase the BR breaker is reclosed and the EV inverter returns to the battery charging operation that can be controlled by the HEMS again.

5. Conclusions

The potential advantages of the V2H services using the off-board inverters are well described in the literature. The novelty of this paper is the reference to the applicable use of on-board 3-phase motor inverter and its control system for the both EV charging and ancillary services—V2H. Such an approach is beneficial since additional infrastructure costs are minimal.

In this paper authors concentrated on creating the accurate test-bed in the PSCAD environment that can be utilized to:

- verify the EV capacity to provide V2H services;
- assess the developed control system for EV on-board motor inverter;

- evaluate the requirements that the home installation should meet to enable V2H services (including operation of protection devices and synchronization system);
- evaluate the operation of the EV motor inverter and protection devices during faults in the supply network.

The conducted tests confirmed that the EV on-board single structure (consisting of motor inverter, its control system, switches/breakers) is sufficient to provide V2H services in the home installation equipped with:

- breaker enabling disconnection of the home installation from the network in case of fault;
- a synchronization system that reconnects the home installation with the supply network.

Since there is no need for additional off-board inverter the cost of V2H services implementation can be minimized. In other words, the V2H services give the EV owner the opportunity to provide emergency power supply (with minimal additional infrastructure expenses) and potentially lower the costs of electrical energy (if a home energy management system is used).

It should be noted that the availability and effectiveness of V2H services (such as peak shaving or uninterruptable power supply) depend on energy stored in the EV batteries. Thus, it is strictly related to the vehicle usage profile (availability of EV battery).

As driving capabilities, the main functionality of EV, should not be limited by additional V2H services, the integration with the home energy management system seems to be advantageous. HEMS can determine the EV battery charging/discharging pattern including restrictions resulting from:

- the necessity to keep the vehicle in readiness for the user;
- energy tariffs;
- other technical aspects referred to the user home installation.

Although the presented simulation results confirm that the EV motor inverter technology is ready for the V2H services, the limited capacity of the EV battery and the impact of charging/discharging cycles might be an issue. On the other hand, intensive research is being carried out into the development of storage technologies, so this should not be a problem in the future.

Author Contributions: Conceptualization, I.W. and R.M. and P.K.; literature review K.W.; formal analysis, R.M., I.W., software, R.M.; investigation and validation R.M., P.K.; writing—original draft preparation, I.W.; writing, visualization and editing P.K.; review and supervision, I.W.; All authors have read and agreed to the published version of the manuscript.

Funding: This research received no external funding.

Conflicts of Interest: The authors declare no conflict of interest.

Appendix A

An instantaneous power p can be expressed as the function of instantaneous values of voltages and currents:

$$p = v_A i_A + v_B i_B + v_C i_C \tag{A1}$$

We apply the known transformations between $\alpha\beta$ and ABC phase coordinates:

$$\begin{cases} v_A = Re\{\boldsymbol{v}\} \\ v_B = -\frac{1}{2}Re\{\boldsymbol{v}\} + \frac{\sqrt{3}}{2}Im\{\boldsymbol{v}\} \\ v_C = -\frac{1}{2}Re\{\boldsymbol{v}\} - \frac{\sqrt{3}}{2}Im\{\boldsymbol{v}\} \\ i_A = Re\{\boldsymbol{i}\} \\ i_B = -\frac{1}{2}Re\{\boldsymbol{i}\} + \frac{\sqrt{3}}{2}Im\{\boldsymbol{i}\} \\ i_C = -\frac{1}{2}Re\{\boldsymbol{i}\} - \frac{\sqrt{3}}{2}Im\{\boldsymbol{i}\} \end{cases} \tag{A2}$$

Applying (A2) to (A1)

$$\begin{aligned} p = {} & Re\,\{\boldsymbol{v}\} \times Re\,\{\boldsymbol{i}\} + \left(-\tfrac{1}{2}Re\,\{\boldsymbol{v}\} + \tfrac{\sqrt{3}}{2}Im\{\boldsymbol{v}\}\right) \\ & \times\left(-\tfrac{1}{2}Re\,\{\boldsymbol{i}\} + \tfrac{\sqrt{3}}{2}Im\{\boldsymbol{i}\}\right) \\ & +\left(-\tfrac{1}{2}Re\,\{\boldsymbol{v}\} - \tfrac{\sqrt{3}}{2}Im\{\boldsymbol{v}\}\right) \\ & \times\left(-\tfrac{1}{2}Re\,\{\boldsymbol{i}\} - \tfrac{\sqrt{3}}{2}Im\{\boldsymbol{i}\}\right) \end{aligned} \quad \text{(A3)}$$

After the simple transformations the instantaneous power equation (as the function of instantaneous voltage and current vectors) is:

$$\mathrm{p} = \frac{3}{2}(Re\,\{\boldsymbol{v}\} \times Re\,\{\boldsymbol{i}\} + Im\{\boldsymbol{v}\} \times Im\{\boldsymbol{i}\}) \quad \text{(A4)}$$

Let us consider now the expression $\frac{3}{2}\boldsymbol{v}^* \times \boldsymbol{i}$ (where $\boldsymbol{v}^*$ stands for conjugate value of complex number $\boldsymbol{v}$):

$$\frac{3}{2}\boldsymbol{v}^* \times \boldsymbol{i} = \frac{3}{2}[(Re\,\{\boldsymbol{v}\} - jIm\{\boldsymbol{v}\}) \times (Re\,\{\boldsymbol{i}\} + jIm\{\boldsymbol{i}\})] \quad \text{(A5)}$$

$$\begin{aligned} \tfrac{3}{2}\boldsymbol{v}^* \times \boldsymbol{i} = {} & \tfrac{3}{2}[Re\,\{\boldsymbol{v}\} \times Re\{\boldsymbol{i}\} + Im\{\boldsymbol{v}\} \times Im\{\boldsymbol{i}\} \\ & + j(Re\{\boldsymbol{v}\} \times Im\{\boldsymbol{i}\} - Im\{\boldsymbol{v}\} \times Re\{\boldsymbol{i}\}] \end{aligned} \quad \text{(A6)}$$

Thus, taking into account (A4) we obtain:

$$\frac{3}{2}\boldsymbol{v}^* \times \boldsymbol{i} = p + jq \quad \text{(A7)}$$

where active power p is expressed by formula (4) and the reactive component is equal to:

$$\mathrm{q} = \frac{3}{2}(Re\,\{\boldsymbol{v}\} \times Im\,\{\boldsymbol{i}\} - Im\{\boldsymbol{v}\} \times Re\{\boldsymbol{i}\}) \quad \text{(A8)}$$

From (A7) the derived equation is:

$$\boldsymbol{i} = \frac{P + jQ}{1.5\boldsymbol{v}^*} \times \frac{\boldsymbol{v}}{\boldsymbol{v}} = \frac{P + jQ}{1.5(|\boldsymbol{v}|)^2} \times \boldsymbol{v} \quad \text{(A9)}$$

References

1. Roadmap to a Single European Transport Area—Towards a competitive and resource efficient transport system—European Environment Agency. 2011. Available online: https://www.eea.europa.eu/policy-documents/roadmap-to-a-single-european (accessed on 6 March 2020).
2. Emadi, A.; Lee, Y.J.; Rajashekara, K. Power Electronics and Motor Drives in Electric, Hybrid Electric, and Plug-In Hybrid Electric Vehicles. *IEEE Trans. Ind. Electron.* **2008**, *55*, 2237–2245. [CrossRef]
3. Chan, C.C.; Bouscayrol, A.; Chen, K. Electric, hybrid, and fuel-cell vehicles: Architectures and modeling. *IEEE Trans. Veh. Technol.* **2010**, *59*, 589–598. [CrossRef]
4. Rajashekara, K. Present status and future trends in electric vehicle propulsion technologies. *IEEE J. Emerg. Sel. Top. Power Electron.* **2013**, *1*, 3–10. [CrossRef]
5. Xu, L.; Marshall, M.; Dow, L. A framework for assessing the impact of plug-in electric vehicle to distribution systems. In Proceedings of the 2011 IEEE/PES Power Systems Conference and Exposition, Phoenix, AZ, USA, 20–23 March 2011.
6. Darabi, Z.; Ferdowsi, M. Aggregated impact of plug-in hybrid electric vehicles on electricity demand profile. *IEEE Trans. Sustain. Energy* **2011**, *2*, 501–508. [CrossRef]
7. Ashtari, A.; Bibeau, E.; Shahidinejad, S.; Molinski, T. PEV charging profile prediction and analysis based on vehicle usage data. *IEEE Trans. Smart Grid* **2012**, *3*, 341–350. [CrossRef]

8. Lopes, J.A.P.; Soares, F.J.; Almeida, P.M.R. Integration of electric vehicles in the electric power system. *Proc. IEEE* **2011**, *99*, 168–183. [CrossRef]
9. De Hoog, J.; Alpcan, T.; Brazil, M.; Thomas, D.A.; Mareels, I. Optimal charging of electric vehicles taking distribution network constraints into account. *IEEE Trans. Power Syst.* **2015**, *30*, 365–375. [CrossRef]
10. Galus, M.D.; Simon Art, G.A. A hierarchical, distributed PEV charging control in low voltage distribution grids to ensure network security. In Proceedings of the 2012 IEEE Power and Energy Society General Meeting, San Diego, CA, USA, 22–26 July 2012; pp. 1–8.
11. Andreotti, A.; Carpinelli, G.; Mottola, F.; Proto, D. A review of single-objective optimization models for plug-in vehicles operation in smart grids part I: Theoretical aspects. In Proceedings of the 2012 IEEE Power and Energy Society General Meeting, San Diego, CA, USA, 22–26 July 2012; pp. 1–8.
12. Timpner, J.; Wolf, L. Design and evaluation of charging station scheduling strategies for electric vehicles. *IEEE Trans. Intell. Transp. Syst.* **2014**, *15*, 579–588. [CrossRef]
13. Moghaddam, Z.; Ahmad, I.; Habibi, D.; Phung, Q.V. Smart Charging Strategy for Electric Vehicle Charging Stations. *IEEE Trans. Transp. Electrif.* **2017**, *4*, 76–88. [CrossRef]
14. Robinson, J.; Brase, G.; Griswold, W.; Jackson, C.; Erickson, L. Business Models for Solar Powered Charging Stations to Develop Infrastructure for Electric Vehicles. *Sustainability* **2014**, *6*, 7358–7387. [CrossRef]
15. Colmenar-Santos, A.; de Palacio, C.; Borge-Diez, D.; Monzón-Alejandro, O. Planning Minimum Interurban Fast Charging Infrastructure for Electric Vehicles: Methodology and Application to Spain. *Energies* **2014**, *7*, 1207–1229. [CrossRef]
16. Pang, C.; Kezunovic, M.; Ehsani, M. Demand side management by using electric vehicles as Distributed Energy Resources. In Proceedings of the 2012 IEEE International Electric Vehicle Conference, Greenville, SC, USA, 4–8 March 2012; pp. 1–7.
17. Xiang, Y.; Liu, J.; Yang, W.; Huang, C. Active energy management strategies for active distribution system. *J. Mod. Power Syst. Clean Energy* **2015**, *3*, 533–543. [CrossRef]
18. Tuttle, D.P.; Fares, R.L.; Baldick, R.; Webber, M.E. Plug-In Vehicle to Home (V2H) duration and power output capability. In Proceedings of the 2013 IEEE Transportation Electrification Conference and Expo (ITEC), Detroit, MI, USA, 16–19 June 2013; pp. 1–7.
19. Monteiro, V.; Exposto, B.; Ferreira, J.C.; Afonso, J.L. Improved Vehicle-to-Home (iV2H) Operation Mode: Experimental Analysis of the Electric Vehicle as Off-Line UPS. *IEEE Trans. Smart Grid* **2017**, *8*, 2702–2711. [CrossRef]
20. Shin, H.; Baldick, R. Plug-In Electric Vehicle to Home (V2H) Operation under a Grid Outage. *IEEE Trans. Smart Grid* **2017**, *8*, 2032–2041. [CrossRef]
21. Yoshimura, Y.; Kondo, T.; Kawanishi, M.; Narikiyo, T.; Sato, A. Model Predictive Control of EV Storage Battery with HEMS based on Particle Swarm Optimization Yuto. In Proceedings of the 2015 IEEE Innovative Smart Grid Technologies-Asia (ISGT ASIA), Bangkok, Thailand, 3–6 November 2015; pp. 1–5.
22. Wu, X.; Hu, X.; Yin, X.; Moura, S. Stochastic Optimal Energy Management of Smart Home with PEV Energy Storage. *IEEE Trans. Smart Grid* **2018**, *9*, 2065–2075. [CrossRef]
23. Brenna, M.; Foiadelli, F.; Longo, M. The exploitation of vehicle-to-grid function for power quality improvement in a smart grid. *IEEE Trans. Intell. Transp. Syst.* **2014**, *15*, 2169–2177. [CrossRef]
24. Saber, A.Y.; Venayagamoorthy, G.K. Plug-in vehicles and renewable energy sources for cost and emission reductions. *IEEE Trans. Ind. Electron.* **2011**, *58*, 1229–1238. [CrossRef]
25. Molina, D.; Hubbard, C.; Lu, C.; Turner, R.; Harley, R. Optimal EV charge-discharge schedule in smart residential buildings. In Proceedings of the IEEE Power and Energy Society Conference and Exposition in Africa: Intelligent Grid Integration of Renewable Energy Resources (PowerAfrica), Johannesburg, South Africa, 9–13 July 2012; pp. 9–13.
26. De Santiago, J.; Bernhoff, H.; Ekergård, B.; Eriksson, S.; Ferhatovic, S.; Waters, R.; Leijon, M. Electrical motor drivelines in commercial all-electric vehicles: A review. *IEEE Trans. Veh. Technol.* **2012**, *61*, 475–484. [CrossRef]
27. Subotic, I.; Bodo, N.; Levi, E.; Jones, M. Onboard Integrated Battery Charger for EVs Using an Asymmetrical Nine-Phase Machine. *IEEE Trans. Ind. Electron.* **2015**, *62*, 3285–3295. [CrossRef]
28. Shi, C.; Tang, Y.; Khaligh, A. A single-phase integrated onboard battery charger using propulsion system for plug-in electric vehicles. *IEEE Trans. Veh. Technol.* **2017**, *66*, 10899–10910. [CrossRef]

29. Standard. IEC 61851-1:2017 Electric vehicle conductive charging system—Part 1: General requirements. International Electrotechnical Commission. 2017. Available online: https://webstore.iec.ch/publication/33644 (accessed on 7 January 2020).
30. Wasiak, I.; Pawelek, R.; Mienski, R. Energy storage application in low-voltage microgrids for energy management and power quality improvement. *IET Gener. Transm. Distrib.* **2014**, *8*, 463–472. [CrossRef]
31. Et-Tolba, E.H.; Maaroufi, M.; Ouassaid, M. Demand side management algorithms and modeling in smart grids a customer's behavior based study. In Proceedings of the 2013 International Renewable and Sustainable Energy Conference (IRSEC), Ouarzazate, Morocco, 7–9 March 2013; pp. 531–536.
32. Wang, J.; Sun, Z.; Zhou, Y.; Dai, J. Optimal dispatching model of smart home energy management system. In Proceedings of the IEEE PES Innovative Smart Grid Technologies, Tianjin, China, 21–24 May 2012; pp. 1–5.
33. Patel, K.; Khosla, A. Home energy management systems in future smart grid networks: A systematic review. In Proceedings of the 2015 1st International Conference on Next Generation Computing Technologies (NGCT), Dehradun, India, 4–5 September 2015; pp. 479–483.
34. Manitoba-HVDC Research Center. User's Guide on the Use of PSCAD v4; *Manitoba-HVDC Research Center*: 2010. Available online: https://hvdc.ca/uploads/ck/files/reference_material/PSCAD_User_Guide_v4_3_1.pdf (accessed on 7 January 2020).

© 2020 by the authors. Licensee MDPI, Basel, Switzerland. This article is an open access article distributed under the terms and conditions of the Creative Commons Attribution (CC BY) license (http://creativecommons.org/licenses/by/4.0/).

energies

Article

Adaptive Smoothing Power Following Control Strategy Based on an Optimal Efficiency Map for a Hybrid Electric Tracked Vehicle

Baodi Zhang [1], Sheng Guo [1], Xin Zhang [1,*], Qicheng Xue [1] and Lan Teng [2]

1 School of Mechanical, Electronic and Control Engineering, Beijing Jiaotong University, Beijing 100044, China; bdzhang@bjtu.edu.cn (B.Z.); shguo@bjtu.edu.cn (S.G.); 17116356@bjtu.edu.cn (Q.X.)
2 School of Economics and Management, Beijing Jiaotong University, Beijing 100044, China; tenglan@bjtu.edu.cn
* Correspondence: zhangxin@bjtu.edu.cn; Tel.: +86-10-51688404

Received: 18 March 2020; Accepted: 9 April 2020; Published: 13 April 2020

Abstract: The series hybrid electric powertrain is the main architecture of the hybrid electric tracked vehicle. For a series tracked hybrid electric bulldozer (HEB), frequent fluctuations of the engine working points, deviation of the genset working points from the pre-set target trajectory due to an insufficient response, or interference of the hydraulic pump consumed torque, will all result in increased fuel consumption. To solve the three problems of fuel economy, an adaptive smooth power following (ASPF) control strategy based on an optimal efficiency map is proposed. The strategy combines a fuzzy adaptive filter algorithm with a genset's optimal efficiency, which can adaptively smooth the working points of the genset and search the trajectory for the genset's best efficiency when the hydraulic pump torque is involved. In this study, the proposed strategy was compared on the established HEB hardware in loop (HIL) platform with two other strategies: a power following strategy in a preliminarily practical application (PF1) and a typical power following strategy based on the engine minimum fuel consumption curve (PF2). The results of the comparison show that (1) the proposed approach can significantly reduce the fluctuation and pre-set trajectory deviation of the engine and generator working points; (2) the ASPF strategy achieves a 7.8% improvement in the equivalent fuel saving ratio (EFSR) over the PF1 strategy, and a 3.4% better ratio than the PF2 strategy; and (3) the ASPF strategy can be implemented online with a practical controller.

Keywords: hybrid electric bulldozer; tracked vehicle; control strategy; adaptive control; power smoothing

1. Introduction

In recent years, hybrid electric vehicle (HEV) technology has been rapidly developed. As a new type of power transmission technology, the HEV is recognized as one of the best solutions for energy saving and emission reduction in the world [1]. Following the success of HEV technology in the automotive field [2–4], the world's major manufacturers and related research institutions tried to apply it in the field of construction machinery. In the research and development of hybrid electric construction machinery (HECM), hybrid technology of the loader and excavator has been previously achieved [5,6]. It was not until 2008 that the Caterpillar company developed a diesel-electric driven bulldozer based on HEV technology, named D7E, which attracted extensive attention worldwide due to its excellent performance of 20% energy saving, 10% productivity improvement, and 50% service life extension [7]. After that, a special project was set up to develop the first hybrid electric bulldozer (HEB) of China in 2011 [8]. Since then, research on the HEB, as a heavy-duty off-road hybrid electric tracked vehicle, has received increasing attention [9–11].

Control strategies, especially energy management strategies, represent a hot topic in HEV technology research. The core purpose of a control strategy is to optimize the fuel economy and emission of the power system, while satisfying the driving purpose, by reasonably controlling the power distribution among multiple energy sources. The technology of the HECM or hybrid special vehicle should be based on the application, transfer, and expansion of HEV technology. Such control strategies are also research hotspots in their fields. Even with the development and introduction of autonomous vehicles, the vehicle control strategy is the most attractive technical research field [12–14]. HEBs mainly adopt the series hybrid electric powertrain due to their violently fluctuating operating resistance. As the amount of research on the control strategy of series HEB is still small, in order to fully draw lessons from related research results, we have reviewed the control strategy of series HEV, the control strategy of other types of HECMs which also face the problem of HEV technology transfer, and some of the recent research on the HEB control strategy.

Research on the control strategy of the series hybrid power system can be simply divided into three types, based on the characteristics: rule-based control [15], optimization-based control [16,17], and intelligent control [18,19]. As shown in Table 1, it can be seen that the three types of control strategies have advantages and challenges in practical engineering applications. The rule-based control strategy is the most commonly used strategy of the series HEV in practical applications. The on-off (thermostat) strategy is the simplest, and permits the genset to always work at the best efficiency point, but the engine is often switched on and off, and the energy loss in start-stop and charge-discharge is large. The power following strategy requires frequent changes in the engine operating points to follow the changing load. However, the engine efficiency is relatively low, the dynamic energy loss is large, and the charge and discharge energy loss of the energy storage unit (ESU) is small [20]. The on-off and power following strategy also are called deterministic rules. The optimization-based control makes use of various optimization algorithms to solve objective functions and obtain the optimal control law. It can be divided to instantaneous and global optimization, according to the time scale of the optimal solution. Two typical instantaneous optimization strategies are the equivalent consumption minimization strategy (ECMS) and its derivative adaptive-ECMS, which calculate the minimum objective function at every moment [16]. Dynamic programming (DP) is a classic and accepted global optimization algorithm used to measure the maximum potential of fuel saving in the whole drive cycle time. Stochastic dynamic programming (SDP) is then formed by considering some uncertainty factors in the load. Between the moment and duration of the optimal time scale, the model predictive control (MPC) converts the solution problem over the entire drive cycle time into a value over a short future period by using sensing devices and methods [21]. An indirect and analytical algorithm employed to solve the global optimal control problem is Pontryagin's minimum principle (PMP), which is also widely adopted and offers an optimal solution close to the DP result by solving a Hamiltonian minimization problem [22]. These optimization-based controls are often segmented into online and offline optimization-based strategies, depending on whether the controls can be applied in real vehicles [1,20]. Fuzzy logic is also mentioned as a kind of rule-based control for fuzzy rules [23], while it is categorized as intelligent control due to its intelligence features of non-model and nonlinear systems in this paper. Intelligent control approaches, including the back propagation neural network (BPNN), genetic algorithm (GA), and machine learning [4,24], which rely on engineering experience and engineering databases, have excellent properties in dealing with uncertain mathematical models, high nonlinearity, and complex task requirements [25,26]. They are also widely adopted in series HEVs and have a good adaptability.

Table 1. Classification and comparison of control strategies of the series hybrid electric powertrain.

Category	Subcategory	Advantages	Challenges
Rule-Based Control	On-off Power following	Simple and reliable, high real-time, widely used in engineering	Poor portability, poor adaptability, low energy optimal efficiency
Optimization-Based Control	Instantaneous optimization Global optimization	Accurate optimization objectives and solutions, instantaneous or global optimal energy distribution, good fuel economy	Large computation and high requirement on processor, poor realizability of global optimization
Intelligent Control	Fuzzy logic Neural network Expert system Machine learning	Unnecessary accurate mathematical model, robust and adaptive, suitable for real-time control	Depend on the engineering experience or database, knowledge base highly required

The control strategies of different types of HECMs have also been studied in terms of the above three aspects. The rule-based control strategies of HECMs have been researched the most. Unlike HEVs, the control strategies of HECMs mainly require the characteristics of operating conditions to be combined. In research on the control strategy of a hybrid excavator, the pressure of the hydraulic pump was often used to identify the working load, and the control rules were made according to the load level [27]. The working process of the excavator and the moving process of each part were analyzed in detail, and the pressure of the operating system was measured in real time to obtain an estimate of the required and recoverable energy [28]. The key point of energy saving control for excavators lies in the design of energy recovery control for moving parts, such as boom and swing systems [29]. Research on the hybrid loader control strategy has also paid attention to the discrimination of operating conditions and to making rules according to the characteristics of the load. The power consumed by a loader's hydraulic system and the impact on the powertrain's dynamic response have often been taken into account to design the strategy [30]. In most studies, the load of the operating system was incorporated into the energy management of the hybrid loader by measuring the outlet pressure of its hydraulic pump [31]. The characteristics of the high transiency and periodicity of construction machinery are both common and specific in control strategy design. The complexity of the working environment of construction machinery, as well as the significant and frequent changes in operating loads, bring new difficulties to energy management and control [6,32]. Research on the HECM control strategy based on an optimization algorithm has been increasing in recent studies. For instance, Nilsson et al. proposed a predictive control approach using SDP under severe disturbances and uncertainties, according to the repetitive pattern of operation of the wheel loaders [33]. SDP control based on prediction also achieved an energy consumption optimization effect on the hybrid excavator [34]. A comparative assessment of ECMS, DP, and thermostat controllers [35]; joint control and parameter optimization by adopting DP control and GA [36]; and a comparison of DP, PMP, and MPC [37], for a hybrid excavator have been conducted. Since intelligent control approaches have a good robustness for nonlinear systems, and HECMs have complex operating conditions and uncertain model parameters, intelligent controls have also been widely studied in this field. Intelligent algorithms such as reinforcement learning [38,39], fuzzy logic [40,41], and the particle swarm optimization (PSO) algorithm [42] have been applied to the energy saving control of HECMs. Therefore, research on other types of HECM control strategies developed according to the characteristics of their operating conditions can provide enlightenment for us to control the HEB: the characteristics of HECMs, such as the periodicity of work, short idle time, and high fluctuation [43], especially the power supply and recovery of the hydraulic actuators, should be taken into account in the control.

At present, only a few institutions have actually developed HEBs, but research articles on control and simulation of the HEB have been increasing in recent years [7,8,44]. In [45,46], the fuel-saving control concept of D7E was briefly introduced; that is, engine control in the series system was not affected by the load torque so that the speed of the generator could be controlled within a narrow range to improve the efficiency, but the detailed control strategy was not given. Song et al. proposed a power following control strategy based on the minimum fuel consumption curve of the engine for an HEB [9]. Wang et al. also proposed a load power following control strategy and adopted three engine speed control methods for a comparative study [11]. However, these studies did not take the efficiency of the generator into account when calculating the optimal efficiency. Wang et al. proposed applying MPC to an HEB, compared the MPC to DP and a rule-based control strategy, and artificially added white noise interference to test the robustness of MPC, which indicated that the energy consumption and robustness of the HEB under the MPC strategy are better than those of the rule-based strategy [10,47–49]. Wang et al. then proposed an improved MPC strategy for an HEB without future driving information [50], and applied the MPC to the analysis of the HEB's energy storage unit [51].

Although various types of algorithms have been applied to the control of the HEB in the existing literature, these studies have basically treated the HEB as a tracked vehicle similar to [52,53], which did not pay enough attention to the difference between HEVs and HEBs. Specifically, in most of these HEB control strategies, the influence of the torque consumed by the hydraulic pump of the working system on the genset operating point has not been considered, even though instantaneous disturbance of the hydraulic pump would increase the fuel consumption and emission [45]. In addition, there has been little discussion about the fact that the transient fuel consumption caused by frequent changes of the engine operating points is higher, in which case the load power of the bulldozer fluctuates violently. Additionally, very little attention has been paid to the fact that the frequent adjustment of the HEB's engine operating points causes an insufficient instantaneous response of the engine and then leads to the points deviating from the optimal target trajectory, before finally resulting in an engine efficiency decline. Therefore, it is necessary to carry out in-depth research on these problems for developing the control strategy of HEBs.

Therefore, the purpose of this paper is to propose an innovative Adaptive Smoothing Power Following (ASPF) control strategy to solve the above mentioned problems. In general, the problems include frequent fluctuations of the engine working points, deviation of the genset working points from the pre-set target trajectory due to an insufficient response, and interference of the hydraulic pump consumed torque, all of which result in the fuel consumption increasing. The control strategy takes the impact of the drastic fluctuation in the bulldozer's load and the abrupt demand torque of the hydraulic pump on the working points of the HEB's genset into consideration. The adaptive smoothing algorithm is used to automatically reduce the transient fuel consumption and the working points' deviation from the high efficiency zone. In addition, the algorithm is combined with an optimal efficiency map of the genset considering the correction for interference of the operating system demand torque. A novel transient fuel consumption model embedded in the HEB model and a hardware in loop (HIL) platform is developed to test the proposed approach. The methodology of the proposed control strategy is a combination of fuzzy control theory, real-based control, and the optimization method. The verification method is based on adopting a validated simulation model and an HIL test platform. The test indicates that the proposed approach could solve the above problems and feature an excellent on-line real-time robustness and adaptability for energy saving of the HEB.

The rest of this article is organized as follows: the HEB model with a novel transient fuel consumption model based on BPNN is developed in Section 2; in Section 3, the ASPF based on an optimal efficiency map strategy is proposed and described; the HEB HIL platform is built in Section 4; Section 5 presents the simulation results and a discussion of the proposed approach; and the last section summarizes the major conclusions.

2. HEB Simulation Model

A simulation model is the basis for recognizing the characteristics of each part of the HEB and analyzing the control strategy. The control strategy performance can be simulated and analyzed after being incorporated into the simulation model. Additionally, the simulation model is essential and primary for the HIL platform test. Therefore, the simulation model of HEB should first be established. The modeling process is based on an analysis of the HEB architecture and component characteristics.

The detailed modeling and validation process information can be found in our previous study [54]. The theoretical foundation of the model is derived from the bulldozer's dynamics and kinetics, the component characteristics, and the modeling method of Advanced Vehicle Simulator (ADVISOR) software (Version 2002, National Renewable Energy Laboratory, Golden, CO, USA) [55].

2.1. HEB Architecture

The object of this research is derived from the first Chinese HEB—a tracked vehicle that has a series hybrid powertrain and a hydraulic operating system. Its architecture schematic diagram can be seen in Figure 1. This powertrain adopts double wheel-side motors to drive tracks on both sides, enabling the HEB to achieve center-based steering. The drive system employs a typical alternating current(AC)-direct current (DC)-AC transmission form to convert mechanical energy into electrical energy and then mechanical energy. The blade operating system is driven by the hydraulic pump, whose power comes from the engine and flows through the transfer case. Table 2 gives the main specifications of the HEB.

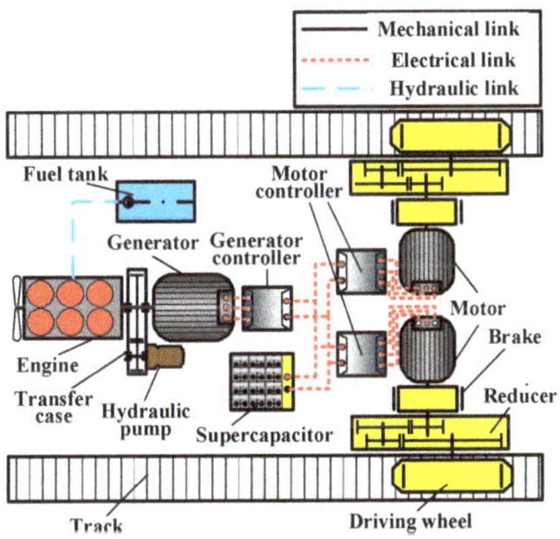

Figure 1. Schematic diagram of the hybrid electric bulldozer (HEB) architecture.

Table 2. Specifications of the HEB.

Component	Parameters	Quantity
Vehicle	Curb weight	27,500 kg
	Dynamic radius	0.468 m
	Track center distance	2.000 m
	Track length	2.730 m
Engine	Rated power	154 kW
	Rated speed	1950 rpm
	Maximum torque	927 Nm
	Maximum power	200 kW
Generator	Rated power	175 kW
	Maximum torque	1123 Nm
	Maximum speed	2200 rpm
Motor	Maximum power	75 kW
	Rated power	105 kW
	Maximum torque	800 Nm
	Maximum speed	6000 rpm
Supercapacitor	Static capacity	5 F
	Energy capacity	0.34 kWh
Hydraulic Pump	Maximum pressure	21 MPa
	Rated flow	200 L/min
Reducer	Speed ratio	93.120

2.2. A Novel Engine Model Based on BPNN

The majority of engine models adopted in HEV simulation research include a kind of simple experimental look-up-table [9,10,48–53], which is acquired from an engine steady-state universal characteristic experiment. This approach can only reflect the static fuel economy, rather than the transient fuel consumption. Therefore, some researchers noticed the difference between steady and transient fuel consumption, began researching transient fuel properties, and attempted to build transient fuel consumption models [56–59]. In [60], it was found that the transient fuel consumption would increase with the increase of the loading rate, as shown in Figure 2.

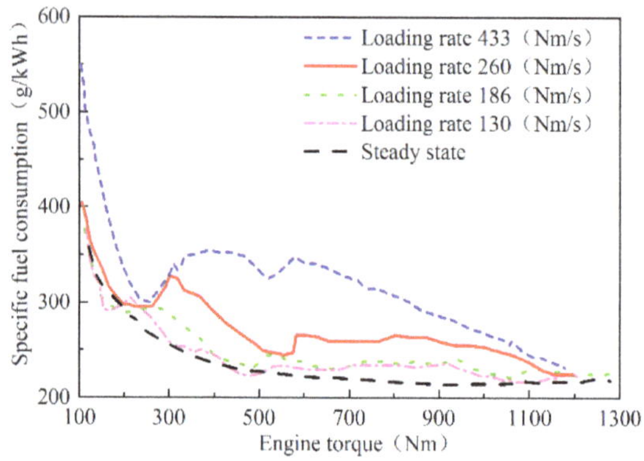

Figure 2. Comparison of the specific fuel consumption of the engine in transient and steady-state conditions. Reproduced from [60], China national knowledge infrastructure, 2015.

Therefore, the increased fuel consumption in a transient state could be more inevitably obvious in the bulldozer's severely fluctuant load. To provide the fuel consumption in this situation with a greater accuracy, a novel engine model with a transient fuel consumption sub-model based on BPNN is proposed in this paper. The modeling approach of key sub-models of the engine model is briefly described in the following.

2.2.1. Transient Fuel Consumption Model

The transient fuel consumption model consists of a steady fuel consumption and a transient correction factor. The steady fuel consumption is based on an engine universal characteristic map and calculated by two-dimensional interpolation, which can be shown as

$$B_s = T_e n_e be(T_e, n_e)/(9549 \cdot \rho_{fu}), \tag{1}$$

where B_s is the steady fuel consumption, T_e is the engine torque, n_e is the engine speed, be is the specific fuel consumption on the basis of the look-up table, and ρ_{fu} is the fuel density.

The BPNN was selected to identify a transient fuel consumption correction factor due to its excellent learning ability. The engine torque T_e, engine speed n_e, and their calculated differentials and product were employed as the input layer of the BPNN, considering the engine transient process, including speed and torque variation. The output layer is the correction factor employed to correct the steady fuel consumption model. The BPNN was first trained by a cluster of data collected in the actual engine operation. Figure 3 shows the structure of the transient fuel consumption model based on BPNN.

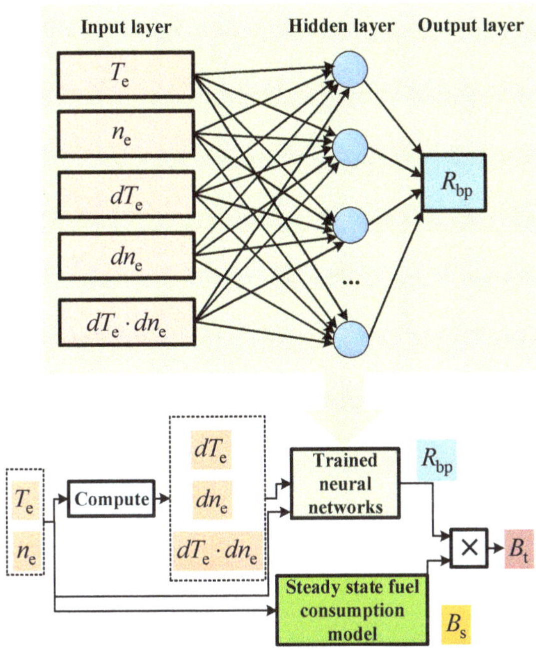

Figure 3. Structure of the transient fuel consumption model based on the back propagation neural network (BPNN).

The transient fuel consumption can be expressed as

$$B_t = B_s(n_e, T_e) \times R_{bp}(n_e, T_e, dn_e, dT_e), \tag{2}$$

where B_t is the transient fuel consumption; R_{bp} is the correction factor based on a trained BPNN calculation; and dn_e and dT_e are the differential of engine speed and torque, respectively.

This BPNN employed the trainlm function as the train function, log-sigmoid as the hidden layer transfer function, and the purelin function as the output layer transfer function. This trained model finally achieved a good simulation performance compared to the experimental data. More details on the model can be found in our previous research [61].

2.2.2. Output Torque Model

The steady output torque can be calculated by interpolation of an engine torque-speed-throttle characteristic map as

$$T_{e,s} = f_{Tn,e}(n_e, \gamma_e), \tag{3}$$

where $T_{e,s}$ is the engine steady torque, $f_{Tn,e}$ is the interpolation function of the torque-speed-throttle characteristic map, and γ_e is the engine throttle.

Considering the transient operation process, a delay of the throttle voltage, injection pressure, and air inflow would lead to a delay of the output torque. For describing the transient process, the dynamic output torque can be expressed as

$$T_e = T_{e,s} \cdot \frac{1}{1 + \tau_e s}, \tag{4}$$

where T_e is the engine dynamic output torque, τ_e is the engine time constant, and s is the Laplace operator.

2.3. Driveline Components' Models

2.3.1. Electric Machine Model

The electric machine model is applicable for modeling a motor or generator. The model uses a first-order inertia link to represent the response delay between the output and the target torque. The dynamic output is also limited by the voltage range, maximum current, and external characteristic torque, which can be denoted by

$$T_{\text{em}} = \begin{cases} \min(T_{\text{em,ul}}, T_{\text{em,cl}}, T_{\text{em,max}}) & \text{motor mode} \\ \max(T_{\text{em,ul}}, T_{\text{em,cl}}, T_{\text{em,max}}) & \text{generator mode} \end{cases}, \tag{5}$$

$$T_{\text{em,ul}} = \begin{cases} T_{\text{em,cmd}} \cdot \frac{1}{1+\tau_{\text{em}}s} & U_{\text{em,min}} \leq U_{\text{em}} < U_{\text{em,max}} \\ 0 & U_{\text{em}} < U_{\text{em,min}} \text{ or } U_{\text{em}} \geq U_{\text{em,max}} \end{cases}, \tag{6}$$

$$T_{\text{em,cl}} = \begin{cases} U_{\text{em}} I_{\text{em,max}} / \omega_{\text{em}} & \text{motor mode} \\ -U_{\text{em}} I_{\text{em,max}} / \omega_{\text{em}} & \text{generator mode} \end{cases}, \tag{7}$$

$$T_{\text{em,max}} = \begin{cases} T^{\text{dr}}_{\text{em,max}}(\omega_{\text{em}}) \cdot \frac{1}{1+\tau_{\text{em}}s} & \text{motor mode} \\ T^{\text{br}}_{\text{em,max}}(\omega_{\text{em}}) \cdot \frac{1}{1+\tau_{\text{em}}s} & \text{generator mode} \end{cases}, \tag{8}$$

where T_{em} is the output torque of the electric machine; $T_{\text{em,ul}}$, $T_{\text{em,cl}}$, and $T_{\text{em,max}}$ are the limited torque of the voltage, current, and external characteristic, respectively; $T_{\text{em,cmd}}$ is the target torque command; τ_{em} is the torque respond time constant; U_{em}, $U_{\text{em,min}}$, and $U_{\text{em,max}}$ are the current, minimum, and maximum voltage of the electric machine, respectively; $I_{\text{em,max}}$ is the maximum current of the electric machine; ω_{em} is the angular speed of the machine; and $T^{\text{dr}}_{\text{em,max}}$ and $T^{\text{br}}_{\text{em,max}}$ are the maximum permissible limit in the driving or braking process, respectively.

The torque balance equation of the electric machine links the relationship of the torque and the angular speed as follows:

$$T_{\text{em}} - T_{\text{em,l}} = J_{\text{em}} \frac{d\omega_{\text{em}}}{dt}, \tag{9}$$

where $T_{\text{em,l}}$ is the load torque and J_{em} is the rotational inertia of the electric machine.

The efficiency module is the main part of the electric machine model. The efficiency is calculated by a look-up table obtained from an electric machine characteristic experiment, which can be described as

$$\eta_{\text{em}} = f_{\eta,\text{em}}(T_{\text{em}}, n_{\text{em}}), \tag{10}$$

where η_{em} is the electric machine efficiency and $f_{\eta,\text{em}}$ is the interpolation function of the electric machine efficiency map.

Figure 4 provides the efficiency maps of a driving motor and generator. Due to the fact that the electric machine could work in four quadrants as a motor, including the driving and braking state of forward and reverse, the efficiency data of the motor spreads over the four quadrants, while the generator can only operate in one quadrant to generate power.

2.3.2. Supercapacitor Model

The supercapacitor model was developed as an equivalent circuit model. The model takes the impact of temperature into account. The resistance and capacitor are the functions of the temperature and current, and described as

$$Re_{\text{sc}} = f_{\text{Re,sc}}(Tem_{\text{sc}}, I_{\text{sc}}), \tag{11}$$

$$C_{sc} = f_{C,sc}(Tem_{sc}, I_{sc}), \tag{12}$$

where Re_{sc} is the resistance of the supercapacitor, Tem_{sc} is the temperature, I_{sc} is the current, C_{sc} is the capacitor, $f_{Re,sc}$ represents the resistance function, and $f_{C,sc}$ represents the capacitor function.

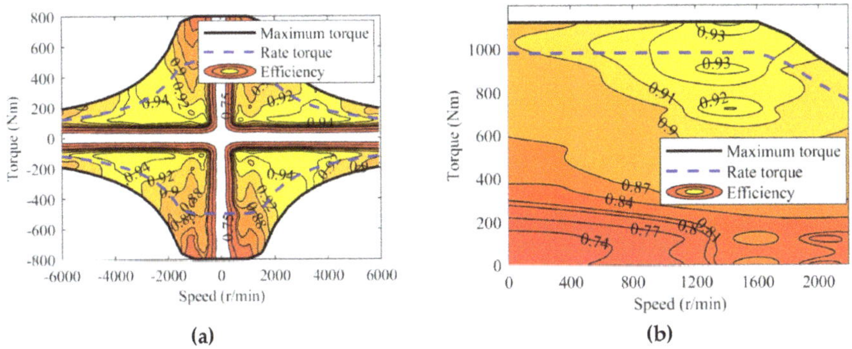

Figure 4. Efficiency map of a (**a**) driving motor and (**b**) generator.

The parameters of the circuit part can be calculated as

$$I_{sc} = \frac{U_{sc} - \sqrt{(U_{sc}^2 - 4Re_{sc}P_{sc})}}{2Re_{sc}}, \tag{13}$$

$$U_{sc}(n+1) = U_{sc}(n) - I_{sc}dt/C_{sc}, \tag{14}$$

$$SOC = (U_{sc}(n+1) - U_{sc,min})/(U_{sc,max} - U_{sc,min}), \tag{15}$$

where U_{sc} is the voltage, and $U_{sc,min}$ and $U_{sc,max}$ are the minimum and maximum voltage of the supercapacitor, respectively.

The temperature of the supercapacitor is calculated according to the power loss of resistance and coulombic efficiency and heat transfer process. The temperature can be shown as

$$Tem_{sc} = Tem_{sc,ini} + \int_0^t \frac{P_{sc,h} - P_{sc,a}}{m_{sc}c_{sc}} dt, \tag{16}$$

$$P_{sc,h} = \begin{cases} I_{sc}^2 Re_{sc}^{dis} & I_{sc} \geq 0 \\ I_{sc}^2 Re_{sc}^{chg} - I_{sc}U_{sc}(1 - \eta_{coul}) & I_{sc} < 0 \end{cases}, \tag{17}$$

$$P_{sc,a} = (Tem_{sc} - Tem_a)/Re_h, \tag{18}$$

where Tem_{sc} is the supercapacitor temperature; $Tem_{sc,ini}$ is the initial temperature of the supercapacitor; Tem_a is the ambient air temperature; $P_{sc,h}$ is the heat power; $P_{sc,a}$ is the heat power transferred to air; m_{sc} is the mass; c_{sc} is the specific heat capacity; Re_{sc}^{dis} and Re_{sc}^{chg} are discharge and charge resistance, respectively; η_{coul} is the coulombic efficiency; and Re_h is the heat resistance.

2.3.3. Hydraulic Pump Model

The hydraulic pump model mainly presents its torque and power consumption and simplifies the hydraulic operation system, which is sufficient for researching the powertrain energy and control. The hydraulic pump model can be denoted by

$$T_{hp} = \frac{p_{hp} q_{hp}}{2\pi \eta_{hp}}, \tag{19}$$

$$P_{hp} = \frac{50 p_{hp} Q_{hp}}{3 \eta_{hp}}, \tag{20}$$

$$Q_{hp} = \frac{q_{hp} n_{hp}}{1000}, \tag{21}$$

where T_{hp} is the torque of the hydraulic pump, n_{hp} is its speed, p_{hp} is the outlet pressure, q_{hp} is its displacement, η_{hp} is its efficiency, P_{hp} is its power consumption, and Q_{hp} is the flow.

2.3.4. Tracked Walking Mechanism Model

The tracked walking mechanism model is represented based on the kinematic relationship and takes the slip ratio as a major factor into consideration. The traction and speed of the tracks can be formulated as

$$F_t = \frac{T_t^{in} \eta_t}{r_{dw}}, \tag{22}$$

$$\omega_t = \frac{v}{(1-\varphi) r_{dw}}, \tag{23}$$

where F_t is the tangential traction of tracks, T_t^{in} is the input torque, η_t is the track efficiency, r_{dw} is the radius of the driving wheel, and φ is the track slip rate.

2.4. Dynamic Model

The HEB has to overcome the resistance of bulldozing, climbing, ground deformation, acceleration, and air friction. The air friction is small and can be neglected due to the velocity of the bulldozer being very low. The dynamic model can be shown as

$$F_t = m_{veh} g \mu \cos\alpha + m_{veh} g \sin\alpha + \xi m_{veh} \frac{dv}{dt} + F_b, \tag{24}$$

where m_{veh} is the HEB mass, g is the gravitational acceleration, α is the slope angle, ξ is the rotary mass coefficient, v is the velocity, μ is the ground deformation resistance coefficient, and F_b is the bulldozing resistance.

In order to accurately calculate the resistance, the ξ was identified and the F_b was extracted in our previous research [7].

Finally, the HEB model was built in the Simulink environment, as shown in Figure 5.

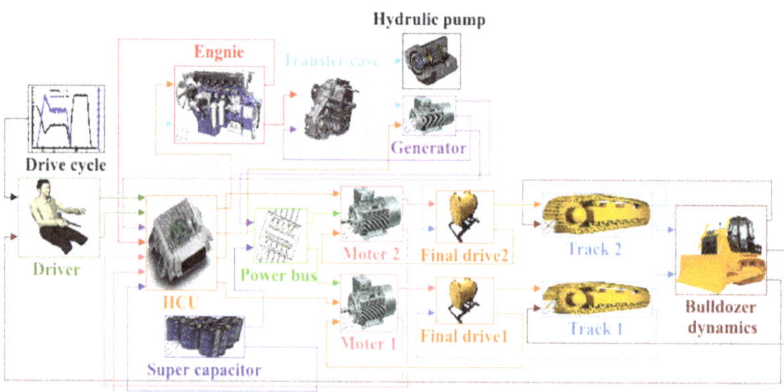

Figure 5. HEB model in Matlab/Simulink.

3. ASPF Control Strategy

The control strategy is the core of this paper and plays an important role in every performance of the HEB. It was developed and optimized according to three current problems that we found in existing research. Therefore, the target problems are presented below, firstly by simulating the operation and control process. The methodology of the control strategy is based on fuzzy control theory, a filter algorithm, optimization, and a rule-based strategy. The detailed structure and composition of the proposed control strategy is then introduced.

3.1. Target Problem

The control improvement goal of this research is to resolve three problems. The first two problems are related to the limitation of genset's output power and the fluctuation of working points. The last is related to re-establishing the working points on the optimal efficiency trajectory according to the involvement of the interference torque.

3.1.1. Frequent Fluctuation of Genset's Working Points

In most studied power following control strategies, the genset output power is conducted to follow the drive system demand power. As the demand power is constantly changing, the operation points of the genset also frequently change. In particular, when the operation points are controlled on a pre-set trajectory, different powers correspond to different speeds and torques, resulting in severe fluctuations of the speed and torque of the genset and further leading to an increase of transient fuel consumption.

3.1.2. Large Change Rate of Output Power

A large change rate of the genset output power will cause its operation points to deviate from the target pre-set trajectory. Due to rapid change of the transient drive system demand power, and switching operation points requiring a certain response time, the points may deviate from the trajectory in the switching process. This deviation will cause the operation points to depart from the minimum fuel consumption curve and spread over a lower efficiency district, resulting in a deterioration of fuel consumption.

The transition processes of different change rates of output power were simulated to illustrate the process, as shown in Figure 6. It indicates that the actual working points deviate from the pre-set minimum fuel consumption curve because of the rapid power change and insufficient responsiveness. However, this phenomenon disappears and the points track the trajectory well if the power change is slow.

3.1.3. Hydraulic Pump's Interference in Target Trajectory

The torque consumed by the hydraulic pump will also lead to the genset operation points deviating from the pre-set target trajectory for torque interference. Most previous studies have ignored the hydraulic pump torque and its interference effect [8–10,50,51], which will finally cause a decrease of the genset efficiency and increase of fuel consumption.

3.2. Development of the ASPF Strategy

An ASPF control strategy is proposed here to provide a comprehensive solution to these problems. Through this strategy, the demand power signal is divided into the trend part and the fluctuation part to make the genset output smooth and stable, and the supercapacitor supply the output power fluctuation. Therefore, the transient fuel consumption and the deviation from the target trajectory can be reduced. Meanwhile, when there is torque consumption of the hydraulic pump in the working system, the optimal efficiency working points of the genset are found for control to avoid interference. The structure and development process are presented below.

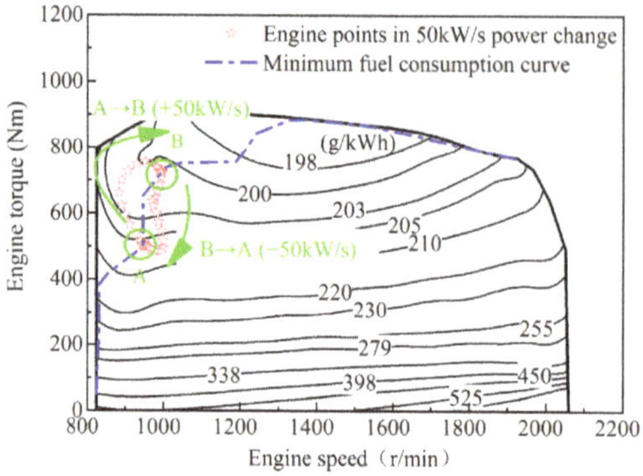

(a)

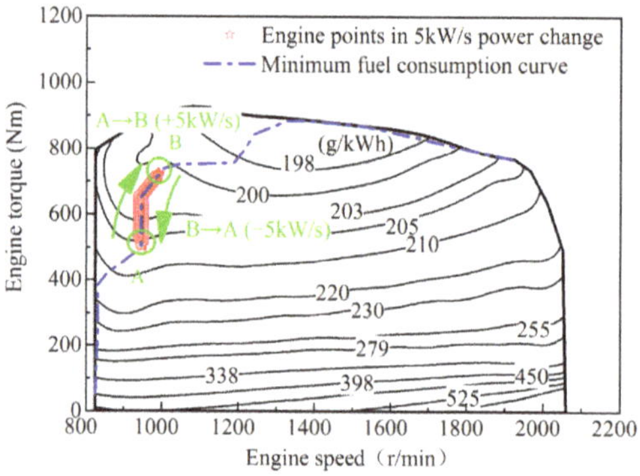

(b)

Figure 6. Effect of different changes of the genset's output power on the engine operating points: (**a**) 50 and (**b**) 5 kW/s power change.

3.2.1. Structure of the ASPF Strategy

The structure of proposed ASPF control strategy is shown as Figure 7. It employs two critical steps, including a fuzzy adaptive filter and optimal efficiency control of the genset module, to solve the problems above and improve the power following control strategy. Utilizing the computed demand power of the drive system as the input, and introducing state of charge (*SOC*) and pressure feedback, the strategy outputs the engine target speed and generator target torque to the subcomponent controller. Energy distribution between the genset and the supercapacitor can be completed through determining the genset output power.

3.2.2. Fuzzy Adaptive Filter

The fuzzy adaptive filter integrates a fuzzy logic controller with a first-order inertia filter. By using the self-adaptability of the fuzzy controller to automatically adjust the filter coefficient, the output power can be smoothed with the *SOC* in a permissive range. Figure 8 shows the smoothing process based on the fuzzy adaptive filter.

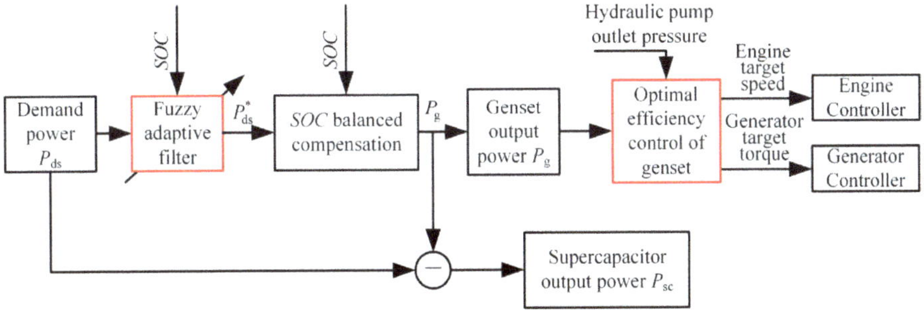

Figure 7. Flow chart of adaptive smoothing power following the energy management strategy.

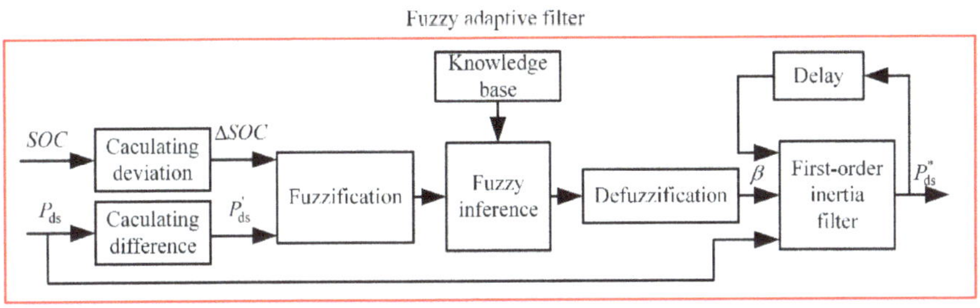

Figure 8. Schematic diagram of smoothing power based on the fuzzy adaptive filter.

The variables in Figure 8 can be expressed as

$$\Delta SOC = SOC - (SOC_{hi} + SOC_{lo})/2, \tag{25}$$

$$P'_{ds} = P_{ds}(t) - P_{ds}(t-1), \tag{26}$$

where ΔSOC is the deviation of SOC; SOC_{hi} and SOC_{lo} are the high and low threshold value, respectively; P'_{ds} is the difference of the demand power; and P_{ds} is the demand power.

The first-order inertia filter can be developed by

$$P^*_{ds}(t) = \beta P_{ds}(t) + (1-\beta)P^*_{ds}(t-1), \tag{27}$$

where P^*_{ds} is the filtered demand power and β is the filter coefficient.

The SOC deviation and the demand power difference are normalized as the fuzzy logic inputs, which can be shown as

$$\Delta SOC_{in} = \Delta SOC/\max(\Delta SOC), \tag{28}$$

$$P'_{ds,in} = P'_{ds}/\max(P'_{ds}), \tag{29}$$

where ΔSOC_{in} and $P'_{ds,in}$ are the normalized SOC deviation and demand power difference, respectively.

The degree of membership (DOM) values of input and output variables of the fuzzy logic controller are defined as shown in Figure 9. Because the filter will introduce a delay effect of input signals, the SOC deviation is entered here in order to prevent the SOC from exceeding its limits. Table 3 shows the fuzzy logic rules in the knowledge base. The basic formulation principle of the rule table is that the demand power is filtered and the SOC is kept within the predetermined range as much as possible.

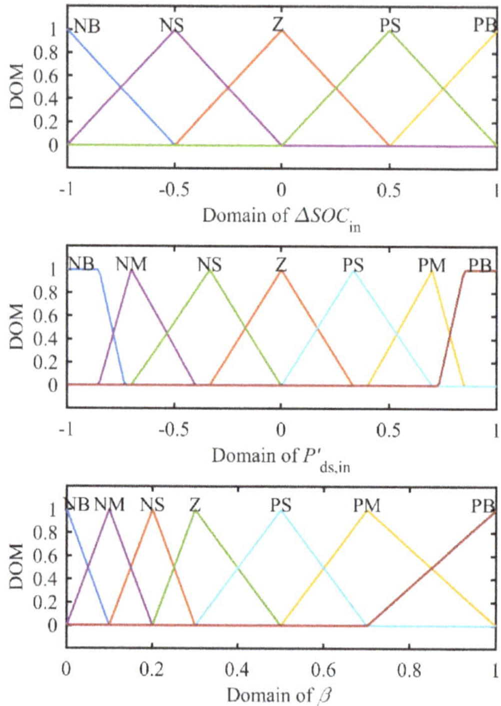

Figure 9. Membership function of input and output variables.

Table 3. Fuzzy logic rules.

β		$P'_{ds,in}$						
		NB	**NM**	**NS**	**Z**	**PS**	**PM**	**PB**
	NB	NB	NB	NM	PB	PB	PB	PB
	NS	NB	NM	NM	Z	PS	PM	PB
ΔSOC_{in}	Z	NS	NM	NM	NB	NM	NM	NS
	PS	PB	PM	PS	Z	NM	NM	NB
	PB	PB	PB	PB	PB	NM	NB	NB

By employing the min-max inference approach of the Mamdani type and the centroid method for defuzzification, the output map of the fuzzy inference system, that is, the fuzzy control table, can be achieved, as shown in Figure 10. The control map is a symmetrical valley type about its center, since the filtering function can be as large as possible within the *SOC* limits when the difference in the demand power and the deviation of *SOC* are opposite in sign.

3.2.3. Correctional Optimal Efficiency Map

The module of optimal efficiency control of the genset is the other crucial part of ASPF, as shown in Figure 11. This module takes the demand power of generation and the pressure of the hydraulic pump as inputs to calculate the generator target torque and the engine target speed as outputs. The input mechanical power of the generator can be calculated by

$$P_{g,m} = \frac{P_g}{\eta_g(T_{g,est}, n_{g,est})}, \tag{30}$$

where $P_{g,m}$ is the input mechanical power of the generator; P_g is the generating electric power; $T_{g,est}$ and $n_{g,est}$ are the estimated torque and speed, respectively; and η_g is the efficiency interpolation function.

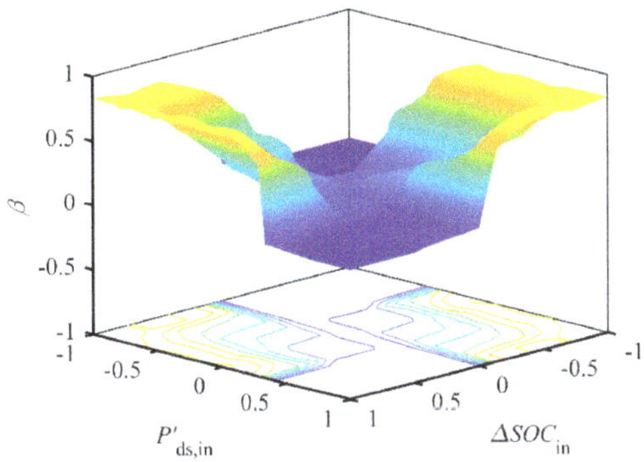

Figure 10. Output map of the fuzzy inference system.

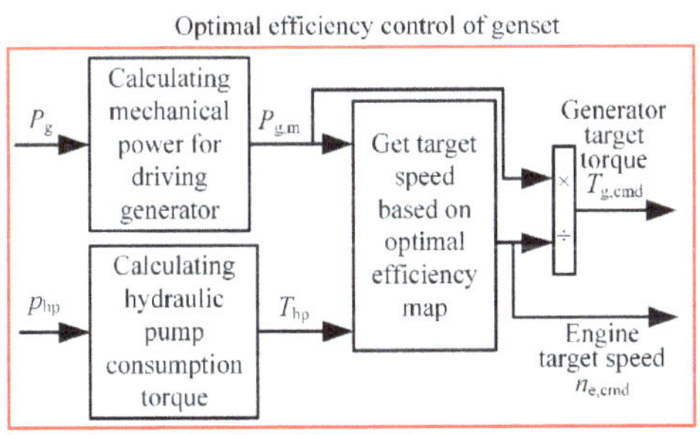

Figure 11. Control principle of the genset based on the optimal efficiency map.

The optimal efficiency map is the core of this module. The problem of searching the optimal efficiency map can be converted into a problem of searching the optimal curves under different input torques of the hydraulic pump by using Equation (30). The consumed torque of the hydraulic pump can be calculated by Equation (19). Through changing the hydraulic pump torque continuously and searching each optimal efficiency curve of the genset at each torque value, the optimal efficiency map can finally be obtained, as shown in Figure 12.

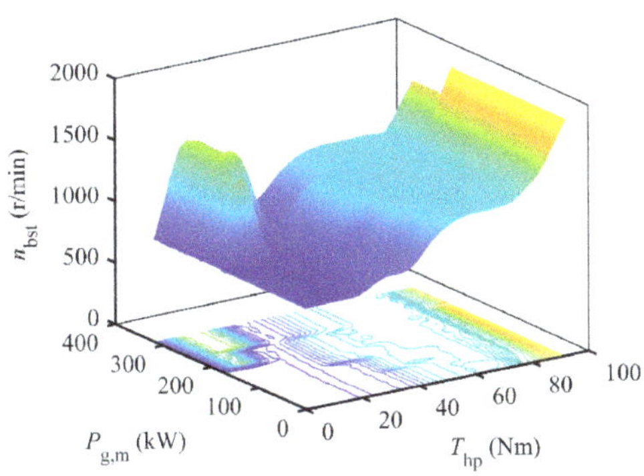

Figure 12. Optimal efficiency map of the genset.

The relationship between the input torque of the generator, the engine torque, and the hydraulic pump torque can be expressed as

$$T_g = T_e - T_{hp}, \tag{31}$$

where T_g is the input torque of the generator.

4. HEB HIL Platform

In order to validate the real-time control effect of the proposed strategy, an HEB controller HIL experimental platform was developed, as shown in Figure 13. The control strategy was converted into an executable C code and integrated with the controller underlying software, and was then downloaded in the vehicle control unit (VCU). Meanwhile, the controlled object model was downloaded to the dSPACE Autobox simulator (DS1103, dSPACE, Paderborn, North Rhine-Westphalia, Germany). The real-time supervisory control interface was developed on the host computer monitor. The communication network among the VCU, the Autobox, and the host computer monitor was established with the means of the control area network (CAN) and transmission control protocol/internet protocol (TCP/IP).

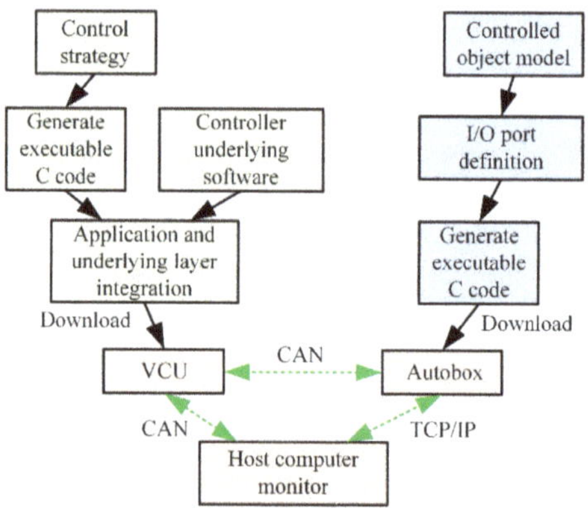

Figure 13. Testing program of hardware in loop (HIL) of the HEB controller.

Figure 14 shows the developed HEB HIL test platform, which could be used to test and compare the real-time control effect of online strategies. The model incorporated in the HIL platform was validated by employing the practical HEB real bulldozing process data, as shown in [61]. Therefore, the combination of the validated model and the real VCU could test the real-time performance and effectiveness of the proposed control strategy to some extent.

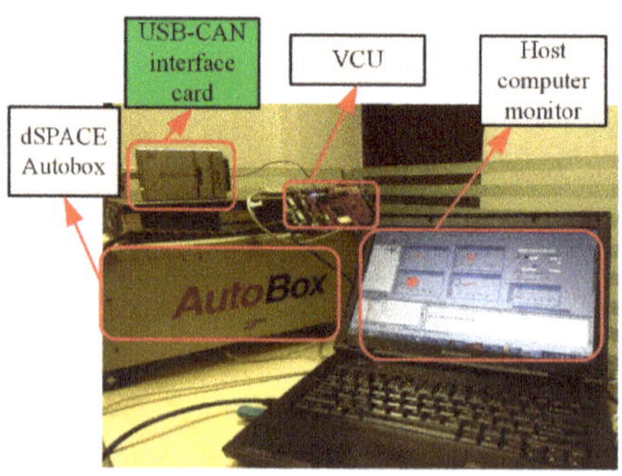

Figure 14. Test platform of HEB HIL.

5. Results and Discussion

Due to the lack of a standard drive cycle of bulldozers like that of automobiles, a representative actual drive cycle of the bulldozer, which was extracted and constructed from a large number of bulldozing experimental data in our previous research [7], was adopted for simulation and comparison. Figure 15 shows the drive cycle, including the bulldozing stage and empty returning stage. The bulldozing stage can be further divided into cutting-soil, transporting-soil, and unloading-soil stages.

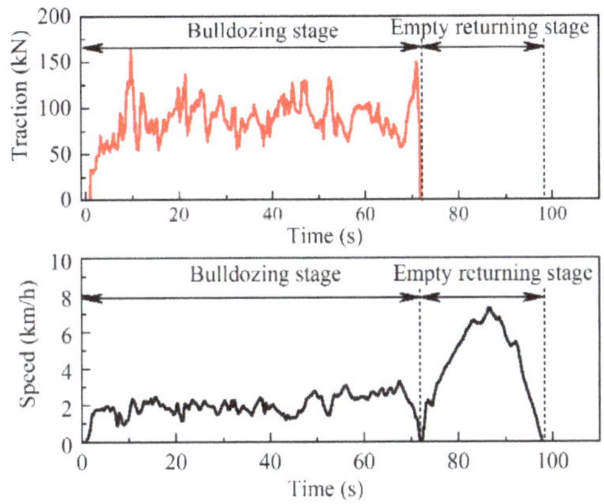

Figure 15. Traction and speed of a representative actual drive cycle.

In order to validate and compare the control effect, three control strategies were compared under the representative actual drive cycle. Table 4 describes the compared strategies: a power following strategy in a preliminarily practical application (PF1), a typical power following strategy based on the engine minimum fuel consumption curve (PF2), and the proposed strategy (ASPF). Figure 16, Figure 17, Figure 18 and Table 5 show the comparison results.

Table 4. Comparison of different control strategies.

Control	Description
PF1	A power following control strategy used in our previous real HEB experiment, where the genset follows the demand power, and the engine speed follows the accelerator and remains within 1300 to 1800 r/min.
PF2	A typical power following control strategy based on the engine minimum fuel consumption curve in the series powertrain.
ASPF	Proposed adaptive smoothing power following control strategy based on the optimal efficiency map.

Table 5. Comparison of fuel consumption.

Vehicle	Control	FC (L)	EFC (L)	EFSR (%)
HMB	-	0.671	0.671	-
HEB	PF1	0.569	0.568	15.4
	PF2	0.537	0.538	19.8
	ASPF	0.517	0.515	23.2

FC: fuel consumption; EFC: equivalent fuel consumption; EFSR: equivalent fuel saving rate.

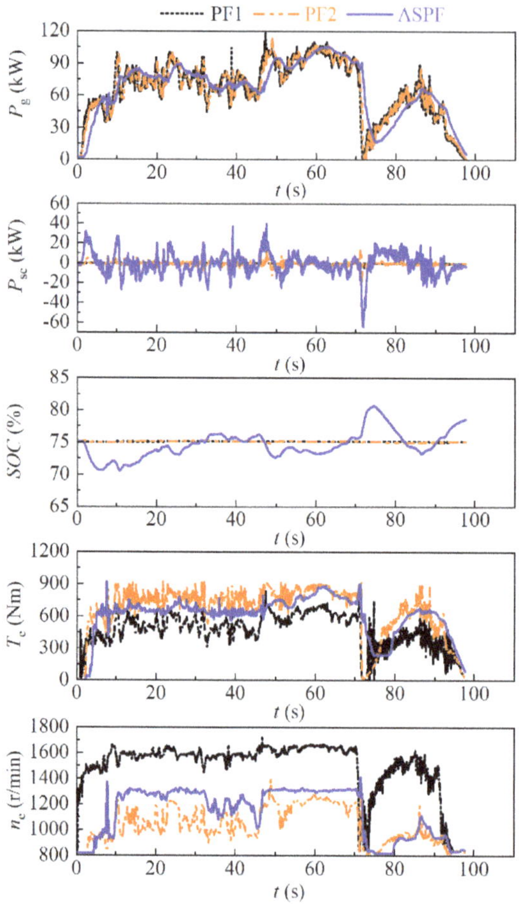

Figure 16. Comparison of operating parameters of key powertrain components.

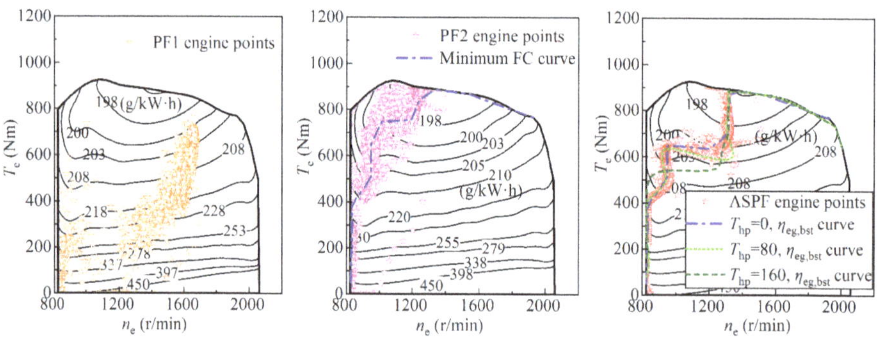

Figure 17. Distribution of engine working points. FC: fuel consumption, $\eta_{eg,bst}$: the best efficiency of the genset.

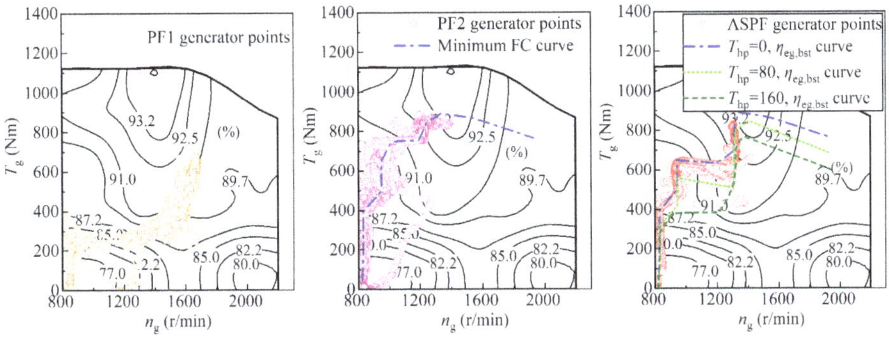

Figure 18. Distribution of generator working points.

Figure 16 shows the comparison of a group of key powertrain parameters, including the generator output power P_g, supercapacitor output power P_{sc}, supercapacitor *SOC*, engine torque T_e, and engine speed n_e. It can be observed that the generator output power fiercely fluctuates and follows the demand power under the PF1 and PF2. However, the change of generator output power is relatively smooth under the ASPF. The supercapacitor power under the ASPF is larger and fluctuates more than that under the PF1 and PF2. Meanwhile, the *SOC* of the ASPF varies within a permissible range. The first three subgraphs of Figure 16 illustrate that the self-adaptive filter algorithm in the ASPF can smoothen the power adaptively and keep the *SOC* within pre-set limits simultaneously by timely adjustment of the filter coefficient, which can prompt higher engagement and take full advantage of the high efficiency of the supercapacitor. The fourth and fifth subgraphs show that ASPF makes the engine speed and torque more stable, especially relative to the PF2 based on the trajectory, through smoothing the genset output power. Therefore, the ASPF can play a positive role in stabilizing the working points of the engine and generator, which can achieve a reduction of the transient energy loss.

Figure 17 compares the engine working points with different strategies under the same representative drive cycle. It can be seen that the engine working point distribution with PF1 is widespread and mainly within the speed range from 1300 to 1800 r/min in different loads, whereas it is far away from the low fuel consumption area. From the middle subgraph, we can see that the engine operating points of PF2 are distributed around the minimum fuel consumption curve. However, they could not coincide with the curve because of their dramatic fluctuation and insufficient response on the timeline shown in Figure 16. The left subgraph shows that the engine working point distribution of ASPF is very close to the pre-set optimal efficiency curves of combining the engine with the generator under different hydraulic pump consumed torque. This relatively concentrated distribution is the result of the effect of the adaptive filter link shown in the above graph.

Figure 18 compares the generator working point distribution with three control strategies under the same drive cycle. It can be seen that the distribution shape of the generator points is similar to that of the engine points on account of the coaxial junction of the engine and the generator. The generator working points of PF1 and PF2 are more widely distributed than those of ASPF for the adaptive filter. The difference between the generator torque below and the engine torque above is the hydraulic pump consumed torque, which is also reflected on the joint optimal efficiency curves in the above and below graphs. The ASPF keeps the generator points along the optimal efficiency curves as much as possible, in which following the routes can result in a greater generator efficiency.

The fuel consumption of the three control strategies and the prototype of the traditional hydro-mechanical bulldozer (HMB) under the same simulated drive cycle is shown in Table 5. The equivalent fuel consumption (EFC) was obtained from balancing the supercapacitor *SOC*. The equivalent fuel saving ratio (EFSR) is the comparison of EFC, reflecting the energy consumption comparison. The HEB equipped with the ASPF strategy can achieve 23.2% EFSR compared with the

HMB. However, it can only achieve 15.4% and 19.8% EFSR with the PF1 and PF2 strategy, respectively. The ASPF strategy can improve EFSR by 7.8% and 3.4% with respect to the PF1 and PF2 strategy, which indicates that the proposed strategy is more effective.

6. Conclusions

An adaptive smoothing power following (ASPF) control strategy based on an optimal efficiency map for a series tracked HEB has been proposed by considering its operation characteristics in this research. The three problems of increased fuel consumption presented in previous control studies to be solved by this strategy, including frequent fluctuation of the engine working points, deviation of the genset working points from the pre-set target trajectory due to a lack of response, and interference of the hydraulic pump consumed torque, were analyzed in detail. The whole HEB system simulation model with a novel transient fuel consumption model based on BPNN was established. The HEB HIL platform was then developed to evaluate the real-time performance and the effect of the strategy in a practical application. The methodology of the proposed control strategy is based on fuzzy control theory, real-based control, and the optimization method. The method of effect verification is incorporated on the basis of the adoption of a validated simulation model and an HIL test platform.

The ASPF strategy was compared with two strategies: a power following strategy in a preliminarily practical application (PF1) and a typical power following strategy based on the engine minimum fuel consumption curve (PF2). The results show that (1) the ASPF strategy can significantly reduce the fluctuation and pre-set trajectory deviation of the engine and generator working points; (2) the EFSR can be improved by about 7.8% and 3.4% with the ASPF strategy compared with the PF1 and PF2 strategy; and (3) the ASPF strategy is able to perform well in a real controller. It can be concluded that the proposed strategy can be effective and efficient in practical online HEB applications. The results also show that the discovered problems indeed exist and the corresponding control strategies are valid in HEB.

Filed tests of the HEB with the ASPF strategy will be carried out to further advance the real application value in our future research. The limitation of the ASPF strategy in real applications relative to previous work is that it should obtain the pressure signal of the hydraulic pump as the input for calculation, so a pressure sensor should be placed in the outlet of the hydraulic pump. Nevertheless, the proposed strategy and the problems solved and focused on in this paper are significant references for other types of HECMs. Moreover, future study will also explore comparisons of the ASPF online strategy and the optimization algorithm-based theoretical offline strategy to evaluate its potential for improvement.

Author Contributions: Conceptualization, B.Z., S.G., and X.Z.; data curation, B.Z., Q.X., and L.T.; formal analysis, B.Z., Q.X., and L.T.; methodology, B.Z., S.G., and X.Z.; project administration, X.Z.; resources, X.Z.; supervision, S.G. and X.Z.; writing—original draft, B.Z.; writing—review and editing, B.Z. All authors have read and agreed to the published version of the manuscript.

Funding: This study is sponsored by the National Key R&D Program of China (2016YFD0700703). The authors would like to thank the anonymous reviewers for their reviews and comments.

Conflicts of Interest: The authors declare no conflict of interest.

Nomenclature

Variables

α	slope angle,
β	filter coefficient
η	efficiency, %
μ	ground deformation resistance coefficient
ξ	rotary mass coefficient
ρ	density, g/L
τ	time constant

φ	track slip rate, %
ω	angular speed, rad/s
be	specific fuel consumption, g/(kWh)
B	fuel consumption rate, L/h
c	specific heat capacity, J/(kg·°C)
C	capacitor, F
f	function
F	force, N
g	gravitational acceleration, m/s^2
I	current, A
J	rotational inertia, kg·m^2
m	mass, kg
n	speed, r/min
p	pressure, Pa
P	power, W
q	displacement, ml/r
Q	flow, L/min
r	radius, m
R	correction factor
Re	resistance, Ω;
s	Laplace operator
SOC	state of charge
t	time, s
T	torque, Nm
Tem	temperature, °C
U	voltage, V
v	velocity, m/s
Subscripts	
a	air
b	bulldozing
bp	back propagation neural network
cl	current limit
cmd	command
coul	coulombic
ds	drive system
e	engine
em	electric machine
est	estimated
fu	fuel
g	generator
h	heat
hi	high
hp	hydraulic pump
in	input
ini	initial
lo	low
m	mechanical
s	steady state
sc	supercapacitor
t	transient state
ul	voltage limit
veh	vehicle

Superscripts

br	braking process
chg	charging process
dis	discharging process
dr	driving process

Abbreviations

ASPF	adaptive smooth power following
BPNN	back propagation neural network
CAN	control area network
DOM	degree of membership
DP	dynamic programming
ECMS	equivalent consumption minimization strategy
EFC	equivalent fuel consumption
EFSR	equivalent fuel saving ratio
ESU	energy storage unit
FC	fuel consumption
GA	genetic algorithm
HEB	hybrid electric bulldozer
HECM	hybrid electric construction machinery
HEV	hybrid electric vehicle
HIL	hardware in loop
HMB	hydro-mechanical bulldozer
MPC	model predictive control
PF	power following
PMP	Pontryagin's minimum principle
PSO	particle swarm optimization
SDP	stochastic dynamic programming
TCP/IP	transmission control protocol/ internet protocol
VCU	vehicle control unit

References

1. Tran, D.D.; Vafaeipour, M.; El Baghdadi, M.; Barrero, R.; Van Mierlo, J.; Hegazy, O. Thorough state-of-the-art analysis of electric and hybrid vehicle powertrains: Topologies and integrated energy management strategies. *Renew. Syst. Energ. Rev.* **2020**, *119*, 109596. [CrossRef]
2. Zhang, Y.; Guo, C.; Li, G.; Liu, Y.; Chen, Z. Cooperative control strategy for plug-in hybrid electric vehicles based on a hierarchical framework with fast calculation. *J. Clean. Prod.* **2020**, *251*, 119627. [CrossRef]
3. Zhang, B.; Yang, F.; Teng, L.; Ouyang, M.; Guo, K.; Li, W.; Du, J. Comparative analysis of technical route and market development for light-duty PHEV in China and the US. *Energies* **2019**, *12*, 3753. [CrossRef]
4. Xu, B.; Rathod, D.; Zhang, D.; Yebi, A.; Zhang, X.; Li, X.; Filipi, Z. Parametric study on reinforcement learning optimized energy management strategy for a hybrid electric vehicle. *Appl. Energy* **2020**, *259*, 114200. [CrossRef]
5. He, X.; Jiang, Y. Review of hybrid electric systems for construction machinery. *Autom. Constr.* **2018**, *92*, 286–296. [CrossRef]
6. Wang, J.; Yang, Z.; Liu, S.; Zhang, Q.; Han, Y. A comprehensive overview of hybrid construction machinery. *Adv. Mech. Eng.* **2016**, *8*. [CrossRef]
7. Zhang, B.; Zhang, X.; Xi, L.; Sun, C. Development of a representative operation cycle characterized by dual time series for bulldozers. *Proc. Inst. Mech. Eng. D J. Aut. Eng.* **2017**, *231*, 1818–1828. [CrossRef]
8. Hong, W.; FengChun, S. Dynamic modeling and simulation on a hybrid power system for dual-motor-drive electric tracked bulldozer. *Appl. Mech. Mater.* **2014**, *494*, 229–233.
9. Song, Q.; Zeng, P.; He, S.; Wang, H. Power flowing control strategy of series hybrid tracked bulldozer at the typical working condition. *J. Mech. Eng.* **2014**, *50*, 136–142. [CrossRef]
10. Wang, H.; Huang, Y.; Khajepour, A.; Song, Q. Model predictive control-based energy management strategy for a series hybrid electric tracked vehicle. *Appl. Energy* **2016**, *182*, 105–114. [CrossRef]

11. Hong, W. *Research on Dynamic Modeling and Control Strategy for Hybrid Tracked Bulldozer*; Beijing Institute of Technology: Beijing, China, 2015.
12. Azmat, M.; Kummer, S. Potential applications of unmanned ground and aerial vehicles to mitigate challenges of transport and logistics-related critical success factors in the humanitarian supply chain. *Asian J. Sustain. Soc. Responsib.* **2020**, *5*, 3. [CrossRef]
13. Azmat, M.; Kummer, S.; Moura, L.T.; Gennaro, F.D.; Moser, R. Future outlook of highway operations with implementation of innovative technologies like AV, CV, IOT and big data. *Logistics* **2019**, *3*, 15. [CrossRef]
14. Wintersberger, S.; Azmat, M.; Kummer, S. Are we ready to ride autonomous vehicles? A pilot study on austrian consumers' perspective. *Logistics* **2019**, *3*, 20. [CrossRef]
15. Shabbir, W.; Evangelou, S.A. Threshold-changing control strategy for series hybrid electric vehicles. *Appl. Energy* **2019**, *235*, 761–775. [CrossRef]
16. Li, H.; Ravey, A.; N'Diaye, A.; Djerdir, A. A novel equivalent consumption minimization strategy for hybrid electric vehicle powered by fuel cell, battery and supercapacitor. *J. Power Sources* **2018**, *395*, 262–270. [CrossRef]
17. Chen, S.Y.; Wu, C.H.; Hung, Y.H.; Chung, C.T. Optimal strategies of energy management integrated with transmission control for a hybrid electric vehicle using dynamic particle swarm optimization. *Energy* **2018**, *160*, 154–170. [CrossRef]
18. Du, G.; Zou, Y.; Zhang, X.; Kong, Z.; Wu, J.; He, D. Intelligent energy management for hybrid electric tracked vehicles using online reinforcement learning. *Appl. Energy* **2019**, *251*, 113388. [CrossRef]
19. Guo, Q.; Zhao, Z.; Shen, P.; Zhan, X.; Li, J. Adaptive optimal control based on driving style recognition for plug-in hybrid electric vehicle. *Energy* **2019**, *186*, 115824. [CrossRef]
20. Enang, W.; Bannister, C. Modelling and control of hybrid electric vehicles (a comprehensive review). *Renew. Sustain. Energ. Rev.* **2017**, *74*, 1210–1239. [CrossRef]
21. Wang, Y.; Wang, X.; Sun, Y.; You, S. Model predictive control strategy for energy optimization of series-parallel hybrid electric vehicle. *J. Clean. Prod.* **2018**, *199*, 348–358. [CrossRef]
22. Lu, X.; Wu, Y.; Lian, J.; Zhang, Y.; Chen, C.; Wang, P.; Meng, L. Energy management of hybrid electric vehicles: A review of energy optimization of fuel cell hybrid power system based on genetic algorithm. *Energy Convers. Manag.* **2020**, *205*, 112474. [CrossRef]
23. Huang, Y.; Wang, H.; Khajepour, A.; Li, B.; Ji, J.; Zhao, K.; Hu, C. A review of power management strategies and component sizing methods for hybrid vehicles. *Renew. Sustain. Energ. Rev.* **2018**, *96*, 132–144. [CrossRef]
24. Wu, Y.; Tan, H.; Peng, J.; Zhang, H.; He, H. Deep reinforcement learning of energy management with continuous control strategy and traffic information for a series-parallel plug-in hybrid electric bus. *Appl. Energy* **2019**, *247*, 454–466. [CrossRef]
25. Wu, J.; Wang, X.; Li, L.; Qin, C.; Du, Y. Hierarchical control strategy with battery aging consideration for hybrid electric vehicle regenerative braking control. *Energy* **2018**, *145*, 301–312. [CrossRef]
26. Li, H.; Ravey, A.; N'Diaye, A.; Djerdir, A. Online adaptive equivalent consumption minimization strategy for fuel cell hybrid electric vehicle considering power sources degradation. *Energy Convers. Manag.* **2019**, *192*, 133–149. [CrossRef]
27. Gong, J.; Zhang, D.; Guo, Y.; Liu, C.; Zhao, Y.; Hu, P.; Quan, W. Power control strategy and performance evaluation of a novel electro-hydraulic energy-saving system. *Appl. Energy* **2019**, *233*, 724–734. [CrossRef]
28. Shi, Y.; Xia, Y.; Zhang, Y.; Yao, Z. Intelligent identification for working-cycle stages of excavator based on main pump pressure. *Autom. Constr.* **2020**, *109*, 102991. [CrossRef]
29. Yu, Y.X.; Ahn, K.K. Optimization of energy regeneration of hybrid hydraulic excavator boom system. *Energy Convers. Manag.* **2019**, *183*, 26–34. [CrossRef]
30. Larsson, L.V.; Ericson, L.; Uebel, K.; Krus, P. Low-level control of hybrid hydromechanical transmissions for heavy mobile working machines. *Energies* **2019**, *12*, 1683. [CrossRef]
31. Li, T.; Liu, H.; Ding, D. Predictive energy management of fuel cell supercapacitor hybrid construction equipment. *Energy* **2018**, *149*, 718–729. [CrossRef]
32. Hippalgaonkar, R.; Ivantysynova, M. Optimal power management of hydraulic hybrid mobile machines-part I: Theoretical studies, modeling and simulation. *J. Dyn. Syst. T Asme* **2016**, *138*. [CrossRef]
33. Nilsson, T.; Froberg, A.; Aslund, J. Predictive control of a diesel electric wheel loader powertrain. *Control. Eng. Pract.* **2015**, *41*, 47–56. [CrossRef]

34. Zhou, H.; Zhao, P.Y.; Chen, Y.L.; Yang, H.Y. Prediction-based stochastic dynamic programming control for excavator. *Autom. Constr.* **2017**, *83*, 68–77. [CrossRef]
35. Kim, H.; Yoo, S.; Cho, S.; Yi, K. Hybrid control algorithm for fuel consumption of a compound hybrid excavator. *Autom. Constr.* **2016**, *68*, 1–10. [CrossRef]
36. Chen, Q.; Lin, T.; Ren, H. Parameters optimization and control strategy of power train systems in hybrid hydraulic excavators. *Mechatronics* **2018**, *56*, 16–25. [CrossRef]
37. Li, T.; Huang, L.; Liu, H. Energy management and economic analysis for a fuel cell supercapacitor excavator. *Energy* **2019**, *172*, 840–851. [CrossRef]
38. Zhu, Q.; Wang, Q.F. Real-time energy management controller design for a hybrid excavator using reinforcement learning. *J. Zhejiang Univ. Sci. A* **2017**, *18*, 855–870. [CrossRef]
39. Zhou, C.; Xu, H.C.; Ding, L.Y.; Wei, L.C.; Zhou, Y. Dynamic prediction for attitude and position in shield tunneling: A deep learning method. *Autom. Constr.* **2019**, *105*, 16. [CrossRef]
40. Zhou, H.; Zhao, P.Y.; Chen, Y.L. Fuzzy logic control for a hydraulic hybrid excavator based on torque prediction and genetic algorithm optimization. *Proc. Inst. Mech. Eng. D J. Aut. Eng.* **2018**, *232*, 983–994. [CrossRef]
41. Liang, X.J.; Wu, W.R. Control strategy of energy saving for power system in hydraulic surface drilling rig. *J. Braz. Soc. Mech. Sci. Eng.* **2018**, *40*, 8.
42. Ye, Y.; Yin, C.B.; Gong, Y.; Zhou, J.J. Position control of nonlinear hydraulic system using an improved PSO based PID controller. *Mech. Syst. Sig. Process.* **2017**, *83*, 241–259. [CrossRef]
43. Truong, D.Q.; Marco, J.; Greenwood, D.; Harper, L.; Corrochano, D.G.; Yoon, J.I. Challenges of micro/mild hybridisation for construction machinery and applicability in UK. *Renew. Sustain. Energ. Rev.* **2018**, *91*, 301–320. [CrossRef]
44. Zhang, B.; Zhang, X.; Zhang, X.; Xi, L. Development of simulation model for series hybrid electric crawler bulldozer (SHECB) based on ADVISOR. *Adv. Mater. Res.* **2013**, *732*, 337–343. [CrossRef]
45. Betz, M.D.; Casey, K.A.; Garnett, S.C.; Liang, C.C.; Tonsor, A.J.; Vanderham, M.E. Electric Powertrain for Machine. U.S. Patent No. 7,950,481, 31 May 2011.
46. Cao, T. *Evaluation of Greenhouse Gas and Criteria Emissions from Conventional and Hybrid Off-Road Equipment*; University of California, Riverside: Riverside, CA, USA, 2014.
47. Hong, W.; Yanjun, H.; Amir, K.; Chuan, H.; Amir, K. *Electrification of Heavy-Duty Construction Vehicles*; Morgan & Claypool: Williston, VT, USA, 2017; p. 1.
48. Wang, H.; Huang, Y.; Khajepour, A.; He, H.; Lv, C. Mpc-Based Power Management Strategy for a Series Hybrid Electric Tracked Bulldozer. In Proceedings of the 2017 IEEE International Conference on Industrial Technology (ICIT), Toronto, ON, Canada, 22–25 March 2017; IEEE: Piscataway, NJ, USA, 2017; pp. 1403–1408.
49. Wang, H.; Huang, Y.; He, H.; Lv, C.; Liu, W.; Khajepour, A. Chapter 5—Energy Management of Hybrid Electric Vehicles. In *Modeling, Dynamics and Control of Electrified Vehicles*; Zhang, H., Cao, D., Du, H., Eds.; Woodhead Publishing: Sawston, UK, 2018; pp. 159–206.
50. Wang, H.; Huang, Y.; Khajepour, A.; He, H.; Cao, D. A novel energy management for hybrid off-road vehicles without future driving cycles as a priori. *Energy* **2017**, *133*, 929–940. [CrossRef]
51. Wang, H.; Huang, Y.; Khajepour, A. Cyber-physical control for energy management of off-road vehicles with hybrid energy storage systems. *IEEE/ASME Trans. Mechatron.* **2018**, *23*, 2609–2618. [CrossRef]
52. Han, X.; He, H.; Wu, J.; Peng, J.; Li, Y. Energy management based on reinforcement learning with double deep Q-learning for a hybrid electric tracked vehicle. *Appl. Energy* **2019**, *254*, 113708. [CrossRef]
53. Wei, S.; Zou, Y.; Sun, F.; Christopher, O. A pseudospectral method for solving optimal control problem of a hybrid tracked vehicle. *Appl. Energy* **2017**, *194*, 588–595. [CrossRef]
54. Zhang, B. *Study on Energy Saving Mechanism and Control Strategy of Hybrid Electric Bulldozer*; Beijing Jiaotong University: Beijing, China, 2017.
55. Markel, T.; Brooker, A.; Hendricks, T.; Johnson, V.; Kelly, K.; Kramer, B.; O'Keefe, M.; Sprik, S.; Wipke, K. ADVISOR: A systems analysis tool for advanced vehicle modeling. *J. Power Sources* **2002**, *110*, 255–266. [CrossRef]
56. Zhou, M.; Jin, H. Development of a transient fuel consumption model. *Transp. Res. Part D Transp. Environ.* **2017**, *51*, 82–93. [CrossRef]
57. Zhang, Y.T.; Claudel, C.G.; Hu, M.B.; Yu, Y.H.; Shi, C.L. Develop of a fuel consumption model for hybrid vehicles. *Energy Convers. Manag.* **2020**, *207*, 112546. [CrossRef]

58. Guang, H.; Jin, H. Fuel consumption model optimization based on transient correction. *Energy* **2019**, *169*, 508–514. [CrossRef]
59. Lindgren, M. A transient fuel consumption model for non-road mobile machinery. *Biosyst. Eng.* **2005**, *91*, 139–147. [CrossRef]
60. Zhang, L. *Investigation of Performance Deterioration and Control Strategy of Automotive Diesel Engine Under Transient Operation Conditions*; Jilin University: Changchun, China, 2015.
61. Zhang, B.D.; Zhang, X.; Zhang, L.; Xi, L.H. Development and validation of a series hybrid electric bulldozer model on whole working condition. *Key Eng. Mater.* **2016**, *693*, 1811–1817. [CrossRef]

© 2020 by the authors. Licensee MDPI, Basel, Switzerland. This article is an open access article distributed under the terms and conditions of the Creative Commons Attribution (CC BY) license (http://creativecommons.org/licenses/by/4.0/).

energies

MDPI

Article

Optimal Location of Fast Charging Stations for Mixed Traffic of Electric Vehicles and Gasoline Vehicles Subject to Elastic Demands

Hong Gao [1], Kai Liu [1,*], Xinchao Peng [1] and Cheng Li [2]

1 School of Transportation and Logistics, Dalian University of Technology, Dalian 116024, China; gaohong9601@mail.dlut.edu.cn (H.G.); pengxc@mail.dlut.edu.cn (X.P.)
2 China Academy of Transportation Sciences, Beijing 100029, China; licheng@motcats.ac.cn
* Correspondence: liukai@dlut.edu.cn; Tel.: +86-411-84706221

Received: 23 March 2020; Accepted: 14 April 2020; Published: 16 April 2020

Abstract: With the rapid development of electric vehicles (EVs), one of the urgent issues is how to deploy limited charging facilities to provide services for as many EVs as possible. This paper proposes a bilevel model to depict the interaction between traffic flow distribution and the location of charging stations (CSs) in the EVs and gasoline vehicles (GVs) hybrid network. The upper level model is a maximum flow-covering model where the CSs are deployed on links with higher demands. The lower level model is a stochastic user equilibrium model under elastic demands (SUE-ED) that considers both demands uncertainty and perceived path constraints, which have a significant influence on the distribution of link flow. Besides the path travel cost, the utility of charging facilities, charging speed, and waiting time at CSs due to space capacity restraint are also considered for the EVs when making a path assignment in the lower level model. A mixed-integer nonlinear program is constructed, and the equivalence of SUE-ED is proven, where a heuristic algorithm is used to solve the model. Finally, the network trial and sensitivity analysis are carried out to illustrate the feasibility and effectiveness of the proposed model.

Keywords: electric vehicles; public charging station; bilevel model; range constraint

1. Introduction

The market share of new energy vehicles without fossil fuels has been increasing rapidly in recent years, especially electric vehicles (EVs), which provide better performance, higher efficiency, and zero emissions [1,2]. To promote the usage of EVs, governments around the world have successively issued a series of active policies to provide subsidies for EV purchases and to deploy public charging infrastructures at convenient locations [2,3]. However, the market acceptance of EVs seems to fluctuate with the decreasing subsidy. On the one hand, limited driving range, insufficient public charging infrastructures, and longer charging time are the main reasons for at-home charging. On the other hand, the deployment of EV charging stations may induce more trips [4,5], that is, demands between each origin to destination (OD) pair that respond to the available fast charging are elastic, so does the route choice. Therefore, it is especially important to study how to effectively and reasonably deploy EV public charging stations within the hybrid network of rapidly increasing EVs and dominated gasoline vehicles (GVs), which should help reduce the range anxiety of EV users and maximize the coverage rate of EVs.

According to previous studies, a variety of factors affect the location of charging stations, including, among others, preference and travel choice behavior of users [6,7], travel demand of users [8], information of the en-route energy consumption of EVs [9–12], information of the remained range of EVs [13,14], charging speed and demand of battery [7,15], road environment [5,16], land supply [17], elastic charging demands [18]. However, it is not practical to build a complicated location model and to solve the location problem by incorporating so many macro and micro factors into the planning and decision making of EV charging stations. This paper mainly studies the uncertainties due to the choice behavior of EV and GV users under elastic demands and uncertain path constraints. A bilevel model is proposed in this paper, where the upper model deals with a maximum flow-covering (MFC) problem, while the lower model is a stochastic user equilibrium model with elastic demands (SUE-ED). It is worth noting that factors, such as road congestion, charging speed, range limit of EVs, and capacity of charging stations, are already considered in users' travel choices. The lower model considers the elastic demands and the distance constraints of EVs, which have a significant influence on route choice and the distribution of link flow. Finally, a generalized Lagrangian function is constructed to prove the equivalence between the stochastic user equilibrium model and the elastic demand model.

The remainder of this paper is structured as follows: Section 2 reviews the relevant literature. Section 3 affirms the problem hypothesis, analyzes three charging paths of the EVs, and gives the symbolic description used in the article. Section 4 establishes a bilevel model. Sections 5 and 6 introduce the algorithms used to solve the model and numerical analysis applying to the Nguyen–Dupuis network, respectively. Finally, the research results and future research directions are discussed in Section 7.

2. Literature Review

The conventional and dominant three types of location optimization models include the point demand model [19,20], the flow demand model [20–23], and the multi-objective optimization model [24–27]. The P-Median model assumes that "charging demand is generated in the road network node" and is widely used as one of the point demand models. The flow demand model is formulated on the basis of the Flow Capturing Location Model (FCLM). For the first time, some researches proposed the Flow Refueling Location Model (FRLM), where mileage limitations were explicitly considered in facility location issues [28–31]. One branch of FRLM aims to maximize the demand coverage by locating a fixed number of charging facilities, which is called the maximum coverage location problem. Although the multi-objective optimization models have the advantage of addressing more complicated experimental requirements, they are not good at dealing with the uncertainty planning problem [32].

In a hybrid network with both EVs and GVs, the layout of EV charging stations and the distribution of EV flow affect mutually. Elastic demands in a hybrid network result in many more uncertainties. Most of the previous studies used the User Equilibrium Model (UE) to locate charging facilities. In these studies, Xu et al. dealt with the user equilibrium problem in a hybrid transport network with battery switching stations and road grade constraints [33]. Jing et al. gave a comprehensive discussion of the equilibrium network model [34]. Jiang et al. introduced the path distance constraints into the UE model [13]. Zheng et al. proposed a bilevel model where the upper layer minimized travel costs, and the lower layer aimed to find the path-constrained EVs equalization flow [35]. The bounded rationality of EV users led to much more complicated energy consumption [7,36,37], and, therefore, route choice behavior reflected more unobserved heterogeneity, which resulted in various elastic demands.

The network design problem with elastic demands, where the induced or transferred OD demand is the subject of responses in traveler itinerary choices to enjoy the improvement of new infrastructure, have several formulations with various motivations. Ge et al. [38] is one of the early attempts to consider both the proportion of EVs and the charge rate of EVs when determining the elastic charging demands from the total number of vehicles on road connections. The elastic demand was formulated based either on feedback of congested travel and congested station on route choices [39] or on the assumption that charge demand between OD pairs follows a nonlinear inverse cost function without considering the pre-generating paths and charging combinations [40].

It should be noted that it is hard to consider all these constraints simultaneously, say, the elastic demand of the road network, the capacity of the charging stations, and the range limit of EVs by using the SUE-ED model in location problems of public EV charging stations. In this study, a novel bilevel public charging station location model that combines a flow-capturing location model and a multi-objective optimization model is proposed.

3. Problem Assumptions and EV Paths Analysis

The proposed location model has two optimization objectives. The upper layer employs an improved maximum coverage model to maximize the coverage of the total EV flow by deploying a given number of charging stations on the links where the EV flow is the largest. The charging stations can serve more EV users by deploying charging facilities on the links where most EV drivers pass, which is an effective way to improve the utilization rate of public charging facilities and alleviate range anxieties of electric vehicle drivers [4]. The lower layer uses an SUE-ED model with incomplete information. The assignment result of SUE is a decisive factor for the placement of the charging stations. There is an inherent difference in the driving behavior of each user (EV users and GV users), in the hybrid network. In particular, range limit, charging time, and location of charging facilities have an important impact on the path choose behaviors of EV users.

3.1. Notations

i: The set of traveler types in the network. i presents an element, mainly including EVs and GVs. When specific instructions are needed, subscripts g and e are used to indicate variables or parameters related to GVs and EVs, respectively;
W: The set of OD pairs. w is an element of the set;
R_w: The set of paths between all OD pairs;
p: The number of charging stations subject to the budget;
U: The utility value of the charging stations, resulting in a reduction in travel cost;
$K{\cdot}U$: The waiting cost in charging stations, where K is a constant coefficient;
l_k^w: Length of path k between OD pair w;
l_k^{ri}: Length of path k between node r and I;
l_k^{ij}: Length of path k between node i and j;
l_k^{js}: Length of path k between node j and s;
R_e: The range limit of EVs;
ε: Charging power. Unit: hour/mile;
t_{ck}^w: The minimum charging time;
v_a: Traffic flow on link a, including EVs and GVs, so $v_a = v_{ae} + v_{ag}$;
$t_a(v_{ag}, v_{ae})$: Actual travel time through link a;
t_a^0: Free travel time through link a;
H_a: Capacity of link a;
x_a: Binary parameter. If the charging facility is on link a, $x_a = 1$, otherwise $x_a = 0$;
δ_{ak}^w: Binary parameter. If the path k crosses the link a, $\delta_{ak}^w = 1$, otherwise $\delta_{ak}^w = 0$;
q_i^w: Elastic travel demand of type-i traveler between OD pair w;
$\bar{q}_i^w$: Maximum potential demand of type-i traveler between OD pair w;
f_{ki}^w: Traffic flow of type-i traveler on path k between OD pair w;
p_{ki}^w: The probability that type-i traveler chooses the path k between OD pair w;
c_k^w: The actual travel time of the path k between OD pair w, and $c_k^w = \sum t_a(v_a)\delta_{ak}^w$;
$\bar{c}_{ki}^w$: The generalized travel cost of type-i traveler on path k between OD pair w;
C_{wi}: The expected perceived travel cost of the type-i traveler between OD pair w;
$D_{wi}(\cdot)$: The traffic demand function of the type-i traveler between OD pair w;
$c^w(\vec{x})$: Vector of the actual travel cost of all paths between OD pair w;
θ_i: a non-negative parameter that characterizes the uncertainty of type-i traveler's understanding of the path travel time.

3.2. Propose Assumptions

This paper focus on a hybrid network, denoted as $G = (N,A)$, where N is the set of nodes, A is the set of links. W is used to represent a set of OD pairs, and w is an element, $w \in W$. R_w represents the set of paths between w. The main differences between GVs and EVs are the limitation of travel distance and the composition of travel costs. Without loss of generality, the following assumptions are made:

(a) There are two types of users in the mixed network: GVs and EVs, where EVs have an identical range limit;
(b) The travel demand of GVs and EVs between each OD pair is elastic. And two types of vehicles have incomplete information about the travel cost;
(c) Each EV is fully charged at its origin;
(d) The level of anxiety and risk-taking behaviors of EV drivers are not considered in this model;
(e) The charging time of electric vehicles is linearly related to mileage;
(f) The charging facilities will be placed in the middle of links whose EV flow ranks p;
(g) Deploying the charging facility on the road/path will increase the attractiveness of the route, which is called the utility of the charging facilities U, reducing the path travel cost;
(h) Charging facilities have a fixed charging capacity. It is assumed that the waiting cost is proportional to the attraction value, expressed by $K \cdot U$.

3.3. Analysis of EV Paths

When performing SUE-ED traffic assignment, the GVs can choose any route because of no range limit. However, the selection of feasible path sets is required before the distribution of EV flow. To describe the three types of EV paths more clearly, the concept of sub-path is described as follows.

Suppose that the origin is r and the destination is s in a pair OD w, and there are two charging stations i, j, which are also regarded as nodes, located in the arc of the path k (Figure 1). When k^{ri}, k^{ij}, k^{js} no longer includes other charging facilities except i, j, they are called sub-paths [41].

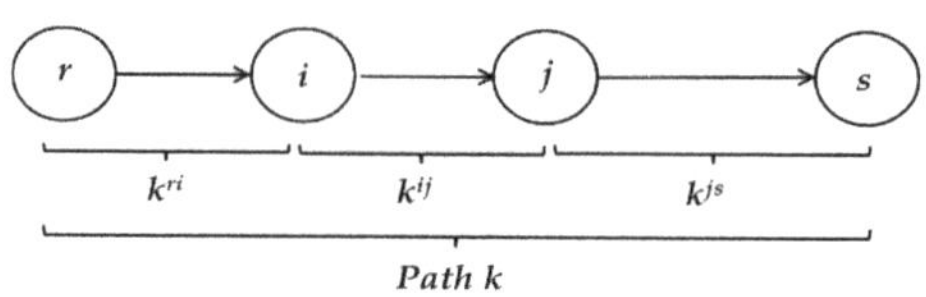

Figure 1. Description of the sub-path.

Based on the relationship between the travel mileage limit R_e and the sub-path distance, the three travel paths of EVs are discussed, as shown in Table 1:

Scenario 1. A travel path that can be completed without charging. When $l_k^w \le R_e$, the EV drivers can reach the destination without charging. The generalized travel cost is $\bar{c}_{ke}^w = c_k^w$ (no charging stations) or $\bar{c}_{ke}^w = c_k^w - U$ (with charging stations).

Scenario 2. A travel path that cannot be completed after charging. When $l_k^{ri} \ge R_e$ or $l_k^{ij} \ge R_e$ or $l_k^{js} \ge R_e$, the path k is an infeasible path. And the generalized path travel cost becomes infinite ($\bar{c}_{ke}^w = \infty$) and the probability that the drivers select the path is zero.

Scenario 3. A travel path that can be completed after charging. The path k is defined as a feasible path, if $l_k^w \ge R_e$ and $l_k^{ri} \le R_e$ and $l_k^{ij} \le R_e$ and $l_k^{js} \le R_e$. The drivers can complete the trip by charging at least once and the minimum charging time is $t_{ck}^w = \varepsilon \ldots (l_k^w - R_e)$. The generalized travel cost is expressed as $\bar{c}_{ke}^w = c_k^w + t_{ck}^w - U + K \cdot U = c_k^w + t_{ck}^w + (K-1)U$, consisting of four parts: travel time, charging time, charging facilities utility, and charging stations' waiting cost.

For GVs, the general path travel cost is $\bar{c}_{kg}^w = c_k^w$.

Table 1. Three travel paths of electric vehicles (EVs).

Scenarios	Conditions	Travel Cost
1. A path that can be completed without charging	$l_k^w \leq R_e$	$\bar{c}_{ke}^w = c_k^w$
2. A path that cannot be completed after charging	$l_k^{ri} \geq R_e$ or $l_k^{ij} R_e$ or l_k^{js} R_e	$\bar{c}_{ke}^w = \infty$
3. A path that can be completed after charging	$l_k^w \geq R_e$ and $l_k^{ri} \leq R_e$ and $l_k^{ij} \leq R_e$ and $l_k^{js} \leq R_e$	$\bar{c}_{ke}^w = c_k^w + t_{ck}^w + (K-1)U$

4. Establish a Double-Layer Model

In this section, a bilevel optimization model for the location of charging facilities is proposed. The upper level model determines the location of the charging facilities by selecting the top p links. The lower level model calculates users' generalized travel cost, and randomly allocates the flow demand between the OD pairs to the filtered paths through known charging facilities location.

4.1. Upper Level Problem

The upper model is designed to maximize the total covered EV link flow by deploying a given number of charging facilities. The EV flow is covered when the charging facility is on the link. That is

$$\text{Max} \sum_a v_{ae} x_a \tag{1}$$

$$\text{Subject to} \sum_{a \in A} x_a = p \tag{2}$$

Equation (1) is the objective function of the upper model, and Equation (2) is the budget constraint indicating the number of charging stations p in a given network.

4.2. Lower Level Problem

The Bureau of Public Road (BPR) function in the lower model is employed [42].

$$t_a(v_{ag}, v_{ae}) - t_a^0 \left\{ 1 + 0.15 \times \left(\frac{v_{ag} + v_{ae}}{H_a} \right)^4 \right\} \tag{3}$$

Since U, t_{ck}^w, k are parameters that are independent of the equilibrium flow of the link, the Jacobian matrix [43] in the lower layer problem is as follows:

$$\frac{\partial \bar{c}_{kg}^w}{\partial v_{ae}} = \frac{\partial \bar{c}_{ke}^w}{\partial v_{ag}} = 0.6 \sum t_a^0 \delta_{ak}^w \frac{(v_{ag} + v_{ae})^3}{{H_a}^4} \tag{4}$$

It proves that the Jacobian matrix is symmetrical, and the lower layer model can be established as a convex function problem. It is assumed that all types of travelers make path selection in a random manner. According to the random utility theory, the probability that type-i traveler chooses the path k between OD pair w is:

$$p_{ki}^w = \frac{exp(-\theta_i \bar{c}_{ki}^w)}{\sum_{k \in R_w} exp(-\theta_i \bar{c}_{ki}^w)} \quad \forall k, w, i \tag{5}$$

Assume that elastic travel demand q_i^w of type-i traveler is a strictly monotonically decreasing function of the expected minimum travel time between OD pair w, and with an upper bound.

$$q_i^w = D_{wi}(C_{wi}) \leq \bar{q}_i^w \quad \forall w, i \tag{6}$$

According to the discrete selection theory [44,45], define C_{wi} as:

$$C_{wi}(c^{w}(\vec{x}))=E\left[\min_{k\in R_w}\{\bar{c}_{ki}^{w}\}|\,c^{w}(\vec{x})\right]=-\frac{1}{\theta_i}\ln\sum_{k\in R_w}exp(-\theta_i\bar{c}_{ki}^{w})\quad\forall w,i \tag{7}$$

For the type-i traveler, its SUE condition can be expressed as:

$$f_{ki}^{w}=q_i^{w}p_{ki}^{w}\;\forall k,w,i \tag{8}$$

In the mixed network, SUE-ED problem can be described by the equivalent mathematical programming model:

$$\begin{aligned}\text{Min } Z(x,f,q)=&\sum_{a\in A}\int_0^{v_a}t_a(w)dw+\sum_{i\in I}\frac{1}{\theta_i}\sum_{w\in W}\sum_{k\in R_w}f_{ki}^{w}(\ln f_{ki}^{w}-1)\\&-\sum_{i\in I}\frac{1}{\theta_i}\sum_{w\in W}\sum_{k\in R_w}q_i^{w}(\ln q_i^{w}-1)-\sum_{i\in I}\sum_{w\in W}\int_0^{q_i^{w}}D_{wi}^{-1}(w)dw\end{aligned} \tag{9}$$

$$\text{Subject to}:\; q_i^{w}=\sum_k f_{ki}^{w},\;\forall w,i \tag{10}$$

$$f_{ki}^{w}\geq 0,\;\forall k,w,i \tag{11}$$

$$q_i^{w}\geq 0,\;\forall w,i \tag{12}$$

$$v_{ai}=\sum_{w\in W}\sum_{k\in R_w}f_{ki}^{w}\delta_{ak}^{w},\;\forall a,i \tag{13}$$

$$R_e-l_k^{ij}\geq 0,\;\forall k \tag{14}$$

Equation (10) is the flow conservation constraint. Equation (11) is the non-negative constraint of path flow for type-i travelers. Equation (12) is the non-negative constraint of type-i traveler's OD flow demand. Equation (13) is the correlation between link flow and path flow. The novelty of this problem is that the introduction of sub-paths in the Equation (14) can generate a feasible set of paths in advance from the finite paths between each OD pair. The superscripts i and j include the origin r and the destination s of all OD pairs in Equation (14).

It is necessary to prove the equivalence between the solution of the proposed program Equation (9) and the solution of the SUE-ED model. The generalized Lagrangian function [46] of the mathematical model is constructed as follows:

$$L=Z+\sum_i\sum_w\lambda_{wi}(q_i^{w}-\sum_{k\in R_w}f_{ki}^{w})-\sum_i\sum_w\sum_k\mu_{ki}^{w}f_{ki}^{w}-\sum_{i\in I}\sum_{w\in W}v_{wi}q_i^{w}-\sum_w\sum_k\alpha_{ke}^{ij}*(R_e-l_k^{ij}) \tag{15}$$

λ_{wi}, μ_{ki}^{w}, v_{wi}, α_{ke}^{ij} are the Lagrange multipliers.

$$\begin{array}{ll}\text{If } l_k^{ij}\leq R_e & \alpha_{ke}^{ij}=0\\ \text{If } l_k^{ij}\geq R_e & \alpha_{ke}^{ij}\geq 0\end{array} \tag{16}$$

According to the Kuhn Tucker conditions [47], Equation (15) must satisfy the following conditions at the extreme point:

$$\frac{\partial L}{\partial f_{ki}^{w}}=0,\;\mu_{ki}^{w}f_{ki}^{w}=0,\;\mu_{ki}^{w}\geq 0\quad\forall k,w,i \tag{17}$$

$$\frac{\partial L}{\partial q_i^{w}}=0,\;v_{wi}q_i^{w}=0,\;v_{wi}\geq 0\quad\forall w,i \tag{18}$$

The partial derivative of f_{ki}^{w} for Equation (15) is derived as follows:

$$\frac{\partial L}{\partial f_{ki}^{w}} = \sum_{a} t_a(v_a)\delta_{ak}^{w} + \frac{1}{\theta_i}\ln f_{ki}^{w} - \lambda_{wi} - \mu_{ki}^{w} = 0 \quad \forall k, w, i \tag{19}$$

Note that if $f_{ki}^{w} = 0$, then $\frac{\partial L}{\partial f_{ki}^{w}}$ does not exist. The above formula is only valid when $f_{ki}^{w} > 0$. So $\mu_{ki}^{w} = 0$.

$$f_{ki}^{w} = \exp\left[-\theta_i(\bar{c}_{ki}^{w} - \lambda_{wi})\right] = \exp\left[-\theta_i\bar{c}_{ki}^{w}\right] \cdot \exp[-\theta_i\lambda_{wi}] \tag{20}$$

$$q_{i}^{w} = \exp(-\theta_i\lambda_{wi}) \sum_{k \in R_w} \exp(-\theta_i\bar{c}_{ki}^{w}) \tag{21}$$

$$p_{ki}^{w} = \frac{exp(-\theta_i\bar{c}_{ki}^{w})}{\sum_{k \in R_w} exp(-\theta_i\bar{c}_{ki}^{w})} \quad \forall k, w, i \tag{22}$$

Equation (22) shows that type-i traveler follows the Logit model to select the travel path, which satisfies the SUE-ED condition described in Equation (8).

Calculate the partial derivative of q_{i}^{w} for Equation (15) as follows:

$$\frac{\partial L}{\partial q_{i}^{w}} = -\frac{1}{\theta_i}\ln q_{i}^{w} - D_{wi}^{-1}(q_{i}^{w}) + \lambda_{wi} - v_{wi} = 0 \ \forall w, i \tag{23}$$

Because $\ln q_{i}^{w}$ should exist, so $v_{wi} = 0$. Then derive from Equation (21):

$$\theta_i\lambda_{wi} = \ln q_{i}^{w} - \ln \sum_{r \in R_w} \exp\left[-\theta_i\bar{c}_{ri}^{w}\right] \tag{24}$$

Comparing Equation (23) and Equation (24), the inverse function of the flow demand function is as follows:

$$D_{wi}^{-1}(q_{i}^{w}) = -\frac{1}{\theta_i}\ln \sum_{r \in R_w} \exp\left[-\theta_i\bar{c}_{ri}^{w}\right] \ \forall w, i \tag{25}$$

Then $q_{i}^{w} = D_{wi}(C_{wi})$. This shows that Equations (9)–(14) can be used to represent a multi-user SUE problem under elastic demand. The BPR function $t_a(v_a)$ is a strictly monotonically increasing function of the link flow v_a. The objective function is a strict convex function about the link flow vector v and the path flow vector f. At the same time, the constraints of Equations (9)–(14) are linear equality constraints and non-negative constraints, so its solution space is a convex set. According to the optimization theory, the strict convex function defined on the convex set only has one optimal solution.

5. Solution Method

The proposed bilevel model is an NP-hard (non-deterministic polynomial hard) problem because the lower level model is a mixed-integer nonlinear program with nonconvex path choosing, subject to uncertainties in elastic demands, where it is difficult to find a polynomial time complexity algorithm [48]. Thus, a heuristic method is employed to solve the problem of SUE-ED and MFC in an iterative manner. The detailed processes in Figure 2 are as follows:

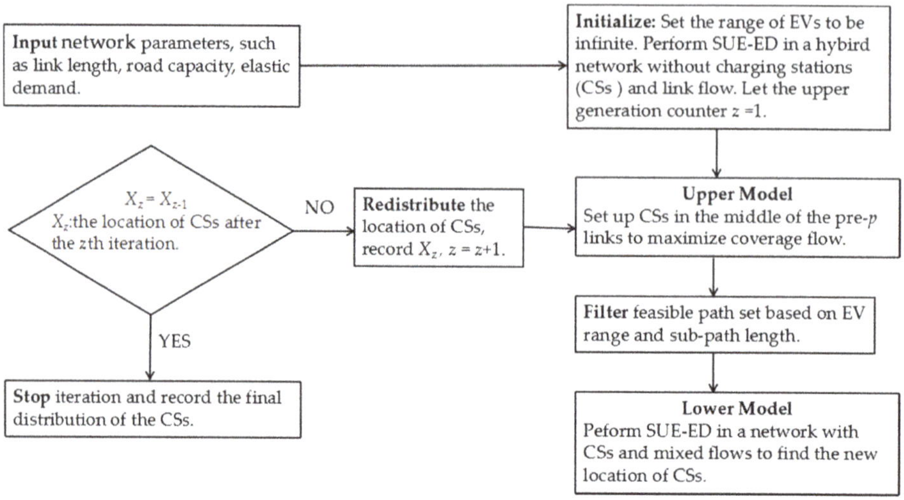

Figure 2. Framework of the bilevel model to solve the optimal location of charging stations (CSs).

Step 1: Input the road network parameters, such as the length of the link and the capacity of the links, and find all the paths between the OD pair w, and record them as the initial path set.

Step 2: Initialize. There is no charging facility in the road network and relax the range limit of EVs. Based on the travel time of the zero-flow link $t_a(0) = t_a^0$, calculate the effective path impedance c_k^w, and obtain the initial link flows $V_{ag}(1)$ and $V_{ae}(1)$. Let the upper generation counter $z = 1$.

Step 3: Sort all the links flow of EVs in ascending order to find the top p ranked links, and arrange the charging facilities in the middle of the p links. Then increase the upper iteration counter by 1.

Step 4: Perform SUE-ED traffic distribution on the network with the location of the charging facilities obtained by step 3. The detailed steps are as follows:

Step 4.1: Check feasible paths and update the path set. If the length of any sub-path is greater than R_e, remove the path from the initial set of paths. The program stops if there is no feasible path between any OD pair. If there is at least one feasible path between each OD pair, proceed to the next step.

Step 4.2: Carry out random loading of traffic flow in the network with charging facilities. Calculate the generalized path travel cost $\bar{c}_{ke}^w$, elastic flow demand between OD pairs, the probability that the path is selected to obtain updated road link flows $V_{ag}(2)$ and $V_{ae}(2)$. Then set the iteration counter $n = 1$.

Step 4.3: Repeat the random loading of step 4.2 to obtain additional link flow $\{y_{ag}\}$, $\{y_{ae}\}$.

Step 4.4: Use the predetermined step size sequence $\{\alpha_n\}$: $_a = 1/n, n = 1,2,\ldots,\infty$.

Step 4.5: Calculate the current flow of links by Method of Successive Average [49].

$V_{ag}^{n+1} = V_{ag}^n + (\frac{1}{n})(y_{ag}^n - V_{ag}^n)$; $V_{ae}^{n+1} = V_{ae}^n + (\frac{1}{n})(y_{ae}^n - V_{ae}^n)$.

Step 4.6: If the convergence condition is met $\frac{\sqrt{\sum_a (\bar{v}_a^{n+1} - \bar{v}_a^n)^2}}{\sum_a \bar{v}_a^n} \le \omega$, where ω indicates the convergence accuracy, then proceed to step 5. $\{V_{ag}^{n+1}\}$, $\{V_{ae}^{n+1}\}$ are the sets of balanced link flow for the GVs and EVs, respectively. Otherwise, set $n = n + 1$ and go to step 4.3.

Step 5: Repeat step 3 and update charging facilities' location. The program stops until the location of the charging facilities is no longer changed; that is, the maximum coverage flow remains stable; otherwise, go to step 4.

6. Numerical Analysis

The model is applied to the Nguyen–Dupuis network [32,45] for a case study in this section, which has been widely used in transportation network researches in the past decades. Due to the small scale of the Nguyen–Dupuis network, the paths can be enumerated to better analyze the relationship between the location of the charging station and the feasible path and the traffic volume of the section. At the same time, the elastic demand between OD pairs, the effects of charging speed, range limitation, charging facilities utility, and waiting cost on the location of charging facilities were evaluated. The test network (shown in Figure 3) consisted of 13 nodes, 19 segments, 25 paths (shown in Table 2), and 4 OD pairs (1, 3), (1, 2), (4, 3), (4, 2). The traveler's OD demand function uses a linear function. The familiarity of the two types of travelers to the network was measured by the parameters θ_e and θ_g, respectively. The parameters were set as follows during the trial calculation. $q_e^w(C_{we}) = 400 - 7C_{we}$, $q_g^w(C_{wg}) = 400 - 7C_{wg}$, $\theta_e = 0.1$, $\theta_g = 0.1$, $U = 5$, $R_e = 20$, $\varepsilon = 1$, $K = 0.5$, $p = 3$. The trial results are shown in Table 3.

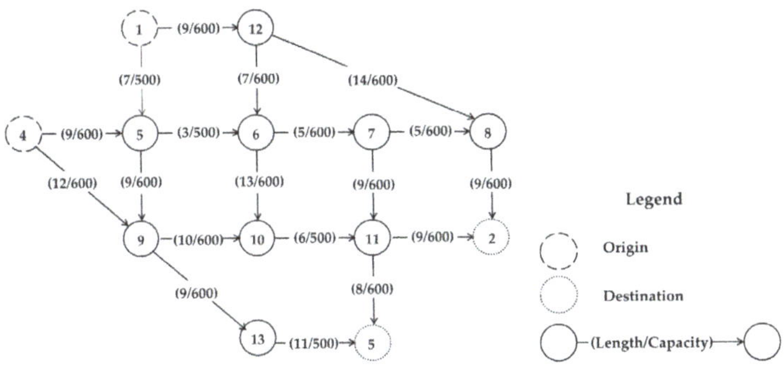

Figure 3. Nguyen–Dupuis network.

Table 2. Path compositions and lengths in the Nguyen–Dupuis network.

OD Pair	Path Number	Path Composition	Length
(1,2)	1	1-5-6-7-8-2	29
	2	1-12-8-2	32
	3	1-5-6-7-11-12	33
	4	1-12-6-7-8-2	35
	5	1-5-6-10-11-2	38
	6	1-12-6-7-11-2	39
	7	1-5-9-10-11-2	41
	8	1-12-6-10-11-2	44
(1.3)	9	1-5-6-7-11-3	32
	10	1-5-9-13-3	36
	11	1-5-6-10-11-3	37
	12	1-12-6-7-11-3	38
	13	1-5-9-10-11-3	40
	14	1-12-6-10-11-3	43
(4,2)	15	4-5-6-7-8-2	31
	16	4-5-6-7-11-2	35
	17	4-9-10-11-2	37
	18	4-5-6-10-11-2	40
	19	4-5-9-10-11-2	43
(4,3)	20	4-9-13-3	32
	21	4-5-6-7-11-3	34
	22	4-9-10-11-3	36
	23	4-5-9-13-3	38
	24	4-5-6-10-11-3	39
	25	4-5-9-10-11-3	42

Table 3. The charging facility locations and Electric Vehicle (EV) flows

Link Node	1		2		3		4		20
	Location	EV Flow	Location	EV Flow	Location	EV Flow	Location	EV Flow	Location
(1,5)	×	367.5	×	219.3	√	197.4	×	172.5	×
(4,5)	×	364.9	×	195.4	×	172	×	143.7	×
(4,9)	×	164.5	×	54.8	×	27.4	×	0	×
(5,6)	×	538.5	√	350.1	√	337.1	√	316.2	√
(5,9)	×	193.9	×	64.6	×	32.3	×	0	×
(6,7)	×	499.7	√	337.3	√	363.7	√	380	√
(7,8)	×	196.7	×	180.5	×	210.3	×	233.2	×
(8,2)	×	248.7	×	197.8	×	219	√	233.2	√
(1,12)	×	196.7	×	100.5	×	83.2	×	63.8	×
(11,3)	×	380.1	×	200.4	×	175.2	×	146.8	×
(12,6)	×	144.8	×	83.2	×	74.6	×	63.8	×
(12,8)	×	51.9	×	17.3	×	8.7	×	0	×
(13,3)	×	157.1	×	52.4	×	26.2	×	0	×
(6,10)	×	183.7	×	95.9	×	48	×	0	×
(7,11)	×	302.9	×	156.8	×	153.4	×	146.8	×
(9,10)	×	201.4	×	67.1	×	33.6	×	0	×
(9,13)	×	157.1	×	52.4	×	26.2	×	0	×
(10,11)	×	385.1	√	163	×	81.5	×	0	×
(11,2)	×	308	×	119.5	×	59.7	×	0	×

Table 3 shows the results of 20 iterations. The total coverage of the charging stations was 0, 906.7, 919.8, 924.9 … 929.5. In the first iteration, only six paths were available among the four OD pairs, thereby causing a change in the probability of path selection. It is not difficult to find that the EV link flow tended to be zero in the infeasible paths, and the EV flows gradually reached equilibrium in the links with charging facilities at last. The final locations of the charging facilities were (5,6); (6,7); (8,2). The changes in the location of charging stations (CSs) and feasible paths are listed in Table 4. After the first iteration, the feasible paths for the EVs were reduced because of range limitations. Among the four OD pairs, only six paths were left to choose, which led to a change in the probability of path selection, so the charging facility location also changed. The reason for the change in CSs from (5,6); (6,7); (1,5) to (5,6); (6,7); (8,2) is mainly due to the fact that there were three paths, including link (8,2), but just one path including link(1,5). In particular, only one feasible path, including the link (8,2), was left in the OD pair (4,2), so all EV users could only select this path.

Table 4. The changes in the location of charging stations (CSs) and feasible paths.

OD Pair	Location 1 (5,6); (6,7); (10,11)	Location 2 (5,6); (6,7); (1,5)	Location 3 (5,6); (6,7); (8,2)
(1,2)	Three paths, including 1,4,5	Two paths, including 1,4	Two paths, including 1,4
(1,3)	Three paths, including 9,11,12	Two paths, including 9,12	Two paths, including 9,12
(4,2)	Two paths, including 15,18	One path, 15	One path, 15
(4,3)	Two paths, including 21, 24	One path, 21	One path, 21

The elastic demand analysis between OD pairs (1,3) is shown in Figure 4. Traffic demands are mainly affected by travel costs. The maximum demand for both electric cars and gasoline were set as 400. At the beginning, there were no CSs on the network, and the range limit of EVs was relaxed, so the flow demands of EVs and GVs were the same, 265.81. In the second iteration, the reduction in travel demand for EVs and GVs was mainly due to the increase in travel costs caused by the increased link flow. For EVs, the feasible paths were less than GVs because of the range limit, and the travel cost was not only affected by the link flow as GVs, but also affected by the utility of charging facilities, charging speed, and waiting cost at charging stations. Therefore, the travel demand of EVs was much lower than GVs, which inspired us to increase the range of EVs and the accessibility of charging facilities to promote the travel demand and development of EVs. The elastic fluctuation of flow demands affect the distribution of link flow and further affect the travel cost and the location of charging facilities, which will react to the travel demand. In the continuous iteration and mutual influence, the final equilibrium was reached. The change trend of elastic demand in other OD pairs was consistent with the OD pair (1,3), which was matched with the actual travel demand.

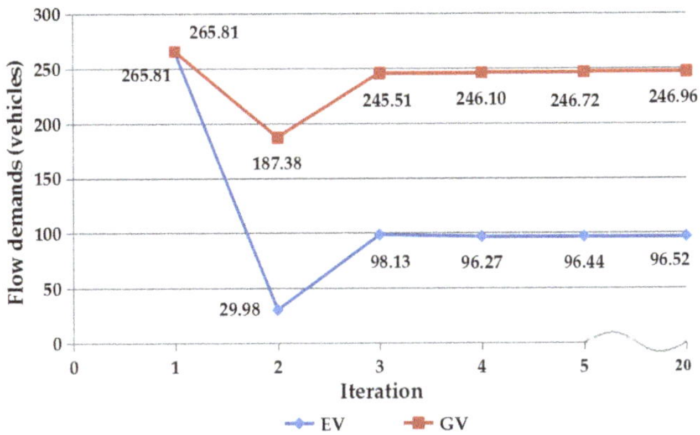

Figure 4. The flow demands of origin-destination (OD) pair (1,3).

The results of sensitivity analysis about range limits, charging speed, charging facilities utility, and charging stations waiting cost are shown in Figures 5–8.

In Figure 5, the impact of EV range limit on the locations of charging facilities is revealed. When the range limits change, the charging facilities were located differently. As the range limit increased, there were more feasible paths within a certain range, and the EV covered flow first increases and then decreased, and finally increases when $R_e = 45$.

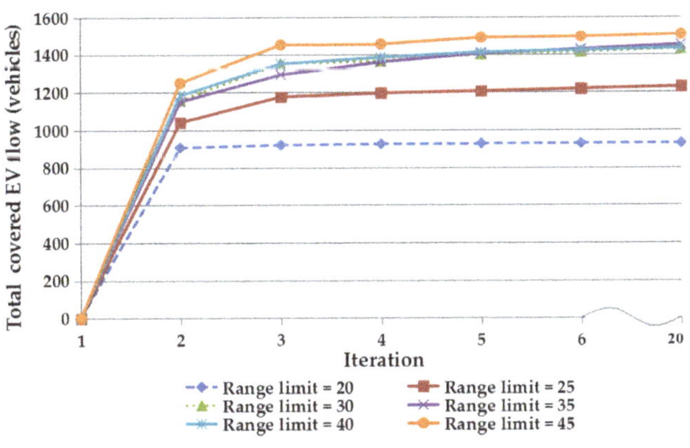

Figure 5. The sensitivity analysis of the range limit.

Then the effect of charging speed on the location of the charging facilities is examined in Figure 6. Formula $t_{ck}^{w} = \varepsilon \cdot (l_k^w - R_e)$ indicates that the charging speed directly affects the charging time and general travel cost, which results in different charging facility layouts. Different values of ε can present different charging methods. $\varepsilon = 0.1$ means fast charging and $\varepsilon = 10$ means slow charging. EV drivers tend to choose a path with a long travel distance without charging facilities, rather than a path with a very slow charging speed.

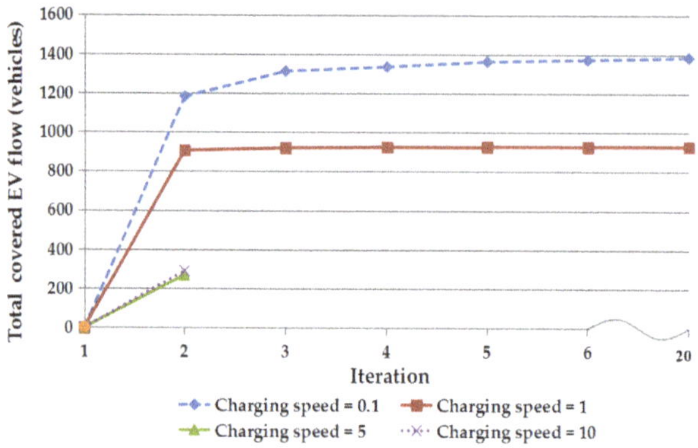

Figure 6. The sensitivity analysis of charging speed.

In Figure 7, the utility of the charging facilities reflects the attractiveness of the link for EV users. The greater the utility value, the higher the level of anxiety for EV users, so they are more likely to choose a path with charging facilities. If there are many types of EVs in the network, the utility value may not be very large for EVs with large battery capacity.

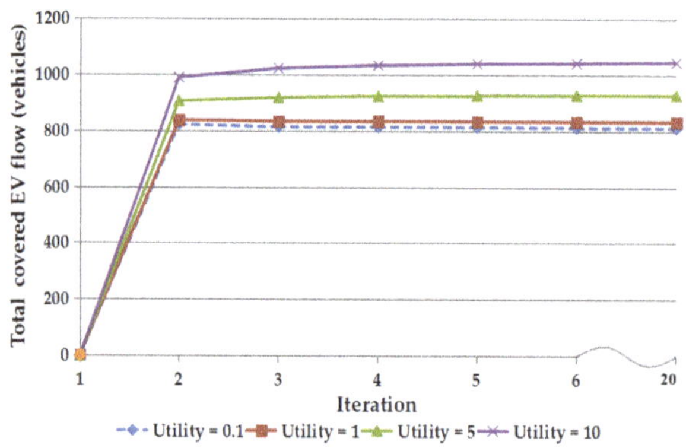

Figure 7. The sensitivity analysis of charging facilities' utility.

The relationship between the charging stations waiting cost coefficient K and the charging station position is shown in Figure 8. When $0 < K < 1$, the charging cost is reduced. When $K > 1$, the charging cost increases, and the total coverage of the electric vehicle begins to decrease. When deploying a fast charging station, a slightly larger capacity should be considered to reduce the occurrence of $K > 1$.

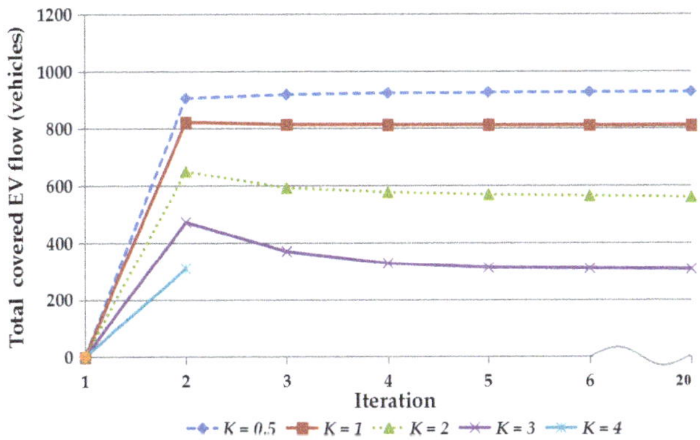

Figure 8. The sensitivity analysis of waiting cost coefficient *K*.

7. Conclusions

This paper studies the deployment of public charging stations in a hybrid network to maximize the service efficiency of the charging facilities. A bilevel model was proposed to depict the interaction between the mixed link flow and the location of the charging stations. The link flow obtained by the lower SUE-ED model is the key to determine the location of the CSs. In the upper level model, the CSs were arranged on the link flow ranking p to achieve the maximum coverage. Four important factors were taken into accounts in the lower level SUE model, including the range limit of EVs that affected the path choice, the elastic travel demand closely related to the distribution of link flow, the road congestion effect, and the capacity of charging facilities in travel costs. These four elements interact with each other and continuously iterate to reach equilibrium, which was more consistent with the actual travel situation. A hybrid integer nonlinear programming method based on the method of successive average (MSA) was constructed to prove the equivalence and the uniqueness of the SUE-ED model with range constraints. Finally, a network trial was conducted to examine the impact of elastic demand between OD pairs, the range limit, charging speed, the charging facilities' utility, and waiting cost on the location problem.

It should be noted that the actual road network systems are very diverse, especially in urban areas, and significantly differ from the example of the Nguyen–Dupuis Network. More realistic factors are required to be considered in the future when designing the location of charging stations, not only by technical and operational factors but also by social factors. And assumptions can be relaxed appropriately in future work. The nonlinearity of the charging time, the uncertainties in EVs energy consumption, as well as the bounded rationality of EV travelers, should be considered.

Author Contributions: Conceptualization, H.G. and K.L.; methodology, H.G. and K.L.; software, H.G. and X.P.; validation, H.G., K.L. and X.P.; formal analysis, H.G., K.L. and X.P.; investigation, H.G., K.L. and C.L.; resources, C.L.; writing—original draft preparation, H.G. and K.L.; writing—review and editing, H.G. and K.L.; visualization, H.G.; supervision, K.L. and C.L.; project administration, K.L.; funding acquisition, K.L. All authors have read and agreed to the published version of the manuscript.

Funding: This research was funded by the National Natural Science Foundation of China (Grant Nos. 51378091 and 71871043) and the open project of the Key Laboratory of Advanced Urban Public Transportation Science, Ministry of Transport, PRC.

Acknowledgments: This work was carried out by the joint research program of the Institute of Materials and Systems for Sustainability, Nagoya University.

Conflicts of Interest: The authors declare no conflicts of interest. The funders had no role in the design of the study; in the collection, analyses, or interpretation of data; in the writing of the manuscript, or in the decision to publish the results.

References

1. Pan, L.; Yao, E.; MacKenzie, D.; Zhang, R. Environmental Effects of BEV Penetration Considering Traffic Status. *J. Transp. Eng. Part A Syst.* **2019**, *145*, 04019048. [CrossRef]
2. Bapna, R.; Thakur, L.S.; Nair, S.K. Infrastructure development for conversion to environmentally friendly fuel. *Eur. J. Oper. Res.* **2002**, *142*, 480–496. [CrossRef]
3. Wang, Y.; Liu, Z.; Shi, J.; Wu, G.; Wang, R. Joint Optimal Policy for Subsidy on Electric Vehicles and Infrastructure Construction in Highway Network. *Energies* **2018**, *11*, 2479. [CrossRef]
4. Dong, J.; Liu, C.; Lin, Z. Charging infrastructure planning for promoting battery electric vehicles: An activity-based approach using multiday travel data. *Transp. Res. Part C Emerg. Technol.* **2014**, *38*, 44–55. [CrossRef]
5. Wang, J.; Liu, K.; Yamamoto, T. Improving Electricity Consumption Estimation for Electric Vehicles Based on Sparse GPS Observations. *Energies* **2017**, *10*, 129. [CrossRef]
6. Xu, M.; Meng, Q.; Liu, K.; Yamamoto, T. Joint charging mode and location choice model for battery electric vehicle users. *Transp. Res. Part B Methodol.* **2017**, *103*, 68–86. [CrossRef]
7. Wang, H.; Zhao, D.; Meng, Q.; Ong, G.P.; Lee, D.-H. Network-level energy consumption estimation for electric vehicles considering vehicle and user heterogeneity. *Transp. Res. Part A Policy Pr.* **2020**, *132*, 30–46. [CrossRef]
8. Jin, F.; Yao, E.; An, K. Analysis of the potential demand for battery electric vehicle sharing: Mode share and spatiotemporal distribution. *J. Transp. Geogr.* **2020**, *82*, 102630. [CrossRef]
9. Liu, K.; Wang, J.; Yamamoto, T.; Morikawa, T. Modelling the multilevel structure and mixed effects of the factors influencing the energy consumption of electric vehicles. *Appl. Energy* **2016**, *183*, 1351–1360. [CrossRef]
10. Liu, K.; Yamamoto, T.; Morikawa, T. Impact of road gradient on energy consumption of electric vehicles. *Transp. Res. Part D Transp. Environ.* **2017**, *54*, 74–81. [CrossRef]
11. Liu, K.; Wang, J.; Yamamoto, T.; Morikawa, T. Exploring the interactive effects of ambient temperature and vehicle auxiliary loads on electric vehicle energy consumption. *Appl. Energy* **2018**, *227*, 324–331. [CrossRef]
12. Wu, X.; Freese, D.; Cabrera, A.; Kitch, W.A. Electric vehicles' energy consumption measurement and estimation. *Transp. Res. Part D Transp. Environ.* **2015**, *34*, 52–67. [CrossRef]
13. Jiang, N.; Xie, C.; Waller, T. Path-Constrained Traffic Assignment. *Transp. Res. Rec. J. Transp. Res. Board* **2012**, *2283*, 25–33. [CrossRef]
14. Tang, T.-Q.; Chen, L.; Yang, S.-C.; Shang, H.-Y. An extended car-following model with consideration of the electric vehicle's driving range. *Phys. A Stat. Mech. Appl.* **2015**, *430*, 148–155. [CrossRef]
15. Wang, H.; Zhao, D.; Cai, Y.; Meng, Q.; Ong, G.P. A trajectory-based energy consumption estimation method considering battery degradation for an urban electric vehicle network. *Transp. Res. Part D Transp. Environ.* **2019**, *74*, 142–153. [CrossRef]
16. Huan, N.; Yao, E.; Fan, Y.; Wang, Z. Evaluating the Environmental Impact of Bus Signal Priority at Intersections under Hybrid Energy Consumption Conditions. *Energies* **2019**, *12*, 4555. [CrossRef]
17. Guo, C.; Yang, J.; Yang, L. Planning of Electric Vehicle Charging Infrastructure for Urban Areas with Tight Land Supply. *Energies* **2018**, *11*, 2314. [CrossRef]
18. Gartner, N.H. Optimal Traffic Assignment with Elastic Demands: A Review Part I. Analysis Framework. *Transp. Sci.* **1980**, *14*, 174–191. [CrossRef]
19. Hakimi, S.L. Optimum Locations of Switching Centers and the Absolute Centers and Medians of a Graph. *Oper. Res.* **1964**, *12*, 450–459. [CrossRef]
20. Daskin, M. Network and Discrete Location: Models, Algorithms and Applications. *J. Oper. Res. Soc.* **1997**, *48*, 763–764. [CrossRef]
21. Hodgson, M.J.; Rosing, K.E. A network location-allocation model trading off flow capturing and p-median objectives. *Ann. Oper. Res.* **1992**, *40*, 247–260. [CrossRef]
22. Hodgson, M.J. A Flow Capturing Location-allocation Model. *Geogr. Anal.* **2010**, *22*, 270–279. [CrossRef]
23. Wang, Y.-W.; Wang, C.-R. Locating passenger vehicle refueling stations. *Transp. Res. Part E Logist. Transp. Rev.* **2010**, *46*, 791–801. [CrossRef]
24. Yi, T.; Cheng, X.-B.; Zheng, H.; Liu, J.-P. Research on Location and Capacity Optimization Method for Electric Vehicle Charging Stations Considering User's Comprehensive Satisfaction. *Energies* **2019**, *12*, 1915. [CrossRef]

25. Kong, W.; Luo, Y.; Feng, G.; Li, K.; Peng, H.; Keqiang, L. Optimal location planning method of fast charging station for electric vehicles considering operators, drivers, vehicles, traffic flow and power grid. *Energy* **2019**, *186*, 115826. [CrossRef]
26. Liu, Q.; Liu, J.; Le, W.; Guo, Z.; He, Z. Data-driven intelligent location of public charging stations for electric vehicles. *J. Clean. Prod.* **2019**, *232*, 531–541. [CrossRef]
27. Wang, H.; Zhao, D.; Meng, Q.; Ong, G.P.; Lee, D.-H. A four-step method for electric-vehicle charging facility deployment in a dense city: An empirical study in Singapore. *Transp. Res. Part A Policy Pr.* **2019**, *119*, 224–237. [CrossRef]
28. Kuby, M.; Lim, S. The flow-refueling location problem for alternative-fuel vehicles. *Socio-Econ. Plan. Sci.* **2005**, *39*, 125–145. [CrossRef]
29. Kuby, M.; Lim, S. Location of Alternative-Fuel Stations Using the Flow-Refueling Location Model and Dispersion of Candidate Sites on Arcs. *Netw. Spat. Econ.* **2006**, *7*, 129–152. [CrossRef]
30. Kuby, M.; Lines, L.; Schultz, R.; Xie, Z.; Kim, J.-G.; Lim, S. Optimization of hydrogen stations in Florida using the Flow-Refueling Location Model. *Int. J. Hydrog. Energy* **2009**, *34*, 6045–6064. [CrossRef]
31. Capar, I.; Kuby, M. An efficient formulation of the flow refueling location model for alternative-fuel stations. *IIE Trans.* **2012**, *44*, 622–636. [CrossRef]
32. Liu, K.; Sun, X. Considering the dynamic refueling behavior in locating electric vehicle charging stations. *ISPRS Ann. Photogramm. Remote. Sens. Spat. Inf. Sci.* **2014**, 41–46. [CrossRef]
33. Xu, M.; Meng, Q.; Liu, K. Network user equilibrium problems for the mixed battery electric vehicles and gasoline vehicles subject to battery swapping stations and road grade constraints. *Transp. Res. Part B Methodol.* **2017**, *99*, 138–166. [CrossRef]
34. Jing, W.; Yan, Y.; Kim, I.; Sarvi, M. Electric vehicles: A review of network modelling and future research needs. *Adv. Mech. Eng.* **2016**, *8*. [CrossRef]
35. Zheng, H.; He, X.; Li, Y.; Peeta, S. Traffic Equilibrium and Charging Facility Locations for Electric Vehicles. *Netw. Spat. Econ.* **2016**, *17*, 435–457. [CrossRef]
36. Tang, T.-Q.; Zhang, J.; Liu, K. A speed guidance model accounting for the driver's bounded rationality at a signalized intersection. *Phys. A Stat. Mech. Appl.* **2017**, *473*, 45–52. [CrossRef]
37. Liu, K.; Liu, D.; Li, C.; Yamamoto, T. Eco-Speed Guidance for the Mixed Traffic of Electric Vehicles and Internal Combustion Engine Vehicles at an Isolated Signalized Intersection. *Sustainability* **2019**, *11*, 5636. [CrossRef]
38. Ge, S.; Feng, L.; Liu, H. The planning of electric vehicle charging station based on Grid partition method. In Proceedings of the 2011 International Conference on Electrical and Control Engineering, Yichang, China, 16–18 September 2011; pp. 2726–2730.
39. Huang, Y.; Kockelman, K.M. Electric vehicle charging station locations: Elastic demand, station congestion, and network equilibrium. *Transp. Res. Part D Transp. Environ.* **2020**, *78*, 102179. [CrossRef]
40. Xu, M.; Meng, Q. Optimal deployment of charging stations considering path deviation and nonlinear elastic demand. *Transp. Res. Part B* **2020**, *135*, 120–142. [CrossRef]
41. Xie, C.; Jiang, N. Relay Requirement and Traffic Assignment of Electric Vehicles. *Comput. Civ. Infrastruct. Eng.* **2016**, *31*, 580–598. [CrossRef]
42. He, N.; Zhao, S. Discussion on Influencing Factors of Free-flow Travel Time in Road Traffic Impedance Function. *Procedia Soc. Behav. Sci.* **2013**, *96*, 90–97. [CrossRef]
43. Gale, D.; Nikaidô, H. The Jacobian matrix and global univalence of mappings. *Math. Ann.* **1965**, *159*, 81–93. [CrossRef]
44. Dial, R.B. A probabilistic multipath traffic assignment model which obviates path enumeration. *Transp. Res.* **1971**, *5*, 83–111. [CrossRef]
45. Daganzo, C.F.; Sheffi, Y. On Stochastic Models of Traffic Assignment. *Transp. Sci.* **1977**, *11*, 253–274. [CrossRef]
46. Di Pillo, G.; Grippo, L.; Pillo, G. A new augmented Lagrangian function for inequality constraints in nonlinear programming problems. *J. Optim. Theory Appl.* **1982**, *36*, 495–519. [CrossRef]
47. Guignard, M. Generalized Kuhn–Tucker Conditions for Mathematical Programming Problems in a Banach Space. *SIAM J. Control.* **1969**, *7*, 232–241. [CrossRef]

48. Cheng, L.; Han, F. Optimal Road Toll Design from the Perspective of Sustainable Development. *Discret. Dyn. Nat. Soc.* **2014**, *2014*, 1–7. [CrossRef]
49. Liu, H.X.; He, X.; He, B. Method of Successive Weighted Averages (MSWA) and Self-Regulated Averaging Schemes for Solving Stochastic User Equilibrium Problem. *Netw. Spat. Econ.* **2007**, *9*, 485–503. [CrossRef]

© 2020 by the authors. Licensee MDPI, Basel, Switzerland. This article is an open access article distributed under the terms and conditions of the Creative Commons Attribution (CC BY) license (http://creativecommons.org/licenses/by/4.0/).

energies

MDPI

Article

Developing Equivalent Consumption Minimization Strategy for Advanced Hybrid System-II Electric Vehicles

Hsiu-Ying Hwang

Department of Vehicle Engineering, National Taipei University of Technology, Taipei 10608, Taiwan; hhwang@mail.ntut.edu.tw

Received: 24 March 2020; Accepted: 17 April 2020; Published: 19 April 2020

Abstract: Compared with conventional vehicles, hybrid electric vehicles (HEVs) have the advantage of high-energy conversion efficiency, which can have better fuel economy and lower emissions. The main issue of HEVs is how to develop an energy management strategy to achieve significantly better fuel efficiency. In this research, the Equivalent Consumption Minimization Strategy (ECMS) was applied to optimize the performance of fuel consumption in the Advanced Hybrid System-II (AHS-II). Based on FTP-75 Test Procedure defined by the U.S. Environmental Protection Agency (EPA), a backward simulation module was established. The baseline simulation module with the rule-based control strategy was validated with the original fuel consumption data. Then, the module with ECMS followed the same control rules of engine on/off and mode selection, and the fuel consumption of ECMS was compared with the simulation results of the baseline model. The fuel economy improvements of ECMS in urban, highway driving pattern, and composite fuel economy were up to 8.5%, 7.7%, and 8.1%, respectively. The simulation results showed that the difference of motors' working efficiency was only 1.2% between ECMS and baseline rule-based control strategies. The main reason of fuel consumption improvement was the engine operation chosen by ECMS, which provided better power distribution.

Keywords: hybrid electric vehicles; equivalent consumption minimization strategy; power-split hybrid

---

1. Introduction

Climate change and the sustainable development of energy are the most serious international issues in the 21st century. The hybrid electric vehicle (HEV) is one of the key technologies for vehicle energy saving. Combining with the internal combustion engine (ICE) and high efficiency electric motor, hybrid electric vehicles have better fuel economy than traditional vehicles. The HEVs are also more achievable than electric vehicles (EVs) under current limitations of battery.

With ICE and electric motors, the operating modes to control engine and motors are necessary designs for the hybrid power system, and the switching between the modes would change its power flow and the operation of the components. A planetary gear set (PGS) is often applied to the configuration of HEV. The most representative design is Toyota Prius released in 1997. In 2005, the Advanced Hybrid System-II (AHS-II), also known as the two-mode hybrid system, was developed by General Motors. AHS-II offers an additional set of electric-continue-variable-transmission (eCVT) mode of operation, and significantly reduce the energy loss in high speed [1]. Arata et al. [2] analyzed two different power-split hybrid-electric vehicle (HEV) powertrains using backward-looking simulations, and compared the Toyota Hybrid System II (THS-II) and the General Motors Allison Hybrid System II (GM AHSII).

Fuel economy and lower pollution emissions are critical issues. Vehicle manufactures are investing in energy management strategy (EMS) to ensure that the system components operate in their safe working range, and, at the same time, to maintain a high energy conversion efficiency to obtain better fuel economy and lower pollution emissions. The EMS can be divided into two categories, rule-based control strategy (RB) and optimization control strategy. Each of these two categories can be further divided into two subcategories. RB is subdivided into fuzzy control and heuristic control, while the optimization strategy is subdivided into global optimization and real-time or online optimization control [3,4]. RB does not require lengthy numerical calculation time [5,6], and can simultaneously monitor a number of parameters, which are usually associated with engine on/off control, and engine and motor operating points [7,8]. However, the fuel consumption is not optimized. For optimization control strategy, the most representative strategy for global optimization is dynamic programming (DP). However, the algorithm requires the information of full driving time, it is difficult to apply for a real vehicle control [9,10]. Chen et al. [11] utilized online control rules but based on offline optimization results of DP for a plug-in HEV to prolong driving range up to 2.86% and reduce the energy consumption up to 5.77%. For real-time optimized control strategy, Equivalent Consumption Minimization Strategy (ECMS) [12] aims at optimal power distribution (between engines and motors) and ensures that the battery pack has sufficient power. Compared to RB, ECMS can have a better fuel economy [13]. Zeng et al. [14] proposed an adaptive simplified-ECMS-based strategy along with particle swarm optimization (PSO) algorithm to optimize PHEV system. The method effectively shortened the calculation time and improved fuel consumption by 16.43%, compared to the Charge Sustaining-Charge Depleting (CS-CD)-based strategy. Dong [15] developed a real-time optimal energy management approach for HEVs and PHEVs using an adaptive coefficient tuning strategy, and validated results using both Model-in-Loop (MIL) and HIL environment. Lu et al. [16] introduced the weighted sum method and no-preference method to solve the multiobjective optimization problem of plug-in electric vehicles and validated with ADVISOR software. Xu et al. [17] developed a fuzzy control strategy for parallel hybrid electric vehicle. The control strategy was adjusted with GA. It was verified that GA could effectively improve the efficiency of the engine and fuel consumption.

This study implemented the AHS-II two-mode system as the transmission structure to establish a Simulink vehicle model, and applied ECMS for EMS to achieve optimal fuel economy.

2. Models of Hybrid Electric Vehicle

Two simulation models with the AHS-II two-mode hybrid system were established. One is the baseline model with the rule-based control which would be validated with manufacture official data to verify the accuracy of vehicle model. The other is the optimized vehicle model with ECMS to achieve a better EMS control and optimize fuel efficiency.

2.1. Introduction of Hybrid Powertrain System

The AHS-II two-mode hybrid powertrain consists of a planetary gear, a compound planetary gear, four clutches, an internal combustion engine, and two motor/generators, MG1 and MG2, as shown in Figure 1. The architecture can produce effects which are similar to the continuously variable transmission (CVT). Therefore, it is also known as electrical continuously variable transmission (eCVT). The symbols R, S, and C represent the ring gear, the sun gear, and the carrier, respectively, and the subscripts 1 and 2 represent the compound planetary gear set and the simple planetary gear set. When the vehicle travels on different road conditions, the powertrain system can operate between two eCVT modes and four fixed-gear modes.

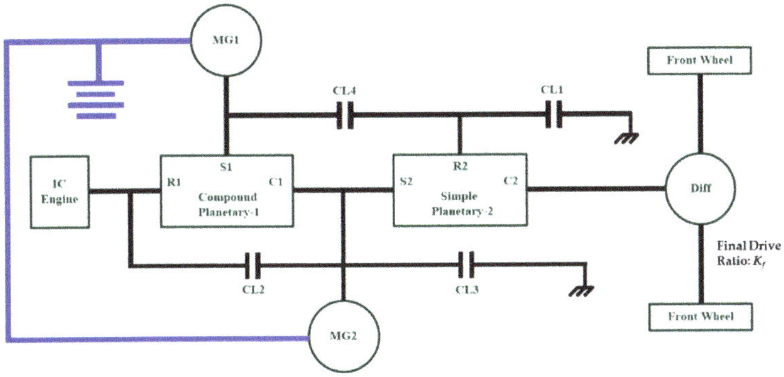

Figure 1. AHS-II two-mode powertrain architecture.

2.2. HEV Simulation Model

The backward calculation dynamic model of the HEV system was implemented using Matlab/Simulink, as shown in Figure 2. The US FTP-75 (EPA Federal Test Procedure) urban and highway drive cycles were applied as the road conditions for simulation, as shown in Figures 3 and 4. According to the known driving speed, the vehicle dynamic model calculated the required vehicle acceleration and driving torque, and through the energy management control module, the operating mode of the system and the output powers of engine and motor/generators, MG1 and MG2, were determined. The speed and torque of the two motor/generators were determined by the gear ratio of the transmission module. According to the speed and torque of the motor/generators, the battery module calculated the state of charge (SOC) of the battery pack. Finally, the fuel consumption was accumulated by ICE module through an engine two-dimensional lookup table.

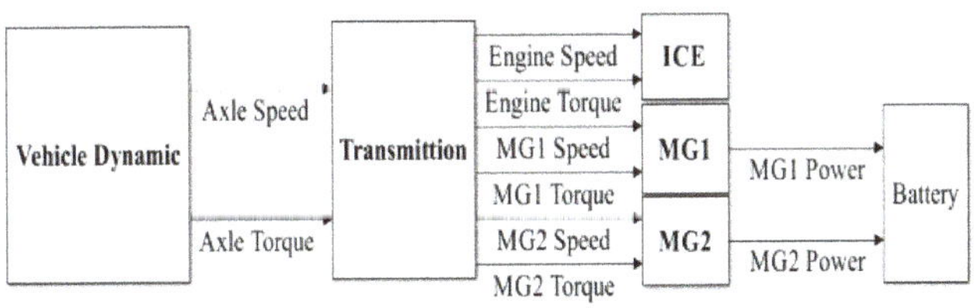

Figure 2. Two-mode powertrain simulation model.

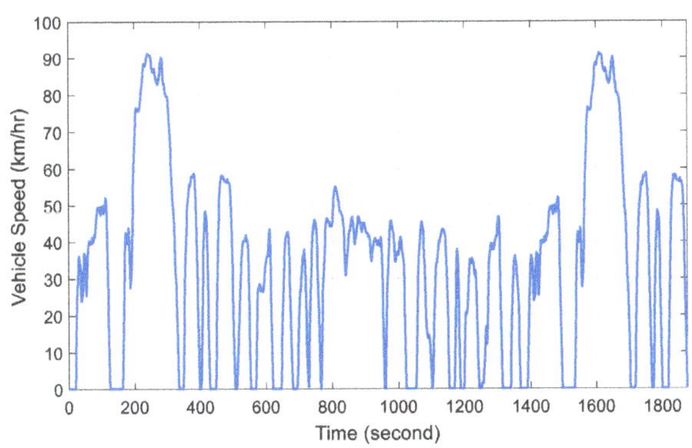

Figure 3. FTP-75 urban driving cycle.

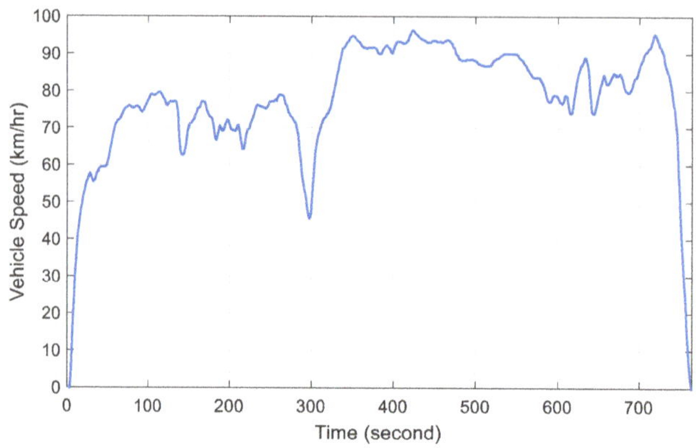

Figure 4. FTP-75 highway driving cycle.

2.3. Vehicle Dynamic Module

According to the driving cycle, the vehicle speed was received, and the road load of vehicle model, including rolling resistance, aerodynamic drag, grade resistance, was calculated by Equations (1)–(4). Further, the AHS-II output shaft required torque was obtained. The vehicle parameters are shown in Table 1.

$$F_{Load} = F_r + F_w + F_g, \tag{1}$$

$$F_r = f_r M g \cos\alpha, \tag{2}$$

$$F_w = \frac{1}{2}\rho A_f C_D V^2, \tag{3}$$

$$F_g = M g \sin\alpha, \tag{4}$$

where F_{Load}, F_r, F_w, and F_g are vehicle road load, rolling resistance, aerodynamic drag, and grade resistance, respectively, and f_r, M, g, α, ρ, A_f, C_D, and V are rolling resistance coefficient, vehicle mass, gravity, road slope, air density, vehicle front area, aerodynamic drag coefficient, and vehicle speed, respectively.

Table 1. Vehicle parameters for dynamic simulation.

Item (Unit)	Value
Mass (kg)	1600
Radius of the tire (m)	0.352
Vehicle frontal area (m^2)	2.642
Rolling resistance coefficient	0.01
Gravity acceleration (m/s^2)	9.81
Aerodynamic drag coefficient	0.386
Air density (kg/m^3)	1.29

2.4. Controller Module

In this study, two sets of controllers were established, the rule-based control module for baseline HEV model and the Equivalent Consumption Minimization Strategy (ECMS) module for the optimized HEV model. These control modules were combined with the engine switch strategy to further determine the operating time of the ICE. The individual controllers were built with Matlab functions.

Based on the torque required for the vehicle driving and the battery SOC, the rule-based controller, heuristic method (if-then-else), determined the speed and torque of ICE. The ESCM with object function developed the working state of the engine. After determining the status of ICE, the mode switch control module would switch between different eCVT and fixed-gear modes.

2.5. Transmission Module

The transmission includes two planetary gear sets, two motor/generators, and four clutches. Based on the vehicle driving condition, the mode switch module would determine the mode of operation, mode 1 for first eCVT mode and mode 2 for second eCVT mode. For mode 1, the motor speeds and torques of MG1 and MG2 were simulated by Equations (5)–(8). Equations (9)–(12) were for mode 2 operation.

$$\omega_{\text{MG1}} = \frac{1}{i_1}\omega_e - \frac{(1-i_1)(1+i_2)}{i_1 i_2}\omega_{out}, \tag{5}$$

$$\omega_{\text{MG2}} = \frac{1+i_2}{i_2}\omega_{out}, \tag{6}$$

$$T_{\text{MG1}} = -i_1 T_e, \tag{7}$$

$$T_{\text{MG2}} = -(1-i_1)T_e + \frac{i_2}{1+i_2}T_{out}, \tag{8}$$

$$\omega_{\text{MG1}} = -\frac{i_2}{1-i_1-i_1 i_2}\omega_e + \frac{(1-i_1)(1+i_2)}{1-i_1-i_1 i_2}\omega_{out}, \tag{9}$$

$$\omega_{\text{MG2}} = \frac{1}{1-i_1-i_1 i_2}\omega_e - \frac{i_1(1+i_2)}{1-i_1-i_1 i_2}\omega_{out}, \tag{10}$$

$$T_{\text{MG1}} = -i_1 T_e + \frac{1}{1+i_2}T_{out}, \tag{11}$$

$$T_{\text{MG2}} = -(1-i_1)T_e + \frac{i_2}{1+i_2}T_{out}, \tag{12}$$

where,

$$i_1 = \frac{R_{\text{S1}}}{R_{\text{R1}}}, \tag{13}$$

$$i_2 = \frac{R_{\text{S2}}}{R_{\text{R2}}}, \tag{14}$$

ω_e, ω_{MG1}, $\omega_{\boldsymbol{MG2}}$, ω_{out}, T_e, T_{MG1}, T_{MG2}, and T_{out} are the rotational speeds and torques of the engine, two motors, and transmission output. R_{R1}, R_{R2}, R_{S1} and R_{S2} are the radii of ring gear 1 and 2 and of sun gear 1 and 2, respectively. i_1, i_2 are the radius ratio of sun gear to ring gear for gear train 1 and 2, respectively.

In the simulation, the rotational inertia of engine, I_e; inertia of ring gear 1 and 2, I_{R1} and I_{R2}; inertia of carrier 1 and 2, I_{C1}, and I_{C2}; inertia of motor/generator 1 and 2, I_{MG1} and I_{MG2}; and inertia of sun gear 1 and 2, I_{S1} and I_{S2}; are all considered. In mode 1 case, the general force-acceleration matrix can be written as shown in Equation (15). Similarly, Equation (16) is the case of mode 2.

$$\begin{bmatrix} \dot{\omega}_e \\ \dot{\omega}_{out} \\ \dot{\omega}_{\text{MG1}} \\ \dot{\omega}_{\text{MG2}} \\ F_1 \\ F_2 \end{bmatrix} = \begin{bmatrix} I_e + I_{\text{R1}} & 0 & 0 & 0 & R_{\text{R1}} & 0 \\ 0 & I_{\text{C2}} + \frac{r_{tire}^2}{K_f{}^2}m & 0 & 0 & 0 & -R_{\text{R2}} - R_{\text{S2}} \\ 0 & 0 & I_{\text{MG1}} + I_{\text{S1}} & 0 & -R_{\text{S1}} & 0 \\ 0 & 0 & 0 & I_{\text{MG2}} + I_{\text{C1}} + I_{\text{S2}} & -R_{\text{R1}} + R_{\text{S1}} & R_{\text{S2}} \\ R_{\text{R1}} & 0 & -R_{\text{S1}} & -R_{\text{R1}} + R_{\text{S1}} & 0 & 0 \\ 0 & -R_{\text{R2}} - R_{\text{S2}} & 0 & R_{\text{S2}} & 0 & 0 \end{bmatrix}^{-1} \begin{bmatrix} T_e \\ -\frac{\sum F_{tire}}{K_f} r_{tire} \\ T_{\text{MG1}} \\ T_{\text{MG2}} \\ 0 \\ 0 \end{bmatrix} \tag{15}$$

$$\begin{bmatrix} \dot{\omega}_e \\ \dot{\omega}_{out} \\ \dot{\omega}_{MG1} \\ \dot{\omega}_{MG2} \\ F_1 \\ F_2 \end{bmatrix} = \begin{bmatrix} I_e + I_{R1} & 0 & 0 & 0 & R_{R1} & 0 \\ 0 & I_{C2} + \frac{r_{tire}^2}{K_f^2} m & 0 & 0 & 0 & -R_{R2} - R_{S2} \\ 0 & 0 & I_{MG1} + I_{S1} + I_{R2} & 0 & -R_{S1} & R_{R2} \\ 0 & 0 & 0 & I_{MG2} + I_{C1} + I_{S2} & -R_{R1} + R_{S1} & R_{S2} \\ R_{R1} & 0 & -R_{S1} & -R_{R1} + R_{S1} & 0 & 0 \\ 0 & -R_{R2} - R_{S2} & R_{R2} & R_{S2} & 0 & 0 \end{bmatrix}^{-1} \begin{bmatrix} T_e \\ -\frac{\sum F_{tire}}{K_f} r_{tire} \\ T_{MG1} \\ T_{MG2} \\ 0 \\ 0 \end{bmatrix} \quad (16)$$

where

$$m = M + \frac{I_{wheel}}{r_{tire}^2}. \quad (17)$$

F_1, F_2, and F_{tire} are the forces acting on the sun gear, ring gear, and tire, respectively. K_f is the final axle ratio, and r_{tire} is the radius of tire. I_{wheel} is the total rotational inertia of the wheels [18].

2.6. Internal Combustion Engine Module

The ICE module of this study was represented by a lookup table. Figure 5 shows the three-dimensional ICE fuel consumption rate. Through the controller module to determine the engine running state, the corresponding engine speed and torque could determine engine fuel consumption rate.

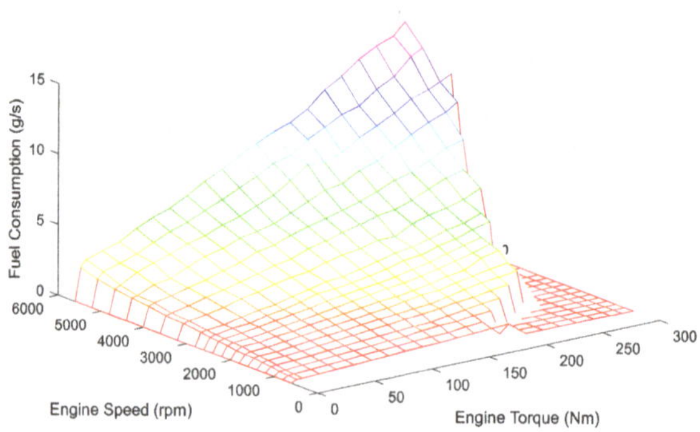

Figure 5. ICE fuel consumption rate.

2.7. Motor/Generator Module

In AHS-II powertrain, there are two electric motor/generators, MG1 and MG2, which have same output power. The motor/generators are 60kW permanent magnet AC motors. In this study, MG1 and MG2 had same specifications. The motor/generators efficiency is a function of speed and output torque, as shown in Figure 6. This module was modeled with a lookup table. The motor/generator power calculation is shown in Equation (18).

$$P_{MG} = \omega_{MG} T_{MG} \eta_{MG}^{K}, \begin{cases} K = 1 \\ K = -1 \end{cases}, \quad (18)$$

where P_{MG}, ω_{MG}, and T_{MG} are motor/generators power, speed, and torque, respectively. If the speed and torque are in the same sign, the motor/generator works as a motor. If the speed and torque are in different sign, the motor/generator works as a generator, which transforms the mechanical energy into electricity and stores in the battery pack. η_{MG} is the efficiency of the motor/generator, and K is the power flow of the motor/generator. $K = 1$ is motoring, and $K = -1$ is generating.

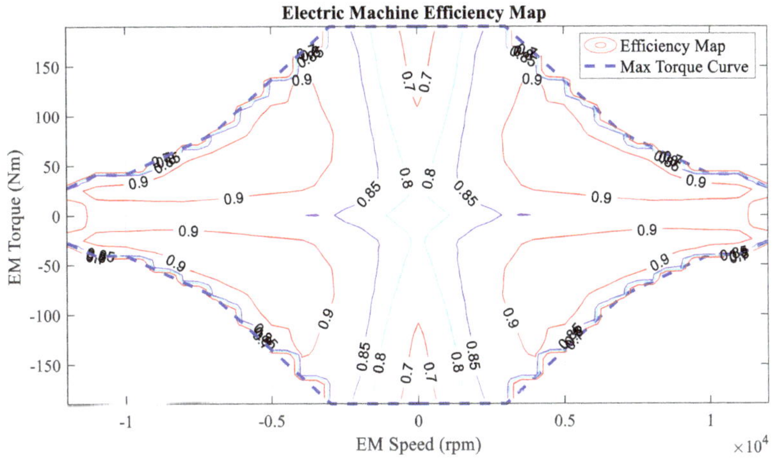

Figure 6. The efficiency of motor/generator.

2.8. Battery Module

This study used a battery equivalent circuit, as shown in Figure 7, to establish the battery module [13]. The model provides the information of the open circuit voltage, output voltage, battery current, required power of battery, battery SOC, SOC changing rate, battery internal resistance, and battery capacity. The battery was mainly to support the power required for the motor/generator in order to keep the system in high fuel efficiency. The proposed model was adequate for the fuel consumption optimization.

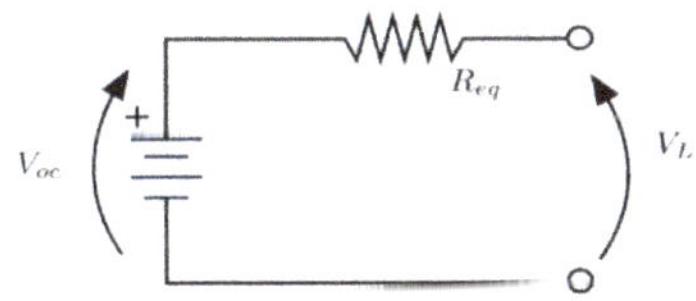

Figure 7. Battery equivalent circuit.

In Figure 7, V_{oc} is the open circuit voltage, R_{eq} is the internal equivalent resistance, and V_L is the output voltage. The power output of battery can be represented in terms of the electric current, I_{batt}, as shown in Equation (19). The total output of battery, $V_{oc}I_{batt}$, includes the power required for the system, P_{em_batt}, and the power consumed by the internal resistance of the battery, $R_{eq}I^2{}_{batt}$. The power required can also represented in terms of motors' power, as shown in Equation (20).

$$P_{em\_batt} = V_{oc}I_{batt} - R_{eq}I_{batt}^2, \quad (19)$$

$$P_{em\_batt} = T_{\mathrm{MG1}}\omega_{\mathrm{MG1}}\eta_{\mathrm{MG1}}^K\eta_{con}^K + T_{\mathrm{MG2}}\omega_{\mathrm{MG2}}\eta_{\mathrm{MG2}}^K\eta_{con}^K, \quad (20)$$

where η_{MG1} and η_{MG2} are the efficiency of the motor/generator 1 and 2, η_{con} is the motor controller efficiency. The battery SOC can be calculated by accumulating the charged and discharged current. The relationship between battery SOC changing rate, battery capacity Q_{max} and current I_{batt} is as follows:

$$\dot{\mathrm{SOC}}(t) = -\frac{I_{batt}}{Q_{\max}}. \quad (21)$$

From Equation (19), the battery current can be derived as follows:

$$I_{batt} = \frac{V_{oc} + \sqrt{V_{oc}^2 - 4R_{eq}P_{em\_batt}}}{2R_{eq}} \quad (22)$$

The open circuit voltage and internal equivalent resistance are function of SOC. From Equations (21) and (22), the SOC rate can be obtained as follows:

$$\dot{\text{SOC}}(t) = -\frac{V_{oc}(\text{SOC}) + \sqrt{V_{oc}^2(\text{SOC}) - 4R_{batt}(\text{SOC})P_{em\_batt}(t)}}{2R_{batt}(\text{SOC})Q_{\max}}. \quad (23)$$

The internal resistance of battery was based on the curve shown in Figure 8. A portion of the battery power output was provided for the driving system, and the other was consumed by the internal resistance. The efficiency of battery can be calculated, as shown in Equation (24).

$$\eta_{batt} = P_{m\_batt}/V_{oc}I_{batt}. \quad (24)$$

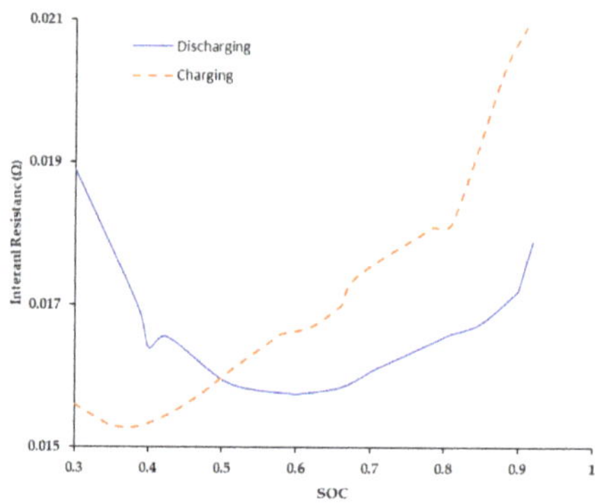

Figure 8. Internal resistance of battery while charging/discharging.

In this study, the SOC of battery was limited in the range between 0.4 and 0.6 since the battery had relatively less energy loss due to the battery internal resistance while considering for both charging and discharging states.

3. Energy Management Strategy

3.1. Optimization

The objective of the optimization problem was to minimize the fuel consumption and satisfy the following requirements for the HEV system: (1) To meet the demand of vehicle driving condition, and (2) to be constrained within the operation limits of the system components, as shown in Equations (25)–(30). The goal of the optimization, the cost function J, is expressed numerically in finite time, as shown in Equation (25). For the hybrid powertrain system with charge-sustaining control, the initial battery SOC and the final state should remain the same. In other words, the power loss of the system must be compensated by the engine. The power required for the vehicle is provided through engine and motors, as shown in Equation (26). Equation (27)–(30) define the SOC controlled limits, battery power output limits, engine power output limits, and motor power output limits, respectively. The optimization problem is defined as the following:

Objective:

$$\min = \left\{ J = \int_{t_0}^{t_f} \dot{m}_{fc}(t)dt \right\}, \tag{25}$$

Subject to

$$P_{req}(t) = P_e(t) + P_{em}(t), \tag{26}$$

$$\mathrm{SOC_{min}} \leq \mathrm{SOC}(t) \leq \mathrm{SOC_{max}}, \tag{27}$$

$$P_{em\_batt\_\min} \leq P_{em\_batt}(t) \leq P_{em\_batt\_\max}, \tag{28}$$

$$P_{e\_\min} \leq P_e(t) \leq P_{e\_\max}, \tag{29}$$

$$P_{em\_\min} \leq P_{em}(t) \leq P_{em\_\max}, \tag{30}$$

where t, J, $m_{fc}(t)$, P_{batt}, P_e, P_{em}, and P_{req} are time, cost function, engine fuel rate, battery power, electric motor power, engine power, and vehicle power required, respectively. In this study, the SOC was limited between 0.4 and 0.6. Battery power and electric motor power were constrained between −60 kW and 60 kW. The engine power was between 0 kW and 157 kW.

3.2. Rule-Based Control Strategy

According to the understanding of the system architecture and the efficiency of each element, the output energy of each driving element is defined based on the different road conditions. The basic principle is to meet the driving force required during vehicle travelling, while the control rule should keep the engine and motor/generators in the high operating efficiency range as long as possible to achieve the best fuel consumption and the lowest emissions. This study applied a rule-based controller for the baseline HEV model, and the heuristic was applied, as shown in Figure 9. The fuel economy of this controller would be compared with the manufacture data to verify the accuracy the HEV model. The rule-based strategy is listed in Table 2. Based on different SOC and required torque output, engine operation conditions are provided.

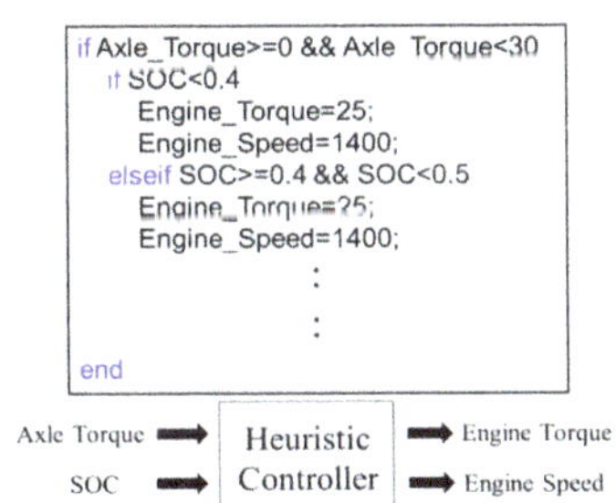

Figure 9. Heuristic controller.

Table 2. Rule-based strategy.

Required Torque \ Engine Output	SOC < 0.4	0.4 ≤ SOC < 0.5	0.5 ≤ SOC ≤ 0.6
0–30 (Nm)	25 Nm 1400 rpm	25 Nm 1400 rpm	25 Nm 1400 rpm
30–50 (Nm)	45 Nm 1500 rpm	45 Nm 1500 rpm	25 Nm 1500 rpm
50–75 (Nm)	140 Nm 1700 rpm	140 Nm 1700 rpm	100 Nm 1700 rpm
75–100 (Nm)	150 Nm 1700 rpm	140 Nm 1700 rpm	130 Nm 1700 rpm
100–125 (Nm)	155 Nm 1800 rpm	145 Nm 1800 rpm	130 Nm 1800 rpm
125–150 (Nm)	170 Nm 1800 rpm	160 Nm 1800 rpm	160 Nm 1800 rpm
150–175 (Nm)	185 Nm 1900 rpm	180 Nm 1900 rpm	170 Nm 1900 rpm
175–200 (Nm)	180 Nm 2000 rpm	160 Nm 2000 rpm	140 Nm 2000 rpm
200–250 (Nm)	180 Nm 2100 rpm	160 Nm 2100 rpm	140 Nm 2100 rpm
250–300 (Nm)	180 Nm 2200 rpm	150 Nm 2200 rpm	130 Nm 2200 rpm
>300 (Nm)		235 Nm 2000 rpm	

3.3. Equivalent Consumption Minimization Strategy

ECMS is one of the best optimization control strategies, and ECMS treats the energy storage system, battery pack/supercapacitor as a buffered energy source. The loss of battery power during travel must be recovered from the brake regeneration or the generator driven by the ICE. The cost function of ECMS is shown in Equation (31), which contains the fuel consumption of the ICE and the electricity energy consumption. Since the electricity power consumption and fuel consumption could not be directly compared, electricity power consumption should be converted into the equivalent fuel consumption by Equations (32) and (33).

$$J(t) = \dot{m}_{fc,eqv} = \dot{m}_{fc}(P_e(t)) + \dot{m}_{eqv}(P_{em}(t)), \tag{31}$$

$$\begin{aligned} \dot{m}_{eqv}(t) = \; & \gamma \cdot s_{dis} \frac{BSFC(t) \cdot P_{em}(t)}{\eta_{batt}(P_{em})\eta_{em}(P_{em})} \\ & +(1-\gamma) \cdot s_{chg} \cdot \eta_{batt}(P_{em})\eta_{em}(P_{em}) \cdot BSFC(t) \cdot P_{em}(t), \end{aligned} \tag{32}$$

$$\gamma = \frac{1 + sign(P_{em}(t))}{2}, \tag{33}$$

where $m_{fc,\, eqv}$ is the summation of instant fuel consumption, $m_{eqv}(t)$ is the equivalent fuel consumption of electricity power, P_{em} is the output power of the motor, and s_{dis} and s_{chg} are the equivalent factors of discharging and charging, respectively. *BSFC* is the fuel consumption per unit ICE output energy. η_{batt} and η_{em} are the working efficiency of battery pack and motor, respectively.

3.4. Engine Switch Control Strategy

In this study, the engine switch was designed to avoid engine operating in high fuel consumption regions and to avoid the overcharging and discharging of the battery pack. Since the battery has less energy loss due to the internal resistance while the SOC is within 0.4 and 0.6, the design strategy uses this interval as SOC limits. In ECMS, the mode of operation should be determined first. Since the mode 2 is applied to the higher speed, if the SOC does not exceed 0.6, the engine will continue to operate to ensure that the battery system has enough power.

For mode 1 operation, when the vehicle speed is less than 20 km/h, the operating efficiency of engine will be in poor condition. If the SOC is not lower than 0.45, the engine will be shut down and vehicle is driven by electric motor to enhance fuel consumption performance. When the vehicle speed is between 20 km/h and 40 km/h, and the SOC is greater than or equal to 0.55, the engine will be shut down. When the vehicle speed is greater than or equal to 40 km/h, and the SOC is greater than or equal to 0.6, the engine will be shut-down. If SOC is greater than 0.6 and engine remains off, the SOC is monitored and checked until SOC is less than 0.55 and the process is reset to the starting block. The control flowchart of the engine switch is shown in Figure 10. This control logic was applied on the energy management strategies of both HEV models in this study.

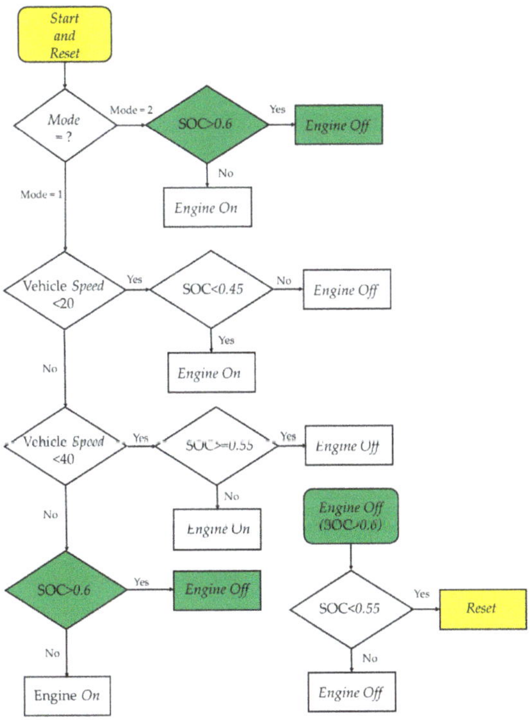

Figure 10. Engine switch flowchart.

4. Simulation Results

4.1. *HEV Baseline Model with Rule-Based Control Strategy*

The rule-based control was applied on the baseline HEV model. With the urban and highway driving simulation, fuel economies were 47 MPG (miles per gallon) and 39 MPG, and composite fuel economy was 43 MPG. The simulation results and the manufacture official data are shown in Table 3. The differences of urban, highway, and composite fuel economy were −2.1%, 5.4%, and 1.6%, respectively, which were acceptable. The simulation model can be used to represent the original vehicle. With this rule-based control model as the baseline, the average efficiency of MG1 and MG2 were monitored as well. The urban and highway efficiency values of MG1 were 0.83 and 0.85, respectively, and those of MG2 were 0.85 and 0.84, as shown in Table 4.

Table 3. Comparison of fuel consumption between rule-based control model (RB) and factory data.

Item	City	Highway	Composite
Rule Based (Baseline) (MPG)	47	39	43
Official Data (MPG)	48	37	42
Difference (Rule Based–Official)/Official × 100%	−2.1%	5.4%	1.6%

Table 4. Motor/generator efficiency in RB.

Item	City	Highway
MG1 (efficiency)	0.83	0.85
MG2 (efficiency)	0.85	0.84

4.2. HEV Optimization Model with ECMS

The ECMS was applied as optimization strategy for fuel economy simulation. The result is shown in Table 5. With ECMS, the fuel economy of urban and highways driving cycles had improvements around 8%, while the efficiencies of MG1 and MG2 were tracked as well, as shown in Table 6. The average of urban and highway efficiency values of MG1 were 0.83 and 0.85, and those of MG2 were 0.85 and 0.83, respectively. The difference between the baseline and ECMS was the MG2 highway efficiency, 84% vs. 83%, which was around 1.2% different. Under two different control strategies, the motor/generator efficiencies were very much the same. That indicates the improvement of fuel economy using ECMS was mainly due to the selection of the engine operating points.

Table 5. Comparison of fuel consumption between the baseline and equivalent combustion minimization strategy (ECMS).

Item	City	Highway	Composite
Baseline (MPG)	47	39	43.0
ECMS (MPG)	51	42	46.5
Improvement, (ECMS-Baseline)/Baseline × 100%	8.5%	7.7%	8.1%

Table 6. Motor/generator efficiency in ECMS.

Item	City	Highway
MG1 (efficiency)	0.83	0.85
MG2 (efficiency)	0.85	0.83

Figure 11 to Figure 12 show the engine operating points of baseline and ECMS models. With ECMS optimization, the distribution of the engine operating points in urban and highway driving cycles was significantly smaller than that of the rule-based control strategy. With less engine power during the driving cycles, the ECMS model had less fuel consumption, as shown in Figures 13 and 14.

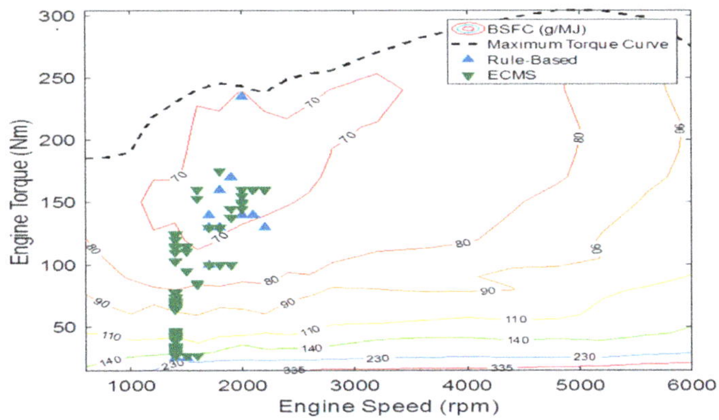

Figure 11. Comparison of engine operating points when driving in urban areas.

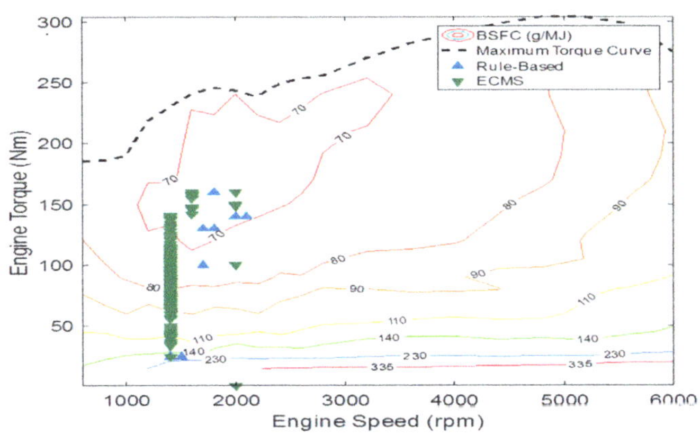

Figure 12. Comparison of engine operating points when driving on expressway.

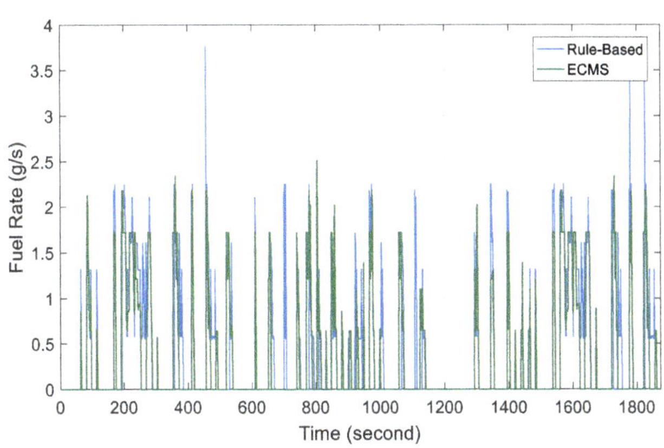

Figure 13. Comparison of fuel consumption rate in urban cycle.

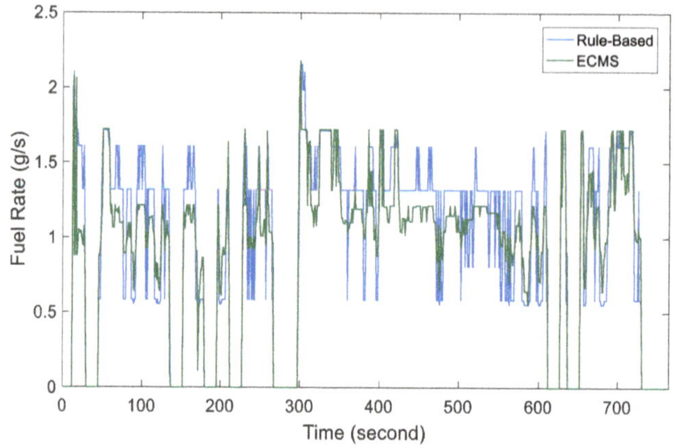

Figure 14. Comparison of fuel consumption rate of highway cycle.

Figures 15 and 16 show the accumulation time of engine operation points in urban driving cycle. The engine load with the rule-based control strategy had up to 42% of the engine running time when operating in the less efficiency range, while the engine load with ECMS had 27% of the engine running time in less efficiency range.

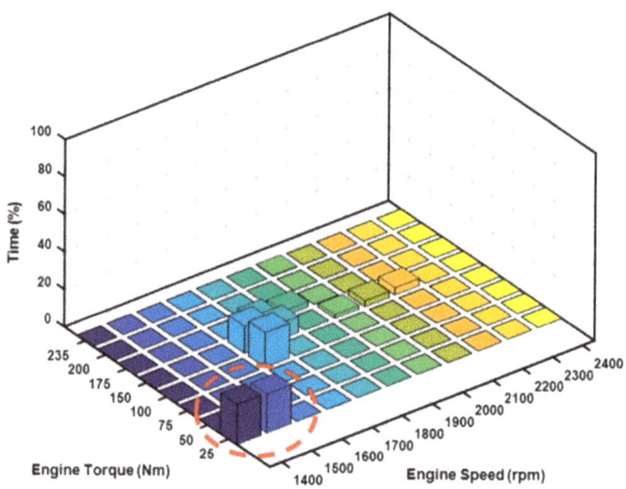

Figure 15. Engine operating time in urban driving cycle with baseline model.

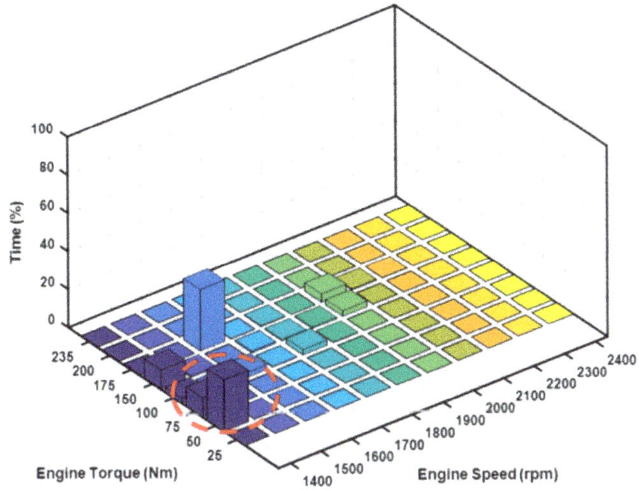

Figure 16. Engine operating time in urban driving cycle with ECMS optimization.

In highway driving cycle, as shown in Figures 17 and 18, the engine load with the rule-based control strategy had a significant operating time ratio in the less working efficiency range, accounting for about 22% of the engine operating time, while ECMS did not operate at all in the less efficiency range. With respect to overall operating time, ECMS optimization had a longer running time when operating in the better efficiency range, so HEV model with ECMS optimization could obtain better fuel economy.

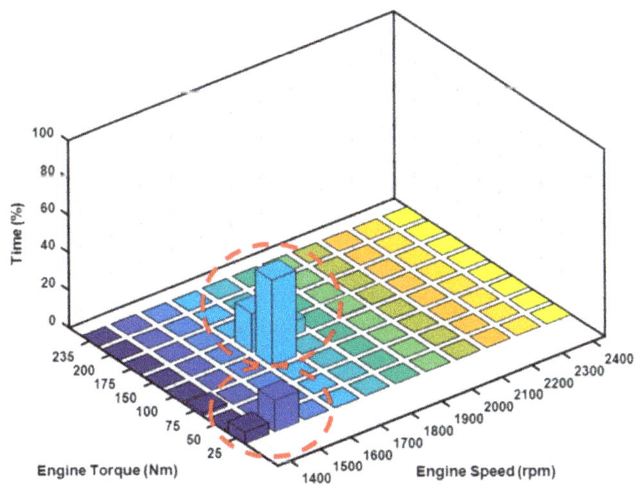

Figure 17. Engine operating time in highway driving cycle with baseline model.

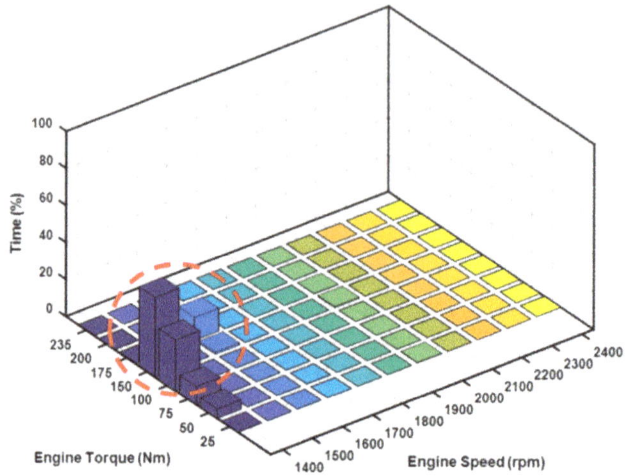

Figure 18. Engine operating time in highway driving cycle with ECMS optimization.

5. Conclusions

Energy management strategy is an important topic for hybrid electric vehicles. In order to effectively improve the performance of fuel consumption, ECMS was selected for instant optimization strategy to optimize the fuel consumption.

The AHS-II two-mode hybrid system was modeled, and a rule-based control was used as baseline model and validated with the official fuel economy data. Then, the ECMS was applied for fuel economy optimization. From the simulation of baseline model and ECMS optimization, the conclusions were drawn as follows:

1. For baseline model with the rule-based control strategy, the fuel economy of the urban, highway, and composite driving cycles were 47 MPG, 39 MPG, and 43 MPG. The official reported fuel economy data of AHS-II are 48 MPG, 37 MPG, and 42 MPG for the urban, highway, and composite driving cycle, respectively. The maximum difference of fuel economy between simulation and vehicle official data is 5.4%, which indicates that the results of simulation model has good correlation with those of the vehicle, and can be used to represent the original vehicle.
2. The fuel economies with ECMS optimization were 51 MPG, 42 MPG, and 46.5 MPG for urban, highway, and composite driving cycles, respectively. Comparing ECMS optimization with the rule-based control strategy, the improvements made were 8.5%, 7.7%, and 8.1%, separately. The proposed ECMS optimization strategy provided a better fuel economy performance.
3. The efficiencies of the motor/generator 1 and 2 in the rule-based model were 0.83 and 0.85 for city, and 0.85 and 0.84 for the highway driving cycles. Those in ECMS were 0.83 and 0.85 for city, and 0.85 and 0.83 for highway, individually. The biggest difference between the baseline and ECMS was only 1.2%. The improvement of fuel economy was mainly due to the selection of the engine operating points which lead to a better fuel performance. ECMS could effectively implement the best engine power distribution to achieve better fuel consumption.
4. In urban driving cycle, there was 42% of time that engine was operated in the less efficiency region for the rule-based control strategy, while there was 27% with ECMS.
5. In highway driving cycle, there was 22% of the time that engine was operated in the less efficiency region for the rule-based control strategy, while there were none for ECMS. Overall, ECMS optimization had engine operated in better efficiency range and provided better fuel economy.

Using vehicle dynamic simulation can help to quickly carry out the preliminary evaluation of various energy management strategies, search for the appropriate optimization of energy management strategies, and reduce R&D process time and cost. The proposed equivalent consumption minimization strategy can be utilized to optimize the performance of fuel consumption.

Funding: This research received no external funding.

Conflicts of Interest: The author declares no conflict of interest.

Nomenclature

A_f	vehicle front area
AHS-II	Advanced Hybrid System-II
C_D	aerodynamic drag coefficient
CVT	continuously-variable-transmission
DP	dynamic programming
ECMS	Equivalent Consumption Minimization Strategy
EMS	energy management strategy
EPA	Environmental Protection Agency
eCVT	electric-continue-variable-transmission
F_1	force acting on sun gear
F_2	force acting on ring gear
F_g	grade resistance
F_{Load}	vehicle road load
F_r	rolling resistance
F_w	aerodynamic drag
f_r	rolling resistance coefficient
F_{tire}	force acting on tire
g	gravity
HEVs	hybrid electric vehicles
I_{batt}	battery current
I_{C1}	rotational inertia of carrier 1
I_{C2}	rotational inertia of carrier 2
I_e	rotational inertia of engine
I_{MG1}	rotational inertia of motor/generator 1
I_{MG2}	rotational inertia of motor/generator 2
I_{R1}	rotational inertia of ring gear 1
I_{R2}	rotational inertia of ring gear 2
I_{S1}	rotational inertia of sun gear 1
I_{S2}	rotational inertia of sun gear 2
I_{wheel}	total rotational inertia of the wheels
ICE	internal combustion engine
J	cost function
K	power flow of the motor/generator (1: motoring; -1: generating)
K_f	final axle ratio
M	vehicle mass
MG	motor/generators
MPG	mile per gallon
$m_{eqv}(t)$	equivalent fuel consumption of electricity power
$m_{fc}(t)$	engine fuel rate
$m_{fc,\,eqv}$	summation of instant fuel consumption
PHEV	plug-in hybrid electric vehicle
P_{batt}	power output of battery
P_e	output power of the engine
P_{em}	output power of the electric motor

P_{req}	vehicle power required
P_{MG}	motor/generators power
PGS	planetary gear set
Q_{max}	battery capacity
R_{eq}	internal equivalent resistance
R_{R1}	radius of ring gear 1
R_{R2}	radius of ring gear 2
R_{S1}	radius of sun gear 1
R_{S2}	radius of sun gear 2
RB	rule-based control strategy
r_{tire}	radius of tire
SOC	state of charge
s_{dis}	equivalent factors of discharging
s_{chg}	equivalent factors of charging
T_e	torques of the engine
T_{MG}	motor/generators torque
T_{MG1}	torques of the motor 1
T_{MG2}	torques of the motor 2
T_{out}	torques of the transmission output
t	time
V	vehicle speed
V_{oc}	open circuit voltage
α	road slope
η_{batt}	working efficiency of battery pack
η_{con}	motor controller efficiency
η_{em}	working efficiency of motor
η_{MG}	efficiency of the motor/generator
η_{MG1}	efficiency of the motor/generator 1
η_{MG2}	efficiency of the motor/generator 2
ρ	air density
ω_e	rotational speeds of the engine
ω_{MG}	motor/generators speed
ω_{MG1}	rotational speeds of the motor 1
ω_{MG2}	rotational speeds of the motor 2
ω_{out}	rotational speeds of the transmission output

References

1. Meisel, J. An Analytic Foundation for the Two-Mode Hybrid-Electric Powertrain with a Comparison to the Single-Mode Toyota Prius THS-II Powertrain. In Proceedings of the 2009 SAE World Congress, Detroit, MI, USA, 20–23 April 2009. SAE Paper No.2009-01-1321.
2. Arata, J.; Leamy, M.J.; Meisel, J.; Cunefare, K.; Taylor, D. Backward-Looking Simulation of the Toyota Prius and General Motors Two-Mode Power-Split HEV Powertrains. In Proceedings of the 2011 the SAE World Congress, Detroit, MI, USA, 12–14 April 2011. SAE Paper No.2011-01-0948.
3. Salmasi, F.R. Control Strategies for Hybrid Electric Vehicles Evolution Classification, Comparison and Future Trends. *IEEE Trans. Veh. Technol.* **2007**, *56*, 2393–2404. [CrossRef]
4. Wirasingha, S.G.; Emadi, A. Classification and Review of Control Strategies for Plug-In Hybrid Electric Vehicles. *IEEE Trans. Veh. Technol.* **2011**, *60*, 111–122. [CrossRef]
5. Mansour, C.J. Trip-Based Optimization Methodology for a Rule-Based Energy Management Strategy Using a Global Optimization Routine: The Case of the Prius Plug-in Hybrid Electric Vehicle. *Proc. Inst. Mech. Eng. Part D J. Automob. Eng.* **2015**, *230*, 1529–1545. [CrossRef]
6. Peng, J.; He, H.; Xiong, R. Rule Based Energy Management Strategy for a Series–Parallel Plug-In Hybrid Electric Bus Optimized by Dynamic Programming. *J. Appl. Energy* **2017**, *185*, 1633–1643. [CrossRef]

7. Moulik, B.; Söffker, D. Optimal Rule-Based Power Management for Online, Real-Time Applications in Hevs with Multiple Sources and Objectives: A Review. *J. Energies* **2015**, *8*, 9049–9063. [CrossRef]
8. Cheng, Y.; Cui, S.; Chan, C.C. Control Strategies for an Electric Variable Transmission Based Hybrid Electric Vehicle. In Proceedings of the 5th IEEE Vehicle Power and Propulsion Conference (VPPC'09), Dearborn, MI, USA, 7–11 September 2009; pp. 1296–1300.
9. Wang, R.; Lukic, S.M. Dynamic Programming Technique in Hybrid Electric Vehicle Optimization. In Proceeding of the 2012 IEEE International Electric Vehicle Conference Electric Vehicle Conference (IEVC), Greenville, SC, USA, 4–8 March 2012; pp. 1–8.
10. Pisu, P.; Rizzoni, G. A comparative study of supervisory control strategies for hybrid electric vehicles. *IEEE Trans. Contr. Syst. Technol.* **2007**, *15*, 506–518. [CrossRef]
11. Chen, Z.; Liu, W.; Yang, Y.; Chen, W. Online Energy Management of Plug-In Hybrid Electric Vehicles for Prolongation of All-Electric Range Based on Dynamic Programming. *Math. Probl. Eng.* **2015**, *2015*, 368769. [CrossRef]
12. Yuan, Z.; Teng, L.; Fengchun, S.; Peng, H. Comparative Study of Dynamic Programming and Pontryagin's Minimum Principle on Energy Management for a Parallel Hybrid Electric Vehicle. *J. Energies* **2013**, *6*, 2305–2318. [CrossRef]
13. Gao, J.P.; Zhu, G.M.; Strangas, E.G.; Sun, F.C. Equivalent Fuel Consumption Optimal Control of a Series Hybrid Electric Vehicle. *Proc. Inst. Mech. Eng. Part D J. Automob. Eng.* **2009**, *223*, 1003–1018. [CrossRef]
14. Zeng, Y.; Cai, Y.; Kou, G.; Gao, W.; Qin, D. Energy Management for Plug-In Hybrid Electric Vehicle Based on Adaptive Simplified-ECMS. *Sustainability* **2018**, *10*, 2060. [CrossRef]
15. Dong, J. Modeling and Real-Time Optimal Energy Management for Hybrid and Plug-in Hybrid Electric Vehicles. Ph.D. Thesis, Tongji University, Shanghai, China, 2009.
16. Lu, X.; Chen, Y.; Wang, H. Multi-Objective Optimization Based Real-Time Control for PEV Hybrid Energy Management Systems. In Proceedings of the IEEE Applied Power Electronics Conference and Exposition (APEC), San Antonio, TX, USA, 4–8 March 2018; pp. 969–975.
17. Xu, Q.; Luo, X.; Jiang, X.; Zhao, M. Research on Double Fuzzy Control Strategy for Parallel Hybrid Electric Vehicle Based on GA and DP Optimization. *IET Electr. Syst. Transp.* **2018**, *8*, 144–151. [CrossRef]
18. Ehsani, M.; Gao, Y.; Emadi, A. *Modern Electric, Hybrid Electric, and Fuel Cell Vehicles: Fundamentals, Theory, and Design*, 2nd ed.; CRC Press Taylor&Francis Group: Boca Raton, FL, USA, 2010.

© 2020 by the author. Licensee MDPI, Basel, Switzerland. This article is an open access article distributed under the terms and conditions of the Creative Commons Attribution (CC BY) license (http://creativecommons.org/licenses/by/4.0/).

energies

MDPI

Article

Comparative Study of Powertrain Hybridization for Heavy-Duty Vehicles Equipped with Diesel and Gas Engines

Ilya Kulikov *, Andrey Kozlov, Alexey Terenchenko and Kirill Karpukhin

National Research Center "NAMI", 125438 Moscow, Russia; a.kozlov@nami.ru (A.K.); terenchenko@nami.ru (A.T.); k.karpukhin@nami.ru (K.K.)

* Correspondence: i.kulikov.mami@gmail.com

Received: 3 April 2020; Accepted: 13 April 2020; Published: 21 April 2020

Abstract: This article describes a study that aimed to estimate the fuel-saving potential possessed by the hybridization of conventional powertrains intended for heavy-duty vehicles based on diesel and natural gas fueled engines. The tools used for this analysis constitute mathematical models of vehicle dynamics and the powertrain, including its components, i.e., the engine, electric drive, transmission, and energy storage system (ESS). The model of the latter, accompanied by experimental data, allowed for an analysis of employing a supercapacitor regarding the selection of its energy content and the interface between the traction electric drive and the ESS (in light of the wide voltage operating range of supercapacitors). The results revealed the influence of these factors on both the supercapacitor efficiency (during its operation within a powertrain) and the vehicle fuel economy. After implementation of the optimized ESS design within the experimentally validated vehicle model, simulations were conducted in several driving cycles. The results allowed us to compare the fuel economy provided by the hybridization for diesel and gas powertrains in different driving conditions, with different vehicle masses, taking into account the onboard auxiliary power consumption.

Keywords: heavy-duty vehicles; diesel engine; natural gas engine; hybrid powertrain; supercapacitor; fuel consumption; mathematical modeling

1. Introduction

Engines fueled with natural gas are considered possible alternatives to diesel engines when used in heavy-duty (HD) vehicles. Diesel engines, although efficient, have complicated issues regarding exhaust emissions, which compel developers to resort to rather sophisticated solutions to conform with regularly tightened emission requirements. Two components of exhaust gases may be highlighted as problematic, namely the nitrogen oxides and the particulate matter. To reduce the emissions of the former component to the level required by the present-day legislation, a selective catalytic reduction system should be employed, which involves a complex chemical reactor with a closed-loop control system and an onboard storage for the additional chemical agent (ammonia) [1]. Reducing the particulate matter requires using filters, which also constitute complex systems that involve an electronic control that regulates the process of regeneration [1]. It should also be noted that diesel engine operating regimes having minima of the two mentioned exhaust components generally do not coincide, which complicates the engine calibration [1]. In contrast, gas-fueled engines have drastically lower raw emissions of particulate matter [2–4] and can operate with the stoichiometric air-to-fuel ratio [5,6], which makes it possible to reduce emissions using less complex three-way catalytic converters [5] that provide more efficient abatement at lower costs. Fuel costs are generally considerably lower for gas (although this is not included in the scope of this work, which is only dedicated to the technical issues). The complicating aspect of replacing diesel engines with gas engines is the necessity of having an onboard storage for the compressed or liquefied gas, which constitutes a complex system that has its own cost, service, safety, and other issues. Therefore, whether to choose a gas-based vehicle or stay conservative with a diesel-based option becomes a trade-off decision.

During the past few years, the National Research Center "NAMI" has been conducting an R&D project in cooperation with one of the country's major producers of heavy-duty vehicles. The project was aimed toward developing a gas-fueled engine family derived from the diesel engine that was newly developed by the aforementioned HD vehicle manufacturer. One can find the details on the project and its results in References [7–9]. The main outcomes of the project were the gas-fueled engines (see example in Figure 1) operating with the Otto and Miller thermodynamic cycles. The engines were installed in vehicles intended for long-haul operations (also shown in Figure 1) and tested in road conditions.

(a) (b)

Figure 1. The developed gas engine at a test bench (**a**) and a heavy-duty (HD) vehicle equipped with the developed liquefied gas system (**b**).

The present work aimed to provide further elaboration of powertrains equipped with the developed gas engines, specifically on the prospects of their hybridization. The primary motivation behind this was to compare the fuel economy improvements that can be achieved by the hybridization of a diesel-based powertrain and a gas-based powertrain. It is assumed that such a powertrain modification could be relevant for those HD vehicles, whose operating schedules include a considerable share of city and suburban haulage since the major effect of HD powertrain hybridization is expected in conditions allowing for intensive use of regenerative braking, which is the most efficient tool for increasing the fuel economy of vehicles with large masses [10].The published studies (see, for example, [11–14]) have demonstrated that hybrid HD vehicles having a gross mass ranging 16 to 36 tons possess a significant fuel-saving potential in city driving conditions, especially in the case of full hybridization, which offers up to 35% fuel consumption reduction. The mild degree of hybridization is able to provide moderate fuel savings of about 10%.

The fuel economy of powertrains, which use different kinds of fuels, cannot be compared directly. One can evaluate the energy content of the consumed fuel [15], or the CO_2 emissions of compared vehicles [12,15], or the economic aspect, i.e., fuel price [12,15,16]. This work employed another approach stemming from the goal of hybridization, namely to reduce the fuel consumption relative to the baseline powertrain. This allows for comparing the percentages of vehicle fuel economy obtained for each considered powertrain rather than the absolute values of fuel consumption. The study was conducted by means of simulations based on mathematical models validated using experimental data.

The layout of the article has the following structure. The next section (Section 2) substantiates the choice of the hybrid powertrain design to be considered in the study. Section 3 describes the mathematical models of the vehicle and the powertrain components. It is followed by a section (Section 4) overviewing the control strategy implemented within the hybrid powertrain model. The driving cycles employed in the simulations are presented in Section 5. The models of conventional vehicles equipped with all the considered engines are validated in Section 6. The model analysis of the powertrain hybridization is presented in Section 7. Finally, Section 8 contains conclusions drawn from the conducted study and outlines further research.

2. Hybrid Powertrain Design

2.1. Selection of the Hybrid Powertrain Topology

Internal combustion engines (ICEs) intended for HD vehicles are remarkably efficient in a wide range of operating modes. Brake-specific fuel consumption (BSFC) becomes significantly higher only at small loads, which, in a hybrid electric vehicle (HEV), may be avoided by electric driving or, in the case of insufficient battery energy, by additional loading of the ICE. The concept of the optimal operating line (OOL), i.e., the locus of operating points having a minimum BSFC for a given power, is undoubtedly efficient in light duty applications where ICEs have a significant BSFC gradient; however, in the case of HD engines, the OOL is not as clear. In the first place, it requires employing a continuously variable transmission (CVT), which in the case of HD vehicles, can be either electric (corresponding to a series HEV topology) or electromechanical (the power-split topology). Both options (frequently called e-CVTs [17]) imply using two electric machines, which complicates the transmission and makes it more expensive. Generally, e-CVTs have a lower efficiency than stepped mechanical transmissions, and using them only makes sense if OOL tracking provides an increase of the ICE efficiency, which is significantly higher than the effect of impairing the efficiency using an e-CVT. In the case of an HD engine, OOL tracking may increase its efficiency only slightly, and considering a relatively low e-CVT efficiency, the overall fuel-saving effect will most likely be small or, in some cases, even negative [13]. This, together with the first aspect (complex and expensive transmission), makes e-CVT-based topologies hardly affordable, especially when it comes to issues of producing and owning such vehicles.

The parallel topology (Figure 2) can be derived from a conventional transmission (with an automated stepped gearbox) through moderate design changes. It implies placing an electric machine (EM) between the ICE shaft and the gearbox input shaft. It is advisable to place the automated clutch between the ICE and the EM to disconnect the former from the transmission and therefore allow the vehicle to be driven solely by the electric machine. The energy storage system (ESS) is connected to the traction electric drive using a DC link shown by the dotted line in Figure 2. The direct mechanical connection established between the ICE and the EM when the clutch is locked allows for the efficient transmission of power from the ICE to the ESS when it is necessary to replenish the latter or to load the former to avoid non-efficient operating regimes.

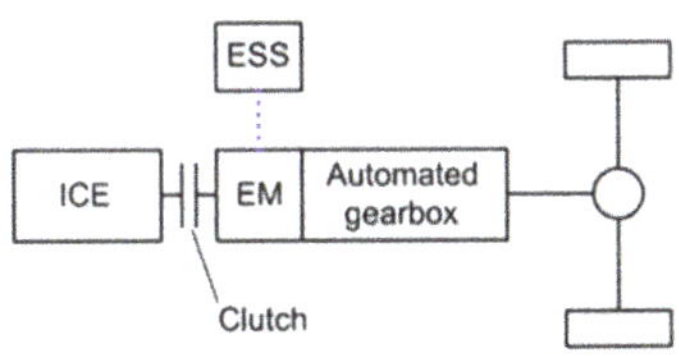

Figure 2. Parallel hybrid powertrain topology with a stepped automated gearbox. EM: Electric machine, ESS: Energy storage system, ICE: Internal combustion engine.

Placing the EM before the gearbox allows for providing a considerable range of electric driving (in terms of both speed and tractive force) without oversizing of the electric machine, which usually takes place in the topologies where the EM is connected to the driving wheels through a reduction gear that has a constant speed ratio. On the other hand, losses in the gearbox reduce the efficiency of the transmitting power from the EM to the driving wheels and backwards (regenerative braking).

Taking into account the above considerations and the previous studies performed by the authors (see, for example [18]) and other researchers [13,14], it was decided that for this study, the parallel topology would be selected for the analysis of powertrain hybridization.

2.2. Selection of the Powertrain Components

While the set of internal combustion engines to be employed in the studied powertrains was defined (i.e., the diesel and two gas engines), the electrical components were to be selected or designed. Considering the available technical resources and the objectives of the study, units available "from the shelf" were preferred over theoretical ones. In particular, the prototype traction electric drive was acquired from one of the commercially available hybrid units intended for HD vehicles. When used in an HEV with the gross weight considered in this work (35–44 tons), the power and torque characteristics of the electric drive (see Section 3.4 for details) correspond to an intermediate solution between the full and mild hybrids.

Two options for the engine control at vehicle stops were considered, namely idling and switching the engine off and starting it back when the vehicle resumes moving (i.e., the start–stop feature). The latter option was assumed to be implemented using an "enhanced" starter rather than the traction electric drive to prevent impairing the vehicle dynamics.

The energy storage system (ESS) is one of the key components in a hybrid powertrain, which defines the amount of regenerated braking energy and the vehicle performance in the pure electric driving mode. The prevailing ESS technology in the field of hybrid and pure electric vehicles are accumulator cells, which are lithium-ion for the most part [19]. Their specific energy capacity and power, which have noticeably increased over the years [20])], along with their moderate prices make them well-suited for passenger and light-duty commercial vehicles. Their known vulnerability to high and low temperatures (capacity and efficiency drop, accelerated degrading) can be alleviated using appropriate temperature management [19], at least in hybrid vehicles, where engine heat can be utilized for that purpose. The cycle life of typical lithium-ion cells, although relatively short (a typical value is around 3000 cycles, but may be lower), is nevertheless admissible for electric vehicles since they employ large battery packs that allow for traveling up to a few hundred kilometers per charge–discharge cycle. Hybrid vehicles, especially non-plugin vehicles, have significantly smaller batteries, which compels developers and producers to resort to more advanced (and expensive) accumulator cells that have higher power densities and longer cycle lives.

When it comes to the hybridization of heavy-duty vehicles, the following specifics of long-haul vehicle applications have to be taken into account while selecting the ESS type. Due to the large mass of such vehicles, the ESS should be able to deliver and consume high amounts of power (over 100 kW) continuously. An HD vehicle usually has a long service life and is operated intensively (on an everyday basis), which entails a large mileage from hundreds of thousands to even millions of kilometers. Throughout the service period, the number of overhauls should be minimal. The replacement of powertrain components, especially expensive ones, is undesirable (the operating life of the powertrain and its major components is to be equal to that of the vehicle). Additional expenses brought about by introducing advanced powertrain technologies should be (at least) compensated for by the economic effect provided by those technologies (e.g., reduced fuel consumption). Climatic operating conditions of the heavy-duty vehicle fleet vary greatly, including both extremely cold and warm regions.

To provide a longer cycle life of accumulator-based solutions, one should use "shallow" charge–discharge cycles. Together with the requirement for higher power, this implies that large batteries are to be used in HD vehicles. As a result, even for hybrid applications, which offer a relatively small pure electric driving range, the battery tends to be oversized due to the necessity in meeting both the power and cycle life requirements.

A known alternative for high-power ESSs are supercapacitors (SCs), which the literature claims as a promising technology for heavy-duty vehicles [11,21–24]. One has to admit that the characteristics of supercapacitors include both substantial advantages and adversities. Among the former, one can provide high operating currents and, consequently, a high power density [25,26]. A single SC cell having a nominal voltage of 3 V can operate continuously with currents up to 300 A and, for a limited time, over 1000 A. The cycle life of supercapacitors is outstandingly higher than that of lithium-ion accumulators and exceeds 1 million cycles [25,26]. Additionally, supercapacitors have a wide operating range of low temperatures, which can reach −40 °C. For example, when used in engine-cranking applications, an SC can start an engine smoothly in a −25 °C ambient temperature [26]. Thus, supercapacitors do not impose the same strict requirements regarding the low-temperature management as accumulators do. On the other hand, the following shortcomings of supercapacitors should be taken into account. Their energy density is far lower (about 95% lower on average) than that of lithium accumulators. The prices for a kWh of supercapacitors are at least 20 times higher than those of the typical lithium cells. For advanced, high-power, long-life accumulators used in compact batteries of hybrid vehicles, this price ratio may be smaller, although not drastically. Like lithium-ion accumulators, supercapacitors are vulnerable to high-temperature degradation; therefore, they require effective cooling systems [25]. Yet another feature may be counted as both positive and negative, namely the proportionality between the supercapacitor voltage and its state of energy. On the one hand, this allows an ESS management system to determine the state of energy using simple voltage monitoring. In contrast, lithium-ion accumulators, especially the $LiFePO_4$-type accumulators, require battery management systems to have sophisticated algorithms for the indirect determination of the state of charge [27] due to its substantial independence from the accumulator voltage. On the other hand, the said proportionality results in a wide operating range of the supercapacitor voltage is much wider than that of lithium-ion accumulators, which raises the question of matching this range with the operating voltage of the traction electric drive. One known solution is employing a DC–DC converter that keeps the input voltage of the electric drive at a reference level, while the supercapacitor voltage travels through its operating range [24,28,29]. This ensures that the traction drive will be able to deliver its rated power independently (to a certain extent) of the ESS voltage. The complication of this approach is, obviously, the need for a high-power buck–boost converter, which entails additional costs, weight, occupied space, and finally, yet importantly, power losses due to voltage transformation. One can expect a 95% peak efficiency at most from such a converter, which is rather noticeable for the powertrain energy balance considering that the major power flow will be transmitted through this converter. Another solution is a direct DC connection between the supercapacitor and the traction drive inverter [29]. This makes the operating characteristics of the drive variable and needs to be addressed in the control algorithm of the powertrain. In the described study, both solutions were compared in terms of the resulting fuel economy and ESS efficiency.

For commercial vehicles, all the mentioned virtues of supercapacitors—high power, long cycle life, and wide operating temperature span—are particularly advantageous. When considering the drawback of low energy density, one can conclude that unlike for pure electric vehicles, for non-plugin hybrids, this is not a critical issue, although the large weight of HD vehicles, of course, does require a proportional energy content of the ESS that can recuperate the bulk of the braking energy. The supercapacitor cost, however, is an issue that raises concerns regarding the additional expenses entailed by hybridization. Since the amount of regenerated energy that can be stored within the ESS is the decisive factor of fuel economy (in the case of HD vehicles), the ESS capacity is a result of a trade-off between lower fuel expenses and higher ESS costs. Therefore, establishing the influence of the ESS energy content on the fuel economy becomes an additional objective of the study.

The prototype of the supercapacitor-based ESS used for this study has been taken from the powertrain of a hybrid city bus that belonged to a limited series produced earlier. The ESS in that bus has an electric capacitance of 28 F and a maximum voltage of 650 V, providing an energy content of 1.6 kWh. The maximum continuous current is 300 A for both discharging and charging. Considering the voltage operating range of 300–640 V, the maximum power delivered by the SC amounts to 90–195 kW. To obtain the characteristics of the prototype supercapacitor, several field experiments were conducted involving the mentioned hybrid bus. During those tests, the supercapacitor current and voltage were being measured and logged while the bus was being driven through predetermined velocity patterns.

3. Mathematical Models

3.1. Vehicle Dynamics

Simulations aimed at the calculation of the fuel economy and assessment of the powertrain operation usually replicate vehicles moving through driving cycles that model real-world driving conditions. Cycle schedules do not include the driving trajectory, and, thus, when modeling, the vehicle motion is considered linear. Furthermore, there were additional assumptions that were made in this work to derive the vehicle dynamics model. The tire slip is neglected due to the assumption that the maximum adhesion coefficient is high, and thus, a sufficient tractive force is exerted with a small slip. The model also neglects the dynamics of wheel vertical forces and variations of the wheel radii. Given these assumptions, the model of the vehicle linear motion can be presented as the dynamics of a single lumped mass:

$$\dot{v} = \frac{T_w/r_w - F_f - F_{air} - F_\alpha}{m_v + n \cdot I_w/r_w^2}, \quad (1)$$

where v is the vehicle velocity; m_v is the vehicle mass; T_w is the torque at the driving wheels; r_w is the wheel radius; I_w is the wheel inertia; n is the number of wheels on the vehicle (including the trailer's wheels); and F_{air}, F_f, and F_α are the resistance forces.

The rolling resistance force is calculated using the following formula: $F_f = m_v \cdot g \cdot f$, where g is the acceleration due to gravity and f is the dimensionless coefficient of tire-rolling resistance. When identifying the model parameters from the coast-down test results, it was found that the function $f(v)$ was satisfactorily approximated using the known quadratic expression: $f = f_0 + k_v \cdot v^2$ [30], where f_0 is the rolling resistance coefficient at near-zero velocity and k_v is the factor of the velocity-dependent growth of the rolling resistance. The rolling resistance force calculated with these formulae was assumed to be equal (i.e., averaged) for all the tires of the vehicle.

A well-known empirical formula is used for calculating the aerodynamic resistance force:

$$F_w = 0.5 \cdot C_x \cdot A \cdot \rho \cdot v^2, \quad (2)$$

where C_x is the air drag coefficient of the vehicle, A is the vehicle frontal area, and ρ is the air density.

The grade resistance force F_β is a projection of the vehicle gravity force vector onto the road plane, which is calculated as follows: $F_\beta = m_v \cdot g \cdot sin(\beta)$, where β is the road inclination angle (positive for an ascending slope).

3.2. Internal Combustion Engines

The powertrain model elaborated in this work used the quasi-static approach for the simulation of the internal combustion engines. The engine characteristics were obtained experimentally in steady operating regimes and then implemented in the form of lookup tables. During simulations, the characteristics were interpolated using dynamic values of the engine rpm and the torque command calculated using the model. The required characteristics of the engines were obtained via laboratory dynamometer tests. In those tests, several parameters were measured and logged, namely the shaft speed, shaft torque, and fuel consumption, which were then employed for modeling purposes.

Figure 3 shows the BSFC map of the diesel engine and its maximum torque curve denoted as "Max. trq." The main performance parameters of this engine confirmed using the tests were as follows: rated power 331 kW (at 1900 rpm) and 1985 Nm maximum torque (at 1300 rpm).

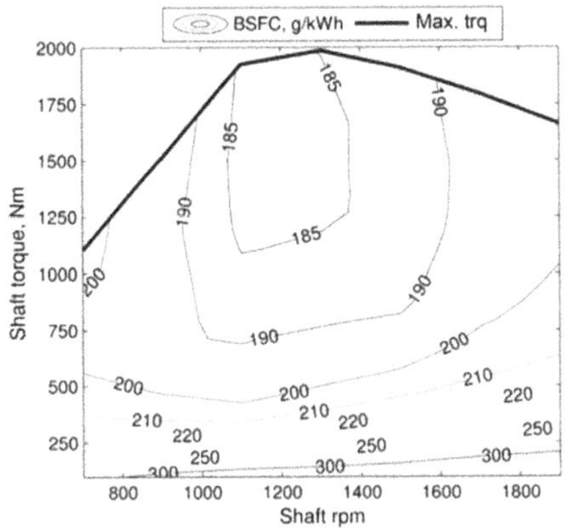

Figure 3. Brake-specific fuel consumption (BSFC) map of the diesel engine.

Similar maps were built from the dynamometer test data for the gas engines (Figure 4). The main performance parameters of these engines were also confirmed: gas Otto's cycle—rated power 326 kW (1900 rpm), maximum torque 1900 Nm (1300 rpm); gas Miller's cycle—rated power 384 kW (1900 rpm), maximum torque 1950 Nm (1400 rpm).

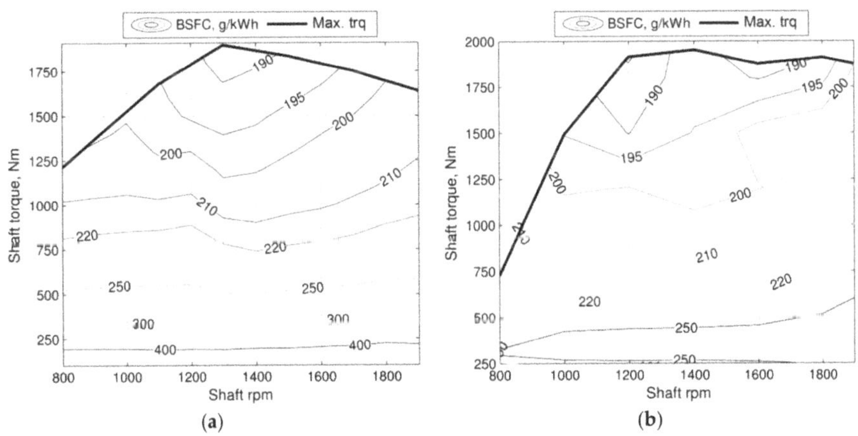

Figure 4. Brake-specific fuel consumption maps of the developed gas engines: (**a**) Otto's cycle and (**b**) Miller's cycle.

One can notice that the engine with the Otto's cycle has a steeper gradient for the BSFC than that of the Miller-cycle engine, especially at low loads. This suggests that a wider part of the engine map should be excluded (if possible) from operating regimes through hybridization in the case of the Otto engine. It can also be seen that the Miller-cycle engine has an apparent torque dip at low shaft angular speeds, which should be compensated for using the additional torque of the electric machine (i.e., the "torque boost" feature).

3.3. Transmission

The transmission model was derived under the assumption that all its shafts were rigid (i.e., shaft torsional stiffness was neglected). Both the conventional stepped transmission and the transmission of the parallel hybrid topology have three operating regimes defined by the automated clutch state:

whether it is disengaged, slipping, or engaged. In the first and the second states, the ICE shaft has a separate degree of freedom, which, for the hybrid transmission, yields the following system of equations:

$$\begin{cases} I_e \cdot \dot{\omega}_e = T_e - T_{clutch} \cdot sgn(\omega_e - \omega_{em}), \\ T_w = \left(T_{clutch} \cdot sgn(\omega_e - \omega_{em}) + T_{em} - I_{em} \cdot \dot{\omega}_{em}\right) \cdot u_{GB} \cdot \eta_{GB}^{sgn(T)} \cdot u_0 \cdot \eta_0^{sgn(T)}, \\ \omega_{em} \cdot r_w = v \cdot u_{GB} \cdot u_0, \end{cases} \quad (3)$$

where ω_e and ω_{em} are the shaft speeds of the ICE and electric machine, respectively; T_e and T_{em} are the shaft torques of the ICE and electric machine, respectively; T_{clutch} is the clutch torque; I_e and I_{em} are the inertias of the ICE and electric machine, respectively; u_{GB} and u_0 are the ratios of the gearbox and final drive, respectively; and $\eta_{GB}^{sgn(T)}$ and $\eta_0^{sgn(T)}$ are the efficiencies of the gearbox and final drive, respectively. The efficiencies are raised to the power of the torque sign function in order to take into account both traction and braking torques entering the transmission.

The term $sgn(\omega_e - \omega_{em})$ introduces a relation between the direction of the clutch torque and the speed differential of the input and output parts of the clutch. When the clutch becomes fully engaged, the separate equation of the ICE shaft dynamics, as well as the torque T_{clutch}, can be excluded from the transmission model, thus eliminating one degree of freedom.

To derive the model of the conventional transmission from the system (Equation (3)), one should exclude the terms associated with the electric machine and replace the angular speed ω_{em} with that of the gearbox input shaft.

3.4. Electric Drive

The electric drive, just like the internal combustion engines, was modeled using experimental static maps. Laboratory tests allowed for obtaining the torque and efficiency characteristics of the traction electric drive (electric machine + inverter) for two voltage levels, namely 650 V and 450 V, which covered the bulk of the prototype supercapacitor operating range. Between these voltages, the characteristics were interpolated, and were extrapolated below 450 V. The main parameters of the electric drive obtained from the experimental data are summarized in Table 1.

Table 1. The main performance parameters of the traction electric drive.

DC Voltage (V)	Motor Mode		Generator Mode	
	Max. Power (kW)	Max. Torque (Nm)	Max. Power (kW)	Max. Torque (Nm)
450	90	1100	100	1100
650	135	1100	165	1100

Figure 5 shows the efficiency map and the maximum torque characteristics of the electric drive in motor mode.

Given the electric machine torque (commanded by the control system) and the shaft speed (calculated by the transmission model), as well as the supercapacitor voltage u_{SC} calculated via its model, one can obtain the electric drive current $i_{em,dc}$ at the DC side:

$$i_{em,dc} = \frac{T_{em} \cdot \omega_{em}}{u_{SC} \cdot \eta_{em}^{sgn(T_{em})}}, \quad (4)$$

where $\eta_{em}^{sgn(T_{em})}$ is the electric drive efficiency, taking into account the torque direction (positive for the motor mode). The current calculated with this formula is used by the SC model as the input signal.

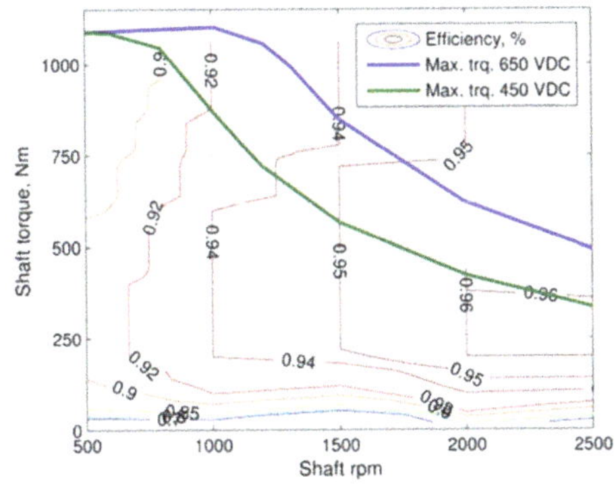

Figure 5. Efficiency map and torque characteristics of the electric drive (motor mode).

3.5. Energy Storage System

The following considerations were taken into account when choosing the approach to mathematical modeling of the supercapacitor. Chemical processes within the SC lie beyond the scope of the research. At the current stage, temperature effects are also neglected under the assumption that the powertrain operates within a moderate range of temperatures, which imposes no substantial effect upon the supercapacitor characteristics. The operating parameters of the SC relevant for the powertrain modeling in this study were the current, voltage, efficiency, and energy content. These considerations suggest using equivalent circuit modeling, which is a widely employed approach for the analysis of ESS not as electrochemical systems but rather electrical components of the powertrain. References [25,31,32] describe an equivalent circuit for the supercapacitor in the form of a "ladder" with each "step" consisting of a capacitor and a resistance. Depending on the required model fidelity, one can use different numbers of "steps," which in the simplest case, would be a single one. The comparison performed in this work with the involvement of experimental data has shown that a reasonable trade-off between model accuracy and complexity is provided by a circuit consisting of three "steps," as shown in Figure 6.

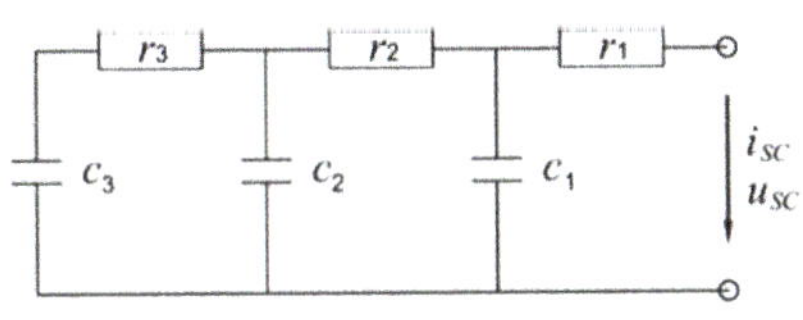

Figure 6. Equivalent circuit for a supercapacitor.

In this scheme, r_1, r_2, and r_3 are the internal resistances; c_1, c_2, and c_3 are the capacitances; and u_{SC} and i_{SC} are the voltage and current at the supercapacitor electric terminals.

From the equivalent circuit, one can derive the following equation system using Kirchhoff's current and voltage laws:

$$\begin{cases} u_{SC} = u_1 = \frac{\int i_{SC}dt - c_2u_2 - c_3u_3}{c_1}, \\ \dot{u}_2 = \frac{u_1 - u_2 - (u_2 - u_3)r_2/r_3}{c_2r_2}, \\ \dot{u}_3 = \frac{u_2 - u_3}{c_3r_3}, \end{cases} \tag{5}$$

where u_1, u_2, and u_3 are the voltages within the sub-circuits corresponding to each "step" of the "ladder."

The initial conditions can be obtained for this model from the assumption of equality between the voltages of all the capacitors, which takes place when no load is applied and transient processes have ended: $u_1 = u_2 = u_3 = u_{SC,init}$. Substituting this condition into the first equation of the system (Equation (5)) yields the initial condition for the integral term:

$$\int i_{SC}dt = u_{SC,init}(c_1 + c_2 + c_3). \tag{6}$$

The term $c_1 + c_2 + c_3$ is the total capacitance of the "ladder." Identification of the model parameters using experimental data showed that the best accuracy for voltage calculation was obtained when the sum $c_1 + c_2 + c_3$ was equal to the rated capacitance c_{SC} of the modeled SC. Furthermore, this provided a correct calculation of the supercapacitor energy content with no need for ad hoc corrections.

Figure 7 shows the results of modeling that simulated two experiments involving the supercapacitor installed within the powertrain of the hybrid city bus mentioned in Section 2. The bus was driven in the pure electric mode through velocity patterns consisting of accelerations and decelerations. The top plots demonstrate the supercapacitor current logged during the experiments (positive for charging). The current was used as the input signal for the equivalent circuit model, which responded with the calculated voltage. The latter is shown in the bottom plots (denoted as "Model") along with the measured voltage ("Experiment"). The identification of the model parameters was performed using the criterion of the minimum root mean square error (RMSE) of voltage calculation. The parameter set satisfying this criterion is listed in Table 2. The RMSEs obtained with these parameters amounted 0.7–3% in about a dozen experiments.

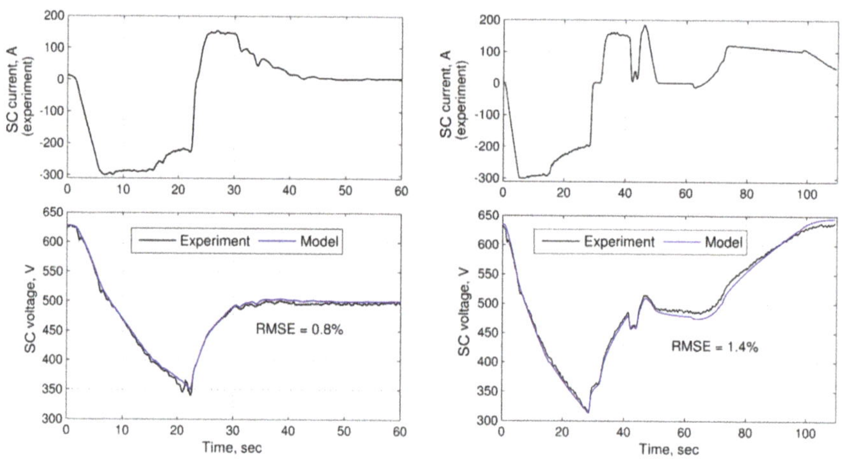

Figure 7. Comparison of the supercapacitor simulation results with the experimental data.

Table 2. Supercapacitor equivalent circuit parameters.

r_1 (Ω)	r_2 (Ω)	r_3 (Ω)	c_1 (F)	c_2 (F)	c_3 (F)
0.17	0.35	1.8	12	8	8

The equivalent circuit includes the internal resistances, which not only provide a voltage drop, but also introduce the dissipation of energy, and therefore, allow for calculating the capacitor efficiency. To do this, it is necessary to determine the power dissipated at the resistances.

The current through the resistance r_1 is equal to that at the SC terminals, i.e., $i_1 = i_{SC}$. While deriving the system (Equation (5)), one can find that the current through the second resistance is equal to $\dot{u}_2 c_2 + \dot{u}_3 c_3$. The current in the third "step" is equal to $\dot{u}_3 c_3$. Therefore, the total power loss is:

$$P_{SC,loss} = r_1 i_{SC}^2 + r_2 \left(\dot{u}_2 c_2 + \dot{u}_3 c_3\right)^2 + r_3 \left(\dot{u}_3 c_3\right)^2. \quad (7)$$

The supercapacitor net power $P_{SC} = i_{SC} \cdot u_{SC}$. The total power drawn from the SC or received by it is a sum of the net power and the loss power, taking into account the signs thereof. These considerations result in the following expression for the SC efficiency:

$$\eta_{SC} = \left(\frac{|P_{SC}|}{|P_{SC}| - sgn(P_{SC}) \cdot P_{SC,loss}} \right)^{-sgn(P_{SC})}. \quad (8)$$

Note that P_{SC} is positive for charging.

Figure 8 demonstrates the simulation results for yet another one of the mentioned field tests, in which the supercapacitor was discharged with the maximum continuous current and then charged back to the initial voltage. Three upper plots show the voltage, current, and net power of the SC, respectively. The fourth plot demonstrates the components of power dissipation associated with the three internal resistances of the equivalent circuit. One can see the clear differences in the dynamics of these components. The first term is the major contributor to the power dissipation. The dynamics of this term is defined by the dynamics of supercapacitor loading (i.e., dynamics of the i_{SC} current). The second and third terms constitute processes with slower dynamics and lower amplitudes. The SC efficiency (the bottom plot) diminished along with the voltage, while the loading current remained constant, as well as the power dissipation at the resistance r_1. However, the power losses at the two other resistances grew due to dynamics of the currents at the capacitors c_2 and c_3. Analysis of the obtained results suggests that when using the considered supercapacitor as the ESS in a hybrid powertrain, one should avoid a deep discharge (i.e., below 450 V), either using the control strategy or via supercapacitor sizing or, most likely, both.

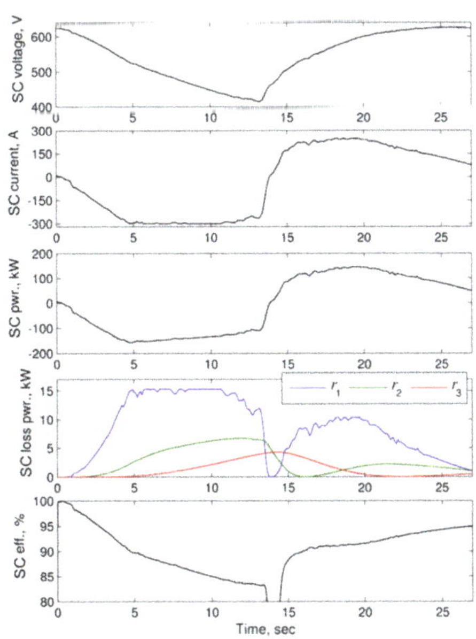

Figure 8. Model analysis of the power losses and efficiency of the supercapacitor.

It is worth noticing the interval of transition between discharging and charging (in the vicinity of the 14th second), where the power dissipation at r_1 drops down to zero while the powers at r_2 and r_3 continue to attenuate at the rates defined by the parameters of the corresponding "steps" of the equivalent circuit. After the load removal, the capacitors c_1, c_2, and c_3 exchange energy between each other until their voltages become equal. This process is accompanied by the dissipation of energy at the resistances r_2 and r_3. If no external load is present during these periods of voltage stabilization, the efficiency of the SC becomes negative. However, in the estimation of the SC mean efficiency, such energy losses may be assigned to the preceding operating interval, which allows for avoiding an unnecessary introduction of the negative efficiency.

An alternative way of determining the supercapacitor efficiency (the average for a certain operating period) stems from the fact that the equivalent circuit model incorporates the power dissipation, and the supercapacitor voltage calculated by it reflects the actual state of energy (at least under no-load conditions). Therefore, one can integrate the net power P_{SC} and multiply it by the average efficiency, which is calculated to bring the integrated energy level to that of the equivalent circuit model by the end of the operating cycle. This can be expressed as the following equation:

$$\widetilde{\eta}_{SC}^{sgn(i_{SC})} \int_0^{t-end} P_{SC} dt = 0.5 c_{SC} \cdot u_{SC,t-end}^2, \quad (9)$$

where $\widetilde{\eta}_{SC}$ is the average efficiency of the SC during the given period, $t-end$ is the end time of the period, and $u_{SC,t-end}$ is the SC voltage at the end of the period.

3.6. Onboard Power Consumers

The main onboard power consumers of an HD vehicle are the air brake compressor, power steering, air conditioner compressor, engine oil and coolant pumps, engine fan, and electrical accessories. In a conventional powertrain, most of these are powered mechanically by drives connected to the engine shaft. Those of them not having disconnecting mechanisms, consume some power even in the idle state. In the literature [13,33], one can find typical ranges of the consumed power for each of these components as a function of the engine shaft speed as well as their operating cycles based on the recommendations of the SAE J1343 standard [34]. An operating cycle is defined by two numbers, namely Δt being the averaged period of the operating cycle and α being the fraction (i.e., percentage) of the operating cycle when the component is engaged. Table 3, based on the information from References [13,33], summarizes the power and duty cycle data for the auxiliaries included in the model of the conventional powertrain. In this table, "A/C" stands for the air conditioning and "Electrical acc." denotes the low-voltage electrical accessories. The columns "Min. power" and "Max. power" contain the ranges of power consumption corresponding to the engine rpm range, which was from 500 rpm (the lower power value) to 2000 rpm (the higher power value).

Table 3. Power and duty cycle parameters of the auxiliary loads in a conventional HD powertrain.

Auxiliary Component	Min. Power (kW)	Max. Power (kW)	Δt (s)	α (%) (Urban Driving)	α (%) (Highway Driving)
Air brake compressor	0.35–1.2	1–3.5	100	30	5
Power steering	0.3–1.2	3–10.5	100	60	10
A/C compressor	0	1–5.3	150	50	50
Engine fan	0	16.7	200	10	5
Electrical acc.	0.6	0.6	-	-	-

The power feeding the electrical accessories is drawn from the engine shaft through the alternator, whose efficiency, obtained from the known literature [35,36], was taken into account in the elaborated model.

The engine coolant and oil pumps are not listed in Table 3 since in this work, they were considered as parts of the engine system. Their power demand is taken into account by the fuel consumption characteristics of the engines since the latter underwent the bench tests with the said auxiliaries installed. The coolant and oil pumps were also considered identical for both the conventional and the hybrid powertrains.

In a hybrid powertrain, the electrified auxiliary components are regulated independently from the engine operating regime, therefore consuming the exact amount of power required for their operation [13]. Table 4, based on the information from [13], shows the average power and operating cycle data for the auxiliaries assumed to be electrified in the hybrid powertrain. Besides the conventional auxiliaries, the cooling system of the high voltage components was taken into account (denoted as HV cooling).

Table 4. Power and duty cycle parameters of the auxiliary loads in the hybrid HD powertrain.

Auxiliary Component	Power (kW) (Urban Driving)	Power (kW) (Highway Driving)	Δt (s)	α (%) (Urban Driving)	α (%) (Highway Driving)
Air brake compressor	1.7	2.2	100	5	5
Power steering	5.5	3	100	60	10
A/C compressor	2.5	3.3	150	50	50
Engine fan	2.4	5.5	200	4	4
Electrical acc.	0.6	0.6	-	-	-
HV cooling	0.3	0.3	100	56	10

A DC–DC converter was considered as the interface between the low voltage auxiliary system and the high voltage traction system.

4. Hybrid Powertrain's Control Strategy

This section gives a brief overview of the control strategy developed for the hybrid powertrain model, highlighting its essential structure and operating principles.

If the modeled HEV is equipped with the start–stop feature, the control algorithm of the latter is quite simple: when the vehicle stops, the engine shuts down; the engine starts again when the vehicle resumes moving (i.e., velocity crosses the minimum threshold). When the ICE is on, the powertrain can operate in one of the two basic modes, namely the electric mode or the hybrid mode, which is illustrated in Figure 9. The former implies for the ICE to be in the idling state having the corresponding torque $T_{e,idle}$, while the electric machine drives and decelerates the vehicle; the EM torque is a function of the accelerator (acc.) and brake pedal signals. In the hybrid mode, both the ICE torque and the electric machine torque are functions of the accelerator and brake signals, as well as the supercapacitor current i_{SC} and voltage u_{SC}.

In Figure 9, several conditions (C1–C6) are assigned to the transition arrows, as well as the logical relations between these conditions (i.e., "and" and "or" operators). The conditions are detailed in Table 5.

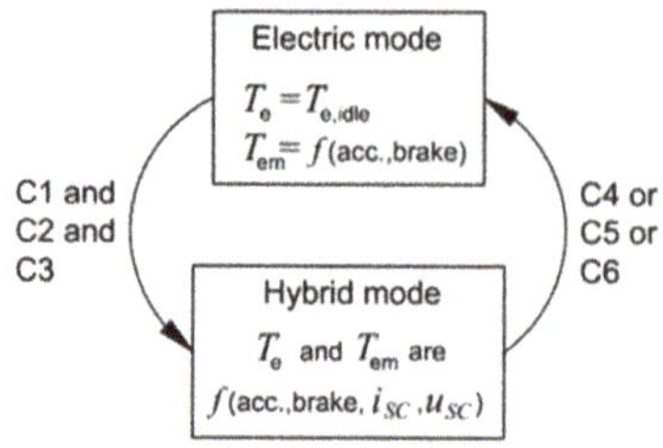

Figure 9. A simplified diagram of the hybrid powertrain operating modes.

Table 5. Basic conditions for the transitions between the hybrid powertrain modes.

Condition No.	C1	C2	C3	C4	C5	C6
Expression	$v > v^{up}$	acc. > acc.^{up}	no gear shift	$v < v^{low}$	acc. < acc.^{low}	gear shift

The parameters v^{up}, v^{low}, acc.^{up}, and acc.^{low} specify the upper and lower thresholds for the vehicle velocity and accelerator signals. The default values of v^{up} and v^{low} are 25 km/h and 15 km/h, respectively. The default values of acc.^{up} and acc.^{low} correspond to 50% and 10% of the maximum ICE torque. These values are used when the ESS voltage stays near its maximum (650 V). When the voltage lowers, the thresholds decrease to engage the ICE earlier and to reduce the electric machine traction torque, preventing an excessive consumption of the supercapacitor energy.

Besides the electric mode, the engine becomes disconnected from the transmission upon gear shifting. During a shift, the electric machine provides gear synchronization.

The hybrid mode implies the torques of the ICE and the electric machine should be combined. When the SC voltage drops below the lower threshold (the default value is in the range of 500–550 V), the ICE begins delivering an excessive power consumed by the electric machine whose torque becomes a function of the SC voltage drop (the deeper the discharge, the higher the torque). When the driver's torque request exceeds the maximum ICE torque, the electric machine provides the "torque boost" in accordance with the map shown in Figure 10 using the example of the diesel-based powertrain (the resulting curve of the hybrid unit's maximum torque is denoted "Hybrid max. trq"). Like the thresholds of the velocity and accelerator signals, the boosting torque constitutes a function of the ESS voltage.

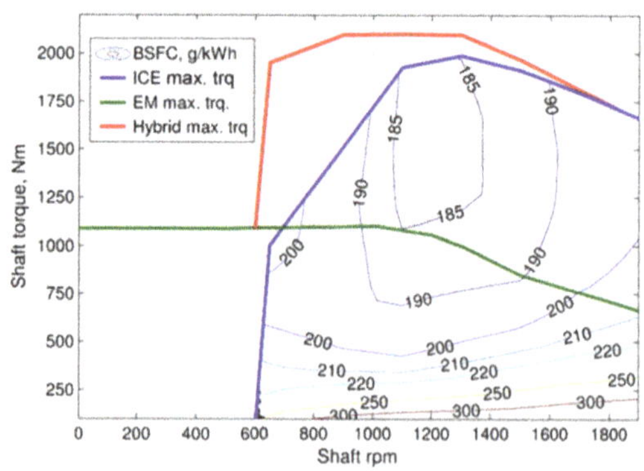

Figure 10. Superimposed torque characteristics of the hybrid unit and its components.

5. Driving Cycles

Several driving cycles intended for heavy-duty vehicles were employed in the simulations to represent the variety of possible driving conditions. These included urban and suburban cycles, which constitute the major interest of the study, as well as highway cycles.

The driving cycles employed in the simulations are shown in Figure 11. The GOST cycles are road-performed (rather than using a chassis-dynamometer) schedules provided by a national standard named the GOST. All the driving cycles, except for the WHVC, imply level roads.

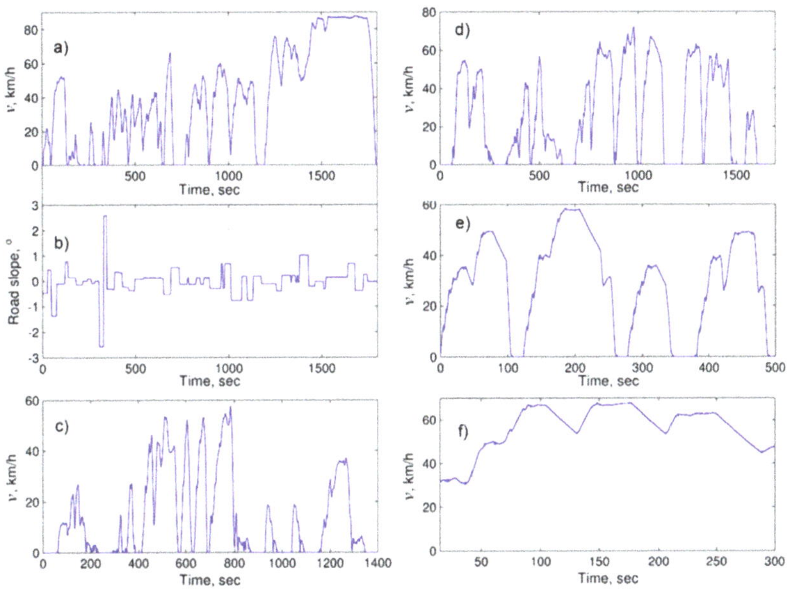

Figure 11. Driving cycles employed in the simulations. (**a**,**b**) velocity and road slope of the World Harmonized Vehicle Cycle (WHVC), (**c**) the West Virginia University city cycle (WVUCITY), (**d**) the West Virginia University suburban cycle (WVUSUB), (**e**) the GOST city cycle (GOST city), (**f**) the GOST highway cycle (GOST highway).

6. Validation of the Conventional Vehicle Model

In the R&D project mentioned in the introduction, several road tests were conducted in the GOST driving cycles involving vehicles equipped with the developed diesel and gas engines. Furthermore, by way of comparison, a vehicle equipped with a production gas-fueled engine was tested in the same driving conditions. The logged parameters were the vehicle velocity, engine rpm, and fuel consumption. The logs allowed for simulating the tests and validating the model of the conventional vehicle.

The baseline vehicle was an articulated road train (tractor-trailer) intended for long haulage. It was equipped with a manual 16-speed mechanical gearbox with two ranges, the high and low, with eight gears each. In normal driving conditions, the former is usually preferred over the latter. The test mass of the vehicle was 44,000 kg (equals to the tractor-trailer gross mass).

Prior to the driving cycle tests, coast-down experiments were performed to estimate the rolling and air resistance forces. Analysis of the coast-down results and the vehicle's technical specification available from the manufacturer allowed for identifying the parameters of the model of vehicle dynamics (Equation (1)), which are specified in Table 6.

Table 6. Parameters of the model of vehicle dynamics.

m_v (kg)	r_w (m)	I_w (kg·m^2)	C_x	A (m^2)	f_0	k_f
44,000	0.492	20	0.58	7.9	0.0045	(1.5–2) × 10^{-6}

Table 7 shows the fuel economy data obtained using the field experiments and simulations, as well as the simulation errors, for the vehicles equipped with the three mentioned engines.

Table 7. Comparison between the experimental and calculated fuel economy for the vehicles equipped with the conventional powertrains.

Driving Cycle (GOST)	Fuel Economy, Diesel (L/100 km)			Fuel Economy, Gas-Otto (m^3/100 km)			Fuel Economy, Gas, Prod. Engine (m^3/100 km)		
	Exp.	Model	Error (%)	Exp.	Model	Error (%)	Exp.	Model	Error (%)
City	58.3	58.5	0.34	74.38	72.4	2.66	61.6	67.38	9.38
Highway	35.8	37.5	4.75	46.98	46.09	1.89	43.3	42.96	0.79

The higher error for the production gas engine in the city cycle resulted from the unknown experimental velocity data, which compelled to use the velocity log from another vehicle's test. Despite the same driving schedule, the velocity profiles most likely had certain differences since the accuracy of road-performed cycles strongly depends on the driver's behavior, which obviously cannot be repeated precisely from test to test.

In general, the model was considered to have sufficient accuracy for performing further simulations with the hybrid vehicles.

7. Model Analysis of the Powertrain Hybridization

Considering the main objective of the study and having analyzed the specifics of the devised hybrid powertrain design, the simulation program was formulated as the following sequence of tasks:

1. Estimate two approaches to match the electric drive voltage to that of the supercapacitor, namely by employing a DC–DC converter and directly interfacing the two devices.
2. Estimate the powertrain characteristics with different ESS energy contents.
3. Having selected the "optimal" solution regarding the ESS energy and the electrical interface, use this solution in comparative simulations estimating the fuel consumption of the conventional and hybrid powertrains. Calculate the fuel economy of the HEVs relative to the conventional vehicles.
4. Estimate the influence of the vehicle mass on the relative fuel economy.

The simulations were conducted in identical driving conditions for all the studied powertrains. The compared variants of HEVs shared the same adjustments of the control system with slight variations due to differences in the characteristics of the engines. The gearbox shift map, which was common for both the conventional and the hybrid powertrains, was defined in a way similar to that employed within the Vecto software tool [36].

Figure 12 shows an example of simulation results obtained in the urban part of the WHVC driving cycle for the HEV equipped with the Otto-type gas engine that had the start–stop feature. ("EM" denotes the variables related to the electric machine).

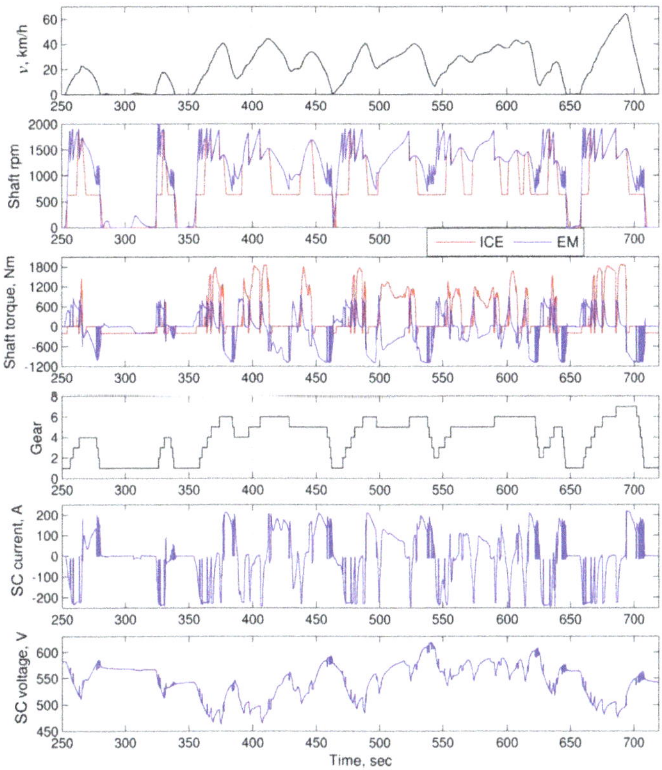

Figure 12. Hybrid powertrain operating variables calculated in the World Harmonized Vehicle Cycle (WHVC) (fragment).

Since the considered hybrid powertrain had no external ESS charge feature, the condition of the energy balance should be pursued in the simulated driving cycles. This means that the ESS energy level at the end of the cycle has to be equal to that observed at the beginning of the cycle. This ensures that no ESS energy is spent without replenishing it and no fuel is consumed to excessively charge the ESS. Originally, since the HEV's control system was not intended to achieve the precise energy balance within a specified time interval, an ad hoc adjustment was introduced, which gradually "pulled" the ESS state of energy toward the initial value when approaching the end of the driving cycle. With the energy balance condition achieved, the fuel economy of the HEV relative to the conventional vehicle was calculated as follows:

$$FE = \frac{FC_{conv.} - FC_{hyb.}}{FC_{conv.}} \times 100\%, \quad (10)$$

where $FC_{conv.}$ and $FC_{hybr.}$ are the fuel consumptions (L/100 km or m^3/100 km) of the conventional and hybrid vehicles, respectively.

The average auxiliary power consumption was calculated from the simulations of all the employed driving cycles. For the conventional vehicle, it amounted 6–7 kW for the urban and suburban conditions and 2–4 kW for the highway schedules. For the hybrid vehicle, it was lower (due to the engine-independent control), with 3.5–4 kW and 1–2 kW, respectively.

Tasks 1 and 2 formulated at the beginning of the section were solved simultaneously. To do this, two sets of simulations were conducted, one of which implemented the direct connection between the ESS and the electric drive, while the other involved a DC–DC converter as the interface. Each set included simulations with different numbers of supercapacitors, ranging from one (1.6 kWh) to four (6.4 kWh). When employing more than one SC unit, the electrical connection between the units was assumed to be parallel. When modeling the variant with a DC–DC converter, the latter maintained the electric drive input voltage at a constant 640 V. The efficiency of the converter was assumed to be 95% (the actual value would most likely be lower).

Figure 13 demonstrates the ESS voltage time histories calculated for the urban/suburban part of the WHVC driving cycle. From the plots, it is clear that using a larger ESS allows for decreasing both the depth of the discharge and the voltage amplitude. Note that this effect is the most pronounced when adding the second SC unit.

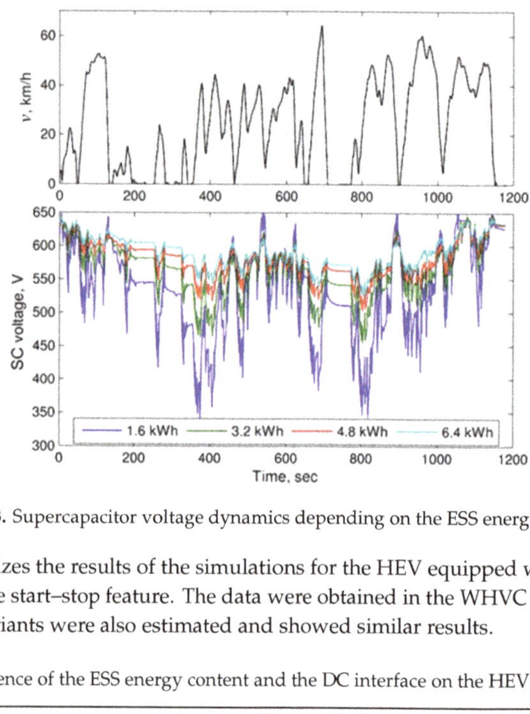

Figure 13. Supercapacitor voltage dynamics depending on the ESS energy content.

Table 8 summarizes the results of the simulations for the HEV equipped with the Otto-type gas engine and having the start–stop feature. The data were obtained in the WHVC urban/suburban cycle. Other powertrain variants were also estimated and showed similar results.

Table 8. Influence of the ESS energy content and the DC interface on the HEV fuel economy.

ESS–Traction Drive Interface	Powertrain Type								
	Conv.	Hybrid, ESS 1.6 kWh		Hybrid, ESS 3.2 kWh		Hybrid, ESS 4.8 kWh		Hybrid, ESS 6.4 kWh	
	FC	FC	FE	FC	FE	FC	FE	FC	FE
Direct link	82.6	65.98	20.12%	62.9	23.8%	61.82	25.16%	61.54	25.50%
DC–DC		70.2	15%	66.95	18.9%	65.7	20.46%	64.98	21.33%

FC in m^3/100 km.

The results show an evident drawback of employing the DC–DC converter, which is a decrease of the fuel economy effect by 4–5%. The lowest fuel savings were obtained when using one SC unit. Although the DC–DC converter prevents the electric drive power from diminishing, therefore keeping the potential of regenerative braking at the maximum level, the SC is discharged below the threshold of the favorable efficiency. Additional losses are introduced by the DC–DC converter itself. As a result, the positive effect of the maximum electric drive power available throughout the driving cycle is not able to outweigh the adverse effect of the lowered system efficiency. However, the simulations with the higher ESS energy contents show that keeping the SC efficiency at a good level does not eliminate the effect of the diminished fuel economy. Moreover, despite the increased ESS efficiency, the fuel economy difference between the two interface options changes negligibly. This suggests that the critical factor is the power loss within the DC–DC converter rather than in the supercapacitor.

Figure 14 shows the average supercapacitor efficiency and the relative fuel economy of the HEV as a function of the ESS energy content, where the plots have resulted from the driving cycle simulations. The efficiency plot denoted "Equiv. circ." was derived using Equation (8), while the second plot, "Const. eff.," was calculated using Equation (9). The difference between the plots stems from the extra-low load regimes where the efficiency of the SC may become negative (see Section 3.5). If those regimes are lumped into the bulk of the power losses, the line "Equiv. circ." lifts toward the second line, eventually coinciding with it. The effect of adding the third and fourth SC units is not so appreciable, while the cost of an ESS having that size may be rather high, outweighing the fuel savings.

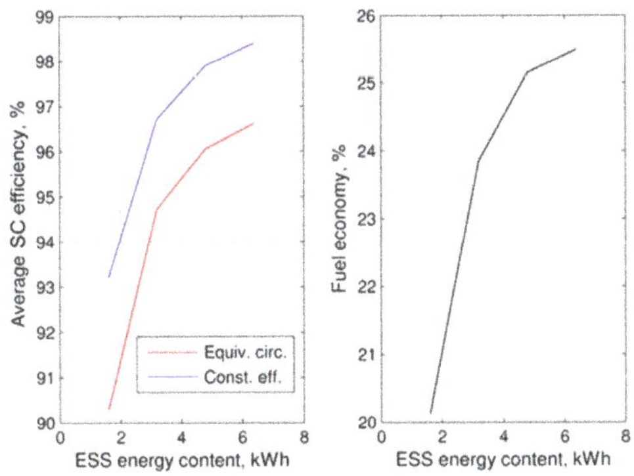

Figure 14. Influence of the ESS energy content on the supercapacitor efficiency and relative fuel economy.

Solving tasks 1 and 2 allowed us to formulate the "optimal" solution regarding the ESS design: it should consist of two SC units having a total energy content of 3.2 kWh and interfacing with the traction electric drive via a direct DC link. This solution was used in the simulations relating to task 3. The results thereof are presented in Figure 15 in the form of histograms showing the relative fuel economy provided by the hybrid powertrains based on all the mentioned ICEs in different driving cycles.

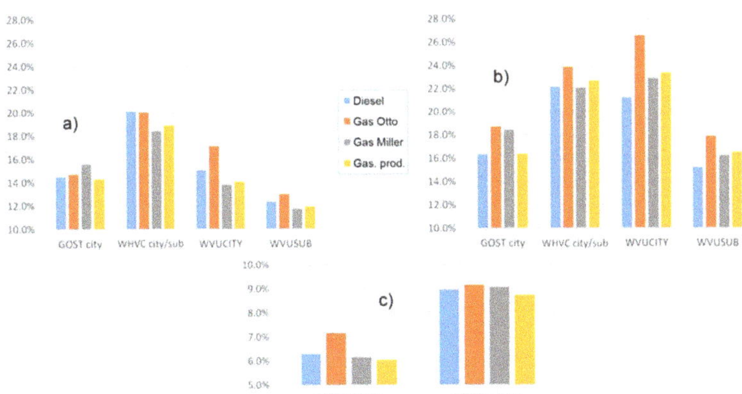

Figure 15. Calculated fuel economy of the HEV relative to the conventional vehicle: (**a**) city and suburban cycles, the start–stop was disabled; (**b**) city and suburban cycles, the start–stop was enabled; and (**c**) highway cycles.

In general, the hybridization provided a similar fuel economy effect for both the diesel-based and gas-based powertrains. The average differences of the relative fuel economy between the gas-based HEVs and the diesel-based HEV are listed in Table 9.

Table 9. Average hybridization effect of the gas-based powertrains relative to the diesel-based powertrain.

Driving Schedule	Start–Stop Feature Status	Gas Engine		
		Otto's Cycle	Miller's Cycle	Production Engine
City/suburban	Disabled	+ 0.7%	−0.6%	−0.7%
City/suburban	Enabled	+ 3%	+ 1.2%	+ 0.7%
Highway	n/a	+ 0.5%	0	−0.5%

The most pronounced effect of the hybridization is observed with the Otto gas engine. In the conventional powertrain, the Otto-cycle engine is up to 5% less effective than the two other gas engines, while the hybrid powertrain allows this engine to reach the absolute fuel economy values of its counterparts.

The fuel economy effect provided by the hybridization mainly depends on two factors: driving schedule and employing the start–stop feature. The former relates to the characteristic called the kinetic factor [10]. It expresses the share of "acceleration–deceleration" driving that directly determines the intensity of using regenerative braking, which in turn, is the major factor of a heavy-duty HEV's fuel economy. From the simulation results without the start–stop feature (i.e., when regenerative braking is the decisive factor of the fuel economy), one can conclude that among the used driving cycles, the city/suburban part of the WHVC has the highest kinetic factor.

The effect of introducing the start–stop feature is particularly pronounced for the gas-based HEVs, especially in the driving cycle containing the highest share of stops, namely the WVUCITY cycle. This is due to the ratio of the idle fuel consumption rate to the maximum fuel rate of the gas engines being approximately two-fold that of the diesel. In other words, the gas engines have a larger share of the idle consumption, which makes its elimination an effective measure to increase the fuel economy. In the absence of the start–stop feature, the same WVUCITY cycle had a 6.2–9.5% lower fuel economy effect, which emphasized the "start–stop" nature of this particular driving schedule.

In the highway cycles, the hybridization provided lower fuel savings, as expected, especially in pure highway regimes represented by the GOST cycle. The highway part of the WHVC cycle is less homogeneous regarding velocity variations, which provide more possibilities to retrieve and reuse the vehicle kinetic energy, resulting in higher fuel savings. It is also worth noting that the work in [14] shows the positive effect of regularly undulated roads with respect to the highway fuel economy due to the larger amount of regenerated energy.

The final task of the study was to estimate the influence of the vehicle's mass on the fuel economy. To do this, the vehicle mass was lowered to 35 tons (with a proportional decrease of the auxiliary power consumption). The resulting variations of the fuel economy relative to that of the initial (44 tons) vehicle are shown in Table 10. In general, the relative fuel economy changes negligibly. An unambiguous effect only shows in the case of the Otto gas-based powertrain, which, again, benefited from the hybridization more than its counterparts did.

Table 10. Variations of the average fuel-saving effect due to the decreased vehicle mass.

Driving Schedule	Start–Stop Feature Status	Engine			
		Diesel	Gas, Otto Cycle	Gas, Miller Cycle	Gas, Prod. Engine
City/suburban	Disabled	−0.1%	+ 0.7%	+ 0.4%	+ 1.2%
City/suburban	Enabled	+ 1%	+ 1.3%	−0.3%	+ 0.2%
Highway	n/a	−0.2%	+ 0.1%	−0.9%	0

8. Conclusions and Future Work

The conducted study allowed for drawing the following conclusions:

- The hybridization of gas-engine-based powertrains provided a fuel economy similar to that of diesel-based powertrains. A pronounced effect of hybridization was obtained in city and suburban driving conditions, as expected, due to a large share of acceleration–deceleration sequences, allowing for the regeneration of a considerable amount of the vehicle's kinetic energy. The relative economy amounted 18–28% using the start–stop feature and 15–20% without it (depending on the driving cycle's kinetic factor and the vehicle's mass). Highway driving does not offer significant opportunities for decreasing the fuel consumption through hybridization. However, an economy of 8–10% can be achieved in the conditions with a sufficiently variable velocity, such as in the highway part of the WHVC.
- The highest potential regarding the hybridization was shown by the gas engine based on Otto's thermodynamic cycle. Eliminating the low load regimes using the hybrid functions allows for this engine to achieve the same level of absolute fuel consumption as its gas-fueled counterparts do have, while in the conventional powertrain, this engine shows the lowest fuel economy of all the considered gas-fueled variants.
- The sufficient ESS energy content for the vehicles having a gross mass of 35–44 tons is 3–3.5 kWh. This allows for operating within the supercapacitor voltage range that ensures high values of the average efficiency, i.e., 95% and above (provided that the HEV control system can keep the voltage within that range).
- If the above conditions concerning the ESS energy content and the control system functionality are satisfied and the traction electric drive performance remains sufficient during ESS voltage transients, one can avoid using a DC–DC converter to interface the ESS and the traction drive. This, in turn, will prevent a significant deterioration of the powertrain efficiency caused by the converter.
- The results of the study have also provided a framework for future research, which includes solving of the following tasks: estimation of the hybridization effect with respect to exhaust emissions; further elaboration of the powertrain design and controls; estimation of the HEV's properties in wider driving conditions, including actual trips logged by GPS-based measurement systems.

Author Contributions: Conceptualization, I.K. and A.K.; Methodology, I.K. and A.K.; Software, I.K.; Validation, I.K., A.K., and A.T.; Formal Analysis, I.K. and A.K.; Investigation, I.K. and A.K., Resources, A.T., A.K., and K.K.; Writing—original draft preparation, I.K.; Writing—review and editing, A.K. and I.K.; Visualization, I.K.; Supervision, A.K., A.T., and K.K.; Project administration, A.T. and A.K.; Funding acquisition, A.T. and K.K. All authors have read and agreed to the published version of the manuscript.

Funding: This research received no external funding. The APC was funded by the National Research Center "NAMI".

Conflicts of Interest: The authors declare no conflicts of interest.

Abbreviations

BSFC	Brake-specific fuel consumption
DC	Direct current
EM	Electric machine
ESS	Energy storage system
FC	Fuel consumption
FE	Fuel economy
HD	Heavy-duty
HEV	Hybrid electric vehicle
ICE	Internal combustion engine
OOL	Optimal operating line
SC	Supercapacitor

References

1. Isermann, R. *Engine Modeling and Control: Modeling and Electronic Management of Internal Combustion Engines*; Springer: Berlin/Heidelberg, Germany, 2014; pp. 573–596.
2. Kamel, M.; Lyford-Pike, E.; Frailey, M.; Bolin, M.; Clark, N.; Nine, R.; Wayne, S. *An Emission and Performance Comparison of the Natural Gas Cummins Westport Inc. C-Gas Plus Versus Diesel in Heavy-Duty Trucks*; SAE Technical Papers 2002-01-2737; SAE: Warrendale, PA, USA, 2002.
3. Giechaskiel, B.; Lähde, T.; Schwelberger, M.; Kleinbach, T.; Roske, H.; Teti, E.; van den Bos, T.; Neils, P.; Delacroix, C.; Jakobsson, T.; et al. Particle Number Measurements Directly from the Tailpipe for Type Approval of Heavy-Duty Engines. *Appl. Sci.* **2019**, *9*, 4418. [CrossRef]
4. Giechaskiel, B. Solid Particle Number Emission Factors of Euro VI Heavy-Duty Vehicles on the Road and in the Laboratory. *Int. J. Environ. Res. Public Health* **2018**, *15*, 304. [CrossRef] [PubMed]
5. Di Maio, D.; Beatrice, C.; Fraioli, V.; Napolitano, P.; Golini, S.; Rutigliano, F.G. Modeling of Three-Way Catalyst Dynamics for a Compressed Natural Gas Engine during Lean–Rich Transitions. *Appl. Sci.* **2019**, *9*, 4610. [CrossRef]
6. Luksho, V.A. Complex Method to Increase the Energy Efficiency of Gas Fueled Engines Having High Compression Ratios with Shortened Intake and Exhaust Strokes. Ph.D. Thesis, National Research Center "NAMI", Moscow, Russia, 2015. (In Russian)
7. Kozlov, A.V.; Terenchenko, A.S.; Luksho, V.A.; Karpukhin, K.E. Prospects for Energy Efficiency Improvement and Reduction of Emissions and Life Cycle Costs for Natural Gas Vehicles. *IOP Conf. Ser. Earth Environ. Sci.* **2017**, *52*, 012096. [CrossRef]
8. Kozlov, A.; Grinev, V.; Terenchenko, A.; Kornilov, G. An Investigation of the Effect of Fuel Supply Parameters on Combustion Process of the Heavy-Duty Dual-Fuel Diesel Ignited Gas Engine. *Energies* **2019**, *12*, 2280. [CrossRef]
9. Kavtaradze, R.Z.; Onishchenko, D.O.; Kozlov, A.V.; Terenchenko, A.S.; Golosov, A.S. Analysis of Local Heat Exchange in Combustion Chamber and Injection Nozzle of Dual-Fuel Engine. *Int. J. Innov. Tech. Exp. Eng.* **2019**, *8*, 2804–2811.
10. O'Keefe, M.; Simpson, A.; Kelly, K.; Pedersen, D. *Duty Cycle Characterization and Evaluation towards Heavy Hybrid Vehicle Applications*; SAE Technical Paper 2007-01-0302; SAE: Warrendale, PA, USA, 2007.
11. Wikström, M.; Folkesson, A.; Alvfors, P. First experiences of ethanol hybrid buses operating in public transport. In Proceedings of the World Renewable Energy Congress, Linkoping, Sweden, 8–13 May 2011; pp. 3653–3660.
12. Zhao, H.; Burke, A.; Zhu, L. Analysis of Class 8 hybrid-electric truck technologies using diesel, LNG, electricity, and hydrogen, as the fuel for various applications. In Proceedings of the IEEE 2013 World Electric Vehicle Symposium and Exhibition (EVS27), Barcelona, Spain, 17–20 November 2013; pp. 1–16.
13. Gao, Z.; Finney, C.; Daw, C.; LaClair, T.; Smith, D. Comparative Study of Hybrid Powertrains on Fuel Saving, Emissions, and Component Energy Loss in HD Trucks. *SAE Int. J. Commer. Veh.* **2014**, *7*, 414–431. [CrossRef]
14. Karbowski, D.; Delorme, A.; Rousseau, A. *Modeling the Hybridization of a Class 8 Line-Haul Truck*; SAE Technical Papers 2010-01-1931; SAE: Warrendale, PA, USA, 2010.
15. Rodman Oprešnik, S.; Seljak, T.; Vihar, R.; Gerbec, M.; Katrašnik, T. Real-World Fuel Consumption, Fuel Cost and Exhaust Emissions of Different Bus Powertrain Technologies. *Energies* **2018**, *11*, 2160. [CrossRef]
16. Topal, O.; Nakir, İ. Total Cost of Ownership Based Economic Analysis of Diesel, CNG and Electric Bus Concepts for the Public Transport in Istanbul City. *Energies* **2018**, *11*, 2369. [CrossRef]
17. Miller, J.M. Hybrid electric vehicle propulsion system architectures of the e-CVT type. *IEEE Trans. Power Electron.* **2006**, *21*, 756–767. [CrossRef]
18. Kulikov, I.A.; Lezhnev, L.Y.; Bakhmutov, S.V. Comparative Study of Hybrid Vehicle Powertrains with Respect to Energy Efficiency. *J. Mach. Manuf. Reliab.* **2019**, *48*, 11–19. [CrossRef]
19. Wang, Q.; Jiang, B.; Li, B.; Yan, Y. A critical review of thermal management models and solutions of lithium-ion batteries for the development of pure electric vehicles. *Renew. Sustain. Energy Rev.* **2016**, *64*, 106–128. [CrossRef]
20. Masias, A.; Snyder, K.; Millet, T. Automaker Energy Storage Need for Electric Vehicles. In Proceedings of the FISITA 2012 World Automotive Congress; Springer: Berlin/Heidelberg, Germany, 2013; pp. 729–741.

21. Serrao, L.; Rizzoni, G. Optimal control of power split for a hybrid electric refuse vehicle. In Proceedings of the American Control Conference, Seattle, WA, USA, 11–13 June 2008; pp. 4498–4503.
22. Barrero, R.; Coosemans, T.; Van Mierlo, J. Hybrid Buses: Defining the Power Flow Management Strategy and Energy Storage System Needs. *World Electr. Veh. J.* **2009**, *3*, 299–310. [CrossRef]
23. Neuman, M.; Sandberg, H.; Wahlberg, B.; Folkesson, A. *Modelling and Control of Series HEVs Including Resistive Losses and Varying Engine Efficiency*; SAE Technical Papers 2009-01-1320; SAE: Warrendale, PA, USA, 2009.
24. Wu, W.; Partridge, J.; Bucknall, R. Development and Evaluation of a Degree of Hybridisation Identification Strategy for a Fuel Cell Supercapacitor Hybrid Bus. *Energies* **2019**, *12*, 142. [CrossRef]
25. Grbović, P.J. *Ultra-Capacitors in Power Conversion Systems. Applications, Analysis and Design from Theory to Practice*; John Wiley & Sons Ltd.: Hoboken, NJ, USA, 2014; pp. 22–76, 149–215.
26. *Electrochemical Capacitors*; Technical Report (In Russian); Elton Inc.: Moscow, Russia, 2012.
27. Plett, G.L. Extended Kalman filtering for battery management systems of LiPB-based HEV battery packs Part 1. Background. *J. Power Sources* **2004**, *134*, 252–261. [CrossRef]
28. Long, B.; Lim, S.T.; Bai, Z.F.; Ryu, J.H.; Chong, K.T. Energy Management and Control of Electric Vehicles, Using Hybrid Power Source in Regenerative Braking Operation. *Energies* **2014**, *7*, 4300–4315. [CrossRef]
29. Zhang, C.; Wang, D.; Wang, B.; Tong, F. Battery Degradation Minimization-Oriented Hybrid Energy Storage System for Electric Vehicles. *Energies* **2020**, *13*, 246. [CrossRef]
30. Genta, G. *Motor Vehicle Dynamics. Modeling and Simulation*; World Scientific: Singapore, 2006; pp. 43–44.
31. Martin, R.; Quintana, J.J.; Ramos, A.; de la Nuez, I. Modeling electrochemical double layer capacitor, from classical to fractional impedance. In Proceedings of the MELECON 2008—The 14th IEEE Mediterranean Electrotechnical Conference, Ajaccio, France, 5–7 May 2008; pp. 61–66.
32. Atcitty, S. Electrochemical Capacitor Characterization for Electric Utility Applications. Ph.D. Thesis, Virginia Polytechnic Institute and State University, Blacksburg, VA, USA, 2006; pp. 20–40.
33. Hendricks, T.; O'Keefe, M. *Heavy Vehicle Auxiliary Load Electrification for the Essential Power System Program: Benefits, Tradeoffs, and Remaining Challenges*; SAE Technical Paper 2002-01-3135; SAE: Warrendale, PA, USA, 2002.
34. SAE Standard J1343. *Information Relating to Duty Cycles and Average Power Requirements of Truck and Bus. Engine Accessories*; SAE: Warrendale, PA, USA, 2000.
35. Andersson, C. On Auxiliary Systems in Commercial Vehicles. Ph.D. Thesis, Lund University, Lund, Sweden, 2004; pp. 37–97.
36. Luz, R.; Rexeis, M.; Hausberger, S.; Jajcevic, D.; Lang, W.; Schulte, L.E.; Steven, H. *Development and Validation of a Methodology for Monitoring and Certification of Greenhouse Gas. Emissions from Heavy Duty Vehicles through Vehicle Simulation*; Technical Report; Technical University Graz: Graz, Austria, 2014.

© 2020 by the authors. Licensee MDPI, Basel, Switzerland. This article is an open access article distributed under the terms and conditions of the Creative Commons Attribution (CC BY) license (http://creativecommons.org/licenses/by/4.0/).

energies

MDPI

Article

Energy Consumption and Lifecycle Cost Analysis of Electric City Buses with Multispeed Gearboxes

Antti Ritari *, Jari Vepsäläinen, Klaus Kivekäs, Kari Tammi and Heikki Laitinen

Department of Mechanical Engineering, School of Engineering, Aalto University, 02150 Espoo, Finland; jari.vepsalainen@aalto.fi (J.V.); klaus.kivekas@aalto.fi (K.K.); kari.tammi@aalto.fi (K.T.); heikki.laitinen@hotmail.com (H.L.)

* Correspondence: antti.ritari@aalto.fi

Received: 29 March 2020; Accepted: 20 April 2020; Published: 24 April 2020

Abstract: This study investigates the potential of improving the energy efficiency and reducing the lifecycle costs of electric city buses with multispeed gearboxes. A two-speed dual clutch gearbox and a continuously variable transmission were studied and compared to a reference fixed gear ratio powertrain. A novel two-level optimization model was introduced. The top level involves an exhaustive search algorithm and quasi-static vehicle dynamic model for optimizing the two-speed gearbox gear ratios, utilizing efficiency maps for the electric motor and the inverter. The second level is an integer programming model, which finds an optimal gear shifting policy subject to constraints on hysteresis and gear shifting induced losses. The model was applied with a standard driving cycle and additionally with three measured cycles acquired from a prototype battery electric city bus operating on a daily schedule on a suburban route in Espoo, Finland. The results showed that a two-speed gearbox reduced energy consumption by 2–3.2%, depending on the driving cycle characteristics. On the other hand, the continuously variable transmission was found to increase consumption by 1.9–4.0% due to large losses of the belt mechanism. It was concluded that the two-speed gearbox is a cost-effective investment for electric city buses characterized by operation profiles with frequent acceleration and braking events.

Keywords: electric vehicle; city bus; gearbox; transmission; optimization; energy efficiency

1. Introduction

In recent years, electric city buses (ECB) have experienced exponentially growing production rates and deployment to the public transportation systems of many cities around the world, as the lifecycle cost has decreased below traditional diesel bus levels in many routes and operating conditions [1]. In 2018, the global ECB fleet grew by 32% and reached 425,000 units worldwide [2]. The major advantage of ECBs is the higher energy conversion efficiency of the powertrain compared to diesel buses. ECBs are also locally emission-free, whereas diesel buses emit harmful particles that stay trapped between buildings, causing significant damage to human health in urban environments [3].

Although battery and charging technologies have developed at a fast pace in recent years, limited range is still a challenge with electric vehicles in general, because energy density of batteries is 25–100 times less than in liquid hydrocarbon fuels [4]. Increasing the battery capacity extends the range, but the size, weight, and cost set upper bound on the battery capacity. This limitation puts an emphasis on efficiency and optimal component selection in ECBs. Improved efficiency translates to an increasing rate at which ECBs displace diesel buses in cities' transportation systems, leading to reduced emissions and cleaner air.

The traction motor is the key component of an electric powertrain, and enabling the motor to operate at its highest efficiency region improves energy efficiency. Traction motors used in ECBs need to provide sufficient torque for acceleration and climbing inclined roads, while the maximum power output needs to be sufficient for high speed cruising. A motor with a wide speed range fulfills these requirements, and there is no absolute need for a multispeed gearbox as in ICE-powered vehicles [5]. However, in a traction application, a motor uses a wide speed range and thus is not operating at optimal efficiency [6]. ECB provides a promising application for a multispeed gearbox, because the vehicle operation consists of frequent braking and acceleration events that use a wide torque and speed range of the motor, which has been shown to result in reduced efficiency with a single gear drivetrain [7]. By contrast, electric vehicles that operate mostly at constant speed, such as long distance buses and trucks, can be equipped with a single speed reduction gear. The gear ratio is selected to operate the motor at the high-efficiency design point at nominal vehicle speed.

1.1. Previous Work

The literature on the advantages of a multispeed gearbox in electric vehicles is extensive, although it is primarily focused on light-duty vehicles. To the knowledge of the authors, there are no studies that investigate multispeed gearbox options for electric city buses.

Wu et al. [8] developed a simulation model in Autonomie to study the impact of a two-speed gearbox (AMT) on the fuel economy of an electric car. Dynamic programming was applied to the optimization problem—A 4.2% reduction in energy consumption in city driving was observed.

Ren et al. [9] reported a simulation study using a backward modeling approach (wheels to motor) and a comparison of different gearbox options in terms of the energy consumption of EVs. The reference was the original single-speed transmission, which was compared to continuous variable transmission (CVT) and multispeed gearbox with two, three, and four speeds. The authors observed that the losses of the multispeed gearbox cancel the improvement in energy efficiency for the NEDC cycle. For the other five cycles studied, the improvement in efficiency varied from 5.3% to 12.4%. The authors also point out that in addition to a possible improvement in energy efficiency, the acceleration performance of the vehicle improves. Specifically, the acceleration time from zero to 100 km/h was reduced from 18.6 s to 12.4 s with a two-speed gearbox.

A similar study of powertrain efficiency comparison was completed by Bottiglione et al. [10]. The gearbox efficiency and increase in vehicle mass were taken into account. The compared gearbox options were infinitely variable transmission (IVT) and CVT with toroidal technology. Single and two-speed transmission were also considered. Both IVT and CVT had two different configurations. The planetary gear in the IVT was assumed to work with 100% efficiency. The simulations were run for the Urban driving cycle (UDC) and the Japanese 10–15 cycle (J10-15). For both cycles, the IVT and CVT were found superior to the two-speed gearbox, with the largest consumption improvement reaching 16.7%. The authors concluded that the energy efficiency of EVs could be significantly improved by optimizing the torque dependent efficiency of CVT and IVT.

Gao et al. [11] focused on the potential of a two-speed gearbox for EVs. The study included simulation and optimization of gear ratios and shift control. The gearbox efficiency was considered equal to the single-gear alternative, and the overall efficiency of the powertrain was assumed to be 96%. The reported results were in agreement with those discussed previously. One notable difference was that the largest energy savings were achieved in the NEDC cycle, which in [9] was found to produce the least savings potential. The authors also mentioned the improved performance of the EV with a two-speed gearbox, i.e., better acceleration and top speed. Their modeling efforts highlight the friction losses due to the clutch in the gear change and the effect of a gear change smoothness to the friction losses. Calm gear shift which is comfortable for the passengers with no torque interrupt generates substantial losses in comparison to more rough and rapid change.

In addition to simulation studies, Spanoudakis et al. [12] discussed a prototype for an experimental test of the potential of a two-speed gearbox for improving the efficiency of an ultralight vehicle for urban environment transportation. The power source in the test vehicle was a hydrogen fuel cell. The gear selection was manual, and the test drive was conducted on a simple oval test track. The results showed that energy savings of 3.4% were reached in comparison to the single-speed gearbox and most of the energy savings were achieved during the acceleration phases of the test run. The energy savings potential is higher in city driving conditions with frequent accelerations and decelerations compared to highway driving.

In the realm of heavy-duty vehicles, Tan et al. [13] reported a study on an iron mine electric dump truck gear ratio optimization using particle swarm optimization algorithm. For a typical working cycle in a mine, energy consumption was found to be reduced by 6.1% with the gearbox and optimized ratios compared to the reference powertrain, which had only a single-speed reduction gear. The study is limited with respect to the conclusions that can be established for ECBs because the operation cycle of the electric dumb truck has significantly less variation than that of a city bus.

1.2. Goals and Scope

Even though the feasibility of multispeed gearboxes has been previously studied for EVs and fuel cell vehicles, the feasibility of this concept for ECBs has not been investigated before. A separate feasibility study is required, since the driving cycles, vehicle structure and powertrain design of an ECB are different by definition from that of an EV. This study seeks to contribute to the literature by providing answers to the following research questions:

- How does the driving cycle impact the benefit of a multispeed gearbox?
- What is the optimal gear shifting policy of a two-speed dual clutch gearbox?
- Does an energy consumption advantage achieved with a multispeed gearbox translate to a lifecycle cost advantage as well?

A simulation model is constructed to study these three aspects of feasibility. The simulation model considers longitudinal dynamics and hence speed, voltage and current signals from real-world operation were measured from a prototype ECB operating in the city of Espoo, Finland. The signals were collected with 1 Hz sampling rate on all driving cycles. The collected dataset presents an opportunity to evaluate the feasibility of a multispeed gearboxes in real-world conditions. The availability of representative driving cycles is considered a valuable contribution in this work, since in previous research standard driving cycles have been used extensively, as described in the previous section. The standard driving cycles are appropriate estimates of driving behavior but cannot capture the complexities and variation of real-world driving. A detailed explanation of the data is provided in Sections 2.4 and 2.5.

Considering new vehicle design, it would be worthwhile to take into account the interaction between the installed two-speed gearbox and the electric motor configuration. The simulation and optimization models developed in this work are in a general form, supporting motors of different sizes, so they can be applied to the motor sizing problem as well although this is not directly addressed in this work.

2. Vehicle Powertrain Model

2.1. Multispeed Gearbox Options for Heavy Duty Vehicles

Multispeed gearbox configurations for EVs constitute automatically shifted multispeed gearboxes, including two-speed gearboxes, and continuously variable transmissions [14]. Hydraulic automatic transmission is considered infeasible for the ECB application due to very low efficiency [15]. Additionally, gearbox designs that are still in the development stage, such as a planetary gear two-speed gearbox presented in [16], are considered out of scope.

CVT permits the electric motor to operate at the ideal point continuously. Power is transmitted by friction via a chain that runs between two axially adjustable taper discs. The diameter of the discs can be varied. Another CVT configuration is a toroidal variator. In this design the variable gear ratio is achieved by swiveling the friction gears.

The major drawback of CVT is the poor efficiency, compared to gear transmission. The traditional CVT that is used with internal combustion engine vehicles has efficiency of less than 85% but with EVs a simpler construction without a torque converter is feasible. Thus, efficiency of over 93% can be achieved [17].

Superior efficiency is achievable with gearboxes that have a discrete number of gear ratios. A greater than 95% efficiency and less than 150 millisecond gearshifts are reported for a two-speed dual clutch gearbox in [18] with 450 Nm maximum input torque and 13,000 rpm maximum speed. A single speed gearbox with the same performance specifications is reported to have two percentage points higher efficiency.

An automatic multispeed gearbox requires actuation for gear shifts. This is commonly implemented with an electrohydraulic actuator. A gear shift loss model for automatic manual transmission is presented in [19]. In this single clutch model, the losses consist of partially lost traction during clutch disengagement and engagement due to slipping, and the total loss of traction during the shifting phase. The clutch slipping is unavoidable, because engagement and disengagement must occur smoothly.

The primary power loss sources are clutch drag, gear churning, gear meshing and bearing seals [14]. Additional losses incur from energy consumption of a pumping cooling and lubricating fluid to the gearbox housing in the case of a wet clutch design. A dry clutch does not have a need for oil pumping, but can sustain lower torques.

The traction loss that occurs in shifting results in the vehicle velocity tracking loss compared to a vehicle equipped with a single speed reduction gear [19]. This loss needs to be compensated by an increased power output after the shifting event.

Based on a preliminary analysis, this work focuses on two gearbox alternatives that are estimated to have the most potential in an electric city bus. The first is a two-speed dual clutch gearbox (Figure 1) and the second is a traditional belt CVT without a torque converter. The model for the two-speed alternative includes both design (gear ratios) and operation (shifting policy) optimization levels. By contrast the CVT model simply selects, at each time step, the gear ratio that represents the highest efficiency operating point of the motor and the inverter.

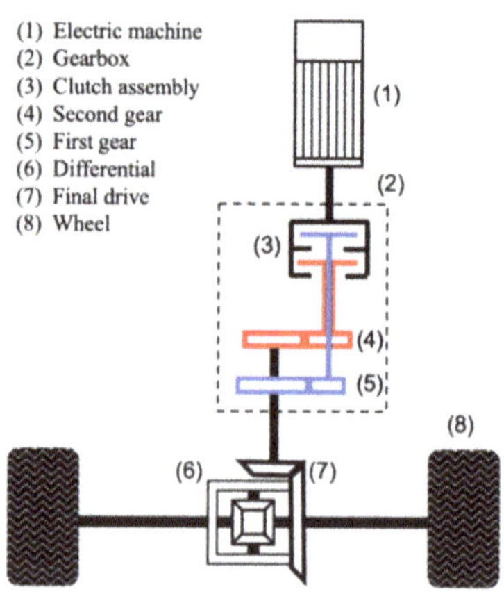

Figure 1. Two-speed dual clutch gearbox layout.

2.2. *Quasi-Static Modeling of Vehicle Dynamics*

EV and ECB powertrain models have been presented in detail in previous work [20–22]. A comprehensive physics-based dynamic ECB model is presented and validated with measurement data in [22]. The model was found accurate but computationally intensive, which limited the usefulness of the model for powertrain design.

Quasi-static, or backward progressing, modeling provides a computationally light alternative to dynamic models that make use of differential equations to describe the system. A quasi-static approach is feasible in research and design applications where fast transients are not the central focus. The backward modeling approach means that the input to the model is a specific driving cycle, which is a presentation of the vehicle speed as a function of time. The powertrain load computation flows from the wheels towards the power source, such as a traction motor.

Discrete time is used instead of continuous time for the numerical computation. In each step, the system is interpreted as a static system. This approach is suitable for slow phenomena, such as energy consumption on a driving cycle. Because this approach is computationally light, a large number of topology and component size options can be explored in a short time. This is fitting to the multispeed gearbox selection problem, which requires a search for optimal gear ratios.

The following assumptions are included in the modeling:

- Energy consumption is computed from the inverter input terminals
- Constant efficiency is applied for the powertrain components other than the motor and the inverter
- The motor is used for regenerative braking within its torque limits while the rest of the braking demand is provided with the mechanical disc brakes if needed
- Regenerative braking is never limited by a fully charged battery
- The motor control electronics provide sufficient response to gear change induced step changes in motor reference speed
- The road profile has zero altitude variation
- Only forward motion and longitudinal forces influence the energy consumption
- Auxiliary power for heat and air conditioning is discarded

The modeled ECB dynamics, gear changing policy, and gear ratio optimization models are introduced in Sections 2.2–2.4. A specific configuration of the model, in which parameter values from the case study are applied, is discussed in Section 3.

2.3. Vehicle Dynamics

The computation starts from the driving cycle $v_{dc}(k)$, which gives the target speed for each time step k. The computation flows from the driving cycle to the inverter (Figure 2).

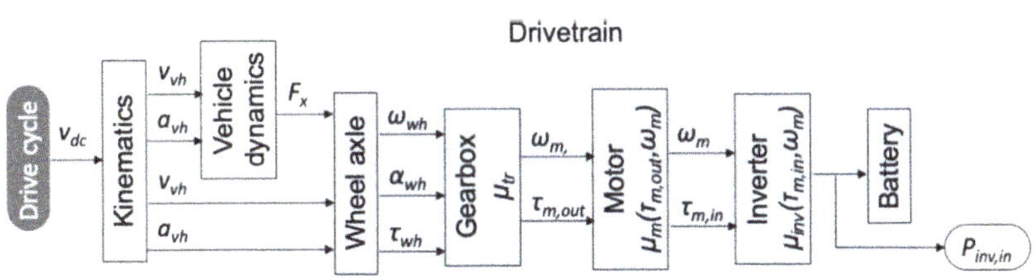

Figure 2. Signal flow in the vehicle model.

Vehicle speed and acceleration are computed by

$$v_{vh}(k) = \begin{cases} \frac{v_{dc}(k)}{2}, & k = 1 \\ \frac{v_{dc}(k)+v_{dc}(k-1)}{2}, & k > 1 \end{cases} \tag{1}$$

$$a_{vh}(k) = \begin{cases} \frac{v_{dc}(k)}{h}, & k = 1 \\ \frac{v_{dc}(k)-v_{dc}(k-1)}{h}, & k > 1 \end{cases} \tag{2}$$

where $v_{vh}(k)$ is the vehicle speed, $a_{vh}(k)$ is acceleration, and h is the time step duration, which is constant for the cycle. Wheel axle angular speed and acceleration at the wheel are given by dividing corresponding vehicle speed and acceleration with the tire radius according to

$$\omega_{wh}(k) = \frac{v_{vh}(k)}{r_{wh}} \tag{3}$$

$$\alpha_{wh}(k) = \frac{a_{vh}(k)}{r_{wh}} \tag{4}$$

where $\omega_{vh}(k)$ is axle angular speed, $\alpha_{vh}(k)$ is the axle angular acceleration, and r_{wh} is the tire radius. The torque on the axle $\tau_{wh}(k)$, required for thrusting the vehicle forward at a given speed and acceleration, is determined by the tractive force $F_x(k)$, which is the sum of the forces acting on the vehicle body.

Rolling resistance, aerodynamic drag and inertia forces as well as the resulting axle torque are given by

$$F_x(k) = Mgc_r + \frac{c_w \rho A v_{vh}(k)^2}{2} + Ma_{vh}(k) \tag{5}$$

$$\tau_{wh}(k) = F_x(k) r_{wh} \tag{6}$$

where M is the vehicle mass, g is gravitational acceleration, c_r is the rolling resistance coefficient, c_w is the aerodynamic drag coefficient, ρ is air density, and A is the frontal area of the vehicle. In the next step of the calculation flow, the torque and speed of the rear wheels are conveyed to the motor. The gearbox gear ratio determines the conversion, and in the case of the motor torque, the rotational inertia must also be taken into account. The motor speed $\omega_m(k)$ and the torque $\tau_m(k)$ are given by

$$\omega_m(k) = \omega_{wh}(k) i \tag{7}$$

$$\tau_m(k) = \frac{\tau_{wh}(k)}{i \eta_{tr} \eta_{gb}} + \alpha_{wh}(k) i I_m \tag{8}$$

where i is the total gear ratio, η_{tr} is the transmission efficiency, η_{gb} is the gearbox efficiency and I_m is the motor moment of inertia. Finally, the motor input power requirement $P_{m,in}$ is computed by

$$P_{m,in}(k) = \frac{\omega_m(k) \tau_m(k)}{\eta_m(\omega_m(k), \tau_m(k)) \eta_{inv}(\omega_m(k), \tau_m(k))} \tag{9}$$

where $\eta_m(\omega_m(k), \tau_m(k))$ is the motor efficiency function, and where the arguments are the motor speed and torque. Likewise, $\eta_{inv}(\omega_m(k), \tau_m(k))$ is the inverter efficiency function. In the case of regenerative braking, the efficiencies in Equations (8) and (9) become multipliers instead of dividers.

2.4. Optimal Shifting Policy

The shifting policy for the two-speed gearbox is optimized with an integer programming model. The objective is to minimize energy consumption for the driving cycle by assigning the gear that translates to the most efficient motor operating point, taking into account the efficiency loss of shifting events and hysteresis. The optimization model is formulated as

$$\underset{y_t, y'_t}{\text{minimize}} \sum_{t=1}^{T} \left(C_t - C_t y_t + C'_t y_t + S y'_t \right) \tag{10}$$

$$y_t - y_{t+1} - y'_t \leq 0, \ \forall t \in \{1, \ldots, T-1\} \tag{11}$$

$$-y_t + y_{t+1} - y'_t \leq 0, \ \forall t \in \{1, \ldots, T-1\} \tag{12}$$

$$\sum_{m=0}^{M} y'_{t+m} - 1 \leq 0, \forall t \in \{1, \ldots, T-M\} \tag{13}$$

$$y_t,\ y'_t \in \{0, 1\}. \tag{14}$$

The objective function in Equation (10) to be minimized is the vehicle energy consumption for the cycle, composed of the sum of the consumptions of each time step t in the set $\{1,..,T\}$. The consumption for each step has three components: the vehicle propulsion, the actuator energy for the shifting procedure, and the friction loss. The first component, motor propulsion energy demand, is represented by the parameters C_t (C'_t), for the low (high) gear. The gearbox has in this case only two gear ratio options to select from, and either one is active at each step, which allows for modeling the gear selection with a single binary variable y_t. The formulation in Equation (10) includes this binary variable to switch on and off the consumption parameter that corresponds to the active gear selection at each step. Both the actuator energy and the friction loss are included in the parameter S_t, which contributes to the objective function at step t. The shifting event at that step is indicated by the binary variable y'_t receiving a value of one.

Equations (11)–(13) define the downshift, upshift, and hysteresis constraints, respectively. The downshift constraint in Equation (11) forces the binary variable y'_t, which indicates a shifting event, to equal one when the gear indicator binary variable y_t changes from one to zero in consecutive time steps. The logic in the constraint Equation (12) is the same as in Equation (11), but the functionality applies to upshifting instead of downshifting.

The hysteresis constraint Equation (13) sets a limit to the shifting frequency. A gear shift is allowed after M time steps have passed since the previous shifting event. The constraint formulation utilizes the binary variable y'_t, which receives a value of one when an upshift or downshift occurs in step t and zero otherwise.

The resulting objective function value is returned to the gear ratio optimizer function, which compares the consumption to the previous lowest value and replaces it in the case that the current run returns the lowest value.

2.5. Gear Ratio Optimization

The two-speed gearbox gear ratios function as an input to the vehicle dynamic model and the shifting policy optimization model. The gear ratios are iterated through such that each combination is covered from a range for both gears. Before the gear ratio optimization is initiated, the search scope is determined by setting the starting and ending gear ratios and the increment. For example, a search between ratios 3.0 and 8.0 with a 0.1 increment is represented by the set {3.0, 3.1, . . . , 7.9, 8.0}. In the following representation, the gear ratio choice set, which contains a sequence of gear ratios, is denoted with list 1 for the first gear and list 2 for the second gear.

Figure 3 shows the exhaustive search process for the gear ratios that produce the lowest consumption for a specific driving cycle. The process consists of an inner loop for the low gear ratio range and an outer loop for the high gear ratio range of the two-speed gearbox. The cycle energy consumption block in the chart constitutes the vehicle dynamics model, described by Equations (1)–(9) and the gear shifting optimization model in Equations (10)–(13). The energy consumption is therefore computed for each combination of gear ratios from the ranges that are selected at the beginning of the process. Finally, the lowest consumption, represented by the stored value, and the corresponding gear ratios are returned.

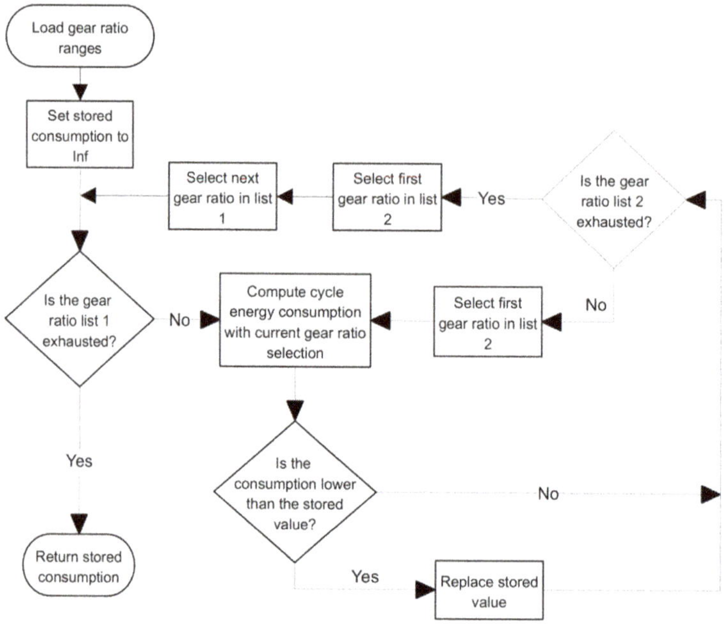

Figure 3. Flow chart of the gear ratio optimization process.

3. Numerical Case Study

The case vehicle examined in this work is a fully electric city bus operating on a predetermined suburban route with a fast-charging station at one end of the route. The vehicle was first deployed for continuous operation at the beginning of 2016. The electric motor is a 180 kW permanent magnet synchronous machine (PMSM) and the energy storage system consists of a lithium-titanate (LTO) battery with 690 V nominal voltage and 55 kWh capacity. The bevel type differential has a ratio of 7.0. The curb weight is 10,500 kg and the maximum allowed payload is 5500 kg, which limits the passenger capacity to 78, assuming that the mean passenger mass is 70 kg. A constant 1500 kg load from 18 passengers and the driver was assumed in the case study. Increased mass from the multispeed gearbox options was considered negligible.

3.1. *Parameter Values*

The parameter values for the case study are presented in Table 1. All the values for the reference single speed transmission are from the work of Vepsäläinen et al. [23], who developed a dynamic simulation model for the case study vehicle. The model has been validated previously based on measurements from multiple electric city buses operating in Finland. In the present work auxiliary power was discarded since it would make no difference for powertrain comparisons.

The motor efficiency map model was based on the work of Mahmoudi et al. [24] for PMSM motors. The map that was modeled based on their work was modified by changing the loss coefficients to match the specifications of the case vehicle, which has 180 kW maximum rating motor with 2000 Nm and 314 rad/s limits for torque and speed. Figure 4a shows the motor efficiency map and Figure 4b shows the inverter map. Appendix A presents the equations and parameters values for drawing the maps.

Table 1. Parameter values for the numerical case study.

Parameter	Description	Value	Unit
M	Vehicle total mass	12,000	kg
r_{wh}	Tire dynamic radius	0.43	m
g	Gravitational acceleration	9.81	m/s^2
c_r	Rolling resistance coefficient	0.01	-
c_w	Aerodynamic drag coefficient	0.6	-
A	Frontal area	6.2	m^2
ρ	Density of air	1.27	kg/m^3
i	Differential gear ratio	7.0	
η_{tr}	Differential efficiency	0.95	-
$\eta_{gb,2s}$	Two-speed gearbox efficiency	0.99	-
$\eta_{gb,CVT}$	CVT efficiency	0.95	-
I_m	Motor moment of inertia	0.63	kg m^2
S	Shifting energy loss	0	J
M	Shifting time delay	1–30	s
h	Simulation step size	1	s

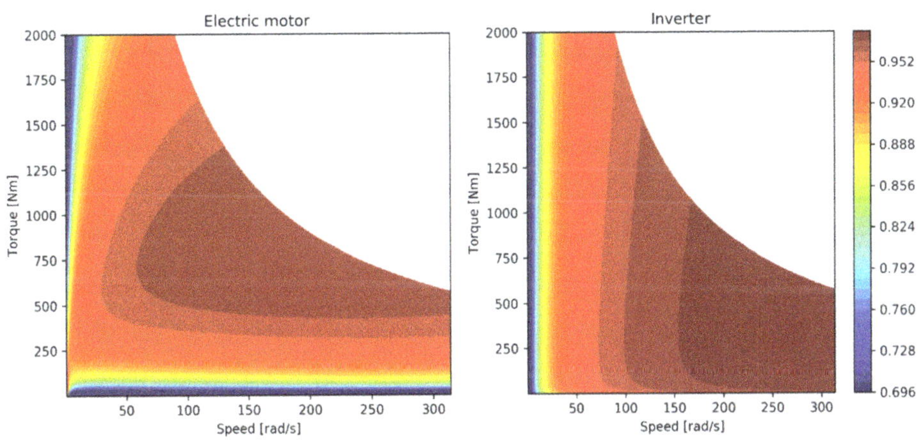

(**a**) (**b**)

Figure 4. (**a**) Motor efficiency map; (**b**) Inverter efficiency map.

The potential of a multispeed gearbox to increase energy efficiency originates from the option to convert between torque and speed, which moves the motor operating point along a power level contour. The instantaneous motor torque and speed are determined by power demand, gear ratios, and gear selection. The advantage that comes from shifting the point along the curve depends on the efficiency characteristics of the motor and the inverter as well as the power level. Figure 5 shows efficiencies on four power levels for the case vehicle motor and inverter combination.

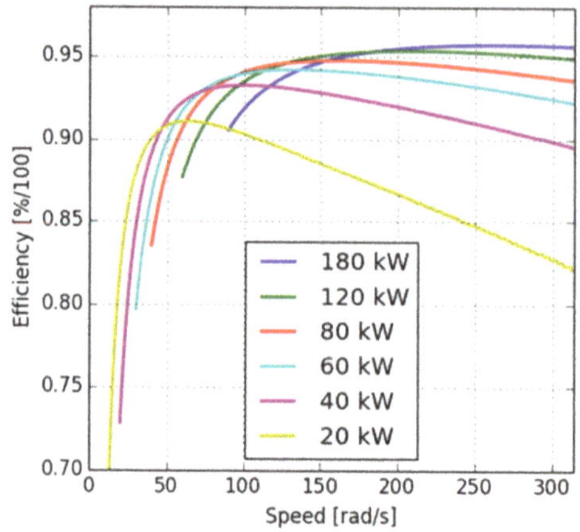

Figure 5. Combined efficiency of the motor and the inverter at different power contours.

The transmission and the gearbox efficiencies were assumed to be constant throughout the study. Efficiency of the reference transmission that was comprised of a differential and a final drive only was 0.95. Efficiencies of 0.99 and 0.95 were used for the two-speed and CVT gearboxes, respectively. The total efficiency of the transmission for the multispeed gearbox configurations was computed by multiplying the gearbox and the differential efficiencies. Due to the lack of experimental data that authors are aware of, the value of the shifting loss parameter S was set to zero.

3.2. Driving Cycle Data Acquisition

The studied vehicle, shown in Figure 6a, was equipped with an online data acquisition system. Speed and GPS coordinate measurements, and the corresponding timestamps, were acquired from this system. The speed signal was obtained from the anti-lock brake system (ABS) sensors with 1 Hz sampling rate.

The suburban bus route examined is line 11 in Espoo, Finland, shown in Figure 6b. This 10.4 km line is a typical suburban bus line consisting of commutes between local business and residential centers, featuring short urban sections and periods with higher speed limits. The route includes a multilevel stack interchange, which causes the route profile and total distance to vary according to the direction of driving. All the cycles in line 11 considered in this work start from Friisilä terminus and end at Tapiola, excluding the drive from the terminus to the fast-charging station nearby. As all the cycles have the same direction, the distances are the same as well.

Suburban bus routes are characterized by a large variation in speed profiles between cycles due to daily and hourly passenger and traffic level variation. To account for this, the speed profile of the bus was measured in three distinctly different traffic intensity levels from the line 11. These cycles in line 11 in Espoo are denoted by the abbreviation E11 and extensions LT, MT, and HT are applied, respectively, for the low, medium, and high traffic instances. Two additional cycles are included for comparison. First, the standard Braunschweig bus cycle is commonly used in academic research and in emission testing. Second, a combined cycle E11-C, which has the three E11 cycles in consecutive order, was constructed. Table 2 shows the cycle features and Figure 7 shows the speed profiles.

(a)

(b)

Figure 6. (**a**) The case study ECB operating in Espoo, Finland; (**b**) The bus route 11 in Espoo. Map data is by OpenStreetMaps.

Table 2. Braunschweig (BR) and three Espoo 11 (E11) cycle characteristics.

Feature	E11-LT	E11-MT	E11-HT	BR
Distance (km)	9.96	9.96	9.96	10.9
Duration (s)	1349	1487	1957	1741
Stops per km	0.70	2.01	2.91	2.57
Max. speed (km/h)	44.4	58.4	53.8	58.26
Avg. driving speed (km/h)	28.4	28.5	28.5	30.48
Max. acceleration (m/s^2)	1.37	1.41	1.41	2.41
Max. deceleration (m/s^2)	1.79	2.00	1.36	3.59
Avg. acceleration (m/s^2)	0.31	0.53	0.43	0.55
Avg. acceleration (m/s^2)	0.30	0.50	0.37	0.72

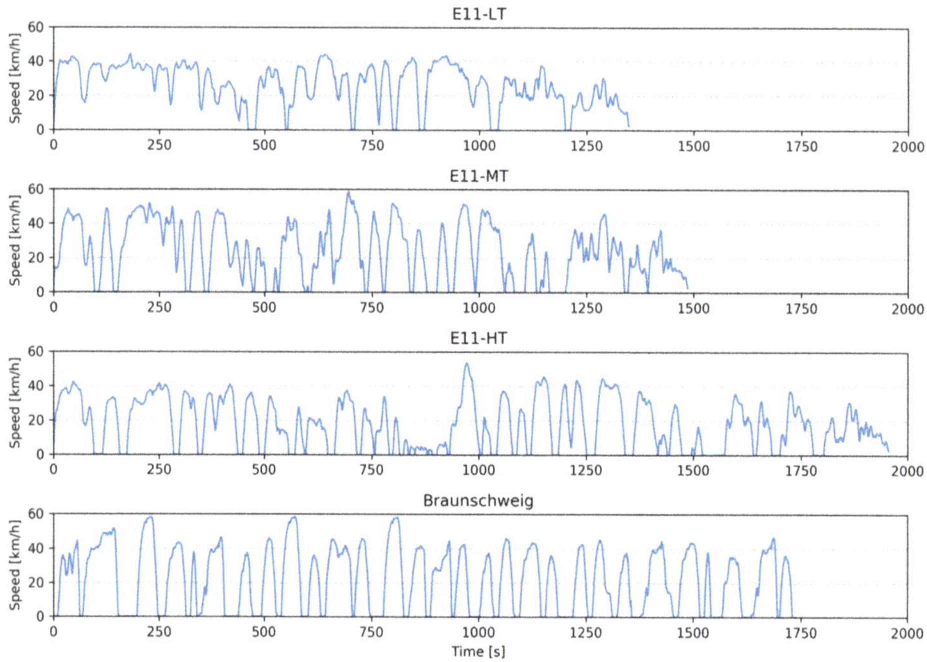

Figure 7. Three measured driving cycles from Espoo line 11 and the Braunschweig standard reference cycle.

The altitude profile of the route E11 included a variation of 14 m between the extreme points, and a road grade change of at most one degree per meter. However, in this work the altitude changes were discarded.

3.3. Lifecycle Costs

Lifecycle costs of electric city buses consist of capital, operation and technology replacement costs [1]. Capital costs consist of only the bus purchase cost, as the salvage value at the end of service life is assumed to be zero and the charging device costs are not included. Operation costs include charged electrical energy and maintenance cost. Technology replacement cost is the cost of replacing the battery system after it has reached the end of its useful life. Table 3 describes the lifecycle costs applied in this work.

Table 3. Lifecycle cost parameters.

Description	Value
Bus purchase cost without a battery and transmission (€)	340,000
Single speed (final drive) transmission (€)	10,000
Two-speed dual clutch transmission (€)	12,000
Battery cost (high power LTO) (€/kWh)	800
Battery useful life (cycles)	10,000
Battery maximum calendar life (years)	10
Maintenance cost (€/km)	0.2
Electricity cost (€/kWh)	0.155
Service life (years)	16
Annual operation (h)	4000

The electricity price is the average price from 28 EU member states for non-household consumers in the first six months of 2019 and includes all taxes and levies [25]. The other cost parameter values, other than the two-speed gearbox, are from the work of Lajunen [1] and Fries et al. [26]. The service life of the bus is 16 years, while the LTO battery maximum calendar life is only ten years. This indicates that the battery must be replaced at least once.

Transmission complexity and cost are related to the maximum input torque, gear ratio and number of gears. Regarding the additional cost to the transmission due to a two-speed gearbox in all electric vehicles, Ruan et al. [17] present an estimate of 20% based on a statistical model built from a comprehensive transmission cost dataset. This translates to a 2000 € cost margin for the electric bus in the case study in the present work.

3.4. Building and Solving the Model

For the model with a two-speed gearbox, the gear ratios were determined. The optimization objective was to minimize the energy consumption of the vehicle. This process was executed for different cycles separately because the optimal gear ratios and shifting policy vary between cycles.

The gear ratio search range of the two-speed gearbox was split at the single speed differential gear ratio of the reference model in order to retain the vehicle acceleration and top speed performance. The reference single speed ratio was 7.0 and, therefore, the first gear ratio in the two-speed gearbox was set to be larger than 7.1 and the second gear smaller than 7.0. In the gear ratio optimization process, ratios from 2.0 to 7.0 and from 7.1 to 20.0 were covered, for the low and high gear, respectively. Increment for the iteration was set to 0.1.

The simulations were carried out for five driving cycles: The standardized city bus cycle Braunschweig, and three measured cycles from Espoo bus line 11. In addition to the measured cycles E11-LT, E11-MT, and E11-HT, the combined cycle E11-C was used.

The vehicle dynamics, powertrain and gear ratio optimization models were formulated with the Python programming language. The integer programming model was formulated algebraically using Gurobi Python interface functions, and solved with Gurobi 8.1 solver with the default settings. The used solver is based on the branch-and-bound algorithm and comes with additional acceleration techniques, such as cutting planes and preprocessing.

4. Results and Discussion

The results of the simulations are first presented graphically for the reference and the two-speed gearbox case to illustrate optimal gear shift scheduling and efficiency improvement by motor and inverter operation point shifting. Then, differences in energy consumption between the reference and the two-speed gearbox and CVT alternatives are presented numerically to address the first research question. In the final section, a comparison of lifecycle costs of the powertrain alternatives shows that the two-speed gearbox is the cost optimal solution on the standard Braunschweig driving cycle.

Figure 8 shows the two-speed gearbox induced efficiency improvement in terms of motor operating points on the E11-MT cycle, which had the highest energy consumption on the Espoo 11 route. The most notable effect of the two-speed gearbox is that the operating points were shifted towards lower motor speeds and higher torques in the positive torque region. In the negative torque region during regenerative braking the full speed range of the motor was utilized. More generally, the operating point pattern variation between propulsion and braking is larger for the two-speed gearbox compared to the reference.

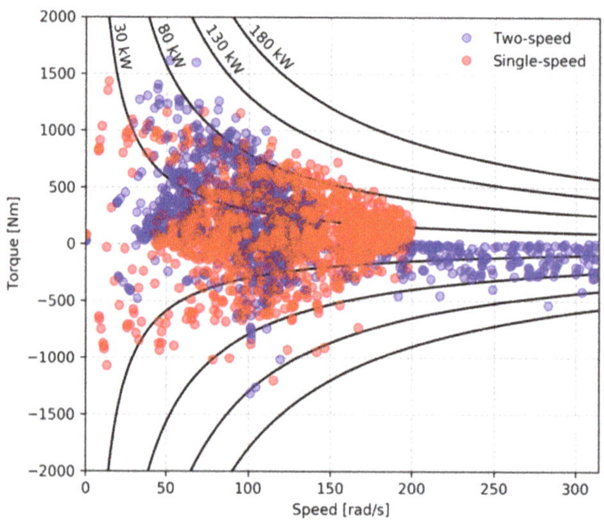

Figure 8. Comparison of operating points of the motor (E11-MT).

The Espoo 11 reference cycle is characterized by a high number of stops and subsequently a high number of acceleration and braking events. With the two-speed gearbox, regenerative braking is frequently executed with the first gear (high gear ratio) resulting in lower torque and higher speed of the motor compared to the second gear. This is evident from the negative torque region in Figure 8. Additionally, Figure 9 shows clearly that high gear ratios are favored for decelerations with both multispeed gearbox options.

Tables 4 and 5 show numerical values of energy consumption for each of the five studied routes, comparing the reference two-speed gearbox and CVT alternatives. The gear shift time limit parameter was set to one second in all the simulations that are reported in the tables.

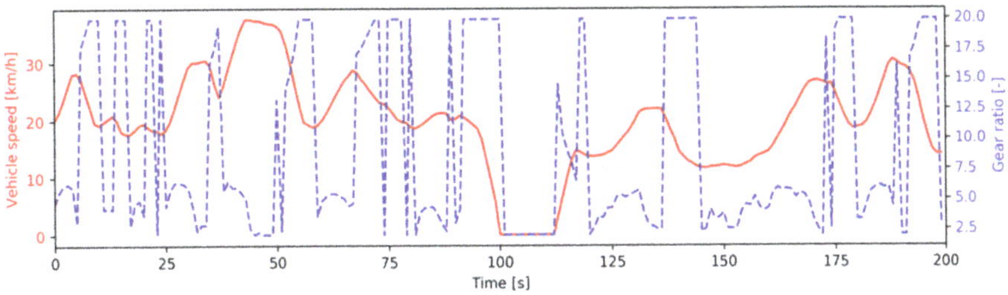

(a)

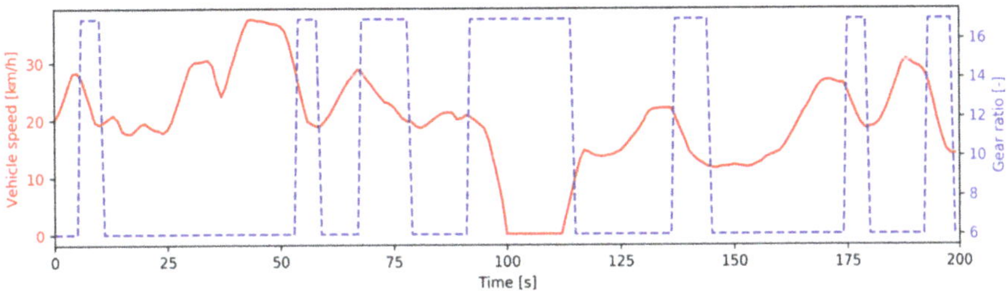

(b)

Figure 9. Illustration of optimal gear ratio selection policy for the CVT (**a**) and the two-speed gearbox with five second minimum gap between changes (**b**). The speed profile is a 200 second section from the E11-LT driving cycle.

Table 4. Energy consumption advantage achieved with the two-speed gearbox in five driving cycles. All energy consumption related numerical values are reported in units of Wh/km.

Cycle	E_{ref}	E_{GB}	ΔE	$\Delta E(\%)$	i_1	i_2
E11-LT	523.1	512.3	−10.8	−2.1	5.0	17.2
E11-MT	596.5	584.6	−11.9	2.0	6.8	19.9
E11-HT	551.6	534.2	−17.4	−3.2	6.4	19.9
E11-C	557.0	545.0	−12.0	−2.2	6.3	19.9
BW	627.4	611.7	−15.7	−2.5	6.8	19.2

Table 5. Energy consumption comparison between the CVT and the reference powertrains. Positive ΔE value represents consumption disadvantage with CVT.

Cycle	E_{ref}	E_{GB}	ΔE	$\Delta E(\%)$
E11-LT	523.1	533.6	10.2	1.9
E11-MT	596.5	620.4	23.9	4.0
E11-HT	551.6	561.9	10.3	1.9
E11-C	557.0	571.8	14.8	2.7
BW	627.4	644.9	17.5	2.8

With the two-speed gearbox the results show increased combined efficiency of the motor, inverter and the gearbox in all the E11 cycles, resulting in lower energy consumption. On the contrary, the CVT results in increased consumption in all the cases. Although the CVT reaches higher efficiency operating points of the motor and the inverter due to continuous gear ratio adjustment, this advantage is undone by the losses in the belt system.

Overall, the two-speed gearbox could improve energy efficiency by 2.0 to 3.2%, depending on the driving cycle and the powertrain characteristics. Energy consumption decreased most on the E11-HT cycle with the two-speed gearbox and inversely increased least with the CVT. This cycle has higher speed variation than the other E11 cycles. The Braunschweig cycle is also characterized by high speed variation and shows a significant 2.5% consumption decrease. On all the cycles the CVT performed worse than the two-speed option.

The difference in optimal gear ratios across all the cycles is small, with the optimal ratios found on the higher ends of the ranges for both the low and high gears on all the E11 cycles. Due to the optimal ratios located at the end of their ranges, the gap between the ratios is wide. Subsequently, the motor control needs to respond to large changes in the reference speed.

The relationship between the two-speed gearbox shifting time delay (gap) parameter M and the energy consumption is presented in Figure 10 for the E11-LT, E11-HT and Braunschweig cycles. The pattern is similar in E11-LT and E11-HT, showing consumption increasing roughly linearly with the shift gap values between 1 and 20 s. For the E11-LT cycle, all of the consumption advantage of the two-speed gearbox is lost when the shift gap value reaches 20 s. On the other hand, increasing the shift gap from one to ten seconds has only a marginal influence on the consumption with the Braunschweig cycle.

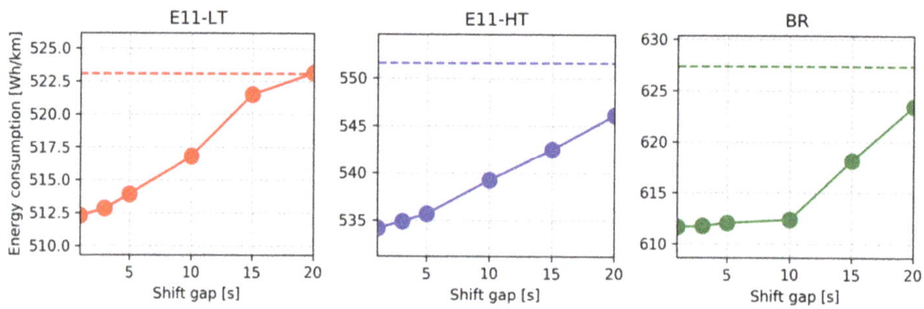

Figure 10. Energy consumption as a function of the minimum shift gap parameter value. The dashed horizontal lines represent the reference consumptions.

Table 6 shows the tradeoff between capital, operation and technology replacement costs. Only comparison between the reference single-speed and the two-speed option are shown, as the CVT didn't show any efficiency improvement. Importantly, the savings in charging cost and battery investment and replacement costs achieved with two-speed gearbox option exceed the increased investment cost of the transmission. The savings are distributed rather evenly between battery cost and charging cost.

Table 6. Comparison of lifecycle costs between the reference and two-speed gearbox vehicles for the Braunschweig cycle. Energy consumption advantage of the two-speed gearbox is translated proportionally to battery capacity and cost.

Cost Item (€)	Reference	Two-Speed	Δ
Purchase cost without battery and transmission	340,000	340,000	0
Transmission	10,000	12,000	2000
Maintenance	192,000	192,000	0
Battery	44,000	42,900	−1100
Battery replacement	44,000	42,900	−1100
Lifetime electricity	90,288	88,031	−2257
Total			−2457

5. Conclusions

The results of this work point towards a marginal potential of a two-speed gearbox utilization to improve ECB energy efficiency. The developed gearbox gear ratio and shifting policy two-level optimization model was found to be a practical and useful tool for vehicle powertrain design. The computational of the two-level optimization time was only a few seconds with a standard desktop computer for a typical 30 min driving cycle with one second time step duration.

The motor and inverter losses were found to depend on the driving cycle characteristics, and the simulation results suggest that powertrain efficiency was increased at higher power, which correlates with higher speeds and frequency of acceleration and deceleration events. On the other hand, utilization of motor and inverter operation point shifting along a power contour brings larger benefits on above average vehicle speeds when power demand is low instead of high. This is most evident in Figure 5, where the slope of the 20 kW power line is the steepest. The energy efficiency of an ECB can be improved with a two-speed gearbox, but the best utilization requires the consideration of the optimal motor-inverter-gearbox combination.

The PMSM motor in the case study vehicle was originally selected because it has high efficiency on a wide speed and torque range, which is useful when the power is transmitted through a single speed gearbox. However, a multispeed gearbox can favor other motor types, such as an induction motor, that have narrow high efficiency region, but lower capital cost than PMSM. Although the focus of the present work was energy efficiency, a two-speed gearbox also improves acceleration and top speed performance. A smaller motor could then be installed to the bus while retaining acceleration and top speed performance with the help of the two-speed gearbox. Further work is needed on the question of how the motor and gearbox interact in terms of selecting the motor design and power rating.

Innovations in gearbox technology, with respect to lower losses, would increase the attractiveness of the multispeed gearbox options for battery electric busses. This is evident from the large increase in motor and inverter efficiency with the CVT, which is then more than canceled out by the losses of the gearbox components. On the other hand, technology improvements in electric machines may lead to a more even efficiency throughout the motor operating region and thus reduce the potential to improve powertrain efficiency by operating point shifting. For instance, a prototype direct liquid cooled PMSM shows variation in efficiency that is at most only five percentage points between the extremes [27].

In translating the reported energy consumption advantages to real vehicles, the limitations of the modeling approach need to be considered. Lower efficiency of the multispeed gearbox options, both during continuous operation and shifting, compared to the single speed reference, influence the attractiveness of the gearbox alternative. Experimental tests, considerations on the increased complexity, and extending the scope of the analysis to reliability, serviceability, and compatibility are advised before final design decisions are made. In the end, the design decision comes down to the relative economic value of the alternatives. The increased efficiency with the multispeed gearbox options translate to lower electricity charging cost and battery investment cost but these advantages may be partially canceled by higher investment and maintenance costs.

Compared to light duty electric vehicles, ECBs have more consistent driving cycles, and the powertrains are dimensioned according to the nominal operation with only a small margin for excess performance. This results in higher overall powertrain efficiency because operating points are close to the motor and inverter maximum power rating, which provides the highest efficiency. Assuming similar main component selection, such as PMSM motor, it can be assumed that the results of the present work can be generalized to ECBs that operate in different, but similar urban routes.

Previous work on light-duty EV multispeed gearbox feasibility has found that the typical power demand in these vehicles is only around 10% of the maximum rating [28]. Optimal gear ratio selection and shifting of operation points in these low-power contours are different from the higher power region, which explains the results in the present work differ from similar work for light-duty vehicles, such as Nissan Leaf. This suggests that a separate analysis for different vehicle types and operating environments is required.

Author Contributions: Conceptualization, A.R., J.V., K.K., H.L. and K.T.; methodology, A.R., H.L. and K.T.; investigation, A.R.; writing—original draft preparation, A.R. and H.L.; writing—review and editing, A.R., J.V., K.K. and K.T.; supervision, K.T.; project administration, K.T. All authors have read and agreed to the published version of the manuscript.

Funding: This research received no external funding.

Acknowledgments: The authors thank Linkker Ltd. for providing data for this study.

Conflicts of Interest: The authors declare no conflict of interest.

Nomenclature

Greek Symbols	
α_{wh}	Angular acceleration
η	Efficiency
ρ	Density of air
τ	Torque
ω	Angular speed
Symbols	
a	Acceleration
A	Vehicle frontal area
c_r	Rolling resistance coefficient
c_w	Aerodynamic drag coefficient
C	Energy consumption with first gear
C'	Energy consumption with second gear
F_x	Tractive force
g	Gravitational acceleration
h	Step size
i	Gear ratio
I_m	Motor output shaft total moment of inertia
k	Index of time step in vehicle dynamic model
M	Vehicle total mass
N	Number of consecutive time steps before gear shift is allowed
P	Power demand
r	Tire dynamic radius
S	Gear shift energy consumption penalty
t	Index of time step in gear change model
T	Number of time steps
v_{dc}	Driving cycle speed
v_{vh}	Vehicle speed
y	Binary variable to indicate which gear is active
y'	Binary variable to indicate if gear change was executed
Subscripts	
dc	Driving cycle
gr	Gearbox
inv	Inverter
m	Electric motor
tr	Transmission
wh	Wheel axle
Abbreviations	
BW	Braunschweig driving cycle
CVT	Continuously variable transmission
E11	Espoo, Finland bus route 11
ECB	Electric city bus
EV	Electric vehicle

HT	High traffic
IVT	Infinitely variable transmission
LT	Low traffic
LTO	Lithium-titanium oxide
MT	Medium traffic
NEDC	New European driving cycle
PMSM	Permanent magnet synchronous machine
UDC	Urban driving cycle

Appendix A

This section presents the motor/generator and the inverter efficiency functions that were applied to compute the maps featured in Figure 4. Modelling of the motor/generator efficiency is based on the work in [24], but with substantially different parameter values due to the different motor characteristics.

Motor/generator efficiency is defined by:

$$P_{m,loss}(\tau_m, \omega_m) = \sum_{i=1}^{I} a_i \left(\frac{|\tau_m|}{\tau_{m,max}}\right)^{m_i} \left(\frac{\omega_m}{\omega_{m,max}}\right)^{n_i} (P_{base\_loss} - b_i) \tag{A1}$$

$$\eta_m(\omega_m, \tau_m) = \max\left\{\eta_{m,min}, \frac{\omega_m \tau_m}{\omega_m \tau_m + P_{m,loss}(\tau_m, \omega_m)}\right\} \tag{A2}$$

where a_i, b_i, m_i, n_i, and P_{base_loss} are parameters in the motor power loss function $P_{m,loss}(\omega_m, \tau_m)$. The max function in (A2) gives a lower limit $\eta_{m,min}$ of the motor/generator efficiency. In the numerical case study, the parameter $\omega_{m,max}$ is 314, $\tau_{m,max}$ is 2000, P_{base_loss} is 11000 and $\eta_{m,min}$ is 0.5. The other parameters that are indexed over $i = 1 \ldots 10$ are given in Table A1.

Table A1. Numerical case study parameter values for the motor power loss function.

i	a_i	b_i	m_i	n_i
1	−0.003	0	0	0
2	0.520	0	0	1
3	0.020	0	1	0
4	−1.022	−4000	1	1
5	1.300	0	2	1
6	0.103	0	2	0
7	−0.334	0	0	2
8	0.450	0	3	0
9	0.500	0	1	2
10	0.120	4000	0	3

The inverter efficiency is computed similarly as the engine/generator, but with only three terms in the weighted sum. The formulation is:

$$P_{m,loss}(\tau_m, \omega_m) = \sum_{i=1}^{I} a_i \left(\frac{|\tau_m|}{\tau_{m,max}}\right)^{m_i} \left(\frac{\omega_m}{\omega_{m,max}}\right)^{n_i} (P_{base\_loss} - b_i) \tag{A3}$$

$$\eta_m(\omega_m, \tau_m) = \max\left\{\eta_{m,min}, \frac{\omega_m \tau_m}{\omega_m \tau_m + P_{m,loss}(\tau_m, \omega_m)}\right\} \tag{A4}$$

where c_1, c_2 and c_3 are parameters of the power loss function (A3) and receive values 5×10^{-4}, 0.5 and 2.0, respectively, in the numerical case study. Additionally, the lower threshold of the inverter efficiency $\eta_{inv,min}$ is set to 0.8.

References

1. Lajunen, A. Lifecycle costs and charging requirements of electric buses with different charging methods. *J. Clean. Prod.* **2018**, *172*, 56–67. [CrossRef]
2. Eckhouse, B. The U.S. Has a Fleet of 300 Electric Buses. China Has 421,000. Available online: https://www.bloomberg.com/news/articles/2019-05-15/in-shift-to-electric-bus-it-s-china-ahead-of-u-s-421-000-to-300 (accessed on 20 June 2019).
3. Anenberg, S.C.; Milled, J.; Henze, D.K.; Minjares, R.; Achakulwisut, P. The global burden of transportation tailpipe emissions on air pollution-related mortality in 2010 and 2015. *Environ. Res. Lett.* **2019**, *14*, 094012. [CrossRef]
4. Besselink, I.J.M.; van Oorschot, P.F.; Meinders, E.; Nijmeijer, H. Design of an Efficient, Low Weight Battery Electric Vehicle Based on a VW Lupo 3L. In Proceedings of the EVS-25 Symposium, Shenzhen, China, 5–9 November 2010.
5. Ehsani, M.; Gao, Y.; Gay, S. Characterization of Electric Motor Drives for Traction Applications. In Proceedings of the IECON 03, Roanoke, VA, USA, 2–6 November 2003.
6. De Santiago, J.; Bernhoff, H.; Ekergård, B.; Eriksson, S.; Ferhatovic, S.; Waters, R.; Leijon, M. Electrical motor drivelines in commercial all-electric vehicles: A review. *IEEE Trans. Veh. Technol.* **2012**, *61*, 475–484. [CrossRef]
7. Ji, F.; Xu, L.; Wu, Z. Effect of driving cycles on energy efficiency of electric vehicles. *Sci. China, Ser. E Technol. Sci.* **2009**, *52*, 3168–3172. [CrossRef]
8. Wu, G.; Zhang, X.; Dong, Z. Impacts of Two-Speed Gearbox on Electric Vehicle's Fuel Economy and Performance. *SAE Tech. Paper* **2013**. [CrossRef]
9. Ren, Q.; Crolla, D.A.; Morris, A. Effect of transmission design on Electric Vehicle (EV) performance. In Proceedings of the VPPC 09, Dearborn, MI, USA, 7–10 September 2009.
10. Bottiglione, F.; De Pinto, S.; Mantriota, G.; Sorniotti, A. Energy consumption of a battery electric vehicle with infinitely variable transmission. *Energies* **2014**, *7*, 8317–8337. [CrossRef]
11. Gao, B.; Liang, Q.; Xiang, Y.; Guo, L.; Chen, H. Gear ratio optimization and shift control of 2-speed I-AMT in electric vehicle. *Mech. Syst. Signal Process.* **2015**, *50–51*, 615–631. [CrossRef]
12. Spanoudakis, P.; Tsourveloudis, N.C.; Koumartzakis, G.; Krahtoudis, A.; Karpouzis, T.; Tsinaris, I. Evaluation of a 2-speed transmission on electric vehicle's energy consumption. In Proceedings of the IEVC 2014, Florence, Italy, 17–19 December 2014.
13. Tan, S.; Yang, J.; Zhao, X.; Hai, T.; Zhang, W. Gear Ratio Optimization of a Multi-Speed Transmission for Electric Dump Truck Operating on the Structure Route. *Energies* **2018**, *11*, 1324. [CrossRef]
14. Dietsche, K.-H. *Automotive Handbook*, 8th ed.; Roberth Bosch: Plochingen, Germany, 2011.
15. Naunheimer, H. *Automotive Transmissions: Fundamentals, Selection, Design and Application*, 2nd ed.; Springer: London, UK, 2011.
16. Shin, J.W.; Kim, J.O.; Choi, J.Y.; Oh, S.H. Design of a 2-speed transmission for an electric commercial vehicle. *Int. J. Automot. Technol.* **2014**, *15*, 145–150. [CrossRef]
17. Ruan, J.; Walker, P.; Zhang, N. A comparative study energy consumption and costs of battery electric vehicle transmissions. *Appl. Energy* **2016**, *165*, 119–134. [CrossRef]
18. *Rimac Automobili Gearbox System*; Rimac Automobili: Sveta Nedelja, Croatia, 2019.
19. Ngo, V.; Hofman, T.; Steinbuch, M.; Serrarens, A. Effect of gear shift and engine start losses on energy management strategies for hybrid electric vehicles. *Int. J. Powertrains* **2015**, *4*, 141–162. [CrossRef]
20. Lajunen, A.; Kalttonen, A. Investigation of thermal energy losses in the powertrain of an electric city bus. In Proceedings of the ITEC 15, Dearborn, MI, USA, 14–17 June 2015.
21. Kivekas, K.; Vepsalainen, J.; Tammi, K. Stochastic Driving Cycle Synthesis for Analyzing the Energy Consumption of a Battery Electric Bus. *IEEE Access* **2018**, *6*, 55586–55598. [CrossRef]
22. Vepsäläinen, J.; Kivekäs, K.; Otto, K.; Lajunen, A.; Tammi, K. Development and validation of energy demand uncertainty model for electric city buses. *Transp. Res. Part D* **2018**, *63*, 347–361. [CrossRef]
23. Vepsäläinen, J.; Ritari, A.; Lajunen, A.; Kivekäs, K.; Tammi, K. Energy Uncertainty Analysis of Electric Buses. *Energies* **2018**, *11*, 3267. [CrossRef]
24. Mahmoudi, A.; Soong, W.L.; Pellegrino, G.; Armando, E. Efficiency maps of electrical machines. In Proceedings of the ECCE 2015, Montreal, QC, Canada, 20–24 September 2015.

25. Eurostat. *Electricity Prices for Non-Household Consumers*; European Commission, Eurostat: Luxembourg, 2020.
26. Fries, M.; Kerler, M.; Rohr, S.; Schickram, S.; Sinning, M.; Lienkamp, M. *An overview of Costs for Vehicle Components, Fuels, Greenhouse Gas Emissions and Total Cost of Ownership 2017*; Institute of Automotive Technology, Technische Universität München: München, Germany, 2017.
27. Lindh, P.; Petrov, I.; Immonen, P.; Pyrhonen, J.; Niemela, M.; Anttila, J.; Paakkinen, M.; Scherman, E. Performance of a Direct-Liquid-Cooled Motor in an Electric Bus under Different Load Cycles. *IEEE Access* **2019**, *7*, 86897–86905. [CrossRef]
28. Laitinen, H.; Tammi, K.; Lajunen, A. Improving electric vehicle energy efficiency with two-speed gearbox. In Proceedings of the VPPC 2017, Belfort, France, 11–14 December 2017.

© 2020 by the authors. Licensee MDPI, Basel, Switzerland. This article is an open access article distributed under the terms and conditions of the Creative Commons Attribution (CC BY) license (http://creativecommons.org/licenses/by/4.0/).

energies

MDPI

Article

Optimized Fuel Economy Control of Power-Split Hybrid Electric Vehicle with Particle Swarm Optimization

Hsiu-Ying Hwang and Jia-Shiun Chen *

Department of Vehicle Engineering, National Taipei University of Technology, Taipei 10608, Taiwan; hhwang@mail.ntut.edu.tw
* Correspondence: chenjs@mail.ntut.edu.tw

Received: 31 March 2020; Accepted: 30 April 2020; Published: 5 May 2020

Abstract: This research focused on real-time optimization control to improve the fuel consumption of power-split hybrid electric vehicles. Particle swarm optimization (PSO) was implemented to reduce fuel consumption for real-time optimization control. The engine torque was design-variable to manage the energy distribution of dual energy sources. The AHS II power-split hybrid electric system was used as the powertrain system. The hybrid electric vehicle model was built using Matlab/Simulink. The simulation was performed according to US FTP-75 regulations. The PSO design objective was to minimize the equivalent fuel rate with the driving system still meeting the dynamic performance requirements. Through dynamic vehicle simulation and PSO, the required torque value for the whole drivetrain system and corresponding high-efficiency engine operating point can be found. With that, the two motor/generators (M/Gs) supplemented the rest required torques. The composite fuel economy of the PSO algorithm was 46.8 mpg, which is a 9.4% improvement over the base control model. The PSO control strategy could quickly converge and that feature makes PSO a good fit to be used in real-time control applications.

Keywords: hybrid electric vehicle; power-split; fuel economy; particle swarm optimization

1. Introduction

Since the industrial era, the demand for fossil fuels has increased and the burning of fossil fuels has led to an increase in global carbon dioxide emissions, which has also increased global warming. The National Oceanic and Atmospheric Administration (NOAA) conducted a network sampling of carbon dioxide concentration based on 40 regions around the world. The survey found that since 1979, the global atmospheric carbon dioxide concentration has risen sharply. With a carbon dioxide concentration of 336 ppm in 1979, the global atmospheric carbon dioxide concentration has reached 414 ppm in March 2020. According to Taiwan's CO2 emissions survey by the Environmental Protection Agency of the Ministry of Administration, the energy sector's CO2 emissions accounted for approximately 10.5% of total fuel combustion emissions, industry accounted for 47.8%, transportation accounted for 14.6%, services accounted for 13.4%, residential emissions accounted for 12.6%, and agriculture accounted for 1.1% [1]. It is clear that transportation emissions are the second-largest source, after industrial emissions.

Due to carbon dioxide emissions, major car manufacturers currently commit to the development of new energy to replace gasoline, including electric energy, solar energy, biomass energy, etc. Many automobile manufacturers are optimistic about electric energy since it can be practically used in mass production. With low fuel consumption and exhaust emissions, hybrid electric vehicles (HEVs) have been attracting widespread public attention in recent years. Hendrickson et al. [2] presented the GM two-mode, front-wheel-drive hybrid powertrain, detailing the mechanical structure and operating mode of this powertrain system. Meisel et al. [3] presented the power distribution of the hybrid transmission in different modes and four fixed-gear ratios in detail. They also compared the differences between the Toyota THS-II gearbox and the GM two-mode gearbox, aiming at energy loss and engine fuel consumption. The advantages of the GM two-mode power system were clearly explained. For the hybrid electric vehicle architecture, the control strategy of the vehicle is critical and is mainly to allocate energy and improve energy efficiency. Many energy management strategies for hybrid electric vehicles have been proposed. They can be divided into two main types: those with 1 a rule-based strategy, and those with 2 an optimization strategy. Torres et al. [4] focused on rule-based controls. The benefits of these controls were rapid rule design and easy implementation. Schouten et al. [5] applied fuzzy logic in HEV control strategies. The rules were judged based on the accelerator and brake pedal signals, battery state of charge (SOC), and motor speed. This optimization method heavily relied on the experience and intuition of the engineers.

The optimization strategy for vehicle fuel economy simulation can be categorized into two areas, global optimization and real-time local optimization. The global optimization algorithm requires completing the whole driving cycle in order to obtain the best fuel economy. This makes it difficult to apply on real road scenarios. Genetic algorithms (GAs) are one of the global optimization algorithms. Montazeri et al. [6] applied GA optimization in parallel HEVs, where the engine torque and battery SOC were design variables with the objective of minimizing fuel consumption and emissions. Dynamic programming (DP) is another typical global optimization algorithm. Wu et al. [7] presented the application of DP on electric buses and used DP to explore energy management strategies for range-extended electric buses (REEBs). Zheng et al. [8] applied stochastic DP in plug-in hybrid vehicles to achieve a global optimization. Wang et al. [9] implemented the DP algorithm in a plug-in HEV (PHEV) and showed a 20% improvement in fuel consumption. DP analyzes the whole driving cycle, searching for minimum fuel consumption to get the global optimum. It provides an improvement for fuel economy; however, it takes a lot of time for the whole simulation and process. An alternative way is to utilize the optimum patterns obtained from pre-calculated DP, such as rule-based (RB) control. RB can be adapted for real-time applications; however, the fuel economy of RB might not be as good as DP's due to the instant change in actual scenarios.

For real-time applications, optimized fuel consumption needs to be carried out quickly for each time step. Local optimization can be suitable for the applications. The equivalent consumption minimization strategy (ECMS) does not require long calculation and is one control algorithm fitting for real-time fuel consumption optimization. Paganelli et al. [10] presented the application of the equivalent consumption minimization strategy (ECMS) in parallel HEVs. They managed the power distribution to minimize fuel consumption, which includes the actual fuel consumed by the engine and an equivalent fuel converted from the electrical energy consumed by motors. Their simulation maintained the battery SOC in a reasonable range by applying a penalty function to ensure battery life. Particle swarm optimization (PSO) is another real-time control algorithm for vehicle fuel consumption. Chen et al. [11] discussed the application of particle swarm optimization (PSO) on HEVs. They optimized the engine output power as an energy management strategy. Wu et al. [12] applied PSO to plug-in HEVs. The main goal of their study was to optimize the control strategy to achieve the best fuel economy. For charge depleting (CD) mode, restrictions were imposed to optimize the PSO. Abido et al. [13] applied PSO to the energy flow control problem of buses. The main goal was to minimize fuel consumption and improve voltage stability. Wang et al. [14] applied a particle-swarm-optimization-based nonlinear model predictive control strategy on a series-parallel hybrid electric bus to optimize the fuel consumption. Chen et al. [15] discussed the application of PSO on plug-in HEVs (PHEVs). In terms of vehicle speed, fuzzy logic judgments were added. Then, the PSO determined the upper and lower limits of the required engine power to obtain the optimal result. Chen et al. [16] implemented improved particle swarm optimization (IPSO) in HEVs. The difference from the original PSO was mainly to add the value of a poor function. This function would cause the PSO particles to speed-up to find the best solution. Beside ECMS and PSO, there are other control algorithms to achieve better fuel economy. Zheng et al. [17] applied Pontryagin's minimum principle to optimize a parallel plug-in hybrid electric bus. Feng et al. [18] combined an artificial neural network model and a fuzzy-logic controller to optimize the fuel consumption of a hybrid electric mining truck. Comparing the differences between PSO and genetic algorithms (GAs), GAs remove the worst position at one time, and PSO keeps the worst particles, judging the best solution according to the position of each particle [18]. In addition, GAs mainly process the replication, mating, and mutation, which requires a large amount of calculation [19]. Relatively speaking, since the information transmission between PSO particles and the interaction mechanism between particles is relatively simple in PSO, the amount of calculation is lower and the delivery time can be shortened. With the advantages of quick convergence, the PSO algorithm is suitable for real-time control applications for vehicle fuel consumption optimization. This research applied PSO for real-time optimization control.

In 2018, global battery electric vehicle (BEV) and HEV sales exceeded 5.1 million, a significant increase of 63% from the previous year. BEVs and HEVs will become the first choice for the public in the future. The two mode hybrid (TMH) system was the power system applied in this research. A basic rule-based control was implemented as the base model with an initial energy management strategy, and the simulation result was compared to the manufacturer data. Then, the PSO control strategy was added to optimize fuel consumption and explore the reasons for improvement. This paper investigated a TMH system that had power-split hybrid functionality.

2. Modeling

In this research, a simulation model was created with Matlab/Simulink that included a vehicle, TMH transmission system, internal combustion engine (ICE), motor/generators (M/Gs), battery, and controller, as shown in Figure 1. With the United States FTP-75 (EPA Federal Test Procedure) of urban and highway driving cycles, the vehicle model could estimate the torque and power required for driving. The controller module (Controller) determined the transmission mode switching and the optimal output torques of the engine and M/Gs. The transmission module simulated the torques and speeds of the two motor/generators. The battery module would simulate the battery's state of charge (SOC) and battery charge and discharge status. These modules are presented in the following sections.

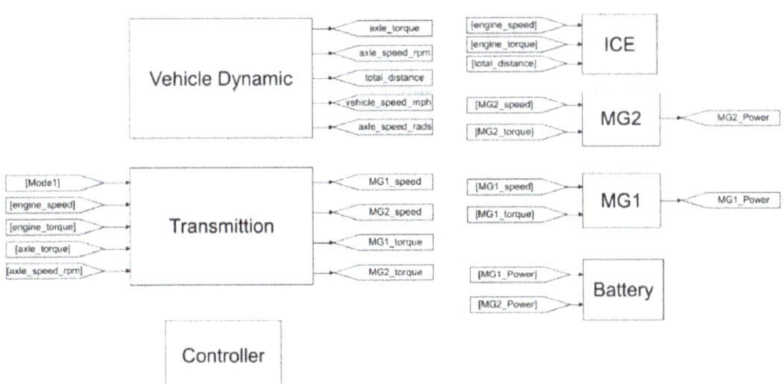

Figure 1. Simulink model of two-mode power-split hybrid system.

2.1. Vehicle Model

U.S. FTP-75 (EPA Federal Test Procedure) was employed in this model. The vehicle model estimated the required driving torque according to the vehicle speed from the driving cycles. The driving resistance forces included aerodynamic resistance, rolling resistance, and grading resistance.

The FTP-75 driving cycles include urban and highway sections. The total distance traveled in the urban cycle is 17.77 km, the total time is 1874 s, the average speed is 34.1 km/h, and the highest speed is 91.2 km/h, as shown in Figure 2. The total distance of the highway cycle is 16.5 km, the total time is 765 s, the average speed is 77.7 km/h, and the maximum speed is 96.4 km/h, as shown in Figure 3.

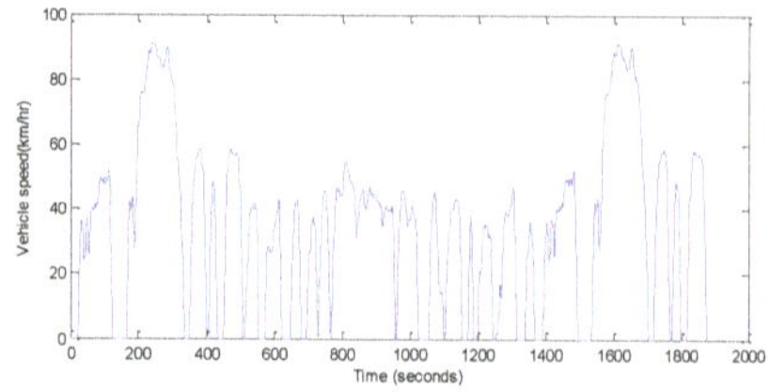

Figure 2. US EPA FTP75 urban driving cycle.

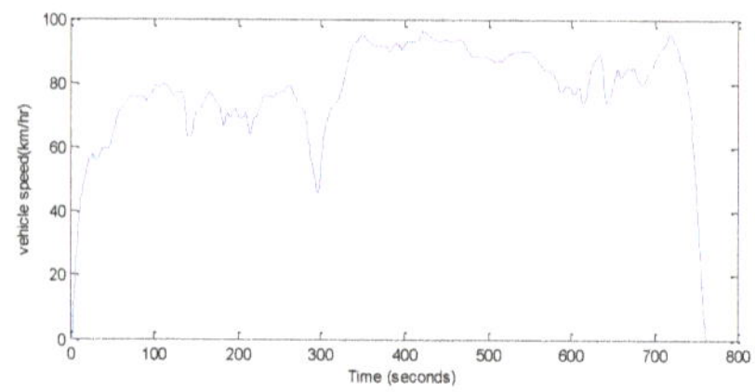

Figure 3. US EPA FTP75 highway driving cycle.

2.2. Transmission Model

In this research, the two-mode hybrid transmission, 2MT70, was implemented as the hybrid electric powertrain system. The entire powertrain was composed of several main components, as shown in Figure 4, which included a simple planetary gear set, a compound planetary gear set, four sets of clutches, two electric motors/generators, an IC engine, and a battery [20].

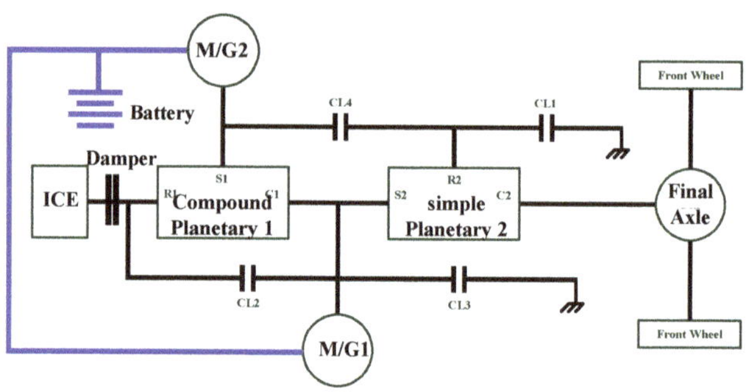

Figure 4. Driveline configuration of two-mode hybrid power system.

2.3. Internal Combustion Engine Model

The internal combustion engine (ICE) module mainly cooperated with the brake-specific fuel consumption (BSFC) lookup table according to the engine speed and torque to find the corresponding fuel rate. The total fuel consumption of the driving cycle was obtained through integration. A V-6 3.6-L engine efficiency chart is shown in Figure 5.

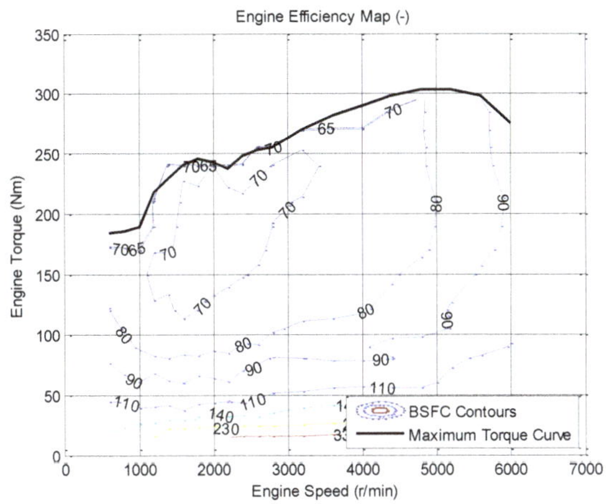

Figure 5. Internal combustion engine's efficiency chart.

2.4. *Electric Motor/Generator Model*

The electric motor/generator (M/G) modules obtain the operating efficiency of the electric motor through a three-dimensional lookup table based on the current speed and torque. Two 60 kW permanent magnet AC motors were constructed. The efficiency diagram is shown in Figure 6. The power of the M/G was calculated as Equation (1).

$$P_{MG} = \omega_{MG} T_{MG} \eta_{MG}^{k} \quad (1)$$

P_{MG}, ω_{MG}, and T_{MG} are the power, speed, and torque of the electric motor/generator, respectively. η_{MG} represented the efficiency of M/G. If the values of the torque and the speed were positive or negative at the same time, the M/G was consuming power and performing as an electric motor. If the values of the speed and torque were of the opposite sign, the M/G was charging and running as a generator. k represents the energy path. If M/G operated as an electric motor, k was −1. If it operated as a generator, k was +1.

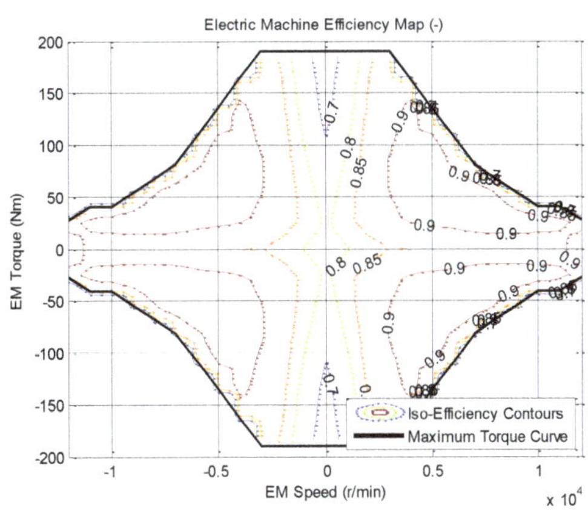

Figure 6. Motor/generator's efficiency chart.

2.5. Battery Model

The electric power required by the M/Gs was provided by the battery, and the total power of the battery was estimated through the motor's power and efficiency, as presented in Equation (2).

$$P_{batt} = T_{MG1}\omega_{MG1}\eta_{MG1}^{k}\eta_{c1}^{k} + T_{MG2}\omega_{MG2}\eta_{MG2}^{k}\eta_{c2}^{k} \quad (2)$$

P_{batt} is the power of the battery, η_c is the efficiency of the electric converter, T_{MG} is the M/G torque, ω_{MG} is the speed of the M/G, η_{MG} is the efficiency of the M/G, and k indicates the energy path. $k = -1$ indicates that the battery was discharged, and $k = 1$ indicates that the battery was charged.

The battery module was composed as an equivalent circuit, including open circuit voltage and a battery pack, as shown in Figure 7. I_{batt} represents the battery current, and P_{batt} is the battery output or input power.

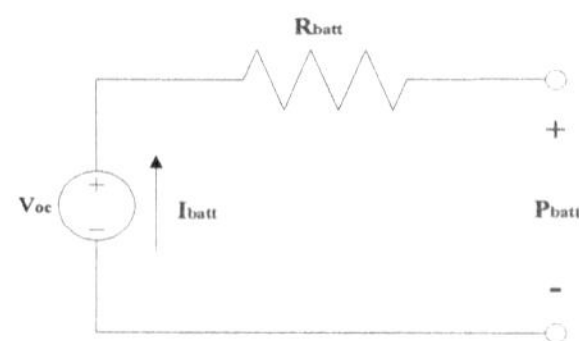

Figure 7. Diagram of battery equivalent circuit.

The relationship between battery SOC and current I_{batt} is as follows.

$$S\dot{O}C = -\frac{I_{batt}}{Q_{\max}} \quad (3)$$

Q_{max} is the ampere-hour capacity of the battery at the current rate I_{batt}. The battery current, open circuit voltage, and internal resistance would vary according to the battery power. The relationship between battery output or input power and current is presented as follows.

$$P_{batt} = V_{OC}I_{batt} - I_{batt}^{2}R_{batt} \quad (4)$$

$$I_{batt} = -\frac{V_{OC} - \sqrt{V_{OC}^{2} - 4P_{batt}R_{batt}}}{2R_{batt}} \quad (5)$$

$$S\dot{O}C = -\frac{V_{OC} - \sqrt{V_{OC}^{2} - 4\left(T_{MG1}\omega_{MG1}\eta_{MG1}^{k}\eta_{c1}^{k} + T_{MG2}\omega_{MG2}\eta_{MG2}^{k}\eta_{c2}^{k}\right)R_{batt}}}{2R_{batt}Q_{\max}} \quad (6)$$

where V_{oc} is the battery open circuit voltage and R_{batt} is the battery's internal resistance.

2.6. Controller Model

2.6.1. Rule-Based Controller

The controller module had three main functions: the first one was the control of switching mode, which decided the timing of switching between the first and the second mode according to the driving cycle; the second one was based on the driving condition and battery SOC to determine the controlled engine speed and engine torque according to the rules established in the controller; the third one was engine state control logic. To achieve fuel saving, the engine would be turned off when it was not required to provide driving power.

2.6.2. PSO Controller

This controller module also had three main functions: the first one was to decide the time of switching, which was the same as the rule-based controller; the second one was to determine the engine speed, which was based on the drive resistance and battery SOC, and the engine torque, which was estimated by the particle swarm algorithm; the third one was the engine state control, which was determined by the PSO controller. The only exception was if the vehicle traveled at very low speed and the engine was turned off—then the rule-based control was used.

3. Energy Management Optimization

3.1. Particle Swarm Algorithm

Particle swarm optimization (PSO) is a group-based optimization search method [21,22]. The advantage is that it has a fast convergence speed and fewer parameter settings. The swarm movement of birds was simulated. Birds usually maintain a specific formation when they travel as a group, and individuals will maintain a certain distance from each other. The search technology of PSO is mainly to simulate the social behavior of birds foraging. When birds are searching for food, individuals will understand the relationship between themselves and food. At the same time, the closest individual to food in the flock is also observed. So, when searching for food, the best path can be found based on these two points. Assume that food is considered as the global best solution in the function and the distance between each individual in the flock and the food is used as the objective degree of the function. Then, the process of each individual in the flock seeking food can be the process to search for the optimized solution. This concept is the foundational idea of the PSO algorithm.

The detailed calculation steps of the PSO algorithm are as follows: ${x_i}^t$ and ${v_i}^t$ are defined as the position and flight speed of the *i*-th particle in the particle group at time *t*, respectively. The model of the PSO algorithm can be divided into speed update function and position update function as follows.

The main process steps of PSO are:

1. Initialize the position and velocity of each particle using random numbers.
2. Use the objective function to calculate the objective value for each particle.
3. Compare the current objective value with the best position of the particle, P_{best}. If the currently searched objective value is better than P_{best}, the new objective value and position will be applied to update the value of P_{best}.
4. Compare the currently searched objective value with the best value in the group, g_{best}. If the currently searched value is better than g_{best}, use the new value and position of the particle to update g_{best}.
5. Change the speed and position of the particles according to Equations (7) and (8).

$$v_i^t = \omega * v_i^{t-1} + c_1 * rand() * \left(p_i - x_i^{t-1}\right) + c_2 * rand() * \left(g_i - x_i^{t-1}\right) \quad (7)$$

$$x_i^t = x_i^{t-1} + v_i^t \quad (8)$$

where ${x_i}^t$ is the position of the *i*-th particle at time t, ${v_i}^t$ is the speed of the *i*-th particle at time *t*, p_i is the best position that the *i*-th particle has traveled to, g_i is the best position that all particles have traveled to, c_1 is the weight of one's own experience to make particles closer to the best solution of the individual, c_2 is the weight of the group's experience to make the particles closer to the best solution of the group, ω is the inertia weight, which affects the range and speed of convergence, and *rand()* is a random value used to maintain the diversity of group movement directions.

3.2. PSO Algorithm Applied to Hybrid Electric Power System

This research implemented PSO to find the minimum instantaneous fuel consumption at each driving moment. The process of the PSO algorithm is described as follows.

1. Initialization: This research applied engine torque as the initial particle position. The initial flying speed was randomly generated. To avoid exceeding the operational range of engine torque during the algorithm search process, the constrained equations were included in the simulation. The engine torque constrained condition is shown in Equation (9). Since the PSO algorithm did not have a practical mechanism to control the speed of the particles, the constrained condition was set for the speed Equation (10). During the particle search process, the battery SOC must be ensured to avoid over-charging and over-discharging. The constrained equation for the SOC upper and lower limits is shown in Equation (11).

$$T_{E,\min} < T_E < T_{E,\max} \tag{9}$$

$$v^t_{i,\min} < v^t_i < v^t_{i,\max} \tag{10}$$

$$SOC^t_{i,\min} < SOC^t_i < SOC^t_{i,\max} \tag{11}$$

where T_E is the engine torque, v_i is the particle speed, and $SOC_i^{\,t}$ is the battery state of charge at time t, $T_{E,min}$ is the lower limit of engine torque, $T_{E,max}$ is the upper limit of engine torque, $v_{i,min}$ is the lower limit of particle speed, $v_{i,max}$ is the upper limit of particle speed, $SOC^t{}_{i,min}$ is the lower limit of the SOC, and $SOC^t{}_{i,max}$ is the upper limit of the SOC.

2. Apply the objective function: In order to improve the performance of the system, a suitable objective function was designed to allow particles to search for the objective value. The lowest instantaneous equivalent fuel consumption was the best solution of this optimization problem. So, instantaneous equivalent fuel consumption was applied as the objective function as in Equation (12).

$$F_{obj} = \dot{m}_{eq} + \beta = \dot{m}_e(T_E, \omega_E) + W(SOC)\dot{m}_b + \beta \tag{12}$$

where $\dot{m}_{eq}$ is the sum of the instantaneous equivalent fuel consumption, $\dot{m}_e(T_E, \omega_E)$ is the instantaneous fuel consumption of the engine, $\dot{m}_b$ is the equivalent fuel consumption of the battery's electrical energy, β is a penalty value to avoid the particle searching process violating restrictions, and *W(SOC)* is the weight factor to prevent the power of the battery from being depleted.

The instantaneous fuel consumption of electric power was calculated by Equation (13). The battery power consumption was converted to the equivalent fuel consumption, $\dot{m}_b$, by the engine BSFC corresponding to the engine torque and speed.

$$\dot{m}_b = \frac{\left(\zeta_{discharge} * \overline{BSFC} * \frac{P_{MG}}{\eta_{batt} * \eta_{MG}} + \zeta_{charge} * \overline{BSFC} * P_{MG} * \eta_{batt} * \eta_{MG}\right)}{1000 * 3600} \tag{13}$$

where $\dot{m}_b$ is the conversion of battery power consumption into equivalent fuel consumption, $\overline{BSFC}$ is the BSFC corresponding to engine torque and speed, P_{MG} is the electric motor power, η_{batt} is the battery operating efficiency, η_{batt} is the electric motor working efficiency. $\zeta_{discharge}$ is the battery discharge factor, and ζ_{charge} is the battery charge factor.

3. Select and memorize: Each particle would remember the corresponding objective value and compare it with the value obtained by the particle at the previous moment. After determining the best value, P_{best}, of each particle, the values of all particles were compared to determine the best value in the group, g_{best}. With the best value, g_{best}, of the group, the corresponding engines best torque could be received.
4. Modify particle's speed and position: Each particle's T_E was moved to the next position according to Equations (7) and (8).
5. End rule: Set the convergence conditions. If the convergence conditions were not met, steps 2–4 until the particle search met the convergence conditions.

3.3. *Formulation of PSO Algorithm*

3.3.1. Initialization and Parameter Setting

Initially, four particles were generated randomly. In this research, vehicle performance was improved by optimizing the instantaneous fuel consumption and the engine torque was applied as the particle's position. The particles' initial position and speed were assigned as follows.

$$\left[\begin{array}{llll} x_1^t & x_2^t & x_3^t & x_4^t \end{array}\right]=\left[\begin{array}{llll} T_{E,1}^t & T_{E,2}^t & T_{E,3}^t & T_{E,4}^t \end{array}\right]=\left[\begin{array}{llll} 0 & 10 & 20 & 30 \end{array}\right] \tag{14}$$

$$\left[\begin{array}{llll} v_1^t & v_2^t & v_3^t & v_4^t \end{array}\right]=\left[\begin{array}{llll} -1 & -2 & 2 & 4 \end{array}\right] \tag{15}$$

The lower speed limit, $v_{i,\min}^t$, and upper speed limit, $v_{i,\max}^t$, were set between (−10–10). If the value of the speed vector was too large, it would cause the particles to jump out of the area with a good solution. If the value of the speed vector was too small, it would cause the particles to fall into the local minimum value. For the better solution, the particle speed was usually set to 10–20% of the search range [18]. The learning factors were both set as 2.

3.3.2. Evaluate Each Particle

The objective value of each particles position was measured. In the initial state, random particles were first moved by the random speed and the updated position of the particles was compared with the initial random position to obtain the local optimal position, P_{best}. Then, based on their own experience and group experience among the particles, the particles moved to the new positions. To avoid unreasonable particle positions and over-charge or over-discharge of the battery, some constraints were set as follows.

$$0 < T_E < 220 \tag{16}$$

$$0.4 < SOC < 0.6 \tag{17}$$

3.3.3. Evaluate the End of Searching

If the position distances between the particles were less than 0.01, the search result was converged (Equations (18)–(21) and the optimal solution was reached. To prevent the searching time from being too long, an extra setting was added. If the particle search time was greater than 0.5 s, the current best value was directly taken as the optimal torque of the engine.

$$\left|T_{E,1} - T_{E,2}\right| < 0.01 \tag{18}$$

$$\left|T_{E,2} - T_{E,3}\right| < 0.01 \tag{19}$$

$$\left|T_{E,3} - T_{E,4}\right| < 0.01 \tag{20}$$

$$\left|T_{E,4} - T_{E,1}\right| < 0.01 \tag{21}$$

4. Simulation Results and Discussion

4.1. *Vehicle Parameters*

The model for this research was implemented with Matlab/Simulink. The parameters of the vehicle are shown in Table 1. A V6 3.6 L engine was applied in this research. This engine had a manufacturer configuration for a midsize power-split HEV, and it was compatible with the power of M/Gs applied in this vehicle. The power-split HEV was designed with this engine in order to maintain the performance as the original ICE version. The additional M/Gs provided the HEV with better acceleration and gradeability.

Table 1. Vehicle parameters.

Engine	
Engine Type	V6, SI
Displacement	3.6 (L)
Peak torque	300 Nm
Peak speed	6000 rpm
EM characteristics	
Type	Permanent magnet motor
Maximum power	60 kW
Maximum speed	10,000 rpm
Peak efficiency	0.92
Peak torque	190 Nm
Battery	
Type	Lithium–ion
Capacity	1.5 kWh
Vehicle characteristics	
Vehicle mass	1600 kg
Radius of tire	0.352 m
Vehicle front area	2642 m^2
Rolling resistance coef.	0.01
Aerodynamic drag coef.	0.386

4.2. Charge and Discharge

In this research, the working range of the battery's SOC was set between 0.4 and 0.6, which could effectively extend the battery's service life. At the same time, the battery's internal resistance was low in this working range, regardless of the state of charge or discharge, as shown in Figures 8 and 9.

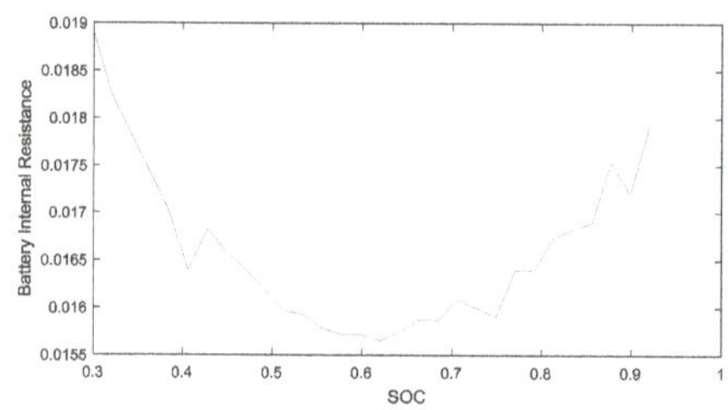

Figure 8. Relation between battery state of charge (SOC) and internal resistance during battery discharge.

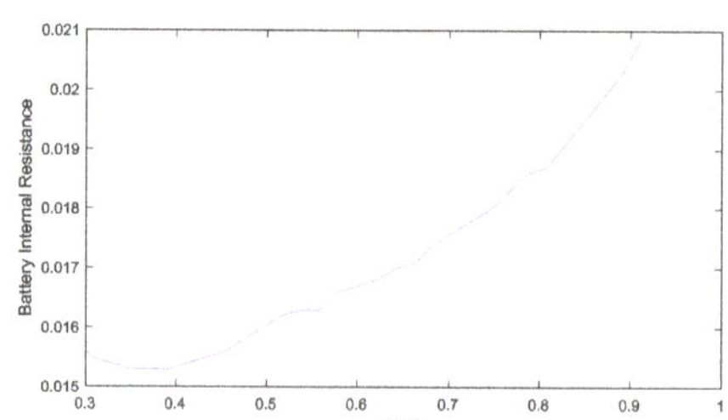

Figure 9. Relation between battery SOC and internal resistance during battery charge.

4.3. Simulation Results

EPA FTP-75 driving cycles were applied in the simulation. The urban and highway driving curves are shown in Figures 2 and 3, respectively. Rule-based simulation results are shown in Table 2. The fuel economy (FE) of urban and highway simulations were 46.28 mpg and 39.11 mpg, respectively, and the composite FE was 42.74 mpg. The formula for composite FE is shown in Equation (22) As shown in Table 2, the difference between the composite FE of the rule-based simulation and that of the manufacturer data was 0.9%, which was within the allowable range. Therefore, the rule-based control model was applied as a base model to evaluate the optimization simulation of the PSO algorithm.

$$Composite\ FE = \frac{1}{\frac{0.55}{CIty\ FE} + \frac{0.45}{Highway\ FE}} \tag{22}$$

Table 2. Vehicle parameters.

Item	Urban	Highway	Composite
Rule-based	46.28 mpg	39.11 mpg	42.74 mpg
Manufacturer	48 mpg	37 mpg	42.34 mpg

Rule-based control and PSO simulation results are presented in Table 3. In the urban driving cycle, the FE of the rule-based simulation was 46.28 mpg, and the FE of the PSO simulation was 51.79 mpg. PSO showed a 12% improvement over rule-based control. In the highway driving cycle, the FE of the rule-based simulation was 39.11 mpg, and the FE of the PSO simulation was 41.85 mpg. PSO showed a 7% improvement over rule-based control. For composite FE, rule-based control was 42.74 mpg and PSO was 46.78 mpg. PSO showed an improvement of 9.4% compared to rule-based control.

Table 3. Comparison of fuel economy.

	Urban	Highway	Composite
Rule-based	46.28 mpg	39.11 mpg	42.74 mpg
PSO	51.79 mpg	41.85 mpg	46.78 mpg
Improved	12%	7%	9.4%

To understand the reasons for the improvements with the PSO algorithm, the instantaneous fuel consumption of the rule-based control and PSO models were compared, as shown in Figures 10 and 11. Figure 10 shows the urban simulation results; the instantaneous fuel consumption of the PSO was generally smaller than that of the rule-based controller. The switching timing for the engine to turn on/off depended on the required engine torque and battery SOC. In rule-based control, the results were obtained from the pre-set rules/tables. In PSO, the algorithm searched for better fuel consumption under the desired engine torque. Therefore, the fuel rate of PSO was smaller and engine-switch timing was different from rule-based control. Figure 11 shows the instantaneous fuel consumption for the highway driving cycle. It can be seen that the maximum instantaneous fuel consumption was 2.6 g/s for rule-based control and 2.3 g/s for PSO. PSO would affect the engine switch timing and engine operating points. The fuel consumption of PSO was better than that of the rule-based control.

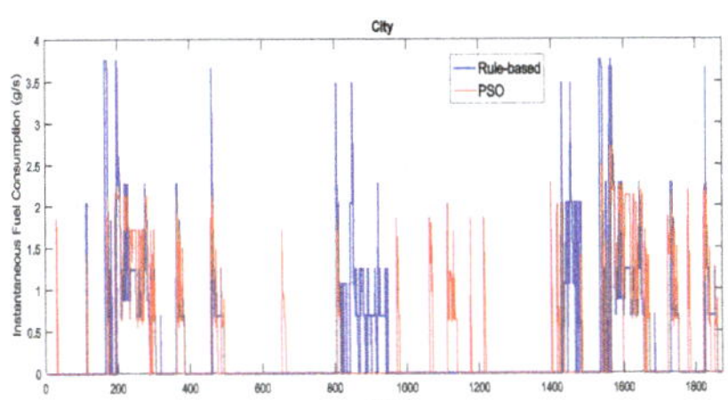

Figure 10. Comparison of instantaneous fuel consumption in the urban driving cycle.

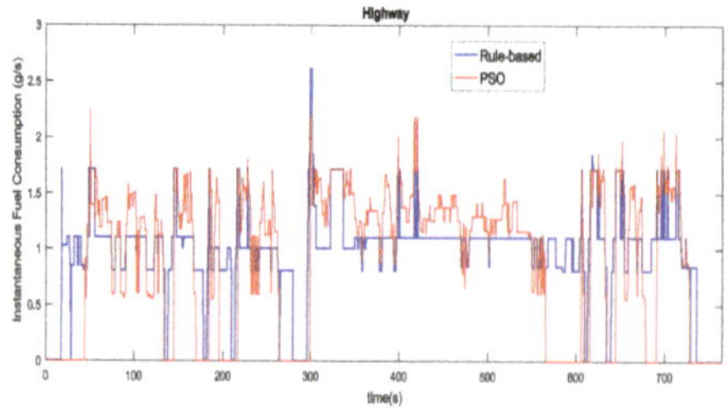

Figure 11. Comparison of instantaneous fuel consumption in the highway driving cycle.

Figures 12 and 13 are comparisons of engine speeds on urban and highway driving cycles, respectively. The engine speeds of PSO and rule-based control were mainly determined by driving resistance and SOC. According to the simulation results of the two methods, most of the engine speeds in urban areas were around 1400 to 2000 rpm. The improvement in fuel consumption was affected by engine torque.

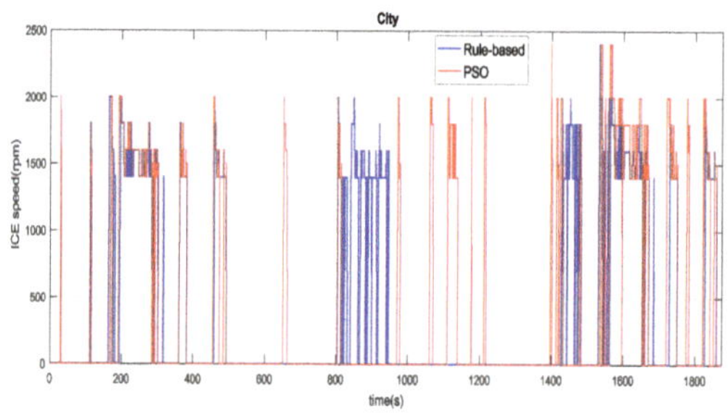

Figure 12. Comparison of engine speed in the urban driving cycle.

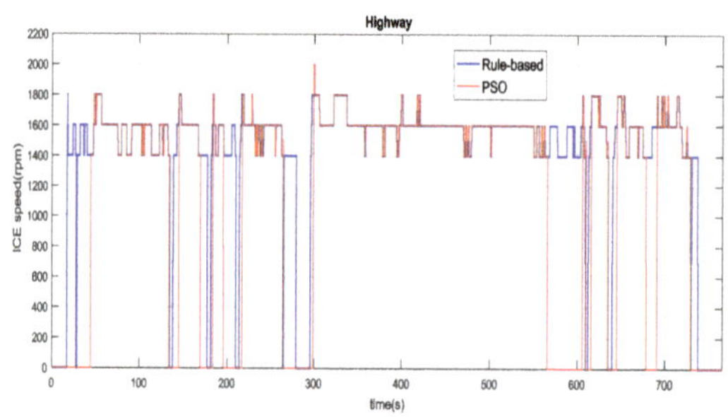

Figure 13. Comparison of engine speed in the highway driving cycle.

Figures 14 and 15 are comparisons of engine torque in urban and highway driving cycles, respectively. It can be seen that the engine torque in PSO was less than that in rule-based control. The peak torque values in urban driving cycles decreased from 250 Nm in rule-based control to 160 Nm in PSO. The reason that PSO algorithm provided a better fuel consumption is mainly due to the engine torque being reduced. In the highway driving cycle, the improvement of fuel consumption was because of different engine operating points and engine running time.

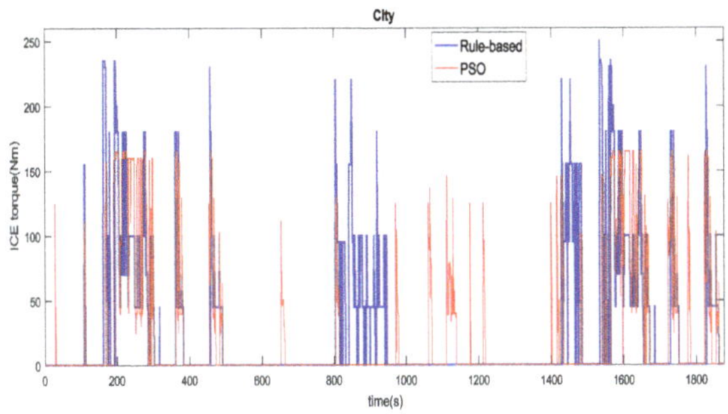

Figure 14. Comparison of engine torque in the urban driving cycle.

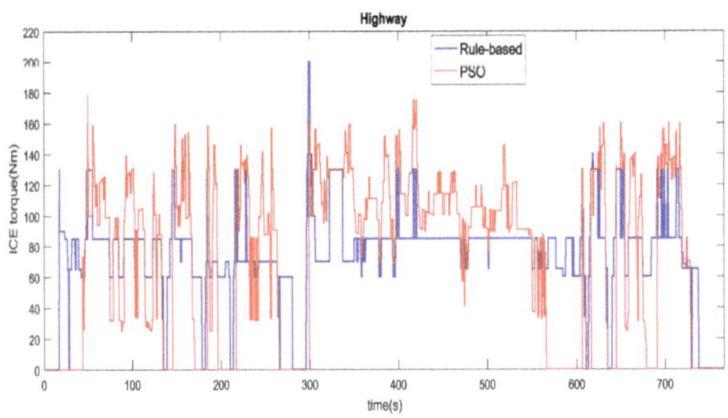

Figure 15. Comparison of engine torque in the highway driving cycle.

Figures 16 and 17 provide the electric conversion loss from the motor/generator to charge the battery. It can be clearly seen that the conversion loss was higher with rule-based control in both driving cycles, which was one of the reasons that PSO could improve the fuel consumption.

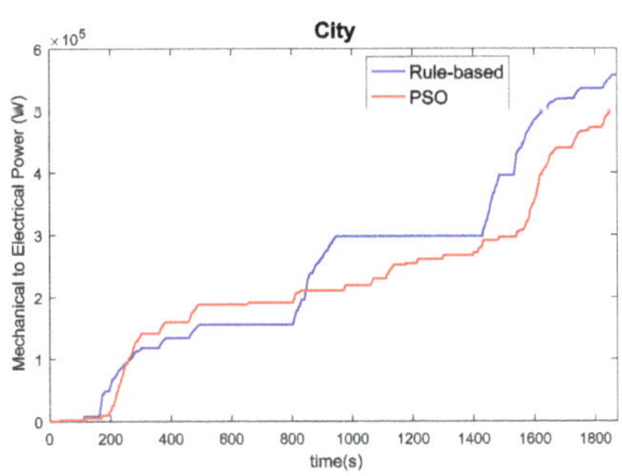

Figure 16. Comparison of electric conversion loss in the urban driving cycle.

Figures 18 and 19 show the engine operating points of rule-based control and PSO in urban areas and the highway driving cycle, respectively. In the urban driving cycle, the engine torque of PSO operating points was mainly in the 50–170 Nm range. Compared to the operating points of rule-based control, the trend of overall PSO engine power decline also led to an improvement of fuel consumption. In rule-based control, some of the engine operating points were around 170–240 Nm, which is in a high-efficiency range. It did not require such a large amount of engine power to drive the vehicle, so the excess engine power would be transferred to the generator to charge the battery. That resulted in an increase of conversion loss. Furthermore, the stored energy in the battery would not stay in the battery very long and would soon be used for driving. This resulted in a second conversion loss and would affect the fuel economy.

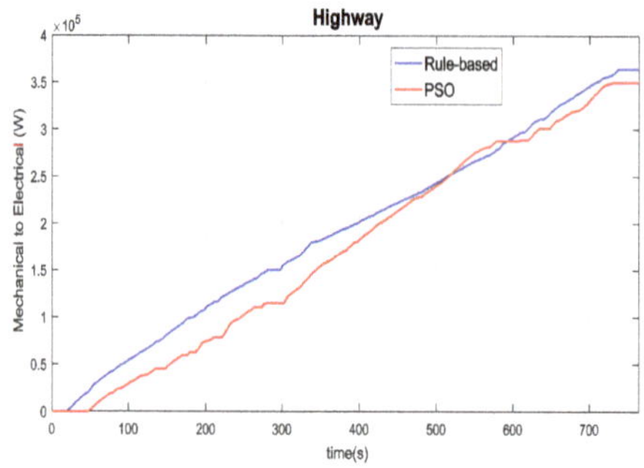

Figure 17. Comparison of electric conversion loss in the highway driving cycle.

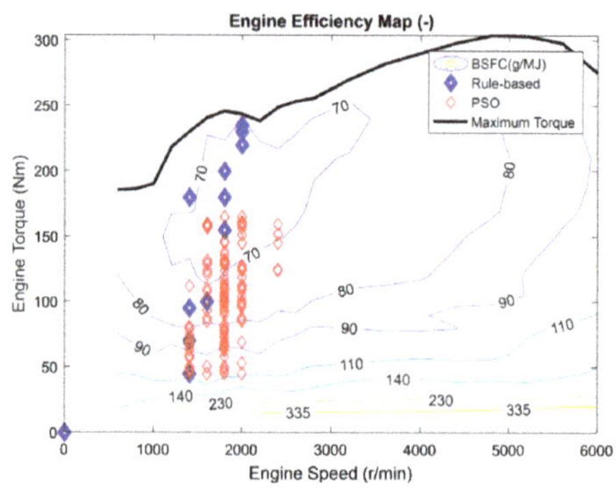

Figure 18. Comparison of urban engine operating points.

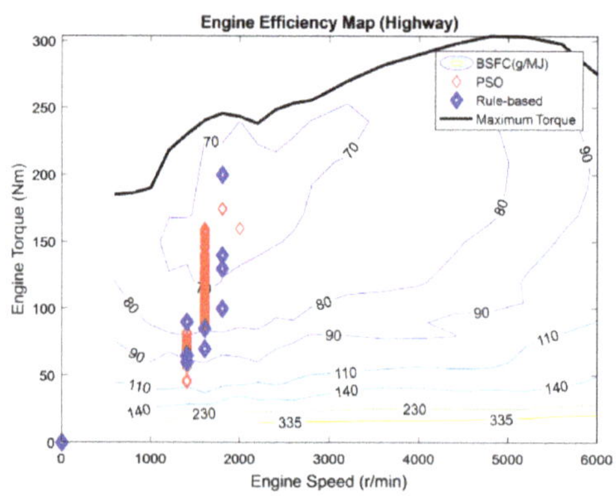

Figure 19. Comparison of highway engine operating points.

On the highway driving cycle, most of the PSO engine operating points were in the range of 75–120 Nm, which is in a better efficiency range of the engine. The engine operating points of rule-based control were scattered throughout a wider range. Some points were around 200 Nm, which is in a high-efficiency region; however, the vehicle did not require such a large amount of engine power. The excess engine power would charge the battery and cause an electric conversion loss. The stored energy in the battery would be applied for driving. This resulted in a second conversion loss. It affected the fuel economy.

Figures 20 and 21 show the comparison of electric motor torque of M/G1 in urban and highway driving cycles, respectively. M/G1 was mainly driven by the engine in mode one. In the urban driving cycle, M/G1 was mostly in the charge condition. In the highway driving cycle, the powertrain mainly stayed on mode two. M/G1 worked as a driving motor and provided power to drive the vehicle.

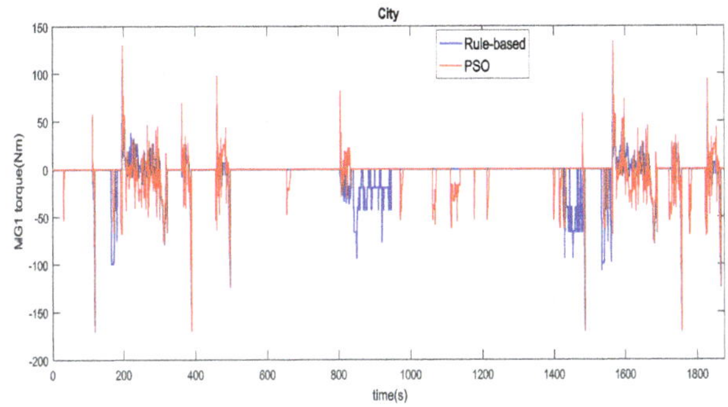

Figure 20. M/G1 torque in the urban driving cycle.

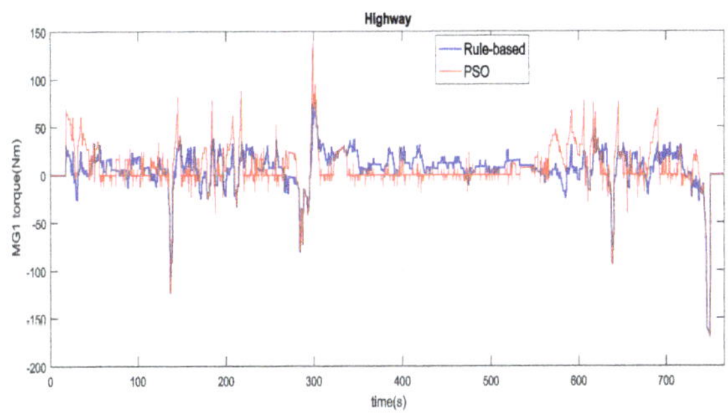

Figure 21. M/G1 torque in the highway driving cycle.

Figures 22 and 23 were the torque of M/G2 in the urban and highway driving cycles, respectively. In the urban driving cycle, the positive torque output time of the PSO algorithm was longer than that of the rule-based control. With the optimization process of the PSO, the vehicle would have more time driven by electric motors to save fuel and improve vehicle fuel economy.

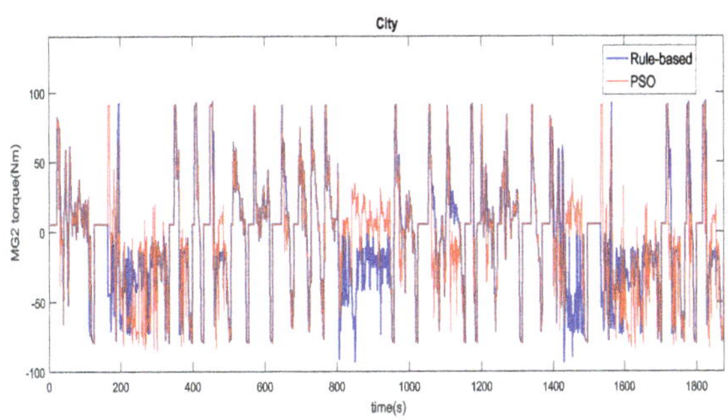

Figure 22. M/G2 torque in the urban driving cycle.

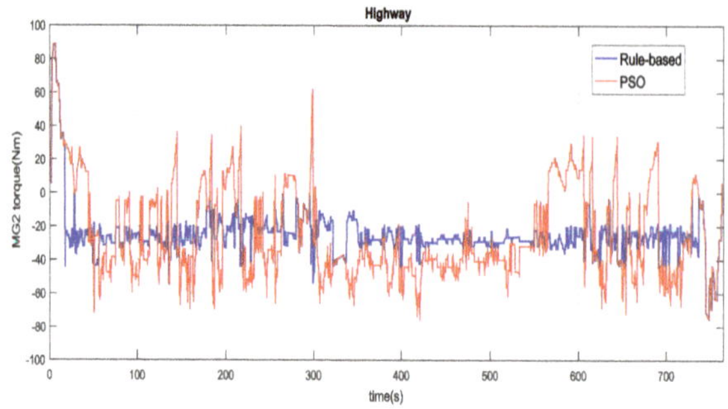

Figure 23. M/G2 torque in the highway driving cycle.

The battery SOC in urban and highway driving cycles are shown in Figures 24 and 25, respectively. Through the driving cycles, the initial SOC and the SOC at the end of the cycles remained very close. During the driving cycle the battery was charged and discharged, and the battery energy at the end of the cycle remained at the same level as at the beginning. All of the driving energy was provided by the engine.

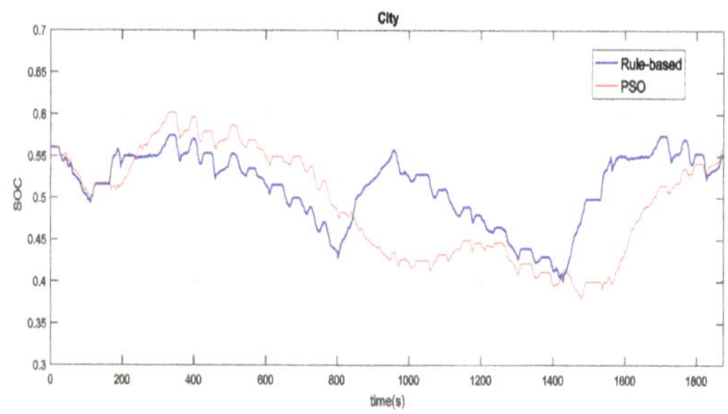

Figure 24. Battery SOC in the urban driving cycle.

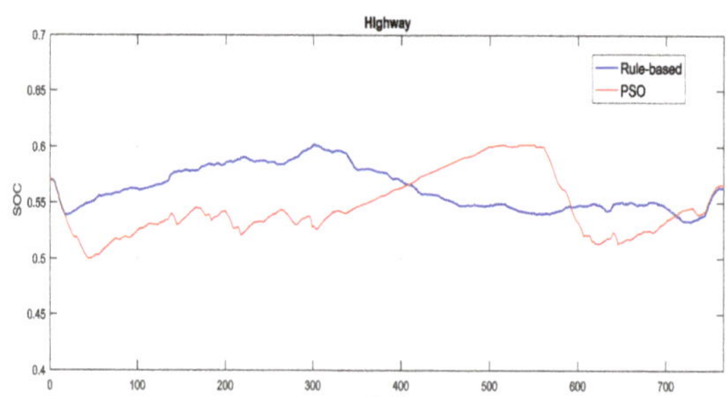

Figure 25. Battery SOC in the highway driving cycle.

5. Conclusions

This research focused on the real-time control algorithm to improve the fuel economy of a two-mode, power-split hybrid electric vehicle. The vehicle model was built with Matlab/Simulink. The fuel economy simulation results of the base model with rule-based control were close to the fuel economy data provided by the original manufacturer, which confirmed the reliability of the vehicle model. Particle swarm optimization (PSO) was implemented as a real-time optimization control with the goal of reducing fuel consumption. The minimum instantaneous fuel consumption was the objective of PSO. The engine torque was the design variable. PSO was to set up a group of sprinkled particles to search for the best solution. The particles were dispersed in a reasonable working area of the engine, and the value of the objective function was calculated for each particle position. The objective function of each particle and the instantaneous fuel consumption was compared, and the particle position was updated based on better results of the group. The above action was repeated until the particles converged, and the objective value of the particle was the current minimum instantaneous fuel consumption. The following conclusions were obtained based on the simulation results.

1. Urban cycle fuel economy: The result of rule-based simulation was 46.28 mpg; the result of PSO simulation was 51.79 mpg; PSO showed a 12% improvement over rule-based control.
2. Highway cycle fuel economy: The result of rule-based simulation was 39.11 mpg; the result of PSO simulation was 41.85 mpg; PSO showed a 7% improvement over rule-based control.
3. Composite fuel economy: The result of rule-based simulation was 42.74 mpg; the result of PSO simulation was 46.78 mpg; PSO showed an improvement of 9.4% compared to rule-based control.
4. Compared with the rule-based control, PSO could more effectively control the engine on/off switch timing and the engine operating points. With the lower engine torque selected by the PSO algorithm, the engine power and fuel consumption were reduced. With less charging and discharging of the battery, the energy conversion loss was smaller with the PSO algorithm and the vehicle would spend more time driven by electric motors to save fuel and improve vehicle fuel economy.
5. The rule-based control requires long-term accumulated experience to set the rule table. Compared with rule-based control, PSO could find the minimum instantaneous fuel consumption at each moment, achieve real-time control, and improve the fuel economy of power-split HEVs.

Author Contributions: Conceptualization, H.-Y.H.; investigation, H.-Y.H.; methodology, H.-Y.H.; project administration, J.-S.C.; software, J.-S.C.; validation, J.-S.C. All authors have read and agreed to the published version of the manuscript.

Funding: There is no external funding for this research.

Conflicts of Interest: The authors declare no conflict of interest.

References

1. Available online: http://www.epa.gov.tw/ct.asp?xItem=10052&ctNode=31352&mp=epa (accessed on 30 April 2018).
2. Hendrickson, J.; Holmes, A.; Freiman, D. *General Motors Front Wheel Drive Two-Mode Hybrid Transmission*; SAE technical paper 2009-01-0508; SAE World Congress & Exhibition: Detroit, MI, USA, 2009.
3. Meisel, J. *An Analytic Foundation for the Two-Mode Hybrid-Electric Powertrain with a Comparison to the Single-Mode Toyota Prius THS-II Powertrain*; SAE technical paper 2009-01-1321; SAE World Congress & Exhibition: Detroit, MI, USA, 2009.
4. Torres, J.L.; Gonzalez, R.; Gimenez, A.; Lopez, J. Energy management strategy for plug-in hybrid electric vehicles. A comparative study. *Appl. Energy* **2014**, *113*, 816–824. [CrossRef]
5. Schouten, N.J.; Salman, M.A.; Kheir, N.A. Energy management strategies for parallel hybrid vehicles using fuzzy logic. *Control Eng. Pract.* **2003**, *11*, 171–177. [CrossRef]
6. Montazeri-Gh, M.; Poursamad, A.; Ghalichi, B. Application of genetic algorithm for optimization of control strategy in parallel hybrid electric vehicles. *J. Frankl. Inst.* **2006**, *343*, 420–435. [CrossRef]
7. Wu, X.; Chen, J.; Hu, C. Dynamic Programming-Based Energy Management System for Range-Extended Electric Bus. *Math. Probl. Eng.* **2015**, *2015*, 1–11. [CrossRef]
8. Zhang, H.; Qin, Y.; Li, X.; Liu, X.; Yan, J. Power management optimization in plug-in hybrid electric vehicles subject to uncertain driving cycles. *eTransportation* **2020**, *3*, 100029. [CrossRef]
9. Wang, X.; He, H.; Sun, F.; Zhang, J. Application Study on the Dynamic Programming Algorithm for Energy Management of Plug-in Hybrid Electric Vehicles. *Energies* **2015**, *8*, 3225–3244. [CrossRef]

10. Paganelli, G.; Delprat, S.; Guerra, T.M.; Rimaux, J.; Santin, J.J. Equivalent Consumption Minimization Strategy for Parallel Hybrid Powertrains. In Proceedings of the IEEE 55th Vehicular Technology Conference, Birmingham, AL, USA, 6–9 May 2002; pp. 2076–2081.
11. Chen, Z.; Xiong, R.; Wang, K.; Jiao, B. Optimal Energy Management Strategy of a Plug-in Hybrid Electric Vehicle Based on a Particle Swarm Optimization Algorithm. *Energies* **2015**, *8*, 3661–3678. [CrossRef]
12. Wu, X.; Cao, B.; Wen, J.; Bian, Y. Particle Swarm Optimization for Plug-in Hybrid Electric Vehicle Control Strategy Parameter. In Proceedings of the 2008 IEEE Vehicle Power and Propulsion Conference, Harbin, China, 3–5 September 2008; pp. 1–5.
13. Abido, M.A. Optimal power flow using particle swarm optimization. *Int. J. Electr. Power Energy Syst.* **2002**, *24*, 563–571. [CrossRef]
14. Wang, Y.; Wang, X.; Sun, Y.; You, S. Model predictive control strategy for energy optimization of series-parallel hybrid electric vehicle. *J. Clean. Prod.* **2018**, *199*, 348–358. [CrossRef]
15. Chen, Z.; Xiong, R.; Cao, J. Particle swarm optimization-based optimal power management of plug-in hybrid electric vehicles considering uncertain driving conditions. *Energy* **2016**, *96*, 197–208. [CrossRef]
16. Chen, S.-Y.; Hung, Y.-H.; Wu, C.-H.; Huang, S.-T. Optimal energy management of a hybrid electric powertrain system using improved particle swarm optimization. *Appl. Energy* **2015**, *160*, 132–145. [CrossRef]
17. Zhang, S.; Hu, X.; Xie, S.; Song, Z.; Hu, L.; Hou, C. Adaptively coordinated optimization of battery aging and energy management in plug-in hybrid electric bus. *Appl. Energy* **2019**, *256*, 113891. [CrossRef]
18. Feng, Y.; Dong, Z. Optimal energy management with balanced fuel economy and battery life for large hybrid electric mining truck. *J. Power Sources* **2020**, *454*, 227948. [CrossRef]
19. Kennedy, J.; Eberhart, R. Particle Swarm Optimization. In Proceedings of the ICNN'95-International Conference on Neural Networks (ICNN), Perth, WA, Australia, 27 November–1 December 1995; pp. 1942–1948.
20. Hwang, H.-Y.; Lan, T.-S.; Chen, J.-S. Optimization and Application for Hydraulic Electric Hybrid Vehicle. *Energies* **2020**, *13*, 322. [CrossRef]
21. Eberhart, R.; Kennedy, J. A New Optimizer Using Particle Swarm Theory. In Proceedings of the Sixth International Symposium on Micro Machine and Human Science Micro Machine and Human Science, Nagoya, Japan, 4–6 October 1995; pp. 39–43.
22. Eberhart, R.C.; Shi, Y. Comparison between Genetic Algorithms and Particle Swarm Optimization. In *Evolutionary Programming VII-7th International Conference 1998*; Springer: Berlin/Heidelberg, Germany, 1998; Volume 1447, pp. 611–616.

© 2020 by the authors. Licensee MDPI, Basel, Switzerland. This article is an open access article distributed under the terms and conditions of the Creative Commons Attribution (CC BY) license (http://creativecommons.org/licenses/by/4.0/).

energies

MDPI

Article

Route Guidance Strategies for Electric Vehicles by Considering Stochastic Charging Demands in a Time-Varying Road Network

Yongxing Wang [1,2], Jun Bi [1,2,*], Chaoru Lu [3] and Cong Ding [1,2]

1 School of Traffic and Transportation, Beijing Jiaotong University, Beijing 100044, China; yx_wang@bjtu.edu.cn (Y.W.); 17120762@bjtu.edu.cn (C.D.)
2 Key Laboratory of Transport Industry of Big Data Application Technologies for Comprehensive Transport, Beijing Jiaotong University, Beijing 100044, China
3 Department of Civil and Environmental Engineering, Norwegian University of Science and Technology, 7033 Trondheim, Norway; chaoru.lu@ntnu.no
* Correspondence: bilinghc@163.com; Tel.: +86-134-8881-2321

Received: 2 February 2020; Accepted: 28 April 2020; Published: 5 May 2020

Abstract: Electric vehicles (EVs) are being increasingly adopted because of global concerns about petroleum dependence and greenhouse gas emissions. However, their limited driving range results in increased charging demands with a stochastic characteristic in real-world situations, and the charging demands should be attributed toward charging stations in time-varying road networks. To this end, this study proposes guidance strategies to provide efficient choice for charging stations and corresponding routes, and it includes the time-varying characteristic of road networks in problem formulation. Specifically, we propose two route guidance strategies from different perspectives based on the charging demand information. The first strategy focuses on the effects of the number of EVs on the charging stations' operation, and the reachable charging stations with the fewest vehicles are selected as the heuristic suggested ones. The other strategy considers the travel cost of individual drivers and selects the charging stations nearest to the destination as heuristic suggested ones. Both strategies ensure that the selected charging stations can be reached in a time-varying road network. In addition, we carry out a simulation analysis to investigate the performance of the proposed route guidance strategies and introduce relevant insights and recommendations for the application of the strategies under various scenarios.

Keywords: electric vehicles; route guidance strategies; stochastic charging demands; time-varying road network

1. Introduction

Dependence on petroleum contributes to a serious environmental and energy problem. The transportation sector is one of the major economic industries that contribute to energy consumption and greenhouse gas emissions. The International Energy Agency found that the transportation sector contributes 28% of global energy consumption and 23% of global greenhouse gas emissions [1]. Electric vehicles (EVs), which are highly energy-efficient, are recognized as a promising solution to alleviate the problem of fossil fuel dependency and increasing greenhouse gas emissions, especially if the energy used for their charging is obtained from a renewable energy source [2]. However, unlike conventional internal combustion engine vehicles, EVs have a relatively short driving range because of their limited battery capacity, thus requiring drivers to recharge their vehicles often to reach their destinations. Insufficient charging infrastructure often causes difficulties in finding charging stations, thus resulting in the range anxiety of EV drivers [3]. Range anxiety can be effectively alleviated by providing guidance information for EV charging based on specific service platforms, such as a smart charging service. Drivers could file charging demands to the charging service provider and receive recommended charging stations through their mobile devices [4]. Guidance strategies should be developed in consideration of charging demand information to determine the recommended charging stations and corresponding travel routes. The traffic condition on a road network also affects the route choice of EVs from departure points to charging stations, because the traffic volume would vary as time progresses in real-world situations, due to the factors such as rush hour, which is regarded as the time-varying characteristic for a road network [5]. Such a characteristic would influence vehicle driving state and should be considered in the route guidance strategies. The stochastic characteristic intrinsic to the charging demands substantially affects the strategies, thus further increasing the difficulty of dealing with charging demands. Large-scale charging behaviors with stochastic characteristics considerably affect the operation efficiency of charging stations. Therefore, given the widespread adoption of EVs, solving stochastic charging demands in complex real-world situations is a critical issue for the current and future global transportation system.

Directing EVs to suitable charging stations is an important and fundamental problem for the adoption of EVs in urban transportation systems. As introduced in Section 2, the traditional methods of route guidance for EVs mainly focus on the problems with deterministic charging demands in a static road network. Although several studies considered the impacts of a time-varying road network on the driving state of vehicles, less attention was paid to the stochastic characteristics intrinsic to charging demands. Therefore, a route guidance method that can be used to deal with stochastic charging demands and that takes into account the time-varying road network is expected. To fill the gap, this study develops route guidance strategies for stochastic charging demands in a time-varying road network from two different perspectives. The performance of the route guidance strategies is explored by considering their effects on the operation efficiency of charging stations. Both strategies can direct EVs with stochastic charging demands to reachable charging stations by considering the time-varying traffic conditions on the routes.

This study makes the following unique contributions: firstly, the stochastic characteristic of charging demands is investigated given the situation with large-scale adoption of EVs. A route guidance problem with stochastic charging demands is formulated by combining the time-varying road network. A dynamic recursive equation is developed to obtain the EV number in charging stations, which varies as time progresses. Secondly, to address stochastic charging demands, two route guidance strategies are established, and the operation efficiency of charging stations and travel cost of individual drivers are considered in the strategies. In actual situations, the operation efficiency of charging stations could be affected by the number of EVs in them, because the sustainable number of EVs for a charging station is limited and the queuing time is increased as vehicle number increases. Therefore, we use the number of EVs in charging stations as the matric to reflect the operation efficiency of charging stations. In addition, the travel cost of individual drivers is generally composed of travel time, energy consumption, and charging cost. These travel cost components are closely correlated with driving distance. Thus, the driving distance is employed to reflect the integration of travel cost components. The reachability of the selected charging stations can be ensured by both strategies in a time-varying road network. Lastly, the proposed strategies are applied in simulation examples to provide guidance for stochastic charging demands in a time-varying road network. The performance of the two strategies is compared in different simulation scenarios, and application recommendations in terms of the strategies are presented based on the simulation results.

The remaining portions of this paper are arranged as follows: in Section 2, the literature review is presented. In Section 3, the route guidance problem is formulated by considering the stochastic charging demands in a time-varying road network. In Section 4, the metrics regarding the charging station selection are analyzed from two different perspectives, and the route guidance strategies are presented. In Section 5, the simulated results are presented to compare the performance of the proposed strategies. Lastly, the conclusions and future studies are discussed in Section 6.

2. Literature Review

EVs are a potential solution to environmental and energy problems because they have high energy efficiency and can be charged by using renewable energy source. Thus, the traveling and charging problems of EVs attracted increasing interest from the scientific community. Given the limited driving range of EVs, several studies attempted to find optimal routes for EVs on the basis of the framework of the constrained shortest path problem [6–8]. However, charging behavior was not involved in the methods. Considering the charging demands incurred by EV travels, Wang et al. used geometric approaches and designed an algorithm for route guidance by considering the charging demand information from drivers [4]. Driving direction and distance were used as choice indicators for optimal charging stations. Sweda et al. proposed two heuristic methods for making adaptive routing and recharging decisions for EVs [9]. Charging costs were involved in the solution. In addition to charging processes, Qin and Zhang and Said et al. considered the effects of queuing time on charging station selection [10,11]. Queuing theory was used to optimize the route guidance. Several studies combined driving time, charging time, and queuing time to discuss charging and route optimization for EVs [12–14]. Wang et al. incorporated energy constraints during travel and proposed an energy-aware routing model for EVs [15]. Cao et al. and Liu et al. considered the effects of charging costs on charging station selection to investigate EV charging problems [16,17]. Yagcitekin and Uzunoglu developed a smart route guidance strategy based on double-layer optimization theory [18]. Sun and Zhou compared the effects of different factors on route selection of EVs by using a cost-optimal algorithm [19]. The trade-off between traveling cost and time consumed was obtained to guide drivers in traveling and charging. Wang et al. integrated drivers' intended traveling and charging choices [20]. A multiobjective model was established to provide guidance for EV charging; the objectives include minimized traveling time, charging costs, and energy consumption. In view of the environmental effects for EV adoption, many studies aimed to search for energy-efficient routes for EVs under different situations [21–25]. However, the aforementioned methods for route guidance are mainly based on problems in a static road network, in which the time or energy consumed in each link is constant. Consequently, the effects of traffic condition on driving state are ignored, which makes the solution unrealistic in complex situations with respect to urban road networks.

In view of this, Alizadeh et al. incorporated time-varying traffic conditions in the traveling and charging problems for EVs to improve the accuracy of route guidance schemes [26]. An extended transportation graph was used to find the optimal routes. Yi and Bauer proposed a model to investigate the effects of traffic condition on energy cost of EVs [27]. The primal–dual interior point algorithm was used to construct the optimal paths. Zhang et al. proposed a multiobjective routing model for EV travel, in which the effects of traffic condition on travel time, driving distance, and energy consumption were considered [28]. The ant colony optimization algorithm was employed to search for optimal routes. Jafari and Boyles incorporated route reliability in the solution for an EV traveling problem under a road network with time-varying traffic condition [29]. Daina et al. explored the EV charging problem by considering uncertain traffic conditions on the basis of random utility theory [30]. The trade-off among driving distance, charging time, and costs for charging selection was analyzed. Huber and Bogenberger utilized real-time traffic information to investigate the time-varying characteristic of traffic conditions and their effects on the EV driving state [31]. Several works introduced network equilibrium theory to explore the optimization models for EV charging and traveling [32–35]. They modeled changing traffic conditions by changing the number of vehicles in each link. However, most existing methods assume that the charging demands of EV drivers are predetermined and overlook their stochastic characteristics. In real-world situations, charging demands with variable information may be made at different periods, and the charging service providers are unable to know the information before they receive the charging demands. Therefore, the previous methods are unable to solve stochastic charging demands in complex real-world situations.

Furthermore, to address stochastic charging demands, Hung and Michailidis proposed a route guidance strategy based on the queuing modeling framework, in which charging demands occur in accordance with a general process during a time period [36]. Nevertheless, the study did not consider the time-varying characteristic of road networks. EVs were assumed to operate with a constant speed in the road network. Energy consumption, which considerably influences the reachability of charging stations, was also ignored in the method. To our knowledge, few studies investigated the route guidance methods of EVs by comprehensively considering stochastic charging demands and a time-varying road network.

Overall, even though the previous studies made achievements in route guidance for EVs, there are still some limitations, as mentioned above. To further clarify the existing studies, we summarize the aforementioned references with respect to their considerations, as listed in Table 1 (considered factors are marked as "√"; otherwise, unconsidered factors are marked as "×"). In view of the limitations, two heuristic-based strategies for route guidance are proposed and introduced in the following sections, which aim to deal with stochastic charging demands in a time-varying road network.

Table 1. Previous studies with respect to their considerations.

References	Considerations		
	Charging Behavior	Time-Varying Road Network	Stochastic Charging Demands
[6–8]	×	×	×
[4,9–25]	√	×	×
[26–35]	√	√	×
[36]	√	×	√

3. Problem Description

During trips, EV drivers often need to recharge their vehicles to reach their destinations, thus resulting in charging demands in a road network with EVs. The charging demand information is assumed to include drivers' travel destinations and the remaining energy of vehicles. Travel destination is one of the critical factors in drivers' travel demand. The remaining energy of EVs can be directly obtained through built-in vehicle dashboards. When EV drivers notice that the remaining energy of their vehicle may be insufficient to reach their destinations, they send the information of charging demands to the charging service provider by using their mobile devices. When receiving charging demands from EV drivers, the charging service provider needs to determine the suitable charging station for each charging demand according to the given information. Note that charging demands in real-world situations have uncertainty and variability by time. That is, the charging demands received in different periods may contain different information about travel destinations of drivers and the remaining energy of EVs. Specifically, both travel destinations and remaining energy have uncertainty from the perspective of the charging service provider, because they derive from the individual travel demands of drivers and operation state of EVs, respectively, thus preventing the charging service provider from predicting the detailed information of charging demands in advance. In situations with large-scale charging demands, multiple charging demands during identical periods often have different information on travel destinations and remaining energy. In actual situations, the charging service provider can obtain the information on EV and charging station locations by using positioning devices. EV drivers do not need to send the location information to the charging service provider, whether the charging demands occur or their detailed information is not predetermined at different periods. Thus, the charging demands in a road network have a significant stochastic characteristic as time progresses. To realize problem formulation, the time is discretized into finite time slots normalized to integral units. Let $\{1, \ldots, t, \ldots, T\}$ denote the set of time slots, where T is the total number of the time slots. With the identical duration for each time slot, the time horizon increases as T increases. The charging demand that occurs in node i and at time slot t is denoted by C_i^t, which is a binary variable in the problem formulation. Its value is equal to 1 if there exists a charging demand that occurs in normal node i $(i = 1, \ldots, m)$ at time slot t; otherwise, it is equal to 0. For each charging demand

$(C_i^t = 1)$, its information includes the travel destination and remaining energy, which are denoted as d_i^t and e_i^t, respectively. The travel destination d_i^t and remaining energy e_i^t from different charging demands may vary. The charging service provider cannot understand or predict C_i^t before time slots t. Thus, the decision-making for all the charging demands needs to be determined based on the traffic condition at corresponding time slots.

The stochastic characteristic of charging demands is one of the challenges in situations with large-scale adoption of EVs. Traffic conditions in a road network also affect the driving speed and energy consumption of EVs, thus affecting their traveling and charging process [37]. Traffic conditions often have time-varying characteristic in real-world situations because of environmental factors [38]. Consequently, the energy and time that are spent while traversing the same links may vary at different time slots. The time-varying characteristic of a road network should be considered to improve the effectiveness of route guidance schemes. The road network structure and EV operating characteristics are combined, and the time-varying road network is defined as $G = (V, A, \tau_a^t, E_a^t)$, where V and A denote the sets of nodes and links, respectively. In set V, two types of nodes, namely, normal and charging station nodes, exist. The latter has the ability to charge EVs. For problem formulation, set V is assumed to consist of m normal nodes and n charging station nodes. τ_a^t and E_a^t in G denote the driving time and energy consumption on link a at time slot t, respectively, where $a \in A$. In every time slot, the values of τ_a^t and E_a^t randomly change within a reasonable range, which reflects the time-varying characteristic of the road network. The conventional optimization problems with time-varying networks often assume that the links remain stable for the duration of a time slot [39]. For the problem of route guidance for EV charging, the assumption signifies the constant driving time and energy consumption during a separate time slot. The assumption conforms to the traffic condition characteristic in an actual road network if the duration for each time slot is relatively short. Therefore, we follow such an assumption in the route guidance problem for EV charging.

The route guidance problem for EV charging is formulated by combining stochastic charging demands and time-varying road networks. The charging demands are assumed to occur in the normal nodes only. The charging station nodes could not generate charging demands. Such an assumption is reasonable, because drivers seek help from the charging service provider only when they have difficulty finding nearby charging stations. The guidance strategies aim to help EV drivers from normal nodes select heuristic suggested charging station nodes based on specific objectives. The binary decision variable in the problem formulation is denoted by x_{ij}^t, which is equal to 1 if the charging demand generated in normal node i $(i = 1, \ldots, m)$ at time slot t is assigned to charging station node j $(j = 1, \ldots, n)$; otherwise, this variable is 0. In every time slot, all the normal nodes in a road network have the potential to generate charging demands. However, whether the charging demands occur or not in every time slot is an uncertain event. In order to reflect such a characteristic of charging demand occurrence, the possibility of the charging demand occurring in node i at each time slot is denoted as $\lambda_i (0 \leq \lambda_i \leq 1)$, as shown in Equation (1).

$$\begin{cases} \Pr(C_i^t = 1) = \lambda_i \\ \Pr(C_i^t = 0) = 1 - \lambda_i \end{cases}, t \in \{1, \ldots, T\}, j \in \{1, \ldots, m\}, \tag{1}$$

where Pr (Λ) represents the possibility of the occurrence of event Λ.

Note that, given the definition of charging demand occurrence as shown in Equation (1), we assume that each normal node is able to generate at most one charging demand following a specific possibility during a time slot. The assumption conforms to the characteristic of charging demand occurrence in actual situations if the duration for each time slot is relatively short. Moreover, to reduce the complexity of problem formulation, we assume that the charging demand occurrence for each node follows a uniform distribution. For example, the possibility of charging demand occurrence for node i has a uniform value λ_i as time progresses, which does not have a time-varying characteristic and is influenced by the node location. Although node i has a constant possibility λ_i for every time slot, the travel destination and remaining energy from the charging demands may vary at different time slots.

When addressing the charging demands at each time slot, the first step is to ensure that the remaining energy can enable the EVs to reach the target charging stations. In a time-varying road network, the energy consumption between normal and charging station nodes may vary at different time slots. Thus, before charging stations are selected, the energy consumption on the routes should be observed, and only the reachable charging stations can be considered as candidates, as shown in Figure 1.

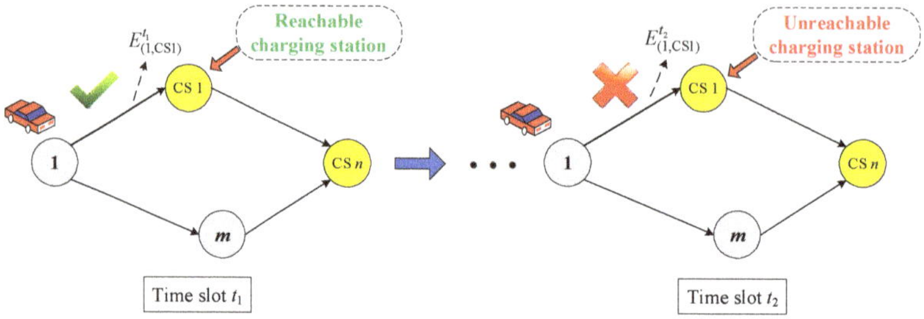

Figure 1. Reachable and unreachable charging stations (CS).

In Figure 1, the energy consumptions between node 1 and CS 1 at time slots t_1 and t_2 are different. The green check mark indicates that the EV can traverse the route, and the red cross indicates that the EV cannot do so because it has insufficient remaining energy. The figure indicates that the reachability of the same charging station may vary at different time slots because of the time-varying traffic conditions on the road network. The driving time on the routes may also change at different time slots, thus determining the suitable time slots as the EVs reach charging stations. The charging service provider is assumed to know the information about the traffic conditions on all the links at the beginning of each time slot, acquiring such information through either real-time traffic information from the transport sector or short-term traffic flow prediction [40]. The basic framework for the stochastic route guidance problem in a time-varying road network is presented in Figure 2.

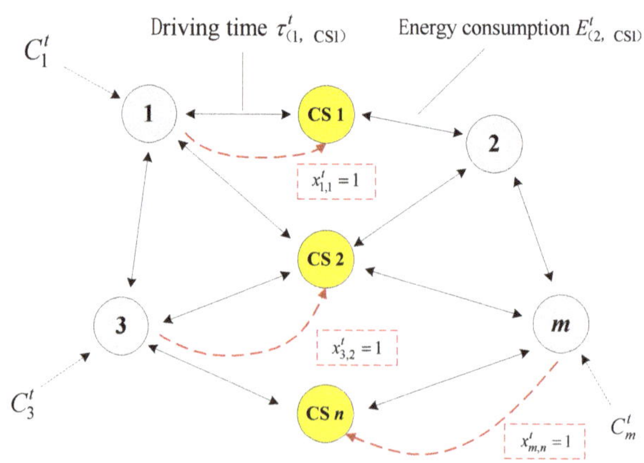

Figure 2. Stochastic route guidance problem in a time-varying road network.

In Figure 2, $\tau^t_{(1,CS1)}$ is the driving time from node 1 to charging station node CS 1 under the traffic condition at time slot t. The energy consumption between node 2 and charging station node CS 1 at time slot t is denoted by $E^t_{(2,CS1)}$. The charging demands that occur in nodes 1, 3, and m at time slot t are denoted as C^t_1, C^t_3, and C^t_m, respectively. The objective of the problem is to provide guidance for every charging demand by considering the traffic condition at time slot t. The recommended charging station nodes would be selected for the charging demands based on specific route guidance strategies. In the figure, the decision for charging station selection is denoted as $x^t_{(1,1)} = 1$, $x^t_{(3,2)} = 1$, and $x^t_{(m,n)} = 1$. For instance, $x^t_{(1,1)} = 1$ indicates that the charging demand that occurs in node 1 is assigned to the charging station node CS 1. How to determine the value of x^t_{ij} at each time slot t is the critical issue to solve the route guidance problem for EV charging in a time-varying road network. This issue should be considered from two aspects. Firstly, the route guidance strategies satisfy the charging demands of EV drivers; that is, an EV should be able to reach the selected charging station under its current remaining energy. For this reason, the relationship between remaining energy and traffic condition needs to be considered. Secondly, the charging behavior has significant effects on the operating state of charging stations, especially in a situation with large-scale adoption of EVs. In every time slot, multiple charging demands may occur in a road network, and the charging stations may have to accept multiple EVs. Given the limited charging rate, the number of EVs in a charging station increases as the time slots pass. However, mass EV charging significantly affects the operating state of charging stations; that is, it may prolong the queuing time and even present a potential burden on local power systems [41–43]. Therefore, aside from drivers' charging demands, the number of EVs at each charging station is another important factor that needs to be considered by route guidance strategies.

We attempt to develop a dynamic recursive equation based on the operation characteristics of charging stations to explore the change trend of EV number in charging stations under the situation with large-scale stochastic charging demands. S^t_j denotes the number of EVs that complete charging and leave charging station j at time slot t. Without loss of generality, the problem assumes that at most one EV can leave a charging station after completing charging at each time slot. The assumption conforms to the actual operating situation in charging stations if the duration for each time slot is relatively short. The possibility of the event that an EV leaves the charging station j after completing charging at each time slot is denoted as $\mu_j (0 \le \mu_j \le 1)$, as shown in Equation (2).

$$\begin{cases} \Pr(S^t_j = 1) = \mu_j \\ \Pr(S^t_j = 0) = 1 - \mu_j \end{cases}, t \in \{1, \ldots, T\}, j \in \{1, \ldots, n\}. \quad (2)$$

The parameter μ_j can reflect the charging levels of the chargers in charging station j. During the actual charging processes, the chargers with different charging levels have different charging rates for EVs [44]. Under the definition of S^t_j and μ_j, the duration between two adjacent events of an EV leaving charging station j follows a geometric distribution [15]. The EV number in charging station j at time slot t is denoted by U^t_j. The dynamic recursive equation for U^t_j is

$$U^t_j = \begin{cases} \varphi_j, & t = 1, j \in \{1, \ldots, n\} \\ \max\{U^{t-1}_j + \sum\limits_{i=1}^{m} \sum\limits_{t''=1}^{t-1} C^{t''}_i x^{t''}_{ij} - S^{t-1}_j, 0\}, & t \in \{2, \ldots, T\}, j \in \{1, \ldots, n\} \end{cases}, \quad (3)$$

where φ_j represents the initial EV number in charging station j within a specific time horizon, and t'' is the time slot when the charging demand from node i occurs. In the equation, the time slots t'' and t satisfy the following relationship:

$$t = t'' + \tau^{t''}_{(i,j)}. \quad (4)$$

Equation (4) indicates that the EV with charging demand $C^{t''}_i$ can reach charging station j after the driving time $\tau^{t''}_{(i,j)}$. Note that, in the equation, the driving time $\tau^{t''}_{(i,j)}$ is defined as the number of time slots with identical duration.

For the problem formulation, the probability variables λ_i and μ_j are introduced to simulate the events of charging demand occurrence and EVs leaving charging stations during the time horizon. However, in real-world situations, the charging service provider could receive the charging demand information and know the number of vehicles that are leaving charging stations at the beginning of each time slot. Therefore, the probability variables λ_i and μ_j do not appear in the dynamic recursive equation. Without loss of generality, the problem assumes that the routes with minimum energy consumption are selected as the travel routes between departure points and charging station nodes. Another assumption is that EV drivers can reach their destinations by charging their vehicles only once, because EVs with a single charge often have sufficient energy to reach their destinations in an urban road network [46].

4. Route Guidance Strategies for Stochastic Charging Demands in a Time-Varying Road Network

Charging station selection decisions at each time slot should be made on the basis of specific strategies to solve the stochastic charging demands in a time-varying road network. In this section, we attempt to develop the heuristic-based route guidance strategies for charging station selection from two different perspectives. Firstly, the strategy considers the effects of stochastic charging demands on charging stations. Secondly, the other strategy aims to select charging stations based on the travel cost of EV drivers. The effectiveness and comparison of the two strategies are discussed in Section 5. Note that charging vehicles by using a renewable energy source, instead of coal power, would improve the energy efficiency of EVs. In this way, the performance of strategies may be influenced if the source of energy to charge EVs is considered, which is not discussed in this study.

4.1. Assumptions

To facilitate problem formulation, several assumptions are made as mentioned in Section 3. In this section, we summarize the assumptions to develop the route guidance strategies as follows:

Assumption 1. *The charging demand information considered in the route guidance strategies is assumed to include drivers' travel destinations and the remaining energy of vehicles.*

Assumption 2. *It is assumed that travel time and energy consumption for each link are constant during a separate time slot. Such an assumption conforms to the traffic condition characteristic if the duration for each time slot is relatively short.*

Assumption 3. *The charging demands are assumed to occur in the normal nodes and the charging station nodes cannot generate charging demands. Such an assumption is reasonable because drivers seek help from the charging service provider only when they have difficulty finding nearby charging stations.*

Assumption 4. *Each normal node is assumed to generate at most one charging demand following a uniform distribution during a time slot. Meanwhile, it is assumed that at most one EV can leave a charging station after completing charging at each time slot. The assumptions conform to the actual situations if the duration for each time slot is relatively short.*

Assumption 5. *The charging service provider is assumed to know the information about the travel time and energy consumption on all the links at the beginning of each time slot.*

Assumption 6. *Without loss of generality, it is assumed that the routes with minimum energy consumption are selected as the travel routes between departure points and charging station nodes.*

Assumption 7. *It is assumed that EV drivers can reach their destinations by charging their vehicles only once, because EVs with a single charge often have sufficient energy to reach their destinations in an urban road network. For the routes from origins to charging stations, EV drivers prefer to focus on reachability rather than distance.*

Assumption 8. *Every charging demand is assumed to have at least one reachable charging station in a road network to guarantee a solution. For the special situation in which no reachable charging station exists, extra cost* *may be incurred in transporting the vehicles, such as through trailer services, which is not discussed in the route guidance strategies.*

4.2. Route Guidance Strategy Based on Vehicle Balance in Charging Stations

In real-world traveling situations, several charging demands may be simultaneously made in road networks. In every time slot, the heuristic suggested charging stations need to be selected for all charging demands. Reachability is the most critical factor for charging station selection in an EV trip. Furthermore, compared with the increasing number of EVs, charging infrastructure is often insufficient. Given the limited charging technology at present and for the foreseeable future [47,48], increasing charging demands and insufficient charging infrastructure may lead to queuing in charging stations. The increasing number of EVs in charging stations affects the charging stations operation. Mass EV charging may increase the operating burden of charging stations, and the queuing time in charging stations may increase as the vehicle number increases. The number of vehicles at each charging station should be considered when addressing stochastic charging demands to ensure the operation efficiency of charging stations. However, in the actual situation, the chance of being selected for charging stations differs if the number of EVs in charging stations is overlooked. For example, the charging stations located in central areas may accept more EVs with charging demands than other stations. Neglecting the number of EVs in charging stations would increase the number of EVs in the charging stations located in central areas.

In this study, we define the charging service system as stable if all charging stations have a sustainable number of EVs in every time slot. In real-world situations, the sustainable number of EVs that could be queued in a charging station is limited due to resource constraints. Thus, balancing the vehicle number at different charging stations effectively reduces the negative influence of mass EV charging on charging stations, which refers to keeping the number of EVs at different charging stations at a relatively similar level. For this reason, the route guidance strategy based on vehicle balance in charging stations is established. An effective method to realize this goal is to direct an EV to the reachable charging station with a minimum number of vehicles during the time slot. In this way, the charging stations with fewer EVs have more chances to accept charging demands. It is also worth noting that such a strategy may increase the travel cost of individual drivers in some cases, especially for the drivers who can select a closer charging station with extra capacity for charging. However, this strategy is focused on the long-term transportation scenario, and it is able to ensure the stability of charging service system in the situation with a long time horizon. In addition, even though the suggested charging station is not the nearest one, its reachability could be ensured based on Assumption 7. The performance of the strategy is discussed in Section 5. To simplify the description, the route guidance strategy based on the vehicle balance in charging stations is represented by the charging station balance (CSB) strategy.

As mentioned in Section 3, the charging service provider would receive charging demands in every time slot. The energy and time consumed to traverse each link, that is, E_a^t and τ_a^t, are known at the beginning of each time slot. The output of the CSB strategy includes heuristic suggested charging stations, corresponding routes and driving time for the charging demands at every time slot. The operating steps of the CSB strategy are detailed as follows:

1. At the beginning of time slot t, on the basis of the information of E_a^t, the minimum energy consumption between all the nodes with C_i^t and charging station nodes is calculated by using shortest path algorithms [49]. The minimum energy consumption between charging demand node i and charging station node j is denoted as $E_{(i,j)}^t$, which is then recorded along with corresponding routes.
2. For each C_i^t at time slot t, $E_{(i,j)}^t$ is compared with e_i^t. If $e_i^t \geq E_{(i,j)}^t$, then charging station j is denoted as reachable charging station j'. Otherwise, charging station j is regarded as unreachable and then deleted from the candidate charging stations.

3. For all the reachable charging stations j' of C_i^t, the EV number in charging station j' at time slot t is checked. The results are denoted as $U_{j'}^t$.
4. For each C_i^t, the EV number $U_{j'}^t$ in all reachable charging stations j' at time slot t is compared. The node with the heuristic suggested charging station is denoted by $j*$, which needs to satisfy the following condition:

$$j^* = \arg \min_{j'} \{U_{j'}^t\}. \quad (5)$$

The values of decision variable x_{ij}^t can be determined as follows:

$$x_{ij}^t = \begin{cases} 1, & j = j^* \\ 0, & j \neq j^* \end{cases}. \quad (6)$$

If multiple charging stations with the same and minimum EV number exist, then one is randomly selected as the heuristic suggested charging station for C_i^t.

5. The driving time on the minimum energy routes between nodes with C_i^t and corresponding heuristic suggested charging station nodes j^* is calculated and recorded. The results are denoted as $\tau_{(i,j*)}^t$.
6. Before the end of time slot t, the heuristic suggested charging stations j^*, driving time $\tau_{(i,j*)}^t$, and corresponding routes for charging demand C_i^t are output.

Figure 3 illustrates the flowchart of the CSB strategy.

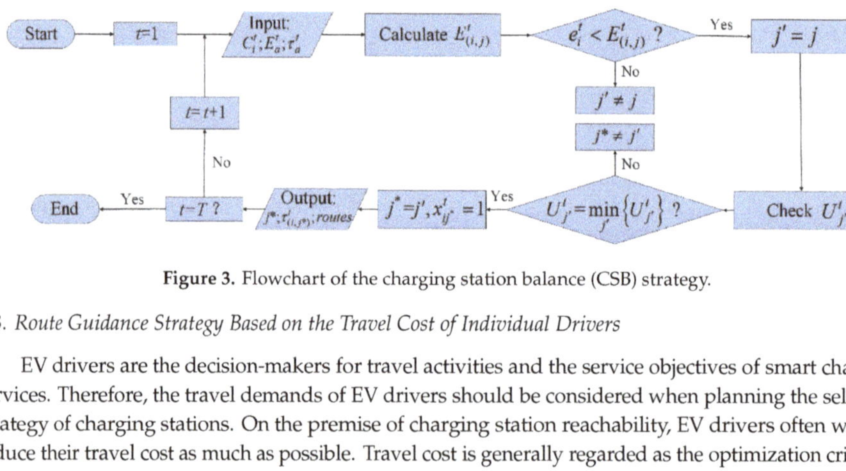

Figure 3. Flowchart of the charging station balance (CSB) strategy.

4.3. Route Guidance Strategy Based on the Travel Cost of Individual Drivers

EV drivers are the decision-makers for travel activities and the service objectives of smart charging services. Therefore, the travel demands of EV drivers should be considered when planning the selection strategy of charging stations. On the premise of charging station reachability, EV drivers often want to reduce their travel cost as much as possible. Travel cost is generally regarded as the optimization criterion for choosing travel routes [50,51]. Travel cost minimization is one of the critical factors for travel demands. Travel cost has multidimensional components during an EV trip, such as travel time, energy consumption, and charging cost [20]. The driving time and energy consumption are closely correlated with the driving distance, with the former being typically proportional to the driving distance with a constant driving speed and the latter having a significant linear relationship with driving distance [52], as well as a significant influence on charging cost; hence, the charging cost would be affected by the driving distance. The driving distance can be used to reflect the integration of travel cost components. Thus, driving distance is minimized to establish the route guidance strategy based on drivers' travel demands.

Unlike driving time and energy consumption, driving distance is a static factor in a time-varying road network. Adopting driving distance as a selection criterion can utilize such an advantage and avoid complicated prediction. The driving distance from charging stations to destinations is considered in the route guidance strategy. For the routes from origins to charging stations, we assume that EV drivers prefer to focus on reachability rather than distance as mentioned in Assumption 7. As a matter of fact, the driving distance from origins to charging stations is unable to fully reflect the travel direction consistency between charging stations and destinations, but it could be reflected by the driving distance from charging stations to destinations to a certain extent [53]. Furthermore, such an assumption could reduce computing burden and ensure the feasibility of solution. Thus, the driving distance between origins and charging stations is not involved in the strategy. To simplify the description, the route guidance strategy based on the travel cost of individual drivers is represented by the shortest driving distance (SDD) strategy.

As mentioned above, the SDD strategy aims to identify all reachable charging stations, and it directs EVs to the ones with nearest to their destinations. Similar to the CSB strategy, the output of the SDD strategy also includes heuristic suggested charging stations, corresponding routes, and driving time for all charging demands at every time slot. The operating steps of the SDD strategy are detailed as follows:

1. At the beginning of time slot t, the minimum energy consumption between all the nodes with C_i^t and charging station nodes is calculated on the basis of the information of E_a^t. The minimum energy consumption between charging demand node i and charging station node j is denoted as $E_{(i,j)}^t$, which is recorded along with corresponding routes.
2. For each C_i^t at time slot t, $E_{(i,j)}^t$ is compared with e_i^t. If $e_i^t \geq E_{(i,j)}^t$, then charging station j is denoted as reachable charging station j'. Otherwise, charging station j is regarded as the unreachable one and is then deleted from the candidate charging stations.
3. For the reachable charging station j' of C_i^t, the driving distance between d_i^t and the node with charging station j' is calculated. The results are denoted as $l_{(j',d_i^t)}$.
4. For each C_i^t, the driving distance between d_i^t and reachable charging station j' is compared. $j*$ denotes the node with the heuristic suggested charging station. The minimum driving distance between d_i^t and heuristic suggested charging station j' needs to satisfy the following condition:

$$l_{(j*,d_i^t)} = \min_{j'}\{l_{(j',d_i^t)}\}. \tag{7}$$

   The values of decision variable can be determined on the basis of Equation (6). If multiple charging stations with the same and minimum driving distance exist between them and d_i^t, then one is randomly selected as the heuristic suggested charging station for C_i^t.
5. The driving time on the minimum energy routes between nodes with C_i^t and corresponding heuristic suggested charging station nodes j^* is calculated and recorded. The results are denoted as $\tau_{(i,j*)}^t$.
6. Before the end of time slot t, the heuristic suggested charging stations j^*, driving time $\tau_{(i,j*)}^t$, and corresponding routes for the charging demand C_i^t are output.

The flowchart of the SDD strategy is given in Figure 4.

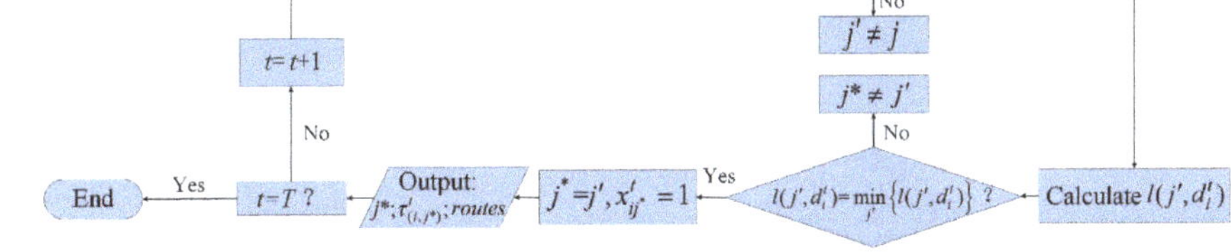

Figure 4. Flowchart of the shortest driving distance (SDD) strategy.

5. Simulation Analysis

5.1. Scenario Description

The simulation examples for a time-varying road network are designed to explore the performance of the SDD and CSB strategies. The structure of the road network is designed based on the Sioux Falls network, which is often adopted to simulate travel optimization problems [54–56]. The network consists of 24 nodes and 76 links, as shown in Figure 5. The road network comprises eight nodes with charging stations, which are marked as CS 1 to CS 8. The other nodes, numbered 1–16, are the normal ones without charging stations, which may generate charging demands in every time slot.

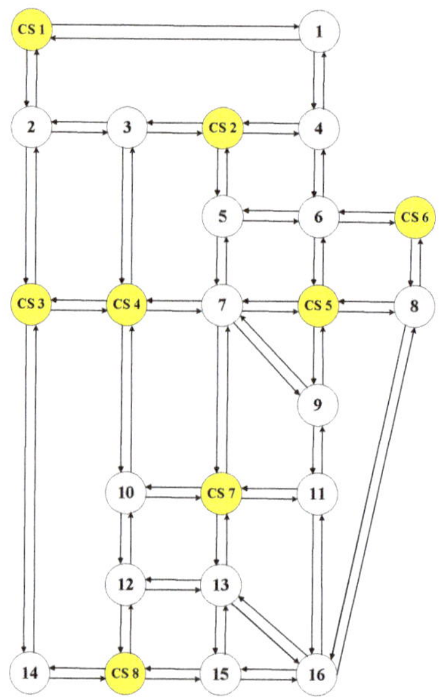

Figure 5. Sioux Falls road network for the simulation example.

Each charging station node in the road network has parameter μ_j to reflect the charging levels of the charging station, where j = {1,2, ... ,7,8}. For the normal nodes, each one has parameter λ_i to reflect the stochastic characteristic of charging demand generation, where I = {1,2, ... ,15,16}. Tables A1 and A2 (Appendix A) list the values of μ_j and λ_i, respectively. The simulation example assumes that the initial number of EVs at each charging station is equal to zero, that is, $\varphi_j = 0$, for each charging station j in the road network.

The information for each charging demand C_i^t includes the travel destination d_i^t and remaining energy e_i^t, which are randomly generated in the simulation example. In every time slot, the travel destination d_i^t is randomly selected from other normal nodes in the road network if a charging demand occurs in a specific normal node. Moreover, the remaining energy e_i^t varies within a given interval, and we suppose that its value ranges from 7.2 kWh to 16.8 kWh by referring to the battery capacity of EVs. In general, EVs with a 24-kWh battery are widely used in unban transportation systems [33]. The energy consumption on each link in a time-varying road network varies as time slot passes. The simulation example randomly determines parameter E_a^t from given intervals for link a at time slot t to reflect such a characteristic. The value intervals of the energy consumption E_a^t on each link a are listed in Table A3 (Appendix A).

Similar to the energy consumption E_a^t, the driving time on each link a also has a time-varying characteristic. Similar to parameter E_a^t, the values of parameter τ_a^t in every time slot are randomly determined based on given intervals for each link a. The value intervals of driving time τ_a^t on each link a are listed in Table A4 (Appendix A). The number of time slots represents the driving time on each link given that the time is slotted into the time slots with identical duration. Without loss of generality, the duration for each time slot is not constrained in the simulation example. In the real-world situation, the duration for time slots could be valued according to actual requirements.

Table A5 (Appendix A) lists the length of each link a, which is denoted by l_a (km), in the road network. Considering the structure characteristic of road networks, the links with a symmetric relation have the same length.

5.2. Simulation Results and Analysis

On the basis of the example scenario, the SDD and CSB strategies are applied in the route guidance problem with stochastic charging demands in a time-varying road network. The total number of time slots is set as $T = 10^2$, $T = 10^3$, $T = 10^4$, $T = 10^5$, and $T = 10^6$ to analyze the performance during different time horizons. The SDD and CSB strategies can ensure the reachability of selected charging stations for the charging demands in every time slot, as mentioned in Section 4. The charging demands of EV drivers can be satisfied by both strategies. Therefore, the simulation example focuses on the effects of the proposed strategies on the operation efficiency of charging stations. The number of EVs in a charging station is a critical factor that reflects the operation state of the charging station. Figure 6 presents the average number of EVs at each charging station during different time horizons T under the proposed strategies, which is computed by averaging over all time slots over the entire number of EVs.

In Figure 6, cases (a)–(h) respectively show the average number of EVs in CS 1–CS 8 during different time horizons. The change trends of the average EV number during different time horizons reflect the stability of charging stations under specific scenarios. Stability is an important criterion for guaranteeing the operation efficiency of charging stations. If the average number of EVs in a charging station has a flat change trend as the time horizon increases, then the charging station would operate stably for the given scenarios [57]; otherwise, the average number of EVs in the charging station would increase rapidly as the time horizon increases. The figure depicts that the SDD and CSB strategies can stabilize the operation states for CS 1–CS 8 under the example scenario, because the number of EVs in all charging stations has flat change trends as time horizon T varies. Although fluctuation trends exist when the time horizon ranges from $T = 10^2$ to $T = 10^4$ for several charging stations under specific strategies, such as CS 2 under the SDD strategy, CS 3 under the CSB strategy, and CS 5 under both strategies, all charging stations could reach a stable state after the time horizon $T = 10^4$. A comparison of the average EV number in CS 1–CS 8 with a stable state shows a difference between the SDD and CSB strategies. For the SDD strategy, the average number of EVs in CS 5 is greater than that in other charging stations because the vehicle balance of charging stations is not considered. The average EV number under the CSB strategy has a similar trend in all charging stations.

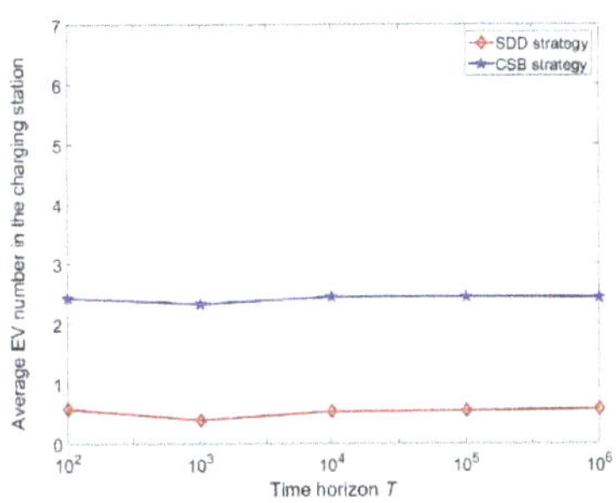

(**a**) CS 1

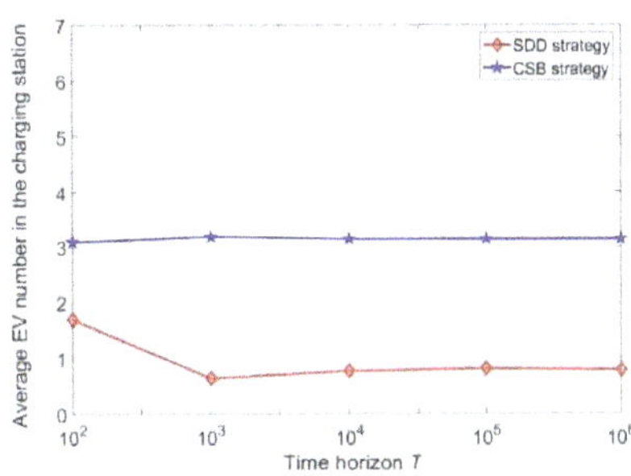

(**b**) CS 2

Figure 6. *Cont.*

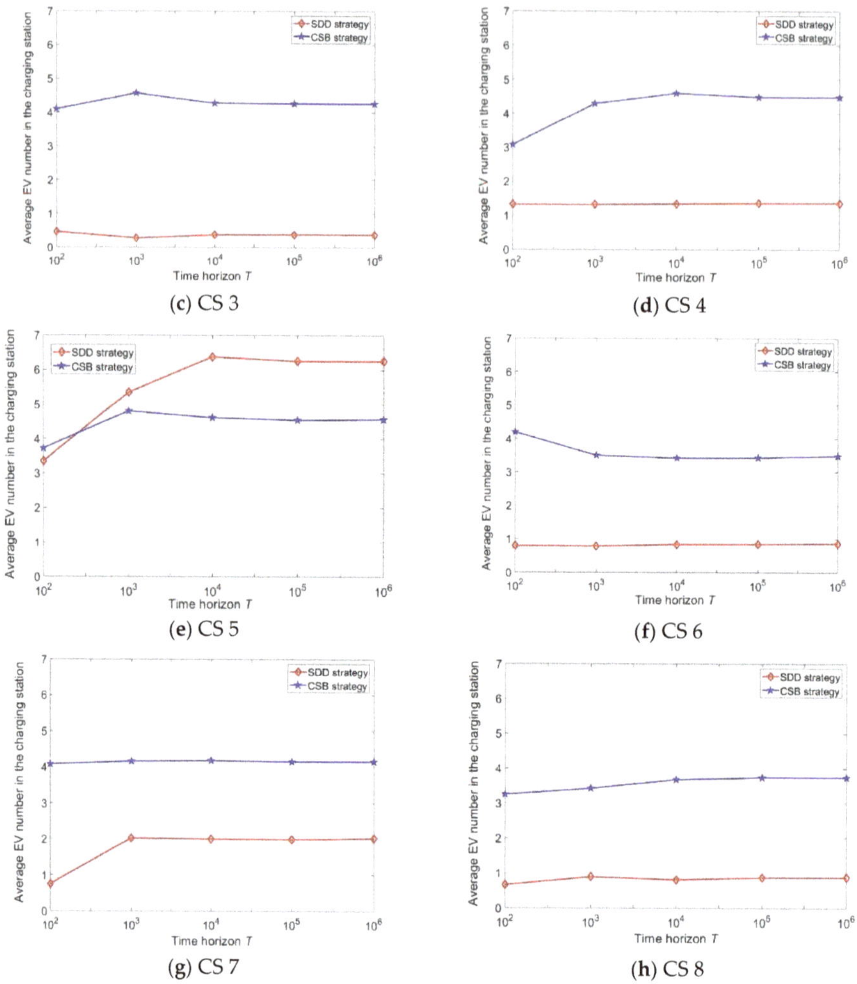

Figure 6. Average electric vehicle (EV) number at each charging station during different time horizons *T*.

Although the average EV number is a critical reflection of the stability of each charging station, it cannot perfectly represent the actual number of EVs in every time slot. The EV number in charging stations at different time slots may vary during the time horizon. As time slots pass, there exists the obvious difference between the maximum and minimum numbers of EVs in a charging station. If the EV number in a selected charging station is relatively large, then the drivers would be reluctant to charge their vehicles by using it at the corresponding time slot. This condition would negatively influence the implementation efficiency of the route guidance service. Therefore, during the time horizon *T*, the maximum number of EVs at each charging station is often regarded as the bottleneck in the application of route guidance strategies under real-world situations. Figure 7 presents the maximum number of EVs at each charging station during different time horizons *T* based on the simulation example to compare the performances of the SDD and CSB strategies.

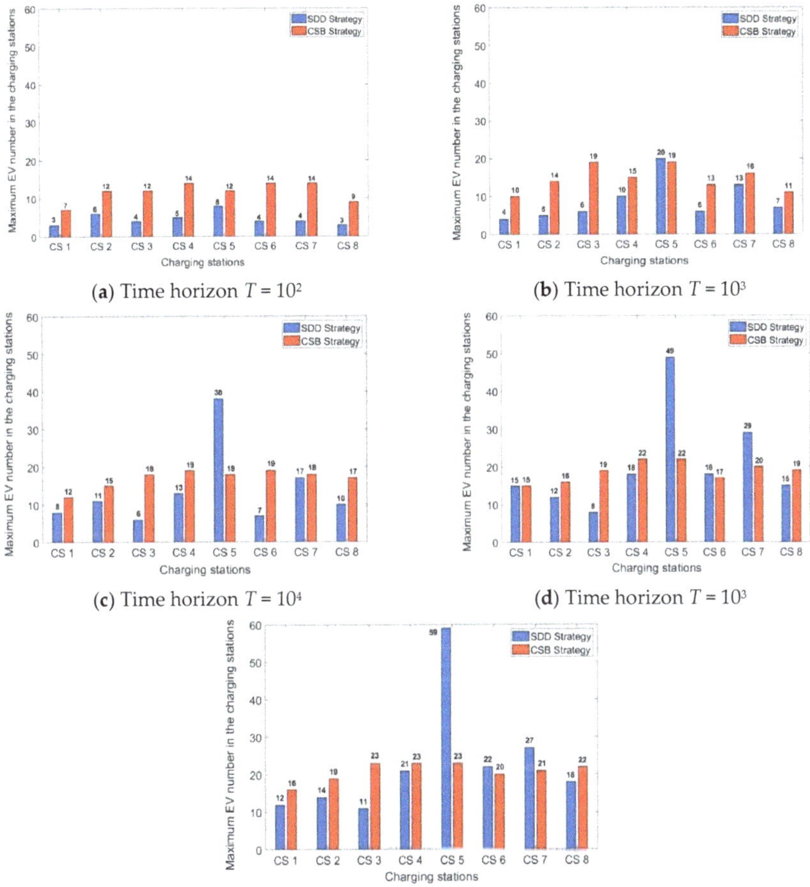

(**a**) Time horizon $T = 10^2$
(**b**) Time horizon $T = 10^3$
(**c**) Time horizon $T = 10^4$
(**d**) Time horizon $T = 10^3$
(**e**) Time horizon $T = 10^6$

Figure 7. Maximum EV number at each charging station during different time horizons *T*.

The maximum number of EVs at each charging station under the SDD and CSB strategies is depicted in Figure 7, where cases (a)–(e) illustrate the results during different time horizons, ranging from $T = 10^2$ to $T = 10^6$, respectively. In case (a), the maximum number of EVs in CS 1–CS 8 under the SDD strategy is less than that under the CSB strategy. However, when the time horizon $T = 10^3$, as shown in case (b), the EV number in CS 5 under the SDD strategy is larger than that under the CSB strategy. In case (c), the SDD strategy enlarges the maximum number of EVs in most charging stations, especially in CS 5, compared with case (b). The extreme gap of maximum EV number among the charging stations is equal to 32. By contrast, the maximum number of EVs under the CSB strategy has a moderate degree of change for all charging stations. The maximum number of EVs in CS 3 and CS 5 in particular has a decreasing trend unlike case (b). The extreme gap of maximum EV number among the charging stations is equal to 7. When the time horizon $T = 10^4$, as shown in case (d), the maximum number of EVs under the SDD strategy increases for all charging stations, and the maximum EV number in CS 5 and CS 7 is larger than that under the CSB strategy. The extreme gap in the maximum EV number among the charging stations under the SDD and CSB strategies is equal to 41 and 7, respectively, thereby indicating a visible difference in vehicle balance among different charging stations between the two strategies. In case (e), the maximum EV number in CS 5, CS 6, and CS 7 under the SDD strategy is larger than that under the CSB strategy. The extreme gap of maximum EV number among the charging stations reaches 48 under the SDD strategy. By contrast, the maximum number of EVs presents a balanced state for different charging stations under the CSB strategy. The extreme gap of maximum EV number among the charging stations is equal to 7.

A comparison of the performances of the SDD and CSB strategies based on the simulation example shows that the CSB strategy has an advantage in vehicle balance among different charging stations, especially in the situation with a long time horizon. Thus, using the CSB strategy would avoid the negative influence of the large number of EVs in a charging station. Unlike the CSB strategy, the SDD strategy would enlarge the gap of the number of EVs at different charging stations as the time horizon increases, thereby affecting the operation efficiency of the charging stations that have relatively more vehicles. However, in the situation with a short time horizon, the SDD strategy, which considers the travel cost of EV drivers, could address stochastic charging demands because of the unobvious difference in the performance of the two strategies in such a situation. In summary, the CSB strategy is suitable to be applied in long-term transportation scenario due to its ability to stabilize the charging service system. By contrast, the SDD strategy fits the short-term transportation scenario to reduce the travel cost of individual drivers.

5.3. Parameter Analysis for Scenario Characteristics

When discussing the route guidance problem for stochastic charging demands, in addition to time horizon, the scenario characteristics have significant effects on the performance of the proposed strategies. For problem formulation, parameter λ_i and μ_j are used to present the stochastic characteristics of charging demands and processes, respectively, as mentioned in Section 3. Such parameters can also reflect the scenario characteristics in terms of the EV scale and charging level. For instance, a large parameter λ_i represents a large EV scale in node i. A large parameter μ_j illustrates a high charging level of the charging station in node j. Parameters λ_i and μ_j are set as different values to explore the performance of route guidance strategies under different scenario parameters. The values of parameter λ_i for all normal nodes i are set as identical value λ to highlight the effects of parameter values on the simulation results. The values of parameter μ_j for all charging station nodes j are also set as identical value μ. The time horizon is set as $T = 10^6$ for all parameter scenarios. Figure 8 presents the maximum number of EVs at each charging station under the SDD strategy as parameters μ and λ vary. The value of λ is set as 0.1, 0.2, 0.3, 0.4, and 0.5. The value of μ is set as 0.6, 0.7, 0.8, 0.9, and 1.0. A parameter scenario consists of a pair of parameters λ and μ. Thus, a total of 25 parameter scenarios are considered.

In Figure 8, cases (a)–(h) respectively illustrate the maximum EV number in CS 1–CS 8 under the SDD strategy for the different parameter scenarios. In several parameter scenarios, the maximum EV number in a specific charging station may exceed its sustainable limit, which indicates that the charging station is unstable. For such a scenario, we let the maximum EV number be equal to zero in the figure. The threshold of sustainable number of EVs for all charging stations is set as 120. In case (a), as parameter μ increases, the maximum EV number in CS 1 presents a decreasing trend. This phenomenon indicates that the maximum EV number reduces as the charging level of the charging station increases. By contrast, as parameter λ increases, the maximum number of EVs in CS 1 has an increasing trend, which indicates that the maximum EV number increases as the EV scale increases in the road network. Among all parameter scenarios, the peak and lowest values of the maximum EV number are equal to 2 and 29, respectively. In case (b), as the scenario parameters change, the change trend of the maximum EV number in CS 2 is similar to that in CS 1. However, unlike case (a), unstable parameter scenarios exist in case (b). Among all stable parameter scenarios, the peak and lowest values of the maximum EV number equal 3 and 82, respectively. In cases (c)–(h), the change trend of the maximum number of EVs in CS 3–CS 8 is also similar to that in CS 1. Similar to CS 2, the unstable state would exist in CS 3–CS 8 under specific parameter scenarios. Among all stable parameter scenarios, the lowest values of the maximum EV number in CS 3–CS 8 are all equal to 4. Comparatively, the peak values of the maximum EV number in CS 3–CS8 are respectively equal to 103, 41, 95, 58, 69, and 67. For a transportation system, the charging service is unstable until all charging stations can reach stability. Therefore, if at least one unstable charging station exists in the road network under a parameter scenario, then the SDD strategy cannot be applied in the parameter scenario. The parameter scenarios that cannot support the SDD strategy can be determined based on such a criterion. The parameter pairs of the unstable scenarios include $(\lambda = 0.3, \mu = 0.6)$, $(\lambda = 0.3, \mu = 0.7)$, $(\lambda = 0.4, \mu = 0.6)$, $(\lambda = 0.4, \mu = 0.7)$, $(\lambda = 0.4, \mu = 0.8)$, $(\lambda = 0.4, \mu = 0.9)$, $(\lambda = 0.5, \mu = 0.6)$,$(\lambda = 0.5, \mu = 0.7)$, $(\lambda = 0.5, \mu = 0.8)$, $(\lambda = 0.5, \mu = 0.9)$, and $(\lambda = 0.5, \mu = 1.0)$. For all the stable parameter scenarios in each case, a significant change trend can be observed as the parameters vary, which indicates that the SDD strategy is sensitive to the change in scenarios.

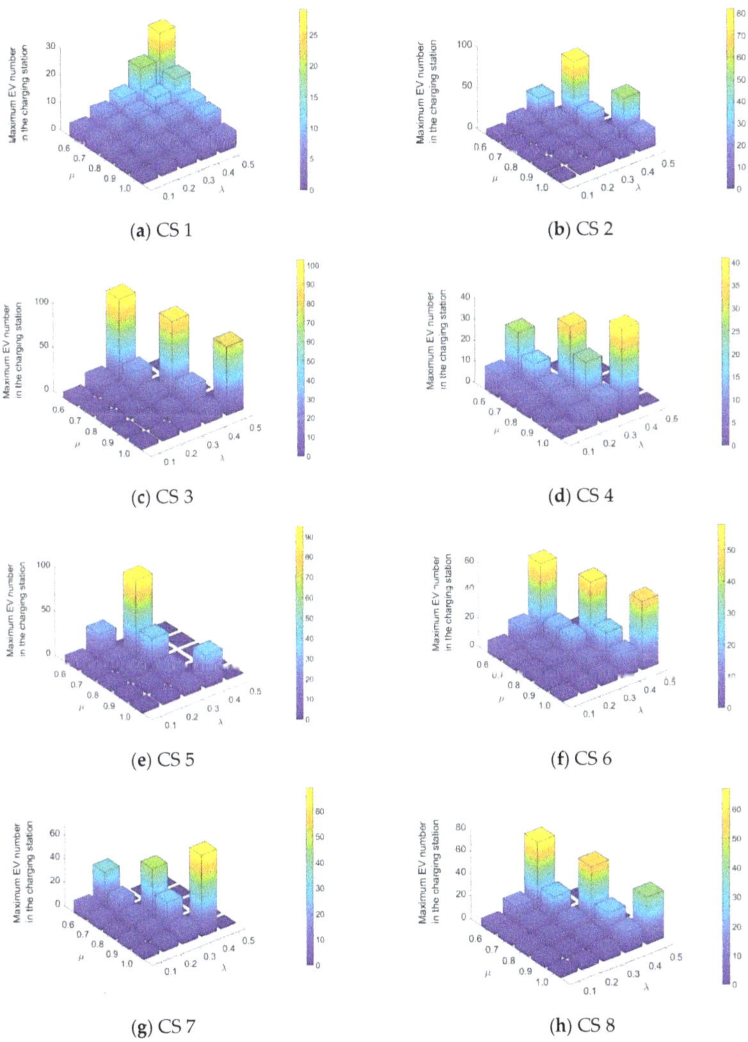

Figure 8. Change trends of the maximum EV number at each charging station under the SDD strategy, where λ is the possibility of charging demand occurrence in normal nodes at each time slot, and μ is the possibility of an EV leaving the charging station after completing charging at each time slot.

On the basis of the parameter scenarios and the time horizon mentioned above, the CSB strategy is further applied in the route guidance problem with stochastic charging demands. As parameters μ and λ vary, the maximum EV number at each charging station is obtained, as shown in Figure 9.

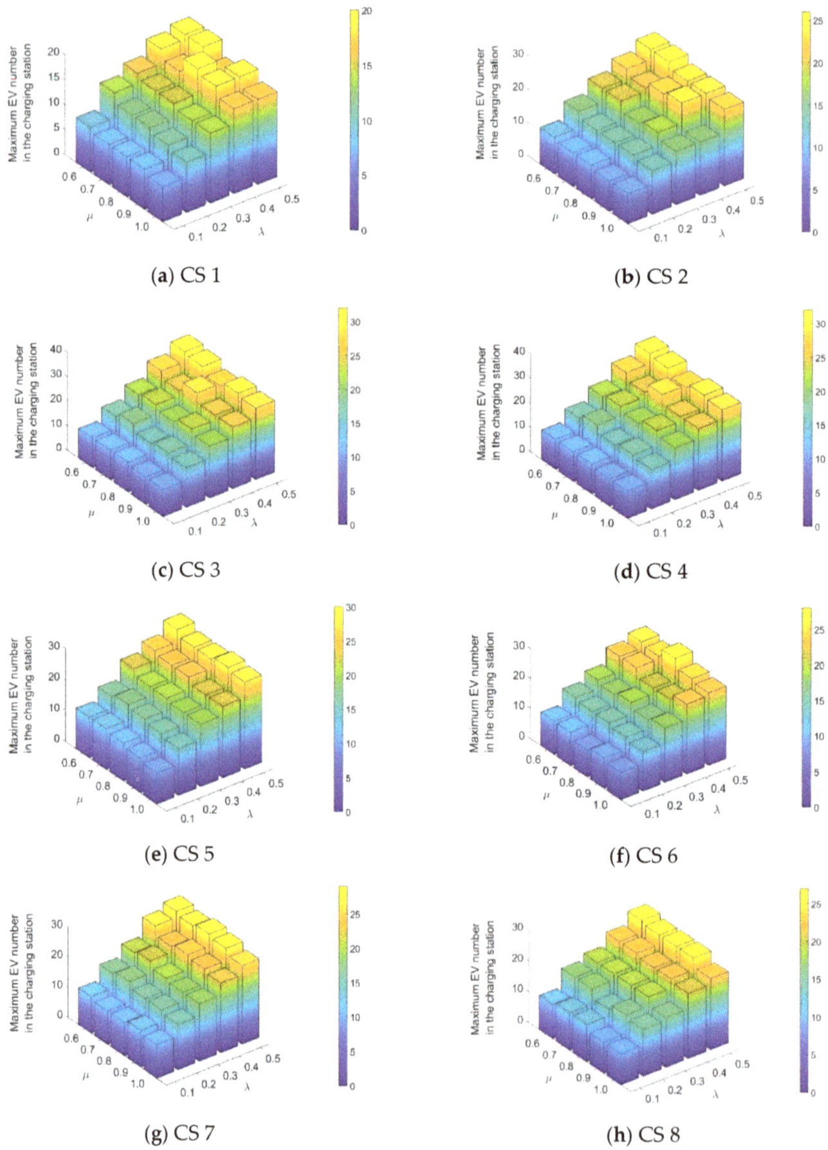

Figure 9. Change trends of the maximum EV number in each charging station under the CSB strategy, where λ is the possibility of charging demand occurrence in normal nodes at each time slot, and μ is the possibility of an EV leaving the charging station after completing charging at each time slot.

In Figure 9, cases (a)–(h) respectively present the maximum number of EVs in CS 1–CS 8 under the CSB strategies for the different parameter scenarios. In case (a), the maximum number of EVs in CS 1 broadly presents a flat increasing trend as parameter μ decreases and parameter λ increases.

For several individual scenarios, a moderate fluctuation occurs as the parameters vary. The results indicate that a change in scenarios has relatively limited effects on the CSB strategy compared with that on the SDD strategy. Among all parameter scenarios, the peak and lowest values of the maximum EV number in CS 1 are equal to 7 and 20, respectively. In cases (b)–(h), the maximum number of EVs in CS 2–CS 8 shows a similar change trend to that in CS 1. Among all parameter scenarios, the lowest values of the maximum EV number in CS 2–CS 8 are respectively equal to 9, 10, 11, 10, 9, 11, and 10. Comparatively, the peak values of the maximum EV number in CS 2–CS 8 are equal to 26, 30, 32, 30, 28, 29, and 27, respectively. Unlike the SDD strategy, the CSB strategy can stabilize the state of CS 1–CS 8 for all parameter scenarios, thus indicating its advantage in charging station stability.

A comparison of the simulation results in Figures 8 and 9 indicates that, for the SDD and CSB strategies, the maximum EV number increases as parameter μ decreases and parameter λ increases, with different change trends. Given the implication of parameters μ and λ, the simulation results conform to the operation state of charging stations in real-world situations.

6. Conclusions

In this study, a route guidance problem was formulated by combining the time-varying road network and stochastic charging demands. To address the problem, we proposed two heuristic-based route guidance strategies from two different perspectives, namely, the SDD and CSB strategies. The SDD strategy uses the driving distance as the optimization criterion based on the travel cost of individual drivers. By contrast, the CSB strategy selects the charging stations with the minimum EV number based on the vehicle balance in charging stations. Despite their differences, both strategies can guarantee the reachability of selected charging stations in a time-varying road network. Simulation experiments were implemented to explore the performances of the SDD and CSB strategies. The results indicate that the CSB strategy has an advantage over the SDD strategy in avoiding the negative influence of mass EV charging on the charging station operation, especially in the situation with a relatively long time horizon.

Furthermore, the parameter analysis was carried out through changing the scenario parameters in the simulation experiments. The results present the different performances of SDD and CSB strategies as parameter scenarios vary. For the SDD strategy, the maximum EV number at different charging stations had a significant gap and ranged from approximately 30 to 100. By contrast, for the CSB strategy, the maximum EV number at different charging station exhibited a moderate degree of change and mainly ranged from 25 to 30. Moreover, the SDD strategy may be unable to stabilize the charging service system in some heavily loaded scenarios (i.e., low μ and high λ), but the CSB strategy can stabilize the charging service system for all parameter scenarios. However, if the charging service is in lightly loaded scenarios (i.e., high μ and low λ), the SDD strategy could be employed due to its similar effects on charging stations with the CSB strategy and consideration of the travel cost of individual drivers.

Each normal node is assumed to generate at most one charging demand in every time slot to simplify the problem formulation. Such an assumption is reasonable if each time slot has a relatively short duration. However, as the scale of EVs increases in the urban transportation system, multiple charging demands may occur simultaneously in the same location on road networks, thereby significantly complicating the solving processes for the stochastic charging demands. On the basis of the proposed strategies, a distribution rule regarding the number of charging demand occurrence will be further explored in future work and considered to improve the route guidance strategies. Furthermore, other evaluation metrics, such as waiting time, trip length extension, and energy efficiency, will be introduced in a future study.

Author Contributions: Conceptualization, Y.W. and J.B.; methodology, Y.W.; validation, Y.W. and C.L.; writing, Y.W.; visualization, Y.W., C.L., and C.D.; supervision, J.B., C.L., and C.D.; funding acquisition, J.B. All authors have read and agreed to the published version of the manuscript.

Funding: This work was funded by the National Natural Science Foundation of China (Nos. 71961137008 and 71621001).

Acknowledgments: The authors gratefully acknowledge fruitful discussions with Prof. Bin Li in the University of Rhode Island, as well as financial support from the China Scholarship Council.

Conflicts of Interest: The authors declare no conflict of interest.

Appendix A

Table A1. Value of parameter μ_j for each charging station node.

Charging Station	μ_j	Charging Station	μ_j	Charging Station	μ_j
CS 1	0.74	CS 4	0.90	CS 7	0.94
CS 2	0.84	CS 5	0.78	CS 8	0.90
CS 3	0.94	CS 6	0.87		

Table A2. Value of parameter λ_i for each normal node.

Normal Node	λ_i	Normal Node	λ_i	Normal Node	λ_i	Normal Node	λ_i
1	0.31	5	0.20	9	0.18	13	0.57
2	0.62	6	0.13	10	0.27	14	0.35
3	0.32	7	0.50	11	0.15	15	0.52
4	0.69	8	0.25	12	0.26	16	0.67

Table A3. Value intervals of energy consumption E_a^t for link a at time slot t.

Link a	E_a^t (kWh)	Link a	E_a^t (kWh)	Link a	E_a^t (kWh)
1–CS 1	(2.64, 5.76)	9–7	(2.16, 4.8)	CS 1–2	(1.68, 4.08)
1–4	(3.6, 5.04)	9–11	(2.16, 5.04)	CS 2–3	(3.6, 6.96)
2–CS 1	(2.4, 4.8)	10–CS 4	(2.64, 5.52)	CS 2–4	(2.88, 6.48)
2–3	(1.92, 4.56)	10–12	(1.44, 4.08)	CS 2–5	(3.6, 4.8)
3–2	(3.12, 4.32)	10–CS 7	(2.64, 4.56)	CS 3–2	(2.88, 6)
3–CS 4	(1.44, 3.84)	11–9	(2.16, 3.6)	CS 3–CS 4	(2.88, 5.28)
3–CS 2	(1.2, 4.32)	11–CS 7	(1.2, 4.56)	CS 3–14	(2.16, 4.8)
4–1	(2.16, 3.6)	11–16	(1.44, 3.36)	CS 4–3	(1.2, 4.8)
4–CS 2	(2.4, 5.28)	12–10	(2.16, 4.8)	CS 4–7	(2.64, 6)
4–6	(3.12, 6.48)	12–13	(2.4, 4.8)	CS 4–10	(3.12, 6.72)
5–CS 2	(2.88, 5.28)	12–CS 8	(1.92, 4.8)	CS 4–CS 3	(2.88, 4.8)
5–6	(2.4, 3.6)	13–CS 7	(3.36, 5.04)	CS 5–6	(1.2, 3.12)
5–7	(1.68, 3.6)	13–12	(2.64, 4.56)	CS 5–7	(1.44, 3.36)
6–4	(1.44.5.04)	13–15	(1.68, 4.56)	CS 5–8	(3.12, 5.04)
6–5	(2.16, 3.6)	13–16	(1.44, 5.04)	CS 5–9	(2.4, 5.04)
6–CS 5	(2.16, 5.28)	14–CS 3	(3.12, 5.04)	CS 6–6	(2.16, 4.08)
6–CS 6	(3.6, 5.52)	14–CS 8	(2.88, 6.24)	CS 6–8	(1.92, 5.52)
7–5	(3.6, 5.52)	15–13	(1.92, 5.28)	CS 7–7	(1.68, 4.56)
7–CS 4	(1.92, 4.32)	15–CS 8	(1.2, 4.56)	CS 7–10	(3.6, 6.72)
7–CS 7	(3.36, 5.52)	15–16	(2.16, 5.28)	CS 7–11	(2.16, 3.84)
7–9	(2.64, 6.24)	16–8	(2.4, 4.08)	CS 7–13	(1.44, 2.88)
7–CS 5	(2.4, 4.32)	16–11	(2.4, 6)	CS 8–12	(2.4, 3.6)
8–CS 6	(3.36, 4.8)	16–13	(3.36, 6.72)	CS 8–14	(1.2, 3.84)
8–CS 5	(3.12, 4.8)	16–15	(3.6, 6.96)	CS 8–15	(2.64, 5.28)
8–16	(2.4, 3.84)	CS 1–1	(2.16, 5.28)	2–CS 3	(2.16, 4.32)
9–CS 5	(1.2, 4.32)				

Table A4. Value intervals of driving time τ_a^t for link a at time slot t.

Link a	τ_a^t	Link a	τ_a^t	Link a	τ_a^t
1–CS 1	(2, 5)	9–7	(1, 2)	CS 1–2	(1, 3)
1–4	(1, 4)	9–11	(1, 2)	CS 2–3	(1, 2)
2–CS 1	(2, 3)	10–CS 4	(1, 2)	CS 2–4	(2, 4)
2–3	(1, 2)	10–12	(1, 2)	CS 2–5	(1, 3)
3–2	(2, 3)	10–CS 7	(2, 4)	CS 3–2	(1, 2)
3–CS 4	(2, 3)	11–9	(1, 2)	CS 3–CS 4	(1, 3)
3–CS 2	(1, 4)	11–CS 7	(2, 4)	CS 3–14	(1, 2)
4–1	(1, 2)	11–16	(1, 2)	CS 4–3	(1, 2)
4–CS 2	(2, 4)	12–10	(2, 3)	CS 4–7	(1, 3)
4–6	(1, 2)	12–13	(1, 2)	CS 4–10	(1, 3)
5–CS 2	(1, 2)	12–CS 8	(1, 4)	CS 4–CS 3	(1, 2)
5–6	(1, 2)	13–CS 7	(1, 3)	CS 5–6	(1, 2)
5–7	(2, 3)	13–12	(1, 2)	CS 5–7	(1, 2)
6–4	(1, 2)	13–15	(2, 3)	CS 5–8	(1, 2)
6–5	(1, 2)	13–16	(1, 2)	CS 5–9	(1, 2)
6–CS 5	(1, 3)	14–CS 3	(2, 4)	CS 6–6	(1, 3)
6–CS 6	(1, 2)	14–CS 8	(2, 3)	CS 6–8	(2, 3)
7–5	(1, 2)	15–13	(2, 3)	CS 7–7	(1, 3)
7–CS 4	(2, 5)	15–CS 8	(2, 5)	CS 7–10	(1, 3)
7–CS 7	(2, 4)	15–16	(1, 2)	CS 7–11	(1, 2)
7–9	(1, 2)	16–8	(1, 3)	CS 7–13	(1, 2)
7–CS 5	(1, 3)	16–11	(1, 2)	CS 8–12	(2, 3)
8–CS 6	(2, 5)	16–13	(1, 3)	CS 8–14	(1, 2)
8–CS 5	(2, 4)	16–15	(2, 3)	CS 8–15	(1, 2)
8–16	(1, 2)	CS 1–1	(1, 2)	2–CS 3	(1, 3)
9–CS 5	(2, 3)				

Table A5. Length l_a of link a in the road network.

Link a	l_a (km)	Link a	l_a (km)	Link a	l_a (km)
1–CS 1	23	9–7	15	CS 1–2	11
1–4	12	9–11	10	CS 2–3	10
2–CS 1	11	10–CS 4	14	CS 2–4	10
2–3	10	10–12	11	CS 2–5	11
3–2	10	10–CS 7	12	CS 3–2	17
3–CS 4	20	11–9	10	CS 3–CS 4	12
3–CS 2	10	11–CS 7	12	CS 3–14	22
4–1	12	11–16	16	CS 4–3	20
4–CS 2	10	12–10	11	CS 4–7	10
4–6	12	12–13	12	CS 4–10	14
5–CS 2	11	12–CS 8	10	CS 4–CS 3	12
5–6	10	13–CS 7	11	CS 5–6	11
5–7	12	13–12	12	CS 5–7	11
6–4	12	13–15	10	CS 5–8	10
6–5	10	13–16	16	CS 5–9	11
6–CS 5	11	14–CS 3	22	CS 6–6	12
6–CS 6	12	14–CS 8	12	CS 6–8	12
7–5	12	15–13	10	CS 7–7	18
7–CS 4	10	15–CS 8	11	CS 7–10	12
7–CS 7	18	15–16	10	CS 7–11	12
7–9	15	16–8	30	CS 7–13	11
7–CS 5	11	16–11	16	CS 8–12	10
8–CS 6	12	16–13	16	CS 8–14	12
8–CS 5	10	16–15	10	CS 8–15	11
8–16	30	CS 1–1	23	2–CS 3	17
9–CS 5	11				

References

1. International Energy Agency. *Global EV Outlook 2017*; IEA: Paris, France, 2017.
2. Rezvani, Z.; Jansson, J.; Bodin, J. Advances in consumer electric vehicle adoption research: A review and research agenda. *Transp. Res. Part D Transp. Environ.* **2015**, *34*, 122–136. [CrossRef]
3. Melliger, M.; Van Vliet, O.; Liimatainen, H. Anxiety vs. reality—Sufficiency of battery electric vehicle range in Switzerland and Finland. *Transp. Res. Part D Transp. Environ.* **2018**, *65*, 101–115. [CrossRef]
4. Wang, Y.; Bi, J.; Zhao, X.; Guan, W. A geometry-based algorithm to provide guidance for electric vehicle charging. *Transp. Res. Part D Transp. Environ.* **2018**, *63*, 890–906. [CrossRef]
5. Gendreau, M.; Ghiani, G.; Guerriero, E. Time-dependent routing problems: A review. *Comput. Oper. Res.* **2015**, *64*, 189–197. [CrossRef]
6. Artmeier, A.; Haselmayr, J.; Leucker, M.; Sachenbacher, M. The shortest path problem revisited: Optimal routing for electric vehicles. In *Annual Conference on Artificial Intelligence*; Springer: Berlin/Heidelberg, Germany, 2010; pp. 309–316.
7. Storandt, S. Quick and energy-efficient routes: Computing constrained shortest paths for electric vehicles. In Proceedings of the 5th ACM SIGSPATIAL International Workshop on Computational Transportation Science, Redondo Beach, CA, USA, 7–9 November 2012; ACM: New York, NY, USA, 2012; pp. 20–25.
8. Neaimeh, M.; Hill, G.; Hübner, Y.; Blythe, P.T. Routing systems to extend the driving range of electric vehicles. *IET Intell. Transp. Syst.* **2013**, *7*, 327–336. [CrossRef]
9. Sweda, T.; Dolinskaya, I.; Klabjan, D. Adaptive routing and recharging policies for electric vehicles. *Transp. Sci.* **2017**, *51*, 1326–1348. [CrossRef]
10. Qin, H.; Zhang, W. Charging scheduling with minimal waiting in a network of electric vehicles and charging stations. In Proceedings of the Eighth ACM International Workshop on Vehicular Inter-Networking, Las Vegas, NV, USA, 19 –23 September 2011; ACM: New York, NY, USA, 2011; pp. 51–60.
11. Said, D.; Cherkaoui, S.; Khoukhi, L. Queuing model for EVs charging at public supply stations. In Proceedings of the 2013 9th International Wireless Communications and Mobile Computing Conference (IWCMC), Sardinia, Italy, 1–5 July 2013; IEEE: Piscataway, NJ, USA, 2013; pp. 65–70.
12. Yang, S.; Cheng, W.; Hsu, Y.; Gan, C.; Lin, Y. Charge scheduling of electric vehicles in highways. *Math. Comput. Model.* **2013**, *57*, 2873–2882. [CrossRef]
13. De Weerdt, M.; Stein, S.; Gerding, E.; Robu, V.; Jennings, N. Intention-aware routing of electric vehicles. *IEEE Trans. Intell. Transp. Syst.* **2015**, *17*, 1472–1482. [CrossRef]
14. Zhang, Y.; Aliya, B.; Zhou, Y.; You, I.; Zhang, X.; Pau, G.; Collotta, M. Shortest feasible paths with partial charging for battery-powered electric vehicles in smart cities. *Pervasive Mob. Comput.* **2018**, *50*, 82–93. [CrossRef]
15. Wang, T.; Cassandras, C.; Pourazarm, S. Energy-aware vehicle routing in networks with charging nodes. *IFAC Proc. Vol.* **2014**, *47*, 9611–9616. [CrossRef]
16. Cao, Y.; Tang, S.; Li, C.; Zhang, P.; Tan, Y.; Zhang, Z.; Li, J. An optimized EV charging model considering TOU price and SOC curve. *IEEE Trans. Smart Grid* **2011**, *3*, 388–393. [CrossRef]
17. Liu, C.; Wu, J.; Long, C. Joint charging and routing optimization for electric vehicle navigation systems. *IFAC Proc. Vol.* **2014**, *47*, 2106–2111. [CrossRef]
18. Yagcitekin, B.; Uzunoglu, M. A double-layer smart charging strategy of electric vehicles taking routing and charge scheduling into account. *Appl. Energy* **2016**, *167*, 407–419. [CrossRef]
19. Sun, Z.; Zhou, X. To save money or to save time: Intelligent routing design for plug-in hybrid electric vehicle. *Transp. Res. Part D Transp. Environ.* **2016**, *43*, 238–250. [CrossRef]
20. Wang, Y.; Bi, J.; Guan, W.; Zhao, X. Optimising route choices for the travelling and charging of battery electric vehicles by considering multiple objectives. *Transp. Res. Part D Transp. Environ.* **2018**, *64*, 246–261. [CrossRef]
21. Wang, Y.; Jiang, J.; Mu, T. Context-aware and energy-driven route optimization for fully electric vehicles via crowdsourcing. *IEEE Trans. Intell. Transp. Syst.* **2013**, *14*, 1331–1345. [CrossRef]
22. Abousleiman, R.; Rawashdeh, O.; Abousleiman, R.; Rawashdeh, O. A Bellman-Ford approach to energy efficient routing of electric vehicles. In Proceedings of the 2015 IEEE Transportation Electrification Conference and Expo (ITEC), Dearborn, MI, USA, 14–17 June 2015; IEEE: Piscataway, NJ, USA, 2015; pp. 1–4.
23. Strehler, M.; Merting, S.; Schwan, C. Energy-efficient shortest routes for electric and hybrid vehicles. *Transp. Res. Part B Methodol.* **2017**, *103*, 111–135. [CrossRef]

24. Fiori, C.; Ahn, K.; Rakha, H. Optimum routing of battery electric vehicles: Insights using empirical data and microsimulation. *Transp. Res. Part D Transp. Environ.* **2018**, *64*, 262–272. [CrossRef]
25. Fernández, R. A more realistic approach to electric vehicle contribution to greenhouse gas emissions in the city. *J. Clean. Prod.* **2018**, *172*, 949–959. [CrossRef]
26. Alizadeh, M.; Wai, H.; Scaglione, A.; Goldsmith, A.; Fan, Y.; Javidi, T. Optimized path planning for electric vehicle routing and charging. In Proceedings of the 2014 52nd Annual Allerton Conference on Communication, Control, and Computing (Allerton), Monticello, IL, USA, 30 September–3 October 2014; IEEE: Piscataway, NJ, USA, 2014; pp. 25–32.
27. Yi, Z.; Bauer, P. Optimal stochastic eco-routing solutions for electric vehicles. *IEEE Trans. Intell. Transp. Syst.* **2018**, *19*, 3807–3817. [CrossRef]
28. Zhang, S.; Luo, Y.; Li, K. Multi-objective route search for electric vehicles using ant colony optimization. In Proceedings of the 2016 American Control Conference (ACC), Boston, MA, USA, 6–8 July 2016; IEEE: Piscataway, NJ, USA, 2016; pp. 637–642.
29. Jafari, E.; Boyles, S. Multicriteria stochastic shortest path problem for electric vehicles. *Netw. Spat. Econ.* **2017**, *17*, 1043–1070. [CrossRef]
30. Daina, N.; Sivakumar, A.; Polak, J. Electric vehicle charging choices: Modelling and implications for smart charging services. *Transp. Res. Part C Emerg. Technol.* **2017**, *81*, 36–56. [CrossRef]
31. Huber, G.; Bogenberger, K. Long-Trip Optimization of Charging Strategies for Battery Electric Vehicles. *Transp. Res. Rec.* **2015**, *2497*, 45–53. [CrossRef]
32. Jiang, N.; Xie, C.; Duthie, J.; Waller, S. A network equilibrium analysis on destination, route and parking choices with mixed gasoline and electric vehicular flows. *EURO J. Transp. Logist.* **2014**, *3*, 55–92. [CrossRef]
33. He, F.; Yin, Y.; Lawphongpanich, S. Network equilibrium models with battery electric vehicles. *Transp. Res. Part B Methodol.* **2014**, *67*, 306–319. [CrossRef]
34. Xie, C.; Jiang, N. Relay requirement and traffic assignment of electric vehicles. *Comput. -Aided Civ. Infrastruct. Eng.* **2016**, *31*, 580–598. [CrossRef]
35. Xu, M.; Meng, Q.; Liu, K. Network user equilibrium problems for the mixed battery electric vehicles and gasoline vehicles subject to battery swapping stations and road grade constraints. *Transp. Res. Part B Methodol.* **2017**, *99*, 138–166. [CrossRef]
36. Hung, Y.; Michailidis, G. Optimal routing for electric vehicle service systems. *Eur. J. Oper. Res.* **2015**, *247*, 515–524. [CrossRef]
37. Bi, J.; Wang, Y.; Sai, Q.; Ding, C. Estimating remaining driving range of battery electric vehicles based on real-world data: A case study of Beijing, China. *Energy* **2019**, *169*, 833–843. [CrossRef]
38. Huang, Y.; Zhao, L.; Van Woensel, T.; Gross, J. Time-dependent vehicle routing problem with path flexibility. *Transp. Res. Part B Methodol.* **2017**, *95*, 169–195. [CrossRef]
39. Neely, M.; Modiano, E.; Rohrs, C. Dynamic power allocation and routing for time-varying wireless networks. *IEEE J. Sel. Areas Commun.* **2005**, *23*, 89–103. [CrossRef]
40. Polson, N.; Sokolov, V. Deep learning for short-term traffic flow prediction. *Transp. Res. Part C Emerg. Technol.* **2017**, *79*, 1–17. [CrossRef]
41. Cheng, L.; Chang, Y.; Huang, R. Mitigating voltage problem in distribution system with distributed solar generation using electric vehicles. *IEEE Trans. Sustain. Energy* **2015**, *6*, 1475–1484. [CrossRef]
42. Xu, Z.; Su, W.; Hu, Z.; Song, Y.; Zhang, H. A hierarchical framework for coordinated charging of plug-in electric vehicles in China. *IEEE Trans. Smart Grid* **2015**, *7*, 428–438. [CrossRef]
43. Esfahani, M.; Yousefi, G. Real time congestion management in power systems considering quasi-dynamic thermal rating and congestion clearing time. *IEEE Trans. Ind. Inform.* **2016**, *12*, 745–754. [CrossRef]
44. Gnann, T.; Funke, S.; Jakobsson, N.; Plötz, P.; Sprei, F.; Bennehag, A. Fast charging infrastructure for electric vehicles: Today's situation and future needs. *Transp. Res. Part D Transp. Environ.* **2018**, *62*, 314–329. [CrossRef]
45. Li, B.; Eryilmaz, A. Non-derivative algorithm design for efficient routing over unreliable stochastic networks. *Perform. Eval.* **2014**, *71*, 44–60. [CrossRef]
46. Franke, T.; Krems, J.F. Understanding charging behaviour of electric vehicle users. *Transp. Res. Part F Traffic Psychol. Behav.* **2013**, *21*, 75–89. [CrossRef]
47. Raslavičius, L.; Azzopardi, B.; Keršys, A.; Starevičius, M.; Bazaras, Ž.; Makaras, R. Electric vehicles challenges and opportunities: Lithuanian review. *Renew. Sustain. Energy Rev.* **2015**, *42*, 786–800. [CrossRef]

48. Bi, J.; Wang, Y.; Sun, S.; Guan, W. Predicting Charging Time of Battery Electric Vehicles Based on Regression and Time-Series Methods: A Case Study of Beijing. *Energies* **2018**, *11*, 1040. [CrossRef]
49. Fu, L.; Sun, D.; Rilett, L.R. Heuristic shortest path algorithms for transportation applications: State of the art. *Comput. Oper. Res.* **2006**, *33*, 3324–3343. [CrossRef]
50. Gao, S.; Frejinger, E.; Ben-Akiva, M. Adaptive route choices in risky traffic networks: A prospect theory approach. *Transp. Res. Part C Emerg. Technol.* **2010**, *18*, 727–740. [CrossRef]
51. Braekers, K.; Ramaekers, K.; Van Nieuwenhuyse, I. The vehicle routing problem: State of the art classification and review. *Comput. Ind. Eng.* **2016**, *99*, 300–313. [CrossRef]
52. Bi, J.; Wang, Y.; Zhang, J. A data-based model for driving distance estimation of battery electric logistics vehicles. *EURASIP J. Wirel. Commun. Netw.* **2018**, *2018*, 251. [CrossRef]
53. Yang, Y.; Yao, E.; Yang, Z.; Zhang, R. Modeling the charging and route choice behavior of BEV drivers. *Transp. Res. Part C Emerg. Technol.* **2016**, *65*, 190–204. [CrossRef]
54. Meng, Q.; Yang, H. Benefit distribution and equity in road network design. *Transp. Res. Part B Methodol.* **2002**, *36*, 19–35. [CrossRef]
55. Chow, J.; Regan, A. Network-based real option models. *Transp. Res. Part B Methodol.* **2011**, *45*, 682–695. [CrossRef]
56. Bell, M.; Kurauchi, F.; Perera, S.; Wong, W. Investigating transport network vulnerability by capacity weighted spectral analysis. *Transp. Res. Part B Methodol.* **2017**, *99*, 251–266. [CrossRef]
57. Hung, Y.; Michailidis, G. Stability and control of acyclic stochastic processing networks with shared resources. *IEEE Trans. Autom. Control* **2012**, *57*, 489–494. [CrossRef]

© 2020 by the authors. Licensee MDPI, Basel, Switzerland. This article is an open access article distributed under the terms and conditions of the Creative Commons Attribution (CC BY) license (http://creativecommons.org/licenses/by/4.0/).

energies

MDPI

Article

Design of a Low Torque Ripple Three-Phase SRM for Automotive Shift-by-Wire Actuator

Grace Firsta Lukman, Xuan Son Nguyen and Jin-Woo Ahn *

Department of Mechatronics Engineering, Kyungsung University, Busan 48434, Korea; gracedr@ks.ac.kr (G.F.L.); nguyenxuanson@ks.ac.kr (X.S.N.)
* Correspondence: jwahn@ks.ac.kr; Tel.: +82-51-663-4773

Received: 6 April 2020; Accepted: 28 April 2020; Published: 7 May 2020

Abstract: The shift-by-wire (SBW) system in vehicles aims to increase performance, safety, and comfort during driving. Switched reluctance motors (SRMs) are simple, resilient, and require only minimum maintenance; these factors make it a good option for use as an SBW actuator. Inherently, SRM generates a higher torque ripple than other AC machines, which can lead to a deterioration of its function. In this paper, a non-uniform air-gap rotor structure combined with the careful positioning of holes near the pole surface is proposed to reduce the torque ripple. The finite element method (FEM) is employed to analyze the electromagnetic characteristics of the design. The proposed motor is manufactured, and experiments are done to verify the performance. The design alone can reduce the torque ripple by 7% and flatten the torque waveform. The experimental result shows that the proposed motor can achieve the desired performance. The estimated torque ripple from the experiment is 26.65%.

Keywords: shift-by-wire (SBW); switched reluctance motor (SRM); non-uniform air-gap; rotor hole placement; torque ripple; rotor structure

---

1. Introduction

Automotive x-by-wire systems have undergone a significant technological development over the last few decades. The "x" refers to the function segment of a car, such as parking, shifting, and braking, and varies widely. In shift-by-wire (SBW), the traditional mechanical connection between the gearshift activator (e.g., shift lever) and transmission is replaced with electric components. There are four common shifting modes—parking (P), reverse (R), neutral (N), and driving (D)—which the driver can switch to easily. When a command is given, it activates the electronic control that rotates the actuator to change the gear. SBW systems are easier to assemble and require less space, so there will is more room for interior design. Moreover, it adds convenience and comfort in driving [1]. According to [2], the three most important aspects in designing and manufacturing vehicle transmissions are their long service life, low repair cost, and low production costs, each of which score 9.00, 4.89, and 4.59 out of 10 points, respectively. In other words, the robustness of the system is the most essential factor.

Switched reluctance motors (SRMs) satisfy the three top requirements. The structure is simple and resilient. The core is doubly-salient with no permanent magnets, and the windings are only wound on the stator in a concentrated way. Moreover, the end-coil length is short, and so the motor is relatively small and compact overall. In terms of its structure, SRM is highly robust. SRM has been reported to be used by Toyota and Lexus as the motor is reliable and highly responsive [3,4].

SRMs are singly-excited and inherently generate a high torque ripple compared to AC machines running on sinusoidal voltage. Figure 1 shows a general 6/4 SRM drive system with an asymmetric half bridge converter and ideal torque generation. As the name suggests, SRM generates only reluctance torque and it pulsates according to the rotor position due to the switching operation and structure.

In other words, the torque generation is based on the tendency of the rotor to be aligned with the excited stator pole pair [5]. Therefore, the reduction of torque ripple is a major topic in SRM research. The methods vary from design to control and simple to complicated approaches. From the design perspective, it is common to change the stator and/or rotor pole shape to form either a non-uniform air-gap or wide arc [6–8]. A unique semi-elliptical core is proposed in [9], in which both sides of each rotor pole are somewhat rounded instead of being straight lines. Rounded rotor pole tips are also presented as a design to reduce torque ripple in [10]. The inclusion of holes on the pole are also viable, as shown in [11–15], but this is usually done to reduce the noise and vibration. Various hole shapes have been studied; the effect of the height, width, and location of rectangular holes can be seen in [12]. Circular holes are preferred in [13], which prevent local saturation in the corners of square-shaped holes. Small diamond-shaped holes are placed strategically in the stator yoke in [14] to enhance the oscillation damping, leading to lower vibration. Triangular shapes are presented in [15] based on the flux path to reduce the stress during excitation. In addition, multi-objective shape optimization using intelligent algorithms (e.g., genetic algorithm, particle swarm, etc.) is usually implemented in the geometry optimization process [16–18]. With this method, various losses and thermal factors can be considered in combination to generate the optimum model. The inherent torque ripple of SRM is due to the nonlinear coupling between the phase current, rotor position, and overlapping angle as well as the complete machine geometry determined in the initial design process [17,19]. Therefore, to effectively reduce the torque ripple, it is best to undertake both design alterations and a control strategy. However, this study solely focuses on the design.

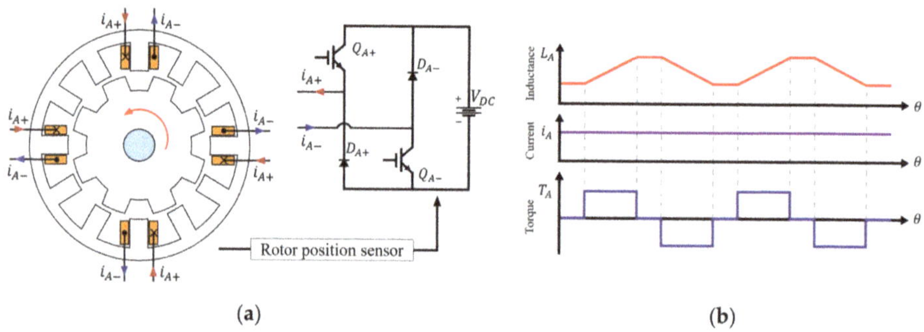

Figure 1. (**a**) General 12/8 switched reluctance motor (SRM) with parallel winding connection and (**b**) constant current excitation.

In this paper, a rotor pole design combining a non-uniform air-gap and circular holes near the tip is proposed to reduce torque ripple. The SBW system actuator is equipped with a cycloidal speed reducer to increase the output torque up to hundreds of Nm. Depending on the gear ratio, many rotations of the motor can only translate to one tooth being moved at the transmission gear. Smooth rotation is necessary to ensure accurate movement and increase the reliability of the SBW. The selection of a non-uniform angle and hole position is described. The finite element method (FEM) is used to analyze the electromagnetic characteristics of the design. The proposed SRM was manufactured and experiments were performed to verify the performance of the motor. Two simple Hall-effect sensors mounted on the back of the motor frame are used as rotation angle sensors to ensure a compact design. The simulation and experiment results show a reduction in torque ripple while maintaining the average torque output.

A similar structure, with holes drilled on the corners of stator and rotor poles, is proposed in [20]. The stator poles have a small shoe-like design, and a conventional uniform air-gap was implemented. The authors also suggested a tapered pole and skewed rotors. However, the rotor positioning method is not clearly described, and the design optimization approach is complicated. A non-uniform air-gap is also used in [21], which also targets the SBW application, but the design is complex. The rotor pole surface is castle-shaped using a plurality of concavities and convexities. A complicated structure is more prone to possible manufacturing errors. The proposed design simply adopts an arched non-uniform air-gap and two holes in the non-uniform air gap section. A hole placement method to modify the torque profile is also presented. Since SRM has no permanent magnet, its torque magnitude depends largely on the length of the air-gap. A non-uniform air-gap results in a larger gap length, and the holes are positioned so that saturation level at the surface can be maintained while still reducing the peak torque, leading to a lower torque ripple. Compared to previous literature, the proposed design is simple and straightforward and can serve as a guideline for modifying the torque profile by design.

2. Design of Proposed SRM

2.1. Design Target and Parameters

For the SBW application, a three-phase 12/8 SRM topology is chosen. A greater pole number is generally associated with higher torque, and a lower number corresponds to a higher rotational speed [22]. In SBW, the motor is not running continuously, but only on command according to the driver. The automotive voltage of 12 V is used as the input supply. The maximum current should be lower than 30 A to reduce the burden of the converter. Detailed specifications of the proposed motor are shown in Table 1. Another important aspect of the application is bi-directional rotation: the SRM must be able to rotate clockwise and counterclockwise to shift the gear back and forth. Most non-uniform air gap SRMs cut a small part of the rotor/stator tip in the same way for every pole. However, this method disables the bi-directional rotating capability since the inductance profile becomes asymmetrical. A solution to this is presented below.

Table 1. Design parameters.

Parameters	Value
Stator outer diameter (mm)	75
Stack length (mm)	18
Air-gap length (mm)	0.25
DC-link voltage (V)	12
Maximum phase current (A)	30
Rated speed (RPM)	1800
Rated torque (Nm)	0.21

2.2. Conventional 12/8 SRM

Common SRMs have a uniform air-gap, as shown in Figure 2a. The notations β_s, β_r, and β_g denote the stator pole arc, rotor pole arc, and the angular distance between the stator and rotor pole tips when fully unaligned. Zero position is as determined in the figure as the position at which the rotor is completely unaligned with the stator. The corresponding torque and inductance profiles can be seen in Figure 2b. The inductance profile depends on the geometry of the motor and determines the torque as follows:

$$T_{ph} = \frac{1}{2} i_{ph}^2 \frac{dL_{ph}}{d\theta} \tag{1}$$

where T_{ph}, i_{ph}, L_{ph}, and θ are the phase torque, current, inductance, and rotor position. L_{ph} can be obtained by one-phase excitation, and this value changes according to rotor position. Considering the linear magnetic characteristic, in SRM, the torque is proportional to the square of current and the change of inductance. The straight line is at 22.5°, where both poles are perfectly aligned and both left and right sides are identical to each other. This symmetry ensures bi-rotational rotation capability. The average torque considering the positive torque area only is 0.324 Nm.

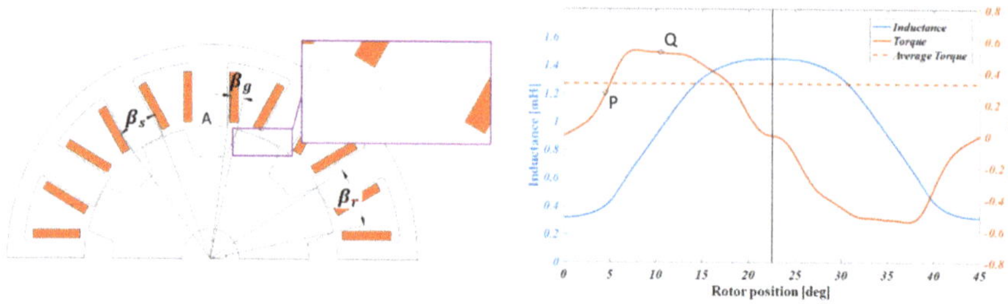

(a) (b)

Figure 2. Conventional 12/8 SRM: (**a**) 2D model and (**b**) characteristics.

The goal of this research is to reduce the torque ripple and shape the waveform close to an ideal profile, as shown in Figure 1b, whereas square-shaped torque is generated with constant current excitation. In Figure 2b, there is a protruding section between P and Q where the torque appears to have some "overshoot" before flattening again at Q. The angular distance between P and Q is 6°.

2.3. Non-Uniform Rotor Stucture

In order to retain the bi-directional rotating capability while maintaining a non-uniform air-gap, a mirrored structure is presented in Figure 3a. The M-type design has the cutting part on the left side, and the N-type design has this on the right side. The distribution is even for the rotor poles, meaning that there will be four "Ms" and four "Ns" for a 12/8 SRM. A similar structure was shown in [21], but instead of cutting/chipping a part of the tip, a castle-shaped pole was introduced. Compared to [21], the proposed rotor will be much simpler. The rotor surface can be divided into two sections: non-uniform and uniform. As can be seen in Figure 3b, the relationship can be expressed by the equation below, where β_{non} and β_{uni} are the arcs of non-uniform and uniform sections, respectively. β_{non} is set to 6° based on previous observation.

$$\beta_r = \beta_{non} + \beta_{uni} \tag{2}$$

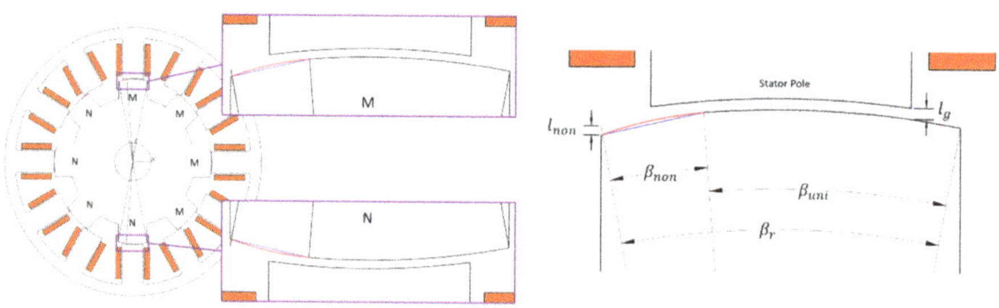

(a) (b)

Figure 3. Bi-directional non-uniform structure: (**a**) pole distribution and (**b**) modeling.

The cutting length, l_{non}, is measured from the pole tip vertically along the side. Therefore, the maximum non-uniform air-gap length is the addition of conventional air-gap l_g and l_{non}.

Inductance is at maximum when the poles are aligned. Considering counterclockwise rotation, the air-gap increases gradually on the M-type pole. Generally, there are two ways to cut the pole: line and arc. The straight (blue) and concave (red) lines in Figure 3b show the difference between these approaches. Nevertheless, both cuts cause a slower increase of inductance compared to the conventional method. Because of this, the torque is also reduced during β_{non}, and instead of having an "overshoot", it appears much rounder. The β_{non} section ends at point Q, and the profile after that is similar using either conventional or non-uniform methods. After alignment, the rotor pole leaves the stator pole, and due to the "N" pole, the air-gap slightly increases, thus creating symmetry in the profiles. This phenomenon is shown and compared in Figure 4. Compared to the straight non-uniform gap, the arched cut preserves some convexity in torque, which results in a greater positive area while still reducing the "overshoot" and creates a more similar feature to the ideal profile. As a result, the average torque is higher than that of the straight cut. The values of the average torque with the same current excitation are 0.324, 0.315, and 0.319 for conventional, straight, and arched cuts, respectively. Because the torque drops by 2.8% in a straight cut and 1.5% in an arched cut, the latter is chosen in this study.

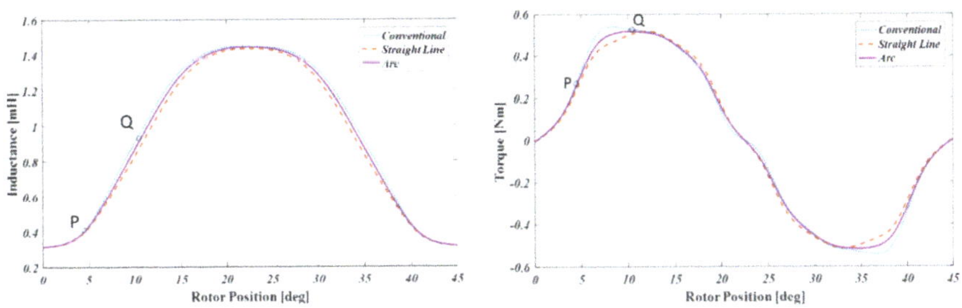

(a) (b)

Figure 4. Comparison of conventional and non-uniform air-gap models: (**a**) torque and (**b**) inductance.

2.4. Proposed Rotor Stucture

After choosing the appropriate non-uniform section, the next step is to flatten the curve to further reduce the ripple. Figure 5 shows the design of the proposed rotor. β_{disR}, β_{disH}, and l_{disH} are the angular distance of the first hole to the nearest pole edge, the angular distance between the holes, and the distance from the pole surface. Note that the l_{disH} is constant regardless of the non-uniform part. Two small holes are placed near the surface because the closer they are to the shaft, the less their effect will be on torque development. Manufacturing accuracy is the only restriction in the model

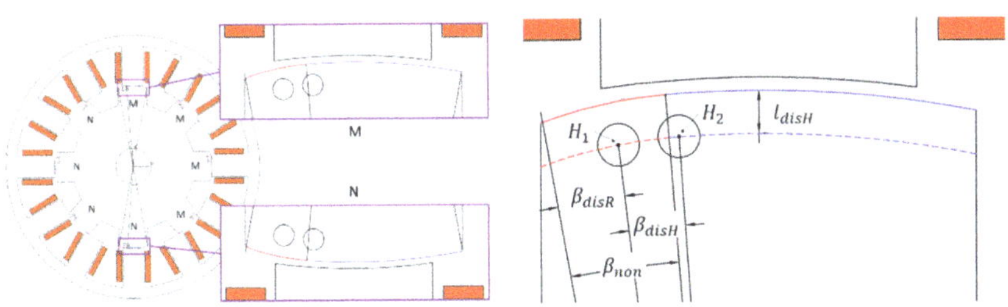

(a) (b)

Figure 5. Proposed rotor structure: (**a**) pole distribution and (**b**) modeling.

In order to investigate this more deeply, the non-uniform, one-hole model is investigated. The comparison of the torque and inductance of the four designs is shown in Figure 6. Two points, R and S, mark the convex area that has to be straightened from the previous non-uniform-only structure to achieve the ideal torque waveform shape. H_1 is located at "R". It can be seen that the simple non-uniform and single hole models do not have distinct differences, as seen in the figure. However, the corner at "R" is slightly lowered in the single-hole design. The second hole and the H_2 placement create the rather square shape, which is discussed in Section 3. In any case, the symmetrical feature is retained, and so bi-directional rotation is possible. In this study, both holes have the same diameter, but it is possible to shape them in any way to create the desired waveform in accordance with the general guidelines given here. The average torque of the one-hole and proposed model for the same current excitation is 0.319 and 0.317 Nm, respectively. In other words, the proposed method has the minimum torque compared to others, with a drop of 2.16% compared the conventional approach.

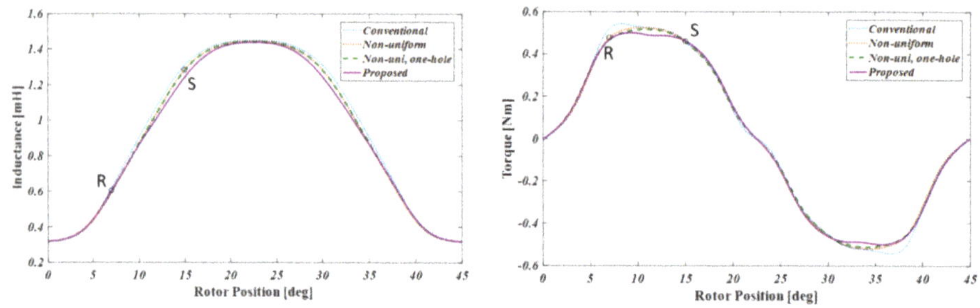

Figure 6. Comparison of the four models: (**a**) inductance and (**b**) torque.

3. Simulation Results

3.1. Flux Distribution

In this section, the flux paths from the stator to the rotor for points P, R, Q, and S are examined. Only the "M" type is used as illustration since the "N" type will work in the opposite manner. Both motors have the same core material, 50PN440, and the current is adjusted to obtain a similar average torque output. The summary of the point positions is given below.

Figures 7 and 8 show the flux distribution of conventional and proposed motors as the rotor rotates along the circumference passing the stator pole. Each figure from (a) to (d) presents points P, R, Q, S, as described Figure 4, Figure 6, and Table 2. The subsection (e), however, is arbitrarily chosen to observe why the proposed motor has a slightly wider torque positive area, as seen in Figure 6b.

Table 2. Selection of point location.

Point	Definition
P	Start of overlap between rotor and stator
R	Start of "torque overshoot" corner
Q	End of "torque overshoot" corner
S	End of convexity of torque waveform

At point P, where the tips of the rotor and stator pole meet, no distinction can be made between the two models other than that the flux is more concentrated at the pole tip for the conventional model. This corresponds to Figures 4b and 6b, in which the value of torque is similar for the four models.

At point R, where the torque convexity corner starts, the stator and rotor overlap in the non-uniform air-gap section. As a result, the flux density of the proposed design is lower, which leads to a lower torque value. In contrast to the conventional type, the sharp increase in torque can be reduced by avoiding high density or saturation in this area, thus obtaining a smoother slope.

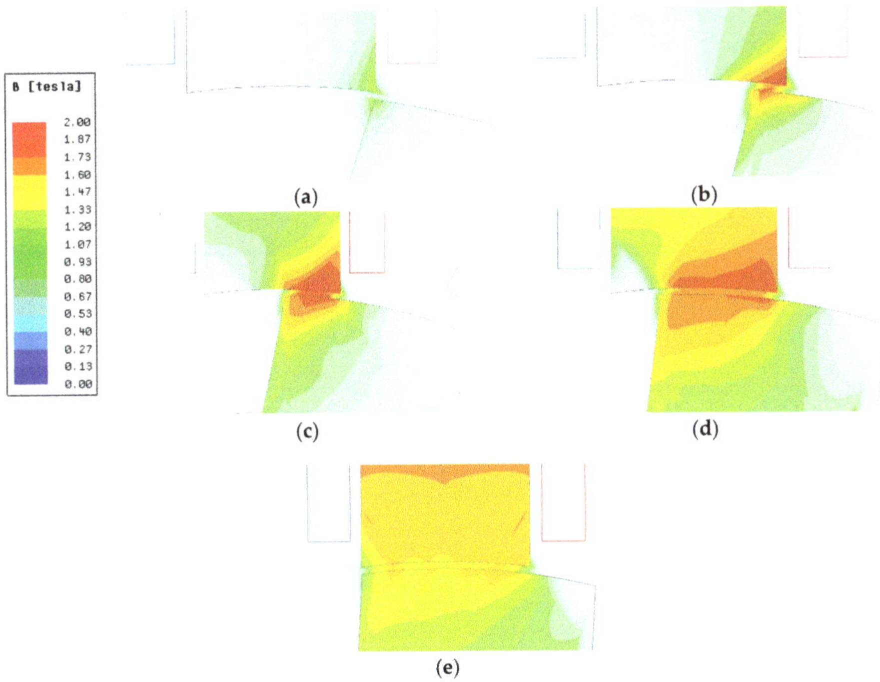

Figure 7. Flux density of conventional type: (**a**) point P, (**b**) point R, (**c**) point Q, (**d**) point S, (**e**) 87% aligned.

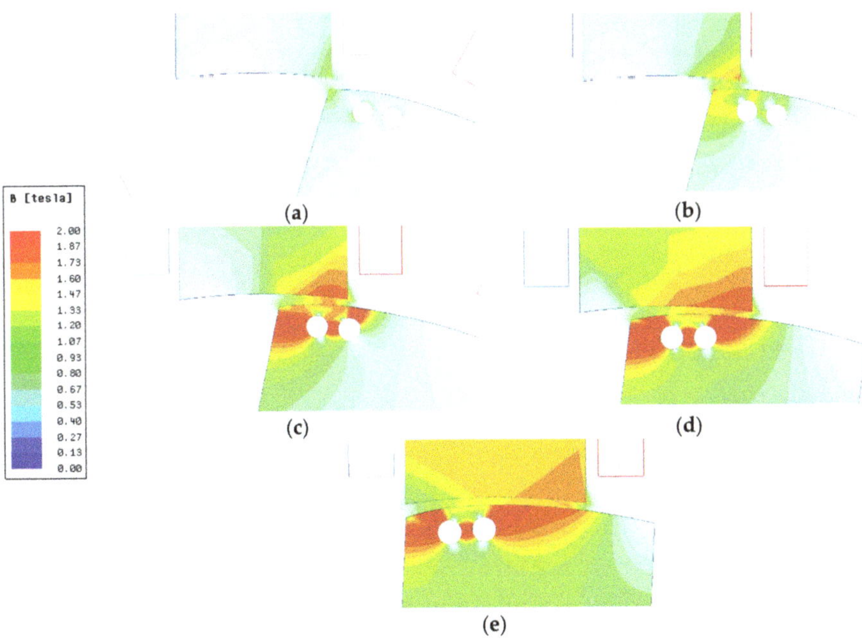

Figure 8. Flux density of proposed motor: (**a**) point P, (**b**) point R, (**c**) point Q, (**d**) point S, (**e**) 87% aligned.

After passing point R, the rotation comes to point Q, where the "torque overshoot" curve in the conventional model ends. Relative to point R, the maximum density of the conventional type is reduced here, and so is the torque value. However, this is not the case in the proposed model. The flux density level on the surface of the overlapping section is only higher than that at point R. Therefore, the torque value can be maintained. It can be seen in the figure that this is where the second hole plays an important role in keeping the rotor surface flux density the same; changing the position and/or size will affect the flatness of the torque.

Finally, the end of the convexity is at point S. Here, the average surface flux density in the overlapping section is similar in both models, which explains the torque values in Figure 6b. After point S, the 87% aligned position is chosen arbitrarily. The proposed model still has slightly higher torque compared to the conventional one, resulting in a wider positive area. The reason for this is shown in Figure 8e, where the average surface flux density of the proposed structure is higher because of the positioning of the second hole. In conclusion, the generated torque is closely related to the flux distribution in the overlapping region of the stator and rotor poles. Maintaining a similar flux density level over the excitation period means keeping a constant torque value and thus lowering the torque ripple.

3.2. Continuous Torque

Single-phase excitation alone cannot verify the reduction of the torque ripple. In this section, continuous torque is observed when all phases are excited sequentially. First, the torque ripple equation is given as a percentage of the average torque as follows:

$$T_{ripple} = \frac{T_{\max} + T_{\min}}{T_{avg}} \tag{3}$$

Figure 9a shows the continuous torque in the proposed SRM. $T_{\min}$ occurs at conduction when the exciting phase is turned off and the next phase is turned on. This switching action inevitably creates the inherent high torque ripple in SRM. Reducing $T_{\max}$ and increasing $T_{\min}$ for the same T_{avg} is an effective way to reduce the torque ripple. This is the main goal of the proposed structure. However, as shown before, the average torque decreases with each modification. For the torque ripple comparison, the current is adjusted so that the four models have the same average torque of 0.45 Nm. The result is presented in Figure 9b. The corresponding torque ripple is shown in Table 3. The torque ripple is reduced by 7%, and the current of the proposed model is merely 0.5 A higher than the conventional approach. It is shown that the torque of the proposed motor is much flatter at the top, with a wider positive torque area, compared to other models. This confirms that the proposed design can reduce the torque ripple with the same output capability.

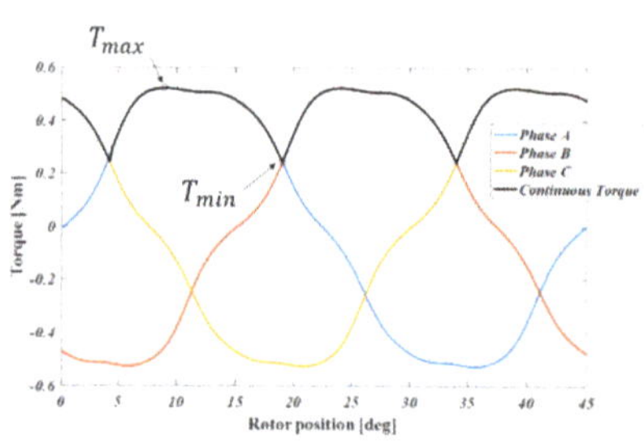

(a)

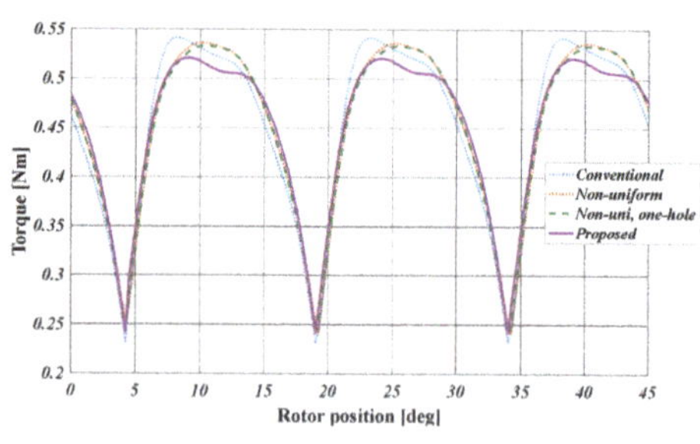

(b)

Figure 9. Continuous torque: (**a**) conventional 12/8 SRM and (**b**) comparison.

Table 3. Torque ripple comparison.

Design	Phase Current (A)	Torque Ripple (%)
Conventional	13.5	68.81
Non-uniform	13.7	65.16
Non-uniform, one-hole	13.8	64.27
Proposed	14.0	61.77

4. Experimental Results

After the verification of the reduced torque ripple, the proposed motor was manufactured to evaluate its performance. The prototype of the proposed 12/8 SRM is shown in Figure 10. The punch-holes were distributed into "M" and "N" types, as previously described. The black plate attached on the rotor in Figure 10b contained small magnets. Two Hall-effect sensors were mounted on the back of the motor frame which detected these magnets for rotor position information. Compared to other encoder types, this method results in a much lighter and more compact system.

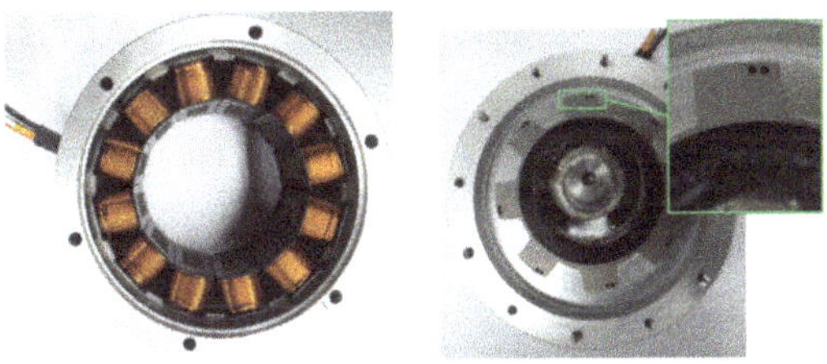

(a) (b)

Figure 10. Manufactured motor: (**a**) stator and (**b**) rotor.

The experimental setup is shown in Figure 11a. The motor is mounted on a zig, and its shaft is coupled with a torque sensor, the other end of which is coupled with an electric clutch. By controlling the power of the electric clutch, the load torque could be induced to the motor. A 12 V power supply—corresponding to the battery supply in vehicles—was used for the motor. Figure 11b shows the overall connection diagram involved in the experiment. A microcontroller, TMS320F28335 from Texas Instruments, was used to control the PWM (Pulse Width Modulation) signal to the asymmetric converter and also read the position signal from the Hall sensors. The clutch model was an ZKG-50AN from Mitsubishi. The power to the clutch was controlled manually by adjusting the voltage of a separate DC power supply.

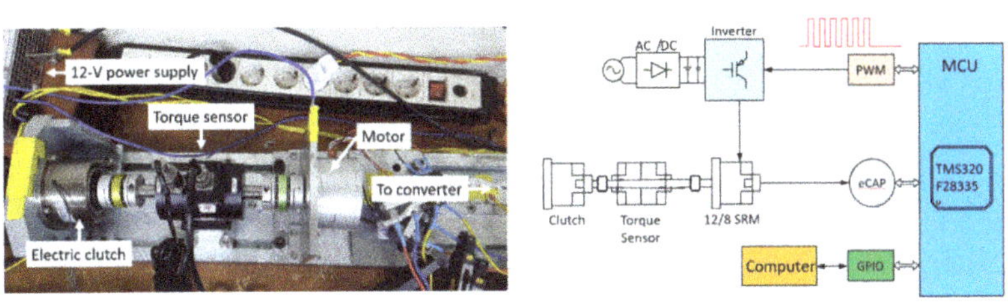

(a) (b)

Figure 11. Experimental setup: (**a**) test bed and (**b**) connection diagram (eCAP—enhanced capture; MCU—microcontroller unit).

The experiment was performed by implementing a 100% duty cycle of PWM, also known as single-pulse control, as shown in Figure 12. The phase currents are also presented in the figure.

The rated value of the proposed motor, which was 0.245 Nm at 1800 RPM, was achieved, as can be seen in Figure 12a. A much higher torque could be obtained at the lower speed level of 600 RPM, at which the motor was able to generate 0.70 Nm of output torque. Nevertheless, the maximum current was under 30 A in any condition. Finally, the speed vs. torque and efficiency curves are shown in Figure 13. The efficiencies at 1800 and 600 RPM were 70.7 and 32.5%, respectively. However, efficiency was not a necessary requirement, since the motor only rotates for a short amount of time with each shift gear command from the driver. The phase RMS (Root-Mean Square) current required to produce the torque is given in Figure 13b. All values were under 15 A and thus met the design target. Therefore, it could be confirmed that the proposed motor satisfies all the requirements presented in Table 1.

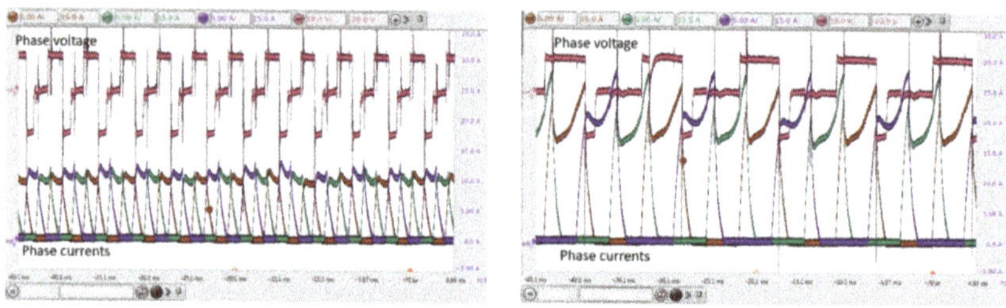

(a) (b)

Figure 12. Experimental result: (**a**) 1800 RPM–0.245 Nm and (**b**) 600 RPM–0.70 Nm.

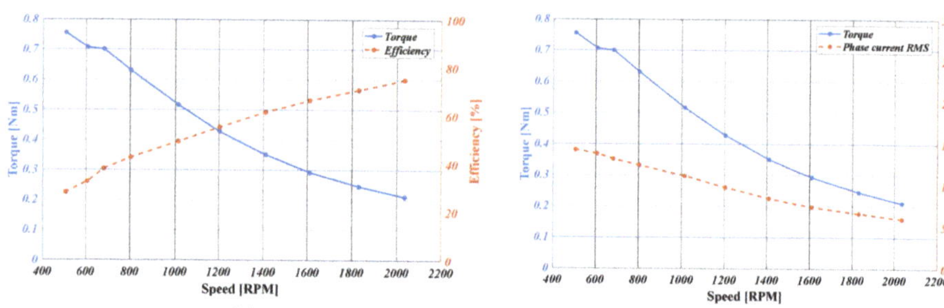

(a) (b)

Figure 13. Performance curves: (**a**) speed vs. torque, efficiency and (**b**) speed vs. torque, phase current RMS.

The calculation of the torque ripple is difficult to perform in real-time during an experiment. Therefore, FEM can be used to estimate the actual torque. A dynamic simulation using the SRM circuit configuration shown in Figure 1a could be set according to the experimental setup. The same control properties (initial position, excitation angle, etc.) were applied in the simulation to obtain a similar phase current to that of the experiment. Thus, the torque output of the experiment could be predicted. The result is shown in Figure 14. The experimental and simulation phase currents are shown in the top and middle plots, respectively. The resulting torque from the middle plot is shown at the bottom. The torque ripple was calculated to be 26.65%.

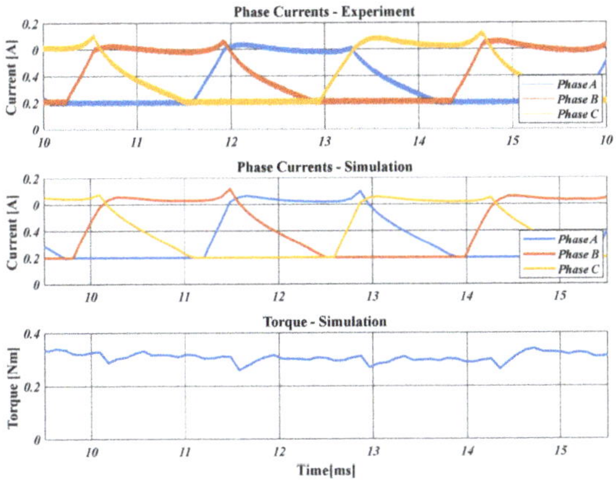

Figure 14. Comparison of phase currents and estimated torque.

5. Discussions

The effectiveness of the proposed design combining a non-uniform air-gap and hole placement and its performance have been confirmed through FEM simulations and experiment. However, an optimization algorithm was not performed, and it is possible that a more optimal model could be designed. The closer the hole is to the surface, the greater the impact on the torque. If the hole is too close, however, then the structure is at risk of breaking during operation. An additional mechanical structural analysis could therefore also be useful. It is also feasible to only use one hole or more than two holes to modify the torque waveform as required. The same principle should be applied, maintaining the flux density on the surface. Another factor to be considered is that it is important to discuss how accurate the motor prototype will be with the corresponding manufacturer; in this case, this relates to the possible size of the hole and air-gap. Nevertheless, the viability of the proposed SRM as an SBW actuator is shown in this study.

SRM has a large number of advantages because of its simple structure, but the implementation of the motor in practical everyday use is still low compared to other machines. In [3,4], the authors presented a 12/8 SRM for an actual SBW application; however, the improvement of the drawbacks of SRM such as torque ripple were not clearly discussed and/or shown. Other than the design method and analysis, it is hoped that this study can guide researchers and industries to utilize and improve SRMs to make it competitive in the market.

Furthermore, various control methods can be applied to increase the performance. As can be seen in the experimental results, the phase currents have a "tail" at the end before decreasing to zero. This is due to the slightly long excitation time allowing more current to flow so that the desired torque can be reached. Even a simple turn-on and turn-off angle control method can be implemented to further reduce the torque ripple; however, this is beyond the scope of this paper.

6. Conclusions

In this paper, a rotor design to reduce torque ripple while maintaining the average torque value and bi-directional rotation is proposed. SRM is a low-cost, robust motor but suffers from relatively high torque ripple, which can hinder its performance. Low torque ripple and smooth operation are necessary for shift-by-wire applications for safety and accuracy in moving the shifting gear.

A non-uniform air-gap is utilized to reduce the "torque overshoot" which occurs in the conventional structure. This is then combined with the positioning of two holes, meaning that the average flux density at the overlapping region of the stator and rotor poles can be kept close to constant during the period of excitation. The rotor pole tip is cut in a mirrored way to maintain a symmetrical property for bi-directional rotation. FEM is used to check the characteristics of the motor, and it is shown that the torque ripple can be reduced by 7% for the same output power by modifying the design alone.

A prototype motor of the proposed design was manufactured to verify its performance. It was shown that the motor satisfied the requirement of 0.245 Nm of torque at 1800 RPM. Moreover, the maximum current was kept below 30 A. The estimated torque ripple by matching the simulation to the experimental result was shown to be low, at 26.65%. The efficiency at rated conditions was 70.7%, but this was not a requirement due to the designated application, which demands the motor only to be run for a short amount of time and not to be in continuous operation. Both simulations and experiments showed the viability of the proposed 12/8 SRM as an actuator for a shift-by-wire system.

SRM has no brushes and commutators, and so there is little—or no—problem with maintenance compared to DC and universal motors. However, compared to permanent magnet motors, the power density and efficiency are generally relatively low. SRM makes up for this with its low manufacturing cost and high robustness (due to its simple structure). When efficiency is not the main issue and a long lifespan and high robustness are prioritized, an SRM with a low torque ripple design is a good actuator candidate that can meet the output requirements, as presented in this paper.

Author Contributions: Conceptualization, X.S.N.; Data curation, G.F.L.; Formal analysis, G.F.L.; Investigation, G.F.L.; Methodology, X.S.N.; Supervision, J.-W.A.; Validation, J.-W.A.; Writing—original draft, G.F.L.; Writing—review & editing, G.F.L. All authors have read and agreed to the published version of the manuscript.

Funding: This work was supported by "Human Resources Program in Energy Technology" of the Korea Institute of Energy Technology Evaluation and Planning (KETEP), and was granted financial resources from the Ministry of Trade, Industry & Energy, Republic of Korea. (No. 20184010201700), and Kyungsung University Research Grants in 2020.

Conflicts of Interest: The authors declare no conflict of interest.

References

1. Smirra, K.; Ferstl, M.; Eiting, T. *Mechatronics for "Shift by Wire"—A Technical Challenge*; SAE Technical Paper 2007-01–1309; SAE International: Warrendale, PA, USA, 2007. [CrossRef]
2. Naunheimer, H.; Bertsche, B.; Ryborz, J.; Novak, W. *Automotive Transmissions: Fundamentals, Selection, Design and Application*, 2nd ed.; Springer-Verlag: Berlin/Heidelberg, Germany, 2011. [CrossRef]
3. Kobiki, Y.; Inoue, Y.; Sekiguchi, S.; Itoh, T.; Kamio, S. *Toyota's New Shift-by-Wire System for Hybrid Vehicles*; SAE Technical Paper 2004-01–1112; SAE International: Warrendale, PA, USA, 2004. [CrossRef]
4. Nakade, Y.; Kamada, A.; Ueno, K.; Kume, M.; Sakaguchi, K. Shift-by-Wire System for Lexus RWD Vehicles. *SAE Int. J. Engines* **2017**, *10*, 689–694. [CrossRef]
5. Ahn, J.-W.; Lukman, G.F. Switched Reluctance Motor Drives. *Modeling Simul. Control Electr. Drives* **2019**, 275–324. [CrossRef]
6. Choi, Y.K.; Yoon, H.S.; Koh, C.S. Pole-Shape Optimization of a Switched-Reluctance Motor for Torque Ripple Reduction. *IEEE Trans. Magn.* **2007**, *43*, 1797–1800. [CrossRef]
7. Hieu, P.T.; Lee, D.-H.; Ahn, J.-W. Design of 2-Phase 4/2 SRM for Torque Ripple Reduction. In Proceedings of the 2012 15th International Conference on Electrical Machines and Systems (ICEMS), Sapporo, Japan, 21–24 October 2012; pp. 1–6.
8. Sheth, N.K.; Rajagopal, K.R. Optimum Pole Arcs for a Switched Reluctance Motor for Higher Torque with Reduced Ripple. *IEEE Trans. Magn.* **2003**, *39*, 3214–3216. [CrossRef]
9. Jing, L.; Cheng, J. Research on Torque Ripple Optimization of Switched Reluctance Motor based on Finite Element Method. *Prog. Electromagn. Res.* **2018**, *74*, 115–123. [CrossRef]
10. Salunke, N.A.; Patel, A.N.; Panchal, T.H. Torque Ripple Reduction of Switched Reluctance Motor by Changing the Rotor Pole Tip Radius. *IJRTE* **2019**, *8*, 4256–4259. [CrossRef]
11. Zhang, H.; Wang, S. Topology Optimization of Rotor Pole in Switched Reluctance Motor for Minimum Torque Ripple. *Electr. Power Compon. Syst.* **2017**, *45*, 905–911. [CrossRef]

12. Nakata, K.; Hiramoto, K.; Sanada, M.; Morimoto, S.; Takeda, Y. Noise Reduction for Switched Reluctance Motor with a Hole. In Proceedings of the Power Conversion Conference-Osaka 2002 (Cat. No.02TH8579), Osaka, Japan, 2–5 April 2002; Volume 3, pp. 971–976. [CrossRef]
13. Faiz, J.; Tahvilipour, F.; Shahgholian, G. Performance Improvement of a Switched Reluctance Motor. In Proceedings of the 31st Progress in Electromagnetics Research Symposium (PIERS), Kuala Lumpur, Malaysia, 27–30 March 2012; pp. 728–732.
14. Tekgun, D. Acoustic Noise and Vibration Reduction on Switched Reluctance Machines through Hole Placement in Stator/Rotor Laminations. Master Thesis, The University of Akron, Akron, OH, USA, 2017.
15. Dickinson, P.G.; Sykes, P.A.; Fulton, N.N.; Durairaj, T.R. Reducing Stress in Electric-Machine Rotors by Optimizing Holes. U.S. 9,246,363, 26 January 2016.
16. Hashemi, Z.; Zohrabi, F.; Mardaneh, M. A Multi-Objective Optimization of Switched Reluctance Motor Using a Hybrid Analytic-ANFIS Model Considering the Vibrations. *Iran. J. Sci. Technol. Trans. Electr. Eng.* **2019**, *43*, 361–371. [CrossRef]
17. Zhang, S.; Li, S.; Harley, R.G.; Habetler, T.G. Performance Evaluation and Comparison of Multi-Objective Optimization Algorithms for the Analytical Design of Switched Reluctance Machines. *CES Trans. Electr. Mach. Syst.* **2017**, *1*, 58–65. [CrossRef]
18. Zhang, J.; Wang, H.; Zhu, S.; Lu, T. Multi-Physics Multi-Objective Optimal Design of Bearingless Switched Reluctance Motor Based on Finite-Element Method. *Energies* **2019**, *12*, 2374. [CrossRef]
19. Schramm, D.S.; Williams, B.W.; Green, T.C. Torque Ripple Reduction of Switched Reluctance Motors by Phase Current Optimal Profiling. In Proceedings of the PESC '92 Record. 23rd Annual IEEE Power Electronics Specialists Conference, Toledo, Spain, 29 June–3 July 1992; Volume 2, pp. 857–860. [CrossRef]
20. Liang, J.; Parsapour, A.; Moallem, M.; Fahimi, B.; Kiani, M. Torque Profile Optimization in Switched Reluctance Motor. In Proceedings of the 2019 IEEE 28th International Symposium on Industrial Electronics (ISIE), Vancouver, BC, Canada, 12–14 June 2019; pp. 414–419.
21. Nakayama, S.; Itoh, T.; Kimura, K. Switched Reluctance Motor. U.S. 7,948,145, 24 May 2011.
22. Krishnan, R. *Switched Reluctance Motor Drives: Modeling, Simulation, Analysis, Design, and Applications*; CRC Press: Boca Raton, FL, USA, 2017.

© 2020 by the authors. Licensee MDPI, Basel, Switzerland. This article is an open access article distributed under the terms and conditions of the Creative Commons Attribution (CC BY) license (http://creativecommons.org/licenses/by/4.0/).

energies

MDPI

Review

A Review on Energy Efficiency in Three Transportation Sectors: Railways, Electrical Vehicles and Marine

Morris Brenna [1], Vittorio Bucci [2], Maria Carmen Falvo [3,*], Federica Foiadelli [1], Alessandro Ruvio [3], Giorgio Sulligoi [2] and Andrea Vicenzutti [2]

[1] Department of Energy, Politecnico di Milano, 20156 Milan, Italy; morris.brenna@polimi.it (M.B.); federica.foiadelli@polimi.it (F.F.)

[2] Department of Engineering and Architecture, University of Trieste, 34127 Trieste, Italy; vbucci@units.it (V.B.); gsulligoi@units.it (G.S.); avicenzutti@units.it (A.V.)

[3] DIAEE Electrical Engineering, University of Rome Sapienza, 00184 Rome, Italy; alessandro.ruvio@uniroma1.it

* Correspondence: mariacarmen.falvo@uniroma1.it

Received: 15 March 2020; Accepted: 8 May 2020; Published: 9 May 2020

Abstract: The present paper is a review on efficiency issues related to three important sectors of the transportation systems: railways, electrical vehicles, and marine. For the three sectors, the authors, in reference of their knowledge and research area, show the results of a wide literature analysis, in order to highlight which are the measures, in terms of technological solutions and management techniques, which are recently investigated and implemented, for improving the three transportation systems, from the point of view of efficiency. In particular, for the railway transportation sector, a wide analysis is presented, detecting which are the main measures adopted for improving the efficiency, related to the power systems for supplying trains and to the train traffic control, with a focus on the storage system integration. For electric road vehicles the analysis is focused on the plug-in electrical vehicles and on the infrastructure for their recharge, with an emphasis on how these vehicles can support the grid, e.g., through Vehicle to Grid (V2G) applications. Finally, for the marine transport service the review is related to the propulsion systems and on how the different solutions can meet the objective of efficiency.

Keywords: electrical transport; electrical vehicle; energy efficiency; marine transport; railway

---

1. Introduction

The guideline suggested by European Commission in 2008 about climate and energy, in a horizon 20-20 context, focuses on the reducing the greenhouse gas (GHG) emissions and increasing the production from renewable energy sources (RES) [1]. The main aspect taken into account, among the first goals reported in [2,3], is the electrification of the transport sector [4].

Nowadays, the current world population of 7.3 billion is expected to reach 8.5 billion by 2030, so the mobility demand for people and goods has already grown in many countries, and it will keep growing, with heavy environmental impacts [5].

From this point of view energy savings solution will be increasingly necessary in order to reduce more and more the energy demand required by the transport systems for the next future. However, energy efficiency has been already increased in the last century thanks to the many technological innovations that have occurred over time in the transport sector [6].

The present paper is a review on efficiency issues related to three important sectors of the transportation systems: railways, electrical vehicles (EVs), i.e., cars, and marine. For the three sectors, the authors, in reference to their knowledge and research area, show the results of a wide literature analysis, in order to highlight which are the measures, in terms of technological solutions and management techniques, which are recently investigated and implemented, for improving the three transportation systems from the point of view of the efficiency, in reference to their peculiarities.

In reference to the railway sector, considering that most of the rail-lines are electrified and use electric trains, the review deals with the measures that can be adopted for improving the efficiency of existing electrical systems, especially in reference to the electrical infrastructure for the supply of the trains and to the management of the train traffic. A focus is reported also on the recent measure of integration of storage systems in the railways system, as action in the direction of efficiency improvement.

In reference to the on-road vehicles the analysis is focused on the plug-in electrical vehicles and on the infrastructure for their recharge, with a focus on how these vehicles can support the grid, e.g., through Vehicle to Grid (V2G) applications.

In reference to the marine transport, the review is related to the propulsion systems and to how the different solutions can meet the objective of efficiency.

It is worth stressing that the final objective is not to find a common metric for the assessment of the efficiency and its improvement in relationship to the different solutions, but to find differences and to cluster the possible measures that can be applied to the three transport sectors for improving the efficiency, in terms of the functions of their peculiarities. It gives an idea of what has been already done and which could be the margin of development to be explored.

This paper is organized in three parts, each dedicated to the single transportation systems investigated.

2. Efficiency in Railway Transportation Sector

Railways are an energy efficient mode of transport for both passengers and freight. Passenger railways cover an average value of 4.1 of toe per million passenger-km; the freights rails cover an average value of 3.49 of toe per million passenger-km. It represents 2% of total transport energy demand [7]. The construction of new railway lines, as the growing of the number of trains on the lines increases, imposes that energy savings solutions are increasingly necessary to reduce the energy demand in the future, considering both urban and suburban systems.

Energy efficiency has already improved in the last century thanks to the many technological innovations that have occurred over time in the railway sector: from the first storage systems on board trains (1903) to cover short distances, until today with the development of different types of electrified lines both DC and AC (single and three-phase). The development of high efficiency electric engines and the increasing progress in power electronics allowed reducing the energy demand by trains, also thanks to braking energy recovery utilization. In electrical substation, from the '60s to today, the replacement of mercury vapours rectifiers with silicon rectifying diodes in DC electrical substations, and the continuous introduction of high performance electric machines for energy conversion, minimizing the losses, have increased the efficiency of the systems. A continuous evolution concerned also the better integration between train and infrastructure (pantograph/overhead lines and wheels/rails). Evaluations about energy recovery, centred on the control of speed profile, taking into account timetable optimization [8–10] and storage design systems [11], offer a best solution to reduce the system's energy consumption, in order to maximize the effectiveness of regenerative braking. The importance of increasing recovering energy, especially when there is no train that can absorb it, or it is not possible to install a reversible substation, is also discussed [11]. Several solutions have been carried out focusing on trains, power plant design and timetable optimization, in order to decrease the energy dissipated on braking rheostat. Regarding the trains driving performance, many solutions are proposed to optimize driving style management and the design of super-caps on board. Several design solutions have been carried out for the siting and sizing of storage systems in railways power plant and at last, some proposal about optimal scheduling process have been formulated for timetable management.

The results of a wide survey on the newest techniques of driving style management with timetable optimization and on the solution of storage systems on board and stationary is summarized in the next paragraph.

2.1. Efficiency Solutions Based on the Driving Style Management and Timetable Optimization

Energy-efficient driving management coupled with the optimization of timetable allows finding optimal train speed profiles and departures to minimize the energy consumption and get the optimal braking trend due to time delay, so as to maximize the recoverable energy in line. Researches start from 1960, with timetable optimization and energy-efficient driving from the first suggested optimal control model in 1968 made by Ishikawa [10].

Several timetable optimization methods have been proposed in recent years [10]. Albrecht in [12] developed a new method based on dynamic programming, to manage running time of trains using an optimal combination of headway and synchronization time, with the task of reducing power peaks and energy consumption. Chen et al. [13] applied the genetic algorithm to optimize train scheduling; particularly the goal was to reduce power consumptions, preventing the synchronous acceleration of many trains. Ramos et al. [14] presented a method to maximize the braking energy recovery, during off-peak hours maximizing the overlapping time between acceleration and braking of the trains. Kim et al. [15] carried out a multi-criteria mixed integer programming, coordinating the train departure times at the starting stations, so as to minimize the peak energy and maximize regenerative energy utilization. Peña-Alcaraz et al. in [16] tested methods to synchronize the movement of trains in the Madrid Metro Line 3 reaching a 3.52% energy saving. In [17,18] a comparison between to real metro-lines in Italy and Spain is made in terms of energy savings that can be obtained by the recovering of the braking energy of the trains. Yang et al. with several studies suggested a cooperative scheduling model to schedule the accelerating and braking phases of nearby trains. In [19] a simulation performed on real data obtained from the Beijing Metro Yizhuang Line shows a great improvement in the overlapping time of around 22%. In [20] a stochastic cooperative scheduling model taking into consideration the randomness of departure delay, for trains considering busy stations, shows a percentage of save energy around 8%, compared with the cooperative scheduling approach reported in [19]. In [21] the same authors offer a model to optimize the timetable, coordinating trains at the same station to maximize the utilisation of recovery energy and reduce waiting time for the passengers. The model reached an 8.86% energy saving, with a waiting time of 3.22% relevant to the current timetable. In [22] a scheduling approach regarding effective speed profiles, to arrange arrivals and departures of all trains, reaches a 6.97% reduction in energy consumption, in comparison with the current timetable.

In [23] a model, based on a multidimensional state vector subspace for train operation, is presented. A smart scheduling methodology useful for multi-train energy saving operation and an optimization procedure based on a genetic algorithm and regenerative kinetic energy, to lowest total energy consumption, is proposed in [23,24].

In [25] the train trajectory optimization is carried out, in order to define a better train target speed profile, to minimize a cost function, including energy consumption and trains arriving on time for all trains. The minimum energy consumption, under different departure headways, is calculated, by using a heuristic algorithm in [26], reaching a reduction in energy consumption up to 19.2%. Furthermore, several studies on driving style are carried out, to define eco-energy driving profile strategies of the trains to couple with timetable optimisation. The motion stage of a train consists in acceleration, cruising, coasting, and braking management. Generally, the speed profile of trains with short travel distance or close intermediate stops, like in tram and metro systems, could not contain the cruising phase.

In [27] a dual speed-curve optimization for energy-saving operation of high-speed trains is proposed using two optimizations. An offline global and online local optimization, demonstrating the increase in energy saving, compared with other well know existing methods that use one-time optimization processes. The main structures of the dual optimization method proposed include: a global optimization to obtain better driving style; after, the speed trajectory is adjusted in real time by local optimization. Additionally, regarding rolling optimization, a closed loop control is integrated with a consistent optimization process that continuously corrects; at least global optimization is reachable using a genetic algorithm, with characteristics and predictive control, with the local optimization characteristics, to compensate for the limitations of a single optimization process [27].

In [28] the authors propose an integrated approach, consisting of both offline and online techniques. The projected framework generates throttle sequences that lead to energy saving under the constraints of trip time and computation time. This work leverages the fast-growing machine learning techniques, so to extract the optimized driving behaviours of human drivers and encode the learned knowledge into a parameter decision tree for fast online optimization. A case study on a given locomotive proved the effectiveness of the proposed framework and an energy saving of 9.84% on different running conditions can be achieved.

In [29] the authors includes a new method for speed curve definition and tracking control, based on a random reinforcement genetic algorithm (GA) to avoid the local optimum and a sliding mode controller developed for speed curve tracking with bounded disturbance.

An improved chicken swarm optimization algorithm for energy-saving for a train, by taking minimum energy-consumption, accurate stopping and punctuality as optimization objectives is in [30] without changing the existing equipment and infrastructure. Chicken swarm optimization is a global optimization algorithm, which integrates the advantages of genetic, particle swarm and bat algorithms.

In [31], the authors introduce an optimization of train speed curve applied in a real case study of the Taipei Mass Rapid Transit System for journeys from "Dingpu Station" to "Yongning Station", showing that operational energy consumption could be reduced up to approximately 58%. A real driving method to reduce the traction energy demand is presented in [32]. In this case the authors carry out theoretical optimal driving solutions thanks to a train simulation using an enhanced Brute Force searching algorithm. A driver practical training system (DPTS) is created to help drivers practice energy-efficient driving controls. A train speed trajectory optimization method associated with a driver practical training system (DPTS) is the main goal. Thanks to the DPTS, traction energy consumption is reduced by around 15%.

The authors in [33] propose a methodology that includes an objective function using cardinality and square of the Euclidean norm functions. The optimization model proposed, allows defining properly the utilization of the regenerative energy. To solve the convex relaxation counterpart of the original NP-hard problem, a two-stage alternating direction method of multipliers is designed. The procedure produces an energy-efficient timetable of trains.

Genetic algorithms have been used for a subway line in Milan and it is reported in [34]. The main goal is to fulfil the transition from a traditional system to a driverless one. It shows an energy saving increase equal to 32.89%.

2.2. *Efficiency Solutions Based on Stationary and on Board Energy Storage Systems*

Many studies about on board and stationary energy storage systems have been developed, especially for DC railway systems, without a reversible substation, where it is not possible to drive the surplus of regenerated energy back to the main AC power supply. Consolidated energy saving solutions using reversible substation focused on different implementations are reported in [35–41]. The innovative technologies used to design energy storage systems are super-capacitor, battery, or flywheel and IEC 62924:2017 standard fixed requirements and test methods. The International Union of Railways (UIC) with the sub commission "Energy Efficiency", creates a database where all relevant railway energy-saving technologies should be analyzed, categorized, and evaluated [42,43].

Regarding on board energy storage systems, they are already in use by some rail transit companies. The main advantages are the reduction of peak power, the stabilization of voltage, the loss reduction and the possibility to operate catenary free [44]. Real applications of on-board storage systems are the Brussels, Madrid metro and Mannheim tramway lines. The percentage of energy saving reported in [45–47] are 18.6% ÷ 35.8%, 24% and 19.4% ÷ 25.6%, respectively. To reach high integration with motor drive control, some research studies are focused on the optimal design, sizing and control of on board energy storage systems [48–52]. Focuses on stationary storage systems, the real implementation of wayside Energy Storage System (ESS), show an increase in energy savings of up to 30%. The percentage of energy saving by ESS moreover is influenced by system features and storage technologies. In [53] it is highlighted that auxiliary battery-based substations could represent a feasible solution to store the required energy for partly powering a train, supporting the electric substation during train accelerations and to compensate for voltage drops. Numerous commercially available stationary systems are available. Sitras SES (Static Energy Storage) system, marketed by Siemens, can reach up to 30% of energy saving using a super-capacitor technology that can offer 1 MW peak power for 20 ÷ 30 s, with 1400 A DC discharging current. This system is in Germany (Dresden, Cologne, Koln and Bochum), Spain (Madrid) and China (Beijing). The EnerGstor of Bombardier Company, based on supercaps, is able to reach 20% ÷ 30% of reduction of energy demand [54]. Another system super-caps based in Hong Kong and Warsaw metro systems [55] is developed, by Meiden and marketed by Envitech Energy, with scalability from 2.8 to 45 MJ of storable energy.

3. Efficiency in Electrical Vehicles Transportation Sector

The growing awareness of environmental issues, social pressures towards a solution to climate change that lead to a progressive disinvestment in fossil fuels [56] and the continuous technological improvements of storage batteries, have now led car manufacturers all over the world [57–59] to invest in new platforms for the construction of electric vehicles (EV). Electric vehicles, unlike those with an internal combustion engine (ICE), have the advantage of avoiding local emissions of greenhouse gases, or eliminating them if powered by renewable sources [60]. In fact, global emissions from electric vehicles vary according to the power generation mode. If coal plants produce the energy, they produce substantial global emissions that determine only local benefits of the use of electric vehicles [60]. Conversely, the use of alternative sources such as wind or photovoltaic allows a significant reduction in global emissions, given their lower carbon intensity [60–64]. Nevertheless, even with electricity generated from coal-fired power plants, the global emissions of an electric vehicle in the well-to-wheel cycle are lower than those ones generated by an ICE vehicle [60]. Electric vehicles are nowadays supported and encouraged by various governments around the world [65], also through measures aimed at reducing the tax burden, setting up free parking lots dedicated to them and equipped with charging infrastructures, the use of preferential lanes, access limited traffic areas, etc. [65]. An electric vehicle, unlike an ICE vehicle, allows it to be recharged from the mains, now available everywhere. However, ICE vehicles can refuel with a method dating back to the early 20th century. This refueling process started from a pharmacy that sold petrol tanks, and it was done completely manually in seconds. Electric vehicles were suffering from much longer charging times and the absence of public infrastructure, which led to their substantial disappearance to the present day [66]. In modern electric vehicles, the recharging process takes place mainly via an on-board charger; this mode requires a very long charging time, which can reach several hours [66]. Most models of electric vehicles nowadays support fast charging through dedicated infrastructures that allow, on average, to reach 80% of the state of charge (SOC) in about 30 min [67]. Nevertheless, the charging power is still not comparable to the refueling of a traditional ICE vehicle. Besides, fast charging, in addition to being more expensive, leads to faster degradation of batteries if used frequently [67].

The impact of electric vehicles on electric transmission and distribution grids is still negligible due to their low diffusion [61,68]. Their growing diffusion will inevitably cause an increase in the demand for electricity which may both have a negative impact, but also have a beneficial effect on the electricity system if well integrated [61,68]. Indeed, a further increase in the demand for electricity at peak times, because it is not restricted [68], could lead to an overload of the electric system and the underutilization of renewable sources [61,68]. It follows that leaving the decision on when to recharge without any coordination to individual users will inevitably lead to the need for further repowering of the transmission and distribution grids, premature aging of the devices and a lower quality of the energy supplied to the users [67,68].

Conversely, if recharges are coordinated among themselves with intelligent logic that also consider the actual production from renewable sources, then the electrical system it will benefit by reducing the percentage of fossil fuels in end uses [61,67,68].

To address these issues, new regulatory rules and management strategies are needed close to new technological advancements for a suitable integration of electric vehicles in the current transmission and distribution grids [69,70]. Some studies, but also common practice, have shown that on average a vehicle is parked for 95% of its life and that the weekly trips are often just the journey homework. It is from these considerations that the EVs can be seen not only as means of transport, but as active elements able to play a role in the management of power lines. The EV, once connected to the electric network, is therefore seen as an integral part of the system capable of supplying energy when demand is high (by discharging the battery) and absorbing the surplus of energy produced when demand is lower (by charging the battery) [71]. This practice is named Vehicle-to-Grid (V2G) and is based on the bidirectional power flow, from the grid to the vehicle but also from the vehicle to the grid [72,73]. When operational and integrated with the network, it can bring significant benefits such as [74]:

- Creation of electrically autonomous islands;
- Lower dependence on foreign countries;
- Better exploitation of renewable energy sources;
- Reduction of emissions for the generation of electricity;
- Increase of power quality in the network;
- Economic benefits for the user who makes their car available for the network.

All this, however, is achievable only with the introduction of the smart grid concept. This requires modernization of the transmission and distribution grids and the introduction of tools for measurement and communication. The definition of a business model, clearly defining the players involved in the value chain and fairly compensating the exploitation of EV batteries, is another crucial factor for the diffusion of such a paradigm.

On a smaller sale, V2G concept can be applied also in the context of a building (vehicle-to-building (V2B)) or of a home (vehicle-to-home (V2H)) [75]; the objective is to benefit from the exploitation of the batteries of EVs when connected to a smart system. Back up and time-shift are two examples of the possible advantages that a bidirectional interaction with EVs can bring. The higher self-consumption rate of renewable energy production achievable thanks to the exploitation of batteries, along with their lower operating costs with respect to conventional cars, represent the main driver for the integration of EVs in smart home environments.

The number of electric vehicles is expected to grow sharply in the incoming decades and the potential impact on the electric grid could be substantial. Of course, this aspect is not related to generation and transmission, in which the effect is relatively small, but it affects the distribution network in a significant way [76]. For the distribution system operator (DSO) the fluctuation of the load due to the plug in of vehicles must be minimized in order to guarantee a good quality of service for the final user. Moreover, these technical problems have to be matched with additional problematics coming from the interaction between the EV user and the grid operator, since the service must be convenient for both the grid and the owner of the vehicle.

The first mention of the vehicle to grid service was proposed by Amory Lovins in 1995, and then developed by William Kempton [76]. The main idea is that the vehicle can be considered as a storage system able to provide energy to the distribution network when parked which can be charged and discharged according to the grid necessities and price of energy fluctuation.

V2G technology can offer a wide range of functions for the grid: load balancing, harmonics suppressing, power quality improvement, peak load shaving, voltage sags reduction and interaction with Renewable Energy Sources (RES) [77]. The V2G concept is still at its first stages and is becoming more and more important as the diffusion of electric vehicles increases. From the literature [76], four key issue areas can be identified: a smart dispatching from the operator point of view, a smart charging management from the vehicle point of view, the bi-directional charger and the effect that the V2G service has on the vehicle's battery. V2G functions have a great potential for electric vehicles to become a tool for the electric grid, to manage power and energy storage applications. The challenge in developing these functions is that it is always mandatory to remember that vehicles have mobility as their primary mission and not storage system for the grid. This means that every time the vehicle's battery is used for providing services to the grid, the battery cannot be charged and discharged in a way that does not guarantee to the vehicle owner a full availability of its own car. Moreover, another critical aspect is the fact that using vehicle's batteries for providing power to and from the network contributes to the degradation of the battery, and so limitations on charge and discharge cycle must be set.

The V2G modelling has to move through the day, simulating hour by hour a real life situation, in which a certain number vehicle is generated according to a specific load profile and interacts with energy requests needed by the photovoltaic park managed by the same aggregator [78]. It is important that the fleet car has to be composed of many vehicles in order to aggregate a total power and energy storage capacity able to deal with the fluctuation of power production during the all day. The figure responsible of managing the vehicles fleet of this region is called aggregator [79].

To be more specific an aggregator is a market participant which aggregates in a unique offer the distributed generation of a certain zone. The reason for the existence of this market figure is that single distributed generation plants are usually of small size, which could be neglected in a big market like the energy one. The aggregator will be in charge of collecting both distributed generation plants and energy storage systems in a defined area. The aggregator can also be responsible for managing electric vehicles charging and discharging processes in order to guarantee to its customers a fully charged vehicle when the parking time is over while guaranteeing to its own aggregated photovoltaic plants the possibility of storing and delivering extra power when requested. The extra power, which could be managed using the fleet batteries, represents the variation in photovoltaic production, or other renewables, for the next hour with respect to the forecasted production of the day ahead.

The reason behind this definition is that the energy market is called Day-Ahead-Market (DAM), since each day the market trades the energy quantities that will be exchange during the next day. The problem is that renewable system production can be predicted but with high errors if we consider time windows of 24 h. From [79] it is possible to see that moving from 24 h ahead time windows to 1 h before the event the error on the prediction is reduced from 24% to 12%. In [80] it is shown that with an artificial neural network it is possible to predict, 24 h in advance, the irradiation during the day with an error as low as 28%. Moreover, being RES market price takers since they bid at 0 €/kWh, they will be always able to enter the market. However, the next day when the real production will change due to the unpredictability of the source there will be a lack or a surplus of energy [81,82]. This means that in the case of extra energy, part of it could be lost since the grid is already balanced thanks to the correct behaviour of the system operator, while in case of a lack of energy, additional energy is requested to traditional plants. Both these solutions act in the opposite way of a smart management of the electric grid. Moreover, when a production plant is not able to provide the energy production set during the day ahead market, it will be responsible for unbalancing the system, being subject to an unbalance forfeiture that the owner has to pay to the transmission system operator.

Some Important Considerations

As previously described, local CO_2 emissions for electric vehicles are negligible, instead they contribute to global CO_2 emissions. The latter depend on the manufacturing process of the batteries and the vehicle, but the most significant component depends heavily on the way electricity is generated. Indeed, for countries that use more renewable sources, CO_2 emissions are significantly lower than the equivalent for an ICE vehicle. On the other hand, for those countries mainly based on fossil fuels, the emissions are comparable to those produced by a very efficient ICE vehicle, as shown in Figure 1.

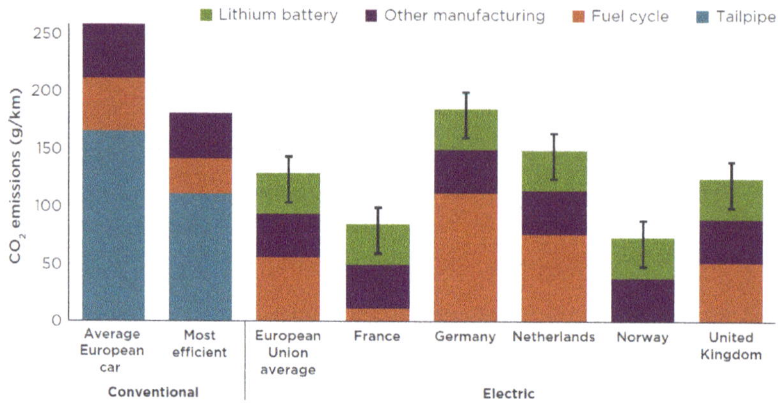

Figure 1. Life-cycle emissions (over 150,000 km) of electric and conventional vehicles in Europe in 2015 [83].

Using electric vehicles batteries for helping the grid, in managing power and energy unbalancing, could be an interesting solution since it occurs without additional costs related to dedicated storage systems.

Thanks to a full communication system between vehicles and their charging poles, it will be possible to elaborate charging strategies in order to use the parked vehicles as a supporting storage system for the grid. However, in order to have a high amount of energy available from the vehicles an aggregator will be needed.

The aggregator will be a market participant that will collect many distributed generation plants and storage systems in a certain area, in order to gather in a single figure a high amount of power increasing its impact on the market.

As a final consideration it is important to consider that if the future vehicles will be provided with batteries able to store a higher amount energy, each single vehicle will be able to exchange more energy during a discharge with a depth of 50%. This means that in that case the number of vehicles required will be even lower with respect to these simulations and the vehicle to grid function will be even more performant.

4. Efficiency in Marine Transportation Sector

International maritime shipping is the most used means to transport goods and people. In fact, global shipping is responsible of moving about 90% of the world trade, while cruise ships in 2017 allowed nearly 25 million of passenger to sail the oceans. Moreover, the availability, the low cost, and the fuel efficiency of the maritime transport made possible the growth of industrial production in emerging economies. These significant results were enabled by the ability of the shipbuilding industry to respond promptly to the market demands, despite the unfavourable economic scenario in recent years caused by the global economic crisis. In particular, at present ship designers and shipbuilders are spending a lot of effort in increasing the performance of ships, by introducing new technologies on board. Nevertheless, it is well known that shipbuilding industry is one of the hardest metal industries, characterized by high levels of raw materials and energy consumption, hazardous materials exposure, and potential risk of sea and air pollution [84]. In addition, shipping greatly contributes to the emission of air pollutants (PM, SOx, NOx), sea pollutants (noise, thermal, discharges), and greenhouse gases. In particular, the contribution of maritime transport to the latter is up to 3% of global emissions.

However, maritime transport is still the most fuel-efficient means of shipping goods, as can be seen from Figure 2. As an example, its CO_2 emissions are nearly 100 times lower than airplanes when considering the tons of goods transported over km, while presenting a similar operative range.

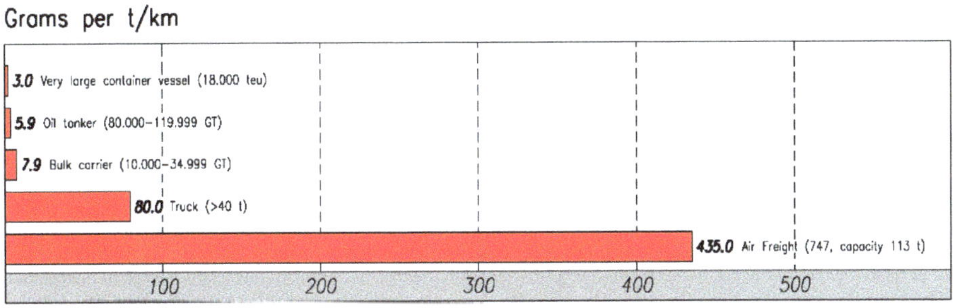

Figure 2. Comparison of typical CO_2 emissions among goods transportation options.

The shipping and shipbuilding industries are governed by global international agreements, promoted by the International Maritime Organization (IMO). In order to face the future environmental challenges caused by the expected growth of world's population and global economy, the IMO decided to set technical and operational measures to reduce ships' emissions, to be applied worldwide. In particular, in July 2011 the IMO delivered a roadmap for the reduction of environmental footprint of ships (Figure 3), by adopting the International Convention for the Prevention of Pollutions from Ships (MARPOL), Annex VI, officially entered into force in January 2013 [85].

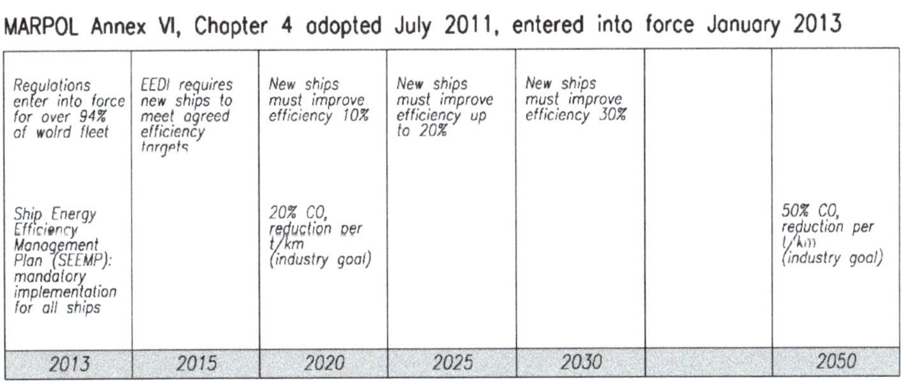

Figure 3. International Maritime Organization (IMO) roadmap for the reduction of ships' environmental footprint.

Such a decision was the result of studies about the mitigation of ship emissions started in 2005 by IMO. The resulting MARPOL annexes collect specific guidelines in this regard. In detail, MARPOL labels specific sea regions as "special areas", considering their oceanographic and ecologic importance, as well as sea traffic and risk of environment contamination. For these areas, MARPOL impose strict limits for air pollutant emissions, including volatile organic compounds (VOCs), nitrogen oxides (NOx), polychlorinated biphenyls (PCBs), sulphur oxides (SOx), chlorofluorocarbons (CFCs) and heavy metals. These substances are specifically addressed since they contribute to the creation of ozone at ground level (VOCs and NOx), rain acidification (NOx and SOx), eutrophication or oxygen depletion of inland and coastal waters (NOx), depletion of atmospheric ozone (CFCs), and the accumulation of PCBs and heavy metals in the food chain. While an emission limit is enforced all over the globe, in special areas stricter limits are imposed depending on their specific characteristics:

- SECAs, Sulphur Emission Control Areas, where there are specific requirements regarding SOx and PM emissions;
- NECAs, Nitrogen oxide Emission Control Areas where there are additional requirements regarding NOx emissions
- PSSAs (Particularly Sensitive Sea Areas), as defined in IMO Resolution A.982(24), are characterized by "unique or rare ecosystem, diversity of the ecosystem, vulnerability to degradation by natural events or human activities, significance of the area for recreation or tourism, biological research value, or historical value".

4.1. Increasing the Efficiency of Ships

The most obvious way of limiting the environmental footprint is the reduction in fuel consumption, which means increasing the overall efficiency of ships. Thus, the shipbuilding industry is continuously working towards improving hull forms, engines, and propellers. Setting aside the hydrodynamic area, the reduction of the ship's pollutant emissions down to a level compliant with the above-depicted goals can be reached by means of three strategies [86]. These are: the switch to a different, cleaner, fuel (e.g., low sulphur fuel oil, natural gas, methanol, etc.); the installation on board of equipment for exhaust gas treatment (e.g., SCR, scrubbers, etc.); and the improvement of a ship's electrification. The use of other fuels can lead to different results, both on the pollutant emissions reduction and the overall ship's fuel consumption, depending on the specific fuel used. The second method is an application of technical solutions already used in other fields, like automotive and land power generation, with some specific modifications due to the particular marine environment. These two strategies have pros and cons, but are both aimed at improving the internal combustion engines (ICEs) performance in terms of pollutant emissions. Conversely, the electrification option implies substituting existing ICE with electrical motors, thus providing: an increase in power output controllability and power density; a reduction in noise, vibrations, heat, and maintenance complexity; more degrees of freedom for arranging the engines in the engine room. Thus, similarly to what is happening in other transportation modes, electrification is one of the most applied methods to improve ships' efficiency and reduce their environmental footprint [87]. Obviously, given the power required by a ship (as an example, the propulsion system of a cruise liner can reach a power of tens of MW [88]), the primary source of electric power is commonly a set of ICE powered generators. In particular, diesel generators are the most used, but generators powered by gas turbines can be also found, typically in naval applications. For these generators, the use of different fuels and the adoption of exhaust gas treatment are viable options to reduce further the ship environmental impact. Other methods to generate electric power are starting to be applied, like fuel cells. Regarding the latter, their use as main power sources is limited to small crafts, while there are some experimental installations on board large ships of fuel cells as auxiliary generators.

However, at present there is no viable option for totally powering a big ship without using fossil fuels. Indeed, the significant energy density of fossil fuels is one of the enabling factors that make big ships feasible, given their scope of work and their typical routes. To give an example, the very large container and bulk carriers that are used to ship goods all around the world require a range of up to 7500 nautical miles to be usable (e.g., the route from the port of Singapore to the port of Trieste is circa 7100 nm long, and must be sailed without intermediate stops to avoid lengthening transport times, and thus losing money). This means that such ships have up to 5000 m^3 of fuel on board, which is usually replenished only at the departure and arrival ports. Conversely, ships for local transport (in a range of a thousand nautical miles) have nearly 1000 m^3 of fuel on board, and the refuelling is done only at the home port due to economic reasons (there are bilateral contracts and national taxation differences among ports, making the refuel in other ports not economically convenient). Besides the energy density of fossil fuels, which is still not in the reach of present energy storage systems and other energy production systems, there is also the issue of the refuelling times. In fact, a ship's refuelling is usually done by means of tanker ships, that have tanks and pumps installed on board, dedicated to such a task. The tankers can refuel a large ship at nearly 500 m^3/h, while smaller ships (the local transport ones) are usually refuelled up to 250 m^3/h, by using in both cases electric pumps that can reach 2–3 MW of power. These figures mean that a complete refuel can be achieved in 10 h for a large ship, and 4 h for a small one (excluding all the logistic and preparation times). By assuming a volumetric energy density of about 36 MJ/l for the fuel (that is nearly 10 MWh/m^3), it is possible to calculate an equivalent refuelling power of 5 GW for a large ship, and 2.5 GW for a small one. Even assuming that it is possible to install on board an energy storage system capacity capable of delivering the required autonomy, the required recharging power level is so high it may be unsustainable for actual port power systems. For the tankers, these are refuelled in their port, with a speed that is very dependent on the specific port operation. Thus, at present, the focus on ships is given to the increase of the overall fuel efficiency of the ship, focusing on the different aspects stated above.

Considering only the propulsion system electrification, several different architectures can be applied. The first is the series configuration, where the propellers are powered by electric motors, and ICEs are used only to generate electric power (Figure 4). The second configuration is the parallel one, also called hybrid propulsion system [89], where electric motors and ICEs are both connected to the propeller (Figure 5).

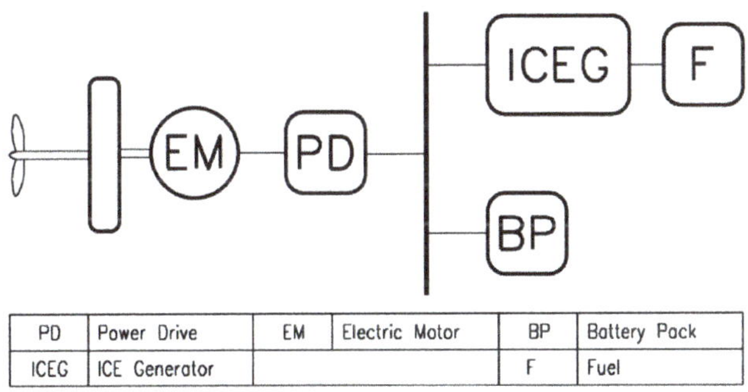

Figure 4. Electric propulsion, series configuration.

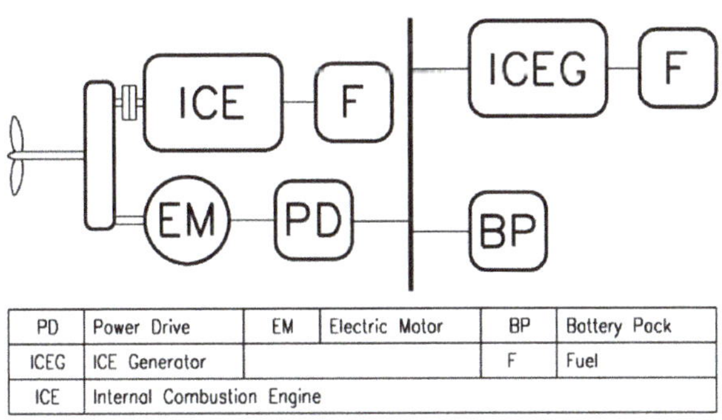

Figure 5. Electric propulsion, parallel configuration.

For ships with ship services loads requiring a significant electric power (e.g., cruise liners), the integrated power system (IPS) configuration is a viable option (Figure 6) [88]. In such a case, the same set of ICE generators is used to power both the propulsion and the other loads through a single power system [90]. Such a configuration allows obtaining a reduction in the total size of on-board generators in respect to the use of separated sets of generators, one for each subsystem. Moreover, the IPS architecture can exploit either the series or the parallel configuration for the propulsion system, depending on the ship operative requirements. Nowadays, the evolution in power electronics is pushing forward the performance levels achievable from the electric drives, leading to the pervasive presence of power electronics converters on board ships. These can be used to control electric motors, manage power flows on the power system, and interface energy storage systems, leading to the so called integrated power and energy system (IPES), shown in Figure 7 [91,92].

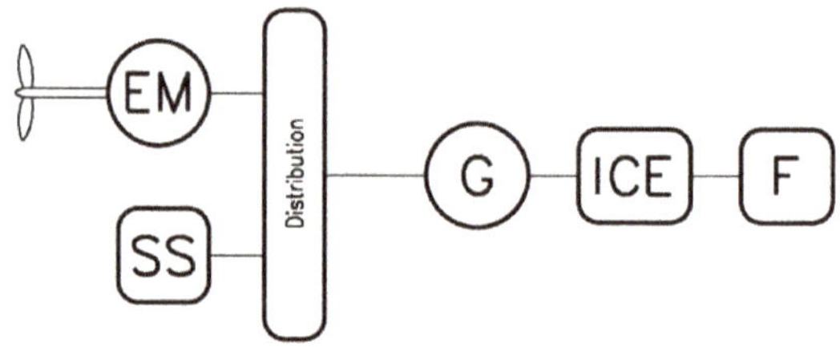

EM	Electric Motor	G	Generator	F	Fuel
SS	Ship Services	ICE	Internal Combustion Engine		

Figure 6. Integrated power system architecture.

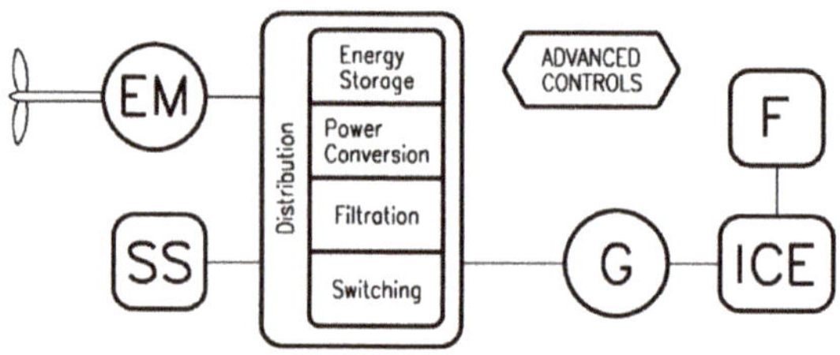

EM	Electric Motor	G	Generator	F	Fuel
SS	Ship Services	ICE	Internal Combustion Engine		

Figure 7. Integrated power and energy System architecture.

The integration of energy storage systems is useful for further increasing the overall ship efficiency. In fact, small ships can sail short routes on stored power only, while big ships can use energy storage systems as a transitional source of power to be used in emergency, avoiding running additional generators in some operative conditions. Moreover, storage systems can be used to perform peak shaving, allowing installing on board smaller generators, with a consequent reduction of weights, volumes, and overall fuel consumption [93].

In addition to the propulsion system electrification, also the replacement of mechanically driven equipment (e.g., pumps, cranes, etc.) with electrically driven ones is a viable option to increase the overall ship efficiency. The coupling of these new electric loads with power electronic converters enables further improvement in efficiency, reliability, and performance, by removing mechanical regulation equipment. Obviously, the introduction of these additional loads requires an increase in the on-board generators' power, further motivating the shift towards IPS or IPES configurations for ships.

4.2. The Issue of Designing Efficient Ships

The design of a ship is a complex process, because it is necessary to integrate different subsystems in a reduced space with several constraints. In fact, each subsystem is needed for the correct operation of the ship, but at the same time, it is in competition with the others for space and weight allocation. Ship designers have to consider both naval architecture issues (e.g., stability, hull form, manoeuvrability, structure, etc.), ship's operative requirements, costs, efficiency, and so on. The ship designers must in fact design a system of systems [94], where the optimal design solution for the overall ship is never the sum of the optimal solutions for each subsystem design. In fact, in the shipbuilding industry the most significant proof of concept is not related to the demonstration of a working technology. Instead, it is given by the demonstration of achievable results in improving ships' key performance indicators (KPI—such as space, weight, safety, and efficiency). Thus, a promising technology which is able to increase the efficiency of a single subsystem may prove to be irrelevant for increasing efficiency in a ship, or may worsen it either due to its effect on other subsystems or due to the ship's specific operative requirements. As an example, electric propulsion allows a significant reduction in fuel consumption compared to mechanical propulsion mechanisms in ships with several different operative speeds, thanks to the small loss of efficiency of electric drives at variable loads. Conversely, ships sailing for long times at constant speed (e.g., tankers) achieve the lowest fuel consumption by using mechanical propulsion, due to the high efficiency of low speed 2-stroke diesel engines when operated at their optimal load point.

In such a context, integrating new technologies dedicated to the efficiency increase into an existing ship is an engineering challenge, possibly being unfeasible or requiring an amount of modifications so high as to make the obtainable gains not justifiable. Likewise, the design of a new ship able to exploit new subsystems to achieve a more efficient operation is a complex task too. Ship designers have to take into account the integration of the new technologies since the first stages of the ship design (i.e., the early stage design), to assure their correct on board exploitation [95–97]. Such an approach is required, since the most impacting decisions about ship design are taken during the first stages of the design process, and cannot be changed on later stages without deeply affecting costs and times. For common ships' designs, the previous knowledge base is, in general, sufficient for making a correct guess about viable design solutions. Conversely, technologies with a significant impact on the ship's KPI may require proceeding with several tests before reaching a feasible design [98].

To overcome the need of making design choices based on uncertain data, tools are being created. In fact, the advancements in information technology led to the creation of software tools, aimed at easing the designers' work [92,99,100]. These tools allow inferring the effect of the design choices overall ship, already during the early stage design. Consequently, designers can compare different solutions in terms of the ship's KPI, possibly leading to the choice of the best overall design. Moreover, such tools provide a means to assure the correct on board integration of innovative technologies, thus ensuring the achievement of the expected efficiency increase, as well as other advantages [101].

Finally, it has to be highlighted that the pursuit of an increase in ships' efficiency may have a significant impact also on other applications. The most obvious case is the need to connect the ship to the port electrical power system (i.e., shore connection [102,103]), for recharging on-board installed energy storage systems or at least shutting off on-board generators. Such a practice can reduce the pollutant emissions in the ports (which are commonly placed inside cities), as well as increase the overall system efficiency (ship's on board generators have a higher CO_2 footprint in respect to the land power grid). However, in order to use a shore connection a specific set of equipment is to be installed on the berth, and the port power system may need to be refitted to support the additional load (e.g., a cruise ship requires up to a MW when at berth). Another example of the global impact of efficiency increase methods can be made referring to the use of cleaner fuels. Indeed, besides the modifications to ship's engines and fuel treatment systems, it is required to create a dedicated supply chain for these new fuels.

5. Conclusions

The paper includes a comprehensive review on efficiency issues related to three important sectors of the transportation systems: railways, electrical vehicles and marine. The measures recently investigated and implemented, for improving three transportation systems from the point of view of efficiency, are reported and analyzed. Many actions deal with the application of devices, apparatus and systems that can improve the efficiency acting on the infrastructure, but other several actions deal with the suitable management techniques for existing systems, also with the aim of integrating them in a wider system. The analysis pointed out that a common way to compare the increased efficiency of the transportation system in the three analyzed sectors is to assess the percentage reduction of the

CO_2 emissions, thanks to the implementation of different electrification solutions of various means. For the rail transport sector, these solutions are related to changes to an already existing electrical infrastructure, that can be integrated with innovative devices (e.g., storage) and management systems (e.g., traffic control), differently from the on road and maritime sectors, where the improvement of the efficiency can be achieved through electrification actions of the vehicles and of the propulsion systems.

However, there are several different factors that make a direct comparison among the different transportation systems a very complex work. First of all, railway, road, and marine sectors can be either integrated or mutually exclusive transportation means. Indeed, for goods transportation at present there is a good integration among them, which implies using ships for long-range transportation, then railways for medium range transportation, and finally road vehicles for short range delivery. This means that a complete assessment of the efficiency of the goods transportation framework includes all of the above described systems, with ratios that depend on the specific start and finish point. Conversely, the people transportation is caused by a different set of needs and aims, depending on the transportation system. Road and railway transportation can be partially overlapped in this regard (excluding locations that are not reachable by train only), while ships are used either for marine routes that cannot be achieved by other means, or for leisure trips. This makes an efficiency comparison among the three systems fully dependent on the specific type of application and route. A second critical point in such a comparison is the strict dependency among the CO_2 footprint of the recharging energy, for grid connected vehicles. As depicted in this paper, different countries have different energy mixes for producing the electrical energy, making it possible to compare these figures for short range transportation only. As an example, the CO_2 emissions of an electric truck that delivers a given set of goods going through different countries is a composition of the amounts of emissions given by the energy mixes used in the places where it stops for recharging. Moreover, given a single starting point and a single arrival point, different routes can be taken, implying different energy consumption and different recharging energy mixes. Thus, a single figure, able to provide a full assessment and comparison of the efficiency among the presented transportation systems cannot be provided.

So, independently from the common metric for the assessment of efficiency and its improvement (in relation to different solutions adopted in the three transport sectors), the main result of the review reported in this paper has been collecting all the possible measures that can be applied to the three transport sectors for improving the efficiency, in function, of their peculiarities. It contributes to the idea of what has been already done within this field and what could be the margin of development to be explored in the future.

Author Contributions: Conceptualization, M.B., M.C.F. and G.S.; methodology, M.B., M.C.F. and G.S.; validation, M.B., M.C.F. and G.S.; investigation, V.B., F.F., A.R., A.V.; resources, V.B., F.F., A.R., A.V.; data curation, V.B., F.F., A.R., A.V.; writing—original draft preparation, V.B., F.F., A.R., A.V.; writing—review and editing, M.B., M.C.F. and G.S.; supervision, M.B., M.C.F. and G.S., M.C.F. and A.R. mainly contributed on Section 2; M.B. and F.F. mainly contributed on Section 3; V.B., G.S., and A.V. mainly contributed on Section 4. All authors have read and agreed to the published version of the manuscript.

Funding: This research received no external funding.

Conflicts of Interest: The authors declare that there is no conflict of interest regarding the publication of this paper.

References

1. European Commission. 20 20 by 2020 Europe's Climate Change Opportunity; COM (2008) 30 Final; Brussels. 2008. Available online: https://eur-lex.europa.eu/LexUriServ/LexUriServ.do?uri=COM:2008:0030:FIN:EN:PDF (accessed on 10 October 2019).
2. Ortmeyer, T.H.; Pillay, P. Trends in Transportation Sector Technology Energy Use and Greenhouse Gas Emissions. *Proc. IEEE* **2001**, *89*, 1837–1847. [CrossRef]
3. Harwey, L.D.D. Global climate-oriented transportation scenarios. *Energy Pol.* **2013**, *54*, 87–103. [CrossRef]
4. Richardson, D.B. Electric Vehicles and the electric grid: A review of modeling approaches, impacts and renewable energy integration. *Renew. Sustain. Energy Rev.* **2012**, *19*, 247–254. [CrossRef]

5. International Energy Agency (IEA). Railway Handbook. 2015. Available online: http://www.uic.org/IMG/pdf/iea-uic_2015-2.pdf (accessed on 7 October 2019).
6. 19 Rail Projects to Watch in 2019. Available online: https://www.railjournal.com/in_depth/19-rail-projects-to-watch-in-2019 (accessed on 1 October 2019).
7. The Future of Rail. Available online: https://www.iea.org/futureofrail/ (accessed on 20 September 2019).
8. Capasso, A.; Lamedica, R.; Ruvio, A.; Giannini, G. Eco-friendly urban transport systems. Comparison between energy demands of the trolleybus and tram systems. *Ing. Ferrov.* **2014**, *69*, 329–347.
9. D'Ariano, A.; Corman, F.; Fujiyama, T.; Meng, L.; Pellegrini, P. Simulation and Optimization for Railway Operations Management. *Hindawi J. Adv. Transp.* **2018**, *2018*, 4896748. [CrossRef]
10. Yang, X.; Li, X.; Ning, B.; Tang, T. A Survey on Energy-Efficient Train Operation for Urban Rail Transit. *IEEE Trans. Intell. Transp. Syst.* **2016**, *17*, 2–13. [CrossRef]
11. Popescu, M.; Bitoleanu, A. A Review of the Energy Efficiency Improvement in DC Railway Systems. *Energies* **2019**, *12*, 1092. [CrossRef]
12. Albrecht, T. *Reducing Power Peaks and Energy Consumption in Rail Transit Systems by Simultaneous Train Running Time Control*; WIT Press: Wessex, UK, 2004; pp. 885–894.
13. Chen, J.F.; Lin, R.L.; Liu, Y.C. Optimization of an MRT train schedule-reducing maximum traction power by using genetic algorithms. *IEEE Trans. Power Syst.* **2005**, *20*, 1366–1372. [CrossRef]
14. Ramos, A.; Pena, M.; Fernández-Cardador, A.; Cucala, A.P. Mathematical programming approach to underground timetabling problem for maximizing time synchronization. In Proceedings of the International Conference on Industrial Engineering and Engineering Management, Madrid, Spain, 5–7 September 2007; pp. 88–95.
15. Kim, K.M.; Kim, K.T.; Han, M.S. A model and approaches for synchronized energy saving in timetabling. In Proceedings of the 9th World Congress on Railway Research—WCRR, Lille, France, 22–26 May 2011; pp. 1–8.
16. Peña-Alcaraz, M.; Fernández, A.; Cucala, A.P.; Ramos, A.; Pecharromán, R.R. Optimal underground timetable design based on power flow for maximizing the use of regenerative-braking energy. *Proc. Inst. Mech. Eng. Part F J. Rail Rapid Transit* **2012**, *226*, 397–408. [CrossRef]
17. Falvo, M.C.; Sbordone, D.; Fernández-Cardador, A.; Cucala, A.P.; Pecharromán, R.R.; López-López, A.J. Energy savings in metro-transit systems: A comparison between an Italian and a Spanish real line. *Proc. Inst. Mech. Eng. Part F J. Rail Rapid Transit* **2016**, *230*, 345–359. [CrossRef]
18. Fernández-Rodríguez, A.; Fernández-Cardador, A.; Antonio, A.; Cucala, P.; Falvo, M.C. Energy efficiency and integration of urban electrical transport systems: EVS and metro-trains of two real European lines. *Energ. J.* **2018**, *12*, 366–385. [CrossRef]
19. Yang, X.; Li, X.; Gao, Z.; Wang, H.; Tang, T. A cooperative scheduling model for timetable optimization in subway systems. *IEEE Trans. Intell. Transp. Syst.* **2013**, *14*, 438–447. [CrossRef]
20. Li, X.; Yang, X. A stochastic timetable optimization model in subway systems. *Int. J. Uncertain. Fuzziness Knowl.-Based Syst.* **2013**, *21* (Suppl. 1), 1–15. [CrossRef]
21. Yang, X.; Ning, B.; Li, X.; Tang, T. A two-objective timetable optimization model in subway systems. *IEEE Trans. Intell. Transp. Syst.* **2014**, *15*, 1913–1921. [CrossRef]
22. Yang, X.; Chen, A.; Li, X.; Ning, B.; Tang, T. An energy-efficient scheduling approach to improve the utilization of regenerative energy for metro rail systems. *Transp. Res. Part C Emerg. Technol.* **2015**, *57*, 13–29. [CrossRef]
23. Zou, B.; Gong, L.; Yu, N.; Chen, J. Intelligent scheduling method for energy saving operation of multi-train based on genetic algorithm and regenerative kinetic energy. *J. Eng.* **2018**, *2018*, 1550–1554. [CrossRef]
24. Zhou, Y.; Bai, Y.; Li, J.; Mao, B.; Li, T. Integrated Optimization on Train Control and Timetable to Minimize Net Energy Consumption of Metro Lines. *Hindawi J. Adv. Transp.* **2018**, *2018*, 7905820. [CrossRef]
25. Zhao, N.; Roberts, C.; Hillmansen, S.; Nicholson, G. A Multiple Train Trajectory Optimization to Minimize Energy Consumption and Delay. *IEEE Trans. Intell. Transp. Syst.* **2015**, *16*, 2363–2372. [CrossRef]
26. Liu, J.; Guo, H.; Yu, Y. Research on the Cooperative Train Control Strategy to Reduce Energy Consumption. *IEEE Trans. Intell. Transp. Syst.* **2017**, *18*, 1134–1142. [CrossRef]
27. Song, Y.; Song, W. A Novel Dual Speed-Curve Optimization Based Approach for Energy-Saving Operation of High-Speed Trains. *IEEE Trans. Intell. Transp. Syst.* **2016**, *17*, 1564–1575. [CrossRef]

28. Huang, J.; Deng, Y.; Yang, Q.; Sun, J. An Energy-Efficient Train Control Framework for Smart Railway Transportation. *IEEE Trans. Comput.* **2016**, *65*, 1407–1417. [CrossRef]
29. Cao, Y.; Wang, Z.; Liu, F.; Li, P.; Xie, G. Bio-Inspired Speed Curve Optimization and Sliding Mode Tracking Control for Subway Trains. *IEEE Trans. Veh. Technol.* **2019**, *68*, 6331–6342. [CrossRef]
30. He, D.; Lu, G.; Yang, A.Y. Research on Optimization of Train Energy-Saving Based on Improved Chicken Swarm Optimization. Special Section on Artificial Intelligence Technologies for Electric Power Systems. *IEEE Access* **2019**, *7*, 121675–121684. [CrossRef]
31. Zhao, X.-H.; Ke, B.-R.; Lian, K.-L. Optimization of train speed curve for energy saving using efficient and accurate electric traction models on the mass rapid transit system. *IEEE Trans. Transp. Electrif.* **2018**, *4*, 922–935. [CrossRef]
32. Tian, Z.; Zhao, N.; Hillmansen, S.; Roberts, C.; Dowens, T.; Kerr, C. SmartDrive: Traction Energy Optimization and Applications in Rail Systems. *IEEE Trans. Intell. Transp. Syst.* **2019**, *20*, 2764–2773. [CrossRef]
33. Luo, Z.; Li, X.; Xiu, N. A Sparse Optimization Approach for Energy-Efficient Timetabling in Metro Railway Systems. *Hindawi J. Adv. Transp.* **2018**, *2018*, 1784789. [CrossRef]
34. Brenna, M.; Foiadelli, F.; Longo, M. Application of Genetic Algorithms for Driverless Subway Train Energy Optimization. *Hindawi Publ. Corp. Int. J. Veh. Technol.* **2016**, *2016*, 8073523. [CrossRef]
35. Khodaparastan, M.; Mohamed, A.A.; Brandauer, W. Recuperation of Regenerative Braking Energy in Electric Rail Transit Systems. *IEEE Trans. Intell. Transp. Syst.* **2019**, *20*, 2831–2847. [CrossRef]
36. Jang, S.J.; Choi, C.Y.; Bae, C.H.; Song, S.H.; Won, C.Y. Study of regeneration power control inverter for DC traction with active power filter ability. In Proceedings of the 31st Annual Conference of IEEE (IECON 2005), Raleigh, NC, USA, 6–10 November 2005.
37. Henning, P.H.; Fuchs, H.D.; Roux, A.D.L.; Mouton, H.A.T. A 1.5-MW seven-cell series-stacked converter as an active power filter and regeneration converter for a DC traction substation. *IEEE Trans. Power Electron.* **2008**, *23*, 2230–2236. [CrossRef]
38. De Jager, W.A.G.; Huizer, M.; van der Pols, E.K.H. Implementation of an active regeneration unit in a traction substation. In Proceedings of the 16th European Conference on Power Electronics and Applications, Lappeenranta, Finland, 26–28 August 2014.
39. Romo, A. Reversible Substation in Heavy Rail. In Proceedings of the Energy Recovery Workshop, Madrid, Spain, 25–29 September 2015; Available online: https://docplayer.net/50090204-Reversible-substation-in-heavy-rail.html (accessed on 6 February 2019).
40. Gray, A.J.; Stinton, D. Designing reversible substations using inverters. In Proceedings of the 7th IET Professional Development Course on Railway Electrification Infrastructure and Systems (REIS 2015), London, UK, 8–11 June 2015.
41. Jefimowski, W.; Szeląg, A. The multi-criteria optimization method for implementation of a regenerative inverter in a 3kV DC traction system. *Electr. Power Syst. Res.* **2018**, *161*, 61–73. [CrossRef]
42. UIC. Energy Efficiency Technologies for Railways. Available online: http://www.railway-energy.org (accessed on 3 October 2019).
43. Hayashiya, H.; Iino, Y.; Takahashi, H.; Kawahara, K.; Yamanoi, T.; Sekiguchi, T.; Sakaguchi, H.; Sumiya, A.; Kon, S. Review of regenerative energy utilization in traction power supply system in Japan: Applications of energy storage systems in d.c. traction power supply system. In Proceedings of the 43rd Annual Conference of the IEEE Industrial Electronics Society (IECON 2017), Beijing, China, 29 October–1 November 2017; pp. 3918–3923.
44. Arboleya, P.; Bidaguren, P.; Armendariz, U. Energy is on board: Energy storage and other alternatives in modern light railways. *IEEE Electrif. Mag.* **2016**, *4*, 30–41. [CrossRef]
45. Barrero, R.; van Mierlo, J.; Tackoen, X. Energy savings in public transport. *IEEE Veh. Technol. Mag.* **2008**, *3*, 26–36. [CrossRef]
46. Barrero, R.; Tackoen, X.; van Mierlo, J. Stationary or onboard energy storage systems for energy consumption reduction in a metro network. *Proc. Inst. Mech. Eng. F J. Rail Rapid Transit* **2010**, *224*, 207–225. [CrossRef]
47. Domínguez, M.; Cucala, A.P.; Fernández, A.; Pecharromán, R.R.; Blanquer, J. Energy efficiency on train control: Design of metro ATO driving and impact of energy accumulation devices. In Proceedings of the 9th World Congress on Railway Research (WCRR), Lille, France, 22–26 May 2011; pp. 1–12.
48. Iannuzzi, D.; Tricoli, P. Speed-based state-of-charge tracking control for metro trains with onboard supercapacitors. *IEEE Trans. Power Electron.* **2012**, *27*, 2129–2140. [CrossRef]

49. Capasso, A.; Ceraolo, M.; Lamedica, R.; Lutzemberger, G.; Ruvio, A. Modelling and simulation of tramway transportation systems. *J. Adv. Transp.* **2019**, *2019*, 1–8. [CrossRef]
50. Ceraolo, M.; Lutzemberger, G.; Poli, D. State-Of-Charge Evaluation of Supercapacitors. *J. Energy Storage* **2017**, *11*, 211–218. [CrossRef]
51. Di Noia, L.P.; Genduso, F.; Miceli, R.; Rizzo, R. Optimal Integration of Hybrid Supercapacitor and IPT System for a Free-Catenary Tramway. *IEEE Trans. Ind. Appl.* **2019**, *55*, 794–801. [CrossRef]
52. Yan, Y.; Li, Q.; Chen, W.; Su, B.; Liu, J.; Ma, L. Optimal Energy Management and Control in Multimode Equivalent Energy Consumption of Fuel Cell/Supercapacitor of Hybrid Electric Tram. *IEEE Trans. Ind. Electron.* **2019**, *66*, 6065–6076. [CrossRef]
53. Graber, G.; Calderaro, V.; Galdi, V.; Piccolo, A.; Lamedica, R.; Ruvio, A. Techno-economic Sizing of Auxiliary-Battery-Based Substations in DC Railway Systems. *IEEE Trans. Transp. Electrif.* **2018**, *4*, 616–625. [CrossRef]
54. *Increasing Energy Efficiency Optimized Traction Power Supply in Mass Transit Systems*; Siemens: Munich, Germany, 2011.
55. Adetel Group. NeoGreen Power System. 2000. Available online: http://www.adetelsolution.com/app/uploads/2017/06/Flyer-NeoGreen.pdf (accessed on 5 January 2019).
56. Healy, N.; Barry, J. Politicizing energy justice and energy system transitions: Fossil fuel divestment and a "just transition". *Energy Pol.* **2017**, *108*, 451–459. [CrossRef]
57. International Energy Agency (IEA). Hybrid and Electric Vehicles—The Electric Drive Captures the Imagination. Annual Report 2011. March 2012. Available online: www.ieahev.org (accessed on 15 January 2020).
58. Jackle, J.A.; Sculle, K.A. *The Gas Station in America*; The Johns Hopkins University Press: Baltimore, MD, USA, 1994.
59. Mills, B.; Andrey, J. Climate Change and Transportation: Potential Interactions and Impacts. In Proceedings of the Potential Impacts of Climate Change on Transportation Conference, Washington, DC, USA, 1–2 October 2002.
60. Poundenx, P. The effect of transportation policies on energy consumption and greenhouse gas emission from urban passenger transportation. *Transp. Res. Part A Pol. Pract.* **2008**, *42*, 901–909. [CrossRef]
61. Mwasilu, F.; Justo, J.J.; Kim, E.K.; Do, T.D.; Jung, J.W. Electric vehicles and smart grid interaction: A review on vehicle to grid and renewable energy sources integration. *Renewable Sustainable Energy Rev.* **2014**, *34*, 501–516. [CrossRef]
62. Commission of the European Communities. Regulation (EEC) No 4064/89Merger Procedure. March 1999. Available online: www.ec.europa.eu (accessed on 17 January 2020).
63. U.S. Energy Information Administration. Electric Power Monthly. February 2013. Available online: www.eia.gov/electricity/monthly (accessed on 15 January 2020).
64. Europe's Energy Portal. March 2013. Available online: www.energy.eu (accessed on 17 January 2020).
65. Crist, P. *Electric Vehicles Revisited: Costs, Subsidies and Prospects*; Discussion Paper No. 2012-O3; International Transport Forum: Paris, France, April 2012. Available online: http://www.bec.mise.gov.it/site/bec/home/contributi/approfondimenti.html (accessed on 21 January 2020).
66. Carley, S.; Graham, J.; Krause, R.M.; Lane, B.W. Intent to purchase a plug-in electric vehicle: A survey of early impressions in large US cities. *Transp. Res. Part D Transp. Environ.* **2012**, *18*, 39–45. [CrossRef]
67. Boulanger, A. Vehicle Electrification: Status and Issues. *Proc. IEEE* **2011**, *99*, 1116–1138. [CrossRef]
68. Lopes, J.A.P.; Soares, F.J.; Almeida, P.M.R. Integration of Electric Vehicles in the Electric Power System. *Proc. IEEE* **2011**, *99*, 168–183. [CrossRef]
69. Lopes, J.A.; Soares, F.J.; Almeida, P.M.R. Identifying management procedures to deal with connection of electric vehicles in the grid. In Proceedings of the 2009 IEEE Bucharest PowerTech, Bucharest, Romania, 28 June–2 July 2009.
70. Ropenus, S.; Skytte, K. Regulatory Review and Barriers for the Electricity Supply System for Distributed Generation in EU-15. In Proceedings of the 2005 International Conference on Future Power Systems, Amsterdam, The Netherlands, 16–18 November 2005.
71. Andersen, P.H.; Mathews, J.A.; Raska, M. Integrating private transport into renewable energy policy: The strategy of creating intelligent recharging grids for electric vehicles. *Energy Pol.* **2009**, *37*, 2481–2486. [CrossRef]

72. Kempton, W.; Tomic, J. Vehicle-to-grid power implementation: From stabilizing the grid to supporting large-scale renewable energy. *J. Power Sources* **2005**, *144*, 280–294. [CrossRef]
73. Kesler, M.; Kisacikoglu, M.; Tolbert, L. Vehicle-to-Grid Reactive Power Operation Using Plug-In Electric Vehicle Bidirectional Offboard Charger. *IEEE Trans. Ind. Electron.* **2014**, *61*, 6778–6784. [CrossRef]
74. Steward, D. *Critical Elements of Vehicle-to-Grid (V2G) Economics*; National Renewable Energy Laboratory: Denver, CO, USA, 2017.
75. Liu, C.; Chau, K.T.; Wu, D.; Gao, S. Opportunities and Challenges of Vehicle-to-Home, Vehicle-to-Vehicle, and Vehicle-to-Grid Technologies. *Proc. IEEE* **2013**, *101*, 2409–2427. [CrossRef]
76. Ma, Y.; Zhang, B.; Zhou, X.; Gao, Z.; Wu, Y.; Yin, J.; Xu, X. An Overview on V2G Strategies to Impacts from EV Integration into Power System. In Proceedings of the 2016 Chinese Control and Decision Conference (CCDC), Yinchuan, China, 28–30 May 2016; pp. 2895–2900.
77. Brenna, M.; Foiadelli, F.; Soccini, A.; Volpi, L. Charging Strategies for Electric Vehicles with Vehicle to Grid Implementation for Photovoltaic Dispatchability. In Proceedings of the 2018 International Conference of Electrical and Electronic Technologies for Automotive, Milan, Italy, 9–11 July 2018.
78. Brenna, M.; Foiadelli, F.; Longo, M.; Zaninelli, D. Energy Storage Control for Dispatching Photovoltaic Power. *IEEE Trans. Smart Grid* **2016**. [CrossRef]
79. Schiavo, L.L.; Delfanti, M.; Fumagalli, E.; Olivieri, V. Changing the regulation for regulating the change: Innovation-driven regulatory developments for smart grids, smart metering and e-mobility in Italy. *Energy Pol.* **2013**, *57*, 506–517. [CrossRef]
80. Voyant, C.; Randimbivololona, P.; Nivet, M.L.; Paoli, C.; Muselli, M. Twenty four hours ahead global irradiation forecasting using multi-layer perceptron. *Meteorol. Appl.* **2014**, *21*, 644–655. [CrossRef]
81. Global Wind Energy Council. Global Wind Report. 2011. Available online: http://gwec.net/wp-content/uploads/2012/06/Annual_report_2011_lowres.pdf (accessed on 6 February 2020).
82. Energinet.dk. Wind Power Production in Denmark. Available online: www.energinet.dk (accessed on 6 February 2020).
83. Hall, D.; Lutsey, N. Effects of Battery Manufacturing on Electric Vehicle Life-Cycle Greenhouse Gas Emissions. ICCT The International Council on Clean Transportation. 2018. Available online: www.theicct.org (accessed on 6 February 2020).
84. Rahman, A.; Karim, M. Green Shipbuilding and Recycling: Issues and Challenges. *Int. J. Environ. Sci. Dev.* **2015**, *6*, 838–842. [CrossRef]
85. IMO. *2017 International Convention for the Prevention of Pollution from Ships*; MARPOL 2017 Consolidated Edition; IMO Publishing: London, UK, 2017.
86. Mauro, F.; Sorrenti, G.; Bonfiglio, L.; Bucci, V. Integrated Design of an Eco-Friendly Luxury Sailing Super-Yacht. In Proceedings of the 2018 SNAME Maritime Convention (SMC 2018), Providence, RI, USA, 24–27 October 2018.
87. McCoy, T.J. Electric Ships Past, Present, and Future. *IEEE Electrif. Mag.* **2015**, *3*, 4–11. [CrossRef]
88. Vicenzutti, A.; Bosich, D.; Giadrossi, G.; Sulligoi, G. The Role of Voltage Controls in Modern All-Electric Ships: Toward the all electric ship. *IEEE Electrif. Mag.* **2015**, *3*, 49–65. [CrossRef]
89. Alexander, D. Hybrid Electric Drive for Naval Combatants. *Proc. IEEE* **2015**, *103*, 2267–2275. [CrossRef]
90. Shen, Q.; Ramachandran, B.; Srivastava, S.K.; Andrus, M.; Cartes, D.A. Power and Energy Management in Integrated Power System. In Proceedings of the 2011 IEEE Electric Ship Technologies Symposium (ESTS 2011), Alexandria, VA, USA, 10–13 April 2011.
91. Markle, S.P. IPES—Harnessing Total Ship Energy & Power, Oral Communication at Sea-Air-Space Exposition. In Proceedings of the Navy League's 2018 Sea-Air-Space Exposition, National Harbor, MD, USA, 9–11 April 2018.
92. Bosich, D.; Bucci, V.; la Monaca, U.; Marinò, A.; Sulligoi, G.; Vicenzutti, A.; Lipardi, G. Early-stage design of integrated power and energy systems for naval vessels electrification: Advanced modeling using CSI. In Proceedings of the 2017 IEEE Transportation Electrification Conference and Expo (ITEC), Chicago, IL, USA, 22–24 June 2017; pp. 387–392.
93. Hebner, R.E.; Davey, K.; Herbst, J.; Hall, D.; Hahne, J.; Surls, D.D.; Ouroua, A. Dynamic Load and Storage Integration. *Proc. IEEE* **2015**, *103*, 2344–2354. [CrossRef]

94. Whitcomb, C.A.; Szatkowski, J.J. Concept Level Naval Surface Combatant Design in the Axiomatic Approach to Design Framework. In Proceedings of the 2000 First International Conference on Axiomatic Design (ICAD), Cambridge, MA, USA, 21–23 June 2000.
95. Trincas, G.; Mauro, F.; Braidotti, L.; Bucci, V. Handling the path from concept to preliminary ship design. In Proceedings of the 13th International Marine Design Conference—IMDC 2018, At Espoo, Finland, 10–14 June 2018; pp. 181–192.
96. Bucci, V.; Marino, A.; Bosich, D.; Sulligoi, G. Inland waterway gas-fueled vessels: CASM-based electrification of a pushboat for the european network. *IEEE Trans. Transp. Electrif.* **2016**, *2*, 607–617. [CrossRef]
97. Chalfant, J. Early-Stage Design for Electric Ship. *Proc. IEEE* **2015**, *103*, 2252–2266. [CrossRef]
98. Sulligoi, G.; Vicenzutti, A.; Menis, R. All-Electric Ship Design: From Electrical Propulsion to Integrated Electrical and Electronic Power Systems. *IEEE Trans. Transp. Electrif.* **2016**, *2*, 507–521. [CrossRef]
99. La Monaca, U.; Bertagna, S.; Marinò, A.; Bucci, V. Integrated ship design: An innovative methodological approach enabled by new generation computer tools. *Int. J. Interact. Des. Manuf.* **2020**, *14*, 59–76. [CrossRef]
100. Vicenzutti, A.; Trincas, G.; Bucci, V.; Sulligoi, G.; Lipardi, G. Early-Stage design methodology for a multirole electric propelled surface combatant ship. In Proceedings of the 2019 IEEE Electric Ship Technologies Symposium (ESTS), Washington, DC, USA, 14–16 August 2019; pp. 97–105.
101. Bucci, V.; la Monaca, U.; Bosich, D.; Sulligoi, G.; Pietra, A. Integrated ship design and CSI modeling: A new methodology for comparing onboard electrical distributions in the early-stage design. In Proceedings of the NAV2018 International Conference on Ship and Shipping Research, Trieste, Italy, 20–22 June 2018; p. 124.
102. Bucci, V.; Marino, A.; Bosich, D.; Sulligoi, G. The design of a slow-cruising superyacht with zero emission navigation and smart berthing modes. In Proceedings of the 2014 9th International Conference on Ecological Vehicles and Renewable Energies, EVER 2014, Monte-Carlo, Monaco, 25–27 March 2014.
103. Sulligoi, G.; Bosich, D.; Pelaschiar, R.; Lipardi, G.; Tosato, F. Shore-to-Ship Power. *Proc. IEEE* **2015**, *103*, 2381–2400. [CrossRef]

© 2020 by the authors. Licensee MDPI, Basel, Switzerland. This article is an open access article distributed under the terms and conditions of the Creative Commons Attribution (CC BY) license (http://creativecommons.org/licenses/by/4.0/).

energies

MDPI

Article

Battery Electric Vehicle Eco-Cooperative Adaptive Cruise Control in the Vicinity of Signalized Intersections

Hao Chen [1] and Hesham A. Rakha [2,*]

[1] Virginia Tech Transportation Institute, 3500 Transportation Research Plaza, Blacksburg, VA 24061, USA; hchen@vtti.vt.edu

[2] Department of Civil and Environmental Engineering, Virginia Polytechnic Institute and State University, 3500 Transportation Research Plaza, Blacksburg, VA 24061, USA

* Correspondence: hrakha@vt.edu

Received: 31 March 2020; Accepted: 10 May 2020; Published: 12 May 2020

Abstract: This study develops a connected eco-driving controller for battery electric vehicles (BEVs), the BEV Eco-Cooperative Adaptive Cruise Control at Intersections (Eco-CACC-I). The developed controller can assist BEVs while traversing signalized intersections with minimal energy consumption. The calculation of the optimal vehicle trajectory is formulated as an optimization problem under the constraints of (1) vehicle acceleration/deceleration behavior, defined by a vehicle dynamics model; (2) vehicle energy consumption behavior, defined by a BEV energy consumption model; and (3) the relationship between vehicle speed, location, and signal timing, defined by vehicle characteristics and signal phase and timing (SPaT) data shared under a connected vehicle environment. The optimal speed trajectory is computed in real-time by the proposed BEV eco-CACC-I controller, so that a BEV can follow the optimal speed while negotiating a signalized intersection. The proposed BEV controller was tested in a case study to investigate its performance under various speed limits, roadway grades, and signal timings. In addition, a comparison of the optimal speed trajectories for BEVs and internal combustion engine vehicles (ICEVs) was conducted to investigate the impact of vehicle engine types on eco-driving solutions. Lastly, the proposed controller was implemented in microscopic traffic simulation software to test its networkwide performance. The test results from an arterial corridor with three signalized intersections demonstrate that the proposed controller can effectively reduce stop-and-go traffic in the vicinity of signalized intersections and that the BEV Eco CACC I controller produces average savings of 9.3% in energy consumption and 3.9% in vehicle delays.

Keywords: eco-driving; battery electric vehicles; signalized intersections; energy-optimized vehicle trajectories; vehicle dynamics model

1. Introduction

The United States is one of the world's prime petroleum consumers, burning more than 20% of the planet's total refined petroleum, and the surface transportation sector alone accounts for around 69% of the United States' total petroleum usage [1]. This presents the transportation sector with three important challenges: availability of fuel to drive vehicles, emissions of greenhouse gases, and vehicular crashes. It is, therefore, important to reduce petroleum consumption to make surface transportation safer, more efficient, and more sustainable [2].

The advent of communication and information technology has enabled vehicle-to-vehicle and vehicle-to-infrastructure connectivity, so that various data, such as signal phase and timing (SPaT), vehicle trajectory, and velocity, can be transmitted and utilized. The advanced communication abilities of connected vehicles ensures that information is updated at a very high rate, which enables researchers to develop connected transportation systems meeting safety, economy, and efficiency challenges [3]. Studies have shown that vehicles have high fuel consumption rates when approaching signalized intersections because of vehicle acceleration/deceleration maneuvers during stop-and-go traffic [2,4]. Over the past few decades, researchers have worked on optimizing traffic signal planning to reduce traffic delay and fuel consumption [5,6]. In recent years, a number of studies have focused on developing eco-driving algorithms to help vehicles approach signalized intersections using connected vehicle technologies. These eco-driving strategies aim to provide, in real-time, recommendations to individual drivers or vehicles so that vehicle maneuvers can be appropriately adjusted to reduce fuel consumption and emission levels [7–9].

Most of the studies in this area have been focused on developing eco-driving strategies for internal combustion engine vehicles (ICEVs). For example, Malakorn and Park proposed to reduce vehicle fuel consumption by minimizing vehicle acceleration maneuvers using a cooperative adaptive cruise control system under a connected environment [10]. Another study [2] developed an optimal control strategy by using dynamic programming and recursive pathfinding techniques, and the control logic was validated by an agent-based modeling approach. In addition, a schedule optimization method was proposed in [11] to search for "green-windows" so that vehicles can traverse multiple signalized intersections by minimizing full stops. A further-improved approach was developed by Guan and Frey to generate a brake-specific fuel map so that the optimized gear ratio can be computed to save fuel levels [12].

In addition to the studies that focused on ICEVs, a few studies have investigated eco-driving strategies for battery electric vehicles (BEVs) near signalized intersections. Using SPaT information passed from connected infrastructure, an energy-optimized speed trajectory can be computed for BEVs while traveling on signalized arterials, thus extending the BEV's range. An eco-driving technique for BEVs was developed in [13]. In that work, the vehicle trajectory control problem was formulated as an optimization problem to minimize the summation of vehicle power, and Bellman's dynamic programming algorithm was used to compute the optimal solution. However, a simple energy model was used in this study by assuming that the recharge efficiency is a constant value. Another BEV eco-driving algorithm was proposed in [14]. A VT-Micro model-based energy consumption model was developed for different BEV operation modes (including acceleration, deceleration, idling, and cruising). Subsequently, an eco-driving model, which used the developed energy model, was proposed for a single signalized intersection. Several example trips in the case study illustrate the proposed eco-driving method's ability to reduce energy consumption efficiently. However, the proposed energy consumption model was a statistical model based on limited collected data, thus the accuracy may not be sufficient for developing an optimal control strategy for dynamic vehicle maneuvers. Moreover, the vehicle dynamics model was not considered in the constraints to compute the acceleration level, so the calculation of the optimal solution may use an unrealistic acceleration level.

The same energy consumption model was used in [15] to develop a connected BEV eco-driving system. A model predictive control logic was considered in the control system to force the vehicle to follow the optimal speed trajectory as closely as possible. A field test with four participants demonstrated an average of 22% energy savings for automated driving with the proposed eco-driving system. However, a 2012 Ford Escape with a hybrid engine was used for the field test in this study, and this vehicle was assumed to be representative of an actual BEV's performance.

An analytical model to calculate a BEV's optimum vehicle trajectory was proposed in [16], with the goal of minimizing electricity usage with consideration of intersection queues. Furthermore, an approximation model was proposed to increase computation efficiency for real-time applications. A 47.5% energy savings was found when evaluating field data from a six-intersection corridor. However, the objective function was the summation of energy consumption from the tractive force only, and the braking force was assumed to be 100% transferable to battery power. In addition, the work in [17] provided a solution to minimize BEVs' energy consumption while traversing a sequence of signalized intersections and always getting a green indication. A simple simulation network (AIMSUN) with five intersections was used in the case study. A sensitivity analysis with different market penetration rates was tested to show a 10% energy savings for a 40% penetration rate. However, the computation of energy consumption in this study did not consider regenerative braking.

There are several issues with the aforementioned studies of BEV eco-driving strategies: a lack of realistic energy consumption models to accurately compute the instantaneous energy consumption when BEVs travel through signalized intersections, and the lack of a vehicle dynamics model to constrain vehicle acceleration maneuvers. In addition, although many previous studies developed eco-driving strategies for ICEVs and BEVs, there is no comparison to demonstrate the differences in the energy-optimal solutions for each. To address these issues, this study develops a connected eco-driving controller for BEVs, called the BEV Eco-Cooperative Adaptive Cruise Control at Intersections (Eco-CACC-I). The developed controller can assist BEVs negotiating signalized intersections by minimizing their energy consumption. The calculation of optimal vehicle trajectory is formulated as an optimization problem subject to the following constraints: (1) vehicle acceleration/deceleration behavior, defined by a vehicle dynamics model; (2) vehicle energy consumption behavior, defined by a BEV energy consumption model; and (3) the relationship between vehicle speed, location, and signal timing, defined by vehicle characteristics and SPaT data shared under a connected vehicle environment. The optimal speed trajectory is computed in real-time by the proposed BEV Eco-CACC-I controller so that a BEV can follow the optimal speed while negotiating a signalized intersection. The proposed BEV controller was tested in a case study to investigate its performance under various speed limits, roadway grades, and signal timings. In addition, a comparison of the optimal speed trajectories for BEVs and ICEVs was conducted to investigate the impact of vehicle engine types on eco-driving solutions. Lastly, the proposed controller was implemented in microscopic traffic simulation software to test its networkwide performance. The test results from an arterial corridor with three signalized intersections demonstrate that the proposed controller can effectively reduce stop-and-go traffic in the vicinity of signalized intersections, and that the BEV Eco-CACC-I controller produces average savings of 9.3% in energy consumption and 3.9% in vehicle delays.

The remainder of this paper is presented as follows. The proposed BEV Eco-CACC-I system and the vehicle dynamics and energy consumption models are described in the next section. Afterward, the details of the proposed system testing on the case study section to investigate the impacts of various factors on system performance are presented. This is followed by implementing the proposed controller into microscopic traffic simulation software to quantify the networkwide impacts. The last section provides conclusions and recommendations for future research.

2. Model Development

2.1. Eco-CACC-I for BEVs

In previous studies, an eco-driving system for gasoline-powered vehicles, named ICEV Eco-CACC-I, was developed and tested under a simulated environment and real-world field tests [18–21]. The ICEV Eco-CACC-I system computes real-time, fuel-optimized speed profiles that vehicles can follow so that they can proceed through signalized intersections while consuming minimum amounts of fuel. In the previous field test study, we implemented the developed eco-driving system into a real-world automated vehicle as an adaptive cruise control system [19]. Note that the developed Eco-CACC-I system does not directly optimize the collaboration between multiple vehicles. Here, the term "cooperative" means the vehicles equipped with the developed system can cooperate with signalized intersections under a connected-vehicle environment. During the previous network-level simulation test, the recommended speed computed by the developed system was used as a variable speed limit, which worked together with other constraints, such as the car following model and collision avoidance constraint, to control vehicle speed [21]. The same control logic of the previously developed algorithm was used in this study to develop an Eco-CACC-I controller that allows BEVs to drive smoothly through signalized intersections with minimal energy consumption and thus extend their range.

The same control environment setup for ICEV Eco-CACC-I was used here to develop the BEV Eco-CACC-I. The interested reader may read about the previous work in [19,21]. The control region was defined as a distance upstream of the signalized intersection (d_{up}) to a distance downstream of the intersection (d_{down}) in which the BEV Eco-CACC-I controller optimizes the speed profiles of vehicles approaching and leaving signalized intersections. Upon approaching a signalized intersection, the vehicle may accelerate, decelerate, or cruise (maintain a constant speed) based on a number of factors, such as vehicle speed, signal timing, phase, distance to the intersection, road grade, and headway distance, etc. [2]. We assumed no leading vehicle ahead of the BEV so that we could compute the energy-optimized vehicle trajectory for the BEV without considering the impacts of other surrounding vehicles. The computed optimal speed was used as a variable speed limit, denoted by $v_e(t)$, which is one of the constraints on the BEV longitudinal motion. When a BEV travels on the roadway, there are other constraints to be considered, including the allowed speed constrained by the vehicle dynamics model, steady-state car following mode, collision avoidance constraint, and roadway speed limit. All of these constraints work together to control the vehicle speed. In this way, the proposed system can also be used in the situation that the BEV follows a leading vehicle and the vehicle speed can be computed by $v(t) = min(v_1(t), v_2(t), v_3(t), v_4(t), v_e(t))$ using the following constraints:

(1) The maximum speed $v_1(t)$ allowed by the vehicle acceleration model for a given vehicle throttle position;
(2) The maximum speed $v_2(t)$ constrained by the steady-state vehicle spacing in the simulation software;
(3) The speed limit of $v_3(t)$ to avoid a rear-end vehicle collision; and
(4) The maximum speed $v_4(t)$ allowed on the road.

Within the control region, the vehicle's behavior can be categorized into one of two cases: (1) the vehicle can pass through the signalized intersection without decelerating or (2) the vehicle must decelerate to pass through the intersection. Given that vehicles drive in different manners for cases 1 and 2, the BEV Eco-CACC-I strategies were developed separately for the two cases.

Case 1 does not require the vehicle to decelerate to traverse the signalized intersection. In this case, the cruise speed for the vehicle to approach the intersection during the red indication can be calculated by Equation (1) to maximize the average vehicle speed during the control region.

$$u_c = min\left(\frac{d_{up}}{t_r}, u_f\right) \tag{1}$$

When the vehicle enters the control region, it should adjust speed to u_c according to the vehicle dynamics model illustrated later in Equations (5) through (7). After the traffic light turns from red to green, the vehicle accelerates from the speed u_c to the maximum allowed speed (speed limit u_f) by following the vehicle dynamics model until it leaves the control region.

In case 2, the vehicle's energy-optimized speed profile is illustrated in Figure 1. After entering the control region, the vehicle with the initial speed of $u(t_0)$ needs to brake at a deceleration level denoted by a, then cruise at a constant speed of u_c to approach the signalized intersection. After passing the stop bar, the vehicle should increase speed to u_f per the vehicle dynamics model, and then cruise at u_f until the vehicle leaves the control region. In this case, the only unknown variables are the upstream deceleration rate a and the downstream throttle f_p. The following optimization problem was formulated to compute the optimum vehicle speed profile associated with the least energy consumption.

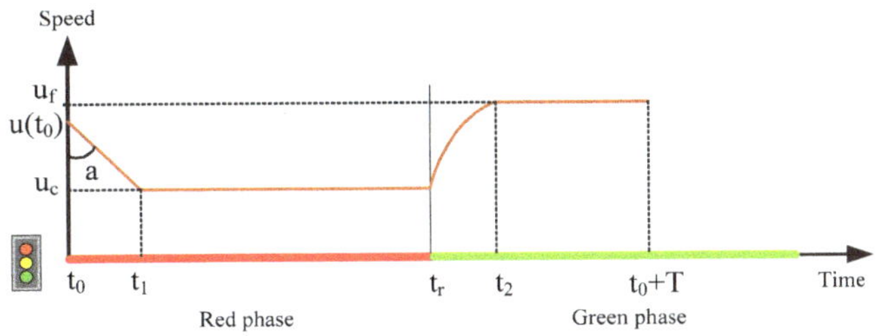

Figure 1. Optimum vehicle speed profile in case 2.

Assuming a BEV enters the Eco-CACC-I control region at time t_0 and leaves the control region at time $t_0 + T$, the objective function entails minimizing the total energy consumption level as:

$$min \int_{t_0}^{t_0+T} EC(u(t)) \cdot dt \tag{2}$$

where EC denotes the electric energy consumption at instant t using Equations (8) through (11). The constraints to solve the optimization problem can be built according to the relationships between vehicle speed, location, and acceleration/deceleration, as presented below:

$$u(t): \begin{cases} u(t) = u(t_0) - at & t_0 \le t \le t_1 \\ u(t) = u_c & t_1 < t \le t_r \\ u(t+t) = u(t) + \frac{F(f_p) - R(u(t))}{m} t & t_r < t \le t_2 \\ u(t) = u_f & t_2 < t \le t_0 + T \end{cases} \tag{3}$$

$$\begin{gathered} u(t_0) \cdot t - \tfrac{1}{2}at^2 + u_c(t_r - t_1) = d_{up} \\ u_c = u(t_0) - a(t_1 - t_0) \\ \int_{t_r}^{t_2} u(t)dt + u_f(t_0 + T - t_2) = d_{down} \\ u(t_2) = u_f \\ a_{min} < a \le a_{max} \\ f_{min} \le f_p \le f_{max} \\ u_c > 0 \end{gathered} \tag{4}$$

In Equation (3), function F denotes the vehicle tractive force calculated by Equation (6), and function R represents all the resistance forces (aerodynamic, rolling, and grade resistance forces) calculated by Equation (7). Note that the maximum deceleration was limited by the comfortable threshold felt by average drivers [2]. The throttle value ranges between 0 and 1. To solve the optimization problem, dynamic programming was used to list all the candidate solutions with the associated electric energy consumption levels. This allowed calculation of optimal parameters for upstream deceleration a and downstream throttle f_p by finding the candidate solution associated with the minimum energy consumption for vehicles passing the control region. To solve the proposed optimization problem in real-time, an A-star search method was selected to ensure fast and efficient computations. The A-star search method is one of the best and most popular path search methods to find the lowest cost path using a heuristic function [22]. The deceleration and throttle levels are considered as constant values in the A-star algorithm when computing the future cost. However, given that the optimal solution is recomputed every decisecond, the acceleration/deceleration level can also be updated every decisecond, thus producing a varying acceleration/deceleration maneuver. In the proposed optimization problem, first, a constant throttle level was assumed (e.g., 0.6) to find the optimal deceleration level, which corresponds to the minimal energy consumption for the entire trip from d_{down} to d_{up}. In this way, the starting speed (cruise speed u_c) and the ending speed (speed limit

u_f) on the downstream roadway are known, so the optimal throttle level which corresponds to the minimal energy consumption for the downstream trip can be located. The details of how the A-star algorithm outperforms other pathfinding algorithms and the steps to implement the A-star algorithm can be found in [22].

2.2. Vehicle Dynamics Model

The proposed BEV Eco-CACC-I system uses a vehicle dynamics model to compute vehicle acceleration behavior. Here, vehicle acceleration is based on the vehicle dynamics model formulated in [23,24], where acceleration level is related to throttle level and vehicle velocity. The vehicle dynamics model is summarized as

$$u(t+t) = u(t) + \frac{F(t) - R(t)}{m}t \tag{5}$$

$$F = \min\left(3600 f_p \beta \eta_D \frac{P_{max}}{u}, m_{ta} g \mu\right) \tag{6}$$

$$R(t) = \frac{\rho}{25.92} C_d C_h A_f u(t)^2 + mg\frac{c_{r0}}{1000}(c_{r1}u(t) + c_{r2}) + mgR_g \tag{7}$$

where F and R are the tractive and resistance forces on the vehicle (N); η_D (unitless) denotes the drivetrain losses; β (unitless) represents the gear reduction factor (light-duty vehicle uses the value of 1.0); f_p (unitless) denotes the throttle input from the driver; m_{ta} (kg) represents the vehicle mass on the tractive axle; P_{max} is the maximum engine power (kW); μ (unitless) denotes the adhesion coefficient between the vehicle tire and road surface; ρ is the sea-level air density (kg/m^3) at 15 °C temperature; A_f denotes the frontal area of the vehicle (m^2); C_d (unitless) represents the drag coefficient; C_h (unitless) is the correction factor for altitude; c_{r0}, c_{r1}, and c_{r2} are rolling resistance constant values (unitless); m (kg) represents the mass of the vehicle; and G (unitless) denotes the road grade.

2.3. Energy Consumption Model for BEVs

The Virginia Tech comprehensive power-based electric vehicle energy consumption model (VT-CPEM), developed in [25], was used in the proposed Eco-CACC-I system to compute instantaneous energy consumption levels for BEVs. The model was selected here for three main reasons: (1) speed is the only required input variable for this model, so it is easy to use to solve the proposed optimization problem; (2) the model has been validated and has demonstrated its ability to produce good accuracy compared to empirical data; and (3) the model can be calibrated to a specific vehicle using publicly available data. The VT-CPEM is a quasi-steady backward highly resolved power-based model, which only requires the instantaneous speed and the electric vehicle characteristics as input to compute the instantaneous power consumed. The VT-CPEM model is summarized in the following equations.

$$EC(t) = \int_0^t P_B(t) \cdot dt \tag{8}$$

$$P_B(t) = \begin{cases} \frac{P_W(t)}{\eta_D \cdot \eta_{EM} \cdot \eta_B} + P_A & \forall\, P_{Wheels}(t) \geq 0 \\ P_W(t) \cdot \eta_D \cdot \eta_{EM} \cdot \eta_B \cdot \eta_{rb}(t) + P_A & \forall\, P_{Wheels}(t) < 0 \end{cases} \tag{9}$$

$$P_W(t) = (ma(t) + R(t)) \cdot u(t) \tag{10}$$

$$\eta_{rb}(t) = \left[e^{\left(\frac{\lambda}{|a(t)|}\right)}\right]^{-1} \tag{11}$$

where EC (kWh) represents the energy consumption from time 0 to t; P_W denotes the power at the wheels (kW); P_B is the power consumed by (regenerated to) the electric motor (kW); P_A is the power consumed by the auxiliary systems (kW); η_D and η_{EM} (unitless) are the driveline efficiency and the efficiency of the electric motor, respectively; η_B (unitless) denotes the efficiency from a battery to an electric motor; η_{rb} represents the regenerative braking energy efficiency (unitless), which can be computed using Equation (11); the parameter λ (unitless) has been calibrated (λ = 0.0411) in [25] using empirical data described in [26]; and $R(t)$ represents the resistance force (N) computed in Equation (7).

3. Case Study

The case study simulated the proposed BEV Eco-CACC-I algorithm to investigate the impact of signal timing, speed limit, and road grade on the optimal solution.

In addition, two electric vehicles (a 2015 Nissan Leaf and a 2015 Tesla Model S) were considered in the simulation test to see if a particular vehicle model with different weight and engine powers affected the optimal solution. To compare BEV and ICEV optimal solutions, two ICEVs (a 2015 Honda Fit and a 2015 Cadillac SRX) were used for the same simulation test.

3.1. Test Eco-CACC-I for BEVs

The simulated test road consisted of a single signalized intersection with a control length starting 200 m upstream and ending 200 m downstream of the intersection (total length of 400 m). An automated connected vehicle, a 2015 Nissan Leaf equipped with the Eco-CACC-I system, was assumed to follow the optimal speed profile calculated by the Eco-CACC-I algorithm in that 400-m distance. Combinations of different speed limits (25, 30, 40, and 50 mph), signal indication offsets (15, 20, 25, and 30 s) and road grades (+3% and −3%) were tested. Given that the test results for the Nissan Leaf under various speed limits were very similar, the test results for the 25-mph speed limit alone are presented in Figures 2 and 3.

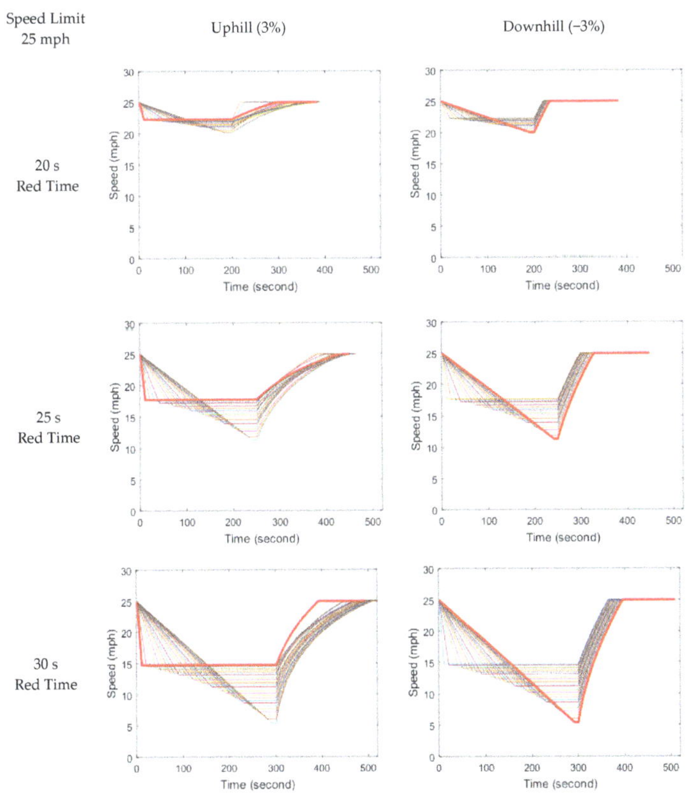

Figure 2. Nissan Leaf speed profile by BEV Eco-Cooperative Adaptive Cruise Control at Intersections (BEV Eco-CACC-I) for a speed limit of 25 mph.

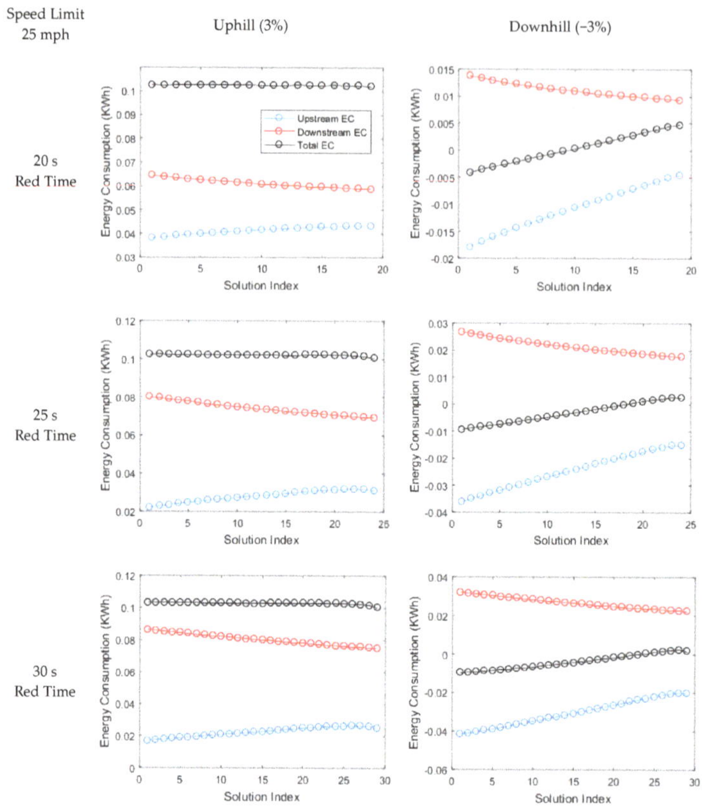

Figure 3. Nissan Leaf BEV Eco-CACC-I energy consumption for a speed limit of 25 mph.

Figure 2 demonstrates the test results under a 25-mph speed limit for different signal timings and road grade values. Note that the vehicle with an initial speed of 25 mph did not need to slow down for the 15-s red indication, so the plots for this case were not included. Each image in Figure 2 presents the sampling of numerous feasible solutions (speed profiles) for each combination of parameters. For instance, the right bottom image in Figure 2 includes 29 curves. Each curve represents a feasible solution when a vehicle approaches a signalized intersection with a certain deceleration level (a_i). The downstream throttle level was the optimal throttle corresponding to the minimal energy consumption given the upstream deceleration level of a_i. Each feasible solution is plotted in a different color, and the optimal solution, which corresponds to the minimal energy consumption trajectory, is presented in a bold red color. It should be noted that all the images in the left column in Figure 2 show that the speed profile associated with the maximum deceleration level was the optimal solution for the uphill direction. Furthermore, all the images in the right column in Figure 2 show that the speed profile associated with the minimum deceleration level was the optimal solution for the downhill direction.

The corresponding energy consumption levels for each feasible solution (speed profile) are presented in Figure 3. Note that the solution index in the x-axis represents the 1st solution, 2nd solution ... nth solution, ordered in ascending order by deceleration levels. All the images in the left column in Figure 3 show that the upstream trip regenerated minimum electric power; much less battery power was regenerated than was consumed. In this case, the cruise speed was the most important factor in identifying the optimal solution, as higher cruise speeds associated with higher deceleration levels result in less energy consumption for the entire trip. Consequently, the maximum deceleration level corresponds to the optimal solution for a BEV driving in the uphill direction. All the images in the right column in Figure 3 illustrate that the upstream trip generated equal or slightly higher electric power than the battery power consumed during the downstream trip due to the impact of gravity in the downhill direction. In this case, the deceleration level was the most important factor in identifying the optimal solution. Lower deceleration levels correspond to longer deceleration times and more regenerative electric power upstream of the intersection, which can result in lower energy consumption for the entire trip. Therefore, the minimal deceleration level corresponds to the optimal solution for a BEV driving in the downhill direction.

The Nissan Leaf is a compact BEV with an 80-HP engine; to investigate the impact of engine size on the optimal control strategy, a 2015 Tesla Model S with a much more powerful 283 HP engine was also tested. The same simulation was conducted assuming a connected and automated Tesla Model S equipped with the Eco-CACC-I controller. The simulation results were very similar to the Nissan Leaf results. There were two main differences. First, downstream of the intersection, the Tesla could accelerate to the maximum allowed speed (speed limit) much more quickly in the downhill direction given that the vehicle is more powerful than the Nissan Leaf. Second, the energy consumption for the Tesla Model S was higher since it weighs more. However, the energy consumption curves across the solutions from minimum to maximum deceleration levels showed the same trends, so the same optimal solution could be found for both vehicles. Given that the test results for the Nissan Leaf are already illustrated, the plots for the Tesla are not presented here. According to the test results for the two BEVs, the optimal solutions for the downhill and uphill directions can be summarized as follows:

- Downhill direction: The optimal speed profile corresponds to the minimum deceleration level in the solution space.
  - Upstream—lower cruise speed produces longer brake time and more regenerative energy.
  - Downstream—lower cruise speed means more energy consumption downstream; however, the benefit of energy regeneration upstream exceeds the additional needs for energy downstream.
- Uphill direction: The optimal speed profile corresponds to the maximum deceleration level in the solution space.
  - Upstream—different from the solution for the downhill direction, the vehicle regenerates minimum energy by decelerating in the uphill direction.
  - Downstream—the vehicle needs the maximum cruise speed while proceeding through the intersection so that the downstream trip requires less energy.

3.2. Eco-CACC-I for ICEVs

The Eco-CACC-I for ICEVs previously developed in [19,20] was considered in comparison with the Eco-CACC-I for BEVs. In this model, the optimization problem was formulated using Equations (2) through (4), and the same vehicle dynamic model used in Equations (5) through (7). Note that the Virginia Tech comprehensive power-based fuel consumption model (VT-CPFM-1) was used in place of the BEV energy model in Equations (8) through (11). More details of the Eco-CACC-I controller for ICEVs can be found in [19,20].

The same simulation was conducted for a 2015 Honda Fit, which has an engine power and weight similar to the 2015 Nissan Leaf. The test results are presented in Figures 4 and 5. Figure 4 shows the test results for a speed limit of 25 mph for different signal timings and roadway grades. All the images in the left column in Figure 4 demonstrate that the speed profile with a deceleration level in the middle area (between the minimum and maximum values) was the optimal solution for the uphill direction. All the images in the right column in Figure 4 demonstrate that the speed profile associated with the maximum deceleration level was the optimal solution for the downhill direction. The corresponding energy consumption levels for each feasible solution (speed profile) in the solution space are presented in Figure 5. Note that the solution index along the x-axis is also ranked and ordered in a descending manner based on the deceleration level. The energy consumption unit is "liters" for ICEVs. In addition, unlike BEVs that regenerate energy while braking, ICEVs always consume fuel during the trip. All the images in the left column in Figure 5 show that the vehicle consumed more energy to reach a higher cruise speed in the uphill direction upstream of the intersection. However, higher cruise speeds resulted in less energy consumption downstream of the intersection. Therefore, the optimal solution for ICEVs driving in the uphill direction is somewhere in the mid-range, depending on the vehicle's specifications and roadway grade. All the images in the right column in Figure 5 demonstrated that different deceleration levels did not change the ICEV's energy consumption while traveling downhill. Therefore, higher cruise speeds resulted in the same level of fuel consumption upstream of the intersection. However, higher cruise speeds resulted in less energy consumption downstream of the intersection. In this case, the deceleration level is the most important factor in locating the optimal solution. Higher deceleration levels corresponded to lower energy consumption for the downstream portion, while energy consumption remained the same for the upstream portion. Therefore, the maximum deceleration level corresponds to the optimal solution for ICEVs driving in the downhill direction.

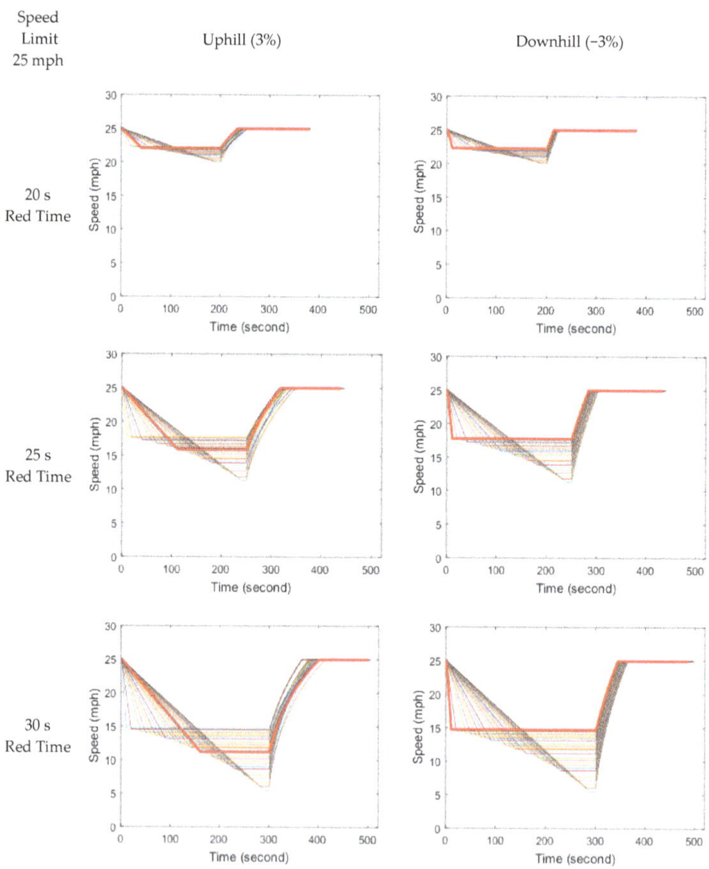

Figure 4. Honda Fit speed profile by ICEV Eco-CACC-I for a speed limit of 25 mph.

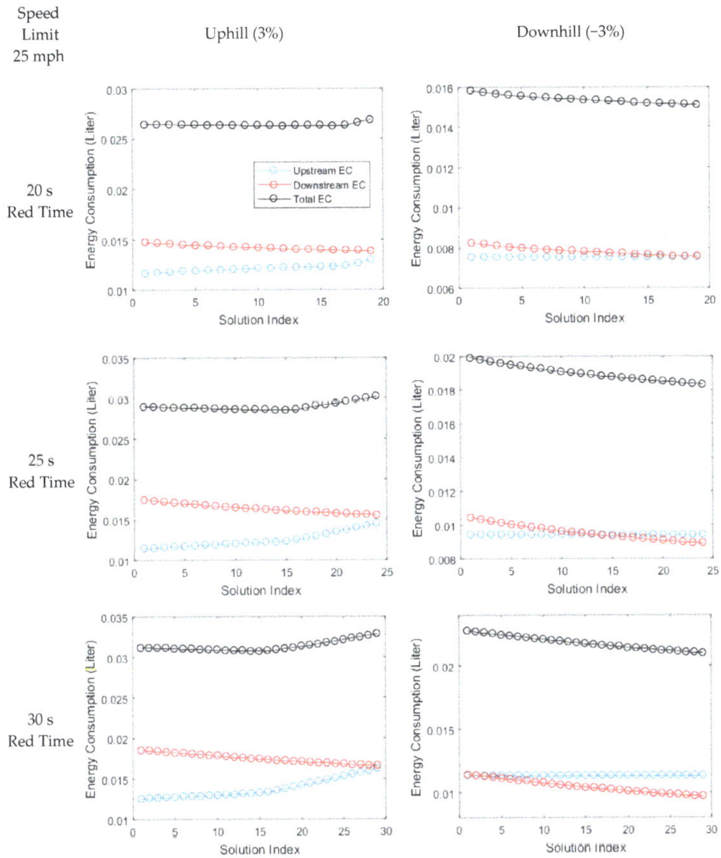

Figure 5. Honda Fit ICEV Eco-CACC-I fuel consumption for a speed limit of 25 mph.

The Honda Fit is a compact gasoline vehicle with a 97-HP engine. To examine whether ICEV optimal solutions are general or engine specific, a 2015 Cadillac SRX with a much more powerful engine of 230 HP was also tested. The same tests were conducted, assuming a connected automated Cadillac SRX equipped with the Eco-CACC-I controller. The simulation results for the two ICEVs were very similar. There were two differences. First, downstream of the intersection, the Cadillac could accelerate to the maximum allowed speed (speed limit) faster in the downhill direction, given that it had more engine power compared to the Honda Fit. Second, the energy consumption for the Cadillac was almost double that of the Honda Fit due to its larger size. However, the energy consumption curves across the solutions from minimum to maximum deceleration levels showed similar trends, demonstrating that the ICEV optimum strategies appear to be general. According to the test results for the two ICEVs, the optimal solutions produced by the Eco-CACC-I system for the downhill and uphill directions can be summarized as follows:

- Downhill direction: The optimal speed profile corresponds to the maximum deceleration level in the solution space.
    - Upstream—different deceleration levels do not change the ICEV's energy consumption during braking, so higher cruise speeds consume a similar amount of fuel.
    - Downstream—higher cruise speeds at the stop bar result in less energy consumption downstream.

- Uphill direction: The optimal speed profile corresponds to the maximum deceleration level in the solution space.
    - Upstream—unlike the downhill direction, the vehicle consumes more energy to reach a higher cruise speed while traveling uphill.
    - Downstream—higher cruise speeds result in less energy consumption downstream. Therefore, the optimal solution sits in the mid-range, depending on the vehicle's weight, engine power, and roadway slope.

3.3. Test Results Analysis and Comparison

The test results indicate that the optimal solutions for BEVs and ICEVs are quite different. The optimal solutions for BEVs and ICEVs when decelerating upstream of an intersection are summarized in Table 1. For downhill roadways, BEVs require longer deceleration times to accumulate more regenerative power to minimize overall energy consumption when traversing the intersection. Conversely, ICEVs need the maximum deceleration level (minimum deceleration time) to minimize overall energy consumption. For uphill approaches, BEVs need to minimize deceleration time to reach the approach stop line at maximum speed, saving energy downstream while accelerating back to the roadway speed limit. Contrarily, the optimum ICEV deceleration level is typically in the mid-range to minimize overall energy consumption. The comparison demonstrates that the energy-optimum solution for BEVs is different from that for ICEVs, as they consume energy differently. The findings in the case study also prove that previous studies, which only considered the optimization of vehicle acceleration and deceleration movements and ignored the specific vehicle energy model, may not correctly compute the energy-optimal eco-driving solutions for different types of vehicles.

Table 1. Optimal solutions for BEVs and ICEVs Eco-CACC-I systems when the vehicle needs to decelerate to traverse a signalized intersection.

Roadway Grade	BEV	ICEV
Uphill	Maximum deceleration	Mid-range deceleration
Downhill	Minimum deceleration	Maximum deceleration

3.4. Test Eco-CACC-I Controllers in Microscopic Traffic Simulation Software

The Eco-CACC-I controllers for BEVs and ICEVs were implemented in the microscopic traffic simulation software INTEGRATION to evaluate their performance. The INTEGRATION software is a trip-based microscopic traffic assignment, simulation, and optimization model that has the capability of modeling networks of up to 3,000,000 vehicle departures. A more-detailed description of INTEGRATION is provided in the literature [27,28].

A simulated traffic network, composed of three signalized intersections, as shown in Figure 6, was used in this test. The major road has a free-flow speed of 40 mph, a speed at a capacity of 30 mph, a saturation flow rate of 1600 veh/h/lane, and a jam density of 160 veh/km/lane. The total length of the main direction roadway is 4000 m. The three traffic signals (1000 m apart) have the same signal timing plan with a 60-s cycle length and 42-s phase length for the main street with 5-s intergreen time (yellow and all-red time indication). The signal offsets are 0 s for all traffic signals. The traffic volume in the main direction is set to be 400 veh/h/lane.

Given that the energy optimum solutions for ICEVs and BEVs are very different for the downhill direction, here we set a −3% grade for the main direction road. We modeled a 2015 Nissan Leaf and a 2015 Honda Fit to represent BEVs and ICEVs in the simulation. Four scenarios, described below, were used to compare the vehicle trajectories.

- Scenario 1 (uninformed drive for ICEVs): All the vehicles were ICEVs, and no Eco-CACC controller was activated. Each vehicle only followed the normal traffic rules (such as vehicle dynamics model, car-following model, collision avoidance) while traversing the network.
- Scenario 2 (uninformed drive for BEVs): All the vehicles were BEVs, and no Eco-CACC controller was activated. Each vehicle only followed the normal traffic rules (such as vehicle dynamics model, car-following model, collision avoidance) while traversing the network.
- Scenario 3 (informed drive by ICEV Eco-CACC-I): All the vehicles were ICEVs, and the ICEV Eco-CACC-I controller was activated when a vehicle was within a 200-m range (both upstream and downstream) of the signalized intersection.
- Scenario 4 (informed drive by BEV Eco-CACC-I): All the vehicles were BEVs, and the BEV Eco-CACC-I controller was activated when a vehicle was within a 200-m range (both upstream and downstream) of the signalized intersection.

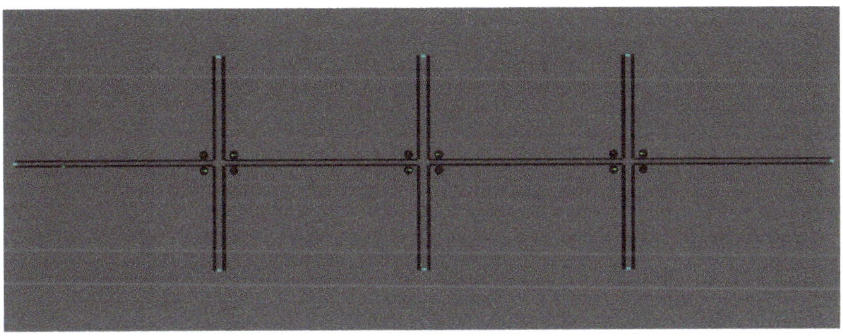

Figure 6. A simulated traffic network with three signalized intersections.

The vehicle trajectories near the first signalized intersection in the four scenarios are presented in Figure 7. Figure 7a presents the speed trajectories of the uninformed drive for ICEVs. We can clearly see that a total of nine vehicles came to a full stop upstream of the first intersection (as demonstrated by the horizontal trajectory lines). Figure 7b presents the speed trajectories of an uninformed drive for BEVs, which were very similar to the trajectories in (a) with nine fully stopped vehicles before the intersection. Figure 7c presents the speed trajectories optimized by ICEV Eco-CACC-I. The findings in Section 3.2 and Table 1 show that an ICEV equipped with an Eco-CACC-I controller quickly reduced speed and then cruised at a constant speed to approach the intersection during red signal indication when traveling in the downhill direction. The INTEGRATION simulation results in scenario 3 were demonstrated to be consistent with our previous findings, and the vehicles produced very smooth trajectories without having to come to a full stop, as shown in Figure 7c. In addition, the findings in Section 3.1 and Table 1 showed that the proposed BEV Eco-CACC-I controller would suggest that the vehicle decelerate mildly to maximize the deceleration time when traversing a signalized intersection on a downhill roadway, which was consistent with the simulation results in scenario 4, as shown in Figure 7d. The test results in the four scenarios proved that the microscopic traffic simulation with the ICEV and BEV Eco-CACC-I controller enabled in the INTEGRATION software produced consistent results to our findings in Table 1. In addition, the comparison of the simulation results in scenarios 2 and 4 showed that the BEV Eco-CACC-I controller produced average savings of 9.3% in energy consumption and 3.9% in vehicle delay.

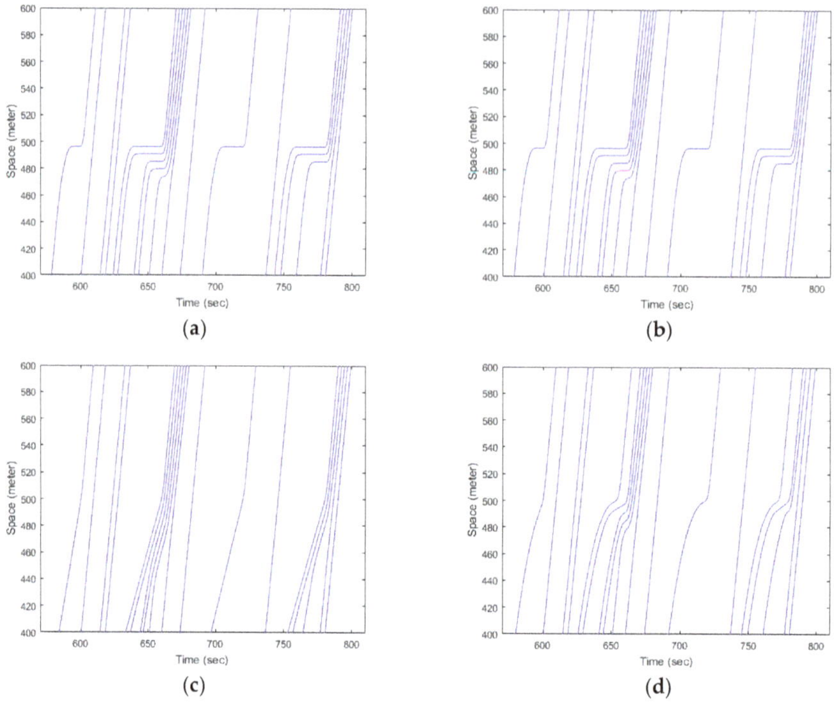

Figure 7. Comparison vehicle speed trajectories by (**a**) uninformed drive for ICEVs; (**b**) uninformed drive for BEVs; (**c**) ICEV Eco-CACC-I; (**d**) BEV Eco-CACC-I.

4. Conclusions and Future Work

A review of the literature shows that there are several issues in the BEV eco-driving strategies developed from existing studies, including a lack of realistic energy consumption models to accurately compute the instantaneous energy consumption when BEVs travel through signalized intersections, and the lack of a vehicle dynamics model to constrain vehicle acceleration maneuvers. In addition, although several previous studies developed eco-driving strategies for ICEVs and BEVs, there is no direct comparison to demonstrate the differences in the energy-optimal solutions for each. To address these issues, this study developed a BEV Eco-CACC-I controller. The developed controller can assist BEVs while traversing signalized intersections with minimal energy consumption. The calculation of the optimal vehicle trajectory was formulated as an optimization problem under the constraints of (1) vehicle acceleration/deceleration behavior, defined by a vehicle dynamics model; (2) vehicle energy consumption behavior, defined by a BEV energy consumption model; and (3) the relationship between vehicle speed, location and signal timing, defined by vehicle characteristics and SPaT data shared under the connected vehicle environment. The optimal speed trajectory was computed in real-time by the proposed BEV Eco-CACC-I controller so that a BEV could follow the optimal speed while negotiating a signalized intersection. The proposed BEV controller was tested in a case study to investigate the performances under various speed limits, roadway grades, and signal timings. In addition, a comparison of the optimal speed trajectories for BEVs and ICEVs was conducted to investigate the impact of vehicle engine types on eco-driving solutions. The comparison results illustrate that previous studies, which only considered the optimization of acceleration/deceleration and ignored the specific vehicle energy model, may not correctly compute the energy-optimal eco-driving solution for different vehicle types. Lastly, the proposed controller was implemented in microscopic traffic simulation software to test its networkwide performance. The test results from an arterial corridor with three signalized intersections demonstrated that the proposed controller can effectively reduce stop-and-go traffic in the vicinity of signalized intersections, and that the BEV Eco-CACC-I controller produced average savings of 9.3% in energy consumption and 3.9% in vehicle delays.

Although the proposed controller was demonstrated to produce very positive energy and delay savings for BEVs from the simulation tests, currently the developed Eco-CACC-I controller can only optimize BEVs or ICEVs separately. In future work, an integrated optimization for different types of vehicles will be considered in developing optimum solutions for mixed traffic conditions. In addition, more simulation tests with various traffic volumes, market penetration rates, signal timings, etc., will be considered to test the proposed controller further.

Author Contributions: Conceptualization, H.C. and H.A.R.; methodology, H.C. and H.A.R.; software, H.C. and H.A.R.; writing, H.C. and H.A.R.; funding acquisition, H.A.R. and H.C. All authors have read and agreed to the published version of the manuscript.

Funding: This work was co-funded by the Department of Energy through the Office of Energy Efficiency and Renewable Energy (EERE), Vehicle Technologies Office, Energy Efficient Mobility Systems Program under award number DE-EE0008209 and the University Mobility Equity Center (UMEC).

Conflicts of Interest: The authors declare no conflict of interest.

References

1. U.S. Energy Information Administration. *Oil: Crude and Petroleum Products Explained*; 2018. Available online: https://www.eia.gov/energyexplained/oil-and-petroleum-products/use-of-oil.php (accessed on 5 May 2020).
2. Kamalanathsharma, R.K. Eco-Driving in the Vicinity of Roadway Intersections—Algorithmic Development, Modeling, and Testing. Ph.D. Thesis, Virginia Polytechnic Institute and State University, Blacksburg, VA, USA, 2014.
3. USDOT.a. *Connected Vehicle Technology*; 2015. Available online: http://www.its.dot.gov/connected_vehicle/connected_vehicle_tech.htm (accessed on 5 May 2020).
4. Rakha, H.; Ahn, K.; Trani, A. Comparison of MOBILE5a, MOBILE6, VT-MICRO, and CMEM models for estimating hot-stabilized light-duty gasoline vehicle emissions. *Can. J. Civil. Eng.* **2003**, *30*, 1010–1021. [CrossRef]
5. Li, X.; Li, G.; Pang, S.-S.; Yang, X.; Tian, J. Signal timing of intersections using integrated optimization of traffic quality, emissions and fuel consumption: A note. *Transp. Res. Part D Transp. Environ.* **2004**, *9*, 401–407. [CrossRef]
6. Stevanovic, A.; Stevanovic, J.; Zhang, K.; Batterman, S. Optimizing traffic control to reduce fuel consumption and vehicular emissions: Integrated approach with VISSIM, CMEM, and VISGAOST. *Transp. Res. Rec. J. Transp. Res. Board* **2009**, *2128*, 105–113. [CrossRef]
7. Saboohi, Y.; Farzaneh, H. Model for optimizing energy efficiency through controlling speed and gear ratio. *Energy Eff.* **2008**, *1*, 65–76. [CrossRef]
8. Saboohi, Y.; Farzaneh, H. Model for developing an eco-driving strategy of a passenger vehicle based on the least fuel consumption. *Appl. Energy* **2009**, *86*, 1925–1932. [CrossRef]
9. Barth, M.; Boriboonsomsin, K. Energy and emissions impacts of a freeway-based dynamic eco-driving system. *Transp. Res. Part D Transp. Environ.* **2009**, *14*, 400–410. [CrossRef]
10. Malakorn, K.J.; Park, B. Assessment of mobility, energy, and environment impacts of IntelliDrive-based Cooperative Adaptive Cruise Control and Intelligent Traffic Signal control. In Proceedings of the IEEE International Symposium on Sustainable Systems and Technology (ISSST) 2010, Arlington, VA, USA, 17–19 May 2010; pp. 1–6.
11. Asadi, B.; Vahidi, A. Predictive cruise control: Utilizing upcoming traffic signal information for improving fuel economy and reducing trip time. *Control. Syst. Technol. IEEE Trans.* **2011**, *19*, 707–714. [CrossRef]
12. Guan, T.; Frey, C.W. Predictive fuel efficiency optimization using traffic light timings and fuel consumption model. In Proceedings of the 2013 16th International IEEE Conference on Intelligent Transportation Systems-(ITSC), The Hague, The Netherlands, 6–9 October 2013; pp. 1553–1558.

13. Miyatake, M.; Kuriyama, M.; Takeda, Y. Theoretical study on eco-driving technique for an electric vehicle considering traffic signals. In Proceedings of the IEEE Ninth International Conference on Power Electronics and Drive Systems (PEDS), Singapore, 5–8 December 2011; pp. 733–738.
14. Zhang, R.; Yao, E. Eco-driving at signalised intersections for electric vehicles. *IET Intell. Transp. Syst.* **2015**, *9*, 488–497. [CrossRef]
15. Qi, X.; Barth, M.J.; Wu, G.; Boriboonsomsin, K.; Wang, P. Energy Impact of Connected Eco-driving on Electric Vehicles. In *Road Vehicle Automation*, 4th ed.; Springer: Berlin/Heidelberg, Germany, 2018; pp. 97–111.
16. Wu, X.; He, X.; Yu, G.; Harmandayan, A.; Wang, Y. Energy-optimal speed control for electric vehicles on signalized arterials. *IEEE Trans. Intell. Transp. Syst.* **2015**, *16*, 2786–2796. [CrossRef]
17. De Nunzio, G.; Wit, C.C.; Moulin, P.; Di Domenico, D. Eco–driving in urban traffic networks using traffic signals information. *Int. J. Robust Nonlinear Control.* **2016**, *26*, 1307–1324. [CrossRef]
18. Almannaa, M.H.; Chen, H.; Rakha, H.A.; Loulizi, A.; El-Shawarby, I. Field implementation and testing of an automated eco-cooperative adaptive cruise control system in the vicinity of signalized intersections. *Transp. Res. Part D Transp. Environ.* **2019**, *67*, 244–262. [CrossRef]
19. Chen, H.; Rakha, H.A.; Almannaa, M.; Loulizi, A.; El-Shawarby, I. Field Implementation of an Eco-cooperative Adaptive Cruise System at Signalized Intersections. In Proceedings of the 94th Annual Meeting Transportation Research Board, Washington, DC, USA, 8–12 January 2017.
20. Chen, H.; Rakha, H.A.; Loulizi, A.; El-Shawarby, I.; Almannaa, M.H. Development and Preliminary Field Testing of an In-Vehicle Eco-Speed Control System in the Vicinity of Signalized Intersections. *IFAC-PapersOnLine* **2016**, *49*, 249–254. [CrossRef]
21. Kamalanathsharma, R.K.; Rakha, H.A.; Yang, H. Networkwide impacts of vehicle ecospeed control in the vicinity of traffic signalized intersections. *Transp. Res. Rec.* **2015**, *2503*, 91–99. [CrossRef]
22. Kamalanathsharma, R.K.; Rakha, H.A. Leveraging connected vehicle technology and telematics to enhance vehicle fuel efficiency in the vicinity of signalized intersections. *J. Intell. Transp. Syst.* **2016**, *20*, 33–44. [CrossRef]
23. Fadhloun, K.; Rakha, H.; Loulizi, A.; Abdelkefi, A. Vehicle dynamics model for estimating typical vehicle accelerations. *Transp. Res. Rec. J. Transp. Res. Board* **2015**, *249*, 61–71. [CrossRef]
24. Yu, K.; Yang, J.; Yamaguchi, D. Model predictive control for hybrid vehicle ecological driving using traffic signal and road slope information. *Control. Theory Technol.* **2015**, *13*, 17–28. [CrossRef]
25. Fiori, C.; Ahn, K.; Rakha, H.A. Power-based electric vehicle energy consumption model: Model development and validation. *Appl. Energy* **2016**, *168*, 257–268. [CrossRef]
26. Gao, Y.; Chu, L.; Ehsani, M. Design and control principles of hybrid braking system for EV, HEV and FCV. In Proceedings of the Vehicle Power and Propulsion Conference, Arlington, TX, USA, 9–12 September 2017; IEEE: Piscataway, NJ, USA, 2007; pp. 384–391.
27. Aerde, M.V.; Rakha, H. *INTEGRATION © Release 2.30 for Windows: User's Guide—Volume II: Advanced Model. Features*; M. Van Aerde & Assoc., Ltd.: Blacksburg, VA, USA, 2007.
28. Aerde, M.V.; Rakha, H. *INTEGRATION © Release 2.30 for Windows: User's Guide—Volume I: Fundamental Model. Features*; M. Van Aerde & Assoc., Ltd.: Blacksburg, VA, USA, 2007.

© 2020 by the authors. Licensee MDPI, Basel, Switzerland. This article is an open access article distributed under the terms and conditions of the Creative Commons Attribution (CC BY) license (http://creativecommons.org/licenses/by/4.0/).

energies

MDPI

Article

The Role of Powertrain Electrification in Achieving Deep Decarbonization in Road Freight Transport

Juan C. González Palencia [1,*], Van Tuan Nguyen [1], Mikiya Araki [1] and Seiichi Shiga [2]

1 Division of Mechanical Science and Technology, Graduate School of Science and Technology, Gunma University, 1-5-1 Tenjincho, Kiryu, Gunma 376-8515, Japan; T17B602@gunma-u.ac.jp (V.T.N.); mikiya.araki@gunma-u.ac.jp (M.A.)

2 Division of Mechanical Science and Technology, Graduate School of Science and Technology, Gunma University, 29-1 Honcho, Ota, Gunma 373-0057, Japan; shiga@gunma-u.ac.jp

* Correspondence: gonzalez@gunma-u.ac.jp; Tel.: +81-277-30-1516

Received: 14 April 2020; Accepted: 10 May 2020; Published: 13 May 2020

Abstract: Decarbonizing road freight transport is difficult due to its reliance on fossil fuel internal combustion engine vehicles (ICEVs). The role of powertrain electrification in achieving deep decarbonization in road freight transport was studied using a vehicle stock turnover model, focusing on Japan. Twelve vehicle types were considered; combining four powertrains, ICEV, hybrid electric vehicle (HEV), battery electric vehicle (BEV) and fuel cell electric vehicle (FCEV); and three vehicle size classes, normal, compact and mini-sized vehicles. A scenario-based approach was used; considering a Base scenario, and three alternative scenarios targeting powertrain electrification. Between 2012 and 2050, tank to wheel CO_2 emissions decrease 42.8% in the Base scenario, due to the reduction of vehicle stock, the improvement of vehicle fuel consumption and the adoption of HEVs. Diffusion of FCEVs in normal vehicles and BEVs in compact and mini-sized vehicles achieves the largest tank to wheel CO_2 emissions reductions, up to 44.6% compared with the 2050 baseline value. The net cash flow is positive over the whole time horizon, peaking at 6.7 billion USD/year in 2049 and reaching 6.6 billion USD/year by 2050. Powertrain electrification is not enough to achieve any of the CO_2 emissions reduction targets in road freight transport.

Keywords: road freight transport; vehicle stock turnover model; deep decarbonization; road freight vehicle; electric-drive vehicle

1. Introduction

1.1. Motivation of the Research

Road freight transport accounts for a significant share of global energy consumption in the transport sector. Almost 45% of global transport energy consumption corresponds to freight transport, with heavy-duty vehicles (HDVs) using more than half of that energy [1]. Furthermore, road freight transport depends heavily on fossil fuels; with medium freight trucks and heavy freight trucks accounting for 24% of global oil-based fuel consumption [2]. Diesel is the main fuel used in road freight transport, accounting for 84% of all oil products used; and corresponding to half of the global diesel demand [3].

Road freight vehicles are difficult to characterize due to the variety of vocational uses that requires a large number of sizes and configurations. However, it is considered that goods delivery from production sites to factories, industries and/or final retailers is one of the essential roles of road freight vehicles [3]. Road freight vehicles consist of light-duty vehicles (LDVs), medium-duty vehicles (MDVs) and HDVs; including but not limited to trucks. One additional difficulty for road freight vehicle characterization is that criteria for vehicle size classification vary depending on the country or region.

Despite the small share in road vehicles, MDVs and HDVs contribute disproportionally to mobile greenhouse gas (GHG) and air pollutant emissions and fossil fuel use, due to high vehicle fuel consumption, large annual traveled distances and long idling times [4]. In the European Union, HDVs account for 30% of on-road CO_2 emissions, despite representing only 4% of the road vehicle stock [5]. Similarly in the United States, MDVs and HDVs account for 26% of transport CO_2 emissions [6]. Additionally, road freight vehicles produce half of particulate matter (PM) emissions and one third of NO_x emissions of the transport sector in cities [7]. In the United States, MDVs and HDVs are responsible for 22%, 28% and 24% of transport sector $PM_{2.5}$, NO_x and CO emissions, respectively [4]. PM and NO_x are associated with adverse health effects in human beings [8]; while exposure to low CO concentrations causes health problems such as headaches and lightheadness [9]. Furthermore, diesel exhaust gas is classified as carcinogenic to humans (Group 1) by the World Health Organization (WHO) [10].

Decarbonization in road freight transport will be difficult to realize [11]. Global road freight activity is expected to grow in the future, driven by economic development. For instance, heavy-duty truck use is expected to increase 2.7%/year between 2000 and 2030 [12]. Increased road freight activity and high dependence on fossil fuels will cause GHG emissions from road freight transport to keep growing in the future; requiring coordinated efforts by shippers, logistics service providers and policymakers to mitigate their growth [2].

Potential for energy consumption and CO_2 emissions reduction through the improvement of internal combustion engine efficiency and aerodynamics is significant and can be cost-effective [13]. However, in the long-term, advanced biofuels and electric-drive vehicles (EDVs) will be the main option to achieve deep decarbonization in road freight vehicles [14]. Additionally, since the time scales involved in road freight vehicle stock turnover are long, it is necessary to consider the dynamics of technology diffusion in the vehicle fleet when assessing the role of powertrain electrification in the decarbonization of road freight transport.

Compared with passenger LDVs, deployment of EDVs in road freight vehicles is still at an early stage. While the global EDV stock in LDVs exceeded 5.1 million vehicles in 2018, EDV stock in the fleet of light commercial vehicles reached 250000 vehicles; and medium-sized trucks sales were in the range of 1000 to 2000 vehicles [15]. Considering that global road freight vehicle stock totaled approximately 186 million vehicles in 2015 [3], penetration of EDVs in road freight transport is almost negligible.

Nevertheless, EDV deployment in road freight vehicles is gaining momentum, with McKinsey projecting that EDVs can account for 15% of global truck sales by 2030 [16]. Currently, several Original Equipment Manufacturers (OEMs) are investing in the development of EDVs for road freight transport. BYD introduced a battery electric class 8 truck and a battery electric refuse truck in 2017 [17]. Isuzu presented the battery electric ELF EV truck at the 2017 Tokyo Motor Show [18]. Tesla announced plans to manufacture battery electric semi-trucks by 2019, with UPS pre-ordering 125 vehicles [19]. Renault plans to start selling battery electric 12–16 ton trucks to complement the line-up of battery electric 4.5 ton trucks in the market since 2010 [20]. Mitsubishi Fuso started mass production of the battery electric eCanter in 2019 [21].

Regarding FCEVs, a Swiss consortium with participation of ESORO announced the development of a 34 ton fuel cell truck [22]. Toyota is working on a fuel cell system for heady-duty trucks to be used at the port of Los Angeles, United States [23]. Ballard and Kenworth have been testing a class 8 drayage fuel cell truck in the United States [24]. Nikola Motors plans to start selling Class 8 fuel cell trucks by 2021 [25].

Compared with passenger LDVs, EDV adoption in road freight transport has several advantages, especially at the early stages of deployment. Under some operating conditions, high annual traveled distances for long-haul vehicles enable fast recovery of the incremental capital costs through lower fuel costs [26]. In the case of BEVs, road freight vehicles follow fixed driving routines based on their delivery routes, making range anxiety concerns less critical [27]. Furthermore, operation according to a predetermined mobility demand makes the requirement for public charging modest [28]. In the case of FCEVs, known daily traveled distance and central refueling station use make fleet-based operation attractive [4].

Nevertheless, EDV deployment in road freight transport also face barriers, such as diversity in terms of vehicle sizes and vocational uses [4], larger weight and longer traveled distances [12], and larger emphasis on economic profitability and reluctance to take risks regarding new automotive technologies [29]. Even though several studies have demonstrated that under some operating conditions EDVs are cost-effective [30–32], road freight vehicle fleet owners often focus on short payback times rather than cost-effectiveness over the vehicle life cycle, which can affect EDV adoption. For example, in the United States, the average payback time for large heavy-duty truck fleets is 24 months [33]; while payback times as short as 6 months are reported for small heavy-duty truck fleets [34].

1.2. *Review of the Existing Literature*

There is a growing interest in research about EDV applications for road freight transport; with most of the research assessing the performance of EDVs compared with internal combustion engine vehicles (ICEVs) from energy consumption, CO_2 emissions and/or cost perspectives.

On a single-vehicle basis, Zhao et al. [35] estimated energy consumption, CO_2 emissions and cost performance for a HEV, a BEV and a FCEV and compared them with a diesel-fueled ICEV for Class 8 trucks in the United States. Lee et al. [36] estimated energy consumption, GHG emissions and total cost of ownership (TCO) for ICEV and BEV urban delivery trucks. Lajunen [37] evaluated energy consumption for HEVs and compared them with ICEVs for HDVs in Finland. Gao et al. [38] assessed several energy consumption reduction technology measures for ICEVs and HEVs for Class 8 trucks in the United States. Lebeau et al. [39] assessed the cost-effectiveness of BEVs for freight transport in Brussels, Belgium, focusing on light commercial vehicles and quadricycles. Zhao et al. [40] assessed energy consumption and GHG emissions for ICEVs, HEVs and BEVs for Class 3–5 delivery trucks in the United States. Kast et al. [4,6] assessed the performance of FCEVs in the MDV and HDV segments in the United States. Sen et al. [31] estimated life cycle GHG and air pollutant emissions, costs and externalities for ICEVs, HEVs and BEVs for Class 8 trucks in the United States. Lee and Thomas [31] evaluated energy consumption, water use, GHG and air pollutant emissions for ICEVs, HEVs and BEVs in medium duty trucks in the United States. Zhao and Tatari [41] evaluated energy consumption and GHG emissions for ICEVs, HEVs and plug-in hybrid electric vehicles (PHEVs) for refuse trucks in the United States. Zhou et al. [42] assessed life cycle GHG emissions and TCO for ICEVs and BEVs for Class 6 trucks in Toronto, Canada.

Other studies assessed the role of EDVs in road freight transport on a fleet basis without considering vehicle stock turnover. Davis and Figliozzi [43] assessed the economic competitiveness of ICEVs and BEVs for MDVs in the United States. Wikström et al. [28] evaluated BEVs and PHEVs for road freight transport in Sweden from technological and social perspectives. Zhao et al. [27] estimated the optimum penetration of HEVs and BEVs in a commercial delivery fleet of MDVs in the United States. Christensen et al. [44] studied the suitability of BEV introduction for road freight transport in LDVs and HDVs in Germany and Denmark. Though valuable, these studies did not capture the dynamics of technology diffusion in the road freight vehicle fleet.

Regarding studies on a fleet basis that considered vehicle stock turnover, Li et al. [45] estimated the most cost-effective pathways to reduce oil consumption in road freight transport in China, considering ICEVs and HEVs; without considering BEVs and FCEVs. Askin et al. [26] analyzed the factors that can influence deployment of advanced technologies in HDVs in the United States, considering natural gas-fueled ICEVs and diesel-fueled HEVs; without considering BEVs and FCEVs. Oshiro and Masui [46] studied the impact of EDV diffusion in road transport in Japan, focusing on HEVs and FCEVs as powertrain options for HDVs; without considering fuel consumption evolution in time and cost. Fridstrøm [47] evaluated the role of EDVs in GHG emissions reduction in road transport in Norway, including freight transport; without considering the impact on cost. Talebian et al. [47] studied the potential of BEVs and FCEVs to reduce GHG emissions in road freight transport in British

Columbia, Canada; without considering the impact on cost. Mullholland et al. [48] assessed the CO_2 emissions reduction potential of HEVs and BEVs in global road freight transport; without considering FCEVs and the impact on cost.

The main characteristics of the most relevant studies regarding powertrain electrification in road freight transport are summarized in Table 1. No research was found in the reviewed literature that assesses the potential of powertrain electrification to reduce CO_2 emissions in road freight transport and the associated cost, considering the dynamics of technology diffusion.

Table 1. Main characteristics of previous studies regarding powertrain electrification in road freight transport on a fleet basis.

Author	Year	EDV			Stock Turnover	Effects Assessed	
		HEV	BEV	FCEV		CO_2 Emissions	Cost
Davis	2013		✓				✓
Wikström	2015		✓				
Zhao	2016	✓	✓			✓	✓
Christensen	2017		✓				
Askin	2015	✓			✓	✓	✓
Li	2015	✓			✓	✓	✓
Oshiro	2015	✓		✓	✓	✓	
Fridstrøm	2017	✓	✓	✓	✓	✓	
Mulholland	2018	✓	✓		✓	✓	
Talebian	2018		✓	✓	✓	✓	
This study	2020	✓	✓	✓	✓	✓	✓

1.3. Contribution

The objective of this research is to estimate the potential of powertrain electrification to reduce road freight transport fossil fuel consumption and CO_2 emissions, and the impact on cost. The road freight vehicle fleet in Japan was chosen as the case study.

To the authors' best knowledge, this is the first research that assesses the CO_2 emissions reduction potential and the associated cost of powertrain electrification in road freight transport on a national scale, considering the dynamics of technology diffusion. The CO_2 emissions reduction potential estimated here corresponds to the upper boundary for CO_2 emissions reductions achievable through EDV deployment in the road freight vehicle fleet. Results from this research can help policymakers understand the maximum contribution of powertrain electrification to reduce CO_2 emissions from road freight transport; and therefore, designing more effective policies for climate change mitigation in the transport sector. Furthermore, results presented here also show the requirements in terms of cost and supply of electricity and hydrogen needed to achieve powertrain electrification in road freight transport.

1.4. Organization of the Article

The rest of the article is organized as follows: the mathematical formulation of the model and the description of the scenarios are presented in Section 2; the results are presented and discussed in Section 3; finally, the conclusions are presented in Section 4.

2. Methods

2.1. Formulation of the Vehicle Stock Turnover Model for the Road Freight Vehicle Fleet

This research focused on energy end-use in road freight vehicles. Other stages of the vehicle cycle and the fuel cycle were not considered. Energy flows in the energy system are shown in Figure 1. The energy system is made of four energy carriers, diesel, gasoline, electricity, and hydrogen; which are used in 12 vehicle types. The 12 vehicle types result from combining four powertrains, ICEV, HEV, BEV, and FCEV; and three vehicle size classes, normal, compact and mini-sized vehicles.

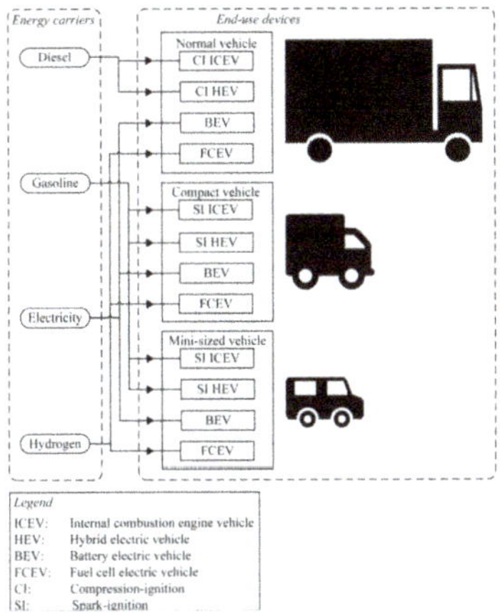

Figure 1. Energy system diagram.

The criteria from the Japan Ministry of Land Infrastructure, Transport and Tourism (MLIT) [49] were used for road freight vehicle size classification. These criteria divide road freight vehicles in normal, compact and mini-sized classes according to the external dimensions and engine displacement. It should be noted that according to MLIT's classification, all mini-sized vehicles are LDVs. However, not all LDVs belong to the mini-sized vehicle size class; and there are LDVs that belong to the compact vehicle size class. MDVs belong to the compact and normal vehicle size classes; while HDVs belong to the normal vehicle size class. The Gross Vehicle Weight (GVW) ranges for road freight vehicles in Japan were extracted by analyzing MLIT data [50]; with GVW for mini-sized vehicles varying between 0.9 and 1.5 ton; compact vehicles between 1.6 and 3.4 ton; and normal vehicles between 2.8 and 59.1 ton.

The Long-range Energy Alternatives Planning system (LEAP) software was used to model the energy system [51]; based on previous work developed by the authors in [52,53]. The model developed can be classified as a dynamic bottom-up accounting energy-economics model. The description of the main components of the model is presented below. For more details, the reader can examine the previous references.

For a given fleet of vehicles type t and vintage v in a calendar year y, the vehicle fleet tank to wheel (TTW) energy consumption is calculated by multiplying the vehicle stock N, the annual traveled distance M and the vehicle fuel consumption R, as indicated in Equation (1):

$$E_{TTW,t,y,v} = N_{t,y,v} M_{t,y,v} R_{t,y,v} \tag{1}$$

The vehicle fleet TTW CO_2 emissions are calculated by multiplying the fleet energy consumption $E_{TTW,t,y,v}$ and the CO_2 emission factor $EF_{TTW,t,y}$ of the fuel used by the vehicle type t in the calendar year y, as indicated in Equation (2):

$$G_{TTW,t,y,v} = E_{TTW,t,y,v} EF_{TTW,t,y} \tag{2}$$

The vehicle fleet well to wheel (WTW) CO_2 emissions are obtained similar to Equation (2), replacing the TTW CO_2 emission factor with the WTW CO_2 emission factor.

The economic assessment was performed using the relative cost of ownership (*RCO*), defined as the sum of the capital cost, the operating and maintenance (O&M) cost and the energy cost, as indicated in Equation (3):

$$RCO_{t,y,v} = S_{t,v=y}c_{cap,t,v=y}CRF + N_{t,y,v}c_{OM,t,y,v} + E_{TTW,t,y,v}c_{ene,t,y} \tag{3}$$

where S is the new vehicle sales, c_{cap} is the vehicle capital cost, CRF is the capital recovery factor, c_{OM} is the annual O&M cost, and c_{ene} is the energy price.

It should be noted that Equations (1) and (2) are built under the assumption that annual traveled distance and vehicle fuel consumption are constant for all vehicles of a given type and vintage during a calendar year. In that sense, differences in vehicle usage across users cannot be captured in the model. Regarding the economic evaluation, it was assumed that costs other than capital cost, O&M cost and energy cost are identical for ICEVs and EDVs [54]; and therefore, excluded in the estimation of the RCO in Equation (3). Road vehicle fleet energy consumption, CO_2 emissions and RCO for a given calendar year y are estimated by summing across all vehicle types and vintages existing in the road freight vehicle fleet.

2.2. Road Freight Vehicles Characteristics

Due to the variety of vocational uses, there is a large number of road freight vehicle types. In Japan, the MLIT reported fuel consumption data for 5272 road freight vehicle types [55], mainly ICEVs. Considering this number of vehicle types in the vehicle stock turnover model is not practical. Therefore, the structure of the road freight vehicle fleet was simplified, assuming there are only 12 vehicle types: ICEVs, HEVs, BEVs, and FCEV, available in three vehicle size classes, normal, compact and mini-sized vehicles. Additionally, based on the analysis of the same data from MLIT, it was assumed that all normal ICEVs use diesel; and all compact and mini-sized ICEVs use gasoline.

In order to calibrate the model against historical data, 2012 was selected as the base year. It was assumed that all road freight vehicles in the Base year were ICEVs. Each vehicle size class was represented only by one ICEV, with vehicle fuel consumption close to the average value reported in data from MLIT [56]. Mini-sized ICEVs were modeled as a Subaru Sambar with a GVW of 1.2 ton; compact ICEVs were modeled as a Toyota Hiace with a GVW of 3.1 ton; and normal ICEVs were modeled as a Fuso Canter with a GVW of 7.9 ton. Vehicle data for road freight EDVs were constructed extracting relationships between ICEVs and EDVs from the existing literature.

Fuel consumption and capital cost for the road freight vehicles in 2012 and 2050 are shown in Figure 2. Fuel consumption data for ICEVs in all vehicle size classes in 2012 were estimated using data reported by MLIT [56]. Using ICEV fuel consumption as a reference, 2012 fuel consumption for EDVs in the normal size class was estimated assuming the ICEV and EDV fuel consumption ratios are identical to values reported in [35] for HEVs, and in [57] for BEVs and FCEVs. Fuel consumption in 2012 for EDVs in the compact size class and the mini-sized BEV was estimated considering the ICEV and EDV fuel consumption ratios are identical to the values reported for normal LDVs in [53]. Fuel consumption in 2012 for the mini-sized HEV and FCEV was estimated assuming the fuel consumption ratios for ICEV and HEV and for ICEV and FCEV are identical to the values reported for LDVs in [58].

The capital costs for the normal ICEV, HEV and BEV were obtained from [40]; while the capital cost for the normal FCEV was estimated as the capital cost for the HEV minus the capital cost of the internal combustion engine plus the capital cost of the fuel cell, using data from [59,60]. Capital costs in 2012 for compact and mini-sized road freight vehicles were assumed equal to values for passenger vehicles reported in [53]. Fuel consumption and capital cost evolution between 2012 and 2050 were assumed identical to the trends reported in the previous reference.

In order to estimate the RCO, vehicle capital costs were annualized over the vehicle service lives using a discount rate of 10%; corresponding to the average value of the range for typical discount rates for trucks reported in [30]. The same reference was used for the O&M costs for the normal ICEV,

HEV and BEV, estimated in 0.124, 0.099 and 0.087 USD/km, respectively. O&M costs for compact and mini-sized vehicles were considered identical, 0.056 USD/km for ICEVs and HEVs, and 0.057 USD/km for BEVs and FCEVs [61,62]. O&M costs for all vehicle types were assumed to remain constant throughout the time horizon.

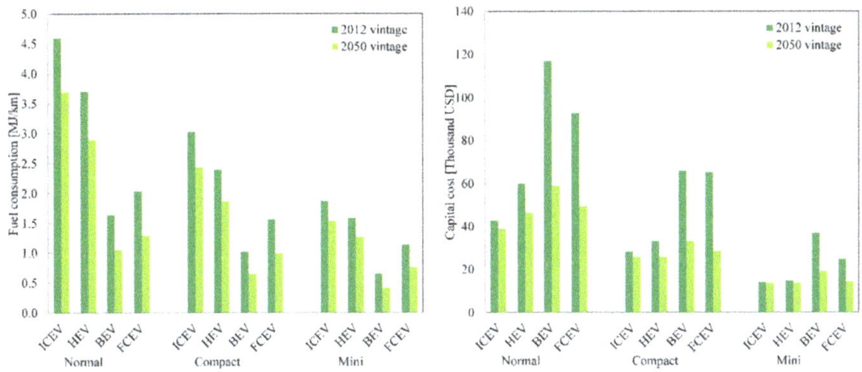

(a) (b)

Figure 2. Main characteristics of the road freight vehicles: (**a**) fuel consumption; (**b**) capital cost.

Vehicle usage characteristics are presented in Figures 3 and 4. Annual traveled distance is reported for public and private vehicles for each vehicle size class by MLIT [63]. The shares of public and private vehicles in each vehicle size class were estimated using data from the Automobile Inspection & Registration Association (AIRIA) [64]. These values were used to calculate the annual traveled distance for each vehicle size class as the weighted sum of the annual traveled distances for public and private vehicles. The median vehicle service life for each vehicle size class corresponds to the average value reported by MOE [65]. Similar to Nishimura [66], survival profiles were estimated using the logistic curve shown in Equation (4):

$$r = 1 - \frac{1}{\alpha + e^{-\beta(a-a_0)}} \quad (4)$$

where r is the vehicle survival rate, a is the vehicle age, a_0 is the vehicle median service life, α is a model parameter set to 1, and β is a growth parameter. The growth parameter β was estimated equal to 0.180, 0.192 and 0.187 for normal, compact and mini-sized vehicles through model calibration against historical data for the road freight vehicle stock.

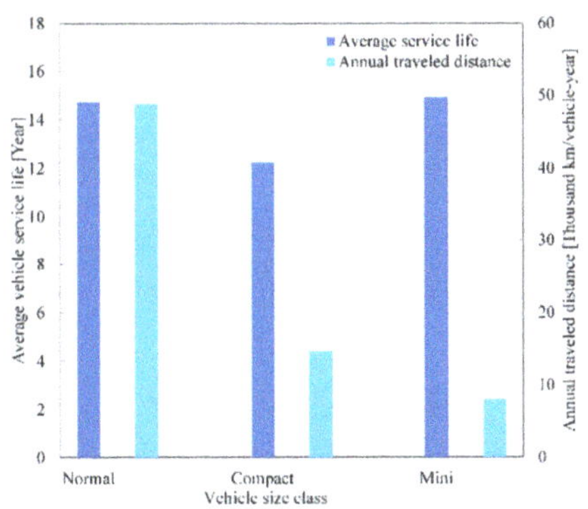

Figure 3. Vehicle service lives and annual traveled distances.

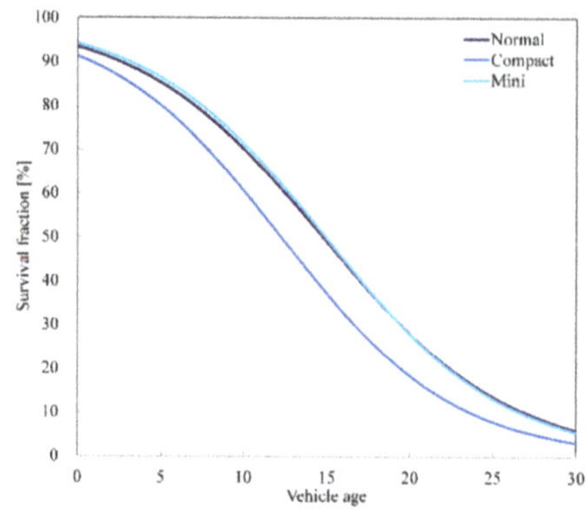

Figure 4. Vehicle survival profiles.

2.3. *Road Freight Transport in Japan and Scenarios for Deep Decarbonization*

Japan was selected as the case study, considering public acceptance of EDVs in the passenger LDV fleet; as well as being home to major OEMs. Historical data for road freight vehicle fleet stock, energy consumption and CO_2 emissions for the years 2012 to 2016 were used to calibrate the model. Due to the long time scales involved in vehicle stock turnover, the time horizon was set between 2012 and 2050.

In the year 2012, road freight vehicle stock totaled 14.8 million vehicles; with normal, compact and mini-sized vehicles accounting for 15.3%, 24.8% and 60.0%, respectively [67]. The road freight vehicle stock distribution by vintage was constructed using data from AIRIA [68]. Data correspond to normal and compact vehicles. Due to data availability constraints, the same vehicle stock distribution was assumed for mini-sized road freight vehicles. Since available data only cover 20 vintages, in contrast with the 30 vintages considered in the model, the aggregated data for vintages 20 or older were redistributed by extrapolating the tendency for vintages newer than 20 years, guarantying that the sum of all vintages totaled 100%. The resulting vehicle stock distribution is shown in Figure 5.

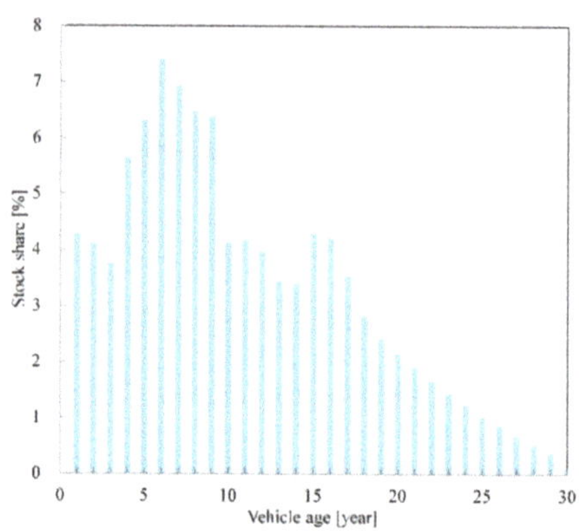

Figure 5. Vehicle stock distribution.

Annual sales for road freight vehicles in 2012 totaled 0.79 million vehicles; with normal, compact and mini-sized vehicles accounting for 17.4%, 28.9% and 53.7%, respectively [67]. Deployment of EDVs in road freight vehicles is still at an early stage, with 25 BEVs and 12204 HEVs in 2012 [69]. For matter of simplicity, it was considered that 2012 road freight vehicle stock and new vehicle sales were made only of ICEVs.

A scenario-based approach was used to assess the role of powertrain electrification in the decarbonization of the road freight vehicle fleet. Four scenarios, the Base scenario and three alternative scenarios for powertrain electrification were considered. All scenarios were defined in terms of the share of the powertrains in the 2050 new vehicle sales, as shown in Table 2.

Table 2. Dominant powertrain in the 2050 new vehicle sales in each scenario.

Scenario	Size Class	Powertrain			
		ICEV	HEV	BEV	FCEV
Base	Normal	△	△		
	Compact	△	△		
	Mini	△	△		
HBB	Normal		○		
	Compact			○	
	Mini			○	
HFF	Normal		○		
	Compact				○
	Mini				○
FBB	Normal				○
	Compact			○	
	Mini			○	

△ Partial dominance; ○ total dominance.

The Base scenario represents the continuation of current trends, with HEVs and ICEVs dominating 2050 new vehicle sales, 38.3% and 55.1%; while BEVs and FCEVs do not achieve significant deployment, each of them accounting for 3.3% of the new vehicle sales [46]. New vehicle sales are shown in Figure 6. The growth trends for the new vehicle sales for each vehicle size class were adjusted using data from the Japan Ministry of Environment (MOE) [65] for future sales forecast, and from the Japanese Automobile Manufacturers Association (JAMA) [70] for historical sales. It was assumed that new vehicle sales do not vary across scenarios.

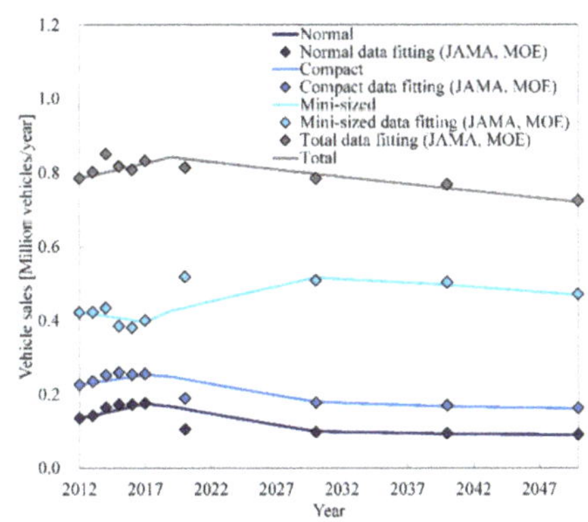

Figure 6. New vehicle sales.

The HBB, HFF and FBB scenarios correspond to alternative scenarios where powertrain electrification is targeted by 2050. These alternative scenarios were built using a 'silver bullet' approach, where only one powertrain technology dominates each vehicle size class. Since normal road freight vehicles are usually used to travel the longest distances, corresponding to trips between cities often located in different prefectures, and are the heaviest, using BEV in this vehicle size class would require large batteries to complete each trip without stopping, or more frequent stops for battery charging; both of which increase the cost. Therefore, only HEVs or FCEVs were considered as candidates for powertrain electrification in normal vehicles. In compact and mini-sized vehicles, used typically for travels within cities that require shorter trips, powertrain electrification using BEVs and FCEVs was assessed. The HBB and HFF scenarios focus on mild decarbonization in normal vehicles using HEVs and deep decarbonization in compact and mini-sized vehicles using BEVs or FCEVs. The FBB scenario focuses on deep decarbonization across all vehicle size classes using FCEVs in normal vehicles and BEVs in compact and mini-sized vehicles.

Market shares for EDVs in the alternative scenarios evolve as indicated in Figure 7. These technology diffusion curves were estimated following the methods described in [71]. It was assumed that 2020 is the initial year for HEV diffusion; while 2025 is the initial year for BEV and FCEV diffusion. EDV diffusion was considered symmetrical, with a technology diffusion span of 30 years [72].

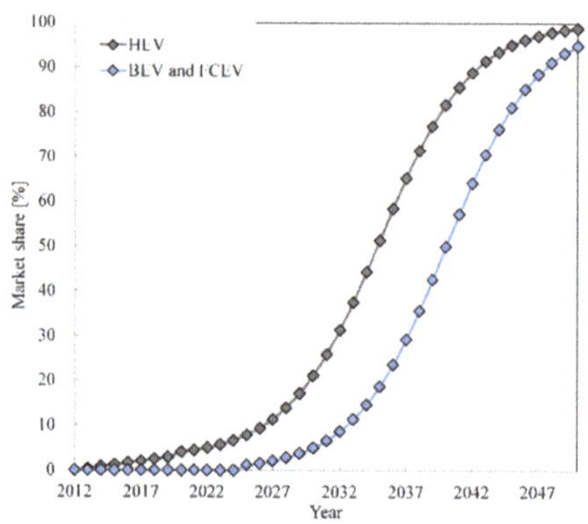

Figure 7. Technology diffusion profiles.

In order to compare the merit of each scenario in the decarbonization of the road freight vehicle fleet, a score analysis was performed. The performance of each scenario in 2050 was assessed in terms of the road freight vehicle fleet TTW energy consumption, WTW CO_2 emissions and RCO. The scores were estimated following the methods developed in [73] using Equation (5):

$$Score = \frac{Maximum - Data_{selected\ scenario}}{Maximum - minimum}100 \quad (5)$$

2.4. Main Assumptions and Limitations

Alternative scenarios for powertrain electrification were built considering that each vehicle size class is dominated only by one powertrain. Even though this assumption is not realistic, it allows to estimate the maximum 'technologically realizable' CO_2 emissions reduction potential of each powertrain electrification strategy. In that sense, CO_2 emissions reduction potential estimated here corresponds to the largest possible CO_2 emissions reductions of each strategy considering the dynamics of technology diffusion. Since there are several barriers that prevent the deployment of EDVs in road freight vehicles, the actual CO_2 emissions reductions that can be achieved through powertrain electrification are smaller. Furthermore, since EDV diffusion is considered independently from vehicle RCO, road freight vehicle fleet energy consumption and CO_2 emissions are not affected by changes in the RCO. This is a limitation of the study, as vehicle RCO affects consumer choices of vehicle type and therefore vehicle fleet energy consumption and CO_2 emissions.

The composition of the Japanese road freight vehicle was simplified, assuming that each vehicle size class-powertrain combination is represented by only one vehicle type. Therefore, the model cannot capture the diversity of vehicle types existing in the real road freight vehicle fleet.

It was assumed that annual traveled distance and vehicle service life are identical for all vehicle types within each vehicle size class. Daily travel patterns were not considered and only the average annual traveled distance was used to characterize road freight transport activity. Vehicle service life and annual traveled distance were assumed to remain identical during the whole time horizon. Based on the model calibration for road freight vehicle fleet energy consumption and CO_2 emissions, it was assumed that fuel consumption deteriorates as vehicles become older at a rate of 0.3%/year

As fuel production is outside of the scope of this research, fuel prices and CO_2 emission factors were considered exogenous to the vehicle stock turnover model. TTW and WTW CO_2 emission factors for gasoline and diesel in 2012 were obtained from [74,75]. These values were assumed to remain constant over the time horizon, considering that TTW CO_2 emissions are determined by the carbon content in the fuel, and gasoline and diesel production processes are mature. Gasoline and diesel prices in 2012 correspond to the values reported in [76]. Gasoline and diesel prices in 2050 were estimated under the assumption that the ratio between fuel price and crude oil price remains constant between 2012 and 2050. Values for the 4DS scenario from the International Energy Agency's (IEA's) Energy Technology Perspectives [77] were assumed for the crude oil price.

Electricity WTW CO_2 emission factor in 2012 was the value reported by the IEA for the electricity generation mix in Japan [78]. Electricity WTW CO_2 emission factor in 2050 corresponds to the value of the Nuclear Phase Out scenario from [79], which assumes electricity generation using 50% fossil fuels and 50% renewable energy. Electricity price in 2012 was the historical value reported in [76]; while the electricity price in 2050 was estimated based on the generation cost from [79] and network, retail and other costs from [80]. Steam Methane Reforming (SMR) share in hydrogen production was assumed to go from 100% in 2012 to 50% in 2050. The remaining hydrogen in 2050 is produced using 25% wind electrolysis and 25% solar photovoltaic electrolysis. Prices in 2012 and 2050 were estimated based on the share of each hydrogen production technology and the near term and future delivered hydrogen cost for each technology from [81]. WTW CO_2 emission factors in 2012 and 2050 were estimated based on the share of each hydrogen production technology and the technology CO_2 emission factor from [75]. Fuel prices and CO_2 emission factors are presented in Figure 8.

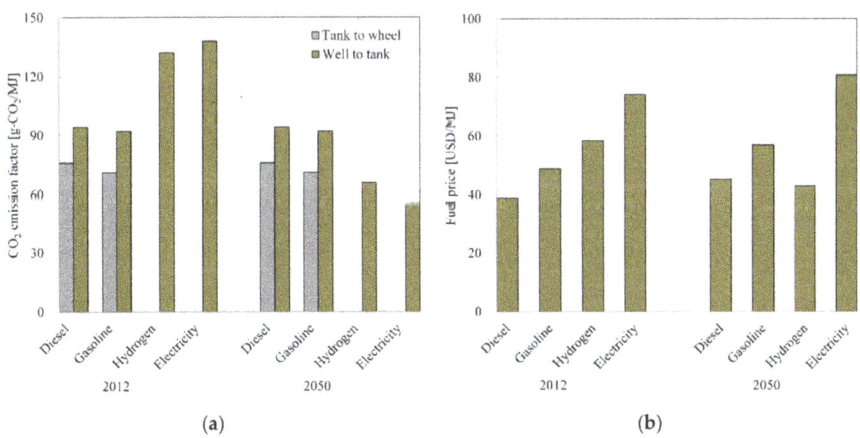

Figure 8. Fuel information: (**a**) Tank to wheel and well to wheel CO_2 emission factors; (**b**) price.

3. Results and Discussion

3.1. Powertrain Electrification of the Road Freight Vehicle Stock

Results for the road freight vehicle fleet stock are presented in Figure 9. Differences between modeling results and historical data for the years 2012 to 2017 [70] are lower than 2%. Vehicle stock decreases in all scenarios from 14.7 to 12.3 million vehicles between 2012 and 2050. Compared with the base year, there is a small shift from normal and compact vehicles to mini-sized vehicles; with the stock share of mini-sized vehicles increasing to 66.6% and the stock shares of normal and compact vehicles decreasing to 13.3% and 20.1%, respectively. In this research, the selection of the vehicle size in road freight transport is considered exogenously. However, opportunities to reduce energy consumption and CO_2 emissions can be unveiled by improving the selection of the vehicle size with a detailed analysis that includes the load capacity utilization rate and the daily travel patterns. This is suggested for future work.

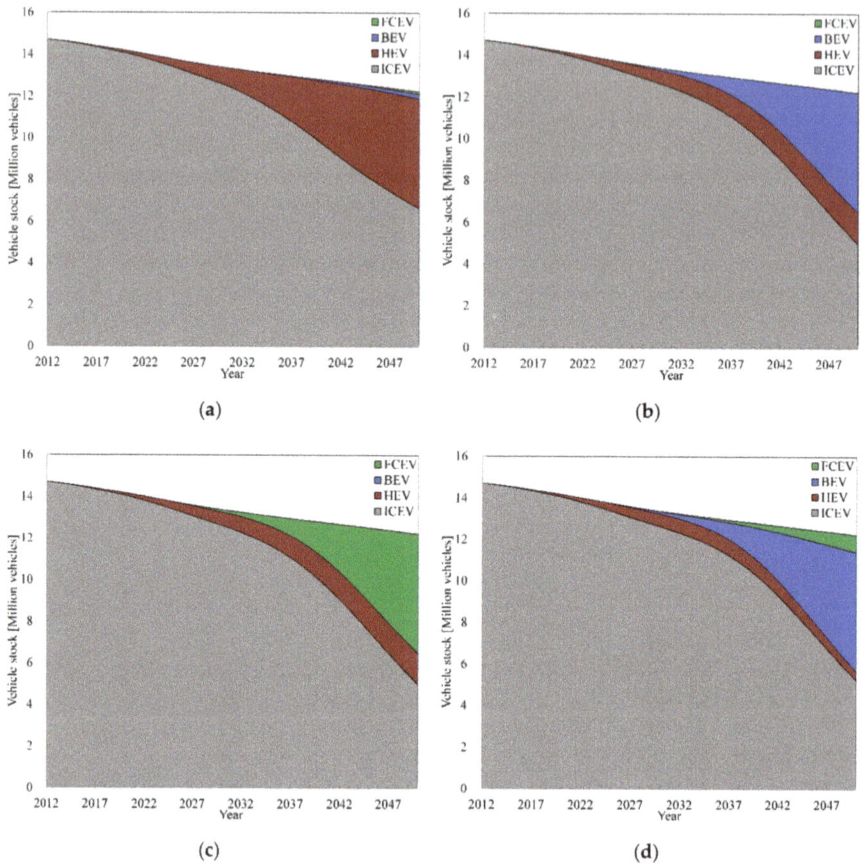

Figure 9. Road freight vehicle fleet stock: (**a**) Base scenario; (**b**) HBB scenario; (**c**) HFF scenario; and (**d**) FBB scenario.

By 2050, the share of ICEVs in the road freight vehicle stock decreases in all scenarios compared with the 2012 values; reaching 54.2% in the Base scenario, 40.9% in the HBB and HFF scenarios, and 42.8% in the FBB scenario. The largest diffusion for HEVs occurs in the Base scenario, reaching 42.9% of the vehicle stock by 2050. HEV diffusion is lower in other scenarios, accounting for 12.2% of the 2050 vehicle stock in the HBB and HFF scenarios, and 3.5% in the FBB scenario. Combining ICEVs and HEVs, diesel- and gasoline-fueled vehicles represent more than 46% of the road freight vehicle stock in all scenarios; evidencing the difficulty of reducing the dependence from fossil fuels in road freight transport.

Compared with EDV diffusion in the new vehicle market share presented in Figure 7, diffusion of EDVs in the road freight vehicle stock shown in Figure 9 is slower due to the time lag effect caused by the vehicle service lives; as most of the new vehicles replace current vehicles in use only after their service lives have finished. Long vehicle service lives are a barrier that prevents powertrain electrification in the road freight vehicle fleet. In the case of Japan, vehicle service lives for road freight vehicles are longer than values reported for other countries. For instance, service lives for heavy-duty trucks, medium-duty trucks, light-duty trucks, and mini-trucks in China are 12, 11, 9, and 8 years [82]; while in the United States, values between 7 and 10 years are reported for heavy-duty trucks [31], and 10 years for medium-duty trucks [27]. Even though reducing the vehicle service life for road freight vehicles might seem as a straight forward measure to accelerate penetration of EDVs in the vehicle stock, EDV manufacturing requires more energy and can produce more CO_2 emissions than manufacturing ICEVs. Therefore, including vehicle cycle in the assessment of strategies for powertrain electrification in road freight transport is recommended for future research.

It should be noted that the road freight vehicle fleet was simplified considering only one vehicle type for each powertrain and vehicle size class combination. This is unrealistic, particularly for normal vehicles, given the broad range of GVWs for normal vehicles according to the MLIT classification. In that sense, a more detailed assessment of the normal road freight vehicle fleet is recommended. Furthermore, it was not possible to consider road freight EDVs as one existing vehicle model due to lack of data. Instead, vehicle data were constructed using different sources in the existing literature. A more realistic characterization of the road freight vehicle fleet can be achieved by modeling each vehicle type using vehicle modeling software such as Autonomie and PAMVEC. This is also suggested for future work.

3.2. Power Electrification Potential for Decarbonization of Road Freight Transport

Results for TTW energy consumption along with historical data [16] are presented in Figure 10. Differences between results for the Base scenario and historical data are lower than 4%. TTW energy consumption in the Base scenario increases, reaching the peak at 1074 PJ/year in 2020, to decrease until reaching 613 PJ/year by 2050. Drivers for TTW energy consumption reduction in the Base scenario are vehicle stock reduction, vehicle fuel consumption improvement and HEV adoption. Since BEV and FCEV diffusion is small, diesel and gasoline account for 99.0% of TTW energy consumption.

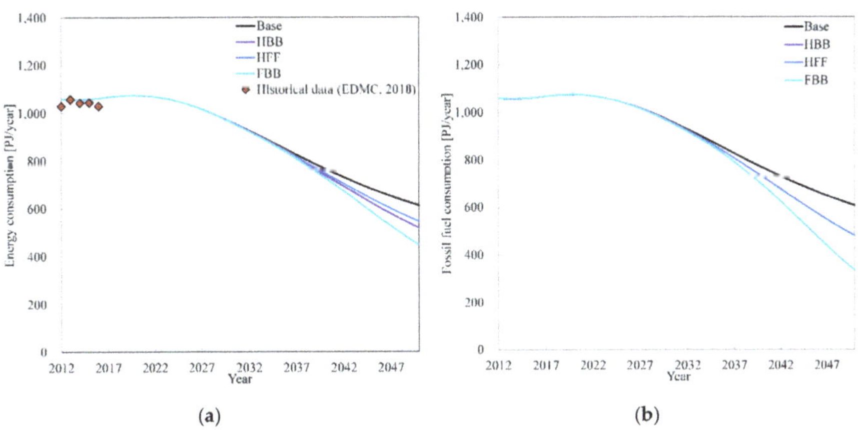

Figure 10. Road freight vehicle fleet tank to wheel energy consumption: (**a**) total energy consumption; (**b**) fossil fuel consumption.

In the alternative scenarios, TTW energy consumption remains identical to the Base scenario until EDV diffusion starts, 2020 in the HBB and HFF scenarios, and 2025 in the FBB scenario. By 2050, TTW energy consumption is reduced 15.1%, 10.7% and 26.7% in the HBB, HFF and FBB scenarios, compared with the baseline value. As fuel shift occurs in the alternative scenarios, reductions for fossil fuel consumption are larger than reductions for energy consumption; up to 20.8% in the HBB and HFF scenarios and 44.7% in the FBB scenario, compared with the 2050 baseline values.

Powertrain electrification increases electricity and/or hydrogen consumption in road freight transport. Electricity consumption and hydrogen consumption reach 3.3 and 2.6 PJ/year in the Base scenario by 2050, respectively. In contrast, electricity consumption increases up to 39.7 PJ/year in the HBB and FBB scenarios; while hydrogen consumption increases up to 66.6 and 73.6 PJ/year in the HFF and FBB scenarios. Since the stock share and vehicle fuel consumption of ICEVs and HEVs are larger than the values for BEVs and FCEVs, diesel and gasoline represent more than 74% of TTW energy consumption in all alternative scenarios by 2050, despite the large increments in electricity and hydrogen consumption.

Even though the fuel cycle is out of the scope of this research, CO_2 emissions are presented both on TTW and WTW basis, with the aim of providing more insights on the impact of powertrain electrification on the road freight vehicle fleet. TTW CO_2 emissions along with historical data [83] are presented in Figure 11a. Differences between the modeling results for the Base scenario and historical data are lower than 3%. In the Base scenario, TTW CO_2 emissions increase until reaching the peak in 2020 at 79.8 Mt-CO_2/year; to decrease thereafter, until reaching 44.9 Mt-CO_2/year by 2050. Since BEV and FCEV diffusion is small and CO_2 emission factors for diesel and gasoline are constant throughout the time horizon, TTW CO_2 emissions reduction is caused by the same drivers that cause TTW energy consumption reduction: vehicle stock reduction, vehicle fuel consumption reduction and HEV adoption.

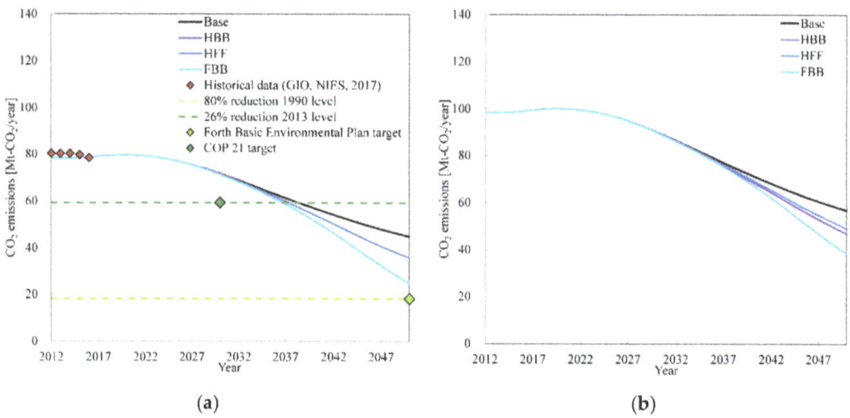

Figure 11. Road freight vehicle fleet CO_2 emissions. (**a**) Tank to wheel CO_2 emissions; (**b**) Well to wheel CO_2 emissions.

TTW CO_2 emissions can be reduced up to 20.1% in the HBB and HFF scenarios and 44.6% in the FBB scenario, compared with the 2050 baseline value. The Japanese government aims to achieve two CO_2 emissions reduction targets: one for the medium-term, proposed at the 21st Conference of the Parties (COP21), corresponding to 26% CO_2 emissions reduction compared with the 2013 values by 2030; and one for the long-term, proposed in the Forth Basic Environmental Plan, corresponding to 80% CO_2 emissions reduction compared with the 1990 values by 2050. Results for the road freight vehicle fleet TTW CO_2 emissions show that none of the CO_2 emissions reductions target can be achieved in any of the alternative scenarios considering powertrain electrification. In that sense, meeting CO_2 emissions reduction targets that involve proportional CO_2 emissions reductions across all sectors in road freight transport requires measures other than powertrain electrification.

WTW CO_2 emissions are shown in Figure 11b. Compared with TTW CO_2 emissions, WTW CO_2 emissions are larger in all scenarios, since CO_2 is emitted during the production of electricity and hydrogen is included. In the Base scenario, WTW CO_2 emissions decrease from 98.8 Mt-CO_2/year in 2012 to 56.9 MT-CO_2/year in 2050. Even including the CO_2 emitted to produce electricity and hydrogen, powertrain electrification yields lower WTW CO_2 emissions than the Base scenario; with maximum CO_2 emissions reductions of 17.3%, 13.4% and 32.6% in the HBB, HFF and FBB scenarios, respectively. Since WTW CO_2 emissions are determined by the energy resources used to produce electricity and hydrogen, it is necessary to focus on the simultaneous decarbonization of electricity and hydrogen production to enhance CO_2 emissions reductions from powertrain electrification in road freight transport.

CO_2 emissions reductions for powertrain electrification estimated in this research represent the maximum 'technologically realizable' CO_2 emissions reduction potential; and they correspond to the upper limit of the CO_2 emissions reductions achievable by replacing ICEVs with EDVs in road freight transport. However, in practice, powertrain electrification in road freight transport will lead to lower CO_2 emissions reductions than estimated here due to barriers that prevent EDV diffusion such as public acceptance, vehicle use diversity, short payback times, and risk aversion. It is recommended to endogenize these barriers in future modeling of the road freight vehicle fleet.

3.3. Economic Impact of Powertrain Electrification on the Road Freight Vehicle Fleet

The evaluation of the economic impact of powertrain electrification on the road freight vehicle fleet was performed in terms of the net cash flow; defined as the difference between the road freight vehicle fleet RCO for a given scenario and the road freight vehicle fleet RCO for the Base scenario. The net cash flows for the alternative scenarios are presented in Figure 12. Since penetration of EDVs in the new vehicle sales is slow during the early stages of deployment, net cash flows for all alternative scenarios remain close to zero until 2030. Differences become larger thereafter. The HBB scenario has the largest net cash flows, with the peak outside the time horizon and reaching 9.6 billion USD/year by 2050. The second largest net cash flows are obtained for the FBB scenario, peaking at 6.7 billion USD/year in 2049, and reaching 6.6 billion USD/year by 2050. The lowest net cash flows are obtained for the HBB scenario, peaking at 4.6 billion USD/year in 2049, reaching 4.5 billion USD/year by 2050. It can be seen that the largest values for the net cash flow are obtained for scenarios with BEV diffusion. Furthermore, if BEVs were to be deployed in compact and mini-sized vehicles, it is more cost-effective to deploy FCEVs than HEVs in normal vehicles.

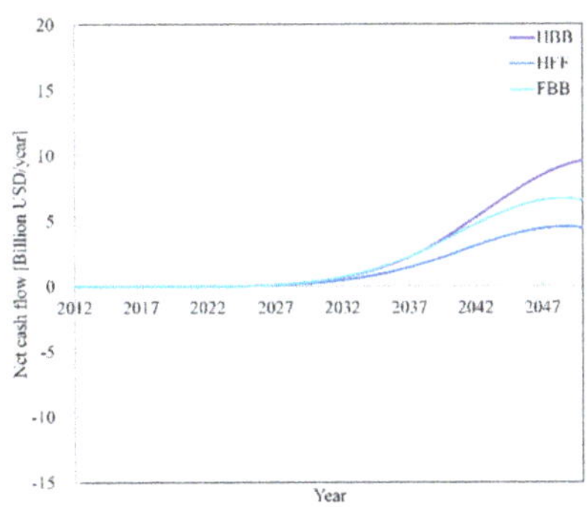

Figure 12. Net cash flow.

Powertrain electrification leads to lower energy consumption in the alternative scenarios than in the Base scenario; with energy savings reaching 4.2, 4.6 and 7.6 billion USD/year by 2050 in the HBB, HFF and FBB scenarios. However, incremental capital and O&M costs increase for all alternative scenarios, reaching 13.8, 9.0 and 14.2 billion USD/year by 2050 in the HBB, HFF and FBB scenarios. Since energy savings cannot outweigh capital cost increments when BEVs and FCEVs are deployed, net cash flows remain positive during the whole time horizon in all alternative scenarios.

3.4. Policy Implications

In order to obtain a complete perspective of the impact of powertrain electrification on the road freight vehicle fleet, the performance of all scenarios was assessed in terms of TTW energy use, WTW CO_2 emissions and RCO. Scores for the Base scenario and the three alternative scenarios in 2050 are shown in Figure 13. Ideally, the goal is to reduce energy consumption and CO_2 emissions while reducing or maintaining the fleet RCO. However, as seen from the net cash flow, powertrain electrification increases the road freight vehicle fleet RCO compared with the Base scenario. The Base scenario is the top performer in terms of cost. Nevertheless, it has the lowest performance in terms of energy use and CO_2 emissions. In contrast, the HFF scenario has better performance in terms of CO_2 emissions; however, cost performance is reduced by half compared to the Base scenario. The HBB scenario can offer larger CO_2 emissions reductions compared with the HFF scenario; nevertheless, it has the lowest cost performance out of all scenarios. The best performance overall is obtained in the FBB scenario, showing the largest energy consumption and CO_2 emissions reductions, with a cost performance between the HFF and HBB scenarios. Therefore, diffusion of FCEVs in normal vehicles and BEVs in compact and mini-sized vehicles is recommended as the best strategy for powertrain electrification in road freight transport in Japan.

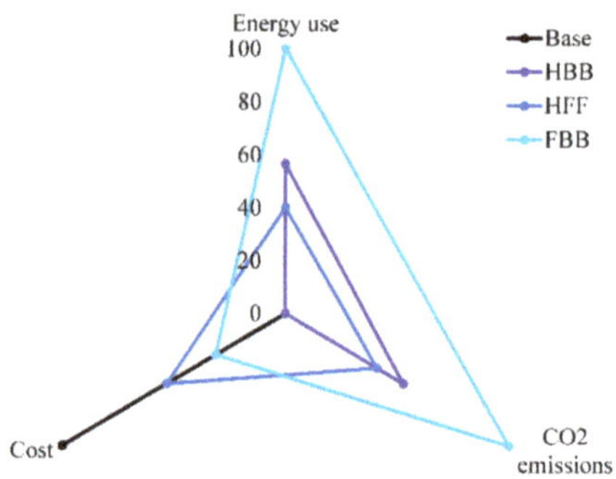

Figure 13. Scenario scores in 2050.

In order to achieve powertrain electrification in road freight transport, it is important to incentivize OEMs to accelerate the development and mass production of road freight EDVs. As road freight vehicle owners often own several vehicles and are more focused on reducing cost than passenger LDV owners [16], policies to incentivize powertrain electrification can have a faster acceptance in road freight vehicles than in passenger LDVs. Therefore, it is also recommended to design measures to help road freight vehicle fleet owners investing in EDVs. Additionally, diffusion of EDVs in road freight transport can help improving social acceptance and developing infrastructure that can benefit EDV diffusion in passenger LDVs. In that sense, the assessment of powertrain electrification strategies that include passenger LDVs and road freight vehicles considering the dynamics of technology diffusion is suggested for future research.

A large asymmetry was found between the vehicle stock distribution by size class and the CO_2 emissions. Normal vehicles account for 13.3% of the road freight vehicle stock in all scenarios. However, they account for more than 61% of TTW and WTW CO_2 emissions. Considering normal vehicle stock is the smallest among all road freight vehicle size classes and it is concentrated in few users, it is recommended to prioritize normal vehicles when designing measures for powertrain electrification in road freight transport.

3.5. Sensitivity Analysis

Since the time scales involved in technology diffusion in road transport are long, a time horizon between 2012 and 2050 was selected to study the role of powertrain electrification in the decarbonization of road freight transport. However, using a long time horizon implies large uncertainties in the parameters used to characterize the elements of the energy system. A sensitivity analysis was performed to understand the impact of those uncertainties on the modeling results. The impact of changes in fossil fuel price, hydrogen price, electricity price, annual traveled distance, new vehicle sales, discount rate, BEV fuel consumption, FCEV fuel consumption, BEV capital cost, FCEV capital cost, and technology diffusion span on the results for the FBB scenario was assessed. A variation of ±20% was considered for all the parameters. Results of the sensitivity analysis for the TTW CO_2 emissions and net cash flow are presented in Figure 14.

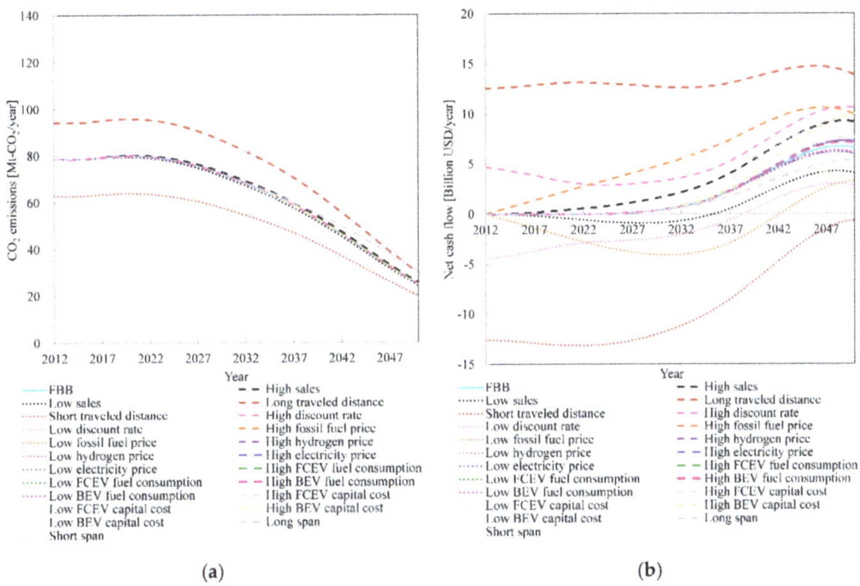

Figure 14. Results of the sensitivity analysis: (**a**) Tank to wheel CO_2 emissions; (**b**) net cash flow.

TTW CO_2 emissions are most sensitive to variations in annual traveled distance and technology diffusion span, as they affect the stock of ICEVs in the road freight vehicle fleet. Other parameters produce small or no variations on TTW CO_2 emissions. This is a consequence of the 'silver bullet' approach used to determine the maximum 'technologically realizable' CO_2 emissions reduction potential of powertrain electrification, which considers EDV diffusion independent of the vehicle fleet RCO. Linking EDV diffusion and cost in the vehicle stock turnover model of the road freight vehicle fleet is recommended for future work.

The net cash flow is most sensitive to variations in annual traveled distance. Changes in the discount rate, fossil fuel price, BEV capital cost, and new vehicle sales also affect significantly the results for the net cash flow. Since the ICEV stock is larger than the EDV stock and fuel consumption is higher for ICEVs than for EDVs, the net cash flow is more sensitive to variations in the fossil fuel price than to variations in electricity and hydrogen price. In that sense, variations in fossil fuel price have a larger impact on the economic competitiveness of EDVs than variations in hydrogen and electricity price. Variations in the capital cost of FCEVs and BEVs change both the value and timing for the net cash flow peak; with the effect of BEV capital cost being larger than the effect of FCEV capital cost.

4. Conclusions

A vehicle stock turnover model of the road freight vehicle fleet was used to assess the role of powertrain electrification in the decarbonization of road freight transport in the case of Japan between 2012 and 2050. Main conclusions are listed below:

1. Driven by the reduction of the vehicle stock, the improvement of vehicle fuel consumption and the adoption of HEVs, energy consumption and CO_2 emissions decrease around 42% between 2012 and 2050 in the Base scenario. In contrast, powertrain electrification can reduce TTW CO_2 emissions up to 20.1% in the HBB and HFF scenarios and 44.6% in the FBB scenario, compared with the 2050 baseline value; while maximum WTW CO_2 emissions reductions are 17.3%, 13.4% and 32.6% in the HBB, HFF and FBB scenarios. Despite the significant reductions in CO_2 emissions, powertrain electrification alone is not enough to achieve any of the CO_2 emissions reduction targets in road freight transport.
2. Despite aggressive deployment of EDVs, diesel- and gasoline-fueled vehicles account for more than 46% of the vehicle stock and more than 74% of the road freight vehicle fleet TTW energy consumption in all scenarios. This evidences the difficulty of reducing the dependence of fossil fuels in road freight transport. In that sense, considering other measures to reduce CO_2 emissions in road freight transport along with powertrain electrification is suggested for future research.
3. The net cash flows are positive for all alternative scenarios considering powertrain electrification during the whole time horizon. The HFF scenario has the lowest net cash flows, peaking at 4.6 billion USD/year in 2049 and reaching 4.5 billion USD/year by 2050. Scenarios where BEVs are deployed have the largest net cash flows by 2050, 9.6 and 6.6 Billion USD/year for the HBB and FBB scenarios, respectively.
4. Deployment of FCEVs for normal vehicles and BEVs for compact and mini-sized vehicles in the Japanese road freight vehicle fleet is recommended. Since road freight vehicles are concentrated in fewer owners than passenger LDVs, it is recommended to design policies to help road freight vehicle fleet owners investing in EDVs. EDV diffusion in road freight transport can help improving social acceptance of powertrain electrification and developing infrastructure that can benefit EDV diffusion in passenger LDVs. Assessing powertrain electrification in passenger LDVs and road freight vehicles considering the dynamics of technology diffusion is recommended for future work.
5. A large asymmetry was found between the vehicle stock distribution by size class and the CO_2 emissions; with normal vehicles accounting for more than 61% of the TTW and WTW CO_2 emissions in all scenarios, despite representing only 13.3% of the vehicle stock. It is therefore recommended to prioritize normal vehicles when designing measures for powertrain electrification in road freight transport.

Author Contributions: The contribution of the authors to this article is as follows: conceptualization, J.C.G.P. and V.T.N.; methodology, J.C.G.P., V.T.N., M.A. and S.S.; software, J.C.G.P. and V.T.N.; validation, J.C.G.P., V.T.N., M.A. and S.S.; formal analysis, J.C.G.P., V.T.N., M.A. and S.S.; investigation, J.C.G.P., V.T.N., M.A. and S.S.; resources, J.C.G.P., M.A. and S.S.; data curation, J.C.G.P. and V.T.N.; writing—original draft preparation, J.C.G.P.; writing—review and editing, M.A. and S.S.; visualization, J.C.G.P.; supervision, M.A. and S.S. All authors have read and agreed to the published version of the manuscript.

Funding: This research received no external funding.

Acknowledgments: The authors appreciate the comments and suggestions made by the reviewers.

Conflicts of Interest: The authors declare no conflict of interest.

Nomenclature

α	Model parameter
β	Growth parameter
a	Vehicle age
a_0	Vehicle median service life
AIRIA	Automobile Inspection & Registration Association
BEV	Battery electric vehicle
c_{cap}	Vehicle capital cost
c_{ene}	Energy price
c_{OM}	Vehicle operating and maintenance cost
COP	Conference of Parties
CRF	Capital recovery factor
E	Road freight vehicle energy consumption
EDV	Electric-drive vehicle
EF	Fuel CO_2 emission factor
FCEV	Fuel cell electric vehicle
G	Road freight vehicle fleet CO_2 emissions
GHG	Greenhouse gas
GVW	Gross Vehicle Weight
HDV	Heavy-duty vehicle
HEV	Hybrid electric vehicle
ICEV	Internal combustion engine vehicle
IEA	International Energy Agency
JAMA	Japanese Automobile Manufacturers Association
LDV	Light-duty vehicle
LEAP	Long-range Energy Alternatives Planning system
M	Annual traveled distance
MDV	Medium-duty vehicle
MLIT	Japan Ministry of Land, Infrastructure, Transport and Tourism
N	Vehicle stock
OEM	Original equipment manufacturer
O&M	Operating and maintenance
PHEV	Plug-in hybrid electric vehicle
PM	Particulate matter
r	Vehicle survival rate
R	Vehicle fuel consumption
RCO	Relative cost of ownership
S	New vehicle sales
SMR	Steam methane reforming
t	Vehicle type
TCO	Total cost of ownership
TTW	Tank to wheel
v	Vehicle vintage
WTW	Well to wheel
y	Calendar year

References

1. Sims, R.; Schaeffer, R.; Creutzig, F.; Cruz-Núñez, X.; D'Agosto, M.; Dimitriu, D.; Figueroa Meza, M.J.; Fulton, L.; Kobayashi, S.; Lah, O.; et al. Transport. In *Climate Change 2014: Mitigation of Climate Change. Contribution of Working Group III to the Fifth Assessment Report of the Intergovernmental Panel on Climate Change*; Edenhofer, O., Pichs-Madruga, R., Sokona, Y., Farahani, E., Kadner, S., Seyboth, K., Adler, A., Baum, I., Brunner, S., Eickemeier, P., Eds.; Cambridge University Press: Cambridge, UK; New York, NY, USA, 2014; pp. 599–670.

2. IEA/OECD (Ed.) *Energy Technology Perspectives 2017: Catalysing Energy Tehnology Transformations*; International Energy Agency and Organization for Economic Cooperation and Development: Paris, France, 2017.
3. IEA/OECD. *The Future of Trucks: Implications for Energy and the Environment*; International Energy Agency and Organization for Economic Cooperation and Development: Paris, France, 2017.
4. Kast, J.; Morrison, G.; Gangloff, J.J.; Vijayagopal, R.; Marcinkoski, J. Designing hydrogen fuel cell electric trucks in a diverse medium and heavy duty market. *Res. Transp. Econ.* **2018**, *70*, 139–147. [CrossRef]
5. ICCT. Overview of the Heavy-Duty Vehicle Market and CO2 Emissions in the European Union; The International Council on Clean Transportation. Available online: http://www.theicct.org/sites/default/files/publications/ICCT_EU-HDV_mkt-analysis_201512.pdf (accessed on 20 February 2020).
6. Kast, J.; Vijayagopal, R.; Gangloff, J.J.; Marcinkoski, J. Clean commercial transportation: Medium and heavy duty fuel cell electric trucks. *Int. J. Hydrog. Energy* **2017**, *42*, 4508–4517. [CrossRef]
7. Dablanc, L. City distribution, a key element of the urban economy: Guidelines for practitioners. In *City Distribution and Urban Freight Transport: Multiple Perspectives*; Macharis, C., Melo, S., Eds.; Edward Elgar Publishing: Cheltenham, UK, 2011; pp. 13–36.
8. WHO. *Review of Evidence on Health Aspects of Air Pollution—REVIHAAP Project; First Results*; World Health Organization, Regional Office for Europe: Copenhagen, Denmark, 2013.
9. HEI. Panel on the Health Effects of Traffic-Related Air Pollution. In *Special Report 17; Traffic-Related Air Pollution: A Critical Review of the Literature on Emissions, Exposure, and Health Effects*; Health Effects Institute: Boston, MA, USA, 2010; Available online: https://www.healtheffects.org/system/files/SR17TrafficReview.pdf (accessed on 4 May 2020).
10. WHO. *Press Release No 213: IARC: Diesel Engine Exhaust Carcinogenic*; World Health Organization: Lyon, France, 2012; Available online: http://www.iarc.fr/en/media-centre/pr/2012/pdfs/pr213_E.pdf (accessed on 30 March 2020).
11. OECD/ITF. The Carbon Footprint of Global Trade; Tackling Emissions from International Freight Transport. Organisation for Economic Co-operation and Development and International Transport Forum: Paris, France, 2016. Available online: http://www.itf-oecd.org/sites/default/files/docs/cop-pdf-06.pdf (accessed on 15 April 2020).
12. Urban Foresight; EVI; IEA/OECD. *EV City Casebook; 50 Big Ideas Shaping the Future of Electric Mobility*; Urban Foresight, Electric Vehicles Initiative of the Clean Energy Ministerial and International Energy Agency and Organization for Economic Cooperation and Development: Newcastle, UK, 2014.
13. ICCT. Transitioning to Zero-Emission Heavy-Duty Freight Vehicles; The International Council on Clean Transportation. Available online: https://www.theicct.org/sites/default/files/publications/Zero-emission-freight-trucks_ICCT-white-paper_26092017_vF.pdf (accessed on 20 April 2020).
14. Den Boer, E.; Aarnik, S.; Kleiner, F.; Pagenkopf, J. Zero Emissions Trucks: An Overview of State-of-the-Art Technologies and Their Potential; CE Delft. Available online: https://www.cedelft.eu/publicatie/zero_emission_trucks/1399 (accessed on 7 April 2018).
15. IEA/OECD. *Global EV Outlook 2019: Scaling Up the Transition to Electric Mobility*; International Energy Agency/Organization for Economic Cooperation and Development and Electric Vehicles Initiative of the Clean Energy Ministerial: Paris, France, 2019.
16. Tryggestad, C.; Sharma, N.; van de Staaij, J.; Keizer, A. New Reality: Electric Trucks and Their Implications on Energy Demand; Energy Insights by McKinsey. Available online: https://www.mckinseyenergyinsights.com/insights/new-reality-electric-trucks-and-their-implications-on-energy-demand (accessed on 21 April 2020).
17. BYD. BYD is Leading the Transition to a Zero-Emissions Future with Advanced Electric Trucks, Buses and Forklifts. Available online: https://en.byd.com/news-posts/press-release-byd-is-leading-the-transition-to-a-zero-emissions-future-with-advanced-battery-electric-trucks-buses-and-forklifts/ (accessed on 10 April 2020).
18. Isuzu Motors Limited. Isuzu Exhibitions in the 45th Tokyo Motor Show 2017. Available online: http://www.isuzu.co.jp/world/press/2017/9_22tms.html (accessed on 24 April 2020).
19. Carey, N. UPS Reserves 125 Tesla Semi-Trucks, Largest Public Pre-Order Yet; Reuters. Available online: https://www.reuters.com/article/us-ups-tesla-orders/ups-reserves-125-tesla-semi-trucks-largest-public-pre-order-yet-idUSKBN1ED1QM (accessed on 17 March 2020).

20. Renault. Renault Trucks will Start Selling Electric Trucks in 2019. Available online: http://corporate.renault-trucks.com/en/press-releases/renault-trucks-will-start-selling-electric-trucks-in-2019.html (accessed on 20 April 2020).
21. Mitsubishi Fuso Truck and Bus Corporation. The World's First Mass-Produced All-Electric Compact Truck "eCanter". Available online: http://www.mitsubishi-fuso.com/content/fuso/jp/truck/ecanter/lp.html (accessed on 13 December 2019). (In Japanese).
22. FCB. Switzerland unveils fuel cell powered heavy truck, and first hydroelectric hydrogen station. *Fuel Cells Bull.* **2016**, *2016*(12), 14–15. [CrossRef]
23. Toyota. Toyota Opens a Portal to the Future of Zero Emission Trucking. Available online: https://media.toyota.ca/releases/toyota-opens-a-portal-to-the-future-of-zero-emission-trucking (accessed on 17 April 2020).
24. FCB. Ballard, Kenworth fuel cell truck validation moves to next stage. *Fuel Cells Bull.* **2018**, *2018*(2), 3.
25. Nikola Motor Company; Bosch. A World-First: The Powertrain for the Electric Long-Haul Truck. Available online: https://www.bosch-presse.de/pressportal/de/en/a-world-first-the-powertrain-for-the-electric-long-haul-truck-126336.html (accessed on 24 March 2020).
26. Askin, A.C.; Barter, G.E.; West, T.H.; Manley, D.K. The heavy-duty vehicle future in the United States: A parametric analysis of technology and policy tradeoffs. *Energy Policy* **2015**, *81*, 1–13. [CrossRef]
27. Zhao, Y.; Ercan, T.; Tatari, O. Life cycle based multi-criteria optimization for optimal allocation of commercial delivery truck fleet in the United States. *Sustain. Prod. Consum.* **2016**, *8*, 18–31. [CrossRef]
28. Wikström, M.; Hansson, L.; Alvfors, P. Socio-technical experiences from electric vehicle utilisation in commercial fleets. *Appl. Energy* **2014**, *123*, 82–93. [CrossRef]
29. Vora, A.P.; Jin, X.; Hoshing, V.; Saha, T.; Shaver, G.; Varigonda, S.; Wasynczuk, O.; Tyner, W.E. Design-space exploration of series plug-in hybrid electric vehicles for medium-duty truck applications in a total cost-of-ownership framework. *Appl. Energy* **2017**, *202*, 662–672. [CrossRef]
30. Lee, D.-Y.; Thomas, V.M. Parametric modeling approach for economic and environmental life cycle assessment of medium-duty truck electrification. *J. Clean. Prod.* **2017**, *142*, 3300–3321. [CrossRef]
31. Sen, B.; Ercan, T.; Tatari, O. Does a battery-electric truck make a difference?—Life cycle emissions, costs, and externality analysis of alternative fuel-powered Class 8 heavy-duty trucks in the United States. *J. Clean. Prod.* **2017**, *141*, 110–121. [CrossRef]
32. Feng, W.; Figliozzi, M.A. Conventional vs electric commercial vehicle fleets: A case study of economic and technological factors affecting the competitiveness of electric commercial vehicles in the USA. *Procedia Soc. Behav. Sci.* **2012**, *39*, 702–711. [CrossRef]
33. Roeth, M.; Kircher, D.; Smith, J.; Swim, R. Barriers to the Increased Adoption of Fuel Efficiency Technologies in the North American On-Road Freight Sector; Report for the International Council on Clean Transportation (ICCT). Available online: https://www.theicct.org/sites/default/files/publications/ICCT-NACFE-CSS_Barriers_Report_Final_20130722.pdf (accessed on 9 April 2020).
34. Schoettle, B.; Sivak, M.; Tunnell, M. A Survey of Fuel Econmy and Fuel Usage by Heavy-Duty Truck Fleets; University of Michigan and American Transportation Reserach Institute. Available online: http://atri-online.org/wp-content/uploads/2016/10/2016.ATRI-UMTRI.FuelEconomyReport.Final_.pdf (accessed on 29 April 2020).
35. Zhao, H.; Burke, A.; Zhu, L. Analysis of Class 8 Hybrid-Electric Truck Technologies Using Diesel, LNG, Electricity, and Hydrogen, as the Fuel for Various Applications. In *Proceedings of the EVS27 International Battery, Hybrid and Fuel Cell Electric Vehicle Symposium*; Institute of Electrical and Electronics Engineers: Barcelona, Spain, 2013; p. 17.
36. Lee, D.Y.; Thomas, V.M.; Brown, M.A. Electric urban delivery trucks: Energy use, greenhouse gas emissions, and cost-effectiveness. *Environ. Sci. Technol.* **2013**, *47*, 8022–8030. [CrossRef] [PubMed]
37. Lajunen, A. Fuel economy analysis of conventional and hybrid heavy vehicle combinations over real-world operating routes. *Transp. Res. Part D Transp. Environ.* **2014**, *31*, 70–84. [CrossRef]
38. Gao, Z.; Smith, D.E.; Stuart Daw, C.; Dean Edwards, K.; Kaul, B.C.; Domingo, N.; Parks, J.E.; Jones, P.T. The evaluation of developing vehicle technologies on the fuel economy of long-haul trucks. *Energy Convers. Manag.* **2015**, *106*, 766–781. [CrossRef]

39. Lebeau, P.; Macharis, C.; Van Mierlo, J.; Lebeau, K. Electrifying light commercial vehicles for city logistics? A total cost of ownership analysis. *Eur. J. Transp. Infrastruct. Res.* **2015**, *15*, 551–569.
40. Zhao, Y.; Onat, N.C.; Kucukvar, M.; Tatari, O. Carbon and energy footprints of electric delivery trucks: A hybrid multi-regional input-output life cycle assessment. *Transp. Res. Part D Transp. Environ.* **2016**, *47*, 195–207. [CrossRef]
41. Zhao, Y.; Tatari, O. Carbon and energy footprints of refuse collection trucks: A hybrid life cycle evaluation. *Sustain. Prod. Consum.* **2017**, *12*, 180–192. [CrossRef]
42. Zhou, T.; Roorda, M.J.; MacLean, H.L.; Luk, J. Life cycle GHG emissions and lifetime costs of medium-duty diesel and battery electric trucks in Toronto, Canada. *Transp. Res. Part D* **2017**, *55*, 91–98. [CrossRef]
43. Davis, B.A.; Figliozzi, M.A. A methodology to evaluate the competitiveness of electric delivery trucks. *Transp. Res. Part E Logist. Transp. Rev.* **2013**, *49*, 8–23. [CrossRef]
44. Christensen, L.; Klauenberg, J.; Kveiborg, O.; Rudolph, C. Suitability of commercial transport for a shift to electric mobility with Denmark and Germany as use cases. *Res. Transp. Econ.* **2017**, *64*, 48–60. [CrossRef]
45. Li, W.; Dai, Y.; Ma, L.; Hao, H.; Lu, H.; Albinson, R.; Li, Z. Oil-saving pathways until 2030 for road freight transportation in China based on a cost-optimization model. *Energy* **2015**, *86*, 369–384. [CrossRef]
46. Oshiro, K.; Masui, T. Diffusion of low emission vehicles and their impact on CO2 emission reduction in Japan. *Energy Policy* **2015**, *81*, 215–225. [CrossRef]
47. Fridstrøm, L. From innovation to penetration: Calculating the energy transition time lag for motor vehicles. *Energy Policy* **2017**, *108*, 487–502. [CrossRef]
48. Mulholland, E.; Teter, J.; Cazzola, P.; McDonald, Z.; Ó Gallachóir, B.P. The long haul towards decarbonising road freight—A global assessment to 2050. *Appl. Energy* **2018**, *216*, 678–693. [CrossRef]
49. MLIT. *Road Transport Vehicle Law Enforcement Act, Ordinance of the Ministry No 74, August 16 of 1951; Last modification in December 28 of 2016*; Ministry of Land, Infrastructure,Transport and Tourism: Tokyo, Japan, 1951. (In Japanese)
50. MLIT. *Vehicle Fuel Consumption List (March 2018)*; Ministry of Land, Infrastructure, Transport and Tourism: Tokyo, Japan, 2018. (In Japanese)
51. SEI. *LEAP User Guide*; Heaps, C., Ed.; Stockholm Environment Institute—SEI: Somerville, MA, USA, 2018; Available online: https://www.energycommunity.org/Help/leap.htm (accessed on 1 March 2020).
52. González Palencia, J.C.; Araki, M.; Shiga, S. Energy consumption and CO2 emissions reduction potential of electric-drive vehicle diffusion in a road freight vehicle fleet. *Energy Procedia* **2017**, *142*, 2936–2941. [CrossRef]
53. González Palencia, J.C.; Sakamaki, T.; Araki, M.; Shiga, S. Impact of powertrain electrification, vehicle size reduction and lightweight materials substitution on energy use, CO2 emissions and cost of a passenger light-duty vehicle fleet. *Energy* **2015**, *93*, 1489–1504. [CrossRef]
54. Teoh, T.; Kunze, O.; Teo, C.C.; Wong, Y.D. Decarbonisation of urban freight transport using electric vehicles and opportunity charging. *Sustainability* **2018**, *10*, 3258. [CrossRef]
55. González Palencia, J.C.; Araki, M.; Shiga, S. CO2 emissions reduction potential of powertrain electrification, lightweight materials use and vehicle size reduction in a passenger light-duty vehicle fleet. In *Proceedings of the International Conference on Power Engineering-15 (ICOPE-15)*; Japan Society of Mechanical Engineers (JSME), American Society of Mechanical Engineers (ASME) and Chinese Society of Power Engineering (CSPE): Yokohama, Japan, 2015; p. 12.
56. MLIT. *Vehicle Fuel Consumption List (March 2013)*; Ministry of Land, Infrastructure, Transport and Tourism: Tokyo, Japan, 2013. (In Japanese)
57. Talebian, H.; Herrera, O.E.; Tran, M.; Mérida, W. Electrification of road freight transport: Policy implications in British Columbia. *Energy Policy* **2018**, *115*, 109–118. [CrossRef]
58. Pagerit, S.; Sharer, P.; Rousseau, A. Fuel Economy Sensitivity to Vehicle Mass for Advanced Vehicle Powertrains. In *Proceedings of the SAE World Congress & Exhibition, SAE paper 2006-01-0665*; Society of Automotive Engineers: Detroit, MI, USA, 2006.
59. Contestabile, M.; Offer, G.J.; Slade, R.; Jaeger, F.; Thoennes, M. Battery electric vehicles, hydrogen fuel cells and biofuels. Which will be the winner? *Energy Environ. Sci.* **2011**, *4*, 3754–3772. [CrossRef]
60. DOE. *DOE Hydrogen and Fuel Cells Program Record # 14012*; Department of Energy United States of America: Washington, DC, USA, 2013. Available online: http://energy.gov/sites/prod/files/14012_fuel_cell_system_cost_2013.pdf (accessed on 10 February 2020).

61. Van Vliet, O.P.R.; Kruithof, T.; Turkenburg, W.C.; Faaij, A.P.C. Techno-economic comparison of series hybrid, plug-in hybrid, fuel cell and regular cars. *J. Power Sources* **2010**, *195*, 6570–6585. [CrossRef]
62. Van Vliet, O.; Brouwer, A.S.; Kuramochi, T.; van den Broek, M.; Faaij, A. Energy use, cost and CO_2 emissions of electric cars. *J. Power Sources* **2011**, *196*, 2298–2310. [CrossRef]
63. MLIT. *Results of the Basic Survey on Motor Vehicle Inspection and Maintenance for the 2004 Fiscal Year (in Japanese)*; Ministry of Land, Infrastructure, Transport and Tourism: Tokyo, Japan, 2004. Available online: http://www.mlit.go.jp/jidosha/iinkai/seibi/6houkokusyo.pdf (accessed on 16 March 2020).
64. AIRIA. 2012 Vehicle stock. Automobile Inspection & Registration Association: Tokyo, Japan, 2012. Available online: http://www.airia.or.jp/publish/file/e49tph00000004t7-att/e49tph00000004te.pdf (accessed on 9 April 2020). (In Japanese).
65. MOE. *The Strategy to Promote Environmentally Friendly Vehicles*; Ministry of Environment: Tokyo, Japan, 2010. Available online: http://www.env.go.jp/air/report/h22-02/index.html (accessed on 18 February 2020). (In Japanese)
66. Nishimura, E. Assessing the Fuel Use and Greenhouse Gas Emissions of Future Light-Duty Vehicles in Japan. Master's Thesis, Engineering System Division, Massachussetts Institute of Technology, Cambridge, MA, USA, 2011.
67. JAMA. *The Motor Industry of Japan 2013*; Japanese Automobile Manufacturers Association: Tokyo, Japan, 2013; Available online: http://www.jama-english.jp/publications/MIJ2013.pdf (accessed on 23 March 2020).
68. AIRIA. *New Vehicle Registration by Year; Light-Duty Vehicle Fleet*; Automobile Inspection & Registration Association: Tokyo, Japan, 2012. (In Japanese)
69. NeV. *Electric Vehicle Stock Statistics (Estimate)*; Next Generation Vehicle Promotion Center: Tokyo, Japan, 2016; Available online: http://www.cev-pc.or.jp/tokei/hanbai.html (accessed on 28 January 2016). (In Japanese)
70. JAMA. *The Motor Industry of Japan 2018*; Japanese Automobile Manufacturers Association: Tokyo, Japan, 2018; Available online: http://www.jama-english.jp/publications/MIJ2018.pdf (accessed on 10 March 2020).
71. Grübler, A. *Technology and Global Change*; Cambridge University Press: Cambridge, UK, 2003.
72. Hollinshead, M.J.; Eastman, C.D.; Etsell, T.H. Forecasting performance and market penetration of fuel cells in transportation. *Fuel Cells Bull.* **2005**, *2005*, 10–17. [CrossRef]
73. Acar, C.; Dincer, I. Review and evaluation of hydrogen production options for better environment. *J. Clean. Prod.* **2019**, *218*, 835–849. [CrossRef]
74. Bandivadekar, A.; Bodek, K.; Cheah, L.; Evans, C.; Groode, T.; Heywood, J.; Kasseris, E.; Kromer, M.; Weiss, M. *On the Road in 2035: Reducing Transportation's Petroleum Consumption and GHG Emissions*; Massachusetts Institute of Technology: Cambridge, MA, USA, 2008.
75. Akerlind, I.B.; Bastani, P.; Berry, I.; Bhatt, K.; Chao, A.; Chow, E.; Karplus, V.; Keith, D.; Khusid, M.; Nishimura, E.; et al. *On the Road toward 2050: Potential for Substantial Reductions in Light-Duty Vehicle Energy Use and Greenhouse Gas Emissions*; Heywood, J., MacKenzie, D., Eds.; Massachusetts Institute of Technology: Cambridge, MA, USA, 2015.
76. EDMC; IEEJ. *EDMC Handbook of Energy & Economic Statistics 2019*; The Energy Data and Modelling Center, The Institute of Energy Economic, The Energy Conservation Center: Tokyo, Japan, 2019. (In Japanese)
77. IEA/OECD (Ed.) *Energy Technology Perspectives 2012; Pathways to a Clean Energy System*; Energy Technology Perspectives; International Energy Agency and Organization for Economic Cooperation and Development: Paris, France, 2012.
78. IEA/OECD (Ed.) *CO2 Emissions from Fuel Combustion Highlights 2013*; International Energy Agency and Organization for Economic Cooperation and Development: Paris, France, 2013.
79. Berraho, D. Options for the Japanese Electricity Mix by 2050. Master's Thesis, School of Industrial Engineering and Management, Division of Energy and Climate Studies, KTH Royal Institute of technology, Stockholm, Sweden, 2012.
80. IEA/OECD. *World Energy Outlook 2012*; International Energy Agency and Organization for Economic Cooperation and Development: Paris, France, 2012.
81. Delucchi, M.A.; Yang, C.; Burke, A.F.; Ogden, J.M.; Kurani, K.; Kessler, J.; Sperling, D. An assessment of electric vehicles: Technology, infrastructure requirements, greenhouse-gas emissions, petroleum use, material use, lifetime cost, consumer acceptance and policy. *Phil. Trans. R. Soc. A* **2014**, *372*, 20120325. [CrossRef]

82. Hao, H.; Wang, H.; Ouyang, M. Fuel consumption and life cycle GHG emissions by China's on-road trucks: Future trends through 2050 and evaluation of mitigation measures. *Energy Policy* **2012**, *43*, 244–251. [CrossRef]
83. GIO; NIES. *Japan's GHG Emissions Data (FY1990-2016, Final Figures)*; Greenhouse Gas Inventory Office of Japan, National Institute of Environmental Studies: Tsukuba, Japan, 2018. Available online: http://www-gio.nies.go.jp/aboutghg/nir/nir-e.html (accessed on 12 February 2020).

© 2020 by the authors. Licensee MDPI, Basel, Switzerland. This article is an open access article distributed under the terms and conditions of the Creative Commons Attribution (CC BY) license (http://creativecommons.org/licenses/by/4.0/).

energies

MDPI

Article

Durability and Reliability of EV Batteries under Electric Utility Grid Operations: Impact of Frequency Regulation Usage on Cell Degradation

George Baure and Matthieu Dubarry *

Hawai'i Natural Energy Institute, SOEST, University of Hawai'i at Mānoa, 1680 East-West Road, POST 109, Honolulu, HI 96822, USA; gbaure@hawaii.edu

* Correspondence: matthieu@hawaii.edu; Tel.: +1-808-9562349

Received: 31 March 2020; Accepted: 13 May 2020; Published: 15 May 2020

Abstract: The usage of electric vehicle batteries to assist the main electric grid for the storage of energy provided by intermittent sources should become an essential tool to increase the penetration of green energies. However, this service induces additional usage on the cells and, therefore, could degrade them further. Since degradation is path-dependent, it is of paramount importance to test the impact of all the different grid applications on the batteries. In this work, we tested the additional usage induced by using electric vehicle batteries for frequency regulation at moderate rates during rest or charge and found no detrimental effect after around 2000 cycles on the cells.

Keywords: lithium-ion; dQ/dV; dV/dQ; frequency regulation; V2G; G2V; electric vehicle

1. Introduction

The Hawai'i Clean Energy Initiative endeavors to gain total oil independence by 2045 [1] for both electricity generation and ground transportation. On the grid side, the energy portfolio will likely include many intermittent resources such as wind and solar and will thus require significant energy storage. The electrification of ground transportation could kill two birds with one stone by providing oil-free vehicles and offer energy storage when the vehicles are not in use. The latter is referred to as vehicle-to-grid (V2G) and grid-to-vehicle (G2V).

With incentives in place [2,3], the number of electric vehicles (EVs) in Hawai'i and the rest of the world continues to rise and will collectively constitute a significant distributed energy storage reservoir for the grid. EV batteries could provide ancillary grid services such as operating reserves, power curtailment, frequency regulation, and voltage smoothing by allowing the network to give (G2V) and take (V2G) energy when necessary [4,5]. The benefits and drawbacks to both the vehicle owner and the energy provider of these strategies have been well-documented [6–18], and the main obstacle was identified to be the additional usage on the cells [19–21], among other challenges [22]. Few experimental studies [23–28] attempted to account for the change in battery degradation resulting from the implementation of these strategies. To accurately account for the change in usage, the path dependence of degradation needs to be considered in the estimation [29]. Each of these ancillary grid services can affect degradation differently, and certain conditions can lead to accelerated capacity loss [30–32]. This accelerated capacity loss, sometimes termed "rollover failure" [33], is a significant safety concern to the battery industry. However, solutions do exist, and it was shown in a previous work [31] that, although this second stage of degradation cannot be predicted from capacity nor the resistance evolution, it might be predicted from the investigation of the voltage response using electrochemical voltage spectroscopies [29,34,35].

This work is a follow-up of our previous studies [25,26,36,37], in which we purchased a batch of commercial cells to test the impact of different aspects of EV battery usage. Part 1 was devoted to the definition of the cell-to-cell variations of the full batch and the emulation of the electrochemical behavior [36]. Part 2 focused on the capacity- and resistance-based analyses of cycle- and calendar-aging experiments to assess the impact of bidirectional charging durability [25]. In Part 3, [26], the cycle and calendar-aging degradation mechanisms were investigated using an incremental capacity analysis and the features of interest (FOI) approach [38]. The analysis was used to quantify the different degradation modes, determine the degradation path dependency, and challenge the Part 2 [25] capacity- and resistance-based forecast. Finally, Part 4 was devoted to the study of the impact of different driving cycles on the degradation mechanisms [37].

The next step of this EV battery degradation research, detailed here, is quantifying the consequences of a distinct grid interaction: frequency response. This investigation elucidates the impact of frequency regulation under several conditions, including a new approach of modulating the charge so that the vehicle can perform grid applications without any additional battery usage.

2. Materials and Methods

In this work, a set of twelve Panasonic cylindrical 3350 mAh NCR 18650B batteries were selected from a batch of 100. More details on these cylindrical 18650-size graphite/$LiNi_xCo_{1-x-y}Al_yO_2$ (NCA) cells, as well as the batch cell-to-cell variations analysis, can be found in previous works [25,26,36,37]. These cells are similar with the type of cells used in some EVs, such as the Tesla Model S [39,40]. The twelve cells chosen were within the outlier boundaries of the cell-to-cell variations distribution [36]. The experimental design for this study consisted of six test scenarios, with common conditioning and reference performance tests (RPT) [41], replicated twice.

As described in previous publications [36,41], all cells were subjected to initial conditioning and characterization tests before the beginning of the cycling experiment to verify the quality of cells. Following the conditioning cycles at C/5 and C/2, and before the start of the duty cycle testing, an RPT consisting of C/35, C/5, and C/3 full cycles was performed on all cells. The RPT procedure included some remnant capacity measurements consisting of a C/50 top-off at the end of each regime, as well as 4-h rests before and after the remnant capacity measurements. The rate capability was calculated by dividing the C/3 capacity by the C/25 capacity at each RPT. The resistance was derived from the ohmic drop associated with the transition from a resting stage to a C/3 rate [41] 0.1 s after the application of current.

Each duty cycle began with a C/2 full charge, as well as a constant voltage step with a 65-mA limiting current followed by a 4-h rest step. The cells were then discharged by around 60% of their typical initial capacity by applying the United States Advanced Battery Consortium (USABC) federal urban driving schedule (FUDS) power profile [3] and scaled to the USABC goal of 400 W/kg. The FUDS profile was shown to replicate driving cell degradation adequately [37]. This usage equated to about three hours of driving per day or about 3 times the typical American roundtrip commute, according to the U.S. Census Bureau. After the discharge, the cells were first rested for half an hour. Subsequently, the cells were subjected to six different scenarios (Figure 1).

In the first scenario (NoFR), no frequency regulation was applied (Figure 1a,b), and the cells rested for 4 h before being charged at a C/4 rate then by a constant-voltage step with a 65-mA limiting current followed by a half-hour relaxation. In scenarios two to four, the cells were used for frequency regulation for 2 h during rest (Figure 1, right column). The frequency regulation profile was extracted from representative real usage data from a grid-scale battery energy storage system located in Hawi, Hawai'i while in frequency regulation mode [42–44] (Figure S1). This frequency regulation profile was normalized to three different maximum currents corresponding to C/5 (Figure 1c, frequency rest (FRr)C/5), C/4 (Figure 1d, FRrC/4), and C/3 (Figure 1e, FRrC/3) rates, but the pulses variations and the depth of the discharge range were kept constant to stay within cell and testing channel availability constraints. The additional usage of the cells added 0.28 Ah (+8.6%), 0.35 Ah (+10.7%), and 0.47 Ah (+14.3%) of exchanged capacity per cycle, respectively. After the 2-h frequency regulation, the cells were rested for an additional one and a half hours, then charged at C/4 followed by a constant-voltage step with a 65-mA-limiting current and a half-hour rest step. In scenarios five and six, the cells were subjected to the same frequency regulation profile, this time added to the C/4 charge instead of during rest (Figure 1f,g). In these two cases, the cells rested for 4 h before being subjected to the amalgamation of the frequency regulation profile and the C/4 charge for 2 h. The frequency regulation was only normalized to two different maximum currents: C/5 (Figure 1f, frequency charge (FRc)C/5) and C/4 (Figure 1g, FRcC/4), so that the current never becomes negative (i.e., discharging) during charge. Since the charge current was always positive while modulated, no additional capacity was exchanged for scenarios five and six compared to not performing the frequency regulation at all. After the frequency regulation, the cells were topped with a charged at C/4 followed by a constant-voltage step with a 65-mA-limiting current and a half-hour rest.

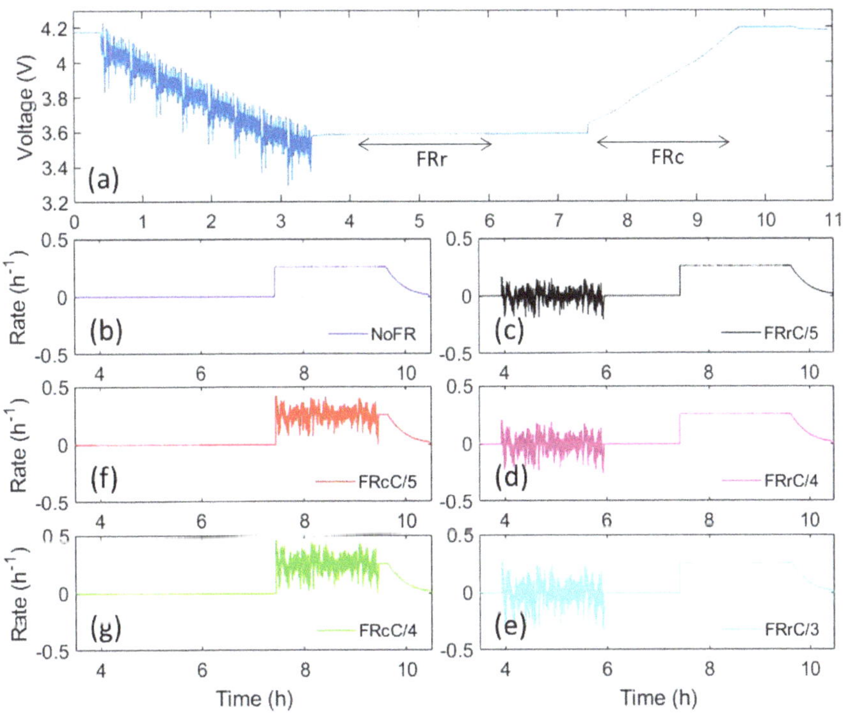

Figure 1. (**a**) Full duty cycle and (**b–g**) charge scenarios for the six duty cycles. noFR: no frequency regulation, FRc: frequency response during charge, and FRr: frequency regulation during rest.

These six duty cycles were typically repeated for four weeks; after which, an RPT was performed. Cycling was interrupted when the cell exhibited a capacity loss of greater than 20% at C/35 for two consecutive RPTs or when the total test time was approximately two years. All the tests were performed in an Amerex IC500-R chamber set at 25 °C. The skin temperature of each of the cells was recorded and was of 24.5 ± 0.5 °C in average. The cut-off voltages were set at 4.2 and 2.5 V. All the tests, except the first RPTs for each cell, were performed by a 40-channel Arbin BT-5HC tester (College Station, TX, USA). The first RPTs were performed using a Biologic BCS-815 battery cycler (Claix, France).

Computer simulations were performed using the mechanistic degradation model proposed in [45] and the 'alawa toolbox. The half-cell data was obtained from previous works [25,26,36,37].

3. Results

Figure 2 plots the normalized capacity (top), normalized resistance (center), and rate capability (bottom) versus exchanged capacity for the twelve cells used in this study. All the cells presented a similar linear ($R^2 > 0.93$) capacity loss for close to 5000 Ah exchanged (1500 equivalent full cycles) at a rate of −0.0024% ± 0.0004% per Ah of exchanged capacity. All but one of the cells exchanged more than 6500 Ah (2000 equivalent full cycles), with capacity losses below 20%. This is a much smaller capacity fade than in our previous work on full discharges [37] but comparable to the one using shallower cycles [25,26]. The most apparent result was that the cells subjected to no frequency regulation (NoFR) degraded slightly faster. These two cells exhibited two of the three highest capacity loss values of the entire experiment: 35% and 17%. In addition, one of these cells (NoFR_1) experienced an accelerated aging stage much earlier than all the others after an exchanged capacity of 4700 Ah and a normalized capacity of 88%. In total, four cells reached a second aging stage defined with more than a 5% loss within an exchanged capacity of 600 Ah. The other three cells that experienced accelerated aging were the second NoFR cell and one each for the cells with the frequency regulation during charge and rest (FRcC/4_1 and FRrC/5_2). These three cells entered the second stage after 5500 Ah of usage. No abnormality in the testing protocol of the affected cells was noticed, and their initial cell-to-cell variations were within the normal boundaries. It can also be observed from Figure 2 that neither resistance nor rate capability could be used to predict the accelerated degradation. The resistance increased linearly and doubled after around 3500 Ah exchanged. The rate capability slowly deceased, also linearly, by less than 1%.

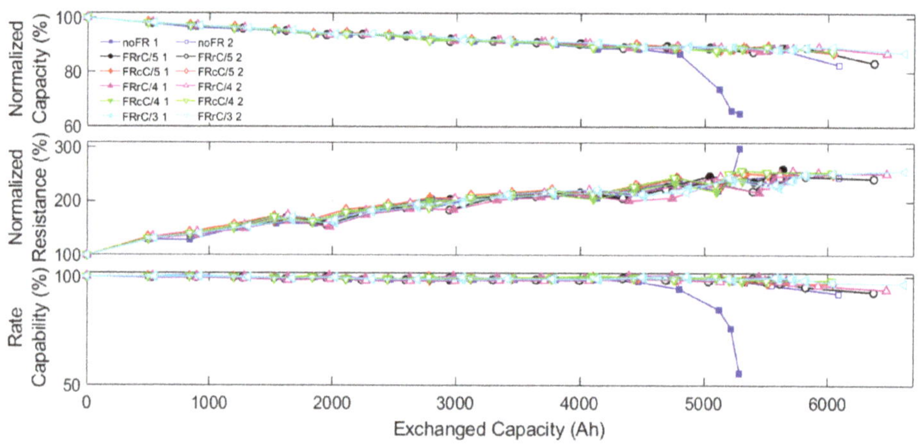

Figure 2. Normalized capacity, resistance, and rate capability versus exchanged capacity.

As discussed in our previous work, capacity loss and degradation must be considered separately as silent degradation (i.e., degradation that was not associated with capacity loss) can be present and play a significant role in the apparition of the accelerated aging phase [29,45]. To assess cell degradations, it was essential to quantify the degradation modes, the loss of lithium inventory, the loss of active material, and the kinetic degradation (RDF) on the positive (PE) and negative (NE) electrodes independently. This separation of the degradation modes can be performed using electrochemical voltage spectroscopies [46], incremental capacity (IC) [35,47], and differential voltage (DV) [34,48,49]. In our previous work on these cells [26,37], we used IC curves to quantify the loss of lithium inventory (LLI,) the loss of active material on the negative electrode (LAM_{PE}), and negative kinetic degradation (RDF_{NE}) directly from the FOIs but had to quantify the LAM_{NE} manually based on the other observations. In this analysis, the technique was refined, and the LAM_{NE} was quantified from the FOI on the DV curves when possible. Figure 3 presents an example of the IC and DV curves for a representative cell with the different FOIs noted. The response for all the cells is remarkably similar and is presented in Figures S2 and S3.

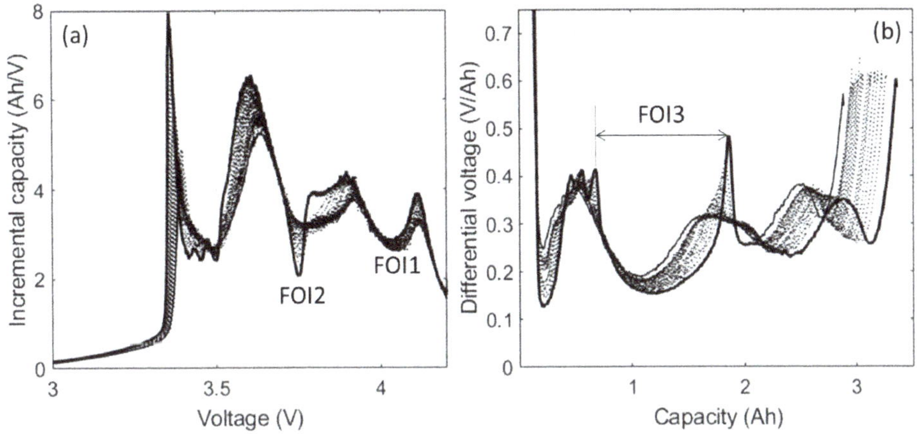

Figure 3. Representative (**a**) incremental capacity and (**b**) differential voltage curve evolutions at C/35 under the aging conditions tested. FOI: features of interest.

Just as in our previous work, and based on the degradation map in [29,45] and the evolution of the experimental curves, the capacity loss was associated entirely to LLI because the high and low-voltage peaks never disappeared completely. Moreover, the intensity of the minimum at 4 V is proportional to the LAM_{PE}, and the intensity of the minimum at 3.75 V is proportional to the RDF_{NE}. Readers are referred to these publications for more details. To quantify the LAM_{NE} automatically, the distance between two graphitic peaks was used as a proxy to the LAM_{NE}, as proposed in [48,50]. The full degradation map for the DV curves showing that the distance between the two peaks varies only for the LAM_{NE} is presented in Figure S4.

Figure 4 presents the evolution of the different FOIs upon aging. Figure 4a shows the evolution of FOI1; the area between 4 and 4.05 V was used in lieu of the intensity at 4 V to minimize the impact of noise on the voltage data. All the cells followed a similar deviation to the pristine behavior, with a decrease comprised between 5% and 6%. In regards to FOI2, as already observed in our previous work [29,45], the evolution of the intensity saturates after the local minimum completely disappeared. This saturation prevented the tracking of further intensity rise and the position of the minimum, (Figure 4c). The variation in the intensity of the minimum is consistent with what was observed in our previous work, but its voltage seems to be rather constant for all cells, which is different from what was found previously [29,45], for some of the tested conditions. Finally, FOI3 (Figure 4d) decreased linearly by between 6% and 10% before the 4000 Ah mark. After that, the peaks disappeared on the DV curves, and the direct estimator was not valid anymore. This change was likely induced by the RDF_{NE} increase (Figure S4).

The quantification of the degradation modes from the FOI variations can be done using the mechanistic modeling approach we pioneered, along with other groups, in the mid-2000s [45,48,51,52]. The approach has been heavily used in recent years within experimental studies using electrochemical voltage spectroscopies (EVS) [46,53], and it has been well-validated [54–56] using independent experimental studies. In the approach, half-cell data from each electrode is matched to emulate the electrochemical behavior of the full cells. Once the initial match is satisfactory, the electrode can be scaled or translated individually to replicate the effect of the different degradation modes [45]. The electrode matching and the voltage changes associated with degradation for these cells have already been heavily discussed in the literature [26,36,37] and will not be repeated here. As in our previous work [26,37], the theoretical variations for the FOIs for each individual degradation mode was computed and compared to the experimental variations to decipher the extent of each degradation mode.

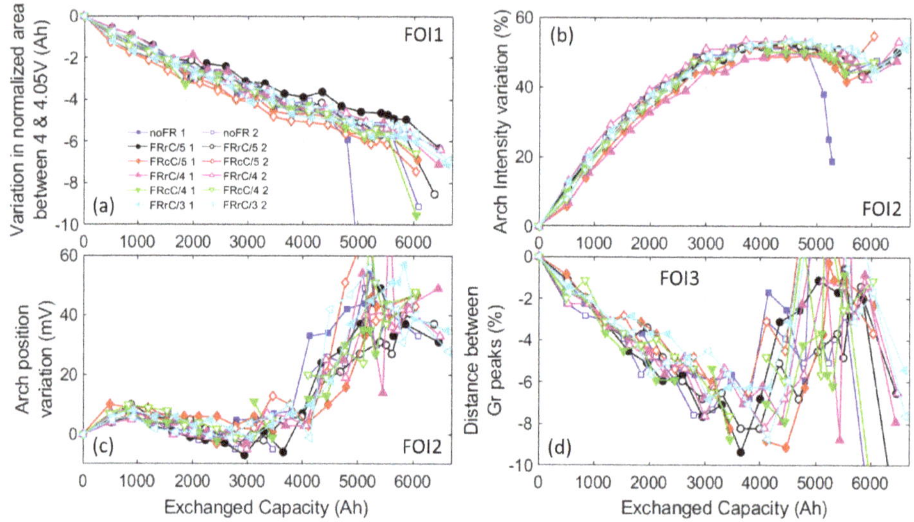

Figure 4. Evolution of (**a**) FOI1, the (**b**) intensity and (**c**) position of FOI2, and (**d**) FOI3.

4. Discussion

Based on Figure 4 and the proportionality between the FOI variations and the main degradation modes, the degradation of the cell up to 3000 Ah can be automatically deciphered using the same strategy that we employed in our previous works [29,45] and the fact that the cells had minimal initial cell-to-cell variations [36]. The results of the analysis are presented in Figure 5. The spider plot in Figure 5a showcases the results of the automatic quantification of the four main degradation modes (LLI, LAM_{PE}, LAM_{NE}, and RDF_{NE}). After 3000 Ah, the cells lost between 7% and 8% of their lithium, 5% and 8% of the NE, and 3% and 4.5% of their PE and their kinetics degraded by a factor comprised between 7 and 8. Overall, as already observable from Figure S2, all the cells degraded in a similar fashion, and no apparent differences were observed between the cells that reached the second stage earlier than the others. Figure 5b displays the same data on a bar plot to elucidate the impact, if any, of the duty cycles. The plotted values are the average between the two cells tested for each condition, and the error bars represent the spread between the individual values. Comparing the six cells that undertook the frequency regulation at different rates during rest (3 FRr x2), the spread of values was similar or smaller than the deviations between cells undergoing the same duty cycles. This is true for the other duty cycles as well. Taking into account the fact that cells under the same duty cycle are well-known to degrade at slightly different paces [57–60], the differences observed in this work are not big enough to conclude that six duty cycles tested in this work induced different degradation on the cells. The spider plot can also be compared to the ones in our previous work that tested the cells under different duty cycles [29,45]. In [26], we diagnosed the cells after a 5% capacity loss, which corresponded to 1500 Ah exchanged in our study. For the duty cycles in this work, we observed a degradation similar to one of the cells that did not experience any constant power (V2G). This outcome was expected, since the duty cycle resembled the NoFR duty cycle, with just a different driving cycle (RD1 in [37]), which we showed had a similar impact as the FUDS cycle used in this work [37]. In [37], where they cycled the cells in the full state of charge (SOC) range, the spider plot compared the degradation at the onset of stage 2 around the same capacity loss (except for RD2 in [37]). All degradation modes are in the same order of magnitude except LAM_{NE}, which was much more pronounced when the full SOC range was used.

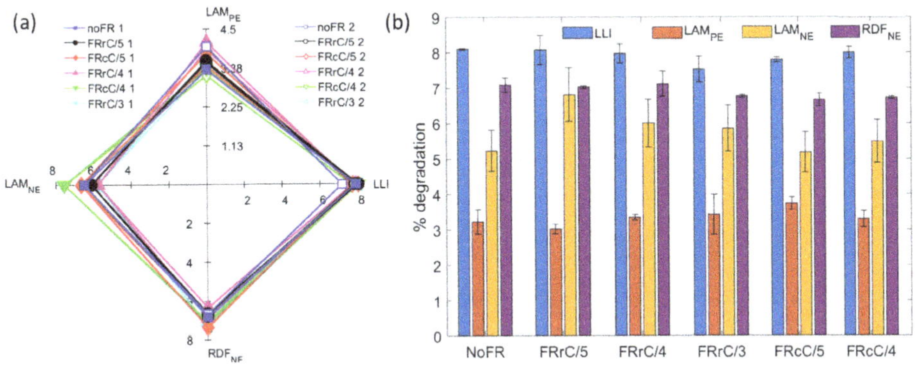

Figure 5. (**a**) Spider plots of the four prominent degradation modes for all the cells at 3000 Ah (~8% capacity loss) and (**b**) bar plot showcasing the impact of the duty cycles. The error bar represents the spread between the two tested cells for each duty cycle, not a distribution of values. LLI: loss of lithium inventory, LAM: loss of active material on the positive (PE) and negative (NE) electrodes, and RDF_{NE}: rate degradation factor on the NE.

Figure 6a presents a validation of the automatic diagnosis performed for Figure 5 for cell NoFR_1. This is important to verify that the automatic quantification of the degradation was correct. If accurate, the changes induced on the electrode half-cell matching by the quantified amount of LLI, LAMs, and RDF should replicate the observed voltage changes between the initial and the aged experimental voltage response. The simulated voltage was nearly overlapping to the experimental one, and this suggests that our automatic quantification of the degradation mode was appropriate, as any other combination of parameters would have led to a different voltage curve. The validation was verified for all the other cells. Since the simulated response was analogous to the experimental one [46], the diagnosis was validated. Hence, our automatic FOI-based methodology managed to perform the diagnosis automatically.

The next point to tackle is the observed differences at which the second stage is appearing. In previous works on similar cells, the second stage apparition was proved to be linked to the LAM_{NE} [37]. It was shown that when the LAM_{NE} reached a threshold, decipherable from the initial conditions and the LLI and LAM_{PE}, lithium plating began. Since the LLI and LAM_{PE} were quantifiable for the entire experiment, the predicted LAM_{NE} necessary for generalized plating to occur was calculated to be around 20% for cell NoFR_1 when the second stage started. However, quantification showed the LAM_{NE} around 11%. This value was lower than the amount of LLI and in good agreement with an extrapolation of the linear trend of LAM_{NE} from the early cycles. The same trend was verified for the other cells that showcased the second stage. Therefore, and unlike our previous study, the second stage of degradation was not induced by a generalized LAM_{NE}. It was also not induced by the LAM_{PE}. This rule out any electrode-wide source for the second stage. A possible explanation is that it might still happen locally because of inhomogeneities in the electrode degradation [61]. The origin of the apparent RDF increase is still unclear. In our approach, it was treated as a kinetic limitation, but, in light of recent literature [62], part of it could also be associated with inhomogeneities in the electrode degradation that could have led to localized plating. This concept is currently under investigation in our laboratory and will be reported in future works.

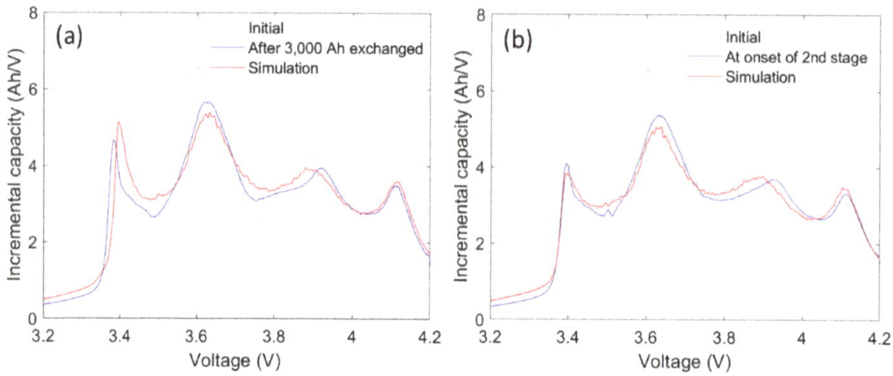

Figure 6. Comparison of experimental and simulated incremental capacity (IC) curves for the NoFR_1 cell (**a**) after 3000 Ah exchanged and (**b**) at the onset of the second stage.

5. Conclusions

The main takeaway from our study was that some frequency regulation V2G usage at moderate rates (from C/5 to C/3 as a maximum current) did not accelerate the cell degradation despite a close to 15% additional usage. In addition, there was no noticeable difference between performing this ancillary service during rest or charge or at maximum current fluctuations up to C/3 during rest and C/4 during charge. These results are extremely positive for the possible application of V2G/G2V strategies. However, it must be noted that our results hold for these specific cells and the duty cycle tested. More research is necessary to generalize the results. Most notably, to test a wide variety of frequency regulation cycles and driving cycles on different cell chemistries. We already verified the results with the same duty cycles on a different batch of commercial cells based on graphite and a blend of lithium cobalt oxide and nickel cobalt aluminum oxide.

Due to the lack of differences in the degradation between duty cycles, the benefits of modulating the charge to eliminate the additional usage on the cells could not be verified. Experiments on other chemistries with more aggressive usages will be launched to test the hypothesis further.

As for the diagnosis for the degradation, a new approach of coupling IC and DV analyses yielded four indicators for automatic diagnosis. This methodology will allow faster diagnosis and bolsters the value of the FOI approach for battery management system implementation. However, our study also showed that, under some conditions—in our case, the apparent kinetic limitations—some of the FOI may become ineffective with time. Therefore, proper validation using full fits and a sensibility analysis to check for the range of efficacy of chosen FOIs is still essential.

Finally, some cells showcased the second stage of aging, and we were not able to predict it from the voltage variations. This suggested that it was not induced by a widespread degradation of the electrodes but was more likely because of localized effects. The cells that did not perform any frequency regulations seemed to be the most affected, but those results need to be moderated by the fact that the differences were probably within what to expect between cells performing the same duty cycles.

Supplementary Materials: The following are available online at http://www.mdpi.com/1996-1073/13/10/2494/s1.

Author Contributions: All the authors designed and performed the experiments, analyzed the results, and wrote the manuscript. All authors have read and agreed to the published version of the manuscript.

Funding: This work was funded by the ONR Asia Pacific Research Initiative for Sustainable Energy Systems, awards No. N00014-16-1-2116 and N00014-17-1-2206. M.D. also has support from the State of Hawai'i. The authors are grateful to the Hawaiian Electric Company for their past support to the operations of the Hawai'i Sustainable Energy Research Facility.

Conflicts of Interest: The authors declare no conflicts of interest.

References

1. Lee, T.; Glick, M.B.; Lee, J.-H. Island energy transition: Assessing Hawaii's multi-level, policy-driven approach. *Renew. Sustain. Energy Rev.* **2020**, *118*, 109500. [CrossRef]
2. Wee, S.; Coffman, M.; La Croix, S. Do electric vehicle incentives matter? Evidence from the 50 U.S. states. *Res. Policy* **2018**, *47*, 1601–1610. [CrossRef]
3. Coffman, M.; Bernstein, P.; Wee, S. Electric vehicles revisited: A review of factors that affect adoption. *Transp. Rev.* **2016**, *37*, 79–93. [CrossRef]
4. Kempton, W.; Letendre, S.E. Electric vehicles as a new power source for electric utilities. *Transp. Res. Part D Transp. Environ.* **1997**, *2*, 157–175. [CrossRef]
5. Raustad, R. The Role of V2G in the Smart Grid of the Future. *Interface* **2015**, *24*, 53–60. [CrossRef]
6. Babrowski, S.; Heinrichs, H.; Jochem, P.; Fichtner, W. Load shift potential of electric vehicles in Europe. *J. Power Sources* **2014**, *255*, 283–293. [CrossRef]
7. Honarmand, M.; Zakariazadeh, A.; Jadid, S. Optimal scheduling of electric vehicles in an intelligent parking lot considering vehicle-to-grid concept and battery condition. *Energy* **2013**. [CrossRef]
8. Lefeng, S.; Qian, Z.; Yongjian, P. The reserve trading model considering V2G Reverse. *Energy* **2013**, *59*, 50–55. [CrossRef]
9. Mwasilu, F.; Justo, J.J.; Kim, E.-K.; Do, T.D.; Jung, J.-W. Electric vehicles and smart grid interaction: A review on vehicle to grid and renewable energy sources integration. *Renew. Sustain. Energy Rev.* **2014**, *34*, 501–516. [CrossRef]
10. Talebizadeh, E.; Rashidinejad, M.; Abdollahi, A. Evaluation of plug-in electric vehicles impact on cost-based unit commitment. *J. Power Sources* **2014**, *248*, 545–552. [CrossRef]
11. Zhou, C. Modeling of the Cost of EV BatteryWear Due to V2G Application in Power Systems. *IEEE Trans. Energy Convers.* **2011**, *26*, 1043–1050. [CrossRef]
12. Zhang, L.; Jabbari, F.; Brown, T.; Samuelsen, S. Coordinating plug-in electric vehicle charging with electric grid: Valley filling and target load following. *J. Power Sources* **2014**, *267*, 584–597. [CrossRef]
13. White, C.D.; Zhang, K.M. Using vehicle-to-grid technology for frequency regulation and peak-load reduction. *J. Power Sources* **2011**, *196*, 3972–3980. [CrossRef]
14. Nykvist, B.; Nilsson, M. Rapidly falling costs of battery packs for electric vehicles. *Nat. Clim. Chang.* **2015**, *5*, 329–332. [CrossRef]
15. Aziz, M.; Oda, T.; Mitani, T.; Watanabe, Y.; Kashiwagi, T. Utilization of Electric Vehicles and Their Used Batteries for Peak-Load Shifting. *Energies* **2015**, *8*, 3720–3738. [CrossRef]
16. Zhao, Y.; Tatari, O. A hybrid life cycle assessment of the vehicle-to-grid application in light duty commercial fleet. *Energy* **2015**, *93*, 1277–1286. [CrossRef]
17. Thompson, A.W.; Perez, Y. Vehicle-to-Everything (V2X) energy services, value streams, and regulatory policy implications. *Energy Policy* **2019**. [CrossRef]
18. Bañol Arias, N.; Hashemi, S.; Andersen, P.B.; Træholt, C.; Romero, R. Assessment of economic benefits for EV owners participating in the primary frequency regulation markets. *Int. J. Electr. Power Energy Syst.* **2020**, *120*. [CrossRef]
19. Uddin, K.; Dubarry, M.; Glick, M. The viability of vehicle-to-grid operations from a battery technology and policy perspective. *Energy Policy* **2018**, *113*, 342–347. [CrossRef]
20. Saldaña, G.; San Martin, J.I.; Zamora, I.; Asensio, F.J.; Oñederra, O. Electric Vehicle into the Grid: Charging Methodologies Aimed at Providing Ancillary Services Considering Battery Degradation. *Energies* **2019**, *12*, 2443. [CrossRef]
21. Guo, J.; Yang, J.; Lin, Z.; Serrano, C.; Cortes, A.M. Impact Analysis of V2G Services on EV Battery Degradation—A Review. In Proceedings of the 2019 IEEE Milan PowerTech, Milan, Italy, 23–27 June 2019.
22. Alshahrani, S.; Khalid, M.; Almuhaini, M. Electric Vehicles Beyond Energy Storage and Modern Power Networks: Challenges and Applications. *IEEE Access* **2019**, *7*, 99031–99064. [CrossRef]
23. Bishop, J.D.K.; Axon, C.J.; Bonilla, D.; Tran, M.; Banister, D.; McCulloch, M.D. Evaluating the impact of V2G services on the degradation of batteries in PHEV and EV. *Appl. Energy* **2013**, *111*, 206–218. [CrossRef]
24. Uddin, K.; Jackson, T.; Widanage, W.D.; Chouchelamane, G.; Jennings, P.A.; Marco, J. On the possibility of extending the lifetime of lithium-ion batteries through optimal V2G facilitated by an integrated vehicle and smart-grid system. *Energy* **2017**, *133*, 710–722. [CrossRef]

25. Dubarry, M.; Devie, A.; McKenzie, K. Durability and reliability of electric vehicle batteries under electric utility grid operations: Bidirectional charging impact analysis. *J. Power Sources* **2017**, *358*, 39–49. [CrossRef]
26. Dubarry, M.; Baure, G.; Devie, A. Durability and Reliability of EV Batteries under Electric Utility Grid Operations: Path Dependence of Battery Degradation. *J. Electrochem. Soc.* **2018**, *165*, A773–A783. [CrossRef]
27. Marongiu, A.; Roscher, M.; Sauer, D.U. Influence of the vehicle-to-grid strategy on the aging behavior of lithium battery electric vehicles. *Appl. Energy* **2015**, *137*, 899–912. [CrossRef]
28. Guo, J.; Yang, J.; Cao, W.; Serrano, C. Evaluation of EV battery degradation under different charging strategies and V2G schemes. In Proceedings of the 8th Renewable Power Generation Conference (RPG 2019), Shanghai, China, 24–25 October 2019.
29. Dubarry, M.; Baure, G.; Anseán, D. Perspective on State-of-Health Determination in Lithium-Ion Batteries. *J. Electrochem. Energy Convers. Storage* **2020**, *17*, 1–25. [CrossRef]
30. Frisco, S.; Kumar, A.; Whitacre, J.F.; Litster, S. Understanding Li-Ion Battery Anode Degradation and Pore Morphological Changes through Nano-Resolution X-ray Computed Tomography. *J. Electrochem. Soc.* **2016**, *163*, A2636–A2640. [CrossRef]
31. Ansean, D.; Dubarry, M.; Devie, A.; Liaw, B.Y.; Garcia, V.M.; Viera, J.C.; Gonzalez, M. Operando lithium plating quantification and early detection of a commercial LiFePO4 cell cycled under dynamic driving schedule. *J. Power Sources* **2017**, *356*, 36–46. [CrossRef]
32. Gering, K.L.; Sazhin, S.V.; Jamison, D.K.; Michelbacher, C.J.; Liaw, B.Y.; Dubarry, M.; Cugnet, M. Investigation of path dependence in commercial lithium-ion cells chosen for plug-in hybrid vehicle duty cycle protocols. *J. Power Sources* **2011**, *196*, 3395–3403. [CrossRef]
33. Burns, J.C.; Kassam, A.; Sinha, N.N.; Downie, L.E.; Solnickova, L.; Way, B.M.; Dahn, J.R. Predicting and Extending the Lifetime of Li-Ion Batteries. *J. Electrochem. Soc.* **2013**, *160*, A1451–A1456. [CrossRef]
34. Bloom, I.; Christophersen, J.; Gering, K. Differential voltage analyses of high-power lithium-ion cells 2. Applications. *J. Power Sources* **2005**, *139*, 304–313. [CrossRef]
35. Dubarry, M.; Svoboda, V.; Hwu, R.; Liaw, B.Y. Incremental capacity analysis and close-to-equilibrium OCV measurements to quantify capacity fade in commercial rechargeable lithium batteries. *Electrochem. Solid State Lett.* **2006**, *9*, A454–A457. [CrossRef]
36. Devie, A.; Dubarry, M. Durability and Reliability of Electric Vehicle Batteries under Electric Utility Grid Operations. Part 1: Cell-to-Cell Variations and Preliminary Testing. *Batteries* **2016**, *2*, 28. [CrossRef]
37. Baure, G.; Dubarry, M. Synthetic vs. Real Driving Cycles: A Comparison of Electric Vehicle Battery Degradation. *Batteries* **2019**, *5*, 42. [CrossRef]
38. Dubarry, M.; Berecibar, M.; Devie, A.; Anseán, D.; Omar, N.; Villarreal, I. State of health battery estimator enabling degradation diagnosis: Model and algorithm description. *J. Power Sources* **2017**, *360*, 59–69. [CrossRef]
39. Berdichevsky, G.K.K.; Straubel, J.B.; Toomre, E. The Tesla Roadster Battery System. *Tesla Mot.* **2007**, *1*, 1–5.
40. Uitz, M.; Sternad, M.; Breuer, S.; Täubert, C.; Traußnig, T.; Hennige, V.; Hanzu, I.; Wilkening, M. Aging of Tesla's 18650 Lithium-Ion Cells: Correlating Solid-Electrolyte-Interphase Evolution with Fading in Capacity and Power. *J. Electrochem. Soc.* **2017**, *164*, A3503–A3510. [CrossRef]
41. Dubarry, M.; Baure, G. Perspective on Commercial Li-ion Battery Testing, Best Practices for Simple and Effective Protocols. *Electronics* **2020**, *9*, 152. [CrossRef]
42. Stein, K.; Tun, M.; Musser, K.; Rocheleau, R. Evaluation of a 1 MW, 250 kW-hr Battery Energy Storage System for Grid Services for the Island of Hawaii. *Energies* **2018**, *11*, 3367. [CrossRef]
43. Stein, K.; Tun, M.; Matsuura, M.; Rocheleau, R. Characterization of a Fast Battery Energy Storage System for Primary Frequency Response. *Energies* **2018**, *11*, 3358. [CrossRef]
44. Dubarry, M.; Devie, A.; Stein, K.; Tun, M.; Matsuura, M.; Rocheleau, R. Battery Energy Storage System battery durability and reliability under electric utility grid operations: Analysis of 3 years of real usage. *J. Power Sources* **2017**, *338*, 65–73. [CrossRef]
45. Dubarry, M.; Truchot, C.; Liaw, B.Y. Synthesize battery degradation modes via a diagnostic and prognostic model. *J. Power Sources* **2012**, *219*, 204–216. [CrossRef]
46. Barai, A.; Uddin, K.; Dubarry, M.; Somerville, L.; McGordon, A.; Jennings, P.; Bloom, I. A comparison of methodologies for the non-invasive characterisation of commercial Li-ion cells. *Progr. Energy Combust. Sci.* **2019**, *72*, 1–31. [CrossRef]

47. Dubarry, M.; Devie, A.; Liaw, B.Y. The Value of Battery Diagnostics and Prognostics. *J. Energy Power Sources* **2014**, *1*, 242–249.
48. Bloom, I.; Jansen, A.N.; Abraham, D.P.; Knuth, J.; Jones, S.A.; Battaglia, V.S.; Henriksen, G.L. Differential voltage analyses of high-power, lithium-ion cells. 1. Technique and Applications. *J. Power Sources* **2005**, *139*, 295–303. [CrossRef]
49. Keil, P.; Jossen, A. Calendar Aging of NCA Lithium-Ion Batteries Investigated by Differential Voltage Analysis and Coulomb Tracking. *J. Electrochem. Soc.* **2017**, *164*, A6066–A6074. [CrossRef]
50. Liu, P.; Wang, J.; Hicks-Garner, J.; Sherman, E.; Soukiazian, S.; Verbrugge, M.; Tataria, H.; Musser, J.; Finamore, P. Aging Mechanisms of LiFePO4 Batteries Deduced by Electrochemical and Structural Analyses. *J. Electrochem. Soc.* **2010**, *157*, A499–A507. [CrossRef]
51. Honkura, K.; Honbo, H.; Koishikawa, Y.; Horiba, T. State Analysis of Lithium-Ion Batteries Using Discharge Curves. *ECS Trans.* **2008**, *13*, 61–73.
52. Dahn, H.M.; Smith, A.J.; Burns, J.C.; Stevens, D.A.; Dahn, J.R. User-Friendly Differential Voltage Analysis Freeware for the Analysis of Degradation Mechanisms in Li-Ion Batteries. *J. Electrochem. Soc.* **2012**, *159*, A1405–A1409. [CrossRef]
53. Pastor-Fernández, C.; Yu, T.F.; Widanage, W.D.; Marco, J. Critical review of non-invasive diagnosis techniques for quantification of degradation modes in lithium-ion batteries. *Renew. Sustain. Energy Rev.* **2019**, *109*, 138–159. [CrossRef]
54. Kassem, M.; Delacourt, C. Postmortem analysis of calendar-aged graphite/LiFePO4 cells. *J. Power Sources* **2013**, *235*, 159–171. [CrossRef]
55. Schmidt, J.P.; Tran, H.Y.; Richter, J.; Ivers-Tiffee, E.; Wohlfahrt-Mehrens, M. Analysis and prediction of the open circuit potential of lithium-ion cells. *J. Power Sources* **2013**, *239*, 696–704. [CrossRef]
56. Birkl, C.R.; Roberts, M.R.; McTurk, E.; Bruce, P.G.; Howey, D.A. Degradation diagnostics for lithium ion cells. *J. Power Sources* **2017**, *341*, 373–386. [CrossRef]
57. Baumhöfer, T.; Brühl, M.; Rothgang, S.; Sauer, D.U. Production caused variation in capacity aging trend and correlation to initial cell performance. *J. Power Sources* **2014**, *247*, 332–338. [CrossRef]
58. Harris, S.J.; Harris, D.J.; Li, C. Failure statistics for commercial lithium ion batteries: A study of 24 pouch cells. *J. Power Sources* **2017**, *342*, 589–597. [CrossRef]
59. Rohr, S.; Müller, S.; Baumann, M.; Kerler, M.; Ebert, F.; Kaden, D.; Lienkamp, M. Quantifying Uncertainties in Reusing Lithium-Ion Batteries from Electric Vehicles. *Procedia Manuf.* **2017**, *8*, 603–610. [CrossRef]
60. Devie, A.; Baure, G.; Dubarry, M. Intrinsic Variability in the Degradation of a Batch of Commercial 18650 Lithium-Ion Cells. *Energies* **2018**, *11*, 1031. [CrossRef]
61. Petz, D.; Mühlbauer, M.J.; Baran, V.; Frost, M.; Schökel, A.; Paulmann, C.; Chen, Y.; Garcés, D.; Senyshyn, A. Lithium heterogeneities in cylinder-type Li ion batteries—fatigue induced by cycling. *J. Power Sources* **2019**. [CrossRef]
62. Fath, J.P.; Dragicevic, D.; Bittel, L.; Nuhic, A.; Sieg, J.; Hahn, S.; Alsheimer, L.; Spier, B.; Wetzel, T. Quantification of aging mechanisms and inhomogeneity in cycled lithium-ion cells by differential voltage analysis. *J. Energy Storage* **2019**, *25*. [CrossRef]

© 2020 by the authors. Licensee MDPI, Basel, Switzerland. This article is an open access article distributed under the terms and conditions of the Creative Commons Attribution (CC BY) license (http://creativecommons.org/licenses/by/4.0/).

energies

MDPI

Article

Research on the Fault Diagnosis of a Polymer Electrolyte Membrane Fuel Cell System

Feng Han, Ying Tian *, Qiang Zou and Xin Zhang

Beijing Key Laboratory of Powertrain for New Energy Vehicle, School of Mechanical, Electronic and Control Engineering, Beijing Jiaotong University, Beijing 100044, China; 17121264@bjtu.edu.cn (F.H.); 17116377@bjtu.edu.cn (Q.Z.); zhangxin@bjtu.edu.cn (X.Z.)

* Correspondence: ytian1@bjtu.edu.cn; Tel.: +86-010-5168-8408

Received: 3 March 2020; Accepted: 10 May 2020; Published: 16 May 2020

Abstract: In this work, the possibilistic fuzzy C-means clustering artificial bee colony support vector machine (PFCM-ABC-SVM) method is proposed and applied for the fault diagnosis of a polymer electrolyte membrane (PEM) fuel cell system. The innovation of this method is that it can filter data with Gaussian noise and diagnose faults under dynamic conditions, and the amplitude of characteristic parameters is reduced to ±10%. Under dynamic conditions with Gaussian noise, the faults of the PEM fuel cell system are simulated and the original dataset is established. The possibilistic fuzzy C-means (PFCM) algorithm is used to filter samples with membership and typicality less than 90% and to optimize the original dataset. The artificial bee colony (ABC) algorithm is used to optimize the penalty factor C and kernel function parameter g. Finally, the optimized support vector machine (SVM) model is used to diagnose the faults of the PEM fuel cell system. To illustrate the results of the fault diagnosis, a nonlinear PEM fuel cell simulator model which has been presented in the literature is used. In addition, the PFCM-ABC-SVM method is compared with other methods. The result shows that the method can diagnose faults in a PEM fuel cell system effectively and the accuracy of the testing set sample is up to 98.51%. When solving small-sized, nonlinear, high-dimensional problems, the PFCM-ABC-SVM method can improve the accuracy of fault diagnosis.

Keywords: fault diagnosis; PEM fuel cell system; PFCM-ABC-SVM

1. Introduction

Hydrogen energy is one of the most important green energy sources. The polymer electrolyte membrane (PEM) fuel cell system can directly convert hydrogen energy into electrical energy through an electrochemical reaction and generate water and heat with minimal pollution [1]. The PEM fuel cell system is a multi-input and-output nonlinear system, and there are some auxiliary elements such as compressors, supply manifolds, return manifolds, compressors, valves, etc. For this reason, the PEM fuel cell system is vulnerable to different sets of faults that can imply its temporal or permanent damage [2]. Therefore, fault diagnosis methods are important to reduce this vulnerability as much as possible.

Considering whether the model is necessary, the diagnosis methods can be classified into two general types, i.e., model- and non-model-based methods [3,4]. The model-based method needs to develop a model to simulate the behavior of the monitored system [4] and, generally, it is performed mostly via residual evaluation, followed by a residual inference for possible fault occurrence detection [5]. Escobet and Feroldi et al. [6,7] proposed a model-based fault diagnosis methodology based on the relative fault sensitivity, and the diagnosis methodology correctly diagnosed the simulated faults in contrast with other methodologies using binary signature matrix of analytical residuals and faults. Rosich et al. [8] designed a subset of consistency relations and residual generators for a fuel cell system. Lira et al. [9] proposed a linear parameter varying (LPV) model-based fault diagnosis methodology based on the relative fault sensitivity. Laghrouche et al. [10] presented an observer-based fault reconstruction method for PEM fuel cells and the method extended the results of a class of nonlinear uncertain systems with Lipschitz nonlinearities. Damiano et al. [11] proposed the Takagi–Sugeno (TS) interval observers to solve the problem of robust fault diagnosis of PEM fuel cells. Kamal et al. [12] proposed a model-based fault detection and isolation (FDI) and found that the residual was sensitive to the fault. Steiner et al. [13] proposed the model-based diagnosis method which was based on a comparison between measured and calculated voltages and pressure drops by an Elman neural network.

A non-model-based method can detect and identify the fault through human knowledge or qualitative reasoning techniques based on a set of input and output data [3,4]. Three types of non-model-based methods include the artificial intelligence method, the statistical method, and the signal processing method. Antoni et al. [14] proposed a fault diagnosis methodology termed visual block fuzzy inductive reasoning and applied it to a fuel cell system. Shao et al. [15] proposed the artificial neural network (ANN) ensemble method based on back-propagating ANN and the Lagrange multiplier method to improve the stability and reliability of the PEM fuel cell systems. Damour et al. [16] proposed a signal-based diagnosis method, based on empirical mode decomposition (EMD). The method did not require any excitation signal or stabilization period as compared with the EIS-based method. Zheng et al. [17] used the electrochemical impedance spectroscopy (EIS) as a basis tool and proposed the double fuzzy method consisting of fuzzy clustering and fuzzy logic to mine diagnostic rules from the experimental data automatically. Ibrahim et al. [18] proposed a diagnosis method using signal-based pattern recognition. All information needed to locate the faults was drawn from the recorded fuel cell output voltage, since certain phenomena leave characteristic patterns in the voltage signal. Pahon et al. [19] used the wavelet transform to identify different patterns or fault signatures and proposed the signal-based pattern recognition approach. Mohammadi et al. [20] used a two-layer feed-forward artificial neural network and developed a reliable fault identification and localization tool for a proton exchange membrane fuel cell. Silva et al. [21] proposed a methodology based on adaptive neuro-fuzzy inference systems (ANFIS) which used, as input, the measures of the fuel cell output voltage during operation. Li et al. [22] proposed a nonlinear multivariable model of a PEM fuel cell system based on support vector regression (SVR) and used an effective informed adaptive particle swarm optimization algorithm to tune the hyper-parameters of the support vector regression (SVR) model. Pei P. et al. [23] reviewed the effect variables of pressure drop and the diagnosis method based on pressure drop was considered to be an online water fault diagnosis. Zhao, [24] proposed a fault diagnosis method based on multi-sensor signals and principle component analysis to improve the fuel cell system performance. Huang, [25] proposed a diagnostic method combining C4.5-based decision tree with a fault diagnosis expert system to solve the fault diagnosis of a fuel cell engine. Liu, [26] proposed a fault diagnosis method which combined an extreme learning machine and the Dempster–Shafer evidence theory to diagnose the faults in a PEM fuel cell system. Bougatef [27] designed the unknown input observer for a delayed LPV model to deal with the fault estimation of actuator fault for a PEM fuel cell. In addition, Wang [28,29] developed a composite support material which possessed intrinsic protonic conductivity and improved electronic conductivity together with the optimization of the microstructure structures. Wilberforce [30–33] researched the effect of humidification of reactive gases and bipolar plate geometry design on the performance of a proton exchange membrane fuel cell.

The PEM fuel cell system is a multi-input and multi-output nonlinear system. The fuel cell stack needs to be integrated with several auxiliary components to form a complete PEM fuel cell system. Therefore, the PEM fuel cell system contains the fuel cell stack, the reactant flow subsystem, heat and temperature subsystem, water management subsystem, power management system and the fuel processor subsystem. The reactant flow subsystem contains the hydrogen supply subsystem and the air supply subsystem. When the source of actual noise is multiplex, Gaussian noise can simulate actual noise well. The probability density of Gaussian noise follows the standard normal distribution. In this paper, Gaussian noise is used to simulate the interference in the PEM fuel cell model, and the fault diagnosis effect of the method can also be verified. In this paper, when the variance of Gaussian noise is 1.0, 0.5, 0.2, 0.1 respectively, the amplitude of characteristic parameters is reduced to ±10%. By simulating the fault scenarios of the PEM fuel cell system, the original dataset is established with eight diagnosis variables. The possibilistic fuzzy C-means clustering artificial bee colony support vector machine (PFCM-ABC-SVM) method is used to diagnose the faults in the PEM fuel cell system.

2. The Relevant Theory of the PFCM-ABC-SVM Method

2.1. PFCM Algorithm

In the fuzzy C-means clustering (FCM) algorithm, the membership value of each sample point must be 1.0, therefore, it is sensitive to noise points and the classification result is not accurate. The PCM algorithm is sensitive to the initial cluster center, and only when the cluster centers are the same can the global optimal solution be obtained, which causes cluster consistency problems [34]. To solve the shortcomings of the above algorithm, Pal proposed the possibilistic fuzzy C-means (PFCM) algorithm based on the above algorithm [35,36]. The PFCM algorithm overcomes the sensitivity of the FCM algorithm to noise and the sensitivity of the PCM algorithm to initial clustering centers. Additionally, the PFCM algorithm improves the accuracy of classification results. The objective function of the PFCM algorithm is as follows:

$$J(U,T,V;X) = \sum_{i=1}^{c}\sum_{j=1}^{n}(au_{ij}^{m}+bt_{ij}^{p})d_{ij}^{2} + \sum_{i=1}^{c}\eta_i\sum_{j=1}^{n}(1-t_{ij})^{p} \tag{1}$$

where $1 \leq i \leq c, 1 \leq j \leq n;\ \sum_{i=1}^{c} u_{ij} = 1$; a and b define the relative importance of fuzzy membership and typicality values in the objective function, $a > 0, b > 0$; m and p are the fuzzy parameters; $d_{ij} = \|x_j - v_i\|$ is the Euclidean distance from sample point x_j to v_i; c is the number of cluster centers; and n is the number of sample points.

The penalty coefficient of the PFCM algorithm is as follows:

$$\eta_i = K\frac{\sum_{j=1}^{n}(au_{ij}^{m}+bt_{ij}^{p})d_{ij}^{2}}{\sum_{j=1}^{n}(au_{ij}^{m}+bt_{ij}^{p})}, K > 0 \tag{2}$$

where η_i is the penalty coefficient; generally, $K = 1$, from the optimal solution of Equation (1), get the following Equations (3)–(5):

$$u_{ij} = \frac{1}{\sum_{k=1}^{c}\left(\frac{d_{ij}}{d_{kj}}\right)^{2/(m-1)}} \tag{3}$$

$$t_{ij} = \frac{1}{1+\left(\frac{bd_{ij}^{2}}{\eta_i}\right)^{1/(p-1)}} \tag{4}$$

$$v_i = \frac{\sum_{j=1}^{n}(au_{ij}^{m}+bt_{ij}^{p})x_j}{\sum_{j=1}^{n}(au_{ij}^{m}+bt_{ij}^{p})} \tag{5}$$

The steps of the PFCM algorithm are as follows:

Step 1 Set the fuzzy parameters, set the terminating threshold ε, set the maximum number of iterations L, set the number of initial iterations l, initialize the cluster center $V^{(0)}$, initialize the membership matrix $U^{(0)}$, and initialize the typicality matrix $T^{(0)}$;

Step 2 According to Formula (2), calculate the penalty coefficient η_i;

Step 3 According to Formula (3), calculate and update the membership matrix $u_{ij}^{(l+1)}$;

$$u_{ij}^{(l+1)} = \frac{1}{\sum_{k=1}^{c} \left(\frac{d_{ij}^{(l+1)}}{d_{kj}^{(l+1)}}\right)^{2/(m-1)}} \tag{6}$$

Step 4 According to Formula (4), calculate and update the typicality matrix $t_{ij}^{(l+1)}$;

$$t_{ij}^{(l+1)} = \frac{1}{1 + \left(\frac{b(d_{ij}^{(l+1)})^2}{\eta_i}\right)^{1/(p-1)}} \tag{7}$$

Step 5 According to Formula (5), calculate and update the cluster center matrix $v_i^{(l+1)}$;

$$v_i^{(l+1)} = \frac{\sum_{j=1}^{n} \left(a(u_{ij}^{(l+1)})^m + (bt_{ij}^{(l+1)})^p\right) x_j}{\sum_{j=1}^{n} \left[a(u_{ij}^{(l+1)})^m + b(t_{ij}^{(l+1)})^p\right]} \tag{8}$$

Step 6 If $\|V^{(l+1)} - V^l\| < \varepsilon$ or $L < l$, output the cluster center, the membership matrix and the typicality matrix; if not, make $l = l + 1$, skip to Step 2. The flow chart of the PFCM algorithm is shown in Figure 1.

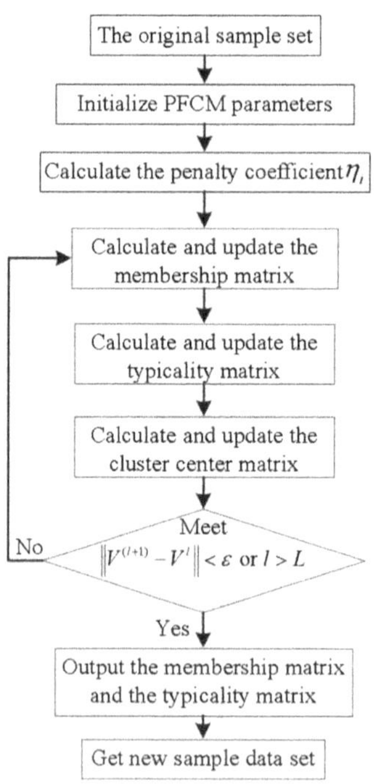

Figure 1. The flow chart of the possibilistic fuzzy C-means (PFCM) algorithm.

2.2. Multi-Parameter Optimization of Support Vector Machine(SVM)

Support Vector Machine (SVM) is a new general machine learning method presented by Vapnik [37]. Traditional statistical research is in the case of sufficient samples or assuming an infinite number of samples, but in actual problems, there are few samples. According to the principle of structural risk minimization and the Vapnik-Chervonenkis(VC) dimension theory, SVM can combine the complexity of the model with the learning ability, find the optimal solution, and obtain better generalization ability. The classification theory of SVM is developed from the problem of linear separable binary classification. In the process of classification, the optimal classification hyperplane is constructed. The training samples are classified correctly according to the principle of least empirical risk, and the maximum classification interval is required to ensure the minimum confidence range. It has advantages in solving small-sized, nonlinear, high-dimensional problems [38–40]. The objective function of the SVM optimization problem is as follows:

$$\begin{cases} max\ L(\alpha) = \sum\limits_{i=1}^{n} \alpha_i - \frac{1}{2} \sum\limits_{i,j=n}^{n} \alpha_i \alpha_j y_i y_j K(x_i, x_j) \\ \sum\limits_{i=1}^{N} \alpha_i y_i = 0\ 0 \le \alpha_i \le Ci = 1, 2 \cdots N \end{cases} \tag{9}$$

where $\forall x_i, x_j \in R^n$; α_i is the Lagrange multiplier; C is the penalty factor; and $K(x_i, x_j)$ is the kernel function which can transform a low-dimensional vector into a high-dimensional inner product.

The corresponding optimal classification function is as follows:

$$f(x) = \text{sgn}[\sum_{i=1}^{n} \alpha_i^* y_i K(x_i, x_j) + b^*] \tag{10}$$

where α^* is the optimal solution; $b^* = y_i - \sum\limits_{i=1}^{N} \alpha_i^* y_i K(x_i, x_j)$.

In the above optimization problem, it is necessary to determine the kernel function $K(x_i, x_j)$. There are four kinds of kernel functions commonly used in SVM as follows: linear kernel function $K(x_i, y_i) = x_i \cdot y_i$; polynomial kernel function $K(x_i, y_i) = [(x_i \cdot y_i) + b]^d$; hyperbolic tangent kernels function $K(x_i, y_i) = \tanh[v(x_i \cdot y_i) + c]$; and radial basis kernel function $K(x_i, y_i) = \exp(-g\|x_i - y_i\|^2)$, where g is the kernel parameter.

Many studies show that radial basis kernel function is a better choice when there is not enough prior knowledge [41]. The radial basis kernel function is used as the kernel function in SVM. After that, the kernel function parameter g and the penalty factor C should be selected which are significant to establish the optimized SVM model.

The artificial bee colony (ABC) algorithm is an intelligent optimization algorithm inspired by biological behaviors proposed by Karaboga [42,43]. It mainly solves practical problems by simulating bees collecting honey. The ABC algorithm finds the global optimal solution through the local optimization behavior of bees. It is often used to solve multi-parameter optimization problems [44]. In the paper, the ABC algorithm is used to obtain the optimal penalty factor C and kernel parameter g. Compared with the genetic algorithm (GA), and particle swarm optimization algorithm (PSO), the ABC algorithm has the advantages of strong global optimization ability and few control parameters.

The multi-parameter optimization of SVM is as follows:

Step 1 Initialize the parameters in the ABC algorithm and SVM, i.e., the number of bee colonies, the number of honey sources, the maximum search number of honey sources (*Limit*), the current search number of honey sources, the maximum number of iterations (MaxIter), the search range of penalty factors C, and the search range of kernel function parameter g.

Step 2 Select the fitness function in the ABC algorithm. The purpose of optimizing the SVM parameters is to improve the accuracy of fault classification. The solution of the optimization problem can be regarded as a process for the bee to find the honey source. The fitness function is as follows:

$$fitness_i = \begin{cases} \frac{1}{1+f_i} & (f_i \geq 0) \\ 1 + \left|f_i\right| & (f_i < 0) \end{cases} \quad (11)$$

where, $fitness_i$ is the fitness value of the i-th parameter, and f_i is the objective function value of the i-th honey source.

Step 3 Employed bees search for the neighborhood of the current honey source according to Formula (12) and calculate the fitness of the new honey source according to Formula (11). If the fitness value of the new honey source is better than that of the original honey source, the new honey source position replaces the original honey source position, otherwise the original honey source remains unchanged.

$$\text{new\_}x_{id} = x_{id} + R_{ij}(x_{id} - x_{kd}) \quad (12)$$

where, new_x_{id} is the value of the d-th dimension in the i-th new honey source; x_{id} is the value of the d-th dimension in the i-th original honey source; R is a random number in [−1, 1]; and k is any honey source except the i-th honey source.

Step 4 After the employed bees complete the global search, onlooker bees select the honey source according to Formula (13), and then search for the neighborhood to get the new honey source according to Formula (12). If the fitness value of the new honey source is better than that of the original honey source, the new honey source position replaces the original honey source position, otherwise the original honey source remains unchanged.

$$P_i = \frac{fitness_i}{\sum\limits_{n=1}^{N} fitness_n} \quad (13)$$

where, P_i is the probability that the i-th honey source is selected, $fitness_i$ is the fitness value of the i-th honey source, and N is the total number of honey sources.

Step 5 Judge whether the current search number of honey sources is bigger than the maximum search number of honey sources. If it is bigger, generate a new honey source according to Formula (14).

$$x_{ij} = \min x_{ij} + R_{ij}(\max x_{ij} - \min x_{ij}) \quad (14)$$

where, x_{ij} is the value of the j-th dimension of the i-th honey source, $j \in \{1, 2\}$.

Step 6 Record the current optimal honey source and judge whether the termination condition is met. If the termination condition is met, skip to Step 7, otherwise skip to Step 3.

Step 7 Get the global optimal honey sources, which are the penalty factor C and kernel parameter g, to establish the optimized SVM model.

3. Fault Simulation of the PEM Fuel Cell System

The PEM fuel cell system can directly convert chemical energy into electricity through electrochemical reaction and produce water and heat at the same time. The PEM fuel cell simulator model uses controller strategies and nonlinear models presented by Pukrushpan et al. [1]. It is assumed that the system is in a constant temperature state, ignoring the influence of the double charge layer, and it is regarded as a rapid dynamic behavior near the electrode/electrolyte. Parameters commonly used in the PEM fuel cell simulator model are described in Table 1.

Table 1. Parameters commonly used in the polymer electrolyte membrane (PEM) fuel cell simulator model.

Symbol	Parameter	Value
n	Number of cells in stack	381
A_{fc}	Fuel cell active area	280 cm^2
d_c	Compressor diameter	0.2286 m
J_{cp}	Compressor and motor inertia	5×10^{-5} kgm^2
V_{ca}	Cathode volume	0.01 m^3
V_{sm}	Supply manifold volume	0.02 m^3
V_{rm}	Return manifold volume	0.005 m^3
$k_{sm,out}$	Supply manifold outlet orifice constant	0.36293×10^{-5} kg/(s·Pa)
$k_{ca,out}$	Cathode outlet orifice constant	0.21773×10^{-5} kg/(s·Pa)
k_v	Motor electric constant	0.0153 V/(rad/s)
k_t	Motor torque constant	0.0225 N·m/A
R_{cm}	Compressor motor resistance	1.2 Ω

The PEM fuel cell simulator model was established by Pukrushpan, J.T. in [1] and some parameters of the PEM fuel cell simulator model are from [45–47] based on actual product parameters. The PEM fuel cell simulator model is widely used for the fault diagnosis of the PEM fuel cell system [7,8,12,14], and represents the75kW fuel cell system with 381 cells. The PEM fuel cell simulator model includes the fuel cell stack model, the compressor model, the supply manifold model, the return manifold model, the air cooler model, and the humidifier model. The PEM fuel cell system block diagram is shown in Figure 2. The five faults are partially quoted from the literature [1,14] and the amplitude of characteristic parameters is reduced to ±10%. The faults in the PEM fuel cell simulator model are described in Table 2.

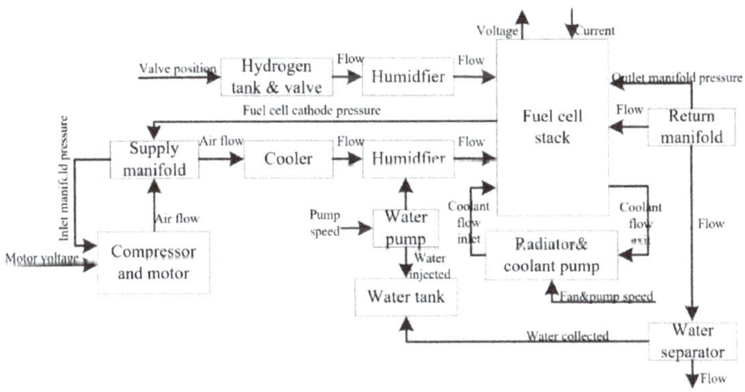

Figure 2. The PEM fuel cell system block diagram [1].

Table 2. Faults in the PEM fuel cell simulator model.

Fault ID	Fault Description	Type	Magnitude
Fault0	Normal state	Parametric unchanged	0
Fault1	There is suddenly an increase of friction in the mechanical component part of the compressor	Parametric abrupt	10% of increment
Fault2	The compressor motor suffers an overheating	Parametric abrupt	10% of increment
Fault3	The fluid resistance increases due to water blocking the channels or flooding in the diffusion layer	Parametric abrupt	10% of reduction of the water flow
Fault4	Air leak in the air supply manifold	Parametric abrupt	10% of reduction of the air flow

The characteristic parameters remain unchanged and Fault0 is in normal state. Equations (15)–(21) [1] are used to simulate Fault1–Fault4. According to the thermodynamic formula, the compressor torque τ_{cp} is expressed as:

$$\tau_{cp} = \frac{C_p}{\omega_{cp}}\frac{T_{atm}}{\eta_{cp}}\left[\left(\frac{p_{sm}}{p_{atm}}\right)^{\frac{\gamma-1}{\gamma}} - 1\right]W_{cp} \tag{15}$$

where, τ_{cp} is the torque needed to drive the compressor, C_p is the specific heat capacity of air, ω_{cp} is the compressor speed, η_{cp} is the compressor efficiency, p_{sm} is the supply manifold pressure, p_{atm} is the pressure of the air, T_{atm} is the temperature of the air, γ is the ratio of the specific heats of the air, and W_{cp} is the air mass flow of compressor.

A lumped rotational parameter model with inertia is used to represent the compressor speed:

$$J_{cp}\frac{d\omega_{cp}}{dt} = (\tau_{cm} - \tau_{cp}) \tag{16}$$

where J_{cp} is the combined inertia of the compressor and the motor, and τ_{cm} is the compressor motor torque input.

The Fault1 state is simulated with the increment Δk_v in the compressor constant k_v. The Fault2 state is simulated with the increment ΔR_{cm} in the compressor motor resistance R_{cm}:

$$\tau_{cm} = \frac{\eta_{cm}k_t}{(R_{cm} + \Delta R_{cm})}\left[v_{cm} - (k_v + \Delta k_v)\omega_{cp}\right] \tag{17}$$

where, η_{cm} is the motor mechanical efficiency, k_t is the motor torque constant, R_{cm} is the compressor motor resistance, ΔR_{cm} is the increment in the compressor motor resistance, k_v is the motor electric constant, and Δk_v is the increment in the motor electric constant.

The maximum mass of the vapor that the gas can hold is calculated from the vapor saturation pressure:

$$m_{v,max,ca} = \frac{p_{sat}V_{ca}}{R_v T_{st}} \tag{18}$$

where, $m_{v,max,ca}$ is the maximum mass of the vapor, p_{sat} is the saturation pressure of the vapor, R_v is the gas constant of the vapor, and T_{st} is the temperature of the stack. If $m_{w,ca} \leq m_{v,\max,ca}$, so $m_{v,ca} = m_{w,ca}$, $m_{l,ca} = 0$; if $m_{w,ca} > m_{v,\max,ca}$,so $m_{v,ca} = m_{v,\max,ca}$,$m_{l,ca} = m_{w,ca} - m_{v,\max,ca}$.

The total cathode pressure is the sum of oxygen, nitrogen, and vapor partial pressure:

$$P_{ca} = P_{O_2,ca} + P_{N_2,ca} + P_{v,ca} = \frac{m_{O_2,ca}R_{O_2}T_{st}}{V_{ca}} + \frac{m_{N_2,ca}R_{N_2}T_{st}}{V_{ca}} + \frac{m_{v,ca}R_v T_{st}}{V_{ca}} \tag{19}$$

where P_{ca} is the cathode pressure; V_{ca} is the cathode volume; $P_{O_2,ca}$, $P_{N_2,ca}$ and $P_{v,ca}$ are the partial pressure of oxygen, nitrogen, and vapor; R_{O_2}, R_{N_2} and R_v are the gas constants of oxygen, nitrogen, and vapor.

Fault3 is simulated with the increment $\Delta k_{ca,out}$ in the cathode outlet orifice constant $k_{ca,out}$:

$$W_{ca,out} = (k_{ca,out} + \Delta k_{ca,out})(p_{ca} - p_{rm}) \tag{20}$$

where, $\Delta k_{ca,out}$ is the increment in the cathode outlet orifice constant, $k_{ca,out}$ is the cathode outlet orifice constant, $W_{ca,out}$ is the air flow in the cathode outlet, p_{ca} is the cathode pressure, and p_{rm} is the return manifold pressure.

Fault 4 is simulated with the increment $\Delta k_{sm,out}$ in the supply manifold outlet orifice constant $k_{sm,out}$:

$$W_{sm,out} = (k_{sm,out} + \Delta k_{sm,out})(p_{sm} - p_{ca}) \tag{21}$$

where, $W_{sm,out}$ is the outlet mass flow, $\Delta k_{sm,out}$ is the increment in the supply manifold outlet orifice constant, and $k_{sm,out}$ is the supply manifold outlet orifice constant.

4. Fault Diagnosis of the PEM Fuel Cell System

In this work, the Gaussian noise with variance of 0.1, 0.2, 0.5, and 1.0 are added to the PEM fuel cell simulator model, respectively. It is difficult to distinguish the Fault 0 to Fault 4 states in Table 2. Signals in a fault state are coupled with signals in other faults. Therefore, the traditional methods cannot diagnose the fault of the PEM fuel cell system effectively.

The Fault 0–4 states are simulated using the PEM fuel cell simulator model in the dynamic condition. Eight diagnostic variables are selected from the PEM fuel cell simulator model, and the eight diagnostic variables are fuel cell current (I_{fc}), fuel cell voltage (V_{fc}), compressor speed (ω_{cm}), compressor outlet pressure ($P_{cm,out}$), compressor motor voltage (V_{cm}), compressor motor current (I_{cm}), hydrogen inlet pressure ($P_{H_2,in}$), and air inlet pressure ($P_{air,in}$). Taking the Fault4 state as an example, Gaussian noise with variance of 1.0 is added to the PEM fuel cell simulator model. The fuel cell current, fuel cell voltage, compressor speed, compressor outlet pressure, compressor motor voltage, compressor motor current, hydrogen inlet pressure, and air inlet pressure change with time, respectively, are shown in Figures 3–10.

In this paper, the Gaussian noise with variance of 0.1, 0.2, 0.5, and 1.0 are added to the PEM fuel cell simulator model, respectively. The PFCM algorithm is used to filter samples with membership and typicality less than 90% and optimize the original dataset. The filtered data is used as the sample dataset. The sample dataset are divided into two groups, one is the training set sample and the other is the testing set sample. The training set sample number is 670, and the testing set sample number is 335. The penalty parameter C and kernel function parameter *g* of SVM are optimized using the ABC algorithm, and then establish the optimized SVM model. The testing set sample is used to test the accuracy of the fault diagnosis method.

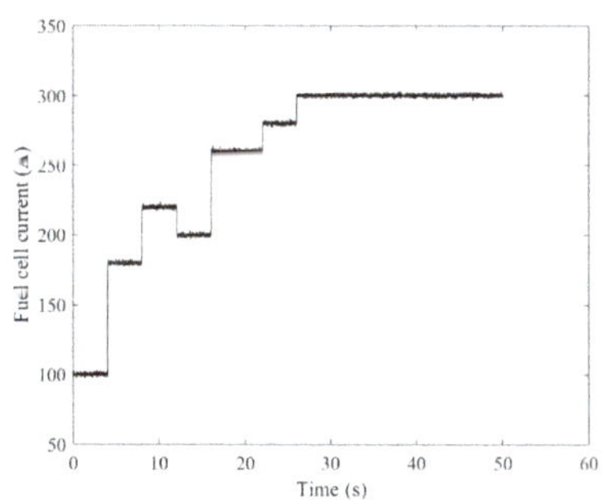

Figure 3. Fuel cell current changes with time.

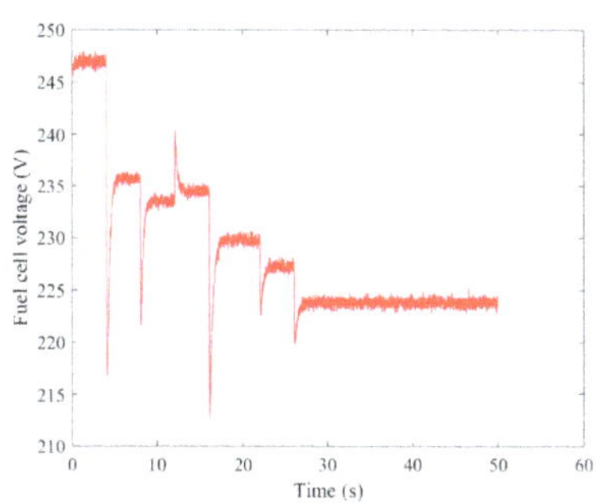

Figure 4. Fuel cell voltage changes with time.

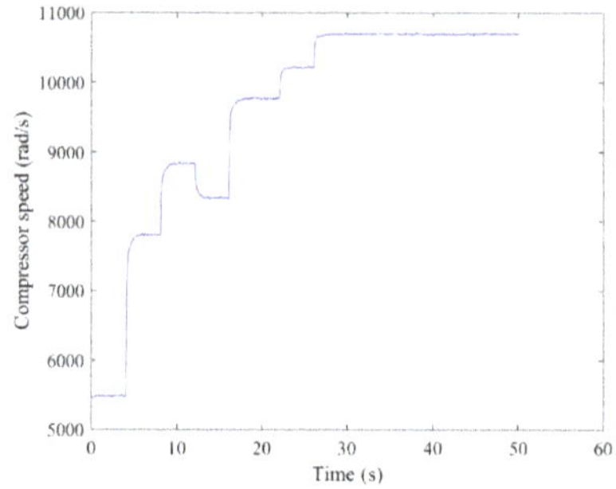

Figure 5. Compressor speed changes with time.

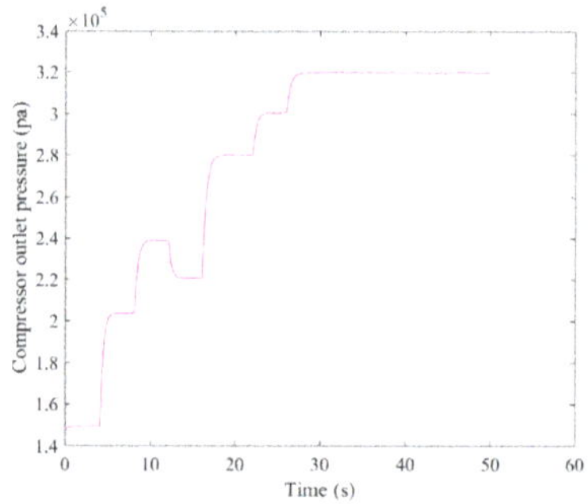

Figure 6. Compressor outlet pressure changes with time.

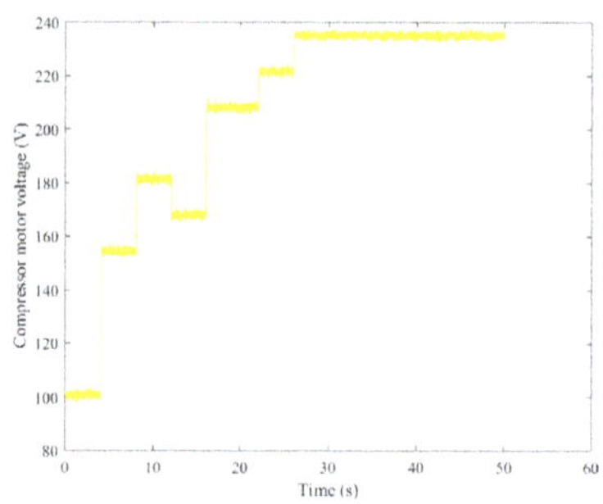

Figure 7. Compressor motor voltage changes with time.

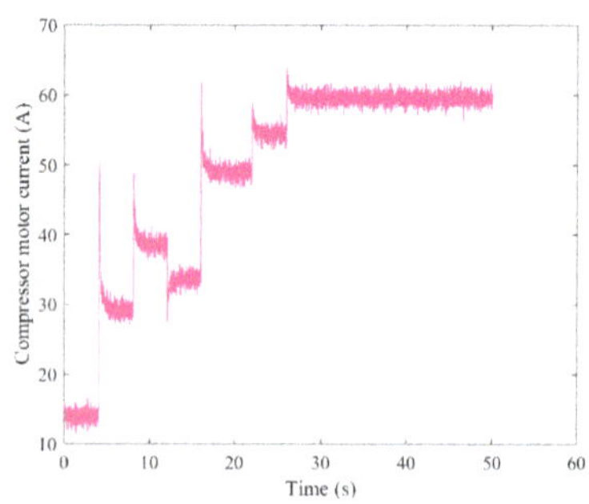

Figure 8. Compressor motor current changes with time.

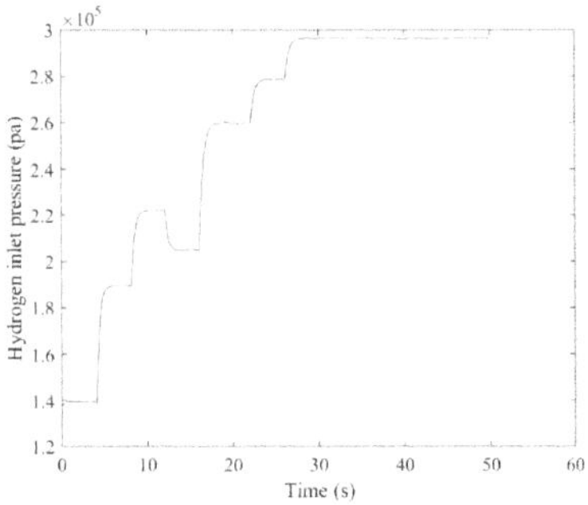

Figure 9. Hydrogen inlet pressure changes with time.

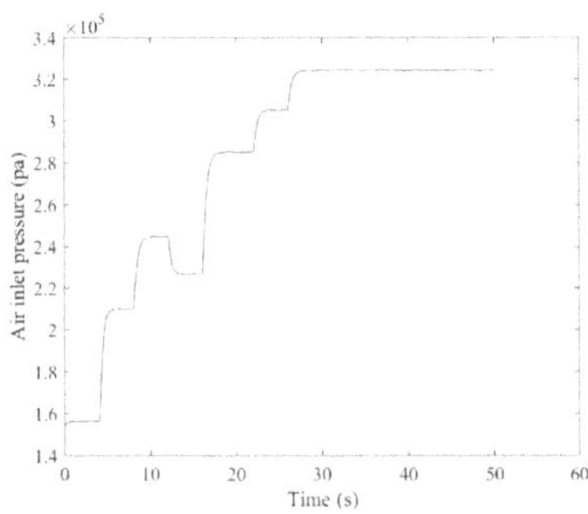

Figure 10. Air inlet pressure changes with time.

The fault diagnosis steps of the fuel cell system based on the PFCM-ABC-SVM method are as follows:

Step 1 Initialize the parameters in the PFCM-ABC-SVM method as follows: set the fuzzy parameters, $m = 2$, $p = 2$; set the terminating threshold $\varepsilon = 10^{-6}$; set the maximum number of iterations $I_t = 100$; set the number of initial iterations $l = 0$; initialize the cluster center $V^{(0)}$, initialize the membership matrix $U^{(0)}$, and initialize the typicality matrix $T^{(0)}$; set the number of bee colonies $n = 20$; set the maximum search number of honey sources $Limit = 100$; set the current search number of honey sources $d = 0$, set the maximum number of iterations $maxIter = 10$; set the search range of penalty factor C: [0.01, 100]; and set the search range of kernel function parameter g: [0.01, 100].

Step 2 Get the original data of the PEM fuel cell system and select eight diagnostic variables. The eight diagnostic variables are fuel cell current, fuel cell voltage, compressor speed, compressor outlet pressure, compressor motor voltage, compressor motor current, hydrogen inlet pressure, and air inlet pressure.

Step 3 Establish the original dataset with eight diagnostic variables and normalize the original dataset using mapminmax Function in Matlab(R2018b).

Step 4 Adapt the PFCM algorithm to eliminate samples with membership and typicality less than 90%, filter the original dataset, and establish the sample dataset.

Step 5 Divide the sample dataset into the training set sample and the testing set sample.

Step 6 Optimize the penalty parameter C and kernel function parameter g of SVM using the ABC algorithm and establish the optimized SVM model.

Step 7 Diagnose faults by the optimized SVM model and obtain the diagnostic result. The fault diagnosis flow chart of the fuel cell system based on PFCM-ABC-SVM method is shown in Figure 11.

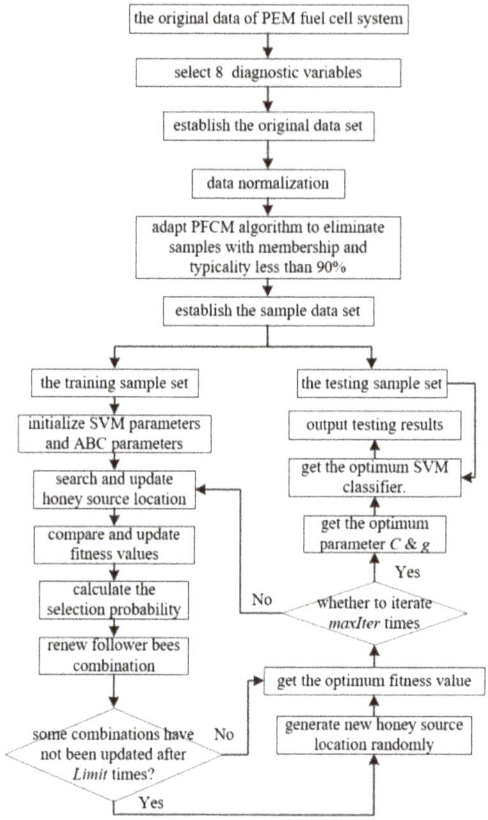

Figure 11. Fault diagnosis flow chart of the fuel cell system based on the PFCM-ABC-SVM method.

When the Gaussian noise variance is 1.0, the PFCM-ABC-SVM method is compared with the GA-SVM, PSO-SVM and ABC-SVM methods. The comparison between the PFCM-ABC-SVM method and the other methods is shown in Table 3. The classification results of the PEMFC-ABC-SVM method when the Gaussian noise variance is 1.0 are shown in Figure 12. For the Fault 0–4 states, the accuracy of the training set sample is 95.67%, and the accuracy of the testing set sample is 92.84% using the PSO-SVM method; the accuracy of the training set sample is 95.82%, and the accuracy of the testing set sample is 94.03% using the ABC-SVM method; the accuracy of the training set sample is 97.46%, and the accuracy of the testing set sample is 97.31% using the PFCM-ABC-SVM method. Therefore, the PFCM-ABC-SVM method can effectively improve the accuracy of fault diagnosis of the PEM fuel cell system. The category label in Figure 12, "0" represents Fault0, "1" represents Fault1, "2" represents Fault2, "3" represents Fault3, and "4" represents Fault4. There are 335 samples in the testing set samples.

Table 3. The comparison between the PFCM-ABC-SVM method and the other methods.

Method.	C	g	Accuracy of the Training Set Sample	Accuracy of the Testing Set Sample
GA-SVM	90.64	41.27	95.97%	92.84%
PSO-SVM	96.84	129.19	95.67%	92.84%
ABC-SVM	82.53	23.31	95.82%	94.03%
PFCM-ABC-SVM	84.28	5.99	97.46%	97.31%

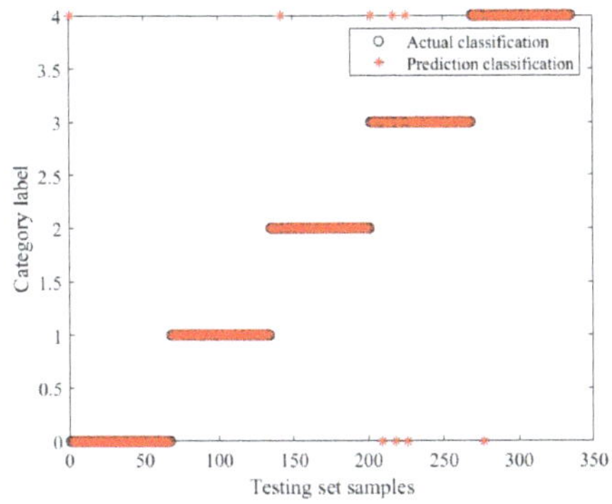

Figure 12. The classification results of the PFCM-ABC-SVM method when the Gaussian noise variance is 1.0.

When the Gaussian noise variance is 0.5, the PFCM-ABC-SVM method is compared with the GA-SVM, PSO-SVM and ABC-SVM methods. The comparison between the PFCM-ABC-SVM method and the other methods is shown in Table 4. The classification results of the PEMFC-ABC-SVM method when the Gaussian noise variance is 0.5 are shown in Figure 13. The accuracy of the training set sample is 98.81% and the accuracy of the testing set sample is 97.91% using the PFCM-ABC-SVM method.

Table 4. The comparison between the PFCM-ABC-SVM method and the other methods.

Method	C	g	Accuracy of the Training Set Sample	Accuracy of the Testing Set Sample
GA-SVM	81.77	39.93	95.67%	92.84%
PSO-SVM	100	128.77	95.52%	93.43%
ABC-SVM	73.02	28.34	95.37%	94.63%
PFCM-ABC-SVM	80.88	15.76	98.81%	97.91%

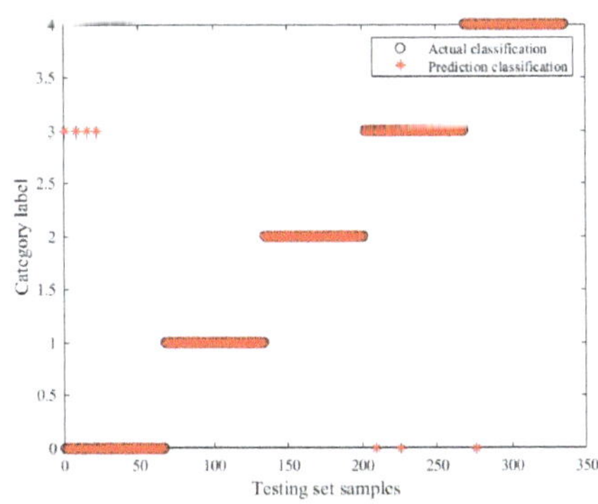

Figure 13. The classification results of the PFCM-ABC-SVM method when the Gaussian noise variance is 0.5.

When the Gaussian noise variance is 0.2, the PFCM-ABC-SVM method is compared with the GA-SVM, PSO-SVM and ABC-SVM methods. The comparison between the PFCM-ABC-SVM method and the other methods is shown in Table 5. The classification results of the PEMFC-ABC-SVM method when the Gaussian noise variance is 0.2 are shown in Figure 14. The accuracy of the training set sample is 98.81%, and the accuracy of the testing set sample is 98.21% using the PFCM-ABC-SVM method.

Table 5. The comparison between the PFCM-ABC-SVM method and the other methods.

Method	*C*	*g*	Accuracy of the Training Set Sample	Accuracy of the Testing Set Sample
GA-SVM	85.50	64.40	96.27%	93.73%
PSO-SVM	100	166.67	96.27%	94.03%
ABC-SVM	92.46	63.77	96.57%	94.33%
PFCM-ABC-SVM	80.14	19.88	98.81%	98.21%

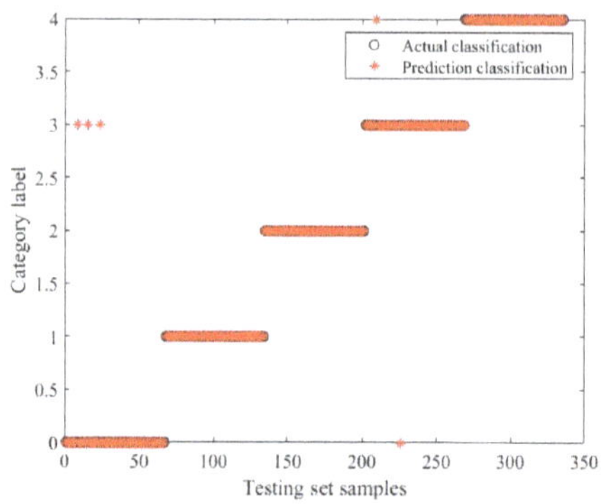

Figure 14. The classification results of the PFCM-ABC-SVM method when the Gaussian noise variance is 0.2.

When the Gaussian noise variance is 0.1, the PFCM-ABC-SVM method is compared with the GA-SVM, PSO-SVM and ABC-SVM methods. The comparison between the PFCM-ABC-SVM method and the other methods is shown in Table 6. The classification results of the PEMFC-ABC-SVM method when the Gaussian noise variance is 0.1 are shown in Figure 15. The accuracy of the training set sample is 98.66%, and the accuracy of the testing set sample is 98.51% using the PFCM-ABC-SVM method.

Table 6. The comparison between the PFCM-ABC-SVM method and the other methods.

Method	*C*	*g*	Accuracy of the Training Set Sample	Accuracy of the Testing Set Sample
GA-SVM	89.03	45.49	95.82%	94.03%
PSO-SVM	100	220.38	96.57%	94.03%
ABC-SVM	99.97	19.36	95.37%	94.63%
PFCM-ABC-SVM	75.18	17.67	98.66%	98.51%

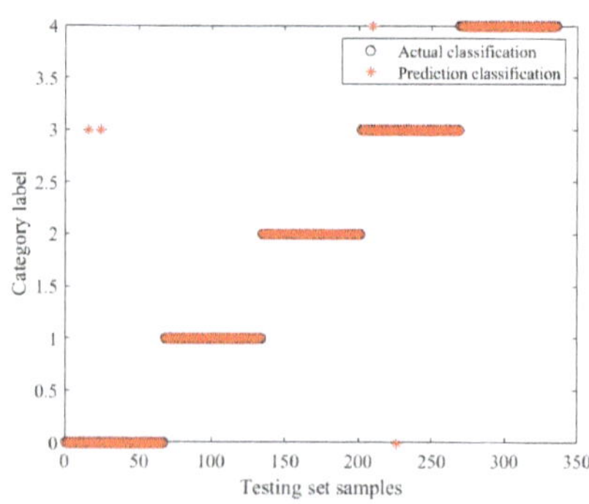

Figure 15. The classification results of the PFCM-ABC-SVM method when the Gaussian noise variance is 0.1.

In order to illustrate the advantages of the PFCM-ABC-SVM method, the GA-SVM, PSO-SVM and ABC-SVM methods are compared with it, in this work. The results of fault diagnosis are shown in Tables 3–6. Under the dynamic conditions with the variance of the Gaussian noise decreasing from 1.0 to 0.1, the accuracy of the testing set sample is as high as 98.51%. Comparing with the other methods, the PFCM-ABC-SVM method has a better effect in fault diagnosis of the PEM fuel cell system.

5. Conclusions

In this work, the PFCM-ABC-SVM method is proposed and verified by the PEM fuel cell simulator model. The Gaussian noise with variance of 0.1, 0.2, 0.5, and 1.0 are added to the PEM fuel cell simulator model, respectively, for fault diagnosis. The PFCM algorithm is used to filter samples with membership and typicality less than 90% and optimize the original dataset. The ABC algorithm is used to optimize the penalty factor C and kernel function parameter g, and the optimized SVM model is used to diagnose the faults of the PEM fuel cell system. The results show that under the dynamic conditions with the variance of the Gaussian noise decreasing from 1 to 0.1, the accuracy of the training set sample increases from 97.46% to 98.81%, and the accuracy of the testing set sample increases from 97.31% to 98.51%. The PFCM-ABC-SVM method is effective to diagnose the faults in the PEM fuel cell system, and it is better than other commonly used methods. The PFCM-ABC-SVM method has an advantage in solving the small-sized, nonlinear, and high-dimensional problems and furthermore, provides references for on-line fault diagnosis of a fuel cell system.

Author Contributions: Methodology, F.H.; writing—original draft preparation, F.H.; software, F.H. and Y.T.; data curation, Y.T.; writing—review and editing, Y.T. and Q.Z.; validation, Q.Z.; supervision, X.Z.; project administration, X.Z.; funding acquisition, X.Z. All authors have read and agreed to the published version of the manuscript.

Funding: This research was funded by the National Key Research and Development Program of China, grant number 2018YFB0105500.

Acknowledgments: The work was sponsored by the National Key Research and Development Program of China—Fuel Cell Bus Electric-Electric Deep Hybrid Power System Platform and Vehicle Development (no. 2018YFB0105500).

Conflicts of Interest: The authors declare no conflicts of interest.

Nomenclature

PFCM	possibilistic fuzzy C-means clustering	ABC	artificial bee colony
SVM	support vector machine	PEM	polymer electrolyte membrane
LPV	linear parameter varying	TS	Takagi–Sugeno
FDI	fault detection and isolation	ANN	artificial neural network
EMD	empirical mode decomposition	EIS	electrochemical impedance spectroscopy
SVR	support vector regression	ANFIS	adaptive neuro-fuzzy inference systems
GS	grid search algorithm	GA	genetic algorithm
PSO	particle swarm optimization algorithm	FCM	fuzzy C-means clustering
PCM	possibilistic C-means clustering	C	penalty factor
g	kernel function parameter	ε	terminating threshold
L	maximum number of iterations	l	the number of initial iterations
η_i	penalty coefficient	αi	Lagrange multiplier
Limit	maximum search number of honey sources	*MaxIter*	maximum number of iterations
τ_{cm}	compressor torque(N·m)	η_{cm}	motor mechanical efficiency
ΔR_{cm}	the increment in the compressor motor resistance(Ω)	Δk_v	the increment in the motor electric constant(V/(rad/s))
t	time(s)	$\Delta k_{ca,out}$	the increment in the cathode outlet orifice constant(kg/(s·Pa))
$W_{ca,out}$	air flow in the cathode outlet(g/s)	p_{ca}	cathode pressure(pa)
p_{rm}	return manifold pressure(pa)	$W_{sm,out}$	outlet mass flow(g/s)

$\Delta k_{sm,out}$	the increment in the supply manifold outlet orifice constant	p_{sm}	supply manifold pressure(pa)
I_{fc}	fuel cell current(A)	V_{fc}	fuel cell voltage(V)
ω_{cm}	compressor speed(rad/s)	$P_{cm,out}$	compressor outlet pressure(pa)
V_{cm}	compressor motor voltage(V)	I_{cm}	compressor motor current(A)
$P_{H_2,in}$	hydrogen inlet pressure(pa)	$P_{air,in}$	air inlet pressure(pa)

References

1. Pukrushpan, J.T.; Stefanopoulou, A.; Peng, H. *Control of Fuel Cell Power Systems: Principles, Modeling, Analysis and Feedback Design*; Springer: Berlin/Heidelberg, Germany, 2004.
2. Benmouna, A.; Becherif, M.; Depernet, D.; Gustin, F. Fault diagnosis methods for proton exchange membrane fuel cell system. *Int. J. Hydrog. Energy* **2017**, *42*, 1534–1543. [CrossRef]
3. Zheng, Z.; Petrone, R.; Pera, M.C.; Hissel, D. A review on non-model based diagnosis methodologies for PEM fuel cell stacks and systems. *Int. J. Hydrog. Energy* **2013**, *38*, 8914–8926. [CrossRef]
4. Petrone, R.; Zheng, Z.; Hissel, D.; Pera, M.C. A review on model-based diagnosis methodologies for PEMFCs. *Int. J. Hydrog. Energy* **2013**, *38*, 7077–7091. [CrossRef]
5. Isermann, R. Supervision, fault-detection and fault-diagnosis methods-an introduction. *Control Eng. Pract.* **1997**, *5*, 639–652. [CrossRef]
6. Escobet, T.; Feroldi, D.; Lira, S.; Puig, V. Model-based fault diagnosis in PEM fuel cell systems. *J. Power Sources* **2009**, *192*, 216–223. [CrossRef]
7. Feroldi, D. Fault Diagnosis and Fault Tolerant Control of PEM Fuel Cell Systems. In *PEM Fuel Cells with Bio-Ethanol Processor Systems: A Multidisciplinary Study of Modelling, Simulation, Fault Diagnosis and Advanced Control*; Springer: Berlin/Heidelberg, Germany, 2012; pp. 185–206.
8. Rosich, A.; Sarrate, R.; Nejjari, F. On-line model-based fault detection and isolation for PEM fuel cell stack systems. *Appl. Math. Model.* **2014**, *38*, 2744–2757. [CrossRef]
9. Lira, S.; Puig, V.; Quevedo, J.; Husar, A. LPV Model-Based Fault Diagnosis Using Relative Fault Sensitivity Signature Approach in a PEM Fuel Cell. In Proceedings of the 18th Mediterranean Conference on Control & Automation Congress Palace Hotel Marrakech, Marrakech, Morocco, 23–25 June 2010.
10. Laghrouche, S.; Liu, J.; Ahmed, F.S.; Harmouche, M. Adaptive second-order sliding mode observer-based fault reconstruction for PEM fuel cell air-feed system. *IEEE Trans. Control Syst. Technol.* **2015**, *23*, 1098–1109. [CrossRef]
11. Rotondo, D.; Fernandez, R.M.; Tornil, S.; Blesa, J. Robust fault diagnosis of proton exchange membrane fuel cells using a Takagi-Sugeno interval observer approach. *Int. J. Hydrog. Energy* **2016**, *41*, 2875–2886. [CrossRef]
12. Kamal, M.M.; Yu, D.W.; Yu, D.L. Fault detection and isolation for PEM fuel cell stack with independent RBF model. *Eng. Appl. Artif. Intell.* **2014**, *28*, 52–63. [CrossRef]
13. Steiner, N.Y.; Hissel, D.; Mocoteguy, P.; Candusso, D. Diagnosis of polymer electrolyte fuel cells failure modes (flooding & drying out) by neural networks modeling. *Int. J. Hydrog. Energy* **2011**, *36*, 3067–3075.
14. Antoni, E.; Àngela, N.; Francisco, M. PEM fuel cell fault diagnosis via a hybrid methodology based on fuzzy and pattern recognition techniques. *Eng. Appl. Artif. Intell.* **2014**, *36*, 40–53.
15. Shao, M.; Zhu, X.J.; Cao, H.F.; Shen, H.F. An artificial neural network ensemble method for fault diagnosis of proton exchange membrane fuel cell system. *Energy* **2014**, *67*, 268–275. [CrossRef]
16. Damour, C.; Benne, M.; Brigitte, G.P.; Bessafi, M. Polymer electrolyte membrane fuel cell fault diagnosis based on empirical mode decomposition. *J. Power Sources* **2015**, *299*, 596–603. [CrossRef]
17. Zheng, Z.; Pera, M.C.; Hissel, D.; Becherif, M. A double-fuzzy diagnostic methodology dedicated to online fault diagnosis of proton exchange membrane fuel cell stacks. *J. Power Sources* **2014**, *271*, 570–581. [CrossRef]
18. Ibrahim, M.; Antoni, U.; Steiner, N.Y.; Jemei, S. Signal-based diagnostics by wavelet Transform for proton exchange membrane fuel cell. *Energy Proced.* **2015**, *74*, 1508–1516. [CrossRef]
19. Pahon, E.; Steiner, N.Y.; Jemei, S.; Hissel, D. A signal-based method for fast PEMFC diagnosis. *Appl. Energy* **2016**, *165*, 748–758. [CrossRef]
20. Mohammadi, A.; Djerdir, A.; Yousfi, S.N.; Khaburi, D. Advanced diagnosis based on temperature and current density distributions in a single PEMFC. *Int. J. Hydrog. Energy* **2015**, *40*, 15845–15855. [CrossRef]

21. Silva, R.E.; Gouriveau, R.; Jeme, S.; Hissel, D. Proton exchange membrane fuel cell degradation prediction based on adaptive neuro-fuzzy inference systems. *Int. J. Hydrog. Energy* **2014**, *39*, 11128–11144. [CrossRef]
22. Li, Q.; Chen, W.; Liu, Z.; Guo, A. Nonlinear multivariable modeling of locomotive proton exchange membrane fuel cell system. *Int. J. Hydrog. Energy* **2014**, *39*, 13777–13786. [CrossRef]
23. Pei, P.; Li, Y.; Xu, H. A review on water fault diagnosis of PEMFC associated with the pressure drop. *Appl. Energy* **2016**, *173*, 366–385. [CrossRef]
24. Zhao, X.W.; Xu, L.F.; Li, J.Q.; Fang, C. Faults diagnosis for PEM fuel cell system based on multi-sensor signals and principle component analysis method. *Int. J. Hydrog. Energy* **2017**, *42*, 18524–18531. [CrossRef]
25. Huang, L.; Zeng, Q.; Zhang, R.M. Fuel Cell Engine Fault Diagnosis Expert System based on Decision Tree. In Proceedings of the IEEE 3rd Information Technology, Networking, Electronic and Automation Control Conference (ITNEC), Chengdu, China, 5–17 March 2019.
26. Liu, J.Y.; Li, Q.; Chen, W.R.; Yan, Y. A Fast Fault Diagnosis Method of the PEMFC System Based on Extreme Learning Machine and Dempster-Shafer Evidence Theory. *IEEE Trans. Transp. Electrif.* **2019**, *5*, 271–284. [CrossRef]
27. Bougatef, Z.; Abdelkrim, N.; Aitouche, A.; Abdelkrim, M.N. Fault detection of a PEMFC system based on delayed LPV observer. *Int. J. Hydrog. Energy* **2020**, *45*, 11233–11241. [CrossRef]
28. Wang, X.D.; Liu, G.Y.; Xu, J.Y.; Jiang, J.M.; Huang, M.; Li, Q.F. Microstructure control and performance of electro-catalysis composite materials for oxygen evolution reaction (OER) in proton exchange membrane water electrolysis. *Chin. Sci. Chem.* **2014**, *44*, 1241–1254.
29. Li, P.P.; Wang, M.; Chen, M.; Yang, Z.Y.; Wang, X.D. Influencing factors of electrochemical test of metal bipolar plate of PEMFC. *Power Technol.* **2018**, *42*, 1679–1681.
30. Wilberforce, T.; Ijaodola, O.; Khatib, F.N.; Ogungbemi, E.O. Effect of humidification of reactive gases on the performance of a proton exchange membrane fuel cell. *Sci. Total Environ.* **2019**, *688*, 1016–1035. [CrossRef]
31. Wilberforce, T.; El Hassan, Z.; Ogungbemi, E.; Ijaodola, O. A comprehensive study of the effect of bipolar plate (BP) geometry design on the performance of proton exchange membrane (PEM) fuel cells. *Renew. Sustain. Energy Rev.* **2019**, *111*, 236–260. [CrossRef]
32. Wilberforce, T.; Khatib, F.N.; Ijaodola, O.S.; Ogungbemi, E. Numerical modelling and CFD simulation of a polymer electrolyte membrane (PEM) fuel cell flow channel using an open pore cellular foam material. *Sci. Total Environ.* **2019**, *678*, 728–740. [CrossRef]
33. Wilberforce, T.; El Hassan, Z.; Khatib, F.N.; Al Makky, A. Development of Bi-polar plate design of PEM fuel cell using CFD techniques. *Int. J. Hydrog. Energy* **2017**, *42*, 25663–25685. [CrossRef]
34. Chang, X.; Sun, J. Application of Fuzzy Clustering Based on Particle Swarm Optimization in Data Processing. *Proc. CSU-EPSA* **2015**, *27*, 78–83.
35. Pal, N.R.; Pal, K.; Keller, J.M.; Bezdek, J.C. A Possibilistic Fuzzy c-Means Clustering Algorithm. *IEEE Trans. Fuzzy Syst.* **2005**, *13*, 517–530. [CrossRef]
36. Pal, N.R.; Pal, K.; Keller, J.M. A New Hybrid C-means Clustering Model. In Proceeding of the IEEE International Conference on Fuzzy System, Budapest, Hungary, 25–29 July 2004; IEEE Press: Piscataway, NJ, USA, 2004; pp. 179–184.
37. Vapnik, V.N. *The Nature of Statistical Learning Theory*, 2nd ed.; Springer: New York, NY, USA, 2000.
38. Xiao, Y.; Wang, Y.J.; Ding, Z.T. The Application of Heterogeneous Information Fusion in Misalignment Fault Diagnosis of Wind Turbines. *Energies* **2018**, *11*, 1655. [CrossRef]
39. Wu, D. Gear box fault diagnosis method based on support vector machine. *J. Vib. Meas. Diagn.* **2008**, *28*, 339–342.
40. Azriel, R.; Wechser, H. Pattern recognition: Historical perspective and future directions. *Int. J. Imaging Syst. Technol.* **2000**, *11*, 101–116.
41. Men, H.; Wu, Y.; Gao, Y. Application of Support Vector Machine to Pattern Classification. In Proceedings of the International Conference on Signal Processing, Beijing, China, 26–29 October 2008; IEEE: Piscataway, NJ, USA, 2008; pp. 1612–1615.
42. Karaboga, D. *An Idea Based on Honey Bee Swarm for Numerical Optimization*; Technical Report-TR06; Erciyes University: Kayseri, Turkey, 2005.
43. Karaboga, D.; Basturk, B. On the performance of artificial bee colony (ABC) algorithm. *Appl. Soft Comput.* **2008**, *8*, 687–697. [CrossRef]

44. Karaboga, D.; Akay, B. A Comparative Study of Artificial Bee Colony Algorithm. *Appl. Math. Comput.* **2009**, *214*, 108–132. [CrossRef]
45. Adams, J.; Yang, W.C.; Oglesby, K.; Osborne, K. *The Development of Ford's P2000 Fuel Cell Vehicle*; SAE Paper: Detroit, MI, USA, 2000.
46. Cunningham, J.; Hoffman, M.; Moore, R.; Friedman, D. Requirements for a flexible and realistic air supply model for incorporation into a fuel cell vehicle (FCV) system simulation. *SAE Trans.* **1999**, 3191–3196.
47. Nguyen, T.; White, R. A water and heat management model for proton-exchange-membrane fuel cells. *J. Electrochem. Soc.* **1993**, *140*, 2178–2186. [CrossRef]

© 2020 by the authors. Licensee MDPI, Basel, Switzerland. This article is an open access article distributed under the terms and conditions of the Creative Commons Attribution (CC BY) license (http://creativecommons.org/licenses/by/4.0/).

energies

Article

Modular Isolated DC-DC Converters for Ultra-Fast EV Chargers: A Generalized Modeling and Control Approach †

Mena ElMenshawy * and Ahmed Massoud

Department of Electrical Engineering, Qatar University, Doha 2713, Qatar; Ahmed.massoud@qu.edu.qa
* Correspondence: MAlmenshawy@qu.edu.qa

† This paper is an extended version of our paper published in 2019 2nd International Conference on Smart Grid and Renewable Energy (SGRE), Doha, Qatar, 19–21 November 2019; pp. 1–6.

Received: 5 April 2020; Accepted: 14 May 2020; Published: 17 May 2020

Abstract: Electric Vehicles (EVs) play a significant role in the reduction of CO_2 emissions and other health-threatening air pollutants Accordingly, several research studies are introduced owing to replacing conventional gasoline-powered vehicles with battery-powered EVs. However, the ultra-fast charging (UFC) of the battery pack or the rapid recharging of the battery requires specific demands, including both: the EV battery and the influence on the power grid. In this regard, advanced power electronics technologies are emerging significantly to replace the currently existing gas station infrastructures with the EV charging stations to move from conventional charging (range of hours) to UFC (range of minutes). Among these power electronics conversion systems, the DC-DC conversion stage plays an essential role in supplying energy to the EV via charging the EV's battery. Accordingly, this paper aims to present possible architectures of connecting multiple Dual Active Bridge (DAB) units as the DC-DC stage of the EV fast charger and study their Small-Signal Modeling (SSM) and their control scheme. These are, namely, Input-Series Output-Series (ISOS), Input-Series Output-Parallel (ISOP), Input-Parallel Output-Parallel (IPOP), and Input-Parallel Output-Series (IPOS). The control scheme for each system is studied through controlling the output filter inductor current such that the current profile is based on Reflex Charging (RC). The main contribution of this paper can be highlighted in providing generalized SSM as well as providing a generalized control approach for the Input-Series Input-Parallel Output-Series Output-Parallel (ISIP-OSOP) connection. The generalized model is verified with three different architectures. The control strategy for each architecture is studied to ensure equal power sharing, where simulation results are provided to elucidate the presented concept considering a three-module ISOS, IPOP, ISOP, and IPOS DC-DC converters.

Keywords: ultra-fast charging; multimodule DC-DC converters; dual active bridge DC-DC converter; full-bridge phase-shift DC-DC converter; input-series output-series; input-series output-parallel; input-parallel output-parallel; input-parallel output-series; input-series input-parallel output-series output-parallel

1. Introduction

Due to high fossil fuel prices, CO2 emissions and other health-threatening emissions, environmental awareness has shown a high interest in Electric Vehicles (EVs) [1]. Replacing conventional gasoline-powered vehicles with battery-powered EVs is expected to reduce fuel consumption significantly [2]. As a response to the ecological issues, a movement towards electrification is taking place in the transport sector. Many research studies have been provided to support this movement in [3–6]. To compete with conventional combustion engine vehicles and ensure the wide adoption of EVs, significant efforts are still needed. One of the main issues limiting the widespread penetration of

EVs is the long charging duration that usually exceeds 30 min [7]. In other words, one of the challenging problems limiting EVs' utilization is the lack of efficient and fast charging capabilities. Accordingly, Ultra-Fast Charging (UFC) is a necessary concept to be investigated to allow massive integration for EVs in the market, particularly when concerning long-distance travel [2]. In this regard, industrial and academic fields have been pushed to explore the Ultra-Fast EV Charging (UF-EVC) concept. This is due to the fact that fast-charging stations allow the recharging process to be done in a few minutes and hence overcoming EVs limitation problem. Studies have shown that the charging process affects EVs' adoption significantly. Therefore, future EV charging stations should have the capability of charging EVs from 10% State-of-Charge (SoC) to 90% SoC in few minutes [8–10]. Nowadays, battery technologies are evolving, enabling the EV's battery to be charged at higher power levels. Currently, offboard fast-charging stations are provided to speed up the charging process [10–14].

Conventionally, the charging methods are categorized based on the charging speed and power into three different levels, which are: level 1, level 2, and level 3 (also termed as the DC fast charging) [15]. Shortly, a new battery technology that reduces the charging time to 3 min is expected to be commercialized [15]. This introduces a new charging level termed the ultra-fast/super-fast charging level, which is different from the present DC fast charging that is commonly used in industrial and academic fields 15. Table 1 shows the electrical specifications for the EV charging stations [15].

Table 1. Electrical vehicle (EV) charging stations' electrical specifications.

Charging Level	Input Voltage	Phase	Power	Time
Level 1	120 VAC	(Single) AC	1.5–2 kW	10–14 h
Level 2	208/240 VAC	(Single) AC	7–20 kW	1–3 h
Level 3 (DC Fast Charging)	480 VAC	(Three) DC	50–100 kW	12–24 min
Ultra-fast/Super-fast Charging	400 VAC or Medium Voltage ranging from 10 − 40 kV	(Three) DC	300–700 kW	< 3 min

To enable UFC for EVs, advanced power electronics converter technologies play an essential role in terms of both flexibility and efficiency. Accordingly, this paper focuses on high-power DC-DC converters that can be employed in UFC stations. As mentioned earlier, one of the main obstacles limiting the broad adoption of EVs is the long charging time. To avoid this problem, UFCs should be designed at a higher power level. Toward this direction, BMW and Porsche, as a part of the Fast Charge Consortium (a group of industrial companies engaged in the "Fast Charge" research project), have demonstrated the latest super-fast EV charger that can considerably charge EVs faster than the present charging methods [16]. Similar systems are demoed by the two companies, where the electric prototype for Porsche is capable of charging at a power level of 400 kW in less than 3 min for the first 100 km with a battery capacity of 90 kWh. While the experimental system for BMW is capable of charging at a power level of 350 kW in 15 min (10% to 80% SoC) [16]. The charging process is done using a DC-DC converter that converts a charging station input voltage of 800 V to a lower voltage of 400 V [17]. The latest prototypes by both companies can work up to 900 and 500 A, meaning that it can charge at a power level of 450 kW, which is three to nine times the capacity of the existing up-to-date DC fast-charging stations [16,17]. This capacity increase significantly reduces the duration of the charging time. Tesla has shown significant efforts to introduce fast charging for EVs. However, the current cars presented by Tesla can only charge at a maximum of 120 kW. Tesla will add enhanced superchargers in 2020. Nonetheless, the charging speed of those cars is boosted to double the charging rate of the current vehicles [16].

The Fast Charge research group is investigating the required technical specifications in terms of the fast charging infrastructure and the EV to match the high charging capacities. Siemens, as a part of the research group, has provided an energy supply system to be used in the project, allowing researchers to examine the fast-charging capacity limits demonstrated by the EV's batteries. The system can handle high voltage that reaches up to 920 V, which is the voltage level predicted for the future EVs [17]. The controller of the charger makes sure that the output is modified automatically according to the EV, allowing different EVs to be charged using a single infrastructure. To illustrate, the UFC station can be used with battery systems rated at 400 and 800 V, where the charging capacity is adapted automatically to the maximum charging capacity on the EV side. The system is characterized by its flexibility and modularity, allowing multiple EVs to be charged at the same time instant [17].

Such fast-charging stations require DC-DC converters of high-power to be designed to achieve high reliability and high efficiency for the system. One of the fast-charging stations' requirements is to develop the DC-DC converters in a modular structure since such configuration can provide redundancy, easier maintenance, as well as scalability, and ride-through capability [18].

To further illustrate, to meet the requirements of the high-voltage high-power fast charger, two approaches are established. The first approach is through semiconductor devices with high-voltage and high-current ratings to be integrated into the two-level converter topologies with series/parallel connections. However, the series connection of power switches results in the unsymmetrical sharing of voltage among the switching devices because of the switches' unequal parameters, such as the switching delays, leakage inductance, and collector-to-emitter capacitance. Accordingly, voltage balancing methods are required to avoid any failure [19].

In the second approach, power electronic converters are usually pursued to be built in a modular manner in fast-charging stations [20]. Modular converters contain several numbers of smaller modules. Building converters in a modular way is a cost-effective solution. Besides, smaller modules can be hot swapped in failure cases, which makes the maintenance of such converters easier. Moreover, the number of modules can be scaled up according to the power rating of the system.

Furthermore, by installing more modules, the concept of redundancy can be established. In other words, in modular structure-based DC-DC converters, each cell handles a small portion of the total input power. Consequently, the selected power switches are of lower voltage ratings, hence, higher switching frequency capability. Therefore, the converter efficiency is improved due to lower losses, and the transformer size is reduced due to the increase in the switching frequency [21,22]. To avoid the demerits associated with the first approach, modular converters' topologies, such as Multimodule Converters and Modular Multilevel Converters, are used to provide the modularity feature and achieve high-voltage and high-power requirements [23]. Nonetheless, the multimodule DC-DC converters are considered in this paper since the modularity concept is not limited only to the power electronics but also is extended to include the magnetics. Moreover, higher switching frequencies can be achieved in the AC link, which results in reduced weight and size of the overall power converter. Although the high-frequency transformer employment allows for a reduction in the overall converter size and weight, the overall system cost is increased, especially with the power increase. Multimodule converters are applicable for EV UFC since scalability, reliability, redundancy, and ride-through capability can be provided. Besides, such converters utilize low-power modules where each module handles only a fraction of the total required power. However, the system complexity increases with the increase in the number of modules. Moreover, the impact of EVs UFC on the grid should be considered. Therefore, coordinated charging should be devised to address challenges that might be encountered with uncoordinated EVs UFC, particularly when multiple batteries are charged through UFC mode. Furthermore, the on-going technology development in the batteries' industry supports the proliferation of EVs UFC through introducing high capacity batteries that can accept such mode of charging.

Multimodule DC-DC converters can offer a bidirectional power flow through utilizing submodules that contain Dual Active Bridge (DAB), Dual Half Bridge (DHB), or series resonant converters, where each configuration has its pros and cons [24,25]. In multimodule converters, soft switching operation, as well as higher switching frequency, can be achieved, resulting in a significant reduction in the component volume without sacrificing the efficiency [24]. The possible architectures of connecting multimodule-based DAB units are classified into four main categories, which are: Input-Series Output-Series (ISOS), Input-Parallel Output-Parallel (IPOP), Input-Series Output-Parallel (ISOP), Input-Parallel Output-Series (IPOS) [26].

A typical fast charger is usually connected to a 400 VAC grid (570 *VDC* rectified) [27]. However, this low voltage results in high current and causes high losses in the magnetic components, bus bars, and switching devices [28]. Accordingly, connecting the fast charger to a higher input voltage level (10–40 kV) will reduce the current and conduction losses in the DC-DC converter. In [29,30], a DC link with a voltage of 1000 V has been proposed to the high power, while, in [31], the fast charger is connected to a medium voltage connection of 2.4 kV. In [28], a multiport converter for EVs' fast chargers has been proposed, where the charging process is done via a 6 to 10 kV supply grid.

In [15], a power architecture for the upcoming super-fast EV charging stations has been proposed. The presented architecture in [15], is interfaced with a medium voltage grid with a voltage level of 4.8 kV. It supports the grid functionality since it can allow bidirectional power flow and can reduce during the EV charging the conduction energy loss of the grid. However, in [7], DC-DC converters for high-power EV UFC stations have been presented. The semiconductor devices voltage ratings are reduced through splitting the DC input voltage. In addition, a modulation scheme named triangular current modulation is applied to achieve Zero Voltage Switching for all the switching devices. This accordingly increases the efficiency of the system. The converter in [7] can achieve an efficiency of 99.5%. To reduce the ripple content of the output current and achieve a high power density, multiple modules are parallel interleaved. In [2], reactive power operation is investigated using an offboard charging station. In [32], two different converter architectures for UF-EVC stations are presented. The first architecture is based on low-frequency isolation (non-isolated DC-DC converter), while the second approach is based on high-frequency isolation (galvanic isolated DC-DC converter). Technical evaluation for the two architectures is carried out where the pros and cons of each topology are highlighted. Besides, simulation results elaborating on the impact of the DC fast charging station on the grid are also provided in [32]. In [1], to realize medium-voltage UF-EVC stations, a multiport power converter has been proposed. A cascaded H-bridge is utilized at the grid side. Besides, to reduce the charging station effect on the grid, the battery energy storage elements are integrated in a split manner on the level of each submodule. Talking about the DC-DC conversion stage presented in [1], multiple Dual Half Bridge units are connected in parallel, where the multiport concept at the output side is achieved by selecting different submodule configurations. This is done to charge multiple EVs at the same time instant without the need for extra chargers. Operation modes, as well as the control techniques, have been addressed in [1].

In [33], a Full-Bridge Phase-Shifted DC-DC converter that combines the features of the double inductor rectifier and the conventional hybrid switching converter has been proposed for EV DC fast chargers. The principle of operation, as well as the characteristics and design specifications, are provided. The presented DC-DC converter in [33], can achieve both: zero voltage switching and zero current switching in the leading and the lagging legs, respectively. In [31], a medium-voltage high-power isolated DC-DC converter for EVs fast chargers have been presented. The employed DC-DC converter is a modular-based structure utilizing silicon carbide switching devices to convert a single-phase rectified input voltage of 2.4 kV to a variable DC output. It is a unidirectional converter that is connected in series from the input side and connected in parallel from the output side.

In [34], the AC-DC and DC-DC stages of an EV charger have been studied. The DC-DC stage utilizes interleaved DC-DC converters to be connected to a high-voltage network (13.8 kV). The interleaved DC-DC converters are accompanied with several advantages that can be highlighted in lower input ripple and lower inductors cost and size. The charger presented in [34] allows for bidirectional power flow so that it can support renewable energy sources and smart grid applications. It is suitable for a high-power fast charger of 400 kW with an EV battery with a rated voltage level of 500 V. However, in [35], a fast charger that is based on a single stage has been proposed. In other words, the power factor correction and zero voltage switching are achieved in one single AC-DC stage. The charger utilizes DAB units that are connected in series and parallel at the input and output sides, respectively. The use of one stage resulted in the DC link capacitor elimination, allowing for higher efficiency and higher power density to be achieved. In [18], a fast-charging system for EVs is proposed. The system consists of a 15 kW multimodule converter that utilizes a three-phase rectifier and a Full-Bridge DC-DC converter. Multiple units are used in a modular manner and connected in parallel at the input side and series or parallel at the output side. To clarify, modules are connected in IPOS to provide a higher voltage that reaches up to (1000 V) or connected in IPOP to provide higher output current at a voltage level of (500 V).

In [36], a power converter that interfaces a three-phase medium-voltage grid with EV batteries has been proposed. The AC-DC stage utilizes a cascaded H-Bridge converter, in which the battery energy storage systems are integrated in a split manner to reduce the effect of the charging station on the grid. Modular DC-DC converters utilizing the Full-Bridge Phase-Shifted topology are employed to charge the EV battery. The DC-DC converters are connected in parallel to provide high currents.

According to the latest technical specifications for fast charging stations, the charging process can be done through a 400 VAC or it can be done through a medium voltage grid supply that ranges from 10 to 40 kV to avoid high current and high losses. Assuming a fast charger rated at 350 kW and EV battery rated at a high voltage level (400–920 V). Different scenarios of the mentioned specifications result in the four different architectures of the multimodule DC-DC converters, which are ISOS, IPOP, ISOP, and IPOS. In this paper, these scenarios are discussed along with the control strategy for each architecture. To support the power converter controller design, the Small-Signal Model (SSM) for the four architectures is studied in detail. Moreover, to ensure equal power sharing among the employed modules, the control scheme for the four architectures is investigated. Moreover, a generalized SSM for any multimodule DC-DC converters' connection, including Input-Series Input-Parallel Output-Series Output-Parallel (ISIP-OSOP), is provided. This is achieved by studying the SSM of Full-Bridge Phase-Shift (FB-PS) DC-DC converter, two-module IPOS, three-module ISOP, and four-module Input-Series Input-Parallel Output-Series (ISIPOS) presented in [37–42], respectively. After deriving the generalized model, the model is verified with the multimodule configurations presented in [38–42].

The main contribution of the paper can be summarized as follows:

- A generalized model for the multimodule ISIP-OSOP DC-DC converter is provided. It is worth mentioning that the work presented in [40] is extended to include the SSM as well as the control schemes for the three other multimodule configurations, which are ISOS, IPOP, and IPOS. In addition to a generalized SSM applicable for all the basic architectures for multimodule DC-DC converters.
- Detailed SSM for the four architectures of the multimodule converter, which are ISOS, ISOP, IPOP, and IPOS, is provided in detail.
- The control strategy to guarantee uniform power distribution among the modules is studied. The strategies are based on current control Reflex Charging (RC) considering high-power-level UFC stations.

This paper is organized such that: Section 2 presents the ISOS circuit configuration as well as the ISOS SSM. Section 3 presents the ISOP multimodule DC-DC converter referring to the work that has been previously published in [40]. Sections 4 and 5 provide the circuit configuration and SSM for the IPOP and IPOS DC-DC converter, respectively. Section 6 addresses the generalized SSM for the ISIP-OSOP multimodule DC-DC converter along with its verification. Section 7 provides the control scheme for the four architectures to achieve equal power sharing among the employed modules. Finally, Section 8 presents the conclusion.

2. Input-Series Output-Series (ISOS) DC-DC Converter

In this section, the ISOS DC-DC converter circuit diagram, as well as the ISOS DC-DC converter SSM, are discussed in detail. The analysis carried out in this section, as well as the following sections, is not restricted to unidirectional power flow and can be applied for bidirectional power flow. In addition, a three-module based DAB topology is considered. Such configurations are employed when galvanic isolation is required.

To reduce the SSM in terms of complexity, it is assumed that all the modules of the ISOS, ISOP, IPOP, and IPOS DC-DC converter have an equal turns ratio and effective duty cycle as well as the same inductor and capacitor values. In other words, the SSM for the four architectures is derived, assuming ideal conditions. The SSM, carried out in this section and the following sections, is used in Section 7 to support the controller design for each configuration.

2.1. ISOS Circuit Diagram

An ISOS DC-DC converter is employed in high input voltage and high output voltage applications [43]. The ISOS converter configuration is shown in Figure 1, where modules are connected in series at the input and the output sides. The ISOS DC-DC converter shown in Figure 1 is needed for a high step-up ratio. This configuration is employed in high-voltage applications where neither the input side voltage nor the output side voltage is within the semiconductor devices voltage ratings. Such configuration enables low-voltage rating switches utilization.

By ensuring equal Input Voltage Sharing (IVS) and Output Voltage Sharing (OVS), the input voltage per module is $\frac{V_{in}}{3}$, and the output voltage per module is $\frac{V_o}{3}$. In which, V_{in} and V_o are the ISOS input voltage and ISOS output voltage, respectively. The features of the ISOS converter can be summarized in reduced voltage stress as well as the reduced power level per module, which facilitates the design of the converter. In the following Section, the SSM of the ISOS converter is studied using the FB-PS DC-DC converter.

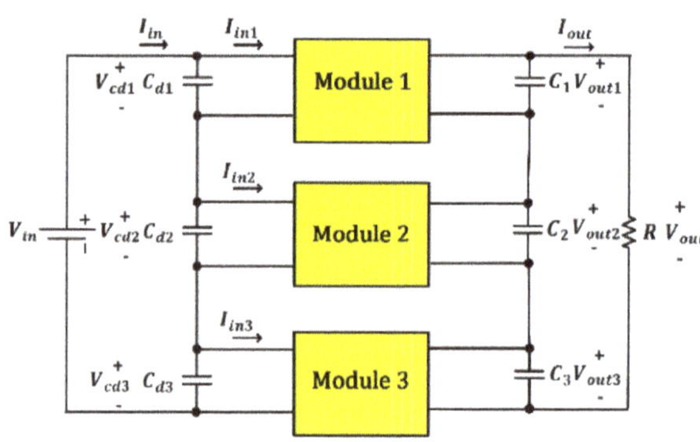

Figure 1. Input-Series Output-Series (ISOS) DC-DC converter circuit diagram.

2.2. ISOS Small-Signal Analysis

The SSM for the ISOS converter shown in Figure 2 is derived, using the SSM presented in [37].

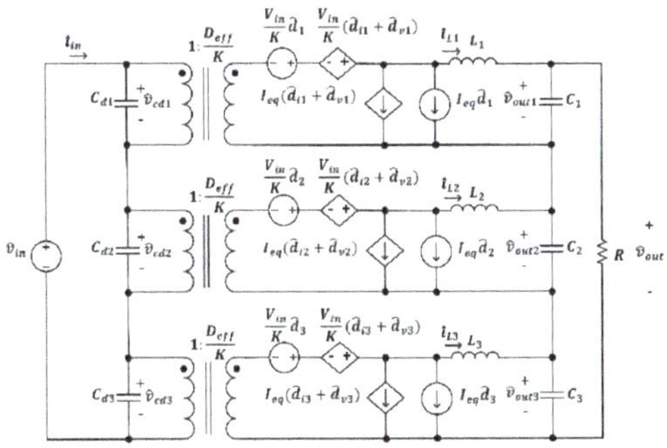

Figure 2. ISOS DC-DC converter small-signal modeling (SSM).

Since the input voltage per module is $\frac{V_{in}}{3}$, and the output voltage per module is $\frac{V_o}{3}$, accordingly, the load resistance per module is $\frac{R}{3}$. Therefore, $\hat{d}_{ij}$ and $\hat{d}_{vj}$, which are the effect of changing the filter inductor current and the effect of changing the input voltage on the duty cycle modulation, as well as I_{eq} presented in Figure 2 can be expressed as follows, where the subscript $j = 1, 2,$ and 3:

$$\hat{d}_{ij} = -\frac{12\, L_{lk} f_s}{KV_{in}} \hat{i}_{Lj} \tag{1}$$

Equation (1) can be rewritten in terms of R_d, where; $R_d = \frac{4L_{lk}f_s}{k^2}$ as:

$$\hat{d}_{ij} = -\frac{3KR_d}{V_{in}} \hat{i}_{Lj} \tag{2}$$

$$\hat{d}_{vj} = \frac{36\, L_{lk} f_s D_{eff}}{k^2 R V_{in}} \hat{v}_{cdj} \tag{3}$$

Similarly, Equation (3) can be rewritten as:

$$\hat{d}_{vj} = \frac{9\, R_d D_{eff}}{RV_{in}} \hat{v}_{cdj} \tag{4}$$

$$I_{eq} = \frac{V_{in}}{KR} \tag{5}$$

The following equations are obtained from Figure 2:

$$\begin{cases} \frac{D_{eff}}{K}\hat{v}_{cd1} + \frac{V_{in}}{3K}\left(\hat{d}_{i1} + \hat{d}_{v1} + \hat{d}_1\right) = sL\hat{i}_{L1} + \hat{v}_{out1} \\ \frac{D_{eff}}{K}\hat{v}_{cd2} + \frac{V_{in}}{3K}\left(\hat{d}_{i2} + \hat{d}_{v2} + \hat{d}_2\right) = sL\hat{i}_{L2} + \hat{v}_{out2} \\ \frac{D_{eff}}{K}\hat{v}_{cd3} + \frac{V_{in}}{3K}\left(\hat{d}_{i3} + \hat{d}_{v3} + \hat{d}_3\right) = sL\hat{i}_{L3} + \hat{v}_{out3} \end{cases} \tag{6}$$

$$\begin{cases} \frac{K}{D_{eff}}\left(\hat{i}_{in} - sC_d\hat{v}_{cd1}\right) = I_{eq}\left(\hat{d}_{i1} + \hat{d}_{v1} + \hat{d}_1\right) + \hat{i}_{L1} \\ \frac{K}{D_{eff}}\left(\hat{i}_{in} - sC_d\hat{v}_{cd2}\right) = I_{eq}\left(\hat{d}_{i2} + \hat{d}_{v2} + \hat{d}_2\right) + \hat{i}_{L2} \\ \frac{K}{D_{eff}}\left(\hat{i}_{in} - sC_d\hat{v}_{cd3}\right) = I_{eq}\left(\hat{d}_{i3} + \hat{d}_{v3} + \hat{d}_3\right) + \hat{i}_{L3} \end{cases} \tag{7}$$

$$\begin{cases} \hat{i}_{L1} = sC\hat{v}_{out1} + \frac{\hat{v}_{out}}{R} \\ \hat{i}_{L2} = sC\hat{v}_{out2} + \frac{\hat{v}_{out}}{R} \\ \hat{i}_{L3} = sC\hat{v}_{out3} + \frac{\hat{v}_{out}}{R} \end{cases} \tag{8}$$

Summing Equations in (8) would result in (9):

$$\sum_{j=1}^{3} \hat{i}_{Lj} = \hat{v}_{out}\left(sC + \frac{3}{R}\right) \tag{9}$$

where

$$\hat{v}_{out1} + \hat{v}_{out2} + \hat{v}_{out3} = \hat{v}_{out} \tag{10}$$

$$\hat{v}_{cd1} + \hat{v}_{cd2} + \hat{v}_{cd3} = \hat{v}_{in} \text{ or } \sum\nolimits_{j=1}^{3} \hat{v}_{cdj} = \hat{v}_{in} \tag{11}$$

2.2.1. Control-To-Output Voltage Transfer Function

The output voltage and the duty cycle relationship is found by summing up the Kirchhoff's voltage law (KVL) Equations in (6), assuming $\hat{v}_{in} = 0$, and $\hat{d}_k = 0$, where $k = 1, 2,$ and 3, and $k \neq j$, and substituting (2), (4), (9)–(11).

$$\frac{V_{in}}{3K}\hat{d}_1 - R_d\left(\hat{v}_{out}\left(sC + \frac{3}{R}\right)\right) = sL\left(\hat{v}_{out}\left(sC + \frac{3}{R}\right)\right) + \hat{v}_{out} \tag{12}$$

Simplifying (12) would result in (13):

$$G_{vd} = \frac{\hat{v}_{out}}{\hat{d}_j} = \frac{\frac{V_{in}}{3K}}{\left(s^2LC + s\left(\frac{3L}{R} + R_dC\right) + \frac{3R_d}{R} + 1\right)} \tag{13}$$

2.2.2. Control-To-Filter Inductor Current Transfer Function

The filter inductor current and the duty cycle relationship is found by substituting $\hat{v}_{out}$ in (12) in terms of $\hat{i}_{Lj}$ using (9), and considering the same assumptions as in Section 2.2.1.

$$\frac{V_{in}}{2K}\hat{d}_1 - R_d\sum_{j=1}^{3} \hat{i}_{Lj} = sL\sum_{j=1}^{3} \hat{i}_{Lj} + \frac{\sum_{j=1}^{3} \hat{i}_{Lj}}{\left(sC + \frac{3}{R}\right)} \tag{14}$$

Simplifying (14) would result in (15):

$$G_{id} = \frac{\hat{i}_L}{\hat{d}_j} = \frac{\frac{V_{in}}{3K}(3 + sRC)}{R\left(s^2LC + s\left(\frac{3L}{R} + R_dC\right) + \frac{3}{R} + 1\right)} \tag{15}$$

2.2.3. Control-To-Module Input Voltage Transfer Function

The module input voltage and the duty cycle relationship is found as follows:
Subtracting the 3rd Equation in (6) from the 1st Equation in (6):

$$\frac{D_{eff}}{K}(\hat{v}_{cd1} - \hat{v}_{cd2}) + \frac{V_{in}}{3K}\left(\hat{d}_{i1} + \hat{d}_{v1} + \hat{d}_1 - \hat{d}_{i2} - \hat{d}_{v2} - \hat{d}_2\right) = sL\left(\hat{i}_{L1} - \hat{i}_{L2}\right) + \hat{v}_{out1} - \hat{v}_{out2} \tag{16}$$

Substituting (2) and (4) in (16):

$$\frac{D_{eff}}{K}\left(1 + \frac{3R_d}{R}\right)(\hat{v}_{cd1} - \hat{v}_{cd2}) + \frac{V_{in}}{3K}\left(\hat{d}_1 - \hat{d}_2\right) = (sL + R_d)\left(\hat{i}_{L1} - \hat{i}_{L2}\right) + \hat{v}_{out1} - \hat{v}_{out2} \tag{17}$$

Subtracting the 2nd Equation in (7) from the 1st Equation in (7), and substituting (5):

$$\frac{K}{D_{eff}}(sC_d\hat{v}_{cd2} - sC_d\hat{v}_{cd1}) = \frac{V_{in}}{KR}\left(\hat{d}_{i1} + \hat{d}_{v1} + \hat{d}_1 - \hat{d}_{i2} - \hat{d}_{v2} - \hat{d}_2\right) + \left(\hat{i}_{L1} - \hat{i}_{L2}\right) \tag{18}$$

Substituting (2) and (4) in (18) and rearranging the equation:

$$\left(\frac{3R_d}{R}-1\right)\left(\hat{i}_{L1}-\hat{i}_{L2}\right)=\frac{V_{in}}{KR}\left(\hat{d}_1-\hat{d}_2\right)+\left(\frac{9R_dD_{eff}}{KR^2}+\frac{sKC_d}{D_{eff}}\right)\left(\hat{v}_{cd1}-\hat{v}_{cd2}\right) \tag{19}$$

Substituting $\hat{v}_{out}$ by $\hat{v}_{out1}+\hat{v}_{out2}+\hat{v}_{out3}$ in (8):

$$\begin{cases}\hat{i}_{L1}=\left(sC+\frac{1}{R}\right)\hat{v}_{out1}+\frac{1}{R}\hat{v}_{out2}+\frac{1}{R}\hat{v}_{out3}\\ \hat{i}_{L2}=\frac{1}{R}\hat{v}_{out1}+\left(sC+\frac{1}{R}\right)\hat{v}_{out2}+\frac{1}{R}\hat{v}_{out3}\\ \hat{i}_{L3}=\frac{1}{R}\hat{v}_{out1}+\frac{1}{R}\hat{v}_{out2}+\left(sC+\frac{1}{R}\right)\hat{v}_{out3}\end{cases} \tag{20}$$

The three Equations in (20) can be represented as:

$$\begin{bmatrix}\hat{i}_{L1}\\ \hat{i}_{L2}\\ \hat{i}_{L3}\end{bmatrix}=\begin{bmatrix}sC+\frac{1}{R} & \frac{1}{R} & \frac{1}{R}\\ \frac{1}{R} & sC+\frac{1}{R} & \frac{1}{R}\\ \frac{1}{R} & \frac{1}{R} & sC+\frac{1}{R}\end{bmatrix}\begin{bmatrix}\hat{v}_{out1}\\ \hat{v}_{out2}\\ \hat{v}_{out3}\end{bmatrix} \tag{21}$$

Solving for the output voltages in (21) would give:

$$\begin{cases}\hat{v}_{out1}=g_1\hat{i}_{L1}-g_2\left(\hat{i}_{L2}+\hat{i}_{L3}\right)\\ \hat{v}_{out2}=g_1\hat{i}_{L2}-g_2\left(\hat{i}_{L1}+\hat{i}_{L3}\right)\\ \hat{v}_{out3}=g_1\hat{i}_{L3}-g_2\left(\hat{i}_{L1}+\hat{i}_{L2}\right)\end{cases} \tag{22}$$

where

$$g_1=\frac{2+sRC}{s^2RC^2+3sC}\ \text{and}\ g_2=\frac{1}{s^2RC^2+3sC}$$

Accordingly;

$$\hat{v}_{out1}-\hat{v}_{out2}=(g_1+g_2)\left(\hat{i}_{L1}-\hat{i}_{L2}\right) \tag{23}$$

Subctituting (23) in (17):

$$\frac{D_{eff}}{K}\left(1+\frac{3R_d}{R}\right)\left(\hat{v}_{cd1}-\hat{v}_{cd2}\right)+\frac{V_{in}}{3K}\left(\hat{d}_1-\hat{d}_2\right)=(sL+R_d+g_1+g_2)\left(\hat{i}_{L1}-\hat{i}_{L2}\right) \tag{24}$$

Substituting (19) in (24):

$$\begin{gathered}\frac{D_{eff}}{K}\left(1+\frac{3R_d}{R}\right)\left(\frac{3R_d}{R}-1\right)\left(\hat{v}_{cd1}-\hat{v}_{cd2}\right)+\frac{V_{in}}{3K}\left(\frac{3R_d}{R}-1\right)\left(\hat{d}_1-\hat{d}_2\right)=\\ \frac{3(sL+R_d+g_1+g_2)}{KR}\left(\hat{d}_1-\hat{d}_2\right)+(sL+R_d+g_1+g_2)\left(\frac{9R_dD_{eff}}{KR^2}+\frac{sKC_d}{D_{eff}}\right)\left(\hat{v}_{cd1}-\hat{v}_{cd2}\right)\end{gathered} \tag{25}$$

Simplifying (25) would result in (26):

$$\left(\hat{v}_{cd2}-\hat{v}_{cd1}\right)=\left(-\frac{D_{eff}\left(R^2\frac{V_{in}}{2}+2RV_{in}(g_1+g_2)+2sLRV_{in}\right)}{\begin{gathered}R^2D_{eff}^2+16R_dD_{eff}^2(g_1+g_2)+16sLR_dD_{eff}^2\\ +K^2R^2C_d\left(s(R_d+g_1+g_2)+s^2L\right)\end{gathered}}\right)\left(\hat{d}_2-\hat{d}_1\right) \tag{26}$$

Equation (26) can be written as:

$$\left(\hat{v}_{cd2}-\hat{v}_{cd1}\right)=A(s)\left(\hat{d}_2-\hat{d}_1\right) \tag{27}$$

where

$$A(s) = -\left(\frac{D_{eff}\left(R^2 \frac{V_{in}}{2} + 2RV_{in}(g_1+g_2) + 2sLRV_{in}\right)}{\begin{array}{c} R^2D_{eff}^2 + 16R_dD_{eff}^2(g_1+g_2) + 16sLR_dD_{eff}^2 \\ +K^2R^2C_d\left(s(R_d+g_1+g_2)+s^2L\right) \end{array}} \right)$$

Rearranging (27) would result in (28):

$$\hat{v}_{cd2} = \hat{v}_{cd1} + A(s)\left(\hat{d}_2 - \hat{d}_1\right) \quad (28)$$

Assuming $\hat{v}_{in} = 0$, hence $\hat{v}_{cd1} + \hat{v}_{cd2} + \hat{v}_{cd3} = 0$,
Equation (28) can be generalized as follows:

$$\hat{v}_{cdj} = \hat{v}_{cd1} + A(s)\left(\hat{d}_j - \hat{d}_1\right) \quad (29)$$

Setting $\hat{v}_{in} = 0$, and substituting (29) in (11) would result in:

$$\sum_{j=1}^{3} \hat{v}_{cd1} + A(s)\left(\hat{d}_j - \hat{d}_1\right) = 0 \quad (30)$$

$$3\hat{v}_{cd1} - 3A(s)\hat{d}_1 + \sum_{j=1}^{3} A(s)\hat{d}_j = 0 \quad (31)$$

Therefore,

$$\hat{v}_{cd1} = A(s)\hat{d}_1 + \frac{A(s)}{3}\sum_{j=1}^{3}\hat{d}_j \quad (32)$$

Substituting (32) in (29) would result in:

$$\hat{v}_{cdj} = A(s)\hat{d}_j + \frac{A(s)}{3}\sum_{j=1}^{3}\hat{d}_j \quad (33)$$

Presenting (33) in a matrix form would give:

$$\begin{bmatrix} \hat{v}_{cd1} \\ \hat{v}_{cd2} \\ \hat{v}_{cd3} \end{bmatrix} = \begin{bmatrix} \frac{2A(s)}{3} & \frac{-A(s)}{3} & \frac{-A(s)}{3} \\ \frac{-A(s)}{3} & \frac{2A(s)}{3} & \frac{-A(s)}{3} \\ \frac{-A(s)}{3} & \frac{-A(s)}{3} & \frac{2A(s)}{3} \end{bmatrix} \begin{bmatrix} \hat{d}_1 \\ \hat{d}_2 \\ \hat{d}_3 \end{bmatrix} \quad (34)$$

2.2.4. Converter Output Impedance

As studied in the SSM presented in [39,40], the ISOS converter output impedance can be found by modifying (8), such that:

$$\begin{cases} \hat{i}_{L1} + \hat{i}_{out} = sC\hat{v}_{out1} + \frac{\hat{v}_{out}}{R} \\ \hat{i}_{L2} + \hat{i}_{out} = sC\hat{v}_{out2} + \frac{\hat{v}_{out}}{R} \\ \hat{i}_{L3} + \hat{i}_{out} = sC\hat{v}_{out3} + \frac{\hat{v}_{out}}{R} \end{cases} \quad (35)$$

Accordingly, Equation (9) is modified as follows:

$$\sum_{j=1}^{3} \hat{i}_{Lj} = \hat{v}_{out}\left(sC + \frac{3}{R}\right) - 3\hat{i}_{out} \quad (36)$$

The output voltage and the output current relationship is found by assuming $\hat{v}_{in} = 0$, and $\hat{d}_j = 0$, $j = 1, 2,$ and 3, summing the KVL Equations in (6), and substituting (2), (4), (10), (11), and (36).

$$\left((sL + R_d)\left(sC + \frac{3}{R}\right) + 1\right)\hat{v}_{out} = 3(R_d + sL)\hat{i}_{out} \tag{37}$$

Simplifying (37) would result in (38):

$$Z_{out} = \frac{\hat{v}_{out}}{\hat{i}_{out}} = \frac{3(R_d + sL)}{s^2LC + s\left(\frac{3L}{R} + R_dC\right) + \frac{3R_d}{R} + 1} \tag{38}$$

2.2.5. Converter Gain

The output voltage and the input voltage relationship is found by assuming $\hat{d}_j = 0$, $j = 1, 2,$ and 3, summing the KVL Equations in (6), and substituting (2), (4), (9)–(11) in the added equation.

$$\frac{D_{eff}}{K}\left(1 + \frac{R_d}{3R}\right)\hat{v}_{in} = \left(\frac{(sL + R_d)(sRC + 1)}{R} + 3\right)\hat{v}_{out} \tag{39}$$

Simplifying (39) would result in (40):

$$G_{vg} = \frac{\hat{v}_{out}}{\hat{v}_{in}} = \frac{\frac{D_{eff}}{K}\left(1 + \frac{R_d}{3R}\right)}{s^2LC + s\left(\frac{L}{R} + R_dC\right) + \frac{R_d}{R} + 3} \tag{40}$$

3. Input-Series Output-Parallel (ISOP) DC-DC Converter

All the related work in for the ISOP DC-DC converter, including system configuration, detailed SSM, and control strategy, has been previously published in [40].

4. Input-Parallel Output-Parallel (IPOP) DC-DC Converter

In this section, the IPOP DC-DC converter circuit diagram, as well as the IPOP DC-DC converter SSM, are discussed in detail.

4.1. IPOP Circuit Diagram

IPOP DC-DC converters are employed in high-current and high-power applications [44]. The IPOP converter configuration is shown in Figure 3, where the modules connected in parallel at the input and the output sides. This architecture is employed when the input side and the output side current are higher than the semiconductor devices' current rating.

By ensuring equal Input Current Sharing (ICS) and Output Current Sharing (OCS), the input current per module is $\frac{I_{in}}{3}$, and the output current per module is $\frac{I_o}{3}$. In which, I_{in} and I_o, are the IPOP input current and the IPOP output current, respectively. The merits of the IPOP DC-DC converter can be highlighted in more accessible semiconductor devices selection due to the low current stress per module. Besides, a higher switching frequency can be achieved and accordingly lowering the size of the magnetic components.

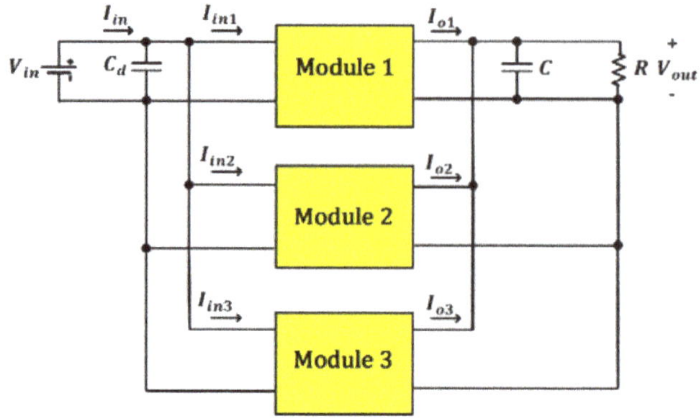

Figure 3. Input-Parallel Output-Parallel (IPOP) DC-DC converter circuit diagram.

4.2. IPOP Small-Signal Analysis

The SSM for the IPOP converter shown in Figure 4 is derived using the SSM presented in [37].

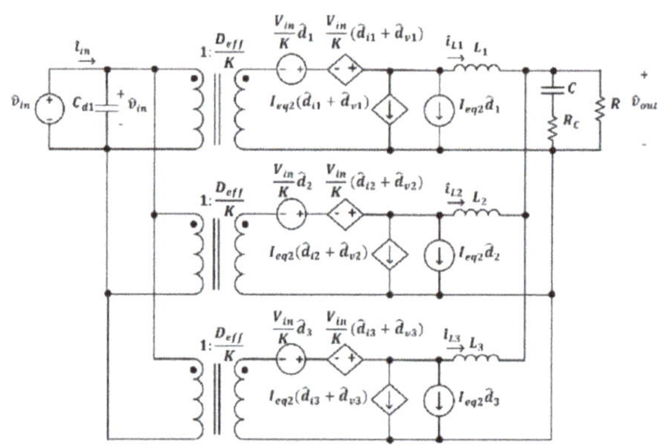

Figure 4. IPOP DC-DC converter SSM.

Since the input current per module is $\frac{I_{in}}{3}$, and the output current per module is $\frac{I_o}{3}$, accordingly, the load resistance per module is $3R$. Therefore, $\hat{d}_{ij}$, $\hat{d}_{vj}$ and I_{eq} presented in Figure 4 can be expressed as follows, where the subscript $j = 1, 2, \text{ and } 3$:

$$\hat{d}_{ij} = -\frac{4\, L_{lk} f_s}{K V_{in}} \hat{i}_{Lj} \tag{41}$$

Rewriting (41) in terms of R_d would result in:

$$\hat{d}_{ij} = -\frac{K R_d}{V_{in}} \hat{i}_{Lj} \tag{42}$$

$$\hat{d}_{vj} = \frac{4 L_{lk} f_s D_{eff}}{3\, k^2 R V_{in}} \hat{v}_{in} \tag{43}$$

Similarly, rewriting (43) in terms of R_d would result in:

$$\hat{d}_{vj} = \frac{R_d D_{eff}}{3\, R V_{in}} \hat{v}_{in} \tag{44}$$

$$I_{eq} = \frac{V_{in}}{3KR} \tag{45}$$

The following equations are obtained from Figure 4:

$$\begin{cases} \frac{D_{eff}}{K}\hat{v}_{in} + \frac{V_{in}}{K}\left(\hat{d}_{i1} + \hat{d}_{v1} + \hat{d}_1\right) = sL\hat{i}_{L1} + \hat{v}_{out} \\ \frac{D_{eff}}{K}\hat{v}_{in} + \frac{V_{in}}{K}\left(\hat{d}_{i2} + \hat{d}_{v2} + \hat{d}_2\right) = sL\hat{i}_{L2} + \hat{v}_{out} \\ \frac{D_{eff}}{K}\hat{v}_{in} + \frac{V_{in}}{K}\left(\hat{d}_{i3} + \hat{d}_{v3} + \hat{d}_3\right) = sL\hat{i}_{L3} + \hat{v}_{out} \end{cases} \tag{46}$$

$$\begin{cases} \frac{K}{3D_{eff}}\left(\hat{i}_{in} - sC_d\hat{v}_{in}\right) = I_{eq}\left(\hat{d}_{i1} + \hat{d}_{v1} + \hat{d}_1\right) + \hat{i}_{L1} \\ \frac{K}{3D_{eff}}\left(\hat{i}_{in} - sC_d\hat{v}_{in}\right) = I_{eq}\left(\hat{d}_{i2} + \hat{d}_{v2} + \hat{d}_2\right) + \hat{i}_{L2} \\ \frac{K}{3D_{eff}}\left(\hat{i}_{in} - sC_d\hat{v}_{in}\right) = I_{eq}\left(\hat{d}_{i3} + \hat{d}_{v3} + \hat{d}_3\right) + \hat{i}_{L3} \end{cases} \tag{47}$$

$$\sum_{j=1}^{3} \hat{i}_{Lj} = \hat{v}_{out}\left(sC + \frac{1}{R}\right) \tag{48}$$

4.2.1. Control-To-Output Voltage Transfer Function

The output voltage and the duty cycle relationship are found by summing up the KVL equations in (46), considering the same assumptions as in Section 2.2.1, and substituting (42), (44), and (48).

$$\frac{V_{in}}{K}\left(-\frac{KR_d}{V_{in}}\left(\hat{v}_{out}\left(sC + \frac{1}{R}\right)\right) + \sum_{j=1}^{3} \frac{R_d D_{eff}}{3\,RV_{in}}\hat{v}_{in} + \hat{d}_1\right) = sL\left(\hat{v}_{out}\left(sC + \frac{1}{R}\right)\right) + 3\hat{v}_{out} \tag{49}$$

Simplifying (49) would result in (50):

$$G_{vd} = \frac{\hat{v}_{out}}{\hat{d}_j} = \frac{\frac{V_{in}}{K}}{s^2LC + s\left(\frac{L}{R} + R_dC\right) + \frac{R_d}{R} + 3} \tag{50}$$

4.2.2. Control-To-Filter Inductor Current Transfer Function

The filter inductor current and the duty cycle relationship is found by substituting $\hat{v}_{out}$ in terms of $\hat{i}_{Lj}$ using (48) in (49), and considering the same assumptions as in Section 2.2.1.

$$\frac{V_{in}}{K}\hat{d}_1 - R_d\sum_{j=1}^{3} \hat{i}_{Lj} = sL\sum_{j=1}^{3} \hat{i}_{Lj} + \frac{3R}{sRC + 1}\sum_{j=1}^{3} \hat{i}_{Lj} \tag{51}$$

Simplifying (51) would result in (52):

$$G_{id} = \frac{\hat{i}_L}{\hat{d}_j} = \frac{\frac{V_{in}}{K}(1 + sRC)}{R\left(s^2LC + s\left(\frac{L}{R} + R_dC\right) + \frac{R_d}{R} + 3\right)} \tag{52}$$

4.2.3. Control-To-Module Filter Inductor Current Transfer Function

The module filter inductor current and the duty cycle relationship is found by substituting (42), (44), and (48) in (46) assuming $\hat{v}_{in} = 0$.

$$\begin{cases} \left(sL + R_d + \frac{R}{sRC+1}\right)\hat{i}_{L1} + \frac{R}{sRC+1}\hat{i}_{L2} + \frac{R}{sRC+1}\hat{i}_{L3} = \frac{V_{in}}{K}\hat{d}_1 \\ \frac{R}{sRC+1}\hat{i}_{L1} + \left(sL + R_d + \frac{R}{sRC+1}\right)\hat{i}_{L2} + \frac{R}{sRC+1}\hat{i}_{L3} = \frac{V_{in}}{K}\hat{d}_2 \\ \frac{R}{sRC+1}\hat{i}_{L1} + \frac{R}{sRC+1}\hat{i}_{L2} + \left(sL + R_d + \frac{R}{sRC+1}\right)\hat{i}_{L3} = \frac{V_{in}}{K}\hat{d}_3 \end{cases} \tag{53}$$

Subtracting the 2nd Equation in (53) from the 1st Equation in (53):

$$\hat{i}_{L2} = \hat{i}_{L1} - \frac{V_{in}}{K(sL + R_d)}\left(\hat{d}_1 - \hat{d}_2\right) \tag{54}$$

Similarly, subtracting the 3rd Equation in (53) from the 1st Equation in (53):

$$\hat{i}_{L3} = \hat{i}_{L1} - \frac{V_{in}}{K(sL + R_d)}\left(\hat{d}_1 - \hat{d}_3\right) \tag{55}$$

Substituting (54) and (55), in the first Equation in (53):

$$\begin{gathered}\left(sL + R_d + \tfrac{3R}{sRC+1}\right)\hat{i}_{L1} = \\ \tfrac{V_{in}}{K}\left[\left(\tfrac{2R}{(sL+R_d)(sRC+1)} + 1\right)\hat{d}_1 - \tfrac{R}{(sL+R_d)(sRC+1)}\hat{d}_2 - \tfrac{R}{(sL+R_d)(sRC+1)}\hat{d}_3\right]\end{gathered} \tag{56}$$

Rearranging (56), the following equation is obtained:

$$\hat{i}_{L1} = A(s)\hat{d}_1 + B(s)\hat{d}_2 + B(s)\hat{d}_3 \tag{57}$$

where

$$A(s) = \frac{V_{in}\left(s^2LCR + s(L + CR_dR) + R_d + 2R\right)}{K(sL + R_d)(s^2LCR + s(L + CR_dR) + 3R)} \tag{58}$$

$$B(s) = \frac{-V_{in}R}{K(sL + R_d)(s^2LCR + s(L + CR_dR) + 3R)} \tag{59}$$

Similar steps can be done for the filter inductor current for module 2 and module 3 such that:

$$\begin{cases} \hat{i}_{L1} = A(s)\hat{d}_1 + B(s)\hat{d}_2 + B(s)\hat{d}_3 \\ \hat{i}_{L2} = B(s)\hat{d}_1 + A(s)\hat{d}_2 + B(s)\hat{d}_3 \\ \hat{i}_{L3} = B(s)\hat{d}_1 + B(s)\hat{d}_2 + A(s)\hat{d}_3 \end{cases} \tag{60}$$

4.2.4. Converter Output Impedance

Similarly, the IPOP converter output impedance can be found by modifying (48) such that:

$$\sum_{j=1}^{3}\hat{i}_{Lj} + \hat{i}_{out} = \hat{v}_{out}\left(\frac{sRC + 1}{R}\right) \tag{61}$$

The output voltage and the output current relationship is found by considering the same assumptions as in Section 2.2.4, summing Equations in (46), and substituting (42), (44), and (61).

$$-R_d\left(\hat{v}_{out}\left(\frac{sRC+1}{R}\right) - \hat{i}_{out}\right) = sL\left(\hat{v}_{out}\left(\frac{sRC+1}{R}\right) - \hat{i}_{out}\right) + 3\hat{v}_{out} \tag{62}$$

Simplifying (62) would result in (63):

$$Z_{out} = \frac{\hat{v}_{out}}{\hat{i}_{out}} = \frac{(R_d + sL)}{s^2LC + s\left(\frac{L}{R} + R_dC\right) + \frac{R_d}{R} + 3} \tag{63}$$

4.2.5. Converter Gain

The output voltage and the input voltage relationship is found by assuming $\hat{d}_j = 0,\ j = 1,2,$ and 3, summing Equations in (46), and substituting (42), (44), and (48) in the added equation.

$$\frac{3D_{eff}}{K}\left(1+\frac{R_d}{3R}\right)\hat{v}_{in} = \left(\frac{(sL+R_d)(sRC+1)}{R}+3\right)\hat{v}_{out} \tag{64}$$

Simplifying (64) would result in (65):

$$G_{vg} = \frac{\hat{v}_{out}}{\hat{v}_{in}} = \frac{\frac{3D_{eff}}{K}\left(1+\frac{R_d}{3R}\right)}{s^2LC + s\left(\frac{L}{R}+R_dC\right)+\frac{R_d}{R}+3} \tag{65}$$

5. Input-Parallel Output-Series (IPOS) DC-DC Converter

In this section, the IPOS DC-DC converter circuit diagram, as well as the IPOS DC-DC converter SSM, are discussed in detail.

5.1. IPOS Circuit Diagram

IPOS DC-DC converters are employed in high input current and high output voltage applications [38]. The IPOS converter configuration is shown in Figure 5, where modules are connected in parallel and series at the input side and the output side, respectively. The IPOS DC-DC converter shown in Figure 5 is required for a high step-up ratio. This architecture is utilized when the output side current is within the semiconductor devices' current rating.

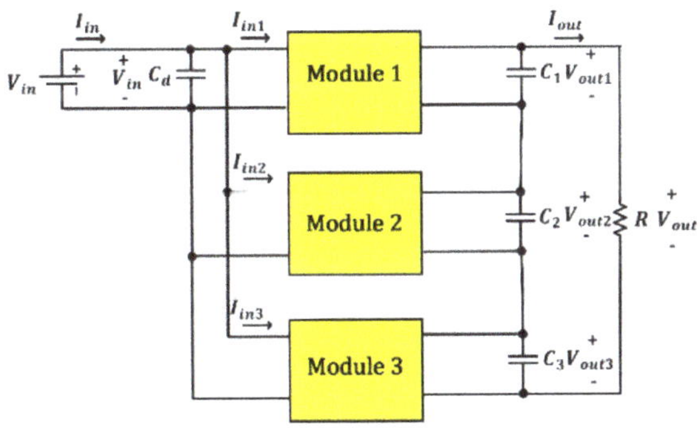

Figure 5. Input-Parallel Output-Series (IPOS) DC-DC converter circuit diagram.

By ensuring equal ICS and OVS, the input current per module is $\frac{I_{in}}{3}$, and the output voltage per module is $\frac{V_o}{3}$. In which, I_{in} and V_o, are the IPOS input current and the IPOS output voltage, respectively.

5.2. IPOS Small-Signal Analysis

The SSM for the IPOS converter shown in Figure 6 is derived using the SSM presented in [37].

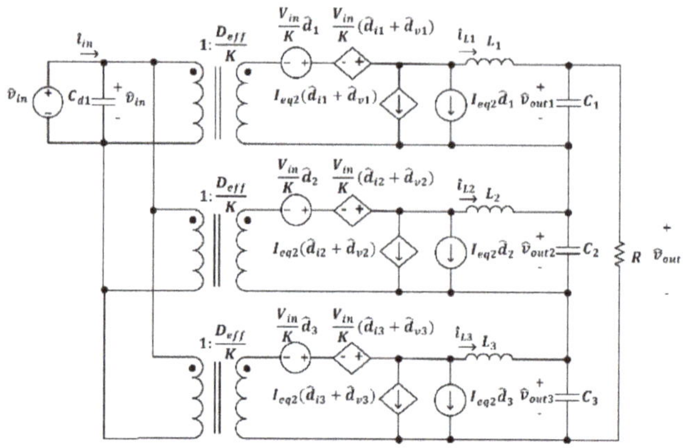

Figure 6. IPOS DC-DC converter SSM.

Since the input current per module is $\frac{I_{in}}{3}$, and the output voltage per module is $\frac{V_o}{3}$, accordingly, the load resistance per module is $\frac{R}{3}$. Therefore, $\hat{d}_{ij}$, $\hat{d}_{vj}$ and I_{eq} presented in Figure 6 can be expressed as follows, where the subscript $j = 1,2,\ and\ 3$:

$$\hat{d}_{ij} = -\frac{4\ L_{lk} f_s}{K V_{in}} \hat{i}_{Lj} \tag{66}$$

Rewriting (66) in terms of R_d would result in:

$$\hat{d}_{ij} = -\frac{K R_d}{V_{in}} \hat{i}_{Lj} \tag{67}$$

$$\hat{d}_{vj} = \frac{12\ L_{lk} f_s D_{eff}}{k^2 R V_{in}} \hat{v}_{in} \tag{68}$$

Similarly, rewriting (68) in terms of R_d would result in:

$$\hat{d}_{vj} = \frac{3\ R_d D_{eff}}{R V_{in}} \hat{v}_{in} \tag{69}$$

$$I_{eq} = \frac{3\ V_{in}}{K R} \tag{70}$$

The following equations are obtained from Figure 6:

$$\begin{cases} \frac{D_{eff}}{K}\hat{v}_{in} + \frac{V_{in}}{K}\left(\hat{d}_{i1} + \hat{d}_{v1} + \hat{d}_1\right) = sL\hat{i}_{L1} + \hat{v}_{out1} \\ \frac{D_{eff}}{K}\hat{v}_{in} + \frac{V_{in}}{K}\left(\hat{d}_{i2} + \hat{d}_{v2} + \hat{d}_2\right) = sL\hat{i}_{L2} + \hat{v}_{out2} \\ \frac{D_{eff}}{K}\hat{v}_{in} + \frac{V_{in}}{K}\left(\hat{d}_{i3} + \hat{d}_{v3} + \hat{d}_3\right) = sL\hat{i}_{L3} + \hat{v}_{out3} \end{cases} \tag{71}$$

$$\begin{cases} \frac{K}{3D_{eff}}\left(\hat{i}_{in} - sC_d\hat{v}_{in}\right) = I_{eq}\left(\hat{d}_{i1} + \hat{d}_{v1} + \hat{d}_1\right) + \hat{i}_{L1} \\ \frac{K}{3D_{eff}}\left(\hat{i}_{in} - sC_d\hat{v}_{in}\right) = I_{eq}\left(\hat{d}_{i2} + \hat{d}_{v2} + \hat{d}_2\right) + \hat{i}_{L2} \\ \frac{K}{3D_{eff}}\left(\hat{i}_{in} - sC_d\hat{v}_{in}\right) = I_{eq}\left(\hat{d}_{i3} + \hat{d}_{v3} + \hat{d}_3\right) + \hat{i}_{L3} \end{cases} \tag{72}$$

$$\begin{cases} \hat{i}_{L1} = sC\hat{v}_{out1} + \frac{\hat{v}_{out}}{R} \\ \hat{i}_{L2} = sC\hat{v}_{out2} + \frac{\hat{v}_{out}}{R} \\ \hat{i}_{L3} = sC\hat{v}_{out3} + \frac{\hat{v}_{out}}{R} \end{cases} \tag{73}$$

Summing Equations in (73) would result in (74):

$$\sum_{j=1}^{3} \hat{i}_{Lj} = \hat{v}_{out}\left(sC + \frac{3}{R}\right) \tag{74}$$

where

$$\hat{v}_{out1} + \hat{v}_{out2} + \hat{v}_{out3} = \hat{v}_{out} \tag{75}$$

5.2.1. Control-To-Output Voltage Transfer Function

The output voltage and the duty cycle relationship is found by summing up Equations in (71), considering the same assumptions as in Section 2.2.1, and substituting (67), (69), and (74).

$$\frac{3D_{eff}}{K}\hat{v}_{in} + \frac{V_{in}}{K}\left(-\frac{KR_d}{V_{in}}\left(\hat{v}_{out}\left(sC + \frac{3}{R}\right)\right) + \sum_{j=1}^{3} \frac{3\,R_d D_{eff}}{RV_{in}}\hat{v}_{in} + \hat{d}_1\right) = sL\left(\hat{v}_{out}\left(sC + \frac{3}{R}\right)\right) + \hat{v}_{out} \tag{76}$$

Simplifying (76) would result in (77):

$$G_{vd} = \frac{\hat{v}_{out}}{\hat{d}_j} = \frac{\frac{V_{in}}{K}}{s^2LC + s\left(\frac{3L}{R} + R_dC\right) + \frac{3R_d}{R} + 1} \tag{77}$$

5.2.2. Control-To-Filter Inductor Current Transfer Function

The filter inductor current and the duty cycle relationship is found by substituting $\hat{v}_{out}$ in terms of $\hat{i}_{Lj}$ using (74) in (76) and considering the same assumptions as in Section 2.2.1.

$$\frac{V_{in}}{K}\hat{d}_1 - R_d\sum_{j=1}^{3} \hat{i}_{Lj} = sL\sum_{j=1}^{3} \hat{i}_{Lj} + \frac{R}{sRC + 3}\sum_{j=1}^{3} \hat{i}_{Lj} \tag{78}$$

Simplifying (78) would result in (79):

$$G_{id} = \frac{\hat{i}_L}{\hat{d}_j} = \frac{\frac{V_{in}}{K}(3 + sRC)}{R\left(s^2LC + s\left(\frac{3L}{R} + R_dC\right) + \frac{3R_d}{R} + 1\right)} \tag{79}$$

5.2.3. Control-To-Module Filter Inductor Current Transfer Function

The filter inductor current and the duty cycle relationship is found by performing the following: Substituting (75) in (76):

$$\begin{cases} \hat{i}_{L1} = \left(sC + \frac{1}{R}\right)\hat{v}_{out1} + \frac{1}{R}\hat{v}_{out2} + \frac{1}{R}\hat{v}_{out3} \\ \hat{i}_{L2} = \frac{1}{R}\hat{v}_{out1} + \left(sC + \frac{1}{R}\right)\hat{v}_{out2} + \frac{1}{R}\hat{v}_{out3} \\ \hat{i}_{L3} = \frac{1}{R}\hat{v}_{out1} + \frac{1}{R}\hat{v}_{out2} + \left(sC + \frac{1}{R}\right)\hat{v}_{out3} \end{cases} \tag{80}$$

Rewriting (80) in terms of the output voltages:

$$\begin{cases} \hat{v}_{out1} = g_1\hat{i}_{L1} - g_2\hat{i}_{L2} - g_2\hat{i}_{L3} \\ \hat{v}_{out2} = g_2\hat{i}_{L1} - g_1\hat{i}_{L2} - g_2\hat{i}_{L3} \\ \hat{v}_{out3} = g_2\hat{i}_{L1} - g_2\hat{i}_{L2} - g_1\hat{i}_{L3} \end{cases} \tag{81}$$

where

$$g_1 = \frac{2 + sRC}{S^2C^2R + 3sC} \quad g_2 = \frac{1}{S^2C^2R + 3sC} \tag{82}$$

Rewriting (81) in a matrix form:

$$\begin{bmatrix} \hat{v}_{out1} \\ \hat{v}_{out2} \\ \hat{v}_{out3} \end{bmatrix} = \begin{bmatrix} g_1 & -g_2 & -g_2 \\ -g_2 & g_1 & -g_2 \\ -g_2 & -g_2 & g_1 \end{bmatrix} \begin{bmatrix} \hat{i}_{L1} \\ \hat{i}_{L2} \\ \hat{i}_{L3} \end{bmatrix} \tag{83}$$

Substituting (67), (69), and (81) in (71) and assuming $\hat{v}_{in} = 0$ would result in:

$$\begin{cases} \frac{V_{in}}{K}\hat{d}_1 = (sL + R_d + g_1)\hat{i}_{L1} - g_2\hat{i}_{L2} - g_2\hat{i}_{L3} \\ \frac{V_{in}}{K}\hat{d}_2 = -g_2\hat{i}_{L1} + (sL + R_d + g_1)\hat{i}_{L2} - g_2\hat{i}_{L3} \\ \frac{V_{in}}{K}\hat{d}_3 = -g_2\hat{i}_{L1} - g_2\hat{i}_{L2} + (sL + R_d + g_1)\hat{i}_{L3} \end{cases} \tag{84}$$

Rearranging (84), the control-to-module filter inductor current can be represented as:

$$\begin{bmatrix} \hat{d}_1 \\ \hat{d}_2 \\ \hat{d}_3 \end{bmatrix} = \begin{bmatrix} g_3 & g_4 & g_4 \\ g_4 & g_3 & g_4 \\ g_4 & g_4 & g_3 \end{bmatrix} \begin{bmatrix} \hat{i}_{L1} \\ \hat{i}_{L2} \\ \hat{i}_{L3} \end{bmatrix} \tag{85}$$

where

$$Sg_3 = \frac{V_{in}(sL + R_d + g_1 - g_2)}{K\left(g_1^2 - g_1g_2 + 2g_1g_5 - 2g_2^2 - g_2g_5 + g_5^2\right)} \tag{86}$$

$$g_4 = \frac{V_{in}g_2}{K\left(g_1^2 - g_1g_2 + 2g_1g_5 - 2g_2^2 - g_2g_5 + g_5^2\right)} \tag{87}$$

$$g_5 = sL + R_d \tag{88}$$

5.2.4. Converter Output Impedance

Similarly, the IPOS converter output impedance can be found by modifying (74) such that:

$$\sum_{j=1}^{3} \hat{i}_{Lj} + \hat{i}_{out} = \hat{v}_{out}\left(sC + \frac{3}{R}\right) \tag{89}$$

The output voltage and the output current relationship is found by considering the same assumptions as in Section 2.2.4, summing Equations in (71), and substituting (67), (69), (75) and (89).

$$-R_d\left(\hat{v}_{out}\left(sC + \frac{3}{R}\right) - \hat{i}_{out}\right) = sL\left(\hat{v}_{out}\left(sC + \frac{3}{R}\right) - \hat{i}_{out}\right) + \hat{v}_{out} \tag{90}$$

Simplifying (90) would result in (91):

$$Z_{out} = \frac{\hat{v}_{out}}{\hat{i}_{out}} = \frac{3(R_d + sL)}{s^2LC + s\left(\frac{3L}{R} + R_dC\right) + 3\frac{R_d}{R} + 1} \tag{91}$$

5.2.5. Converter Gain

The output voltage and the input voltage relationship is found by assuming $\hat{d}_j = 0,\ j = 1,2,$ and 3, summing Equations in (71) and substituting (67), (69), and (74) in the added equation.

$$\frac{3D_{eff}}{K}\left(1 + \frac{3R_d}{R}\right)\hat{v}_{in} = \left(\frac{(sL + R_d)(sRC + 3)}{R} + 1\right)\hat{v}_{out} \tag{92}$$

Simplifying (92) would result in (93):

$$G_{vg} = \frac{\hat{v}_{out}}{\hat{v}_{in}} = \frac{\frac{3D_{eff}}{K}\left(1 + \frac{3R_d}{R}\right)}{s^2LC + s\left(\frac{3L}{R} + R_dC\right) + \frac{3R_d}{R} + 1} \tag{93}$$

6. Generalized Small-Signal Analysis for Dual Series/Parallel Input-Output (ISIP-OSOP) DC-DC Converter

In this section, the SSM for the four architectures is expanded to include a generalized SSM applicable for ISIP-OSOP configuration. In addition, after generalizing the model, the equations derived in this section are validated with the IPOS, ISOP, and ISIPOS presented in [38–42], respectively.

6.1. ISIP-OSOP Generic DC-DC Converter Circuit Diagram

The ISIP-OSOP generic DC-DC converter configuration shown in Figure 7 consists of n modules that are connected in series and/or parallel at the input side, and in series and/or parallel at the output side, as shown in Figure 7.

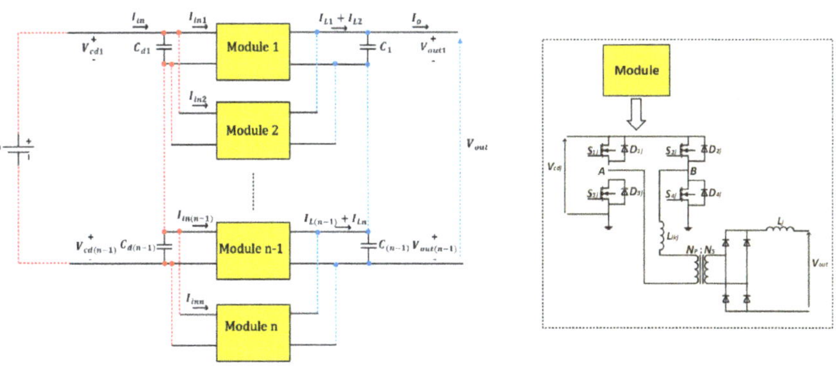

Figure 7. Generalized multimodule DC-DC converter circuit diagram.

By ensuring ICS and IVS, the input current per module is $\frac{I_{in}}{\alpha}$, and the input voltage per module is $\frac{V_{in}}{\beta}$. In which, I_{in} is the input current and V_{in} is the input voltage of the ISIP-OSOP DC-DC converter, and α is the number of modules connected in parallel, and β is the number of modules connected in series at the input side. Similarly, by ensuring OCS and OVS, the output current per module is $\frac{I_o}{a}$, and the output voltage per module is $\frac{V_o}{b}$. In which I_o is the output current and V_o is the output voltage of the ISIP-OSOP DC-DC converter, and a is the number of modules connected in parallel, and b is the number of modules connected in series at the output side.

6.2. ISIP-OSOP Generic DC-DC Converter Small-Signal Analysis

The SSM for the ISIP-OSOP converter shown in Figure 8 is derived using the SSM presented in [37], and expanding the study of the presented multimodule DC-DC converters presented in [38–42].

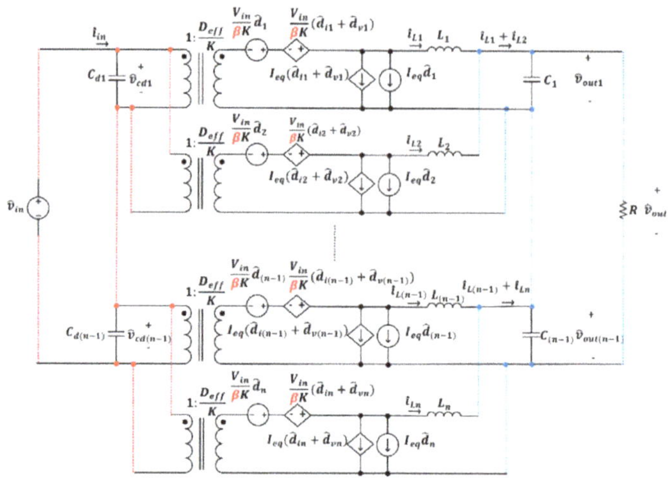

Figure 8. n module Input-Series Input-Parallel Output-Series Output-Parallel (ISIP-OSOP) DC-DC converter SSM.

Since the input current and voltage per module are $\frac{I_{in}}{\alpha}$ and $\frac{V_{in}}{\beta}$, respectively, and the output current and voltage per module is $\frac{I_o}{a}$ and $\frac{V_o}{b}$, respectively. Therefore, the load resistance per module is $\frac{a}{b}R$. Accordingly, $\hat{d}_{ij}$ and $\hat{d}_{vj}$, which are the effect of changing the filter inductor current and the effect of changing the input voltage on the duty cycle modulation, as well as I_{eq} presented in Figure 8 can be expressed as [39,42]:

$$S\hat{d}_{ij} = -\frac{4\beta L_{lk} f_s}{KV_{in}}\hat{i}_{Lj}, \quad j = 1, 2, \ldots, n \tag{94}$$

Equation (94) can be re-written as:

$$\hat{d}_{ij} = -\frac{\beta K R_d}{V_{in}}\hat{i}_{Lj}, \quad j = 1, 2, \ldots, n \tag{95}$$

where $R_d = \frac{4L_{lk} f_s}{k^2}$.

$$\hat{d}_{vj} = \frac{4\beta b L_{lk} f_s D_{eff}}{a k^2 R V_{in}}\hat{v}_{cdj}, \quad j = 1, 2, \ldots, n \tag{96}$$

Equation (96) can be re-written as:

$$\hat{d}_{vj} = \frac{\beta b R_d D_{eff}}{a R V_{in}}\hat{v}_{cdj}, \quad j = 1, 2, \ldots, n \tag{97}$$

$$SI_{eq} = \frac{bV_{in}}{\beta a K R} \tag{98}$$

The following equations are obtained from Figure 8:

$$\begin{cases} \frac{D_{eff}}{K}\hat{v}_{cd1} + \frac{V_{in}}{\beta K}\left(\hat{d}_{i1} + \hat{d}_{v1} + \hat{d}_1\right) = sL\hat{i}_{L1} + \hat{v}_{out1} \\ \frac{D_{eff}}{K}\hat{v}_{cd2} + \frac{V_{in}}{\beta K}\left(\hat{d}_{i2} + \hat{d}_{v2} + \hat{d}_2\right) = sL\hat{i}_{L2} + \hat{v}_{out2} \\ \vdots \\ \frac{D_{eff}}{K}\hat{v}_{cdn} + \frac{V_{in}}{\beta K}\left(\hat{d}_{in} + \hat{d}_{vn} + \hat{d}_n\right) = sL\hat{i}_{Ln} + \hat{v}_{outn} \end{cases} \tag{99}$$

$$\begin{cases} \hat{i}_{L11} + \hat{i}_{L21} + \cdots + \hat{i}_{La1} = \frac{sC}{sR_cC+1}\hat{v}_{out1} + \frac{\hat{v}_{out}}{R} \\ \hat{i}_{L12} + \hat{i}_{L22} + \cdots + \hat{i}_{La2} = \frac{sC}{sR_cC+1}\hat{v}_{out2} + \frac{\hat{v}_{out}}{R} \\ \vdots \\ \hat{i}_{L1b} + \hat{i}_{L2b} + \cdots + \hat{i}_{Lab} = \frac{sC}{sR_cC+1}\hat{v}_{outn} + \frac{\hat{v}_{out}}{R} \end{cases} \tag{100}$$

Based on the feature of modularity, it is assumed that all the employed modules are ideal. Moreover, the Equivalent Series Resistance (ESR) of the output capacitance is considered in this model.

Summing Equations in (100):

$$\sum_{i=1}^{a}\sum_{j=1}^{b}\hat{i}_{Lij} = \frac{sC}{sR_cC+1}\hat{v}_{out} + \frac{b\hat{v}_{out}}{R} \tag{101}$$

Equation (101) can be written as:

$$\sum_{i=1}^{a}\sum_{j=1}^{b}\hat{i}_{Lij} = \hat{v}_{out}\left(\frac{sRC + s\mathbf{b}R_cC + b}{R(1+sR_cC)}\right) \tag{102}$$

Defining the summation terms of the module's input and output voltage appearing after summing up Equations in (99):

$$\sum_{j=1}^{n}\hat{v}_{cdj} = \gamma\hat{v}_{in} \tag{103}$$

where:

- $\gamma = 1$, if all the modules at the input side are connected in series.
- $\gamma = \alpha$, if all the modules at the input side are connected in parallel.
- $\gamma = \left(1 + \frac{\alpha}{\beta}\right)$, if the modules at the input side are connected in series and parallel.

$$\sum_{j=1}^{n}\hat{v}_{outj} = c\hat{v}_{out} \tag{104}$$

where:

- $c = 1$, if all the modules at the output side are connected in series.
- $c = a$, if all the modules at the output side are connected in parallel.
- $c = \left(1 + \frac{a}{b}\right)$, if the modules at the output side are connected in series and parallel.

6.2.1. Control-to-Output Voltage Transfer Function

The output voltage and the duty cycle relationship is found by summing up Equations in (99), assuming $\hat{v}_{in} = 0$, and $\hat{d}_k = 0$, where $k = 1, 2, \ldots, n$ and $k \neq j$, and substituting (95), (97), (102), (103), and (104).

Summing Equations in (99):

$$\frac{D_{eff}}{K}\sum_{j=1}^{n}\hat{v}_{cdj} + \frac{V_{in}}{\beta K}\left(\sum_{j=1}^{n}\hat{d}_{ij} + \sum_{j=1}^{n}\hat{d}_{vj} + \sum_{j=1}^{n}\hat{d}_{j}\right) = sL\sum_{j=1}^{n}\hat{i}_{Lj} + \sum_{j=1}^{n}\hat{v}_{outj} \tag{105}$$

$$\frac{D_{eff}}{K}\sum_{j=1}^{n}\hat{v}_{cdj} + \frac{V_{in}}{\beta K}\left(\sum_{j=1}^{n} -\frac{\beta K R_d}{V_{in}}\hat{i}_{Lj} + \sum_{j=1}^{n}\frac{\beta b R_d D_{eff}}{aRV_{in}}\hat{v}_{cdj} + \hat{d}_1\right) = sL\sum_{j=1}^{n}\hat{i}_{Lj} + \sum_{j=1}^{n}\hat{v}_{outj} \tag{106}$$

$$\frac{D_{eff}}{K}\gamma\hat{v}_{in}+\frac{V_{in}}{\beta K}\left(\sum_{j=1}^{n}-\frac{\beta KR_d}{V_{in}}\hat{i}_{Lj}+\frac{\beta bR_dD_{eff}}{aRV_{in}}\gamma\hat{v}_{in}+\hat{d}_1\right)=sL\sum_{j=1}^{n}\hat{i}_{Lj}+c\hat{v}_{out} \tag{107}$$

Simplifying (107) would result in (108).

$$G_{vd}=\frac{\hat{v}_{out}}{\hat{d}_j}=\frac{\frac{V_{in}}{\beta K}(1+sR_cC)}{s^2LC\left(1+\frac{bR_c}{R}\right)+s\left(\frac{bL}{R}+R_dC\left(1+\frac{bR_c}{R}\right)+cR_cC\right)+\frac{bR_d}{R}+c} \tag{108}$$

6.2.2. Control-To-Filter Inductor Current Transfer Function

The filter inductor current and the duty cycle relationship is found by using (102) to find the following equation:

$$\hat{v}_{out}=\frac{R(1+sR_cC)}{sRC+sbR_cC+b}\sum_{i=1}^{a}\sum_{j=1}^{b}\hat{i}_{Lij} \tag{109}$$

Substituting (109) in (107) and considering the same assumptions as in Section 7.2.

$$\frac{V_{in}}{\beta K}\hat{d}_1-R_d\sum_{j=1}^{n}\hat{i}_{Lj}=sL\sum_{j=1}^{n}\hat{i}_{Lj}+\frac{cR(1+sR_cC)}{sRC+sbR_cC+b}\sum_{j=1}^{n}\hat{i}_{Lj} \tag{110}$$

Simplifying (110) would result in (111).

$$G_{id}=\frac{\hat{i}_L}{\hat{d}}=\frac{\frac{V_{in}}{\beta K}(b+sRC+sbR_cC)}{R\left(s^2LC\left(1+\frac{bR_c}{R}\right)+s\left(\frac{bL}{R}+R_dC\left(1+\frac{bR_c}{R}\right)+cR_cC\right)+\frac{bR_d}{R}+c\right)} \tag{111}$$

6.2.3. Output Impedance

Similarly, as studied in the SSM presented in [39,40], the ISIP-OSOP converter output impedance can be found by modifying (100), such that:

$$\begin{cases}\hat{i}_{L11}+\hat{i}_{L21}+\cdots+\hat{i}_{La1}+\hat{i}_{out}=\frac{sC}{sR_cC+1}\hat{v}_{out1}+\frac{\hat{v}_{out}}{R}\\ \hat{i}_{L12}+\hat{i}_{L22}+\cdots+\hat{i}_{La2}+\hat{i}_{out}=\frac{sC}{sR_cC+1}\hat{v}_{out2}+\frac{\hat{v}_{out}}{R}\\ \vdots\\ \hat{i}_{L1b}+\hat{i}_{L2b}+\cdots+\hat{i}_{Lab}+\hat{i}_{out}=\frac{sC}{sR_cC+1}\hat{v}_{outn}+\frac{\hat{v}_{out}}{R}\end{cases} \tag{112}$$

Summing Equations in (112):

$$\sum_{i=1}^{a}\sum_{j=1}^{b}\hat{i}_{Lij}=\frac{sC}{sR_cC+1}\hat{v}_{out}+\frac{b\hat{v}_{out}}{R}-b\hat{i}_{out} \tag{113}$$

Accordingly, (100) is modified as follows:

$$\sum_{i=1}^{a}\sum_{j=1}^{b}\hat{i}_{Lij}=\hat{v}_{out}\left(\frac{sRC+sbR_cC+b}{R(1+sR_cC)}\right)-b\hat{i}_{out} \tag{114}$$

The output voltage and the output current relationship is found by considering the same assumptions as in Section 7.2, summing Equations in (99), and substituting (95), (97), (103), (104), and (114).

$$\frac{V_{in}}{\beta K}\left(-\frac{\beta KR_d}{V_{in}}\right)\sum_{j=1}^{n}\hat{i}_{Lj}=sL\sum_{j=1}^{n}\hat{i}_{Lj}+c\hat{v}_{out} \tag{115}$$

$$-R_d\left(\hat{v}_{out}\left(\frac{sRC+sbR_cC+b}{R(1+sR_cC)}\right)-b\hat{i}_{out}\right)=sL\left(\hat{v}_{out}\left(\frac{sRC+sbR_cC+b}{R(1+sR_cC)}\right)-b\hat{i}_{out}\right)+c\hat{v}_{out} \quad (116)$$

Rearranging (116) would result in (117).

$$Z_{out}=\frac{\hat{v}_{out}}{\hat{i}_{out}}=\frac{b(R_d+sL)(1+sR_cC)}{s^2LC\left(1+\frac{bR_c}{R}\right)+s\left(\frac{bL}{R}+R_dC\left(1+\frac{bR_c}{R}\right)+cR_cC\right)+\frac{bR_d}{R}+c} \quad (117)$$

6.2.4. Converter Gain

The output voltage and the input voltage relationship is found by assuming $\hat{d}_j=0$, $j=1,2,\ldots,n$, summing Equations in (99), and substituting (95), (97), (102)–(104) in the added equation.

$$\frac{D_{eff}}{K}\sum_{j=1}^{n}\hat{v}_{cdj}+\frac{V_{in}}{\beta K}\left(\sum_{j=1}^{n}-\frac{\beta KR_d}{V_{in}}\hat{i}_{Lj}+\sum_{j=1}^{n}\frac{\beta bR_dD_{eff}}{aRV_{in}}\hat{v}_{cdj}\right)=sL\sum_{j=1}^{n}\hat{i}_{Lj}+\sum_{j=1}^{n}\hat{v}_{outj} \quad (118)$$

$$\frac{D_{eff}}{K}\gamma\left(1+\frac{bR_d}{aR}\right)\hat{v}_{in}=(sL+R_d)\left(\hat{v}_{out}\left(\frac{sRC+sbR_cC+b}{R(1+sR_cC)}\right)\right)+c\hat{v}_{out} \quad (119)$$

Rearranging (119) would result in (120).

$$G_{vg}=\frac{\hat{v}_{out}}{\hat{v}_{in}}=\frac{\frac{D_{eff}}{K}\gamma\left(1+\frac{bR_d}{aR}\right)(1+sR_cC)}{s^2LC\left(1+\frac{bR_c}{R}\right)+s\left(\frac{bL}{R}+R_dC\left(1+\frac{bR_c}{R}\right)+cR_cC\right)+\frac{bR_d}{R}+c} \quad (120)$$

6.3. ISIP-OSOP DC-DC Converter SSM Verification

In this section, the generalized multimodule DC-DC converter SSM is verified using three different models presented in [38–42].

6.3.1. Generalized Model Verification with a Two-Module IPOS DC-DC Converter

The proposed configuration in [38] consists of two modules that are connected in parallel at the input side, and series, at the output side, where the modules are FB-PS-based DC-DC converters. In this model, the effect of the ESR R_c is ignored, accordingly, when verifying the model presented in [34] with the generalized model, R_c should be equated to zero.

By ensuring ICS and OVS, the input current per module is $\frac{I_{in}}{2}$. However, the input voltage per module is V_{in}. Regarding the output side, the output voltage per module is $\frac{V_o}{2}$. However, the output current per module is I_o. In which, V_{in} is the IPOS input voltage, I_{in} is the IPOS input current, V_o is the IPOS output voltage, and I_o is the IPOS output current. Accordingly, the values for the six parameters that are previously defined are known and shown in Table 2.

Table 2. Values for the six parameters for a two-module Input-Parallel Output-Series (IPOS) power converter.

Defined Variables		
Input Side	α	2
	β	1
	γ	2 (same as α)
Output Side	a	1
	b	2
	c	1

Substituting the six parameters with their values in Equations (109), (112), (117), and (120) and substituting the ESR R_c with zero in the generalized transfer functions would result in the transfer functions presented in Table 3.

Table 3. Generalized model verification with the two-module IPOS power converter.

Transfer Functions for Two-Module IPOS DC-DC Converter	
G_{vd}	$\dfrac{\frac{V_{in}}{K}}{s^2LC+s\left(\frac{2L}{R}+R_dC\right)+\frac{2R_d}{R}+1}$
G_{id}	$\dfrac{\frac{V_{in}}{K}(2+sRC)}{R\left(s^2LC+s\left(\frac{2L}{R}+R_dC\right)+\frac{2R_d}{R}+1\right)}$
Z_{out}	$\dfrac{2(R_d+sL)}{s^2LC+s\left(\frac{2L}{R}+R_dC\right)+\frac{2R_d}{R}+1}$
G_{vg}	$\dfrac{\frac{2D_{eff}}{K}\left(1+\frac{2R_d}{R}\right)}{s^2LC+s\left(\frac{2L}{R}+R_dC\right)+\frac{2R_d}{R}+1}$

As can be seen from Table 3, the derived generalized model matches the model presented in [38], where the R_d term in the above transfer functions is substituted with its equivalence $\frac{4\,L_{lk}\,f_s}{K^2}$.

6.3.2. Generalized Model Verification with a Three-Module ISOP DC-DC Converter

In this Section, the generalized model is validated with the three-module ISOP power converter model-derived and presented in [39,40]. The three-module ISOP configuration consists of three modules that are connected in series at the input side, and parallel at the output side. In this model, the effect of the ESR R_c is considered, accordingly, when verifying the model presented [39] with the generalized model, R_c should not be ignored.

By ensuring IVS and OCS, the input voltage per module is $\frac{V_{in}}{3}$. However, the input current per module is I_{in}. Regarding the output side, the output current per module is $\frac{I_o}{3}$. However, the output voltage per module is V_o. Accordingly, the values for the six parameters that are previously defined are known and shown Table 4.

Table 4. Values for the six parameters for a three-module Input-Series Output-Parallel (ISOP) power converter.

Defined Variables		
Input Side	α	1
	β	3
	γ	1
Output Side	a	3
	b	1
	c	3

Substituting the six parameters with their values in Equations (109), (112), (117), and (120) would result in the transfer functions presented in Table 5.

Table 5. Generalized model verification with the three-module ISOP power converter.

Transfer Functions for Three-Module ISOP DC-DC Converter	
G_{vd}	$\dfrac{\frac{V_{in}}{3K}(1+sR_cC)}{s^2LC\left(1+\frac{R_c}{R}\right)+s\left(\frac{L}{R}+R_dC\left(1+\frac{R_c}{R}\right)+3R_cC\right)+\frac{R_d}{R}+3}$
G_{id}	$\dfrac{\frac{V_{in}}{3K}(1+sRC+sR_cC)}{R\left(s^2LC\left(1+\frac{R_c}{R}\right)+s\left(\frac{L}{R}+R_dC\left(1+\frac{R_c}{R}\right)+3R_cC\right)+\frac{R_d}{R}+3\right)}$
Z_{out}	$\dfrac{(R_d+sL)(1+sR_cC)}{s^2LC\left(1+\frac{R_c}{R}\right)+s\left(\frac{L}{R}+R_dC\left(1+\frac{R_c}{R}\right)+3R_cC\right)+\frac{R_d}{R}+3}$
G_{vg}	$\dfrac{\frac{D_{eff}}{K}\left(1+\frac{R_d}{3R}\right)(1+sR_cC)}{s^2LC\left(1+\frac{R_c}{R}\right)+s\left(\frac{L}{R}+R_dC\left(1+\frac{R_c}{R}\right)+3R_cC\right)+\frac{R_d}{R}+3}$

As can be seen from Table 5, the derived generalized model matches the model presented in [39,40].

6.3.3. Generalized Model Verification with a Four-Module ISIPOS DC-DC Converter

In this section, the generalized model is validated with the four-module ISIPOS power converter model presented in [41,42].

The four-module ISIPOS configuration consists of four modules that are connected in series and parallel at the input side, and series at the output side. In this model, the effect of the ESR R_c is ignored, accordingly, when verifying the model presented in [41] with the generalized model, R_c should be equated to zero.

By ensuring IVS, ICS, and OVS, the input voltage per module is $\frac{V_{in}}{2}$ and the input current per module is $\frac{I_{in}}{2}$. Regarding the output side, the output voltage per module is $\frac{V_o}{4}$. However, the output current per module is I_o. Accordingly, the values for the six parameters that are previously defined are known and shown in Table 6.

Table 6. Values for the six parameters for a four-module Input Series Input-Parallel Output-Series (ISIPOS) power converter.

Defined Variables		
Input Side	α	2
	β	2
	γ	$2=\left(1+\frac{\alpha}{\beta}\right)$
Output Side	a	1
	b	4
	c	1

Substituting the six parameters with their values in Equations (109), (112), (117), and (120) would result in the transfer functions presented in Table 7.

Table 7. Generalized model verification with the four-module ISIPOS power converter.

	Transfer Functions for Four-Module ISIPOS DC-DC Converter
G_{vd}	$\dfrac{\frac{V_{in}}{2K}}{s^2LC + s\left(\frac{4L}{R} + R_dC\right) + \frac{4R_d}{R} + 1}$
G_{id}	$\dfrac{\frac{V_{in}}{2K}(4 + sRC)}{R\left(s^2LC + s\left(\frac{4L}{R} + R_dC\right) + \frac{4R_d}{R} + 1\right)}$
Z_{out}	$\dfrac{4(R_d + sL)}{s^2LC + s\left(\frac{4L}{R} + R_dC\right) + \frac{4R_d}{R} + 1}$
G_{vg}	$\dfrac{\frac{2D_{eff}}{K}\left(1 + \frac{4R_d}{R}\right)}{s^2LC + s\left(\frac{4L}{R} + R_dC\right) + \frac{4R_d}{R} + 1}$

As can be seen from Table 7, the derived generalized model matches the model presented in [41,42].

7. Power Balancing in ISIP-OSOP DC-DC Converters

To guarantee stable operation in the presence of parameter variations, a control strategy that ensures power balancing is essential to provide equal power distribution among the modules. In this section, the control strategy for each multimodule configuration is studied to ensure IVS, ICS, OVS, and OCS if needed.

The possible control strategies for the four configurations have been discussed in [45]. According to the study presented in [45], for input parallel-connected systems, OCS and OVS controllers are required for IPOP and IPOS DC-DC converters, respectively. However, for input series-connected systems, IVS and OVS are required for ISOS DC-DC converters, and IVS and OCS are required for ISOP DC-DC converters. To further illustrate, active power sharing control for input parallel-connected systems is achieved by designing ICS controllers that achieve OCS and OVS for IPOP and IPOS systems or by designing OCS controllers for IPOP systems and designing OVS controllers for IPOS systems that also achieves ICS between the employed modules. However, for input-series connected systems, IVS controllers are necessary to achieve OVS and OCS for ISOS and ISOP DC-DC converters. However, in [46], Cross Feedback OCS (CFOCS) has been proposed for ISOP systems to achieve OCS without the need for IVS controllers, hence simplifying the overall control design. This control strategy has been tested for a three-module ISOP DC-DC converter considering RC in [40] to achieve IVS and OCS.

Since this paper mainly focuses on multimodule DC-DC converters designed for EV fast chargers, the control scheme designed for the four DC-DC converters is current-controlled considering an RC technique that is termed as burp charging or negative pulse charging. This charging technique is based on applying a short negative pulse or a short discharge pulse during the charging cycle. Such an algorithm offers significant advantages that can be highlighted in shortening the charging time and lowering the rise in temperature. Generally, the RC technique consists of three charging sequences, which are: a positive charging pulse, a rest period where no charging occurs, and a negative charging pulse or a discharge pulse [47]. Accordingly, the designed control schemes are based on controlling the filter inductor current of the DC-DC converters such that the output current profile is based on RC. Consequently, for IPOP and IPOS DC-DC converters, active power sharing is achieved through controlling the filter inductor currents, as shown in Figure 9a, that accordingly achieves equal output current distribution for IPOP systems and equal voltage distribution for IPOS systems. However, active power sharing in the ISOP DC-DC converter is achieved through the CFOCS presented in [40,46] and shown in Figure 9b, which eliminates the need for IVS controllers and control the output currents of the converters. The difference between Figure 9a and b is that, in Figure 9b, the current feedback for the individual module is the summation of the other two output currents and not its output current. Unlike the three systems ISOP, IPOP, and IPOS, active power sharing in the ISOS system is achieved through the use of both IVS and OVS controllers. In other words, the control scheme for the ISOS DC-DC converter combines the control scheme presented in Figure 9a with an IVS control, resulting in Figure 9c. To clarify, three IVS controllers are designed for each module to achieve equal input voltage distribution between the employed modules. In addition, the filter inductor currents are controlled using a reference current with an RC profile to achieve equal OVS between the modules of the ISOS converter. Therefore, it can be said that the control scheme presented in Figure 9a is applicable for the four architectures with slight differences presented in Figure 9b,c.

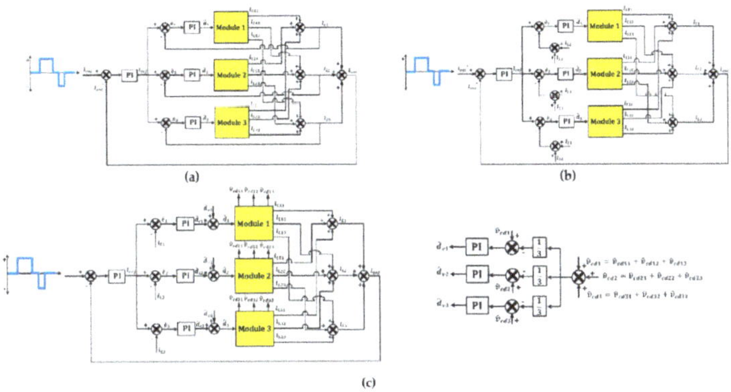

Figure 9. Control schemes for the four DC-DC converters; (**a**) control scheme used for input parallel-connected systems; (**b**) control scheme used for ISOP DC-DC converter; (**c**) control scheme used for ISOS DC-DC converter.

7.1. ISOS Control Strategy

In this Section, the ISOS DC-DC converter overall control scheme is examined with the system parameters presented in Table 8. To test the controller's power balancing capability when handling uncertainties, the component parameters for each module are purposely assumed to be different, as presented in Table 8.

Table 8. Input-Series Output-Series (ISOS) converter parameters.

Parameters	First Module	Second Module	Third Module 3
Overall converter rated power		350 kW	
Rated power per module		117 kW	
Overall input voltage		10 kV	
Input voltage per module		3.3 kV	
Overall output voltage		920 V	
Output voltage per module		307 V	
Modules Number		3	
Turns ratio	8.696 : 1	8.152 : 1	7.609 : 1
Leakage inductance	1142.72 μH	1004.35 μH	874.89 μH
Effective duty cycle	0.8	0.75	0.7
Input Capacitance	50 μF	80 μF	60 μF
Output filter inductor	50 mH	60 mH	60 mH
Output capacitance	300 μF	350 μF	300 μF
Load resistance		2.418 Ω	
Switching frequency		10 kHz	

Talking about the simulation results, the control scheme presented in Figure 9c for the ISOS DC-DC converter is examined considering a reference current with a reflex-current profile to the total current flowing in the filter inductance. The charging pulse is applied, where the charging cycle begins at 0.2 s and ends at 0.6 s. After that, a rest period for 0.1 s is applied to the total filter inductor current reference signal. The results shown in Figure 10 demonstrate that the power-sharing controller presented in Figure 9c compensates for the negative influences resulting from the systems' parameters mismatch. In which, the individual filter inductor currents are equally shared between the three modules. Moreover, the total filter inductor current of the ISOS system follows the reference signal applied based on RC. It is worth mentioning that the results in this section, as well as the following sections, involve the number of signals presented in the legend. However, the signals are coincident, meaning that power balance control is achieved. Consequently, it can be concluded that the control scheme is reliable and achieves equal power distribution between the modules.

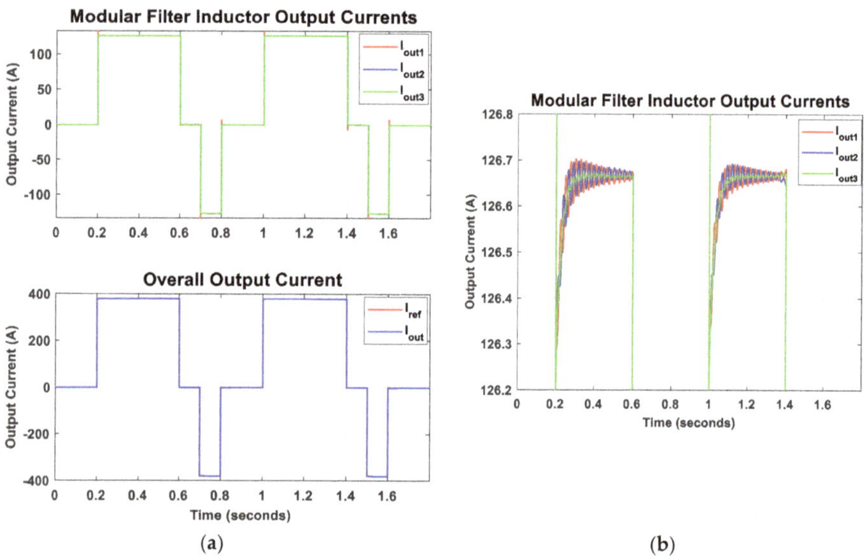

Figure 10. Simulation results for ISOS DC-DC converter: (**a**) modular filter inductor output currents and overall output current; (**b**) a zoomed-in illustration for the modular filter inductor output currents shown in (**a**).

7.2. ISOP Control Strategy

The overall control strategy for the ISOP system is examined and presented in [40] considering parameter mismatch for the three modules and a reference output current profile based on RC.

7.3. IPOP Control Strategy

In this Section, the IPOP DC-DC converter overall control scheme is examined with the system parameters presented in Table 9. To test the controller's power balancing capability when handling uncertainties, the component parameters for each module are purposely assumed to be different, as presented in Table 9.

Table 9. Input-Parallel Output-Parallel (IPOP) converter parameters.

Parameters	First Module	Second Module	Third Module 3
Overall converter rated power		350 kW	
Rated power per module		117 kW	
Overall input voltage		570 V	
Input voltage per module		570 V	
Overall output voltage		400 V	
Output voltage per module		400 V	
Modules Number		3	
Turns ratio	1.14 : 1	1.069 : 1	1.003 : 1
Leakage inductance	3.712 μH	3.262 μH	2.871 μH
Effective duty cycle	0.8	0.75	0.7
Input Capacitance	50 μF	80 μF	60 μF
Output filter inductor	50 mH	60 mH	60 mH
Output capacitance	300 μF	350 μF	300 μF
Load resistance		0.457 Ω	
Switching frequency		10 kHz	

For parallel connected systems, active power-sharing control is achieved through directing the control efforts toward the output filter inductor currents without the need for any additional controllers. After designing the control scheme for the output filter inductor currents, the input currents are monitored to ensure that ICS is achieved through OCS. The same reference signal based on RC is applied to assess the controller scheme shown in Figure 9a. It can be observed from Figure 11 that the control scheme for the three-module IPOP system can achieve equal power distribution among the modules under parameter mismatch. In which, the modular input currents, as well as the modular output currents, are equally shared between the three modules. In addition, the overall filter inductor current of the IPOP power converter follows the reference signal applied based on RC.

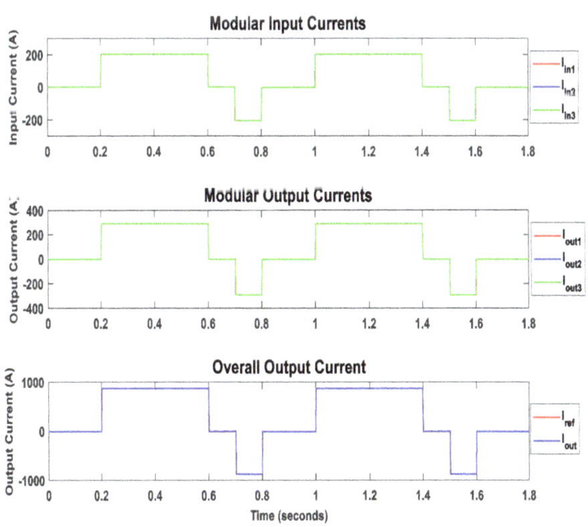

Figure 11. Simulation results for IPOP DC-DC converter.

7.4. IPOS Control Strategy

In this Section, the IPOS DC-DC converter overall control scheme is examined with the system parameters presented in Table 10. To test the controller's power balancing capability when handling uncertainties, the component parameters for each module are purposely assumed to be different, as presented in Table 10.

Table 10. IPOS converter parameters.

Parameters	First Module	Second Module	Third Module 3
Overall converter rated power		350 kW	
Rated power per module		117 kW	
Overall input voltage		570 V	
Input voltage per module		570 V	
Overall output voltage		920 V	
Output voltage per module		307 V	
Modules Number		3	
Turns ratio	1.485 : 1	1.393 : 1	1.299 : 1
Leakage inductance	33.383 μH	29.341 μH	25.559 μH
Effective duty cycle	0.8	0.75	0.7
Input Capacitance	50 μF	80 μF	60 μF
Output filter inductor	50 mH	60 mH	60 mH
Output capacitance	300 μF	350 μF	300 μF
Load resistance		2.421 Ω	
Switching frequency		10 kHz	

In IPOS, OVS is achieved through controlling the filter inductor currents of each module. This accordingly achieves ICS between the three modules. The reference signal for the overall filter inductor current is also based on RC. It can be observed from Figure 12 that the control scheme for the three-module IPOS DC-DC converter presented in Figure 9a can achieve equal ICS and equal OVS among the employed IPOS modules. In which, the modular input currents, as well as the modular output voltages, are equally shared between the three modules. In addition, the overall filter inductor current follows the same current profile of the reference signal.

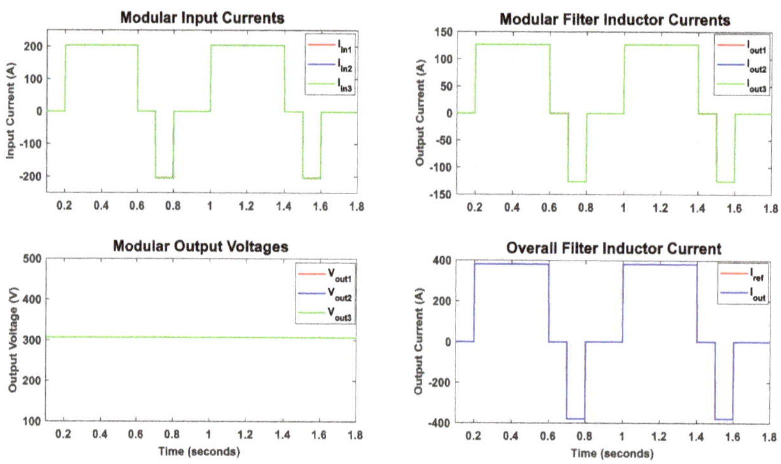

Figure 12. Simulation results for IPOS DC-DC converter.

8. Conclusions

In recent years, EVs have gained attention because of their significant role in reducing CO_2 emissions, minimizing noise pollution, avoiding the high price of fuel, and providing consumers with more efficient and environmentally friendly EVs. To allow for broader adoption of such EVs, diminishing the charging process would significantly increase the EV's efficiency as well as their utilization. Towards this direction, advanced power electronics technologies are emerging significantly to achieve high efficiency as well as ensuring extended battery lifetime. Among these power electronics conversion systems, the DC-DC conversion stage plays a vital role in charging the EV's battery.

Consequently, this paper aims to present the four basic architectures for the modular DAB DC-DC converter with their SSM and control strategy to guarantee uniform power sharing among the modules. In addition, the main contribution of this paper can be summarized in providing generalized SSM for any multimodule connection, including ISIP-OSOP configurations to support in the design of the DC-DC stage control. In other words, the dynamic behavior of such multimodule converters is studied using the SSM to develop a generalized dual series/parallel input-output dual active bridge SSM applicable for EVs. The provided model is verified with three different models presented in the literature, which are: two-module IPOS converter, three-module ISOP converter, and four-module ISIPOS converter. Moreover, to assess the effectiveness of the control schemes, simulation results are presented where parameter mismatch is introduced between the modules to ensure power balancing in the overall converter system.

Author Contributions: For research articles with several authors, a short paragraph specifying their individual contributions must be provided. M.E. and A.M. contributed to the whole research work and analysis tools; M.E. wrote the paper. This work was performed under the supervision with regular and continuous feedback of A.M. All authors have read and agreed to the published version of the manuscript.

Funding: This work was supported by NPRP grant NPRP (10-0130-170286) from the Qatar National Research Fund (a member of Qatar Foundation).

Acknowledgments: This work was supported by NPRP grant NPRP (10-0130-170286) from the Qatar National Research Fund (a member of Qatar Foundation). The statements made herein are solely the responsibility of the authors.

Conflicts of Interest: The authors declare no conflict of interest.

Abbreviations

EVs	Electric Vehicles
UFC	Ultra-Fast Charging
UF-EVC	Ultra-Fast EV Charging
SoC	State-of-Charge
DAB	Dual Active Bridge
DHB	Dual Half Bridge
ISOS	Input-Series Output-Series
IPOP	Input-Parallel Output-Parallel
ISOP	Input-Series Output-Parallel
IPOS	Input-Parallel Output-Series
SSM	Small-Signal Model
ISIP-OSOP	Input-Series Input-Parallel Output-Series Output-Parallel
FB-PS	Full-Bridge Phase-Shift
ISIPOS	Input-Series Input-Parallel Output-Series
RC	Reflex Charging
IVS	Input Voltage Sharing
OVS	Output Voltage Sharing
ICS	Input Current Sharing
OCS	Output Current Sharing
ESR	Equivalent Series Resistance
CFOCS	Cross Feedback Output Current Sharing

References

1. Vasiladiotis, M.; Rufer, A. A Modular Multiport Power Electronic Transformer with Integrated Split Battery Energy Storage for Versatile Ultrafast EV Charging Stations. *IEEE Trans. Ind. Electron.* **2015**, *62*, 3213–3222. [CrossRef]
2. Kesler, M.; Kisacikoglu, M.C.; Tolbert, L.M. Vehicle-to-Grid Reactive Power Operation Using Plug-In Electric Vehicle Bidirectional Offboard Charger. *IEEE Trans. Ind. Electron.* **2014**, *61*, 6778–6784. [CrossRef]

3. Hu, X.; Zou, C.; Tang, X.; Liu, T.; Hu, L. Cost-Optimal Energy Management of Hybrid Electric Vehicles Using Fuel Cell/Battery Health-Aware Predictive Control. *IEEE Trans. Power Electron.* **2020**, *35*, 382–392. [CrossRef]
4. Teng, F.; Ding, Z.; Hu, Z.; Sarikprueck, P. Technical Review on Advanced Approaches for Electric Vehicle Charging Demand Management, Part I: Applications in Electric Power Market and Renewable Energy Integration. *IEEE Trans. Ind. Appl.* **2020**. [CrossRef]
5. Seth, A.K.; Singh, M. Resonant controller of single-stage off-board EV charger in G2V and V2G modes. *IET Power Electron.* **2020**, *13*, 1086–1092. [CrossRef]
6. Zeb, M.Z.; Imran, K.; Khattak, A.; Janjua, A.K.; Pal, A.; Nadeem, M.; Zhang, J.; Khan, S. Optimal Placement of Electric Vehicle Charging Stations in the Active Distribution Network. *IEEE Access* **2020**, *8*, 68124–68134. [CrossRef]
7. Christen, D.; Tschannen, S.; Biela, J. Highly efficient and compact DC-DC converter for ultra-fast charging of electric vehicles. In Proceedings of the 15th International Power Electronics and Motion Control Conference (EPE/PEMC), Novi Sad, Serbia, 2–4 September 2012; pp. LS5d.3-1–LS5d.3-8.
8. Albert, G.B.; Andrew, C.C.; Suzanne, M.; David, L.W. Vehicle Electrification: Status and Issues. *Proc. IEEE* **2011**, *99*, 1116–1138.
9. Pierre, M.; Jemelin, C.; Louvet, N. Driving an Electric Vehicle: A sociological analysis on pioneer users. *Proc. Energy Effic.* **2011**, *4*, 511–522. [CrossRef]
10. Hidrue, M.; Parsons, G.; Kempton, W.; Gardner, M. Willingness to pay for electric vehicles and their attributes. *Resour. Energy Econ.* **2011**, *33*, 687–705. [CrossRef]
11. Yilmaz, M.; Krein, P.T. Review of battery charger topologies, charging power levels and infrastructure for plug-in electric and hybrid vehicles. *IEEE Trans. Power Electron.* **2013**, *28*, 2151–2169. [CrossRef]
12. Khaligh, A.; Dusmez, S. Comprehensive topological analysis of conductive and inductive charging solutions for plug-in electric vehicles. *IEEE Trans. Veh. Technol.* **2012**, *61*, 3475–3489. [CrossRef]
13. Hartmann, M.; Friedli, T.; Kolar, J.W. Three-phase unity power factor mains interfaces of high power EV battery charging systems. In Proceedings of the Workshop ECPE Power Electronics for Charging Electric Vehicles, Valencia, Spain, 21–22 March 2011; pp. 1–66.
14. Intel. *Revolutionizing Fast Charging for Electric Vehicles*; Intel: Santa Clara, CA, USA, 2013.
15. Wang, S.; Crosier, R.; Chu, Y. Investigating the power architectures and circuit topologies for megawatt superfast electric vehicle charging stations with enhanced grid support functionality. In Proceedings of the 2012 IEEE International Electric Vehicle Conference, Greenville, SC, USA, 4–8 March 2012; pp. 1–8.
16. Whitwam, R. BMW, Porsche Demo Super-Fast Electric Car Charger. *ExtremeTech* **2018**. Available online: https://www.extremetech.com/extreme/282364-bmw-porsche-demo-super-fast-electric-car-charger (accessed on 9 March 2019).
17. BMW. Ultra-Fast Charging Technology Ready for Future of Electric Vehicles. *Electron. Compon. News* **2018**. Available online: https://www.ecnmag.com/news/2018/12/ultra-fast-charging-technology-ready-future-electric-vehicles (accessed on 9 March 2019).
18. Yuan, Z.; Xu, H.; Chao, Y.; Zhang, Z. A novel fast charging system for electrical vehicles based on input-parallel output-parallel and output-series. In Proceedings of the 2017 IEEE Transportation Electrification Conference and Expo, Asia-Pacific (ITEC Asia-Pacific), Harbin, China, 2–5 August 2017; pp. 1–6.
19. Beldjajev, V. Research and Development of the New Topologies for the Isolation Stage of the Power Electronic Transformer. Master's Thesis, Tallinn University of Technology, Tallinn, Estonia, 2013.
20. Engel, S.P.; Stieneker, M.; Soltau, N.; Rabiee, S.; Stagge, H.; de Doncker, R.W. Comparison of the Modular Multilevel DC Converter and the Dual-Active Bridge Converter for Power Conversion in HVDC and MVDC Grids. *IEEE Trans. Power Electron.* **2015**, *30*, 124–137. [CrossRef]
21. Sari, H.I. *DC/DC Converters for Multi-terminal HVDC Systems Based on Modular Multilevel Converter*; Norwegian University of Science and Technology: Kongeriket, Norway, 2016.
22. Yang, H. Modular and Scalable DC-DC Converters for Medium-/High-Power Applications. Master's Thesis, Georgia Institute of Technology, Atlanta, GA, USA, 2017.
23. Fan, H.; Li, H. A high-frequency medium-voltage DC-DC converter for future electric energy delivery and management systems. In Proceedings of the 8th International Conference on Power Electronics—ECCE Asia, Jeju, Korea, 30 May–1 June 2011; pp. 1031–1038.
24. Papadakis, C. Protection of HVDC Grids Using DC Hub. Master's Thesis, Delft University of Technology, Delft, the Netherlands, 2017.

25. Carrizosa, M.J.; Benchaib, A.; Alou, P.; Damm, G. DC transformer for DC/DC connection in HVDC network. In Proceedings of the 15th European Conference on Power Electronics and Applications (EPE), Lille, France, 2–6 September 2013; pp. 1–10.
26. Davidson, C.C.; Trainer, D.R. Innovative concepts for hybrid multi-level converters for HVDC power transmission. In Proceedings of the 9th IET International Conference on AC and DC Power Transmission (ACDC 2010), London, UK, 19–21 October 2010; pp. 1–5.
27. Alatalo, M. *Module Size Investigation on Fast Chargers for BEV*; Chalmers University of Technology: Gothenburg, Sweden, 2018.
28. Jalakas, T.; Roasto, I.; Vinnikov, D. Electric vehicle fast charger high voltage input multiport converter topology analysis. In Proceedings of the 2013 International Conference-Workshop Compatibility and Power Electronics, Ljubljana, Slovenia, 5–7 June 2013; pp. 326–331.
29. Rivera, S.; Wu, B. Electric Vehicle Charging Station with an Energy Storage Stage for Split-DC Bus Voltage Balancing. *IEEE Trans. Power Electron.* **2017**, *32*, 2376–2386. [CrossRef]
30. Rivera, S.; Wu, B.; Kouro, S.; Yaramasu, V.; Wang, J. Electric Vehicle Charging Station Using a Neutral Point Clamped Converter with Bipolar DC Bus. *IEEE Trans. Ind. Electron.* **2015**, *62*, 1999–2009. [CrossRef]
31. Srdic, S.; Liang, X.; Zhang, C.; Yu, W.; Lukic, S. A SiC-based high-performance medium-voltage fast charger for plug-in electric vehicles. In Proceedings of the 2016 IEEE Energy Conversion Congress and Exposition (ECCE), Milwaukee, WI, USA, 18–22 September 2016; pp. 1–6.
32. Aggeler, D.; Canales, F.; Zelaya, H.; La Parra, D.; Coccia, A.; Butcher, N.; Apeldoorn, O. Ultra-fast DC-charge infrastructures for EV-mobility and future smart grids. In Proceedings of the 2010 IEEE PES Innovative Smart Grid Technologies Conference Europe (ISGT Europe), Gothenburg, Sweden, 10–13 October 2010; pp. 1–8.
33. Cui, T.; Liu, C.; Shan, R.; Wang, Y.; Kong, D.; Guo, J. A Novel Phase-Shift Full-Bridge Converter with Separated Resonant Networks for Electrical Vehicle Fast Chargers. In Proceedings of the 2018 IEEE International Power Electronics and Application Conference and Exposition (PEAC), Shenzhen, China, 4–7 November 2018; pp. 1–6.
34. Justino, J.C.G.; Parreiras, T.M.; Filho, B.J.C. Hundreds kW Charging Stations for e-Buses Operating Under Regular Ultra-Fast Charging. *IEEE Trans. Ind. Appl.* **2016**, *52*, 1766–1774.
35. Tian, Q.; Huang, A.Q.; Teng, H.; Lu, J.; Bai, K.H.; Brown, A.; McAmmond, M. A novel energy balanced variable frequency control for input-series-output-parallel modular EV fast charging stations. In Proceedings of the 2016 IEEE Energy Conversion Congress and Exposition (ECCE), Milwaukee, WI, USA, 18–22 September 2016; pp. 1–6.
36. Vasiladiotis, M.; Rufer, A.; Béguin, A. Modular converter architecture for medium voltage ultra fast EV charging stations: Global system considerations. In Proceedings of the 2012 IEEE International Electric Vehicle Conference, Greenville, SC, USA, 4–8 March 2012; pp. 1–7.
37. Vlatkovic, V.; Sabate, J.A.; Ridley, R.B.; Lee, F.C.; Cho, B.H. Small-signal analysis of the phase-shifted PWM converter. *IEEE Trans. Power Electron.* **1992**, *7*, 128–135. [CrossRef]
38. Lian, Y.; Adam, G.; Holliday, D.; Finney, S. Modular input-parallel output-series DC/DC converter control with fault detection and redundancy. *IET Gener. Transm. Distrib.* **2016**, *10*, 1361–1369. [CrossRef]
39. Ruan, X.; Chen, W.; Cheng, L.; Tse, C.K.; Yan, H.; Zhang, T. Control Strategy for Input-Series–Output-Parallel Converters. *IEEE Trans. Ind. Electron.* **2009**, *56*, 1174–1185. [CrossRef]
40. ElMenshawy, M.; Massoud, A. Multimodule ISOP DC-DC Converters for Electric Vehicles Fast Chargers. In Proceedings of the 2nd International Conference on Smart Grid and Renewable Energy (SGRE), Doha, Qatar, 19–21 November 2019; pp. 1–6.
41. Lian, Y.; Adam, G.P.; Holliday, D.; Finney, S.J. Medium-voltage DC/DC converter for offshore wind collection grid. *IET Renew. Power Gener.* **2016**, *10*, 651–660. [CrossRef]
42. ElMenshawy, M.; Massoud, A. Multimodule DC-DC Converters for High-Voltage High-Power Renewable Energy Sources. In Proceedings of the 2nd International Conference on Smart Grid and Renewable Energy (SGRE), Doha, Qatar, 19–21 November 2019; pp. 1–6.
43. Sha, D.; Guo, Z.; Luo, T.; Liao, X. A General Control Strategy for Input-Series–Output-Series Modular DC-DC Converters. *IEEE Trans. Power Electron.* **2014**, *29*, 3766–3775. [CrossRef]
44. Cheng, J.; Shi, J.; He, X. A novel input-parallel output-parallel connected DC-DC converter modules with automatic sharing of currents. In Proceedings of the 7th International Power Electronics and Motion Control Conference, Harbin, China, 2–5 June 2012; pp. 1871–1876.

45. Chen, W.; Ruan, X.; Yan, H.; Tse, C.K. DC/DC Conversion Systems Consisting of Multiple Converter Modules: Stability, Control, and Experimental Verifications. *IEEE Trans. Power Electron.* **2009**, *24*, 1463–1474. [CrossRef]
46. Sha, D.; Guo, Z.; Liao, X. Cross-Feedback Output-Current-Sharing Control for Input-Series-Output-Parallel Modular DC–DC Converters. *IEEE Trans. Power Electron.* **2010**, *25*, 2762–2771. [CrossRef]
47. Lee, C.S.; Lin, C.H.; Lai, S.-Y. Development of Fast Large Lead-Acid Battery Charging System Using Multi-state Strategy. *Int. J. Comput. Consum. Control. (IJ3C)* **2013**, *2*, 56–65.

© 2020 by the authors. Licensee MDPI, Basel, Switzerland. This article is an open access article distributed under the terms and conditions of the Creative Commons Attribution (CC BY) license (http://creativecommons.org/licenses/by/4.0/).

energies

Article

Optimal Energy Management of Plug-In Hybrid Electric Vehicles Concerning the Entire Lifespan of Lithium-Ion Batteries

Zeyu Chen *, Jiahuan Lu, Bo Liu, Nan Zhou and Shijie Li

School of Mechanical Engineering and Automation, Northeastern University, Shenyang 110819, China; jhlu@stu.neu.edu.cn (J.L.); Boliu@stu.neu.edu.cn (B.L.); nanzhou@mail.neu.edu.cn (N.Z.); shijie02@stu.neu.edu.cn (S.L.)

* Correspondence: chenzy@mail.neu.edu.cn; Tel.: +86-(024)-8369-1095

Received: 31 March 2020; Accepted: 10 May 2020; Published: 17 May 2020

Abstract: The performance of lithium-ion batteries will inevitably degrade during the high frequently charging/discharging load applied in electric vehicles. For hybrid electric vehicles, battery aging not only declines the performance and reliability of the battery itself, but it also affects the whole energy efficiency of the vehicle since the engine has to participate more. Therefore, the energy management strategy is required to be adjusted during the entire lifespan of lithium-ion batteries to maintain the optimality of energy economy. In this study, tests of the battery performances under thirteen different aging stages are involved and a parameters-varying battery model that represents the battery degradation is established. The influences of battery aging on energy consumption of a given plug-in hybrid electric vehicle (PHEV) are analyzed quantitatively. The results indicate that the variations of capacity and internal resistance are the main factors while the polarization and open circuit voltage (OCV) have a minor effect on the energy consumption. Based on the above efforts, the optimal energy management strategy is proposed for optimizing the energy efficiency concerning both the fresh and aging batteries in PHEV. The presented strategy is evaluated by a simulation study with different driving cycles, illustrating that it can balance out some of the harmful effects that battery aging can have on energy efficiency. The energy consumption is reduced by up to 2.24% compared with that under the optimal strategy without considering the battery aging.

Keywords: battery aging; plug-in electric vehicles; energy management; global optimization; state of health; particle swarm algorithm; genetic algorithm

1. Introduction

With the deepening of environmental deterioration and energy crisis issues, developing high efficient and clean automobiles has been recognized as a matter of global significance [1]. In the recent years, electric vehicles (EVs) are widely recognized as the development tendency of automobile industries all over the world [2,3]. The benefits of EVs highly depend upon the onboard high-capacity battery pack that can be recharged by the power grid. The frequently discharge/charge cycles during the vehicular utilizations will inevitably cause the degradation of the power battery [4]. Battery aging, which implies a complex electrochemical evolution process of the gradually loss of lithium inventory and active material, has been fully discussed in many literatures [5,6]. In order to monitor the battery health condition and improve the battery performance, many efforts have been exerted on estimation of the battery state of health (SOH) or remaining useful life (RUL) [7–9]. However, battery aging not only influences the performance of the battery itself, but it also has an impact on the vehicle performance, like reducing maximum power, driving range, energy economy, etc., of the vehicle. Especially in hybrid electric vehicles (HEVs), battery aging will seriously affect the overall energy efficiency since the engine has to contribute more power than expected, resulting in the increased energy consumption and emissions. Therefore, the energy management strategy should be adjusted to maintain the optimality during the entire lifespan of lithium-ion batteries.

Energy management strategy (EMS) is integral part of improving the fuel economy of both the traditional HEVs and plug-in hybrid electric vehicles (PHEVs), which have drawn attentions from many researchers [10–13]. Nevertheless, the current studies mainly focus on the optimization methods towards how to maximize the hybrid system's advantages, without enough concerns of the impacts of battery aging. Normally, the existing methods of EMS can be divided into two categories: the rule-based method and the optimization-based method. The rule-based strategies mainly depend upon some predefined control rules, containing deterministic rules and fuzzy logic rules, to operate the power units at high efficiency [14]. For example, Gao et al. [15] proposed a deterministic rule-based energy management strategy for PHEV focused on all electric range and charge depletion range operations, which has been verified by an example passenger car in a typical urban driving cycle; Schouten et al. [16] presented a fuzzy logic-based energy management strategy to improve the fuel economy of the parallel hybrid electric vehicle; Ali et al. [17] proposed a fuzzy logic control for electric vehicles, the presented method can achieve an efficient and fast-charging of the lithium-ion batteries. The rule-based strategies have been widely used in real-time control because they are simple, easy to be online implemented, and have good robustness.

The optimization-based strategies are designed to achieve the optimal control performance by using advanced optimization algorithms, such as dynamic programming (DP) [18,19], genetic algorithm (GA), particle swarm optimization (PSO) [20], etc. These algorithms adopt a common cost function, namely, to minimize the fuel consumption (or maximum the fuel conversion efficiency) of the vehicle during a certain time horizon. For example, Larsson et al. [21] investigated the DP-based energy management strategy to minimize the fuel consumption of a hybrid electric vehicle and discussed how much computational demand can be reduced. The drawback of these global optimization algorithm-based strategies is that they can barely be implemented in real-time control due to their dependence on an a priori known speed profile. Therefore, they are often implemented offline as a reference or a benchmark for other algorithms [22]. In addition, there is another kind of optimization approach, namely, the instantaneous minimization algorithm, which is to minimize the cost function at each time step. Most representative one of this kind is equivalent consumption minimization strategy (ECMS). Although ECMS can only provide a near-optimal solution, it can be implemented online because it does not rely on an a priori known speed profile. The specific descriptions about ECMS can be found in References [23–26].

The above investigations have achieved a great progress of resolving the energy management issues; however, most of these studies are based on the characteristics of the fresh battery. Although the battery aging induces a significant impact on energy consumption, it is still unknown how much the extra energy consumption can be caused by the battery aging in PHEVs and there are few studies to deal with battery aging from the perspective of energy management. In this study, we expect to reveal the maximum influence that battery aging can produce on the vehicle energy consumption and to present a global optimal control strategy over the entire lifespan of onboard batteries. The main target of the presented strategy is to maintain the optimal energy efficiency even after the serious aging of the battery and partially compensate for the negative impact of battery aging from the system level. The remainder of this paper is organized as follows: the model, energy management scheme and optimization method for PHEVs concerning the impacts of battery aging are described in Section 2; the battery model and the mathematical expression of the aging characteristics are proposed in Section 3; the impacts of the battery aging on energy consumption and the results of the EMS are illustrated in Section 4 while the conclusions are summarized in Section 5.

2. Energy Management Strategy

2.1. Problem Formulation

Figure 1 shows the power system of PHEV with a serious topology, in which the battery pack is the main onboard energy storage system (ESS) to provide the power load of the electric motor and an engine generator group is used as the assistance power unit (APU). The electric motor drives the front axle of the vehicle through an automatic mechanical transmission (AMT). The electrical power demand of the motor is split between the ESS and APU, as commanded by the energy management system. The parameters of these main components and the modeling have been described in our previous research presented in Reference [27,28], that is the foundation of this paper.

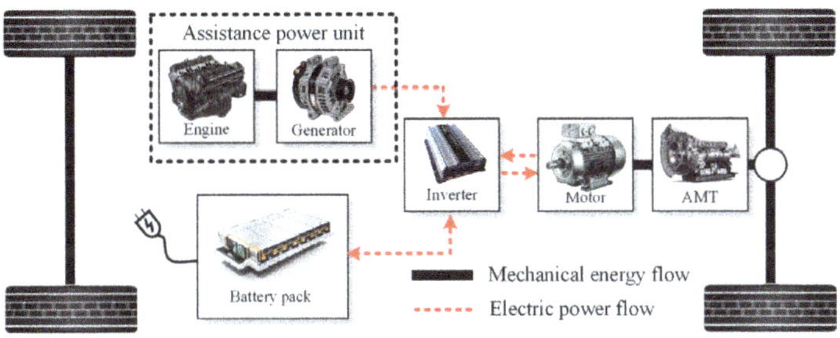

Figure 1. The structure diagram of plug-in hybrid electric vehicles (PHEV) power system.

The energy management issue is described as finding the optimal power allocation for minimizing the cost function while meeting the constraints. The objective function is minimizing the energy consumption during a certain driving cycle, described by

$$\text{Minimize} : J(u) = \Phi_B + \sum_{k=1}^{N}\left\{v_{fuel}P_E(k)\Delta t\right\} \tag{1}$$

where u denotes the control policy, Φ_B is the cost of electricity, N is the total number of steps, P_E is the engine power, v_{fuel} is the price of fuel (CNY-¥/kWs), t_0 and t_f denote the time at the beginning and ending the driving cycle, respectively, t denotes the discrete time instant, $t = t_0 + k\Delta t$, $k = 1,2,\ldots,N$.

For a certain driving condition, the electricity consumption is directly calculated according to the total battery power consumption, described by

$$\begin{aligned}\Phi_B &= v_{elec} \cdot \Delta E_{Batt} \\ &= v_{elec} \cdot \frac{\sum_{k=1}^{N}\{P_{Batt}(k)\Delta t\}}{\eta_b};\end{aligned} \tag{2}$$

in which

$$\eta_b = \begin{cases} \eta_{discha} & \text{at discharging} \\ \frac{1}{\eta_{cha}} & \text{at charging} \end{cases}$$

where ΔE_{Batt} is the battery electricity consumption, v_{elec} is the price of electricity (CNY-¥/kWs), P_{Batt} is power of the battery, η_b is the efficiency of the battery, η_{cha} and η_{discha} denote the efficiencies at charging and discharging, respectively.

If the battery aging is not considered, the cost function is only impacted by the power allocation. Thus, the control variable can be defined as $u = [P_E(k)\ P_{Batt}(k)]^T$. The optimization problem is to find out the optimal u^* to minimize the above cost function. To ensure the optimization results conform to feasible solutions, the control variable is subject to some constraints below.

$$P_{Batt,\min}(z(k)) \le P_{Batt}(k) \le P_{Batt,\max}(z(k)) \tag{3}$$

$$0 \le P_E(k) \le P_{E,\max} \tag{4}$$

$$\varphi_1 \le z(k) \le \varphi_2 \tag{5}$$

where $P_{Batt,\max}$ and $P_{Batt,\min}$ are power limitations of the battery pack, $P_{E,\max}$ is the maximum power of the engine, φ_1 and φ_2 are limitations of battery state of charge (SOC).

Based on the power balance relation, equality constraints are given as:

$$P_{\mathrm{Req}}(k) = P_E(k)\eta_{APU} + P_{Batt}(k)\eta_{Batt} \tag{6}$$

in which

$$\eta_{Batt} = \begin{cases} \eta_{dis}\eta_{Inv}, & if\ P_{Batt}(k) \ge 0 \\ \frac{1}{\eta_{chg}\eta_{Inv}}, & if\ P_{Batt}(k) < 0 \end{cases}$$

where P_{Req} is the power demand of the electric motor, η_{APU} and η_{Batt} are efficiencies of APU and battery system, respectively, η_{dis} is the discharging efficiency of battery, P_{Batt} is positive while the battery is discharging and negative while the battery is charging.

In addition, the battery SOC at the initial time of the optimization horizon should be pre-set by

$$z(t_0) = z_0 \tag{7}$$

where z_0 is initial SOC of the battery pack.

2.2. Control Strategy

The above model describes the basic mathematical problems of energy management issue, but in practice, we still need to consider the influence of aging on the model. Therefore, in the design of control strategy, battery aging characteristics are taken into account. For convenience of description, here we simply use the SOH (defined as the ratio of the current maximum capacity and the nominal capacity) to describe the different state of battery aging. The more specific parameters variations and the mathematical expression of the battery aging model will be further discussed in Section 3. Figure 2 shows the flow chart of the presented EMS. The blended control strategy [29] is used for power allocation, where the total power demand is split between the lithium battery pack and APU according to the control rules described below. To clearly illustrate the control algorithm, two thresholds are given at first, namely δ_1 and δ_2, where the δ_1 denotes the high SOC level threshold and δ_2 denotes the low SOC threshold. If the battery SOC $\ge \delta_1$, the battery will provide as much power to supply the load requirement; in this case, the optimization is not required. When battery SOC drops below the δ_1, the control algorithm is given as below.

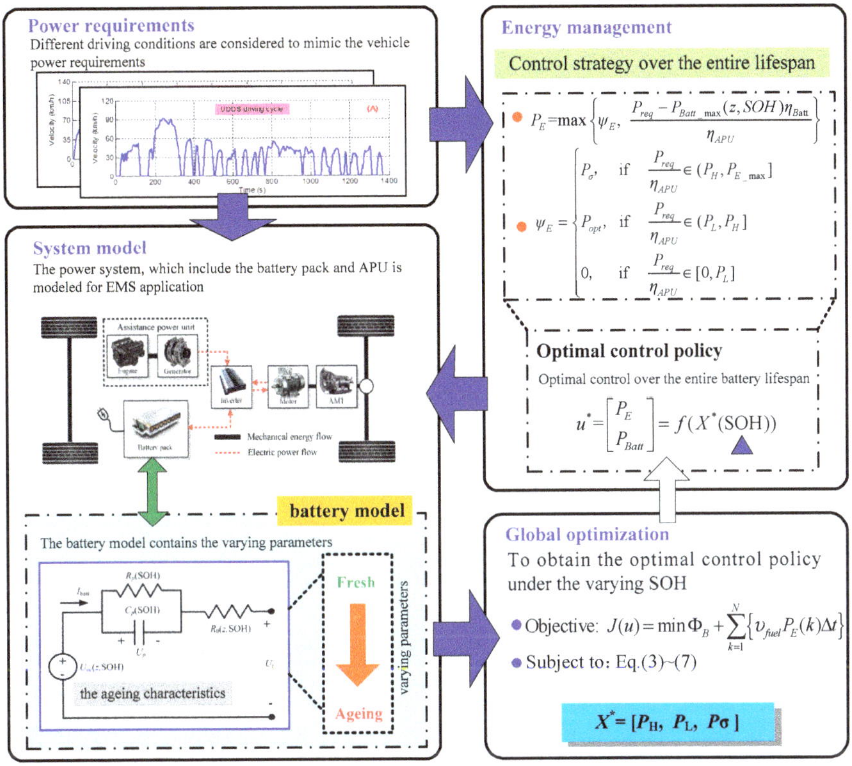

Figure 2. The schematic diagram of energy management strategy.

(1) When battery SOC is higher than the low threshold ($SOC \geq \delta_2$),

$$P_E = \max\left\{\psi_E, \frac{P_{req} - P_{Batt\_max}(z, SOH)\eta_{Batt}}{\eta_{APU}}\right\} \tag{8}$$

in which

$$\psi_E = \begin{cases} P_\sigma, & \text{if } \frac{P_{req}}{\eta_{APU}} \in (P_H, P_{E\_max}] \\ P_{opt}, & \text{if } \frac{P_{req}}{\eta_{APU}} \in (P_L, P_H] \\ 0, & \text{if } \frac{P_{req}}{\eta_{APU}} \in [0, P_L] \end{cases} \tag{9}$$

where P_{opt} denotes the power of engine at its highest efficiency point, P_L and P_H are two thresholds to define a high efficiency range of the engine. P_σ denotes the power demand of the engine that needs to be optimized.

The maximum output power of the battery pack is treated as a function of battery SOC and SOH, and is calculated by a discrete solving process [30]:

$$P_{Batt\_max,k}(z, SOH) = n_{Batt} \cdot U_{t\min}\left\{\frac{OCV(z_{k-1}, SOH_{k-1}) - U_{p,k} - U_{t\min}}{\frac{\Delta t \eta_{batt}}{Q_{Batt}(SOH_{k-1})}\frac{\partial OCV(z)}{\partial z}\Big|_{z=z_{k-1}} + R_0(z_{k-1}, SOH_{k-1})}\right\} \tag{10}$$

where n_{Batt} is the number of the cells that contained in the lithium battery pack, $U_{t\min}$ is lower cut-off voltage, z denotes the battery SOC.

(2) When SOC is quite low (SOC < δ_2), the battery pack stop discharging and the engine provides the power demand:

$$P_E = \max\left\{0, \frac{P_{req}}{\eta_{APU}}\right\} \tag{11}$$

2.3. Optimization Algorithm

The power demands from the typical driving cycles are used as the input of the program, which are the foundations of the model simulation and the energy management implementation. The optimization is conducted by the PSO algorithm. There are three control coefficients (two thresholds and one power demand) that required to be determined, expressed by

$$X = [P_H \ P_L \ P_\sigma] \tag{12}$$

where X denotes the particle position in PSO algorithm.

The PSO algorithm is offline implemented with varying battery aging condition to do the optimization. The numerical processing of the PSO algorithm is described in Reference [27]. The scale of the particle swarm $M = 60$, the maximum iteration steps $N_p = 1000$. For each particle i, the velocity and position are updated according to the following expression:

$$\begin{cases} V^i(k+1) = wV^i(k) + c_1r_1\left(P^i(k) - X^i(k)\right) + c_2r_2\left(G^i(k) - X^i(k)\right) \\ X^i(k+1) = X^i(k) + V^i(k) \end{cases} \tag{13}$$

where w is inertia factor, r_1 and r_2 denote two random values, r_1, $r_2 \in (0, 1)$, c_1 and c_2 are weight coefficients, P^i denotes the best position of the particle i amongst the historical iterations, G^i denotes the best position within a certain neighborhood at the current iteration step.

Once the optimal X is obtained, the control policy $u = [P_E, P_{Batt}]^T$ can be further deduced based on previous blended strategy, denoted as $u = f(X)$. It should be noticed that all these coefficients in X are treated as functions of SOH since the parameters-varying battery aging model has been adopted to replace the conventional battery model.

Thus, the optimal control policy is obtained by

$$u^* = \begin{bmatrix} P_E \\ P_{Batt} \end{bmatrix} = f(X^*(\mathrm{SOH})) \tag{14}$$

It should be noted that the presented algorithm is using PSO to offline optimize the control policy under different aging condition and can be used in online implementation by a look-up table method. However, to implement the presented algorithm in real-time control, we need the battery management system (BMS) hardware to provide the online estimation of the current SOH and SOC information. Accurate estimations [31–34] are the premise guarantee for this method improving the energy economy improvement.

3. Impacts of Battery Aging

The battery performance parameters will be notably changed after the battery is seriously aged, resulting in the influences of the total hybrid power sources and the optimal control policy. In order to establish a global optimal strategy, it is necessary to dynamically adjust the control parameters as the battery ages. Physical methods like X-ray diffraction and scanning electron microscopy are very useful to analyze the aging mechanism of the battery [35], but they are not suitable for onboard energy management application. In this section, the mathematical expression of the battery aging characteristics is presented and the parameters-varying aging model of lithium battery is used to incorporate the battery aging into the EMS design.

3.1. Modeling

Here an equivalent circuit model, namely the first-order RC model, is employed to mimic the basic electrical behavior of the battery, as shown in Figure 3. The parameters in the model are treated as functions of battery SOC and SOH. In this study, the inconsistency of single cells is neglected and the battery SOC is supposed to be known correctly. SOH is described by

$$SOH = \frac{Q_{Batt}}{Q_{Batt\_new}} \times 100\% \tag{15}$$

where Q_{Batt} is the maximum capacity of the battery at current, Q_{Batt_new} is the nominal capacity of the battery.

The parameters in this model are considered as functions of both SOH and SOC, described as

$$U_t(k) = U_{oc}(z, \mathrm{SOH}) - \varphi_B(k) R_0(z, \mathrm{SOH}) - U_p(k) \tag{16}$$

$$U_p(k) = U_p(k-1) e^{\frac{-\Delta t}{\tau(z,\mathrm{SOH})}} + \varphi_B(k-1) R_p(z, \mathrm{SOH}) (1 - e^{\frac{-\Delta t}{\tau(z,\mathrm{SOH})}}) \tag{17}$$

$$\varphi_B(k) = \frac{P_B(k)}{U_t(k)} \tag{18}$$

where U_t is the terminal voltage, U_{oc} is the open circuit voltage (OCV), U_p is the voltage of the RC network, R_p and τ are resistance and time constant of RC network, respectively, P_B is the battery output power, φ_B is the battery current, R_0 is the internal resistance.

The maximum output power of the battery pack is treated as a function of battery SOC and SOH, and is calculated by a discrete solving process:

$$\begin{cases} I_{\mathrm{max}}^{\mathrm{dis}} = \dfrac{U_{oc}(z_k) - U_{p1,k+1} - U_{p2,k+1} - U_{t,\mathrm{min}}}{\frac{\eta_c \Delta t}{C_{\mathrm{aged}}} \frac{\partial U_{oc}(z)}{\partial z}\big|_{z=z_k} + R_{dis}} \\ I_{\mathrm{min}}^{\mathrm{chg}} = \dfrac{U_{oc}(z_k) - U_{p1,k+1} - U_{p2,k+1} - U_{t,max}}{\frac{\eta_c \Delta t}{C_{\mathrm{aged}}} \frac{\partial U_{oc}(z)}{\partial z}\big|_{z=z_k} + R_{chg}} \\ SoP_{\mathrm{dis}} = n_{\mathrm{s}} n_{\mathrm{p}} \left(I_{\mathrm{max}}^{\mathrm{dis}} U_{t,\mathrm{min}} \right) \\ SoP_{\mathrm{chg}} = n_{\mathrm{s}} n_{\mathrm{p}} \left(I_{\mathrm{max}}^{\mathrm{chg}} U_{t,\mathrm{max}} \right) \end{cases} \tag{19}$$

where n_p and n_s are the parallel number and series number of the cells that contained in the lithium battery pack, $U_{t\mathrm{min}}$ is lower cut-off voltage.

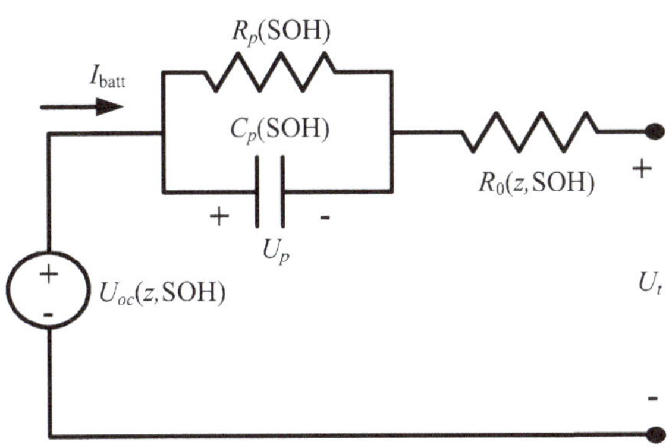

Figure 3. Diagram of the first order RC model.

3.2. Exprimental Study

The above model includes parameters that are related to battery SOH. In this study, we totally use thirteen 18650-type $LiFePO_4$ lithium cells to conduct the aging tests, in which each cell is cycled to a different aging stage under room temperature. Then we test the cell characteristics under urban dynamometer driving schedule (UDDS) to provide dataset of the battery behaviors in different aging conditions. The UDDS voltage/current profile is provided in Section 5. In this section, we use the test data of four typical cells (denoted as No.1~4) to illustrate the method and propose the mathematical expression. The capacity and the corresponding SOH of these four cells are summarized in Table 1. The test battery parameters are given in Table 2.

Table 1. The aging conditions of the four cells.

Battery Number	1	2	3	4
Capacity	1.299	1.217	1.158	1.071
SOH	96.21%	90.13%	85.76%	79.34%

Table 2. Specifications of the test cells.

Parameters	Value
Nominal capacity/Ah	1.35
Nominal voltage/V	3.2
Temperature range/°C	−20–60
Charge cut-off voltage/V	3.65
Discharge cut-off voltage/V	2.5
Internal resistance/mΩ	33

3.3. Mathematical Expression

For better application in EMS, we further establish the mathematical expression of the model parameters. The method contains the following two steps.

- **Step 1:** Test data are divided into many small data segments ranging from SOC = 0.1 to SOC = 1.0; GA is implemented to optimize the model parameters at each data segment. The programming of GA has been introduced in Reference [28], so it is not reproduced here for brevity. Optimization objective is to find the best parameters $\rho^{(j)}= [U_{oc}{}^{(j)}, R_0{}^{(j)}, R_p{}^{(j)}, \tau^{(j)}]$ to minimize the model error at each segment j. The results are shown in Figure 4.
- **Step 2: Mathematical expression.** From the results, we found U_{oc} and R_0 have clear correspondences with SOC under each SOH condition, but R_p and τ show fluctuate with some certain value. Thus, U_{oc} and R_0 at the entire SOC range are further fitted by the following continuous polynomials while R_p and τ are replaced by their mean value

$$\begin{cases} U_{oc}(z) = a_1 z^3 + a_2 z^2 + a_3 z + a_4 + a_5 \exp(-\frac{a_6}{z}) \\ R_0(z) = b_1 z^4 + b_2 z^3 + b_3 z^2 + b_4 z + b_5 \\ \overline{R}_p = \frac{\sum R_p{}^{(j)}}{n} \\ \overline{\tau} = \frac{\sum \tau^{(j)}}{n} \end{cases} \tag{20}$$

where a_1–a_6 and b_1–b_5 are coefficients, n is the total number of segments.

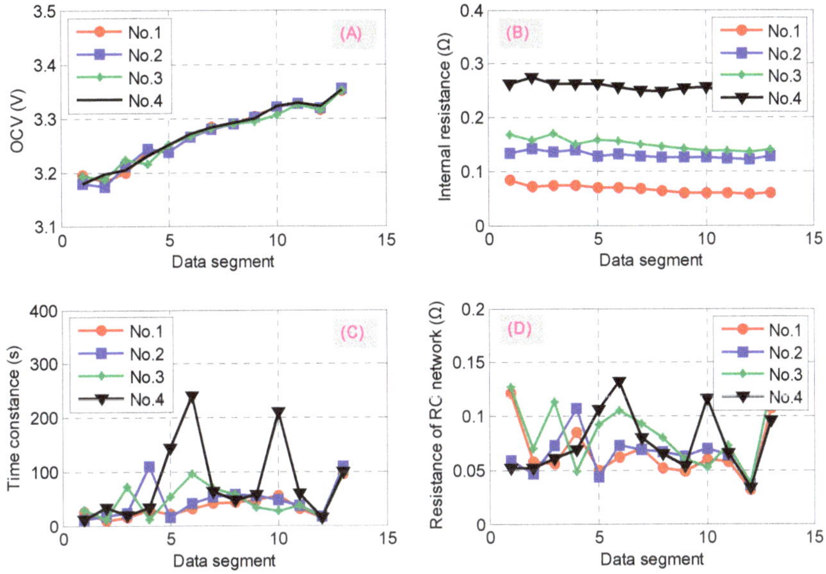

Figure 4. Identification results on data segments of four cells: (**A**) open circuit voltage (OCV); (**B**) internal resistance; (**C**) time constant; (**D**) resistance of RC network.

Based on the above expressions, GA is carried out once again to perform the identification of parameters on the entire SOC range under each aging condition. The coefficients for GA to optimize are rewritten as

$$\rho(SOH) = [a_1, a_2, \ldots, a_6, b_1, b_2, \ldots, b_5, \overline{R}_p, \overline{\tau}] \tag{21}$$

in which a_1,..., a_6, b_1,..., b_5 are subject to

$$\begin{cases} a_i \in [a_{i,\min}, a_{i,\max}],\ i = 1, 2, \ldots, 6 \\ b_j \in [b_{j,\min}, b_{j,\max}],\ j = 1, 2, \ldots, 5 \end{cases}$$

where $a_{i,\min}$, $a_{i,\max}$, $b_{j,\min}$, and $b_{j,\max}$ are limitations of the coefficients' boundary.

4. Results and Analysis

4.1. The Impacts of Battery Aging

Figure 5 shows the results of electrical behaviors and the parameter variations of the batteries under thirteen aging stages (from the fresh to deep aged battery). It can be seen that the internal resistance increases significantly with the deep aging of the battery, however, the polarization internal resistance and polarization capacitance do not show obvious change regulations. The varying battery parameters are used to design the control parameters in optimal EMS; therefore, the EMS can be adaptive to the battery aging process according to the current SOH value. The results of the parameters in aging expression are provided in Table 3. The driving cycles of UDDS and Extra Urban Driving Cycle (EUDC) are used to assess the increased energy consumption induced by the battery aging. Compared with the fresh battery, the deep aged battery (SOH = 79.34%) cause extra energy cost of 15.19% and 14.28% under UDDS and EUDC, respectively.

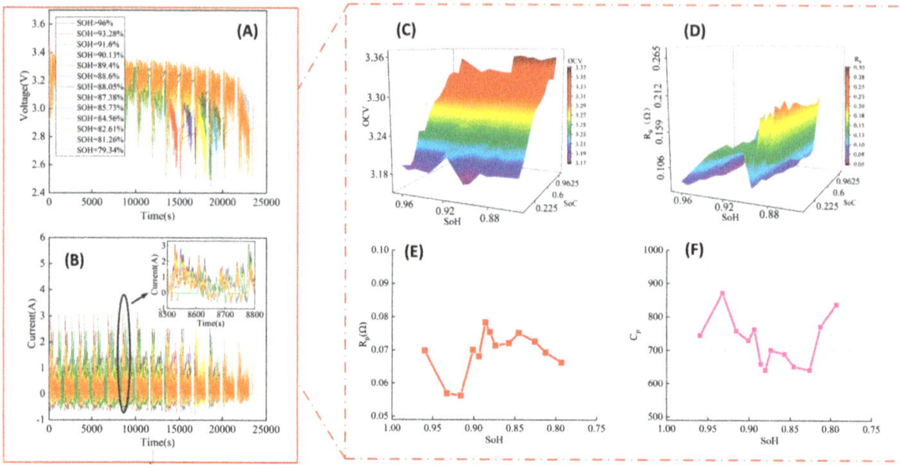

Figure 5. Aging characteristics of batteries under thirteen different states of health (SOH) conditions: (**A**,**B**): test results of battery voltages and currents under urban dynamometer driving schedule (UDDS) driving cycles; (**C–F**): the varations of the battery parameters caused by the aging.

Table 3. Battery model parameters under the varying SOH phases.

SOH	τ	R_p(mΩ)	b1	b2	b3	b4	b5	a1	a2	a3	a4	a5	a6
96.21%	52.1	69.9	0.1	−0.4	0.55	−0.31	0.12	−3.1	50.0	3.14	0.61	−0.84	0.44
93.28%	49.7	57.0	1.8	−3.9	2.88	−0.89	0.22	26.8	73.0	3.12	0.69	−0.99	0.52
91.60%	42.7	56.3	1.4	−2.8	1.89	−0.56	0.20	−475	475	3.17	0.54	−0.85	0.51
90.13%	51.2	70.2	0.3	−0.8	0.69	−0.29	0.18	301.9	310.1	3.12	0.74	−1.09	0.59
89.40%	52.1	68.2	0.6	−1.5	1.29	−0.51	0.14	−498.0	116.7	3.12	0.70	−1.0	0.53
88.60%	51.7	78.4	0.3	−0.7	0.53	−0.22	0.22	479.9	138.0	3.13	0.67	−0.99	0.55
88.05%	48.5	75.5	0.2	−0.6	0.55	−0.27	0.16	−475	139.9	3.13	0.65	−0.94	0.51
87.38%	50.1	71.5	0.2	−0.6	0.62	−0.31	0.25	483.5	145.9	3.13	0.64	−0.92	0.49
85.73%	49.8	72.2	0.2	−0.3	0.14	−0.07	0.12	10.5	72.3	3.12	0.69	−1.01	0.55
84.56%	49.1	75.3	0.4	−1.0	0.90	−0.37	0.17	4.7	54.6	3.1	0.79	−1.18	0.63
82.61%	46.7	72.7	0.6	−1.4	1.24	−0.49	0.19	0.8	41.5	3.1	0.78	−1.14	0.61
81.26%	53.6	69.5	0.4	−0.9	0.73	−0.27	0.24	1.2	51.0	3.12	0.69	−1.02	0.55
79.34%	55.7	66.4	−0.3	0.4	0.07	−0.18	0.29	270.3	356.8	3.12	0.71	−1.03	0.54

For ease of designing the adaptive control algorithm, we wish to find out the main parameters (among the multiple battery parameters) that contribute the most to energy consumption. Therefore, the specific impacts of each battery aging parameters on increased consumption cost are further analyzed, in which the percentage of cost increment is calculated. The results are shown in Figure 6. The results illustrate that the capacity loss and resistance increase are the main factors leading to the increase of energy consumption. The capacity degradation accounts for up to 10.24% of the aging-caused energy cost while the internal resistance accounts for up to 6.42% of the aging-caused energy cost. On the contrary, the influences of OCV and RC items are minor (less than 0.29%) and can be neglected.

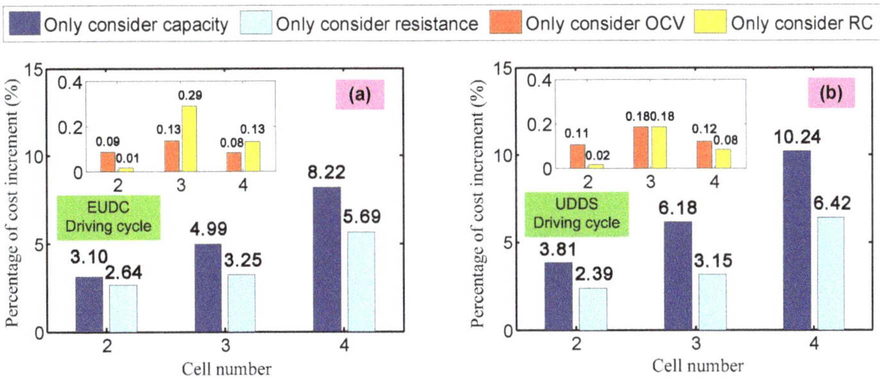

Figure 6. Specific impacts of battery aging on vehicle energy cost: (**a**) simulation at Extra Urban Driving Cycle (EUDC); (**b**) simulation at UDDS driving cycle.

4.2. Energy Consumption

To evaluate the effectiveness of the proposed energy management strategy (denoted as Strategy A), another energy management strategy (denoted as Strategy B) is employed in the following to make a comparison study. Two strategies are implemented with same driving conditions. The two strategies are based on the same control algorithm and optimization process. The only difference is that the impacts of battery aging are not considered in Strategy B while the varying control parameters enable the Strategy A to be adaptive to the battery aging process according to the current SOH value. The simulation conditions are provided as follow. The prices of fuel and electricity are 5.86 CNY-¥/L and 0.82 CNY-¥/kWh, respectively, the thresholds δ_1 and δ_2 are set as 0.5 and 0.2, respectively. Since the optimization is not implemented when SOC $\geq \delta_1$, the initial SOC in this study is set as 0.5. The impact of temperature is neglected in this simulation.

The comparison results of the energy cost from two strategies are summarized in Tables 4 and 5. The optimal energy cost is enhanced by up to 15.19% in UDDS driving cycle and 14.28% in EUDC driving cycle when SOH changes from 96.21% to 79.34% if the battery aging is not concerned in energy management (Strategy B). Without replacing the aging battery pack, using the presented energy management strategy can reduce the effect of battery aging to some certain extent. The energy cost is decreased by up to 2.24% both in UDDS driving cycle and in EUDC driving cycle compared with the strategy without considering the battery aging. The detailed simulation results of power allocation in UDDS driving cycle are shown in Figure 7.

Table 4. Comparison on cost functions of two strategies: with the case of UDDS.

SoH	Driving Distance: 50 km			Driving Distance: 100 km		
	Strategy A	Strategy B	Reduction	Strategy A	Strategy B	Reduction
96.21%	13.16	13.16	–	32.95	32.95	–
90.13%	13.75	13.90	1.08%	33.63	33.83	0.59%
85.76%	14.13	14.34	1.46%	34.05	34.38	0.96%
79.34%	14.82	15.16	2.24%	34.90	35.38	1.36%

Table 5. Comparison on cost functions of two strategies: with the case of EUDC.

SoH	Driving Distance: 50 km			Driving Distance: 100 km		
	Strategy A	Strategy B	Reduction	Strategy A	Strategy B	Reduction
96.21%	13.30	13.30	–	33.26	33.26	–
90.13%	13.86	14.02	1.14%	33.84	34.08	0.70%
85.76%	14.23	14.41	1.25%	34.22	34.52	0.89%
79.34%	14.86	15.20	2.24%	34.91	35.40	1.38%

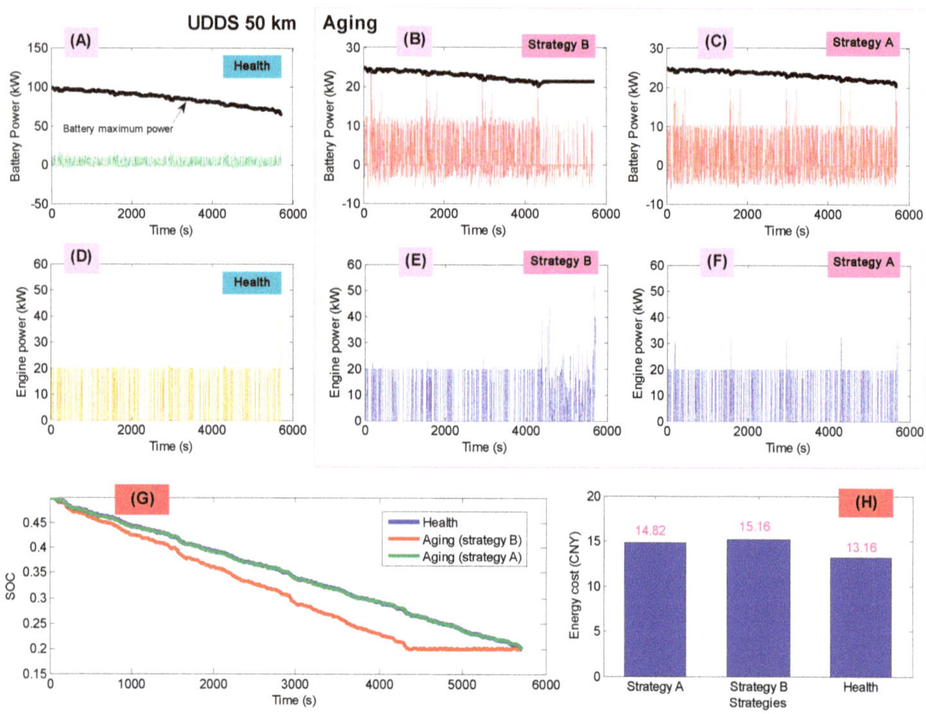

Figure 7. Simulation results in UDDS (50 km): (**A**) Battery power of healthy battery; (**B**) battery power of aging battery with Strategy B; (**C**) battery power of aging battery with Strategy A; (**D**) engine power of healthy battery; (**E**) engine power of aging battery with Strategy B; (**F**) engine power of aging battery with Strategy A; (**G**) SOC trajectory; (**H**) energy cost.

Figure 7A–C show the apparently fade of maximum battery power. When battery is deeply aging, the battery performance is influenced, resulting in an improper energy management result. Therefore, adjusting the energy management parameters at varying battery SOH is necessary. The battery power in Strategy A is properly reduced and well-distributed compared with that in Strategy B. From Figure 7D–F show the engine power distribution. The engine power in Strategy B becomes very large near the terminal of the trip leading in a low efficient performance. This is because the battery energy is insufficient near the end of the trip, see Figure 7G. The battery SOC in Strategy B drops to bottom earlier than that in Strategy A, and then the higher cost power from the engine supply the power demand. On the contrary, the SOC trajectory in Strategy A has a good agreement with that of health battery. The optimal cost of Strategy A at the terminal of trip is decreased effectively in comparison with Strategy B. Similar analysis results can also be found in Figure 8. When trip length is long, the impact of battery aging tends to be minor because the energy cost from engine system (gasoline) occupies a larger proportion than that of short trip length. Although the battery aging cannot be eliminated through the energy management approach, a proper design of energy management strategy can partly reduce its negative impacts on energy cost of the PHEV.

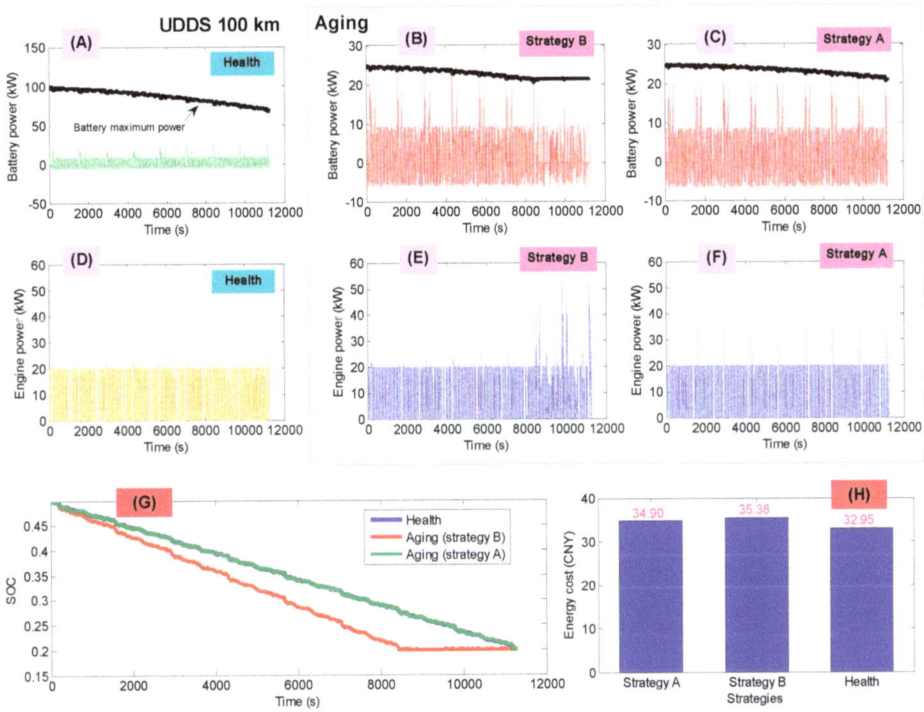

Figure 8. Simulation results in UDDS (100 km). (**A**) Battery power of healthy battery; (**B**) battery power of deeply aged battery with Strategy B; (**C**) battery power of deeply aged battery with Strategy A; (**D**) engine power of healthy battery; (**E**) engine power of deeply aged battery with Strategy B; (**F**) engine power of deeply aged battery with Strategy A; (**G**) results of SOC; (**H**) energy cost.

5. Conclusions

In this study, the battery behaviors under 13 different aging conditions are investigated experimentally, based on which, an aging-conscious battery model is proposed for energy management application. The optimal control strategy is then proposed for PHEVs energy management against the impact of battery aging. The presented control strategy can achieve the optimal control performance over the entire battery lifespan based on the PSO algorithm. The quantitative impact of battery aging on the energy consumption has been revealed, indicating that the capacity and internal resistance are the main factors that cause the extra energy cost. The presented energy management strategy is evaluated and analyzed by a simulation study under two typical driving cycles. The results indicate that the energy cost of PHEV can be increased by up to 15.19% due to the battery aging. The aging-conscious energy management can balance out some of the harmful effects that battery aging can have on energy efficiency. Compared with the strategy without considering the battery aging, the presented strategy can reduce the aging-induced energy consumption by up to 2.24% at certain driving condition.

Author Contributions: Conceptualization and methodology are provided by Z.C. and J.L.; software and data analysis are conducted by B.L.; validation is presented by N.Z. and S.L. All authors have read and agreed to the published version of the manuscript.

Funding: This research was funded by "the National Natural Science Foundation of China, grant number 51977029, 51607030" and "the Fundamental Research Funds for the Central Universities, grant number N2003002".

Conflicts of Interest: The authors declare no conflict of interest. The funders had no role in the design of the study; in the collection, analyses, or interpretation of data; in the writing of the manuscript, or in the decision to publish the results.

References

1. Khayyam, H. A Bab-Hadiashar. Adaptive intelligent energy management system of plug-in hybrid electric vehicle. *Energy* **2014**, *69*, 319–335. [CrossRef]
2. Karplus, V.; Paltsev, S.; Reilly, J. Prospects for plug-in hybrid electric vehicles in the United States and Japan: A general equilibrium analysis. *Transp. Res. Part A Policy Pract.* **2010**, *44*, 620–641. [CrossRef]
3. Weiller, C. Plug-in hybrid electric vehicle impacts on hourly electricity demand in the United States. *Energy Policy* **2011**, *39*, 3766–3778. [CrossRef]
4. Peterson, S.; Apt, J.; Whitacre, J. Lithium-ion battery aging resulting from realistic vehicle and vehicle-to-grid utilization. J Power Sources. *J. Power Sources* **2010**, *195*, 2385–2392. [CrossRef]
5. Dubarry, M.; Liaw, B.; Chen, M. Identifying battery aging mechanisms in large format Li ion cells. *J Power Sources* **2011**, *196*, 3420–3425. [CrossRef]
6. Vetter, J.; Novák, P.; Wagner, M.; Veit, C.; Moller, K.; Besenhard, J.; Winter, M.; Wohlfahrt-Mehrens, M.; Vogler, C.; Hammouche, A. Ageing mechanisms in lithium-ion batteries. *J Power Sources* **2005**, *147*, 269–281. [CrossRef]
7. Nuhic, A.; Terzimehic, T.; Guth, T.; Buchholz, M.; Dietmayer, K. Health diagnosis and remaining useful life prognostics of lithium-ion batteries using data-driven methods. *J. Power Sources* **2013**, *239*, 680–688. [CrossRef]
8. Kim, I. A Technique for Estimating the State of Health of Lithium Batteries Through a Dual-Sliding-Mode Observer. *IEEE Trans. Power Electron.* **2010**, *25*, 1013–1022.
9. Zhang, Y.; Xiong, R.; He, H.; Michael, P. Long short-term memory recurrent neural network for remaining useful life prediction of lithium-ion batteries. *IEEE Trans. Veh. Technol.* **2018**, *67*, 5695–5705. [CrossRef]
10. Shams-Zahraei, M.; Kouzani, A.; Kutter, S.; Bäker, B. Integrated thermal and energy management of plug-in hybrid electric vehicles. *J. Power Sources* **2012**, *216*, 237–248. [CrossRef]
11. Hou, C.; Ouyang, M.; Xu, L.; Wang, H. Approximate Pontryagin's minimum principle applied to the energy management of plug-in hybrid electric vehicles. *Appl. Energy* **2015**, *115*, 174–189. [CrossRef]
12. Li, L.; You, S.; Yang, C.; Yan, B.; Song, J.; Chen, Z. Driving-behavior-aware stochastic model predictive control for plug-in hybrid electric buses. *Appl. Energy* **2016**, *162*, 868–879. [CrossRef]
13. Wirasingha, S.; Emadi, A. Classification and review of control strategies for plug-in hybrid electric vehicles. *IEEE Trans. Veh. Technol.* **2011**, *60*, 111–122. [CrossRef]
14. Peng, J.; He, H.; Xiong, R. Rule based energy management strategy for a series–parallel plug-in hybrid electric bus optimized by dynamic programming. *Appl. Energy* **2017**, *185*, 1633–1643. [CrossRef]
15. Gao, Y.; Ehsani, M. Design and control methodology of plug-in hybrid electric vehicles. *IEEE Trans. Ind. Electron.* **2010**, *57*, 633–640.
16. Schouten, N.; Salman, M.; Kheir, N. Energy management strategies for parallel hybrid vehicles using fuzzy logic. *Control Eng. Pract.* **2003**, *11*, 171–177. [CrossRef]
17. Li, S.; Sharkh, S.; Walsh, F.; Zhang, C. Energy and battery management of a plug-in series hybrid electric vehicle using fuzzy logic. *IEEE Trans. Veh. Technol.* **2011**, *60*, 3571–3585. [CrossRef]
18. Zhang, S.; Xiong, R. Adaptive energy management of a plug-in hybrid electric vehicle based on driving pattern recognition and dynamic programming. *Appl. Energy* **2015**, *155*, 68–78. [CrossRef]
19. Patil, R.; Filipi, Z.; Fathy, H. Comparison of supervisory control strategies for series plug-in hybrid electric vehicle powertrains through dynamic programming. *IEEE Trans. Control Syst. Technol.* **2014**, *22*, 502–509. [CrossRef]
20. Chen, S.; Hung, Y.; Wu, C.; Huang, S. Optimal energy management of a hybrid electric powertrain system using improved particle swarm optimization. *Appl. Energy* **2015**, *160*, 132–145. [CrossRef]
21. Larsson, V.; Johannesson, L.; Egardt, B. Analytic Solutions to the Dynamic Programming sub-problem in Hybrid Vehicle Energy Management. *IEEE Trans. Veh. Technol.* **2015**, *64*, 1458–1467. [CrossRef]
22. Chen, B.; Wu, Y.; Tsai, H. Design and analysis of energy management strategy for range extended electric vehicle using dynamic programming. *Appl. Energy* **2014**, *113*, 1764–1774. [CrossRef]

23. Nüesch, T.; Cerofolini, A.; Mancini, G.; Cavina, N.; Onder, C.; Guzzella, L. Equivalent consumption minimization strategy for the control of real driving NOx emissions of a diesel hybrid electric vehicle. *Energies* **2014**, *7*, 3148–3178. [CrossRef]
24. Geng, B.; Mills, J.; Sun, D. Energy management control of microturbine-powered plug-in hybrid electric vehicles using the telemetry equivalent consumption minimization strategy. *IEEE Trans. Veh. Technol.* **2011**, *60*, 4238–4248. [CrossRef]
25. Sezer, V.; Gokasan, M.; Bogosyan, S. A novel ECMS and combined cost map approach for high-efficiency series hybrid electric vehicles. *IEEE Trans. Veh. Technol.* **2011**, *60*, 3557–3570. [CrossRef]
26. Chen, C.; Xiong, R.; Shen, W. A lithium-ion battery-in-the-loop approach to test and validate multi-scale dual H infinity filters for state of charge and capacity estimation. *IEEE Trans. Power Electron.* **2017**. [CrossRef]
27. Chen, Z.; Xiong, R.; Cao, J. Particle swarm optimization-based optimal energy management of plug-in hybrid electric vehicles considering uncertain driving conditions. *Energy* **2016**, *96*, 197–208. [CrossRef]
28. Lu, J.; Chen, Z.; Yang, Y.; Lv, M. Online estimation of state of power for lithium-ion batteries in electric vehicles using genetic algorithm. *IEEE ACCESS.* **2018**, *6*, 20868–20880. [CrossRef]
29. Padmarajan, B.; Mcgordon, A.; Jennings, P. Blended rule based energy management for PHEV: System Structure and Strategy. *IEEE Trans. Veh. Technol.* **2016**, *65*, 8757–8762. [CrossRef]
30. Xiong, R.; Sun, F.; He, H.; Nguyen, T. A data-driven adaptive state of charge and power capability joint estimator of lithium-ion polymer battery used in electric vehicles. *Energy* **2013**, *63*, 295–308. [CrossRef]
31. Ali, M.; Zafar, A.; Nengroo, S.; Hussain, S.; Hussain, S.; Alvi, M.; Kim, H. Towards a smarter battery management system for electric vehicle applications: A critical review of Lithium-Ion battery State of Charge Estimation. *Energies* **2019**, *12*, 446. [CrossRef]
32. Tian, J.; Xiong, R.; Yu, Q. Fractional-Order Model-Based Incremental Capacity Analysis for Degradation State Recognition of Lithium-Ion Batteries. *IEEE Trans. Ind. Electron.* **2019**, *66*, 1576–1584. [CrossRef]
33. Andre, D.; Appel, C.; Guth, T.; Sauer, D. Advanced mathematical methods of SOC and SOH estimation for lithium-ion batteries. *J. Power Sources.* **2013**, *224*, 20–27. [CrossRef]
34. Zhang, Z.; Zhang, L.; Hu, L.; Huang, C. Active cell balancing of lithium-ion battery pack based on average state of charge. *Int. J. Energy Res.* **2020**, *44*, 2535–2548. [CrossRef]
35. Han, X.; Ouyang, M.; Lu, L.; Li, J.; Zheng, Y.; Li, Z. A comparative study of commercial lithium ion battery cycle life in electrical vehicle: Aging mechanism identification. *J. Power Sources* **2014**, *251*, 38–54. [CrossRef]

© 2020 by the authors. Licensee MDPI, Basel, Switzerland. This article is an open access article distributed under the terms and conditions of the Creative Commons Attribution (CC BY) license (http://creativecommons.org/licenses/by/4.0/).

energies

MDPI

Article

A Publicly Available Simulation of Battery Electric, Hybrid Electric, and Gas-Powered Vehicles

Lawrence Fulton

Department of Health Administration, Texas State University, San Marcos, TX 78666, USA; lf25@txstate.edu

Received: 3 April 2020; Accepted: 12 May 2020; Published: 19 May 2020

Abstract: Volatility in energy markets has made the purchase of battery electric vehicles (BEV) or hybrid vehicles (HEVs) attractive versus internal combustion engine vehicles (ICEVs). However, the total cost of ownership (TCO) and true environmental effects, are difficult to assess. This study provides a publicly available, user driven simulation that estimates the consumer and environmental costs for various vehicle purchase options, supporting policymaker, producer, and consumer information requirements. It appears to be the first to provide a publicly available, user interactive simulation that compares two purchase options simultaneously. It is likely that the first paper to simulate the effects of solar recharging of electric vehicles (EV) on both cost-benefit for the consumer and environmental benefit (e.g., carbon dioxide, oxides of nitrogen, non-methane organic gasses, particulate matter, and formaldehyde) simultaneously, demonstrating how, as an example, solar-based charging of BEVs and HEVs reduces carbon emissions over grid-based charging. Two specific scenarios are explicated, and the results of show early break-even for both BEV and Plug-in HEV (PHEV) options over ICEV (13 months, and 12 months, respectively) with CO_2 emissions about 1/2 that of the gasoline option (including production emissions.) The results of these simulations are congruent with previous research that identified quick break-even for HEVs versus ICEV.

Keywords: simulation; electric vehicles; battery electric vehicles; plug-in hybrid

1. Introduction

This study provides an analysis of both, environmental and total cost of ownership (TCO) estimates for vehicle capabilities identified by user input in a publicly available simulation (available in Supplementary materials). The simulation focuses on capability, not necessarily an existing vehicle type. Located here https://rminator.shinyapps.io/Vehicles/, the simulation was built to address the necessity for publicly-available TCO tools that compare different vehicle types, as well as environmental conditions. The next sections highlight the necessity for this analysis, environmental motivations for vehicle purchases, previous TCO analyses of EVs, consumer decision-making regarding EVs, modeling considerations, and this study's significance, as well as location in the literature.

1.1. *Necessity for Total Cost of Ownership Analysis*

With the volatility of gasoline and electrical prices [1], the purchase of battery electric vehicles (BEV) or hybrid vehicles has become an attractive option to some [2], but understanding the actual total costs of ownership (TCO) associated with such a purchase, as well as environmental concerns requires analysis The metrics "miles per gallon" or "kilometers per liter" are not appropriate for electric vehicles (EVs), which underscores that the consumer understanding of EVs and energy costs is often incomplete [3]. When conducting a cost-benefit analysis of potential options, the analysis of EVs must include both cost to the consumer and cost to the environment, as well as performance and re-charging [4]. The automobile industry must evaluate and compare the costs and the demand trade-offs from a consumer perspective to ensure that the engineering of sustainable products meets consumer requirements [5,6]. The bottom line is that there is a necessity for consumers to understand the TCO as well as environmental considerations.

1.2. Environmental Considerations

EVs may provide a partial solution to the growing efforts to reduce carbon emissions and improve the environment [7]. One study suggested that they could reduce Global Warming Potential (GWP) by 15%, carbon monoxide (CO) by 37%, and carbon dioxide by 14% [8]. As part of any energy efficiency analysis, emissions, costs, and demand trade-offs [9] and geographic location need to be evaluated, because the efficacy of energy-efficient technology varies across regions [10], just as emissions vary with economic growth, population density, urbanization, fixed capital investment, and technological adoption [11]. The adoption of a BEV powered by photovoltaic systems (PVS) may be cost-effective and would reduce greenhouse gas (GHG) emissions [12]. Of course, TCO is just as important to consumers as environmental impact.

1.3. TCO of EVs

Most ownership and total cost of ownership (TCO) studies of BEV and hybrids are relatively recent, due to the emergence of the relevant technology [7]. One study developed a model of the performance, energy use, manufacturing cost, retail cost, and lifecycle cost of BEVs versus comparable gasoline-powered vehicles [13]. This study found that BEVs were cost-competitive, but the battery life and manufacturing costs were concerns. In another study, a vehicle cost simulation was used to analyze the manufacturing costs, retail prices, and lifecycle costs of hybrid gasoline-electric vehicles, conventional vehicles, electric-drive vehicles, and other alternative-fuel vehicles [14]. However, the research is dated. This study uses methods like the previous studies but updates the analysis with newer data in a publicly available format. Another study indicated that hybrid vehicles were better than gas-powered vehicles in terms of life-cycle costs and high travel miles [15]. Palmer et al. used panel regression to compare life-cycle costs for four geographic location. The authors did not forecast energy costs or consider seasonality. The vehicle data were also dated [16].

One study focused on the battery performance impact on TCO. This study found that TCO for EVs and PHEVs was 24 to 36% higher than HEVs [17]. Another study found that HEVs were associated with lower TCO in comparison with other EVs [18]. The purchase of hybrids is correlated with their TCO, and cost parity is reached relatively quickly (e.g., 16 months in the United Kingdom) [16]. Further, consumers prefer PHEVs to HEVs, with only a few percent opting for pure EVs [19].

A 2018 study demonstrated that, with proper policy incentives, TCO was better for EVs than than HEVs [20]. This study suggests that even without subsidies, EVs will gain market share by 2025 [20]. EVs are cost competitive without incentives now. However, incentives for purchase should be tailored based on battery development [21,22]. Further, grid-powered EVs (and PHEVs) may have deleterious effects on the grid [23,24] and produce unwanted pollution due to the reliance on grid power [25]. A solution for handling this problem is the use of solar charging for EVs [26]. Such a solution has been used in both industry [26] and in residences [27]. Fulton [28] compared specific BEV and non-plugin hybrids via simulation estimating that both were reasonable options. None of these studies provided a publicly available simulation with flexible parameters that compared break-even costs for gasoline vehicles, BEV, and non-plug hybrid vehicles.

Another, somewhat dated German study comparing ownership cost of BEV versus gas vehicles concluded a break-even cost of six years with 4 kW vehicles [29]. However, studies all depend on geography and climate assessments, as BEVs and HEVs experience different decay rates for the lithium ion batteries, based on climatology and driving experience [30]. Based on this discussion, the evidence for BEVs versus PHEVs is mixed. What motivates consumers to buy either is another concern altogether.

1.4. Consumer Decision Making in Vehicle Purchasing

Marketing studies have investigated consumer preferences and BEV viability [31]. Shin et al. evaluated how consumers would change habits based on the adoption of a BEV and how heterogeneity of vehicles might affect the market [32]. He, Chen and Conzelmann evaluated vehicle usage and consumer profile attributes, in order to assess vehicle usage versus consumer choices of hybrid vehicles [33]. Another study indicated how utility prices versus replacement costs are factors in consumer decision making [34]. All these studies provide evidence that given proper capabilities and price, EVs may be acceptable to consumers.

1.5. Modeling Considerations

From their unique perspective, Ozdemir and Hartmann estimated the optimal electric driving range for different oil price levels [35]. Some researchers have helped refine calculation of fuel consumption and emission factors based on driving styles [36]. This type of work provides a good framework for cost estimation. Focusing on driving range and gasoline costs, one study compared the lifecycle costs of electric cars to similar gasoline-powered vehicles [37]. In their study, the authors concluded that electric cars with 150 km range are viable and meet most consumer needs. The effects of charging behaviors (i.e., time of day and location) on electricity demand was studied by Weiller [38]. The study estimated a consumption of 1.5–2.0 kWh per day with home electric chargers. These studies provide the framework for simulation analysis.

1.6. Study Research Question and Significance

This study provides a user-interactive simulation to estimate the consumer costs and environmental costs for various vehicle configurations. The research question for this study is straightforward: What consumer trade-off considerations make purchasing a BEV/HEV versus a gasoline-powered vehicle reasonable in terms of acquisition costs, operations and maintenance costs, disposal/residual costs, and environmental costs? The study addresses these questions through a freely available and online simulation located here: https://rminator.shinyapps.io/Vehicles/. This simulation that compares two vehicle options simultaneously.

Unlike much of the other work in this field, this study does not focus on a particular vehicle or vehicle set. Instead, the focus is on a capability set that is customizable based on current and emerging technology (e.g., the improvement of miles per kWh in emerging vehicle lines). This differentiates the work from other simulations which are based on fixed capability values.

This study appears to be the first to provide a publicly available simulation accessible by any interested party that compares two purchase options simultaneously. It is likely the first paper to simulate the effects of solar recharging of EV vehicles on both cost-benefit for the consumer and environmental benefit (e.g., CO_2, NO_x + NMOG, PM, and HCHO) simultaneously, demonstrating how solar-based charging of electrical vehicles reduces CO_2 emissions (as an example). No other study appears to estimate acquisition costs for solar panels necessary to support a residential EV that is to be charged by such means. This work appears to provide the only user-available sensitivity analysis for various options of natural gas, electricity, or solar residential power options coupled with solar power generation capability estimates that vary by state and Error, Trend, and Seasonality (ETS) forecasts of utility costs based on Energy Information Administration data. The study is novel in that it combines user-based input with time series forecasting and simulation to generate interesting TCO and environmental assessments. The study also addresses the primary theme of the journal's special issue: Energy savings and reduced environmental impact associated with solar versus natural gas versus coal-powered grid systems.

2. Materials and Methods

2.1. *Method, Software, and Flowchart*

This study leverages Monte Carlo simulation [39] in R Statistical Software [40] and an R Shiny web application [41]. The simulation follows the flowchart in Figure 1. Each step A through S is now discussed in different sections. Feasible ranges of user-based, simulation parameters along with their default values are shown in Table 1, while variable distributions are discussed in subsequent sections.

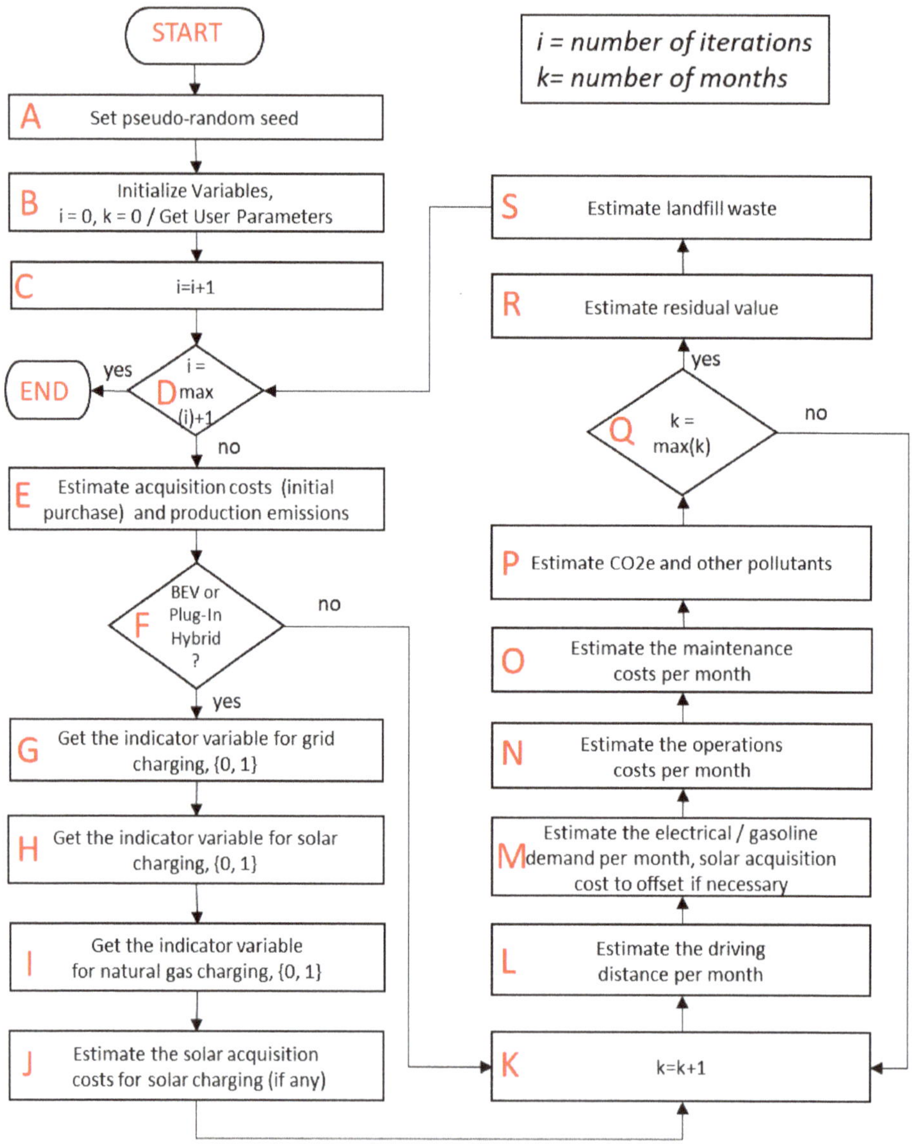

Figure 1. Simulation flowchart.

Table 1. Parameter Options in the Simulation.

Parameter/Variable	Options	Default
Month of purchase	January, February, ... December	April
Year of purchase	2020, 2021, 2022	2020
State of use	50 states	Texas
Vehicle Type	Gas, NPHEV, PHEV, BEV	Gas
Residential Power for EV	Electric, Natural Gas, Solar	Electric
Monthly Driving Distance	100 to 20,000 miles	1100 miles
Years of Ownership	1 to 10	8 years
% of Vehicle Power by Battery	0 to 100	0%
Vehicle Purchase Cost	$10,000 to $20,000	$30,000
Miles Per Gallon (Gas)	0 to 120	30 MPG
Tax Credit	0 to 30%	0%
EV kWh/mile	0.2 to 0.5	0.34 EV kWh/mile
Charging Station Cost for Home	$0 to $3000	$0
Cost of Solar Panels: $/Watt	$2 to $5	$3.3

2.2. Simulation Initialization

The simulation starts by setting a pseudo-random number seed for replicability (step A). This seed ensures that statistical differences are not due to the selection of random numbers, as they are identical between models. Step B initializes the indices and variables (e.g., the indices for iterations, *i*, and months, *k*, are set equal to zero) and gathers the input parameters from the user interface for use in future steps (e.g., user estimated lifespan or retention of vehicle). The iterations then increase by 1 (step C).

2.3. Acquisition Costs, Environmental Effects, and Additional Simulation Parameters

If the user selected number of iterations is not met (step D), then the vehicle acquisition costs from the user interface including, tax credits and home charging station costs are estimated. These are user-entered parameters for the simulation. In addition, production emissions, a value derived from a previous study [42], are calculated based on vehicle selection type, a user parameter (step E, Table 2).

Table 2. Vehicle Emissions Analysis from Previous Study.

Vehicle Type	Estimated Lifecycle Emissions (Tonnes CO_2e)	Proportion of Emissions from Production	Estimated Emissions in Production (Tonnes CO_2e)
ICEV	24	23%	5.6
NPHEV	21	31%	6.5
PHEV	19	35%	6.7
BEV	19	46%	8.8

2015 vehicle after 150 k KM with 10% ethanol blend and 500 g/kWh grid electricity.

Step F evaluates whether the vehicle is a BEV or PHEV. Residential grid electric charging (step G), residential solar charging (step H), and the residential natural gas charging (step I) are pulled from the user interface. Solar charging is used to estimate solar acquisition costs for the panels necessary to charge the vehicle. Step J estimates the associated solar acquisition cost per panel watt required for the percent solar charging specified in the user interface. Cost per solar panel watt was user parameter with values between $2 and $5, representative of the range $3.10 and $4.50 found in [43] with the possibility of growth and shrinkage provided. Total cost of solar acquisition to charge a BEV or PHEV is then calculated by Equation (1) (author-proposed). In this equation, solar acquisition cost is estimated based on and indicator variable for a BEV or PHEV (*EV*) times an indicator variable for solar charging (*Solar*) times a state-based geographic system-sizing multiplier [44] times the miles driven (*Miles*) times the BEV kWh per mile (*kWh/Mile*) (from the user interface, defaulted to 0.34 based on a Nissan Leaf [28]) times the dollars per kWh (*$/kWh*) times the percentage of power provided by the battery (*% Battery*, which is 100% for BEV but less than 100% for PHEVs). Tax credits, if any, are assigned 12 months after vehicle acquisition:

$$\text{Solar Acquisition Cost} = EV \times Solar \times \text{Geographic Photovoltaic Size} \times Miles \times kWh/Mile \times \$/kWh \times \%\ Battery. \quad (1)$$

After estimating the acquisition costs in steps E through J, the month of the simulation is then incremented (step K). Driving distance per month is based on user interface (step L), with a default value of the average monthly driving distance of 1123 miles based on the United States Department of Transportation [45].

2.4. Operations and Maintenance Costs/Solar Acquisition Costs (If Needed)

2.4.1. Miles Driven

Electrical, natural gas, and gasoline monthly demands are estimated based on average monthly driving distance (per user specifications) and user input parameters for vehicle mpg and mpkWh. Step M leverages forecasts from publicly available information of the Energy Information Administration [46] to assign costs based on miles driven, mpg, and mpkWh (see author-proposed Equation (2):

$$\begin{aligned}\text{Variable Fuel Costs} &= Grid \times \%\ Battery \times (\$/kWh\ Electricity) \times (kWh/mile) \times miles + \\ &(1\text{-}GRID)\ \%\ Natural\ Gas \times \%\ Battery \times (\$/kWh\ Natural\ Gas) \times (kWh/mile) \times miles + \\ &\%\ Solar\ x\ 0 + (1 - \%\ Battery) \times (\$/gl) \times (gl/mile) \times miles.\end{aligned} \quad (2)$$

In Equation (2), *GRID* is an indicator variable identifying that the residence relies on an electrical grid for recharging the vehicle rather than natural gas. *% Battery* is the percent of vehicle power generated by the battery for PHEVs and BEVs. *% Natural Gas* is the percentage residential battery recharge from natural gas. Electric car costs are then calculated by taking the estimated cost per kWh, multiplying by kWh per mile and the distance driven in miles. *Solar* charging assumes away gray power, power produced when solar is not active, and is set to zero. This is an artificial simplification. Cars that are not fully electric generate costs based on dollars per gallon times gallons per mile and distance in miles for the percentage, not powered by batteries.

2.4.2. Natural Gas, Electricity, and Regular Gasoline Prices

Monthly natural gas residential price data from the Energy Information Administration (EIA) in dollars per 1000 cubic feet (converted to dollars per kWh) were used to estimate costs for BEV and PHEV re-charging from residences using. The data were state-dependent with Hawaii being an obvious outlier. To account for seasonality in natural gas costs, simple error, trend, seasonality (ETS) models implemented using the *fpp2* library in R were used [47]. These models proved reasonable versus ARIMA (auto-regressive, integrated, moving average models), as well as random walks for this data in previous research [5].

Grid electricity costs from the EIA in dollars per kWh have trended slowly upwards since 2001 and vary largely by state, with Hawaii having the most expensive residential costs [46]. Due to the large cost variability, ETS models by state were used to forecast costs over the vehicles' lifespans. A previous study also found that ARIMA and ETS models performed nearly identically for this variable [5].

The EIA does not publish gasoline prices for each of the states but rather only U.S. averages, U.S. petroleum regions, and data for nine selected states and cities. All states in petroleum production regions were then assigned the regional cost for regular gasoline, unless the state had its own estimates from the EIA.

Figure 2 compares the costs of natural gas, electricity, and regular gasoline per kWh for the United States based on the publicly available EIA data discussed and shows the values used for forecasting from ETS modeling. About 1000 cubic feet of natural gas is 293.07 kWh, and 1 gallon of regular gasoline is equivalent to about 36 kWh, so for comparison, the *y*-axis is shown in dollars per kWh. The data exhibit significant seasonality and illustrate that natural gas is the cheapest alternative of the three.

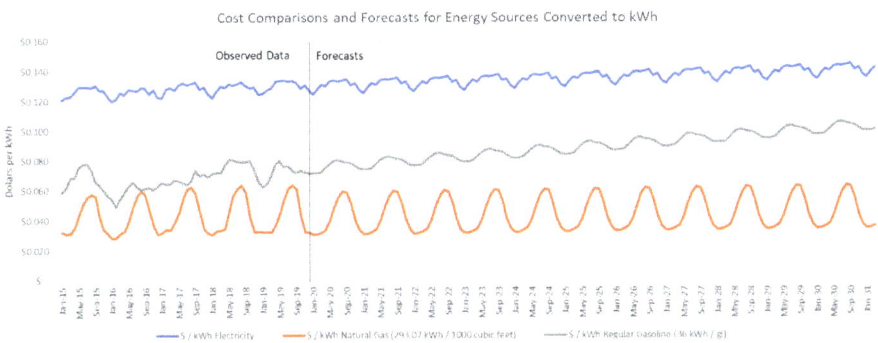

Figure 2. Actual energy costs and Error, Trend, Seasonality forecasts by source.

2.4.3. Insurance and Maintenance Costs

Insurance costs depend on vehicle type and selection, so a fixed cost of 5% of the initial vehicle price was initially assigned as in [5]. However, a recent study concluded that insurance rates for EVs are generally 21% higher, due to the higher repair costs [27], congruent with previous research [48]. To account for this factor, insurance rates for BEVs and PHEVs were inflated to 6% of the initial vehicle cost and assigned monthly.

In step O, maintenance costs for BEV, hybrid, and gasoline vehicles are estimated. The consensus from the literatures is that EVs are cheaper to maintain than ICEVs [21]. From [28,49], EVs average about 3.5 cents per mile for maintenance compared to 6 cents per mile for hybrids. As PHEVs include both traditional gasoline engines, as well as BEV components, they may experience the maintenance profile of ICEVs, although some less conservative studies have argued that maintenance is less expensive [50]. Given the paucity of data, PHEV and EV's are assigned maintenance costs of an average of 3.5 cents per mile while NPHEVs and ICEVs are assumed to experience an average of 6 cents per mile. These values are used as the center for triangular distributions with minimum and maximum set at ±2 cents from the center values.

2.4.4. Environmental Costs and Residual Values

Environmental costs are estimated in step P, and these include carbon dioxide emissions (CO_2e) [51] as well as estimates of nitrogen oxides (NO_x)/non-methane organic gases (NMOG), carbon monoxide (CO), and particulate matter (PM). Formaldehyde (HCHO) is a constant, based on Tier 3 standards of the Environmental Protection Agency (EPA) [52]. BEV cars are non-emissive (although there is carbon emission in the production phase). Other vehicles are assigned emission standards based on estimates from [53]. Table 3 illustrates the estimates by vehicle type.

Table 3. Simulation emissions estimates by selected vehicle type.

Vehicle Type	NO_x + NMOG, mg/mile	CO_2, g/mile	CO, g/mile	PM, mg/mile
ICEV	160	475	4.2	10
NPHEV *	128	323	3.0	10
PHEV	95	170	2.1	10
BEV	0.07	0	0.0	0

* imputed as mean of ICEV and PHEV.

In step Q, the simulation evaluates whether the vehicles have reached their usable lifespan. If not, steps K through P are repeated until that lifespan has been reached. Then, in step R, residual value for the selected vehicles is estimated based on Moody's Analytics *Electric Vehicle Residual Value Outlook* [54]. Gasoline cars retained on average 45% of their value over 4 years. Electric vehicles retained barely more than 25% of their value after adjustment for tax credits [54]. The geometric residual rate per year for gasoline and non-gasoline cars is then 82%, and 71%, respectively. In other words, $0.82^4 = 0.45$ and $0.71^4 = 0.25$. These geometric values are applied to the user-selected life span (not to exceed 10 years). PHEVs are estimated to be 0.75 (closer to BEV), while non-plug-in hybrids are estimated at 0.78 (closer to gasoline vehicles).

Finally, landfill lithium ion battery (LIB) waste is estimated in step S [55,56]. The concern for LIB waste is that; (1) it is highly toxic to humans, (2) it requires cobalt, which is often extracted irresponsibly with child labor, and (3) it may be recycled [55]. Waste in pounds is estimated by a uniform distribution between 100 and 544 pounds, the minimum and maximum specified in Figure 1 of [56].

2.5. Verification, Validation, and Iterations

R's native debugging ensured proper coding, while a comparison of a priori and posterior distributions, through hypothesis testing, ensured distributional integrity. To be valid for model comparison, pseudo-random number streams were identical across experimental. To do so, the function "*set.seed*" was used in R, a function which randomly generates a series of numbers so that the differences in simulation results are not to selection of random values.

3. Results

There are an infinite number of scenarios for break-even analysis based on the simulation user inputs. For analysis, this study analyzes two basic scenarios, the purchase of a BEV versus ICEV and the purchase of a PHEV versus an ICEV. All scenarios leverage the mean monthly driving distance from [45], which is about 1100 miles, and the state selected for all scenarios is Texas. Miles per gallon for gasoline vehicles is set to 30. All other variables were left at the initialization default unless specified.

3.1. Scenario 1: Gasoline versus Electrical Vehicle Charged via Solar

Scenario 1 evaluated the default values for a gasoline vehicle (termed the "baseline" simulation) against a BEV priced at the same value that receives a 30% tax credit and is completely charged by solar panels (termed the "comparison" simulation). An additional $1000 is also charged for a home-charging station.

The break-even is on month 13, after the 30% tax credit is assigned (Figure 3), despite the assigned PVS acquisition costs in month 1. The simulation demonstrates the slower growth of total expenditures, but an initial increase in expenses associated with solar panel acquisition required to charge the BEV. Figure 4 provides cost breakouts for maintenance, energy, and insurance. While lifetime insurance costs are more expensive for the BEV, lifetime maintenance costs are less expensive and energy costs are nominal, as the BEV recharges through a PVS in this simulation.

The environmental analysis (also depicted in Figure 4) shows the value of the BEV most clearly. The BEV CO_2 emission is largely in the production process and less than 1/6 of the ICEV emissions over its lifecycle. CO, PM, and NO_x/NMOG emissions are negligible for the BEV option, whereas they continuously increase over the lifecycle of the ICEV. There is a clear advantage to the BEV in terms of the environment.

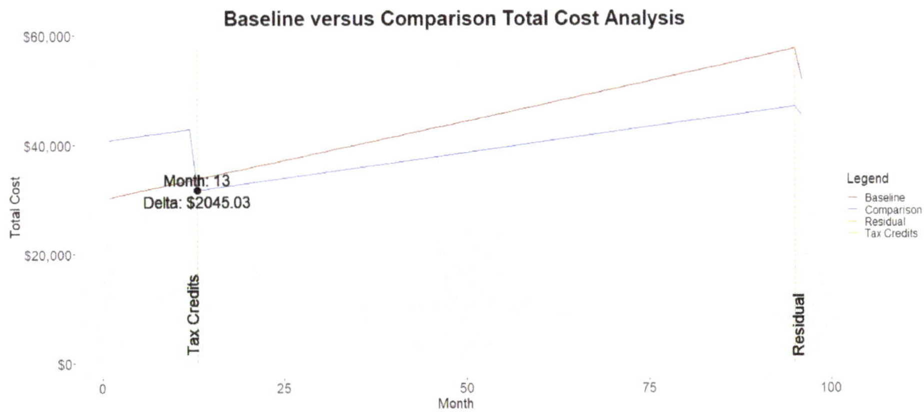

Figure 3. Break-even for Scenario 1 (curved line is a smoothed error curve estimate). Cost and environmental metrics for Scenario 1.

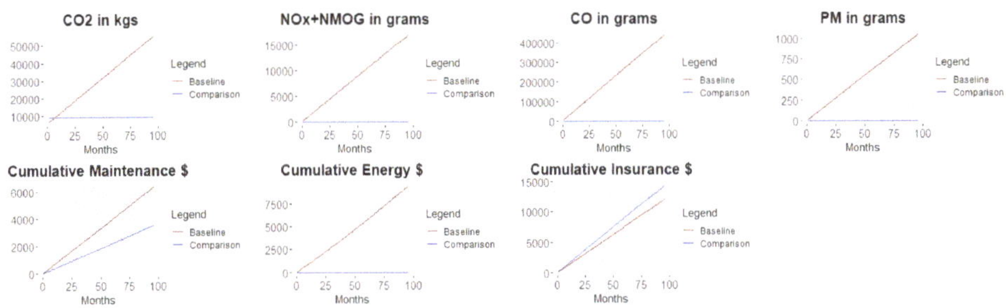

Figure 4. Cost and environmental metrics for Scenario 1.

3.2. *Scenario 2: Gasoline vs. PHEV*

In Scenario 2, the baseline simulation is again the traditional, gasoline-based vehicle priced at $30,000 with all default simulation values. The comparison option is an equally-priced PHEV (no tax credit) recharged through the electrical grid and obtaining 110 mpg (both electric and gas) with a battery offsetting 30% of vehicle power consumption. Again, a home charging station cost of $1000 is included.

The break-even analysis (Figure 5) demonstrates that the PHEV and ICEV are approximately equal at month 16. Since the up-front cost differences was the home charging station, the analysis suggests about $1000 savings every year for the adoption of a PHEV (not including residual analysis). The residual is smaller for the PHEV option. However, the TCO is lower (although not strikingly). Once again, insurance costs are clearly higher over the vehicle life cycle, maintenance and energy costs are lower (see Figure 6).

Figure 6 also depicts, the PHEV CO_2 emissions are much lower than gasoline vehicles. Comparing Figure 6 to Figure 4 demonstrates that BHEV is superior in this category, as well as in all other emissions categories. PM is tied for PHEV and ICEV but zero post-production for BEVs.

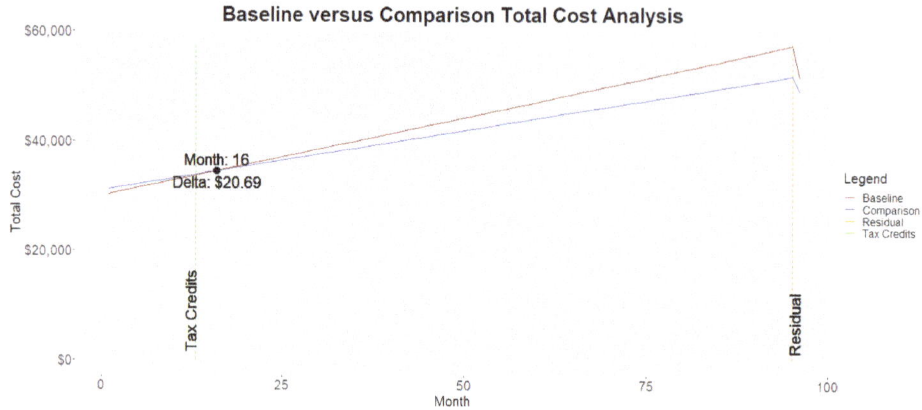

Figure 5. Break-even for Scenario 2.

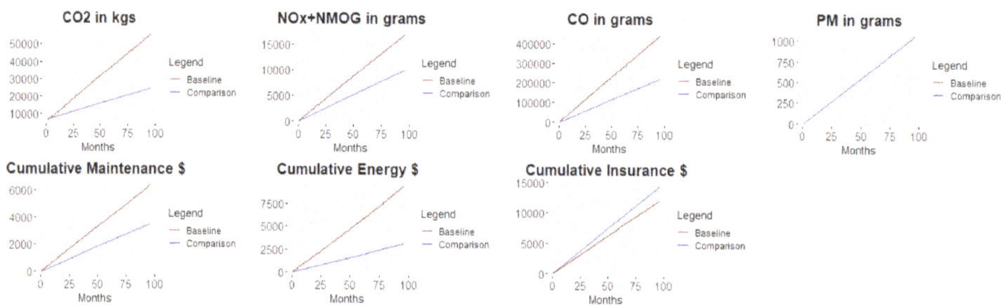

Figure 6. Cost and environmental metrics from Scenario 2.

4. Discussion

4.1. Findings and Relationships to Previous Studies

This study provides a comparative simulation for various vehicle options, based on real-world data. The results of the two scenarios provided demonstrate that BEV and PHEV options may be both green for the pocketbook and the environment. BEVs have a much lower GHG emission profile, as found in previous studies [57]. BEVs rely on tax credits for an earlier break-even point as in Scenario 1, and these tax credits are likely to incentivize purchases [58] and allow for lower TCO [48]. Any initial up-front costs may be offset by savings depending on geographic considerations and other factors.

In Scenario 1, the results show an initial upfront cost increase for the BEV due to the purchase of requisite solar panels capable of powering the vehicle and the purchase of a home charging station. BEV lifecycle costs of maintenance are about half of that experienced by ICEV ($3.5 K versus $6.5 K), while energy costs are near zero after solar panel acquisition versus $10 K for the ICEV. Insurance costs are higher for the BEV ($14 K versus $12.5 K ICEV), which is congruent with simulation input and previous research [48]. By eliminating grid use, the upfront cost reduces GHG emissions as in [12]. The production phase produces the majority of the lifecycle waste for BEVs.

In Scenario 2, the unincentivized purchase of a PHEV resulted in a 16-month break-even point congruent with previous research [28]. The higher initial cost due to the home installation of a charger was offset quickly due to lower maintenance costs similar to the BEV (arising from reduced wear on the gasoline engine) and lower energy costs ($10 K versus $3.5 K ICEV), although insurance costs were higher. The GHG emissions are higher than the solar-charged BEV option in Scenario 1, as the scenario involves use of an electrical grid. This finding is also congruent with previous research [59].

Both scenarios reinforce previous research showing that EVs can reduce GHGs [60] and have a lower TCO [7].

A significant concern of the consumer might be the residual value of the vehicle upon resell. Both scenarios illustrate that residual values are lower for BEV and PHEV. ICEV appear to retain their value much better. High-end BEVs have seen some improvement in this area with Tesla Model S holding its value better than any vehicle regardless of type [61].

Utility market changes are also consideration for purchasing of a vehicle, as they effect fuel-cost savings and thus the TCO [3]. By including forecasting models for various energy options, the simulation models these effects for inclusion in the analysis.

4.2. Energy Savings and Reduced Environmental Impact of Electrical Power Systems for Transportation

One of the findings of this study is that GHGs for solar-powered BEVs are nominal. Battery-electric power systems powered by PVS produce an environmental impact that is almost negligible. When electrical grid power is used for charging EVs under the conditions of Scenario 2, the simulation depicts reductions in CO_2, NO_x + NMOG, and CO. For CO_2, that lifecycle reduction is from about 60,000 kg (ICEV) to 25,000 g (PHEV). For NO_x + NMOG, lifecycle reductions are estimated to be from 17.5 kg (ICEV) to 10 kg (PHEV). CO estimates were 450 kg for ICEV versus 225 kg for PHEV. Under both scenarios, EV adoption resulted in a reduced environmental impact.

The verdict on true GHGs, however, may be mixed. A recent study has proposed a more robust method for capturing grid energy CO_2 emissions in the electricity production phase [62]. The results suggest that under some conditions, ICEVs might be preferred. This result is surprising and underscores the value in PVS offsets for electrical charging requirements [23].

The total kWh for the BEV operating at 0.34 kWh/mile for 1100 miles × 12 months × 8 years is 35,904 lifecycle kWh, about 4408 kWh monthly. (The 0.34 kWh/mile is a variable parameter in the simulation.) With approximately 1 million BEVs on the road today [63], the additional monthly kWh demand is 4408 GW at the rate of consumption. EV loads are now placing pressure on power grids [64]. Using PVS for recharging these vehicles could offset that increased demand. Given that EVs may increase global consumption of electricity between 11–20% by the year 2040 [65], this type of solution along with smart grid and others may prove vital.

4.3. Implications

This simulation has policy implications for both the government and the manufacturers. First, the use of tax credits visibly accelerates break-even analysis in the online simulation. Tax credits for hybrid and BEV should be considered an investment in the environment [58]. Second, the expansion of EVs is likely to result in an 11–20% increase in grid demand by 2040 [65]. The solutions to this problem might include smart grid development acceleration or incentives for renewable recharging.

4.4. Limitations

One of the limitations of this study is that it focuses on the break-even costs and environmental effects. Consumer vehicle type selection (e.g., gasoline versus hybrid versus fully electric) is often based on other criteria such as driving range, charging time, and performance [66]. A BEV may not be pragmatic for exceedingly long trips.

Another limitation of this simulation is that it assumes zero gray-power costs for solar charging of BEVs. This simplification belies the fact that many solar power systems do not have battery back-up. This is an important limitation when considering environmental effects.

A third limitation is that the simulation is based on data and distributions generated from the United States. There are valid reasons to assume that future costs of gasoline, electricity, and natural gas are unlikely to be the same among countries. Further, other factors are likely to be different, including driving range, availability of recharging stations, and energy policy.

4.5. Future Improvements

This is the first publicly available simulation for this type of analysis. Future work will include modeling of solar gray-power considerations and include additional parameters for user input. Updated data distributions will be included as available, and additional variables will be added.

5. Conclusions

The research question in this study addressed the consumer trade-off considerations for purchasing a BEV, a hybrid, or a gasoline-powered car using a publicly available simulation, the first of its type. The findings of the paper show early break-even points for the two alternative scenarios: A BEV re-charged by PVS and a PHEV re-charged through the grid. In both cases, the break-even was about a year (13, and 12 months, respectively). However, the BEV benefited from tax credits to achieve that early break-even. Due to other considerations that affect buying a BEV, such as a driving range [66], the continuation of tax credits should be a policy focus.

The environmental impacts associated with the PVS-BEV scenario are nominal. Further, such a scenario avoids the 11–20% increase in demand on the grid that may occur by 2040 [65]. Good policy would consider incentivizing the use of the PVS-BEV relationship, while developing smart grid capability. For the PHEV, the GHG are estimated to be much less than for the ICEV; however, the true impact is possibility underestimated [23].

This study provides a foundation for sensitivity analysis in relation to both cost and environmental effects associated with car purchasing decisions. It also highlights the differences in terms of GHG and time. It illustrates how tax credits and residuals values affect TCO and how the choice of re-charging sources affects total costs. Further, the use of ETS forecasts for utility costs provides a reasonable method for estimating energy costs. Finally, it highlights an issue that may require a policy solution: Grid demand increases based on increased EV demand. The solution might include PVS-BEV incentives.

Supplementary Materials: The following are available online at https://rminator.shinyapps.io/Vehicles/.

Author Contributions: Conceptualization through writing, L.F. The author has read and agreed to the published version of the manuscript.

Funding: This research received no external funding.

Conflicts of Interest: The author declares no conflict of interest.

References

1. Baruník, J.; Křehlík, T. Combining high frequency data with non-linear models for forecasting energy market volatility. *Expert Syst. Appl.* **2016**, *55*, 222–242. [CrossRef]
2. Sheldon, T.L.; DeShazo, J.R.; Carson, R.T. Electric and plug-in hybrid vehicle demand: Lessons for an emerging market. *Econ. Inq.* **2017**, *55*, 695–713. [CrossRef]
3. Coffman, M.; Bernstein, P.; Wee, S. Electric vehicles revisited: A review of factors that affect adoption. *Transp. Rev.* **2017**, *37*, 79–93. [CrossRef]
4. DeLuchi, M.; Wang, Q.; Sperling, D. Electric vehicles: Performance, life-cycle costs, emissions, and recharging requirements. *Transp. Res. Part A Gen.* **1989**, *23*, 255–278. [CrossRef]
5. Fulton, L.; Bastian, N. A Fuel Cost Comparison of Electric and Gas-Powered Vehicles. In Proceedings of the 2012 AutumnSim Conference on Energy, Climate and Environmental Modeling & Simulation, San Diego, CA, USA, 28–31 October 2012.
6. Beggs, S.; Cardell, S.; Hausman, J. Assessing the potential demand for electric cars. *J. Econom.* **1981**, *17*, 1–19. [CrossRef]
7. Hagman, J.; Ritzén, S.; Stier, J.J.; Susilo, Y. Total cost of ownership and its potential implications for battery electric vehicle diffusion. *Res. Transp. Bus. Manag.* **2016**, *18*, 11–17. [CrossRef]

8. Yu, A.; Wei, Y.; Chen, W.; Peng, N.; Peng, L. Life cycle environmental impacts and carbon emissions: A case study of electric and gasoline vehicles in China. *Transp. Res. Part D Transp. Environ.* **2018**, *65*, 409–420. [CrossRef]
9. Hackney, J.; De Neufville, R. Life cycle model of alternative fuel vehicles: Emissions, energy, and cost trade-offs. *Transp. Res. Part A Policy Pract.* **2001**, *35*, 243–266. [CrossRef]
10. Park, C.; Xing, R.; Hanaoka, T.; Kanamori, Y.; Masui, T. Impact of Energy Efficient Technologies on Residential CO_2 Emissions: A Comparison of Korea and China. *Energy Procedia* **2017**, *111*, 689–698. [CrossRef]
11. Xu, B.; Lin, B. Regional differences of pollution emissions in China: Contributing factors and mitigation strategies. *J. Clean. Prod.* **2016**, *112*, 1454–1463. [CrossRef]
12. Coffman, M.; Bernstein, P.; Wee, S. Integrating electric vehicles and residential solar PV. *Transp. Policy* **2017**, *53*, 30–38. [CrossRef]
13. Delucchi, M.A.; Lipman, T.E. An analysis of the retail and lifecycle cost of battery-powered electric vehicles. *Transp. Res. Part D Transp. Environ.* **2001**, *6*, 371–404. [CrossRef]
14. Lipman, T.E.; Delucchi, M.A. A retail and lifecycle cost analysis of hybrid electric vehicles. *Transp. Res. Part D* **2006**, *11*, 115–132. [CrossRef]
15. Ahmadi, P.; Cai, X.M.; Khanna, M. Multicriterion optimal electric drive vehicle selection based on lifecycle emission and lifecycle cost. *Int. J. Energy Res.* **2018**, *42*, 1496–1510. [CrossRef]
16. Palmer, K.; James, E.T.; John, N.; Zia, W. Total cost of ownership and market share for hybrid and electric vehicles in the UK, US and Japan. *Appl. Energy* **2018**, *209*, 108–119. [CrossRef]
17. Prevedouros, P.; Mitropoulos, L. Impact of Battery Performance on Total Cost of Ownership for Electric Drive Vehicle. In Proceedings of the 2018 21st International Conference on Intelligent Transportation Systems (ITSC), Maui, HI, USA, 4–7 November 2018; pp. 1155–1160.
18. Mitropoulos, L.K.; Prevedouros, P.D.; Kopelias, P. Total cost of ownership and externalities of conventional, hybrid and electric vehicle. *Transp. Res. Procedia* **2017**, *24*, 267–274. [CrossRef]
19. Axsen, J.; Kurani, K.S. Hybrid, plug-in hybrid, or electric—What do car buyers want? *Energy Policy* **2013**, *61*, 532–543. [CrossRef]
20. Danielis, R.; Giansoldati, M.; Rotaris, L. A probabilistic total cost of ownership model to evaluate the current and future prospects of electric cars uptake in Italy. *Energy Policy* **2018**, *119*, 268–281. [CrossRef]
21. Weldon, P.; Morrissey, P.; O'Mahony, M. Long-term cost of ownership comparative analysis between electric vehicles and internal combustion engine vehicles. *Sustain. Cities Soc.* **2018**, *39*, 578–591. [CrossRef]
22. Moon, S.; Lee, D.-J. An optimal electric vehicle investment model for consumers using total cost of ownership: A real option approach. *Appl. Energy* **2019**, *253*, 113494. [CrossRef]
23. Khan, S.; Ahmad, A.; Ahmad, F.; Shafaati Shemami, M.; Saad Alam, M.; Khateeb, S. A Comprehensive Review on Solar Powered Electric Vehicle Charging System. *Smart Sci.* **2018**, *6*, 54–79. [CrossRef]
24. Kelly, J.C.; MacDonald, J.S.; Keoleian, G.A. Time-dependent plug-in hybrid electric vehicle charging based on national driving patterns and demographics. *Appl. Energy* **2012**, *94*, 395–405. [CrossRef]
25. Das, H.S.; Rahman, M.M.; Li, S.; Tan, C.W. Electric vehicles standards, charging infrastructure, and impact on grid integration: A technological review. *Renew. Sustain. Energy Rev.* **2020**, *120*, 109618. [CrossRef]
26. Chandra Mouli, G.R.; Bauer, P.; Zeman, M. System design for a solar powered electric vehicle charging station for workplaces. *Appl. Energy* **2016**, *168*, 434–443. [CrossRef]
27. Fulton, L.; Beauvais, B.; Brooks, M.; Kruse, C.S.; Lee, K. Green for the Environment and Green for the Pocketbook: A Decade of Living Sustainably. *Preprints* **2020**. [CrossRef]
28. Fulton, L. Ownership Cost Comparison of Battery Electric and Non-Plugin Hybrid Vehicles: A Consumer Perspective. *Appl. Sci.* **2018**, *8*, 1487. [CrossRef]
29. Ernst, C.-S.; Hackbarth, A.; Madlener, R.; Lunz, B.; Uwe Sauer, D.; Eckstein, L. Battery sizing for serial plug-in hybrid electric vehicles: A model-based economic analysis for Germany. *Energy Policy* **2011**, *39*, 5871–5882. [CrossRef]
30. Jaguemont, J.; Boulon, L.; Dubé, Y. A comprehensive review of lithium-ion batteries used in hybrid and electric vehicles at cold temperatures. *Appl. Energy* **2016**, *164*, 99–114. [CrossRef]
31. Lieven, T.; Mühlmeier, S.; Henkel, S.; Waller, J.F. Who will buy electric cars? An empirical study in Germany. *Transp. Res. Part D* **2011**, *16*, 236–243. [CrossRef]
32. Shin, J.; Hong, J.; Jeong, G.; Lee, J. Impact of electric vehicles on existing car usage: A mixed multiple discrete–continuous extreme value model approach. *Transp. Res. Part D* **2012**, *17*, 138–144. [CrossRef]

33. He, L.; Chen, W.; Conzelmann, G. Impact of vehicle usage on consumer choice of hybrid electric vehicles. *Transp. Res. Part D* **2012**, *17*, 208–214. [CrossRef]
34. He, H.; Fan, J.; Li, Y.; Li, J. When to switch to a hybrid electric vehicle: A replacement optimisation decision. *J. Clean. Prod.* **2017**, *148*, 295–303. [CrossRef]
35. Özdemir, E.D.; Hartmann, N. Impact of electric range and fossil fuel price level on the economics of plug-in hybrid vehicles and greenhouse gas abatement costs. *Energy Policy* **2012**, *46*, 185–192. [CrossRef]
36. Silva, C.; Farias, T.; Ross, M. Evaluation of energy consumption, emissions and cost of plug-in hybrid vehicles. *Energy Convers. Manag.* **2009**, *50*, 1635–1643. [CrossRef]
37. Werber, M.; Fischer, M.; Schwartz, P.V. Batteries: Lower cost than gasoline? *Energy Policy* **2009**, *37*, 2465–2468. [CrossRef]
38. Weiller, C. Plug-in hybrid electric vehicle impacts on hourly electricity demand in the United States. *Energy Policy* **2011**, *39*, 3766–3778. [CrossRef]
39. Fulton, L.; McMurry, L.T.C.P.; Kerr, C.O.L.B. A Monte Carlo Simulation of Air Ambulance Requirements During Major Combat Operations. *Mil. Med.* **2009**, *174*, 610–614. [CrossRef]
40. R Core Team. *R: A Language and Environment for Statistical Computing*; R Core Team: Vienna, Austria, 2018.
41. Chang, W.; Cheng, J.; Allaire, J.J.; Xie, Y.; McPherson, J. Shiny: Web Application Framework for R. 2019. Available online: https://shiny.rstudio.com/reference/shiny/1.4.0/shiny-package.html (accessed on 1 May 2020).
42. LowCVP. LowCVP Study Demonstrates the Increasing Importance of Measuring Whole Life Carbon Emissions to Compare Vehicle Performance. Available online: https://d1v9sz08rbysvx.cloudfront.net/ricardo/media/media/news%20assets/lowcvp%20study%20demonstrates%20importance%20of%20whole%20life%20co2%20emissions.pdf (accessed on 27 April 2020).
43. Barbose, G.; Darghouth, N. Tracking the Sun|Electricity Markets and Policy Group. 2019. Available online: https://emp.lbl.gov/tracking-the-sun (accessed on 27 April 2020).
44. Fulton, L.; Bradley, B.; Matthew, B.; Clemens Scott, K.; Lee, K. A Publicly Available Cost Simulation of Sustainable Construction Options for Residential Houses. *Sustainability* **2020**, *12*, 2873. [CrossRef]
45. Average Annual Miles per Driver by Age Group. Available online: https://www.fhwa.dot.gov/ohim/onh00/bar8.htm (accessed on 27 April 2020).
46. U.S. Energy Information Administration (EIA). Available online: https://www.eia.gov/ (accessed on 27 April 2020).
47. Hyndman, R.A.G. *Forecasting: Principles and Practice*; OTexts: Melbourne, Australia, 2019.
48. Breetz, H.L.; Salon, D. Do electric vehicles need subsidies? Ownership costs for conventional, hybrid, and electric vehicles in 14 U.S. cities. *Energy Policy* **2018**, *120*, 238–249. [CrossRef]
49. Total Cost of Ownership of an Electric Car|PluginCars.com. Available online: https://www.plugincars.com/eight-factors-determining-total-cost-ownership-electric-car-127528.html (accessed on 21 April 2020).
50. Xia, Y.; Yang, J.; Liu, Z.; Dong, J. Cost-Effectiveness Analysis of Plug-In Hybrid Electric Vehicles using Vehicle Usage Data Collected in Shanghai, China. *Transp. Res. Rec.* **2019**, *2673*, 251–261. [CrossRef]
51. Alternative Fuels Data Center: Emissions from Hybrid and Plug-In Electric Vehicles. Available online: https://afdc.energy.gov/vehicles/electric_emissions.html#wheel (accessed on 21 April 2020).
52. E.P.A. Federal and California Light-Duty Vehicle Emissions Standards for Air Pollution, EPA-420-B-19-043. 2019. Available online: https://nepis.epa.gov/Exe/ZyPDF.cgi?Dockey=P100XCIV.pdf (accessed on 21 April 2020).
53. Randolph, J.; Masters, G.M. Transportation Energy and Efficient Vehicles. In *Energy for Sustainability: Foundations for Technology, Planning, and Policy*; Randolph, J., Masters, G.M., Eds.; Island Press/Center for Resource Economics: Washington, DC, USA, 2018; pp. 389–428. [CrossRef]
54. Moody's. Electric Vehicle Residual Value Outlook. Available online: https://www.moodysanalytics.com/-/media/presentation/2017/electric-vehicle-residual-value-outlook.pdf (accessed on 28 April 2020).
55. Harper, G.; Sommerville, R.; Kendrick, E.; Driscoll, L.; Slater, P.; Stolkin, R.; Walton, A.; Christensen, P.; Heidrich, O.; Lambert, S.; et al. Recycling lithium-ion batteries from electric vehicles. *Nature* **2019**, *575*, 75–86. [CrossRef] [PubMed]
56. Berjoz, D.J.I. Influence of batteries weight on electric automobile performance. *Eng. Rural Dev.* **2017**, *24*, 1388–1394.
57. Ma, S.X.; Junping, J.; Xiao, M. Comparative Life Cycle Energy and GHG Emission Analysis for BEVs and PhEVs: A Case Study in China. *Energies* **2019**, *12*, 834. [CrossRef]

58. Narassimhan, E.; Johnson, C. The role of demand-side incentives and charging infrastructure on plug-in electric vehicle adoption: Analysis of US States. *Environ. Res. Lett.* **2018**, *13*, 074032. [CrossRef]
59. Falahati, B.; Shahverdi, M.; Mohajeryami, S.; Fajri, P. Examining the impact of PHEVs on GHG emissions based on various objectives. In Proceedings of the 2017 IEEE Conference on Technologies for Sustainability (SusTech), Phoenix, AZ, USA, 12–14 November 2017; pp. 1–5.
60. Li, J.; Yang, B. Analysis of greenhouse gas emissions from electric vehicle considering electric energy structure, climate and power economy of ev: A China case. *Atmos. Pollut. Res.* **2020**. [CrossRef]
61. Guo, Z.; Zhou, Y. Residual value analysis of plug-in vehicles in the United States. *Energy Policy* **2019**, *125*, 445–455. [CrossRef]
62. Manjunath, A.; Gross, G. Towards a meaningful metric for the quantification of GHG emissions of electric vehicles (EVs). *Energy Policy* **2017**, *102*, 423–429. [CrossRef]
63. EEI Celebrates 1 Million Electric Vehicles on U.S. Roads. Available online: https://www.eei.org/resourcesandmedia/newsroom/Pages/Press%20Releases/EEI%20Celebrates%201%20Million%20Electric%20Vehicles%20on%20U-S-%20Roads.aspx (accessed on 30 April 2020).
64. Falahati, S.; Taher, S.A.; Shahidehpour, M. A new smart charging method for EVs for frequency control of smart grid. *Int. J. Electr. Power Energy Syst.* **2016**, *83*, 458–469. [CrossRef]
65. Kapustin, N.O.; Grushevenko, D.A. Long-term electric vehicles outlook and their potential impact on electric grid. *Energy Policy* **2020**, *137*, 111103. [CrossRef]
66. Hidrue, M.K.; Parsons, G.R.; Kempton, W.; Gardner, M.P. Willingness to pay for electric vehicles and their attributes. *Resour. Energy Econ.* **2011**, *33*, 686–705. [CrossRef]

© 2020 by the author. Licensee MDPI, Basel, Switzerland. This article is an open access article distributed under the terms and conditions of the Creative Commons Attribution (CC BY) license (http://creativecommons.org/licenses/by/4.0/).

energies

MDPI

Article

Multidispatch for Microgrid including Renewable Energy and Electric Vehicles with Robust Optimization Algorithm

Ruifeng Shi [1,2], Penghui Zhang [1], Jie Zhang [1], Li Niu [3,*] and Xiaoting Han [4]

1 School of Control and Computer Engineering, North China Electric Power University, Beijing 102206, China; shi.ruifeng@ncepu.edu.cn (R.S.); penghui_zhang@ncepu.edu.cn (P.Z.); jie_zhang@ncepu.edu.cn (J.Z.)
2 China Institute of Energy and Transportation Integrated Development, Beijing 102206, China
3 School of Information Resource Management, Renmin University of China, Beijing 100872, China
4 School of Economics and Management, Beihang University, Beijing 100083, China; hanxiaoting@buaa.edu.cn
* Correspondence: rucniuli@ruc.edu.cn; Tel.: +86-10-6251-1461

Received: 25 March 2020; Accepted: 21 May 2020; Published: 2 June 2020

Abstract: With the deterioration of the environment and the depletion of fossil fuel energy, renewable energy has attracted worldwide attention because of its continuous availability from nature. Despite this continuous availability, the uncertainty of intermittent power is a problem for grid dispatching. This paper reports on a study of the scheduling and optimization of microgrid systems for photovoltaic (PV) power and electric vehicles (EVs). We propose a mathematical model to address the uncertainty of PV output and EV charging behavior, and model scheduling optimization that minimizes the economic and environmental cost of a microgrid system. A semi-infinite dual optimization model is then used to deal with the uncertain variables, which can be solved with a robust optimization algorithm. A numerical case study shows that the security and stability of the solution obtained by robust optimization outperformed that of stochastic optimization.

Keywords: electric vehicle; microgrid; photovoltaic; robust optimization; stochastic optimization

1. Introduction

Energy shortages and environmental pollution are now problems that cannot be ignored. To build an environment-friendly and resource-saving society, renewable energy generation such as wind and photovoltaic (PV) power have received increasing attention [1–3]. Renewable energy accounted for approximately 18.1% of total final energy consumption in 2017 and was mainly concentrated in the field of electricity generation. According to the "Renewables (2019) Global Status Report" estimate, by the end of 2019 more than 26% of global electricity generation had come from renewable sources [4]. Renewable energy has the advantage of being inexhaustible, but its intermittent and volatile nature poses new challenges to the safe and stable operation of the power grid.

Renewable energy generation is influenced by climate and weather, and is not controlled by human factors. The electricity generation and the electricity consumption cannot match well; there will often be no electricity available at the peak of electricity consumption, or no electricity production surplus at the trough of the situation. With the increasing number of electric vehicles, the two-way energy dispatching between electric vehicles and the power grid provides the solution for renewable energy power generation and large-scale application of electric vehicles. Electric vehicles consume renewable energy when electricity consumption demand is low, and even discharge electricity into the grid when demand is high. Reasonable and flexible scheduling strategies can ensure the coordinated operation of electric vehicles and renewable energy generation.

To solve the problem of grid-connected distributed generation, the microgrid came into being. A microgrid manages the power generation units and power load dispatching within a certain range.

The key to the safe and stable operation of a microgrid is to reduce the uncertainty of renewable energy generation. Currently, the main methods to solve the uncertainty problem are stochastic optimization (SO), fuzzy programming, and robust optimization (RO) [5]. The probability distribution function is used by SO to express the output of renewable energy. This method must usually collect much scene data, which increases the difficulty of the problem [6–8]. Fuzzy programming uses fuzzy variables to represent uncertainty and fuzzy sets to describe constraints. The satisfaction of constraints is represented by a membership function. This method usually depends on the activities of people, so some deviation is unavoidable [9–11]. RO uses a set approach to represent uncertainty. This method does not need to know the probability distribution of renewable energy output, nor does it depend on human activity. The results of RO can deal with worst-case scenarios. In view of this characteristic, RO is increasingly favored [12,13].

RO has good advantages in tolerating uncertainties in dispatch problems [14]. To alleviate the risk of microgrid energy trading under the uncertainty of renewable energy and transaction price, Luhao Wang et al. proposed a risk avoidance method based on RO and created a two-stage energy-trading RO model. Their results show that the model can not only achieve the optimal operating cost of the microgrid system but also ensure the robustness of energy trading between systems at various prices [15]. Yan Cao et al. used RO techniques to study the scheduling problem of electric vehicle (EV) aggregators with price uncertainty and to participate in the market with the goal of maximizing the benefits of aggregators. When they modeled market price uncertainty, they used the upper and lower limits of the upstream grid price instead of the estimated price. The output of the algorithm was used to construct various charging and discharging strategies for operators to use for robust scheduling of EV aggregators under upstream grid price uncertainty. The results show that compared with a deterministic strategy, the total profit of the EV aggregators under the optimistic strategy had increased by 69.78% [16]. Carlos D. Rodríguez-Gallegos et al. proposed an optimized configuration method for solar panels and batteries, forming a PV hybrid system. A multi-objective optimization problem was established by considering economic goals, environmental goals, and grid quality. The worst weather conditions were considered, and RO algorithm was applied. The results showed that PV hybrid power systems have the advantages of reducing costs and improving the environment and the grid quality in remote areas [17].

Considering the uncertainty of the renewable energy output and EV charging behavior, in this study the RO theory was applied to a microgrid system containing a PV power station and EVs. To reduce the running cost and environmental cost, the goal was to build a multi-objective scheduling model. Then, the scheduling objective function was optimized by minimizing the economic and environmental costs of the microgrid system. Based on the duality principle, the semi-infinite problem containing uncertain variables was converted into an easier dual problem. Finally, SO and RO were compared. RO has a high cost, but its security and stability are most important for a microgrid. The main contributions of this paper are as follows:

1. The uncertainty of PV power output and EV charging behavior is expressed as a set, which contains all possible values of uncertain parameters. The results of robust optimization ensure that the system can run safely and stably even under the worst condition.
2. In order to take the various practical constraints into consideration, a multi-objective robust scheduling optimization model is proposed to address these constraints. The model is a semi-infinite optimization problem, which is difficult to solve directly. According to the duality theory, the model is transformed into a definite linear programming model, which can be easily solved with Lagrange relaxation algorithm.
3. By numerical analysis, SO and RO are compared. The results show that RO is higher in cost, while it is more robust than SO. System operators can select appropriate optimization methods in balancing between economy and safety.

The paper is organized as follows: the models of PV power output and EV charging behaviors are described in Section 2. Section 3 introduces the multi-objective scheduling system and actual constraints. The theory and application of RO are presented in Section 4. A case study is used in Section 5 to show the performance of the proposed models. Conclusions are drawn in Section 6.

2. Uncertainty Modeling

In this paper, the microgrid system connected to the main power grid includes PV power stations, microturbines (MTs), diesel engines (DEs), and EVs. The uncertainties of PV output power and EV charging behavior are expressed as a set.

2.1. Photovoltaic Output Model

Solar power is the use of battery modules to directly convert solar energy into electricity. Solar energy is an inexhaustible source of clean energy, and a solar PV power station is safe, reliable, and unaffected by the energy crisis. However, it is difficult to accurately predict the power generation of a PV system [18]. Due to the periodicity and randomness of the output of a PV power station, the output is considered as [19]

$$\begin{gathered} PV^{G}_{l,t} = \overline{PV_{l,t}} + P\hat{V}_{l,t} \\ \text{s.t.}\underline{P\hat{V}_{l,t}} \le P\hat{V}_{l,t} \le \overline{P\hat{V}_{l,t}} \end{gathered} \tag{1}$$

In other words, the uncertainty of the PV can be expressed as $[\overline{PV_{l,t}} + \underline{P\hat{V}_{l,t}}, \overline{PV_{l,t}} + \overline{P\hat{V}_{l,t}}]$, where $PV^{G}_{l,t}$ is the output of the lth PV power station at period t, $\overline{PV_{l,t}}$ is the forecasted output of the lth PV power station at period t, $P\hat{V}_{l,t}$ is the deviation, and $\underline{P\hat{V}_{l,t}}$, $\overline{P\hat{V}_{l,t}}$ are the lower and upper limits of $P\hat{V}_{l,t}$, respectively.

2.2. Electric Vehicle Charging Model

The probability of an individual EV traveling a distance d can be represented by the logarithmic normal distribution function [20]

$$h(d, \mu, \sigma) = \frac{1}{d\sqrt{2\pi\sigma^2}} e^{-\frac{(\ln d-\mu)^2}{2\sigma^2}} \tag{2}$$

where μ and σ are the mean and standard deviation, respectively.

According to the traveling distance d of the EV, the remaining capacity can be calculated:

$$SOC = (1 - \frac{d}{d_m}) \times 100\% \tag{3}$$

where d_m is the maximum travel distance.

According to [21], the charging start time can be modeled as a normal distribution function with specified parameters, and we represent the time when the EV starts charging as a set.

$$T^{k}_{start} = [\underline{T^{k}_{start}}, \overline{T^{k}_{start}}] \tag{4}$$

where T^{k}_{start} is the charging start time of the nth EV, and $\underline{T^{k}_{start}}$ and $\overline{T^{k}_{start}}$ are the lower and upper limits, respectively.

In this study, we considered that all EV batteries have the same capacity E and the same charging power P_c. Thus, the charging end time T^k_{end} can be obtained:

$$T^k_{end} = T^k_{start} + \frac{(1 - SOC^k) \times E}{P_c} \tag{5}$$

The total charging power of EVs at each moment is the sum of the charging power of an individual EV:

$$PEV^G_t = \sum_{k=1}^{K} PEV^G_{k,t} \tag{6}$$

where $PEV^G_{k,t}$ is the charging power of the kth EV at period t. In addition, K is the number of EVs dispatched.

The uncertainty of the charging power of the EV is still expressed in the form of a set:

$$\begin{aligned} PEV^G_t &= \overline{PEV_t} + P\hat{E}V_t \\ \text{s.t.} \underline{P\hat{E}V_t} &\leq P\hat{E}V_t \leq \overline{P\hat{E}V_t} \end{aligned} \tag{7}$$

where PEV^G_t is the total charging power of EVs at period t, $\overline{PEV_t}$ is the forecasted value of the charging power of EVs at period t, $P\hat{E}V_t$ is the deviation, and $\underline{P\hat{E}V_t}$, $\overline{P\hat{E}V_t}$ are its lower and upper limits, respectively.

3. Multi-Objective Dispatch System

The microgrid system includes PV power stations, MTs, DEs, and EVs. The optimization objective is to minimize total costs, including the cost of operating the microgrid and the cost of environmental protection. In this section, various practical constraints are also considered. The mathematical model is described in Sections 3.1 and 3.2 [22,23].

3.1. Objective Functions

3.1.1. Objective Function 1: Minimum Operating Cost

The total operating costs include fuel costs for the DEs and MTs; operation and maintenance costs for DEs, MTs, and the PV power station; and transmission costs between the microgrid and the main grid. The battery degradation cost is disregarded [24,25].

$$C_1 = \sum_{t=1}^{T} [C_f(P_{i,t} + PMT_{j,t}) + C_{OM}(P_{i,t} + PV_{l,t} + PMT_{j,t}) + C_{grid,t}] \tag{8}$$

where $P_{i,t}$ is the output power of the ith DE at period t, $PMT_{j,t}$ is the output power of the jth MT at period t, $PV_{l,t}$ is the output power of the lth PV power station at period t. $C_f(\cdot)$ is the fuel costs of the DEs and MTs [26]; $C_{OM}(\cdot)$ is the operation and maintenance costs of the DE, MTs, and the PV power station [27]; and $C_{grid,t}$ is the cost of transmission between the microgrid and the main power grid.

$$\begin{aligned} C_f(P_{i,t}) &= [c_1 P^2_{i,t} + c_2 P_{i,t} + c_3]_{DE} + [y \frac{PMT_{j,t}}{\eta(PMT_{j,t})}]_{MT} \\ C_{OM}(P_{i,t} + PV_{l,t} + PMT_{j,t}) &= K_{OM}(P_{i,t} + PV_{l,t} + PMT_{j,t}) \\ C_{grid,t} &= P_{grid,t} M_t \Delta t \end{aligned} \tag{9}$$

where c_1, c_2, c_3 represent fuel cost parameters of DEs, y is the cost parameter of MTs, $\eta(PMT_{j,t})$ is the work efficiency of the jth MT at period t, K_{OM} is the OM cost parameter, Δt is scheduling interval, and M_t is the price of electricity.

3.1.2. Objective Function 2: Minimum Environmental Cost

The traditional output units and the power transmission process of the grid will cause environmental pollution problems, which incurs the cost of environmental protection. Three important pollutants, sulfur dioxide (SO_2), carbon dioxide (CO_2), and nitrogen oxide (NO_X), are considered in this paper [26].

$$C_2 = \sum_{t=1}^{T}\sum_{p=1}^{P}\sum_{h=1}^{H}(C_h\mu_{p,h})P_{p,t} + \sum_{t=1}^{T}\sum_{h=1}^{H}(C_h\mu_{grid})P_{grid,t} \tag{10}$$

where C_h is the treatment cost of the hth pollutant; $\mu_{p,h}$ is the hth pollutant emission coefficients of the pth type power source including DE, MT, and PV; $P_{p,t}$ is the output power of the pth power source; $P_{grid,t}$ is the transmission power between the microgrid and the main power grid at period t, and μ_{grid} is the pollutant emission coefficients of the main power grid.

3.1.3. Total Cost Function

The objective of the scheduling system proposed in this paper is to minimize the system operating cost and the environmental protection cost. Therefore, the total cost function (C_{total}) can be expressed by

$$\min C_{total} = \min(C_1 + C_2) \tag{11}$$

3.2. Constraints

3.2.1. Power Balance Constraint

Electricity supply and demand should be balanced in the microgrid system:

$$\sum_{i=1}^{I} P_{i,t} + P_{grid,t} + \sum_{j=1}^{J} PMT_{j,t} + \sum_{l=1}^{L} PV_{l,t} - PEV_t = P_{load,t} \tag{12}$$

where $P_{load,t}$ is the total load demand at period t.

3.2.2. Generation Capacity Constraints

The output power of DEs and MTs should be within a certain range:

$$\begin{gathered} P_{i,\min} \le P_{i,t} \le P_{i,\max} \\ PMT_{j,\min} \le PMT_{j,t} \le PMT_{j,\max} \end{gathered} \tag{13}$$

where $P_{i,\min}$ and $P_{i,\max}$ are the lower and upper limits of the power output of the ith DE, respectively. Similarly, $PMT_{j,\min}$ and $PMT_{j,\max}$ are the lower and upper limits of the power output of the jth MT, respectively.

3.2.3. Ramp Rate Limits

Ramp rate refers to the increase or decrease of output power per unit of time from traditional power sources.

$$\begin{gathered} P_{i,down} \le P_{i,t} - P_{i,t-1} \le P_{i,up} \\ P_{j,down} \le PMT_{j,t} - PMT_{j,t-1} \le P_{j,up} \end{gathered} \tag{14}$$

where $P_{i,down}$ and $P_{i,up}$ are the lower and upper limits of the ramp rate of the ith DE, respectively, and $P_{j,down}$, $P_{j,up}$ are the lower, upper limits of the ramp rate of the jth MT, respectively.

3.2.4. Capacity Constraints of PV

The output of PV is affected by the predicted value at period t.

$$0 \leq PV_{l,t} \leq PV_{l,t}^{G} \tag{15}$$

3.2.5. Constraints of Transmission Capacity

The transmission power between the microgrid and the main grid should not exceed the limit.

$$P_{down} \leq P_{grid,t} \leq P_{up} \tag{16}$$

where P_{down} and P_{up} are the lower and upper limits, respectively, of the main grid transmitting power.

3.2.6. Charge Constraints of the EV Battery

To slow the aging of the battery, the charging power should not exceed the maximum:

$$0 \leq PEV_{k,t} \leq PEV_{k,t}^{\max} \tag{17}$$

3.2.7. Spinning Reserve Constraint

The spinning reserve constraint is to ensure a reliable power supply:

$$\sum_{i=1}^{I} P_{i,t}^{\max} + \sum_{j=1}^{J} PMT_{j,t}^{\max} + P_{grid,t}^{\max} + \sum_{l=1}^{L} PV_{l,t}^{G} \geq (1 + L_t)(P_{load,t} + \sum_{k=1}^{K} PEV_{t}^{G}) \tag{18}$$

where L_t is the spinning reserve rate.

4. Robust Optimization Model

4.1. Robust Optimization Algorithm

RO is designed to deal with uncertainty problems. Unlike stochastic optimization, it represents uncertain variables in interval form. It can be said that the uncertainty of variables is fully considered in modeling. The result of RO is the most conservative result. Any value in the set of uncertain variables can be satisfied. Therefore, it is especially suitable for microgrid systems that are very important for security.

RO first models the uncertain variables, then transforms the uncertain variables into deterministic variables according to the robust equal conversion, and finally solves the robust equal conversion model to obtain the robust optimal solution [28,29].

4.2. Robust Equal Conversion

On the basis of the sets of uncertainties introduced in Sections 2.1 and 2.2, the spinning reserve constraint (Equation (18)) can be transformed to

$$\sum_{i=1}^{I} P_{i,t}^{\max} + \sum_{j=1}^{J} PMT_{j,t}^{\max} + P_{grid,t}^{\max} + \sum_{l=1}^{L} (\overline{PV_{l,t}} + P\hat{V}_{l,t}) \geq (1 + L_t)[P_{load,t} + \sum_{k=1}^{K} (\overline{PEV_{k,t}} + PE\hat{V}_{k,t})] \tag{19}$$

RO deals with uncertain data in the worst-case scenario. The worst-case scenario can be defined as

$$
\begin{gathered}
F = \max[\sum_{l=1}^{L} P\hat{V}_{l,t} - (1+L_t)\sum_{k=1}^{K} PE\hat{V}_{k,t}] \\
\underline{P\hat{V}_{l,t}} \leq P\hat{V}_{l,t} \leq \overline{P\hat{V}_{l,t}} \\
\underline{PE\hat{V}_{k,t}} \leq PE\hat{V}_{k,t} \leq \overline{PE\hat{V}_{k,t}}
\end{gathered} \tag{20}
$$

The schedule objective function is monotonically increasing, strictly convex, and differentiable. Its dual problem is also feasible and bounded, and the objective values coincide according to strong duality [19]. Therefore, the dual problem becomes

$$
\begin{gathered}
\min(-\sum_{l=1}^{L} \underline{P\hat{V}^{G}_{l,t}}\alpha_t + \sum_{l=1}^{L} \overline{P\hat{V}^{G}_{l,t}}\beta_t - \sum_{k=1}^{K} \underline{PE\hat{V}_{k,t}}\gamma_t + \sum_{k=1}^{K} \overline{PE\hat{V}_{k,t}}\delta_t) \\
-\alpha_t + \beta_t \geq 1 \\
-\gamma_t + \delta_t \geq -1 - L_t \\
\alpha_t, \beta_t, \gamma_t, \delta_t \geq 0
\end{gathered} \tag{21}
$$

where $\alpha_t, \beta_t, \gamma_t, \delta$ are the dual coefficients. Then the original spinning reserve constraint is converted to

$$
\begin{gathered}
\sum_{i=1}^{I} P_{i,t}^{\max} + \sum_{j=1}^{J} PMT_{j,t}^{\max} + P_{grid,t}^{\max} + \sum_{l=1}^{L} \overline{PV_{l,t}} - (1+L_t)\sum_{k=1}^{K} \overline{PEV_{k,t}} - \sum_{l=1}^{L} \underline{P\hat{V}_{l,t}}\alpha_t \\
+\sum_{l=1}^{L} \overline{P\hat{V}_{l,t}}\beta_t - \sum_{k=1}^{K} \underline{PE\hat{V}_{k,t}}\gamma_t + \sum_{k=1}^{K} \overline{PE\hat{V}_{k,t}}\delta_t \geq (1+L_t)P_{load,t}
\end{gathered} \tag{22}
$$

4.3. Robust Economic Dispatch Model

The purpose of this study was to minimize system operating costs and environmental protection costs while meeting the load on the microgrid system:

$$
\begin{gathered}
\min\left\{ \sup_{PV_{l,t}, PEV_{k,t}} C_1 + C_2 \right\} \\
s.t.\ (12)\text{–}(17),\ (20)\text{–}(22)
\end{gathered} \tag{23}
$$

The probability of spinning reserve constraint violated (POV) is expressed as,

$$
P_r\left[\sum_{i=1}^{I} P_{i,t}^{\max} + \sum_{j=1}^{J} PMT_{j,t}^{\max} + P_{grid,t}^{\max} + \sum_{l=1}^{L} PV_{l,t}^{G} < (1+L_t)(P_{load,t} + \sum_{k=1}^{K} PEV_{k,t}^{G})\right] \leq P_r\left[\sum_{m\in V} \eta_{m,t}\omega_{m,t} \geq \Gamma_t\right] \tag{24}
$$

where Γ_t is the number of the uncertain variables. Furthermore, according to literature [29], the robustness of the system can be expressed by Formula (25):

$$
P_r\left[\sum_{m\in V} \eta_{m,t}\omega_{m,t} \geq \Gamma_t\right] \leq \exp\left[-\frac{\Gamma_t^2}{2|J_t|}\right] \tag{25}
$$

where $\eta_{m,t}, \omega_{m,t}, J_t$ are the coefficients of POV.

5. Case Study

5.1. Problem Description

In this study, we verified the validity of the model by selecting a typical grid-connected microgrid in an area where people work. Figure 1 is a diagram of the initial state of the scheduling system. After the simulation of the model built in Section 2, Figure 1 is obtained. PV-down and PV-up represent the lower and upper bounds of PV power output at each moment, respectively. EV charging-up and EV charging-down represent the upper and lower bounds of the charging power of electric vehicles, respectively. As shown in Figure 1, when people come to the work area in the morning, they start charging; the charging behavior is concentrated between 8:00 a.m. and 10:30 a.m. At this time, the demand for electricity in the office area increases, but the output of the PV power station is still relatively low. To compensate for the disordered charging behavior of EVs, DEs have to increase their output, which results in an increase in the total cost of the system. At noon, people have a rest time and the system electricity demand is reduced, but PV power stations have the most output at that time. If there is no electrical equipment to consume the electrical energy, this will cause a discard phenomenon. Therefore, an unoptimized system is neither stable nor economical.

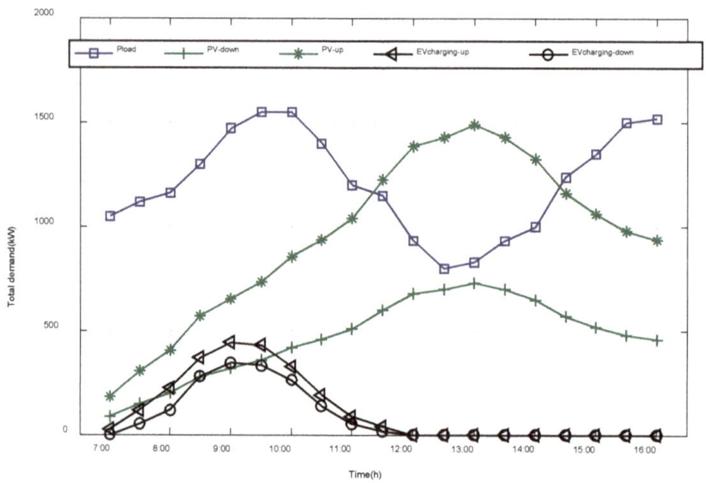

Figure 1. Initial status of the microgrid of an unoptimized system.

5.2. Parameter Setting

In this study, the microgrid consisted of a DE with a maximum output of 1500 kW, two MTs with a maximum power of 250 kW, and two PV power stations with a maximum power of 800 kW. The maximum deviation of actual output and predicted output in the dispatch period was set at ±30% [19]. Additionally, the system considered that 100 EVs participated in the dispatch. We assumed that the battery capacity of the EVs was 60 kWh, the microgrid was connected to the main grid in real time, the microgrid spinning reserve ratio was set at 0.2, and the simulation time was set at 7:00 a.m. to 4 p.m. The simulation time interval was set at 30 min. The detailed parameters of each unit are shown in Table 1 [26]. The electricity price is shown in Figure 2 [26]. The operation and maintenance costs are shown in Table 2, and the environmental treatment cost is shown in Table 3 [26]. The Lagrange relaxation algorithm with faster convergence speed is used to solve the model on MATLAB/CPLEX software.

Table 1. Maximum capacity of each scheduling unit.

Type	Maximum (kW)
Microturbines	250
Diesel engine	1500
Photovoltaic power stations	800
Charging power of electric vehicle	6
Main grid transmission power	300

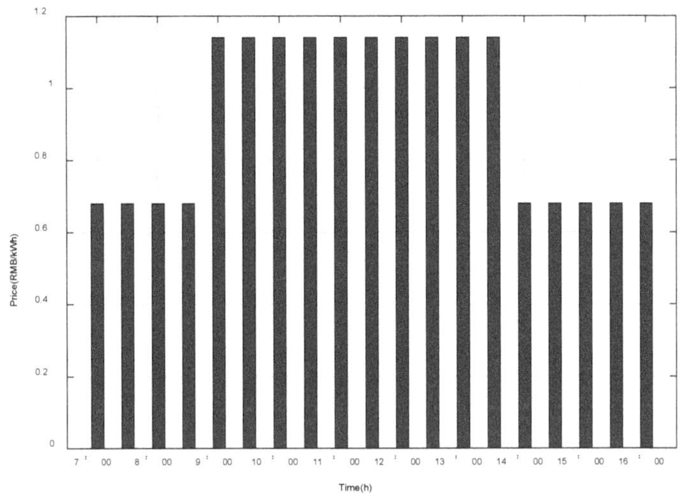

Figure 2. Time of use electricity prices.

Table 2. Power supply operation and maintenance cost parameters.

Type	Microturbine	Diesel Engine	Photovoltaic
K_{OM} (RMB/kWh)	0.04	0.08	0.005

Table 3. Environmental protection cost parameters.

Type	Source	CO_2	SO_2	NOx
Pollution discharge (g/kW)	MT	724	0.0036	0.2
	DE	680	0.306	10.09
	PV	0	0	0
	Main grid	889	1.8	1.6
Governance costs (RMB/kg)		0.21	6	8

5.3. *Simulation Result*

5.3.1. Stochastic Optimization

The purpose of this study was to reduce system operating costs and environmental pollution as much as possible while ensuring electricity demand is met. As shown in Figure 3, the EVs are charged at 12:00. That is not the peak period of microgrid power consumption, and although the electricity price of the main grid is higher at this time, the system has enough light energy to be used. Therefore, the renewable energy efficiency can be maximized by the microgrid system. The remaining electricity demand is supplemented by DEs, MTs, and the main grid. However, the system does not consider the uncertainty of PV power output and EV charging behavior, it is less robust, and it operates in a nonoptimal, nonstable state.

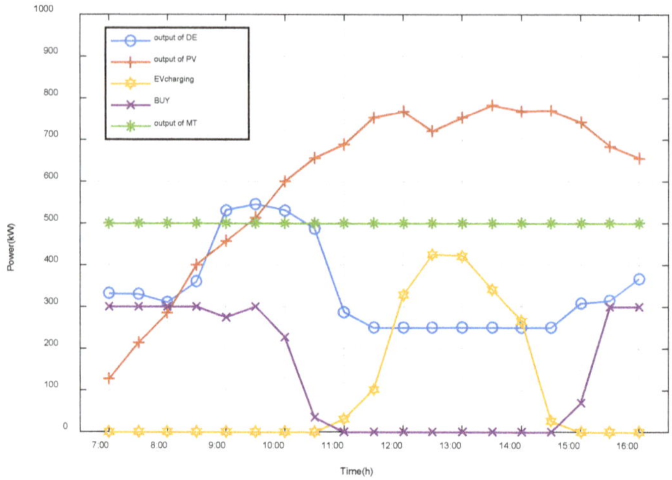

Figure 3. Stochastic optimization result.

5.3.2. Robust Optimization

RO fully considers the uncertainty of the system, and can guarantee stable operation of the system even in a worst-case scenario. The worst-case scenario means the least PV output and the most EV charging. When this happens, the microgrid system must increase the output of the DE to supplement the demand for electricity. Due to the relatively high price of electricity at noon, the microgrid does not purchase electricity from the main grid. Figure 4 shows that the PV output power is fully used, and the EVs are effectively charged at noon to achieve peak load shifting. RO guarantees the stability of the system, but the total cost of the system must increase.

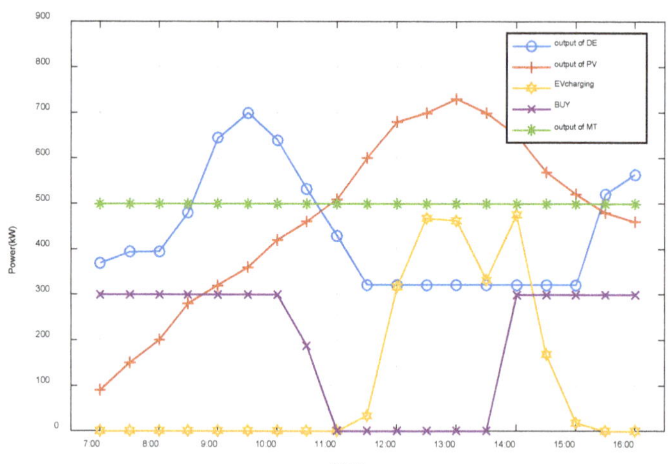

Figure 4. Result of robust optimization in a worst-case scenario.

Figure 5 shows the simulation result of RO in a best-case scenario, where PV power generation accounts for a large proportion in the system, and the EV charging capacity is significantly reduced. Because the system is basically self-sufficient, the amount of electricity purchased from the main grid has also dropped markedly.

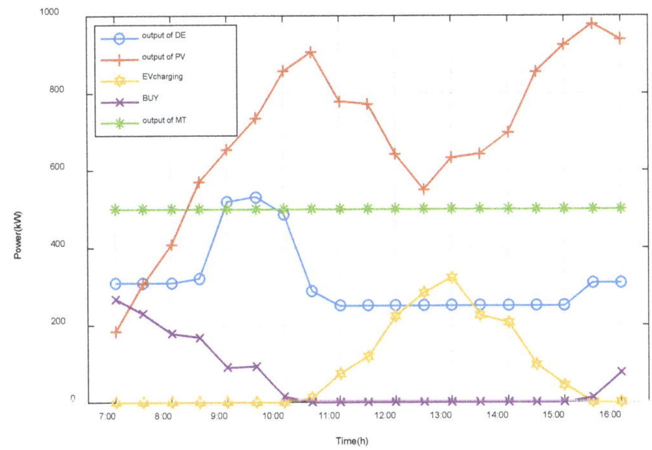

Figure 5. Result of robust optimization in the best-case scenario.

5.3.3. Comparison of Stochastic Optimization and Robust Optimization

In the three results, EVs are charged near 12:00 and renewable energy is used preferentially. Compared with SO, RO fully considers the uncertainty of the system and can operate safely even in the worst case. Table 4 shows that in a worst-case scenario the robust-optimized PV output is smaller, but RO meets more EV charging requirements, and the output of the DE as a rotating standby unit also increases. The total system cost is higher than that of the SO. On the contrary, when RO is in a best-case scenario, the PV output greatly increases while the EV charging capacity decreases, so the system reduces the output of the diesel generator and the power purchase from the main grid, which leads to a reduced total system cost.

Table 4. Comparison of stochastic optimization and robust optimization.

Type	DE (kW)	PV (kW)	MT (kW)	Pgrid (kW)	EV (kW)	Total Cost (RMB)	POV
RO worst-case	8243.9	8880	9500	3787.7	2280	22,508	0%
RO best-case	5943.5	13,035.6	9500	1133.2	1614	20,611	—
SO	6460.3	11,345.3	9500	2706.5	1947	21,456	100%

6. Conclusions and Discussion

The participation of electric vehicles in microgrid dispatching provides a new solution for grid connection of renewable energy power generation. Finding a method to dispatch the electric vehicle to play its energy storage unit role is a problem that must be solved. An optimal scheduling strategy is very important to the economy, environmental protection, and safety of a microgrid. This paper first analyzes the uncertainty of PV power output and EV charging behavior and expresses this uncertainty in the form of a set. Then, to enhance the economic and environmental benefits of the microgrid, a multi-objective optimization scheduling model is established, and an RO theory is applied to the scheduling model. Finally, the simulation results verify the validity of the proposed model. Compared with SO, RO considers the most conservative situation of the system. The system's economic cost is lowered, and the stable operation of the microgrid system is better guaranteed. The scheduling strategy proposed in this paper is a conservative method, which is suitable for the microgrid system with high security requirements. The case study of this paper can help decision-makers to find corresponding solutions for different microgrid systems.

Next, the following three aspects of research can be summarized.

1. Robust optimization guarantees the safe and stable operation of the system under the worst conditions, leaving the system in an overly conservative state. Adjustable coefficients can be introduced to adjust the robustness of the system by changing the number of uncertain variables in the system, and to coordinate the relationship between the robustness and economics of the system according to the needs of the actual situation.
2. On the dispatching problem with electric vehicles, we assume that all electric vehicles are involved in dispatching. Future research can consider the psychological willingness of electric vehicle users to participate in dispatching, and explore the impact of user psychological factors on the total dispatchable power of electric vehicles to make them more in line with actual conditions.
3. The electric vehicles in this article only charge. As the V2G technology of electric vehicles matures, the way in which electric vehicles participate in dispatching will become more flexible. It can be considered that electric vehicles not only consume photovoltaic power when the photovoltaic output is strong, but also discharge at the peak of electricity consumption, alleviating the pressure on the grid, making the electric vehicle a mobile energy storage unit.

Author Contributions: Conceptualization, R.S. and L.N.; methodology, R.S. and P.Z.; software, P.X. and J.Z.; validation, P.X., and J.Z.; formal analysis, R.S. and P.Z.; investigation, X.H.; resources, L.N.; data curation, J.Z. and X.H.; writing—original draft preparation, R.S., P.Z. and J.Z.; writing—review and editing, R.S., P.Z. and J.Z.; visualization, P.Z. and J.Z.; supervision, R.S.; project administration, L.N.; funding acquisition, R.S. and L.N. All authors have read and agreed to the published version of the manuscript.

Funding: This research was funded by the National Natural Science Foundation of China, grant numbers 71801224 and 61203100, the National Social Science Fund of China, grant numbers 19BTQ098 and by the Fundamental Research Funds for the Central Universities, grant number 16MS42.

Conflicts of Interest: The authors declare that they have no conflicts of interest in this work.

Nomenclature

A. Nouns and numbers

RO	Robust optimization
SO	Stochastic optimization
EV	Electric vehicle
PV	Photovoltaic
SOC	State of charge
SO_2	Sulfur dioxide
CO_2	Carbon dioxide
NO_X	Nitrogen oxide
MT	Microturbine
DE	Diesel engine
F	The worst-case scenario
POV	The probability of spinning reserve constraint violated
i	The ith DE
j	The jth MT
l	The lth PV
k	The kth PV
h	The hth pollutant
p	The pth power source
I	The number of DEs
J	The number of MTs
L	The number of PVs
K	The number of EVs dispatched
H	The number of pollutants
P	The number of power sources
Γ_t	The number of uncertain variables

B. Uncertain sets

$PV_{l,t}^{G}$	The output of the lth PV power station at period t (kW)
$\overline{PV_{l,t}}$	The forecasted output of the lth PV power station at period t (kW)
$\hat{PV}_{l,t}$	The deviation of the charging power of the EV at period t (kW)
$\underline{\hat{PV}_{l,t}}, \overline{\hat{PV}_{l,t}}$	The lower and upper limits of $\hat{PV}_{l,t}$ (kW)
μ,σ	The mean and standard deviation of function $h(\cdot)$ (m)
d,d_m	The traveling distance and the maximum travel distance of the EV (m)
T_{start}^{k}	The charging start time of the kth EV (h)
$\underline{T_{start}^{k}}$	The lower limits of charging start time of the kth EV (h)
$\overline{T_{start}^{k}}$	The upper limits of charging start time of the kth EV (h)
E	The capacity of EV batteries (kWh)
P_c	The charging power of EV (kW)
T_{end}^{k}	The charging end time of the kth EV (h)
$PEV_{k,t}^{G}$	The charging power of the kth EV at period t (kW)
PEV_{t}^{G}	The total charging power of EVs at period t (kW)
$\overline{PEV_t}$	The forecasted value of the charging power of EVs at period t (kW)
$\hat{PEV}_t$	The deviation of $\overline{PEV_t}$ (kW)
$\underline{\hat{PEV}_t}$	The lower limits of $\hat{PEV}_t$ (kW)
$\overline{\hat{PEV}_t}$	The upper limits of $\hat{PEV}_t$ (kW)

C. Function parts

$h(\cdot)$	The daily traveling distance of an individual EV (%)
$C_f(\cdot)$	The fuel costs of DEs and MTs (RMB)
$C_{OM}(\cdot)$	The operation and maintenance costs of DEs, MTs and PV power station (RMB)
C_1	The operating cost (RMB)
C_2	The environmental cost (RMB)
C_{total}	The total cost of system (RMB)
$C_{grid,t}$	The cost of transmission between the microgrid and the main power grid (RMB)
c_1,c_2,c_3	The fuel cost parameters of DEs (RMB/kW)
y	The cost parameter of MTs (RMB/kW)
$\eta(\cdot)$	The work efficiency of MTs (%)
C_h	The treatment cost of the hth pollutant (RMB/kg)
$\mu_{p,h}$	The hth pollutant emission coefficients of the pth type power source (g/kW)
$P_{p,t}$	The output power of the pth power source (kW)
$P_{grid,t}$	The transmission power between the microgrid and the main power grid (kW)
μ_{grid}	The pollutant emission coefficients of the main power grid (g/kW)
Δt	Scheduling interval (h)
K_{OM}	The OM cost parameter (RMB/kW)

D. Variables and constants

$P_{i,t}$	The output power of the ith DE at period t (kW)
$PMT_{j,t}$	The output power of the jth MT at period t (kW)
$PV_{l,t}$	The output power of the lth PV power station at period t (kW)
$P_{load,t}$	The total load demand at period t (kW)
$P_{i,\min}$	The lower limits of the power output of the ith DE (kW)
$P_{i,\max}$	The upper limits of the power output of the ith DE (kW)
$PMT_{j,\min}$	The lower limits of the power output of the jth MT (kW)
$PMT_{j,\max}$	The upper limits of the power output of the jth MT (kW)
$P_{i,down}$	The lower limits of the ramp rate of the ith DE (kW)
$P_{i,up}$	The upper limits of the ramp rate of the ith DE (kW)
$P_{j,down}$	The lower limits of the ramp rate of the jth MT (kW)
$P_{j,up}$	The upper limits of the ramp rate of the jth MT (kW)
P_{down}	The lower limits of the main grid transmitting power (kW)
P_{up}	The upper limits of the main grid transmitting power (kW)
M_t	The price of electricity (RMB/kWh)
$\alpha_t, \beta_t, \gamma_t, \delta_t$	Lagrange coefficients
$\eta_{m,t}, \omega_{m,t}, J_t$	The coefficients of POV
L_t	The spinning reserve rate (%)

References

1. Zhou, E.; Cole, W.; Frew, B. Valuing variable renewable energy for peak demand requirements. *Energy* **2018**, *165*, 499–511. [CrossRef]
2. Sharvini, S.R.; Noor, Z.Z.; Chong, C.S.; Stringer, L.C.; Yusuf, R.O. Energy consumption trends and their linkages with renewable energy policies in East and Southeast Asian countries: Challenges and opportunities. *Sustain. Environ. Res.* **2018**, *28*, 257–266. [CrossRef]
3. Zhang, D.; You, P.; Liu, F.; Zhang, Y.; Zhang, Y.; Feng, C. Regulating cost for renewable energy integration in power grids. *Global Energy Interconnect.* **2018**, *1*, 544–551.
4. Xia, T.; Ji, Q.; Zhang, D.; Han, J. Asymmetric and extreme influence of energy price changes on renewable energy stock performance. *J. Clean Prod.* **2019**, *241*, 118338. [CrossRef]
5. Li, X.; Jiang, C. Short-term operation model and risk management for wind power penetrated system in electricity market. *IEEE T Power Syst.* **2010**, *26*, 932–939. [CrossRef]
6. Zakaria, A.; Ismail, F.B.; Lipu, M.H.; Hannan, M. Uncertainty models for stochastic optimization in renewable energy applications. *Renew. Energ.* **2019**, *145*, 1543–1571. [CrossRef]
7. Meng, N.; Xu, Y.; Huang, G. A stochastic multi-objective optimization model for renewable energy structure adjustment management–A case study for the city of Dalian, China. *Ecol. Indic.* **2019**, *97*, 476–485. [CrossRef]
8. Yu, J.; Ryu, J.; Lee, I. A stochastic optimization approach to the design and operation planning of a hybrid renewable energy system. *Appl. Energ.* **2019**, *247*, 212–220. [CrossRef]
9. Hocine, A.; Kouaissah, N.; Bettahar, S.; Benbouziane, M. Optimizing renewable energy portfolios under uncertainty: A multi-segment fuzzy goal programming approach. *Renew. Energ.* **2018**, *129*, 540–552. [CrossRef]
10. Daim, T.U.; Kayakutlu, G.; Cowan, K. Developing Oregon's renewable energy portfolio using fuzzy goal programming model. *Comput. Ind. Eng.* **2010**, *59*, 786–793. [CrossRef]
11. Heo, E.; Kim, J.; Boo, K. Analysis of the assessment factors for renewable energy dissemination program evaluation using fuzzy AHP. *Renew. Sustain. Energ. Rev.* **2010**, *14*, 2214–2220. [CrossRef]
12. Roberts, J.J.; Cassula, A.M.; Silveira, J.L.; da Costa Bortoni, E.; Mendiburu, A.Z. Robust multi-objective optimization of a renewable based hybrid power system. *Appl. Energ.* **2018**, *223*, 52–68. [CrossRef]
13. Gazijahani, F.S.; Salehi, J. Reliability constrained two-stage optimization of multiple renewable-based microgrids incorporating critical energy peak pricing demand response program using robust optimization approach. *Energy* **2018**, *161*, 999–1015. [CrossRef]
14. Bertsimas, D.; Brown, D.B.; Caramanis, C. Theory and applications of robust optimization. *Siam Rev.* **2011**, *53*, 464–501. [CrossRef]
15. Wang, L.; Li, Q.; Cheng, X.; He, G.; Li, G.; Wang, R. A robust optimization approach for risk-averse energy transactions in networked microgrids. *Energy Procedia* **2019**, *158*, 6595–6600. [CrossRef]
16. Cao, Y.; Huang, L.; Li, Y.; Jermsittiparsert, K.; Ahmadi-Nezamabad, H.; Nojavan, S. Optimal scheduling of electric vehicles aggregator under market price uncertainty using robust optimization technique. *Int. J. Electr. Power* **2020**, *117*, 105628. [CrossRef]
17. Rodríguez-Gallegos, C.D.; Yang, D.; Gandhi, O.; Bieri, M.; Reindl, T.; Panda, S. A multi-objective and robust optimization approach for sizing and placement of PV and batteries in off-grid systems fully operated by diesel generators: An Indonesian case study. *Energy* **2018**, *160*, 410–429. [CrossRef]
18. Fu, L.; Yang, Y.; Yao, X.; Jiao, X.; Zhu, T. A Regional Photovoltaic Output Prediction Method Based on Hierarchical Clustering and the mRMR Criterion. *Energies* **2019**, *12*, 3817. [CrossRef]
19. Peng, C.; Xie, P.; Pan, L.; Yu, R. Flexible robust optimization dispatch for hybrid wind/photovoltaic/hydro/thermal power system. *IEEE T Smart Grid* **2015**, *7*, 751–762.
20. Shi, R.; Sun, C.; Zhou, Z.; Zhang, L.; Liang, Z. A robust economic dispatch of residential microgrid with wind power and electric vehicle integration. In Proceedings of the 2016 Chinese Control and Decision Conference, Yinchuan, China, 28–30 May 2016; pp. 3672–3676. Available online: https://ieeexplore.ieee.org/document/7531621 (accessed on 23 March 2020).
21. Tian, L.; Shi, S.; Jia, Z. A statistical model for charging power demand of electric vehicles. *Power Syst. Technol.* **2010**, *34*, 126–130.
22. Moghaddam, A.A.; Seifi, A.; Niknam, T.; Pahlavani, M.R.A. Multi-objective operation management of a renewable MG (micro-grid) with back-up micro-turbine/fuel cell/battery hybrid power source. *Energy* **2011**, *36*, 6490–6507. [CrossRef]

23. Ma, Z.; Zou, S.; Ran, L.; Shi, X.; Hiskens, I.A. Efficient decentralized coordination of large-scale plug-in electric vehicle charging. *Automatica* **2016**, *69*, 35–47.
24. Kheshti, M.; Ding, L.; Nayeripour, M.; Wang, X.; Terzija, V. Active power support of wind turbines for grid frequency events using a reliable power reference scheme. *Renew. Energ.* **2019**, *139*, 1241–1254. [CrossRef]
25. Wang, Y.; Wang, Y.; Huang, Y.; Li, F.; Zeng, M.; Li, J.; Wang, X.; Zhang, F. Planning and operation method of the regional integrated energy system considering economy and environment. *Energy* **2019**, *171*, 731–750.
26. Lu, X.; Zhou, K.; Yang, S. Multi-objective optimal dispatch of microgrid containing electric vehicles. *J. Clean Prod.* **2017**, *165*, 1572–1581. [CrossRef]
27. Mohamed, F.A.; Koivo, H.N. System Modelling and Online Optimal Management of MicroGrid Using Multiobjective Optimization. In Proceedings of the2007 International Conference on Clean Electrical Power, Capri, Itlay, 21–23 May 2007; pp. 148–153. Available online: https://ieeexplore.ieee.org/abstract/document/4272373 (accessed on 23 March 2020). [CrossRef]
28. Bai, J.; Ding, T.; Wang, Z.; Chen, J. Day-Ahead Robust Economic Dispatch Considering Renewable Energy and Concentrated Solar Power Plants. *Energies* **2019**, *12*, 3832. [CrossRef]
29. Bertsimas, D.; Sim, M. The price of robustness. *Oper. Res.* **2004**, *52*, 35–53.

© 2020 by the authors. Licensee MDPI, Basel, Switzerland. This article is an open access article distributed under the terms and conditions of the Creative Commons Attribution (CC BY) license (http://creativecommons.org/licenses/by/4.0/).

energies

Article

Battery Management Systems—Challenges and Some Solutions

Balakumar Balasingam [1,*], Mostafa Ahmed [1] and Krishna Pattipati [2]

[1] Department of Electrical and Computer Engineering, University of Windsor, 401 Sunset Ave., Office#3051, Windsor, ON N9B3P4, Canada; ahmed168@uwindsor.ca
[2] Department of Electrical and Computer Engineering, University of Connecticut, 371 Fairfield Rd, Office#350, Storrs, CT 06269, USA; krishna.pattipati@uconn.edu
* Correspondence: singam@uwindsor.ca; Tel.: +1-(519)-253-3000 (ext. 5431)

Received: 23 April 2020; Accepted: 27 May 2020; Published: 2 June 2020

Abstract: Electric vehicles are set to be the dominant form of transportation in the near future and Lithium-based rechargeable battery packs have been widely adopted in them. Battery packs need to be constantly monitored and managed in order to maintain the safety, efficiency and reliability of the overall electric vehicle system. A battery management system consists of a battery fuel gauge, optimal charging algorithm, and cell/thermal balancing circuitry. It uses three non-invasive measurements from the battery, voltage, current and temperature, in order to estimate crucial states and parameters of the battery system, such as battery impedance, battery capacity, state of charge, state of health, power fade, and remaining useful life. These estimates are important for the proper functioning of optimal charging algorithms, charge and thermal balancing strategies, and battery safety mechanisms. Approach to robust battery management consists of accurate characterization, robust estimation of battery states and parameters, and optimal battery control strategies. This paper describes some recent approaches developed by the authors towards developing a robust battery management system.

Keywords: battery management systems; battery fuel gauge; state of charge, state of health; power fade; capacity fade; robust estimation; predictive control

1. Introduction

Automobiles powered by gasoline engines account for nearly 25% of the global energy consumption [1]. Rechargeable batteries promise a way to replace them by electric vehicles (EVs) in the near future. In addition to EVs, rechargeable batteries have been widely adopted in portable electronic equipment, household appliances, power tools, aerospace equipment and renewable energy storage systems. A battery management system (BMS) ensures the safety, efficiency and reliability of a battery powered system. Research on BMS has been very intense in the last two decades and significant improvements were achieved in the safety, efficiency and reliability of battery systems [2,3]. However, there are challenges remaining and in this paper we describe a list of challenges and outline possible solutions.

Two schools of approaches for battery management systems have emerged over time; one models the battery through *electrical equivalent circuit model (ECMs)* [2,3] and the other seeks to model it through *electrochemical models* [4]. However, most practical systems adopt the electrical ECM based approaches due to their simplicity. The research challenges faced by the present day BMS are three pronged: *safety, efficiency* and *reliability*. Lithium ion batteries are susceptible to *thermal runaway* which is an irreversible chemical process triggered by several conditions including over-voltage and high temperature. The need to fast-charge the battery, which is important in electric vehicle applications, increases the possibility of thermal runaway and safety issues [5,6]. There are wide ranging issues affecting the efficiency of energy storage in batteries; particularly, electric vehicle applications strive to improve efficiency in every possible way. For example, charging efficiency is the percentage of the total energy needed during charing [7]; fast charging requirements results in significant energy waste in the form of heat. BMS algorithms attempt to enhance efficiency of batteries in multiple ways; optimal charging algorithms aim to reduce the amount of heat waste and the degradation of state of health; precise SOC estimation algorithms will help to improve the efficiency by helping to design minimal battery-pack configurations based on specific needs. Individual cells in a battery-pack are known to become imbalanced over time causing safety and reliability issues; short circuited cells are another common cause of safety and reliability issues in Li-ion batteries [8–10].

An emerging challenge for battery management systems comes in the form of battery reuse [11,12]. It is predicted that the electric vehicle sales are about to grow by nearly 500% in the next 10 years [13]. The state of the art BMS algorithms heavily depend on prior characterization carried out in laboratories [2,3]; Consequently, they are only effective for first time use of batteries. Considering the fact that the first use of the battery alters its electrochemical characteristics in unique ways, traditional BMS approaches that rely on empirical modeling, under the assumption that batteries of the same chemistry and size have similar characteristics, will be inadequate to manage used batteries.

The present manuscript is written in the form of an *expository paper* detailing the many solutions developed by the authors in the recent past in order to address specific challenges in battery management systems. Section 2 describes in more details about the specific goals of a state of the art battery management systems and the challenges it needs to overcome. Section 3 describes some specific solutions developed by the authors in order to address the challenges faced by the present day battery management system. Finally, the paper is concluded in Section 4.

2. Battery Management System: Goals and Challenges

In this section, some of the challenges faced in designing battery BMS are briefly described.

2.1. State of Charge Estimation

Coulomb counting is the easiest approach to estimate the state of charge (SOC) of a battery [2,3]. Figure 1a gives the approximate Coulomb counting equation that is used to compute SOC in a recursive manner. However, Coulomb counting method suffers from the following sources of errors:

1. *Initial SOC error.* Since it is a recursive integration, any errors in the initial SOC assumption will remain as a bias.
2. *Current measurement error.* Current sensors are corrupted by measurement noise; simple, inexpensive current sensors are likely to be more noisy and possibly biased.
3. *Current integration error.* Coulomb counting methods employ a simple, rectangular approximation for current integration. Such an approximation results in errors that increase with sampling interval as the load changes rapidly.
4. *Uncertainty in the knowledge of battery capacity [14].* Coulomb counting method assumes perfect knowledge of the battery capacity, which is known to vary with temperature, usage patterns and time (age of the battery) [15,16].
5. *Timing oscillator error.* Timing oscillator provides the clock for (recursive) SOC update, that is, the *measure of time* comes from the timing oscillator. Any error/drift in the timing oscillator will have an effect on the measured Coulombs.

Alternatively, the open circuit voltage (OCV) can be modeled as a function of the SOC of the battery. This OCV-SOC model [17] can be exploited to estimate the SOC based on voltage measurements. However, measuring the OCV in real-time during battery operation is not feasible because the battery needs to be rested for several hours before the OCV can be measured. While the battery is operational a measure of OCV can be obtained by estimating the voltage across the battery ECM; this requires the estimation of the ECM parameters as well. Once the OCV is estimated, the SOC can be looked-up [17] using the OCV-SOC characterization parameters. Figure 1b summarizes the voltage based approach to SOC estimation. The following errors are encountered by the OCV-SOC based state of charge estimation approach:

1. Errors in the parameters estimated for the electrical ECM of the battery.
2. Voltage and current measurement error.

Most of the advanced BFG's use a *fusion based approach* where both the Coulomb counting method and the OCV-lookup method a combined in an efficient manner.

$$s(k) = s(k-1) + \frac{1}{3600C_{\text{batt}}} \int_{t(k-1)}^{t(k)} i(\tau)d\tau$$

$$s(k) \approx s(k-1) + \frac{\Delta_k i(k)}{3600C_{\text{batt}}}$$

(**a**) Current based approach

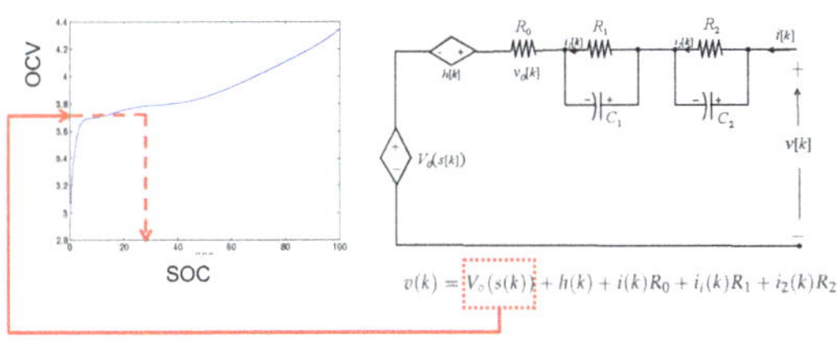

(**b**) Voltage based approach

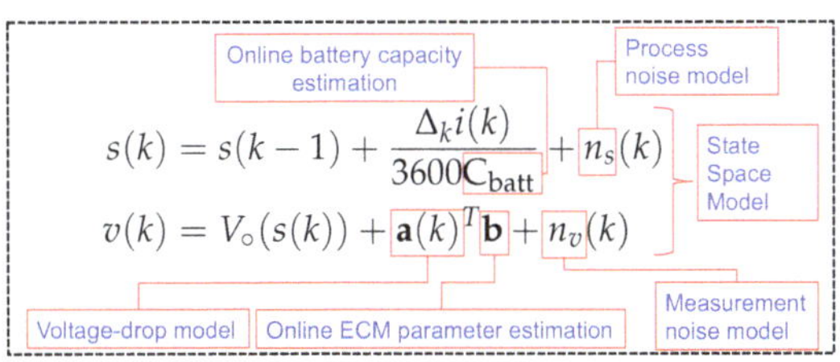

(**c**) Fusion based approach

Figure 1. State of charge estimation. The fusion based approach is one of the most robust approaches to accurate battery SOC estimation.

Th fusion approach to SOC estimation (more appropriately, SOC tracking) is modelled as a recursive Bayesian estimation problem and by employing a nonlinear filtering approach (such as an extended Kalman filter) for online SOC tracking [2,3]. A complete SOC tracking solution involves the following:

(i) *Estimation of the OCV parameters that form part of the state space model through offline OCV characterization:* The OCV-SOC characterization is stable over temperature changes and aging of the battery. Once estimated, these parameters form part of a state-space model with known parameters.
(ii) *Estimation of the dynamic ECM parameters:* These parameters can change depending on the battery age, temperature, and SOC, therefore, they must be estimated in real time.
(iii) *Estimation of battery capacity:* Even though the the manufacturer provides the nominal capacity of the battery, it changes over time. Some important factors that cause *capacity fading* are, elevated temperature, cycling (usage), depth of discharge patterns, and calendar aging. Due to this, the battery capacity needs to be estimated in real-time for an accurate BFG. Capacity estimation is still being actively investigated in the literature [14].
(iv) *Model parameter-conditioned SOC tracking:* As soon as the model parameters are estimated, a filtering approach can be used to track the SOC using the state-space model discussed above. In order to do this, numerous filtering approaches, including extended Kalman filter, Unscented Kalman filter and particle filter, were experimented in the literature. However, it is observed that the resulting state-space model contains correlated process and measurement noise processes. Properly addressing the effect of these correlations will yield better SOC tracking accuracy.

Figure 1c illustrates the fusion based approach to SOC estimation. The fusion based approach needs to have the knowledge of OCV parameters, battery capacity, ECM parameters as well as the sensitivity of the voltage and current measurement sensors. Section 3 briefly describes the approaches to estimate them.

2.2. Real-Time State of Health Estimation

Today's BMS technology is inadequate to accurately predict the state of health (SOH) of a battery. The available choices are either to prematurely replace the battery or to wait until an explicit failure event occurs. Both of these choices have undesirable consequences: premature replacement will result in increased cost to the end user and excessive waste to the environment; waiting out will negatively impact the safety and quality of experience to the end user.

Many of the methods proposed in the literature for SOH estimation are data driven methods. The existing approach to SOH estimation differ in terms of the features used to train and the machine learning topology used. For instance, both approaches presented in References [18,19] used neural networks; in Reference [18], the following features were used: change in the SOC, current, temperature and the internal circuit parameters; in Reference [19] the voltage curve was used as a feature. In References [20–22], support vector machines approach was employed for SOH classification. Feng et al. [20] used partial charging voltage curves as there feature to estimate the SOH online while [21] identified the charging time and capacity to be the features in order to estimate the SOH. In Reference [23], a sample entropy algorithm is used identify the measured terminal voltage under hybrid pulse power characterization current profile; this is then used as a feature to estimate the SOH using a sparse Bayesian predictive modeling. Yun and Qin [24] proposed the use of the time required for the terminal voltage to drop from and to a certain value as the feature to train.

Real-time SOH estimation remains one of the open problems in battery management system research.

2.3. Optimal Charging

The state of the art in battery charging is primitive: time consuming, less efficient and less safe compared to gasoline refueling. Research on optimal charging algorithms (OCA) received significant attention in the recent past. One of the most common method of charging is the constant trickle current based charge strategy. Because a low charging current is used, it requires a long charging time (around 10 h) [25]; charging time can be reduced by increasing the charging current, however, as the batter OCV increases due to charging, this will cause the battery terminal to reach a voltage that is above the safety threshold. Hence, the higher current that is applied at the initial stage needs to be reduced when the terminal voltage reaches a certain threshold value. Consequently, the *constant-current constant-voltage (CC-CV)* [25] strategy has become one of the widely used approaches to fast charging. In order to shorten the charging time and perpetuate the cycle life of the battery a multi-step constant-current charging is used in References [26,27]. The Taguchi-based methods for battery charging [28,29] uses orthogonal arrays to put forward a systematic method to find the optimal solution with guidelines for choosing the design parameters. Another strategy to use is the boost charging strategy, where a very high current is applied to close-to-fully discharged batteries [30]. In pulse-charging methods [31–35], the battery is exposed to very short rest or even deliberate discharging periods during the charging process. Soft-computing approaches can also be used to optimize the battery charging profile. In Reference [36], optimal charging is achieved by simplifying the problem to be in the form of an optimization problem with the objective function of maximizing the charge within 30 min. Through the use of a multistage constant current charging algorithm, the optimal solution can be obtained by using an ant-colony approach. The authors of Reference [37] proposed a universal voltage protocol, its goal is to enhance the charging efficiency and cycle life by applying a certain charging profile, this charging profile is determined based on the SOH of the battery, which is estimated during the optimization process [38]. Recently, in Reference [39], to find the optimal charging strategy, an optimization approach with cost function of time-to-charge and energy loss is used. However, an analytical solution has not been presented; rather a numerical solution is given to the problem. Many other approaches are presented in the literature for battery charging, such as data mining [40,41], genetic algorithm and neural network based strategies [42], Grey-predicted charging system [43].

2.4. Fast Characterization

Two important offline characterizations required in a BMS are the SOC and SOH characterizations. In SOC characterization, the SOC is modelled against the OCV by collecting one full cycle of data (fully charged battery $\rightarrow$ fully discharged battery $\rightarrow$ fully charged battery) whereas the state of the art SOH characterization is done against the number of cycle requiring hundreds and even thousands of cycles of data. This makes SOC ans SOH characterization a time-consuming process. Hence, it is important to find ways to reduce characterization time.

One approach to reduce characterization time is to do it in real-time while the battery is in use. Some approaches for real-time SOC characterization were proposed in the past [44,45]. One of the drawbacks of these approaches is due to the fact that simpler OCV-SOC models need to be employed (due to computational bottlenecks in the BMS) for online estimation of parameters; this will lead to loss of accuracy [46]. Secondly, different sources of error can accumulate from other estimated parameters, that will be incorporated during OCV estimation [47]. Lastly, based on the SOC range that the battery goes through, the estimated OCV model will cover that SOC portion—which depends on the battery usage pattern. In orderfor OCV-SOC model to cover the entire SOC range, the battery has to undergo a complete discharge/charge profile—this cannot be guaranteed. In Reference [48], an approach that uses the data pieces-based parameter identification was proposed to estimate the entire OCV-SOC model. However, this approach has its own drawbacks where the modelling error can be high at the initial stages and the the convergence is not always insured.

Compared to SOC characterization, SOH characterization is nearly impossible to do in real-time locally for a particular battery pack. However, the abundance today's connectivity offers an alternative solution for real-time SOC characterization. Figure 2 depicts how a *cloud assisted BMS* can collect data from numerous batteries to estimate crucial parameters for real-time management of battery packs.

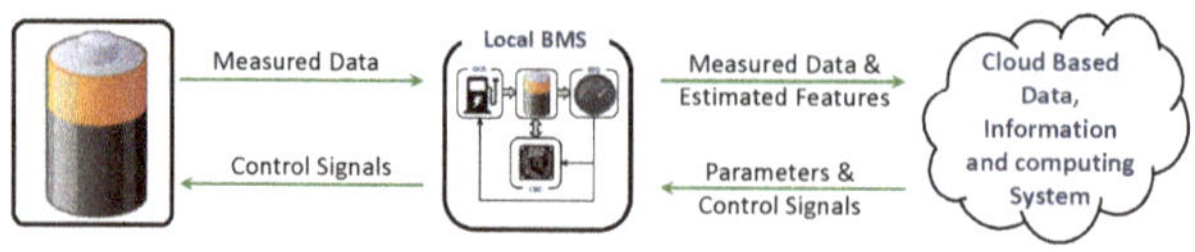

Figure 2. Cloud assisted battery management system. The abundance of todays connectivity allows crucial parameters related to SOC and SOH characteristics of battery packs in real-time. (image from Reference [49] being reprinted with permission from IEEE).

2.5. Battery Reuse

As counties race towards decreasing their green house gas emissions, the public is being encouraged to use EVs by offering various incentives. As a result, the manufacturing of Lithium-ion (Li-ion) batteries is expected to increase very rapidly in the next few decades due to their expected use in electric vehicles [50]. Battery packs used in electric vehicles are expected to be replaced when they reach about 80% of their original capacity [11], since range is an important quality in EVs. Research on BMS algorithms so far has predominantly focused on the first use of the battery-pack. The batteries retired from the EVs are still an excellent medium of renewable energy storage in other applications, such as renewable energy storage systems [11]. However, it is still not well understood as to how usage affects the SOC and SOH characterizations of a battery pack. Environmental and usage conditions affect battery characteristics; based on how, when and where an EV was predominantly in use, its battery-pack might have significantly different reliability, efficiency and safety compared to another battery-pack that was made by the same manufacturer during the same assembly process. In other words, even though two batteries were identical twins out of the assembly line, after their first retirement, they would possess two different characteristics based on the patterns of environmental and usage conditions that they experienced. Hence, there is a need to invest in research and to develop BMS that ensure safe, reliable and efficient operation of EV batteries during their second use as renewable energy storage systems. Even though the electric vehicle production is expected to grow exponentially in the next few decades [13], research on battery reuse is still in early stages [51,52]. Figure 3 demonstrates the overall block diagram of a BMS during battery reuse. One of the important challenges here is that each used battery pack is different from one another.

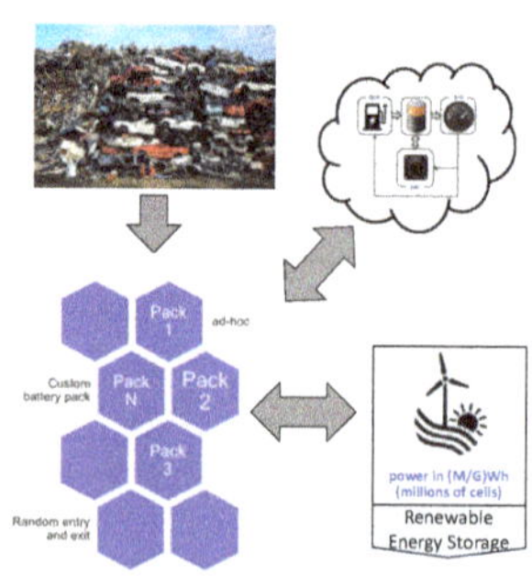

Figure 3. Battery reuse: from scrapyard to powering living rooms. Millions of vehicles are scrapped each year due to accidents. In the case of electric vehicles, the batteries could be reused store renewable energy. However, more research needs to be done about managing used batteries of various size, chemistry and manufacturers.

2.6. Universality

Existing BFG algorithms depend on prior characterization for accurate estimation of SOC and SOH [2,3]; as a result, their application is limited to certain type of batteries to which they have the parameters for. The state of the art BMS is constrained to a particular chemistry, manufacturer, and size of the battery to which it is characterized for, that is, the present-day BMS is not universal; this restricts battery selection and results in increased cost; also, such a restrictive BMS does not allow one to repurpose used battery packs for energy storage. In smaller, household, applications, custom battery chargers generate excessive electronic clutter and environmental waste.

The first ones to think about the universal battery systems were the battery charger designers who had to address the huge number of different chemistry and types of batteries that in each application requires its own customized charger; this increases the amount of electronic wastage and adds to the cost of the device. Hence, the problem of *universal battery charger* received attention in the literature [7,53–55]. Earlier versions of universal battery chargers are programmed to look for appropriate voltage to terminate the charging process. Most of them used a look-up table of incremental voltage in response to charging by a certain number of Coulombs [7,53–55].

A preliminary achievement regarding the universality objective is reported in Reference [52] where a probabilistic data association filter [56,57] was employed to associate the online measurements from batteries to their model parameters, thus, resulting in a *chemistry-adaptive BFG*. Further research needs to be done on this topic so that reliable algorithms can be developed to extend adaptivity for load-range, size, temperature, nominal voltage and age as well. This would require large computing power that the traditional battery management systems are not allocated for, for example, portable electronics. Cloud computing [58] allows one to outsource intense computing to external sources; that is, by combining information fusion with cloud computing, a greater deal of universality can be achieved in battery management systems, paving the way for optimal battery reuse (see Section 2.5) and reduced electronic clutter in households and work places.

2.7. Self Evaluation

Battery management system evaluation is a very challenging research problem since there are no proven mathematical models to represent the complex features of a Li-ion battery, these features include power fade (PF), capacity fade (CF), temperature effects on parameters, aging, hysteresis and relaxation effects.

There is little literature focusing on BFG algorithm evaluation under realistic usage conditions. The importance of BMS evaluation is discussed in Reference [59]; in Reference [60], the need to minimize power dissipation and extend battery run-time for portable devices is discussed; the advantages of hardware-in-the-loop (HIL) testing to validate a BMS under various failure conditions was motivated in Reference [61]; and a HIL test to validate the BFG using a multi-cell battery pack was proposed in References [62,63].

Evaluating BMS algorithms is a time consuming task [16] that requires research to find efficient solutions. Particularly, the following aspects needs to be studied further:

- The state of the art BMS evaluation is done in a lab setting. Real-time self-evaluation through data driven approaches need to be developed.
- Majority of the existing research experiments are done at a constant temperature. BMS evaluation in the presence of gradual and rapid temperature changes is needed.

3. Solutions Through Model Based Algorithms

Figure 4 shows an overall block diagram of a BMS that consists of three important components [49]: BFG, OCA, and cell-balancing circuitry (CBC). The BFG is considered as the the primary component of a BMS since the BFG output is required in both the OCA and CBC. The BFG estimates the SOC and SOH of the battery-pack based on three measurements: voltage, current, and temperature. The OCA is responsible for regulating the battery charging by generating charging waveforms. The charing waveforms vary in complexity; at the simplest level, a charger applies a constant voltage across the battery terminal; in constant-current constant voltage (CC-CV) charging, the battery SOC (and hence the OCV) rises fast due to the relatively high current; then the charing is switched to CV in order to safeguard the battery from overcharging. Complex charging strategies closely monitor critical battery parameters and adaptively alter the charging pattern. The ultimate goal of an OCA is to charge the battery faster without negatively affecting its SOH [64,65]. When new, individual cells in the battery-packs have similar battery capacity and impedance. However, it is well know that after many charge/discharge cycles these parameters can deviate away from one other causing cell imbalance. Cell-imbalance has many drawbacks from reduced power output, reduced cycle life to catastrophic failures, including fire. The CBC helps to maintain the battery-pack balanced. In addition to this, CBC is also responsible for thermal balancing [66] of the battery-pack. In the remainder of this section, recent contributions to some of the BMS components are described.

The BMS consists of several smaller modules that are critical for improving its safety, efficiency and reliability. Many of todays research is focused on improving these individual modules. In the remainder of this section, we brief the details of the several BMS modules that were developed as improvements to the state of the art.

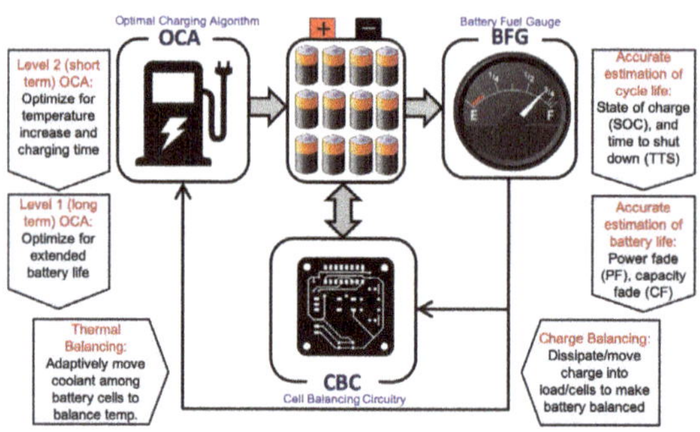

Figure 4. Functional block diagram of a battery management system. Three important components of a BMS are battery fuel gauge, optimal charging algorithm and cell balancing circuitry.

3.1. *Normalized Open Circuit Voltage Characterization*

Open circuit voltage characterization is one of the most important elements of any BMS, as it allows one to estimate the SOC based on a given OCV. Earlier approaches to OCV modelling suggested to store different OCV-SOC parameters at different temperatures. It was shown in Reference [17] that the *normalized approach* to OCV characterization results in a single set of parameters for all temperatures. Further, various models for OCV characterization were evaluated in Reference [17]. Figure 5 shows the results of applying the normalized OCV modelling approach [17] for OCV-SOC characterization. The important advantage of the normalized modelling approach is that the OCV-SOC characterization does not need to be repeated at multiple temperatures. Just one characterization at room temperature is shown to be enough to cover typical operational temperatures experienced by batteries.

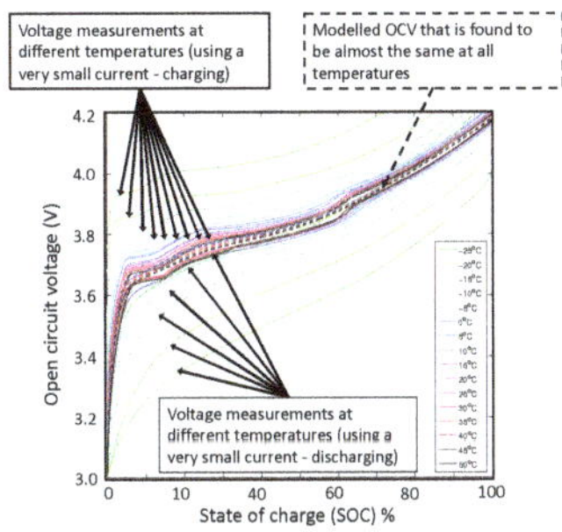

Figure 5. Normalized open-circuit voltage modelling. It is shown that the OCV-SOC parameters obtained through the proposed normalized OCV modelling approach in Reference [17,67] showed little variations with temperature.

The combined model and its variations, such as the combined+3 model [17], remain one of the most used approaches to OCV modelling. Many existing OCV models, including combined model and its variants, suffer from the fact they are not defined at the extreme limits of the SOC, that is, (SOC = 0% and SOC = 100%). For example, let us consider the combined model equation where the OCV ($V_o(s)$ relates to the SOC (s) as follows

$$V_o(s) = \kappa_0 + \frac{\kappa_1}{s} + \kappa_2 s + \kappa_3 \ln(s) + \kappa_4 \ln(1-s) \tag{1}$$

where it can be noticed that the function is undefined when $s = 0$ (SOC = 0%) and when $s = 1$ (SOC = 100%). Existing approaches to the above problem not very optimal. In References [67,68], the effect of not scaling on the performance of SOC estimation is formally quantified and an approach was presented to find the optimal scaling factor; further, it was shown in Reference [67] that the optimal scaling factor remained the same across different battery chemistries and temperatures. Figure 6 summarizes the results of scaling

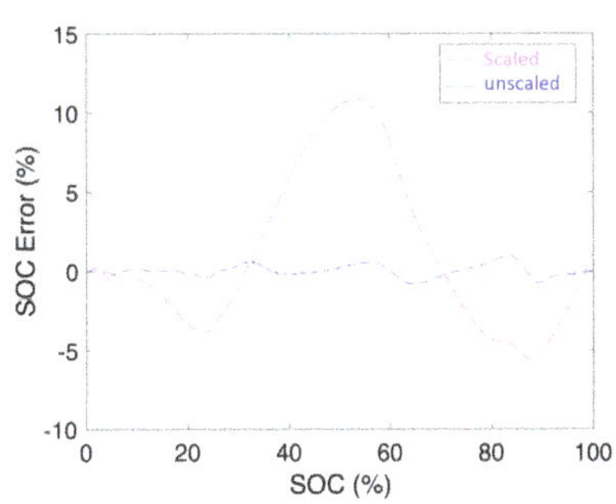

Figure 6. Scaling. The scaling approach [67] reduces the worst case error in OCV modelling.

3.2. Equivalent Circuit Model Identification

Li-ion batteries are powered through chemical reactions. modelling such chemical reactions using physics and chemistry result in very complex models that are challenging to solve. In contrast, ECM provided a simplistic, albeit adequately approximate representation of batteries and battery packs. Figure 7 shows a generalized ECM of a battery.

It was shown in Reference [69] that different approximations of ECM in Figure 8 can be used based on the battery load. Using the appropriate ECM can reduce computation time and complexity and give accurate results. Figure 8a shows the ECM when there is a constant low current load. In this mode the hysteresis can be neglected due to the small effect it has and the capacitors can be ommited due to the constant current. The remaining resistances can be lumped together to form the output resistance (R_0).

The second mode is shown in Figure 8b. This model is used when there is a high current for extended periods of time. Due to the high current, hysteresis cannot be neglected anymore, and must be incorporated in the model. However, the capacitors can still be omitted due to the constant current. The remaining resistances can still be lumped together in one output resistance (R_0).

Figure 8c shows model 3 that can be used when there is a dynamic load with a constant average load. In this case, the hysteresis cannot be ignored, along with the capacitor/resistor component (C_1/R_1). On the other hand, model 4 is shown in Figure 8d where there is a dynamic current with varying average load. Therefore, a second capacitor/resistor (C_2/R_2) need to be used for accurate battery modelling.

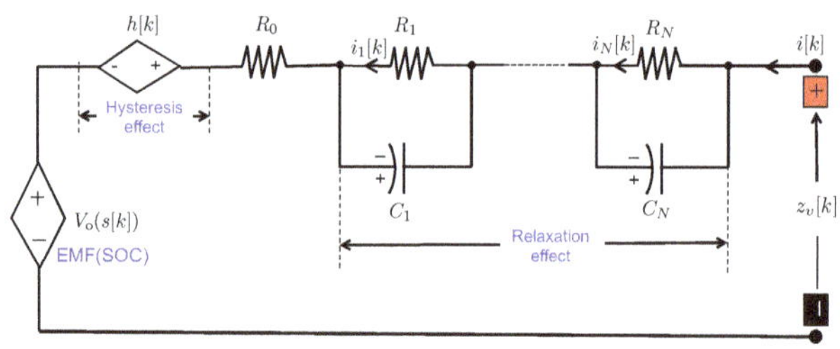

Figure 7. Equivalent circuit model of a battery. Identifying the battery model and estimating its parameters are crucial steps for all aspects of a battery management system, from state of charge estimation to optimal charging to charge and thermal balancing. In practice, reduced models, shown in Figure 8, are employed; In Reference [69] a unified approach to ECM model parameter estimation is developed.

The authors of Reference [69] proposed an approach that is based on weighted least squares method to identify the battery parameters online, which is inexpensive and has high accuracy. This method has the ability to switch between the battery models easily based on the current profile. Furthermore, instead of modelling the hysteresis as a function of the SOC, which can be very complex and inaccurate, the authors proposed a way to model the hysteresis as an error in the OCV, which has the added benefit of fast recovery when the initial SOC is inaccurate.

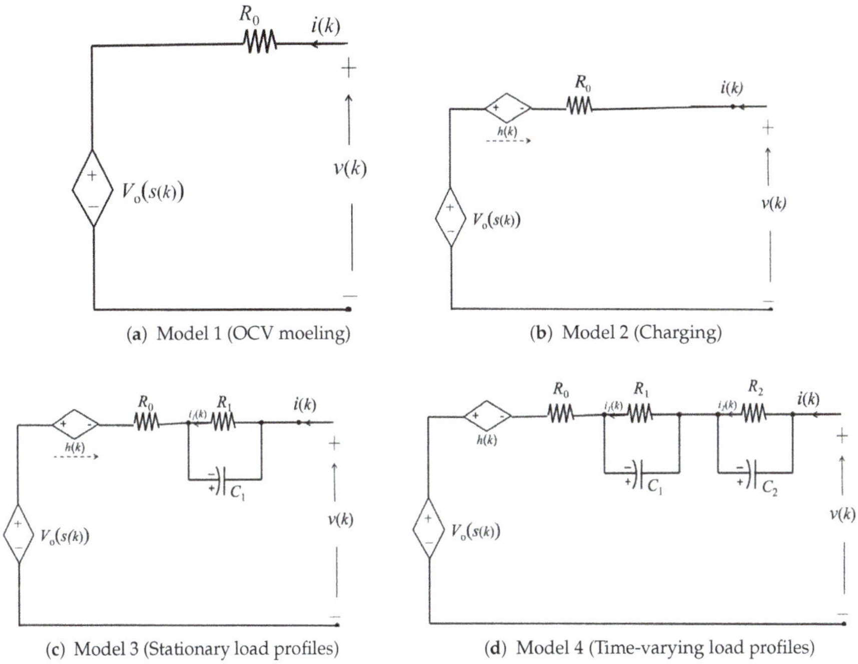

Figure 8. Reduced equivalent circuit models of a battery. Each model is appropriate for different types of loading condition as indicated.

3.3. Real-Time Battery Capacity Estimation

Real-time battery capacity estimation is a very important factor to achieve a universal BMS. It is also one of the ways to improve the accuracy of SOC estimates. The work done in Reference [14] aims to establish an approach that can estimate the battery capacity in real-time. In this paper, two approaches to estimate the battery capacity were investigated; the first approach uses total least squares, while the second approach used the rest states and models the hysteresis as an error in the OCV. Furthermore, both approaches are fused together to estimate the battery capacity with a high accuracy. Finally, HIL approach was used to validate the estimation algorithm; the results showed that it is accurate within 1% of the true value.

3.4. Optimized Charging

Optimal battery charging is one of the most active research areas of BMS. Figure 9 outlines the level-2 and level-1 charging goals.

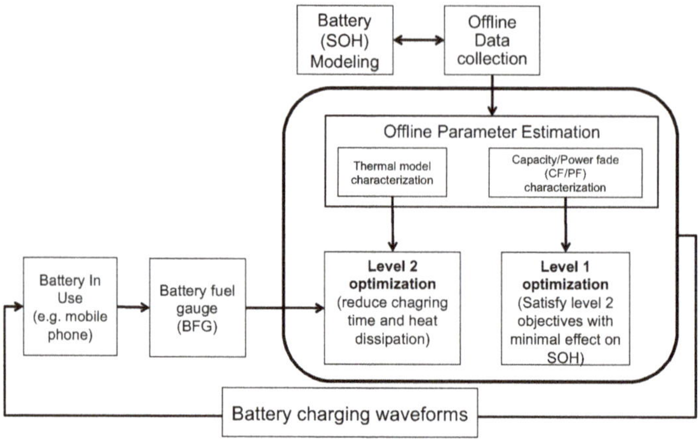

Figure 9. Elements of a smart (optimal) charger.

A closed-form solution to the problem of optimally charging a Li-ion battery was presented in Reference [64] by considering a combination of three cost functions: time-to-charge (TTC), energy losses (EL), and a temperature rise index. It was theoretically shown in Reference [64] that the optimal charging strategy for the simple equivalent model case reduces to the well-known CC-CV policy with the value of the current in the CC stage being a function of the ratio of weighting on TTC and EL and of the resistance of the battery.

In Reference [65], two models were presented for normalized battery capacity: the LAR-$\alpha\eta\gamma$ model and the control variable dependent model. The first model is based on the number of cycles and the latter is a function of the number of cycles as well as two other charge control parameters, viz., maximum terminal voltage of the battery ($v_{\max}$) and maximum charge current ($i_{\max}$). In order to evaluate the accuracy of these models experimental data were gathered from aging experiments performed on Samsung GS4 battery. The results show that these methods are far more superior to the bi-exponential capacity model [70].

3.5. *Adaptive Algorithms for Universality*

Developing a generalized BFG that is independent of battery chemistry can be broken down into two categories. The first category is to simply compile a library of all possible OCV parameters and to select the most suitable OCV model for fuel gauging through online deteciton. In other words, this first approach seeks to resolve the association ambiguity between several possible OCV parameters and the battery being monitored in a supervised way (e.g., employing nearest neighbor or any of the machine learning-based classifiers).

The second category seeks to use online data to estimate the OCV parameters [71–73]; an iterative process is used to keep the OCV parameters and battery capacity up to date. Since users can swap the battery at any time this can cause an issue where the BFG has to be aware of this change and adapt accordingly by restarting the OCV parameter estimation process. Additionally, this routine should only be applied when required. Further, the iterative estimation of SOC, OCV & ECM parameters and the battery capacity can lead to loss of robustness and instability for the BFG algorithm.

One of the first few approaches towards achieving chemistry adaptive BFG was reported in Reference [52] where the probabilistic data association (PDA) methodology was used to achieve this goal.

Here the ultimate goal is to be able to mange an arbitrary battery (present day BMS rely on parameters that are obtained from the same battery type). Figure 10 shows a demonstration of the chemistry adaptivity reported in Reference [52]. Chemistry adaptivity is a desired feature in the secondary applications of used batteries, for example, used EV batteries used in power grid.

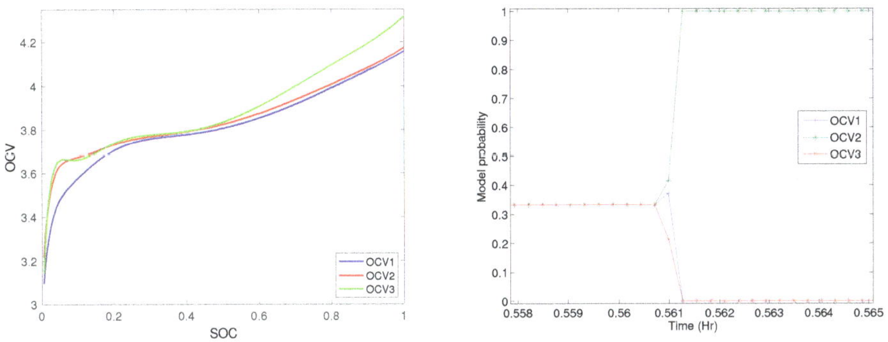

(a) (b)

Figure 10. Chemistry adaptive BMS. The proposed chemistry probabilistically selects the battery parameters based on the measured data from the battery (voltage and current). In the above demonstration, all three models were initialized with equal probability (1/3); within few samples of measured data, the PDA algorithm was able converge to the correct model. (**a**) OCV curves of different chemistries; (**b**) Model probabilities of the PDA algorithm [52].

3.6. Approaches to BMS Evaluation

In References [15,16], systematic approaches were presented to validate BFG algorithms. Particularly, the following three BFG evaluation metrics were proposed and analyzed in References [15,16]:

- *CC-metric.* The CC-metric is used to evaluate the accuracy of the SOC estimates of a BFG. It was known that the Coulomb counting method is an error prone approach to SOC estimation. However, if the battery capacity and initial SOC are known, the Coulomb counting approach will provide a very accurate estimate of SOC. The CC-metric proposes to use special *BFG validation load profiles* [16] such that the initial SOC and the battery capacity can be accurately estimated in order to evaluate the SOC estimate of a BFG. It must be noted that the CC-metric is a *laboratory based metric*, that is, it cannot be implemented in real-time when the battery is being operated by the end user.
- *OCV-metric:* The OCV-SOC metric proposes to employ the OCV curve [17] in order to find the true SOC which can then be used to validate the SOC estimate given by a BFG. Similar to the CC-metric, the OCV-SOC metric is also a *laboratory based metric* because the battery needs to be rested before the OCV can be directly measured.
- *Time to voltage (TTV) Metric.* The TTV metric [16] is the most rigorous way to test the accuracy of a BFG algorithm. This metric tests several features of a BFG at once. Let us consider an example: the BFG in an EV predicts the remaining milage as 100 miles. The most accurate way to validate this prediction is to actually drive the EV until it reaches end of charge; by subtracting the actual distance travelled from the prediction, the true BFG error can be computed. Now, instead of miles,

consider this in voltage: A BFG can predict the time it takes to reach a certain voltage, given a constant load or constant charging current. Similar to how an EV can be driven to check the accuracy of the mileage prediction, the TTV metric is computed based on the predicted vs. actual time it took for the battery to reach a certain terminal voltage. One drawback of the TTV metric is that it requires a constant current to implement the metric. Most battery chargers employ constant current charging for a certain amount of time—this provides an opportunity to implement the TTV metric in real-time. It must be re-emphasized that the TTV metric is used to quantify the accuracy of the following BFG estimates at once: such as, SOC, battery capacity and ECM parameter estimates

In order to compute the *BFG evaluation metrics* (CC-metric, OCV-metric and TTV-metric), the battery needs to undergo a specific load profile that is named *BFG evaluation profile* in References [15,16]. Figure 11 shows a BFG evaluation profile that is designed to compute the above three metrics for a battery with a nominal capacity of 1.5 Ah. The evaluation profile requires to start the experiment with a fully charged battery and apply various discharge profiles and rest periods until the battery is given a final rest period of about 2 h before constant current profile of moderate to low magnitude is applied until the battery becomes empty; this procedure ensures the total capacity [17] of the battery can be accurately estimated—the estimated total capacity in return ensures the computation of a rigorous CC-metric. The constant current discharge at the end also allows to compute the TTV metric; this is particularly significant because remaining time prediction is crucial when the battery nears it end of charge.

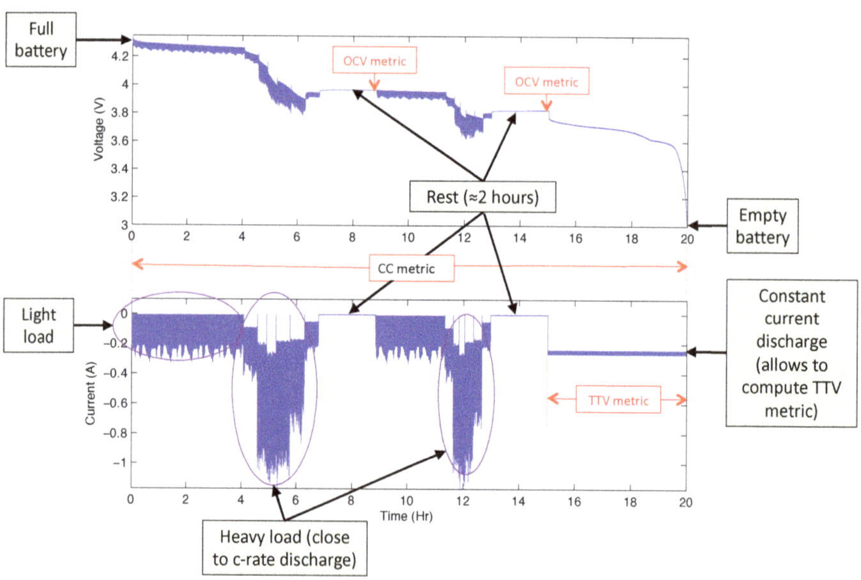

Figure 11. BFG evaluation profile. The battery fuel gauge evaluation profile is a specially designed load current profile that allows to implement all three BFG evaluation metrics: *CC-metric*, *OCV-metric* and the *TTV-metric*. The profile starts with a full battery and ends when the battery becomes empty; this allows to compute the battery capacity accurately. Intermittent rest periods within the profile allows to compute the OCV metric and the constant current load profile allows to compute the TTV metric.

4. Conclusions

In this paper, we detailed the challenges involved in developing a novel battery operating system that is suitable in future applications and described the details of some solutions that we developed. Particularly, details of the following elements of a robust battery management system are described:

- *Open circuit voltage modelling:* It is demonstrated how careful modelling and optimization can result in parameters that are applicable to a wide range of temperatures. The need for careful modelling is demonstrated using *scaling*, a strategy, when ignored, results in up to 90% higher SOC errors.
- *Battery impedance estimation:* Battery impedance changes with temperature and other battery states; real-time impedance estimation is required for effective battery management. In this paper, we summarize a real-time approach to battery impedance estimation.
- *Battery capacity estimation:* Accurate knowledge battery capacity is crucial for all aspects of a battery management system.
- *Adaptive strategies for universal battery management:* Newer versions of batteries come in slightly different chemical compositions. How to develop a battery management system that can stay relevant with ever changing battery types? This paper offers a glimpse into futuristic solutions based on probabilistic data and information fusion.
- *Optimal charging strategies:* Battery chargers have two competing objectives; one seeks to charge fast and the other attempts to minimize capacity fade and temperature rise due to charging. This paper offers high-level summary of 'level-1' and 'level-2' optimal charging algorithms designed to satisfy the above goals.
- *Strategies to evaluate battery management systems:* We describe the challenges involved in evaluating a battery management system and present several guidelines.

Further, we provided insights into the remaining challenges that needs to be addressed in the domain of battery management systems research.

Author Contributions: Conceptualization, B.B.; Methodology, B.B.; Validation, B.B., M.A. and K.P.; Resources, B.B. and K.P.; Writing—Original Draft Preparation, B.B. and M.A.; Writing—Review & Editing, K.P.; Visualization, B.B. and M.A.; Supervision, B.B.; Project Administration, B.B. and K.P.; Funding Acquisition, B.B. and K.P. All authors have read and agreed to the published version of the manuscript.

Funding: B. Balasingam acknowledges the support of the Natural Sciences and Engineering Research Council of Canada (NSERC) for financial support under the Discovery Grants (DG) program [funding reference number RGPIN-2018-04557]. Research of K. Pattipati was supported in part by the U.S. Office of Naval Research and US Naval Research Laboratory under Grants #N00014-18-1-1238, #N00173-16-1-G905, #HPCM034125HQU and by a Space Technology Research Institutes grant (#80NSSC19K1076) from NASA's Space Technology Research Grants Program.

Conflicts of Interest: The authors declare no conflict of interest. The founding sponsors had no role in the design of the study; in the collection, analyses, or interpretation of data; in the writing of the manuscript, and in the decision to publish the results.

Abbreviations

The following abbreviations are used in this manuscript:

BFG	Battery fuel gauge
BMS	Battery management system
CBC	Cell-balancing circuitry
CF	Capacity fade
ECG	Electrocardiography
ECM	Equivalent circuit model
EL	Energy loss

EV	Electric vehicle
HIL	Hardware-in-the-loop
Li-ion	Lithium ion
OCA	Optimal charging algorithm
OCV	Open circuit voltage
PF	Power fade
SOC	State of charge
SOH	State of health
TTC	Time to charge
TTV	Time to voltage

References

1. Rodrigue, J.P.; Comtois, C.; Slack, B. *The Geography of Transport Systems*; Taylor & Francis: New York, NY, USA, 2016.
2. Plett, G.L. *Battery Management Systems, Volume I: Battery Modeling*; Artech House: Norwood, MA, USA, 2015.
3. Plett, G.L. *Battery Management Systems, Volume II: Equivalent-Circuit Methods*; Artech House: Norwood, MA, USA, 2015.
4. Thomas, K.E.; Newman, J.; Darling, R.M. Mathematical modeling of lithium batteries. In *Advances in Lithium-Ion Batteries*; Springer: Boston, MA, USA, 2002; pp. 345–392.
5. Abada, S.; Marlair, G.; Lecocq, A.; Petit, M.; Sauvant-Moynot, V.; Huet, F. Safety focused modeling of lithium-ion batteries: A review. *J. Power Sources* **2016**, *306*, 178–192. [CrossRef]
6. Taylor, W.; Krithivasan, G.; Nelson, J.J. System safety and ISO 26262 compliance for automotive lithium-ion batteries. In Proceedings of the 2012 IEEE Symposium on Product Compliance Engineering Proceedings, Portland, OR, USA, 5–7 November 2012; pp. 1–6.
7. Hussein, H.H.; Batarseh, I. A review of charging algorithms for nickel and lithium battery chargers. *IEEE Trans. Veh. Technol.* **2011**, *60*, 830–838. [CrossRef]
8. Dey, S.; Mohon, S.; Pisu, P.; Ayalew, B. Sensor fault detection, isolation, and estimation in lithium-ion batteries. *IEEE Trans. Control. Syst. Technol.* **2016**, *24*, 2141–2149. [CrossRef]
9. Kang, Y.; Duan, B.; Zhou, Z.; Shang, Y.; Zhang, C. Online multi-fault detection and diagnosis for battery packs in electric vehicles. *Appl. Energy* **2020**, *259*, 114170. [CrossRef]
10. Wang, Z.; Hong, J.; Liu, P.; Zhang, L. Voltage fault diagnosis and prognosis of battery systems based on entropy and Z-score for electric vehicles. *Appl. Energy* **2017**, *196*, 289–302. [CrossRef]
11. Ahmadi, L.; Fowler, M.; Young, S.B.; Fraser, R.A.; Gaffney, B.; Walker, S.B. Energy efficiency of Li-ion battery packs re-used in stationary power applications. *Sustain. Energy Technol. Assess.* **2014**, *8*, 9–17. [CrossRef]
12. Ahmadi, L.; Yip, A.; Fowler, M.; Young, S.B.; Fraser, R.A. Environmental feasibility of re-use of electric vehicle batteries. *Sustain. Energy Technol. Assess.* **2014**, *6*, 64–74. [CrossRef]
13. EV Sales Forecasts. Available online: https://evadoption.com/ev-sales/ev-sales-forecasts/ (accessed on 10 February 2019).
14. Balasingam, B.; Avvari, G.; Pattipati, B.; Pattipati, K.; Bar-Shalom, Y. A robust approach to battery fuel gauging, part II: Real time capacity estimation. *J. Power Sources* **2014**, *269*, 949–961. [CrossRef]
15. Balasingam, B.; Avvari, G.; Pattipati, K.; Bar-Shalom, Y. Performance analysis results of a battery fuel gauge algorithm at multiple temperatures. *J. Power Sources* **2015**, *273*, 742–753. [CrossRef]
16. Avvari, G.; Pattipati, B.; Balasingam, B.; Pattipati, K.; Bar-Shalom, Y. Experimental set-up and procedures to test and validate battery fuel gauge algorithms. *Appl. Energy* **2015**, *160*, 404–418. [CrossRef]
17. Pattipati, B.; Balasingam, B.; Avvari, G.; Pattipati, K.; Bar-Shalom, Y. Open circuit voltage characterization of lithium-ion batteries. *J. Power Sources* **2014**, *269*, 317–333. [CrossRef]
18. Eddahech, A.; Briat, O.; Bertrand, N.; Delétage, J.Y.; Vinassa, J.M. Behavior and state-of-health monitoring of Li-ion batteries using impedance spectroscopy and recurrent neural networks. *Int. J. Electr. Power Energy Syst.* **2012**, *42*, 487–494. [CrossRef]

19. Wu, J.; Wang, Y.; Zhang, X.; Chen, Z. A novel state of health estimation method of Li-ion battery using group method of data handling. *J. Power Sources* **2016**, *327*, 457–464. [CrossRef]
20. Feng, X.; Weng, C.; He, X.; Han, X.; Lu, L.; Ren, D.; Ouyang, M. Online State-of-Health Estimation for Li-Ion Battery Using Partial Charging Segment Based on Support Vector Machine. *IEEE Trans. Veh. Technol.* **2019**, *68*, 8583–8592. [CrossRef]
21. Deng, Y.; Ying, H.; E, J.; Zhu, H.; Wei, K.; Chen, J.; Zhang, F.; Liao, G. Feature parameter extraction and intelligent estimation of the State-of-Health of lithium-ion batteries. *Energy* **2019**, *176*, 91–102. [CrossRef]
22. Quin, T.; Zeng, S.; Guo, J.; Skaf, Z. A Rest Time-Based Prognostic Framework for State of Health Estimation of Lithium-Ion Batteries with Regeneration Phenomena. *Energies* **2016**, *9*, 896. [CrossRef]
23. Hu, X.; Jiang, J.; Cao, D.; Egardt, B. Battery Health Prognosis for Electric Vehicles Using Sample Entropy and Sparse Bayesian Predictive Modeling. *IEEE Trans. Ind. Electron.* **2016**, *63*, 2645–2656. [CrossRef]
24. Yun, Z.; Qin, W. Remaining Useful Life Estimation of Lithium-Ion Batteries Based on Optimal Time Series Health Indicator. *IEEE Access* **2020**, *8*, 55447–55461. [CrossRef]
25. Cope, R.C.; Podrazhansky, Y. The art of battery charging. In Proceedings of the Conference (Cat. No.99TH8371), Fourteenth Annual Battery Conference on Applications and Advances, Long Beach, CA, USA, 12–15 January 1999; pp. 233–235.
26. Ikeya, T.; Sawada, N.; Takagi, S.; ichi Murakami, J.; Kobayashi, K.; Sakabe, T.; Kousaka, E.; Yoshioka, H.; Kato, S.; Yamashita, M.; et al. Multi-step constant-current charging method for electric vehicle, valve-regulated, lead/acid batteries during night time for load-levelling. *J. Power Sources* **1998**, *75*, 101–107. [CrossRef]
27. Ikeya, T.; Sawada, N.; ich Murakami, J.; Kobayashi, K.; Hattori, M.; Murotani, N.; Ujiie, S.; Kajiyama, K.; Nasu, H.; Narisoko, H.; et al. Multi-step constant-current charging method for an electric vehicle nickel/metal hydride battery with high-energy efficiency and long cycle life. *J. Power Sources* **2002**, *105*, 6–12. [CrossRef]
28. Liu, Y.; Luo, Y. Search for an Optimal Rapid-Charging Pattern for Li-Ion Batteries Using the Taguchi Approach. *IEEE Trans. Ind. Electron.* **2010**, *57*, 3963–3971. [CrossRef]
29. Vo, T.T.; Chen, X.; Shen, W.; Kapoor, A. New charging strategy for lithium-ion batteries based on the integration of Taguchi method and state of charge estimation. *J. Power Sources* **2015**, *273*, 413–422. [CrossRef]
30. Notten, P.; het Veld, J.O.; van Beek, J. Boostcharging Li-ion batteries: A challenging new charging concept. *J. Power Sources* **2005**, *145*, 89–94. [CrossRef]
31. Purushothaman, B.; Landau, U. Rapid Charging of Lithium-Ion Batteries Using Pulsed Currents. *J. Electrochem. Soc.* **2006**, *153*, A533–A542. [CrossRef]
32. Zhang, J.; Yu, J.; Cha, C.; Yang, H. The effects of pulse charging on inner pressure and cycling characteristics of sealed Ni/MH batteries. *J. Power Sources* **2004**, *136*, 180–185. [CrossRef]
33. Chen, L. A Design of an Optimal Battery Pulse Charge System by Frequency-Varied Technique. *IEEE Trans. Ind. Electron.* **2007**, *54*, 398–405. [CrossRef]
34. Chen, L. Design of Duty-Varied Voltage Pulse Charger for Improving Li-Ion Battery-Charging Response. *IEEE Trans. Ind. Electron.* **2009**, *56*, 480–487. [CrossRef]
35. Li, J.; Murphy, E.; Winnick, J.; Kohl, P.A. The effects of pulse charging on cycling characteristics of commercial lithium-ion batteries. *J. Power Sources* **2001**, *102*, 302–309. [CrossRef]
36. Liu, Y.H.; Teng, J.H.; Lin, Y.C. Search for an optimal rapid charging pattern for lithium-ion batteries using ant colony system algorithm. *IEEE Trans. Ind. Electron.* **2005**, *52*, 1328–1336. [CrossRef]
37. Guo, Z.; Liaw, B.Y.; Qiu, X.; Gao, L.; Zhang, C. Optimal charging method for lithium ion batteries using a universal voltage protocol accommodating aging. *J. Power Sources* **2015**, *274*, 957–964. [CrossRef]
38. Guo, Z.; Qiu, X.; Hou, G.; Liaw, B.Y.; Zhang, C. State of health estimation for lithium ion batteries based on charging curves. *J. Power Sources* **2014**, *249*, 457–462. [CrossRef]
39. Hu, X.; Li, S.; Peng, H.; Sun, F. Charging time and loss optimization for LiNMC and LiFePO4 batteries based on equivalent circuit models. *J. Power Sources* **2013**, *239*, 449–457. [CrossRef]
40. Aliev, R.; Aliev, R.; Guirimov, B.; Uyar, K. Dynamic data mining technique for rules extraction in a process of battery charging. *Appl. Soft Comput.* **2008**, *8*, 1252–1258. [CrossRef]

41. Guo, G.; Xu, P.; Bai, Z.; Zhou, S.; Xu, G.; Cao, B. Optimization of Ni-MH Battery Fast Charging in Electric Vehicles Using Dynamic Data Mining and ANFIS. In *Advanced Intelligent Computing Theories and Applications. With Aspects of Artificial Intelligence*; Springer: Berlin/Heidelberg, Germany, 2008; pp. 468–475.
42. Petchjatuporn, P.; Khaehintung, N.; Sunat, K.; Sirisuk, P.; Kiranon, W. Implementation of GA-trained GRNN for Intelligent Fast Charger for Ni-Cd Batteries. In Proceedings of the 2006 CES/IEEE 5th International Power Electronics and Motion Control Conference, Shanghai, China, 14–16 August 2006; Volume 1, pp. 1–5.
43. Chen, L.; Hsu, R.C.; Liu, C. A Design of a Grey-Predicted Li-Ion Battery Charge System. *IEEE Trans. Ind. Electron.* **2008**, *55*, 3692–3701. [CrossRef]
44. Waag, W.; Sauer, D.U. Adaptive estimation of the electromotive force of the lithium-ion battery after current interruption for an accurate state-of-charge and capacity determination. *Appl. Energy* **2013**, *111*, 416–427. [CrossRef]
45. Petzl, M.; Danzer, M.A. Advancements in OCV Measurement and Analysis for Lithium-Ion Batteries. *IEEE Trans. Energy Convers.* **2013**, *28*, 675–681. [CrossRef]
46. Hu, X.; Li, S.; Peng, H. A comparative study of equivalent circuit models for Li-ion batteries. *J. Power Sources* **2012**, *198*, 359–367. [CrossRef]
47. Wei, Z.; Tseng, K.J.; Wai, N.; Lim, T.M.; Skyllas-Kazacos, M. Adaptive estimation of state of charge and capacity with online identified battery model for vanadium redox flow battery. *J. Power Sources* **2016**, *332*, 389–398. [CrossRef]
48. Gao, W.; Zou, Y.; Sun, F.; Hu, X.; Yu, Y.; Feng, S. Data pieces-based parameter identification for lithium-ion battery. *J. Power Sources* **2016**, *328*, 174–184. [CrossRef]
49. Balasingam, B.; Pattipati, K. Elements of a Robust Battery-Management System: From Fast Characterization to Universality and More. *IEEE Electrif. Mag.* **2018**, *6*, 34–37. [CrossRef]
50. Smokers, R.; Verbeek, M.; van Zyl, S. EVs and post 2020 CO2 targets for passenger cars. *World Electr. Veh. J.* **2013**, *6*, 1068–1078. [CrossRef]
51. Heymans, C.; Walker, S.B.; Young, S.B.; Fowler, M. Economic analysis of second use electric vehicle batteries for residential energy storage and load-levelling. *Energy Policy* **2014**, *71*, 22–30. [CrossRef]
52. Avvari, G.; Balasingam, B.; Pattipati, K.; Bar-Shalom, Y. A battery chemistry-adaptive fuel gauge using probabilistic data association. *J. Power Sources* **2015**, *273*, 185–195. [CrossRef]
53. Barth, H.; Schaeper, C.; Schmidla, T.; Nordmann, H.; Kiel, M.; Van der Broeck, H.; Yurdagel, Y.; Wieczorek, C.; Hecht, F.; Sauer, D.U. Development of a universal adaptive battery charger as an educational project. In Proceedings of the IEEE Power Electronics Specialists Conference, Rhodes, Greece, 15–19 June 2008; pp. 1839–1845.
54. Hussein, H.H.; Pepper, M.; Harb, A.; Batarseh, I. An efficient solar charging algorithm for different battery chemistries. In Proceedings of the IEEE Vehicle Power and Propulsion Conference, Dearborn, MI, USA, 7–10 September 2009; pp. 188–193.
55. Park, S.Y.; Miwa, H.; Clark, B.T.; Ditzler, D.; Malone, G.; D'souza, N.S.; Lai, J.S. A universal battery charging algorithm for Ni-Cd, Ni-MH, SLA, and Li-Ion for wide range voltage in portable applications. In Proceedings of the IEEE Power Electronics Specialists Conference, Rhodes, Greece, 15–19 June 2008; pp. 4689–4694.
56. Bar-Shalom, Y.; Li, X.R.; Kirubarajan, T. *Estimation with Applications to Tracking and Navigation: Theory, Algorithms, and Software*; John Wiley & Sons: Hoboken, NJ, USA, 2004.
57. Bar-Shalom, Y.; Willett, P.K.; Tian, X. *Tracking and Data Fusion*; YBS Publishing: 2011. Available online: http://isif.org/sites/isif.org/files/web-files/documents/TDFBKPUBF.pdf (accessed on 1 June 2020).
58. White, T. *Hadoop: The Definitive Guide*; O'Reilly Media, Inc.: Newton, MA, USA, 2012.
59. Stolitzka, D. An electronic fuel gauge accuracy study. In Proceedings of the Twelfth Annual Battery Conference on Applications and Advances, Long Beach, CA, USA, 14–17 January 1997; pp. 211–213.
60. Chen, M.; Rincon-Mora, G.A. Accurate electrical battery model capable of predicting runtime and IV performance. *IEEE Trans. Energy Convers.* **2006**, *21*, 504–511. [CrossRef]

61. Li, Y.; Sun, Z.; Wang, J. Design for battery management system hardware-in-loop test platform. In Proceedings of the 2009 9th International Conference on Electronic Measurement & Instruments, Beijing, China, 16–19 August 2009; pp. 3–399.
62. He, Y.; Liu, W.; Koch, B.J. Battery algorithm verification and development using hardware-in-the-loop testing. *J. Power Sources* **2010**, *195*, 2969–2974. [CrossRef]
63. Wu, H. Hardware-in-loop verification of battery management system. In Proceedings of the 2011 4th International Conference on Power Electronics Systems and Applications, Hong Kong, China, 8–10 June 2011; pp. 1–3.
64. Abdollahi, A.; Han, X.; Avvari, G.; Raghunathan, N.; Balasingam, B.; Pattipati, K.; Bar-Shalom, Y. Optimal battery charging, Part I: Minimizing time-to-charge, energy loss, and temperature rise for OCV-resistance battery model. *J. Power Sources* **2016**, *303*, 388–398. [CrossRef]
65. Abdollahi, A.; Han, X.; Raghunathan, N.; Pattipati, B.; Balasingam, B.; Pattipati, K.; Bar-Shalom, Y.; Card, B. Optimal charging for general equivalent electrical battery model, and battery life management. *J. Energy Storage* **2017**, *9*, 47–58. [CrossRef]
66. Karimi, G.; Li, X. Thermal management of Lithium-ion batteries for electric vehicles. *Int. J. Energy Res.* **2013**, *37*, 13–24. [CrossRef]
67. Ahmed, M.S.; Raihan, S.A.; Balasingam, B. A scaling approach for improved state of charge representation in rechargeable batteries. *Appl. Energy* **2020**, *267*, 114880. [CrossRef]
68. Ahmed, M.; Balasingam, B. A Scaling Approach for Improved Open Circuit Voltage Modeling in Li-ion Batteries. In Proceedings of the 2019 IEEE Electrical Power and Energy Conference, Montreal, QC, Canada, 16–18 October 2019.
69. Balasingam, B.; Avvari, G.; Pattipati, B.; Pattipati, K.; Bar-Shalom, Y. A robust approach to battery fuel gauging, part I: Real time model identification. *J. Power Sources* **2014**, *272*, 1142–1153. [CrossRef]
70. He, W.; Williard, N.; Osterman, M.; Pecht, M. Prognostics of lithium-ion batteries based on Dempster–Shafer theory and the Bayesian Monte Carlo method. *J. Power Sources* **2011**, *196*, 10314–10321. [CrossRef]
71. He, H.; Zhang, X.; Xiong, R.; Xu, Y.; Guo, H. Online model-based estimation of state-of-charge and open-circuit voltage of lithium-ion batteries in electric vehicles. *Energy* **2012**, *39*, 310–318. [CrossRef]
72. Xiong, R.; Sun, F.; Gong, X.; Gao, C. A data-driven based adaptive state of charge estimator of lithium-ion polymer battery used in electric vehicles. *Appl. Energy* **2014**, *113*, 1421–1433. [CrossRef]
73. Chiang, Y.H.; Sean, W.Y.; Ke, J.C. Online estimation of internal resistance and open-circuit voltage of lithium-ion batteries in electric vehicles. *J. Power Sources* **2011**, *196*, 3921–3932. [CrossRef]

© 2020 by the authors. Licensee MDPI, Basel, Switzerland. This article is an open access article distributed under the terms and conditions of the Creative Commons Attribution (CC BY) license (http://creativecommons.org/licenses/by/4.0/).

energies

MDPI

Review

Life Cycle Assessment of Electric Vehicle Batteries: An Overview of Recent Literature

Andrea Temporelli *, Maria Leonor Carvalho * and Pierpaolo Girardi *

Ricerca Sistema Energetico—RSE SpA, 20134 Milan, Italy

* Correspondence: andrea.temporelli@rse-web.it (A.T.); marialeonor.carvalho@rse-web.it; (M.L.C.); pierpaolo.girardi@rse-web.it (P.G.)

Received: 30 March 2020; Accepted: 28 May 2020; Published: 4 June 2020

Abstract: In electric and hybrid vehicles Life Cycle Assessments (LCAs), batteries play a central role and are in the spotlight of scientific community and public opinion. Automotive batteries constitute, together with the powertrain, the main differences between electric vehicles and internal combustion engine vehicles. For this reason, many decision makers and researchers wondered whether energy and environmental impacts from batteries production, can exceed the benefits generated during the vehicle's use phase. In this framework, the purpose of the present literature review is to understand how large and variable the main impacts are due to automotive batteries' life cycle, with particular attention to climate change impacts, and to support researchers with some methodological suggestions in the field of automotive batteries' LCA. The results show that there is high variability in environmental impact assessment; CO_2eq emissions per kWh of battery capacity range from 50 to 313 g CO_2eq/kWh. Nevertheless, either using the lower or upper bounds of this range, electric vehicles result less carbon-intensive in their life cycle than corresponding diesel or petrol vehicles.

Keywords: battery electric vehicles; environmental impacts; life cycle assessment; review

1. Introduction

In recent years, the introduction of electric vehicles, and in particular electric passenger cars, has been seen as a great opportunity to reduce both urban air pollution and greenhouse gas emissions from the transport sector [1]. In particular, for what concerns urban air quality, the absence of tailpipe emissions from electric vehicles (EVs) justify this idea, confirmed by a recent study that estimates 500,000 premature deaths every year due to pollutants in the European Union, where transportation represents the main air pollutants source, especially in urban areas [2]. Regarding the reduction of greenhouse gases, EVs can rely on an overall higher efficiency [3] and, in countries where it is relevant, on the penetration of renewable energy sources in the national electric generation mix [4]. However, these considerations do not allow us to state that electric vehicles are better than Internal Combustion Engine Vehicles (ICE Vehicles), since it is not possible to compare EVs and ICE Vehicles considering only emissions that occur during vehicles use phase. In order to properly compare ICE vehicles and EVs, researchers should consider impacts related to electric energy production, fossil fuels production, vehicle and battery production and end of life phases in the LCA of EVs. In other words, an LCA approach, which allows analyzing of the environmental impacts occurring during vehicles entire life cycle, should be adopted [5]. Of course, many comparative LCAs of EVs vs. ICE Vehicles and also some literature reviews on the topics have been published in the last decade [3], but in this paper, we want to focus our attention on a particular component of the EVs: the battery. Batteries in fact are a central element in electric vehicles, and one of the most relevant distinctive elements (together with the powertrain) between EVs and ICE Vehicles. Moreover, batteries' production generates energy consumption and environmental impacts which have the potential to negatively affect the electric vehicles' benefits due to the use phase, with particular reference to climate change emissions. In order to investigate this issue and also to support researchers that are going to perform new LCA studies on EVs batteries with methodological suggestions, a review of recent LCA studies is presented in this paper. Far from being comprehensive and exhaustive, our study focuses its attention on studies performed in the last decade, since the traction battery sector is characterized by a continuous and rapid technological evolution [6]. The analysis of the selected studies has been carried out following the scheme of an ISO 14040 compliant LCA study: Goal and Scope, Inventory (Life Cycle Inventory—LCI), Life Cycle Impact Assessment (LCIA), Conclusions as summarized in Table 1, where the description of each section is reported in italic. In the following paragraphs, besides a brief description of the selected studies, for each of the LCA steps recommended in ISO 14040 standard, we analyze the main methodological differences among the studies, trying to draw useful conclusions for future traction batteries LCA studies.

Table 1. Analysis scheme considered to evaluate literature review documents.

	Title	
Bibliographic reference		Bibliographic reference following the APA style (American Psychological Association).
Goal and Scope	Target of the study	Specify the target of the study; specify if the LCA is attributional or consequential (if possible).
	Functional unit	Specify the functional unit considered and evaluate whether it is suitable to represent the service analyzed.
	System boundaries	Specify the system boundaries and the phases of the analysis considered. Possibly, state the omitted phases and the reason for their exclusion.
	Allocation system	Indicate any allocation system used (mass, economic, etc.,) and how the end of life is managed (cut-off, default, system expansion).
	Cut-off rules	Specify any cut-off rules and the parameters considered.
	Impact categories and methods	Indicate impact categories and methods used in the study.
Inventory—LCI	Data source	Report data source, specifying if primary or secondary data are considered.
LCIA	Results	Summarize the results of the study.
Sensitivity and uncertainty analysis	Considered parameters and techniques	Specify parameters and techniques considered to realize sensitivity and uncertainty analysis (if present).
Conclusions	Main conclusions of the study	Summarize the main conclusions of the document.

2. The Assessed Documents

The literature review was realized by searching in Google Scholar for the following keywords: Automotive batteries life cycle assessment, Automotive batteries life cycle, Battery life cycle, Electric vehicle batteries environmental impacts.

From the search results, we selected only the works that presented the following features:

- LCA of batteries used in automotive applications, rejecting all the papers that analyze batteries in other contexts (e.g., for stationary use).
- Documents assessing a specific battery life cycle phase, for example, production or end of life and recycling phase, and identifying materials and operations with relevant environmental impacts.
- Studies comparing different battery models, characterized by different chemistry, power, energy density and storage capacity.
- Automotive batteries literature reviews.
- Studies assessing a single battery model and identifying the major impacts due to materials and operations, suggesting sustainable alternatives.

According to these criteria, seventeen documents were suitable for the current literature review (Table 2).

Table 2. Documents analyzed within this bibliographic review.

Authors	Title	Year	Type of Document
Cusenza, M, A; Bobba, S; Ardente, F; Cellura, M; Di Persio, F	Energy and environmental assessment of a traction lithium-ion battery pack for plug-in hybrid electric vehicles [7]	2019	Journal paper
Helmers, E; Weiss, M	Advances and critical aspects in the life-cycle assessment of battery electric cars [8]	2017	Journal paper
Ioakimidis, C, S; Murillo-Marrodàn, A; Bagheri, A; Thomas, D; Genikomaskis, K	Life Cycle Assessment of a Lithium Iron Phosphate (LFP) Electric Vehicle Battery in Second Life Application Scenarios [9]	2019	Journal paper
Ellingsen, L, A, W; Majeau-Bettez, G; Singh, B; Srivastava, A, K; Valøen, L, O; Strømman, A, H	Life Cycle Assessment of a Lithium-Ion Battery Vehicle Pack [10]	2014	Journal paper
Notter, D, A; Gauch, M; Widmer, R; Wager, P; Stamp, A; Zah, R; Althaus, H, J	Contribution of Li-ion batteries to the environmental impact of electric vehicles [11]	2010	Journal paper
Romare, M; Dahllöf, L	The Life Cycle Energy Consumption and Greenhouse Gas Emissions from Lithium-Ion Batteries [12]	2017	Technical report
Dunn, J, B; Gaines, L; Barnes, M; Sullivan, J	Material and Energy Flows in the Materials Production, Assembly, and End-of-Life Stages of the Automotive Lithium-Ion Battery Life Cycle [13]	2014	Technical report
Amarakoon, S; Smith, J; Segal, B	Application of Life-Cycle Assessment to Nanoscale Technology: Lithium-ion Batteries for Electric Vehicles [14]	2013	Technical report
ReCharge	PEFCR - Product Environmental Footprint Category Rules For High Specific Energy Rechargeable Batteries for Mobile Applications [15]	2018	Technical guide
Nordelöf, A; Messagie, M; Tillman, A. M; Söderman, M. L; Van Mierlo, J	Environmental impacts of hybrid, plug-in hybrid, and battery electric vehicles—what can we learn from life cycle assessment? [4]	2014	Journal paper
Richa, K; Babbitt, C. W; Nenadic, N, G; Gaustad, G	Environmental trade-offs across cascading lithium-ion battery life cycles [16]	2017	Journal paper
Faria, R; Marques, P; Garcia, R; Moura, P; Freire, F; Delgado, J; de Almeida, A, T	Primary and secondary use of electric mobility batteries from a life cycle perspective [17]	2014	Journal paper
Oliveira, L; Messagie, M; Rangaraju, S; Sanfelix, J, Rivas, M, H; Van Mierlo, J	Key issues of lithium-ion batteries–from resource depletion to environmental performance indicators [18]	2015	Journal paper
Liu, C; Lin, J; Cao, H; Zhang, Y; Sun, Z	Recycling of spent lithium-ion batteries in view of lithium recovery: A critical review [19]	2019	Journal paper
Peters, J, F, Baumann, M, Zimmermann, B; Braun, J; Weil, M	The environmental impact of Li Ion batteries and the role of key parameters–A review [20]	2017	Journal paper
Majeau-Bettez, G; Hawkins, T, R; Strømman, A, H	Life Cycle Environmental Assessment of Lithium-Ion and Nickel Metal Hydride Batteries for Plug-In Hybrid and Battery Electric Vehicles [21]	2011	Journal paper
Dai, Q; Kelly, J, C; Gaines, L.; Wang, M	Life Cycle Analysis of Lithium-Ion Batteries for Automotive Applications [22]	2019	Journal paper

Most of these documents were published between 2014 and 2019, and only three of them between 2009 and 2013. Thirteen documents are papers published in scientific journals, three documents are research centers' technical reports and one work is a technical guide.

3. Goal and Scope

The goal and scope phase describes and defines the LCA study and the processes involved in the life cycle: targets, functional unit, system boundaries, impact categories, possible allocation procedures and cut-off rules. In the present paragraph, we analyze the main methodological choices made by different authors for this relevant part of an LCA study and we make some recommendations for future LCAs application on traction batteries.

3.1. Functional Unit

ISO14040 [23] and ISO14044 [24] standards define the LCA functional unit as the quantified performance of a product system, to be used as a reference unit. The functional unit has to be consistent with the goal and scope of the study and must provide a reference for normalizing the input and the output data. It is important to remind that thanks to the choice of a proper functional unit, it is possible to compare different systems and products offering similar services. In our analysis, despite the relatively limited number of analyzed papers, we found several different functional units. In three studies, [7,9,16] the functional unit is the battery pack. This kind of functional unit does not seem to be particularly appropriate, since it does not refer to the service offered by the systems (as requested by the ISO 14040 norm) and does not allow us to easily compare the environmental performances of different batteries. Although in reference [15], the suggested functional unit is 1 kWh of delivered energy over the service life of batteries, in five works [10,12,18,21,22], the functional unit is the battery unit storage capacity (e.g., 1 kWh, except for [21] where it is 50 MJ). As the assessed batteries are housed in electric vehicles, the authors in [11] and [4] decided to choose 1 km traveled as the functional unit, whereas in references [14] and [17], the distance traveled by the vehicle during its entire lifetime is considered. The functional unit is not clearly defined only in references [8,13,19] due to the nature and contents of these documents. As discussed, 1 kWh of battery capacity is the most used functional unit as it allows the comparison of different batteries' systems in an easy way [7]. Nevertheless, this functional unit seems more suitable for cradle to grave studies, and other functional units relating to the distance travelled, such as 1 km or the distance travelled in the battery life time, can be used, as far as transport is the service provided by the system. However, assumptions concerning the life of the battery itself must be well clarified. Although sometimes it is useful to express results also per unit of mass (kg of battery) [10], mass-based functional units are not related to the performance of the analyzed systems and we do not recommend to use them in traction batteries LCA studies.

3.2. System Boundaries

In the LCA methodology, the system boundaries definition is a way to identify which processes within the entire life cycle of the involved systems need to be analyzed or, for the sake of simplification, can be neglected. The system should be modelled to have only input and output elementary flows. System boundaries define process phases that need to be included within the LCA and their choice has to be consistent with the target of the study. Except for [8,13,19], which do not define clearly the system boundaries of their work, all the other documents explain their LCAs system boundaries in a clear way. Eight studies out of seventeen [7,9,11,14,16–18,20] analyze all batteries phases (cradle to grave): raw materials extraction and manufacturing, batteries production, transportation, use phase, end of life with material recycling. Three studies [10,12,22] analyze only batteries production, by a cradle to gate assessment, due to the lack of reliable information for modelling the use and the end of life phases. In reference [21], system boundaries include the production and the use phases, not considering the end of life impacts. Whenever possible, as suggested in references [15] and [4], we recommend to go through cradle to grave analysis, considering all phases of a battery during its life cycle.

3.3. Allocation System

The allocation process splits the input and output flows of a multiple product process between the analyzed system product and one or more other system products. The inputs and outputs have to be allocated to the different products following clear rules, defined at the beginning of the analysis. Most of the assessed documents do not specify any allocation rules, whereas only two studies mention the allocation: [11] and [15]. The first document [11] declares that there is no allocation regarding recycled products in the end of life. As a consequence, all the charges that may arise from material production are assigned to the first life of the product, even if the same can be reused, for example in a stationary domestic storage system. The second document [15] states that no accurate indication for a possible allocation is provided, as there are no co-product cases identified during batteries' production phase. However, if it is necessary to allocate impacts of any co-products linked to the battery manufacturing process, to solve multi-functionality problems, authors recommend to apply a predefined hierarchic approach: division or system expansion; allocation based on a relevant physical relationship; allocation based on some other relationships. In batteries' LCAs, allocation seems relevant only when considering recycling and second life scenarios.

3.4. *Cut-Off Rules*

Cut-off rules define material or energy flows, associated with the process unit, which are excluded from the study. Eight studies use a cut-off system, whereas nine studies do not use it or it is not possible to deduce the following criteria [4,7,8,13,14,16,18,19,21]. In five studies [9,10,12,20,22], authors do not include in their analyses impacts and benefits linked to material recycled during end of life phase. This choice is due to a high uncertainty of data and information about recycling, reuse and substitution of primary raw material with secondary material. In reference [11], materials and processes are excluded when their potential contribution is negligible. This choice is based on a mass, energy demand and expected impacts per unit of mass or energy. Of course, given the growing importance of recycling materials in the framework of the circular economy and the use of critical materials in battery production, we encourage the inclusion of end of life and recycling in future LCA studies. As in reference [15], which uses a 1% cut-off rule, neglecting all phases which have impacts lower than this threshold, we suggest that negligible phases are: batteries distribution during the end of life, infrastructure and equipment for batteries assembly and recycling.

3.5. *Impact Categories and Methods*

As described in the European Commission ILCD Handbook recommendations [25], impact categories selection must be consistent with the goal of the study. Furthermore, impact categories choice has to be complete, and should cover all the main environmental issues related to the system. In order to compare the results from different studies, it is certainly needed that studies use the same metrics: i.e., the same impact categories and the same impact method for their quantification. In our analysis, only eight studies [7,9,10,17,18,20–22] clearly explain the impact method used for quantifying midpoint impact categories [25], while six studies [4,11,12,14–16] declare the impact categories used but they do not specify the impact method followed. Finally, three studies [8,13,19] do not report any results evaluable with usual LCA impact categories or methods proposed by the ILCD Handbook [25]. In Figure 1, the impact categories used in the assessed studies are resumed.

The most used impact category is global warming (fourteen studies out of seventeen), followed by acidification (ten out of seventeen) and eutrophication (nine out of seventeen). Seven works use the impact categories ozone depletion and particulate matter, whereas six documents use the impact categories CED—cumulated energy demand, abiotic depletion, human toxicity and ecotoxicity. Other categories considered less frequently are: photo oxidant formation and resource depletion (five studies), fossils depletion (four studies), ionizing radiation (three studies), land use and water use (two studies). On the basis of these results, considering the impact categories used in almost 40% of the assessed studies, and taking into account the lesson learnt from reference [20], in an automotive battery LCA, it is suggested to use the following impact categories: global warming, acidification, eutrophication, ozone depletion, particulate matter, abiotic depletion, human toxicity, ecotoxicity and CED (Cumulated Energy Demand).

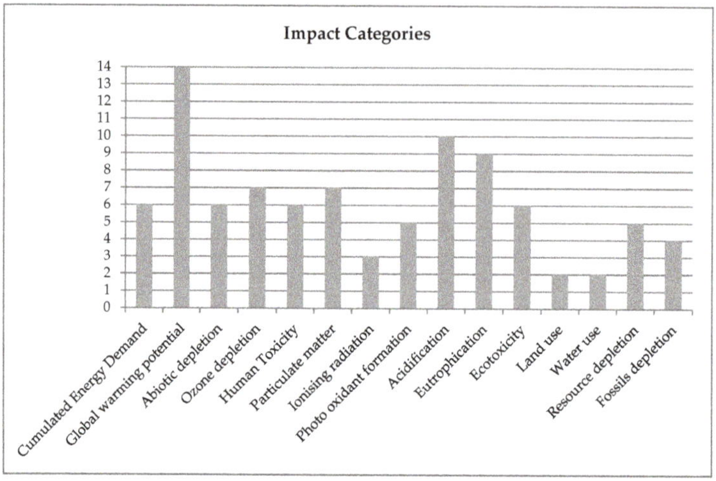

Figure 1. Impact categories used in the assessed studies.

4. Life Cycle Inventory—LCI

Inventory includes data collection and calculation procedures to quantify relevant inputs and outputs in the assessed system. Data collection includes their validation, data and process units' relationship, the relationship between data and reference flow and functional unit. In our work, we focus our attention on data quality. Data for the analysis can be divided in two categories: primary data, directly collected from producers and users of the systems, and secondary data, derived from the existing literature (including databases). Although technical guidelines in reference [15] and two literature reviews realized by the authors in [12] and [20] strongly recommend using primary data, most of the existing studies use secondary information. Only six of the analyzed works [7,10,11,14,18,22] use primary data obtained thanks to direct collaboration with batteries manufacturers, which provided information about the amount of material for each component, energy consumption for battery production, waste, percentage of recycled material used and battery maintenance operations. One study, reference [17], uses primary data only to evaluate battery energy consumption during the use phase, while [9,16,21] consider only secondary data, obtained from the available literature and from the Ecoinvent database. In general, it is possible to observe a lack of primary data that either are absent or cannot be presented in the studies due to industrial confidentiality reasons. Although this critical issue is justified by the high rate of competition and innovation of the sector, the lack of information related to primary data affects the transparency and replicability of many studies. Furthermore, it is difficult to update studies based on outdated databases (for example, reference [10]) or to check the results against different energy mixes.

5. Life Cycle Impact Assessment—LCIA

In an LCA, the impacts evaluation phase (Life Cycle Impact Assessment—LCIA) allows the assessment of potential impacts extent using data collected in the LCI. This operation links inventory data with specific impact categories and indicators, in order to better evaluate these impacts. The LCIA phase gives important information for life cycle results interpretation. Since different studies rely on different hypotheses, make use of different databases for background data and, above all, use different Life Cycle Impact Assessment Methods with their own unit, results cannot be compared easily with each other [20]. Nevertheless, some general conclusions may be drawn. First of all, for almost all impact categories, results show that the environmental major impacts of batteries life cycle occur during the production phase [7] and are due to energy consumption during materials and component production [11,16]. In particular, anode production process is responsible for the greatest impacts for impact categories such as eutrophication and acidification, whereas the cathode has major impacts for global warming and abiotic depletion [17]. Coming to the amount of the environmental impacts, results show great variability. Variability is due to, as mentioned, the use of different hypotheses and databases, but it is also linked to the different batteries' chemistry. As discussed, global warming is the most investigated impact category, since EV market penetration is mainly driven by transport sector decarbonization. Figure 2 summarizes results variability linked to greenhouse gas emissions per kWh of batteries capacity, relating to batteries production phase. These values are extracted or inferred by the assessed studies in this literature review. Depending on the different technologies and on the age of the studies, greenhouse gas emissions per kWh batteries capacity can range from 53 kg CO_2eq/kWh to more than 300 kg CO_2eq/kWh.

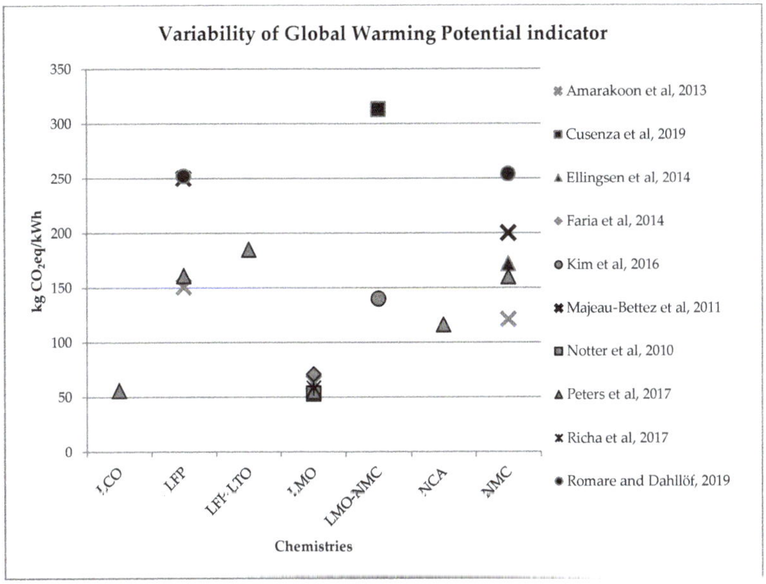

Figure 2. Variability of the global warming potential indicator (kg CO_2eq/kWh) for batteries production phase (LCO: Lithium Cobalt Oxide; LFP: Lithium iron phosphate; LFP-LTO: Lithium iron phosphate-Lithium Titanate; LMO: Lithium Manganese Oxide; LMO-NCM: Lithium Manganese Oxide-Lithium Nickel Cobalt Manganese; NCA: Lithium Nickel Cobalt Aluminum Oxide; NCM: Lithium Nickel Cobalt Manganese Oxide).

Considering a modern EV equipped with a 40 kWh battery lasting for 210,000 km [3], the lower and the upper values in Figure 2 correspond to an emission per km ranging from less than 10 g CO_2eq/km to almost 60 g CO_2eq/km. Nevertheless, despite this high variability, if we consider for the other vehicles life cycle phases, the CO_2eq/km reported in reference [3], the total CO_2eq/km life cycle emissions of an average middle size EV equipped with a 40 kWh battery, are lower than those of similar diesel or petrol cars, no matter if we consider the upper or the lower bound of battery CO_2eq emission variability (see Table 3).

A similar range of variability can be found for other, less investigated, environmental impact categories (see Figures 3–6). Again, if we consider a 40 kWh battery lasting for 210,000 km, and we consider the results from reference [1] for the other life stages, we can see that while for some impact categories for which EV perform worst, like eutrophication [1], the variability of the impacts associated with battery production does not affect the environmental ranking among EV and the corresponding ICE Vehicle. For categories like acidification, the use of the lower bound value implies that EV performs better than ICE Vehicle, while the use of the upper bound value implies that the ICE Vehicle is the best performer (see Table 4).

Table 3. Effects of the variability of CO_2eq emission per kWh of battery on the life cycle comparison among a middle size electric, diesel and petrol car. Battery CO_2eq emission per km derives from Figure 2, considering 40 kWh of capacity and 210,000 km of life. CO_2eq emission per km of remaining life cycle phases are taken from reference [1].

	Vehicle Production	Battery Production		Maintenance	Road	Fuel/Electricity Production	Use	Total	
g CO_2eq/km	(w/out battery)	Min	Max			IT marg. Mix	Urban Cycle	Min	Max
Diesel	38.2	0.0	0.0	7.7	0.6	41.1	198.5	286.1	286.1
Electric	37.7	9.5	59.6	6.2	0.6	92.5	0.0	146.6	196.7
Petrol	41.5	0.0	0.0	7.4	0.5	59.1	221.6	330.0	330.0

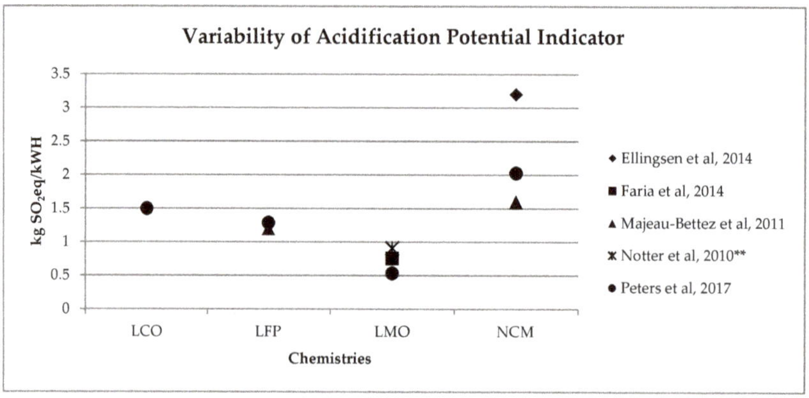

Figure 3. Variability of acidification potential (kg SO2eq/kWh) for batteries production phase (LCO: Lithium Cobalt Oxide; LFP: Lithium iron phosphate; LMO: Lithium Manganese Oxide; NCM: Lithium Nickel Cobalt Manganese Oxide); ** data from reference [11] have been updated using Ecoinvent v 3.5.

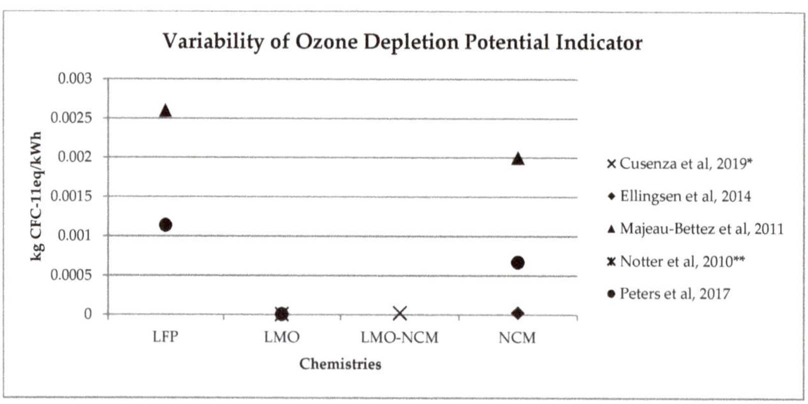

Figure 4. Variability of ozone depletion potential (kg CFC11eq/kWh) for batteries production phase (LFP: Lithium iron phosphate; LMO: Lithium Manganese Oxide; LMO-NCM: Lithium Manganese Oxide-Lithium Nickel Cobalt Manganese; NCM: Lithium Nickel Cobalt Manganese Oxide); * data from reference [7] are calculated on the basis of the total amount and the percentage for battery production, ** data from reference [11] have been updated using Ecoinvent v. 3.5.

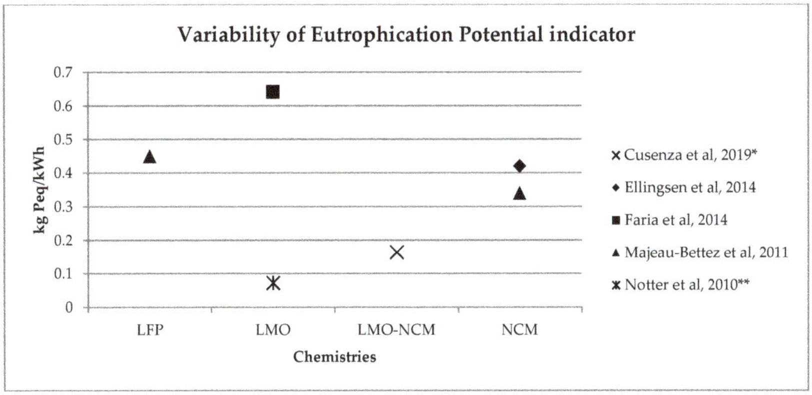

Figure 5. Variability of eutrophication potential (kg Peq/kWh) for batteries production phase (LFP: Lithium iron phosphate; LMO: Lithium Manganese Oxide; LMO-NCM: Lithium Manganese Oxide-Lithium Nickel Cobalt Manganese; NCM: Lithium Nickel Cobalt Manganese Oxide); * data from reference [7] are calculated on the basis of the total amount and the percentage for battery production, ** data from reference [11] have been updated using Ecoinvent v 3.5.

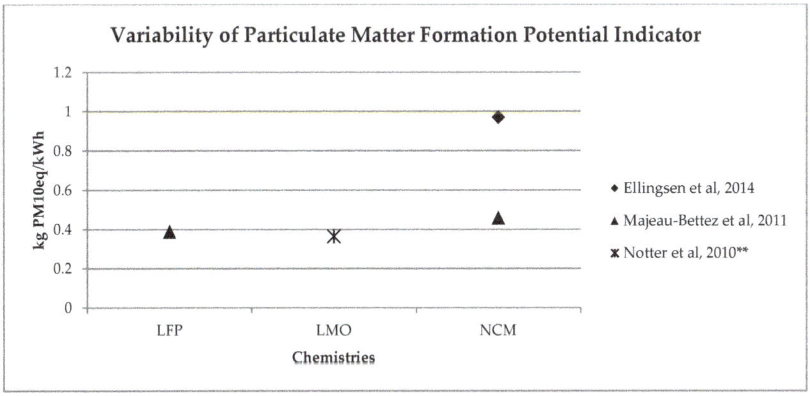

Figure 6. Variability of particulate matter formation potential (kg PM10eq/kWh) for batteries production phase (LFP: Lithium iron phosphate; LMO: Lithium Manganese Oxide; NCM: Lithium Nickel Cobalt Manganese Oxide); ** data from reference [11] have been updated using Ecoinvent v 3.5.

Table 4. Effects of the variability of acidification potential (g SO_2eq) and eutrophication potential (g PO_4eq) per kWh of battery on the life cycle comparison among a middle size electric, diesel and petrol car. Battery Emissions per km derives from Figures 3 and 5, considering 40 kWh of capacity and 210,000 km of life. Emissions per km of remaining life cycle phases are taken from reference [1].

	Vehicle	Vehicle Production & Disposal	Battery Production & Disposal		Fuel/Electricity Production & Supply	Use & Maintenance	Total	
Impacts/km	Type		Min	Max			Min	Max
Acidification Potential	EV	0.10	0.10	0.70	0.23	0.02	0.46	1.00
g SO_2 eq	ICE	0.14	-	-	0.55	0.11	0.79	0.79
Eutrophication Potential	EV	0.06	0.01	0.12	0.06	0.01	0.27	0.25
g PO_4 eq	ICE	0.06	-	-	0.07	0.03	0.16	0.16

Of course, many of the impacts associated with battery production could be lowered by recycling battery components and using recycled materials for battery production. Recycling may reduce material production energy demand up to 50% and can help to decrease environmental impacts for all the impact categories assessed [4]. Although there are a number of technologies and combinations of technologies being developed for batteries recycling (hydrometallurgy is close at hand, and can potentially extract more materials than pyrometallurgy) [12], battery recycling options are not always included in the analysis, due to the lack of relevant and reliable information [20]. Recently, a very careful recycle phase analysis has been realized in reference [7]. This work states that the environmental credits associated with materials recovered through battery recycling processes exceed the environmental impacts associated with recycling processes in all the impact categories examined, with the exception of ozone depletion, ionizing radiation and freshwater ecotoxicity. The environmental credits are particularly relevant for some impact categories such as: marine eutrophication (−27%), human toxicity (about −20% for human toxicity no cancer effect and −40% for human toxicity cancer effect), particulate matter (−17%) and abiotic depletion (−16.4%). In particular, the environmental credits related to cobalt, nickel and manganese sulphates, copper and steel are really significant and rise up to almost 80% for an important category such as abiotic depletion. Moreover, the environmental benefits linked to recycling could be increased if other cell components/materials, such as graphite, electrolyte and aluminum, are recovered, i.e., by designing battery cells to make disassembling and separating the cell components easier and more secure [7]. Additionally, for climate change impact, recycling can gain relevant positive effects, and the saved emission can be in a range of 16–32 kg CO_2eq/kWh [26]. For what concerns lithium recycling, instead, further research is still needed [19].

6. Sensitivity and Uncertainty Analysis

In general, because of the lack of primary and reliable data from industry, several assumptions have to be made in LCA studies. For this reason, sensitivity analysis has an important role, especially in a traction battery LCA, where some data and information are difficult to be found or cannot be declared by battery manufacturers due to their confidentiality. Moreover, in comparative LCAs, sensitivity analysis is requested by the ISO 14040 standard. However, sensitivity analysis is realized only in eight studies out of seventeen, and the related parameters can be organized in three categories: energy, distance driven, battery components materials and their recycling rate. The first category refers mainly to the energy mix consumed during the use phase and in the battery manufacturing phase [4,7,10,11,14,21]. The energy mixes consumed can considerably affect the final results, especially if these mixes are characterized by a high rate of non-renewable energy sources. For this reason, it is important to consider an appropriate energy mix [1] or simulate different mixes or different daily charging period. Sensitivity analysis should also be applied to the amount of energy used for battery production and to the composition of the energy mix used in this life cycle stage, given its relevance in traction battery LCA [7,10]. The second category is linked with the total distance driven, during their entire lives, by the e-cars where batteries are deployed [7,10,11,14,17]. This parameter can influence the final results of the studies and for this reason, different distances can be considered in order to verify the robustness of the results. Finally, the third category refers to battery components' materials and their recycling rate during the end of life phase [7,11,14,16,21], which could represent a relevant parameter in an LCA study. The sensitivity analysis related to this parameter could help to identify the materials with higher environmental impacts and if material recovery can help to reduce environmental impacts or if recycling operations generate more impacts than components disposal.

7. Conclusions

The review analyzed seventeen recent studies on automotive batteries LCA. This analysis is realized to give useful information to carry out new LCAs of automotive batteries and to provide a more complete picture of electric vehicle batteries LCAs. Almost all the assessed works have a good degree of compliance with the indications given by ISO 14040 [23] and ISO 14044 [24] international standards, but some documents do not fully comply with these two standards when they analyze batteries' impacts. We found that the functional unit definition is very heterogeneous and not always appropriate, and many studies consider different functional units, basing their choice on the analysis they have to realize. Consequently, the assessed works suggest many functional units: the whole battery pack, 1 kWh of storage capacity, 1 kg of battery, the distance travelled by the electric vehicle (equipped with the batteries) during its lifetime or 1 km. In view of the foregoing, the most suitable functional units for LCA of traction batteries seem to be 1 km of travel distance along the entire battery life cycle or, for sake of comparability with the existing literature, 1 kWh of battery storage capacity (specifying the battery's number of charging cycles during its lifetime). Many studies, except for [8,13,19], clearly define the system boundaries of their analysis. Only eight out of seventeen studies consider all impacts generated by the batteries during their life, realizing a cradle to grave assessment. The evaluated phases are: raw materials extraction and manufacturing, batteries production, transportation, use phase, and end of life with material recycling. Only few studies rely on primary data, while many of the assessed studies use secondary data, obtained from available literature documents or from the Ecoinvent LCA database. In general, we register a lack of primary data and of transparency both on bills of material and on energy consumption during the battery production phases. It seems important to encourage new automotive battery LCA using updated and reliable primary data, since using old data in a sector where technologies are evolving rapidly can lead to wrong conclusions and wrong decisions. For what concerns impact categories, there is a very heterogeneous situation, even if some impact categories are more frequently used (global warming, acidification, eutrophication) whereas others are used only in few studies (e.g., water use, land use, ionizing radiation). Basing on this literature review, to realize an automotive battery LCA authors suggest to consider the following: global warming, acidification, eutrophication, ozone depletion, particulate matter, abiotic depletion, human toxicity, ecotoxicity and CED (Cumulated Energy Demand). The review also underlines the importance of carrying out sensitivity analysis on some key parameters such as: battery lifetime, recycling/second life scenarios, energy mixes in production and use phase, percentage of recycled material used during the production phase.

As regards LCIA results, there is a great variability in all the impact categories that were comparable among different studies (global warming, ozone depletion, acidification, eutrophication and particulate matter formation). For global warming, one of the main impact categories under the spotlight, this variability ranges from 53 to 313 g CO_2eq/kWh of battery capacity. Our analysis shows that no matter the value considered within this range, the EVs show lower impact in their life cycle when compared to diesel or petrol cars. For other analyzed impact categories we found similar variability but if for eutrophication EVs perform worse than ICE Vehicles for any value within the variability range, for impact categories such as acidification and particulate matter formation, the use of the lower or upper bound of the variability range completely change the comparison among EVs and ICE Vehicles. This confirms that other impact categories than global warming should be investigated in LCA of traction batteries. Moreover, our review shows that batteries components which generate the greatest impacts during the production phase are the cathode active material and the anode copper and aluminum. Key aspects that could be improved to reduce these impacts are: battery lifetime extension, increase in battery efficiency and energy density. In addition, energy mix considered during the battery different life phases could be very important to decrease impacts: an energy mix with an important contribution of renewable energy sources can reduce dramatically battery overall impacts. Many studies underline that battery second life, that is battery use in stationary storage systems after their use in the automotive field, can help to reduce storage systems overall impacts. Finally, although investigated by a relatively small number of studies, it appears that material recycling, especially cobalt and nickel, could represent another useful solution to further reduce batteries' overall impacts, avoiding virgin material use during storage devices' production.

Author Contributions: Conceptualization, A.T. and P.G.; Methodology, A.T. and P.G.; Software; Validation, M.L.C. and P.G.; Formal Analysis A.T. and M.L.C.; Investigation, A.T., M.L.C. and P.G.; Resources; Data Curation, A.T.

and M.L.C.; Writing—Original Draft Preparation, A.T.; Writing—Review and Editing, A.T., M.L.C. and P.G.; Visualization, A.T. and M.L.C.; Supervision, P.G.; Project Administration, P.G.; Funding Acquisition. All authors have read and agreed to the published version of the manuscript.

Funding: This work has been financed by the Research Fund for the Italian Electrical System in compliance with the Decree of 16 April 2018.

Conflicts of Interest: The authors declare no conflict of interest.

References

1. Girardi, P.; Gargiulo, A.; Brambilla, P.C. A comparative LCA of an electric vehicle and an internal combustion engine vehicle using the appropriate power mix: The Italian case study. *Int. J. Life Cycle Assess.* **2015**, *20*, 1127–1142. [CrossRef]
2. Valverde, V.; Clairotte, M.; Bonnel, P.; Giechaskiel, B.; Carriero, M.; Otura, M.; Suarez-Bertoa, R. *Joint Research Centre 2018 Light-Duty Vehicles Emissions Testing*; JRC117625; Publications Office of the European Union: Luxembourg, 2019.
3. Girardi, P.; Brambilla, P.C.; Mela, G. Life Cycle Air Emissions External Costs Assessment for Comparing Electric and Traditional Passenger Cars. *Integr. Environ. Assess. Manag.* **2020**, *16*, 140–150. [CrossRef] [PubMed]
4. Nordelöf, A.; Messagie, M.; Tillman, A.M.; Söderman, M.L.; Van Mierlo, J. Environmental impacts of hybrid, plug-in hybrid, and battery electric vehicles—what can we learn from life cycle assessment? *Int. J. Life Cycle Assess.* **2014**, *19*, 1866–1890. [CrossRef]
5. Gao, L.; Winfield, Z.C. Life cycle assessment of environmental and economic impacts of advanced vehicles. *Energies* **2012**, *5*, 605–620. [CrossRef]
6. Feng, S.; Magee, C.L. Technological development of key domains in electric vehicles: Improvement rates, technology trajectories and key assignees. *Appl. Energy* **2020**, *260*, 114264. [CrossRef]
7. Cusenza, M.A.; Bobba, S.; Ardente, F.; Cellura, M.; Di Persio, F. Energy and environmental assessment of a traction lithium-ion battery pack for plug-in hybrid electric vehicles. *J. Clean. Prod.* **2019**, *215*, 634–649. [CrossRef] [PubMed]
8. Helmers, E.; Weiss, M. Advances and critical aspects in the life-cycle assessment of battery electric cars. *Energy Emiss. Control Technol.* **2017**, *5*, 1. [CrossRef]
9. Ioakimidis, C.S.; Murillo-Marrodán, A.; Bagheri, A.; Thomas, D.; Genikomsakis, K.N. Life Cycle Assessment of a Lithium Iron Phosphate (LFP) Electric Vehicle Battery in Second Life Application Scenarios. *Sustainability* **2019**, *11*, 2527. [CrossRef]
10. Ellingsen, L.A.W.; Majeau-Bettez, G.; Singh, B.; Srivastava, A.K.; Valøen, L.O.; Strømman, A.H. Life cycle assessment of a lithium-ion battery vehicle pack. *J. Ind. Ecol.* **2014**, *18*, 113–124. [CrossRef]
11. Notter, D.A.; Gauch, M.; Widmer, R.; Wager, P.; Stamp, A.; Zah, R.; Althaus, H.J. Contribution of Li-ion batteries to the environmental impact of electric vehicles. *Environ. Sci. Technol.* **2010**, *44*, 6550–6556. [CrossRef] [PubMed]
12. Romare, M.; Dahllöf, L. The life cycle energy consumption and greenhouse gas emissions from lithium-ion batteries. *Stockholm Zugriff Am* **2017**, *23*. Available online: https://www.ivl.se/download/18.5922281715bdaebede9559/1496046218976/C243+The+life+cycle+energy+consumption+and+CO2+emissions+from+lithium+ion+batteries+.pdf (accessed on 30 March 2020).
13. Dunn, J.B.; Gaines, L.; Barnes, M.; Sullivan, J.L.; Wang, M. *Material and Energy Flows in the Materials Production, Assembly, and End-of-Life Stages of the Automotive Lithium-Ion Battery Life Cycle*; Argonne: Lemont, IL, USA, 2014.
14. Amarakoon, S.; Smith, J.; Segal, B. *Application of Life-Cycle Assessment to Nanoscale Technology: Lithium-Ion Batteries for Electric Vehicles*; No. EPA 744-R-12-001; Environmental Protection Agency: Washington, DC, USA, 2013.
15. ReCharge. *PEFCR—Product Environmental Footprint Category Rules for High Specific Energy Rechargeable Batteries for Mobile Applications*; Recharge: Brussels, Belgium, 2018.
16. Richa, K.; Babbitt, C.W.; Nenadic, N.G.; Gaustad, G. Environmental trade-offs across cascading lithium-ion battery life cycles. *Int. J. Life Cycle Assess.* **2017**, *22*, 66–81. [CrossRef]
17. Faria, R.; Marques, P.; Garcia, R.; Moura, P.; Freire, F.; Delgado, J.; de Almeida, A.T. Primary and secondary use of electric mobility batteries from a life cycle perspective. *J. Power Sources* **2014**, *262*, 169–177. [CrossRef]

18. Oliveira, L.; Messagie, M.; Rangaraju, S.; Sanfelix, J.; Rivas, M.H.; Van Mierlo, J. Key issues of lithium-ion batteries–from resource depletion to environmental performance indicators. *J. Clean. Prod.* **2015**, *108*, 354–362. [CrossRef]
19. Liu, C.; Lin, J.; Cao, H.; Zhang, Y.; Sun, Z. Recycling of spent lithium-ion batteries in view of lithium recovery: A critical review. *J. Clean. Prod.* **2019**, *228*, 801–813. [CrossRef]
20. Peters, J.F.; Baumann, M.; Zimmermann, B.; Braun, J.; Weil, M. The environmental impact of Li-Ion batteries and the role of key parameters–A review. *Renew. Sustain. Energy Rev.* **2017**, *67*, 491–506. [CrossRef]
21. Majeau-Bettez, G.; Hawkins, T.R.; Strømman, A.H. Life cycle environmental assessment of lithium-ion and nickel metal hydride batteries for plug-in hybrid and battery electric vehicles. *Environ. Sci. Technol.* **2011**, *45*, 4548–4554. [CrossRef] [PubMed]
22. Dai, Q.; Kelly, J.C.; Gaines, L.; Wang, M. Life cycle analysis of lithium-ion batteries for automotive applications. *Batteries* **2019**, *5*, 48. [CrossRef]
23. ISO—The International Organization for Standardization. *ISO 14040:2006: Environmental Management—Life Cycle Assessment—Principles and Framework*; ISO—The International Organization for Standardization: Geneva, Switzerland, 2006.
24. ISO—The International Organization for Standardization. *ISO 14044:2006: Environmental Management—Life Cycle Assessment—Requirements and Guidelines*; ISO—The International Organization for Standardization: Geneva, Switzerland, 2006.
25. European Commission—Joint Research Centre—Institute for Environment and Sustainability. *International Reference Life Cycle Data System (ILCD) Handbook—Recommendations for Life Cycle Impact Assessment in the European Context*, 1st ed.; EUR 24571 EN; Publications Office of the European Union: Luxembourg, 2011.
26. Ellingsen, L.; Hung, C.; Strømman, A. Identifying key assumptions and differences in life cycle assessment studies of lithium-ion traction batteries with focus on greenhouse gas emissions. *Transp. Res. Part D Transp. Environ.* **2017**, *55*, 82–90. [CrossRef]

© 2020 by the authors. Licensee MDPI, Basel, Switzerland. This article is an open access article distributed under the terms and conditions of the Creative Commons Attribution (CC BY) license (http://creativecommons.org/licenses/by/4.0/).

energies

Article

Optimization of Vehicle Braking Distance Using a Fuzzy Controller

Peter Girovský *, Jaroslava Žilková and Ján Kaňuch

Department of Electrical Engineering and Mechatronics, Faculty of Electrical Engineering and Informatics, Technical University of Košice, Letná 9, 042 00 Košice, Slovakia; jaroslava.zilkova@tuke.sk (J.Ž.); jan.kanuch@tuke.sk (J.K.)

* Correspondence: peter.girovsky@tuke.sk; Tel.: +421-55-602-2270

Received: 31 March 2020; Accepted: 6 June 2020; Published: 11 June 2020

Abstract: The paper presents the study of an anti-lock braking system (ABS) that has been complemented by a fuzzy controller. The fuzzy controller was used to improve the braking performance of the vehicle, particularly in critical situations, for example, when braking a vehicle on wet road. The controller for the ABS was designed in the MATLAB/Simulink program. The designed controller was simulated on a medium-size vehicle model. During testing, three braking systems were simulated on the vehicle model. We compared the performance of a braking system without an ABS, a system with a threshold-based conventional ABS, and a braking system with the proposed ABS with a fuzzy controller. These three braking systems were simulation tested during braking the vehicle on a dry straight road and on a road with combined road adhesion. A maneuverability test was conducted, where the vehicle had to avoid an obstacle while braking. The results of each test are provided at the end of the paper.

Keywords: anti-lock braking system; fuzzy logic system; controller; vehicle; simulation

1. Introduction

The increasing number of cars on the roads and the related rising accident rate have brought the concept of road safety to the forefront. The primary safety objective is to protect the health and lives of passengers and minimize the consequences of an accident. In order to achieve this goal, various security features are being applied in vehicles. They can be divided into two groups:

(1) Active safety features.
(2) Passive safety features.

One of the most important active safety features includes the braking system of the vehicle, which is considered to be the most important system in the vehicle, as safe stopping or slowing down of the vehicle is one of the ways of accident prevention. Since its introduction, the braking system has undergone numerous improvements, one of them being the application of the ABS. The anti-lock braking system (ABS), which has been developed and implemented from the late 1970s, prevents the wheels from locking during braking of the vehicle, and thereby, allows the driver to maintain control over steering [1]. Although this vehicle safety system has been used for decades, it is constantly being improved using either conventional or intelligent control methods [2].

Several studies have been published that focus on optimizing the vehicle's braking system using various strategies, both for vehicles with internal combustion engines and electric vehicles. For electric vehicles, the most commonly applied method is regenerative braking control. The authors of paper [3] present regenerative braking control strategy intended to improve braking performance while maximizing braking energy recovery. Application of fuzzy logic rules enables the optimization of regenerative braking in order to achieve better braking performance. This issue is addressed in paper [4], where the rules of fuzzy logic were used to optimize the slip parameter of a sliding mode controller and to thereby achieve a shortening of the vehicle's braking distance and an increase in the energy efficiency of regenerative braking.

As shown in several studies [5–12], the use of fuzzy logic rules in different ways of controlling the ABS has improved the braking performance of the vehicle in simulations on different road surfaces. Another possible approach to optimizing the braking performance of a vehicle is to use fuzzy logic to optimize the parameters of conventional methods of controlling ABSs. In paper [13], the authors present the achievement of optimal slip rate parameters using fuzzy logic. The presented simulation results prove the improved maneuverability and stability of the vehicle. In paper [14], the authors deal with the optimization of braking properties and shortening of vehicle braking distance using a PID-fuzzy controller with parameter adaptation.

Using fuzzy logic is not the only way to improve the vehicle ABS. The application of artificial neural networks to optimize the rules of the fuzzy controller [15], or as observers, or the implementation of neuro-controllers as such, seem to be suitable alternatives. Comparison of various strategies [16] and selection of the most appropriate method or combination of methods can result in optimization of the ABS performance, and thus, in the improvement of vehicle safety.

Mathematical modeling and simulation are powerful tools in the initial stages of research and in the verification of hypotheses, which is sometimes not possible to implement in practice, be it for safety, for physical or other reasons [17,18]. Therefore, in our design, we used a combination of two simulation tools, MATLAB/Simulink and CarSim, which allowed detailed setting of the vehicle parameters and simulation in different environments. Suitable setting and interlinking with other software, such as MATLAB/Simulink, provides the opportunity of testing vehicle performance in real time. Using the CarSim simulation software enables the verification of vehicle performance in situations where failure could have destructive consequences or even result in loss of life.

2. ABS—Anti-Lock Braking System

In terms of control, the ABS represents a feedback control system that controls brake pressure based on measuring wheel deceleration, and thereby, maintaining wheel slip at optimum value. The basic idea is the use of friction properties of tires on the road surface. As a result of the transmitted longitudinal friction force during braking, deformation and slipping of the individual parts of the tire tread on the road will occur. The result is that during braking, the wheel rotates more slowly than would correspond to the vehicle speed, so it rotates with specific slips. The maximum braking force that a tire is able to transfer onto the concrete road surface is highly dependent on this very specific slip of the wheel. The braking force reaches maximum at the specific value of slip and a further increase in slip reduces the braking force. At maximum value of brake force on normal road surfaces, using normal tires, the specific slip is in the range of 15–20%.

The ABS ensures that even at maximum braking (maximum pressure of the driver on the brake pedal), the specific slip of the wheel does not exceed the said range of 15–20%. This ensures that the vehicle is braked with maximum efficiency, i.e., on the shortest path, plus the transferable lateral forces are in the area of specific slip and are even large enough to allow the correction of direction during braking, in case the vehicle has to get round an obstacle [19–21].

2.1. Tire Slip Characteristics

Tire slip characteristics are described by curves that characterize the longitudinal and lateral forces that are transmitted onto the road by the tire at various slips of the tire. The tire slip characteristic shown in Figure 1 presents a typical course of the longitudinal and lateral forces.

The longitudinal forces Fx are the braking and driving forces that a tire can transfer onto the road under various slip values. The lateral forces Fy are forces that a tire transfers onto the road when the vehicle is changing direction. Figure 1 demonstrates the size of the braking (driving) force and the lateral force that a tire can transfer onto the road at various slips [22,23].

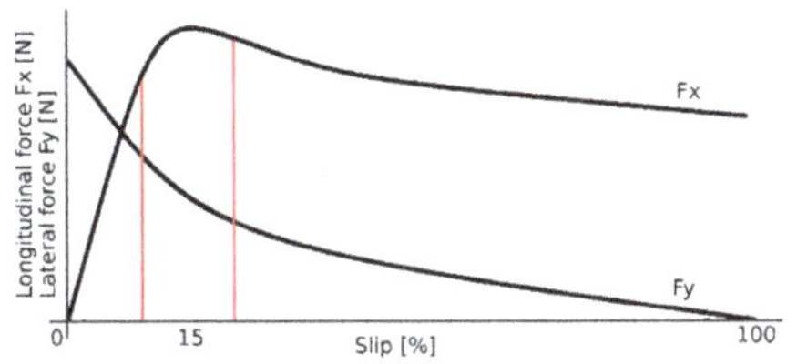

Figure 1. Slip characteristics of tires with radial construction.

Slip λ is calculated according to the equation:

$$\lambda = \frac{v - \omega \times R_\omega}{v} = 1 - \frac{\omega \times R_\omega}{v} \tag{1}$$

where:

v is vehicle velocity,
ω is angular velocity of tire,
R_ω is tire radius.

2.2. Vehicle Braking System

The principle scheme of the braking system used for simulation is shown in Figure 2.

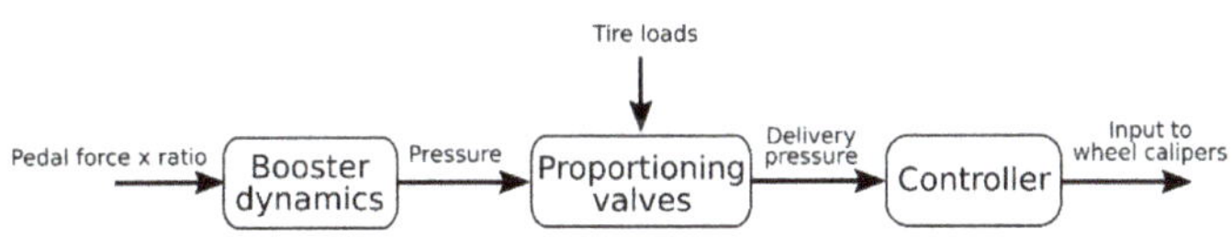

Figure 2. Basic scheme of the vehicle brake system.

The input variable for the vehicle braking system is the force exerted on the brake pedal by the driver the pedal force. The force that enters the master cylinder is multiplied by the mechanical transmission brake pedal—ratio.

$$F_{in} = F_{pedal} \times ratio \tag{2}$$

where:

F_{pedal} is the pedal force,
$ratio$ is the mechanical transmission brake pedal,
F_{in} is the force entering the brake booster.

The pressure of the master cylinder is calculated according to the equation:

$$P_{MC} = \frac{F_{MC}}{D_{MC}^2 \times \frac{\pi}{4}} \tag{3}$$

where:

F_{MC} is the master cylinder output force,
D is the diameter.

The braking torque depends on the pressure in the brake cylinders and the temperature of the brake disc. The calculation of brake disc temperature uses the input energy from the braking torque, the rotational speed and the energy loss due to cooling. Input energy is dependent on brake torque and wheel speed. Output energy loss due to cooling is dependent on the cooling coefficient as a function of vehicle speed, brake disc weight, specific heat of the wheel as a function of the temperature of the wheel, brake disc temperature and air temperature.

The output heat change (ΔT) is calculated according to:

$$\Delta T = \frac{(E_{in} - E_{out})}{M_d \times C_p(T_d)} \tag{4}$$

where:

ΔT is output heat change,
E_{in} is input energy,
E_{out} is output energy,
M_d is brake disc weight,
$C_p(T_d)$ is specific heat of the wheel as a function of the temperature of the wheel.

First order delay is applied to the brake booster. It is a specific time constant that depends on whether the brake pedal is pressed or released. Consequently, the brake pressure is transferred through proportioning valves to the brake cylinders of the wheels as the supply pressure, which together with other parameters generates the mechanical braking torque, as demonstrated in Figure 3. This type of brake system was used for simulation in the CarSim program.

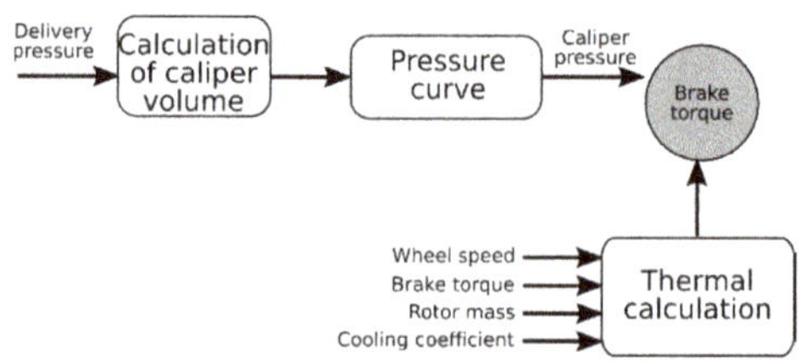

Figure 3. Brake system—basic scheme of generating mechanical braking torque.

2.3. *Basic Model of ABS Control*

The input to the ABS controller is the vehicle speed and the equivalent speeds of the individual wheels. These speeds are processed in the controller and slip is evaluated on their basis. The output of the controller are then values from −1 to 1 depending on the speed and slip. This dependence of the output values is shown in Table 1.

Table 1. ABS controller logic.

Output Values	−1	0	1
Slip	>0.2	0.05–0.2	<0.05
Vehicle speed (km/h)	>5	>5	>5

The basic block diagram of the ABS control model used in the CarSim program is shown in Figure 4.

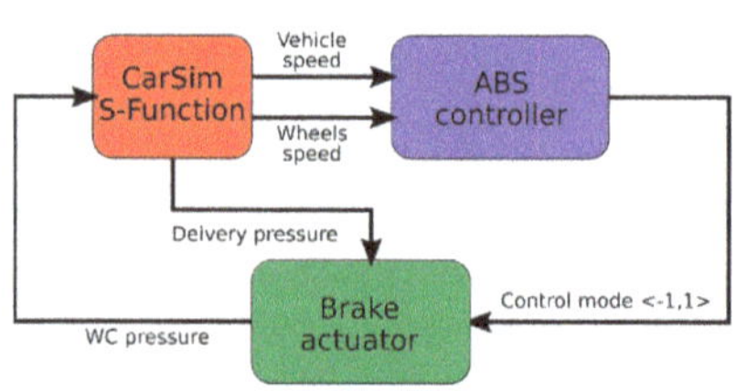

Figure 4. Block scheme of ABS control model.

3. Fuzzy Control of ABS

The fuzzy control of the ABS is based on a conventional ABS complemented by a fuzzy controller. The design and the subsequent verification by simulation were carried out in CarSim and MATLAB/Simulink. The designed fuzzy ABS control system consists of three parts:

(1) Speed controller.
(2) Fuzzy controller.
(3) Brake pressure control device.

The block model of the proposed fuzzy control is shown in Figure 5.

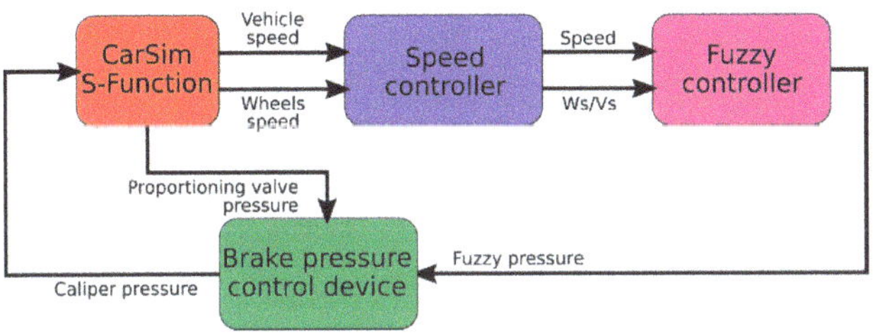

Figure 5. Block scheme of ABS control with fuzzy controller.

The basic parameters for speed controller input were the values of vehicle speed and the speed values of individual wheels. As shown in Figure 6, the output of the speed controller is the value of the vehicle speed multiplied by a constant that modifies the speed value, and this then, becomes the input value to the fuzzy controller. Constant 1/200 was selected on the basis of the defined maximum speed 200 km/h. For tests involving a higher speed, this constant has to be modified. The second output is the ratio of the wheel speed and the vehicle speed. The ratio of vehicle speed and wheel speed can be in the range of values between 0 and 1, where value 0 represents an unblocked wheel (slip 0), and value 1 represents a blocked wheel (slip 1).

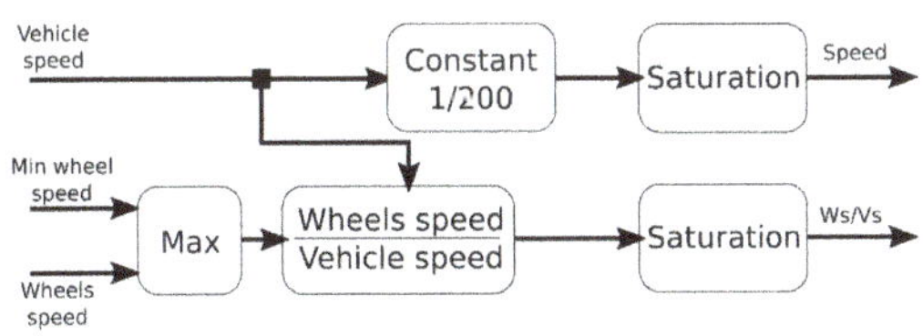

Figure 6. Block scheme of the speed controller.

The input for the fuzzy controller, shown in Figure 7, is the modified vehicle speed and speeds of the individual wheels.

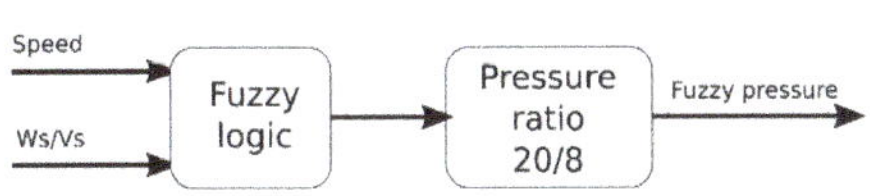

Figure 7. Block scheme of the fuzzy controller.

The range of values for the first input is from 0 to 1. Zero value represents zero vehicle speed and 1 is the highest speed. The range of values for the second input can also be from 0 to 1. In this case, the 0 value represents zero slip and value 1 represents maximum slip value.

The fuzzy controller output value is in the range −1 to 1. Value 1 represents full brake pressure applied to the wheels and −1 represents the application of negative pressure, which means that the wheel is released. The constant output pressure ratio is 20/8 for the front/rear wheels. The constant ratio in favor of the front wheels was chosen in view of the fact that during braking, the vehicle weight is transferred forward and the rear wheels are relieved, which at full braking force would cause blocking of the wheels and subsequent loss of control over the vehicle. This ratio value has been taken as ideal from the CarSim brake pressure control actuator model. Since a fuzzy controller is used to control the braking force, a ratio of 20/8 was chosen as ideal for achieving the largest possible parameter of relative deceleration. The maximum pressure for the simulation was limited to 10 MPa. Table 2 shows the decision rules for the fuzzy controller setting.

Table 2. Table of rules for the fuzzy controller.

Fuzzy Rules		**Speed**				
		VS	**S**	**M**	**H**	**VH**
Ws/Vs	VS	VN	VN	VN	VN	VN
	S	N	N	N	N	VN
	M	Z	Z	Z	Z	Z
	H	P	P	P	P	P
	VH	VP	VP	VP	VP	VP

VS—very small, S—small, M—middle, H—high, VH—very high, VN—very negative, N—negative, Z—zero, P—positive, VP—very positive.

Values of the individual variables for the FIS matrix shown in Figure 8 were selected on the basis of tire slip, which was set at 20%. At such slip, the vehicle is still operable even under heavy braking. A graph showing the dependence between vehicle speed, ratio of vehicle and wheel speed and the resulting pressure is presented in Figure 9.

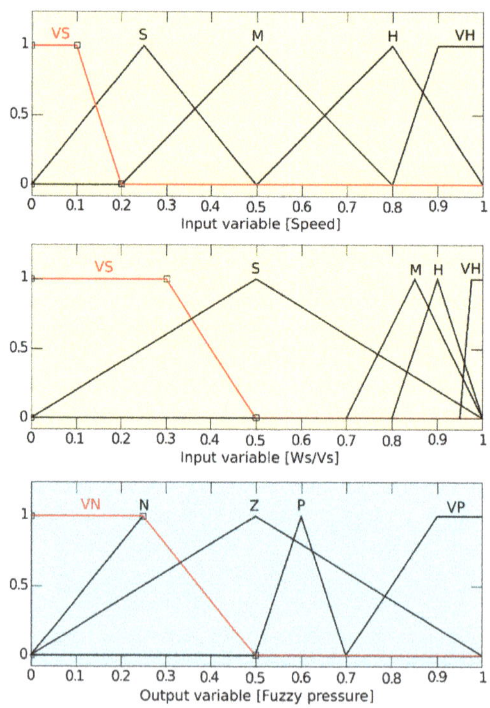

Figure 8. Setting rules of FIS matrix for fuzzy controller.

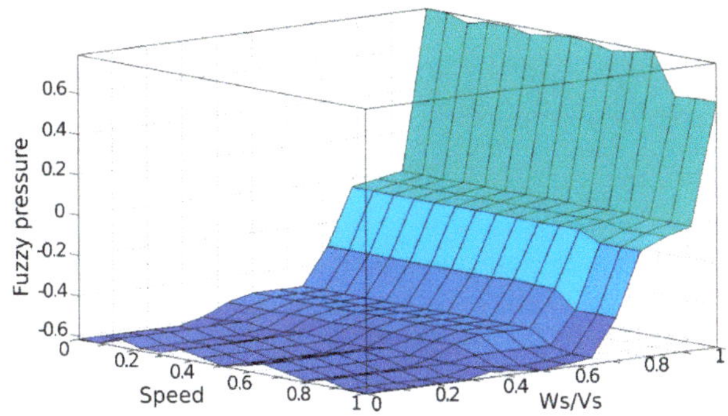

Figure 9. Graph showing the dependence of output variable on the input variables in the fuzzy controller.

4. Braking Systems Comparison by Simulation

In order to verify the properties of the proposed ABS with the fuzzy controller, simulation tests were performed in the CarSim and MATLAB/Simulink programs. The simulation tests were performed on a medium-size category vehicle model with independent front and rear suspension. The vehicle had a 7-speed automatic transmission and the maximum engine torque was 534 Nm. During the simulation, the vehicle had 225/60/R18 radial tires.

The tests were performed for a given vehicle in three different combinations. A comparison was made of the vehicle's braking behavior with a conventional threshold-based ABS, with the proposed ABS with a fuzzy controller designed to slip around 0.2, and with braking without an ABS, where the wheels were locked during braking.

The tests of these systems were carried out for the following operating states:

(1) Dry road;
(2) Road with combined adhesion;
(3) Avoiding an obstacle on the road.

The results of each simulation test are presented below.

4.1 Vehicle Braking Systems Test on Dry Road

In this test, we observed the performance of individual braking systems when the vehicle was braking on a straight dry road. The test was performed for the initial vehicle speeds of 100 and 140 km/h. The results of the braking test for the individual braking systems from speed 100 km/h are shown in Figure 10 and Table 3.

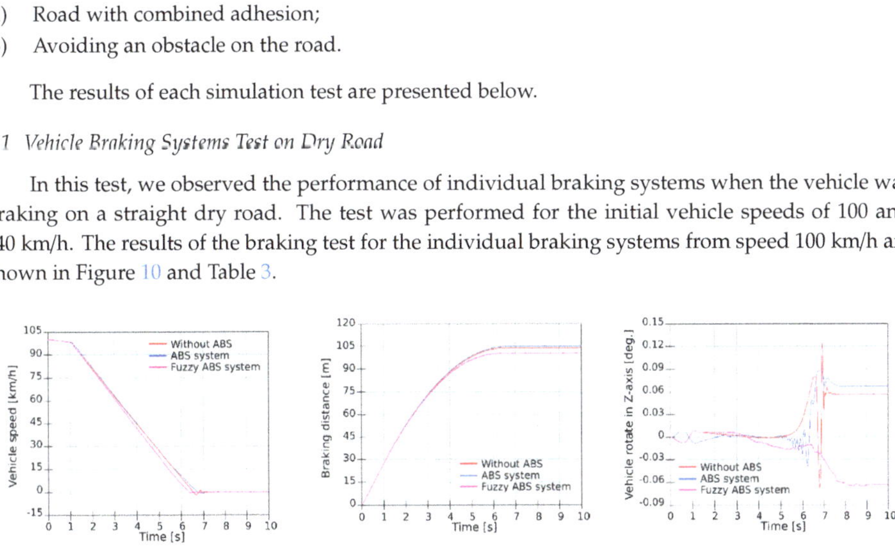

(a) (b) (c)

Figure 10. Test on dry road at speed 100 km/h—(**a**) Speed curve; (**b**) Braking distance; (**c**) Vehicle rotation.

Table 3. Test results on dry road at speed 100 km/h.

Monitored Value	Without ABS	ABS	Fuzzy ABS
Braking time (s)	6.53	6.73	6.35
Braking distance (m)	103.66	104.80	100.7
Max. rotation in *z*-axis (deg.)	0.124	0.088	−0.067

Figure 11 and Table 4 show the results of braking test runs at the speed of 140 km/h.

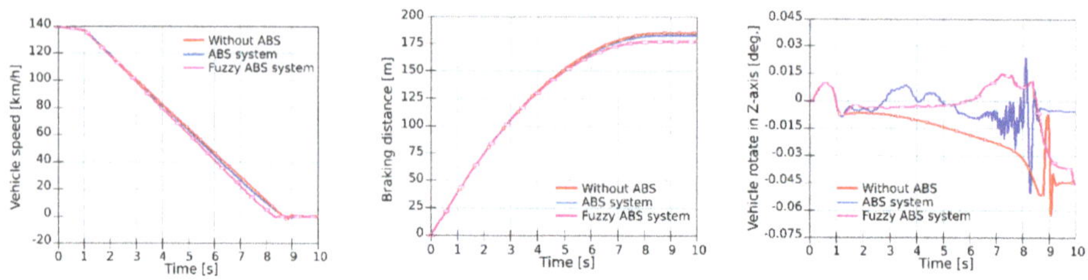

(a) (b) (c)

Figure 11. Test on dry road at speed 140 km/h—(**a**) Speed curve; (**b**) Braking distance; (**c**) Vehicle rotation.

Table 4. Test results on dry road at speed 140 km/h.

Monitored Value	Without ABS	ABS	Fuzzy ABS
Braking time (s)	8.65	8.73	8.30
Braking distance (m)	185.04	182.83	177.08
Max. rotation in *z*-axis (deg.)	−0.063	−0.052	−0.045

As seen from the test results, all vehicles were kept in their lane and the vehicle with the fuzzy ABS in both cases had the shortest braking distance.

4.2. Vehicle Braking Systems Test on Road with Combined Adhesion

This test was performed on a straight road with different adhesion values for the right and left sides. The left side of the vehicle was on the road with adhesion value 0.2, and the right side of the vehicle on the road with adhesion value 0.5. In general, such combination of surface adhesion represents an icy road. The test was performed for the initial vehicle speeds of 100 and 140 km/h. The results of the braking test for the individual braking systems for speed 100 km/h are shown in Figure 12 and Table 5.

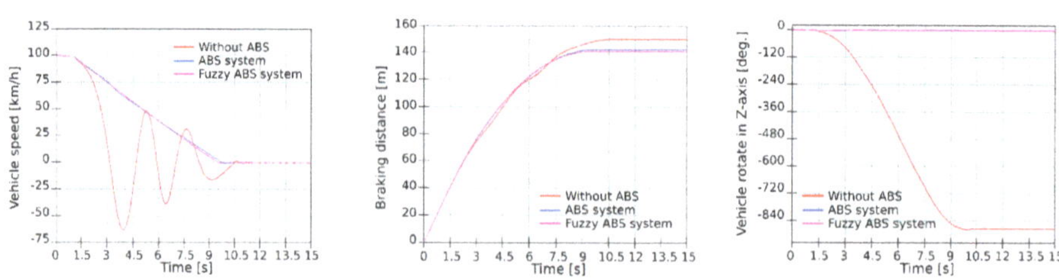

(a) (b) (c)

Figure 12. Test on road with combined adhesion at speed 100 km/h—(**a**) Speed curve; (**b**) Braking distance; (**c**) Vehicle rotation.

Table 5. Test results on road with combined adhesion at speed 100 km/h.

Monitored Value	Without ABS	ABS	Fuzzy ABS
Braking time (s)	10.5	9.73	9.6
Braking distance (m)	150.52	143.23	141.78
Max. rotation in z-axis (deg.)	−873.24	−3.21	−2.14

Figure 13 and Table 6 show the results of braking test runs at speed 140 km/h.

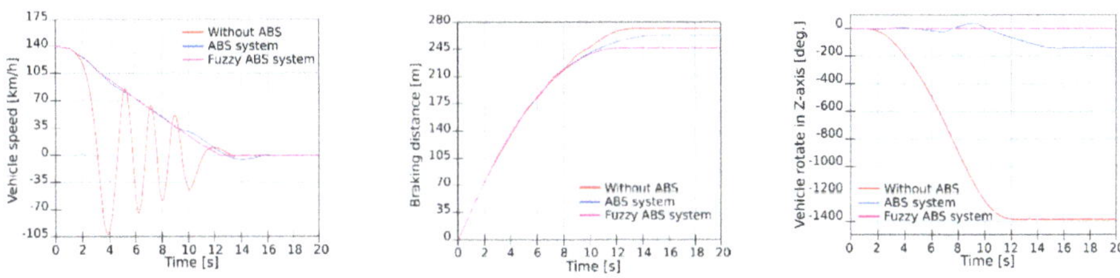

(a) (b) (c)

Figure 13. Test on road with combined adhesion at speed 140 km/h—(**a**) Speed curve; (**b**) Braking distance; (**c**) Vehicle rotation.

Table 6. Test results on road with combined adhesion at speed 140 km/h.

Monitored Value	Without ABS	ABS	Fuzzy ABS
Braking time (s)	13.23	15.55	12.43
Braking distance (m)	272.21	263.83	247.56
Max. rotation in z-axis (deg.)	−1372.30	−141.29	−2.83

As demonstrated by the test results, a vehicle without an ABS slipped into skidding on the icy road and at both speeds ended off the road. The vehicle with a conventional ABS also skidded at the speed of 140 km/h and finished off turned in the opposite direction. The vehicle with the fuzzy ABS was the only one able to remain on the road and had the shortest braking distance at both speeds.

4.3. Vehicle Braking Systems Test When Avoiding an Obstacle on the Road

In this test, we observed the ability of the vehicle to maneuver on a dry road during braking, employing the individual braking systems. During this test, the vehicle was avoiding an obstacle during braking. The obstacle was simulated by delimiting the route using cones, as illustrated in Figure 14.

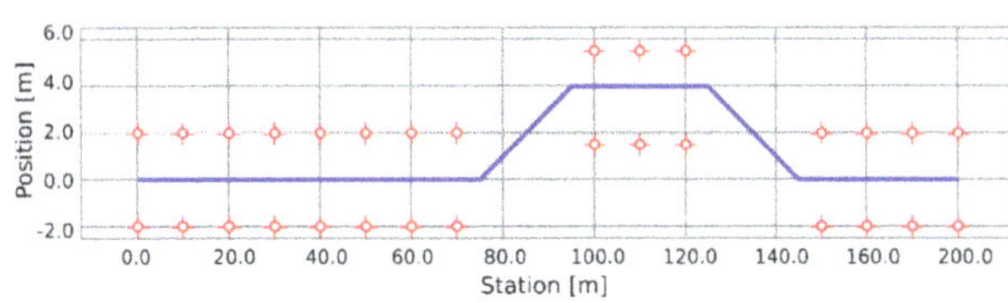

Figure 14. Road during maneuverability test.

The test was performed for the initial vehicle speeds of 100 and 130 km/h. Because it was important to avoid an obstacle in this test, instead of vehicle rotation in z-axis, the shift of the vehicle from the defined route was under observation. The results of this test for individual braking systems at speed 100 km/h are shown in Figure 15 and Table 7.

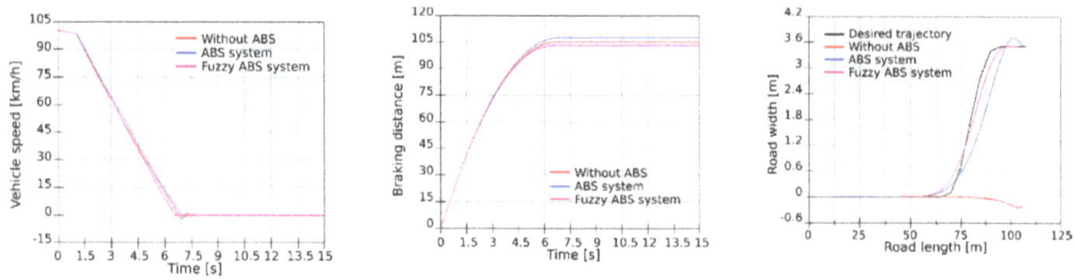

(a) (b) (c)

Figure 15. Maneuverability test on dry road at speed 100 km/h—(**a**) Speed curve; (**b**) Braking distance; (**c**) Vehicle trajectory.

Table 7. Maneuverability test on dry road at speed 100 km/h.

Monitored Value	Without ABS	ABS	Fuzzy ABS
Braking time (s)	6.78	6.95	6.58
Braking distance (m)	105.10	107.52	103.03
Max. deviation of the vehicle (m)	3.76	0.2	0.01

As seen from the vehicle maneuverability tests during braking at speed 100 km/h, the vehicle with the fuzzy ABS was able to copy the desired trajectory with most accuracy. A vehicle without an ABS would not be able to maneuver when braking, it would just move straight forward with locked wheels and would hit the obstacle.

Figure 16 and Table 8 show the results of test runs at a speed of 130 km/h.

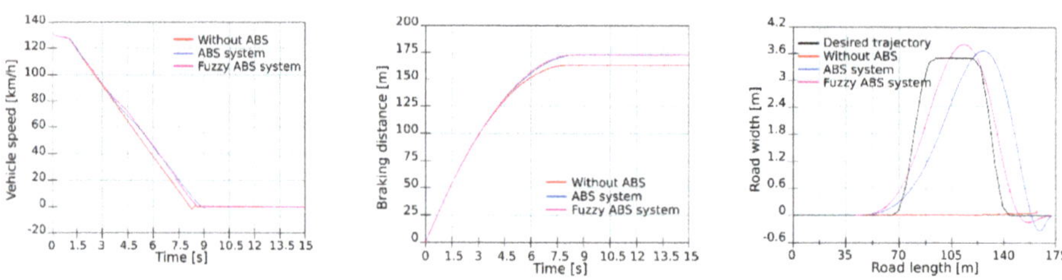

(a) (b) (c)

Figure 16. Maneuverability test on dry road at speed 130 km/h—(**a**) Speed curve; (**b**) Braking distance; (**c**) Vehicle trajectory.

Table 8. Maneuverability test on dry road at speed 130 km/h.

Monitored Value	Without ABS	ABS	Fuzzy ABS
Braking time (s)	8.15	8.8	8.58
Braking distance (m)	162.93	172.66	172.43
Max. deviation of the vehicle (m)	3.52	0.34	0.19

As demonstrated in the vehicles' maneuverability test during braking at speed 130 km/h, the vehicle with the fuzzy ABS was able to copy the desired trajectory most accurately of all the vehicles. The vehicle with the conventional ABS showed problems in copying the desired trajectory at this speed. The vehicle without an ABS would not be able to maneuver when braking, it would just move straight forward with locked wheels and would hit the obstacle the same as before.

5. Discussion

The CarSim and MATLAB/Simulink programs were used to verify the design of the vehicle's fuzzy ABS controller. The main goal of this work was to improve the braking properties of the vehicle by using fuzzy logic for the brake pressure regulator. We have chosen three types of operating states for testing the fuzzy ABS. The first was to test the vehicle's braking behavior on dry roads at speeds of 100 and 140 km/h.

In this test, the shortest braking distance was the main criterion for assessing the quality of the proposed fuzzy ABS controller. The value of the braking distance of a vehicle with a classic ABS was taken as a reference value for comparison, as we wanted to improve the properties of this ABS by using a fuzzy logic.

As can be seen from the comparison in Table 9 and Figure 17, the ABS with the fuzzy controller achieved better properties with all monitored parameters, i.e., directional stability, braking time and braking distance than when using ABS without a fuzzy controller.

Table 9. Comparison of parameters when braking on dry road.

Test on Dry Road	Max. Rotation in Z-Axis (deg.)	Braking Time (s)	Braking Distance (m)
100 km/h			
ABS	0.088	6.730	104.800
Fuzzy ABS	0.067	6.350	100.700
Without ABS	0.124	6.530	103.660
140 km/h			
ABS	0.052	8.730	182.830
Fuzzy ABS	0.045	8.300	177.080
Without ABS	0.063	8.650	185.040

reference value, better value, worse value.

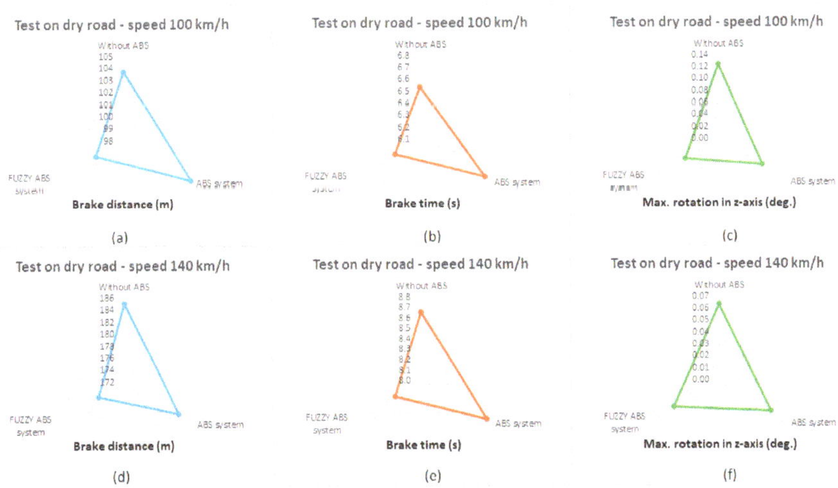

Figure 17. Graphical comparison of the test on dry road.(**a**): Graphical comparison of the braking distance during a dry road test at a vehicle speed of 100 km/h; (**b**): Graphical comparison of the braking time during a dry road test at a vehicle speed of 100 km/h; (**c**): Graphical comparison of the vehicle rotation in z-axis during a dry road test at a vehicle speed of 100 km/h; (**d**): Graphical comparison of the braking distance during a dry road test at a vehicle speed of 140 km/h; (**e**): Graphical comparison of the braking time during a dry road test at a vehicle speed of 140 km/h; (**f**): Graphical comparison of the vehicle rotation in z-axis during a dry road test at a vehicle speed of 140 km/h.

In the second test, we observed the braking behavior of the vehicle on the road with different adhesion on the right and left side. In this test, in addition to monitoring the shortest braking distance, the ABS's ability to maintain the vehicle in a straight line when braking was also monitored. Again, as in the previous test, the reference values were obtained by simulating a conventional ABS. As can be seen from the comparison shown in Table 10 and Figure 18, the ABS with the fuzzy controller achieved better results also in this case.

Table 10. Comparison of parameters when braking on road with combined adhesion.

Test on Dry Road	Max. Rotation in *Z*-Axis (deg.)	Braking Time (s)	Braking Distance (m)
100 km/h			
ABS	3.210	9.730	143.230
Fuzzy ABS	2.140	9.600	141.780
Without ABS	873.240	10.500	150.520
140 km/h			
ABS	141.290	15.550	263.830
Fuzzy ABS	2.830	12.430	247.560
Without ABS	1372.300	13.230	272.210

reference value, better value, worse value.

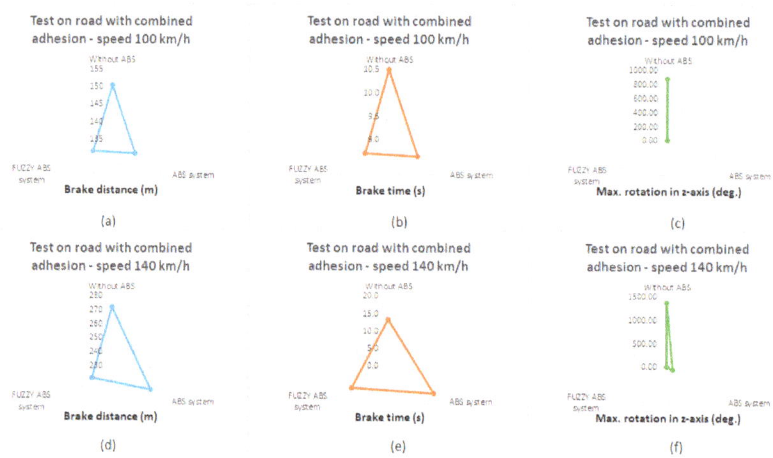

Figure 18. Graphical comparison of the test on road with combined adhesion. (**a**): Graphical comparison of the braking distance during a test on a road with combined adhesion at speed of 100 km/h; (**b**): Graphical comparison of the braking time during a test on a road with combined adhesion at speed of 100 km/h; (**c**): Graphical comparison of the vehicle rotation in z-axis during a test on a road with combined adhesion at speed of 100 km/h; (**d**): Graphical comparison of the braking distance during a test on a road with combined adhesion at speed of 140 km/h; (**e**): Graphical comparison of the braking time during a test on a road with combined adhesion at speed of 140 km/h; (**f**): Graphical comparison of the vehicle rotation in z-axis during a test on a road with combined adhesion at speed of 140 km/h.

In the last test, we observed the behavior of the vehicle during braking and simultaneously avoiding an obstacle in the roadway. In this test, we monitored the vehicle's ability to maintain within the prescribed path, and the reference value in this test was the value of the maximum deviation from the prescribed path for a vehicle with a conventional ABS. As Table 11 and Figure 19 demonstrates, although the vehicle without ABS had the shortest braking distance at speed 130 km/h, this vehicle was not able to avoid the obstacle and continued straight on, which would eventually cause the vehicle to collide with the obstacle.

Table 11. Comparison of parameters during avoiding an obstacle on the road.

Test on Dry Road	Max. Deviation of the Vehicle (m)	Braking Time (s)	Braking Distance (m)
100 km/h			
ABS	0.200	6.950	107.520
Fuzzy ABS	0.010	6.580	103.030
Without ABS	3.760	6.780	105.100
140 km/h			
ABS	0.340	8.800	172.660
Fuzzy ABS	0.190	8.580	172.430
Without ABS	3.520	8.150	162.930

reference value, better value, worse value.

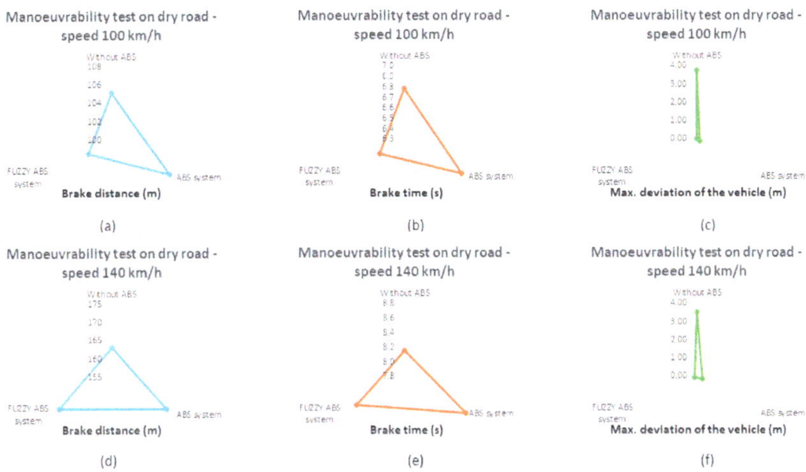

Figure 19. Graphical comparison of the maneuverability test on road. (**a**): Graphical comparison of the braking distance during a manoeurability test on a dry road at speed of 100 km/h; (**b**): Graphical comparison of the braking time during a manoeurability test on a dry road at speed of 100 km/h; (**c**): Graphical comparison of the vehicle deviation from the prescribed road during a manoeurability test on a dry road at speed of 100 km/h; (**d**): Graphical comparison of the braking distance during a manoeurability test on a dry road at speed of 140 km/h; (**e**): Graphical comparison of the braking time during a manoeurability test on a dry road at speed of 140 km/h; (**f**): Graphical comparison of the vehicle deviation from the prescribed road during a manoeurability test on a dry road at speed of 140 km/h.

Based on the previous results, we made a percentage comparison of the improvement in the performance of the ABS with the fuzzy controller compared to the classic ABS. The result of this comparison is shown in Table 12.

Table 12. Percentage comparison of the improvement of key indicators.

Monitored Value	Improvement of the Measured Parameter (%)	
	100 km/h	140 km/h (130 km/h)
Test on dry road—Braking distance	3.91	3.14
Test on road with combined adhesion—Braking distance	1.01	6.17
Test on road with combined adhesion—Max. rotation in z-axis	33.33	98.00
Maneuverability test on dry road—Max. deviation of the vehicle	95.00	44.12

It is shown in Table 11 that by using a fuzzy controller for the vehicle's ABS, we can achieve up to a 6% reduction in braking distance when braking on a road with combined adhesion. In the simulations, however, we achieved much more significant results in the directional stability of the vehicle or when avoiding an obstacle.

6. Conclusions

In this paper, we presented the design of an ABS with a fuzzy controller. Integrating fuzzy control into a conventional ABS has resulted in achieving improved maneuverability of the vehicle when braking on different surfaces and in different situations at high speed. As can be seen from the simulation results presented in the discussion, in all tests, the ABS with a fuzzy controller achieved better results than the conventional ABS. Comparing the key parameters in Table 12, we can see that by using a fuzzy controller, we achieved a reduction in the braking distance compared to the conventional ABS by 6% in the test on the road with combined adhesion. We achieved significantly better results mainly in the directional stability of the vehicle during the test on a road with a combined surface and the test of maneuverability.

Author Contributions: P.G. conceived and wrote the paper; P.G., J.Ž. and J.K. methology; P.G. designed the experiments; P.G., J.Ž. and J.K. analysed the experimental data; P.G., J.Ž. and J.K. proposed the theory. All authors have read and agreed to the published version of the manuscript.

Funding: This work was supported by the Slovak Research and Development Agency under the contract No. APVV-15-0750. This work was supported by the Slovak Research and Development Agency under the contract No. APVV-16-0206. The authors wish to thank the project VEGA 1/0187/18 for its support.

Conflicts of Interest: The authors declare no conflicts of interest.

References

1. Minh, V.; Oamen, G.; Vassiljeva, K.; Teder, L. Development of Anti-lock Braking System (ABS) for Vehicles Braking. *Open Eng.* **2016**. [CrossRef]
2. Aly, A.; Zeidan, E.; Hamed, A.; Salem, F. An Antilock-Braking Systems (ABS) Control: A Technical Review. *Intell. Control Autom.* **2011**, 186–195. [CrossRef]
3. Liu, H.; Lei, Y.; Fu, Y.; Li, X. An Optimal Slip Ratio-Based Revised Regenerative Braking Control Strategy of Range-Extended Electric Vehicle. *Energies* **2020**, *13*, 1526. [CrossRef]
4. Guo, J.; Jian, X.; Lin, G. Performance Evaluation of an Anti-Lock Braking System for Electric Vehicles with a Fuzzy Sliding Mode Controller. *Energies* **2014**, *7*, 6459–6476. [CrossRef]
5. Sun, J.; Xue, X.; Cheng, K.W.E. Fuzzy Sliding Mode Wheel Slip Ratio Control for Smart Vehicle Anti-Lock Braking System. *Energies* **2019**, *12*, 2501. [CrossRef]
6. Cabrera, J.A.; Ortiz, A.; Castillo, J.J.; Simon, A. A fuzzy logic control for antilock braking system integrated in the imma tire test bench. *IEEE Trans. Veh. Technol.* **2005**, *54*, 1937–1949. [CrossRef]
7. Keshmiri, R.; Shahri, A.M. Intelligent ABS Fuzzy Controller for Diverse Road Surfaces. *World Acad. Sci. Eng. Technol.* **2007**, *29*, 292–297.
8. Layne, J.R.; Passino, K.M.; Yurkovich, S. Fuzzy learning control for antiskid braking systems. *IEEE Trans. Control Syst. Technol.* **1993**, *1*, 122–129. [CrossRef]
9. Lennon, W.K.; Passino, K.M. Intelligent control for brake systems. *IEEE Trans. Control Syst. Technol.* **1999**, *7*, 188–202. [CrossRef]
10. Mauer, G.F. A fuzzy logic controller for an ABS braking system. *IEEE Trans. Fuzzy Syst.* **1995**, *3*, 381–388. [CrossRef]
11. Ayush, A.K.; Amitosh, K.; Sridevi, S.; Venkateswaran, K. ABS using Fuzzy Logic in MATLAB and Its Hardware Implementation. *Int. J. Recent Technol. Eng.* **2019**, *8*, 1007–1010.
12. Modi, D.; Padia, Z.; Patel, K. Fuzzy logic Anti-lock brake system. *Int. J. Sci. Eng. Res.* **2012**, *3*, 1–8.
13. Yin, G.; Wang, S.; Jin, X. Optimal Slip Ratio Based Fuzzy Control of Acceleration Slip Regulation for Four-Wheel Independent Driving Electric Vehicles. *Math. Probl. Eng.* **2013**, 1–7. [CrossRef]
14. Chen, C.K.; Shih, M.C. PID-Type Fuzzy Control for Anti-Lock Brake Systems with Parameter Adaptation. JSME International Journal Series C Mechanical Systems. *Mach. Elem. Manuf.* **2004**, *47*, 675–685. [CrossRef]

15. Raesian, N.; Khajehpour, N.; Yaghoobi, M. A New Approach in Anti-lock Braking System (ABS) Based on Adaptive Neuro-Fuzzy Self-tuning PID Controller. In Proceedings of the 2nd International Conference on Control, Instrumentation and Automation (ICCIA), Shiraz, Iran, 27–29 December 2011; pp. 530–535. [CrossRef]
16. Pretagostini, F.; Ferranti, L.; Berardo, G.; Ivanov, V.; Shyrokau, B. Survey on Wheel Slip Control Design Strategies, Evaluation and Application to Antilock Braking Systems. *IEEE Access* **2020**, *8*, 10951–10970. [CrossRef]
17. Lonkwic, P.; Syta, A. Nonlinear analysis of braking delay dynamics for the progressive gears in variable operating conditions. *J. Vibroeng.* **2016**, *18*, 4401–4408. [CrossRef]
18. Wolszczak, P.; Lonkwic, P.; Cunha, A.; Litak, G.; Molski, S. Robust optimization and uncertainty quantification in the nonlinear mechanics of an elevator brake system. *Meccanica* **2019**, *54*, 1057–1069. [CrossRef]
19. Day, A. *Braking of Road Vehicles*; Elsevier: Oxford, UK, 2014.
20. Ribbens, W. *Understanding Automotive Electronics—An Engineering Perspective*, 7th ed.; Elsevier: Oxford, UK, 2012.
21. Stone, R.; Ball, J.K. *Automotive Engineering Fundamentals*; SAE International: Warrendale, PA, USA, 2004.
22. Aksjonov, A.; Vodovozov, V.; Augsburg, K.; Petlenkov, E. Design of regenerative anti-lock braking system controller for 4 in-wheel-motor drive electric vehicle with road surface estimation. *Int. J. Automot. Technol.* **2018**, *19*, 727–742. [CrossRef]
23. Pacejka, H. *Tire and Vehicle Dynamics*, 3rd ed.; Elsevier: Oxford, UK, 2012.

© 2020 by the authors. Licensee MDPI, Basel, Switzerland. This article is an open access article distributed under the terms and conditions of the Creative Commons Attribution (CC BY) license (http://creativecommons.org/licenses/by/4.0/).

energies

MDPI

Article

Impact of Intersection Control on Battery Electric Vehicle Energy Consumption

Kyoungho Ahn [1], Sangjun Park [2] and Hesham A. Rakha [3,*]

[1] Virginia Tech Transportation Institute, 3500 Transportation Research Plaza, Blacksburg, VA 24061, USA; kahn@vt.edu
[2] Department of Civil Engineering, Chosun University, 309 Pilmun-Daero, Dong-Gu, Gwangju 61452, Korea; spark@chosun.ac.kr
[3] Department of Civil and Environmental Engineering, Virginia Polytechnic Institute and State University, 3500 Transportation Research Plaza, Blacksburg, VA 24061, USA
* Correspondence: hrakha@vt.edu; Tel.: +540-231-1500

Received: 22 May 2020; Accepted: 16 June 2020; Published: 19 June 2020

Abstract: Battery electric vehicle (BEV) sales have significantly increased in recent years. They have different energy consumption patterns compared to the fuel consumption patterns of internal combustion engine vehicles (ICEVs). This study quantified the impact of intersection control approaches—roundabout, traffic signal, and two-way stop controls—on BEVs' energy consumption. The paper systematically investigates BEVs' energy consumption patterns compared to the fuel consumption of ICEVs. The results indicate that BEVs' energy consumption patterns are significantly different than ICEVs' patterns. For example, for BEVs approaching a high-speed intersection, the roundabout was found to be the most energy-efficient intersection control, while the two-way stop sign was the least efficient. In contrast, for ICEVs, the two-way stop sign was the most fuel-efficient control, while the roundabout was the least efficient. Findings also indicate that the energy saving of traffic signal coordination was less significant for BEVs compared to the fuel consumption of ICEVs since more regenerative energy is produced when partial or poorly coordinated signal plans are implemented. The study confirms that BEV regenerative energy is a major factor in energy efficiency, and that BEVs recover different amounts of energy in different urban driving environments. The study suggests that new transportation facilities and control strategies should be designed to enhance BEVs' energy efficiency, particularly in zero emission zones.

Keywords: battery electric vehicles; signalized intersections; traffic signal control

1. Introduction

The objective of this study is to quantify battery electric vehicles' (BEVs') energy consumption for different intersection control types, including a roundabout, a traffic signal, and a two-way stop sign using a microscopic traffic simulation model. The findings of this study can be used in developing vehicle-specific routing strategies that favor or avoid specific types of intersections depending on the vehicle technology. The market penetration rate of BEVs has significantly increased in recent years. The International Energy Agency reported that in 2018, 1.98 million BEVs were sold worldwide [1]. During the first 6 months of 2019, global sales of BEVs increased by 92% [2].

Compared to ICEVs, BEVs have significant benefits, including low maintenance costs, zero emissions, energy savings, and the ability to use electricity generated from renewable resources. Furthermore, BEVs can produce regenerative braking energy to improve energy efficiency. Specifically, in a BEV's regenerative braking system, the electric motor works as a generator by sending energy from the vehicle's wheels to the electric motor and storing it in the battery system.

Previous studies demonstrated that BEVs have different energy consumption patterns than ICEVs. An ICEVs' fuel efficiency is typically maximized under constant speed highway driving conditions, whereas BEVs are more energy efficient in intermittent driving conditions, as these conditions allow the regenerative braking system to recover more energy [3]. A previous study also investigated the fuel and energy consumption associated with various driving cycles for both BEVs and ICEVs [4]. For a test ICEV (a Nissan Versa), the "LA92" and "New York" cycles were the worst fuel economy driving cycles, whereas the "Freeway Level of Service A-C" and "Freeway Level of Service D" cycles were the best fuel economy cycles. In contrast, for the test BEV (a Nissan Leaf), the "Area" and "Ramp" cycles utilized more electricity than the other driving cycles, and the "Freeway Level of Service E" and "Freeway Level of Service F" cycles were the most energy efficient. These results suggest that BEVs' energy consumption patterns should be systematically compared to ICEVs' patterns.

A number of studies have investigated the effects of traffic controls on ICEVs and developed optimum traffic signal controls to reduce fuel consumption and emissions. Varhelyi investigated the effects of small roundabouts on emissions and fuel consumption [5]. Coelho et al. [6] utilized a vehicle-specific power method to evaluate fuel consumption and emissions impacts on roundabout intersections in Raleigh, North Carolina and in Lisbon, Portugal. Researchers from Kansas State University utilized video-recorded traffic flow through the roundabout to extract speed, acceleration, deceleration, and other characteristics, and evaluated the emission impacts of roundabouts using SIDRA software (SIDRA SOLUTIONS, Balwyn, Australia) [7]. Ahn et al. [8] investigated the impacts of fuel consumption and emissions on a high-speed roundabout using a traffic simulation model. Kwak et al. [9] investigated the effects of traffic signal optimization on fuel and greenhouse gas emissions in an urban corridor and optimized traffic signal timing plans based on vehicle fuel consumption using a genetic algorithm. Park et al. [10] introduced a sustainable traffic signal control system and demonstrated that the system could reduce fuel consumption and pollutants with moderate increases in vehicle delays and number of stops. Stevanovic et al. [11] developed a fuel optimum traffic control system by integrating the Comprehensive Modal Emission Model and Vissim-based genetic algorithm optimization of signal timings with the Vissim traffic simulation model. Michel et al. [12] investigated the potential impacts of connected and automated vehicles on fuel consumption and found that the connectivity removed some stops and strong accelerations and increased overall vehicle speed, concluding that the best fuel consumption levels for hybrid electric vehicles and BEVs are achieved with connectivity.

While the above studies investigated the effects of traffic control strategies on ICEV fuel consumption and emissions, most did not consider BEVs. Thus, a systematic analysis of the effects of various intersection controls and signal plans on BEVs' energy consumption is needed. To address this gap in knowledge, this study aimed to quantify BEVs' energy consumption for different intersection controls—a roundabout, traffic signal, and two-way stop sign—and compare BEV's energy and ICEV's fuel consumption patterns. The study also investigated the effects of various traffic signal coordination planned BEVs' energy consumption. In attempt to reduce greenhouse gas emissions, more BEVs will be utilized in the near future and some cities may adopt zero emission zones where only BEVs may be operated. The results of this study can be utilized by researchers, transportation practitioners, and politicians to make decisions related to alternative intersection control strategies for BEVs in the near future.

According to the definition given by the US Department of Energy, plug-in electric vehicles include BEVs (e.g., Nissan Leaf) and plug-in hybrid electric vehicles (PHEVs). The latter include blended/parallel vehicles (e.g., Toyota Prius Plug-In) along with extended-range electric vehicles (EVs) (e.g., Chevy Volt) which have a variety of configurations. In this study, EVs refer to battery-only EVs (i.e., BEVs), since a Nissan Leaf was utilized as a test vehicle.

The contributions of the study include the following: (a) we attempted to identify the impacts of various traffic controls on ICEV and BEV fuel/energy consumption; (b) we used a microscopic analysis tool to evaluate instantaneous ICEV and BEV fuel/energy consumption; and (c) while most other studies have utilized simplified BEV energy models which consider an average regenerative braking energy efficiency or a regenerative braking factor that depends on vehicle speed, this study utilized a BEV energy model that can estimate the instantaneous energy consumed (kWh) and the instantaneous energy regenerated (kWh) required to accurately capture the regenerative braking energy.

The remainder of this paper is organized as follows. The next sections present the methodology utilized in the study; the energy model for BEVs and the fuel consumption model for ICEVs; and the simulation model and results. The discussion and future works are summarized in the final section.

2. Methodology

This study evaluated the effects on energy/fuel efficiency of two factors. First, the effects of different intersection controls (i.e., a roundabout, a traffic signal, and a stop sign) were investigated for both BEV and ICEV energy/fuel consumption. Second, the effects of fixed-time traffic signal control strategies with various signal coordination plans were assessed by evaluating BEV and ICEV energy/fuel consumption levels in a corridor with three intersections using three signal coordination plans: a well-coordinated plan, a partially coordinated plan, and a poorly coordinated plan.

The study required two main tasks: (1) the simulation or measurement of vehicle driving patterns for different intersection control types; and (2) measurements and/or estimates both BEV and ICEV energy/fuel consumption. Typically, field measurements and simulation are used to identify vehicle driving patterns. The use of a probe vehicle is a popular method to record vehicle driving patterns; however, it is difficult to collect the speed profiles of all approaching vehicles for different intersection control strategies. Thus, this study used the INTEGRATION software (Hesham Rakha, Blacksburg, VA, USA), a microscopic traffic simulation software, to obtain realistic representations of individual vehicles for different traffic control types. INTEGRATION software was validated against standard traffic flow theory and has also been used to evaluate various intelligent transportation system applications [13].

The energy consumption of BEV and the fuel consumption of ICEV can be measured with on-board measurement devices or dynamometer testing. While field measurements are relatively accurate in measuring the BEV's battery state-of-charge (SOC) and ICEV fuel consumption, data collection is limited to test vehicles equipped with measurement devices. This limits the usefulness of this method, because it is difficult to recruit test vehicles, install the data collection equipment, and collect sufficient field data. Thus, this study employed a microscopic BEV energy model to estimate BEV energy consumption and a microscopic fuel consumption model to estimate ICEV fuel consumption.

A number of BEV energy models and ICEV fuel consumption models use an average speed as an input variable. However, this approach is not suitable for the purposes of this study. In particular, the average speed-based models cannot identify the effects of transient changes in a vehicle's speed and acceleration which have significant effects on energy/fuel consumption as demonstrated in earlier studies [14]. Significant and frequent speed changes are observed while approaching and traversing intersection controlled approaches including roundabouts, traffic signals, and stop signs [8]. Thus, this study utilized microscopic energy/fuel consumption models to estimate the vehicle energy/fuel consumption for various scenarios. These models are described in more detail below.

2.1. BEV Energy Model

The study utilized the Virginia Tech Comprehensive Power-Based EV Energy Consumption Model (i.e., VT-CPEM), to estimate BEV energy consumption [4]. The VT-CPEM model is a microscopic, power-based EV energy model developed to estimate BEVs' instantaneous energy consumption using vehicle operational variables, including instantaneous speed, acceleration, and grade information. The model estimates the instantaneous energy consumed (kWh), the instantaneous energy regenerated (kWh), and the final SOC of the electric battery (%). The VT-CPEM has a simple structure that allows it to be implemented in other modeling tools, including in-vehicle/smartphone applications and microscopic traffic simulation models. One of the major advantages of using VT-CPEM is that it can capture instantaneous braking energy regeneration, which is not available in most BEV energy models. Most other BEV energy models instead use an average regenerative braking energy efficiency [15] or a regenerative braking factor that depends on the vehicle's speed [16]. However, these models cannot accurately capture the instantaneous regenerative braking energy required in this study.

Using VT-CPEM, the power at the electric motor was computed from the power at the wheels, assuming a driveline efficiency of 92% and an electric motor with an efficiency of 91%. The model also considered the power consumed by the auxiliary systems (700 W) [17]. The VT-CPEM was validated against independent data, producing an average error of 5.9% [18]. A detailed description of VT-CPEM is found in Reference [4].

2.2. ICEV Fuel Consumption Model

The study utilized the VT-Micro model to estimate ICEVs' fuel consumption. The VT-Micro model uses instantaneous speed and acceleration as explanatory variables and estimates instantaneous vehicle fuel consumption and emissions. The model consists of the following form:

$$MOE_e = \begin{cases} \exp\left(\sum_{i=0}^{3} \sum_{j=0}^{3} L^e_{i,j} \times u^i \times a^j \right) & \forall\, a \geq 0 \\ \exp\left(\sum_{i=0}^{3} \sum_{j=0}^{3} M^e_{i,j} \times u^i \times a^j \right) & \forall\, a < 0 \end{cases} \tag{1}$$

where $L^e{}_{i,j}$ and $M^e{}_{i,j}$ denote coefficients for measure of effectiveness (MOE) including fuel consumption, HC, CO, NOx, and CO_2 emissions at speed exponent i and acceleration exponent j; u is instantaneous speed; and a is instantaneous acceleration rate.

The VT-Micro model utilizes a number of data sources, including the Environmental Protection Agency and Oak Ridge National Laboratory (ORNL). Due to the simplicity and accuracy of the model, the VT-Micro model is one of the most popular microscopic fuel and emission models. A detailed description of the model is found in Reference [19].

3. Microscopic Traffic Simulation Modeling and Results

INTEGRATION software, which has been utilized and validated for a number of traffic control applications, was used in the simulations in this study. The model provides a reasonable assessment of how the studied intersection functions.

3.1. Analysis of Alternative Intersection Control Strategies: Roundabout, Traffic Signal, and Stop Sign Control

Driving behavior and deceleration/acceleration events were modeled at a roundabout, a two-way stop sign, and a signalized intersection using INTEGRATION software. The intersection of Ariane Way and Virginia 606 in Loudoun County, Virginia—which is adjacent to the Washington Dulles Airport (Figure 1)—was used as the case study site.

The test intersection is typically used as a substitute route for airport travelers. The speed limit is 88 km/h for eastbound and westbound and 40 km/h for northbound and southbound. Northbound and southbound traffic is controlled by two-way stop signs. The average afternoon peak hour traffic volumes are shown in Figure 1. Virginia 606 is a four-lane corridor with an extra left-turn lane, and Ariane Way has two lanes. While traffic volume is low during non-peak hours, travel time and delays are significantly increased at the two-way, stop-controlled intersection during peak hours. The typical queue length at Ariane Way is from 10 to 15 vehicles and those vehicles make for unsafe gap-acceptance maneuvers through the high-speed approaching vehicles from eastbound and westbound directions.

We developed simulation models using parameters derived from field data, including speed, traffic volume, saturation flow rate, jam density, number of lanes, and lane striping data. We modeled a base saturation flow rate and a jam density of 1800 veh/h/lane and 120 veh/km/lane, respectively, for all links except the northbound approaches. Due to the aggressive driving behavior of the northbound approaches, we used a saturation flow rate of 2000 veh/h/lane. We set the base gap between 3 and 4 s to simulate the driver gap behaviors. We calibrated and validated the simulation model against collected field data. For signalized intersection scenarios, we designed two phase movements with a 35 s cycle length, which was estimated based on the traffic demand. For roundabout scenarios, we used an entry speed of 50 km/h and a diameter of 60 m, which were recommended in *Roundabouts: An Information Guide* [20].

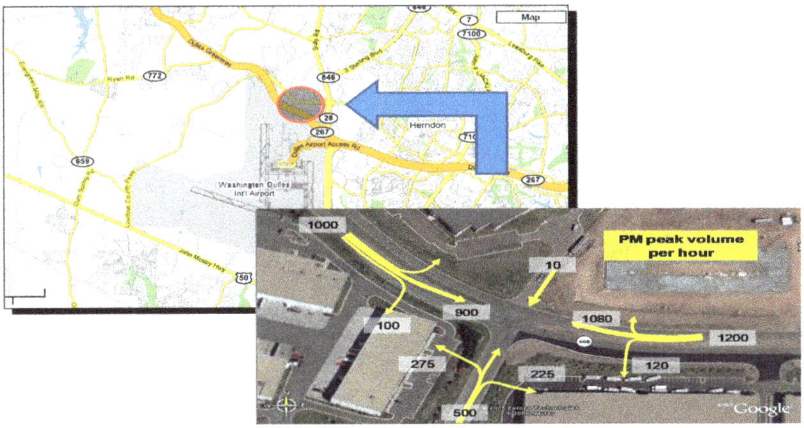

Figure 1. Map and aerial view of the modelled intersection (Source: Google Maps).

The BEVs' energy consumption was estimated using VT-CPEM from individual second-by-second profiles of vehicle speed that were generated using the INTEGRATION simulation. The ICEV fuel consumption was generated directly from the INTEGRATION software, since INTEGRATION includes the VT-Micro model. Three approach speeds, 56 km/h (35 mph), 72 km/h (45 mph), and 88 km/h (55 mph) were evaluated to identify the effects of different approach speeds.

Figure 2 compares the simulated ICEV fuel consumption and BEV energy consumption at a roundabout, a signalized intersection, and a two-way stop-sign-controlled intersection for an approach speed of 88 km/h. BEVs and ICEVs clearly had different energy/fuel consumption patterns. We converted ICEV fuel consumption to a comparable energy consumption unit, kWh based on Reference [21]. Based on the report, one liter of gasoline contains the energy equivalent to 8.9 kWh of electricity. The fuel consumption of ICEV increased by 126% and 79% when the stop-sign-controlled intersection was replaced with a roundabout and a signalized intersection, respectively. In contrast, for BEVs, the roundabout reduced energy consumption by 2.3% and 8.4% compared to the stop-sign-controlled and signalized intersections, respectively. The roundabout decreased total delay by 68% (from 18.81 to 6.07 s/veh) compared to the stop-sign-controlled intersection and by 45% (from 8.78 to 6.07 s/veh) compared to the signalized intersection. These results demonstrate that, for the modeled intersection, a roundabout would improve BEVs' energy efficiency, whereas ICEVs would have better fuel economy at a two-way stop-controlled intersection.

Figure 3 illustrates BEV and ICEV energy/fuel consumption for different intersection controls with an approach speed of 72 km/h. For ICEVs, the results were similar to those obtained at the approach speed of 88 km/h; the stop-sign-controlled intersection was the most fuel efficient, while the roundabout led to the worst fuel economy. However, for BEVs, the stop-sign-controlled intersection was slightly more energy efficient than the other two control strategies, although the energy consumption was similar for all three control types (0.15, 0.15, and 0.14 kWh/veh for the roundabout, signalized intersection, and stop-sign-controlled intersection, respectively). The roundabout considerably reduced vehicle delay by 46% and 78% compared to the signalized and stop-sign-controlled intersections, respectively.

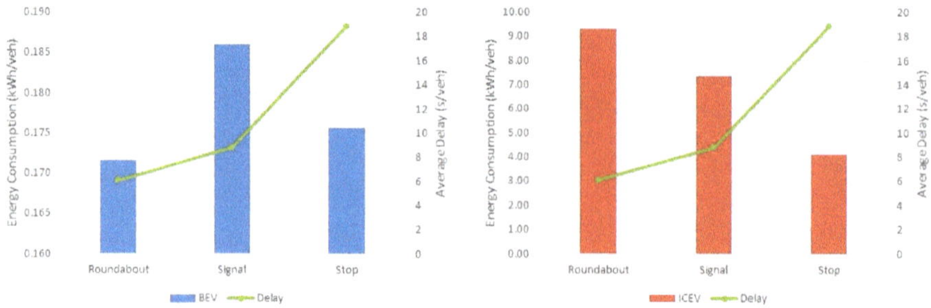

Figure 2. Comparison of battery electric vehicle (BEV) and internal combustion engine vehicle (ICEV) energy/fuel consumption for different intersection control strategies (approach speed = 88 km/h).

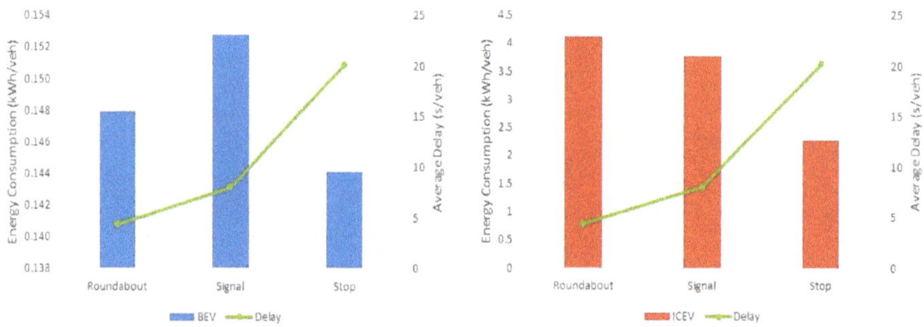

Figure 3. Comparison of BEV and ICEV energy/fuel consumption for different intersection control strategies (approach speed = 72 km/h).

The BEV and ICEV energy/fuel consumption levels at the different intersections are shown in Figure 4 for an approach speed of 56 km/h. For ICEVs, the stop-sign-controlled intersection was the most fuel efficient, while the signalized intersection was the least fuel efficient. For BEVs, the energy efficiencies were nearly identical (0.12 kWh/veh) for all three control strategies. As for the other approach speeds, the roundabout was the most efficient intersection control type in terms of vehicle delay.

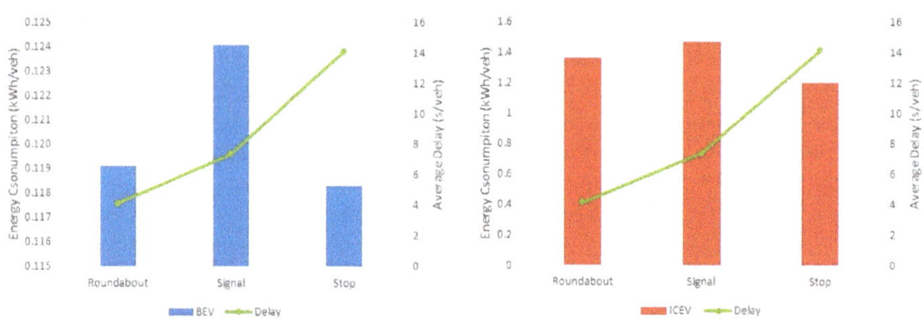

Figure 4. Comparison of BEV and ICEV energy/fuel consumption for different intersection control strategies (approach speed = 56 km/h).

The simulation results for all evaluated approach speeds show that the roundabout significantly reduced vehicle delay compared to the signalized and stop-sign-controlled intersections. For ICEVs, the stop-sign-controlled intersection significantly reduced fuel consumption compared to the other two control methods at all three approach speeds, and the roundabout resulted in the worst fuel efficiency at approach speeds of 88 and 72 km/h. The ICEVs' higher fuel consumption in the roundabout resulted from the acceleration of vehicles leaving the roundabout. A vehicle entering a roundabout must yield or stop before entering the roundabout; thus, after negotiating the roundabout, the vehicle must accelerate to full speed upon exiting, increasing the rate of fuel consumption. Acceleration rate is one of the most important contributors to vehicle fuel consumption [19,22].

Compared to ICEVs, BEVs' energy consumption patterns were significantly different. For example, at an approach speed of 88 km/h, the roundabout was the most energy-efficient intersection, whereas the stop-sign-controlled intersection resulted in the lowest electricity usage at the approach speed of 72 km/h. Nonetheless, the differences in BEV energy consumption between the different intersection types were relatively small compared to the fuel consumption of ICEVs. These results can be explained by the energy BEVs regenerate during deceleration.

Figure 5 illustrates the average recovered energies corresponding to the three intersection controls at different approach speeds. The regenerative energy produced increased with increasing approach speed. For the roundabout, BEVs generated 193% more regenerative energy at an approach speed of 88 km/h compared to an approach speed of 56 km/h. Figure 5 also demonstrates that BEVs recovered more energy at the roundabout compared to the signalized and stop-sign-controlled intersections. Table 1 shows the percentage of regenerated energy compared to total BEV energy consumption. For example, BEVs recovered 32.9% of the total energy through regenerative energy at the roundabout with an approach speed of 88 km/h. These results indicate that regenerative energy is critical to BEVs maintaining their energy efficiency at roundabouts.

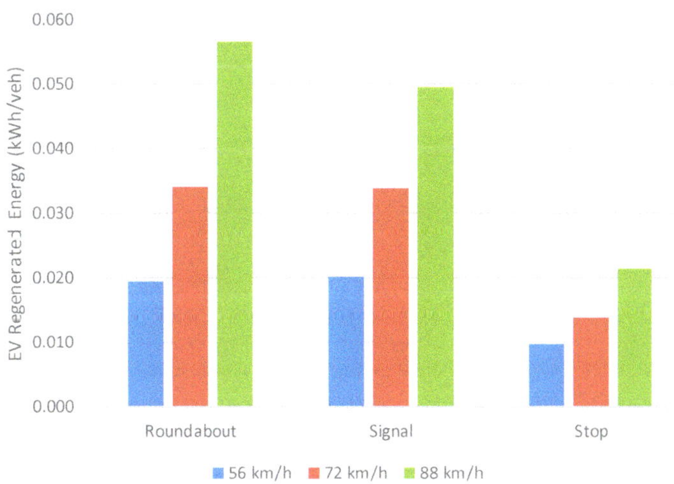

Figure 5. BEV's regenerative energy generated for different intersection control strategies and approach speeds.

Table 1. Percentage of regenerated energy compared to total BEV energy consumption.

	Roundabout	**Signalized**	**Stop-Sign-Controlled**
56 km/h approach speed	16.2%	16.3%	8.0%
72 km/h approach speed	22.9%	22.1%	9.6%
88 km/h approach speed	32.9%	26.6%	12.2%

3.2. Analysis of Various Traffic Signal Coordination Plans

The effects of various signal coordination plans on BEV and ICEV energy/fuel consumption were also investigated. A previous study found that good traffic signal coordination can significantly reduce fuel consumption and ICEV emissions [14]. Figure 6 illustrates a sample corridor with three intersections and four links. Each link was 0.5 km in length which is a reasonable intersection length in an urban area. The demand from the start node to the end node was 1200 veh/h. The last vehicle injected into the simulation departed 15 min from the beginning of the simulation. A free speed of 80 km/h was applied to the entire corridor. This scenario was meant to show how a signal coordination plan can affect the energy/fuel consumption in a specific corridor. To evaluate the effects of signal coordination in this study, three scenarios were adopted: (1) a poorly coordinated fixed-time signal plan, (2) a partially coordinated fixed-time signal plan, and (3) a well-coordinated fixed-time signal plan. Each traffic signal had a 60 s cycle length, 30 s effective green time, and a 2.5 s effective lost time.

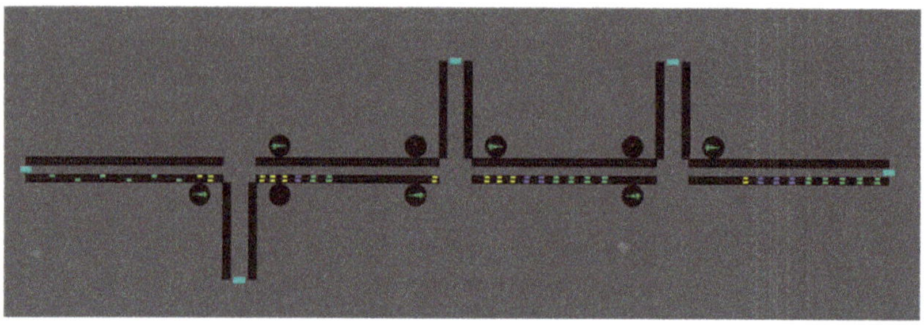

Figure 6. A sample coordinated corridor.

Figure 7 illustrates the effects of different signal coordination plans on BEV and ICEV energy/fuel consumption and vehicle delay. As expected, the well-coordinated signal plan significantly reduced vehicle delay from 72.64 s to 15.51 s compared to the poorly coordinated signal plan. The well-coordinated signal plan also reduced vehicle delay by 26% compared to the partially coordinated signal plan. Figure 7 also shows that the well-coordinated signal plan improved energy/fuel efficiency in both ICEVs and BEVs. In particular, ICEVs' fuel consumption was reduced by 28% and 35% compared to the partially coordinated and poorly coordinated signal plans, respectively, while BEVs' energy efficiency was improved by 17% and 20%, respectively. These results indicate that BEVs' energy consumption was less sensitive to signal coordination than ICEVs' fuel consumption, since BEVs produced more regenerative energy on the corridors with partially and poorly coordinated signal plans. The BEVs recovered 21%, 40%, and 49% of total energy through regenerative braking for the well-coordinated, partially coordinated, and poorly coordinated signal plans, respectively, as illustrated in Figure 8.

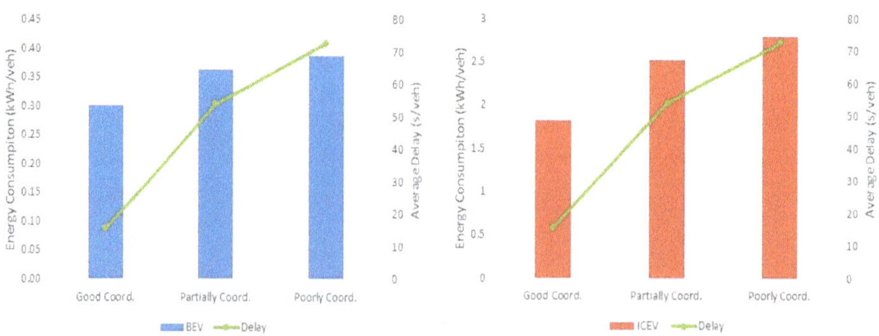

Figure 7. BEV and ICEV energy/fuel consumption for different signal coordination plans.

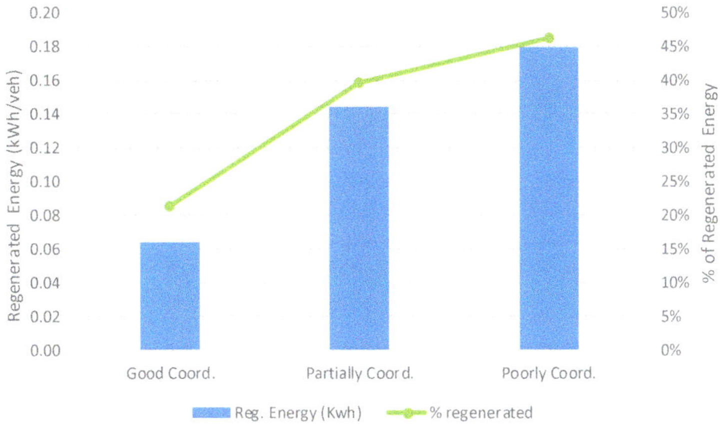

Figure 8. Regenerated energy of BEV for different signal coordination plans.

4. Conclusions and Recommendations for Future Research

This study investigated the effects of different intersection controls (i.e., a roundabout, traffic signal, and stop sign) and signal coordination plans (i.e., well-coordinated, partially coordinated, and poorly coordinated) on BEV and ICEV energy/fuel consumption. The second-by-second speed profiles of individual vehicles were derived from a traffic simulation model, and the speed profiles were utilized as inputs to estimate ICEV fuel consumption and BEV energy consumption using microscopic fuel consumption and energy models.

The results indicate that the most energy/fuel efficient traffic controls are different for BEVs and ICEVs. For BEVs approaching an intersection at a speed of 88 km/h, the roundabout was the most energy-efficient intersection control, while the stop sign was the least energy efficient. In contrast, for ICEVs at the same approach speed, the two-way stop sign was the most fuel-efficient control, while the roundabout was the least fuel efficient. The BEV and ICEV energy/fuel consumption patterns also differed at other approach speeds (72 and 56 km/h). For BEVs, the energy consumption at all intersection control types was similar, while for ICEVs, stop-sign-controlled intersections significantly reduced fuel consumption compared to the other two intersection types.

The ICEVs' higher fuel consumption levels at roundabouts resulted from acceleration leaving the roundabout, as acceleration rate is a major contributor to vehicle fuel consumption. The BEVs' energy consumption levels were less affected by the intersection control strategy than were ICEVs'. This can be attributed to the energy regenerated by BEVs during deceleration; BEVs recovered 32.9% of the total energy through regenerative braking at the roundabout for an approach speed of 88 km/h, thereby improving overall energy efficiency.

The effects of different signal coordination plans on energy/fuel consumption were also investigated for BEVs and ICEVs. Both BEV and ICEV energy/fuel efficiency was improved by the well-coordinated signal plan. Signal coordination had a weaker effect on energy efficiency in BEVs compared to ICEVs, because BEVs produced more regenerative energy on corridors with partially and poorly coordinated signal plans. The study found BEVs recovered 21%, 40%, and 49% of total energy through regenerative braking on the corridors for well-coordinated, partially coordinated, and poorly coordinated signal plans, respectively.

In summary, this study demonstrated that regenerative energy in BEVs is a critical factor in energy efficiency. Among the tested intersection controls, BEVs recovered the largest amount of energy at the roundabout for the studied intersection site. The market penetration rate of BEVs has increased sharply in recent years, and the number of BEVs on the road is expected to continue to increase in the near future. Recently, a number of transportation researchers and engineers have developed fuel-efficient transportation facilities and control strategies to improve fuel economy and reduce ICEV emissions. The findings of this study suggest that transportation facilities and control strategies should be designed to enhance BEVs' energy efficiency, particularly in zero-emission zones [23], where only BEVs can be utilized. Furthermore, the findings of this study can be used in the development of energy-efficient routing strategies. For example, the routing of BEVs might entail sending the vehicles through roundabouts, whereas the routing for ICEVs might send the vehicles through two-way stop sign intersections.

Future research will investigate the effects of combinations of various vehicle types, different demand levels, and more facility types on BEVs' energy consumption. The development of a multimodal traffic signal control system that can improve both BEV and ICEV energy/fuel efficiencies is also recommended.

Author Contributions: The authors confirm contributions to the paper as follows: study conception and design, K.A., S.P.; simulation model development, K.A., S.P.; analysis and interpretation of results, K.A., H.A.R.; draft manuscript preparation, K.A., H.A.R. All authors reviewed the results and approved the final version of the manuscript.

Funding: This work was funded by the Department of Energy through the Office of Energy Efficiency and Renewable Energy (EERE), Vehicle Technologies Office, Energy Efficient Mobility Systems Program under award number DE-EE0008209.

Conflicts of Interest: The authors declare no conflict of interest.

Abbreviations

BEV	Battery Electric Vehicle
EV	Electric Vehicle
ICEV	Internal Combustion Engine Vehicle
PHEV	Plug-In Hybrid Electric Vehicle
SOC	State-of-Charge
VT-CPEM	Virginia Tech Comprehensive Power-Based Energy Model

References

1. International Energy Agency (IEA). Tracking Transport. 2019. Available online: https://www.iea.org/reports/tracking-transport-2019 (accessed on 1 May 2020).
2. Munoz, F. Global sales of pure electric vehicles soar by 92% in H1 2019. 2019. Available online: https://www.jato.com/global-sales-of-pure-electric-vehicles-soar-by-92-in-h1-2019/ (accessed on 1 May 2020).
3. De Gennaro, M.; Paffumi, E.; Scholz, H.; Martini, G. GIS-driven analysis of e-mobility in urban areas: An evaluation of the impact on the electric energy grid. *Appl. Energy* **2014**, *124*, 94–116. [CrossRef]
4. Fiori, C.; Ahn, K.; Rakha, H.A. Power-based electric vehicle energy consumption model: Model development and validation. *Appl. Energy* **2016**, *168*, 257–268. [CrossRef]
5. Várhelyi, A. The effects of small roundabouts on emissions and fuel consumption: A case study. *Transp. Res. Part D Transp. Environ.* **2002**, *7*, 65–71. [CrossRef]
6. Coelho, M.C.; Farias, T.L.; Rouphail, N. Effect of roundabout operations on pollutant emissions. *Transp. Res. Part D Transp. Environ.* **2006**, *11*, 333–343. [CrossRef]
7. Mandavilli, S.; Rys, M.J.; Russell, E.R. Environmental impact of modern roundabouts. *Int. J. Ind. Ergon.* **2008**, *38*, 135–142. [CrossRef]
8. Ahn, K.; Kronprasert, N.; Rakha, H.A. Energy and Environmental Assessment of High-Speed Roundabouts. *Transp. Res. Rec. J. Transp. Res. Board* **2009**, *2123*, 54–65. [CrossRef]
9. Kwak, J.; Park, B.; Lee, J. Impacts of urban traffic signal timing optimization on energy and GHG emissions. In Proceedings of the 2010 IEEE International Symposium on Sustainable Systems and Technology, Arlington, VA, USA, 17–19 May 2010; Institute of Electrical and Electronics Engineers (IEEE): Piscataway, NJ, USA, 2010; p. 1.
10. Park, B.B.; Yun, I.; Ahn, K. Stochastic Optimization for Sustainable Traffic Signal Control. *Int. J. Sustain. Transp.* **2009**, *3*, 263–284. [CrossRef]

11. Stevanovic, A.; Stevanovic, J.; Zhang, K.; Batterman, S. Optimizing Traffic Control to Reduce Fuel Consumption and Vehicular Emissions Integrated Approach with VISSIM, CMEM, and VISGAOST. *Transp. Res. Rec.* **2009**, *2128*, 105–113. [CrossRef]
12. Michel, P.; Karbowski, D.; Rousseau, A. Impact of Connectivity and Automation on Vehicle Energy Use. In Proceedings of the SAE 2016 World Congress and Exhibition, Detroit, MI, USA, 12–14 April 2016.
13. Van Aerde, M.; Rakha, H. *INTEGRATION © Release 2.40 for Windows: User's Guide – Volume I: Fundamental Model Features*; M. Van Aerde & Assoc., Ltd.: Blacksburg, VA, USA, 2013.
14. Rakha, H.; van Aerde, M.; Ahn, K.; Trani, A.A. Requirements for evaluating traffic signal control impacts an energy and emissions based an instantaneous speed and acceleration measurements. *Energy Air Qual. Fuels* **2000**, *1738*, 56–67. [CrossRef]
15. Fleurbaey, K.; Omar, N.; El Baghdadi, M.; Timmermans, J.-M.; Van Mierlo, J. Analysis of Hybrid Rechargeable Energy Storage Systems in Series Plug-In Hybrid Electric Vehicles Based on Simulations. *Energy Power Eng.* **2014**, *6*, 195–211. [CrossRef]
16. Zhou, M.; Gao, Z.; Zhang, H. Research on regenerative braking control strategy of hybrid electric vehicle. In Proceedings of the 2011 6th International Forum on Strategic Technology, Harbin, China, 22–24 August 2011; Volume 1, pp. 300–303.
17. Szadkowski, B.; Chrzan, P.J.; Roye, D. A study of energy requirements for electric and hybrid vehicles in cities. In Proceedings of the 2003 International Conference on Clean, Efficient and Safe Urban Transport, Gdansk, Poland, 4–6 June 2003.
18. U.S. Department of Energy (USDOE). Advanced Vehicle Testing Activity (AVTA) of the Idaho Nation Laboratory (INL). 2013. Available online: http://avt.inel.gov/pdf/fsev/fact2013nissanleaf.pdf (accessed on 30 June 2015).
19. Ahn, K.; Rakha, H.A.; Trani, A.; Van Aerde, M. Estimating Vehicle Fuel Consumption and Emissions based on Instantaneous Speed and Acceleration Levels. *J. Transp. Eng.* **2002**, *128*, 182–190. [CrossRef]
20. Robinson, B.; Rodegerdts, L.; Scarborough, W.; Kittelson, W.; Troutbeck, R.; Brilon, W.; Bondzio, L.; Courage, K.; Kyte, M.; Mason, J.; et al. *Roundabouts: An Information Guide*; Kittelson & Associates Inc.: McLean, VA, USA, 2000; p. 284.
21. Canada, N.R. Understanding the tables. 2020. Available online: https://www.nrcan.gc.ca/energy-efficiency/energy-efficiency-transportation/personal-vehicles/choosing-right-vehicle/buying-electric-vehicle/understanding-tables/21383 (accessed on 15 June 2020).
22. Rakha, H.A.; Ahn, K.; Trani, A. Development of VT-Micro model for estimating hot stabilized light duty vehicle and truck emissions. *Transp. Res. Part D Transp. Environ.* **2004**, *9*, 49–74. [CrossRef]
23. Müller, J.; Petit, Y.L. *Low-Emission Zones are A Success -But They Must Now Move to Zero-Emission Mobility*; Transport & Environment: Brussels, Belgium, 2019.

© 2020 by the authors. Licensee MDPI, Basel, Switzerland. This article is an open access article distributed under the terms and conditions of the Creative Commons Attribution (CC BY) license (http://creativecommons.org/licenses/by/4.0/).

energies

MDPI

Article

Total Cost of Ownership Model and Significant Cost Parameters for the Design of Electric Bus Systems

Anders Grauers [1,*], Sven Borén [2] and Oscar Enerbäck [3]

1 Department of Electrical Engineering, Chalmers University of Technology, SE-412 96 Gothenburg, Sweden
2 Blekinge Institute of Technology, TISU, SE-37179 Karlskrona, Sweden; sven.boren@bth.se
3 Rise Research Institutes of Sweden, Lindholmspiren 3A, SE-41756 Gothenburg, Sweden; oscar.enerback@ri.se
* Correspondence: anders.grauers@chalmers.se; Tel.: +46-31-772-3729

Received: 14 May 2020; Accepted: 17 June 2020; Published: 24 June 2020

Abstract: Without experiences of electric buses, public transport authorities and bus operators have faced questions about how to implement them in a cost-effective way. Simple cost modelling cannot show how costs for different types of electric buses differ between different routes and timetables. Tools (e.g., HASTUS, PtMS, and optibus) which can analyse such details are complicated, time consuming to use, and provide insufficient insights into the mechanisms that influence the cost. This paper therefore proposes a method for how to calculate total cost of ownership, for different types of electric buses, in a way which can predict how the cost varies based on route and timetable. The method excludes factors which cause minor cost variations in an almost random manor, in order to better show the fundamental mechanisms influencing different costs. The method will help in finding ways to reduce the cost and help to define a few cases which deserve a deep analysis with more complete tools. Testing of the method in a Swedish context showed that the results are in line with other theoretical and practical studies, and how the total cost of ownership can vary depending on the variables.

Keywords: electric bus; cost model; total cost of ownership; TCO; charging strategy; public transport; sustainability

1. Introduction

Electric cars and buses has been proposed by several studies and authorities as a long-term solution for the sustainable development of transport systems, mainly because of their high efficiency, very low emissions when being driven, low noise levels, and the possibility of using renewable sources for their electricity (e.g., [1–3]). Several predictions estimate that electric vehicles will dominate the sales within the next decade (e.g., BloombergNEF believe electric cars will dominate after 2036, and electric buses after 2030 [4] and IEA believe that sales of EVs will be 70% in 2030 [5]) as the price for batteries is likely to decrease and governmental incentives are likely to increase to support such development of the transport sector. A commonality in countries with incentives to tackle climate-change issues related to transport (e.g., taxes on fossil fuels) is that the current approximately 50% higher purchase price for electric cars can be compensated with a much lower price for electricity per km. For example, in Sweden in 2019, the VW e-Golf had a lower total cost of ownership (TCO) after three years (accumulated TCO after five years) than a comparable VW Golf powered by fossil fuels with a mileage of 15 km per year [6]. However, TCO is different for cars and buses, because buses usually have a higher use rate and longer mileage (a bus is normally driven 4–8 times more kilometres than a car). However, bus operators have been reluctant to use electric buses in their operations, mainly because of higher purchase costs and a lack of knowledge regarding how to design, operate, and maintain electric bus systems [7]. Contradictory, a study by Borén [1] summarized several studies about electric buses in Sweden and showed that electric buses can reduce the total cost of ownership for bus operators by more than 10% over a 10-year period, as well as societal costs due to low life-cycle emissions and noise levels when compared to buses powered by diesel and gas (methane). This depends on country-specific conditions, and a recent study in Texas (USA) show that electric buses can become cost competitive in about 5 years [8], while a study in India show that electric buses can be cost competitive within 25 years [9], and a study in Turkey showed that electric buses have twice as long pay-back time (almost six years) than diesel buses [10]. Since the greenhouse gas emissions from electricity production varies a lot from country to country, electric buses could actually emit more greenhouse gas emissions in total than diesel buses if the electricity used for electric buses is produced with high carbon intensity, e.g., electricity mix in Malta, Poland, Latvia, and Estonia [11]. Some of these emissions can be linked to the production of batteries if there is an extensive use of energy that stem from fossil resources (e.g., oil, coal, and natural gas) [12]. This can, however, be compensated by subscriptions or shares in (or establishment of) new facilities for electricity produced from renewable sources. However, the efforts to reduce climate change and other sustainability impacts cannot only address the transport system, leaving the electricity sector to continue to use fossil fuels. The increase in renewable electricity production and a decline in the cost of it will likely make it possible to achieve a rather quick transition towards sustainable electricity production. Electric buses will then have a minor contribution to climate change and other emissions.

While there can be many environmental and health-related advantages to switch from fossil-fuelled buses to electric, it is important to make that transition relatively easy and cost-efficient to get bus-operators (and taxpayers in the long run) onboard. There have been analyses completed and models/tools designed that focus on charging systems' design and costs [13], location of charging infrastructure [14], costs and sustainability for electric buses [1], life-cycle environmental impacts [12], and procurement processes [15]. There are also several commercial tools (e.g., HASTUS, PtMS, and Optibus [16–18]) used by public transport authorities and bus operators for the calculation of costs related to bus traffic, but without the integration of electric bus systems. Based on that, the authors of this paper have identified a need to focus on modelling and analysis of the total cost of operating electric buses that are charged either at the bus depot or/and along the bus route. What complicates the search for the most cost-effective electric bus system is that the cost of different types of bus systems changes with route properties and timetables.

1.1. Different Types of Electric Buses

Electric buses have a very low operating cost compared with conventional buses, but a higher investment cost of the battery and chargers, and sometimes additional cost of the driver waiting during charging and for extra buses required due to the charging time. It is not possible to minimize all these cost at the same time, so there is a need to find a cost-effective compromise, which depend on the timetable and bus route properties, and that is why several different charging strategies are relevant for analysis.

Charging at the end stops means that the buses can have smaller batteries than buses charged at a bus depot, as they can charge after each trip. This can be cost effective as long as there are many departures per hour, allowing the chargers to be frequently used. The chargers are used very little if the bus route has low bus traffic density, and a low utilization of the chargers increases the cost per trip kilometre. One of the drawbacks of charging at the end stop after each trip is that there is a need for extra buses to have time to charge.

The operators want to minimize the number of buses for a bus line since the investment in buses is a major cost driver. That is why the second charging strategy in this paper, end-stop charging off-peak, is included. All buses then drive during peak times without charging. Between the morning and afternoon peak times, typically 09:00 to 14:00, all buses are not needed in traffic, so then they can charge. This will reduce the number of buses, but it will also require bigger batteries onboard since the buses need to be able to drive for about three hours without charging. It is not obvious which timetables and bus routes that could make one or the other charging strategy more cost effective.

1.2. Aim of the Paper

This paper addresses the needs and challenges described earlier. There is not one solution of electric buses which can always be assumed to be the best, and there are many factors which influence when a certain type of electric bus is cost effective or not. The goal of this paper is to present a cost model which can model the main mechanisms that influence the costs of electric buses when different routes and timetables are compared. Besides calculating costs, the model can also help explain why and how different factors influence the costs, and thus make it easier to find ways to reduce costs. The model can be used for many types of electric buses, but in this paper, it is only used to analyse electric buses mainly charged at the end stops. The costs for electric buses are also compared with buses powered by biofuels (gas and diesel) to find which type of routes of electric buses with end-stop charging are most cost effective.

1.3. Limitations

To better serve the purpose of identifying the underlying mechanisms, rather than giving an exact conclusion of a specific route, several simplifications are made. For example, the Total Cost of Ownership (TCO) model can calculate the cost of a non-integer number of buses. Such simplifications make it much easier to identify some general trends in cost changes and explain what causes them, but they also mean that the model is not intended to use when conducting the final and detailed cost analysis on a route. Rather, its purpose is to help determine which types of buses are interesting to investigate for a specific route, while the final analysis should be made with a more detailed bus-planning and cost analysis tool. The focus of the paper is to present the cost model, while analysis of bus routes is included only as examples of how the model can be used. The cost results presented should therefore not be seen as representative for all bus routes. The parameter values used in the examples are for a Swedish context and may need to be changed when analysing other bus routes.

The paper analyses the two charging strategies for electric buses: end-stop charging—when they are charged at the end stops on a route (included in the opportunity charging concept); and end-stop off-peak charging—buses charged at the end stops for the whole day except during the peak hours. The buses are also charged in the depot during night in order to be fully charged when they start operations. Both the electric bus charging strategies are compared with buses powered by biomethane or Hydrogenated Vegetable Oil (HVO), which is a biodiesel that can be used as a drop-in fuel in conventional diesel engines. In Sweden almost all buses run on biofuels, so we have not included diesel buses in the comparisons in Section 5. Diesel buses will have exactly the same costs as HVO buses, excluding the fact that the price of diesel fuel differs to that of HVO.

1.4. Structure of the Paper

Section 2 explains what the model calculates and in what steps. First, the method and core assumptions are presented. The modelling starts from a formula for calculating the TCO, and that formula is used to identify which intermediate variables influences the total cost. In Section 3, the parameters which are used to describe the bus route and timetable are presented, as are the basic cost parameters. Section 4 then shows how the intermediate variables can be calculated from the parameters for the route and timetable. The explanations of the charging strategies, and some assumptions related to them, are also found in Section 4, as deriving the model is closely linked to the explanation of the charging strategies. In Section 5, an analysis of end-stop-charged buses is presented, with the aim of explaining the mechanisms which influence the cost of running a route and how different types of routes and timetables influences their TCO. This analysis is intended to demonstrate what the model can be used for, rather than providing cost results which can be used for any bus route. Finally, Section 6 summaries the main findings in the paper, followed by a critical assessment and comparison with other studies. The very last part then focuses on how the paper contributes to the research community, and on recommended further work.

2. Model for Total Cost of Ownership

2.1. Method

The TCO model is built in Matlab [19], which is a software that integrates computation, visualization, and programming in an environment where problems and solutions are expressed in mathematical notation. The model is based on a direct step by step calculation of the results starting only from the parameters describing the route and the timetable, as well as the bus and cost parameters. This direct calculation method is possible since the calculations have been broken down in steps which can calculate their respective outputs while only knowing input parameters and intermediate results calculated in previous steps. All the equations have been solved analytically before creating the TCO calculation program. Furthermore, the order of the calculations has been carefully selected so that there is no need to feed results back to previous calculation steps, avoiding the need for numerical solvers in the program. This leads to a quick program that is easy to follow, despite having a significant number of steps in the calculations.

Another key part of the method is to avoid calculating more details than needed. This is done by, as far as possible, directly calculate energies and bus time for the whole fleet of buses, rather than for each individual bus. Furthermore, the smallest step in the calculation of a whole day of bus traffic is the single bus trip. No finer time step is analysed, which means that there is no need to simulate the individual buses, which also ensures a quick calculation.

Since the purpose is to find and model the general mechanisms which influence the TCO, it is important to only include the important factors and exclude all minor effects that will mainly show up as a noise in the calculation of the cost. The way we do this is to start from the end result, the TCO equation, and derive the model backwards from there. This ensures that we only include things that can influence the TCO, while all other aspects of buses will automatically be excluded. Later in this section, we also discuss how the model avoids factors which can cause small variations up and down in cost when timetable and route parameters only slightly change. These are therefore seen as noise, which is not a part of the overall trend in the cost variations.

2.2. Output of the Total Cost of Ownership Model

In this paper, the TCO will be presented either as a total cost per year for the investigated route, or as the route's total cost per trip kilometre. The cost per trip kilometre is the TCO per year divided by the total distance all the buses drive in service. Cost of driving outside the timetable, such as driving to or from the depot and driving between different routes, is included in both cost measures, but when the specific cost per km is calculated, the cost is only divided on the kilometers driven during the trips. This is important since driving to and from the depot adds nothing to the value of the route, while it adds to the cost of operating the route.

The total cost per year is obviously a good measure of how cost effective a certain bus type is for a specific route. However, it is not so useful for general conclusions and comparisons of different bus routes which have very different bus traffic density and different lengths. Then, the total yearly cost will be very different and does not clearly illustrate which system is more cost-effective. When comparing different routes and different bus traffic volume, it is often better to compare cost per trip (km) instead. Even if bus lines can have total yearly costs which are different by an order of magnitude, a comparison of their cost per trip (km) will be revealing. The difference in costs per trip (km) then indicates why one of the routes is more cost-effective than the other.

2.3. Parameters Used to Determine Total Cost of Ownership

The TCO is calculated as the sum of operating cost and investment-related costs. In this model, the operating cost is the sum of driver cost, energy cost, maintenance and insurance and electric grid fees. The cost of the investment is determined from the depreciation of the chargers, batteries and buses. The final calculation steps of the TCO are shown in Figure 1.

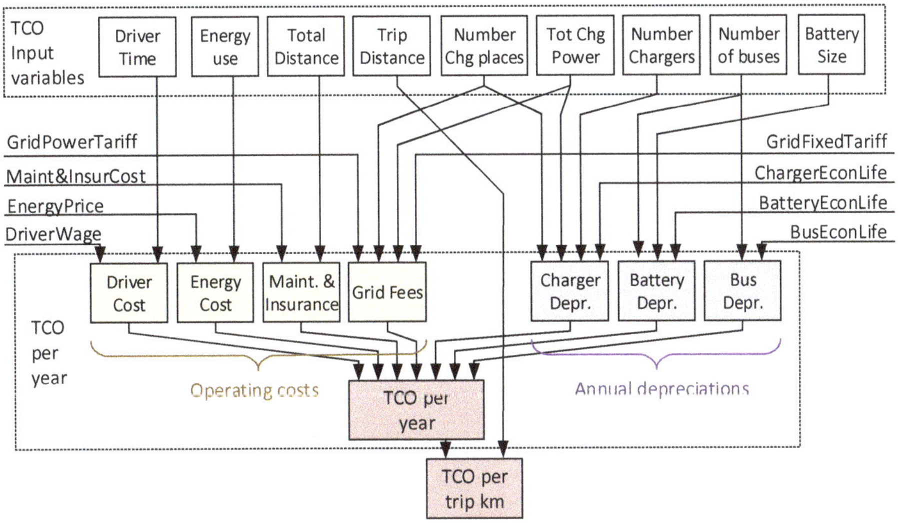

Figure 1. The calculation steps of the TCO model, its nine input variables in the white boxes at the top and its eight cost parameters as inputs from the left and right side of the model.

There are, of course, other costs for a route, such as the cost of depots (excluding chargers), ticket systems, etc. In this analysis it is assumed that such costs are the same for all the investigated bus types and therefore will not be important when comparing different bus types with each other. However, when cost differences between the alternatives are small, minor variations in these other costs can very well be what tips the scale in favour of one bus type or another, and that is why a final decision on what type of buses to use for a route should always be conducted based on a detailed bus-planning and cost analysis tool.

To determine these seven costs, the model use nine intermediate variables which are calculated for the analysed routes and timetables. These are:

- Driver time per year;
- Energy use per year;
- Total driven distance per year;
- Trip distance per year;
- Number of places with chargers (i.e., number of grid connections);
- Total combined power of all chargers;
- Number of chargers;
- Number of buses;
- Bus battery size.

These nine variables, together with eight cost parameters, determine the seven parts that make up the TCO. The TCO per year is the sum of the seven costs, and the TCO per trip kilometre is the TCO per year divided by the number of trip kilometres per year.

2.4. *Simplifications Aimed to Find General Trends Rather than Route-Specific Results*

There are some costs which vary only in steps, and these steps can make it difficult to see the general trend in the costs. For example, when the headway is varied, the number of buses needed for a route changes in steps of one. The exact headway at which the number of buses changes is not necessarily the same for the compared charging strategies. Comparing two types of buses, it may sometimes look as if one is more cost effective than the other, but with just a slight adjustment of the headway, the result can be the opposite. To avoid this, the cost model is defined to calculate as if it is possible to buy a non-integer number of buses, and by extension, a non-integer number of depot chargers. Normally several routes are driven by the same operator, and then the possibility to use a non-integer number of buses for a route is even more reasonable since buses and drivers can be shared between different routes, making it possible to plan the bus schedules so that buses can operate on several routes.

The number of end-stop chargers are, however, an integer in the model, as the step in the number of chargers will be much more important for the cost effectiveness of the different charging strategies. End-stop chargers are sometimes less well-utilized than depot chargers, and it is important that the model includes that effect, as it is a reason why the TCO for end-stop charging significantly varies between different routes. Furthermore, end-stop chargers can only support the routes that use the bus stop at which they are placed and can therefore not be shared as easily between routes as the buses can.

When planning bus schedules for a route, there is often a need for buses to be inactive and wait for the next departure. The need for such waiting time can vary in a very random way with changes in route properties and timetables. To avoid such "noise" influences on the TCO calculations, the TCO model does not create real bus schedules. Instead, the model just estimates the total number of buses needed in traffic, how many will be driving to and from the depot, and how many will be charging at different times during the day. This simplification is a feature and not a bug, since it allows for a clearer illustration of the system effects when analysing the total amount of buses occupied by different tasks rather than focusing on analysing the buses individually.

Another simplification is that the model does not keep track of the State of Charge (SoC) of individual batteries, but instead, the SoC of the buses are ensured by a few conditions regarding the size of the required batteries, and by determining how much charging is required in total to achieve energy balance of the fleet over the day. To allow for this simplification, the model assumes that the batteries are sized to handle some worst-case energy use that individual buses can experience between charges. Optimizing the battery size can further reduce costs, but this is not included in this version of the cost model.

2.5. Cost of Conventional Combustion Engine Buses

The TCO model is mainly developed to analyse electric buses, but it can also analyse conventional buses since they are less complex. As there is no need for conventional buses to be fuelled during the day, they only need to meet the requirement of driving their designated amount and the minimum number of trips to and from the depot. They need no extra time to charge or extra time to drive to and from the depot, as may be needed by electric buses. The TCO for them is calculated using the same formula, but with slightly different cost parameters, as shown in Table 1. Rather than calculating the volume of fuel consumed, the cost of the conventional buses fuel is calculated from required traction energy, the fuel cost per litre and the average fuel efficiency of the powertrain.

Table 1. The cost parameters for buses powered by HVO, biomethane, and electricity when charging at the end stop, and at the end stop only during off-peak time. The values are relevant for Sweden 2019, and based on results from pilot projects.

Cost Parameters	HVO	Biomethane	Electricity	
			End-Stop	End-Stop Off-Peak
Price (Million SEK)	2.2	2.5	3 (excl. battery)	
Battery capacity (kWh)	-	-	100	200
Max energy used between charging (kWh)	-	-	25	75
Maintenance including chargers (SEK/km)	3	3.6	3.3	3.3
Bus Economic Life (year)	10	10	10	10
Battery Economic Life (year)	-	-	7	7
Battery Price (SEK/kWh)	-	-	4000	4000
Energy Cost (SEK/kWh)	3.5	4	0.82	0.82

3. Model Input Parameters and Variables

The TCO model has many input variables which we use to describe the route and the timetable. Several of these are, later in this paper, varied to analyse how they influence the TCO. The parameters are factors which we do not vary in this analysis, but they are still needed to determine the TCO. The parameters describe important values which influence the cost of buses, batteries, chargers, drivers, and the electricity grid.

3.1. Route Variables

We do not need to know all details of the route but must know any property which influences the nine TCO variables. In this TCO model we selected to base the route description on the time it takes to drive a trip rather than how long the route is, since the required number of buses and driver time are both directly determined by the time required, rather than the distance driven. The route distance is also an input parameter, but most calculations are made based on analysing time. Then, only a few results are translated into driven distance when it is needed to calculate the TCO variables. The route properties are independent of the used timetable.

The main route variable is the net time it takes to drive one trip, which can vary over the day. In our model, we have different trip times in the off-peak period ($T_{\text{TripNetOffPeak}}$), in the peak period ($T_{\text{TripNetPeak}}$), and during the evening ($T_{\text{TripNetEvening}}$). We also need to know the time to drive from the depot to the route or back from the end stop to the depot ($T_{\text{PullInOut}}$). For simplicity reasons, we assume it to be the same time for both end stops. This is often not the case, but the given value can then be the average time for the two end stops. To determine the driven distance, we also need parameters for the trip distance (l_{Trip}), and the distance from depot to the route's end stops (l_{Depot}). If needed, the trip length and net trip time can be used to calculate the average speed of the bus.

3.2. Timetable Variables

In this cost model, we have assumed that the route and timetable are identical in both directions. A simple way of describing the timetable is shown in Figure 2. It shows a generic timetable description as a curve showing the number of departures per hour from the end stops and its variation during the day. In the diagram we can see that the timetable can be described by seven time-dependent variables and three variables for number of departures per hour. By changing these 10 variable values, the timetable can be altered in our TCO analysis. Since the buses often run into the night, beyond midnight, the calculation uses time values beyond 24 h, as this simplifies the calculations. In Figure 2, the last bus departs from the end stop at 01:00 in the night, and this is coded as 25.0 h in our model.

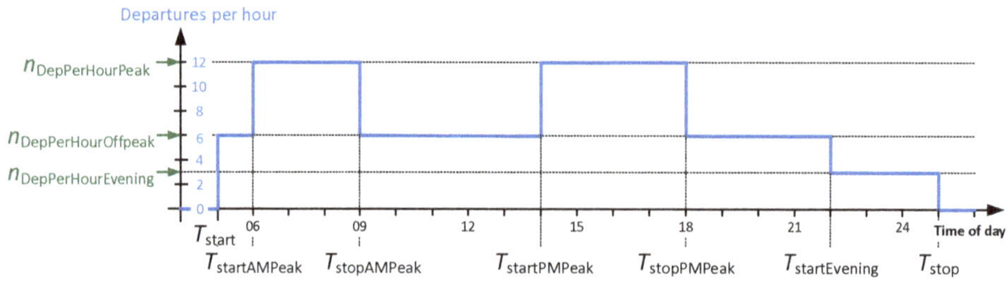

Figure 2. The generic way of describing the timetable which is used in the TCO model.

In addition to the variables in Figure 2, there are also variables to define the layover time needed off-peak ($T_{\text{LayoverOffPeak}}$), during peak times ($T_{\text{LayoverPeak}}$), and in the evening ($T_{\text{LayoverEvening}}$). The layover time is the time between when the bus arrives at the end stop from one trip, and the time it departs for the return trip. The layover time is mainly used as a buffer time so that a delayed incoming bus shall still mostly be able to depart on time on its next trip. The layover time can also provide some breaks for the driver between trips.

Instead of defining different timetables for the different types of days (weekdays, weekends, holidays etc.) the total traffic during a whole year is instead calculated as the defined typical day, multiplied with the number of effective traffic days which give an estimate of the total yearly traffic. In this paper we use $N_{\text{TrafficDays}} = 313$.

3.3. Bus, Driver and Battery Parameters

In this section, the parameters used to determine the cost of buses, driver, and batteries are described. In the examples in this paper, we assume 12 m buses, and they have the values given in Table 1. The different size of batteries for the two types of electric buses depends on different charging strategies. Why different battery sizes are used depends on the longest time a bus can be in traffic without charging, and this will be explained more in Section 4.1. A conservative estimation of battery cost has been used. The battery price is assumed to be constant, so it is the same after seven years when the batteries are replaced. The residual battery value after seven years has also been set to zero. Table 1 shows parameter values which are relevant for Sweden 2019, mainly based on data from pilot projects. The parameter values are included to allow the reader to check and interpret the numeric results in this paper. It shall, however, be noted that development of electric buses is rapid, and the production volumes are growing fast; therefore, these parameter values can change and should not be seen as generally applicable.

For all the buses, the average power during the trip has been assumed to be 25 kW, which is based on measured energy consumption of electric buses in Sweden. It includes auxiliary loads, heating and cooling, and is a typical average value over the year. The effect that the worst-case consumption will be higher has to be considered when sizing the batteries. The economic life of the bus (depreciation period) has been set to 10 years, while the batteries have been assumed to last seven years. In the near future, batteries will most likely last 10 years, but the first generation of bus batteries may live a little shorter.

The driver wage is set to 300 SEK/h, and the driver schedules are assumed to be planned so that 90% of the driver's time can be used to drive the bus and wait during layover time or wait during charging. This means that the effective driver cost will be 333 SEK/h that there is a driver in the bus. It is also assumed that the driver must be paid during the time the bus charge at end stops, but not when charging at the depot.

3.4. Electric Grid Parameters

There is a need to invest in one grid connection at each charger location, irrespective of how many chargers are in the same location. It is assumed that the chargers will require so much power from the grid that one new transformer or substation will have to be built for each charger location, including some new distribution lines. The cost of a new substation will depend on the total power of all the chargers in that location. The initial cost of building a substation is set to 1 Million (M) SEK and the total cost includes an additional 1000 SEK/kW. This means that a 500-kW substation will cost SEK 1.5 M SEK and a 2 MW substation 3 MSEK. These cost levels are based on dialogue with the local electric utility company in Gothenburg, Sweden, and assume that end-stop chargers are normally not built in the city centre where cost is often much higher. When calculating the depreciation of the grid investment, the economic life for the substations and grid connection has been set to 20 years.

There is also an annual fee for using the grid. This can be very different in different regions, and in this paper, it includes one fixed annual fee per substation of 5000 SEK per year plus an annual fee depending on the installed peak power of the chargers, which is 500 SEK/kW per year.

3.5. Charger Parameters

The chargers are assumed to have a base cost of 5000 SEK per charger plus a size-dependent cost of 3000 SEK/kW. The low base cost means that the cost is almost only proportional to the total installed power of the chargers. The charger depreciation is calculated using an economic life of 10 years. This is similar to a normal contract period with a bus operator in Sweden. These cost levels are based on data from pilot projects and estimates how that cost will be reduced when building many new bus chargers at once for a contract with many bus lines. The cost has also been found to be consistent with the cost of high-power charges for electric cars, which are based on the same technology.

The assumed charger power $P_{ChgEndstop}$ is 300 kW for the end-stop-charged buses, and $P_{ChgEndstopNight}$ is 11 kW per bus for the night chargers for buses with 100 kWh battery and 22 kW for buses with 200 kWh battery.

There is also the factor of how much of the layover time which, on average over several trips, can be used for charging at end-stop chargers ($k_{LayoverChargingFactor}$), and in this paper it is assumed to be 50%. Since the layover time is required to avoid delays, there are situations in which there will be some trips which are delayed so that there is no layover time to charge. However, the bus batteries are big enough so that one missed charge is not be a problem as long as the bus later during the day can compensate for that missed charging. That is a reason why it is assumed that some of the layover time may, on average, be used for charging.

3.6. Other Parameters

The TCO calculation also requires some other parameters. The interest rate is used to calculate the capital cost of the investments and it has been set to 3%. This is a low interest rate, but a city or government can often borrow at such low rate.

4. Calculating TCO Input Variables from Timetable and Bus Route Parameters

Figure 1 shows the nine variables needed to calculate the TCO of the buses. However, these in turn have to be calculated from the route and timetable, and this section shows the steps in which this is done. In Figure 3, the main steps are shown, and they are then described in the following sections.

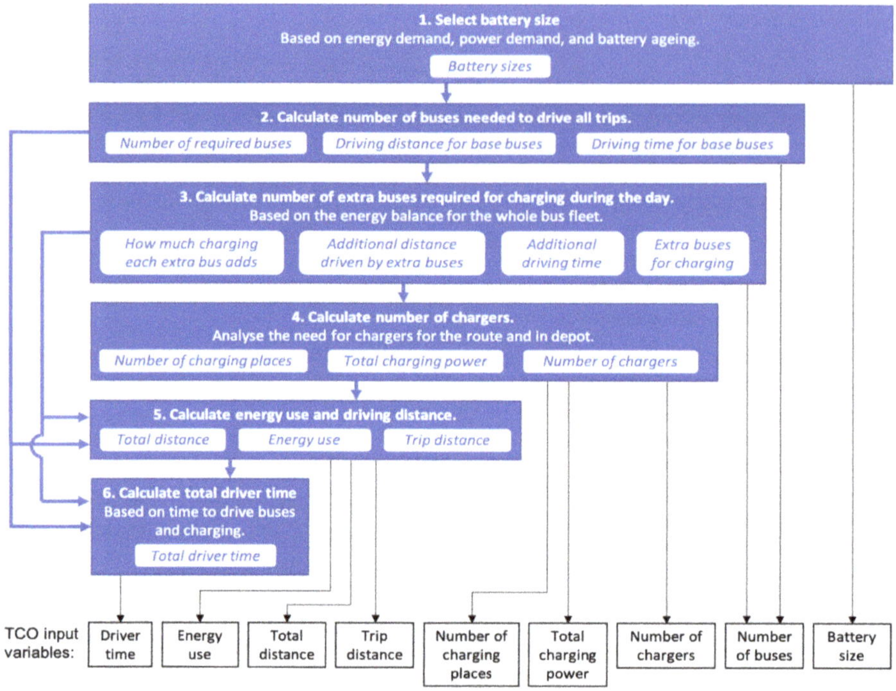

Figure 3. Steps 1–6 for how to calculate the nine TCO input variables, and information flow between these steps (to the left). The white boxes show some of the key variables calculated in different steps and the output variables.

4.1. Battery Size and Need to Charge during the Day

Depending on charging strategy, the buses will need to have different battery sizes. The size of the battery must be chosen to meet several requirements. The battery must have enough capacity (kWh) to supply the energy needed for the most demanding bus schedule and should also have some margin to handle disturbances, which may sometimes lead to a shortened or a completely missed charging.

Another size-related criterion is that the battery must have enough capacity to be able to deliver the necessary traction power and to handle the charger power. It is not possible to have a very small battery and discharge or charge at very high power. The fact that batteries in buses with end-stop charging must be capable of charging at high power is a reason for why they are assumed to be more expensive per kWh of stored energy. The higher cost is due to the battery cells being more expensive per kWh when they are optimized for high charging power and the battery system needing a more effective cooling system.

Finally, the battery must not wear out to quickly, and have margin so that it can still meet all the requirements for energy and power also when the battery has aged. Typically, the capacity of the battery is reduced by up to 20% when it reaches its end of life, but the maximum discharge and charge power will also be reduced when the battery ages and, as such, this also needs to be included when deciding the battery size for a bus.

For the end-stop-charged buses, the charge power and number of charge cycles will be the critical factors, and therefore it is assumed that a 100-kWh power-optimized battery is required, despite the fact that a trip typically only requires 25 kWh if it is one hour long.

For the buses which use end-stop charging but only off-peak, a 200-kWh power-optimized battery has been assumed. The peak traffic periods are up to about three hour long which requires about 75 kWh of energy.

The battery size could be optimized for each route and timetable, but it is deemed likely that the market will settle on a few battery sizes as this simplifies moving buses between different contracts and the offers possibility for the buses to have a second life if a contract is not renewed. Therefore, it is not likely that buses will be optimized for the route they operate on, rather, there will be a few standard battery sizes which the operator selects from. The optimal sizing of a battery is a very complex task and is not included in this paper.

4.2. Determining the Number of Buses Needed to Drive the Trips

The number of buses needed is calculated in two steps. First, the number of buses required to drive the trips are calculated. This will be determined by the highest number of buses in traffic during the peak periods, and it will be equal to the number of conventional buses required. After that, there is a calculation of how many extra buses are needed in order to have time to charge electric buses. This number can be zero or larger, depending on the timetable and charging strategy. The number of extra buses is calculated in the next section.

We start by looking in detail at each bus needed to drive the trips from one of the two end stops, and later we derive the formulas needed to calculate the number of buses from the detailed analysis. The use of the buses is illustrated in Figure 4. There, we can see that bus 1 starts the first trip at time T_{start} according to the timetable, and before that, it has used some time driving from the depot to the start of the route, illustrated by the light blue bar. bus 1 drive the first trip during the time shown by the green bar, and there is a need for layover time at the end of it. One headway time after bus 1, bus 2 starts the second trip, followed by bus 3 and 4 after each additional headway time. Thus, the number of buses initially increases by one bus for each headway time that passes. The increase in number of buses stops after the gross trip time, because at this point, the buses which have been driving the route in the other direction have arrived and had their layover time, and they are ready to drive the next trip as a return trip. Therefore, after the gross trip time, the number of buses in traffic does not need to be increased as long as the headway is constant. Figure 4 shows that the buses alternate driving the route in both directions, as indicated by the green and blue bars.

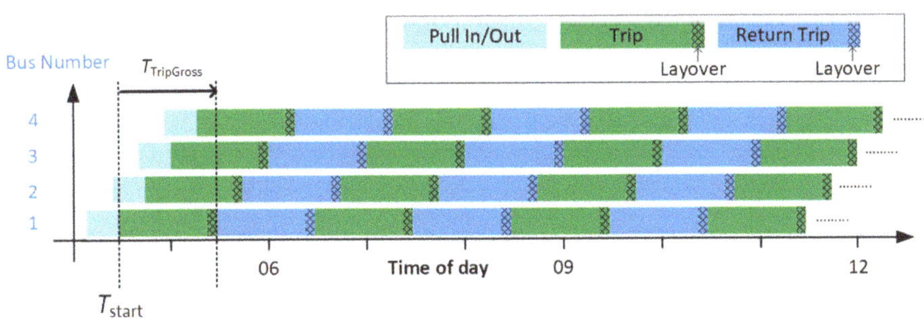

Figure 4. Example of buses needed to drive the trips during early morning traffic.

The headway is reduced during the morning rush hours after some time in the early morning. There are no longer enough of buses returning from earlier trips to start all the trips. If the headway during rush hour is half of the headway during the early morning, the number of buses will need to increase, as shown in Figure 5, in which the morning rush hours starts at 06:00 and ends at 09:00. Note that these times show when the headway changes for the departures from the end stop. Further down the line, the reduced headway will occur later, as it takes some time for the buses to drive from the end stop. Just like at the start of the traffic in early morning, there will be a need for more buses at the beginning of the rush hours. In this example, every second bus starting a trip from the end stop must be an additional bus coming from the depot. As before, the number of buses increases, now by one every second headway time. This continues for a time equal to the gross trip time when enough buses arrive from the other direction of the route.

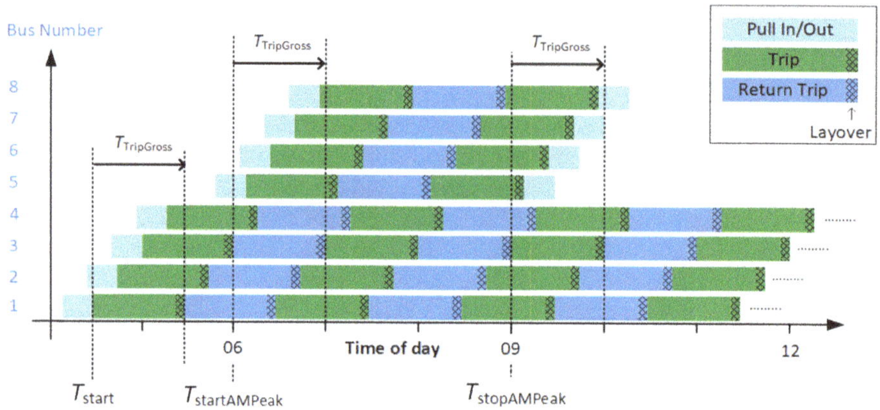

Figure 5. Example of buses needed to drive the trips during early morning traffic and the morning rush hours.

At the end of the rush hour, when the headway is increased, not all buses arriving from the other direction are needed, so after the end of the morning peak, some of the buses are taken out of traffic and return to the depot.

Based on the previous analysis, we can determine the number of buses needed in traffic during the whole day. Note that we previously showed which bus is driving which trip, so that each row in the diagram is the schedule for one particular bus during the day. In the following analysis, we will derive a diagram that looks very similar, but it only shows how many buses are occupied by different activities during the day, without showing which bus is doing what. This way, we can simplify the analysis a lot, and do not need to plan the schedules of the buses. On the other hand, this analysis cannot capture all the small details involved in planning bus schedules, and some of the details in the scheduling are instead included as factors to take into account that it is not possible to plan bus schedules completely without slack for the bus and drivers.

The number of buses required for the traffic will vary during the day, as shown in the diagram in Figure 6, and it is derived from the timetable and data regarding driving time and layover time for the route. We will later use this diagram to determine how much time is available for charging during different parts of the day. Right now, we only need to know the number of buses required to drive during the off-peak period, during the peak times and in the evening. Note that despite being similar to the timetable diagram in Figure 2, this shows the total number of busses in traffic, while the timetable diagram shows the frequency of departures. How many buses are needed will not only depend on the timetable but also on the time it takes to drive the route. A short route of course needs fewer buses to follow a certain timetable than a longer bus route with the same timetable.

The number of buses required for driving all trips during peak traffic:

$$N_{\text{busPeak}} = 2 \times T_{\text{TripGrossPeak}} \times n_{\text{DepPerHourPeak}}, \tag{1}$$

where the gross trip time in the peak is:

$$T_{\text{TripGrossPeak}} = T_{\text{TripNetPeak}} + T_{\text{LayoverPeak}} \tag{2}$$

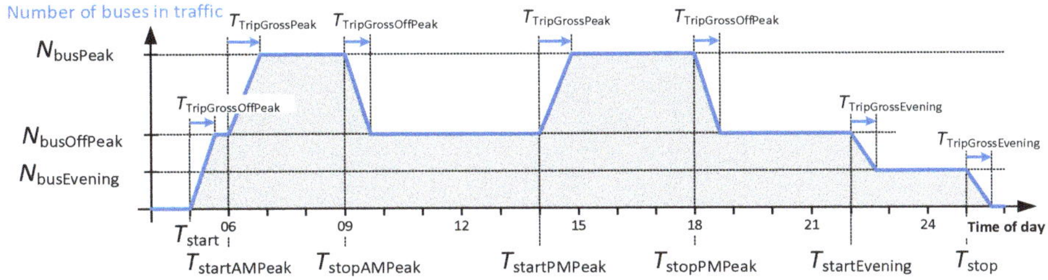

Figure 6. Number of buses in traffic during the day. Derived from the timetable and route parameters.

As stated earlier we do not round this off to the nearest higher integer, but instead analyse the TCO based on a non-integer number of buses. This way of calculating the number of buses assumes that the gross trip time is shorter than the peak periods. That is the case for most routes in cities, at least in Sweden, since the peak period in the morning and afternoon are typically 2 h and 3 h or more, respectively, while very few routes have more than a 2-h trip time. The number of buses in traffic during the midday off-peak period can be calculated in the same way:

$$N_{\text{busOffPeak}} = 2 \times T_{\text{TripGrossOffPeak}} \times n_{\text{DepPerHourOffPeak}}, \tag{3}$$

where the gross trip time off-peak is:

$$T_{\text{TripGrossOffPeak}} = T_{\text{TripNetOffPeak}} + T_{\text{LayoverOffPeak}} \tag{4}$$

Finally, the number of buses in traffic during the evening is:

$$N_{\text{busEvening}} = 2 \times T_{\text{TripGrossEvening}} \times n_{\text{DepPerHourEvening}}, \tag{5}$$

where the evening gross trip time is:

$$T_{\text{TripGrossEvening}} = T_{\text{TripNetEvening}} + T_{\text{LayoverEvening}} \tag{6}$$

4.3. *Determining the Number of Extra Buses to Provide Time to Charge*

We now know the number of base buses needed to drive the traffic, but there is also a need to provide time for the buses to charge, and that may require extra buses. Thus, in this section, we determine the number of extra buses needed for charging. Note that the biogas buses and HVO buses do not require any extra buses beyond the base buses. Besides night charging, which is assumed to allow the buses to start each day fully charged, we divide the charging in three categories to make it easier to build and understand the model. The categories of daily charging are:

- At the end stop between trips, aiming at restoring the state of charge to what it was before the last trip (red colour in Figures 7–10 below);
- Extra charging at the end stops, aiming at increasing the state of charge to a higher level than what it was before the last trip (purple colour in Figures 8 and 9 below).
- At the depot during the day. (green colour in Figures 8 and 9 below);

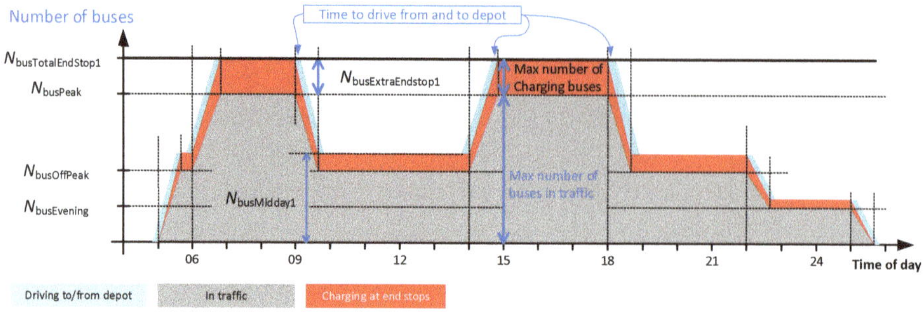

Figure 7. Number of buses, including buses charging at end stops and driving to and from the depot.

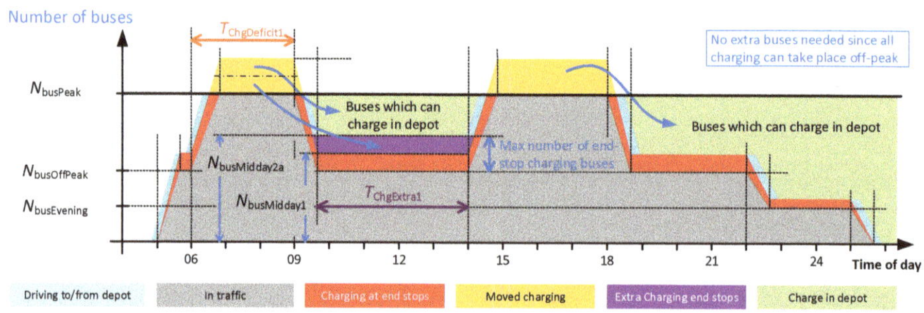

Figure 8. How the number of buses during peak traffic can be reduced by moving charging to other times.

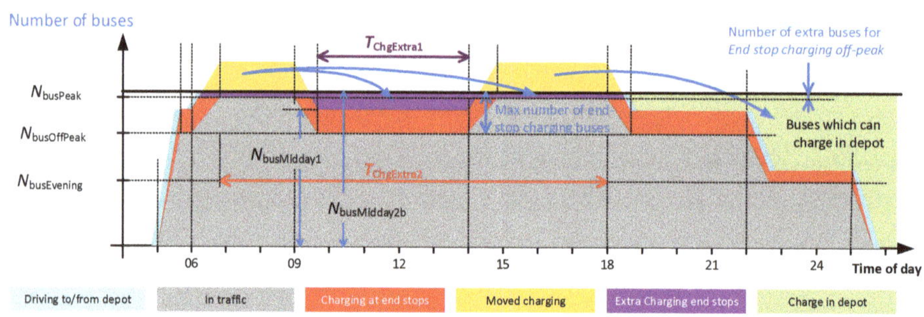

Figure 9. How the number of buses during peak traffic can be reduced by moving charging to other times, when the off-peak traffic is almost as high as the peak traffic.

4.3.1. Extra Buses for End-Stop Charging for a Whole Day (EndStop1)

With this charging strategy, the buses are charged after each trip, and it is called **EndStop1** in the calculations. The energy charged equals the energy used during the last trip, which means that the buses always starts each trip with the same battery state-of-charge. The number of extra buses required is determined by calculating how much time is required to charge the bus after each trip. The calculation is made for the peak periods, as that is when the greatest number of buses will be charging simultaneously. The amount of energy that the bus must charge at each end stop is:

$$W_{\text{TripPeak}} = T_{\text{TripNetPeak}} \times P_{\text{BusAv}} \tag{7}$$

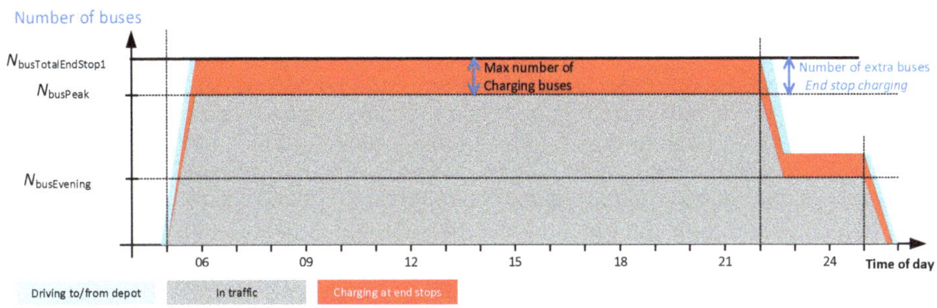

Figure 10. How the number of buses during peak traffic cannot be reduced if the off-peak traffic is the same as the peak traffic.

The charge time is:

$$T_{ChgPeakEndStop} = \frac{W_{TripPeak}}{P_{ChgEndStop}} \tag{8}$$

This charge time can be used to calculate a factor for how many buses need to be charging per bus in traffic:

$$k_{busChgPeakEndStop} = \frac{T_{ChgPeakEndStop}}{T_{TripGrossPeak}} \tag{9}$$

Now the number of extra buses required to allow for charging at end stops during peak times can be determined:

$$N_{BusExtraEndstop} = N_{BusPeak} \times k_{busChgPeakEndStop} \tag{10}$$

The number of buses charging off-peak and in the evening can be calculated with Equations (7)–(10) using the net trip time off-peak and in the evening. The extra buses required for charging at the end stops are illustrated by the red area in Figure 7, where the grey area is the buses in traffic from Figure 6. In the diagram, the time to drive from, or back to the depot is also shown as light blue segments. The longer the distance between depot and bus route, the longer the light blue segments will be.

From the diagram in Figure 7, we can see that the maximum number of buses used during the day will be during peak traffic:

$$N_{BusTotalEndStop1} = N_{BusPeak} + N_{BusExtraEndstop1} \tag{11}$$

Since the diagram also shows the highest number of buses which are simultaneously charging, it can also be used to determine the number of end-stop chargers required, which will be completed in a later section.

4.3.2. Extra Buses for End-Stop Charging during Off-Peak Time Only (EndStop2)

Since the highest number of buses during the day will determine how many buses must be bought, there is a possibility to save on the bus investment if it is possible to change when the buses charge so that fewer buses are needed during the peak times. This will occupy more buses off-peak, but that does not influence the investment in buses if it does not exceed the bus number in the peak times. The lowest number of buses required is the number of buses in traffic during the peak times, so the best we can do, in terms of reducing the number of buses, is to limit them to the highest number of buses in traffic during the peak times. If we do that, it means that no buses can charge during the peak periods.

Figure 8 shows a charging strategy which adds to the charging off-peak in order to compensate for the elimination of the charging in the peak periods, and it is called EndStop2 in the calculations below. The yellow colour shows the charging which cannot be done during the peak since the number of buses have been reduced. This will be compensated for in two ways. Some of the buses which needs extra charging will be driven to the depot and can charge there, illustrated by the green colour in Figure 8. The other buses, which remains in traffic in the midday period, will need to further charge at the end-stop chargers. That extra charging at the end stops is illustrated by the purple colour in Figure 8, and it allows the buses to charge the battery so that it is full before the afternoon peak starts. The charging shown in red is the charging which is needed between the trips just to keep battery state-of-charge the same from trip to trip, and it is the same as the charging shown in Figure 7.

There is a charge deficit also from the afternoon peak, but it will not be necessary to charge the batteries to full capacity again after that peak. The reason why this deficit does not need to be restored is that there will be enough time during the rest of the day to charge at the end stops so that the battery state-of-charge can remain constant from the start of one trip to the start of the next. This deficit can be compensated for before the next day during the night charging in the depot. Thus, we do not need to analyse any extra charging after the second peak. This does not mean that the buses cannot charge a little extra after that peak; it simply means that such charging is not necessary to consider when sizing the number of buses and chargers. In general, there will be possibilities for additional charging also after the afternoon peak, but it will not be necessary to do that.

The buses which are in traffic between the peak times, besides needing to charge in order to keep the batteries from draining, will need to stay for additional time at the end stops to further charge the battery after each trip to ensure that it is full before the second peak. How much extra time is required for the charging depends on the ratio between how long the first peak was, and how long the period between the peak times are. A conservative (high) estimate of how long the buses drained their batteries for during the first peak can be seen in Figure 8:

$$T_{\text{ChgDeficit1}} = T_{\text{stopAMPeak}} - T_{\text{startAMPeak}}. \tag{12}$$

Based on that, a worst-case charging energy deficit can be determined from the number of departures per hour, the trip time, and the average power consumption:

$$W_{\text{ChgDeficit1}} = T_{\text{ChgDeficit1}} \times n_{\text{DepPerHourPeak}} \times T_{\text{TripNetPeak}} \times P_{\text{BusAverage}}. \tag{13}$$

A conservative (low) estimate of the time they can charge up again can also be seen in Figure 8:

$$T_{\text{ChgExtra1}} = T_{\text{startPMPeak}} - \left(T_{\text{stopAMPeak}} + T_{\text{TripGrossOffPeak}}\right). \tag{14}$$

If we assume that all the charging can be completed at midday, without having to add any extra buses, we can determine the number of buses needed to charge as the sum of the buses needed for normal end-stop charging, plus the extra buses needed to be charged:

$$N_{\text{busMidday2a}} = \frac{W_{\text{ChgDeficit1}}}{T_{\text{ExtraChg1}} \times P_{\text{ChgEndStop}}} + N_{\text{busMidday1}}. \tag{15}$$

Note that this number in some extreme cases can become higher than the number of buses in traffic during the peak times. That is taken care of in Equation (22) when this number of buses is compared with a calculation of $N_{\text{busMidday2b}}$ made for the case when we need to add buses to allow some charging also during the peak times.

If the timetable is such that the number of buses in traffic between the peak times is not significantly lower than in the peak times, we obtain the second case where there may not be enough buses available to charge off-peak in order to compensate for the charging deficit from the peak. It will still be possible to move charging from the peak times to off-peak time, but not fully, so some extra buses will be needed to allow for additional charging. Those extra buses do not only increase the possibility of charging between the peak times but will also allow charging during the peak times. Such a case is illustrated in Figure 9, and it can be seen that the number of buses is higher than the highest number of buses in traffic during the peak times, but it is still lower than what would have been required if the buses were charged to full capacity after each trip as well as also during the peak times.

Thus, if the number of buses required midday, $N_{busMidday2a}$, is higher than the number of buses in traffic during the peak times, the charge balance equation must be altered to include the need for some buses that can charge both during the two peak times as well as in the midday period. In this case we also know that no buses will drive to the depot in the midday period.

Thus, it can be calculated how much extra charging is possible during the midday period for all the buses which are not needed in traffic during the peak times, $N_{busPeak}$

$$W_{ExtraMiddayMax} = \left(N_{busPeak} - N_{busMidday1}\right) \times P_{ChgEndStop} \times T_{ChgExtra1}. \quad (16)$$

There will be a remining energy deficit if this maximum midday charging is not sufficient:

$$W_{Remain} = W_{ChgDeficit1} - W_{ExtraMiddayMax}. \quad (17)$$

The extra buses that need to charge this energy have to do it during the morning peak, the midday period and during the afternoon peak. It may seem strange that the charging can be done also during the afternoon peak, since we earlier stated that the buses should be fully charged before the second peak starts, but that statement was based on the assumption that no charging could take place during the afternoon peak. It will not be important that all buses are fully charged at the beginning of the second peak if some of them avoid draining too much by charging a little also during the peak. The critical factor is that they shall not be at their minimum charge level before the end of the afternoon peak, not that they are fully charged at the beginning of that peak.

Note that it is not the added buses themselves which charge all this energy, as they will have a full battery when the morning peak starts. Instead, the added buses take over the task of driving the trips in order to relieve the other buses so that they can have more time to charge up. The time which one extra bus can be used to relieve other buses is:

$$T_{ChgExtra2} = T_{stopPMPeak} - T_{startAMPeak}. \quad (18)$$

It will take some time for the extra buses to relieve the other buses, and there will always be some waiting time for a bus which has been charging before it can start driving trips again. Therefore, it is not realistic to assume that all of the time added by the extra buses can be used for the charging of buses. The fraction of the added time which can be used for charging is:

$$k_{UtilExtraBusChg} = 50\%. \quad (19)$$

The charge balance for the added buses lets us calculate the number of extra buses required:

$$N_{BusExtraEndStop2} = \frac{W_{Remain}}{T_{ChgExtra2} \times P_{ChgEndStop} \times k_{UtilExtraBusChg}}. \quad (20)$$

Since the equations do not check that the remaining energy is a positive value, this number of extra buses can become negative. This never happens in reality, but according to Equation (22), it will not be a problem since it is compared with the number for the first case to find out what the right number of buses is. The total number of buses needed for the route at midday, for this second case, is:

$$N_{BusMidday2b} = N_{BusExtra} + N_{BusPeak}. \quad (21)$$

We can now use the number of buses determined for the two cases to decide what the actual need of the buses will be in the midday period:

$$N_{BusMidday2} = min\left[N_{BusMidday2a}\,,\ max\left(N_{BusPeak}\,,\ N_{BusMidday2b}\right)\right]. \quad (22)$$

A third and even more extreme case is if the timetable has the same number of departures per hour during the whole day. This case is illustrated in Figure 10, and we can see that there is no possibility to move any charging to the midday period, so the number of extra buses will be higher. In this case, the strategy to charge off-peak will no longer add any benefit, and the resulting number of buses becomes the same as the number of buses needed for the normal end-stop-charging strategy.

The TCO variable "Number of buses", can now be determined for the charging strategy EndStop2:

$$N_{\text{busTotal}} = max\left(N_{\text{busPeak}}\ ,\ N_{\text{BusMidday2}}\right). \tag{23}$$

4.4. Number of Chargers

There is a need in the depot for one charger per bus for all types of electric buses, and the power demand will depend on the size of the buses' batteries. Thus, there is always a need to build one new substation and pay for one connection to the grid at the depot. For end-stop-charged buses, there is additionally a need for building chargers at both end-stops of the route, so in total three substations and grid connections are needed. Thus, for end-stop-charged buses, the total number of chargers will equal the total number of buses plus the number of end-stop chargers.

The number of end-stop chargers is typically one per end stop for one route, but if the number of buses driving on that route becomes very high, there will be a need to add more chargers if more than one bus at a time need to charge at each end stop. Therefore, the number of end-stop chargers can be determined from the maximum number of buses simultaneously charging at the end stops. This is represented by the red and purple parts in Figures 7–10.

For the EndStop1 charging strategy, the number of charging buses is always highest during the peak periods, while the number is highest between the peak times for EndStop2 charging strategy. It is not realistic to assume that a charger can be used 100% of the time, so we assume that a charger should on average not be used more than $k_{\text{MaxUtilEndStopChg}}$ of the time, and the maximum utilization is therefore set to 50% in this paper. This will provide a margin to allow for buses to be delayed without having a big influence on other buses' ability to charge. A system with such a margin will also be able to work even if one charger is out of order for a limited time. Thus, the number of end-stop chargers for a bus route with end-stop charging for the whole day will be:

$$N_{\text{ChgEndStop1}} = max\left(2,\ \text{ceil}\left(\frac{N_{\text{BusExtraEndstop1}}}{k_{\text{MaxUtilEndStopChg}}}\right)\right), \tag{24}$$

where ceil is a function rounding upwards to the nearest integer number.

Note that this means that we can never have less than two end-stop chargers but can have both odd and even number of chargers from three and up. It is possible to have a different number of chargers at the different end stops since bus schedules can be planned so that the buses charge longer on one side of the route than the other. However, we do not allow a system with only one charger per route, since such a system will not be able to maintain service if that single charger fails for longer than a short time.

The strategy to charge between the peak times rather than during the peak times will require another number of end-stop chargers. Still, the number of chargers can be determined by the highest number of simultaneously charging buses; it is just that the highest number of charging buses will occur in the midday period and may require a different number of chargers.

$$N_{\text{ChgEndStop2}} = max\left(2,\ \text{ceil}\left(\frac{N_{\text{BusMidday2}} - N_{\text{BusOffPeak}}}{k_{\text{MaxUtilEndStopChg}}}\right)\right). \tag{25}$$

We can now determine the required number of chargers:

$$N_{\text{chg}} = N_{\text{BusTotal}} + N_{\text{ChgEndStop}}. \tag{26}$$

The total power of all the chargers is also needed in order to determine the cost of the chargers:

$$P_{chgTotal} = N_{busTotal} \times P_{ChgNight} + N_{ChgEndStop} \times P_{chgEndstop}. \quad (27)$$

The end stop charger power is $P_{ChgEndStop} = 300$ kW and the depot charger power $P_{ChgNight}$ is 11 kW for EndStop1 and 22 kW for EndStop2.

4.5. Calculating Energy Use and Driving Distance

Energy use can be determined from all the driving by all the buses. For the cost analysis, we do not need to know which bus is driving where, just the sum of all the driving. This is the sum of driving the trips plus driving to and from the depot. The number of trips during a day can be determined from the timetable parameters. All the trip numbers are multiplied by a factor of two since the route has the same departures from both directions. First, it is determined how long during the day the bus route operates at different numbers of departures per hour:

$$T_{Peak} = \left(T_{stopAMPeak} - T_{startAMPeak}\right) + \left(T_{stopPMPeak} - T_{startPMPeak}\right), \quad (28)$$

$$T_{OffPeak} = T_{startEvening} - T_{start} - T_{Peak}, \quad (29)$$

$$T_{Evening} = T_{stop} - T_{startEvening}, \quad (30)$$

The number of trips are:

$$N_{TripOffPeak} = 2 \times T_{OffPeak} \times n_{BusPerHourOffPeak}, \quad (31)$$

$$N_{TripPeak} = 2 \times T_{Peak} \times n_{BusPerHourPeak}, \quad (32)$$

$$N_{TripEvening} = 2 \times T_{Evening} \times n_{BusPerHourEvening}, \quad (33)$$

which results in a total trip number:

$$N_{TripTotal} = N_{TripOffPeak} + N_{TripPeak} + N_{TripEvening}, \quad (34)$$

From this, the total trip distance can be calculated:

$$d_{TripTotal} = N_{TripTotal} \times l_{Trip}. \quad (35)$$

The total energy used during that distance is:

$$W_{TripTotal} = T_{TripNetTotal} \times P_{Average}, \quad (36)$$

where the total net trip time, excluding the layover time, is:

$$\begin{aligned} T_{TripNetTotal} &= N_{TripOffPeak} \times T_{TripNetOffPeak} \\ &+ N_{TripPeak} \times T_{TripNetPeak} + N_{TripEvening} \times T_{TripNetEvening}. \end{aligned} \quad (37)$$

The distance driven during the trips and the energy consumed for it are the same, irrespective of charging strategy. However, the distance driven to and from the depot will differ between the strategies, since the number of buses and the number of the buses which have to drive to the depot in the midday period vary. The number of times a bus has driven to or from the depot during a day is represented by the light blue areas in Figures 7–9. All the blue areas have the same length, $T_{PullInOut}$ so we need to determine the number of times a bus drives from the depot per day for each charging strategy. The number of times a bus drives to the depot will of course always be the same.

$$N_{PullInOut} = 2 \times N_{BusTotal} + 2 \times \left(N_{BusTotal} - N_{BusMidday}\right) \quad (38)$$

Based on the number of times a bus drives to and from the depot, the distance driven to and from the depot can be determined:

$$d_{\text{PullInOutTotal}} = N_{\text{PullInOut}} \times l_{\text{PullInOut}}. \tag{39}$$

The energy used when driving to and from the depot:

$$W_{\text{PullInOutTotal}} = N_{\text{PullInOut}} \times T_{\text{PullInOut}} \times P_{\text{Average}}, \tag{40}$$

$$W_{\text{Total}} = W_{\text{TripTotal}} + W_{\text{PullInOutTotal}}, \tag{41}$$

$$d_{\text{Total}} = d_{\text{TripTotal}} + d_{\text{PullInOutTotal}}. \tag{42}$$

We have now determined the TCO variables of "Total Energy", "Trip distance" and "Total distance".

4.6. Calculating Total Driver Time

It is important to calculate how much driver time is needed for the different charging strategies since the driver wage is a high cost in many countries. A driver is of course needed for the whole time that the bus is in motion, but also for the layover time, and for some, but not all, charging. Charging at the end stops between each trip is typically only a few minutes at a time, for which the driver will have to wait. In contrast, charging in the depot is completed while the bus is parked, hence, drivers do not need to be in duty during this type of charging. The extra charging at the end stops can sometimes be completed during a driver's break. This will not lead to any extra driver cost, but in this paper, we assume that a driver must be paid also during all the extra charging at the end stops. The total time drivers must be in duty during a day can be calculated as:

$$T_{\text{Driver}} = \frac{T_{\text{Driving}} + T_{\text{LayoverTotal}} + T_{\text{ChgTotal}}}{k_{\text{DriverUtilFactor}}}, \tag{43}$$

where $k_{\text{DriverUtilFactor}}$ takes into account that it is not possible to plan driver schedules so that all of the driver's time is spent in active duty. In this example, it is assumed that 90% of the driver's time can be deemed as active duty. The same equation can be used for all types of buses, but the charging time will be different between them, and the driving time will differ due to different amount of driving to and from the depot. The time the buses drive is:

$$T_{\text{Driving}} = T_{\text{TripNetTotal}} + T_{\text{PullInOut}}, \tag{44}$$

where:

$$T_{\text{TripNetTotal}} = N_{\text{TripOffPeak}} \times T_{\text{TripNetOffPeak}} + N_{\text{TripPeak}} \times T_{\text{TripNetPeak}} + N_{\text{TripEvening}} \times T_{\text{TripNetEvening}}. \tag{45}$$

The charging time is different for the different types of buses. For EndStop1, it is:

$$T_{\text{ChgTotal}} = T_{\text{ChargePeakEndStop1}} \times N_{\text{TripPeak}} + T_{\text{ChargeOffPeakEndStop1}} \times N_{\text{TripOffPeak}} + T_{\text{ChargeEveningEndStop1}} \times N_{\text{TripEvening}}, \tag{46}$$

where the charging time per trip during different parts of the day for EndStop1, $T_{\text{ChargeXXXEndStop1}}$, can be determined according to Equation (8).

For EndStop2, the corresponding value is:

$$T_{\text{ChgTotal}} = T_{\text{ChargeOffPeakEndStop1}} \times N_{\text{TripOffPeak}} + T_{\text{ChargeEveningEndStop1}} \times N_{\text{TripEvening}} + (N_{\text{BusTotal}} - N_{\text{BusPeak}}) \times T_{\text{ChgExtra2}} + \left(N_{\text{BusMidday2}} - N_{\text{BusMidday1}}\right) \times T_{\text{ChgExtra1}}, \tag{47}$$

where the first two terms represent the charging according to EndStop1, but only off-peak and in the evening. This charging is shown in red in Figures 8 and 9. The third part corresponds to the extra charging by the busses which are not in traffic during the midday period, and the fourth part is the charging by extra buses (if there are any).

4.7. TCO for Combustion Engine Buses

We have now defined equations for the nine TCO parameters for two types of electric buses. In order to better interpret the variations in the TCO for these two types of electric buses, they will be compared with the TCO for combustion engine buses. We use the same way of determining the TCO of the electric buses based on the nine TCO parameters, with different values of most parameters. Since there is no need for combustion engine buses to charge, we can analyse them only based on the trips they drive and driving to and from the depot, as illustrated in Figure 11.

$$N_{\text{BusTotalICE}} = N_{\text{BusPeak}}, \tag{48}$$

$$T_{\text{ChgTotalICE}} = 0, \tag{49}$$

$$W_{\text{BatteryICE}} = 0, \tag{50}$$

$$N_{\text{ChgTotalICE}} = 0, \tag{51}$$

$$P_{\text{ChgTotalICE}} = 0, \tag{52}$$

$$N_{\text{ChgPlacesICE}} = 0. \tag{53}$$

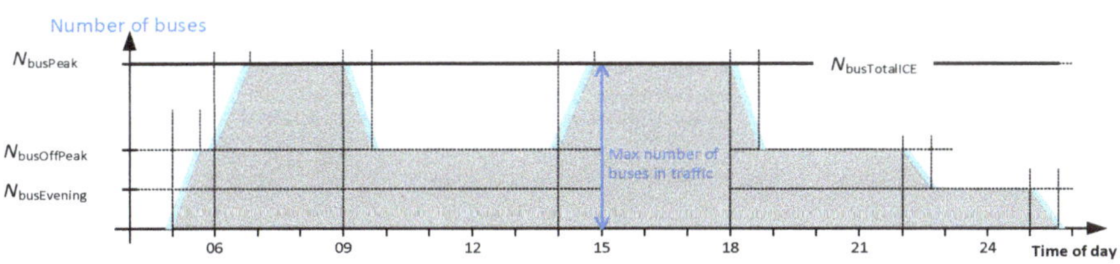

Figure 11. How the use of buses with combustion engines varies during the day.

Trip distance is the same for all types of buses, as it is determined only by the timetable and route. The total driving distance, traction energy use, and trip distance are calculated in the same way for combustion engine buses as electric buses, but with the difference of being based on a different number of buses during the peak time and midday.

Instead of calculating the cost of electricity, the energy cost of the combustion engine buses will be the fuel cost, which is determined from the required traction energy, just like for the electric buses. The cost parameter for the fuel takes into account both the fuel cost per litre and the fuel consumption per kWh of traction energy, i.e., there is no need to calculate the fuel consumption in litres explicitly. The fuel cost of HVO buses used in this paper is 3.5 SEK/kWh for traction energy, and 4.0 SEK/kWh for biogas. They are based on prices from Reference [1], which indicates a HVO price of 1.46 SEK/kWh and a biogas price of 1.25 SEK/kWh, and this paper has assumed an average powertrain efficiency of 36%.

Chargers are replaced by one fuel station at the depot. However, many bus routes share the cost of the fuel station, so we assume the investment cost of the fuel station to be included in the fuel cost per litre.

5. TCO Analysis

In this section, the TCO model is used to explore how the cost of the two end-stop-charging strategies vary for different bus routes and different timetables. As a reference, the costs will be compared to the cost of a bus for compressed biogas (CBG) and one bus running on biodiesel (HVO). The main purpose of the paper is to develop the TCO model and explain how different charging strategies influence the TCO, so the analysis in this section is mainly included to demonstrate how the TCO model can be used to understand how the timetable, charging strategy and bus route influence the TCO.

The parameter values used in this paper are relevant for Swedish bus operators in 2019. The technology for buses, batteries and chargers are under rapid development, so the results discussed below should only be considered as an example. Conclusions for other regions in the world or for costs from 2020 and onwards should always be based on updated parameter values.

5.1. Cost Comparison for Different Bus Types

The cost of different types of buses are compared in Figure 12 for a specific bus route and timetable, where a bar-chart shows how the TCO varies between different types of buses. Besides the bar-chart, there is another diagram showing the investigated timetable as the number of departures per hour over a full day. We start by comparing the TCO for the electric buses with two different charging strategies. Due to reduced number of buses, the cost of the bus depreciation is lower for the charging strategy that does not charge during the peak times (EndStop2) when compared to charging during the whole day (EndStop1). A second order effect is that less driver hours are needed for EndStop2 since fewer buses are driven to and from the depot. However, the higher cost of bigger batteries outweighs the savings from fewer buses, and the total cost is slightly higher for the EndStop2 strategy than for EndStop1, at least in this case.

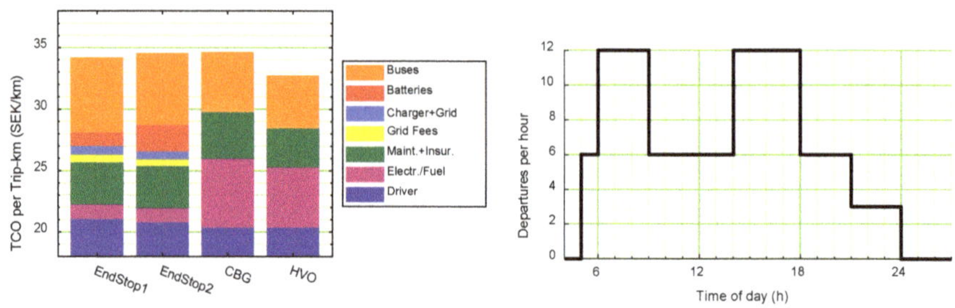

Figure 12. The cost of different types of buses on the reference bus route and with the reference timetable. EndStop1 and 2 buses are powered by electricity, while the others are powered by biogas (CNG) or biodiesel (HVO). (Note that the y-axis starts at 18 SEK/km, so the driver cost is more than half the TCO).

Electric buses have almost the same total cost as buses powered by biogas. Electricity is much cheaper than biogas, but the buses, batteries and chargers are still much more expensive than the biogas buses. The extra driver time for charging also adds to the cost of these electric buses. HVO buses are cheaper than biogas buses, due to a lower bus price, less maintenance and, for now, cheaper fuel. However, it will be difficult to find enough supply of HVO with a high reduction in greenhouse gas emissions in large quantities. Therefore, HVO buses are not seen as a long-term, large-scale solution for buses in Sweden.

However, only comparing the bus types for one route and one timetable can be misleading as it can give the impression that the cost difference between different types of buses are always the same. This is not the case, since the TCO can vary differently for the different type of buses when the bus route and timetable are changed.

5.2. TCO Variations for Different Timetables

The TCO model can help analyse how the timetable influences the costs of different types of buses. In Figure 13, the cost of three types of buses are compared for one bus route, but with three different timetables of varying bus traffic density. In this analysis, the bus traffic density is varied in the same proportions over the whole day. Thus, the ratio between peak and off-peak traffic remains the same, and the start and end of the traffic periods also remain the same in the three compared cases.

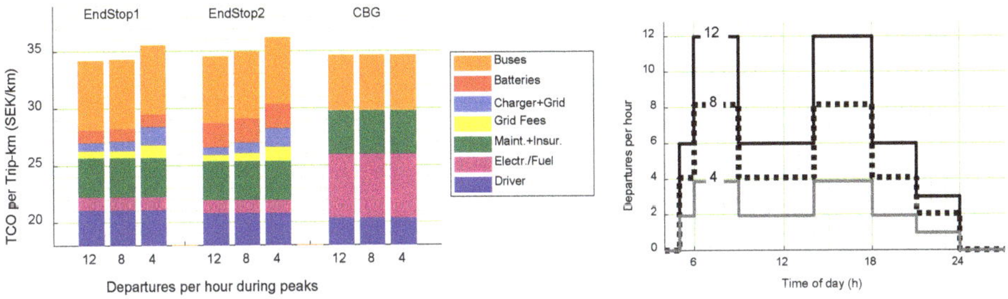

Figure 13. The cost per km for end-stop-charged and biogas (CBG) buses with varying bus traffic density for a whole day. (Note that the *y*-axis starts at 18 SEK/km, so the driver cost is more than half the TCO).

In Figure 13, the cost per km for biogas buses is found to be the same, irrespective of the number of departures per day. Since conventional buses do not share any infrastructure on the bus route, doubled bus traffic density results in all costs increasing by a factor of two. Thus, the cost per trip (km) for biogas buses is not changed when the bus traffic density is changed by the same factor over the whole day. Electric buses with chargers at the end-stops instead become increasingly expensive per trip (km) as the bus traffic density is lowered. This is caused by the investment in grid connection and chargers being shared by fewer and fewer buses as the bus traffic density decreases. This can be seen as the yellow and blue parts in the bar-chart growing as the bus traffic density is lowered. The increase in total cost per km is only in the order of 5%, but it represents an important difference compared to conventional buses, as the electric buses are relatively more cost effective at a higher bus traffic density. The two different end-stop-charging strategies both change cost in a similar way, as the bus traffic density influences their respective TCO in a similar way.

Another important timetable parameter is the ratio of departures per hour in the peak and off-peak periods. In the following, we analyse the effect of keeping the peak bus traffic density constant and varying the off-peak traffic. This will give a different result to that when we vary the bus traffic density over the whole day, since the number of buses is mainly determined by the bus traffic density during the peak times and not so much by the off-peak traffic. Therefore, an increased traffic off-peak can be expected to reduce the cost per km. The results of varying the off-peak bus traffic density is shown in Figure 14. For all three cases, the timetable has 12 departures per hour in the peak times. The difference is in the number of departures off-peak which are either the same (12) two thirds (8) or one third (4) of the number of departures in the peak times. The number of departures during the evening has been set to 50% of the off-peak departures per hour.

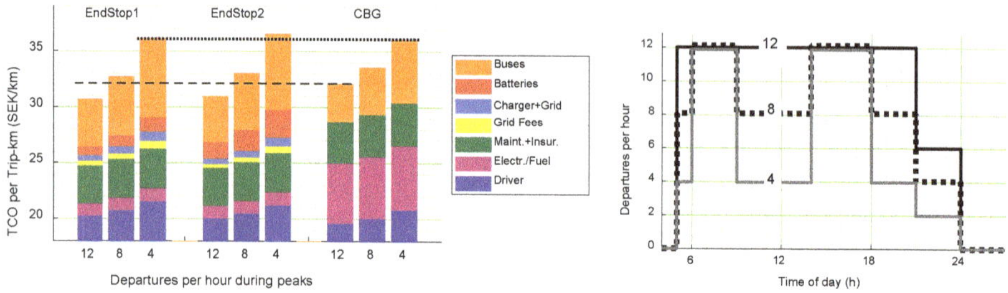

Figure 14. The cost per km for end-stop-charged and biogas (CBG) buses with varying bus traffic density off-peak. (Note that the y-axis starts at 18 SEK/km, so the driver cost is more than half the TCO).

According to Figure 14, the cost per km for biogas buses is found to be lower the higher the bus traffic density is off-peak. The main reason for this reduction is that the same number of buses are driving more trips and, thus, the bus depreciation is divided among more trips (km). Another contributing mechanism is that the number of times the bus drives to and from the depot decrease as more and more of the buses drive the whole day without a midday break in the depot. This is reflected as a lower driver cost per trip (km) and lower fuel consumption per trip (km).

The electric buses also reduce their cost per trip (km) if the departures per hour off-peak is increased. They have the same reason for reducing the costs as the biogas buses, but on top of that, they have charger depreciation and grid fees which are also divided by more and more km in traffic. Therefore, the electric buses have even more of a reduction to their TCO when the off-peak traffic is increased than the biogas buses have. In Figure 14, the dotted line shows that biogas buses have lower or the same TCO as the electric buses at four departures per hour off-peak. However, the dashed line shows that the electric buses are significantly cheaper than the biogas buses at 12 departures per hour during the whole day. Generally, we can conclude that electric buses with end-stop charging will be most cost effective for routes which have a high amount of and constant traffic during the whole day, but it is not possible to determine one type of bus which is always the cheapest.

Now that we have looked at the TCO with some different timetables, we can see that it seems that the whole idea of not charging during the peak times in order to reduce the number of buses manages to achieve the expected reduction in the number of buses; but in these cases, this does not lead to a lower TCO than charging for the whole day, mainly due to a more expensive battery. The authors of this paper hypothesized that the costs for these two charging strategies would change differently with timetable variations, leading to EndStop2 having a lower TCO for some timetables, but so far it seems that the TCO for EndStop1 and EndStop2 follow each other well. This illustrates the value of having a rather detailed TCO model, as there are often several second order effects which may influence the TCO.

5.3. *TCO with Future Cost Levels*

Since electric buses are new products, their cost is still high. Once they are produced in high volumes, the cost of an electric bus, excluding the battery, is likely to be the same as for a conventional bus. This means that the cost of the electric powertrain is assumed to be the same as the cost of a diesel powertrain. Furthermore, increasing production volumes of batteries for cars has led to rapidly falling battery prices, and it is likely that the cost of power-optimized batteries for buses can fall to about 2500 SEK/kWh and to 1500 SEK/kWh for energy-optimized batteries. The TCO for our compared bus types under these assumptions are shown in Figure 15, which reveals that electric buses might have a lower TCO than conventional diesel buses that are powered by HVO (or diesel).

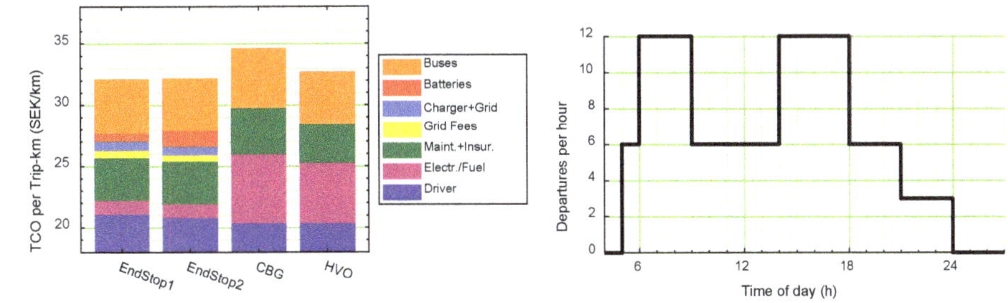

Figure 15. The cost per km for end-stop-charged and biogas (CBG) buses with cost parameters estimated for high production volumes. (Note that the y-axis starts at 18 SEK/km, so the driver cost is more than half the TCO).

The TCO in Figure 15 is not a lower limit for electric buses. There are at least three more ways in which the cost may be further reduced. The service life of an electric bus is likely to be longer than that of a combustion engine bus, due to less vibrations and fewer parts that can be worn out. The service life of the battery is also likely to increase, as it is currently extensively researched by universities and companies globally due to the high economic value in such improvements. Finally, the maintenance cost can be expected to become lower for electric buses than for combustion engine buses. When these improvements also occur, electric buses are likely to become even cheaper than conventional buses. Looking at the size of the different parts in the TCO bar-chart, it can be seen that even though the battery is often said to be a very expensive part, an increase in the service life of electric buses and a reduction in the maintenance cost may be just as important for reducing the TCO as future reductions of battery prices.

6. Concluding Discussions

6.1. Main Findings

This paper explains and shows the main mechanisms that influence the costs of electric buses when different routes and timetables are compared. The most significant results of this paper can be summarized as:

- A new model that demonstrates how to calculate the TCO for electric buses that depends on the nine most significant input variables. The calculations result in four operating and three annual depreciation cost parameters that forms the TCO.
- Testing of the method in a Swedish context from 2019 showed that the TCO for electric buses is generally in line with buses powered by biomethane and slightly higher than buses powered by HVO. However, the TCO can be both higher or lower depending on cost variations related to departures per hour, electric grid connections, the distance to the depot, and the length of the route. It is likely that future TCOs will be lower for electric buses when compared to buses powered by biomethane or HVO, mainly due to lower prices for batteries and buses and costs related to maintenance.

This paper has presented a TCO model which fills a gap between the very simple cost comparisons which present one TCO for each type of bus, and the very complex bus-planning tools which can calculate the exact TCO for one specific route with all its details and its exact timetable. The presented TCO model is aimed at describing the main mechanisms which make the TCO for different type of buses vary in different ways when route properties and timetables vary. It ignores some details on purpose to make it easier to understand, for example, the reasons why one type of bus can have the lowest cost of one timetable and at the same time be more expensive than other types of buses for another timetable.

By explaining these mechanisms, this TCO model can help build a general understanding of the cost structure of electric buses. Such knowledge is important to determine what type of buses to investigate with the more detailed bus-planning tools. The TCO model can also be used to find modifications to a charging strategy which can reduce costs, or to estimate how future cost reductions in different parts of the system will influence the determination of which type of bus will be most cost effective.

The results from the TCO analysis in Section 5 demonstrate how useful this type of model is, as it shows, for example, that it is not at all sufficient to analyse the cost effectiveness of charging off-peak by only looking at its influence on the number of buses required. Reducing the number of buses was the primary motivation for such a strategy, but other effects such as a bigger battery and less driver hours will also influence how the TCO changes using this strategy.

6.2. Critical Assessment and Comparisons with Other Studies

The results in this paper are theoretical but based on experiences where the authors have been involved in or lead earlier projects with electric buses (e.g., [1,7,15,20,21]). Results from testing the proposed methodology for the TCO of electric buses is well in line with these previous experiences, as well as other recent studies with comparable economic prerequisites (e.g., the Nordic countries) [22]. There are, however, some differences in results regarding the TCO of other buses when compared to studies from countries other than those in northern Europe (e.g., [23,24]). As mentioned in Section 1, electric buses can become cost competitive in about 5 years in Texas [8] and 25 years in India [9], and have twice as long pay-back time (almost six years) than diesel buses in Turkey [10]. These differences are mainly related to incentives for fossil fuels, but also the cost of drivers and maintenance personnel, as well as regional fluctuations in prices for busses and batteries. As shown in this paper, such variations in input data can, in the end of a procurement period, make a difference in terms of revenue (or loss) for a bus operator. The authors of this paper therefore stress the importance of using the model with updated prices and other data for the routes(s) if the model is used in a study for the procurement of electric bus traffic.

As for all models, it is important to know this model's limitations, and its purpose. The model should be used for more strategic investigations, such as when making general comparisons between different charging strategies or analysing what bus routes are especially good to focus on when introducing electric buses. The final decision on how to operate a certain bus route should always be based on the results of tools that are created for that purpose.

6.3. Conclusions

The authors of this paper believe the results have contributed to the research community, as there is currently a lack of theoretical TCO models based on experiences from real-life public transport tests of electric buses. Both the model and the data received from testing in a Swedish environment are considered to be useful for public transport authorities, bus operators, and other stakeholders involved in public transport planning that have intentions to move towards sustainability. The model and results from testing are adapted to conditions in Sweden and northern Europe, but the model could be adopted to other regional conditions as well.

For further work, the model could be tested further with data from several real-life cases, which could lead to a database of TCOs for different routes, buses, and regional prerequisites that could be useful for future investment analysis for public transport authorities and bus operators. The TCO model could also be complemented with societal costs of emissions of air and noise, and possible social sustainability-related costs in line with earlier studies (e.g., [1,23]). The model could also be complemented with scenarios that includes electric buses charged only at bus depots to find out when it would be most cost competitive to invest in depot- or end-stop-charged buses, and also in comparison with other buses with low climate impacts.

Author Contributions: Conceptualization, A.G., S.B., and O.E.; methodology, A.G., S.B., and O.E.; software, A.G.; validation, A.G., and S.B.; formal analysis, A.G.; investigation, A.G., and S.B.; resources, A.G.; data curation, A.G.; writing—original draft preparation, A.G., S.B., and O.E.; writing—review and editing, A.G., and S.B.; visualization, A.G., and S.B.; supervision, A.G., and S.B.; project administration, S.B.; funding acquisition, S.B. All authors have read and agreed to the published version of the manuscript.

Funding: This research was until June 2018 funded by the Swedish Energy Agency and in-kind co-funding organizations in the project "Decision support for implementing electric buses in public transport", with grant number 41411-1. After that date, the Swedish universities Blekinge Institute of Technology and Chalmers University of Technology have funded the research.

Acknowledgments: Organizations involved in the project "Decision support for implementing electric buses in public transport" contributed to this research by providing real-time data and experiences from using electric buses in public transport in Gothenburg, Karlstad, Umeå, Västerås, and Ängelholm, or looking into possibilities to use electric buses in other Swedish cities. In addition, participating organizations in earlier projects regarding energy transfer solutions for electrified bus systems (EAEB), and electric bus studies in the GreenCharge project can be considered to have contributed to this research, as it builds on findings of these earlier projects.

Conflicts of Interest: The authors declare no conflict of interest. The funders had no role in the design of the study; in the collection, analyses, or interpretation of data; in the writing of the manuscript, or in the decision to publish the results.

References

1. Borén, S. Electric buses' sustainability effects, noise, energy use, and costs. *Int. J. Sustain. Transp.* **2019**, 1–16. [CrossRef]
2. Johansson, T.B.; Kågesson, P.; Johansson, H.; Jonsson, L.; Westin, J.; Hejenstedt, H.; Hådell, O.; Holmgren, K.; Wollin, P. *Fossilfrihet på väg*; Ministry of Enterprise, SOU: Stockholm, Sweden, 2013.
3. UK Department for Transport. *The Road to Zero*; Department of Transport: London, UK, 2018.
4. BloombergNEF Electric Vehicle Outlook 2020—Executive Summary. Available online: https://bnef.turtl.co/story/evo-2020/page/1 (accessed on 2 June 2020).
5. IEA. *Global EV Outlook 2019*; IEA: Paris, France, 2019.
6. Rask, K. *Allt om elbil*. 2019. Available online: https://alltomelbil.se/elbilen-blir-lonsam-allt-snabbare-jamfort-med-fossilalternativ/ (accessed on 2 June 2020).
7. Borén, S.; Nurhadi, L.; Ny, H. Preferences of Electric Buses in public Transport; Conclusions from Real Life Testing in Eight Swedish Municipalities. *Int. J. Environ. Ecol. Eng.* **2016**, *10*, 259–268.
8. Quarles, N.; Kockelman, K.M.; Mohamed, M. Costs and Benefits of Electrifying and Automating Bus Transit Fleets. *Sustainability* **2020**, *12*, 3977. [CrossRef]
9. Sheth, A.; Sarkar, D. Life cycle cost analysis for electric vs diesel bus transit in an Indian scenario. *Int. J. Technol.* **2019**, *10*, 105–115. [CrossRef]
10. Topal, O.; Nakir, İ. Total Cost of Ownership Based Economic Analysis of Diesel, CNG and Electric Bus Concepts for the Public Transport in Istanbul City. *Energies* **2018**, *11*, 2369. [CrossRef]
11. Moro, A.; Lonza, L. Electricity carbon intensity in European Member States: Impacts on GHG emissions of electric vehicles. *Transp. Res. Part D Transp. Environ.* **2018**, *64*, 5–14. [CrossRef] [PubMed]
12. Nordelöf, A.; Romare, M.; Tivander, J. Life cycle assessment of city buses powered by electricity, hydrogenated vegetable oil or diesel. *Transp. Res. Part D Transp. Environ.* **2019**, *75*, 211–222. [CrossRef]
13. Grauers, A.; Olsson, O.; Carlsson, N.; Ohlin, G.; Rosengren, L.-G. *Analys av Energiförsörjning för Elektrifierade Bussystem*; Chalmers University of Technology: Gothenburg, Sweden, 2016.
14. Xylia, M.; Leduc, S.; Patrizio, P.; Kraxner, F.; Silveira, S. Locating charging infrastructure for electric buses in Stockholm. *Transp. Res. Part C Emerg. Technol.* **2017**, *78*, 183–200. [CrossRef]
15. Borén, S.; Grauers, A. Stakeholder collaboration models for public transport procurement of electric bus systems. *Int. J. Sustain. Policy Pract.* **2019**, *15*, 19–29. [CrossRef]
16. GIRO Inc. HASTUS Software. Available online: https://www.giro.ca/en-ca/our-solutions/hastus-software/ (accessed on 1 June 2020).
17. Optibus Optibus—The Future of Mass Transit. Today. Available online: https://www.optibus.com (accessed on 1 June 2020).
18. Automated Business Solutions PtMS for Windows. Available online: http://ptmsforwindows.com/ (accessed on 1 June 2020).

19. MathWorks MATLAB. Available online: https://uk.mathworks.com/products/matlab.html (accessed on 9 June 2020).
20. Nurhadi, L.; Borén, S.; Ny, H. Advancing from Efficiency to Sustainability in Swedish Medium-sized Cities: An Approach for Recommending Powertrains and Energy Carriers for Public Bus Transport Systems. *Procedia-Soc. Behav. Sci.* **2014**, *111*, 1218–1225. [CrossRef]
21. Ntemiris, S.; Grauers, A. *EAEB Project - Role Playing Workshop*; Chalmers University of Technology: Gothenburg, Sweden, 2017.
22. Mahmoud, M.; Garnett, R.; Ferguson, M.; Kanaroglou, P. Electric buses: A review of alternative powertrains. *Renew. Sustain. Energy Rev.* **2016**, *62*, 673–684. [CrossRef]
23. Tong, F.; Hendrickson, C.; Biehler, A.; Jaramillo, P.; Seki, S. Life cycle ownership cost and environmental externality of alternative fuel options for transit buses. *Transp. Res. Part D Transp. Environ.* **2017**, *57*, 287–302. [CrossRef]
24. Meishner, F.; Uwe Sauer, D. Technical and economic comparison of different electric bus concepts based on actual demonstrations in European cities. *IET Electr. Syst. Transp.* **2020**, *10*, 144–153. [CrossRef]

© 2020 by the authors. Licensee MDPI, Basel, Switzerland. This article is an open access article distributed under the terms and conditions of the Creative Commons Attribution (CC BY) license (http://creativecommons.org/licenses/by/4.0/).

energies

MDPI

Article

Development of Enhancing Battery Management for Reusing Automotive Lithium-Ion Battery

Wen-Poo Yuan [1], Se-Min Jeong [2], Wu-Yang Sean [3,*] and Yi-Hsien Chiang [4]

[1] Wistron Corporation, Hsichih 22181, Taiwan; patrick19941211@gmail.com
[2] Department of Naval Architecture and Ocean Engineering, Chosun university, Gwangju 61452, Korea; smjeong@chosun.ac.kr
[3] Center for Environmental Risk Management, Chung Yuan Christian University, ChungLi 32023, Taiwan
[4] Vigourpack Co., Ltd., Taichung 408, Taiwan; acloud.c@gmail.com
* Correspondence: wysean@cycu.org.tw; Tel.: +886-3-2654905

Received: 29 May 2020; Accepted: 23 June 2020; Published: 28 June 2020

Abstract: In this study, a battery management system (BMS) is developed for reused lithium-ion battery (RLIB). Additional enhancing functions of battery management are established, i.e., estimation of life-sensitized parameters and life extension. Life-sensitizing parameters mainly include open-circuit voltage (OCV) and internal resistances (IRs). They are sensitized parameters individually relative to state of charge (SOC) and state of health (SOH). For estimating these two parameters, an adaptive control scheme is implemented in BMS. This online adaptive control approach has been extensively applied to nonlinear systems with uncertainties. In two experiments, OCV and IRs of reused battery packs are accurately extracted from working voltage and discharge current. An offline numerical model using a schematic method is applied to verify the applicability and efficiency of this proposed online scheme. Furthermore, a solution of actively extending life by using an ultracapacitor to share peak power of RLIB through adjusting duty ratio is also proposed. It is shown that this enhancing battery management for RLIB can properly estimate OCV and IRs, and actively extend the life of the RLIB in two experiments.

Keywords: reused battery; adaptive control theory; battery management system (BMS); internal resistances; open-circuit voltage

1. Introduction

Well-known greenhouse gas (GHG) emissions, due to extensive use of diesel and gasoline, threaten our environment [1]. A large amount of CO_2 from transportation is gradually decreasing [2,3]. Meanwhile, electric vehicles (EVs) have been identified as being a key technology in reducing emissions and energy consumption in the transport sector [4]. Most countries are stepping up the introduction of electric vehicles, and are trying to progressively replace traditional fuel vehicles with EVs. However, with the rapidly increasing popularity of various EVs, the demand for lithium-ion batteries (LIBs) increases annually [5,6]. Sales of pure electric cars rose by nearly 14% in 2018 compared to 2017, while hybrid and plug-in hybrid sales rose by more than 20%. Now, more than 1.15 million EVs are on the roads today in the world [7]. Consequently, the quantity of wasted LIBs is also quickly increasing [8]. However, wasted battery packs are still available for energy balancing in thermal power plants, due to their residual electrical capacity [9]. Reuse of LIBs has become crucial in recent years [10], since renewable energy sources such as solar energy and wind energy are intermittent in nature. They have to keep continuous and reliable supply [11–13]. One of the best ways of stabling these renewable resources is to be stored in batteries. Among them, LIB is widely accepted due to its high energy density, long lifespan, and high efficiency [8,14,15]. However, the lifecycles of reused LIBs (RLIBs) become short, because the geometrical structure of the battery is possibly damaged by cycling use [16]. Therefore, an effective battery management system (BMS) for RLIB is compulsory, so that batteries can operate safely and reliably, preventing any physical damages and cell unbalancing [17]. Besides this, the state of charge (SOC) in BMS is considered as one of the critical and important basic parameters [14,18–20], which indicates how much remaining capacity is left inside a battery. Accurate estimation of SOC not only helps to provide information about the charge current and remaining performance of the battery, but also gives assurance of reliable and safe operation of the battery. However, the performance of the battery is highly affected by aging, temperature variation, and charge/discharge cycles, which make the task of estimating accurate SOC very challenging [21]. Therefore, we apply an adaptive control algorithm [16] for online estimation of key parameters, which are known as simple and easy to implement in BMS. For reused batteries, online estimations of electrochemical related parameters, such as internal resistances, are capable of real-time monitoring of the safety of reused batteries, rather than normal battery management systems for new batteries. Besides this, for prolonging a RLIB's life, one ultracapacitor (UC) is connected and controlled by BMS to share peak load power. One simple pulse-width modulation (PWM) module is adopted to adjust the duty ratio of RLIB. Verification of this battery management for RLIB is conducted by systematic testing on bench by monitoring online results of estimation. It is expected that this study will contribute to promoting the systematic reuse of wasted LIB in industry before the chemical method.

2. Enhancing Battery Management of Extending Life for RLIB

Generally, the states of LIB monitored by battery management systems include working temperature, voltage and current, and safety functions to avoid any overcharge/over-discharge. Relative functions of BMS can be categorized into hardware side and software side, as shown in Figure 1. In the hardware side, in case of centralized circuit BMS, cell voltage and temperature are measured by sensors in the external board, MMU (module monitoring unit), as shown in Figure 2a,b. Total data from the slave MMU are collected in the master BCU (battery control unit) for estimating SOC and SOH (state of health) through SPI (serial peripheral interface) protocol, as shown in Figure 2a. The analog-to-digital converter (ADC) measures the working current of the battery. This work is implemented based on Renesas Electronics Corporation ISL78600, as shown in Figure 2b. ISL78600 IC supervises up to 12 series-connected cells in the MMU. For keeping the high-voltage safety of RLIB, one insulation-resistance-measure device is attached on the BCU, as shown in Figure 2a. Besides this, controller area network (CAN) and universal asynchronous receiver transmitter (UART) protocols are also adopted. Through these functions of communication, the online real-time states of the battery, such as working current, SOC, SOH, and temperature, are gathered and transferred to the display or upper-level controller. In the software side, an algorithm embedded in the BCU is developed for estimating SOC and SOH by using basic battery parameters. Cell balancing is conducted in the MMU. A real-time simulator is supported to achieve automatic data acquisition and processing, to further verify the algorithm [22].

In this study, additional enhancing battery management is studied in reusing a large-scale automotive LIB. Functions of life estimation and life extension are highlighted for an RLIB. The schematic hardware of life extension is listed in Figure 2a. One proper ultracapacitor (UC) is parallelly connected with the BMS under the PWM (pulse-width modulation) control, which is used to switch the duty ratio of the IGBT (insulated gate bipolar transistor) in unit time. The application of the ultracapacitor achieves sharing of the RLIB's peak power so as to decrease the depth of discharge (DOD), and hence extends the lifecycle (i.e., SOH) of the RLIB. The duty ratio is a stationary value in every control case. A proper duty ratio can be estimated by calculating from the demand power, sustained time, and maximum power from the battery. The operation board of the BCU is shown in Figure 2c.

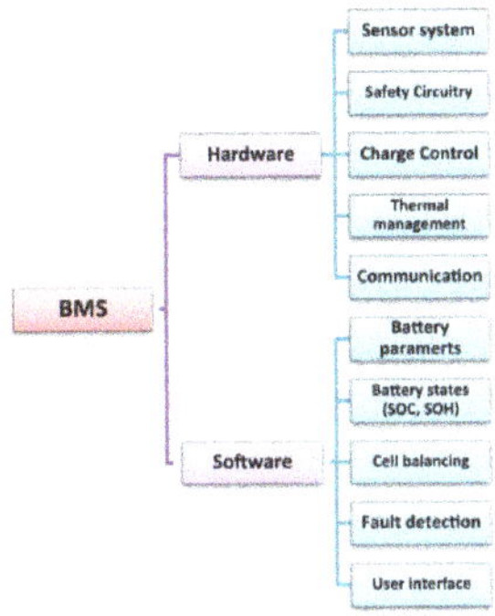

Figure 1. Basic framework of the battery management system (BMS) [17].

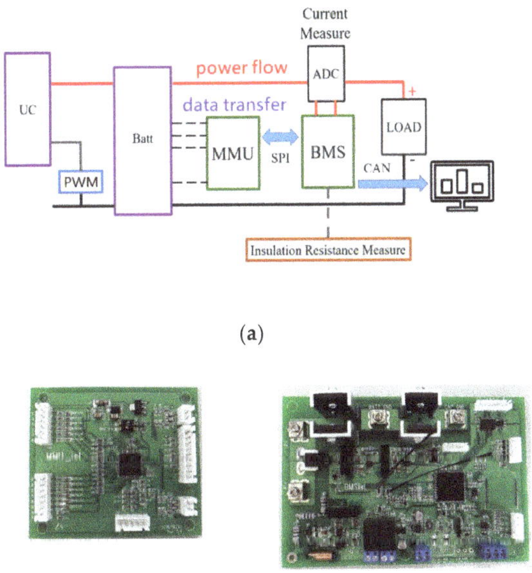

(a)

(b) (c)

Figure 2. In-house battery management system: (**a**) basic diagram; (**b**) board of module monitoring unit (MMU), based on chip ISL78600; (**c**) operation board of BMS.

3. Adaptive Control Scheme for Estimating OCV and IRs

Online estimation of parameters open-circuit voltage (OCV) and internal resistances (IRs) of new battery is illustrated in [16]. These two parameters are relative to SOC and SOH [23,24]. The adaptive control algorithm guarantees estimating error converges to zero, by applying Lyapunov stability criteria. As a result, OCV and IRs can be precisely estimated from input parameters, i.e., working voltage and current. The detailed mathematical modeling in Simulink/MATLAB, according to an adaptive control scheme, is shown in Figure 3. The procedure is briefly summarized as below. First, the battery voltage can be expressed as:

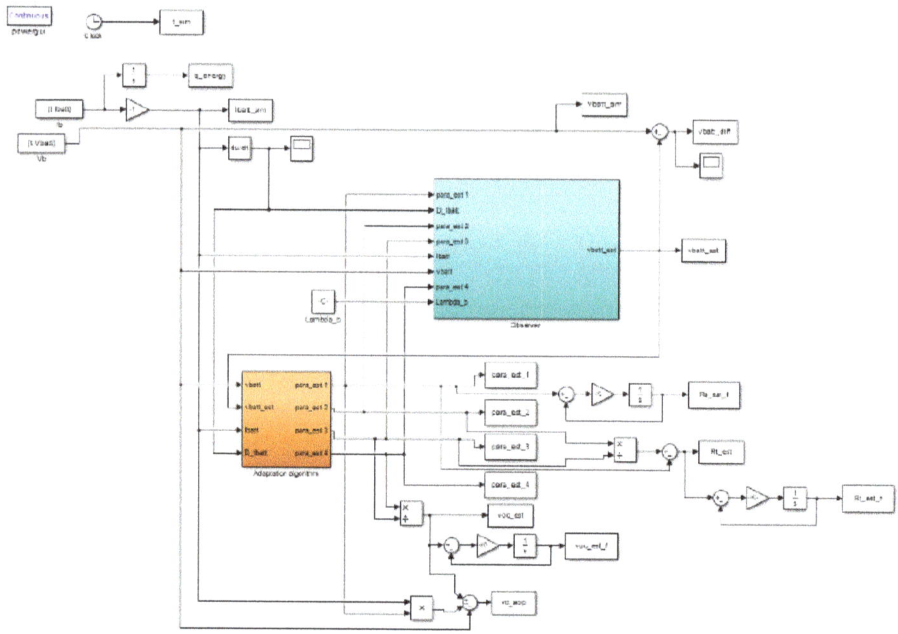

Figure 3. Adaptive control scheme in Simulink/MATLAB.

$$\begin{aligned} \dot{v_b} &= \dot{v}_{oc} - \dot{R}_s i_b - R_s \dot{i_b} - \dot{v}_c \\ &= \sigma(v_{oc})x - \sigma(R_S)x i_b - R_s \dot{i_b} - \dot{v}_c \end{aligned} \tag{1}$$

Here,

$$\sigma(p) = \left[\frac{\partial p}{\partial SoC} \; \frac{\partial p}{\partial T} \; \frac{\partial p}{\partial h} \right] \tag{2}$$

$$x^T = \left[\frac{\partial SoC}{\partial t} \; \frac{\partial T}{\partial t} \; \frac{\partial h}{\partial t} \right] \tag{3}$$

Here, v_b, v_{oc}, R_S, R_T, $\dot{i_b}$, and v_c represent voltage of battery, open-circuit voltage, first-order and second-order IRs, battery current, and voltage drop across RC circuit, respectively, based on the electrical circuit model (ECM) shown in Figure 3. The projected parameters v_b and R_S are assumed as functions of *SOC*, *T*, and *h*, and change with time. In addition, Equation (1) can be simplified with the following assumptions:

(1) Small deviation of every battery's discharge is assumed, compared to rated useful capacity, thus $\partial SoC/\partial t \approx 0$.

(2) In normal operating conditions, deviation of cell temperature, *T* is slow by using a fan, thus $\partial T/\partial t \approx 0$.

(3) A long-time usage history, h, is performed, thus $\partial h/\partial t \approx 0$.

Accordingly, it follows that Equation (3) ≈ 0; furthermore, Equation (1) can be rewritten as:

$$\begin{aligned} \dot{v_b} = &\left(\frac{\partial v_{oc}}{\partial SoC}\frac{\partial SoC}{\partial t} + \frac{\partial v_{oc}}{\partial T}\frac{\partial T}{\partial t} \right) - \left(\frac{\partial R_S}{\partial SoC}\frac{\partial SoC}{\partial t} + \frac{\partial R_S}{\partial T}\frac{\partial T}{\partial t} \right) i_b \\ &- \frac{1}{C_t R_t} v_b - R_s \dot{i_b} - \frac{R_t + R_s}{C_t R_t} i_b + \frac{v_{oc}}{C_t R_t} \end{aligned} \tag{4}$$

Equation (4) can also be written as a vector form.

$$\dot{v_b} = \theta^T X \tag{5}$$

$$\theta^T = [\theta_1\ \theta_2\ \theta_3\ \theta_4]^T = \left[R_s\ \frac{R_s + R_t}{C_t R_t}\ \frac{1}{C_t R_t}\ \frac{v_{oc}}{C_t R_t}\right] \tag{6}$$

$$X^T = \left[-\dot{i_b} - i_b - v_b\ 1\right] \tag{7}$$

Equation (6) is rewritten as Equation (8) with respect to every estimated state $\dot{\hat{v}_b}$

$$\dot{\hat{v}_b} = \hat{\theta}^T \hat{X} + u \tag{8}$$

where $\hat{X}^T = [-\dot{i_b} - i_b - \hat{v}_b\ 1]$, u is adjustable for input parameters. θ is estimated results of target parameters.

The adaptive control algorithm is based on tracking input signals to modify the target parameters in control states, so that the convergent criteria are satisfied as below:

$$\lim_{t \to \infty} e = \lim_{t \to \infty} (v_b(t) - \hat{v_b}(t)) = 0 \tag{9}$$

Here, Rt and Ct represent first-order IR and capacitance, respectively, based on ECM. If the thermal effect of temperature is not considered, $\partial T / \partial t \approx 0$ is obtained.

The adaptive control algorithm shown in Figure 3 is introduced to optimize the unknown target parameters of Rs, Rt, and OCV in Equation (4). A filtering process is used to improve the measured noise and enhance the estimation reliability. In the estimation process, IR is a sensitized parameter of lifecycle, hence it is used to indicate the deviation of lifecycle. An OCV curve related to the battery's useful capacity is applied to estimate the remained capacity of SOC. The algorithm is discretized, and embedded in BMS.

4. Setup of Test Bench

There are limitations of an electrochemical battery's performance. The charge/discharge response of a lithium-ion battery is much slower than a UC. Therefore, for reusing a lithium-ion battery, a high-power UC can support sudden peak current, and extend the life of an LIB. To take a simple parallel combination between an RLIB and a UC, the ultracapacitor is operated as a dc-side buffer for supporting peak current. In case of using a DC-DC converter, it allows more flexible management between an RLIB and an ultracapacitor. However, it is not competitive in reusing cost. Consequently, a simple hybrid parallel connection is established in this study. PWM control is applied for adjusting the duty ratio of an RLIB. A UC connected with the BMS directly is to improve the life of the LIB, by restricting the voltage drop or DOD. RLIB packs with rated voltage of 52 V and 50.4 V are employed, respectively, as shown in Figure 4a. The first pack of 52 V uses a commercialized BMS for reference, and the other one of 50.4 V uses an in-house BMS. Figure 4b shows the developed circuit of the BMS with a PWM control in this study. If the duty ratio per unit time of battery is selected as 40%, then the UC's load becomes 60%.

(a) (b)

Figure 4. (**a**) Two battery packs in test bench (52 V and 50.4 V); (**b**) circuit of BMS with pulse-width modulation (PWM) control.

5. Results

5.1. Life Extension Test by Using An Ultracapacitor

An ultracapacitor is a link between source and load. When used in parallel with the battery, it smooths out the load on the battery whilst improving the source impedance seen by the load. This may be viewed as the battery supplying the energy and the ultracapacitor supplying the short term power. Furthermore, to use an ultracapacitor matters in a battery's life extension, due to the constrained voltage drop. Here, the duty ratio controls the loading between battery and ultracapacitor. Figure 5a,b shows voltage drop of 52 V/10 Ah battery pack, versus three duty ratios. For the first case, with a duty ratio of 20%, it presents the duty ratio of the battery's loading as 80%, and on the contrary, the UC's loading is 20%. In cases of duty ratios of 20%, 60%, and 80%, the terminal voltages were 47.4 V, 47.7 V, and 47.9 V, respectively. It shows the increased effect of constrained voltage drops of battery when increasing the duty ratio of ultracapacitor. Similar cases of 50 V battery pack were measured and are shown in Figure 6a,b. In Figure 6a,b, the blue line is the battery's discharge, and the red line is the battery in parallel connection with UC, by setting to 75% duty ratio. The UC effectively reduces the voltage drop and decreases the DOD of the battery.

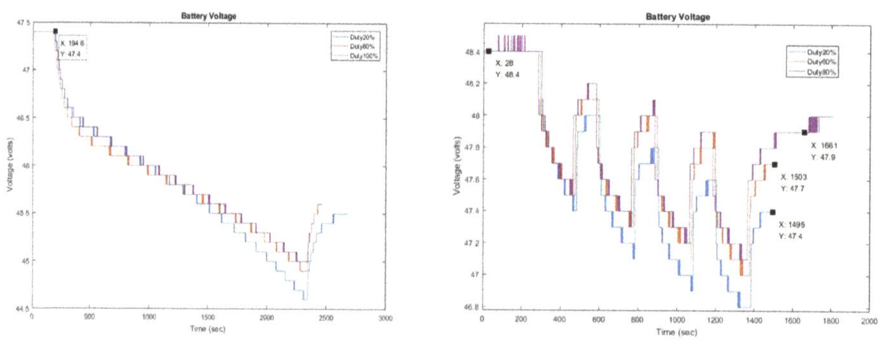

(a) (b)

Figure 5. (**a**) Voltage drop vs. duty ratio in constant-current discharge: (**b**) voltage drop vs. duty ratio in pulse discharge.

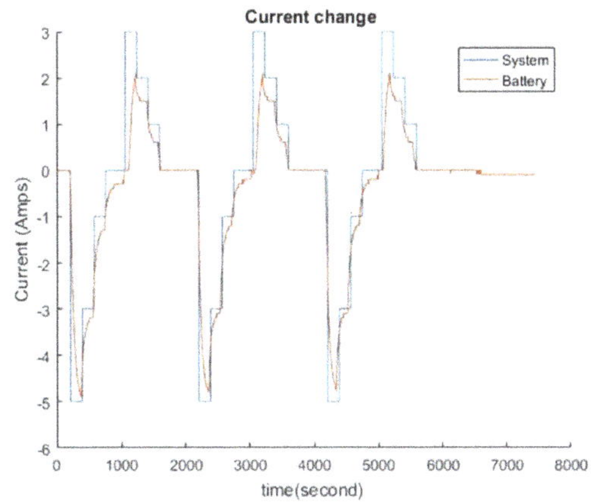

(a)

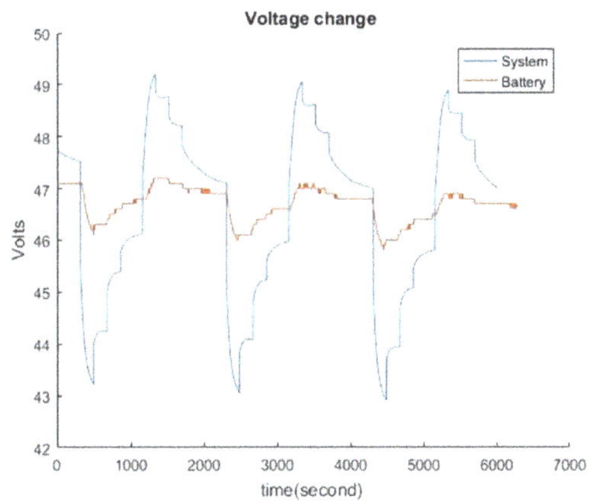

(b)

Figure 6. (**a**) Discharge current of battery and system (battery/ultracapacitor (UC)); (**b**) voltage drop of battery and system.

5.2. Estimation of OCV and IRs

Two experiments in this section were carried out for confirming the achievability for developing an adaptive control scheme on monitoring of a reused battery's parameters, e.g., IRs and OCV simultaneously. Here, lithium-ion batteries were selected and integrated in nominal voltages of 52 V and 50.4 V modules to simulate reused batteries. Working voltage v_b and current i_b of a reused battery were inputs required. According to the electrical circuit model (ECM), the estimated battery voltage was formulated through estimating model parameters, $\hat{\theta}_{i=1\sim4}$. The target estimating parameters were v_{oc}(OCV) and internal resistances (R_s and R_t). v_{oc}, R_s, and R_t can be extracted from online estimation algorithms.

5.2.1. Experiment 1

The objective of this experiment was to verify the accuracy of the proposed method of estimating IRs and OCV. One module of 12.6 V in a 50.4 V battery pack was used and composed by a random discharge current. Each discharge cycle was lower than 5 s so as to simulate a random load. Figure 7a shows voltage drop, battery current, and estimating error separately in a, b, and c. The estimating error tends to zero after 150 s.

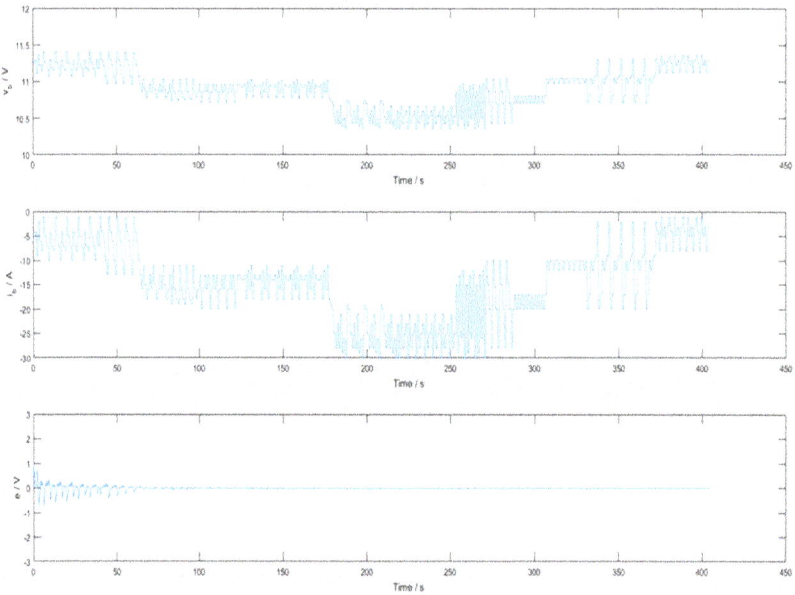

(a)

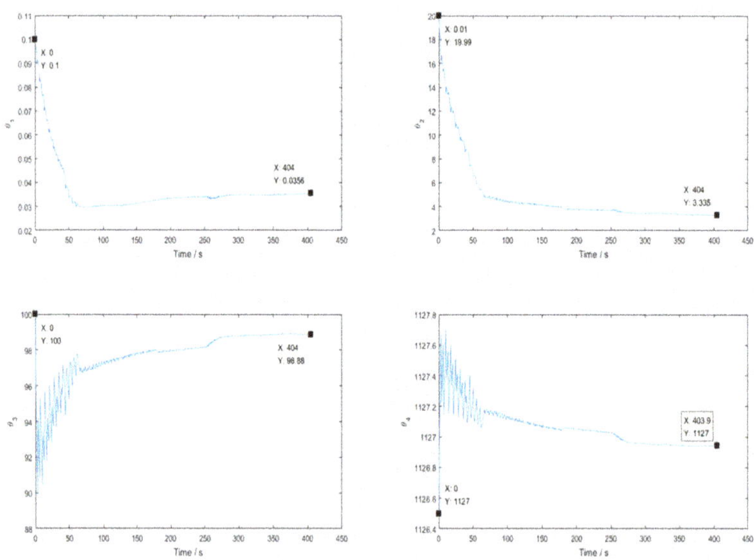

(b)

Figure 7. *Cont.*

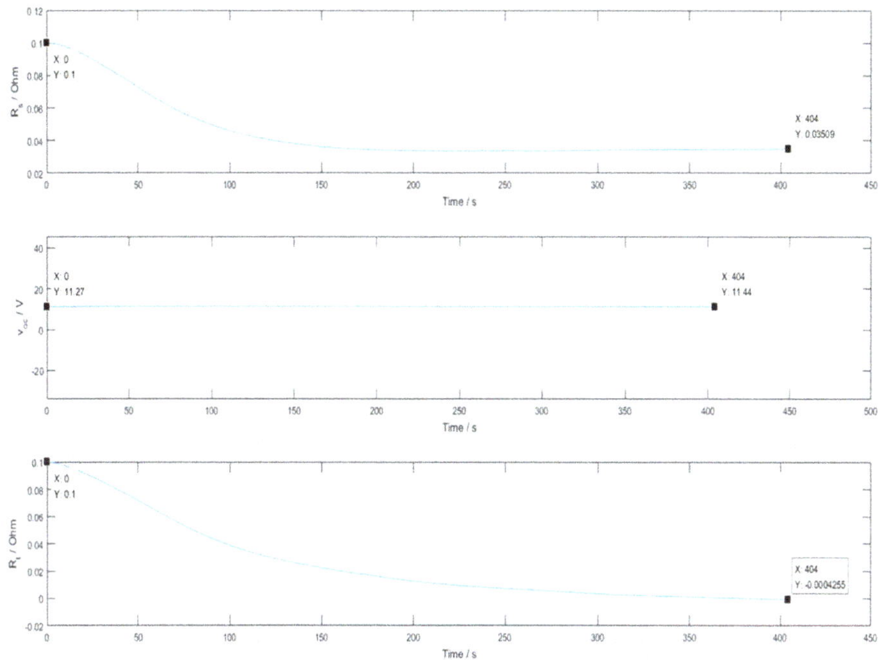

(**c**)

Figure 7. (**a**) Terminal voltage, discharge current, and estimating error; (**b**) estimating model parameters, $\hat{\theta}_i$, i = 1~4; (**c**) estimated Rs, Rt, and voc in Experiment 1.

The trajectories of estimating parameters are illustrated in Figure 7b, and the parameters, $\hat{\theta}_1$,$\hat{\theta}_2$, $\hat{\theta}_3$, and $\hat{\theta}_4$ approach values of 0.1, 19.99, 100, and 1127, respectively. R_s, R_t, and OCV are listed in Figure 7c. The trajectory of ohmic resistance relates the parameter, $\hat{\theta}_1$. The polarized resistance converges to 0.03 Ω, and the OCV converges gradually from 11.27 to 11.44 V.

5.2.2. Experiment 2

To validate the compatibility of adaptive control scheme in an EV level's voltage, a 50.4 V battery pack was applied in Experiment 2. The working voltage, current, and estimating error are shown in Figure 8a. The estimating error has converged to near zero in 150 s. The trajectories of the estimating parameters are listed in Figure 8b, the steady parameters, $\hat{\theta}_1$,$\hat{\theta}_2$, $\hat{\theta}_3$, and $\hat{\theta}_4$ converge to 0.078 Ω, 7.45 F^{-1}, 99.55 S^{-1}, and 4950 V^{-1}, respectively. The target parameters of R_s, R_t, and OCV are plotted in Figure 8c.

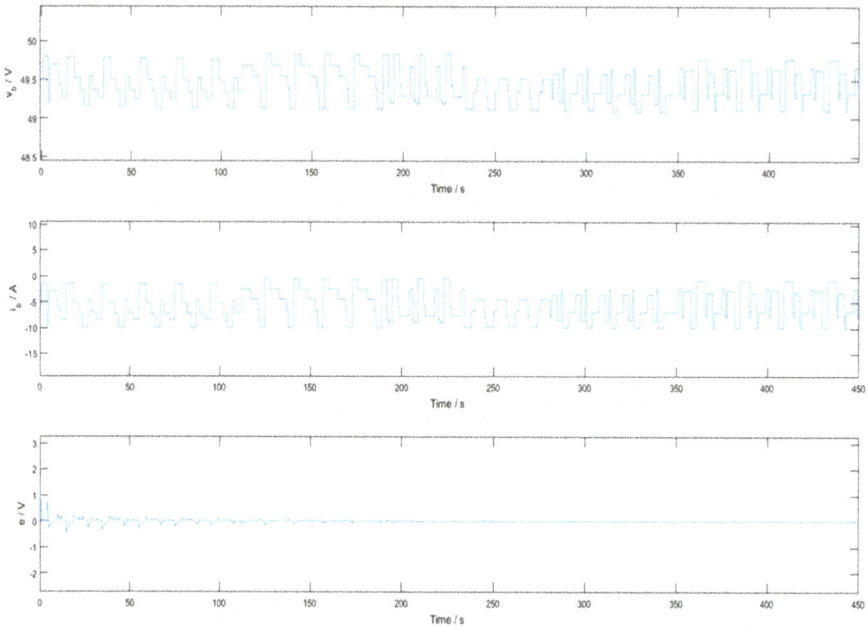

(**a**)

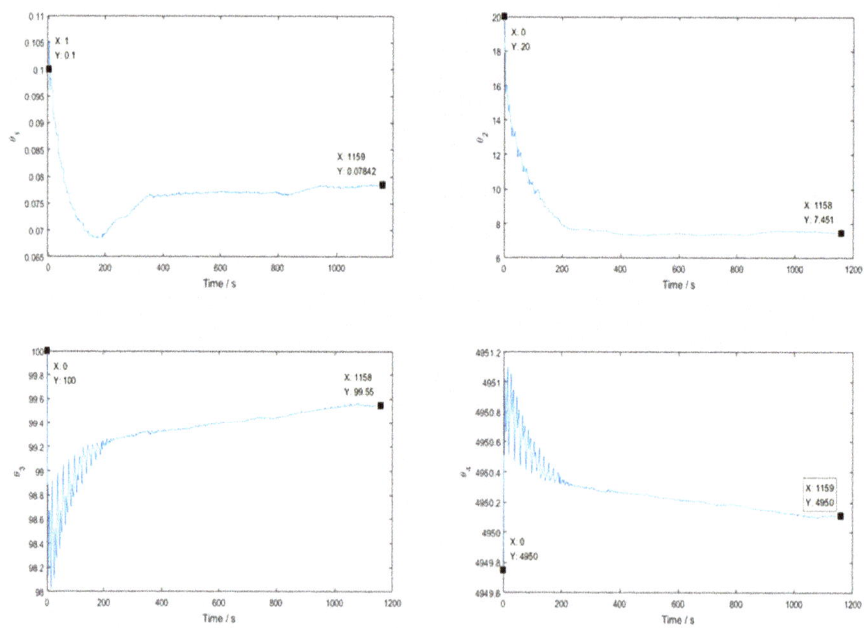

(**b**)

Figure 8. *Cont.*

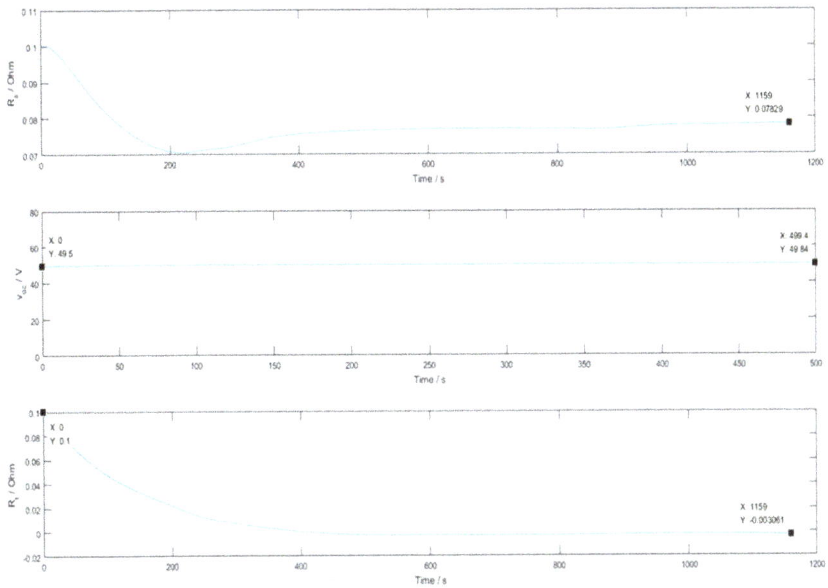

(c)

Figure 8. (**a**) Terminal voltage, discharge current, and estimating error; (**b**) estimating model parameters, $\hat{\theta}_i$, i = 1~4; (**c**) estimated Rs, Rt, and voc in Experiment 2.

For checking online battery management for a battery, an offline numerical model based on a schematic method [16] was established, and is shown in Figure 9. From sudden voltage drop and voltage rise in one discharge, it is possible to roughly estimate R_s, R_t, and C_t in the figure. Then, the analytical solution of voltage drop can be derived, as listed in Figure 9. In experiment 2, battery discharges 5 A for 12 min and rests for 1 h, then repeats this pattern for 12 h, as shown in Figure 10a. In Figure 10b, the red and blue lines represent the offline numerical and online estimated results, respectively. Trends of estimating OCV and IRs estimation are close to the offline numerical model. The deviation of OCV is roughly within 2%. Furthermore, the battery pack is composed of continuous discharging patterns, as shown in Figure 11a. For OCV, the estimated values, in the green line, are compared with offline numerical results, in the red line, as shown in Figure 11b. The deviation between online and offline values in the left-hand side of Figure 11b might be due to the heat accumulated in the battery, but the trend is identical. In this study, the algorithm of OCV and internal resistance assumes the battery pack is well-ventilated and isothermal. Temperature is not considered in the calculation. As for R_t, the online values are almost overlapped with offline results, as expected.

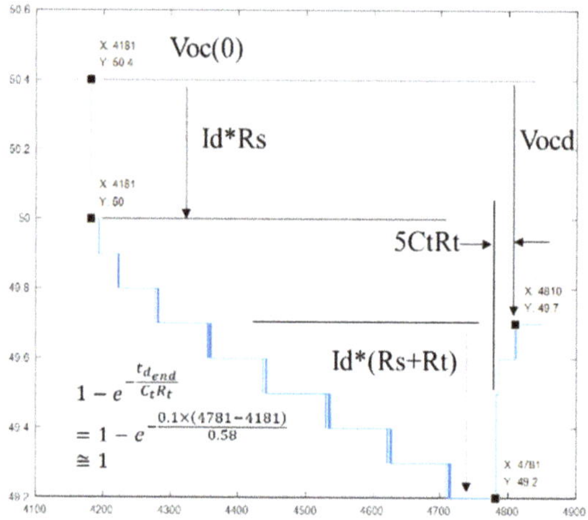

Figure 9. An offline numerical model based on the schematic method [16] (Id: discharge current; td: discharge time).

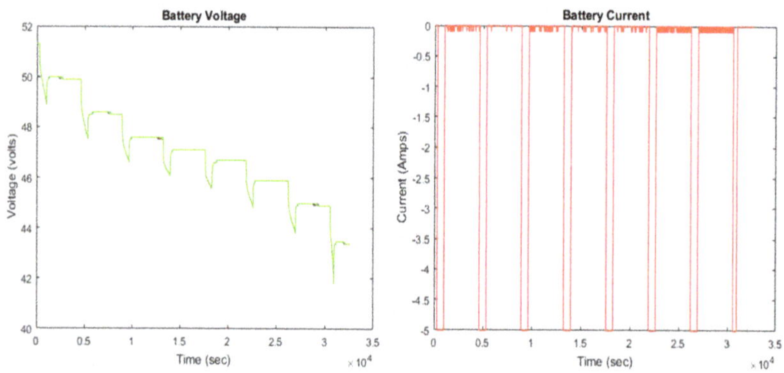

(a)

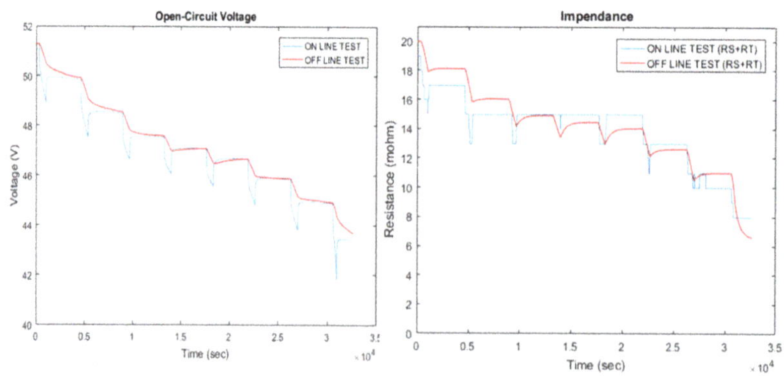

(b)

Figure 10. (**a**) Voltage drop and discharge current; (**b**) comparison of online estimating open-circuit voltage (OCV) and internal resistances (IRs) with the offline numerical model.

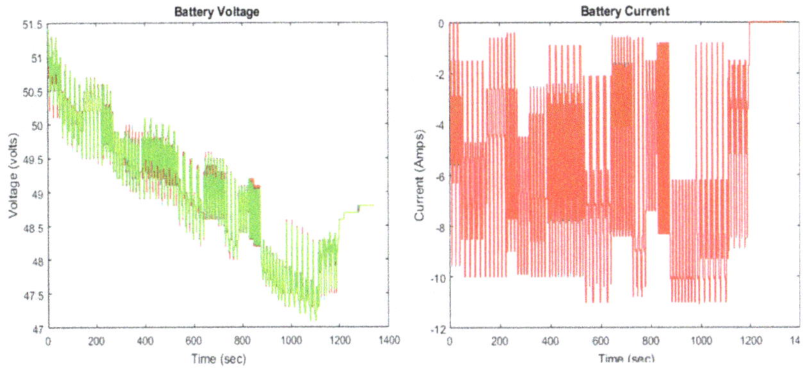

(**a**)

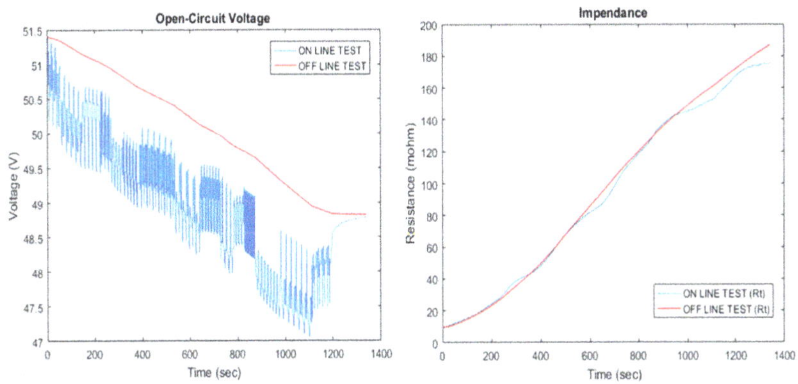

(**b**)

Figure 11. (**a**) Voltage drop and discharge current; (**b**) comparison of online estimating OCV and Rt with the offline numerical model.

6. Conclusions

In this study, an in-house BMS is developed for RLIB. Additional enhanced battery management is established. Adaptive control schemes in the BMS are established for estimating a battery's IRs and OCV relative to key parameters of RLIB, e.g., SOC and SOH, respectively. Here, IRs and OCV of a battery pack are accurately extracted from working voltage and discharge current in two experiments in this study. An offline numerical model using the schematic method is applied to verify the results of the online proposed scheme. In experiment 2, results of online estimations regarding OCV and IRs show good agreement with offline numerical model. The deviation of OCV is roughly within 2%. Furthermore, a hybrid battery pack using a UC is proposed to share peak power of RLIB by adjusting the duty ratio in the BMS. It shows more constrained voltage drops of battery when increasing the duty ratio of UC. The UC effectively reduces the voltage drop and decreases the DOD of the battery in the life extension test. It is shown that enhancing battery management for an RLIB can properly estimate OCV and IRs, and actively extend the life of the battery. For a new battery, offline estimation of parameters such as OCV and IRs meet the requirement. However, enhancing online management is indispensable on safety. This study proves the achievability of this managing solution for RLIBs. In the next phase, an RLIB integrated with this in-house BMS will be arranged, for storage of the intermittent renewable energies in site, in order to evaluate the performance.

Author Contributions: Conceptualization: W.-Y.S. and Y.-H.C.; methodology: Y.-H.C.; software: W.-P.Y.; validation: W.-P.Y., W.-Y.S., and Y.-H.C.; formal analysis: W.-P.Y., W.-Y.S., and Y.-H.C.; writing—original draft preparation: W.-P.Y.; writing—review and editing: W.-Y.S. and S.-M.J.; supervision: Y.-H.C. and S.-M.J.; project administration: W.-Y.S.; funding acquisition: W.-Y.S. All authors have read and agreed to the published version of the manuscript.

Funding: This research was financially supported by the Chung Yuan Christian University (Project No: 109609432).

Conflicts of Interest: The authors declare no conflicts of interest.

Nomenclature

DOD	depth of discharge, %
ECM	electrical circuit model
IGBT	insulated gate bipolar transistor
v_b	voltage of battery, A
v_{oc}	open-circuit voltage (OCV), V
PWM	pulse-width modulation
RLIB	reused lithium-ion battery
R_s	first-order, ohm
R_t	second-order, ohm
i_b	battery current, A
v_c	voltage across RC circuit, V
I_{RS}	current across RC curcuit based on ECM, A
SoC	state of charge, %
T	temperature, K
h	historical data
u	adjustable for input parameters
UC	ultracapacitor
θ	estimated results of target parameters
C_t	first-order capacitance based on ECM, C

Subscripts: b, c, oc, s, t: battery, capacitance, open-circuit, first-order, second-order parameter, based on ECM.

References

1. Van der Hoek, J.P.; Mol, S.; Giorgi, S.; Ahmad, J.I.; Liu, G.; Medema, G. Energy recovery from the water cycle: Thermal energy from drinking water. *Energy* **2018**, *162*, 977–987. [CrossRef]
2. Bala, B.K.; Alam, M.S.; Debnath, N. Energy Perspective of Climate Change: The Case of Bangladesh. *Strateg. Plan. Energy Environ.* **2014**, *33*, 6–22. [CrossRef]
3. Wai, R.; Jhung, S. Design of energy-saving adaptive fast-charging control strategy for Li-FePO4 battery module. *IET Power Electron.* **2012**, *5*, 1684–1693. [CrossRef]
4. Omar, N.; Daowd, M.; Hegazy, O.; Mulder, G.; Timmermans, Je.; Coosemans, T.; van den Bossche, P.; van Mierlo, J. Standardization Work for BEV and HEV Applications: Critical Appraisal of Recent Traction Battery Documents. *Energies* **2012**, *5*, 138. [CrossRef]
5. Song, X.; Hu, S.; Chen, D.; Zhu, B. Estimation of waste battery generation and analysis of the waste battery recycling system in China. *J. Ind. Ecol.* **2017**, *21*, 57–69. [CrossRef]
6. Provazi, K.; Campos, B.A.; Espinosa, D.C.R.; Tenório, J.A.S. Metal separation from mixed types of batteries using selective precipitation and liquid–liquid extraction techniques. *Waste Manag.* **2011**, *31*, 59–64. [CrossRef] [PubMed]
7. Chiang, Y.-H.; Sean, W.-Y.; Wu, C.-H.; Huang, C.-Y. Development of a converterless energy management system for reusing automotive lithium-ion battery applied in smart-grid balancing. *J. Clean. Prod.* **2017**, *156*, 750–756. [CrossRef]
8. Richa, K.; Babbitt, C.W.; Gaustad, G.; Wang, X. A future perspective on lithium-ion battery waste flows from electric vehicles. *Resour. Conserv. Recycl.* **2014**, *83*, 63–76. [CrossRef]

9. Ahmed, R.; Sayed, M.E.; Arasaratnam, I.; Tjong, J.; Habibi, S. Reduced-Order Electrochemical Model Parameters Identification and SOC Estimation for Healthy and Aged Li-Ion Batteries Part I: Parameterization Model Development for Healthy Batteries. *IEEE J. Emerg. Sel. Top. Power Electron.* **2014**, *2*, 659–677. [CrossRef]
10. Shareef, H.; Islam, M.M.; Mohamed, A. A review of the stage-of-the-art charging technologies, placement methodologies, and impacts of electric vehicles. *Renew. Sustain. Energy Rev.* **2016**, *64*, 403–420. [CrossRef]
11. Zhang, Y.; Sivakumar, M.; Yang, S.; Enever, K.; Ramezanianpour, M. Application of solar energy in water treatment processes: A review. *Desalination* **2018**, *428*, 116–145. [CrossRef]
12. Bendary, A.F.; Ismail, M.M. Battery Charge Management for Hybrid PV/Wind/Fuel Cell with Storage Battery. *Energy Procedia* **2019**, *162*, 107–116. [CrossRef]
13. Ibrahim, H.; Ilinca, A.; Perron, J. Energy storage systems—Characteristics and comparisons. *Renew. Sustain. Energy Rev.* **2008**, *12*, 1221–1250. [CrossRef]
14. Ng, K.S.; Moo, C.-S.; Chen, Y.-P.; Hsieh, Y.-C. Enhanced coulomb counting method for estimating state-of-charge and state-of-health of lithium-ion batteries. *Appl. Energy* **2009**, *86*, 1506–1511. [CrossRef]
15. Speirs, J.; Contestabile, M.; Houari, Y.; Gross, R. The future of lithium availability for electric vehicle batteries. *Renew. Sustain. Energy Rev.* **2014**, *35*, 183–193. [CrossRef]
16. Chiang, Y.H.; Sean, W.Y.; Ke, J.C. Online estimation of internal resistance and open-circuit voltage of lithium-ion batteries in electric vehicles. *J. Power Sources* **2011**, *196*, 3921–3932. (In English) [CrossRef]
17. Hannan, M.A.; Lipu, M.S.H.; Hussain, A.; Mohamed, A. A review of lithium-ion battery state of charge estimation and management system in electric vehicle applications: Challenges and recommendations. *Renew. Sustain. Energy Rev.* **2017**, *78*, 834–854. [CrossRef]
18. Shuo, P.; Farrell, J.; Jie, D.; Barth, M. Battery state-of-charge estimation. In Proceedings of the 2001 American Control Conference (Cat. No.01CH37148), Arlington, VA, USA, 25–27 June 2001; pp. 1644–1649.
19. Salkind, A.J.; Fennie, C.; Singh, P.; Atwater, T.; Reisner, D.E. Determination of state-of-charge and state-of-health of batteries by fuzzy logic methodology. *J. Power Sources* **1999**, *80*, 293–300. [CrossRef]
20. Farag, M. Lithium-Ion Batteries: Modelling and State of Charge Estimation. Master's Thesis, McMaster University, Hamilton, ON, Canada, 2013.
21. Hu, C.; Jain, G.; Tamirisa, P.; Gorka, T. Method for estimating capacity and predicting remaining useful life of lithium-ion battery. *Appl. Energy* **2014**, *126*, 182–189. [CrossRef]
22. Wu, C.-H.; Chiang, Y.-H.; Sean, W.-Y.; Lo, S.-M.; Ke, J.-C.; Hung, Y.-H. Optimal designs and experimental verification for a hybrid energy storage system. In Proceedings of the 2010 International Symposium on Computer, Communication, Control and Automation (3CA), Tainan, Taiwan, 5–7 May 2010.
23. Available online: http://www.mpoweruk.com/performance.htm (accessed on 1 May 2020).
24. Maretat, H.; Jalalmaab, M.; Azad, N.L. Energy management of battery electric vehicles hybridized with supercapacitor using stochastic dynamic programming. In Proceedings of the 2018 SICE International Symposium on Control Systems (SICE ISCS), Piscataway, NJ, USA, 9–11 March 2018; pp. 199–205.

© 2020 by the authors. Licensee MDPI, Basel, Switzerland. This article is an open access article distributed under the terms and conditions of the Creative Commons Attribution (CC BY) license (http://creativecommons.org/licenses/by/4.0/).

energies

MDPI

Article

Comparative Analysis of National Policies for Electric Vehicle Uptake Using Econometric Models

Jia Yao [1], Siqin Xiong [1,2,*] and Xiaoming Ma [1,2]

[1] School of Environment and Energy, Shenzhen Graduate School, Peking University, Shenzhen 518055, China; yao.jia@pku.edu.cn (J.Y.); xmma@pku.edu.cn (X.M.)

[2] College of Environmental Sciences and Engineering, Peking University, Beijing 100871, China

* Correspondence: xiongsiqin@pku.edu.cn

Received: 30 May 2020; Accepted: 6 July 2020; Published: 13 July 2020

Abstract: As electric vehicles (EVs) have been widely discussed as a promising way to mitigate the effect of climate change, various policies have been implemented across the world to promote the uptake of EVs. Policymakers also paid attention to the density of public charging points. In this paper, we examined the impact of policies on EV markets in the post subsidy era with multiple linear regression analysis using panel data on 13 countries from 2015 to 2018. Five of the independent variables showed significantly positive effects on the 1% level in different regression models: fast/slow charger density, mandate, purchasing restriction and waiver. Subsidies showed significance only on 5% level for battery electric vehicles (BEVs). Financial stimulates have experienced a declining marginal effect, whereas a high density of fast chargers has the most significantly positive effect on EV uptake. This paper suggests policymakers can invest more in completing the public infrastructures of EVs, especially on fast charging points.

Keywords: electric vehicle; policy incentive; public charging infrastructure

1. Introduction

In order to lower greenhouse gas (GHG) emissions and to help achieve the goal of climate change mitigation, electric vehicle (EV) adoption has been discussed worldwide. Policies aiming to incentivize consumers have been widely adopted by governments in North America, Europe, and Asia. The effectiveness of policies to increase the market share of EVs has always been discussed in the literature.

We all know that the higher acquisition cost of electric vehicles compared to conventional vehicles is one of the important barriers to adoption, due to R&D investment and fixed asset investment of the car manufactures. For a new environment-friendly technology, the barriers also include the lack of knowledge for consumers and low consumer risk tolerance [1], leading to an inefficient allocation of goods and services, known as a market failure in economics. Thus, adopting policy incentives is reasonable to fix the market.

As for electric vehicles, policy incentives include financial incentives and non-financial incentives. Table 1 is adapted from reports of International Energy Agency (IEA), listing the existing policies in some selected countries in 2018/2019 [2]. Financial incentives consist of direct purchasing subsidy, registration/emission/tax fee exemption, etc., which are most widely used worldwide to lower the initial purchasing cost and cost in daily use. The authors of [3] studied financial and tax incentive policies in Norway based on a survey and found that tax exemptions were significant motivators for more than 80% of the respondents. Moreover, [4] concluded that the markets with high EV penetration, such as Norway, the Netherlands, and the State of California, were mainly attributed to supportive incentive policies. However, existing literatures have conflicting results regarding the effectiveness of financial incentives. The authors of [5] found that financial incentives were significantly correlated to

EV uptake, whereas [6] showed a very weak correlation between purchase subsidies and consumers' willingness to buy EVs. There are literatures show that repealing incentives may result in a sharp drop in sales. For example, the U.S. state Georgia experienced a drop in EV sales by over 80% after the state government repealed the tax credit [7] and [8] found that when subsidies and tax exemption for EVs are abrogated with no change to other policies, the market share of EVs will suffer a sharp fall by 42%. Thus, it is important to explore the crucial factors influencing EV adoption.

Table 1. Electric vehicle (EV)-related policies in selected regions.

	Policies	Canada	China	European Union	India	Japan	United States
Regulations (vehicles)	ZEV (zero-emissions vehicle) mandate	√ *	√				√ *
	Fuel economy standards	√	√	√	√	√	√
Incentives (vehicles)	Fiscal incentives	√	√	√	√		√
Industrial policies	Subsidy	√	√			√	
Regulations (chargers)	Hardware standards **	√	√	√	√	√	√
	Building regulations	√ *	√ *	√	√		√ *
Incentives (chargers)	Fiscal incentives	√	√	√		√	√ *

* Indicates that the policy is only implemented at a state/province/local level. ** Standards for chargers are a fundamental prerequisite for the development of EV supply equipment. All regions listed here have developed standards for chargers.

As substitutes and complements of financial incentives, non-financial incentives and public charging infrastructures also matter a lot to EV adoption. Non-financial incentive policies are designed for the convenience of EV users and vary across countries, such as free parking policy, toll tax exemption, highway lane excess, etc. The authors of [3] supported that convenience measures such as free parking are very effective for an increase in EV sales using stated preference (SP) data. Based on a large number of EV consumers, free access to bus lanes and exemption from toll taxes were the crucial factors for EV adoption. Furthermore, [9] attributed about 25% of California's EV sales to high-occupancy vehicle (HOV) lane access policy. However, [10] demonstrated that access to bus lanes and road toll waivers are not statistically significant, whereas charging infrastructures are a strong predictor to EV adoption. The authors of [5] drew a similar conclusion regarding the relationship between the market share of electric vehicles and the number of charging stations per capita on national level.

Besides, government regulations are important factors to facilitate EV sales, targeting both consumers and automobile manufacturers, usually including setting goals for EV uptake, restricting the fuel used and mandating the EV production. A few examples of such regulations are GHG standards, a zero-emission vehicle mandate, and a low-carbon fuel standard. The authors of [11] revealed that corporate average fuel economy (CAFE) regulation could help promote the market penetration of EV, especially if executed along with other incentives. The authors of [12] studied dual-credit policy in China and found that the Corporate Average Fuel Consumption rules alone may stimulate more plug-in electric vehicle (PEV) sales than the dual-credit policy; however, the dual-credit policy could stimulate more battery electric vehicles (BEVs) in the market compared to other policy scenarios. Additionally, more economic benefits could be achievable for fleet owners once the carbon tax is introduced [13].

Battery electric vehicles (BEVs) operating solely on electricity are not the only alternative for the electrification of passenger cars. Plug-in hybrid electric vehicles (PHEV) running on both gasoline and electricity, usually with a lower cruising range, have the potential to replace fuel used with electricity and thus lower the impact on the environment, without compromising the range of the vehicle [14]. Since PHEVs do not suffer from range limitation, they bring higher flexibility for drivers and can be well suited for diversified driving needs. For which type of EVs to become a major option on the private car market, the total economics of the EV type would presumably have to be favorable compared to the alternatives, since the currently dominant battery technology (Li-ion) is still relatively expensive. Earlier studies have commonly focused on total battery cost and some recent studies have discussed the marginal cost and its effect on cost-effective battery sizing. For example, [15] found the battery range has a small impact on the total cost of ownership (TCO) for PHEVs, while [16] concluded that short-range PHEVs would reduce more gasoline consumption than PHEVs with a larger range. These different characteristics caused different market shares for EVs and PHEVs even in the same country. It will be interesting to study the effects of incentives on BEVs and PHEVs separately.

The trend of the relevant studies in recent five years is summarized in Table 2, adapted from a review by [13], selecting researches in field of Business Management and policy, as well as the field of Transportation and Environment science together. The adoption of electric vehicles is most frequently discussed in the USA and China, since the they have the largest EV markets. Studies in the USA are the most over the years, while studies based on other countries have been increasing, with China being the second, especially from 2017 onwards. The more prominent research design used in literature is a quantitative method compared to a qualitative method. Survey-based methodology is the most predominant method used for EV adoption, followed by simulation, optimization techniques, and secondary data analysis. Most cross-country studies used survey-based analysis and secondary data analysis, which indicates that heterogeneity, causality, and locational disparity are important indicators in the area [17]. Noticeably, the stated preference (SP) data of consumers may be subjective and are usually not be in accordance with reality [8]; thus, quantitative research using fact analysis may be preferred.

Table 2. Trend of the relevant studies in recent years [1] [13].

Research Categories	2015	2016	2017	2018	2019	Total
Country wise stratification (top 5)						
USA	8	6	10	17	9	50
China	0	0	8	10	6	24
Across countries	4	3	2	9	3	21
Germany	2	1	0	5	0	8
United Kingdom	0	2	3	2	2	9
Research design						
Qualitative	1	1	1	3	5	11
Quantitative	25	19	41	62	30	177
Methodology used (top 5)						
Survey based	4	5	12	24	7	52
Simulation based	8	4	10	5	3	30
Optimization techniques	2	3	7	9	5	26
Secondary data analysis	2	2	5	11	3	23
Mixed (combination of above)	2	2	1	2	5	12

[1] Selecting only Scopus Q1 Journal ranking in 2017 in field of Business Management and policy, as well as field of Transportation and Environment science together.

Though plenty of researchers have done a considerable amount of work on EV-related policies, few papers have discussed the diminishing effect of subsidies for EVs using quantitative analysis on cross-country data. This scenario is also well known as the post subsidy era, referring to the stage when subsidies retreat and have less impact on consumers' purchasing decisions. This paper focuses on this recent subtle change of subsidies and fills the gap by using multiple linear regression and panel data on 13 countries from 2015 to 2018 to examine the impact of policies and infrastructures on electric vehicle market.

The rest of the paper is organized as follows. Section 2 gives the descriptive statistics of the data in 13 countries and introduces the methodology. Section 3 demonstrates the impact of the policy incentives and infrastructures on EV uptake, and explanations for the findings are given in this section. Section 4 draws conclusions and provides policy implications.

2. Data and Methods

2.1. Data Description

EV market share and sales data were collected from International Energy Agency [2], containing 13 countries over 4 years (2015–2018). These countries are Canada, China, France, Germany, India, Japan, Korea, the Netherlands, Norway, Portugal, Sweden, the United Kingdom and the United States. Electric vehicles in the 13 countries accounted for over 90% worldwide in recent years. The macroeconomic data were collected from the website of World Bank. Details of variables and data sources are provided in Appendix A.

Policies we selected in the paper are considered as important ones to EV adoption or have conflicting results in related literatures. All these countries mentioned above are adopting targeted zero emission vehicle (ZEV) regulations and incentives to accelerate the rate of deployment. Mandate and purchasing restriction are dummy variables, indicating whether it is implemented. The other non-financial policies are categorical variables evaluated by policy intensity. The wider the policy is applied in the country, the higher the scores it gets. Details about how we quantify the effect of non-financial policies by assigning different values are provided in Appendix B.

Table 3 provides the financial and non-financial policies we have considered in our model. Fuel standard is the level of fuel standards/regulations drivers need to meet and has been raised over the years to meet air quality standards and greenhouse gas emission reduction goals. ZEV (Zero emission vehicle) mandates on manufacturers can also result in increased model availability in the market. Insufficient model options can deter consumers from purchasing EVs even after adequate emphasis on consumer incentives and charging infrastructures.

Among the non-financial policy incentives, the variable target evaluates the governments' ambition to promote EVs. Different countries released a different target year to ban internal combustion engine (ICE) sales and achieve 100% ZEV sales. For example, Norway took the lead and promised to replace all fuel cars by 2025 [18], followed by some other European countries, such as Denmark and Iceland, announcing 100% ZEV by 2030. The Netherlands and the UK promised to replace all traditional vehicles later than 2040. The earlier the 100% ZEV target year is, the higher scores the variable target will get. We expected a positive relationship between the goal and EV uptake. [19] has highlighted the importance of policy goals in the UK and Germany to decrease GHG emissions; however, [20] found that the EV climate mitigation strategy was not effective in the United Kingdom. Despite the conflicting conclusions, the policy goal will be considered and tested in our model.

Table 3. Financial and non-financial variables.

Financial Variables		Non-Financial Variables
Purchase phase	Use phase	Waivers on access restrictions
Subsidies	Circulation tax exemptions	Access to HOV lanes
Sales tax exemptions (excl. VAT)	Waivers on fees (e.g., tolls, parking, ferries)	Access to restricted traffic zones
VAT exemptions	Electricity supply reductions/exemptions	Access to bus lanes
Tax credits		
Fuel Standard: the level of fuel standards/regulations in the countries		
Mandate: ZEV (Zero emission vehicle) mandate level in the countries		
Target: the target year when the 100% ZEV sales goal is achieved (Heidrich et al., 2017)		

For better accuracy, after using dummy variables to describe the policies, we collected numerical dollar values to describe the purchasing incentives (usually are tax credits or purchasing subsidies). The numerical incentives are used in the extended analysis part where BEVs and PHEVs are discussed, respectively. BEVs, which totally use electricity were able to receive higher subsidies because the level of subsidies was usually decided by battery capacity across the countries. In this study, we assigned the highest level of money (tax credits or subsidies) a passenger vehicle can receive from the central government to the variable subsidy. Though we did not use median or average values because of data unavailability, the maximum value can still reflect the intensity of the subsidy, since both manufacturers and consumers will strive for the maximum subsidy. It is also a reason that a government giving a higher upper limit for subsidy standard, tends to provide a higher medium level of subsidies, that is to say, the maximum value of standard should be highly related to the average subsidy consumers get. Furthermore, we used linear regression for analysis; thus, we care much more about the relative relationship between countries and the difference between years, than the absolute value of subsidies. Thus, using the highest level of subsidy standard is reasonable. Besides, the subsidy itself is based on many parameters, and each country or even each province has different standards. For a global scope research involving 13 countries and 4 years, it is too complicated to refine and calculate subsidies and we only divided it into two categories: BEV and PHEV subsidies, which are considered as two variables and will be applied to BEV and PHEV sales, respectively.

For BEVs, Korea and China offered the highest subsidy in Asia. The subsidy in Korea was up to 10,900 dollars and was the highest of all countries over the years, although the standard of getting total subsidy was not easy to meet. China offered a subsidy up to 8900 dollars before 2018, slightly higher than the federal tax credits (USD 7500) in the United States. Norway was the most generous country in Europe to subsidize the promotion of electric vehicles with the highest level of 10,300 dollars over the 4 years. Except for France and Sweden, the other European countries, such as Germany and Portugal, gave lower than average subsidies (USD 6000) to EV consumers. Among all the countries listed in our study, India gave the least subsidies of up to 2000 dollars for BEVs, lower than Portugal (USD 2400), which ranked the second last.

For PHEVs, subsidies were much lower. The UK with USD 7100 subsidies ranked first and Norway with USD 7000 ranked second. Though Korea offered the highest subsidy for BEVs, PHEVs could only get the maximum of USD 860 dollars. When BEVs and PHEVs are considered as a whole, we used the weighted average of subsidies by the sales volume. Appendix B shows the value of some non-financial policy and numerical variables, such as subsidies and charger density, in 2018.

Table 4 below is the summary statistics, which contains the characteristics of all variables. The variables fast/slow chargers per million population means the number of chargers per million people in the country, which can also be presented by chargers' density. EV share is the proportion of electric vehicle sales in total vehicle sales. We implemented a logit transformation to normalize the distribution of EV share because the data are skewed to the right [21]. After the transformation, the data showed a normal distribution and validated the ordinary least squares (OLS) regression we plan to use. This is one of the necessary conditions for a good estimation result, not a sufficient condition. OLS regression also needs to satisfy the assumptions of homoscedasticity of the error term, no multicollinearity, etc. We have tested them below in the study and all results are satisfactory. To solve traffic congestion and air pollution problem, China adopts administrative orders to control the supply of vehicle licenses, while the restrictions is not applicable for EVs [22]. To identify this effect, the variable purchasing restriction is considered in the pooled regression in the basic results part.

Table 4. Summary statistics.

Variables	(1) N	(2) Mean	(3) Sd	(4) Min	(5) Max
EV share	52	0.0451	0.0895	0.0002	0.464
EV sales	52	80.98	175.5	0.450	1079
Subsidies	52	5.150	2.575	1.717	11.66
BEV subsidy	52	6.008	2.695	2	12
PHEV subsidy	52	3.209	2.374	0	9
Waiver	52	0.981	0.779	0	2
Mandate	52	0.135	0.345	0	1
Target	52	0.346	0.814	0	3
Fuel standard	52	0.808	0.445	0	2
Purchasing restriction	52	0.0769	0.269	0	1
Fast chargers per million population	52	35.24	42.32	0	230.7
Slow chargers per million population	52	375.1	569.8	0	2097
GDP per capita	52	41,686	22,498	1752	92,121
Population density	52	215.2	184.7	3.926	529.7

In Table 5, correlation coefficients are provided. Between a pair of independent variables, the correlation between slow chargers and fast chargers shows the largest cross-correlation coefficient of 0.804. Thus, there should be no severe linear correlations during regressions.

Table 5. Correlation coefficients of variables.

Variables	Subsidies	Waiver	Mandate	Target	Fuel Standard	Ln (Fast-Charger Density)	Ln (Slow-Charger Density)	Ln (GDP per Capita)	Ln (Population Density)
Subsidies	1								
Waiver	0.222	1							
Mandate	0.0968	0.141	1						
Target	0.0519	0.341 *	−0.177	1					
Fuel standard	0.417 **	0.513 ***	0.293 *	0.168	1				
ln(Fast-charger density)	0.342 *	0.335 *	0.0382	0.197	0.431 **	1			
ln(Slow-charger density)	0.0684	0.392 **	−0.0493	0.351 *	0.369 **	0.804 ***	1		
ln(GDP per Capita)	0.183	0.118	−0.128	0.152	0.279 *	0.643 ***	0.802 ***	1	
ln(Population density)	−0.0236	0.202	−0.262	0.355 *	−0.217	−0.131	−0.241	−0.353 *	1

* $p < 0.05$, ** $p < 0.01$, *** $p < 0.001$.

For an overview of the electric vehicle markets, we used cross-sectional data in 2018 to draw a worldwide map of the sales and market share of EVs as shown in Figure 1. Though both crediting to the governments' effort in promotion, market shares and sales give different information on the countries' EV market. EV sales is more related to the market volume given that China and United States made remarkable achievements, while market shares showed the overall recognition of the citizens for EVs; thus, European countries showed higher EV share, especially Norway, the highest of all. Later in the basic result part, we will interpret the difference with statistical models.

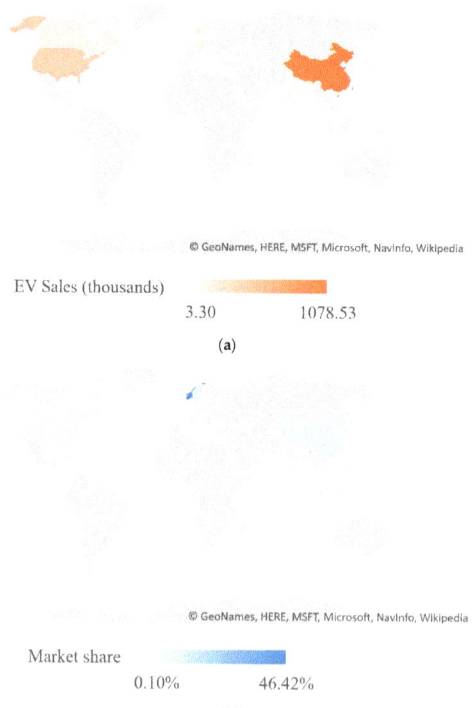

Figure 1. EV sales and market share worldwide overview. (**a**) EV sales in 2018; (**b**) market share of EVs in 2018.

2.2. Model

We used panel data regression to analyze EV uptake to reduce time-invariant heterogeneity resulting from some unobservable variables that affect the EV market share in different countries. The multiple linear regression model with panel data is shown as follows:

$$\begin{aligned} ln(y_{it}) = \alpha + \beta_1 \quad & (Subsidies_{it}) + \beta_2(Waiver_{it}) + \beta_3(Mandate_{it}) + \beta_4(Fuel\ Standard_{it}) \\ & + \beta_5\ ln\left(Fast\ Charger\ Density_{it}\right) + \beta_6\ ln\left(Slow\ Charger\ Density_{it}\right) \\ & + \beta_7\ ln\left(GDP\ per\ Capita_{it}\right) + \beta_8\ ln\left(Population_{it}\right) \\ & + \beta_9\ ln(Purchasing\ Restriction_{it}) + u_j + \mu_t + \varepsilon_{it} \end{aligned} \quad (1)$$

where subscripts i and t represent the i-th country and the t-th year, respectively. The dependent variable is $ln(y_{it})$ represent a logit transformation of y for country i in year t. y represents the EV share or EV sales, respectively, in two different regressions. u_j is the fixed effects for individual countries and μ_t is the time fixed effects. Using this model, we evaluated the effectiveness of policies and charging infrastructures with macroeconomic factors controlled. To reduce the level of heteroscedasticity, some variables are taken natural logarithm along with EV shares [23].

Moreover, a simplified model of this is used in the basic results part for a pooled regression to look at the effectiveness of different policies and other influential factors, shown in Equation (2):

$$\begin{aligned} ln(y_i) = \alpha + \beta_1 \quad & (Subsidies_i) + \beta_2(Waiver_i) + \beta_3(Mandate_i) + \beta_4(Fuel\ Standard_i) \\ & + \beta_5\ ln\left(Fast\ Charger\ Density_i\right) + \beta_6\ ln\left(Slow\ Charger\ Density_i\right) \\ & + \beta_7\ ln\left(GDP\ per\ Capita_i\right) + \beta_8\ ln\left(Population_i\right) \\ & + \beta_9\ ln(Purchasing\ Restriction_i) + \varepsilon_i \end{aligned} \quad (2)$$

The pooled regression model is one type of model that has constant coefficients, referring to both intercepts and slopes. For this model, we can pool all of the data and run an ordinary least squares regression model without considering difference across countries and years.

Two types of models are usually considered for panel data regressions: the fixed effects and random effects model, different in dealing with endogeneity. Before deciding on the best regression method, we first have to figure out if our predictor variables are endogenous. The Hausman specification test was used to detect endogenous regressors in a regression model [24,25]. Details of the tests will be provided in the result part.

Besides, we used two statistical tests for our data and model in Section 3: the White test [26] (verifying the data conforming to the OLS homoscedasticity assumption) and the variance inflation factor test [27] (making sure that there is no multicollinearity problem in the pooled model). The tests showed that our data meet the requirements/assumptions of OLS regression.

For further analysis, we implemented the vector autoregression (VAR) model as an alternative method, in order to take time lag into consideration. Ordinarily, regressions reflect "mere" correlations, but [28] argued that causality in economics could be tested for by measuring the ability to predict the future values of a time series using prior values of another time series. The VAR model proposed by [29] is used to capture the linear interdependencies among multiple time series, with the lagged values of all endogenous variables to estimate the reverse impact of them [30]. Moreover, it allows us to consider both long-run and short-run restrictions justified by economic considerations [31]. Consequently, the VAR model can be used to capture the dynamic impacts of the influencing factors on EV sales. The mathematical expression of the general VAR model is given as follows:

$$y_t = c + A_1 y_{t-1} + A_2 y_{t-2} + \cdots + A_p y_{t-p} + e_t \quad (3)$$

where the observation y_{t-i} (i periods back) is called the i-th lag of y, c is a vector of constants (intercepts), A_i is a time-invariant matrix and e_t is a vector of error terms. In this paper, we used one period lag for VAR, which is *i* equals to 1, and we explored the relationship between subsidies, infrastructures and EV sales as three variables. We showed the results of Granger causality test, which is a statistical hypothesis test for determining whether one time series is useful in forecasting another, first proposed in 1969 and widely used for explaining the results of VAR models [28].

3. Results and Discussion

3.1. Descriptive Results

For further analysis of governments' effect on EV uptake, we drew pictures of the relationship between EV uptake and policy incentives in the 13 countries and then showed the correlation between EV uptake and the density of public charging infrastructures. The aim of the descriptive analysis here is to provide a vivid understanding of the effectiveness of the variables before we used econometric models; however, they are probably not accurate and pervasive enough because of the existence of other influential factors which will be eliminated to a large extent in further discussion.

Figure 2 shows the correlation between policy incentives and the market share in the countries in 2018. The bars refer to the value of incentives and the red triangles represent the level of market share. EV market share in European countries are usually higher than in other countries. Norway had the highest level of EV uptake in the 13 countries, up to about 46% with highest policy incentives value; however, the relationship between uptake and policies is not always right for other countries. Korea and the U.S. had nearly the same scores for policy incentives as Norway; however, the uptake was not as good. Some European countries may not perform outstandingly in policies while the EV uptake was a lot higher than average. Overall, the descriptive analysis shows a relatively weak dependence between policy incentives and EV uptake.

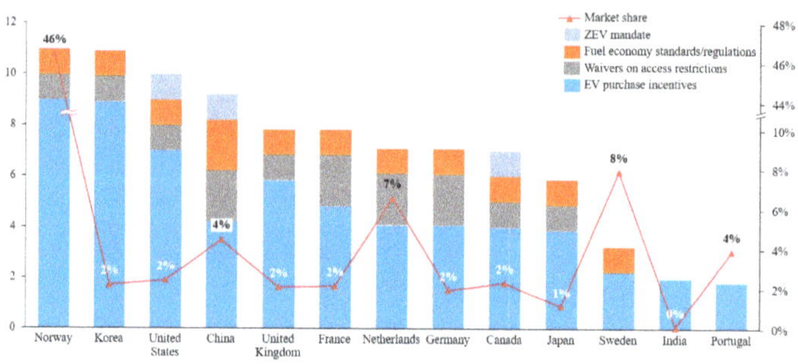

Figure 2. EV market share and policy incentives in 2018.

The relationship between market share and density of public chargers is depicted in Figures 3 and 4. The vertical axis and horizontal are the numbers of fast chargers and slow chargers per million population, respectively. The size of the bubble represents market share for the particular country. Norway performs best in both the density of fast and slow chargers. The Netherlands has a higher level in slow chargers than fast chargers. In China, fast chargers are developing relatively faster in China than slow chargers. Overall, it is consistent with the assumption that the larger the bubble size, the higher the country ranks in both axes.

3.2. Basic Results

We conducted a simple pooled regression model of all the countries over the years to describe the effectiveness of different policies and other influential factors. After performing the White test, we verified that the data conform to the OLS homoscedasticity assumption, that is, the variance of the errors in the regression models is constant, which means that the model was well defined. Due to the relatively large number of explanatory variables, in order to assess the possible effects of multicollinearity, the model was tested for variance inflation factor (VIF) [27]. The results are shown in Table 6. The maximum value of the variance expansion factor is 5.12 and the average value is 2.54, all less than the commonly used threshold of 10. Thus, it can be considered that there is no multicollinearity problem in the pooled model.

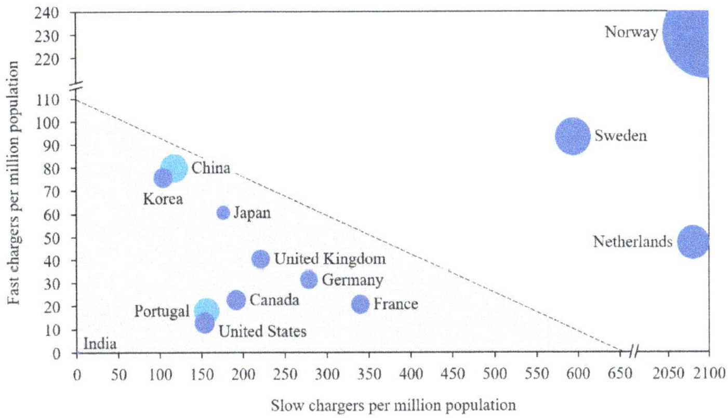

Figure 3. Market share and numbers of slow and fast chargers per million population in 2018.

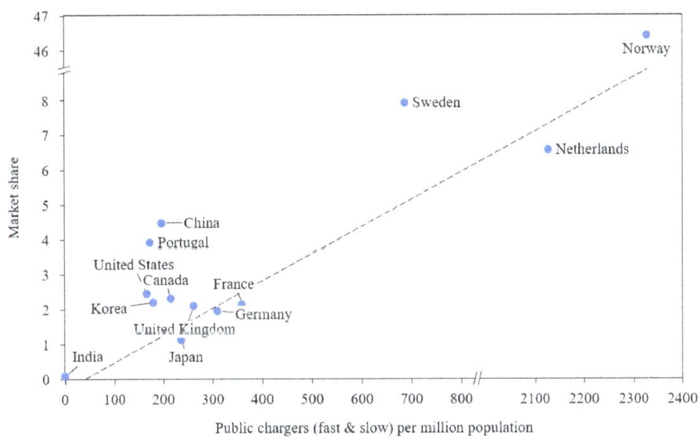

Figure 4. Market share and numbers of all public chargers per million population in 2018.

Table 6. Results of variance inflation factor.

Variable	VIF	1/VIF
ln(Slow-charger density)	5.12	0.20
ln(Fast-charger density)	4.65	0.21
Fuel standard	1.88	0.53
Waiver	1.75	0.57
Subsidies	1.64	0.61
Purchasing restriction	1.61	0.62
Target	1.32	0.76
Mandate	1.22	0.82
Mean VIF	2.54	

Using market share along with sales volume can partly eliminate the interference of external macro factors, such as income growth and economic growth [22]. We still controlled two macro factors in our regression model in order to measure the impact of macro factors. In the regression with EV share as dependent variable (the left column in Table 7), only slow-charger density was positively significant while in the regression of EV sales, mandate, waiver, fast-charger density and purchasing restriction policies showed a significant positive impact. The significant variables vary a lot if changing the dependent variable, but could be explained as follows.

For EV promotion, Europe usually started early in the history for higher citizens' environmental awareness and thus showed a higher EV uptake. In the early days, only slow chargers other than fast chargers were available, so the more public slow chargers, the more it showed that the countries took earlier actions, which is highly related to the high market share nowadays. On the other hand, EV sales are more related to the size of the market, as is shown in Figure 1, meaning that China and the U.S. performed best. Purchasing restriction for traditional cars is the policy used only in China and is proved to be significantly effective [22]. Mandate policies were also used mainly in China and North America, and the large market scale makes the mandate policy remarkable. Waivers on access restrictions are the convenience EVs brought to the driver, especially in densely populated cities. It is good to know that the policy is effective in the history of promoting EV sales volume.

An explanation for the insignificancy of subsidies and other non-financial policies is that most of the countries offered high subsidies and other preferential treatments over the years; thus, those do not make a significant difference now. We will use panel data to analyze the impact thoroughly in the next section.

Table 7. Policy impact on total EV market using pooled regression.

Variables	(1) Total EV Share	(2) Total EV Sales
Subsidies	0.0813	0.0136
	(0.0607)	(0.0515)
Waiver	−0.219	0.489 **
	(0.208)	(0.176)
Mandate	0.308	1.627 **
	(0.381)	(0.323)
Target	−0.0813	−0.188
	(0.168)	(0.142)
Fuel standard	0.135	0.393
	(0.393)	(0.333)
ln(Fast-charger density)	−0.00940	0.305 *
	(0.149)	(0.126)
ln(Slow-charger density)	0.704 **	0.00334
	(0.153)	(0.130)
Purchasing restriction	0.531	1.471 **
	(0.559)	(0.474)
Constant	−7.844 **	1.302 **
	(0.557)	(0.472)
Observations	50	50
R-squared	0.709	0.798
Adjusted R-squared	0.652	0.759

** $p < 0.01$, * $p < 0.05$.

After performing the Hausman test mentioned in Section 2, the result of test showed that the p-value is small (less than 0.05) and rejected the null hypothesis, which is that the preferred model is random effects. The result shows significant differences between the coefficients for the fixed effects and random effects model. Therefore, we used the fixed effects model in the study with EV market share and EV sales as dependent variables, respectively.

In Table 8, regression coefficients are tabulated based on four models, in which seven and nine independent variables are utilized, respectively. Compared to pooled regression, the panel model controlled the country fixed effect; thus, we deleted purchasing restriction for highly multicollinearity with the dummy variable China, the only country conducted the restriction policy. Moreover, we will not include target in the regression since the target set does not change over time, meaning that it is not suitable for panel models.

Table 8. Policy impact on total EV market.

Variables	(1) Total EV Share	(2) Total EV Share	(3) Total EV Sales	(4) Total EV Sales
Subsidies	−0.177	−0.187	−0.155	−0.161
	(0.0964)	(0.0966)	(0.0830)	(0.0839)
Waiver	0.273	0.308	0.0638	0.0943
	(0.201)	(0.201)	(0.173)	(0.175)
Mandate	−0.207	0.0817	0.0983	0.296
	(0.350)	(0.412)	(0.302)	(0.357)
Fuel standard	−0.0232	−0.185	−0.174	−0.330
	(0.375)	(0.412)	(0.323)	(0.358)
ln(Fast-charger density)	0.649 **	0.630 **	0.387 *	0.363 *
	(0.186)	(0.189)	(0.160)	(0.164)
ln(Slow-charger density)	0.186	0.316	0.0944	0.191
	(0.254)	(0.264)	(0.218)	(0.230)
ln(GDP per Capita)		−2.806		−1.237
		(4.951)		(4.299)
ln(Population density)		−22.18		−17.14
		(13.79)		(11.98)
Constant	−5.868 **	54.62	2.164	38.88
	(1.354)	(63.07)	(1.167)	(54.76)
Observations	50	50	50	50
R-squared	0.958	0.962	0.970	0.972
Macro Factors	NO	YES	NO	YES
Country FE	YES	YES	YES	YES
Year FE	YES	YES	YES	YES
Adjusted R-squared	0.927	0.928	0.947	0.948

** $p < 0.01$, * $p < 0.05$.

By gradually introducing a socioeconomic factor in the later model, the possible impact of the factor can be observed more clearly. Referring to Table 8, the adjusted R square value for each model is above 0.90. That is, the confidence in explaining the cross-country EV share/sales is above 90% for the variables tested.

Most variables are not significant referring to Table 8. Only the density of fast chargers is significantly positive for all models. The parameter means that 1% increase in density of fast chargers can cause a 0.63% increase for EV uptake or a 0.36% increase for EV sales. Our study uses data from 2015–2018, when fast-charging infrastructures rather than slow chargers were constructed rapidly worldwide and contributed considerably to the global EV development. Besides, the result that other policy incentives are not significant may also due to the time period we studied. In the post subsidy era, policies have experienced a declining marginal effect. Taking the waiver on access restrictions as an example, we see from the pooled regression that the policy did show some positive impact on EV sales in the history; however, the effect was declining over the years, because HOV and bus lanes got crowded as the number of electric vehicles increased. In other words, the benefits diminished across time for some non-financial policies. According to the literature, in post subsidy era, it is chargers' density and fuel price that really counts [8]. In this study, we will shed more light on the effect of charging infrastructures.

3.3. *Heterogeneity: Considering Different EV Types*

After the basic model analysis, we examined the effects of the policy incentives and infrastructures by separating PHEVs from BEVs. After the separation, the fast charging infrastructures have more apparent impacts for BEVs than those for PHEVs, as illustrated in Table 9 with macroeconomic factors controlled. The parameter means that a 1% increase in density of fast chargers can cause a 0.246% increase for BEV sales, while the effect for PHEV sales is not significant enough. The reason may be that PHEV drivers suffer much less from range anxiety.

The variable subsidy is significant for BEVs but not significant for PHEVs, which may due to the low subsidy PHEV get for its relatively low battery density. The average subsidies for BEVs (USD 6000) are twice the average for PHEVs (USD 3200). The models separating PHEV from BEV also further indicates the robustness for analysis [32].

Table 9. Policy impact on the battery electric vehicle (BEV) market and plug-in hybrid electric vehicle (PHEV) market.

Variables	(1) BEV Share	(2) BEV Sales	(3) PHEV Share	(4) PHEV Sales
BEV/PHEV subsidy	0.0989	0.126 *	−0.0575	−0.0739
	(0.0588)	(0.0576)	(0.131)	(0.126)
Waiver	−0.0500	0.0120	0.405	0.413
	(0.137)	(0.134)	(0.360)	(0.345)
Mandate	0.0395	0.186	0.516	0.656
	(0.272)	(0.267)	(0.752)	(0.721)
Fuel standard	−0.186	−0.220	0.301	0.261
	(0.276)	(0.270)	(0.741)	(0.710)
ln(Fast-charger density)	0.226	0.256 *	0.579	0.642
	(0.124)	(0.122)	(0.352)	(0.337)
ln(Slow-charger density)	0.0665	0.0286	0.651	0.616
	(0.170)	(0.167)	(0.460)	(0.441)
Constant	−63.07	−57.62	177.3	192.8
	(41.62)	(40.77)	(127.3)	(122.0)
Observations	50	50	48	48
R-squared	0.980	0.986	0.892	0.906
Adjusted R-squared	YES	YES	YES	YES
Macro Factors	YES	YES	YES	YES
Country FE	YES	YES	YES	YES
Year FE	0.962	0.974	0.798	0.823

** $p < 0.01$, * $p < 0.05$.

3.4. Extended Analysis: Vector Autoregression

Infrastructures for facilitating EV sales could be enhanced over years, and factors such as subsidies may influence upon others like chain reactions. Financial aid influences upon the BEV market; meanwhile, the reverse impact could occur. If the feedback relationships exist, vector autoregression should be used to reflect the dynamic relationships [30,33,34]. In this part, we are going to explore if there's reverse causality in EV markets worldwide.

The authors of [8] develop a system dynamics model of China's EV adoption to analyze the effectiveness of EV policies. In the dynamic model, the relationship of government incentives, customers' behavior and infrastructure providers showed complex causality, inspiring us to predict the values of a time series using prior values of another time series. We run a vector autoregression (VAR) model using the three variables mentioned. For the data only contain four periods, we used the first lag of each variable.

We showed the results of the Granger causality test below in Table 10. The results show that all the lagged items do not show significance, even on the 10% level, which means that all lagged variables have no explanatory power for the other variables. The reversed causal relationship is not significant referring to EV adoption; thus, the pooled and panel data regressions we used above are sufficient to explain the causal effect.

Table 10. Results of the Granger causality test.

Variables	(1) ln(EV Sales)	(2) ln(Charger Density)	(3) Subsidies
L. ln(EV sales)	53.64	−1.708	54.14
	(206.1)	(21.73)	(262.2)
L. ln(Charger density)	−257.4	11.34	−263.9
	(0)	(91.29)	(546.6)
L. Subsidies	−247.8	10.76	−252.6
	(193.6)	(88.60)	(542.4)
Observations	25	25	25

4. Conclusions and Policy Implications

4.1. Conclusions

The purpose of this research is to explore the relationship between government incentives and other related factors to electric vehicle adoption across the main countries with EVs. Using panel data from 2015 to 2018, this paper studies the EV uptake among 13 countries. An econometric model for the uptake is established with eight independent variables and two macro control variables. Five of them showed significantly positive effects on 1% level in different regression models: fast/slow charger density, mandate, purchasing restriction and waiver. Subsidies showed significance only on the 5% level for BEVs. The zero–emission vehicle (ZEV) target set did not have apparent impacts. Descriptive analysis drew the same conclusion that charger infrastructure density predicts best for electric vehicle uptake on the national level.

Fast–charge infrastructures were positively related to EV uptake all the time. In the panel data regression, a 1% increase in the density of fast chargers can cause a 0.63% increase for EV uptake or a 0.36% increase for EV sales. It is reasonable that increasing the number of charging stations contributes to EV adoption, while from another perspective, fast–charger density is a sign that the country invested in public infrastructures in recent years. Slow charge infrastructures were positively significant when using pooled regression, that is to take all years into comparation. 1% increase in the density of slow chargers can cause 0.7% increase for EV uptake. Technology always evolves gradually [35] and even though fast chargers have been common in recent years, slow chargers occupied the market in the primary stage. Those countries with high slow charger density, such as Norway and the Netherlands [18], usually have developed electric vehicle markets for years. Thus, citizens have a higher acceptance for the new mobility tools, which was shown in the national EV uptake.

On the other hand, waiver, mandate and purchasing restriction (for fuel cars) were significantly effective to sales volume though they seemed to have little impact on EV market share, in pooled regression. They all showed significance in the 1% level, and purchasing restriction has the best effect for EV sales, that is a country with the fuel car restriction would have 1.47% more sales than others, with other factors controlled. We believe the three policies were directly intended for the sales volume and because of the mandatory, it should be effective in the short run, thus not enough to effect EV share. Market share is a better indicator for the overall acceptance of EV, requiring the governments' persistent work for years, while EV sales are more sensitive to some powerful mandatory policy in the short term.

4.2. Policy Implications

The significance of the policies in the panel regression can be explained well by the logic behind consumers' behaviors. If the policy makers can understand what people really care about, the policy is likely to be effective. For example, we can use financial policies to encourage the adoption of EVs; however, if we cannot relieve the range anxiety of EV drivers, it will not work well. In addition to increasing the energy density of the battery, which has already been encouraged in various ways, building more fast chargers along the highway is a good choice compared to other incentives. That is why the density of fast chargers was positive and significant in predicting EV adoption rates, especially for BEVs compared with PHEVs, because PHEVs can get refueling easily as traditional cars.

Besides, financial incentives have showed a declining marginal effect in promoting EV. Only the subsidies to BEV showed positive significance at a low level. The time we studied was the post subsidy era (2015–2018) [36], financial policies have been generous and used for years. Almost all the counties mentioned in the study had high subsidies or tax incentives from central government, not accounting for the local subsidies. The increase in subsidies will not bring as much benefit as before, just as the law of diminishing marginal utility indicates. The marginal effect is higher for the increase in infrastructures. According to the literature, chargers' density needs more attention paid during this time [8].

In the future, we should shed more light on the effect of charging infrastructures. Policymakers should focus more on the infrastructures of electric vehicles, especially fast public charging points, which is also consistent with some of the literatures [5]. Shown in a recent review summarizing the top antecedents in related studies, the first was charging infrastructure [13]. Many studies also highlighted the importance of fast chargers, especially in densely populated areas [37,38]. Apart from public chargers, private chargers have also played an important role in creating a favorable EV ecosystem [39,40]. Thus, policies related to vehicle–to–grid, wireless charging, and the use of technology can further increase the likelihood of EV adoption [41,42].

Besides, use–based incentive policies designed for the convenience of EV users, such as free parking policy, toll tax exemption and highway lane excess, all showed significance in the pooled regression in this study. The zero–emission mandate is also proved to be relatively effective both in this study and in related literatures. Even policies like the purchasing restriction of fuel cars in China showed a certain level of significance for EV adoption. These policies are not the most widely discussed topics in the literatures, while they are quite effective in reality and worth considering by policymakers.

Author Contributions: Conceptualization, J.Y., S.X. and X.M.; Data curation, J.Y.; Formal analysis, J.Y.; Funding acquisition, S.X. and X.M.; Investigation, J.Y.; Methodology, J.Y.; Resources, S.X.; Supervision, S.X. and X.M.; Visualization, J.Y.; Writing—original draft, J.Y.; Writing—review & editing, S.X. All authors have read and agreed to the published version of the manuscript.

Funding: This research received no external funding.

Acknowledgments: We would like to express our gratitude to all those who helped us during the writing of this paper, especially Ma, our supervisor, for his constant encouragement and guidance.

Conflicts of Interest: The authors declare no conflict of interest.

Appendix A

Table A1. Description of variables and sources.

Variables	Data Descriptions	Sources
EV share	National market share of electric vehicles as percentage of all car sales	[2]
EV sales	National EV car sales	[2]
Subsidies	Direct purchasing subsidies (BEV & PHEV) available to consumers, using the exchange rate of domestic currency against US$ on May 20th 2020	[13] National websites
BEV subsidy	Direct subsidies available to consumers when purchasing BEVs, using the exchange rate of domestic currency against US$ on May 20th 2020	[13] National websites
PHEV subsidy	Direct subsidies available to consumers when purchasing PHEVs, using the exchange rate of domestic currency against US$ on May 2020	[13] National websites

Table A1. *Cont.*

Variables	Data Descriptions	Sources
Waiver	Waivers on access restrictions, such as access to HOV/bus lanes and access to restricted traffic zones	[2,18,43–45]
Mandate	ZEV (Zero emission vehicle) mandate policy	[2,18,43–45]
Target	Year set by the government when 100% ZEV sales goal should be achieved	[2,18,43–45]
Fuel standard	Level of fuel standards/regulations in the country	[2,18,43–45]
Purchasing restriction	Restriction referring to purchasing traditional cars, used only in China	[2,18,22,43–45]
Fast chargers per million population	Number of fast public chargers per million population. Data of chargers are collected from IEA and population of countries is collected from world bank.	[2] World Bank
Slow chargers per million population	Number of slow public chargers per million population. Data sources are the same as above.	[2] World Bank
GDP per capita	Gross domestic product per capita, in current US dollars	World Bank
Population density	People per sq. km of land area, in national level	World Bank

Appendix B

This appendix provides the details of how we quantify the effect of non–financial policies. We assign different values to the policies. Mandate and purchasing restriction are dummy variables, indicating whether it is implemented. The other non-financial policies are categorical variables evaluated by policy intensity. The wider the policy is applied in the country, the higher the scores it gets. For example, if the policy was implemented nationwide other than just in some target cities, it will get higher value. For waiver, we added up the value of three variables to get the total scores, which are shown in Table A2 below. Table A3 gives the value of some non-financial policy in 2018 and Table A4 provides the value of numerical variables. Data are adapted from IEA reports.

Table A2. Summary of policy value for *waiver* in 2018 [2,43].

Country	Waiver on Access Restrictions		
	Access to Bus Lanes	Access to HOV Lanes	Access to Restricted Traffic Zones
Canada	0 *	1	0
China	0	1	1
France	1 **	0	1
Germany	1	0	1
India	0	0	0
Japan	0	1	0
Korea	0	0	1
Netherlands	2 ***	0	0
Norway	1	0	0
Portugal	0	0	0
Sweden	0	0	0
UK	1	0	0
USA	0	1	0

* No policy; ** Only in target cities; *** Nationwide policy.

Table A3. Summary of policy value for the variable in selected countries in 2018.

Country	Waiver	Mandate	Target	Fuel Standards	Purchasing Restriction
Canada	1	1	0	1	0
China	2	1	0	2	1
France	2	0	1	1	0
Germany	2	0	0	1	0
India	0	0	0	0	0
Japan	1	0	0	1	0
Korea	1	0	0	1	0
Netherlands	2	0	2	1	0
Norway	1	0	3	1	0
Portugal	0	0	0	0	0
Sweden	0	0	1	1	0
UK	1	0	1	1	0
USA	1	1	0	1	0

Table A4. Summary of numerical variables in 2018.

Country	BEV Subsidy ($)	PHEV Subsidy ($)	Weighted Average Subsidy ($)	Market Share of BEVs	Market Share of PHEVs	Public Slow Chargers per Million Population	Public Fast Chargers per Million Population
Canada	4.0	4.0	4.0	1.2%	1.1%	191.6	22.7
China	4.6	3.0	4.2	3.4%	1.1%	117.5	79.9
France	6.7	1.1	4.8	1.4%	0.7%	339.4	20.8
Germany	4.5	3.6	4.1	1.1%	0.9%	278.7	31.5
India	2.0	0.0	2.0	0.1%	0.0%	0.2	0.0
Japan	5.6	1.9	3.9	0.6%	0.5%	176.1	60.7
Korea	10.0	0.9	8.9	2.0%	0.3%	104.5	75.7
Netherlands	4.4	2.0	4.1	5.7%	0.9%	2080.7	47.5
Norway	9.0	9.0	9.0	29.5%	17.0%	2097.2	230.7
Portugal	2.4	1.2	1.8	2.1%	1.8%	155.8	17.9
Sweden	6.0	1.0	2.2	2.0%	6.0%	594.1	93.3
UK	3.9	6.7	5.8	0.7%	1.4%	221.6	40.5
USA	7.5	6.0	7.0	1.6%	0.8%	153.6	13.0

References

1. Jaffe, A.B.; Stavins, R.N. The energy–efficiency gap what does it mean? *Energy Policy* **1994**, *22*, 804–810. [CrossRef]
2. IEA. Global EV Outlook 2019: Scaling–Up the Transition to Electirc Mobility. Available online: https://www.iea.org/reports/global-ev-outlook-2019 (accessed on 5 May 2019).
3. Bjerkan, K.Y.; Nørbech, T.E.; Nordtømme, M.E. Incentives for promoting battery electric vehicle (BEV) adoption in Norway. *Transp. Res. Part D Transp. Environ.* **2016**, *43*, 169–180. [CrossRef]
4. Melton, N.; Axsen, J.; Goldberg, S. Evaluating plug–in electric vehicle policies in the context of long–term greenhouse gas reduction goals: Comparing 10 Canadian provinces using the "PEV policy report card". *Energy Policy* **2017**, *107*, 381–393. [CrossRef]
5. Sierzchula, W.; Bakker, S.; Maat, K.; Van Wee, B. The influence of financial incentives and other socio–economic factors on electric vehicle adoption. *Energy Policy* **2014**, *68*, 183–194. [CrossRef]
6. Zhang, X.; Wang, K.; Hao, Y.; Fan, J.-L.; Wei, Y.-M. The impact of government policy on preference for NEVs: The evidence from China. *Energy Policy* **2013**, *61*, 382–393. [CrossRef]
7. Badertscher, N. *Electric Car Sales Hit the Brakes as Tax Credit Axed and Fee Added*; Politifact: Washington, DC, USA, 2015.
8. Wang, N.; Tang, L.; Pan, H. A global comparison and assessment of incentive policy on electric vehicle promotion. *Sustain. Cities Soc.* **2019**, *44*, 597–603. [CrossRef]
9. Sheldon, T.L.; DeShazo, J.R. How does the presence of HOV lanes affect plug–in electric vehicle adoption in California? A generalized propensity score approach. *J. Environ. Econ. Manag.* **2017**, *85*, 146–170. [CrossRef]
10. Mersky, A.C.; Sprei, F.; Samaras, C.; Qian, Z.S. Effectiveness of incentives on electric vehicle adoption in Norway. *Transp. Res. Part D Transp. Environ.* **2016**, *46*, 56–68. [CrossRef]
11. Sen, B.; Noori, M.; Tatari, O. Will Corporate Average Fuel Economy (CAFE) Standard help? Modeling CAFE's impact on market share of electric vehicles. *Energy Policy* **2017**, *109*, 279–287. [CrossRef]
12. Ou, S.; Lin, Z.; Qi, L.; Li, J.; He, X.; Przesmitzki, S. The dual–credit policy: Quantifying the policy impact on plug–in electric vehicle sales and industry profits in China. *Energy Policy* **2018**, *121*, 597–610. [CrossRef]
13. Kumar, R.R.; Alok, K. Adoption of electric vehicle: A literature review and prospects for sustainability. *J. Clean. Prod.* **2020**, *253*, 119911. [CrossRef]
14. Björnsson, L.-H.; Karlsson, S. Electrification of the two–car household: PHEV or BEV? *Transp. Res. Part C Emerg. Technol.* **2017**, *85*, 363–376. [CrossRef]
15. Neubauer, J.; Brooker, A.; Wood, E. Sensitivity of plug–in hybrid electric vehicle economics to drive patterns, electric range, energy management, and charge strategies. *J. Power Sources* **2013**, *236*, 357–364. [CrossRef]
16. Peterson, S.B.; Michalek, J.J. Cost–effectiveness of plug–in hybrid electric vehicle battery capacity and charging infrastructure investment for reducing US gasoline consumption. *Energy Policy* **2013**, *52*, 429–438. [CrossRef]
17. Lieven, T. Policy measures to promote electric mobility–A global perspective. *Transp. Res. Part A Policy Pract.* **2015**, *82*, 78–93. [CrossRef]
18. IEA. Global EV Outlook 2020: Entering the Decade of Electric Drive? Available online: https://www.iea.org/reports/global-ev-outlook-2020 (accessed on 5 June 2020).

19. Mazur, C.; Contestabile, M.; Offer, G.J.; Brandon, N.P. Assessing and comparing German and UK transition policies for electric mobility. *Environ. Innov. Soc. Transit.* **2015**, *14*, 84–100. [CrossRef]
20. Heidrich, O.; Hill, G.A.; Neaimeh, M.; Huebner, Y.; Blythe, P.T.; Dawson, R.J. How do cities support electric vehicles and what difference does it make? *Technol. Forecast. Soc. Chang.* **2017**, *123*, 17–23. [CrossRef]
21. Levine, M.D.; Koomey, J.G.; McMahon, J.E.; Sanstad, A.H.; Hirst, E. Energy efficiency policy and market failures. *Annu. Rev. Energ. Environ.* **2003**, *20*, 535–555. [CrossRef]
22. Ma, S.-C.; Fan, Y.; Feng, L. An evaluation of government incentives for new energy vehicles in China focusing on vehicle purchasing restrictions. *Energy Policy* **2017**, *110*, 609–618. [CrossRef]
23. Jenn, A.; Springel, K.; Gopal, A.R. Effectiveness of electric vehicle incentives in the United States. *Energy Policy* **2018**, *119*, 349–356. [CrossRef]
24. Wooldridge, J.M. A unified approach to robust, regression–based specification tests. *Econom. Theory* **1990**, 17–43. [CrossRef]
25. Hausman, J.A. Specification tests in econometrics. *Econom. J. Econom. Soc.* **1978**, 1251–1271. [CrossRef]
26. White, H. A heteroskedasticity–consistent covariance matrix estimator and a direct test for heteroskedasticity. *Econom. J. Econom. Soc.* **1980**, 817–838. [CrossRef]
27. Craney, T.A.; Surles, J.G. Model–dependent variance inflation factor cutoff values. *Qual. Eng.* **2002**, *14*, 391–403. [CrossRef]
28. Granger, C.W.J. Long memory relationships and the aggregation of dynamic models. *J. Econ.* **1980**, *14*, 227–238. [CrossRef]
29. Sims, C.A. Comparison of Interwar and Postwar Business Cycles: Monetarism Reconsidered. *Am. Econ. Rev.* **1980**, *70*, 250–257.
30. Xu, B.; Lin, B. Carbon dioxide emissions reduction in China's transport sector: A dynamic VAR (vector autoregression) approach. *Energy* **2015**, *83*, 486–495. [CrossRef]
31. Magkonis, G.; Tsopanakis, A. Exploring the effects of financial and fiscal vulnerabilities on G7 economies: Evidence from SVAR analysis. *J. Int. Financ. Mark. Inst. Money* **2014**, *32*, 343–367. [CrossRef]
32. Li, X.; Chen, P.; Wang, X. Impacts of renewables and socioeconomic factors on electric vehicle demands–Panel data studies across 14 countries. *Energy Policy* **2017**, *109*, 473–478. [CrossRef]
33. Choi, H. Technology–push and demand–pull factors in emerging sectors: Evidence from the electric vehicle market. *Ind. Innov.* **2018**, *25*, 655–674. [CrossRef]
34. Zhang, Y.; Zhong, M.; Geng, N.; Jiang, Y. Forecasting electric vehicles sales with univariate and multivariate time series models: The case of China. *PLoS ONE* **2017**, *12*, e0176729. [CrossRef] [PubMed]
35. Jenn, A.T. *Advanced and Alternative Fuel Vehicle Policies: Regulations and Incentives in the United States*; Carnegie Mellon University: Pittsburgh, PA, USA, 2014.
36. Brown, D.; Hall, S.; Davis, M.E. Prosumers in the post subsidy era: An exploration of new prosumer business models in the UK. *Energy Policy* **2019**, *135*, 110984. [CrossRef]
37. Neaimeh, M.; Salisbury, S.D.; Hill, G.A.; Blythe, P.T.; Scoffield, D.R.; Francfort, J.E. Analysing the usage and evidencing the importance of fast chargers for the adoption of battery electric vehicles. *Energy Policy* **2017**, *108*, 474–486. [CrossRef]
38. Bunsen, T.; Cazzola, P.; Gorner, M.; Paoli, L.; Scheffer, S.; Schuitmaker, R.; Tattini, J.; Teter, J. *Towards Cross–Modal Electrification*; Global EV Outlook: Paris, France, 2018.
39. Helveston, J.P.; Liu, Y.; Feit, E.M.; Fuchs, E.; Klampfl, E.; Michalek, J.J. Will subsidies drive electric vehicle adoption? Measuring consumer preferences in the US and China. *Transp. Res. Part A Policy Pract.* **2015**, *73*, 96–112. [CrossRef]
40. Wu, X. Role of workplace charging opportunities on adoption of plug–in electric vehicles–Analysis based on GPS–based longitudinal travel data. *Energy Policy* **2018**, *114*, 367–379. [CrossRef]
41. Liu, H.; Wang, D.Z. Locating multiple types of charging facilities for battery electric vehicles. *Transp. Res. Part B Methodol.* **2017**, *103*, 30–55. [CrossRef]
42. Xiong, Y.; Wang, B.; Chu, C.-C.; Gadh, R. Vehicle grid integration for demand response with mixture user model and decentralized optimization. *Appl. Energy* **2018**, *231*, 481–493. [CrossRef]
43. IEA. Global EV Outlook 2016: Beyond One Million Electric Cars. Available online: https://www.oecd.org/publications/global-ev-outlook-2016-9789264279469-en.htm (accessed on 6 July 2016).

44. IEA. Global EV Outlook 2017: Two Million and Counting. Available online: https://webstore.iea.org/global--ev--outlook--2017 (accessed on 5 June 2017).
45. IEA. Global EV Outlook 2018: Towards Cross–Modal Electrification. Available online: https://www.oecd.org/publications/global--ev--outlook--2018--9789264302365--en.htm (accessed on 5 June 2018).

© 2020 by the authors. Licensee MDPI, Basel, Switzerland. This article is an open access article distributed under the terms and conditions of the Creative Commons Attribution (CC BY) license (http://creativecommons.org/licenses/by/4.0/).

energies

MDPI

Article

Synergies between e-Mobility and Photovoltaic Potentials—A Case Study on an Urban Medium Voltage Grid

Julia Vopava [1,*], Ulrich Bergmann [2] and Thomas Kienberger [1]

1 Chair of Energy Network Technology, Montanuniversitaet Leoben, Franz-Josef Straße 18, A-8700 Leoben, Austria; thomas.kienberger@unileoben.ac.at
2 Verkehrplus—Prognose, Planung und Strategieberatung GmbH, Elisabethinergasse 27a, A-8020 Graz, Austria; Ulrich.Bergmann@verkehrplus.at
* Correspondence: Julia.Vopava@unileoben.ac.at

Received: 15 June 2020; Accepted: 22 July 2020; Published: 23 July 2020

Abstract: To reduce CO_2 emissions, it is necessary to cover the increasing energy demand of e-mobility with renewable energy sources. Therefore, the influence of increasing e-mobility and synergy effects between e-mobility and renewable energy sources need to be investigated. The case study presented here shows results from the analysis of grid-side and energetic synergy effects between e-mobility charged only at work and photovoltaic (PV) potentials. The basis of the grid study is a simplified cell-based grid model. Following the determination of synthetic charging profiles for e-mobility, PV potential profiles, load and production profiles, we perform load flow calculations for different scenarios and a simulation period of one year using the grid model. After the grid study, the energy analyses are carried out using four key performance indicators. The grid study shows that line overloads caused by PV production are only reduced and not avoided by increasing e-mobility and vice versa. The increase in the power peak of e-mobility, by shifting the charging processes into the peak of PV potentials, leads to a reduction of the production surplus in summer, while in winter the line utilisation increases. By modelling PV potentials on real irradiation and temperature data, the investigation of key performance indicators can identify not only seasonal fluctuations but also daily fluctuations.

Keywords: electric vehicles; power grid; grid side effects; photovoltaic potentials; controlled charging

1. Introduction

To contribute to the reduction of CO_2 emissions, it is necessary to change over to alternative drive systems such as electric vehicles (EV) in the transport sector. The change to e-mobility leads to an increasing electricity demand, which can lead to a reduction of grid stability and security of supply [1–5]. The additional electricity demand of electric vehicles must be provided by renewable energy sources (RES) to achieve decarbonisation targets. The impact of the EVs electricity demand together with the fluctuating RES production can lead to additional challenges compared to a single consideration of e-mobility [6–9]. Munkhammar et al. [6] focused on home charging and investigated the interaction between the household power consumption, the electric vehicle home charging and the photovoltaic (PV) power production using a probability model. The investigation shows on the one hand the increasing peak load caused by EV charging and on the other hand, the time-shift between EV charging and PV power production. This time-shift is based on the fact that the PV power production and the resulting peak is available during the day, while the peak caused by charging at home occurs in the evening. To reduce this shift and thereby increase power self-consumption, a change in the user behaviour of EV drivers or the application of demand side measures becomes necessary. For a better use of the PV power production, stays during the day should be used for charging the EVs, such as charging at work and charging commercially. [10] Since the load profiles of e-mobility and thus the required charging infrastructure are strongly dependent on respective user behaviour of EV drivers [1,11], load shift potentials of electric vehicles for different available charging infrastructures (e.g., for charging at home vs. for charging at work) are determined in the references [10,12]. Even with an uncontrolled charging, the PV power production can be better utilised when charging at work compared to charging at home. This is caused by the fact that charging at work is usually performed during the day, while charging at home is generally performed in the evening-hours and at night. [10] Babrowski et al. [12] conclude that the greatest potential for controlled charging is in the area of workplace charging using PV power production.

Despite the large number of presented approaches and results on the interaction between PV power production and uncontrolled and controlled EV charging, to the best of our knowledge, the impact of this interaction on the distribution grids have not yet been sufficiently studied. Therefore, the case study presented in this paper will focus on charging at work and the impact of different penetrations of work-charged EVs and PV potentials on an urban medium-voltage grid. For this purpose, we have developed a grid model for the medium-voltage grid of a medium-sized city based on a cellular approach, using the example of Leoben in Austria. The application of the cellular approach simplifies the complex grid structure and allows reduced calculation times. The approach is therefore suitable for usually time-consuming time resolved load flow calculations, e.g., with annual load profiles. [11] Besides the modelling of existing consumer and producer profiles, we determine production profiles of PV potentials and synthetic charging load profiles of e-mobility. While Su et. al. [13], for example, characterise PV production profiles by averages of "rainy days", "cloudy days" and "sunny days", the case study presented here determines the time-resolved production profiles of PV potentials (15-minute mean value) for each calendar day on the basis of irradiation and temperature data [14] and a solar roof register depicting the actual roof areas [15]. The determination of synthetic charging load profiles for e-mobility is based on traffic analysis, mobility pattern and statistical data. [11] The method for this determination, we have already shown in [11], has been extended in this work, among others, by the function of taking controlled charging into account. In comparison to Gnann et al. [10], where summer charging is shifted to midday and in winter the focus is on peak-shaving by charging at night, here a combination of the summer and winter strategy for controlled charging is applied, regardless of the season. The method distinguishes between shifting the charging process into the peak of the PV production profile or, distributing the charging processes over a defined period of time. For each charging process, it is decided separately if it can be shifted under the defined parameters (for example duration of stay and distance travelled) into the intended period of controlled charging. By comparing the scenarios for uncontrolled charging with those of controlled charging, we investigate in addition to the energetic benefits for the direct use of PV production, whether controlled charging can prevent the negative effects of uncontrolled charging on the power grid. The focus of this analysis is therefore to determine the benefits of controlled charging on the existing infrastructure considering today's mobility behaviour, while Zhang and Chen [16] discuss smart charging management based on mobility behaviour and regional energy prices. Using the cell-based grid model and the determined annual load and production profiles, we perform time resolved load flow calculations for different scenarios. The scenarios differ between the penetrations of e-mobility and PV potentials, the charging power as well as controlled and uncontrolled charging. Based on the results of the load flow calculations, we present grid-side synergy effects and energy-related key performance indicators between e-mobility and PV potentials for the different scenarios.

2. Data Description

In this section, all data required for the case study and its preparation is presented. Starting with the description of the medium-voltage grid of the city of Leoben, the data required to model the synthetic charging load profiles for e-mobility are presented. This includes traffic analyses, mobility patterns and vehicle-related specifications. Finally, the fundamental data for modelling the PV potential profiles are described.

2.1. Medium-Voltage Grid

The original medium-voltage grid for the city of Leoben within the framework of the Austrian research funding agency (FFG) projects "Smart Exergy Leoben" [17] and "Move2Grid" [18] has been anonymised and made available by the local grid operator Energienetze Steiermark GmbH, in compliance with the data protection regulations. The data was provided in the form of a detailed database of the electrical equipment for the year 2014 and a power grid plan in the commercially available software NEPLAN 5.5.6 [19]. The software NEPLAN is a high-end power system analysis tool for electricity, gas, water, and district heating networks. It is used in transmission, distribution, generation, industry, renewable energy systems and smart grid applications. The medium-voltage grid of the city of Leoben consists of a 5 kV and a 30 kV voltage level. While the 30 kV voltage level is designed as a ring topology, the 5 kV voltage level is split into a mesh topology in the inner-city area and into stub lines in rural areas. The database contains, for example, the annual energy consumption of the 194 grid stations, which represent about 16000 customers with an annual energy consumption of ~181 GWh, in an anonymous form. Each grid station contains the number of customers and the corresponding annual energy consumption or production for all standard load profiles according to the „Federal Association of the German Energy and Water Industry" (BDEW) [20] and the synthetic load profiles from the Austrian regulation agency (e-control) [21]. These profiles are normalised to a consumption/production of 1000 kWh per year and are given as 15-minute average power values over a day. Based on these annual energy consumptions, annual load and production profiles are modelled as described in Section 3.2. A hydroelectric power plant (10 MW), a biomass cogeneration plant (0.5 MW) and eight small PV plants feed into the medium-voltage grid. In addition, the database provides all technical data and the geographical location of the local grid transformers and the electrical equipment (transformers, lines, grid nodes, busbars, etc.). To ensure the allocation of the equipment to the energy cells in the context of the development of the simplified cell-based grid model, the geographic coordinates are visualised using a geographic information software system (GIS), QGIS [22]. Due to data protection guidelines, we cannot provide this original data. However, this has no negative influence on the general statement of this work.

2.2. Traffic Analysis, Mobility Pattern and Traffic Model

Traffic analysis enables firstly the identification of potential locations and their respective connection points to the power grid of charging stations. They provide secondly information for the energy demand at these charging stations, which can be derived based on vehicles driving distances. In addition, they provide the necessary data for modelling synthetic charging load profiles for e-mobility.

The identification of potential locations for charging stations is based on the so-called traffic demand. To support the identification of potential locations, two basic parameters are defined: the number of motor vehicles at one location and the duration of stay at the location. In addition to the analysis of the mobility behaviour of people (individual person-related consideration), the field of traffic planning also deals with the location-related calculation of traffic volumes.

Traffic is caused by the fact that people travel to different locations in order to pursue certain activities. These activities can be summed up for the "purpose of the stay" of the people on site. The required change of location is assigned to the "purpose of stay", also called user group. Therefore, the traffic analysis differs between seven various user groups: a trip home, a trip to work with a private or company car, a trip for shopping, a trip for execution (e.g., visit to the doctor), a trip to leisure activities and a trip to education. [23–25]

The mobility pattern of each user group includes original destination matrices according to Bosserhoff [23]. These matrices describe the relative proportion of arriving and departing vehicles in each hour of the day in relation to the total amount of vehicles of one day (24 h). Finally, cumulative distribution functions of arriving and departing vehicles are generated for these daily distributions. Two examples of original daily distributions according to Bosserhoff for the purpose "trip home" and "trip to work-shift operation" as well as the respective cumulative distributions are shown in Figure 1 [11]. Furthermore, by using the daily distributions of arrivals and departures, as step 1, relative distributions of the present vehicles can be derived for every hour: from the vehicles present at the beginning of an hour, the departing vehicles are added and the departing vehicles are subtracted. In step 2, the average duration of stay at the considered location will be calculated. From the daily distribution of the vehicles present, the sum of the duration of stay of all vehicles can be calculated. This sum divided by the number of arrived vehicles gives the average duration of stay per vehicle.

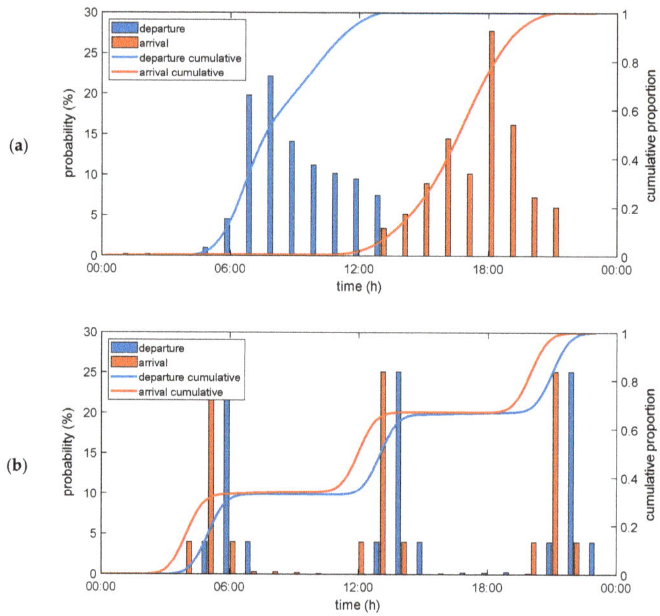

Figure 1. Original destination matrices according to Bosserhoff [23] and cumulative distribution function (**a**) trip home; (**b**) trip to work—shift operation. Reproduced from [11].

The distances travelled are needed for the charging station design (number of charging points, charging power) as well as for the determination of the energy demand. They are the result of the individual behaviour of persons (choice of destination) and the distance between the locations. This means that they cannot be derived directly from the location-related information. Therefore, surveys on mobility behaviour are carried out to obtain such average data like the distance travelled. Such surveys usually deal with the mobility behaviour of the inhabitants of a geographical unit (municipality, city, etc.). While the private sector is well represented in such surveys, commercial traffic is not treated so intensively. In addition, the commercial traffic differs very clearly between the various sectors. A determination of the distances travelled for commercial traffic is made possible by the evaluation of sector-specific "driving profiles" [24]. While the consideration of "driving profiles" is especially important for commercial traffic, as this is the only way to describe its characteristics and dimensions, this is of negligible interest in the private sector for an average working day. This is due to the fact that in the private sector there are usually fixed residences and thus trip purposes, for example "living - working - shopping – living". Such trip chains are used in this paper to illustrate charging at workplaces.

In traffic modelling, the division of an area usually takes place in smaller geographical units, so-called traffic cells. The formation of traffic cells is geographically based on contiguous road network sections and administrative borders. Traffic cells combine several (postal) addresses into one larger unit. In a traffic model, the traffic between the cells (and not within the cells) is essentially calculated. Since the boundaries of the traffic cells do not necessarily match with those of the energy cells, which are created during the development of the simplified cell-based grid model, an approach had to be found to transfer the information from the traffic cells into the energy cells. For this approach, the "grid point's method" is used in this work. Grid point data are available in Austria in the smallest unit in a 100 by 100 meter square. For the case study Leoben, the traffic model is therefore built based on this 100 by 100 meter grid unit. The cell boundaries of the energy cells can therefore be defined taking into account the grid topology and the guidelines described in [11] to achieve the best possible accuracy of the cell-based grid model. As shown in Figure 2, each grid unit has a grid point, in which the statistical data such as number of trips, average duration of stay and average distance travelled are stored for the seven various user groups. A more detailed breakdown of the data can be found in Appendix A. All data is created for each hour of the day in an average weekday. The grid units enable each energy cell to incorporate the specific grid points within its boundaries, with the result that the statistical data and the resulting traffic data can be clearly allocated to the energy cells. All grid points within an energy cell are finally aggregated to an energy node.

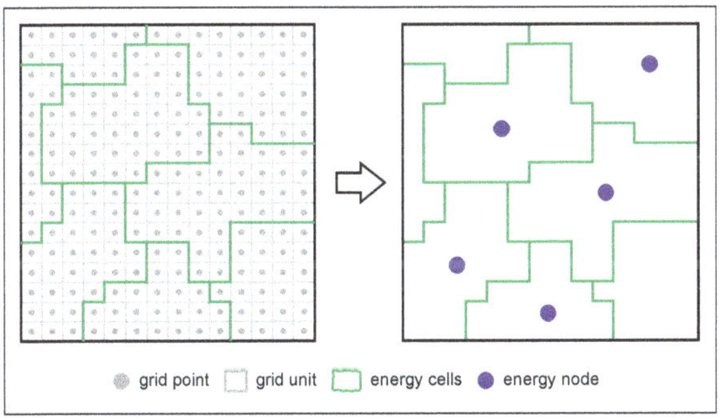

Figure 2. Visualization grid point, energy cell and energy node.

However, the statistics stored in the grid points do not fully cover the need for data to determine the energy demand for charging. The missing data is supplemented by comprehensive reports on mobility behaviour, such as "Österreich unterwegs 2013/2014" [25] and "Mobilität in Deutschland" [26].

In addition to the traffic analyses and mobility patterns, vehicle-related specifications such as battery capacity, average energy consumption and charging efficiency are required to model the synthetic charging load profiles for e-mobility. Therefore, based on the EV-models registered in Germany [27], 15 different EV models for passenger traffic are identified and a distribution function of the EV type is derived. An overview of the considered 15 different EV models is given in Appendix B, Table A2. Subsequently, the specific vehicle parameters mentioned above are assigned to each EV type based on the ADAC eco-test [28]. The specified average energy consumption is defined for an ambient temperature of 20 °C [29]. In order to take into account the seasonal variations in temperature and average energy consumption, the average energy consumption for the seasons are calculated based on average ambient temperatures for summer, winter and transition [30].

2.3. PV Potential Profiles

To determine the production profiles of the PV potentials, we use irradiation and temperature data from 2014 of the Austrian Central Institute for Meteorology and Geodynamics (ZAMG) [14] in combination with the area of suitable roofs we gained from the Styrian solar roof register from the regional government of Styria [15]. The Styrian solar roof register contains all roof areas that are suitable for the generation of solar energy and exceed a size of 20 m^2. Based on Airborne laser scanning (ALS) the suitable roof areas were identified and divided into "good" and "very good" areas as shown in Figure 3. The division into "good" and "very good" suitable roof areas is carried out during the data collection of the solar roof register and takes into account the size, azimuth, tilt and shading of the roof areas [31]. For each suitable roof area, detailed information such as the area in m^2 and the potential annual production in kWh is available in addition to the geographical location.

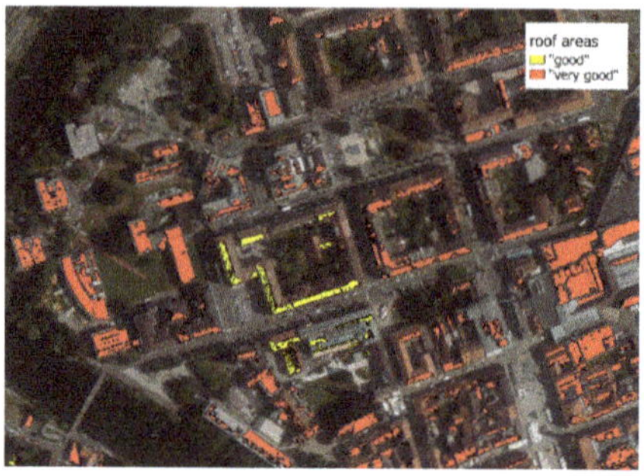

Figure 3. Example of suitable roof areas, including the classification "very good" and "good" based on data from the solar roof register [15].

3. Methods

The basis of the presented case study is a simplified cell-based grid model. Therefore, the methodology used for model development is explained below. This is followed by a description of the determination and grid-implementation of time-resolved load profiles, e.g., for households, as well as for the production profiles of already existent plants (hydroelectric power, biomass cogeneration and PV, as already mentioned in Section 2.1). We also explain the determination of the synthetic charging load profiles for e-mobility and the profiles of the PV potentials in this section. Furthermore, the necessary steps for the evaluation of the grid-side effects and the energy analyses are presented. Finally, we describe the scenarios we investigate during the case study.

3.1. Cell-based Grid Model

The simplified cell-based grid model presented here is developed in five steps as shown in Figure 4: (i) cell classification, (ii) cell allocation, (iii) energy aggregation, (iv) development of the grid model and (v) accuracy check. If a grid model containing several voltage levels is to be developed, steps 2 to 4 must be repeated for each voltage level [11].

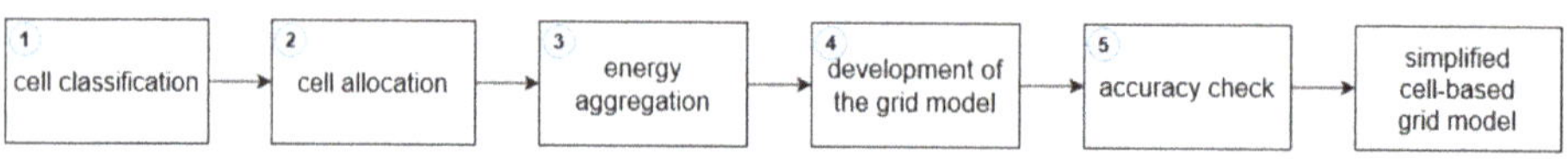

Figure 4. Overview: development of a simplified cell-based model.

In step 1, the original medium-voltage grid is divided into energy cells, considering the grid topology and the topography of the depicted area. As already mentioned in Section 2.1, the cell boundaries of these cells are adapted to the grid units of the traffic analysis. In step 2, the electrical equipment (transformers, lines, grid nodes, busbars, etc.), the consumers and producers are allocated to the individual energy cells. In step 3, the assigned annual energy consumption of the consumers and annual energy production of the producers within an energy cell are aggregated into a fictitious energy node, representing virtual busbars. This aggregation is performed separately for the respective categories of standard load profiles from the BDEW [20] and the synthetic load profiles from the e-control [21]. In step 4, the simplified cell-based grid model is generated by using the software NEPLAN 5.5.6 [19]. For this, a busbar is inserted into the grid model for each energy node. After the implementation of the electrical equipment, it is connected to the energy node. Subsequently, the energy nodes are connected according to the grid topology of the original medium-voltage grid. This connects those cells with each other that enable energy export or import beyond their cell boundaries. Lines within a cell in the original medium-voltage grid structure are taken into account by implementing E-RLC modules in the grid model. In step 5, the accuracy is checked by comparison of the load flows of the simplified cell-based grid model with those of the original medium-voltage grid. We describe this approach of grid simplification in detail in [11].

3.2. Determination Load and Production Profiles

Following the model development, annual time-resolved (15-min mean values) load and production profiles are modelled for the aggregated annual energy consumption and production. Therefore, we use the standard load profiles from the BDEW [20] and the synthetic load profiles from the e-control [21]. These profiles represent normalised load profiles for categories like household, commercial, agricultural, interruptible power supplies, and public lighting. Based on these normalised profiles (1000 kW/a), we calculate the time-resolved annual load profiles.

For the determination of the synthetic charging load profiles for the e-mobility we apply as shown in Figure 5, four steps: (i) data preparation, (ii) determination of each charging process, (iii) modelling charging curve and (iv) aggregation of the load profiles [11].

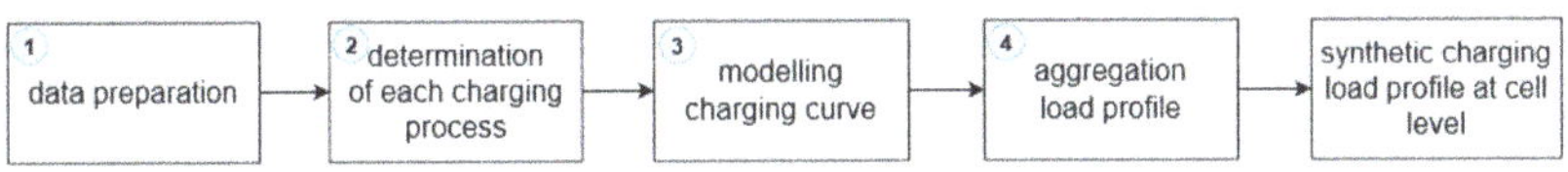

Figure 5. Overview: determination of synthetic charging load profiles of electric vehicles.

In step 1, the traffic grid points and the mobility behaviour presented in Section 2.2 are prepared for the further procedure. In addition to the aggregation of all grid points and the data contained therein (e.g., number of trips) within a cell, this includes the preparation of the distribution functions (arrival/departure/EV type), already mentioned in Section 2.2. Based on the average distances travelled at grid point level, a distribution function for the distance travelled per user group and cell is derived by aggregating all distances travelled within a cell. In step 2, each charging process (CP) within a user group and cell is determined based on the distribution functions by applying probabilistic approaches and the selected charging strategy (e.g., uncontrolled/controlled charging). Each charging process is described by time of arrival, time of departure, EV type and distance travelled. The number of charging processes per cell and user group on a weekday is derived from the aggregation of grid points performed in step 1. Based on the number of trips per weekday, empirical factors from statistical data are used to determine the number of trips for Saturday and Sunday for each user group and cell. The number of trips per day corresponds to today's mobility behaviour. A change in mobility behaviour is difficult to estimate according to traffic planners, therefore it is not taken into account. Hence, we assume that in a scenario with an EV penetration of 100%, each distance travelled represents a trip with an electric vehicle. In step 3, the charging curve for each charging process is modelled according to the state-of-the-art constant current (CC) – constant voltage (CV) charging process. In step 4, all charging processes within a cell are aggregated to a synthetic charging load profile. We describe the general procedure of determining synthetic cell-based charging load profiles for electric vehicles in detail in [11].

For this work we have expanded the method we have described in [11]: Firstly to consider only selected user groups like "trip to work with a private car" for modelling the synthetic charging load profiles. Secondly to enable "intelligent controlled charging", called charging strategy 2.

The extension regarding the selection of the considered user groups consists of creating a trip chain for the selected user groups, which takes place in step 2. The creation of such a trip chain is necessary because the selection of user groups means that there is no longer a charging process after each trip. Therefore, the trips of the non-selected user groups are cumulated into the selected ones, so that the distances travelled by the non-selected user groups can be considered. For this purpose, a trip of the selected user groups consists of several "sub trips", e.g., the trip chain: "working - living - shopping - working" of the user group "trip to work with a private car" consists of three "sub-steps". For each trip of the selected user groups resulting from step 1, the total distance travelled is determined as shown in Figure 6. For this, the number of "sub trips" is defined by applying the probabilistic approach to the statistical distribution of the number of trips made per day and vehicle. Subsequently, the distances travelled are determined for each "sub trip" and summed up, so that the charging curve is modelled, based on the total distance travelled.

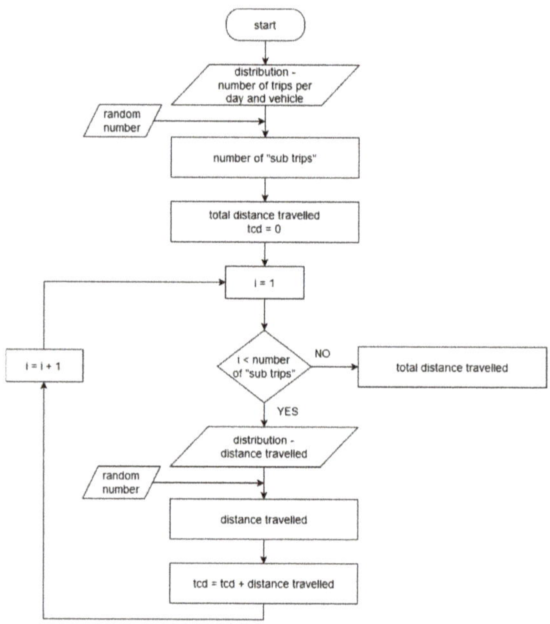

Figure 6. Determination of total distance travelled for each trip of the selected user groups.

The introduction of charging strategy 2, "intelligent controlled charging", also takes place in step 2. Table 1 shows the comparison of charging strategy 1 and 2: As shown, in charging strategy 1 the arrival of the EV starts the charging process, while in charging strategy 2, a time shift of the start of the charging process is considered.

Table 1. Comparison of charging strategy 1 and charging strategy 2.

Charging Strategy	Type of Charging	Start of the Charging Process (CP)	Further Information
1	Uncontrolled	Time of arrival	Basis for charging strategy 2
2	Intelligent controlled	Depending on the time span in which the CP can be shifted	Two approaches: Shifting CPs into PV Peak or sequencing of CPs

The charging strategy 2 takes into account two different approaches, which depend on the mobility behaviour. For each peak in the synthetic charging load profile, a time span is defined in which the charging processes can be shifted depending on the occurrence (morning, midday, afternoon, evening) of the peak. Charging processes which occur during the morning and midday peak are shifted into the peak of the PV production. For the charging processes that take place during the afternoon and evening peaks, the aim is to reduce the charging load by sequencing the charging processes with the boundary condition that the EVs are fully charged in the morning. This does not involve a reduction of the charging power of the individual charging process.

The defined time span is used to decide for each charging process whether it can be shifted or not, see Figure 7. As a generic example, it is not possible to shift the charging processes CP 1 to CP 4 (red) into the time span, because their time of arrival and their time of departure are before or after the defined time span. This means that this charging process takes place at the original time as defined in charging strategy 1. In principle, the charging processes CP 5 to CP 9 can be shifted into or within the time span. During the shifting and sequencing of the charging process, it must be ensured that this is only possible within the duration of stay of the EV, with the boundary condition that the EV is fully charged at the time of departure. For example, if charging process CP5 is shifted to the start of the time span and the remaining time is not sufficient for a fully charged EV, the charging process will also take place at the original time as defined in charging strategy 1.

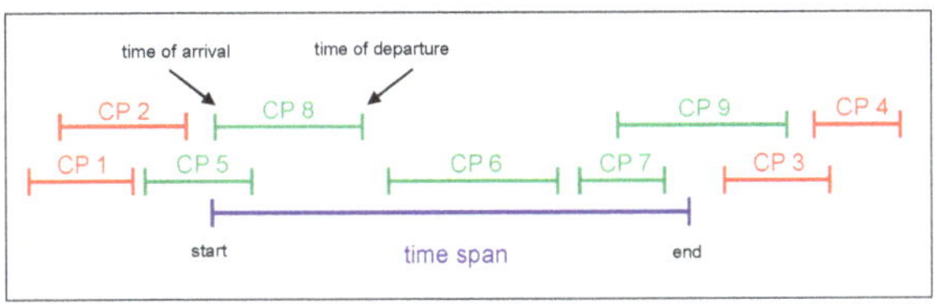

Figure 7. Generic examples of charging processes, which can be shifted in or within the time span in green and those, which cannot in red.

Following the identification of those charging processes that can be shifted into or within the time span, the shifting or sequencing of these charging processes is carried out according to defined criteria. These criteria represent the duration of stay, a priority factor, start of the charging process and the duration of the charging process. The priority factor describes the ratio between duration of stay and duration of the charging process. This means that the closer this ratio is to one, the shorter the period of time in which this charging process can be shifted within the duration of the vehicle's stay. Such charging processes are assigned with new starts of the charging process as soon as possible. Within the framework of this charging strategy, only the start of the charging processes is changed. All other data (e.g., EV type, charging power, distance travelled) for modelling the charging curves remain unchanged. Therefore, after applying charging strategy 2, the methodology described above can be continued in step 3.

3.3. Determination of PV Potential Profiles

The modelling of the PV potential profiles is based on an irradiation model according to Perez et al. [32–34], which is described in more detail in Duffie et al. [35] and Pretschuh et al. [36]. Based on the irradiation and temperature data as well as data from the described solar roof register, the model calculates the direct and diffuse share of the radiation on the inclined surface and converts it into electricity. As a result, annual time resolved PV potential profiles are generated for each cell, which take into account seasonal conditions based on real irradiation and temperature data. A penetration of 100% PV potential requires that all potential roof areas classified as "good" and "very good" in the solar roof register are considered during the determination of the PV potential profiles.

3.4. Simulation, Calculation and Evaluation

To analyse the grid-side synergy effects between e-mobility and PV potentials, load flow calculations with annual load profiles (15-min mean values) are performed for different scenarios.

These scenarios differ between the penetration of EV, the penetration of PV, charging power and charging strategy (controlled and uncontrolled charging). The results from the load flow calculations are analysed with regard to the utilisation of the equipment (e.g., transformer, lines). This is followed by the identification of overloaded equipment, the percentage value, the duration and time of occurrence of these overloads and thus the identification of worst-case weeks. In addition, the influence of PV and EV interactions on these overloads is investigated. Besides the grid-side synergy effects, energy-based analyses are carried out for the same scenarios. For this purpose, the following four key performance indicators are defined: (i) the residual load P_{Res}, (ii) the degree of self-generation (DSG), (iii) the degree of self-sufficiency (DSS) and (iv) the self-consumption ratio (SCR). The residual load describes the difference between the power of consumers and e-mobility $P_{Consumer+EV}$ and the power of the photovoltaic potential P_{PV}. If in a certain 15-min time step the PV production exceeds the total load, the residual load becomes negative and vice versa.

$$P_{Res} = P_{Consumer+EV} - P_{PV} \tag{1}$$

While the residual load is determined for each 15-min time step of the annual total load profile and the annual PV potential profile, the defined time frame for the determination of DSG, DSS and SCR is defined as one day (96 15-min time steps). The calculation of daily values, in comparison to the determination of an annual value, enables the seasonal consideration of the interaction between the e-mobility and PV potentials.

To explain the DSG, DSS and SCR, Figure 8 shows a generic load profile (consumer and EV) and a generic PV production profile as an example.

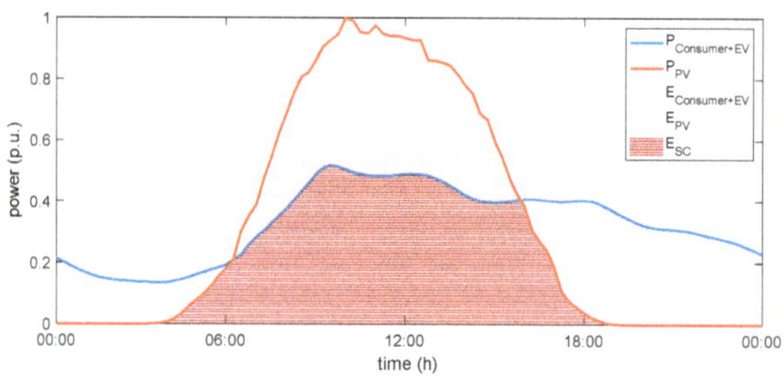

Figure 8. Relevant areas for calculating the DSG, DSS and SCR.

The degree of self-generation (DSG) describes the locally produced energy in relation to the locally consumed energy for a defined time frame:

$$DSG = \frac{E_{PV}}{E_{Consumer+EV}} = \frac{\int_{t_{start}}^{t_{end}} (P_{PV}(t))\, dt}{\int_{t_{start}}^{t_{end}} (P_{Consumer+EV}(t))\, dt} \tag{2}$$

where E_{PV} is the total locally produced energy and $E_{Consumer+EV}$ is total energy demand for the defined time frame.

The degree of self-sufficiency (DSS) considers the ratio between the directly consumed part of the locally produced energy E_{SC} and the total energy demand. This means that this criterion describes the proportion of the total load required that can be covered directly by the PV potential:

$$DSS = \frac{E_{SC}}{E_{Consumer+EV}} = \frac{\int_{t_{start}}^{t_{end}} (\min\{P_{Consumer+EV}(t),\, P_{PV}(t)\})\, dt}{\int_{t_{start}}^{t_{end}} (P_{Consumer+EV}(t))\, dt} \tag{3}$$

The self-consumption ratio (SCR) relates the directly consumed part of the locally produced energy to the total locally produced energy:

$$SCR = \frac{E_{SC}}{E_{PV}} = \frac{\int_{t_{start}}^{t_{end}} (\min\{P_{Consumer+EV}(t),\ P_{PV}(t)\})\ dt}{\int_{t_{start}}^{t_{end}} (P_{PV}(t))\ dt} \tag{4}$$

3.5. Scenario Definition

To identify which of the seven user groups from the traffic analyses are particularly relevant for the investigation of the synergy effects between e-mobility and PV potential, the potential for the direct use of PV production per user group is examined. For the case study presented here, the user groups "trip for shopping", "trip for execution" and "trip to work" have the highest energy demand in relation to the total energy demand of e-mobility based on the traffic analysis. Since the user groups "trip for shopping" and "trip for execution" have the same characteristic load profile, only the user groups "trip for shopping" and "trip to work" are shown as examples in Figure 9. The figure is used to visualize the determination of the potential for the direct use of PV production for the user groups "trip for shopping" and "trip to work". For the user group "trip to work", the user groups "trip to work with a private car" and "trip to work with a company car" are aggregated due to their similar behaviour. The characteristic of the synthetic charging load profile for the user group "trip to work" is strongly dependent on the sector-specific jobs within the considered area. Therefore, the evening peak, which occurs due to an existing shift operation, can be omitted when looking at another area. The number of peaks for the user group "trip for shopping" is independent of the area under investigation. This means that as soon as there are shopping facilities in the area under consideration, and thus a shopping user group exists, a peak at noon and in the afternoon or evening occurs. In the example shown, around 80% of the energy demand of the user group "trip to work" can be directly covered by PV production, while only 60% can be directly supplied for the user group "trip for shopping".

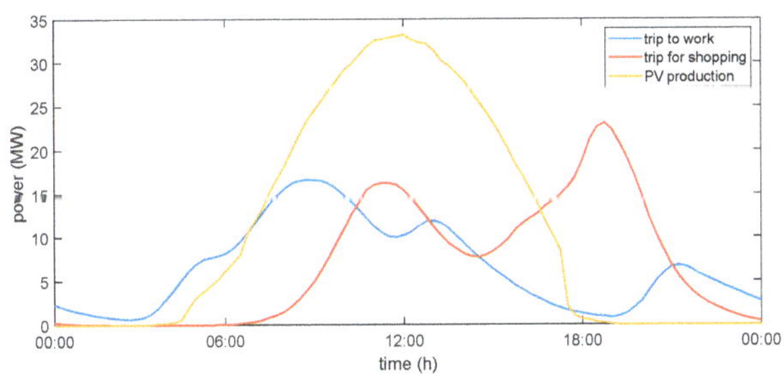

Figure 9. Generic example of a load profiles of the user groups "trip to work" and "trip for shopping" as well as a PV production profile.

The analysis of all user groups indicates that with an increasing number of charging processes during the day, the direct use of the PV production increases and thus the potential. Based on the user behaviour and the share of PV use, the user group "trip to work" can be identified as the user group with the highest potential. Therefore, we have focused on the user groups "trip to work with a private car" and "trip to work with a company car" in combination with a trip chain model to investigate the synergy effects between e-mobility and PV potentials in the case study presented here. Using the trip chain model, the charging energy of the other five user groups is accumulated into these two user groups. Since the number of weekend trips for these user groups is decreasing significantly and it should be ensured that all distances travelled at the weekend are also taken into account, the user group "trip home" is also considered at the weekend.

As already mentioned, the scenarios differ among each other in the penetration of EV and PV. For this purpose, the penetration of EV and PV is varied by 0%, 20%, 40%, 60%, 80% and 100% respectively. In a scenario, all cells have always the same penetration of EV and PV, this means for example that all cells have a penetration of 20% PV and a penetration of 80% EV. This assumption was made, because there are no statistical distributions for urban areas that allow an individual penetration for each cell. Furthermore, two different charging powers (3.7 kW and 11 kW) are considered. For both charging powers, a distinction is made between charging strategy 1 (uncontrolled charging) and charging strategy 2 (controlled charging).

The reference scenario describes the status quo (without PV Potential and future e-mobility) of the medium-voltage grid to be examined. The overview of all simulated scenarios is given in Appendix B, Table A1.

4. Results and Discussion

In this section the developed cell-based grid model, which forms the basis of the case study, is presented. Afterwards, the results of the load flow calculations are analysed with regard to grid-side synergy effects. Finally, the results from the energy analyses are discussed.

4.1. Cell-Based Grid Model

The original medium-voltage grid of the city of Leoben is divided into 26 energy cells. Figure 10 illustrates the developed simplified cell-based grid model, where the 30 kV voltage level is shown in blue and the 5 kV voltage level in green. To clearly distinguish the two voltage levels, the 30 kV level is marked with the ending _030 and the 5 kV level with the ending _005. Since the 30 kV voltage level is not designed to cover the complete area of Leoben, not every cell has a connection to the 30 kV voltage level. The 110 kV voltage level of the Leoben substation was chosen as the slack node, so that the load flows between the 30 kV and 110 kV voltage level can also be analysed. The slack node thus corresponds to the energy node N10_110. Since there are two 110 kV/30 kV transformers in the substation due to redundancy, both are also shown in the grid model.

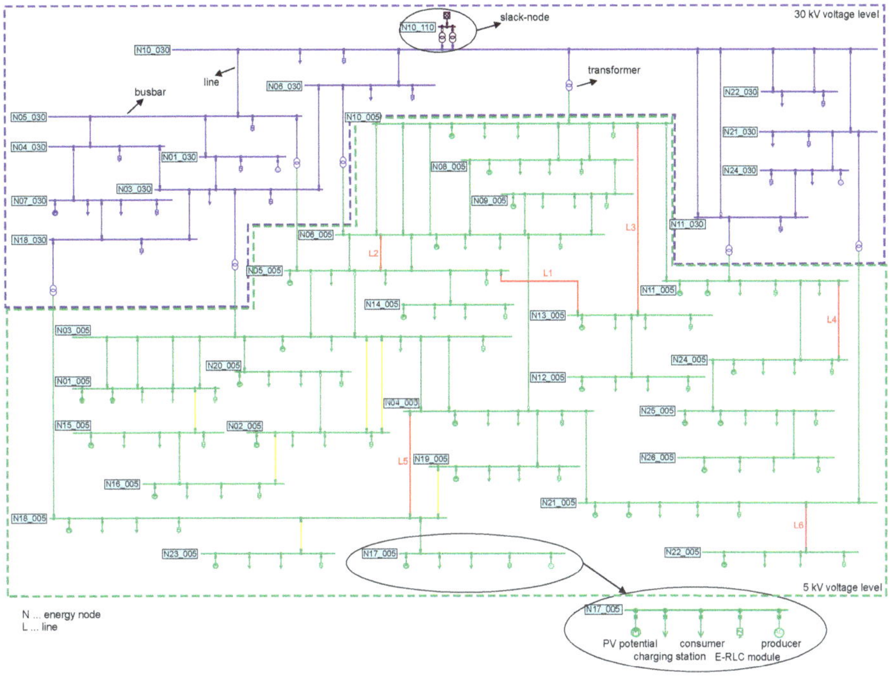

Figure 10. Simplified cell-based grid model for the city of Leoben.

4.2. *Grid-side Synergy Effects*

Based on the simplified cell-based grid model and the annual load and production profiles, load flow calculations are performed for the different scenarios, described in Section 3.5. The results of the load flow calculations are analysed with regard to grid-side synergy effects between e-mobility and PV potential. The results presented here focus on the utilisation of the lines of the 5 kV voltage level.

4.2.1. Charging Strategy 1 – Uncontrolled Charging

The number of overloaded lines of the 5 kV voltage level with corresponding penetration of EV and PV for the charging capacities 3.7 kW and 11 kW and charging strategy 1 is shown in Figure 11. Since a penetration of 0% EV only takes into account the demand for e-mobility that is already present in the status quo of the power grid, no additional charging power is assigned to these scenarios and they are therefore referred to as without EV. With increasing EV and PV penetration as well as charging power, the number of overloaded lines increases. The use of PV potentials without e-mobility leads to line overloads at penetration rates of 40% PV and higher. E-mobility without PV potential, too, only causes line overloads when the penetration rate is over 40% and the charging power is 11 kW. With a penetration of 20% PV, the line overload caused by a penetration of 40% EV and a charging power of 11 kW can be avoided.

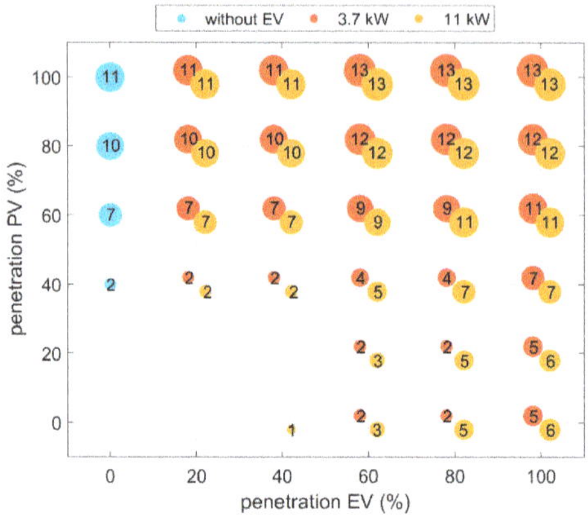

Figure 11. Number of overloaded lines of the 5 kV voltage level for charging strategy 1.

In addition to the number of overloaded lines, the model also determines which lines are affected and where they are located in the power grid model and geographically. Figure 12 shows the cell division for the energy cells of the city of Leoben and the geographical location of the 13 overloaded lines, with the six most stressed lines in red. The identification of the most stressed lines is based on the maximum utilisation of the overloaded lines and the duration of the overload for all scenarios. These overloaded lines are also illustrated in Figure 10, which shows the grid plan of the cell-based grid model.

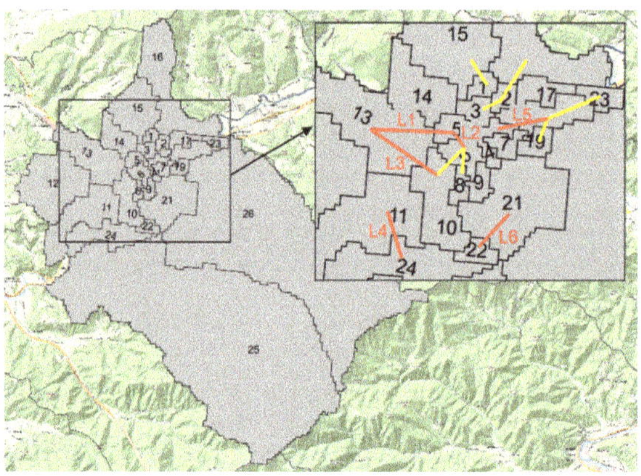

Figure 12. Cell division of the city of Leoben and detailed view with geographical location of the identified overloaded lines - in red: most stressed lines.

With the exception of the already mentioned avoidance of line overload at a penetration of 40% EV with a penetration of 20% PV, the different combinations of the penetrations of PV and EV do not lead to a reduction of the number of overloaded lines. Therefore, according to Figure 11 there are no obvious positive synergies between e-mobility and PV potential for charging strategy 1 for the number of overloaded lines. To investigate the interaction between e-mobility and PV potential in more detail, we determine the maximum line utilisation and the duration of overloads. For this purpose, we examine when the maximum utilisations of the lines occur and whether this is caused by the penetration of EV or PV. As illustrated for the most stressed lines in Figure 13, line overloads occur mainly in winter as well as in April and June. While the overloads in winter are due to load peaks of e-mobility, those in April and June are mainly due to production peaks of PV potentials. Furthermore, the worst-case weeks are identified based on the time of occurrence of the maximum load. Those worst-case weeks, for example, are used to compare scenarios with uncontrolled and controlled charging (charging strategy 1 vs. charging strategy 2).

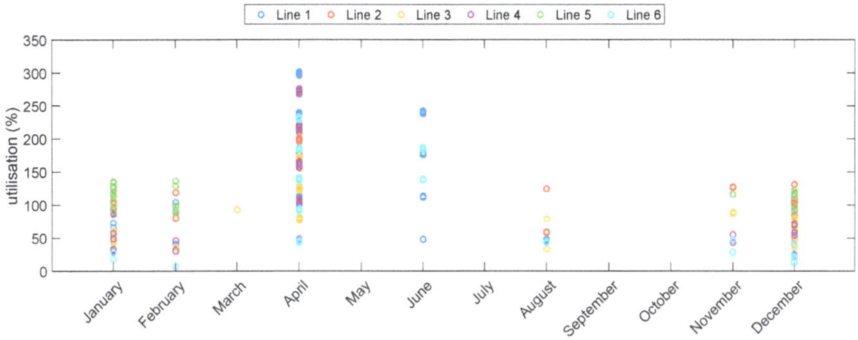

Figure 13. Month of occurrence for the maximum line utilisation of all scenarios with charging strategy 1 during the simulation period of one year.

Two of the six most stressed lines are selected to analyse the duration of the line overloads in detail. For this purpose, annual duration curves are modelled for each line and scenario based on the sorted utilisation of the line. While line 1 is overloaded for the first time at a penetration of 40% PV, the overload on line 5 occurs first at a penetration of 40% EV and a charging power of 11 kW. Since the duration of overload is less than 10 h for both lines, the next higher penetration of PV and EV is chosen for demonstration purposes of the interaction between e-mobility and PV potential. Table 2 shows the comparison of the duration of the line overloads of lines 1 and 5 depending on the charging power. For line 1 we investigate the influence of an increasing e-mobility penetration at a fixed penetration of 60% PV. With a penetration of 60% PV and the current demand, the duration of overload is 506.5 h per year. By integrating a penetration of 100% EV, the duration of overload can be reduced to 258.5 h per year at a charging power of 3.7 kW and 206.75 h per year at a charging power of 11 kW respectively. The comparison of the duration of overloads for a penetration of 100% EV shows that at a charging power of 11 kW, the reduction in the duration of the overload is greater by around 50 h per year than at a charging capacity of 3.7 kW. This higher reduction is due to the increased power requirement of e-mobility at a charging power of 11 kW compared to 3.7 kW. As a result of this increase, the production surplus at the peak of the PV potential is reduced and thus the duration of the line overload.

Table 2. Duration of overloads (utilisation > 100%) for the lines 1 and 5 – charging power 3.7 kW and 11 kW, using charging strategy 1.

Penetration (%)		Duration of Overload (h) Line 1		Penetration (%)		Duration of Overload (h) Line 5	
PV	**EV**	**3.7 kW**	**11 kW**	**PV**	**EV**	**3.7 kW**	**11 kW**
60	0	506.50	506.50	0	60	154.50	155.75
60	20	454.00	450.50	20	60	107.75	132.25
60	40	396.75	392.00	40	60	84.25	106.00
60	60	349.50	346.25	60	60	62.75	92.50
60	80	300.75	294.75	80	60	47.50	79.00
60	100	258.5	206.75	100	60	39.00	67.00

However, an increased power requirement of e-mobility due to the higher charging power can also lead to an increasing grid load. This increase in grid load can be seen in Table 2 for line 5. While the duration of the line overload at a fixed penetration of 60% EV and a charging power of 3.7 kW is reduced from 154.50 h per year without PV potential to 39.00 h per year with a penetration of 100% PV, the duration of the overload at a charging power of 11 kW is reduced from 155.75 h per year only to 67.00 h per year.

The duration of the overload shown in Table 2 is graphically illustrated in Figure 14 using the sorted annual duration curves of the line utilisation for the lines 1 (a) and 5 (b) for the range 0 to 600 h at a charging power of 3.7 kW. In addition to the already described reduction of the duration of the line overload, the decrease of the utilisation can also be seen. For the shown range (0 to 600 h) of the sorted annual duration curves, the comparison of the utilisation of line 1 at a penetration of 0% EV with the utilisation at a penetration of 100% EV shows a reduction of the utilisation between 6% and 19%. For line 5, a decrease in line utilisation of up to 15% is achieved for the range of 0 to 600 h of the sorted annual duration curve with an increase in penetration from 0% PV to 100% PV.

(a)

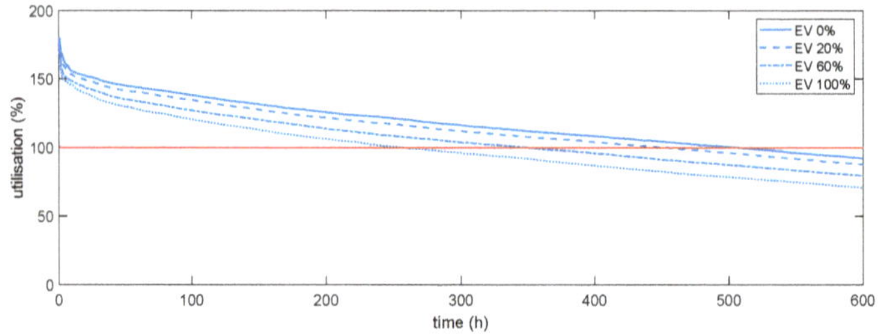

(b)

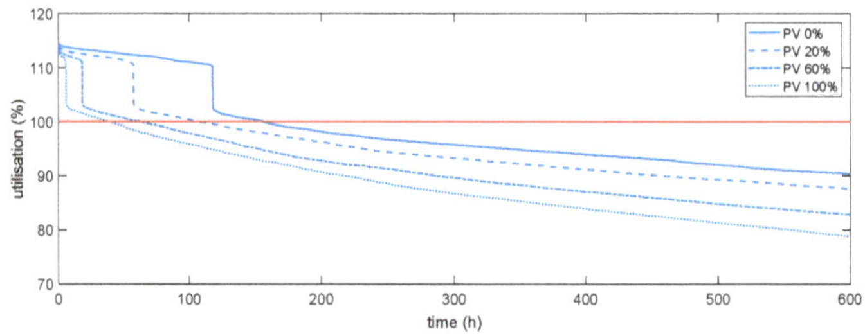

Figure 14. Annual duration curve of the utilisation for the period 0 to 600 h for 3.7 kW charging power and charging strategy 1 (**a**) line 1 and a fixed penetration of 60% PV (**b**) line 5 and a fixed penetration of 60% EV.

As demonstrated by the example of the lines 1 and 5, the higher the penetration of EV with fixed penetration of PV, the lower the duration of the line overloads as well as the maximum utilisation and vice versa. The charging power also has a considerable influence on the line utilisation and thus on the overload of the lines. While a higher charging power in summer further reduces the power surplus caused by PV production and thus relieves the grid, in winter it can lead to higher and longer overloads.

In the scenarios presented here, the same penetration of EV and PV is considered for all cells, therefore in some cells large differences between PV production and the power requirements of e-mobility occur. For further reduction or possibly complete avoidance of overloads, it is necessary to find the optimal balance between PV production and the power requirements of e-mobility for each cell.

4.2.2. Charging Strategy 2—Controlled Charging

A comparison of the scenarios with charging strategy 1 and 2 based on two worst-case weeks is carried out to determine the influence of controlled charging on the grid utilisation compared to uncontrolled charging and the resulting synergy effects between e-mobility and PV potential. For this purpose, one week is in winter (20 January to 26 January) and one in summer (2 June to 8 June). The most stressed days of this week are 24 January and 2 June, respectively. The comparison of the line utilisation in Figure 15 shows for line 1 for a penetration of 40% EV and 60% PV and a charging power of 3.7 kW that controlled charging can lead to both, a reduction but also to an increase in the utilisation of the line. By applying charging strategy 2, the charging processes during the day are shifted into the peak of the PV production; thereby an increase of the power requirement of e-mobility can occur within the period of the peak of the PV potentials as shown in Figure 16.

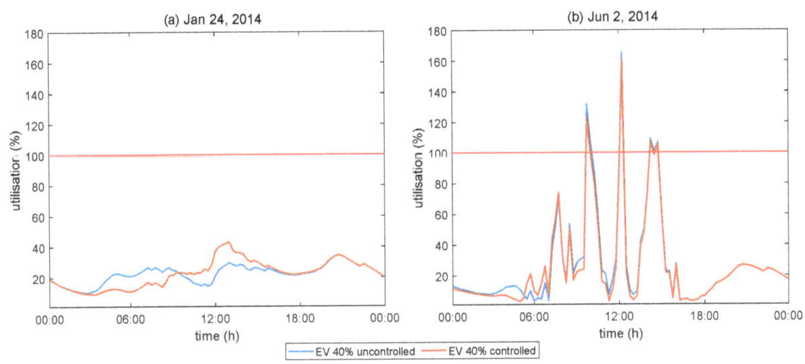

Figure 15. Comparison of the utilisation of the line 1 for a penetration of 40% EV and 60% PV with 3.7 kW charging power and charging strategy 1 and 2 for (**a**) 24 January (**b**) 2 June.

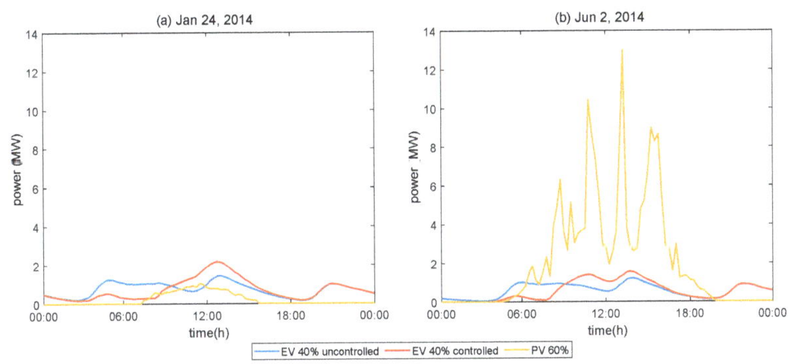

Figure 16. PV potential profile and synthetic charging load profiles for the e-mobility with 3.7 kW charging power, charging strategy 1 and 2 for cell 13 (**a**) 24 January (**b**) 2 June.

This increase in power demand leads to an increased line utilisation if the PV potential is not sufficient, as demonstrated in Figure 15a for the 24 January. On 2 June the PV potential is significantly higher than the power demand from EV (see Figure 16b), which means that the line utilisation can be reduced at noon. As shown in Figure 15b, however, it is by far not possible to avoid the overloads. The difference of about 12 MW between the maximum peak value of the PV production on the 2 June and that on the 24 January, shown in Figure 16, does not allow any conclusions concerning other days and their maximum peak values due to the PV potentials modelled in dependence on irradiation and temperature data.

The utilisation of line 1 cannot be directly compared with the difference between the EV charging load and the PV potential of cell 13 (Figure 16). On the one hand, the utilisation of the line depends on the balance of the total demand (e-mobility and other consumption) as well as the PV potential and the resulting power requirement or power surplus of cell 13. On the other hand, this line is in a closed ring system and the line utilisation is therefore dependent on several other cell balances, see Figure 10 (cell-based grid model).

The described increase or reduction of the utilisation of line 1 by controlled charging can be transferred to other lines. The amount of this positive as well as negative effect is strongly dependent on the demand of e-mobility and the PV potential of the respective cell as well as the grid topology (mesh, ring, radial) in which the investigated line is located.

Figure 17 shows the sorted duration curves of the line utilisation of line 1 for different scenarios for both worst-case weeks. In addition to the charging strategy, the scenarios differ between the penetration of 40% and 100% EV. Furthermore, a fixed penetration of 60% PV and a charging power of 3.7 kW is specified. As shown in Figure 17a, in winter with a fixed penetration of 60% PV, the utilisation on line 1 for a penetration of 100% EV is higher than for 40% EV regardless of the charging strategy. The increase in line utilisation is caused by the lower PV potential in winter compared to summer (see Figure 16). The increase in the power requirement of e-mobility within the peak of the PV potential resulting from controlled charging leads to an increase in the line utilisation at a penetration of 100% EV compared to uncontrolled charging. Due to the higher PV potential in summer, a higher production surplus is produced in the worst-case week in summer with a penetration of 40% EV, compared to a penetration of 100% EV. This surplus leads to an increase in line utilisation, as shown in Figure 17b. The duration of the overload is increased by about one hour at a penetration of 40% EV by controlled charging, while controlled charging has no effect on the duration of the overload at a penetration of 100%.

(a)
(b)

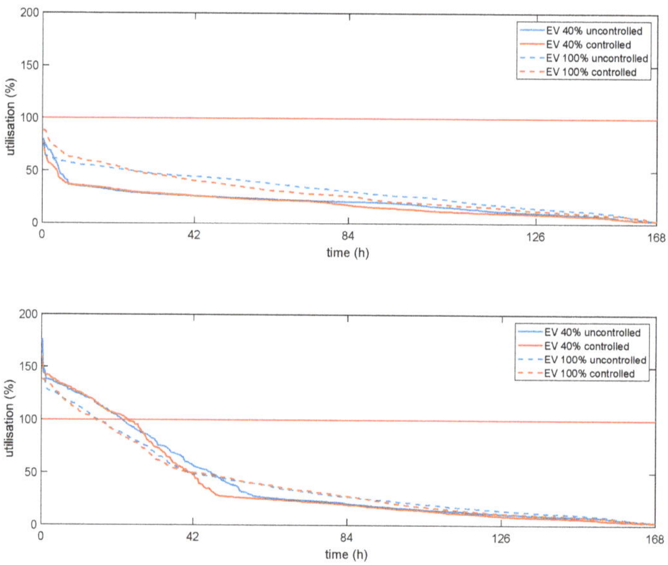

Figure 17. Sorted duration curve of the utilisation of line 1 at a fixed penetration of 60% PV and 3.7 kW charging power for the selected worst-case weeks (**a**) winter (**b**) summer.

Regardless of the charging strategy, the comparison of the line utilisation of a worst-case week in winter with one in summer clearly shows the need to consider seasonal effects. A fixed penetration of PV generates significantly more power in summer than in winter (Figure 16). The high penetrations of EV, which are used to reduce line overloads (caused by PV potentials), can lead to increased grid load in winter due to insufficient PV production. In order to avoid increased loads and to ensure sufficient reserves in the grid, the seasonal effect should therefore always be taken into account. Instead of completely expanding e-mobility in these cells, reducing the penetration of the PV potential, for example, could be the better solution.

Although the influence of controlled charging is small for the worst-case weeks presented here, controlled charging is most effective for the use of PV production in summer. However, it is important to find the optimal ratio between e-mobility and PV potential in order to make best use of the effect of reducing line utilisation in summer. In winter, controlled charging can also lead to higher line utilisation by shifting the charging processes into the peak of the PV, so the seasonal effect between PV production in summer and winter should be considered more closely.

4.3. Energy Analyses

In the energy analysis depicted here, we show the interaction of the energy flows between the cells and their significance for line overloads on the basis of the determined energy demand of e-mobility and the potential of PV production. Furthermore, we present the results on the energetic synergy effects between e-mobility and PV potential for different scenarios, which are based on the determination of the key performance indicators explained in Section 3.3. The key performance indicators are determined and analysed for each cell and the overall system regardless of grid parameters.

4.3.1. Interaction of Energy Surpluses between the Cells

The annual energy demand for each cell is determined on the basis of the synthetic charging profiles for e-mobility. Figure 18a illustrates the annual energy demand spatially resolved for a penetration of 100% EV. Depending on the cell, the annual energy demand is in the range of 20 MWh (cell 26) to 12,500 MWh (cell 13). Figure 18b shows the determined annual energy production of the PV potentials for a penetration of 100% spatially resolved. Depending on the cell, these vary between 290 MWh (cell 19) and 22,500 MWh (cell 13).

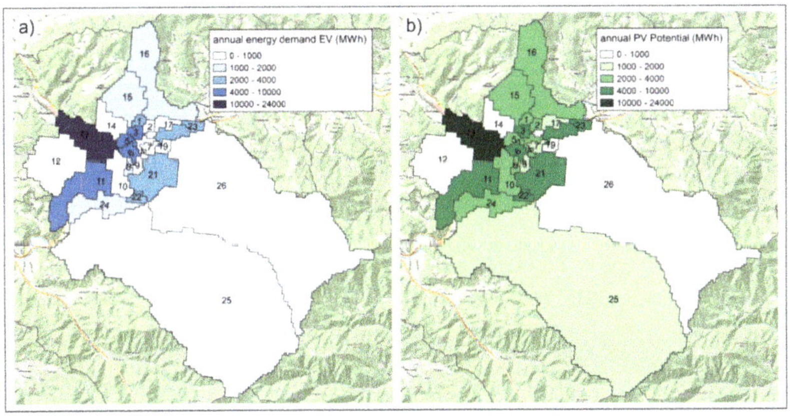

Figure 18. (**a**) Annual energy demand e-mobility and (**b**) annual PV potential at cell level for the city of Leoben.

In addition to the highest PV potential and the highest demand for work-related e-mobility, cell 13 is a very industry intensive cell. Furthermore, two of the three lines connecting cell 13 with its neighbour cells are among the most stressed lines (line 1 and line 3). The overloads of the lines 1 and 3 are mainly caused by the PV potentials. Moreover, lines 1 and 3 belong to a closed ring system in which another most stressed line (line 2) is located. Figure 19 shows a schematic section of the cell-based grid model shown in Figure 10 and the relevant load flows for the closed ring structure. The directions of the load flows correspond to those at the time of the maximum utilisation of line 1 for a penetration of 60% PV and 100% EV, a charging power of 3.7 kW and charging strategy 1. For this time step, the directions of the load flows indicate that the unused PV potential of cell 12 covers the electricity demand of the cells 5, 6 and 14 (ratio PV to EV < 1, see Figure 20) and leads to a feed into the 30 kV level.

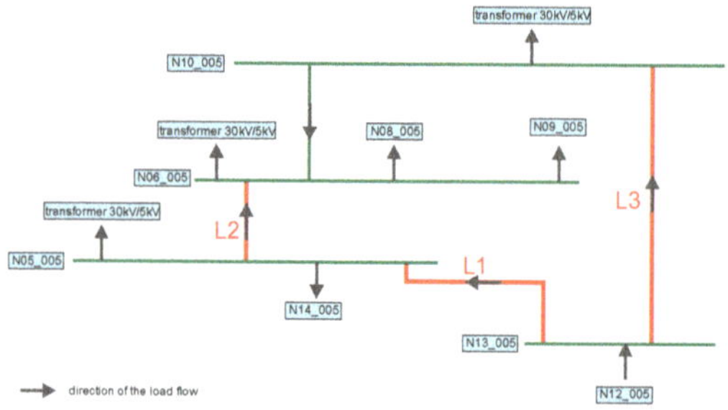

Figure 19. Schematic grid section - illustration of the direction of load flows for a penetration of 60% PV and 100% EV, 3.7 kW charging power and charging strategy 1 for the time of maximum utilisation of line 1.

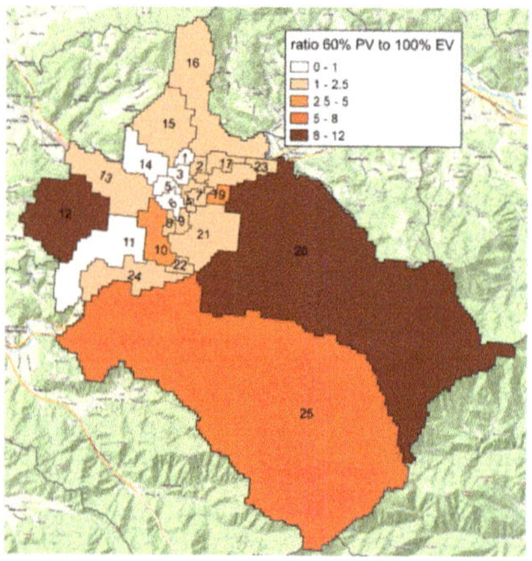

Figure 20. Energetic ratio of the annual energy production of 60% PV potential to the annual energy demand of 100% EV.

As already mentioned, the optimal ratio between e-mobility and PV potential is important to avoid overloads. For this purpose, Figure 20 shows the energetic ratio of the annual energy production of 60% of the PV potential to the annual energy demand for 100% e-mobility. The ratio varies between 0.42 (cell 1) and 11.45 (cell 26). While cell 13 has a ratio between e-mobility and PV potential of 1.07 for these penetrations, the neighbour cell (cell 12) with a ratio of 8.43 has a significant production surplus. This surplus is transported from cell 12 to cell 13 and from there via lines 1 and 3 to cells 5 and 10, where this surplus leads to overloads. To reduce overloads caused by the PV potential that cannot be avoided at a penetration of 100% EV, it is sometimes necessary to reduce the PV potential of these cells from a penetration of 60% to 40%.

An optimal energetic ratio between e-mobility and PV potential is, however, not sufficient to avoid overloads, since the annual energy demand of e-mobility does not allow conclusions to be drawn about the load curve and thus about the power ratio of the individual time step. This means that with an energetic optimum, there can be significant differences in the peak values between e-mobility and PV potential on the one hand, and a time shift between the two peaks on the other. Cell 13, for example, is characterised by shift operation and office workplaces. This characteristic results in a division of the energy demand into four load peaks. This means that the optimum balance between e-mobility and PV potential must be analysed and determined in terms of both energy and power.

The self-consumption of the cells, for example from households, also has an important influence on the cell balance and the load flows between the cells. For this reason, the self-consumption of the cells is taken into account in our further energy analyses.

4.3.2. Charging Strategy 1 – Uncontrolled Charging

The calculation of the residual load enables the determination of the maximum power surpluses and power demand. While the energetic ratio between 60% PV potential and 100% e-mobility is 1.07, Figure 21 shows that the power profile of the PV potential is significantly higher than the load demand (e-mobility and consumption) of cell 13. Due to this high difference, the PV production profile during the day can be seen from the residual load curve in Figure 21. The significant fluctuations in the negative residual load are due to the use of real irradiation and temperature data during the modelling of the PV potentials. The profiles therefore consider, for example, seasonal effects or shading by clouds. Especially at night, there is a positive residual load and thus an increased power demand. As shown in Figure 21a, depending on the weather, negative residual loads of up to 5 MW also occur in the worst-case week in winter. The residual load for the represented worst-case week in summer (identified in the grid study) indicates that June 5 is a cloudy and rainy day due to the predominantly positive residual load also during the day (see Figure 21b). With increasing penetration of EV and a fixed penetration of 60% PV, the power surplus of the PV potential decreases and the load demand at night increases, regardless of the season.

(a)
(b)

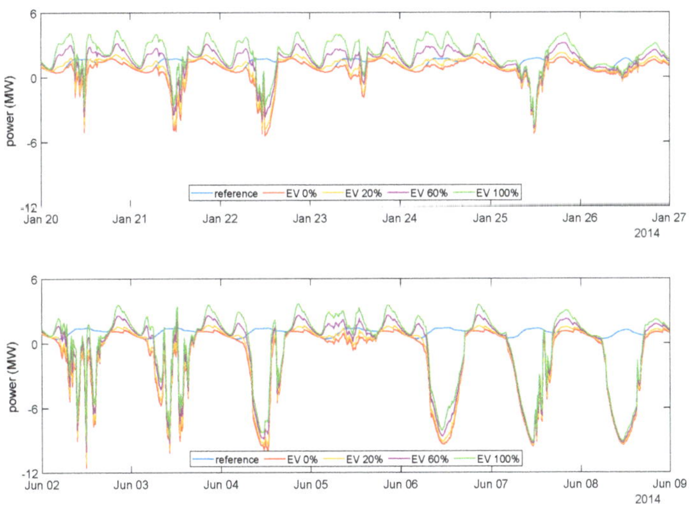

Figure 21. Residual load of cell 13 for the selected worst-case weeks for a fixed penetration of 60% PV and a variation of the penetration of EV for 3.7 kW charging power and charging strategy 1 as well as for the reference scenario (without PV and EV) (**a**) winter (**b**) summer.

The sorted annual duration curve of the residual load of cell 13 in Figure 22 shows that in an observation period of one year, about 2/3 of the year show a positive residual load. In the last third, there is a rapid increase to a maximum power surplus of 11.8 MW at a penetration of 60% PV and no e-mobility. The influence of increasing e-mobility at a fixed penetration of 60% PV can be seen clearly from the annual duration curve. The curves are shifted in the direction of positive residual loads. In the case of a fixed penetration of EV and increasing penetration of PV, the annual duration curves behave in exactly the opposite way.

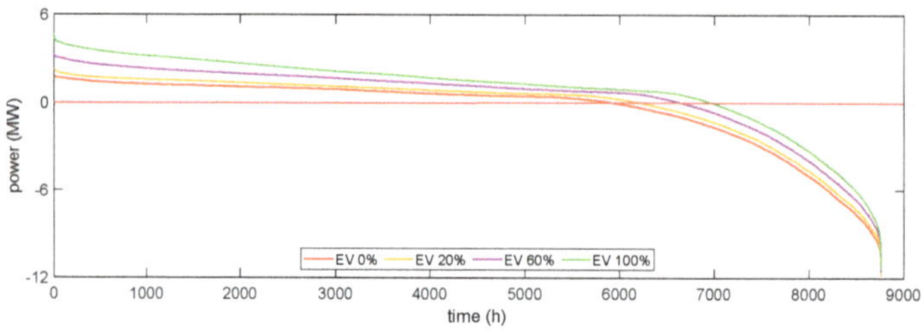

Figure 22. Sorted duration curve of the residual load of cell 13 for a fixed penetration of 60% PV and a variation of the penetration of EV for 3.7 kW charging power and charging strategy 1 and a simulation period of one year (8760 h).

The key performance indicators DSG, DSS and SCR are determined for each day for the simulation period of one year and are presented in Figure 23 for cell 13 at a fixed penetration of PV and increasing penetration of EVs.

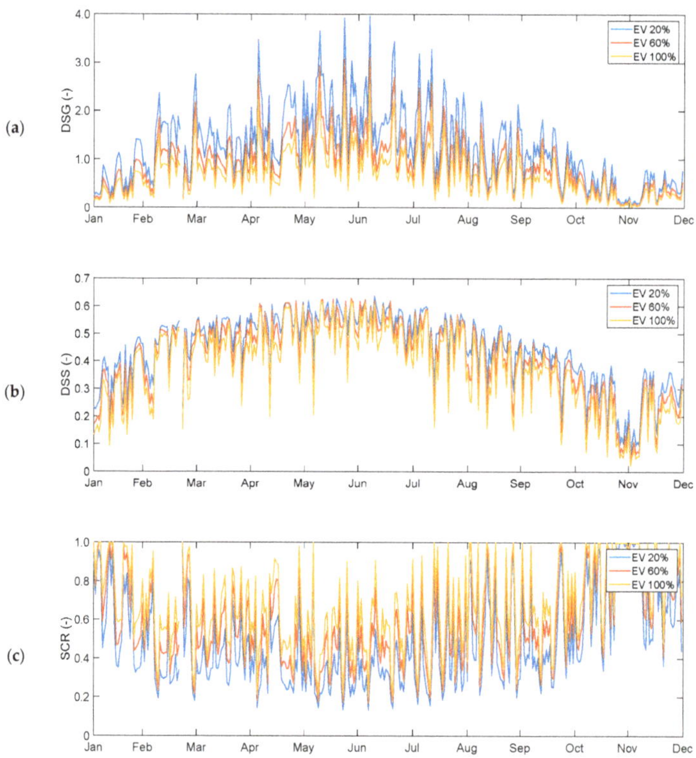

Figure 23. Key performance indicator of cell 13 for the simulation period of one year and for a fixed penetration of 60% PV and a variation of the penetration of EV for 3.7 kW charging power and charging strategy 1 (**a**) DSG (**b**) DSS (**c**) SCR.

The DSG decreases with increasing penetration of EV at a fixed penetration of PV (Figure 23a) and vice versa. Due to the high-power profile of the PV potential compared to the synthetic charging load profile of cell 13, maximum DSG of up to four are possible (without EV). The DSS has its maximum in summer with up to 0.65. This means that up to 65% of the locally produced PV Potential can be consumed directly. The DSS behaves like the DSG, which means that the DSS decreases with increasing penetration of EV at a fixed penetration of PV (Figure 23b) and vice versa. The seasonal dependence is less significant for SCR compared to DSG and DSS. In contrast to the DSG and DSS, the SCR increases with increasing penetration of EV at a fixed penetration of PV (Figure 23c) and vice versa.

4.3.3. Charging Strategy 2 – Controlled Charging

To determine the influence of controlled charging on the key performance indicators, two scenarios are compared, which differ only in the charging strategy. The basis for the results presented below is based on scenarios, which assume a fixed penetration of 60% PV, an increasing penetration of EV and a charging power of 3.7 kW. For the analysis, we take the cells 4, 5, 13 and 18 into account. Figure 24 shows the maximum and minimum relative deviation of selected cells that occur within the comparison of DSG, DSS and SCR of the respective scenarios. The relative deviation varies between −0.16 and 0.31 and is calculated for the respective key performance indicators (KPI) as follows:

$$relative\ deviation = \frac{KPI_{charging\ strategy\ 2} - KPI_{charging\ strategy\ 1}}{KPI_{charging\ strategy\ 1}} \tag{5}$$

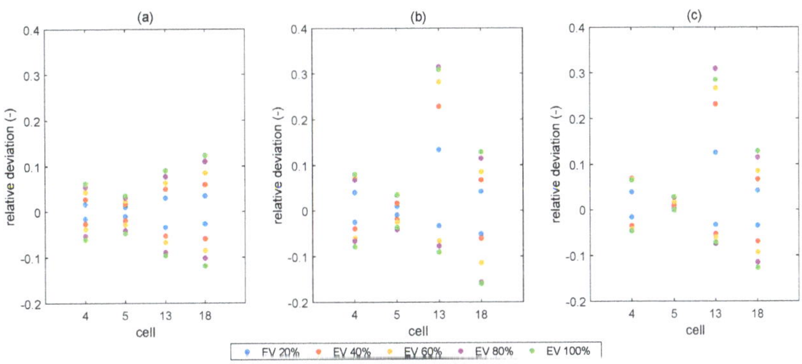

Figure 24. Maximum and minimum relative deviation of selected cells for a fixed penetration of 60% PV and a variation of the penetration of EV for 3.7 kW charging power (**a**) DSG (**b**) DSS (**c**) SCR.

Since the focus of the presented case study is on the user group "charging at work" and these charging processes take place mainly during the day, the daily energy demand of e-mobility remains almost constant on weekdays. On weekends, the user group "trip home" is also selected, where, as already mentioned, charging takes place mainly in the evening or at night. To avoid load peaks of the e-mobility in the evening, in charging strategy 2 these charging processes are distributed over the night, this can lead to shifts in energy demand from one day to the other. Among the other effects, this shift leads to the relative deviations of the DSG as shown in Figure 24a. Since the DSS and SCR calculate the directly consumed part of the locally produced energy on the basis of the total energy demand or the total locally produced energy, the positive relative deviations shown in Figure 24b,c are based on the time shift of the charging processes into the peak of the PV potential. The negative relative deviations, which mean, that controlled charging has a negative effect compared to uncontrolled charging, can be caused by both methods, shifting of the charging processes into the peak of the PV and sequencing the charging processes overnight. For instance, through the distribution of charging processes overnight, the evening peak of e-mobility is reduced. Since in summer there is often sufficient PV production until the evening hours, this reduction of the evening peak can lead to a lower use of the PV potential and therefore to a decrease of DSS and SCR. As shown in Figure 24 for different cells, the benefit of controlled charging (charging strategy 2) depends on the one hand on the user behaviour of the sector-specific jobs within the cells and on the other hand from the balance between energy demand

and the PV potential. This means that the more charging processes take place in the early morning hours and can be shifted into the peak of the PV potential, the higher the DSS and SCR (positive relative deviation) with sufficient PV potential. However, if the ratio of EV to PV is very high, as for example in winter, the shifting of the charging processes has a minor influence on the DSS and SCR. Therefore, in addition to considering the optimal ratio between e-mobility and PV potential, the actual benefits of controlled charging should also be evaluated. However, applying controlled charging leads to a minor positive effect.

5. Conclusions

This paper presents results from a case study for the city of Leoben (Austria), which investigates the synergy effects between e-mobility charged only at work and photovoltaic potentials. The basis for the determination of the grid-side synergy effects is a cell-based grid model. Taking into account the spatial cell division of the grid model, load and production profiles, synthetic charging load profiles for e-mobility and PV potential profiles are determined. For different scenarios (variation of the penetration of EV and PV, charging power, charging strategy), load flow calculations are performed for a simulation period of one year. The results of the load flow calculations are investigated regarding grid-side synergy effects. Additionally, four key performance indicators (residual load, DSG, DSS and SCR) are used to analyse energetic synergy effects between e-mobility and PV potentials.

By using the cellular approach to develop a grid model, there is always a compromise between accuracy, resolution and calculation time. By aggregating the original power grid into energy nodes, this grid model is not suitable for detailed planning, but provides an overview of the effects caused by increasing e-mobility and PV potential. The possibility of carrying out simulations on the grid model for one year means that annual trends, worst-case days and weeks as well as countermeasures can be derived and analysed. As a result, these findings can be used as a basis for detailed studies on the original power grid.

The analyses of the grid-side effects show that with increasing penetration of EV and PV as well as increasing charging power, the grid utilisation and thus the number of line overloads increases. Although line overloads cannot be avoided completely, the maximum line utilisation and the duration of the overloads can be reduced by the correct ratio between e-mobility and PV potential. By using controlled charging (charging strategy 2), the charging processes are shifted to the peak of the PV potential, therefore the power requirement of e-mobility can increase. On the one hand, this increase relieves the grid because it reduces the excess power of the PV potential and on the other hand, it leads to an increased grid load if there is not sufficient PV production available. The comparison of a worst-case week in winter with one in summer shows the need to consider seasonal effects during grid-side and energy analysis. For example, with fixed penetration of PV and increasing penetration of EV, production surpluses are reduced in summer, while increasing e-mobility in winter leads to an increase in line utilisation due to insufficient PV potential. However, if the PV potential is much higher than the demand for e-mobility, controlled charging has little impact on the reduction of line utilisation and thus on overloads caused by PV potentials. Furthermore, the four key performance indicators of the energy analysis demonstrate the importance of modelling PV potentials using real irradiation and temperature data. Besides the residual load, which indicates the fluctuating power surplus caused by the PV potential, DSG, DSS and SCR vary significantly between two consecutive days. The investigation of the residual loads of different scenarios shows that high negative residual loads, which lead to overloads mainly in summer, are reduced by an increasing penetration of EV while the penetration of PV remains the same. However, since the overloads cannot be avoided completely, further possibilities to reduce these high negative residual loads, like a reduction of the PV potential or the use of controlled charging need to be considered. A reduction of the PV potential as well as the increasing e-mobility lead to increasing positive residual loads, especially in winter on PV production weak days. The use of controlled charging can reduce both, negative and positive residual loads.

While negative residual loads are reduced by shifting the charging processes into the peak of the PV potential, positive residual loads are reduced by sequencing the charging processes.

The analysis of grid and energy-side effects has shown the importance of balancing e-mobility and PV potential. The same penetration of EV and PV over all cells for the scenarios considered leads to large differences in the ratios between the energy demand of e-mobility and the PV potential of the individual cells. With penetrations of 60% PV and 100% EV, these ratios are in the range of 0.42 (cell 1) to 11.45 (cell 26) based on the annual energy production of the PV Potential and the annual energy demand of e-mobility. Due to this uneven distribution, for example, a penetration of 100% EV can lead to a line overload, whereas this penetration would be necessary to reduce a power surplus of the PV potential in another cell.

In summary, the key findings of the grid-side and energetic synergy effects between e-mobility charged only at work and PV potentials for the presented case study are as follows:

- To avoid line overloads, spatial planning is important in addition to the optimal energy and power ratio between e-mobility and PV potential.
- The optimal ratio between e-mobility and PV potential is limited by the seasonal component. This means, for example, while increasing e-mobility in summer counteracts negative effects of the PV potential, it leads to higher line utilisations in winter.
- Controlled charging for charging at work has only a minor positive effect on line overloads and key performance indicators.

Although this study shows first trends for the synergies between e-mobility charged only at work and PV potentials, it is also subject to some limitations. Besides the limitations already mentioned, the lack of the possibility of detailed planning and constant penetration across all cells, the data basis is a dominating factor. The real irradiation and temperature data [14] and database for the electric power grid including consumption and generation data are from 2014. In addition, the traffic analyses and statistical data [23–25] used refer to today's mobility behaviour and thus mainly to vehicles with combustion engines. New technologies open up new possibilities, which influence mobility behaviour and therefore change it. To model the synthetic charging load profiles, the vehicles registered in Germany up to 2018 [27] (battery capacity, average energy consumption and charging efficiency) are used. Furthermore, the focus of the presented case study is on a grid-side and energetic consideration, this means that no ecological and economic factors are taken into account. In the area of controlled charging, the trends of line overloads identified in this case study and the associated necessary grid expansion should be analysed in combination with economic and ecological strategies for controlled charging. Furthermore, the use of storage systems at strategic grid points to avoid overloads should be investigated in further studies. Due to the limiting factor of a constant penetration across all cells, following studies should analyse and determine the optimal ratio between e-mobility and PV potentials at cell level to define the scenarios. Based on this optimum per cell, the impact on the power grid should be analysed to verify the hypothesis that overloads can be avoided. The possibility of increasing the determined optimal penetration rate of EV or PV by controlled charging or by the use of storages should also be investigated.

Author Contributions: Conceptualization, J.V.; methodology, J.V., T.K.; software, J.V.; validation, J.V.; formal analysis, J.V.; investigation, J.V.; data curation, J.V., U.B.; writing—original draft preparation, J.V., U.B.; writing—review and editing, J.V., T.K.; visualization, J.V.; supervision, T.K.; funding acquisition, T.K.; All authors have read and agreed to the published version of the manuscript.

Funding: This research was funded by the Austrian Federal Ministry of Transport, Innovation and Technology via the FFG program "Energie der Zukunft", grant number 854637 and by the "Klima- und Energiefonds" via the FFG program "Leuchttürme eMobilität", grant number 865447.

Conflicts of Interest: The authors declare no conflict of interest.

Appendix A

The following direct statistical data and derived variables in relation to the residence or trip purposes were summarised in grid points or stored as a calculation result:

- A trip home
    - number of residents
        - number of residents arrivals and departures
        - number of simultaneous residents vehicles
        - average duration of stay
        - average distance travelled
    - number of visitors
        - number of visitors arrivals and departures
        - number of simultaneous visitor vehicles
        - average duration of stay
        - average distance travelled
    - number of delivery trips
        - number of delivery trips
- A trip to work with private or official car
    - number of employees
        - number of employee arrivals and departures
        - number of simultaneous employee vehicles
        - average duration of stay
        - average distance travelled
    - number of delivery trips
        - number of delivery trips
    - employers
        - sector/industry
    - number of business trips
        - number of company vehicles
        - number of company vehicle arrivals and departures
        - number of simultaneous company vehicles
        - average duration of stay for company vehicles
        - average ride-distance company vehicles
- A trip for shopping
    - Sales area per sector
    - number of employees
        - number of employee arrivals and departures
        - number of simultaneous employee vehicles
        - average duration of stay
        - average distance travelled

- number of customers and visitors
  - number of customers and visitor vehicles arrivals and departures
  - number of simultaneous customers and visitors vehicles
  - average duration of stay
  - average distance travelled
- number of delivery trips
  - number of delivery trips

- A trip for execution (e.g., doctor's visit)
  - number of employees
    - number of employee arrivals and departures
    - number of simultaneous employee vehicles
    - average duration of stay
    - average distance travelled
  - number of customers and visitors
    - number of customers and visitor vehicles arrivals and departures
    - number of simultaneous customers and visitors vehicles
    - average duration of stay
    - average distance travelled
  - number of delivery trips
    - number of delivery trips
- A trip to leisure activities
  - size (area)
  - number of employees
    - number of employee arrivals and departures
    - number of simultaneous employee vehicles
    - average duration of stay
    - average distance travelled
  - number of visitors per day
    - number of customers and visitor vehicles arrivals and departures
    - number of simultaneous customers and visitors vehicles
    - average duration of stay
    - average distance travelled
  - number of delivery trips
    - number of delivery trips
- A trip to education
  - number of employees
    - number of employee arrivals and departures
    - number of simultaneous employee vehicles
    - average duration of stay

        - average distance travelled
    - number of pupils/students
        - number of employee arrivals and departures
        - number of simultaneous employee vehicles
        - average duration of stay
        - average distance travelled
    - number of delivery trips
        - number of delivery trips

Appendix B

Table A1. Overview of all simulated scenarios during the presented case study.

Penetration (%)		CP (kW)	CS	Penetration (%)		CP (kW)	CS	Penetration (%)		CP (kW)	CS	Penetration (%)		CP (kW)	CS
PV	EV			PV	EV			PV	EV			PV	EV		
0	0*	0	0												
0	20	3.7	1	0	20	11	1	0	20	3.7	2	0	20	11	2
0	40	3.7	1	0	40	11	1	0	40	3.7	2	0	40	11	2
0	60	3.7	1	0	60	11	1	0	60	3.7	2	0	60	11	2
0	80	3.7	1	0	80	11	1	0	80	3.7	2	0	80	11	2
0	100	3.7	1	0	100	11	1	0	100	3.7	2	0	100	11	2
20	0*	0	0												
20	20	3.7	1	20	20	11	1	20	20	3.7	2	20	20	11	2
20	40	3.7	1	20	40	11	1	20	40	3.7	2	20	40	11	2
20	60	3.7	1	20	60	11	1	20	60	3.7	2	20	60	11	2
20	80	3.7	1	20	80	11	1	20	80	3.7	2	20	80	11	2
20	100	3.7	1	20	100	11	1	20	100	3.7	2	20	100	11	2
40	0	3.7	1	40	0	11	1	40	0	3.7	2	40	0	11	2
40	20	3.7	1	40	20	11	1	40	20	3.7	2	40	20	11	2
40	40	3.7	1	40	40	11	1	40	40	3.7	2	40	40	11	2
40	60	3.7	1	40	60	11	1	40	60	3.7	2	40	60	11	2
40	80	3.7	1	40	80	11	1	40	80	3.7	2	40	80	11	2
40	100	3.7	1	40	100	11	1	40	100	3.7	2	40	100	11	2
60	0*	0	0												
60	20	3.7	1	60	20	11	1	60	20	3.7	2	60	20	11	2
60	40	3.7	1	60	40	11	1	60	40	3.7	2	60	40	11	2
60	60	3.7	1	60	60	11	1	60	60	3.7	2	60	60	11	2
60	80	3.7	1	60	80	11	1	60	80	3.7	2	60	80	11	2
60	100	3.7	1	60	100	11	1	60	100	3.7	2	60	100	11	2
80	0	0	0												
80	20	3.7	1	80	20	11	1	80	20	3.7	2	80	20	11	2
80	40	3.7	1	80	40	11	1	80	40	3.7	2	80	40	11	2
80	60	3.7	1	80	60	11	1	80	60	3.7	2	80	60	11	2
80	80	3.7	1	80	80	11	1	80	80	3.7	2	80	80	11	2
80	100	3.7	1	80	100	11	1	80	100	3.7	2	80	100	11	2
100	0*	0	0												
100	20	3.7	1	100	20	11	1	100	20	3.7	2	100	20	11	2
100	40	3.7	1	100	40	11	1	100	40	3.7	2	100	40	11	2
100	60	3.7	1	100	60	11	1	100	60	3.7	2	100	60	11	2
100	80	3.7	1	100	80	11	1	100	80	3.7	2	100	80	11	2
100	100	3.7	1	100	100	11	1	100	100	3.7	2	100	100	11	2

CP … charging power/CS … charging strategy; 0* … Since a penetration of 0% EV only takes into account the demand for e-mobility that is already present in the status quo of the power grid, no charging power and charging strategy is assigned to these scenarios. Therefore, the simulation of scenarios with a penetration of 0% EV for different charging power and charging strategy is not necessary.

More than 30 different EV models are registered in Germany [27]. However, since 15 vehicle types account for 96% of all registered vehicles, the other vehicle models are not included due to their small share. In Table A2 an overview of the considered 15 different EV models is given, including the battery capacity and the average energy consumption from the ADAC eco-test [28].

Table A2. Overview of the 15 different considered EV types.

Brand of Car	Model	Battery Capacity [28] (kWh)	Average Energy Consumption [28] (kWh/100 km)
Audi	E-tron	95.0	23.7
BMW	i3	27.2	18.4
Hyundai	Ioniq Elektro	28.0	14.7
Hyundai	Kona	64.0	19.5
Nissan	e-NV 200	40.0	28.1
Nissan	Leaf	40.0	22.1
Renault	Zoe	41.0	20.3
VW	e-Golf	34.9	17.3
VW	e-up!	18.7	17.7
Jaguar	I-PACE	90.0	27.6
Kia	Soul EV	27.0	19.1
Smart	fortwo	17.6	18.3
Smart	forfour	17.6	18.3
Tesla	Model S	90.0	24.0
Tesla	Model 3	75.0	20.9

References

1. Vopava, J.; Kienberger, T. Impact of increasing electric mobility on a distribution grid at the medium voltage level. In Proceedings of the 2nd E-Mobility Power System Integration, Stockholm, Sweden, 15 October 2018.
2. Clement-Nyns, K.; Haesen, E.; Driesen, J. The Impact of Charging Plug-In Hybrid Electric Vehicles on a Residential Distribution Grid. *IEEE Trans. Power Syst.* **2010**, *25*, 371–380. [CrossRef]
3. Deb, S.; Tammi, K.; Kalita, K.; Mahanta, P. Impact of Electric Vehicle Charging Station Load on Distribution Network. *Energies* **2018**, *11*, 178. [CrossRef]
4. Leemput, N.; Geth, F.; van Roy, J.; Olivella-Rosell, P.; Driesen, J.; Sumper, A. MV and LV Residential Grid Impact of Combined Slow and Fast Charging of Electric Vehicles. *Energies* **2015**, *8*, 1760–1783. [CrossRef]
5. Osório, G.; Shafie-khah, M.; Coimbra, P.; Lotfi, M.; Catalão, J. Distribution System Operation with Electric Vehicle Charging Schedules and Renewable Energy Resources. *Energies* **2018**, *11*, 3117. [CrossRef]
6. Munkhammar, J.; Widén, J.; Rydén, J. On a probability distribution model combining household power consumption, electric vehicle home-charging and photovoltaic power production. *Appl. Energy* **2015**, *142*, 135–143. [CrossRef]
7. Chaouachi, A.; Bompard, E.; Fulli, G.; Masera, M.; De Gennaro, M.; Paffumi, E. Assessment framework for EV and PV synergies in emerging distribution systems. *Renew. Sustain. Energy Rev.* **2016**, *55*, 719–728. [CrossRef]
8. Su, S.; Hu, Y.; Yang, T.; Wang, S.; Liu, Z.; Wei, X.; Xia, M.; Ota, Y.; Yamashita, K. Research on an Electric Vehicle Owner-Friendly Charging Strategy Using Photovoltaic Generation at Office Sites in Major Chinese Cities. *Energies* **2018**, *11*, 421. [CrossRef]
9. Chandra Mouli, G.R.; Bauer, P.; Zeman, M. System design for a solar powered electric vehicle charging station for workplaces. *Appl. Energy* **2016**, *168*, 434–443. [CrossRef]
10. Gnann, T.; Klingler, A.L.; Kühnbach, M. The load shift potential of plug-in electric vehicles with different amounts of charging infrastructure. *J. Power Sources* **2018**, *390*, 20–29. [CrossRef]
11. Vopava, J.; Koczwara, C.; Traupmann, A.; Kienberger, T. Investigating the Impact of E-Mobility on the Electrical Power Grid Using a Simplified Grid Modelling Approach. *Energies* **2020**, *13*, 39. [CrossRef]
12. Babrowski, S.; Heinrichs, H.; Jochem, P.; Fichtner, W. Load shift potential of electric vehicles in Europe. *J. Power Sources* **2014**, *255*, 283–293. [CrossRef]
13. Su, W.; Wang, J.; Zhang, K.; Chow, M.Y. Framework for investigating the impact of PHEV charging on power distribution system and transportation network. In Proceedings of the IECON 2012, 38th annual conference on IEEE Industrial Electronics Society, Montreal, QC, Canada, 25–28 October 2012; pp. 4735–4740, ISBN 978-1-4673-2421-2.
14. ZAMG Zentralanstalt für Meteorologie und Geodynamik. *Einstrahlungsmessdaten und Temperaturmesswerte des Jahres 2014 für Kapfenberg*; ZAMG Zentralanstalt für Meteorologie und Geodynamik: Wien, Austria, 2014.

15. Amt der Steiermärkischen Landesregierung. Solardachkataster Steiermark. Available online: https://www.landesentwicklung.steiermark.at/cms/beitrag/11864478/142970647/ (accessed on 3 July 2020).
16. Zhang, T.Z.; Chen, T.D. Smart charging management for shared autonomous electric vehicle fleets: A Puget Sound case study. *Transp. Res. Part D* **2020**, *78*, 102184. [CrossRef]
17. Steinmüller, H.; Tichler, R.; Kienberger, T.; Gawlik, W.; Lehner, M.; Muggenhumer, G.; Kriechbaum, L.; Böckl, B.; Winter, A.; Biegger, P.; et al. *Smart Exergy Leoben—Final Report. Exergetische Optimierung der Energieflüsse für eine smarte Industriestadt Leoben*; Klima- und Energiefonds: Wien, Austria, 2017.
18. Kienberger, T.; Hammer, A.; Vopava, J.; Thormann, B.; Kriechbaum, L.; Sejkora, C.; Hermann, R.; Watschka, K.; Bergmann, U.; Frewein, M.; et al. *Move2Grid—Final Report. Umsetzung regionaler Elektromobilitätsversorgung durch hybride Kopplung*; Bundesministerium für Verkehr, Innovation und Technologie: Wien, Austria, 2019.
19. *NEPLAN AG*; NEPLAN: Küsnacht, Switzerland, 2015.
20. BDEW—Budesverband der Energie- und Wasserwirtschaft e.V. Standardlastprofile Strom. Available online: https://www.bdew.de/energie/standardlastprofile-strom/ (accessed on 3 April 2019).
21. E-CONTROL. Sonstige Marktregeln. Kapitel 6: Zählwerte, Datenformate und Standardisierte Lastprofile. Available online: https://www.e-control.at/recht/marktregeln/sonstige-marktregeln-strom#p_p_id_56_INSTANCE_10318A20066_ (accessed on 3 April 2019).
22. QGIS Development Team. QGIS Geographisches Informationssystem. Available online: https://qgis.org/en/site/ (accessed on 3 July 2020).
23. Bosserhoff, D. *Integration von Verkehrsplanung und Räumlicher Planung. Teil2: Abschätzung der Verkehrserzeugung Durch Vorhaben der Bauleitungsplanung*; Hessisches Landesamt für Straßen- und Verkehrswesen: Wiesbaden, Germany, 2000.
24. Fraunhofer-Institut für System- und Innovationsforschung ISI. *Codebook REM 2030 Driving Profiles Database*; Fraunhofer-Institut für System- und Innovationsforschung ISI: Karlsruhe, Germany, 2015.
25. Infas—Institut für angewandte Sozialwissenschaft GmbH; TRICONSULT—Wirtschaftsanalytische Forschung Gesellschaft m.b.H; HERRY Consult GmbH; Sammer und Partner Zivilingenieur GmbH, ZIS+P Verkehrsplanung; Institut für Verkehrswesen der Universität für Bodenkultur. *Österreich Unterwegs 2013/2014. Ergebnisbericht*; Bundesministerium für Verkehr, Innovation und Technologie: Wien, Austria, 2016.
26. Infas-Institut für angewandte Sozialwissenschaften GmBH; Deutsches Zentrum für Luft und Raumfahrt e.V.; Institut für Verkehrsforschung; IVT Research GmbH. *Mobilität in Deutschland—MiD. Ergebnisbericht*; Bundesministerium für Verkehr und digitale Infrastruktur: Bonn, Germany, 2018.
27. Kraftfahrt-Bundesamt. Neuzulassungen—Deutschland. Available online: https://www.kba.de/DE/Statistik/Fahrzeuge/Neuzulassungen/neuzulassungen_node.html (accessed on 15 July 2019).
28. Allgemeiner Deutscher Automobil-Club e.V. ADAC Ecotest. Available online: https://www.adac.de/infotestrat/tests/eco-test (accessed on 1 July 2019).
29. Tober, W. *Praxisbericht Elektromobilität und Verbrennungsmotor. Analyse elektrifizierter Pkw-Antriebskonzepte*; Springer: Wiesbaden, Germany, 2016; ISBN 9783658136017.
30. World Meteorological Organization. Mean Daily Temperature Provided for Each Month by "World Weather Information Service". Available online: https://www.wwis.dwd.de/de/home.html (accessed on 3 July 2020).
31. Kapfenberger-Pock, A. *Grazer Solardachkataster. Solarthermie bzw. Photovoltaik*; Stadtvermessungsamt: Graz, Austria, 2013.
32. Perez, R.; Seals, R.; Ineichen, P.; Stewart, R.; Menicucci, D. A new simplified version of the perez diffuse irradiance model for tilted surfaces. *Sol. Energy* **1987**, *39*, 221–231. [CrossRef]
33. Perez, R.; Stewart, R.; Seals, R.; Guertin, T. *The Development and Verification of the Perez Diffuse Radiation Model*; Sandia National Labs: Livermore, CA, USA, 1988.
34. Perez, R.; Ineichen, P.; Seals, R.; Michalsky, J.; Stewart, R. Modeling daylight availability and irradiance components from direct and global irradiance. *Sol. Energy* **1990**, *44*, 271–289. [CrossRef]
35. Duffie, J.A.; Beckman, W.A. *Solar Engineering of Thermal Processes*, 3rd ed.; Wiley: Hoboken, NJ, USA, 2006; ISBN 9780471698678.
36. Pretschuh, P. *Solares Energiepotential Kleiner und Mittlerer Städte*; Montanuniversität Leoben: Leoben, Austria, 2016.

© 2020 by the authors. Licensee MDPI, Basel, Switzerland. This article is an open access article distributed under the terms and conditions of the Creative Commons Attribution (CC BY) license (http://creativecommons.org/licenses/by/4.0/).

energies

MDPI

Article

The Simultaneous Impact of EV Charging and PV Inverter Reactive Power on the Hosting Distribution System's Performance: A Case Study in Kuwait

Heba M. Abdullah [1,2,*], Rashad M. Kamel [2,3], Anas Tahir [1], Azzam Sleit [4] and Adel Gastli [1]

1 Electrical Engineering Department, Qatar University, Doha 2713, Qatar; at1404302@qu.edu.qa (A.T.); adel.gastli@qu.edu.qa (A.G.)
2 Electrical Engineering Department, Kuwait University, Kuwait City 72860, Kuwait; rmkamelm@gmail.com
3 Electrical Engineering Department, Assuit University, Assuit 71515, Egypt
4 KINDI Center for Computing Research, Doha 2713, Qatar; azzamibrahim@qu.edu.qa
* Correspondence: ha1808917@qu.edu.qa; Tel.: +965-9751-9996

Received: 29 March 2020; Accepted: 21 May 2020; Published: 26 August 2020

Abstract: Recently, electric vehicles (EVs) have become an increasingly important topic in the field of sustainable transportation research, alongside distributed generation, reactive power compensation, charging optimization, and control. The process of loading on existing power system infrastructures with increasing demand requires appropriate impact indices to be analyzed. This paper studies the impact of integrating electric vehicle charging stations (EVCSs) into a residential distribution network. An actual case study is modeled to acquire nodal voltages and feeder currents. The model obtains the optimal integration of solar photovoltaic (PV) panels with charging stations while considering reactive power compensation. The impact of EV integration for the case study results in two peaks, which show a 6.4% and 17% increase. Varying the inverter to the PV ratio from 1.1 to 2 decreases system losses by 34% to 41%. The type of charging is dependent on the maximum penetration of EVCSs that the network can install without system upgrades. Increasing the number of EVCSs can cause an increase in power system losses, which is dependent on the network architecture. Installing PV reduces the load peak by 21%, and the installation of PV with consideration of reactive power control increases system efficiency and power delivery.

Keywords: electric power systems; transportation; efficiency; sustainability; PV; inverter

1. Introduction

Under the Paris Agreement signed in 2016, nine countries pledged to cut the world's emissions by 2030. Transportation is the second-highest contribution to worldwide emissions, and it is expected to increase 60% by 2050 in the absence of mitigation measures [1]. To meet emission reduction goals, it is essential to expedite the adaptation of electric vehicles (EVs), which are currently being promoted to replace conventional vehicles, thus taking the EV market to a new level [2]. The success of this transition depends on the ability to provide adequate EV charging stations, which, in turn, must provide adequate power for the charging demands of EVs' batteries. It is anticipated that the increasing rate of EVs will lead to a high electricity demand burden on the power grid.

1.1. Impact of EV Charging on Power Distribution Systems

The overall impact of EVs on the power distribution system was reviewed by [3]. This impact varies according to EV penetration, EV battery characteristics, charging demand, load characteristics, driving behavior, driving distance, demand response strategy, and electricity tariffs. The main identified problems were the increase in peak demand, power quality problems, power loss, transformer heating, and system overloading.

Connecting EV charging stations to the grid can negatively affect the power quality and efficiency of the network by introducing losses and voltage violations [4]. The optimal location and size of charging stations and renewable energy sources help to reduce this negative effect. The charging process has a direct effect on distribution system voltage levels; therefore, EV impact assessments are conducted for worst-case scenarios while modeling for EV electrical loads. Authors of [5,6] developed a model for EV load in MATLAB SIMULINK, which represented the behavior of the battery charge and discharge loading characteristics, and concluded that integrating EV fast-charging reduces the steady-state voltage stability limit of the power system.

The battery charge demand of an EV defines the impact on the power system. For instance, uncontrolled charging will cause the charging demand to require an upgrade to the power system, while controlled charging takes into account the distribution network's constraints and does not require further upgrades to the network [7]. In other words, uncontrolled charging requires adding new equipment like transformers, cables, and protection devices to the existing distribution system in order to successfully host and deal with EV charging. Modeling the EV load can be done through a car trip destination based on the household and the person present, such as the Markov chain [8]. A probabilistic approach using the probability density function in modeling EV, grid loads, and PV outputs was reported by [9]. A stochastic modeling-implemented approach in the load flow based on the Monte Carlo method was discussed in [10]. The load demand of the EV fleet in [11] considers socio-economic, technical, and spatial factors. Deterministic EV load estimation based on actual measurements requires traffic data and surveys. The study in [12] suggested considering the degradation of EV batteries when sizing for renewable energy sources' integration for EVCSs to avoid high-capacity sizing.

1.2. PV Implementation with EVCSs

Improving the distribution voltage levels affected by EV charging demand was suggested in [10] by combining PV systems with EV chargers, which proved effective in reducing system losses and in improving load profile and voltage levels. The modeled EV was modeled as a constant load, which was preliminary and did not reflect the actual load characteristics as proven in [6]. The authors of these papers did not incorporate the reactive power involved in EV chargers reported in [13].

To reduce the demand for EV charging of the electric grid, wind-powered EVCSs were suggested by [14]. A PV-based charging station with historical PV prediction was implemented in [15], which showed optimal performance. The author of [16] considered renewable energy technologies, including a diesel generator, concluded that a charging station with PV can benefit most when grid-connected.

Renewable energy sources, such as PV systems, have energy benefits regarding the grid, but may cause disturbances if not analyzed closely [17]. The increased penetration of distributed generation can lead to power system violations or problems such as voltage violations, line capacity or transformer overloading, and extreme line losses [18]. It is suggested in [15] that the most accurate hosting capacity limit for distributed generators (DGs) must be based on assessments of historical limits, performance limits, perception, and enhancement techniques. In other words, uncontrolled charging requires adding new equipment like transformers, cables, and protection devices so the existing distribution system can successfully host and deal with the EV charging.

1.3. Inverter Reactive Power

Modern inverters have the capability of providing reactive power support to improve voltage profiles and to minimize power losses, and therefore were investigated for the application of EV charging in [19]. According to the IEEE global community published standard for Interconnection and Interoperability of Distributed Energy Resources with Associated Electric Power Systems, namely IEEE 1574, there is no requirement for power factors [20]. On the other hand, in the international standard

IEC 61727, the average lagging power factor of PV inverters is stated to be 0.9 when PV generation is 90% [21].

The charger also has the capability of allowing reactive power flow to the grid by utilizing the DC-link capacitor. Several authors have addressed this topic using the bi-directional EV battery charger enabled with the vehicle-to-grid (V2G) concept [22].

A control scheme in [23] implemented PV inverter reactive power compensation to improve network power quality, to reduce losses and overvoltage, and to increase renewable generation capacity. A method in [24] proposed an improved control response for grid inverter reactive power compensation, which controls the power flow between distributed generation and utility. An experimental study in [25] utilized the PV smart inverter as a Volt-Ampere reactive (VAr) controller to regulate the distribution voltage without the installation of additional devices.

1.4. Paper Contribution and Organization

Generally, the literature defines the impact of EV stations' integration into the grid; however, most of the above-mentioned impact studies were performed at a unity power factor. Nowadays, grid-tie inverters can regulate their power factor by shifting their output voltage phase angle to supply or draw reactive power [19]. Another concern is that the increased demand for renewable energy sources (RESs) may have a negative effect on the distribution network [17], and enhancement techniques such as reactive power compensation will be considered alongside other factors while sizing for RES capacity [26]. According to a survey of the literature, most of the research concentrates predominantly on the impact of EV alone on power quality or with renewable energy sources, with less focus on the combined managed impact of EV with PV systems on the grid with reactive power control at non-unity operation.

This paper proposes analyzing the impact of EV integration on the distribution network by considering PV generation with reactive power compensation. This study suggests sizing the PV holding capacity based on enhancement techniques, such as system power loss minimizing. This capacity is based on considering the inverter size compared to the PV size, which is referred to as the inverter-to-PV ratio throughout this paper.

The summary of the proposed work is as follows: The maximum EV penetration for the distribution network is obtained for different scenarios. Subsequently, the optimal integration of solar PV with charging stations is obtained based on reactive power consideration and the inverter-to-PV ratio (power factor). The impact of the EV penetration level on the distribution system's constraints is analyzed. The effect of reactive power generation on the grid on the impact study is considered. A forward/backward load flow model is used based on daily load profiles. Finally, modeling and simulation of a real-life residential area case study in Kuwait (Ahmadi) is carried out to obtain actual results.

The paper is structured as follows: In Section 2, the proposed approach for assessing the electric vehicle's impact is explained; this section defines how to obtain EV charging demand, the PV capacity with its effect on the EV charging station, and the reactive power consideration from the PV inverter. Section 3 presents the models for the distribution network, the system load model, and the load flow problem. In Section 4, the case study explains the location of the test area, and the assumptions, considerations, and load profiles are defined. The results are presented and discussed in Section 5; this section considers the EVCS demand for an IEEE 33 bus, the EVCS demand of the Ahmadi residential area, the effect of PV inverter penetration on line losses with reactive power control, the effect of PV inverter power factors on PV optimum size, and a daily assessment of the power system for the Ahmadi distribution network with PV generation and EVCSs. The paper ends with a final summary of the findings and the conclusion in Section 6.

2. Proposed Approach for Assessing EV Impact

In this paper, different scenarios are considered to analyze the impact on the demands of EVCSs on the electric grid. The proposed approach considers EVCS to be the main load connected to the power system, rather than considering individual EVs as loads. Load flows are run to study the performance of distribution networks to assess certain performance and impact indices, such as nodal voltage variation and feeder overloading [26]. An approach for optimal PV integration is based on obtaining the optimal inverter size for power loss minimizing, considering the reactive power control capacity of the PV inverter. This method is based on the enhancement technique sizing of renewable sources in [15]. Finally, the effectiveness of this approach is assessed by analyzing the EVCS impact study combined with PV and reactive power control.

2.1. EV Charging Demand

The installation of EVCSs requires power system adequacy assessments. The concept of extra effective available energy determines the number of EV loads that can be connected to the system without infrastructure expansion [27]. The same concept was applied in this study to obtain the EVCSs' maximum demand. The capacity of the chargers was obtained by load flow simulation, as explained in Figure 1a.

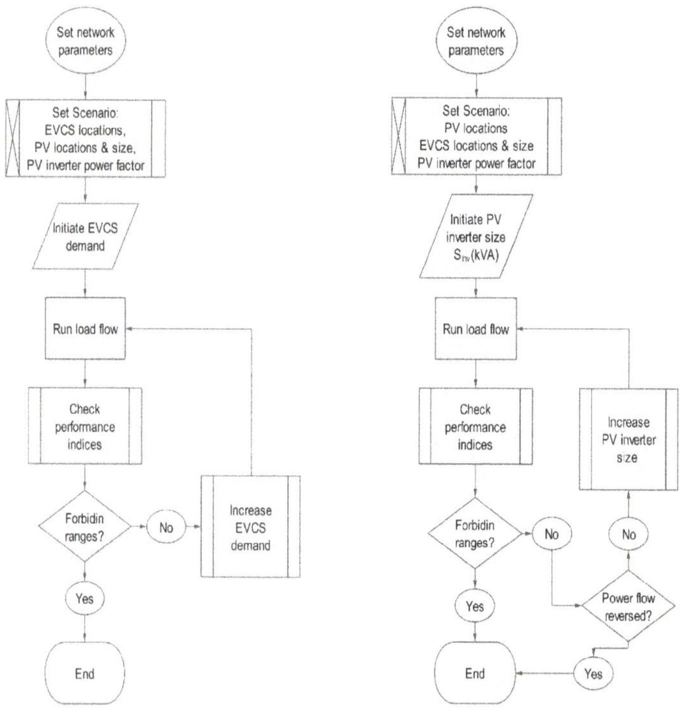

(a) (b)

Figure 1. Flow chart for (**a**) electric vehicle charging stations (EVCSs)' maximum demand. (**b**) Maximum PV penetration. PV, photovoltaic; PF, power factor.

The load flow iteration obtains the maximum demand of EVCSs that can be added to the existing distribution system. The power system's physical boundaries impose constraints on the voltage magnitudes ($V_{\min} = 0.95\ p.u.$, $V_{\max} = 1.05\ p.u.$) and phase angles ($\delta_{\min} = -30^{\circ}$ and $\delta_{\max} = +30^{\circ}$) for all bus voltages in the power network.

$$V_{\min} \leq V_i \leq V_{\max} \quad (1)$$

$$\delta_{\min} \leq \delta_i \leq \delta_{\max} \quad (2)$$

The currents in the power lines must not exceed the rated value (1 p.u.), as in Equation (3).

$$I_{line} \leq I_{Rated} \quad (3)$$

To obtain the maximum demand for EVCSs, the power system is studied for worst-case scenarios when the loading is during peak hours. The actual maximum load demand for the case study was considered in the load flow.

2.2. PV Power with EVCSs

Overloading the power system with EVCSs impacts power system voltage profiles, line currents, and total losses, according to the system topology and physical characteristics. Through the paper; P_T and Q_T refer to the power system's total loading, while $P_{T,i}$ and $Q_{T,i}$ refer to the power loading of a specific bus. Adding PV-distributed generation (PVDG) with EVCSs affects the overall loading of the power system. The main objectives of adding PVDG are:

- Line loss minimization
- Voltage profile improvement
- Increased power reliability
- Peak shaving
- Operation cost reduction

The total system power is therefore sum of the load (P_L and Q_L), EV load (P_{EVCS} and Q_{EVCS}), line losses (P_{Loss} and Q_{Loss}), and PV inverter output power ($-P_{inv}$ and $\pm$ Q_{inv}). Therefore, the total active and reactive powers can be calculated as follows:

$$P_T - P_L + P_{EVCS} + P_{Loss} - P_{inv} \quad (4)$$

$$Q_T = Q_L \pm Q_{EVCS} + Q_{Loss} \pm Q_{inv} \quad (5)$$

where, the negative sign at the inverter active power represents the direction of generated power (−) towards the grid. Also, the reactive power output of the inverter can either generate (−) or absorb (+) reactive power to or from the grid.

The maximum PV penetration is the capacity of the PV plant that can be added to the existing distribution network without the need for upgrading the infrastructure. The values of the maximum PV penetration are obtained by the optimum load flow, and the optimum solution is based on minimizing the line losses of the system without violating the physical constraints. The locations of the PV panels are predefined according to the available space.

2.3. Maximum PV Power Penetration

The PVDG connected to the distribution network acts as a negative load, as given in Equation (4). Increasing PV generation reduces line currents until the power flow is reversed. As the PV continues to increase, line currents start to increase, contributing to power system line losses, as calculated by (6):

$$P_{lineloss} = I_{line}^2 R_{Line} \quad (6)$$

The maximum active power the inverter can deliver from the PV plant at unity power factor is limited by the inverter's size and efficiency. Otherwise, non-unity power factor operation allows the inverter the operation of reactive power injection or consumption.

This paper suggests obtaining the maximum PV penetration from the maximum apparent power of the inverter (S_{inv}) at the worst-case (lowest) power factor operation using Equations (7)–(9).

$$Q_{inv} = S_{inv}\sin(\theta) \tag{7}$$

$$P_{inv} = \sqrt{S_{inv}^2 - Q_{inv}^2} \tag{8}$$

$$P_{\mathrm{PV}} = \frac{P_{inv}}{\eta_{inv}} + P_{Loss}^{\mathrm{DC}} + P_{Loss}^{\mathrm{PV}} \tag{9}$$

where Q_{inv} is the reactive power injected bythe EV charging inverter; P_{inv} is the active power injected by the inverter; θ is the angle between them; S_{inv} is the complex (active and reactive) power injected by the EV charging inverter; PF is the inverter power factor; η_{inv} is the inverter efficiency; P_{PV} is the PV power; P_{Loss}^{DC} refers to the losses in the DC side (before inverter); and P_{Loss}^{PV} refers to the power losses in the PV.

DC system losses are taken into consideration when designing PV size, the calculation for which is as shown in Equation (9). Finally, the inverter-to-PV ratio is obtained according to this approach. The inverter size compared to the PV size is referred to as the inverter-to-PV ratio throughout this paper.

2.4. PV Inverter Reactive Power

The available reactive power at the inverters of EVCSs and PV can be utilized for reactive power compensation in order to improve power loss reduction and voltage profile. Line losses are a function of line current, as in Equation (6). Reducing line losses can be achieved by the PV inverter's active and reactive power generation

Before adding the PV to the power system, the line current can be calculated as follow:

$$I_{line} = \frac{R_{line}P_L + X_{line}Q_L}{V|Z_{line}|} \tag{10}$$

After adding PV to the distribution system, the line current can be calculated by Equation (11) as follow:

$$I_{line}^{new} = \frac{R_{line}(P_L - P_{\mathrm{PV}}) + X_{line}(Q_L + Q_{\mathrm{PV}})}{V|Z_{line}|} \tag{11}$$

where V is the voltage between the two buses, and I_{line} is the cable line in between. The available reactive power at the inverter is calculated by:

$$Q_{inv} = \sqrt{S_{inv}^2 - P_{inv}^2} \tag{12}$$

where P_{inv} can be the PV power at the PV inverter or the power consumed by the EV, which is considered an instantaneous value. Therefore, reactive power compensation application for voltage profile improvement can be realized when the EVCS is not receiving power [22].

3. System Modeling

3.1. The Distribution Network

Two distribution systems were studied. The first network is the IEEE 33 radial bus system, and the second network is an actual urban case study. Both systems are radial low-voltage (LV) systems, modeled in MATLAB code by defining busloads and line impedances. EVCSs were added to the connected loads at defined buses. PV solar power generation was considered for optimal integration at defined locations. The studies were repeated for multiple scenarios.

3.2. System Load Model

The different types of load models have a direct impact on distributed system planning [28]. The type of load models that are voltage-dependent, as defined in [29], are residential, commercial, and industrial loads. Voltage-dependent loads are calculated using Equations (13) and (14) and the values in Table 1.

$$P = P_0 V^{\alpha} \tag{13}$$

$$Q = Q_0 V^{\beta} \tag{14}$$

where α is an active power exponent, β is the reactive power exponent, V is the bus voltage in per unit (p.u.), P_o is the active power at rated voltage, and Q_o is the reactive power at rated voltage.

Table 1. Exponent values for residential loads.

Parameter	α	β
Summer day	0.72	2.96
Summer night	0.92	4.04

3.3. The Load Flow Problem

The basic tool for electrical system analysis is the load flow, which was used to determine the performance of the system. The load flow involves finding the node voltages, line currents, and system losses that are necessary for optimizing network planning, the process of which involves repeating the load flow for multiple iterations. In the application of optimization, the efficiency of the load flow technique was considered.

The classification and comparison of load flow techniques were addressed by [30]. The popular backward-forward sweep (BFS) approach was used to determine the performance indices in the proposed study [31].

A distribution line illustrated in Figure 2 shows the effective active power P_i and reactive power Q_i flowing in branch j through line resistance r_j and reactance x_j from node i to node $i + 1$.

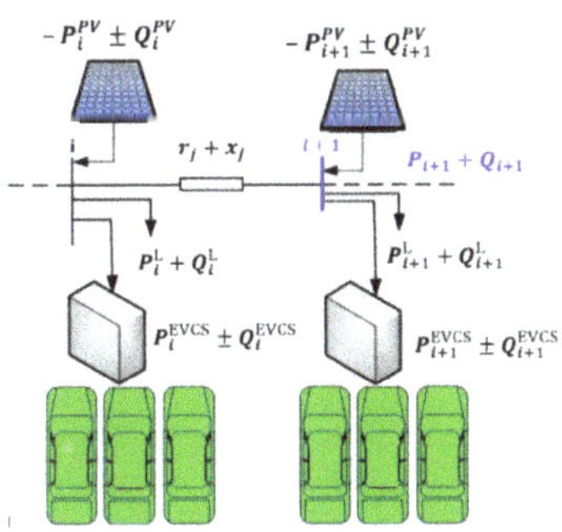

Figure 2. Series impedance line and bus model.

The active and reactive powers can be calculated by Equations (15) and (16):

$$P_i = P_{T,i+1} + r_j \frac{(P_{T,i+1}^2 + Q_{T,i+1}^2)}{V_{i+1}^2} \tag{15}$$

$$Q_i = Q_{T,i+1} + x_j \frac{(P_{T,i+1}^2 + Q_{T,i+1}^2)}{V_{i+1}^2} \tag{16}$$

where $P_{T,i+1}$ and $Q_{T,i+1}$ are the total active and reactive power at node $i + 1$, as formulated in Equations (17) and (18) without including both the EVCS and PV.

$$P_{T,i+1} = P_{i+1} + P^L_{i+1} \tag{17}$$

$$Q_{T,i+1} = Q_{i+1} + Q^L_{i+1} \tag{18}$$

Considering both EVCS and PV implementation in the system, the total power equations are modified into Equations (19) and (20).

$$P_{T,i+1} = P_{i+1} + P^L_{i+1} + P^{\mathrm{EVCS}}_{i+1} - P^{\mathrm{PV}}_{i+1} \tag{19}$$

$$Q_{T,i+1} = Q_{i+1} + Q^L_{i+1} \pm Q^{\mathrm{EVCS}}_{i+1} \pm Q^{\mathrm{PV}}_{i+1} \tag{20}$$

The magnitudes and phase angles of the voltages at each bus are calculated using Equations (21) and (22).

$$V_{i+1} = \sqrt{\left[V_i^2 - 2\left(P_i r_j + Q_i x_j\right) + \left(r_j^2 + x_j^2\right)\frac{P_i^2 + Q_i^2}{V_i^2}\right]} \tag{21}$$

$$\delta_{i+1} = \delta_i - \tan^{-1}\left(\frac{Q_i r_j - P_i x_j}{V_i^2 - \left(P_i r_j + Q_i x_j\right)}\right) \tag{22}$$

The total load S_i on the main bus i is calculated using the bus voltage V_i and the outgoing feeder currents I_i by:

$$S_{i=} V_i I_j^* \tag{23}$$

4. Case Study

4.1. The Test Area Description

To study the proposed assessment approach, a distribution network in an urban area located in Ahmadi, Kuwait, was considered. This section describes this base case study for the power system analysis. Kuwait has an area of 17,818 km^2 with borders of about 195 km located along the Arabian Gulf (Figure 3).

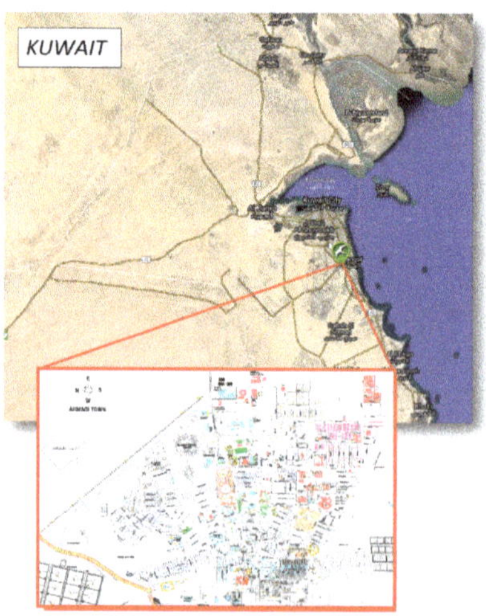

Figure 3. Illustration of Kuwait and Ahmadi city.

Ahmadi city is located 42 km south of Kuwait City and covers an area of 60 km^2. This is a residential area for Kuwait Oil Company employees. The company provides housing, along with facilities and vehicles, and the residents of this area have a higher than average income. Therefore, the possibility of their owning EVs in the future may be higher than others in the country, which may be provided by the company or purchased using their own income. Therefore, analyzing the impact of EVs on this area is essential to promoting the first transition to EVs in Kuwait in the near future.

4.2. Assumptions and Considerations

The Ahmadi low-voltage (LV) residential network delivers power to domestic customers and operates on a medium-voltage to low-voltage (MV/LV) transformer, which supplies 40 houses with five outgoing pillars or feeders (Figure 4). The main transformer of the substation is loaded at 80% of its rated capacity.

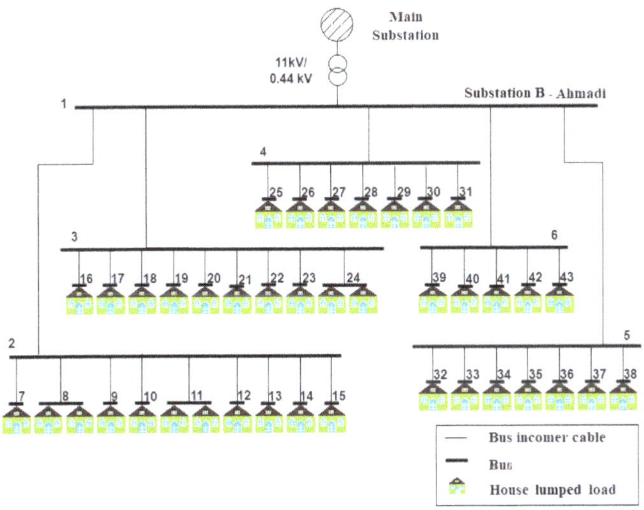

Figure 4. Ahmadi residential network-typical.

All networks are assumed to be three-phase with balanced loads. Residential customers were modeled individually as lumped loads according to the main feeder maximum demand with an average power factor of 0.85.

Both the loading and the characteristic data of the network's buses and lines were considered in this study. The cables feeding the customers were modeled according to their characteristics, sizes, and lengths. As shown in Table 2, the maximum current demand was measured during peak hours in Kuwait at 14:00 on 6 June.

Table 2. Max current demand (6 June 2018 at 14:00).

Bus	Max Demand (A)	No. Customers
2	210	11
3	210	10
4	240	7
5	195	7
6	105	5

The total number of EVCSs was chosen according to the required placement locations in the distribution network (see Figure 4 and Table 3).

Table 3. Scenarios for the EV and PV locations for each case study.

Case Study	EVCS Number of Buses/Locations	Terminology
IEEE 33 Bus	33, 16, 8, 4	33 EVCS, 16 EVCS, 8 EVCS, 4 EVCS
Ahmadi Residential Network	All residence 100%, 50%, 25%	40 EVCS, 20 EVCS, 10 EVCS

It was assumed that only one EVCS is placed at each defined location/bus. The maximum capacity for the EVCS to be added to the network without violating the grid constraints was first obtained for every case study. Then, the optimal location of the PV was obtained for every EVCS location scenario. This was repeated for different values of the power factor. The reactive power at the inverter was taken into consideration for these scenarios by varying the PV inverter power factor (PF = 1, 0.9, and 0.8).

For the IEEE 33 bus system, the PV location was assumed to be at the EV charger bus. The impact assessment of EV charging on the Ahmadi residential area as performed when the chargers were at the residential buses as a case study. Another case was considered when charging was centralized at upper buses. The cases for the PVDG locations are defined in Table 4

Table 4. Cases for PV locations.

Case Study	PVDG Locations
K1	At EV charging locations (all buses)
K2	Centralized PV, main buses (2–6) for Ahmadi residential area

4.3. PV and Load Profiles

After considering the impact of EVCSs on the grid during maximum demand (peak hours), a daily assessment of the power system was performed over a period of 24 h. during a summer peak day. The power flow was run with daily EVs, loads, and PV profiles to obtain the total power system load profile (see Table 5). The load profile was constructed from IEEE Reliability Test System (RTS) system [32]; see Table 5.

Table 5. Daily variation % of peak load [32], EVCS hourly peak load in % daily peak load [27], and PV hourly generation for a 5 kWp plant in Kuwait on 11 June PVSYST© [33].

Hour	Residential Loads %	Residential Charging %	PV Generation kW
0–1	64	54.41	0
1–2	60	39.71	0
2–3	58	35.29	0
3–4	56	32.35	0
4–5	56	26.47	0
5–6	58	17.65	0
6–7	64	10.29	0.55
7–8	76	11.03	1.2
8–9	87	11.76	1.89
9–10	95	11.62	2.55
10–11	99	11.76	3.06
11–12	100	13.97	3.39
12–13	100	16.18	3.51
13–14	100	22.06	3.44
14–15	100	26.47	3.18
15–16	97	28.68	2.74
16–17	96	36.76	2.14
17–18	96	51.47	1.46
18–19	93	73.53	0
19–20	92	85.29	0
20–21	92	88.24	0
21–22	93	100	0
22–23	87	95.59	0
23–24	72	75	0

The hourly variation of the peak demand is provided as a percentage for every hour. The EVCS profile is based on the hourly peak loads as a percentage of daily peak loads. The individual house peak demand on the residential Ahmadi network is considered to be 22 kW. This is based on the data obtained from the site survey from the Kuwait Oil Company (KOC) housing team [34]. A rooftop PV power generation of 5 kW is obtained from the PVSYST© results in Table 5 and confirmed by capacity factors from the real-life PV rooftop projects installed in Kuwait [35]. The estimated PV generation is based on satellite data obtained for site location coordinates N 28°49′29.12″ (Latitude: 28.97), E 47°45′36.81″ (Longitude: 47.62).The actual meteorological data of the location is taken from satellite measures provided by SolarGIS [36]. The software used PVSyst© considers the location's real temperatures and insulation on an hourly bases [33]. The electrical components are based on a typical PV rooftop system installed on the rooftop of a house in Kuwait City obtained from the contractor. According to the collected weather data, the peak irradiance in Kuwait occurred on 11 June. The hourly daily max demand load profile was constructed by PVSYST©; the selected PV panel type was Solar Frontier SF 155-S [37], and the inverter was PVI-6000-OUTD [38].

5. Results and Discussion

This section discusses the impact of EVCS penetration with PVDG and reactive power compensation on power system performance indices. The base case scenario is when there is no EVCS connected to the network system.

5.1. EVCSs' Demand for IEEE 33 Bus

The IEEE 33 bus distribution system was used to investigate the model under study. The power system load flow results, represented as node voltages, line currents, and losses, were obtained for the different demands of the EVCS connected to the system. The results were assessed by the performance indices shown in Table 6. For all cases, the integration of the EVCS into the network system was limited by the capacity of the incoming feeder of the main bus 1, as shown in Figure 5.

Table 6. Maximum EVCS penetration-demand obtained by load flow.

Case	Scenario	EVCS Size (kW)	Total (kW)	I_{max} (p.u.)	V_{max} (p.u.)	Total Line Losses (kW)
IEEE 33 Bus Network	33 EVCS	14	462	0.997	0.913	244.1
	16 EVCS	29	464	0.999	0.903	244.5
	8 EVCS	58	464	0.999	0.905	243
	4 EVCS	119	476	1	0.906	238.5

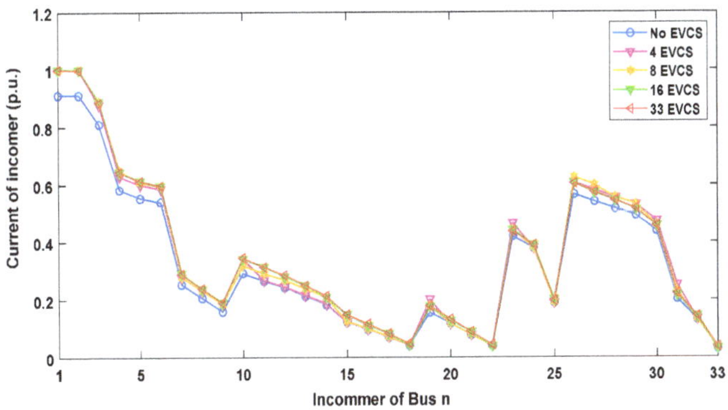

Figure 5. Current in the incomer buses of 33 bus systems.

The limitation factor is defined by the incomer capacity, which should not exceed 1 p.u. Another limiting parameter is the voltage profile, which was taken into consideration while obtaining the EVCSs' demand. The voltage profile at each bus was checked for all the scenarios, and also plotted in Figure 6.

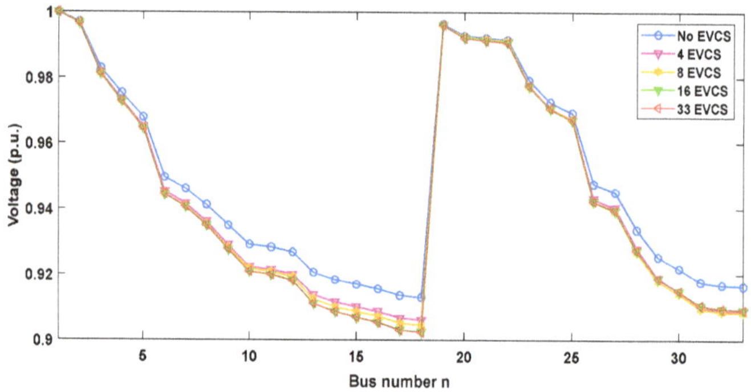

Figure 6. Voltage profile for 33 bus distribution systems.

For the base case, the minimum voltage of 0.913 p.u. occurs at bus 18. This result was benchmarked and validated with the same bus voltage values published in [39,40]. The worst-case scenario was obtained for the maximum overloading of EVCSs, for which a voltage drop violation occurred.

For the IEEE 33 bus system with 33 integrated EVCSs, the voltage drop is approximately 1% compared only to the base case. In all cases, there is no voltage violation while integrating the maximum EV charger demand.

Figure 7 compares the line losses in the incoming lines (incomers) to the buses for all studied cases. It is observed that the incomers with most line losses are those for buses 2–6. It is also observed that the fewer the line losses, the greater the EVCSs' installed capacity.

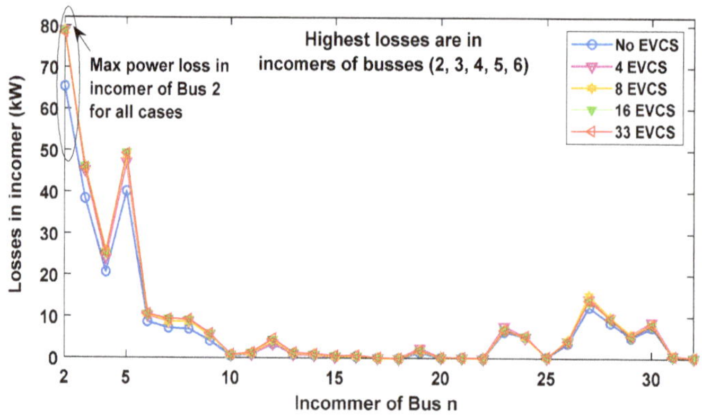

Figure 7. Losses in the incoming lines of the buses in 33 bus systems.

The charging station's maximum penetration for this IEEE 33 bus was obtained for the cases without PV. According to the results in Table 6, the capacity of the charging station is related to the number of total EVCSs distributed in the network.

For the case of four EVCSs in the network, the max capacity is 119 kW per charger station. This results in a higher total charging capacity of 476 kW in the overall system and lower losses of approximately 238 kW. Installing 33 EVCSs limits each charger to 14 kW. This case shows higher losses

(244 kW) than the previous. The limitation factor in all cases is the incomer line capacity. For the IEEE 33-bus system, installing fewer EVCSs is more efficient than having higher numbers in the distribution network.

5.2. EVCSs' Demand for the Ahmadi Residential Network

The maximum penetration of EVCSs at the Ahmadi Residential Network was obtained through load flow simulation, and the results are listed in Table 7. According to Figure 8, the maximum demand is limited by the maximum capacity of the incoming feeder to main bus 1.

Table 7. Maximum EVCS penetration—demand obtained by load flow.

Case	Scenario	EVCS Size (kW)	Total (kW)	I_{max} (p.u.)	V_{max} (p.u.)	Total Line Losses (kW)
Ahmadi Residential Network	40 EVCS	4	160	0.997	0.949	7.262
	20 EVCS	7	140	0.988	0.948	7.044
	10 EVCS	15	150	0.999	0.948	7.541

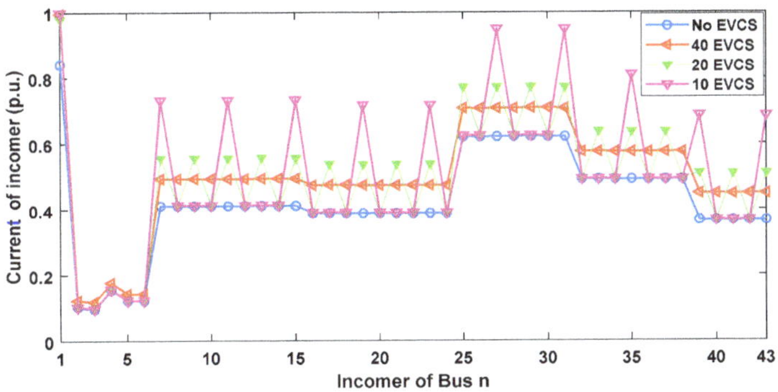

Figure 8. Current in incomer buses of the Ahmadi distribution system.

Besides, the line currents are not uniformly drawn from the main bus because not all customers have charging stations installed at their bus/house. The load profile is less affected by these variations with the increase of customers with EV chargers integrated into the system.

Figure 9 shows the voltage profile at each bus. For the base case, the minimum voltage is found at buses 25 and 31 with a value of 0.95 p.u., which is within the acceptable limits. Notably, the maximum voltage drop for all cases under study is less than 1%.

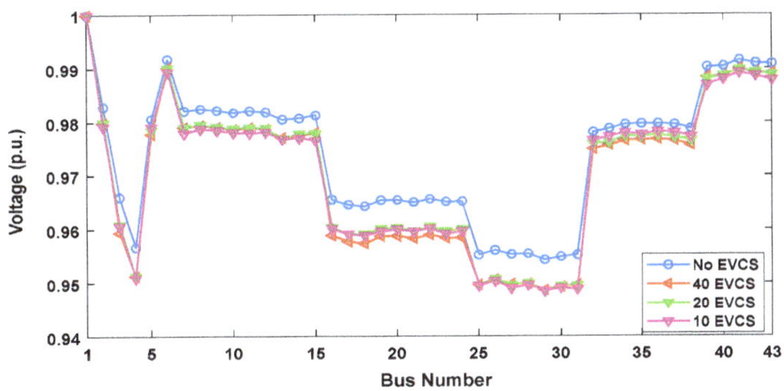

Figure 9. Voltage profile for the Ahmadi distribution system.

In the case when all houses are equipped with EV chargers, which is equivalent to 100% EVCS integration, if the demand per charger is 4 kW, then the total charging demand is 160 kW. The analysis of this case scenario also shows less power loss in the network system. Figure 10 shows that the feeders of main buses 2–6 have the highest losses; these are the feeders of the main buses that feed multiple houses in parallel. Furthermore, the sizes of these feeders are 240 mm^2 and run for approximately 300–700 m. The maximum ampacity (amperage) of the incomer of each feeder pillar at these buses is 400 A. The incomers to the rest of the buses are the residential buses, 185 mm^2 in size, and run for approximately 30–200 m. For all cases, the currents in these lines reach maximum cable capacity because of the high charger loading current, in addition to the losses in the feeder line.

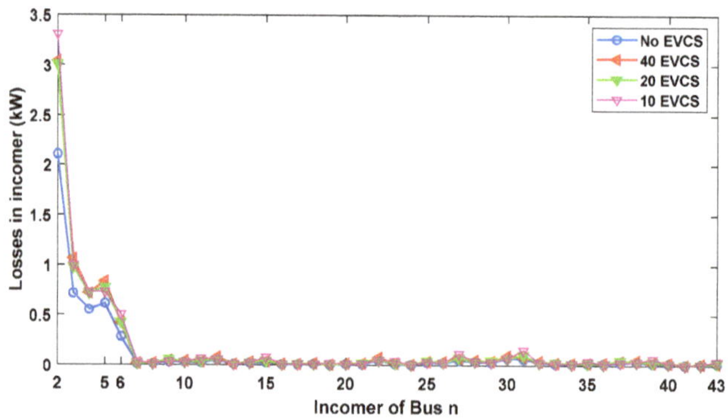

Figure 10. Losses in incomers of the buses in the Ahmadi distribution system.

The limitation of this design is that fast charging cannot be implemented under the circumstances of only a 4 kW charging capacity. For the case with 10 EVCSs connected to the system (25% EVCS penetration), the capacity per charger is 15 kW, and the total charring demand is 150 kW. This case introduces higher line losses compared with the previous. As the number of EVCSs is reduced in this network, the individual charger capacity becomes larger. Consequently, the overall charging demand is decreased because of the increased line losses. Indeed, the line losses are directly affected by the EVCSs' loading demand, location, number, and power line characteristics.

The plot in Figure 11 shows the overall active and reactive powers on the main bus (Bus 1) for different sizes of connected PV and EVCs. For the case of 100% EVCSs (40 EVCS) without integrating any PV power, the bus active power is 1354 kW and reactive power is 691 kVAr (blue), which is the transformer's capacity 1.5 kVA.

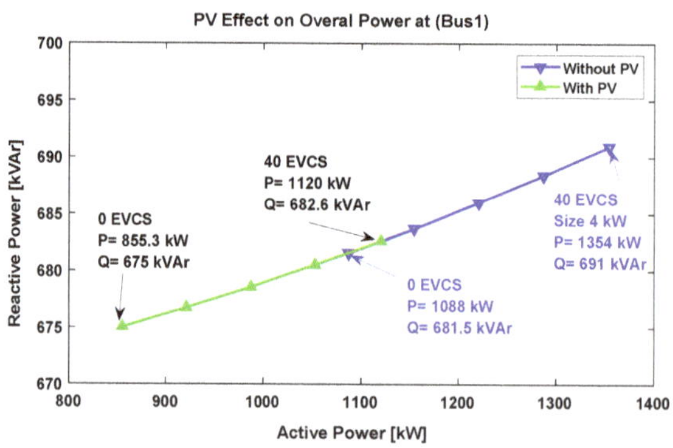

Figure 11. System load at main bus (Bus 1) at 100% EVCS penetration (PV = 20 kW).

Notice that adding 20 kW PV power to the system reduces the total active load kW by 17.3% and kVA load by 14.5% (green). The following section studies the effect of PV penetration and inverter reactive power on power system line losses.

5.3. *Effect of PV Inverter Penetration on Line Losses with Reactive Power Control*

This paper proposes PV generation application with EV charging stations to minimize power system losses. The effect of PV penetration levels on power system losses was obtained through multiple iterations of the load flow simulation, with the objective to minimize power losses in the feeder lines. The results are presented in Figure 12, which are the losses obtained by the load flow for the Ahmadi network cases. The locations of the PV are mentioned in detail in Table 4.

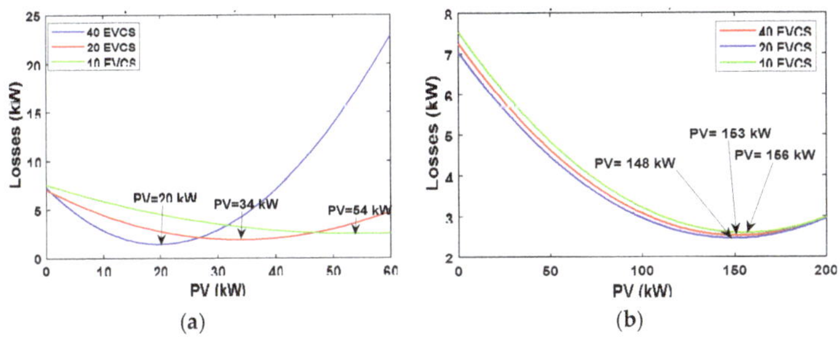

Figure 12. Optimum PV size at minimum line loss for PV locations K1 and K2 for unity power factor: (**a**) Optimum PV at K1 locations; (**b**) optimum PV at K2 locations.

Line losses start decreasing with the increasing size of PV generation, and then start increasing at the point when PV generation exceeds the power at the bus and line losses. This is the point when the current is reversed. Therefore, the point of minimum power loss is the suggested optimal point of operation.

The maximum PV penetration for every case was obtained by the proposed load flow technique. The results are shown in Tables 8 and 9. For the unity power factor, the PV rating was considered the inverter's apparent power for the case study. For the IEEE 33 bus system, the PV locations were considered connected directly to the EVCS. PV generation introduces current to the network load, which in turn reduces losses in the lines by around 60–70% (see Table 8). This is the case when the PV inverter is operating at a unity power factor. As the PF decreases, the inverter compensates for the system's reactive power. Setting the inverter power factor to 0.9 or 0.8 reduces line losses by approximately 90%. The maximum PV inverter penetrations for the Ahmadi distribution network are shown in Table 9. The results in Table 9 refer to the cases when EVCSs are connected at different capacities, and are as follows:

Table 8. Effect of PV penetration on losses for the IEEE 33 bus system.

	PF		1			0.9			0.8		
Case	Size/EVCS (kW)	PV	S_{inv} (kVA)	P_{Loss} (kW)	P_{Loss} Reduction (%)	S_{inv} (kVA)	P_{Loss} (kW)	P_{Loss} Reduction (%)	S_{inv} (kVA)	P_{Loss} (kW)	P_{Loss} Reduction (%)
4 EVCS	118	K1 *	243	68.29	72.1	276	138	43.5	276	17.91	92.6
8 EVCS	58	K1	243	68.29	72.1	276	138	43.5	276	17.91	92.6
16 EVCS	29	K1	243	68.29	72.1	276	138	43.5	276	17.91	92.6
33 EVCS	14	K1	117	69.43	71.5	134	18.07	92.5	134	19.2	92.1

* K1: PV location at EVCS.

Table 9. Effect of PV penetration on losses for the Ahmadi residential network.

Case	Size/EVCS (kW)	PV	PF = 1 S_{inv} (kVA)	P_{Loss} (kW)	P_{Loss} Reduct. (%)	PF = 0.9 S_{inv} (kVA)	P_{Loss} (kW)	P_{Loss} Reduct. (%)	PF = 0.8 S_{inv} (kVA)	P_{Loss} (kW)	P_{Loss} Reduct. (%)
40 EVCS	4	K1 *	20	1.455	80.0	22	0.104	98.6	22	0.335	95.4
		K2 **	153	2.522	65.3	169	1.407	80.6	166	1.591	78.1
20 EVCS	7	K1	34	1.918	72.8	38	0.729	89.7	37	0.934	86.7
		K2	148	2.46	65.1	166	1.347	80.9	165	1.513	78.5
10 EVCS	15	K1	54	2.519	66.6	59	1.545	79.5	58	1.817	75.9
		K2	156	2.591	65.6	173	1.475	80.4	169	1.684	77.7

* K1: PV location at EVCS. ** K2: PV location at the upper bus (feeder pillar).

Case 1: For 40 EVCSs in the system, the maximum PV inverter capacity obtained is 20 kW at the EVCS location and 153 kW if located at the main buses (2–6) for the unity power factor. The inverter size is increased when the power factor is reduced in order to obtain the optimum active and reactive power the system requires to minimize line losses. Therefore, at 0.9 power factor operation, the PV inverter size is 22 kW at the EVCS buses and 169 kW at the main buses. At 0.8 power factor operation, the change in the maximum size of the inverter is not significant.

Case 2: For 20 EVCSs in the system, the maximum PV inverter size obtained is 34 kW at the EVCS location (higher than Case 1) and 148 kW if located at the main buses (lower than Case 1) for the unity power factor. At 0.9 power factor operation, the PV inverter size is 38 kW at the EVCS buses and 166 kW at the main buses. At 0.8 power factor operation, the change in the maximum size is reduced. The reactive power generated by the inverter exceeds the load reactive power consumed by the power system load when the inverter operates at a power factor of 0.8.

Case 3: The maximum inverter size increases compared to Cases 1 and 2. The increase is more for the cases when the PV is installed directly at the main buses.

It is observed that, for the three cases, the maximum PV penetration is higher when the PV is installed at the EVCS buses when fewer EV chargers with higher capacity are connected to the system. For the Ahmadi distribution system, the line loss reduction for unity, 0.9, and 0.8 power factors is in the range of 60–80%, 80–90%, and 70–90%, respectively. Varying the inverter's power factor results in a noticeable decrease in the line losses of both networks under study. In the load flow simulation, the value of apparent power was increased after each iteration, and the real and reactive power was calculated and added to the network. Thus, the numerical results can be used to develop a performance indicator that can be utilized for optimizing the inverter-to-PV ratio in the distribution network, along with EVCS planning.

5.4. *Effect of PV Inverter Power Factor on PV Optimum Size*

As the inverter power factor is reduced from unity, the optimum PV size kVA is affected. According to the results presented in Figures 13 and 14, it is observed that the maximum PV penetration into the power system varies with the inverter power factor. For non-unity power factor operation, the optimum size of PV is chosen according to the optimum inverter-to-PV ratio for each power factor. This section did not consider the available area when calculating the maximum PV penetration. The inverter-to-PV ratios considered were 1 for PF = 1, 1.11 for PF = 0.9, and 1.25 for PF = 0.8. As the power factor decreases, the power system's losses are lessened by reducing the reactive power load until the point where power losses start to increase, as for the case when PF = 0.8. The optimum inverter-to-PV ratio is 1.1 for this system. Changing the power factor setting of the inverter allows an increase in PV installation. Varying the inverter-to-PV ratio can affect the maximum connection that the system can allow. Therefore, the PV connection is more effective for systems with high line losses, as is the case when the PV is installed at the main buses (case in Figure 14).

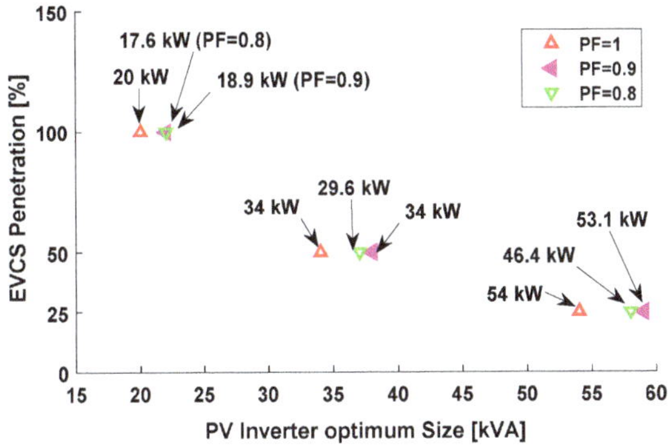

Figure 13. Effect of PV inverter PF on PV optimum size—PV at EVCSs.

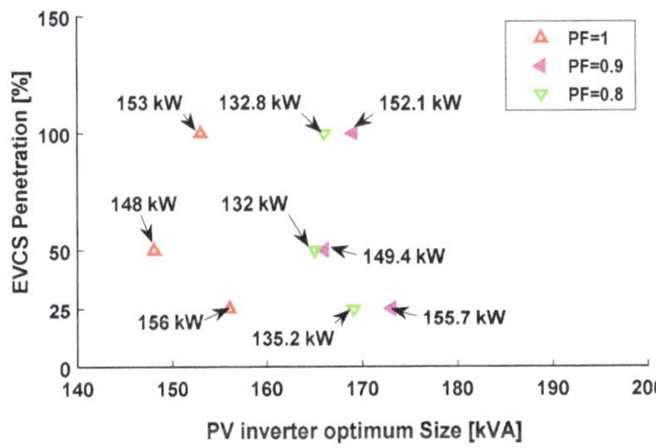

Figure 14. Effect of PV inverter PF on PV optimum size—PV at main buses.

According to the Ahmadi distribution, the rooftop area of the residence can allow an installation of no more than 5 kW per house, which refers to the case study in the next section.

5.5. Daily Assessment of the Power System for the Ahmadi Distribution Network with EVCSs and PV

The daily load flow was obtained according to EVCSs' demand, PV power generation, and PV inverter reactive power. The case under study was conducted with the following (obtained from previous results and assumptions):

- EVCS penetration 100% located at residential buses.
- EVCS capacity size 4 kW per house (see Table 9).
- PV plant at residential buses and capacity of 5 kW.
- The load on buses and EV demand from Table 5.
- PV generation from PVSYST© (see Table 5).

Figure 15 shows the load profile for the Ahmadi distribution network, which is the overall active power drawn by the network at bus 1.

Case 1: The base case with zero EVCSs connected. The load profile shows an increasing peak around noon. The daily load profile is generated by the network's maximum demand [32]. This case was studied according to the available data and the future development of the actual daily load demand required. The peak demand is approximately 1 MW and occurs between 11:00 and 15:00.

Case 2: This case studies the network at 100% EVCS penetration with no PV power. The EVCSs' demand is reflected in the overall power profile. The increase in the load occurs at night between 16:00 and 07:00, which refers to the time during the weekdays where people return to their homes after work.

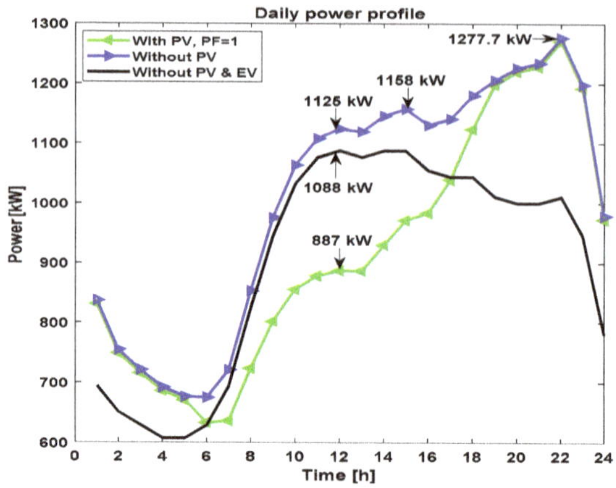

Figure 15. Daily load profile at bus 1.

The noon peak increases by approximately 3.4%, and the new peak at 15:00 is 1.15 MW, which increases by 6.4%, with a higher peak at 22:00 of 1.27 MW (an increase of 17.4%). The load starts to reduce exponentially, which reflects the final behavior of the charging process of the electric vehicles' batteries. Adding an EVCS to the load demand affects daily power losses by an increase of 17% (see Table 10).

Table 10. Inverter power factor effect on system line losses.

EV	PV	PV Inverter Power Factor	System Losses (kWh/day)	Change in Power Loss (%)
×	×	N/A	97.18	0
√	×	N/A	114.05	+17.36%
√	√	1	76.81	−32.65%
√	√	0.9	74.89	−34.33%
√	√	0.8	72.46	−36.47%
√	√	0.6	69.46	−39.10%
√	√	0.5	67.03	−41.22%

Similar effect of EV charging on the load profile is found in the literature. The report in [41] shows two types of load factors (LFs) that of the load profiles. The system peak (SP) load factor and the non-coincident peak (NCP). The SP load factor is the ratio of the average EV demand to the EV peak demand, obtained here as 1.08. The NCP is the ratio between the average EV demand to the EV NCP demand, obtained here as 0.198.

Case 3: PV generation is the minimum where every house has an installation of only 4–5 kW, which operates around 2–3 kW per house. The total maximum PV generation occurs at noon, with approximately 130 kW, and load demand is reduced from 1.12 to 0.88 MW (a reduction of 21.4%). The power losses reduce by approximately 41% with PV generation (at 12:00 pm), as shown in Figure 16. At night, the PV inverter is used for reactive power compensation. The line losses are reduced by 16.6% during the night peak load (at 10 pm), and the daily power losses are reduced by 20% (see Table 10). This case considers the inverter-to-PV ratio to be 1; the available reactive power at the PV does not reach the installed capacity. Therefore, there is always available reactive power at the inverter to compensate for the system losses.

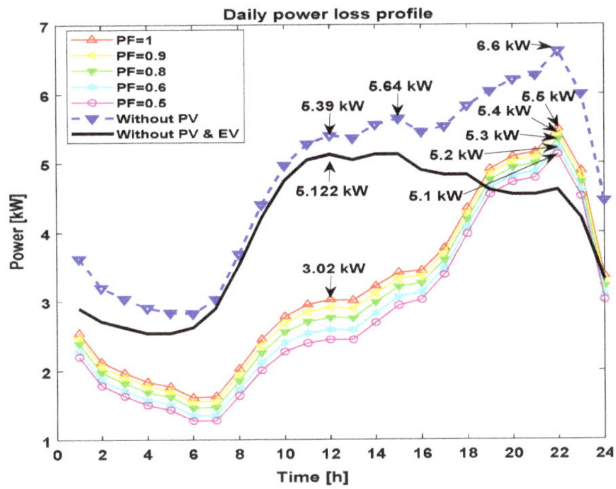

Figure 16. Daily losses.

Case 4: The non-unity power factor of the PV inverter is considered. The effect of reactive power compensation is reflected in the line losses. PV generation is not affected by the power factor because the inverter-to-PV ratio is 1.11 for this case (PF = 0.9). The daily power losses are reduced by 2.5% compared with the case involving unity power factor operation.

Case 5: The inverter-to-PV ratio of this case is 1.25. The effect of having the inverter compensate at PF = 0.8 causes a 5.67% reduction in daily line losses compared to the system at a unity power factor.

Furthermore, other cases were considered for PF = 0.6 and 0.5, with an inverter-to-PV ratio of 1.5 and 2, respectively, with further daily loss reductions of 7.26% and 12.73%, respectively.

6. Conclusions

In this study, the impact of EV charging demand was analyzed for two case study power distribution networks with different characteristics. The power flow was validated by the standard IEEE 33 network. Different load flow scenarios were introduced to analyze the performance indices. Instead of varying the number of EVs, the number of EV charging stations was varied, and the maximum charging load capacity was obtained using the load flow analysis technique.

This paper suggests operating EVCS with PV power generation and reactive power compensation. The optimal PV size was considered by obtaining the optimum inverter-to-PV ratio for minimum power loss. The inverter optimum size was obtained at 0.9 power factor for the Ahmadi network without considering rooftop area limitations. The maximum PV size is not always the optimal solution when PVs are located at the same EVCS buses. Other cases show that reactive power increases the PV maximum penetration when the PV is located at the main feeder buses. The limitation factor refers to both cable ampacity and line losses at PV locations.

It was observed that PV integration with EV charging stations can decrease 21% of the load profile at noon in Kuwait. A high penetration of EV chargers among the houses in the Ahmadi case study causes a 17% increase in power demand at night. The increase does not exceed the network power system limitations for the worst-case scenario, which is the summer peak load. The capability for fast charging at residential areas is limited to the physical characteristics of the network. The maximum demand of an EVCS that can be provided for all customers is 4 kW, which limits the DC fast charging to be centralized at the network's main buses only. Deciding whether or not to have central or distributed charging stations depends on both network characteristics and customer needs. Varying the Inverter to PV ratio from 1.1 to 2 can decrease the system losses by 34% to 41%, which shall be considered based on an economic assessment. Considering reactive power compensation with PVDG when integrating EVCS into the power system has triggered improvements in system efficiency and power delivery.

Author Contributions: Conceptualization, H.M.A.; formal analysis, H.M.A.; investigation, H.M.A. and R.M.K.; methodology, H.M.A.; project administration, A.G.; software, H.M.A. and A.T.; supervision, A.S. and A.G.; validation, R.M.K.; visualization, H.M.A. and A.T.; writing—original draft, H.M.A.; writing—review and editing, R.M.K and A.G. All authors have read and agreed to the published version of the manuscript.

Funding: This research received no external funding.

Acknowledgments: The technical support given by the Housing team of Kuwait Oil Company, which provided all of the necessary data required for modeling Ahmadi Residential Network in this study.

Conflicts of Interest: The authors declare that there is no conflict of interest.

Nomenclature

Acronyms

DC	Direct Current
DG	Distributed generator
EV	Electric vehicle
EVCSs	Electric vehicle charging stations
PV	Photovoltaic
PF	Power Factor
PVDG	Photovoltaic distributed generation
RES	Renewable Energy Sources

Greek symbols

δ	Angle of the bus voltage (degree)
η_{inv}	Inverter efficiency
α	Active power exponent
β	Reactive power exponent

Variables

I_{line}	The line current (p.u.)
P_T	Total active power of power system(p.u.)
P_L	Total load active power (p.u.)
P_{EVCS}	Total EVCS active power (p.u.)

Variables

P_{inv}	Power injected by PV inverter (p.u)
P_{Loss}	Total active power losses (p.u.)
P_o	Active power at rated voltage (p.u.)
P_{PV}	PV active power (p.u.)
Q_T	Total reactive power (p.u.)
Q_L	Total load reactive power (p.u.)
Q_{EVCS}	Total EVCS reactive power (p.u.)
Q_{inv}	Reactive power of PV inverter (p.u.)
Q_{loss}	Total reactive power losses (p.u.)
Q_o	Reactive power at rated voltage (p.u.)
R_{line}	Line resistance (ohm)
S_{inv}	PV inverter apparent power (p.u.)
S_i	Apparent power at certain bus (p.u.)
V_i	Voltage of bus i (p.u.)
V_{min}	Minimum limit of bus voltage (p.u.)
V_{max}	Maximum limit of bus voltage (p.u.)
X_{line}	Line reactance (ohm)
Z_{line}	Line impedance (ohm)

References

1. Transport Outlook 2017.Enw. Available online: http://read.oecd-ilibrary.org/transport/itf-transport-outlook-2017_9789282108000-en#page1 (accessed on 23 May 2020).
2. Wikström, M.; Eriksson, L.; Hansson, L. Introducing Plug-in Electric Vehicles in Public Authorities. *Res. Transp. Bus. Manag.* **2016**, *18*, 29–37. [CrossRef]
3. Shareef, H.; Islam, M.M.; Mohamed, A. A review of the stage-of-the-art charging technologies, placement methodologies, and impacts of electric vehicles. *Renew. Sustain. Energy Rev.* **2016**, *64*, 403–420. [CrossRef]
4. Moradi, M.H.; Abedini, M.; Hosseinian, S.M. Improving operation constraints of microgrid using PHEVs and renewable energy sources. *Renew. Energy* **2015**, *83*, 543–552. [CrossRef]
5. Ma, G.; Jiang, L.; Chen, Y.; Dai, C.; Ju, R. Study on the impact of electric vehicle charging load on nodal voltage deviation. *Arch. Electr. Eng.* **2017**, 66. [CrossRef]
6. Dharmakeerthi, C.H.; Mithulananthan, N.; Saha, T.K. Impact of electric vehicle fast charging on power system voltage stability. *Int. J. Electr. Power Energy Syst.* **2014**, *57*, 241–249. [CrossRef]
7. De Hoog, J.; Alpcan, T.; Brazil, M.; Thomas, D.A.; Mareels, I. Optimal charging of electric vehicles taking distribution network constraints into account. *IEEE Trans. Power Syst.* **2015**, *30*, 65–375. [CrossRef]
8. Munkhammar, J.; Widén, J.; Rydén, J. On a probability distribution model combining household power consumption, electric vehicle home-charging and photovoltaic power production. *Appl. Energy* **2015**, *142*, 135–143. [CrossRef]

9. Shariful Islam, M.; Mithulananthan, N.; Quoc Hung, D. Coordinated EV charging for correlated EV and grid loads and PV output using a novel, correlated, probabilistic model. *Int. J. Electr. Power Energy Syst.* **2019**, *104*, 335–348. [CrossRef]
10. Tovilović, D.M.; Rajaković, N.L.J. The simultaneous impact of photovoltaic systems and plug-in electric vehicles on the daily load and voltage profiles and the harmonic voltage distortions in urban distribution systems. *Renew. Energy* **2015**, *76*, 454–464. [CrossRef]
11. Hu, Q.; Li, H.; Bu, S. The prediction of electric vehicles load profiles considering stochastic charging and discharging behavior and their impact assessment on a real uk distribution network. *Energy Procedia* **2019**, *158*, 6458–6465. [CrossRef]
12. Atia, R.; Yamada, N. More accurate sizing of renewable energy sources under high levels of electric vehicle integration. *Renew. Energy* **2015**, *81*, 918–925. [CrossRef]
13. Haidar, A.M.A.; Muttaqi, K.M. Behavioral characterization of electric vehicle charging loads in a distribution power grid through modeling of battery chargers. *IEEE Trans. Ind. Appl.* **2016**, *52*, 483–492. [CrossRef]
14. Lee, Y.; Hur, J. A simultaneous approach implementing wind-powered electric vehicle charging stations for charging demand dispersion. *Renew. Energy* **2019**, *144*, 172–179. [CrossRef]
15. Ismaela, S.M.; Abdel-Aleemb, S.H.E.; Abdelazizc, A.Y.; Zobaad, A.F. State-of-the-art of hosting capacity in modern power systems with distributed generation. *Renew. Energy* **2019**, *130*, 1002–1020. [CrossRef]
16. Hafez, O.; Bhattacharya, K. Optimal design of electric vehicle charging stations considering various energy resources. *Renew. Energy* **2017**, *107*, 576–589. [CrossRef]
17. Kyritsis, A.; Voglitsis, D.; Papanikolaou, N.; Tselepis, S.; Christodoulou, C.; Gonos, I.; Kalogirou, S.A. Evolution of PV systems in Greece and review of applicable solutions for higher penetration levels. *Renew. Energy* **2017**, *109*, 487–499. [CrossRef]
18. Gevers, D.N.; Aoki, A.R.; Impinnisi-Cleverson, P.R.; da Pinto, S.L.; da Damasceno, V.P.M.R. Evaluation of the impacts of renewables sources and battery systems in distribution feeders with different penetration levels. In Proceedings of the 2019 IEEE PES Innovative Smart Grid Technologies Conference-Latin America, Gramado, Brazil, 15–18 September 2019.
19. Hung, D.Q.; Dong, Z.Y.; Trinh, H. Determining the size of PHEV charging stations powered by commercial grid-integrated pv systems considering reactive power support. *Appl. Energy* **2016**, *183*, 160–169. [CrossRef]
20. IEEE 1547 Standard for Interconnecting Distributed Resources with Electric Power Systems. Available online: http://grouper.ieee.org/groups/scc21/1547/1547_index.html (accessed on 22 April 2020).
21. IEC Standard. Photovoltaic (PV) Systems -Characteristics of the Utility Interface. Available online: https://www.iecee.org/dyn/www/f?p=106:49:0::::FSP_STD_ID:5736 (accessed on 22 April 2020).
22. Yong, J.Y.; Ramachandaramurthy, V.K.; Tan, K.M.; Mithulananthan, N. Bi-directional electric vehicle fast charging station with novel reactive power compensation for voltage regulation. *Int. J. Electr. Power Energy Syst.* **2015**, *64*, 300–310. [CrossRef]
23. Collins, L.; Ward, J.K. Real and reactive power control of distributed pv inverters for overvoltage prevention and increased renewable generation hosting capacity. *Renew. Energy* **2015**, *81*, 464–471. [CrossRef]
24. Sampaio, L.P.; de Brito, M.A.G.; de Melo, A.G.; Canesin, C.A. Grid-tie three-phase inverter with active power injection and reactive power compensation. *Renew. Energy* **2016**, *85*, 854–864. [CrossRef]
25. Howlader, A.M.; Sadoyama, S.; Roose, L.R.; Sepasi, S. Distributed voltage regulation using volt-var controls of a smart pv inverter in a smart grid: An experimental study. *Renew. Energy* **2018**, *127*, 145–157. [CrossRef]
26. SIslam, M.S.; Mithulananthan, N. PV based EV charging at universities using supplied historical PV output ramp. *Renew. Energy* **2018**, *118*, 306–327. [CrossRef]
27. Kamruzzaman, M.; Benidris, M. Effective accessible energy to accommodate load demand of electric vehicles. In Proceedings of the 2018 IEEE Industry Applications Society Annual Meeting (IAS), Portland, OR, USA, 23–27 September 2018; pp. 1–8. [CrossRef]
28. Singh, D.; Misra, R.K. Load type impact on distribution system reconfiguration. *Int. J. Electr. Power Energy Syst.* **2012**, *42*, 583–592. [CrossRef]
29. Price, W.W.; Casper, S.G.; Nwankpa, C.O.; Bradish, R.W.; Chiang, H.D.; Concordia, C.; Wu, G. Bibliography on load models for power flow and dynamic performance simulation. *IEEE Trans. Power Syst.* **1995**, *10*, 523–538. [CrossRef]
30. Ghatak, U.; Mukherjee, V. An improved load flow technique based on load current injection for modern distribution system. *Int. J. Electr. Power Energy Syst.* **2017**, *84*, 168–181. [CrossRef]

31. Augugliaro, A.; Dusonchet, L.; Favuzza, S.; Ippolito, M.G.; Sanseverino, E.R. A backward sweep method for power flow solution in distribution networks. *Int. J. Electr. Power Energy Syst.* **2010**, *32*, 271–280. [CrossRef]
32. Laengen, Ø.S. Application of Monte Carlo Simulation to Power System Adequacy Assessment. Master's Thesis, Norwegian University of Science and Technology, Trondheim, Norway, 2018.
33. PVSyst. Available online: https://www.pvsyst.com/ (accessed on 22 April 2020).
34. Al-Sabti, A. Meeting with the housing team in Kuwait oil company (KOC). In Proceedings of the SPE Kuwait Oil & Gas Show and Conference, Mishref, Kuwait, 13–16 October 2019.
35. Life Energy Projects. Available online: http://www.thelifeenergy.com/projects/ (accessed on 22 April 2020).
36. SolarGIS. Available online: https://solargis.com/ (accessed on 22 April 2020).
37. Solar Frontier SF155-S (155W) Solar Panel. Available online: http://www.solardesigntool.com/components/module-panel-solar/Solar-Frontier/2294/SF155-S/specification-data-sheet.html (accessed on 22 April 2020).
38. PVI-5000/6000-TL-OUTD-Legacy Solar Inverters (ABB Solar inverters). Available online: https://new.abb.com/power-converters-inverters/solar-old/legacy/pvi-5000kw-6000kw (accessed on 22 April 2020).
39. Rupa, J.A.M.; Ganesh, S. Power flow analysis for radial distribution system using backward/forward sweep method. *Int. J. Electr. Comput. Electron. Commun. Eng.* **2014**, *8*, 5.
40. Ahmed, A.H.; Hasan, S. Optimal allocation of distributed generation units for converting conventional radial distribution system to loop using particle swarm optimization. *Energy Procedia* **2018**, *153*, 118–124. [CrossRef]
41. *Electric Vehicle Charging Station*; Xcel Energy: Minneapolis, MN, USA, 2015.

© 2020 by the authors. Licensee MDPI, Basel, Switzerland. This article is an open access article distributed under the terms and conditions of the Creative Commons Attribution (CC BY) license (http://creativecommons.org/licenses/by/4.0/).

MDPI
St. Alban-Anlage 66
4052 Basel
Switzerland
Tel. +41 61 683 77 34
Fax +41 61 302 89 18
www.mdpi.com

Energies Editorial Office
E-mail: energies@mdpi.com
www.mdpi.com/journal/energies

www.ingramcontent.com/pod-product-compliance
Lightning Source LLC
Chambersburg PA
CBHW061447220926
51850CB00004B/30